Little, Brown's Paperback Book Series

Basic Medical Sciences

Colton	Statistics in Medicine
Hine & Pfeiffer	Behavioral Science
Levine	Pharmacology
Peery & Miller	Pathology
Selkurt	Physiology
Sidman & Sidman	Neuroanatomy
Siegel & Albers	Basic Neurochemistry
Snell	Clinical Anatomy for Medical Students
Snell	Clinical Embryology for Medical Students
Valtin	Renal Function
Watson	Basic Human Neuroanatomy

Clinical Medical Sciences

Clark & MacMahon	Preventive Medicine
Eckert	Emergency-Room Care
Grabb & Smith	Plastic Surgery
Green	Gynecology
Judge & Zuidema	Methods of Clinical Examination
Keefer & Wilkins	Medicine
MacAusland & Mayo	Orthopedics
Nardi & Zuidema	Surgery
Thompson	Primer of Clinical Radiology
Ziai	Pediatrics

Nursing Sciences

DeAngelis	Basic Pediatrics for the Primary Health Care Provider
Sana & Judge	Physical Appraisal Methods in Nursing Practice
Selkurt	Basic Physiology for the Health Sciences

Manual and Handbooks

Arndt	Manual of Dermatologic Therapeutics
Children's Hospital Medical Center, Boston	Manual of Pediatric Therapeutics
Condon & Nyhus	Manual of Surgical Therapeutics
Friedman & Papper	Problem-Oriented Medical Diagnosis
Gardner & Provine	Manual of Acute Bacterial Infections
Massachusetts General Hospital	Manual of Nursing Procedures
Neelon & Ellis	A Syllabus of Problem-Oriented Patient Care
Papper	Manual of Medical Care of the Surgical Patient
Shader	Manual of Psychiatric Therapeutics
Spivak & Barnes	Manual of Clinical Problems in Internal Medicine: Annotated with Key References
Wallach	Interpretation of Diagnostic Tests
Washington University Department of Medicine	Manual of Medical Therapeutics
Zimmerman	Techniques of Patient Care

Little, Brown and Company
34 Beacon Street
Boston, Massachusetts 02106

Physiology

Physiology

Fourth Edition

Edited by
Ewald E. Selkurt, Ph.D.
Professor and Chairman, Department
of Physiology, Indiana University
School of Medicine, Indianapolis

Little, Brown and Company   Boston

Contributing Authors

The contributors to *Physiology* are all affiliated with the Department of Physiology, Indiana University School of Medicine, Indianapolis.

Ewald E. Selkurt, Ph.D., Professor and Chairman; Editor

William McD. Armstrong, Ph.D., Professor

Julius J. Friedman, Ph.D., Professor

Kalman Greenspan,* Ph.D., Professor

William V. Judy, Ph.D., Associate Professor

Leon K. Knoebel, Ph.D., Professor

Thomas C. Lloyd, Jr.,† M.D., Professor

Richard A. Meiss,‡ Ph.D., Assistant Professor

Ward W. Moore, Ph.D., Professor

Sidney Ochs, Ph.D., Professor

Carl F. Rothe, Ph.D., Professor

*Professor of Physiology and Medicine, Terre Haute Center for Medical Education.
†Joint Appointment in Department of Medicine.
‡Joint Appointment in Department of Obstetrics and Gynecology.

Preface

The fourth edition of this text adheres to the format of a compact, yet comprehensive handling of facts essential to a good foundation in mammalian physiology particularly suited to the needs of professional courses in medicine and dentistry. Development of so-called core curricula has made this style of text particularly useful. In previous editions, it has met the needs of advanced undergraduate college courses and graduate courses, particularly in the biological sciences. It is a useful foundation text for emerging disciplines in biomedical engineering.

Established facts are emphasized, and complementary subject matter taught by other basic science disciplines—for example, anatomy and biochemistry—is limited to a review of pertinent essentials. Although basic principles are stressed, sufficient clinical application is included to provide a foundation for subsequent development of the subject matter in a clinical setting. We are indebted to our clinical colleagues for many helpful suggestions in this direction.

The fourth edition has undergone thoroughgoing revision, but several subdisciplines have received particularly comprehensive revision and expansion in the light of developments in the field: neurophysiology, muscle physiology, cardiovascular, respiratory, renal, and electrolyte physiology, acid-base regulation, and endocrinology.

We are indebted to Terri Spradlin, Pauline Mayfield, and Helen Glancy for secretarial help and to Phil Wilson and Joe Demma of the Department of Medical Illustrations of Indiana University School of Medicine for their invaluable services; and we thank Helène Marer for the preparation of the index.

E. E. S.

Indianapolis

Contents

xi

Physiology

1. The Cell Membrane and Biological Transport

William McD. Armstrong

Physiology is concerned with the integrated functioning of the tissues and organs of which the body is composed. Essentially, these tissues and organs are *organized* assemblies of large numbers of cells. Thus, in a very real sense, the cell is a fundamental functional unit in relation to the physiology of the body as a whole.

Living cells are complex and highly organized. In addition to a fluid phase (intracellular fluid or cytoplasm), the interior of a typical cell contains a number of *inclusions* or *organelles* (nucleus, mitochondria, lysosomes, etc.). Modern research has shown that these subcellular particles are themselves highly organized structures that play a vital role in the overall activity of the cell. For present purposes, however, we shall be concerned mainly with the exchanges of material that take place between the cell as a whole and its external environment.

Cytoplasm or intracellular fluid differs markedly in composition from the external medium (blood or interstitial fluid) bathing the cell. In intact, normally functioning cells some of the differences in composition remain virtually constant over long periods of time. At the same time, there is a continuous exchange of material between the cell interior and the external environment, with some substances entering the cell and other substances leaving it. Evidently cells must possess a mechanism or mechanisms by which the entry and exit of dissolved solutes are regulated. Two alternative kinds of mechanism have been postulated. One invokes the idea that the cytoplasm as a whole is capable of selectively accumulating certain substances while excluding others. The other assigns the control of the transfer of materials between the cell and its environment to a special region at or near the periphery of the cell, the *cell membrane* or *plasma membrane*. At present, although the idea of functional selectivity by the cytoplasm as a whole is vigorously championed by some workers (Ling, 1962), most physiologists accept the special role of the membrane in the regulation of cellular composition as the most satisfactory interpretation of the available experimental evidence. Therefore, in the present chapter, cellular transport mechanisms will be discussed from this point of view.

STRUCTURE OF THE CELL MEMBRANE

At present, the existence of the plasma membrane as a discrete entity is not seriously challenged, even by those who dispute its importance in controlling the exchange of material between the cell interior and the external environment. Moreover, during recent years there has been a rapidly growing awareness of the importance of internal membranous structures (e.g., the *sarcoplasmic reticulum* of muscle) as regulators of overall cellular function. It is a truism of cellular biology that a satisfactory understanding of the function of any system requires a detailed knowledge of its molecular architecture. The problem of the molecular structure of cell membranes, particularly as it relates to their functional properties, is, therefore, of crucial importance to contemporary physiology.

Unfortunately, we are still a long way from a complete understanding of the fine details of the structure of the cell membrane. Nevertheless, recent theoretical and experimental research has reached a stage where the disposition of its major components can be inferred with some confidence. In this section an attempt will be made to review some current concepts of membrane structure. In particular, the fluid mosaic model of Singer (Singer and Nicolson, 1972) will be discussed since it appears, in this

1

writer's opinion, to represent a convincing synthesis of theoretical concepts and experimental data. However, some cautionary comments are appropriate. First, there is by no means unanimous agreement on even the grosser aspects of membrane structure. A number of models of varying ingenuity, complexity, and plausibility have been proposed (Hendler, 1971). Second, as a result of vigorous research utilizing a variety of physical, chemical, and biological techniques, our insights into the molecular structure of membranes are expanding so rapidly that any model of the membrane which can be proposed at present must be regarded as tentative. Third, in considering any generalization about membrane structure, one must bear in mind that cell membranes exhibit considerable diversity in composition. Thus there may be, and probably are, significant differences in structural detail between the membranes of different cells and between different membranous structures within the same cell. This phenomenon becomes especially apparent when one considers highly specialized membranous organelles such as the *myelin sheath* of nerve.

The principal structural components of cell membranes are lipids (mainly cholesterol and phospholipids), proteins, and oligosaccharides. In addition, membranes contain water (some of which is probably in a more or less highly *ordered* state because of its close association with the ionized and polar groups of phospholipids and proteins) and small amounts of low-molecular-weight species such as inorganic ions. Of the three major components listed above, the most abundant is protein. Chemical analysis of membrane fractions obtained from a variety of cells indicates that the protein/lipid weight ratio ranges from about 1.5 to 4.0 (Singer and Nicolson, 1972). Oligosaccharides, though less abundant than the other two, can be of crucial importance in certain properties of cell surfaces (e.g., immunological responses) (Stein, 1967). Thus the basic problem of cell membrane structure is the organization of membrane lipids and proteins. In general, one may assume that the factors which determine the organization of oligosaccharides within the membrane are not very different from those which apply to membrane proteins. Lipids and proteins contain both polar and nonpolar groups and are, therefore, *amphiphilic* substances—that is, substances that display both *hydrophilic* and *lipophilic* or *hydrophobic* properites. Their polar groups tend to interact strongly with water or other polar solvents (hydrophilic reactions), and at the same time their nonpolar residues tend to interact strongly with nonpolar solvents or other nonpolar moieties (hydrophobic reactions). Thus if proteins or lipids are spread as a layer one molecule thick (*monolayer*) at the interface between water and a nonpolar phase (e.g., oil or air), they tend to orient themselves so that their polar groups are in contact with the polar phase while their nonpolar portions are in contact with the nonpolar phase (Fig. 1-1A). In aqueous solutions, polar-nonpolar lipids are likely to clump together to form *micelles* (Fig. 1-1B), in which the polar "heads" are oriented outward toward the water while the nonpolar "tails" are tucked inside the micelle where their interactions with the polar water molecules are minimized and their interactions with each other are maximized. Similarly, proteins in aqueous solution are oriented so that as many as possible of their ionic residues are in contact with water (mostly at the outside of the protein molecule) and as many as possible of their hydrophobic residues are folded inside the molecule away from the external aqueous environment.

This tendency of amphiphilic substances to orient themselves so that, as far as possible, both hydrophilic and hydrophobic interactions are maximized is a reflection of a very general principle which is embodied in the second law of thermodynamics: The state of maximum stability for any system is that state in which the free energy of the system is at a minimum. Looked at in another way, the state of minimum free energy is the most probable state for the system to achieve as a result of spontaneous changes; or, conversely, one can say that spontaneous changes in any system always proceed in the direction of decreasing free energy. Applying this principle to the problem of the orientation of lipids and proteins in cell membranes, which may be regarded essentially as a thin layer, about 75 to 100 Å in thickness, bounded on both sides by an aqueous solution, one can predict that, within the constraints imposed by the overall geometry of

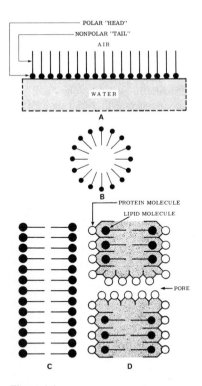

Figure 1-1
A. Orientation of monolayer of amphiphilic lipid at the interface between a polar and a nonpolar phase. B. Phospholipid micelle in a polar liquid. C. Phospholipid bilayer. D. Model of cell membrane proposed by J. F. Danielli.

the membrane, both these molecular species will dispose themselves in such a way that their capacity for hydrophilic and for hydrophobic interaction with neighboring molecules or groups is maximized. A conformation which permits phospholipids to achieve this condition within the cell membrane is the *bilayer* (Fig. 1-1C). In this structure the polar head groups of the phospholipids make contact with the aqueous media on both sides of the membrane while the predominantly hydrocarbon tails are sequestered inside the bilayer.

The suggestion that the lipid moiety of the cell membrane is organized as a bilayer was initially put forward about 50 years ago on the basis of studies with lipids extracted from red blood cell "ghosts," i.e., red blood cells from which the internal contents have escaped following *hemolysis* (see p. 17).* Since then the concept not only has remained viable but has been strongly reinforced by recent experimental work (Hendler, 1971; Singer and Nicolson, 1972). Thus, despite the fact that a number of models for cell membrane structure have been proposed in which the majority of membrane lipids are supposed to be organized as micelles, either alone or as mixed lipoprotein micelles (Lucy, 1968; Hendler, 1971), the existence of a phospholipid bilayer as an essential structural component of the cell membrane seems now well established. In particular, artificially produced phospholipid bilayers have been shown to possess many properties that are similar to those of naturally occurring cell membranes (Christensen, 1975). One

*A detailed account of the early experimental and theoretical work which led to the evolution of the bilayer concept is given by Davson (1970).

may conclude (Singer and Nicolson, 1972) that the greater part (at least 70 percent) of the membrane phospholipid is in the bilayer form, though some fraction (about 30 percent) may be organized in a different form, either alone or in combination with protein. It is important to note that recent studies, utilizing such techniques as electron spin resonance, strongly indicate that the interior of the bilayer is essentially fluid in nature; i.e., there appears to be little constraint within the bilayer on rotational and vibrational movements of the hydrocarbon moieties of the phospholipid molecules. Consequently the nonpolar phase of the phospholipid bilayer can be considered to approximate a liquid oil layer in its solvent properties, for it permits small lipid-soluble molecules to move rather freely within and across it. A rigid or quasi-crystalline lipid phase would not be expected to possess the "lipid-solvent" properties exhibited by most cell membranes.

An early suggestion concerning the organization of membrane proteins was that of Danielli (Fig. 1-1D; see also Davson, 1970). According to this model the lipid "core" of the membrane was surrounded on both sides by layers of adsorbed protein, forming a kind of "sandwich" or *trilamellar* structure. In the model, the bimolecular lipid leaflet could be considered to be interrupted at intervals, thus permitting the formation of hydrophilic pores. The existence of such pores appears to be virtually required to explain certain aspects of the permeability of cell membranes. The Danielli model enjoyed a long period of dominance in cell physiology and seemed for a time to be strongly supported by the appearance, in electron micrographs, of a trilamellar structure close to the surface of cells which was thought to be consistent at least with the proposed structure. However, the existence, in electron micrographs, of this structure by no means requires that the membrane itself have the trilamellar nature assigned to it in Danielli's model. Further, it has been found that the fixation procedures used to prepare specimens for the electron microscope can markedly alter the conformation of membrane proteins. In addition, modern research into the conformational states of protein molecules in aqueous and nonaqueous media strongly suggests that the structure proposed by Danielli is highly improbable on thermodynamic grounds. First, the Danielli model requires considerable unfolding of the proteins at the surface of the membrane and thus forces a large fraction of the nonpolar residues of these molecules to be in direct contact with water. Second, the protein layer to a large extent sequesters the ionic and polar groups of the phospholipid bilayer from the aqueous environment and frequently requires that they be contiguous with nonpolar protein residues. Thus neither hydrophilic nor hydrophobic interactions can be maximized in terms of this model. It has also been found that the conformation of membrane proteins is partly α-helical and partly random coil. This does not indicate extensive unfolding.

These objections are largely overcome in the model proposed by Singer and Nicolson (Fig. 1-2). Here the membrane is regarded as a *mosaic* of lipids and globular proteins, the proteins being inserted as discrete "plugs" into the bilayer matrix in more or less random fashion. A three-dimensional representation of the concept is given in Figure 1-2B. Several features of this model deserve comment. First, as already mentioned, the hydrophobic core of the phospholipid matrix is considered to be essentially fluid in nature; i.e., lateral adhesion between neighboring molecules is relatively weak. Thus the whole ensemble of lipid and protein is a fluid mosaic or two-dimensional solution. The membrane, therefore, is a highly labile, dynamic entity which can readily alter its fine structure in response to a variety of stimuli—for example, hormones. Some hormones are known to act directly on the membranes of their *target cells.**

The model illustrated in Figure 1-2 may be looked upon as representing the essential core of the living cell membrane and, as discussed below (p. 8), provides a plausible

---

*An interesting discussion of the fluid mosaic model in relation to physiological and pathological alterations in membrane structure, including changes in surface properties, which are associated with the transformation of cells from normal to malignant states, is given by Singer and Nicolson (1972).

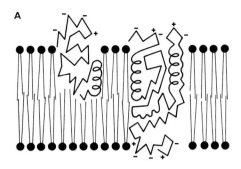

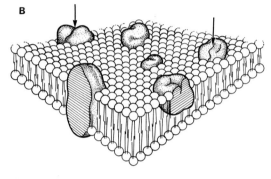

Figure 1-2
A. Cross-sectional, and B. three-dimensional representation of the cell membrane as a fluid mosaic of lipid and (globular) protein. Note disposition of ionic groups in membrane proteins (A) and occurrence of oligomers (arrow in B).

structural basis for many aspects of membrane transport. However, the high overall protein/lipid ratio found in most membranes, together with the fact that cell membranes (in addition to functioning as barriers which control the movement of water and solutes between the cell interior and the outside medium) are the site of a rich variety of enzymatic reactions, suggests that additional proteins, bound to the membrane by forces other than the predominantly hydrophobic lipid-protein interactions suggested by Figure 1-2, may be essential functional components of living membranes. Such proteins have been designated *peripheral* proteins by Singer (Singer and Nicolson, 1972). In contrast to the *integral* proteins shown schematically in Figure 1-2, peripheral proteins may be thought of as rather loosely adsorbed to the surface of the membrane (e.g., by electrostatic interactions between their ionized groups and the ionized groups of the lipid bilayer). Evidence for two different classes of membrane-bound proteins has been obtained during the separation and characterization of the membrane fractions of different cell species. Some proteins can be dissociated from the membrane by very mild treatment, and when dissociation occurs, the protein comes off the membrane in lipid-free form. A good example is the *cytochrome c* of mitochondrial membranes, which can be caused to dissociate by high salt concentrations. Such proteins are the peripheral proteins and account for about 30 percent of the total protein of most membranes. The remaining 70 percent (the integral proteins) require more drastic treatment (detergents, bile salts, organic solvents, etc.) to separate them from the membrane matrix, and, when separation does occur, the extracted proteins are usually contaminated with greater or lesser amounts of lipid.

THE TRANSPORT OF SMALL MOLECULES AND IONS ACROSS THE CELL MEMBRANE

Mechanisms of Membrane Transport: General Principles

As generally employed in physiology, the term *membrane transport* has come to have a rather arbitrarily restricted meaning. Ordinarily, any general discussion such as that presented in this chapter is largely confined to transmembrane movements of water and low-molecular-weight solutes, the latter being assumed to be in the monomeric state. However, many cells of the body subserve important transport functions which are not subject to the rules to be discussed in detail below. For example, the epithelial cells lining the lumen of the small intestine absorb fats. Multimolecular aggregates such as micelles (Fig. 1-1B) play an important role in this process. Again, the transmembrane transport of large molecules such as proteins is of vital importance to the normal functioning of the body. One may cite as examples the intestinal absorption of immunoglobulins by the newborn and the transfer of salivary and pancreatic enzymes from the cells in which they are synthesized to the lumen of the alimentary canal. It is generally believed that the process of *exocytosis* and its converse *endocytosis* are notable factors in the transport of relatively large molecules and aggregates. In this process a segment of the cell membrane folds around the transported species to form a vesicle which then separates either externally (exocytosis) or internally (endocytosis) from the rest of the membrane. The membranous "wall" of the vesicle is then degraded enzymatically, thus releasing the transported species to the external or internal medium. These processes, though of importance in a number of physiological situations, will not be further considered.

Even within the restrictions noted above, the characterization of membrane transport processes is complex, particularly because certain substances (notably Na^+ and K^+ ions) are frequently transported at the same time across cell membranes by at least two different mechanisms. Figure 1-3 is an attempt to summarize, in schematic form, what are believed at present to be the major routes of membrane penetration. Let us first consider the processes labeled I, II, and IIa in the diagram. In these instances the transported species crosses the membrane by simple diffusion, either through the membrane matrix itself (I) or through polar pores (II and IIa). The walls of these pores can be considered to consist primarily of ionic groups (IIa) or of neutral polar groups (II). In reality, one supposes that most pores in the membrane are a combination of the types suggested by II and IIa. One notes that penetration by diffusion does not require any specific interaction between the transported species and the cell membrane. Indeed, it might be predicted that, in the case of a substance crossing the membrane by a process analogous to I in Figure 1-3, the major determinant of the ease with which it penetrates (i.e., its permeability) would be its solubility in the matrix of the membrane. Similarly, for penetration via the route designated II or IIa in Figure 1-3, the size of the penetrating molecule would be expected to be of prime importance.

These predictions are confirmed by two generalizations for molecular species which cross cell membranes by a process analogous to diffusion. These generalizations, first proposed by the German physiologist E. Overton toward the end of the last century, are frequently known as "Overton's rules." Countless studies have confirmed their essential validity.

Overton's first rule states that for small, predominantly nonpolar molecules ease of penetration or permeability is *directly proportional to lipid solubility*. The term *lipid solubility* refers to the *partition* or *distribution* coefficient for a given substance between water and a fat solvent (e.g., chloroform or benzene) which is immiscible with water. That is,

$$K = C_1/C_2 \tag{1}$$

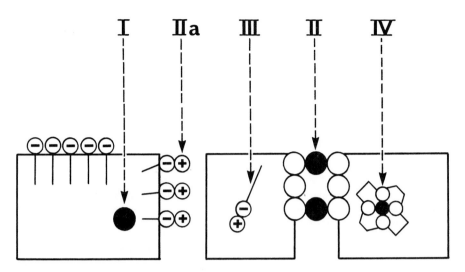

Figure 1-3
Mechanisms of membrane permeation. ⊕, cation; ⊖, anionic group; ●, neutral (nonpolar) molecule; O, oxygen.

where K is the partition coefficient and C_1 and C_2 are the equilibrium concentrations of the solute in question in the nonaqueous medium and in water respectively. Thus, for nonpolar solutes, high lipid solubility is associated with high permeability and vice versa.

The role of lipid solubility in cell penetration is further underlined by the behavior of weak electrolytes—i.e., electrolytes which undergo *reversible dissociation* of the type

$$HA \rightleftharpoons H^+ + A^-$$ (2)

or

$$BH^+ \rightleftharpoons H^+ + B$$ (2a)

within the physiological pH range. It is found that, whereas weak electrolytes (e.g., acetic acid) penetrate cells readily in the nonionized state, they fail to do so when *ionized*. This finding is in keeping with the fact that the lipid solvent : water partition coefficients of these compounds are much greater in the nonionized than in the ionized state.*

The implications for drug therapy are of considerable practical importance since many drugs are weak electrolytes. Consider the absorption from the alimentary canal into the bloodstream of two orally administered drugs, one of which (A) dissociates according to equation 2 while the other (B) dissociates in conformity with equation 2a. In any given situation, the degree of ionization of both of these will be determined by the relationship between the pH of the surrounding medium and the pK (pH of half dissociation) of the drug. For drug A the degree of ionization will decrease as the ambient pH decreases. The opposite will be true for drug B. Let us suppose that both our hypothetical drugs have a pK of 6 (and that neither is chemically altered by the enzymes of gastrointestinal juice). In the stomach, where the pH of the gastric juice is

*A much fuller discussion of the role of lipid solubility, molecular size, and dissociation in the permeability of nonelectrolytes and weak electrolytes is given by Davson (1970).

ordinarily about 2 to 3, drug A will be virtually all in the nonionized form and thus will pass freely through the membranes of the gastric mucosa, whereas drug B will be almost completely ionized and will not be absorbed to any great extent during its passage through the stomach. On the other hand, farther down the intestine where the alkaline secretions of the pancreas increase the luminal pH to 7 or 8, drug B will be largely converted to the nonionized form and readily absorbed. In consequence, if both A and B are administered together, one would expect to find an effective level of A in the bloodstream before B appeared in quantity. Such considerations (in addition to intrinsic therapeutic potency, toxicity, etc.) underlie the many chemical variants of a basic therapeutic agent which are frequently available to the physician.

Overton's second rule is concerned with predominantly polar or hydrophilic molecules. It states that, for these molecules, permeability is *inversely related to molecular size;* i.e., the smaller the molecule, the more readily it penetrates the cell and vice versa. Thus, small polar molecules such as urea (and water) and small inorganic ions which have virtually zero lipid solubility penetrate cells by simple diffusion. It must be remembered, however, that, compared to substances such as urea and (nonionized) acetic acid, the diffusion of ions across cell membranes is normally quite slow and in certain permeability experiments of brief duration (e.g., in studying the hemolytic effects of rapidly penetrating solutes on red blood cells—see p. 17) can be neglected.

The model of the cell membrane shown in Figure 1-2 is easily reconciled with Overton's rules. Since, in this model, the fluid matrix of the membrane is lipid in nature, one can easily see how lipid solubility can be the major determinant of permeability for nonpolar molecules. Similarly, the model readily permits the existence of discrete aqueous pores through which small water-soluble molecules and ions (in addition to water itself) might be expected to move. Although single protein molecules do not contain such pores, it is reasonable to suppose that many of the integral proteins of membranes are *oligomers* consisting of two or more *subunits* (Fig. 1-2B). Aggregates of this type frequently contain narrow water-filled channels near their central axes. A good example (though admittedly it is not a membrane protein) is the internal protein hemoglobin of red blood cells. This normally exists as a *tetramer* (four subunits) and has been shown by x-ray crystallography to have a central pore some 10 Å in diameter, close to the apparent average pore diameters (about 4 to 8 Å) that have been estimated for a number of cell membranes.

The Kinetics of Diffusion—The Permeability Coefficient

Diffusion results from the spontaneous tendency for any substance (e.g., a gas or a solute in aqueous solution) to distribute itself uniformly throughout the whole space available to it. For example, if a solute is present initially at a higher concentration in one region of a solution than in another, with the passage of time the concentration difference disappears; the solute concentration becomes uniform throughout the solution. As long as a solute concentration difference or *concentration gradient* exists between two different regions of the solution, there will be a spontaneous net movement of solute from the region of higher to the region of lower concentration. When all concentration gradients have been dissipated, *net* solute movement between different regions of the solution will cease. The solution is then said to be in a state of equilibrium with respect to the solute.

The rate of solute diffusion (V) between two regions of a solution containing unequal concentrations of solute is obviously equal to $-dS/dt$, the amount of solute passing from the region of higher to that of lower solute concentration in unit time. At constant temperature, this quantity is given by Fick's equation:

$$V = -\frac{dS}{dt} = DA(S_1 - S_2) \tag{3}$$

In this equation S_1 and S_2 are the solute concentrations in the regions of higher and lower concentration respectively, A is the cross-sectional area of the boundary between these regions (the area across which diffusion is taking place), and D is the *diffusion coefficient* or *diffusivity* of the solute. It is apparent from equation 3 that D is numerically equal to the rate of diffusion across unit area (1 cm²) when the concentration difference $(S_1 - S_2)$ across the boundary layer is unity. Thus D is a measure of the inherent ability of the solute molecules to move through the solution. It is also clear from equation 3 that the net rate of diffusion at constant temperature depends on the concentration gradient $(S_1 - S_2)$ and on the magnitude of D. D has been found to depend on temperature, becoming larger as the temperature is increased. It also varies with the molecular weight of the dissolved substance, becoming smaller as the molecular weight increases. D likewise depends to some extent on molecular shape as well as on molecular size. With large molecules in particular, the extent and strength of their interactions with other molecules, and hence the overall resistance to their diffusion in aqueous solution, are governed by shape as well as size. In these circumstances, the relationship between D and molecular weight can be rather complex.

In thermodynamic terms, the tendency of a solution to move spontaneously in the direction of equilibrium can be viewed as a reflection of the general tendency of natural systems to strive to attain the condition of minimum free energy, or maximum *entropy* (p. 2). It can be shown by the methods of statistical thermodynamics that this condition corresponds to the most probable state of the system, which is, in turn, the state of *maximum disorder* or *randomness.*

When a concentration gradient (ΔS) for a solute exists across a cell membrane, the rate of solute diffusion across the membrane is given by

$$V = \frac{PA}{x} \cdot \Delta S \tag{4}$$

Equation 4 is analogous to equation 3 except for the term x, which represents the thickness of the membrane, and the term P, the *permeability coefficient* for the solute, which replaces the diffusion coefficient D in equation 3. The substitution of P for D merely reflects the fact that, in these conditions, it is the rate of diffusion of the solute within the membrane, rather than its diffusivity in free solution, that is the rate-determining step in the penetration process.

Equation 4 states that, for a diffusive process, the overall rate of penetration V is a linear function of ΔS. This is illustrated graphically in Figure 1-4A. It is evident that the

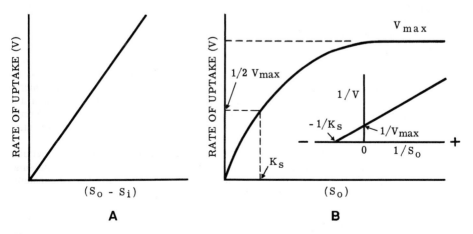

Figure 1-4
Kinetic characteristics of transport processes. A. Diffusion type. B. Carrier-mediated transport. (From H. Stern and D. L. Nanney. *The Biology of Cells.* New York: Wiley, 1965. P. 326.)

slope of the line relating V to ΔS in the graph is equal to PA/x. Thus, if one knows A and x, P is readily determined. In practice, A, the area of the membrane, and x may be difficult to determine with accuracy. However, if these parameters are assumed to remain constant, the *relative* permeabilities of a given cell species for a number of solutes can be derived from uptake studies such as the one illustrated in Figure 1-4A.

It is important to note that in the present discussion and elsewhere in this chapter *net* rates of transport are considered. If the concentration of a permeable solute is finite on both sides of a membrane, there will be, at one and the same time, an *influx* of solute into the cell and an *efflux* of solute out of the cell. The *net flux* is obviously the difference between these two *unidirectional fluxes*. Now, in general, a flow or flux of material in any system implies a conjugate *driving force*. In thermodynamic terms, the concentration gradient (ΔS of equation 4) is the force driving the flow V in a diffusive process. The term PA/x is an example of a coupling coefficient that expresses the relationship between S and V. A familiar example of an equation describing a linear relationship between a flow and its conjugate driving force is Ohm's law, $\Delta E = IR$, which relates the current or flow of electricity (I) between two points in a conductor to the driving force ΔE, the difference in electrical potential, between these two points. The electrical resistance R is an expression of the coupling coefficient between E and I. Systems of linear equations of this kind form the basis of Onsager's analysis of the thermodynamics of nonequilibrium systems. In recent years Onsager's theory has been extensively applied to the theoretical analysis of membrane transport (Katchalsky and Curran, 1965).

It is plain from equation 4 that, in diffusive transport, the net flow is always in the direction of the concentration gradient; i.e., net flow is *downhill* in an energetic sense. Hence diffusion is an example of what is frequently called *passive* transport. When ΔS becomes zero, net flux is also zero. Clearly, this is the result when the concentrations of the solute inside and outside the cell are equal.*

Carrier-Mediated Transport

With many substances of physiological importance (e.g., sugars, amino acids, and ions), the experimental curve relating the rate of entry to the outside concentration is frequently of the type shown in Figure 1-4B. At first, as S_0 is increased, the rate of entry also increases. Ultimately, however, an external substrate concentration is reached at which the rate of entry becomes maximal and is not further increased in the presence of higher external substrate concentrations. Kinetic behavior of this type is referred to as *saturation kinetics* and is often exhibited by systems in which the kinetic process involves the reversible combination of the substrate with a receptor site (e.g., enzyme reactions).

In the case of membrane processes, it is postulated that the transported solute or *substrate* enters the cell by combining reversibly with a specific membrane component or *carrier* at the outer surface of the membrane. The carrier-substrate complex so formed moves across to the inner surface of the membrane, where it dissociates, releasing free substrate to the cell interior. The carrier then moves back across the membrane to the outer boundary, where it combines with a second molecule of substrate and the cycle begins again. Thus a relatively small number of carrier molecules operating cyclically can transport large amounts of substrate. On this basis the occurrence of a maximum in the rate of transport with increasing substrate concentration is readily explained as being due to saturation of the available carrier sites when the external substrate concentration reaches a sufficiently high level. The processes

---

*This discussion assumes that the transported solute is *uncharged.* For charged species (ions), the conditions of zero net driving force across the membrane are more complex (p. 19). In this state influx and efflux of solute may both be vigorous but they will be equal. The phenomenon of solute exchange across a membrane without net transport is called *exchange diffusion.* It is readily demonstrated by the use of radioactive isotopes.

labeled III and IV in Figure 1-3 are schematic representations of carrier-mediated processes. In IV the carrier molecule is supposed to be neutral, as one might expect for a carrier that participates in the transfer of uncharged molecules such as sugars. However, recent work with model systems such as artificial bilayers has shown that neutral carrier molecules can also participate in the transport of ions (see Pressman and Haynes, 1969, and p. 29). In III a charged (anionic) molecule is envisaged as a carrier for cations. Combination of a cation with an anionic carrier could result in an electrically neutral complex capable of moving freely within the lipid core of the membrane.

Kinetics of Carrier-Mediated Transport
A model for a simple carrier process is diagrammed in Figure 1-5A. Applying the law of mass action to the reversible combination of substrate with carrier, one may write

$$S + C \rightleftharpoons CS \quad \text{and} \quad \frac{[C] \times [S]}{[CS]} = K_s \tag{5, 6}$$

where K_s is the *dissociation constant* for the complex CS. The brackets denote concentration. K_s may be calculated from the equation

$$\frac{V}{V_{max}} = \frac{[S]}{[S] + K_s} \tag{7}$$

where V is the rate of transport at a substrate concentration (S) and V_{max} is the maximal or saturation rate of transport. (This equation is identical with the well-known Michaelis-Menten equation of enzyme kinetics.)

As shown in Figure 1-4B, K_s is numerically equal to the substrate concentration when $V = \frac{1}{2} V_{max}$. If the association-dissociation steps on both sides of the membrane (equation 5) are *rapid* compared to the movement or *translocation* of the carrier substrate complex across the membrane, then, as indicated by equation 6, $1/K_s$ gives a measure of the *affinity* of the carrier for the transported substrate. Note that this affinity is *reciprocally* related to the magnitude of K_s.

Both K_s and V_{max} are more conveniently determined from a plot of $1/V$ against $1/[S]$, as shown in the insert in Figure 1-4B. With this plot, a straight line is obtained which,

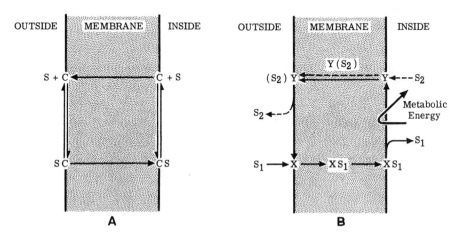

Figure 1-5
Diagram of two types of carrier-mediated transport. A. A sỹmmetrical carrier of the equilibrating type. B. A carrier capable of accumulative transport.

when extrapolated to cut the axes of the graph, gives an intercept equal to $1/V_{max}$ on the $1/V$ axis and an intercept equal to $-1/K_s$ on the $1/[S]$ axis.

A symmetrical carrier of the type illustrated in Figure 1-5A, which has the same affinity for substrate on both sides of the membrane, can lead only to equalization of the substrate concentration across the membrane. As in the case of simple diffusion, the direction of *net* transport will depend upon the direction of the substrate concentration gradient. When $S_{inside} = S_{outside}$, the same amount of substrate will be carried across the membrane in each direction and no net transport of S will occur. Equilibrating processes of this kind are frequently referred to as *facilitated diffusion*. They do not, in the strict thermodynamic sense, require an external source of energy. However, the production of carrier molecules by the cell may be dependent on metabolism. Consequently, equilibrating carrier processes may show some sensitivity to metabolic inhibitors, in which case they are occasionally described erroneously as "active" transport processes. The entry of sugars and amino acids into human red blood cells occurs by way of equilibrating carrier processes.

The transport of sugars and amino acids by cells such as those of kidney tubule epithelia and intestinal mucosa also shows saturation kinetics. Unlike red blood cells, however, these cells accumulate sugars and amino acids against a concentration gradient. The accumulation process depends on metabolic energy and is inhibited by metabolic poisons—cyanide and dinitrophenol, for example.

A model of a carrier system capable of *accumulative* or *active transport* is shown in Figure 1-5B. An important difference between this model and the one illustrated in Figure 1-5A lies in the fact that, in the model shown in Figure 1-5B, the carrier is asymmetrical; i.e., it undergoes a change at the inner surface of the membrane from a form X, with a relatively high affinity for the substrate, to a form Y, with a relatively low substrate affinity. The reverse takes place at the outer surface of the membrane, where the X form of the carrier is regenerated.

Certain other features of the carrier model shown in Figure 1-5B may be noted. Since the transport process effected by the carrier is an "uphill" or active process, it requires a continuous supply of energy. In the reversible transformation of the carrier between the X and Y forms (which might be, for example, a configurational or other intramolecular change), one step at least must be energy requiring. As indicated in Figure 1-5B, it is rather generally believed that the energy-requiring step is the one that occurs at the inner surface of the membrane. This provides a mechanism by which metabolic energy can be continuously fed into the system.

Again, as indicated by the solid arrows in Figure 1-5B, an accumulating carrier process or "pump" can serve to transport a single substance S_1 across the cell membrane, the carrier returning across the membrane in the free or uncombined form during the second half of the cycle.* Alternatively, as indicated by the dotted arrows, such a mechanism can act as a so-called exchange pump in which the Y form of the

---

*This statement is of course an approximation. Since there are finite though different affinities between carrier and substrate on both sides of the membrane, there will always be some "reverse" flux of S from the low-affinity to the high-affinity side providing there is a finite concentration of S in the compartment bounded by the former. Using primes to denote the kinetic parameters pertaining to the low-affinity side (e.g., V′ is the unidirectional flux of solute from this side to the high-affinity side), one can describe the kinetics of the system by the following equation:

$$V_{net} = (V - V') = V_{max} \left[\frac{[S]}{[S] + K_s} - \frac{[S']}{[S'] + K'_s} \right] \tag{8}$$

This equation shows that when $V = V'$ (i.e., $V_{net} = 0$), $[S']/[S] = K'_s/K_s$. The ratio K'_s/K_s is thus the *maximum accumulation ratio* of the system and can be used to predict the extent to which a metabolically linked carrier of the type under discussion can accumulate a transported solute on one side of the membrane.

carrier combines with a second substrate S_2, which is transported out of the cell in an amount equivalent to the amount of S_1 transported inward. The essential requirement for an exchange pump is that there be a reciprocal relationship between the affinities of two forms of the carrier for S_1 and S_2. The transport of sodium and potassium ions by red blood cells is a well-known example of an exchange pump (p. 29).

Biological transport processes that show saturation kinetics also exhibit a number of other characteristics which are easily explained on the basis of carrier theory. For example, among a series of closely related compounds such as sugars and amino acids, some may be transported readily while others may be transported very slowly or not at all. This *substrate specificity* may be regarded as arising from the possession of a highly specific structure by the combining site on the carrier molecule, requiring an equally *specific structure* on the part of the substrate and the carrier-substrate complex if transport is to be easily effected. Note the contrast with the more general properties such as lipid solubility and molecular size that appear to determine transport rates in purely diffusive permeation. A further consequence of substrate specificity is *competition,* in which two or more structurally similar substrates compete for the same carrier sites. In this situation the relative amount of each substrate that passes through the membrane will depend on its relative affinity for the carrier and on its relative concentration in the vicinity of the carrier sites. The existence of substrate specificity and competition has been amply demonstrated in many cell species.

Finally, as already stated (p. 6), many physiologically important substances can move across cell membranes by both diffusive and carrier-mediated pathways. These pathways appear to be independent of one another, and they frequently subserve different physiological functions. A particularly striking example is the transport of Na^+ and K^+ ions across the membranes of excitable cells such as muscle and nerve. This is discussed more fully below (pp. 25–29). For the present we may note that diffusive transport is passive or downhill in character. Active or uphill transport, in every case for which its existence has been unequivocally demonstrated, requires the participation of a carrier mechanism, although, since equilibrating carrier-mediated transport does occur, the demonstration that a given transport process is mediated by a carrier does not alone serve to characterize it as an active transport process.

Water Movement—Osmosis and Osmotic Pressure

Virtually all cells are freely permeable to water. The regulation of the water content of cells and tissues is an important factor in the maintenance of the overall *fluid and electrolyte balance* of the body. The latter is crucial in the context of human health and disease. Indeed, one of the most striking contributions of physiology to modern clinical practice lies in the sophisticated system of fluid and electrolyte therapy that has been and is still being developed to deal with imbalances in body water and electrolytes resulting from pathological conditions or from accidental or surgically induced trauma. It is therefore not surprising that the movement of water across cell membranes and the forces governing this movement have been of prime interest to physiologists for many years.

Osmotic Phenomena in Dilute Solutions

Consider a vessel divided into two parts by a membrane which is permeable to water but impermeable to dissolved solutes (a *semipermeable* membrane). If the compartment on one side of the membrane contains pure water and the other compartment contains an aqueous solution, water will flow spontaneously from the side containing pure water to the side containing the solution. This spontaneous net flow of water is called *osmosis* or *osmotic flow*. Its origin may be explained as follows. As already noted (p. 2), all natural processes tend to proceed spontaneously in the direction of equilibrium. In the system under consideration, there are, initially, two concentration gradients across the membrane: a solute concentration gradient from the side containing the solution to the side containing pure water, and (because of the diluting effect of the solute on the water in the side containing the solution) a concentration gradient for water in the opposite direction. In the absence of external restraints both water and solute would diffuse freely in the direction of their respective concentration gradients until mixing was complete. Because of the restraint imposed on the system by the semipermeable membrane, the solute cannot diffuse into the side containing pure water. The only net movement of material that can take place across the membrane is a flow of water into the side containing the solution. If, instead of pure water and a solution, two solutions containing unequal concentrations of solutes were used, a similar osmotic flow of water would take place. In this case osmosis would occur from the more dilute to the more concentrated solution.

Clearly, unless some additional restraint is imposed, water will continue to move in the direction of its concentration gradient as long as the gradient exists—i.e., until the concentrations of water on both sides of the membrane become equal. In the case of two solutions, equilibrium would be achieved when the solute concentrations on both sides of the membrane became equal. In a system containing pure water and an aqueous solution, osmotic equilibrium cannot, in principle at least, be realized in this way at any finite solute concentration. However, it is evident that if a force or pressure equal and opposite to the force generated by the concentration gradient for water across the membrane could be applied to the side containing the solution, osmotic flow could be prevented. The pressure required to prevent osmotic flow of water into a given solution is called the *osmotic pressure* of that solution.

There are many ways in which osmotic flow can be prevented by the application of an external force. One of the simplest—and it also permits the osmotic pressure to be determined directly—is utilization of the force of gravity. The solution whose osmotic pressure is to be measured is placed inside a thin semipermeable bag, which is then immersed in a vessel containing water. One end of a fine capillary tube is inserted into the bag, and the apparatus (called an *osmometer*) is adjusted so that the capillary tube is vertical. Initially the apparatus is adjusted so that the liquid levels in the capillary and in the outer vessel, apart from the slight rise due to capillary action, are the same. As water enters the bag by osmosis, the level of the liquid in the capillary rises until the hydrostatic pressure developed is just sufficient to balance the osmotic driving force across the wall of the bag. If the volume of the column of solution in the capillary is very small compared to the total volume enclosed in the bag (so that the total amount of water which enters the bag and its diluting effect on the solution within the bag are negligible), the osmotic pressure of the original solution is given by

$$\pi = h\rho g \tag{9}$$

where π is the osmotic pressure, h is the height of the column of solution in the capillary necessary to balance the osmotic driving force, ρ is the density of the solution, and g is the acceleration due to gravity. For moderately dilute solutions ρ does not differ significantly from the density of water; hence the height in centimeters of the column in the capillary will give the osmotic pressure directly in centimeters of water. Since 1034 cm water is equivalent to 76 cm mercury or 1 standard atmosphere, π is readily obtained in either of these units.

Osmotic Pressure and Solute Concentration: Units of Osmotic Concentration
It is clear from the preceding discussion that the magnitude of the osmotic pressure in any given solution depends only on the difference between the concentration of water in that solution and its concentration in the pure liquid. This dependence, however, is of little practical use in physiology. Useful and meaningful relationships can be obtained if osmotic pressure is expressed in terms of solute concentration. Fortunately, despite the fact that the solute, aside from its diluting effect on the concentration of water in the solution, is not a fundamental factor in the generation of osmotic pressure, a simple

relationship between osmotic pressure and solute concentration does exist. The reason is that, in moderately dilute solutions, there is a simple complementary relationship between the concentration of water and that of solute.

Consider a solution containing n_1 moles of water and n_2 moles of solute per unit volume. The *molar fractions* of water and solute, respectively, are $X_1 = n_1/(n_1 + n_2)$ and $X_2 = n_2/(n_1 + n_2)$. Hence, $(X_1 + X_2) = 1$, and $X_2 = (1 - X_1)$. Evidently it is the factor $(1 - X_1)$ that determines the osmotic pressure of a solution in contact with pure water. Therefore, the osmotic pressure is directly proportional to X_2 or solute concentration. For a dilute solution the quantitative relationship between osmotic pressure and solute concentration is given by the van't Hoff equation

$$\pi = CRT \tag{10}$$

where C is the solute concentration, R is the gas constant, and T is the absolute temperature.

It is apparent from equation 10 that, if the concentration of solute in a given solution is known, its osmotic pressure can readily be calculated. Conversely, given the osmotic pressure, the solute concentration can be obtained from this equation. It is important to realize that there is a difference between concentrations as they relate to osmotic activity and ordinary chemical concentrations. Since, as has already been pointed out, the solute has no intrinsic effect on osmotic pressure, the osmotic pressure of a solution is independent of the nature of the solute particles and depends only on their number per unit volume. In other words, in a given volume of solution, equal numbers of dissolved particles will contribute equally to osmotic pressure whether the particles are large molecules, small molecules, or ions. For this reason the osmotic activites of solutions containing equal *chemical* concentrations of different solutes will not necessarily be identical. Consider, for example, two solutions, one containing $0.1M$ sucrose and the other containing $0.1M$ NaCl. Although the concentrations of these two solutions are equal in chemical terms (moles per liter), their osmotic pressures are not the same because NaCl exists in solutions as Na^+ and Cl^- ions. Consequently the osmotic pressure of a $0.1M$ solution of NaCl is approximately twice that of a $0.1M$ sucrose solution. Thus it is apparent that, with electrolyte solutions, if one wishes to relate the osmotic pressure or effective osmotic concentration of solute to its chemical concentration, one must multiply the term C in equation 10 by a factor G (the osmotic coefficient), where G is the number of ions produced by one molecule of electrolyte.

The situation is further complicated by the fact that this simple relationship between the number of ions formed by an electrolyte and its osmotic activity holds only for very dilute solutions. Because of the attractive forces between ions of opposite charge and between individual ions and water molecules, the value of G for a given electrolyte varies with concentration. Also, the concentration dependence of G is different for different electrolytes. At physiological concentrations the divergence of G from its limiting value in dilute solution is sufficiently great to affect appreciably the accuracy of results calculated on the basis of equation 10. Further, the physiologist is frequently confronted with solutions, such as blood or urine, that contain complex mixtures of solutes, both electrolytes and nonelectrolytes, in widely different concentrations. In this situation the necessity for a practical unit of osmotic concentration which is independent of the variation of G with concentration for different individual solutes will readily be appreciated. The *osmole* (Osm) is such a unit. A solution having an osmotic pressure of 22.4 standard atmospheres is said to have an effective osmotic concentration of one osmole per liter or to be an *osmolar* solution, regardless of its chemical composition. For any individual substance the osmole is defined as the weight in grams which gives rise to an osmotic pressure of 22.4 standard atmospheres when dissolved in 1 liter of solution. For osmotic purposes the concentration of any given solution can be expressed directly in terms of *osmoles per liter* or *osmolarity*.* In physiological work the milli-osmole (mOsm) is usually employed as the unit of osmotic concentration (1 Osm = 1000 mOsm). In the case of dilute solutions one may write for nonelectrolytes (e.g., sucrose, urea), *milliosmoles = millimoles,* and for electrolytes, *milliosmoles = G × millimoles,* where G has its ideal value—i.e., the number of ions formed by each molecule of the electrolyte. For solutions containing a single solute, these "ideal" relationships may be used for approximate purposes at physiological concentrations.

The Determination of Osmotic Concentrations

In principle, the osmolarities of physiological solutions can be determined by directly measuring their osmotic pressure and converting it to osmoles or milliosmoles. Because of technical complications, however, this method is unsuitable for routine physiological or clinical investigations, in which rapid determination of the osmolarities of large numbers of samples is usually required. An alternative method of determining osmolarities which is at once simpler in practice, more rapid, and more accurate than all but the most elaborate instruments for the direct measurement of osmotic pressure is based on the relative freezing points of solutions.

Osmotic pressure is one of the so-called *colligative* properties of dilute solutions; i.e., its magnitude is related to the concentration (number per unit volume) of dissolved particles and is not affected (in very dilute solutions at least) by such factors as their size, shape, or chemical composition. Other colligative properties of dilute solutions are vapor pressure lowering, depression of the freezing point, and elevation of the boiling point. These four properties of solutions are very simply related so that if one of them is known for a given set of circumstances the others may readily be calculated.

In the case of freezing point depression it can be shown that, in very dilute solutions, the amount by which a nondissociating solute like sucrose lowers the freezing point of pure water is 1.86°C per mole. This factor, the *cryoscopic constant,* is the same for all nondissociating solutes providing they can be considered to behave ideally. Under similar conditions one would expect an electrolyte like NaCl, which gives rise to 2 ions per molecule, to lower the freezing point of water by 2 × 1.86 or 3.72°C per mole, and so on. In practice it is found that the factor 1.86°C per mole is subject to the same kind of concentration dependence as the osmotic coefficient G. Therefore, for practical purposes an osmolar (or osmolal) solution may be defined as one that lowers the freezing point of water by 1.86°C regardless of the precise conditions, or of its exact chemical composition. Since, as already pointed out (p. 15), such a solution has an osmotic pressure of 22.4 atmosphere (atm), the following relations between osmolarity, freezing point lowering, and osmotic pressure are at once apparent:

$$\text{milliosmoles} = \frac{\Delta T_f \times 1000}{1.86}$$

and

$$\text{osmotic pressure (atm)} = \frac{\Delta T_f \times 22.4}{1.86}$$

*From a physicochemical point of view, *osmolality* (number of osmoles per kilogram of water) is preferable to *osmolarity* (number of osmoles per liter of solution), the osmole being defined as the weight in grams of solute which, when dissolved in 1 kg water, yields a solution having an osmotic pressure of 22.4 standard atmospheres. However, in the present discussion, osmolarities rather than osmolalities will be employed for two reasons. First, at normal physiological concentrations, the error introduced by using osmolarities instead of osmolalities is sufficiently small to be neglected. Second, in dealing with physiological fluids such as blood plasma and urine it is often more practicable to refer one's results to a volume of solution rather than to an amount of water as a standard. It is assumed that the solutions in question are not subject to wide variations in temperature.

where ΔT_f is the difference between the freezing point of the solution and that of pure water (0°C).

Nowadays a number of instruments (osmometers) are available with which the freezing points of solutions can be rapidly and accurately measured and which give a direct readout in milliosmoles of the results obtained. Thus, freezing point lowering (cryoscopy) is at present the usual method of choice for the determination of osmolarities in physiological and clinical investigations.

Osmotic Behavior of Red Blood Cells: Osmotic Hemolysis

Red blood cells have been a frequent choice in osmotic and permeability studies for a number of reasons, including the following: the ready availability in quantity of these cells (1 mm³ of human blood contains about 5 million red cells); their ease of handling; and the fact that, since they are free cells whose whole surface is exposed to the external medium, the volume changes they undergo in response to changes in their osmotic environment and the rates of entry of various substances into them are easily determined. Although there are undoubtedly differences in detail in the osmotic behavior of different cells, the red cell may be considered a fairly typical representative, in this respect, of animal cells in general.

The total osmotic concentration of mammalian blood plasma is about 300 mOsm. Present estimates indicate that in most cases, including red cells, the intracellular osmolarity in vivo is essentially the same as that of plasma. Intracellular osmolarity is principally due to the presence inside the cell of K^+, Na^+, and Cl^- ions, together with smaller amounts of HCO_3^-, phosphates, amino acids, etc. In red cells as in many other cells the sum of Na^+ and K^+ ions greatly exceeds the total amount of small anions present. To preserve electroneutrality it is necessary to postulate the existence within the cell of large nondiffusible or "fixed" anions. These are partly the ionized anionic groups (e.g., $-COO^-$ groups) of cellular proteins and also include smaller anionic molecules (e.g., adenosine triphosphate: ATP) which are "nondiffusible" in the sense that they do not readily move across the cell membrane.

In addition, cell membranes are not semipermeable. Rather, they are selectively permeable or *permselective;* that is, they are impermeable to some solutes. Other solutes penetrate the membrane but do so more slowly than water. Other solutes (e.g., urea) penetrate as fast or almost as fast as water. Variability in the rate of solute penetration is reflected in the permeability coefficient of the membrane for various solutes (cf. equation 4) but can, for osmotic purposes, be conveniently expressed in terms of a related parameter, the *reflection coefficient* (σ). σ ranges from a value of unity for a completely impermeable solute to a value of zero for a solute whose permeability approximates that of water. The osmotic response of a cell to a given solution therefore depends not only on the total osmolality of the solution but also on the specific solute or solutes which it contains.

Osmotic Response of Red Blood Cells to Solutes for Which $\sigma \rightarrow 1$

We will consider a suspension of human red blood cells in a relatively large volume of a solute such as NaCl for which we can assume $\sigma \sim 1$.* Because the volume of the outside solution is very large compared to the total water content of the cells, any movement of water from the cells to the medium or vice versa will not appreciably affect the solute concentration in the latter.

If our solution contains 300 mOsm (approximately 0.15M) per liter of NaCl (i.e., if it has the same osmolarity as the cell interior or is *isosmotic* with the latter), the volume of the cells will not change since there is no osmotic gradient across the membrane. Such

---

*Although it is now known that Na^+ and K^+ ions can exchange fairly freely across the red cell membrane (p. 29), net transmembrane movements of salt are slow. For this reason NaCl may be regarded, for practical purposes, as a nonpenetrating solute in osmotic experiments of relatively short duration.

a solution is said to be *isotonic* with the cells. If the cells are suspended in a solution containing more than 300 mOsm NaCl per liter, they shrink because of osmotic loss of water. In such shrunken cells the surface is often roughened or *crenated,* and the solution that caused them to shrink is said to be *hypertonic.* Note that it is also *hyperosmotic* because it has a greater osmotic pressure than the cell contents. Conversely, if the solution bathing the cells contains less than 300 mOsm NaCl per liter (i.e., is hyposmotic with respect to the cell contents), the cells swell because of osmotic entry of water from the external medium. A solution that causes swelling of the cells is said to be *hypotonic.* Swelling in red blood cells is accompanied by a change in shape. Normally the cells are biconcave discs with an average major diameter of about 7.2 μm. When water enters them as the result of an osmotic gradient, the cells become progressively spherical. In this way they can increase their volume by a maximum of about 67 percent without any appreciable change in surface area. According to present ideas, the red blood cell membrane possesses very little tensile strength. Thus, if the cells cannot, by becoming spherical, neutralize the osmotic gradient imposed on them, they cannot further increase their volume without damage to the membrane. Entry of water beyond the point at which the cells become fully spherical results in damage to the membrane and escape of the characteristic red protein *hemoglobin* from the cell interior. This phenomenon is called *hemolysis.* The residual cells left following hemolysis are called *ghosts* or *stroma.* The cell volume beyond which hemolysis occurs is the *critical hemolytic volume.* Within the limits imposed by hemolysis, swelling and shrinking of red cells are reversible.

Fragility

The resistance of red blood cells to hypotonic hemolysis is an important diagnostic parameter in the clinic. For this purpose the term *fragility* is often employed. Fragility is determined as follows. Since all the cells in a suspension do not hemolyze simultaneously, one determines the external concentration of NaCl which induces a definite amount of hemolysis (say, 50 percent) in the blood sample under investigation. The reciprocal of this concentration is the fragility. In certain diseases (e.g., pernicious anemia) fragility is significantly less than normal. In others, such as hemolytic jaundice, fragility may be considerably greater than in normal cells.

Osmolarity and Tonicity—Osmotic Responses to Solutes for Which $\sigma < 1$

Any two solutions having the same osmotic pressure are *isosmotic.* Solutions which have a greater or smaller osmotic pressure than a given reference solution are *hyperosmotic* and *hyposmotic,* respectively, compared to the reference solution. Thus these terms refer only to the properties of solutions. Tonicity, on the other hand, is determined by the permeability properties of the cell membrane as well as by the osmolarity of the bathing medium. For nonpenetrating solutes ($\sigma \rightarrow 1$) this distinction is, admittedly, somewhat academic.

With penetrating solutes such as urea, the situation is quite different. A solution containing 300 mOsm urea (approximately 0.3M) is *isosmotic* with the fluid contents of the red cell. Nevertheless, if red cells are placed in that solution, they rapidly hemolyze because urea readily penetrates the cell membrane. Since the cell is virtually impermeable, over short periods of time, to the osmotically active substances it initially contains, the entry of urea cannot be balanced by an equivalent exit of osmotically active material and must therefore bring about an increased intracellular osmolarity. This sets up an inwardly directed osmotic gradient for water, as a result of which water flows into the cell. Urea is normally present in only very small amounts in red cells; therefore, unless the volume of external fluid is exceedingly small, the cells cannot compensate for the osmotic gradient resulting from the entry of urea and will hemolyze. Since water moves across the cell membrane very rapidly, the osmotic response to the entry of urea will be correspondingly rapid. There will be a virtually simultaneous movement of water and urea into the cells, and the time required for hemolysis to occur will depend on the rate at which urea moves across the membrane. An equivalent situation will arise if any other penetrating solute is substituted for urea. Thus, the time taken for hemolysis to occur in an isosmotic solution of a given solute is an index of the rapidity with which that solute penetrates the cell, and the rate of hemolysis has been widely used to determine relative rates of solute penetration.

The above considerations concerning hemolysis by a penetrating solute are not limited to the case in which the initial external solute concentration is osmotically equal to that of the cell contents but can be extended to other initial concentrations of the penetrating solute. Hence it can be concluded that a solution containing a penetrating solute only, or a mixture of penetrating solutes, whether or not it is isosmotic with the cell interior, cannot be isotonic. In fact, from the definition of tonicity given above, *all* solutions of penetrating solutes must be regarded as *hypotonic,* whatever their actual osmolarity may be.

The mechanism of osmotic water flow across biological membranes is complex. It appears to depend largely upon the structure of the membrane and specifically on the size of its aqueous pores or channels. If these are relatively large (as, for example, in the endothelial lining of the blood capillaries), osmosis occurs primarily by *bulk flow* of water. Bulk flow is an ordered or laminar streaming such as occurs when water is allowed to flow slowly through a pipe under a hydrostatic pressure head. On the other hand, if the pores in the membrane are small (their diameter is not much greater than that of a water molecule), only diffusion can take place through them. The apparent pore radii of a number of animal cells range from about 0.9 to 7.4 Å (Dick, 1966). The water molecule has a diameter of about 1.5 Å. For a cell such as the human red blood cell for which the apparent pore radius is about 4.0 Å, one would predict that bulk flow and diffusion would each account for roughly half of the total osmotic water flow across the membrane. This supposition has been experimentally confirmed (Dick, 1966).

Nonosmotic Hemolysis

In addition to hemolysis resulting from an osmotic gradient across the membrane or from the presence of a penetrating solute in the external solution (osmotic or hypotonic hemolysis), hemolytic injury to red cells may stem from a number of causes, among them excessive heat or mechanical agitation and a variety of "hemolytic agents"—that is, substances which impair or destroy the selective permeability of the membrane by specific interaction with one or more of its components. Hemolytic agents of this kind include fat solvents such as chloroform and ether, surface-active agents (detergents, bile salts, saponin, etc.), heavy-metal ions, and certain snake venoms.

The Transport of Ions

The Energy Gradient for Ion Transport: The Resting Potential

If a microelectrode (tip diameter 1 μm or less) is inserted into a cell (Fig. 1-6), a *potential difference* is observed between it and a reference electrode immersed in the external medium. This potential difference, which in resting cells is usually such that the intracellular electrode is *negative* compared to the external reference electrode, is the *resting potential* (E_R) or *membrane potential* of the cell. By convention, the potential of the external electrode is taken as zero at all times. Typical values for the resting potential are about 90 mv for skeletal muscle and about 70 mv for nerve.

The existence of a potential difference across a membrane implies that a charged particle such as an ion moving through the membrane will be subjected to an electrical force which may, according to the direction of movement and sign of charge of the ion, either assist or hinder its passage. In the case of the cell membrane, the potential difference is oriented in such a way as to assist the movement of cations into the cell and oppose their outward movement from the cell interior. The opposite is true for anions. Thus, it is clear that the net energy gradient for ions across the cell membrane cannot be

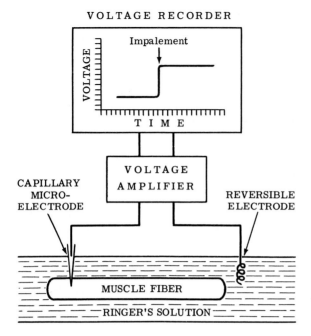

Figure 1-6
Diagram of the measurement of a resting membrane potential. An open-tip capillary micro-electrode filled with KCl solution is inserted into the cell, and the potential difference between it and a reversible half-cell (e.g., calomel or silver–silver chloride) immersed in the bathing medium is recorded. At the moment of impalement there is a sharp deflection in the voltage trace.

specified in terms of concentration differences alone but must also take into account the effect of electrical forces.

The Origin of Membrane Potentials: Resting Potentials of Muscle and Nerve

Living cells are characterized by an unequal distribution of ions between the cell interior and the external environment. The most important quantitative inequalities relate to Na^+, K^+, and Cl^-. The interior of a typical mammalian cell contains about 20 to 30 times as much K^+ as Na^+. In blood plasma or interstitial fluid this ratio is reversed. Chloride is also unequally distributed between the cell and its environment, being present in lower concentration in the cell interior than in the external fluid (Table 1-1).

This asymmetry in ionic distribution raises two interrelated questions which are of fundamental importance in cell physiology: (1) To what extent can the resting potential of cells be accounted for by the unequal distribution of Na^+, K^+, and Cl^- across the membrane? (2) What in turn is the contribution of the resting potential to the maintenance of this asymmetry in ionic distribution? In discussions of the origin of cell membrane potentials, two principles must always be borne in mind. First, the development of an electrical potential in any system depends on separation of charge within that system. That is to say, somewhere within the system there must be an excess of positive charges and, consequently, somewhere else there must be an excess of negative charges. These regions must be spatially separate, and further, if a potential difference between them is to be maintained for any length of time, there must be a restraint on the flow of electrical charge; i.e., current should not flow too freely from the positive to the negative region. Otherwise the potential difference between them will be quickly dissipated. In the case of aqueous solutions, ions are the charged entities in question. A potential difference in an aqueous system therefore implies the existence of a region containing an excess of cations and another region containing an excess of anions.

On the other hand, it is a necessary consequence of the laws of thermodynamics that in any *finite* volume of an electrolyte solution, whatever inequalities may exist in the concentrations of individual ions, the total numbers of positive and negative charges must be equal. This is the principle of *electroneutrality*. If cytoplasm and extracellular fluid are regarded as two essentially electroneutral solutions separated by a membrane, the question of the origin of membrane potentials can be stated very simply as follows: How can charge separation be achieved in such a system without violating the condition of electroneutrality?

Granted an asymmetrical distribution of individual ionic species across a membrane, and regardless of how this asymmetry came about, two kinds of electrical potentials may be expected: *equilibrium potentials* and *diffusion potentials*. Both have been used as a basis for interpreting cell membrane potentials, but diffusion potentials offer a more satisfactory and complete explanation of the observed behavior of cellular systems.

In the simplest case, an equilibrium potential arises when a membrane is permeable to one ionic species only. For example, if a cell membrane is permeable to K^+ only and there is an unequal distribution of K^+ ions across it, the following situation may be visualized. Because of their unequal distribution across the membrane, K^+ ions will tend to diffuse from the inside of the cell, where their concentration is high, to the outside. This diffusion of K^+, unaccompanied by any compensatory movement of anions, would, if unchecked, result in a net transfer of positive charge to the outside of the cell in violation of the electroneutrality principle. Therefore, the tendency for K^+ to diffuse across the membrane is immediately countered by the development of a potential difference of a size and an orientation sufficient to prevent any further movement of K^+. The end result, in the system under discussion, will be a slight charge separation, in the *immediate region of the membrane* only. This will involve no more than a minute quantity of K^+ ions, will cause the outside of the membrane to become positively charged with respect to the inside, but will not significantly affect the overall electroneutrality of the solutions on each side of it. The magnitude of the membrane potential generated in this situation is given by the Nernst equation

$$E_M = E_K = \frac{2.303 \, RT}{zF} \log \frac{[K_i^+]}{[K_o^+]} \qquad (11)$$

where R is the gas constant, T the *absolute* temperature, F the Faraday (96,500 coulombs), and $[K_i^+]$ and $[K_o^+]$ denote the K^+ concentrations (strictly speaking the *activities*, but in most physiological investigations concentrations may be substituted without undue error) inside and outside the cell respectively; z is the valence of the ionic species involved, and for K^+ and other *monovalent* ions, $z = 1$.

This equation leads to certain theoretical predictions about the behavior of the membrane potential which can be tested by actual measurements on the systems involved. First, according to equation 11, it is obvious that at any given temperature E_M will depend only on the ratio $[K_i^+]/[K_o^+]$. Further, equation 11 predicts that the relationship between E_M, expressed in *millivolts* (mv), and $\log [K_i^+]/[K_o^+]$ should be a straight line with a slope (2.303 RT/F) of 58, at 18°C, or 61, at 37°C, for a 10-fold change in the ratio $[K_i^+]/[K_o^+]$.* In other words, equation 11 states that a membrane which obeys it behaves as an ideal K^+ electrode. In muscle and nerve there is evidence that the fiber membrane, under certain conditions, approximates closely the behavior of

---

*In this chapter the experimentally observed potential difference across a cell membrane will be called the resting potential (E_R). The theoretical membrane potential calculated from a specific equation or set of equations will be denoted by E_M. E_K, E_{Na}, etc., will be used to indicate the equilibrium (Nernst) potentials of individual ions. For simplicity a value of 60 is assigned to the factor 2.303 RT/F in subsequent equations of the Nernst and similar types.

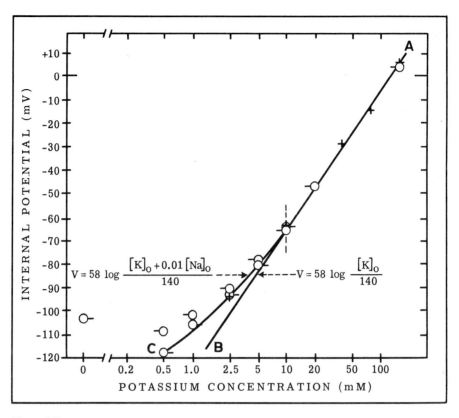

Figure 1-7
Effect of external potassium on the membrane potential of a single fiber taken from the frog sartorius muscle. (From Hodgkin and Horowicz, 1959.)

a K^+ electrode. Figure 1-7, taken from an investigation of the resting potential in frog sartorius muscle (Hodgkin and Horowicz, 1959), illustrates the point. In the experiment shown in this figure $[K_o^+]$ was altered under conditions in which $[K_i^+]$ remained virtually constant (140mM). In this situation E_M, according to equation 11, should be a linear function of log $[K_o^+]$. The straight line AB in Figure 1-7 is the theoretical plot obtained using equation 11. The circles are the experimental values of E_R under the same conditions. It is clear from Figure 1-7 that when the external K^+ concentration is greater than 10mM there is excellent agreement between the observed values of E_R and the calculated values of E_M; thus, when the external K^+ concentration is relatively high, the resting potential of frog muscle is satisfactorily accounted for by equation 11. When the external K^+ concentration is less than 10mM, the observed potential deviates rather markedly from the values predicted by equation 11. This is discussed later (p. 24).

Apart from the fact that the observed behavior of the resting potential in skeletal muscle deviates noticeably from the predictions of equation 11 for $[K_o^+]$ values below 10mM, which is, of course, the situation of greatest relevance to normal physiological conditions (an essentially similar type of behavior is seen with nerve fibers), attempts to explain E_R as a Nernst equilibrium potential encounter other serious difficulties. In addition to the restriction that the membrane should be permeable to just one ionic species, the Nernst equation is valid only when there is no flow of current across the membrane. At present, evidence against both of these assumptions is overwhelming.

Following the introduction of radioactive tracers into biological research, it soon became clear that cell membranes are permeable to Na^+ and Cl^-, though, in general, their permeabilities to these ions are considerably less than their permeabilities to K^+. Also, in many instances current flow across cell membranes can be demonstrated. These and other findings have led to an alternative formulation of the membrane potential in terms of diffusion potentials.

Diffusion potentials arise where a boundary such as a membrane separates two solutions containing different concentrations (activities) of individual ionic species and where the individual ionic permeabilities of the membrane also differ. A simple system illustrating the development of a diffusion potential is illustrated in Figure 1-8A. In this illustration, a membrane permeable to both K^+ and Cl^- separates a KCl solution from water. Because of the concentration gradient, both K^+ and Cl^- diffuse across the membrane. Since the membrane is assumed to be more permeable to K^+ than to Cl^- (as indicated by the lengths of the arrows labeled P_{K^+} and P_{Cl^-} respectively), K^+ initially diffuses more rapidly than Cl^- from side 1 to side 2. Thus a potential difference is quickly generated across the membrane such that side 2 is positive with respect to side 1. This potential has a retarding effect on K^+ ions and an accelerating effect on Cl^- ions, and, once a very small amount of K^+ has leaked across the membrane, the potential stabilizes at a value which permits K^+ and Cl^- ions to diffuse from side 1 to side 2 at the same rate.

Several important differences exist between diffusion potentials and equilibrium potentials. Diffusion potentials permit a flow of current across the membrane. Also, in diffusion systems such as that illustrated in Figure 1-8A, unless the volume of solution on each side of the membrane is infinite, the membrane potential will dissipate with time as the salt concentration on side 2 approaches that on side 1. By contrast, an equilibrium potential does not decay with time and, in fact, represents an extreme case of a diffusion potential in which all permeability factors except one are zero. Thus in fluid systems of finite volume such as the cells of the body together with the extracellular and vascular fluid spaces, the maintenance of steady-state diffusion potentials requires the operation of additional mechanisms to conserve the ionic concentration differences which give rise to these potentials. The mechanisms are discussed below (pp. 23–31).

To return to a consideration of the resting potential as a diffusion potential: It is evident that, since cell membranes show a wide spectrum of permeabilities to different ions, and since both cytoplasm and extracellular fluid contain a large number of diffusible ionic species, all of these could, in principle, contribute to the potential difference observed across the membrane. Fortunately, there is ample evidence to show that, quantitatively, K^+, Na^+, and Cl^- are the only ions that need to be considered. On this basis, and with the assumption that the potential gradient across the membrane is

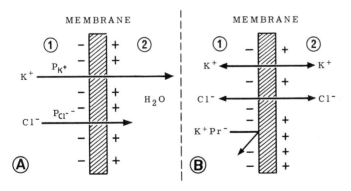

Figure 1-8
A. Origin of a diffusion potential. B. Diagram of a Gibbs-Donnan equilibrium.

linear (the *constant field* assumption), the following equation for the membrane potential can be derived (Goldman, 1943; Hodgkin and Katz, 1949):

$$E_M = 60 \log \frac{P_K [K_i^+] + P_{Na} [Na_i^+] + P_{Cl} [Cl_o^-]}{P_K [K_o^+] + P_{Na} [Na_o^+] + P_{Cl} [Cl_i^-]} \qquad (12)$$

In this equation, usually called the *Goldman* or *constant field* equation, the expressions in brackets denote concentrations as before (equation 11), and P_K, P_{Na}, and P_{Cl} are the permeability coefficients for K^+, Na^+, and Cl^-. According to equation 12, the contribution of each ion to the overall membrane potential is determined by its concentration ratio across the membrane and by the magnitude of its permeability coefficient.

Equation 12 has proved to be exceptionally useful in studying the relationship of the resting potential to transmembrane ionic concentration differences and in determining the relative permeabilities of K^+, Na^+, and Cl^- in many cell types. As an example of its use, let us return to the data shown in Figure 1-7. In this experiment Cl^- was replaced by sulfate (a virtually impermeant anion) so that only K^+ and Na^+ need be considered. As already mentioned, when $[K_o^+]$ was greater than 10mM, the resting potential followed the predictions of equation 11 for a potassium equilibrium potential. In terms of equation 12 this is explained by postulating that in skeletal muscle $P_K \gg P_{Na}$ so that when $[K_o^+] > 10$mM the contribution of Na^+ to the resting potential becomes negligible, and equation 12 reduces to equation 11. Thus, for these conditions, there appears to be little to choose between the equilibrium approach and the diffusion approach (in fact, the former seems to be a great deal simpler!). However, equation 11 offers no explanation for the fact that when $[K_o^+]$ is reduced below 10mM the resting potential ceases to be a linear function of log $[K_i^+]/[K_o^+]$ and becomes progressively less than the value predicted by this equation. In terms of the Goldman equation this discrepancy can be explained simply as being due to an increasing effect of Na^+ on the resting potential as $[K_o^+]$ is progressively lowered. As already mentioned, Na^+ is usually present in much lower concentrations inside the cell than in the external medium. Since the membrane is permeable to Na^+, one would expect a continuous net diffusion of Na^+ ions into the cell (which must be continuously removed from it under steady-state conditions—p. 28). This inward "leak" of Na^+ would, on the basis of equation 12, be expected to affect the resting potential. For the conditions shown in Figure 1-7 one would, in fact, predict that the membrane potential would obey an equation of the type

$$E_M = 60 \log \frac{[K_i^+] + \alpha [Na_i^+]}{[K_o^+] + \alpha [Na_o^+]} \qquad (13)$$

where $\alpha = P_{Na}/P_K$. The curve AC in Figure 1-7 is drawn to fit such an equation, α being taken as 0.01—that is, assuming a K^+ permeability 100 times greater than the Na^+ permeability. It is clear that, when $[K_o^+]$ is less than 10mM, this curve fits the observed values much more closely than the line AB, which is drawn on the basis of equation 11.

Thus, equation 12 can be used to determine the relative permeabilities of the membrane to any two ions providing the conditions are such that E_R depends essentially on these two ions only. Similar methods to those illustrated by Figure 1-7 have been used to determine P_{Na}/P_{Cl} in skeletal muscle (Hodgkin and Horowicz, 1959) and hence the ratio $P_K : P_{Na} : P_{Cl}$ in this and other tissues. Finally, it should be reemphasized that in any situation in which the permeability of one ionic species greatly exceeds that of all others, equation 12 reduces to the appropriate form of the Nernst equation. In other words, under these conditions, one would expect E_R to approximate closely the equilibrium potential of the highly permeant ion. The application of the Nernst equation to the resting potentials of cells is therefore frequently justified in practice even though one is not dealing with an equilibrium situation of the kind for which it is theoretically valid.

The Regulation of Intracellular Ionic Composition
We have seen that the analysis of resting potentials in terms of ionic diffusion potentials gives a reasonably satisfactory answer to the first of the two questions posed on page 20; i.e., one can say, on this basis, that the resting potentials of living cells can in large measure be accounted for by the asymmetry in ionic distribution between the cytoplasm and the external medium. The second question, the role of the resting potential in the maintenance of this asymmetry, will now be examined in the broader context of the overall regulation of the electrolyte content of cells.

As already pointed out, the maintenance of steady-state diffusion potentials across a membrane requires that mechanisms be brought into play which prevent the dissipation of the ionic gradients giving rise to these potentials. In the case of living cells, the question is: To what extent can the gradients be maintained by purely physicochemical or "passive" forces and to what extent must active, metabolism-linked transport processes be invoked to explain the normal steady-state distribution or "intracellular homeostasis" with respect to inorganic ions?

The Distribution of K^+ and Cl^-: The Gibbs-Donnan Equilibrium
A situation in which an unequal distribution of ions across a membrane can arise as a result of purely physicochemical forces and which is of particular relevance to the behavior of cell membranes as well as to a number of other physiological functions (Chaps. 4A, 22, 23, and 24) is that shown diagrammatically in Figure 1-8B. Two solutions, one containing KCl only and the other containing KCl together with KPr (the potassium salt of a large anion Pr^-), are separated by a membrane which is permeable to water, K^+, and Cl^-, but not to Pr^-. If the membrane were permeable to Pr^- as well as to K^+ and Cl^-, all three ions would diffuse freely until they were equally distributed throughout the whole solution on both sides of the membrane. By making the membrane impermeable to one ionic species (Pr^-) one imposes a restriction on free diffusion which gives rise to a number of important consequences. Of particular interest in the present context is the fact that the impermeability of the membrane to Pr^- has a marked effect on the equilibrium distribution of the diffusible ions K^+ and Cl^- across the membrane. This distribution is given by the Gibbs-Donnan relationship, which states that *at equilibrium the products of the concentrations of the diffusible ions on each side of the membrane are equal.* For the conditions illustrated in Figure 1-8B, this may be written:

$$[K_1^+] \times [Cl_1^-] = [K_2^+] \times [Cl_2^-] \tag{14}$$

or

$$[K_1^+] / [K_2^+] = [Cl_2^-] / [Cl_1^-] \tag{14a}$$

Certain consequences of this relationship will be immediately apparent. Clearly, since the solution on each side of the membrane must be, as a whole, electrically neutral, $[K_2^+]$ must be equal to $[Cl_2^-]$. Also, since, in the solution containing the nondiffusible ion Pr^-, the condition of overall electroneutrality requires that there be sufficient K^+ ions present to balance the negative charges on the Pr^- ions as well as those on the Cl^- ions, $[K_1^+]$ must be greater than $[Cl_1^-]$. Again, equation 14 requires that the sum of $[K_1^+]$ and $[Cl_1^-]$ be greater than the sum of $[K_2^+]$ and $[Cl_2^-]$.* Thus for the situation shown in Figure 1-8B, $[K_1^+] > [K_2^+]$ and $[Cl_1^-] < [Cl_2^-]$.

---

*This follows from the fact that if $ab = c^2$, where a and b are unequal quantities, then $(a + b) > 2c$.

A well-known geometrical illustration of this relationship is the fact that the sum of any two adjacent sides of a rectangle is greater than the sum of any two sides of a square of equal area (Davson, 1970).

These considerations lead to the important generalization that in a Gibbs-Donnan system at equilibrium there is a concentration gradient across the membrane for each species of diffusible ion present. This gradient (the Gibbs-Donnan ratio) is identical in magnitude for each diffusible ionic species but is opposite in direction for cations and anions. Thus, the equilibrium condition cannot be specified in terms of chemical concentrations alone since the individual concentration gradients would, if unopposed, result in an equal distribution of diffusible ions across the membrane. In the Gibbs-Donnan situation the concentration gradients for diffusible ions which exist at equilibrium must therefore be neutralized by an equal and opposite driving force. This force is an equilibrium potential, generated by the unequal distribution of diffusible ions across the membrane and, for the system illustrated in Figure 1-8B, is given by the Nernst equation in the following form:

$$E_M = 60 \log \frac{[K_1^+]}{[K_2^+]} = 60 \log \frac{[Cl_2^-]}{[Cl_1^-]} \tag{15}$$

Note that the side of the membrane in contact with the solution containing the nondiffusible anion is negative with respect to the opposite side.

The forces giving rise to this potential are similar to those already discussed in the case of potassium equilibrium potentials (p. 21). The main point to note is that the Gibbs-Donnan equilibrium represents a state of balance between two opposing tendencies: the tendency for each diffusible ion to achieve an equal chemical concentration throughout the system, and the tendency for the system as a whole to remain electrically neutral.* This has major implications for the distribution of K^+ and Cl^- ions between many types of cell and the external environment.

Often the observed ratios of intracellular to extracellular concentrations of K^+ and Cl^- are close to those predicted by the Gibbs-Donnan relationship. This observation, together with the fact that at the intracellular pH values which prevail in vivo the cell interior is known to contain relatively large quantities of nondiffusible anions, raises the question of the extent to which the Gibbs-Donnan equilibrium can account for the distribution of K^+ and Cl^- across the cell membrane without the intervention of other forces. The matter will be examined in detail for two types of cell, skeletal muscle fibers and red blood cells. These may be considered to represent two extreme situations with respect to the maintenance of intracellular K^+ concentrations. Other cells of the body show intermediate types of behavior. Some, like cardiac muscle and nerve fibers, approximate skeletal muscle in behavior. Others (e.g., smooth muscle, epithelia, liver cells) diverge rather widely from the pattern seen in skeletal muscle.

The essence of the situation is as follows: The chemical force tending to cause net movement of an ion into or out of a cell is its intracellular/extracellular concentration ratio. It is frequently possible, by direct chemical analysis, to obtain fairly accurate estimates of this ratio for Na^+, K^+, and Cl^-. The electrical force tending to drive ions across the membrane in one direction or the other is the resting potential. This too can often be measured accurately. The two forces can be in the same or in opposite directions, but in either case their *algebraic* sum (taking into account their orientation with respect to the membrane) is the net force or *electrochemical potential gradient* which tends to displace the distribution of a given ion from the steady-state condition

---

*It is important to realize that the term *equilibrium* in a Gibbs-Donnan system applies only to the diffusible ions present and not to the system as a whole. Obviously, in the case shown in Figure 1-8B, there cannot be an equilibrium in the case of Pr⁻, which is confined to one side of the membrane. Also, the presence of Pr⁻ on one side only of the membrane, together with the presence on the same side of an *excess* of total diffusible ions, gives rise to an osmotic gradient which, unless it is opposed, will cause water to move into the side containing the nondiffusible ion. This osmotic pressure difference is called the *colloid osmotic pressure* and is of importance in such physiological functions as transcapillary fluid movement (Chap. 12).

Table 1-1. Ionic Concentrations, Measured Resting Potentials, and Calculated Equilibrium Potentials in Frog Skeletal Muscle and Human Red Blood Cells

	Na^+	K^+	Cl^-	E_R	E_{Na}	E_K	E_{Cl}
Frog Skeletal Muscle							
Fibers	13	140	~3	−90	+56	−105	−86
Plasma	110	2.5	90				
Human Red Blood Cells							
Cells	19	136	78	−(7 to 14)	+55	−86	−9
Plasma	155	5	112				

Note: Potentials in millivolts (minus sign indicates inside negative). Concentrations for cells and fibers in milliequivalents per liter cell water. Plasma concentrations in milliequivalents per liter.

and which must be overcome if a steady state is to be conserved. The analysis of ionic distribution thus resolves itself into the determination of the magnitude and orientation of the electrochemical gradients involved and the identification and measurement of the restoring forces which maintain the steady-state distribution of ions across the cell membrane.

The electrochemical potential gradient for a given ion can be determined as follows: By inserting its observed intracellular/extracellular concentration ratio into the Nernst equation one can calculate its equilibrium potential. This is the electrical potential required to balance exactly the driving force arising from its concentration gradient across the membrane. The electrochemical potential gradient is the difference between this equilibrium potential and the measured resting potential.* Obviously, if E_R is equal in size and opposite in direction to the calculated equilibrium potential for any ion, that ion is in electrochemical equilibrium across the membrane and there is no net force acting upon it.

In Table 1-1 this type of analysis is applied to frog skeletal muscle and human red blood cells. The Na^+, K^+, and Cl^- concentrations listed are average figures from the published literature (Harris, 1960). The data shown for skeletal muscle make it clear that, within the limits of experimental error, $E_R = E_{Cl}$. Therefore one may conclude that in this tissue Cl^- ions are in electrochemical equilibrium ("passively" distributed) across the fiber membrane and that no direct energy expenditure is required to maintain a steady state with respect to Cl^- in the living muscle fiber.† In other words, the inwardly directed concentration gradient for Cl^- is balanced by an equal and opposite outwardly directed electrical gradient.

With regard to potassium, it is seen that in skeletal muscle fibers, under normal physiological conditions, E_K is significantly greater than E_R, though both are oriented in the same direction. This means that an E_R value some 10 to 15 mv greater than that observed would be required to balance the outwardly directed concentration gradient for this ion. In other words, K^+ is present within the fibers in excess of the amount corresponding to electrochemical equilibrium, although the excess is relatively small. Because of this net outwardly directed electrochemical potential gradient, K^+ must leak continuously out of the fibers and the loss must continuously be made good. Thus, there is a necessity in skeletal muscle for an inwardly directed energy-requiring potassium pump which can transport K^+ against an electrochemical gradient. The nature of this pump will be discussed later in connection with Na^+ transport (p. 28).

* In effect, this procedure is equivalent to converting the observed chemical (concentration) driving force into an equivalent electrical force, which can then be compared with the measured electrical force across the membrane.
† Inasmuch as metabolism is necessary to the life of cells, there is, of course, an *indirect* dependence of all membrane properties on metabolism.

As for the distribution of K^+ and Cl^- between red blood cells and plasma, it is seen from Table 1-1 that the situation is qualitatively similar to but quantitatively different from that in muscle. In recent years several workers have succeeded in penetrating red blood cells with microelectrodes and have recorded potentials in the range 7 to 14 mv (inside negative). Once again, these figures are in essential agreement with the calculated value of E_{Cl} (Table 1-1), indicating an equilibrium distribution of Cl^- across the red blood cell membrane.

By contrast, the calculated value of E_K (Table 1-1) shows that the concentration of K^+ ions inside the red blood cell is far in excess of the amount required for electrochemical equilibrium. An inwardly directed active potassium pump must therefore be a major component of the potassium-transporting machinery in these cells. Such a pump, showing saturation kinetics and depending on a continuous direct supply of metabolic energy, has, in fact, been shown to exist and has been studied in considerable detail (Hoffman, 1966).

The Distribution of Sodium: The Sodium Pump
Inspection of Table 1-1 shows clearly that the situation with respect to intracellular Na^+ is not only quantitatively but qualitatively different from that of K^+ and Cl^-. In both muscle and red blood cells, E_{Na} is different from E_R both in magnitude and in orientation; i.e., E_{Na} requires that the inside surface of the membrane be *positive* with respect to the outside whereas in reality the opposite is true.* That is, the intracellular Na^+ concentration in muscle and red blood cells is much *lower* than that required for electrochemical equilibrium, and both the chemical and electrical gradients tend to "push" Na^+ into these cells. Further, the same appears to be true for virtually all cells, not only those of man and other animals, but also higher plant cells and microorganisms. Since cells are permeable to Na^+ and since many cell species exist in an environment which is rich in Na^+ ions, there is an almost universal requirement for a mechanism which removes Na^+ from the cell interior as fast as it enters. Removal of Na^+ is usually accomplished against an electrochemical potential gradient. Consequently, the mechanism involved must be an active process. It is generally referred to as the *sodium pump.*

The precise nature of the sodium pump in a number of cell species has been the subject of much discussion. In general, the movement of Na^+ into or out of cells is found to be associated with a converse movement of K^+. For example, when the metabolism of isolated muscles or red cells is blocked by low temperature or metabolic inhibitors, the cells accumulate Na^+ and lose K^+ in approximately equivalent amounts. When normal metabolic activity is restored, these changes are reversed, Na^+ ions being extruded and K^+ ions being taken up by the cells. Since the active nature of the Na^+ extrusion process is clearly established on energetic grounds alone for virtually all animal cells, much interest has centered upon the nature of the associated inward movement of K^+.

There is little doubt that the underlying reason for the occurrence of Na^+-K^+ exchange in association with the operation of the Na^+ pump is the necessity for maintaining overall electroneutrality in the cell and its surroundings. The movement, in any finite quantity, of a charged species such as Na^+ across the membrane would, if unaccompanied by compensatory movements of other charged particles, quickly generate a potential gradient which would effectively prevent further movement of the charged species. In the case of the Na^+ pump, electroneutrality is very simply achieved by a counterflow of K^+ ions. Thus each Na^+ ion leaving the cell is balanced by a K^+ ion entering it. The exchange is effected without altering the overall osmotic equilibrium of the cell and also serves to replace K^+ lost by outward diffusion.

Two mechanisms have been proposed by which K^+ ions could move into the cell in

---

*This "inversion" of sign of E_{Na} is of fundamental importance in relation to the phenomenon of "overshoot" in the action potentials of muscle and nerve (Chaps. 2, 3B, and 14).

exchange for Na^+ ions. The first postulates the existence of a *linked* or *coupled pump* in the cell membrane; that is, for each Na^+ ion pumped out of the cell a K^+ ion is pumped in. It has been suggested that this exchange pump operates by way of an asymmetrical carrier such as that shown in Figure 1-5B. In this case the X form of the carrier molecule as illustrated in the figure is supposed to have a high affinity for K^+ and a low affinity for Na^+. In the Y form of the carrier the relative affinities for K^+ and Na^+ are reversed.* It should be noted that a coupled pump of this kind does not give rise to a potential difference across the membrane and thus would not be expected to influence E_R. The reason is that in such a system there is a close correspondence between the rate of outward movement of Na^+ and the rate of inward K^+ movement. Pumps of this type are said to be *nonelectrogenic* or *electroneutral*.

A second mechanism proposed for Na^+-K^+ exchange is one in which the outward movement of Na^+ is the only active process. According to this concept K^+, on account of its much greater permeability, is taken up in preference to Na^+ by simple diffusion from the external medium to balance the excess negative charges inside the cell that would otherwise result from the outward movement of Na^+. Since there is no reason to suppose that the rate of outward Na^+ movement will correspond exactly to the rate of inward diffusion of K^+, such a mechanism can generate a diffusion potential which can appear as a component of the resting potential. Mechanisms of this kind, therefore, are frequently called *electrogenic* pumps.

In red blood cells convincing evidence for the existence of a coupled Na^+-K^+ pump has been obtained (Hoffman, 1966; Glynn, 1968). Part of this evidence may be summarized as follows: In the presence of glucose, K^+ influx in the red blood cell shows two components, a component which follows diffusion kinetics and a saturable component. In the absence of glucose or in the presence of glycolytic inhibitors, only the diffusion component of K^+ influx is observed. The saturable component of K^+ influx has a K_s with respect to $[K_o^+]$ of $2mM$.

Na^+ efflux likewise shows a diffusion component and a metabolism-dependent component which is saturable with respect to $[K_o^+]$; that is, as $[K_o^+]$ is increased, Na^+ efflux reaches a maximal value. The K_s in terms of $[K_o^+]$ for the saturable component of Na^+ efflux is virtually the same as the K_s for K^+ influx ($2mM$). These facts point to a very tight obligatory coupling of Na^+ efflux to K^+ influx in the red cells, although recent evidence indicates that the coupling ratio is not one to one but is probably 3 Na^+ out to 2 K^+ in (Glynn, 1968).

In nerve and muscle the active extrusion of Na^+ alone could, in principle, account for the maintenance of normal steady-state Na^+ and K^+ concentrations within the cell. However, because certain metabolic poisons (e.g., cyanide, azide, and dinitrophenol) markedly inhibit K^+ influx in nerve under conditions in which Na^+ efflux is not much affected, and because Na^+ efflux in muscle and nerve has been shown to depend strongly on external K^+, it has been suggested that part at least of the influx of K^+ in these tissues is an active process linked to Na^+ efflux. Studies with denervated skeletal muscles from the frog and the rat, previously allowed to accumulate Na^+ by immersion in a cold K^+-free solution, indicate that, during the early stages of extrusion of this accumulated Na^+, E_R is considerably higher than E_K. This finding suggests a relatively large contribution from an electrogenic Na^+ pump to the overall resting potential under these conditions. In muscles with an intact nerve supply, evidence for a more tightly coupled Na^+-K^+ exchange has been obtained (Dockry et al., 1966).

*In recent years this interpretation of Na^+-K^+ exchange across the cell membrane has received dramatic and unexpected support from the discovery of a series of naturally occurring antibiotics which have remarkable effects on ion transport in isolated mitochondria and which can facilitate the transfer of ions across artificial lipid bilayers. In the latter systems they have been shown to function as cyclic "carriers." The possibility that these substances can be regarded as artificial models of naturally occurring membrane carriers has opened up new and exciting perspectives in the molecular biology of membrane transport. A discussion of this topic is outside the scope of the present chapter. The reader is referred to Pressman and Haynes (1969) for a fuller account.

The Sodium Pump and Metabolism: The Role of Adenosine Triphosphate (ATP)
The dependence of the Na^+ pump on a supply of metabolic energy has been amply demonstrated in numerous investigations. In all cases studied it has been found that interference with the normal metabolic activity of the cell, whether this proceeds by way of oxidative pathways or through glycolysis, inhibits the active transport of ions. The nature of the precise coupling of metabolic energy to the task of driving Na^+, and in some cases K^+, ions across the cell membrane in opposition to an electrochemical gradient is, as yet, an unsolved problem in membrane physiology. However, recent studies have indicated that the phosphate bond energy of ATP, which is known to be the immediate energy source for the majority of cellular functions, plays a vital role in the active transport of Na^+ by several kinds of cell.

Na^+ efflux and K^+ influx in squid nerve are inhibited when cyanide or dinitrophenol is incorporated in the bathing solution. Removal of the inhibitors from the bathing solution results in restoration of the normal fluxes. The changes in Na^+ efflux are paralleled by changes in the ATP and arginine phosphate content of the nerves. In the presence of the inhibitors these compounds disappear from the nerve fibers. When the inhibitors are removed, ATP and arginine phosphate are resynthesized. Further, it has been shown that Na^+ efflux can be restored to normal by microinjection of ATP or arginine phosphate into fibers poisoned with cyanide or dinitrophenol.

Red blood cell ghosts prepared by careful hypotonic hemolysis (p. 18) can be "reconstituted" by immediate reimmersion in isotonic saline solution. The membranes of the reconstituted cells recover their normal permeability properties for some hours following reconstitution. During reconstitution certain substances, including ATP, to which the membrane is normally impermeable can be incorporated in the cells. Cells to which ATP has been administered in this way show pump activity for Na^+ and K^+. In the absence of ATP the reconstituted red blood cell ghosts do not pump these ions.

The discovery of a Na^+ and K^+ specific ATPase (which splits ATP to form adenosine diphosphate and inorganic phosphate and uses the magnesium salt of ATP as a substrate) in the membranes of a number of cell species has shed further light on the stimulation of active ion transport by ATP (Hoffman, 1966; Glynn, 1968; Skou, 1974). This enzyme, which was first discovered in the membrane of crab nerve but which has now been detected in the membranes of many kinds of cell, including red blood cells, shows some remarkable analogies with the Na^+ and K^+ transporting system in these cells. The most extensive studies to date in this area have been made with red blood cells, and it seems appropriate to conclude this section with a brief description of some of the more striking results obtained (Hoffman, 1966; Glynn, 1968).

First, the enzyme found in the membrane fraction of red blood cells behaves in many ways like the Na^+-K^+ exchange pump in these cells. Both the enzyme and the pump require Na^+ and K^+ together for operation, neither ion alone being effective. NH_4^+ ions can substitute for K^+ but not for Na^+ in both systems. Both systems are inhibited by the cardiac glycoside ouabain. Finally, the concentrations at which Na^+, K^+, NH_4^+, and ouabain exert their half-maximal effects are the same for both the pump and the enzyme.

Second, two kinds of red blood cell are found in sheep. One is a normal (HK) cell containing a high concentration of K^+ and a relatively low concentration of Na^+. The other (LK) has a relatively high Na^+ and a relatively low K^+ content. The difference between the two kinds of cell is apparently due to a single gene mutation. It has been found that the failure of LK sheep cells to accumulate K^+ to the normal extent is due to the fact that in them the Na^+-K^+ pump is only about one-fourth as active as it is in HK cells. Both types of cell possess an identical Na^+ and K^+ requiring ATPase, but in LK cells the ATPase activity is only one-fourth as great as it is in HK cells.

Third, experiments by I. M. Glynn and his associates (Garrahan and Glynn, 1967; Glynn and Lew, 1970) have provided a fascinating additional demonstration of the tight coupling between ATP metabolism and cation transport in red blood cells. When the amount of energy available from ATP hydrolysis is compared with the energy required to move 3 Na^+ out of the cell and 2 K^+ into it under physiological conditions, it is found that the energy "surplus" available to the pump is only about 4 kcal per mole ATP hydrolyzed. Thus, under extremely "unfavorable" conditions (i.e., a very high outward concentration gradient for K^+ and a similarly high inward gradient for Na^+), it might be possible to make the sodium pump "run backward" and synthesize ATP. Garrahan and Glynn (1967) achieved such conditions by suspending reconstituted ghosts (made rich in K^+ and inorganic phosphate during reconstitution) in high-Na^+ media. They were then able to demonstrate ATP synthesis during the entry of Na^+ into the cells. This synthesis was inhibited by cardiac glycosides. Glynn and Lew (1970) have shown that the ouabain-sensitive K^+ efflux from starved intact human red blood cells incubated in high-Na^+, K^+-free media (associated with a ouabain-sensitive Na^+ entry and believed to be due to a reversal of the pump mechanism which normally carries Na^+ out of and K^+ into the cells) is accompanied by ATP synthesis. Further, the stoichiometric ratio involved (1M ATP synthesized/2 to 3 equivalents of K^+ leaving the cell) is of the order expected from an ATP-driven Na^+-K^+ pump of the type believed to exist in the red blood cell membrane.

The Sodium Pump and the Regulation of Cell Volume
Unlike plant cells, which possess a tough outer wall capable of withstanding large pressure differences, animal cells, as illustrated by the above discussion of red blood cell hemolysis (p. 18), have little mechanical ability to resist volume changes due to osmotic gradients. If all the diffusible ions present in the body were distributed across cell membranes in accordance with the Donnan relationship (as Cl^- frequently is and K^+ sometimes is approximately), the resulting colloid osmotic pressure would cause cell swelling and damage on account of osmotic entry of water. By keeping Na^+ "off balance" relative to its electrochemical equilibrium, the Na^+ pump successfully counteracts this tendency for osmotic swelling to occur. Colloid osmotic swelling of cells has been demonstrated under conditions in which the Na^+ pump is inhibited, and it has been found that where the inhibition of the pump is reversible swelling can also be reversed (Tosteson, 1964).

Ion Transport in Epithelial Cells
So far this chapter has been concerned with transport phenomena in "symmetrical" cells—i.e., cells in which the membrane (aside from small localized special areas such as the end-plate zones in skeletal muscle fibers) is homogeneous all around the cell periphery. Such cells may be said to discriminate between two spatial regions only, the cell interior and the external milieu. In other words, once a molecule or ion enters the cell, there is no reason why it should emerge in any one specific direction rather than another. Thus the cell *as a whole* has no vectorial properties in relation to membrane transport. A schematic summary of ion transport in a symmetrical cell is shown in Figure 1-9A.

Several groups of cells in the body, however, show a completely different behavior because the membrane bounding one part of these cells has different properties from the membrane surrounding the remainder. Such cells can effect net *transcellular* movement of ions and other solutes from the solution bathing one side of them to the solution bathing the other, often against an energy gradient.

These are *epithelial cells*. They usually occur in sheets or layers in which individual cells are joined together by discrete junctional areas. Examples of highly functional epithelia (from the point of view of membrane transport) in the human body are the gastric and intestinal mucosa, the tubular epithelium of the kidney, and the cells lining the gall bladder and the ducts of exocrine glands. One should also mention two amphibian preparations, the isolated surviving frog skin and the toad urinary bladder. These have been widely used as experimental models for the investigation of epithelial function in general. In particular, many fundamental features of epithelial ion transport were first discovered through experiments with the isolated frog skin (Ussing, 1960). For

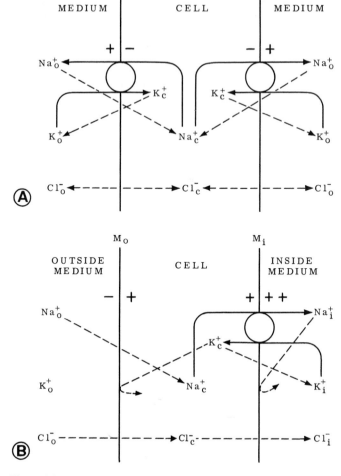

Figure 1-9
A. Symmetrical cell—regulation of ionic content. M (membrane): Na^+ leaks in, pumped out; K^+ leaks out, pumped in (possibly as coupled Na^+-K^+ pump); Cl^- in equilibrium across membrane. B. Asymmetrical cell (frog skin epithelium)—transcellular transport. M_o, M_i (outer and inner membranes): M_o: $P_{Na} \gg P_K$—no pump; M_i: $P_K \gg P_{Na}$—Na^+ pump. Na^+ leaks in through M_o, pumped out through M_i. A potential is generated which can drag a permeable anion (e.g., Cl^-) across the cell.

this reason, the electrophysiological and transport properties of this tissue are discussed in some detail below (p. 33).

It should be emphasized that ion transport across epithelial cell layers (*transepithelial transport*) is now known to take place via two distinct routes, the transcellular route and an *intercellular* or *paracellular* route. The latter is frequently referred to as the *shunt pathway*. Anatomically, it is believed to correspond to the intercellular junction–intercellular space complex.* In addition, it should be noted that active ion transport

*A delightful and completely painless summary of the development of the concept that the shunt pathway is an important regulatory factor in epithelial transport and electrophysiology will be found in a recent article by J. M. Diamond (1974).

across epithelial layers occurs only through the transcellular pathway. The shunt pathway is essentially a simple conductive route which permits a flow of ions to occur only when an appropriate transepithelial electrochemical gradient exists. In some epithelia (e.g., frog skin, toad bladder, gastric mucosa) the conductance of the shunt pathway may be relatively low compared to that of the transcellular pathway. Thus the transport and electrophysiological properties of the tissue may be largely determined by transcellular events. These epithelia are sometimes referred to as "tight" epithelia (Frömter and Diamond, 1972). In other epithelia (the "leaky" epithelia—e.g., the small intestine, gall bladder, and proximal kidney tubule) the conductance of the shunt pathway is high relative to that of the transcellular route. Consequently, the physiological behavior of such epithelia may depend to a marked degree on the conductance and permeability of the shunt pathway (Schultz et al., 1974; Armstrong, 1975).

Transcellular Transport: The Isolated Frog Skin

It has been known for over a century that isolated frog skin when mounted between two identical physiological solutions maintains an electrical potential difference between its outer and inner surfaces. This potential difference is normally oriented so that the inside surface is positive with respect to the outside. In addition, the skin has a remarkable capacity for net Na^+ transport from the outer to the inner solution.

Net Na^+ transfer can be effected against both electrical and chemical potential gradients and therefore constitutes a clear and unequivocal example of active transport according to Rosenberg's definition (Wilbrandt and Rosenberg, 1961). As might be expected, the Na^+ transport process is extremely sensitive to a number of metabolic inhibitors (Ussing, 1960).

Not surprisingly, the view that these two processes, the maintenance of an electrical potential difference and active Na^+ transport, are somehow interrelated gained wide acceptance among physiologists. The clarification of the mechanisms underlying the interrelationship is due largely to the work of H. H. Ussing and his collaborators (Ussing, 1960). In particular, their introduction of the so-called short-circuit technique provided an experimental tool which proved to be extremely useful not only in the study of isolated amphibian skin but in the investigation of the physiological behavior of a number of other isolated epithelial tissues (Schoffeniels, 1967). The principle of this technique is illustrated in Figure 1-10. The skin (S) is mounted between two identical Ringer solutions. The potential difference across it is measured by a pair of reversible electrodes (A, A′ in Fig. 1-10) placed as close as possible to its outer and inner surfaces without actually touching them. The potential is recorded by the millivoltmeter (P). By means of a source of electromotive force (D) and a potential divider (W), a variable current can be passed through the skin using a second pair of electrodes (B, B′). This current can be applied in such a way as to oppose the spontaneous skin potential and can gradually be increased until the potential becomes zero. Under these conditions the current flowing in the external circuit and measured by the microammeter (M) is the *short circuit current.*

There are, therefore, two fundamental electrical parameters to be interpreted in the analysis of this system, the short circuit current and the spontaneous or open circuit potential across the skin. The latter is of course the potential difference (PD) observed when no external current is drawn from the battery (D). Let us consider each of them in turn.

First, what is the meaning of the short circuit current? It is evident that under short circuit conditions all external electrical, chemical, osmotic, etc., gradients which could cause a passive net flow of ions across the skin have been eliminated. In this situation the short circuit current must represent the algebraic sum of all *active net ionic fluxes* across the skin. Under normal circumstances the short circuit current is found to correspond to a net transfer of positive charge from the solution bathing the outside surface to the solution bathing the inside.

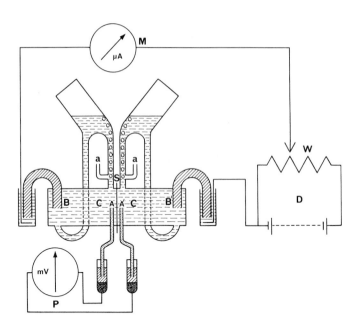

Figure 1-10
Apparatus for the determination of open circuit potential difference, short circuit current, and isotopic fluxes across isolated frog skin. See text for explanation of apparatus. (From Ussing and Zerahn, 1951.)

How can one identify which ions contribute to this current flow? The amounts of ions involved are so minute, compared to the total ionic content of the bathing solutions, that direct chemical estimates would be extremely difficult. The use of radioactive tracers permits the unidirectional fluxes of ions from outside the skin to inside (influx) and from inside to outside (efflux) to be determined accurately. Net flux is obviously the difference between these. The tracer method is particularly powerful when two different isotopes can be used simultaneously as in the case of Na^+. In this case one can label the solution on one side of the skin with ^{24}Na and the solution on the other side with ^{22}Na and determine influx and efflux simultaneously in the same preparation.

The application of tracer methods to the isolated frog skin has shown that, under normal conditions, short circuit current and net Na^+ flux from outside to inside are equal. Thus, one may conclude that in this preparation Na^+ is the *only* ion which shows *net* movement across the skin under short circuit conditions and is therefore the only ion which normally is actively transported across the tissue. Under open circuit conditions the situation is different. The electrical PD across the skin constitutes a driving force for *all* ions which are free to move. Thus, in this situation Cl^-, to which the skin is permeable, will also move (in the net sense) from the outside bathing solution to the inner one though less rapidly than Na^+.* The capacity for net transcellular Na^+ transport found in isolated frog skin has been shown to be a property of epithelia in general (Schoffeniels, 1967). Frequently, however, the situation under short circuit conditions is more complex than that exhibited by this tissue. In addition to active Na^+ transport there may be active transport of other ions (e.g., Cl^-), particularly in short-circuited preparations of the small intestine both under normal conditions and under the influence of pathological agents such as cholera toxin (Schultz et al., 1974).

*This is of considerable importance to the frog! The sodium-transporting mechanism of the skin enables the intact animal to absorb salt from dilute solutions such as pond water and thus plays a vital role in its overall electrolyte regulation.

The Origin of Transepithelial Potential Difference (PD)

Tight Epithelia—The Isolated Frog Skin. Concerning the question of the origin of the electrical potential difference (PD) across the skin, let us first consider some of the experimental data obtained by Ussing and his collaborators (Ussing, 1960). When Cl^- in the bathing solutions is replaced by an impermeant anion such as sulfate, the PD across the skin is increased. (In Cl^- media this potential is frequently in the range 50 to 70 mv. Replacement of Cl^- with sulfate increases it by some 20 to 30 mv.) Under these conditions the potential across the skin shows a linear dependence on the logarithm of the Na^+ concentration in the medium bathing its outer surface, and the slope of this line is close to the value predicted by the Nernst equation.*

On the other hand, the K^+ content of the outside bathing medium has very little effect on the skin potential. The opposite is found when the K^+ and Na^+ concentrations of the medium bathing the inner surface are changed. Here the skin potential shows a linear dependence on $\log [K_i^+]$ which is reminiscent of the situation found with skeletal muscle (Fig. 1-8), but Na^+ has virtually no effect. In addition, specific Na^+ pump inhibitors such as the cardiac glycoside ouabain when applied to the inner surface of the skin abolish the PD across it but have little effect when applied to the outer surface (Herrera, 1971). Thus in the frog skin the Na^+ pump is confined to the serosal or lateral-serosal border of the transporting cells (as indicated in Fig. 1-9B). A similar restriction of the Na^+ pump to the lateral-serosal cell membrane is believed to hold for other transporting epithelia (e.g., small intestine).

Ussing (1960) has combined these and other observations in the model of transcellular transport illustrated in Figure 1-9B. Before discussing this in detail, however, we offer a word of caution. The model illustrated assigns the task of net Na^+ transport across the whole skin to a single layer of cells (one of which is shown schematically in Figure 1-9B). At one time it was thought that a single epithelial cell layer (the *stratum germinativum*) did in fact control net Na^+ transport in this tissue. Recent electron microscope and other studies (Schoffeniels, 1967) have shown that frog skin epithelium is a highly complex tissue, so that the "single cell layer" model is an oversimplification. It is nevertheless valuable because it illustrates the principles involved in a clear and simple fashion and also because it can, with suitable modifications, be applied to other epithelia (e.g., toad bladder, intestine, kidney tubules) in which transcellular Na^+ transport can with confidence be thought of as being effected by a single layer of cells.

In the model shown in Figure 1-9B the outer membrane of the epithelial cell is regarded as being virtually a Na^+ electrode ($P_{Na} \gg P_K$) and the inner membrane is considered a virtual K^+ electrode ($P_K \gg P_{Na}$). In the absence of a penetrating anion, Na^+ diffuses into the cell from the outside medium, causing a diffusion potential which is given by the equation

$$E_o = 60 \log \frac{[Na_o^+]}{[Na_c^+]} \tag{16}$$

Similarly, K^+ diffuses outward across the inner membrane of the cell, giving rise to a second diffusion potential:

$$E_i = 60 \log \frac{[K_c^+]}{[K_i^+]} \tag{17}$$

(in these equations o, c, and i refer to the outside bathing medium, cell interior, and inside medium, respectively). In the model illustrated the Na^+ pump in the inner

---

*Note that in this case the PD increases with increasing external Na^+ as one would expect since the intracellular Na^+ in this tissue (as in virtually all tissues) is low and raising the external Na^+ concentration will raise the ratio $[Na_o^+]/[Na_c^+]$ (see equation 16).

membrane is assumed to be a coupled (nonelectrogenic) pump. Therefore the total PD across the skin is given by

$$E_T = E_o + E_i \tag{18}$$

Recent studies have indicated that the Na^+ pump in frog skin and other tight epithelia is not always a tightly coupled Na^+-K^+ exchange pump as suggested by Ussing's original model. In other words, it is now thought that the lateral-serosal Na^+ pump is electrogenic and contributes to the overall skin potential. Thus, one should write

$$E_T = E_o + E_i = E_p \tag{18a}$$

for the latter where E_p is the electrogenic pump potential.

In summary, one can say that the transepithelial PD in tight epithelia such as the frog skin and toad bladder is due in part to the asymmetrical diffusion potentials E_o and E_i and in part to an electrogenic pump component E_p. The exact quantitative contribution of each of these elements to the overall transepithelial potential is not always easy to assess, but, in any event, equations 18 and 18a predict that the total potential across a single cell should be the sum of two membrane potentials in series and have the "staircase" form shown in Figure 1-11A; the cell interior is positive with respect to the outside solution and negative with respect to the inside bathing solution. This prediction has been experimentally confirmed.

In the presence of an ion such as Cl^- to which both outer and inner membranes are permeable, both E_o and E_i will be reduced by the effect of Cl^- permeability (cf. equation 12).

Leaky Epithelia—The Small Intestine. In contrast to tight epithelia, the transmural PD across a leaky epithelium, as in the small intestine, is usually small (5 to 10 mv). Further, when one examines the individual mucosal and serosal membrane potentials in such an epithelium, one finds that they are both oriented so that the cell interior is

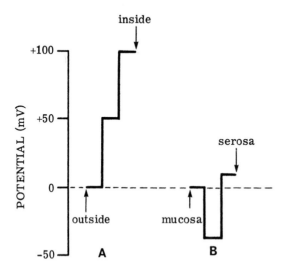

Figure 1-11
Potential profiles across (A) frog skin and (B) small intestine. The outside (A) or mucosal (B) bathing solution is considered to be at zero (ground) potential.

negative with respect to the appropriate bathing solution (Armstrong, 1975). The potential profile across the epithelial cell thus has the "well" configuration illustrated in Figure 1-11B. This has been interpreted as indicating that the mucosal and lateral-serosal cell membranes (unlike those of the frog skin) have very similar, perhaps identical, ionic permeabilities. Hence, the diffusional components of the total trans-epithelial potential may virtually cancel each other out, and this potential may be due, almost exclusively, to the operation of an electrogenic Na^+ pump. The intrinsic PD generated by the pump is probably much larger than the rather low potentials normally recorded across a leaky epithelium such as that of the small intestine. The low measured values for the transepithelial PD are probably due to shunting, across the relatively highly conducting junctional pathways, of a large fraction of the PD generated by ion pumping.

In cells of tight and leaky epithelia, as in other types of cell, the Na^+ pump plays a vital role in the maintenance of the steady-state electrochemical gradients for Na^+ and K^+ across both the mucosal (or outside) and serosal (or inside) cell membranes. When the Na^+ pump is inhibited (e.g., with ouabain), epithelial cells are found to accumulate Na^+ and lose K^+.

Co-Transport

The movement of Cl^- ions across the isolated open-circuited frog skin is an example of the phenomenon of co-transport. Co-transport takes place where the energy gradient created by the active transport of an ion (in the present case the electrical gradient generated by active Na^+ transport) is used to drive another ion or molecule in the same direction across a membrane. Co-transport has been much studied in recent years, particularly in the small intestine, where some of its most striking manifestations have been found (Armstrong and Nunn, 1970). In intestinal epithelia net transcellular Na^+ transport from lumen to blood is thought by many to provide the energy gradient necessary for the active transport of sugars and amino acids as well as for passive transfer of water and of other ions. Some aspects of intestinal co-transport are discussed in Chapter 26. At the present time the mechanisms by which energy coupling between intestinal Na^+ transport and intestinal transport of sugars and amino acids is achieved are not known. However, this is one of the most active fields in current membrane research.

REFERENCES

Armstrong, W. McD. Electrophysiology of Sodium Transport by Epithelial Cells of the Small Intestine. In T. Z. Csaky (Ed.), *Intestinal Absorption and Malabsorption.* New York: Raven, 1975.

Armstrong, W. McD., and A. S. Nunn, Jr. (Eds.). *Intestinal Transport of Electrolytes, Sugars, and Amino Acids.* Springfield, Ill.: Thomas, 1970.

Christensen, H. N. *Biological Transport* (2nd ed.). London: Benjamin, 1975.

Davson, H. *A Textbook of General Physiology* (4th ed.). Baltimore: Williams & Wilkins, 1970.

Diamond, J. M. Tight and leaky junctions of epithelia: A perspective on kisses in the dark. *Fed. Proc.* 33:2220–2224, 1974.

Dick, D. A. T. *Cell Water.* Washington: Butterworth, 1966.

Dockry, M., R. P. Kernan, and A. Tangney. Active transport of sodium and potassium in mammalian skeletal muscle and its modification by nerve and by cholinergic and adrenergic agents. *J. Physiol.* (Lond.) 186:187–200, 1966.

Frömter, E., and J. M. Diamond. Route of passive ion permeation in epithelia. *Nature* [*New Biol.*] 235:9–13, 1972.

Garrahan, P. J., and I. M. Glynn. The incorporation of inorganic phosphate into adenosine triphosphate by reversal of the sodium pump. *J. Physiol.* (Lond.) 192:237–256, 1967.

Glynn, I. M.   Membrane ATP-ase and cation transport. *Br. Med. Bull.* 24:165–169, 1968.

Glynn, I. M., and V. L. Lew.   Synthesis of adenosine triphosphate at the expense of downhill cation movements in intact human red cells. *J. Physiol.* (Lond.) 207:393–402, 1970.

Goldman, D. E.   Potential, impedance and rectification in membranes. *J. Gen. Physiol.* 27:37–60, 1943.

Harris, E. J.   *Transport and Accumulation in Biological Systems* (2nd ed.). New York: Academic, 1960.

Hendler, R. W.   Biological membrane ultrastructure. *Physiol. Rev.* 51:66–97, 1971.

Herrera, F. C.   Frog Skin and Toad Bladder. In E. E. Bittar (Ed.), *Membranes and Ion Transport.* London: Wiley-Interscience, 1971. Vol. 3.

Hodgkin, A. L., and P. Horowicz.   The influence of potassium and chloride ions on the membrane potential of single muscle fibers. *J. Physiol.* (Lond.) 148:127–160, 1959.

Hodgkin, A. L., and B. Katz.   The effect of sodium ions on the electrical activity of the giant axon of the squid. *J. Physiol.* (Lond.) 108:37–77, 1949.

Hoffman, J. F.   The red cell membrane and the transport of sodium and potassium. *Am. J. Med.* 41:666–680, 1966.

Katchalsky, A., and P. F. Curran.   *Nonequilibrium Thermodynamics in Biophysics.* Cambridge, Mass.: Harvard University Press, 1965.

Ling, G. N.   *A Physical Theory of the Living State.* Boston: Blaisdell, 1962.

Lucy, J. A.   Ultrastructure of membranes: Bimolecular organization. *Br. Med. Bull.* 24:127–134, 1968.

Pressman, B. J., and D. H. Haynes.   Ionophorous Agents as Mobile Ion Carriers. In D. C. Tosteson (Ed.), *The Molecular Basis of Membrane Function.* Englewood Cliffs, N.J.: Prentice-Hall, 1969.

Schoffeniels, E.   *Cellular Aspects of Membrane Permeability.* Elmsford, N.Y.: Pergamon, 1967.

Schultz, S. G., R. A. Frizzell, and H. R. Nellans.   Ion transport by mammalian small intestine. *Physiol. Rev.* 36:51–91, 1974.

Singer, S. J., and G. L. Nicolson.   The fluid mosaic model of the structure of cell membranes. *Science* 175:720–731, 1972.

Skou, J. C.   The ($Na^+ + K^+$) activated enzyme and its relationship to transport of sodium and potassium. *Q. Rev. Biophys.* 7:401–434, 1974.

Stein, W. D.   *The Movement of Molecules Across Cell Membranes.* New York: Academic, 1967.

Tosteson, D. C.   Regulation of Cell Volume by Sodium and Potassium Transport. In J. F. Hoffman (Ed.), *The Cellular Functions of Membrane Transport.* Englewood Cliffs, N.J.: Prentice-Hall, 1964.

Ussing, H. H.   The Alkali Metal Ions in Isolated Systems and Tissues. In O. Eichler and A. Farah (Eds.), *Handbuch der Experimentellen Pharmakologie.* Berlin: Springer, 1960. Vol. 13, pp. 1–195.

Ussing, H. H., and K. Zerahn.   Active transport of sodium as the source of electric current in the short-circuited isolated frog skin. *Acta Physiol. Scand.* 23:110–127, 1951.

Wilbrandt, W., and T. Rosenberg.   The concept of carrier transport and its corollaries in pharmacology. *Pharmacol. Rev.* 13:109–183, 1961.

2. General Properties of Nerve

Sidney Ochs

One of the fundamental properties of cells is excitability, defined as the capacity of cells to respond to changes in the external environment. Excitation is demonstrated even in a unicellular organism such as the ameba. When this cell is subjected to either mechanical or chemical stimulation on part of its surface, the response is the formation of pseudopods and movement of the ameba. Much evidence suggests that these reactions are initiated at the surface membrane where special physicochemical properties relate ions to membrane potential, as has been discussed in Chapter 1. Here we will discuss those special electrical changes present in the membrane which, while found in the single-cell organism, become more highly developed in the multicellular organism. In the course of evolutionary specialization, cells whose chief function is concerned with excitation and conduction arose, namely, the *neurons*. Other cells which also show excitation and conduction, the *effectors,* became specialized for contraction and movement of the organism. In higher forms these are the muscles (Chap. 3B).

Several representative neurons found in the vertebrate are shown in Figure 2-1. The sensory neuron with its cell body in the dorsal root ganglion has one long fiber passing outward to innervate the skin, muscle, and other parts of the body. At its termination in the periphery the afferent fiber ending is often combined with other cells to form a sensory receptor organ. Impulses from the body pass in the usual direction from the receptor ending to the nerve cell body in the dorsal root ganglion, and another fiber process of the cell carries the impulses into the central nervous system (CNS), where synapses are made on nerve cells within the spinal cord or other parts of the CNS. A motoneuron, such as is typically found in the central portion of the gray matter of the spinal cord, may have a large, wide, ramifying system of the treelike branches known as the *dendrites.* These arise from the cell body of the neuron, which is also referred to as the *soma* or *perikaryon.* Many synapses take place on the dendrites as well as on the soma. As can be observed in electron microscope studies, there is no merging of the membrane of one cell with another at the synapse. The presynaptic cell excites the postsynaptic cell by the release of a special neurotransmitter substance (see Chaps. 3A and 5).

Most neurons have all their portions within the CNS. The cerebral cortex (see Chaps. 8, 9), the thin layer of gray matter over the surface of the brain, contains as one of its characteristic cells the *pyramidal cell* (Fig. 2-1). This cell takes its name from the pyramidal form of the cell body. The apex of the pyramid is pointed radially outward to the surface, from which a large dendrite takes its origin, the *apical dendrite.* The apical dendrite passes upward for some distance before dividing into smaller and smaller branches, thus making available a relatively large area of dendritic surface receiving synapses from other axons terminating on the branches. Thus the neuron is influenced by many other cells of the cortex or of other regions in the brain. Its axon passes to other parts of the cortex or to other regions of the CNS. Another type of cortical neuron is the *stellate* cell. It has a short axon and performs an integrative function. Regions of the CNS where cell bodies and dendrites and their connections predominate are *gray* in color. Regions where axon tracts are compact have a whiter appearance and are referred to as *white* areas.

Most of what is known of nerve excitability has been obtained from studies made on peripheral nerve. As shown in Figure 2-2, a nerve trunk is composed of a large number

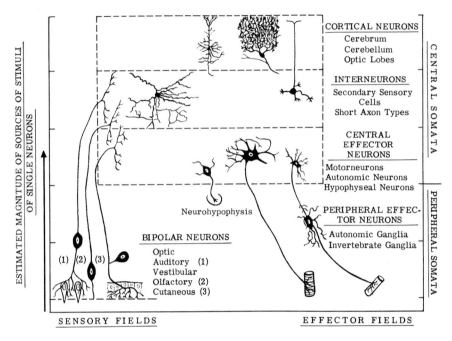

ESTIMATED MAGNITUDE OF SOURCES OF STIMULI OF SINGLE NEURONS

CORTICAL NEURONS
Cerebrum
Cerebellum
Optic Lobes

INTERNEURONS
Secondary Sensory Cells
Short Axon Types

CENTRAL EFFECTOR NEURONS
Motorneurons
Autonomic Neurons
Hypophyseal Neurons

Neurohypophysis

PERIPHERAL EFFEC-TOR NEURONS
Autonomic Ganglia
Invertebrate Ganglia

BIPOLAR NEURONS
Optic
Auditory   (1)
Vestibular
Olfactory (2)
Cutaneous (3)

(1) (2) (3)

CENTRAL SOMATA
PERIPHERAL SOMATA

SENSORY FIELDS   EFFECTOR FIELDS

Figure 2-1

Varieties of neurons. Sensory neurons in the higher animal forms are typically bipolar cells, shown to the left. Both the fiber portion receiving sensory input and the portion synapsing centrally are long. Effector neurons often have a long terminal axon, except for those of the autonomic nervous system. Shown in the dashed box are cells with very great arborizations of dendrites. (From Bodian, 1952.)

of fibers of varying size, with diameters ranging from a fraction of $1\,\mu$ up to approximately $20\,\mu$. Fibers above $2\,\mu$ in diameter have a myelin sheath around them and are known as *myelinated nerve fibers.* Myelin has a large content of lipid, apparent when it is stained with dyes having an affinity for it (e.g., osmium tetroxide). These show the fiber to be covered by a sheath of myelin (Fig. 2-2). Fibers below $2\,\mu$ in diameter are usually not observed with light microscopy and are referred to as *nonmyelinated nerve fibers.* Electron microscope studies of the myelin sheath of the myelinated fibers have revealed that the myelin is composed of layers which are regularly arranged in the form of lamellae (Fig. 2-3). The myelin layers are laid down by the Schwann cells positioned at intervals of 1 to 2 mm along the length of the fibers. The axon is thus left uncovered with myelin at the *nodes,* where, as will be discussed later, excitation occurs. The myelinated portions between the nodes are referred to as the *internodes.* Electron microscope studies of growing nerves during embryogenesis have revealed that the myelin is laid down by a jelly-roll wrapping process of membrane around the axon to produce the typical lamellated appearance (Fig. 2-3). The nerve fibers in the CNS are similarly myelinated by their analogous satellite cells—the *glial cells.* The thinner nonmyelinated fibers have been shown in electron micrographs to have one layer of myelin around them.

The giant fibers of squid and cuttlefish are a special type of nonmyelinated nerve fiber. Under the electron microscope they are seen to have several irregular myelin-like layers around them, supplied by satellite cells analogous to the Schwann cell. The giant nerve fibers attain diameters of 0.5 to 1.0 mm. The use of such large-diameter fibers has

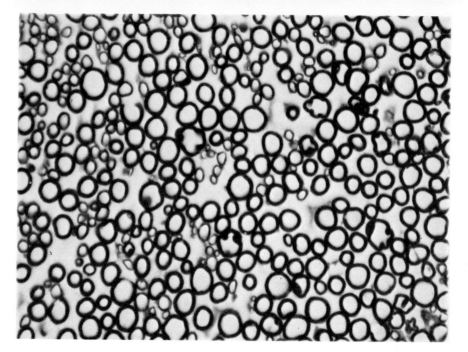

Figure 2-2

Cross section of myelinated nerve fibers. The dense bands around the fibers are myelin sheaths. The smallest groups of fibers are the nonmyelinated fibers, which typically are found in bundles, and are not visible here. ×425.

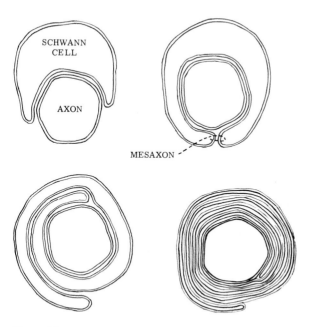

SCHWANN CELL

AXON

MESAXON

Figure 2-3

Embryological development of the myelin sheath. Early in development the Schwann cell becomes approximated to the axon. The membranes of the Schwann cell meet to form the mesaxon. One surface pushes forward and in so doing wraps a double membrane around the nerve. Myelin is thus laid down in a jelly-roll fashion. (Adapted from Geren, 1954.)

been of considerable service in the development of the modern theory of nerve excitation, as will be discussed later in this chapter.

EXCITATION AND RECORDING OF THE ACTION POTENTIAL

The frog sciatic nerve has long been favored for the study of the basic properties of excitation and conduction because it can survive for many hours after it has been removed from the animal. The isolated sciatic nerve is placed in a closed chamber containing moist air where electrical contact can be made to it through electrodes at various points along its length. As indicated in Figure 2-4, two stimulating electrodes are used to convey a brief pulse of current through a portion of the nerve at one end. When a current of adequate strength is used to excite the fibers, a nerve impulse passes along the nerve. If the nerve is innervating a muscle, a brief contraction (twitch) of the muscle innervated by the motor fibers results. Some simple observations can be made using the muscle twitch as an index of effective nerve excitation. Stimulation below a strength effective to excite is called a *subthreshold stimulus.* A strength of stimulation causing a

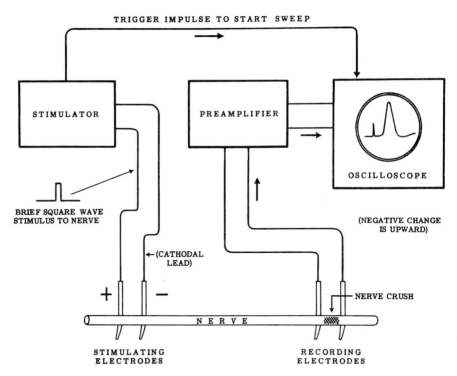

Figure 2-4

Excitation and recording of action potentials. The stimulator starts the horizontal sweep of the oscilloscope beam with a trigger pulse, and at some set time thereafter a brief pulse of current is delivered through the stimulating electrodes on which the nerve is placed. Farther along the nerve, electrodes pick up the propagated action potential after a conduction delay. After amplification, the action potential is displayed on the vertical plates of the oscilloscope. The combined time movement in the horizontal direction and voltage changes in the vertical direction give an X-Y graphic display.

just discernible motor response is referred to as a *threshold stimulation.* A further increase of stimulation causes more fibers to respond, and the twitch increases in size. When all nerve fibers are excited, the result is a *maximal response.* An additional increase in strength, a *supramaximal stimulation,* does not further increase the twitch size.

More accurate studies of excitability can be made by recording the electrical changes that constitute the nerve impulse. Two recording electrodes placed along the sciatic nerve are connected through an amplifier to an oscilloscope, as shown in Figure 2-4. The nerve impulse is a small brief negative voltage change which is picked up by these electrodes. The response must be considerably amplified before it can be displayed upon the screen of an oscilloscope. A brief description of this instrument will be necessary to explain the display of the response. A small beam of electrons is focused on the inner face of the oscilloscope tube and visualized by the fluorescence it produces. Voltages applied to the horizontal and vertical plates arranged along the path of the beam cause it to be displaced horizontally and vertically on the face of the screen. A special sweep circuit is used to apply appropriate voltages on the horizontal plates to make the beam move uniformly in the horizontal direction from the left to the right side of the tube screen. This constitutes the *time axis.* After a horizontal sweep is completed, the beam is shut off, and it returns to the starting position until a trigger pulse starts the movement once again. With the beam moving in the horizontal (X) direction, voltages applied to the vertical (Y) plates cause the beam to rise or fall, giving rise to a time display of fast-changing voltages.

The brief electrical signal seen on the oscilloscope trace at the time of the brief stimulating pulse applied to the nerve is an *artifact* caused by the adventitious electrical coupling to the recording electrodes. It serves as a convenient indication of the time when the stimulus is applied. After the stimulus, a *latency* period is seen as the oscilloscope beam moves along with no further voltage change until the beginning of the action potential is recorded. The latency depends on the speed at which the nerve impulse is propagated along the nerve from the stimulating electrodes to the recording electrodes, and on the distance from the cathode of the stimulating electrode to the first recording electrode (Fig. 2-4). There is some variation in the site of origin of excitation under the cathode lead of the stimulating electrode. The excitation may, because of spread of current in the perineural sheath, be initiated a millimeter or more away, but this usually is not too large an error. As the action potential passes under the first electrode, it becomes negative relative to the other electrode, and the beam rises vertically on the oscilloscope (Fig. 2-5A). The action potential continues past the first electrode until it reaches an intermediate position between the recording electrodes so that no difference in potential exists and the beam returns to the baseline level. Then, as the action potential passes under the second recording electrode, it becomes negative with respect to the first electrode, and the beam is directed downward. As the action potential moves on, the beam returns to the baseline, thus displaying the *diphasic action potential.*

To convert the diphasic action potential to the more useful *monophasic action potential,* the nerve is crushed between the two recording electrodes. The nerve impulse can then pass under the first electrode as before but not under the second one, leaving only the first negative deflection to be recorded on the oscilloscope (Fig. 2-5B).

The action potential is said to be *all-or-none,* analogous to a burning fuse which obtains its energy from the amount and condition of the powder along its length. If, for example, a damp or deficient region is encountered, the burning will be slowed or blocked. But if this area is passed, burning resumes in the next region at its usual rate. Similarly in the nerve, conduction of the action potential does not decrease in amplitude down the length of the fibers but develops its full amplitude in all the regions encountered. It can recover its full amplitude after passing through a narcotized region of reduced amplitude.

The monophasic action potential is followed by a smaller, longer-lasting potential,

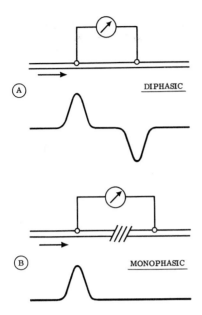

Figure 2-5
Diphasic and monophasic action potentials. The arrows show the direction of an action potential as it moves along a nerve in A past the first and then the second recording electrode. As it passes each electrode, that electrode becomes relatively negative with respect to the other, accounting for the deflection first one way, then the other, giving rise to the diphasic action potential. In B a monophasic action potential is seen because the crush of the nerve prevents the action potential from passing to the second recording electrode.

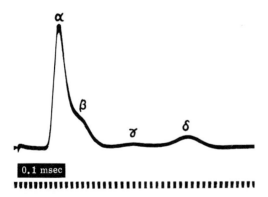

Figure 2-6
Compound action potential. The A fibers are subdivided into subgroups labeled with Greek letters having successively higher stimulation thresholds and slower conduction velocities. (Note that a large γ group is present only in motor nerves.) (From H. S. Gasser. *Association for Research in Nervous and Mental Disease*. Proceedings [1942] 23:48, 1943.)

the *negative afterpotential,* merging with a much-longer-lasting *positive afterpotential.* These are usefully described in the giant nerve axon, where analysis is easier.

The fibers of a mixed nerve such as the frog sciatic nerve have different diameters (Fig. 2-2). The action potential excited at the lower stimulus strengths comes from the most excitable fibers and, as found by Erlanger and Gasser (1937), from the group of fibers with the largest diameters. These fibers are classified as the *A fiber group.* It is divided into subgroups of successively smaller sizes, designated by Greek letters: *alpha, beta, gamma,* and *delta.* The alpha fibers, ranging in diameter from about 15 to 18 μ in the frog sciatic nerve, are the most excitable subgroup of the A group. With just threshold stimulus strengths, only the most excitable fibers of the alpha group are excited, and the response is small in size. As the stimulation strength is increased, more fibers of the alpha group are added to the responding population until the action potential response reaches a maximum. Further increase in stimulation strength does not increase the size of the alpha group response. With greater strengths a hump appears on the descending portion of the action potential. This potential is due to the excitation of the next small group—beta fibers (Fig. 2-6).

Still greater stimulation strength similarly excites later-appearing responses because the smaller-fiber groups have a lower conduction velocity as well as having higher thresholds. Advantage is taken of this property to separate the action potentials of the different subgroups by increasing the distance between the stimulating and recording electrodes. Just as in a race the faster and slower runners are lined up at the start but in

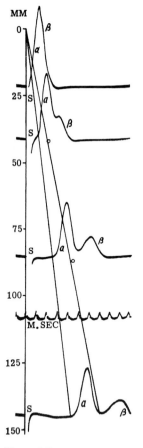

Figure 2-7

Separation of alpha (α) and beta (β) groups. If the distance between stimulated and recording sites along a nerve is successively increased, the α and β groups become easier to distinguish. The velocities of the two groups are indicated by the differing slopes of the two lines drawn through the beginning (the foot) of their action potentials. At a greater distance the foot of the β group is clearly separated from that of the α group. (From Erlanger and Gasser, 1937.)

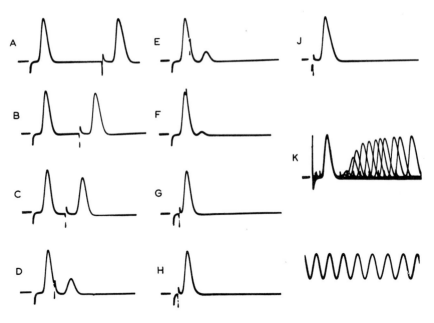

Figure 2-8
Two monophasic responses at decreasing time intervals from A-J give rise to a decrease of the second test response in C-F and its absence in G-J (relative and absolute refractoriness). In K, a series of double-shock traces is overlayed to show in a more compact form the decreasing amplitude of the test response during refractoriness. 1000 cps calibration in the lowermost trace. (From Ochs, 1965.)

the course of the race spread out, so in nerve the different action potentials representing the subgroups traveling at their various speeds spread out. The spread is best seen when the distance between stimulating and recording electrodes is increased, as is shown for the alpha and beta groups in Figure 2-7.

Another designation of fiber sizes useful for mammalian nerve fibers is that of Lloyd. Fibers from 20 to 12 μ are designated as type I; those from 12 to 6 μ, type II; and those from 6 to 2 μ, type III.

If a second stimulus is initiated too soon after a first response, it does not excite an action potential and the nerve is said to show *absolute refractoriness*. It is inexcitable no matter how strong a stimulus is used. In the alpha group of A fibers of the frog sciatic nerve, absolute refractoriness lasts a little longer than 1 msec and is followed by a period of *relative refractoriness*. In this state some response is possible if the strength of stimulus is increased, and as the fibers recover their excitability, the amplitude of response increases along an S-shaped curve until the action potential equals that of the control response (Fig. 2-8). The duration of the relative refractory period for the alpha group of the frog sciatic nerve is approximately 3 to 5 msec.

STIMULATING CURRENTS AND EXCITABILITY
Electrical conduction in electrolyte solutions takes place by the movement of ions. As discussed in Chapter 1, the major ions of intercellular fluids in mammalian tissue are Na^+ and Cl^-. The positively charged Na^+ moves toward the cathode, and the negative Cl^- moves toward the anode when a current is impressed on tissues. Nerves and body tissues in general can be considered as electrolyte solutions of various ionic composi-

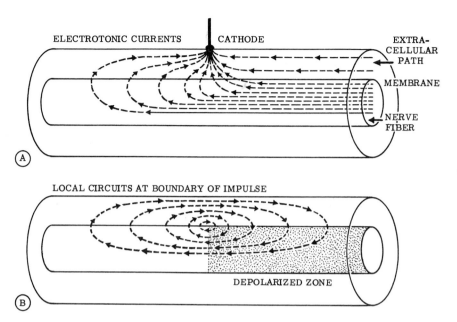

Figure 2-9
Paths of applied current and local circuits resulting from an action potential. In A the lines of current from an applied source are shown passing from the nerve to the cathode. Part of the current passes between the fibers, and part inside. The latter current is the effective portion, and excitation occurs in the region of the cathode as current passes outward across the nerve membranes and depolarizes the membranes. In B a similar direction of outward current in the region in front of an active part of the fibers depolarizes and can excite that region of nerve.

tions, with mainly K$^+$ and the larger molecular anions present within the nerve and muscle fibers. Ions moving between electrodes because of an imposed potential constitute a current flow; the number of ions flowing is directly related to the quantity of current. The proportion of the current carried by any given ion species also depends on its charge and mobility. Lines drawn between the electrodes signify the direction, and the number of lines the density, of the current flow (Fig. 2-9A). By convention, current flows from the anode through the nerve trunk and then back to the cathode.

As a general rule, nerve cells are excited in the region close to or under the cathode electrode. The current flowing between the nerve fibers is ineffective as regards excitation; that part of the current flowing inside and then leaving the fibers at the cathode constitutes the effective portion. The lines of current pictured in Figure 2-9 show only the form of current flow with no particular designation of the ion species. The ion species which carries current varies in the cells and in the extracellular fluid.

The membranes of nerve fibers (and other excitable cells such as muscle) offer a high resistance to the passage of electrical current. This relatively high resistance can be pictured by a suitable resistance element in an electrical model of the nerve (Fig. 2-10). Excitation occurs when the current flowing outward across the membrane reduces its *resting membrane potential;* i.e., it causes a *depolarization.* When the depolarization reaches a critical level, the impedance changes and an action potential is excited. In a later section of this chapter the ionic events related to excitation and the production of an action potential will be discussed in more detail.

Because of the high resistance of the membranes and their electrical *capacitance,* an application of electrical currents gives rise to a distribution of potential along the length of the fiber to which the term *electrotonus* is applied. Electrotonus refers not only to

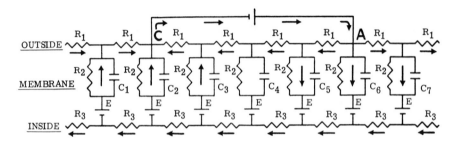

Figure 2-10
Electrical model of the membrane. The arrows in this model show the direction of current through an electrical network representing the nerve membrane. The model is composed of resistance and capacitance elements with a battery representing resting membrane potential.

distributed voltages along the nerve resulting from applied currents but also to the potentials produced by the current flows resulting from action potentials in the nerve. The distributed currents and resulting voltage changes affect the nearby membrane to change its excitability. This event will be described later on in discussion of local currents, and may be noted in Figure 2-9B.

The duration of a pulse of exciting current related to the strength needed to excite is given by the *strength-duration curve* (Fig. 2-11). The curve is determined by the strength of stimulating current required to reach threshold with different durations of stimulating current. As the duration of the pulse is shortened, the strength required to reach threshold rises. With the long stimulus durations the stimulating current required decreases to reach an asymptote called the *rheobase*. The term *chronaxie* is defined as that duration of stimulus required to excite at a strength twice the rheobase. The more excitable tissues generally have shorter chronaxies. Nerve, for example, has a shorter chronaxie than does muscle. Thus excitation is more likely with stimulation of certain

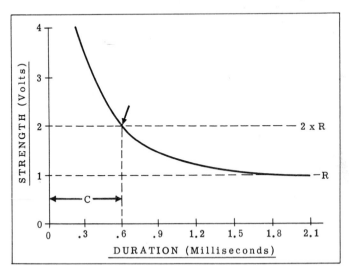

Figure 2-11
Strength-duration curve. The curve is derived from the strength required to attain a given level of response when various durations of stimulation are used. The longer pulse durations approach the rheobase (R). Chronaxie (C) is the duration of the stimulus required for excitation at a strength of stimulation twice rheobase (arrow).

points over the surface of muscles in the body than with stimulation at other points. These *motor points* are places where there is a confluence of motor nerves entering the muscle. The strength-duration curve over the motor point shows the short chronaxie and strength-duration curve typical of nerve excitation. This property is made use of in testing for nerve degeneration in cases of nerve lesion and for the recovery of function after nerve regeneration takes place (see below).

IONIC BASIS OF MEMBRANE POTENTIALS

Resting Potentials

For many years the presence of an electrical potential across the membrane of nerve and muscle was known. It was demonstrated by the negative potential seen at a crushed or injured site on one part of the nerve or muscle relative to the intact portion. This *injury potential* or *demarcation potential* is due to current passing from the *polarized* intact part of the fiber to injured *depolarized* portion. The source of current is the asymmetry of ions across the normal membrane and the resting membrane produces in the manner described in Chapter 1. In the Nernst equation

$$E = \frac{RT}{F} \ln \frac{[K^+]_i}{[K^+]_o}$$

where E represents the electrical driving force in equilibrium with the chemical driving force due to the ratio of $[K^+]$ across the membrane (Chap. 1). An important point to note is that ion concentration differences of only several picamoles ($10^{-12}M$) across the membrane can produce a voltage of some -90 mv, the sign referring to the inside of the cell with the outside taken as zero.

The giant nerve axons obtained from squid and cuttlefish have diameters up to 0.5 mm or more, and in these fibers the *transmembrane* potential can be readily measured by inserting a microelectrode into the fiber from one end with another electrode placed in the outside medium. Potentials of approximately 50 to 60 mv are found with the inside negative to the outside. The transmembrane potential or *resting membrane potential* is accounted for by the Nernst equation. Additionally, micro-electrodes with tips measuring 0.5 μ or less can be inserted through the membrane of smaller nerve fibers and nerve cell bodies as well as into muscle cells. By this means resting membrane potentials of the order of 50 to 90 mv have been found in a wide variety of tissues in various species including man.

On changing the concentration of K^+ in the external medium, the potential changes were seen to conform to the Nernst relationship, except that at the lower concentrations of K^+ in the outside medium a small deviation from the expected voltage was found. This is due to a small leakage of Na^+ into the cell causing a reduction of the resting membrane potential a little below that expected of the K^+ ratio. In muscle fibers there is also a large contribution of Cl^- to the resting membrane potential which acts as a "battery" in parallel with the K^+ "battery" and with the same orientation of potential.

It is possible to squeeze the axoplasm out of the giant axons and replace it with fluids of known composition. In such *perfused axons,* when K^+ was present at its usual concentration, the usual resting potential was seen. When K^+ was diluted with a nonelectrolyte such as sucrose to lower its internal concentration, the membrane potential was reduced according to the expectations of the Nernst equation. With equal concentration of K^+ across the membrane, the resting potential was close to zero as expected from the Nernst equation.

Action Potentials

With a micropipette inserted into a giant fiber and a resting membrane potential of 60 mv recorded, stimulation of the fiber gives rise to action potentials having amplitudes

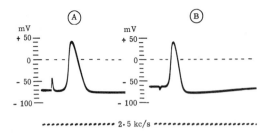

Figure 2-12

Action potentials from giant nerve fibers. In A the action potential is recorded from a giant nerve fiber in situ with a microelectrode. The resting membrane potential is 70 mv, inside negative. An action potential shows an overshoot of approximately 35 mv. In B, an internal capillary electrode inserted from one end measures a smaller resting membrane potential of approximately 60 mv. A hyperpolarization follows the action potential—the positive afterpotential. (From A. L. Hodgkin. *Proc. R. Soc.* [*Biol.*] 148:5, 1958.)

of 90 mv (Fig. 2-12). An action potential larger in amplitude than the resting membrane potential was a new and unexpected finding. The excess potential approximately 30 mv beyond that of a simple depolarization was called the *overshoot.* The explanation of the overshoot resulted in a new theory of nerve action, the *sodium hypothesis,* now generally accepted as the basis of the action potential. Sodium ion is high in concentration outside the nerve fiber and low inside and in effect constitutes a "battery" in opposition to the resting membrane potential. The reason is that the Na^+ concentration difference across the membrane is so directed that it tends to make Na^+ enter the fiber. Also, the negative charge inside the cell attracts the positively charged Na^+ ion adding an electrical driving force. Radioactive tracer studies have shown that Na^+ in actuality continuously enters the cells. This process would soon result in the entrance of sufficient Na^+ into the cells to bring the resting membrane potential to zero. It is counteracted by the operation of the *sodium pump* (Chap. 1 and below), whereby Na^+ is pumped out of the cell interior and in effect reduces the effect of Na^+ on the resting membrane potential to a negligible factor.

The action potential is explained by the sodium hypothesis. When the resting potential across the membrane of the nerve is reduced (depolarized) to a *critical level,* the permeability of the membrane to the Na^+ rapidly increases, allowing Na^+ to move into the cell within a fraction of a millisecond. The entry of Na^+ accounts for the 30-mv overshoot of the action potential. This occurs until the voltage across the membrane moves toward the *equilibrium potential for Na$^+$ (E_{Na^+}).* The equilibrium potential can be computed from the Nernst equation and the concentration ratio of Na^+ across the membrane, and it is that transmembrane voltage at which a further Na^+ entry is prevented.

The increase of Na^+ permeability does not continue indefinitely because of a process called *inactivation.* As the voltage across the membrane approaches E_{Na^+}, inactivation shuts off further entry of Na^+. Additionally, after a brief lag, K^+ permeability increases, and K^+ leaves the interior of the fiber, bringing its positive charge out of the cell. Thus, the membrane voltage returns to the resting membrane potential at the termination of the action potential. The increase in K^+ permeability continues for some time, causing the *after-hyperpolarization* or *positive afterpotential* (Fig. 2-12). The size of the after-potential is related to the level of the resting membrane potential and the deviation existing from the *equilibrium potential for K$^+$ (E_{K^+}).* The hyperpolarization is larger when the nerve has been depolarized below its normal resting membrane potential.

It must be emphasized that only a few picamoles of Na^+ and K^+ are exchanged across the membrane to give rise to the action potential. The net effect of these ion shifts is that the fiber has gained a tiny bit of Na^+ (3 to 6pM) and lost an equivalent amount of K^+

for each action potential. The sodium pump will later redress the imbalance of the gain of Na+ and loss of K+ after a number of action potentials to restore the original ionic asymmetry. The small ionic change occurring for each discharge allows a great many action potentials to be generated before any significant effect on ionic concentration and resting membrane potential is seen.

The analysis of the sodium hypothesis was accomplished by Hodgkin, Huxley, and Katz using electrodes placed inside the length of giant nerve axons. By means of a control circuit completed through the outside electrodes, the nerve could be "clamped" at a fixed level of potential. The entire axon membrane could thus be depolarized (to excite it) and the relatively large flow of current through the whole of the membrane measured without the complication of the conduction of the action potential. Upon depolarization, an early brief inflow of current appeared which was related to the entry of Na+. Later, the current became reversed as a result of the outward flow of K+. This sequence was shown in experiments in which the nerve was immersed in solutions from which Na+ was removed—for example, by replacing Na+ with sucrose or choline. The early inward portion of the current was eliminated while the outward current—i.e., that due to K+ outflow—remained.

The computed permeability changes to Na+ and K+ underlying these ionic flows are shown in Figure 2-13. They may be conceived of as the successive opening of selective *channels* in the membrane, first to Na+ and then to K+, allowing the Na+ to flow in and the K+ to flow out through the channels. The behavior of nerve in response to various other ions shows in general that these channels are apparently separate, though not perfectly selective to these ion species.

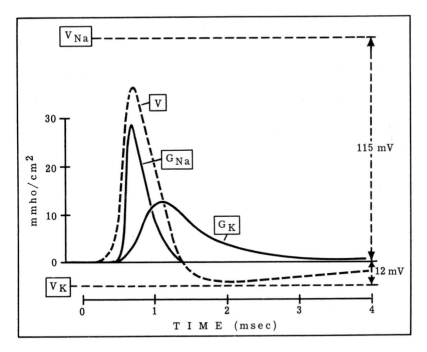

Figure 2-13
Computed conductance change in milliohm (mmho) per square centimeter to Na (G_{Na}) and to K (G_K) accounting for the action potential (V). Notice that the result of Na+ entry is to drive the membrane potential to the Na equilibrium (V_{Na}). The membrane at rest is close to the equilibrium potential for K+ (V_K). (From A. L. Hodgkin. *Proc. R. Soc. [Biol.]* 148:5, 1958.)

The electrical resistivity of the membrane is high. As shown in electron micrographs, the membrane thickness is of the order of 75 Å. Physicochemical studies show it to be composed of a bilayer of lipoproteins with globular proteins embedded in the lipid. These proteins appear to be the channels for ion movement. Just at the onset of the action potential, the electrical resistivity of the membrane falls drastically, to about one-fortieth of its original value. As shown in Figure 2-13, there is a *Na+ conductance* (G_{Na+}) increase which is soon followed by a *K+ conductance* (G_{K+}) increase. Conductance is the inverse of resistance. Conductance increases may be viewed as being due to a changed conformation of globular proteins embedded in the membrane, opening a channel through which the ions move. It is possible that each ion has a specific channel. To describe the conductance changes related to permeability Hodgkin and Huxley developed a set of kinetic formulations. In their formulation K+ conductance is controlled by a variable raised to the fourth power. Intuitively we can consider that the increased conductance comes about when four associated ions are grouped at some critical site within the K+ channel. A more complicated situation controls Na+ permeability; three ions are considered to occupy a critical site to bring about an increase in Na+ conductance. An opposite process controlled by another variable, *inactivation,* can be viewed as a closing of Na+ channels.

The underlying idea is that the control of Na+ and K+ is brought about by the voltage difference existing across the membrane. The crucial event is a quick depolarization to reach the threshold or *critical level* for the initiation of increased Na+ conductance. If the voltage change is too slow, the opposite phenomenon of inactivation prevents Na+ from entering the cell to cause excitation.

Support for the sodium hypothesis has been gained in several ways. Radioactive tracers of Na+ and K+ were used to show an extra uptake of Na+ with a corresponding loss of K+ in nerve or muscle fibers following a repetitive series of action potentials. The amounts conformed to the expected relation of charge moved across the membrane, of the order of several picamoles per action potential.

Specific toxins acting on the giant axon membranes have helped reveal the properties of the active Na+ channel and the permeability increase on depolarization. A toxin obtained from the puffer fish, *tetrodotoxin* (TTX), acts in very low concentrations to reduce Na+ permeability and to block the activation of the Na+ permeability increase during excitation, apparently by binding the Na+ channels at or just within the outer face. *Batrachotoxin* (BTX), another toxin which blocks in very small concentrations, appears to act on the Na+ channels by keeping them open and thus making the nerve inexcitable. Local anesthetics, such as procaine, block excitability by depressing permeability of both Na+ and K+ channels. *Tetraethylammonium* (TEA) blocks K+ channels by acting at the inner surface.

Calcium is associated in some important way with excitability, as is shown with regard to the property of rhythmicity. Nerve axons do not usually discharge rhythmically, but do so in low Ca++ media. In clinical cases of *hypocalcemia,* abnormal excitability of nerve and muscle is observed, often as a period of sustained rhythmic nerve discharge following an excitation.

LOCAL CURRENTS AND CONDUCTION— CONTINUOUS AND SALTATORY

The conduction of the action potential is explained on the basis of the electrical properties of the membrane: its high resistance and the electrical function capacity distributed all along the length of the fiber (Fig. 2-10). At the region where an action potential has been excited, either by an applied pulse of electrical current or naturally at nerve endings (Chap. 3A), the voltage change across the membrane which constitutes the action potential can be considered as a "battery." This battery causes current to spread out into the adjoining membrane, depolarizing it and bringing it to the critical level for excitation. After initiation of an action potential, the process is repeated again and again, constituting the propagated action potential (Fig. 2-9B).

The ionic difference between the local potential changes due to local currents and the currents involved in the action potential cannot be stressed too much. The local currents are passive, with no specificity of the species of ions carrying the current. In the nerve fiber K^+ and anions carry the major part of the current while Na^+ and Cl^- carry the major part of the current outside the fiber. These current flows follow the general law of electrical conductance in an electrolyte medium. In contrast, the active action potential is, as described above, the result of specific ionic processes, with first Na^+ and then K^+ moving through their respective specific channels in the membrane when excitation occurs.

We can account for conduction velocity by considering the resistances inside the fiber and across the membrane, neglecting, for the time being, the electrical capacitance of the membrane (Fig. 2-10). Considering the voltage change due to an action potential across the membrane as a battery, we may say that the current flowing in the nearby membrane falls off with distance from that point. The fall-off is expressed by the length constant (λ) given in the resulting potential recorded along the length of the fiber. At the point where the voltage falls to $1/e$ of its original level

$$\frac{V}{V_o} = e^{-(x/\lambda)}$$

where V = measured voltage at different distances, x, from the source voltage, V_o. When $x = \lambda$, $V = 1/e \cdot V_o$. This formula is used to determine the length constant, λ. It can be seen intuitively that the greater the λ, the farther the electrotonic voltage reaches out in the membrane. In effect, excitation will then occur at a more distant site and be directly related to a faster conduction velocity.

A much faster conduction velocity is seen in the myelinated nerve fibers (Fig. 2-3). Single myelinated nerve fibers can be isolated for a study of their physiological properties, which requires care and patience. The fiber is shown to be more excitable to cathodal current at the nodes, where there is no myelination, than at the internodes (Fig. 2-14). This finding follows from the theory of *saltatory conduction,* which takes into account the fact that the myelin acts as an electrical insulator. Electrical currents that flow as a result of an action potential can enter or leave only at the nodes. Excitation takes place at the node when the membrane is depolarized to the critical level, as was described for the giant nerve membrane. The Na^+ permeability then rapidly increases, and Na^+ enters, causing a *local current* to pass down inside the axon and out the neighboring nodes. The outward current across the nodal membrane causes excitation when the critical level of depolarization is reached. Sodium ions enter that node, and the same process is repeated. Excitation, therefore, instead of occurring continuously along the membrane, as is the case in the giant axon fiber and in the small non-myelinated axons, occurs at the nodes and is thus discontinuous, or *saltatory.* As a result, the velocity is higher than for a correspondingly sized nonmyelinated fiber. Velocities of up to 120 meters per second are attained in the large 20 to 22 μ myelinated fibers of animals and man.

The current flowing from node to node within a fiber does not normally excite the nerve fibers lying alongside them, but it is not wholly without effect. Under special conditions, small alterations in the excitability of nearby fibers can be shown. In some diseases—e.g., multiple sclerosis—when the myelin is lost, an abnormal excitation of neighboring fibers may take place.

METABOLISM RELATED TO EXCITABILITY

After generating an action potential, the nerve has gained several picamoles of Na^+ and lost an equivalent amount of K^+. A complete restitution of ionic asymmetry after nerve activity requires the pumping out of the extra Na^+ gained and the recapture of the lost K^+. Otherwise, the ionic composition of the fiber would eventually change after a long time, with a loss of resting potential and excitability. Sodium pumping requires the

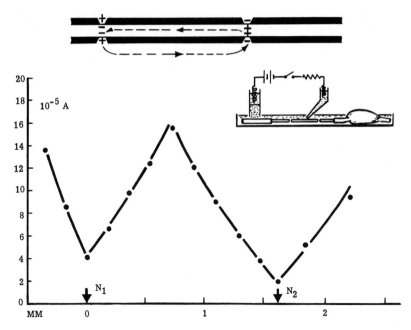

Figure 2-14
Transmission in a myelinated nerve occurs via currents passing down along the inside of the fiber and out the adjacent nodes, as shown in the insert above. The internodal portion covered with myelin is less excitable than the nodes. Threshold to electrical currents for a microelectrode placed at various points along a single fiber, as shown in the insert above, is lowest at the two nodes, N_1 and N_2, and is high in the internodal region. (From I. Tasaki, *Nervous Transmission*, 1953. Courtesy of Charles C Thomas, Publisher, Springfield, Illinois.)

operation of an enzyme located in the membrane, *sodium potassium activated ATPase* (Na K ATPase). This enzyme is polarized so that its high Na^+-requiring side faces internally and its K^+-requiring portion faces externally. *Adenosine triphosphate* (ATP) is required by the Na^+ pump, the ATP supplying energy through its energy-rich phosphate bond ($\sim$P). Oxygen is utilized by the nerve to generate ATP through oxidative phosphorylation in the mitochondria found all along the inside of the axons.

   If metabolic blocking agents, such as cyanide or dinitrophenol, are applied to the giant nerve fiber or injected into it, the resulting block of oxidative phosphorylation causes a fall in ATP and a consequent cessation of the outward flux of Na^+. Na^+ then accumulates inside the fiber. When the Na^+ pump of the giant nerve fiber is poisoned by these agents, the resting potential and action potentials can continue for a relatively long time, 90 minutes or more, before a change of Na^+ and K^+ sufficient to block excitation takes place. The reason is the large volume of axoplasm and the small ionic changes which result from the relatively small inward flux of Na^+. In comparison, much faster concentration changes occur in the smaller fibers of myelinated nerves, where potentials fail within a matter of 15 minutes.

   When ATP is injected into metabolically poisoned giant axons, Na^+ pumping resumes. In the giant axon *arginine phosphate* can also cause a resumption of pumping because the $\sim$P is transferred from arginine phosphate to ATP to restore its level. In mammalian tissues this storage function is performed by *creatine phosphate (CP)*.

   The nerve, like other tissues in the body, utilizes glucose to supply the energy needed to perform its function. The glucose undergoes *glycolysis* to give rise to pyruvate, which then enters the *citric acid cycle* present in the mitochondrion. In the operation of the citric acid cycle, the oxygen taken up by the terminal end of the *electron transfer chain* is utilized, and ATP is generated. Various toxic agents are known to block glycolysis, the citric acid cycle, and the electron transfer chain of mitochondria and thus produce a block of metabolism. Diseases such as diabetes or thiamine deficiency which alter energy metabolism can cause neurological disturbances through a defect of ATP supply resulting in ionic changes and changed excitability. As described below, transport of materials within the fibers is also affected.

Protein Metabolism and Axoplasmic Transport

Electron microscope and biochemical studies have revealed that the nerve cell body has a high level of protein synthesis. The small bodies within the cytoplasm of the neuron soma, the *Nissl bodies,* when stained a deep blue with aniline dyes, are seen in electron microscope studies to be made up of a system of channels, the *endoplasmic reticulum,* along which small particulate structures, the *ribosomes,* are associated. Protein chain formation is known to occur on the ribosomes. In recent years, proteins, polypeptides, and other materials have been shown to move outward in the fibers by a process called *axoplasmic transport.*

A swelling of nerve fibers dammed above a site of ligation or constriction has suggested an outflow of material. Recently, with isotopically labeled amino acid used in sufficiently long lengths of nerve such as are available in the cat sciatic nerve, the rate and characteristics of downflow were directly measured. In such experiments the amino acid ^{3}H-leucine is injected into the seventh lumbar (L7) dorsal root ganglia or into the L7 motoneuron region of the spinal cord. The precursor is taken up by the nerve cell bodies and is rapidly incorporated into proteins and polypeptides, which then move down the sciatic nerve fibers at a rate of 410 mm per day (Fig. 2-15). The rate of the advancing crest of radioactivity is linear and is independent of nerve fiber size or function, either sensory or motor. The same rate is seen in a variety of animal species and it is probably the same in man. Specific substances, such as norepinephrine in sympathetic nerve fibers and acetylcholinesterase (AChE) in cholinergic nerves, are also transported at a fast rate.

A hypothesis proposed to account for axoplasmic transport is that a *transport filament* binds the various materials transported which move down the microtubules found in the nerve fibers by cross-bridges (Fig. 2-16). This model is based on the sliding-filament theory of muscle (Chap. 3B). The cross-bridges require an energy supply, and fast axoplasmic transport was shown to be closely dependent on oxidative metabolism. Nitrogen, anoxia, azide, cyanide, and dinitrophenol, all of which block oxidative phosphorylation, also block axoplasmic transport. When transport is blocked in mammalian nerves, membrane excitability is also lost; both functions are gone in about 15 minutes. This loss indicates a common supply of ATP to both the Na^+ pump and the transport mechanism present in the nerve fibers.

The anoxic block of nerve action potentials and axoplasmic transport is reversible. After periods of anoxia in vivo produced by the application of high pressures in cuffs around the hind limb of cats to block blood flow, electrical responses and fast axoplasmic transport can recover after *compression ischemia* lasting up to five or six hours. With ischemic compression times longer than six hours, irreversible block and wallerian degeneration of the nerve fibers occur (see below).

The microtubules, considered to be the stationary elements in the nerve along which transport of material occurs, are found in various other cells and are probably present in all cells. They are associated with the translocation of materials and the maintenance of cellular form. The microtubules are sensitive to *mitotic blocking agents,* such as *colchicine* and *Vinca* alkaloids (*vinblastine,* for example), which arrest and block cell division. These agents act on the microtubules in the spindles of dividing cells causing them to disassemble and thus block cell division. These agents can also block axoplasmic transport. Electron microscopy of nerve fibers exposed to colchicine or vinblastine shows that these agents can cause a complete disassembly of microtubules at

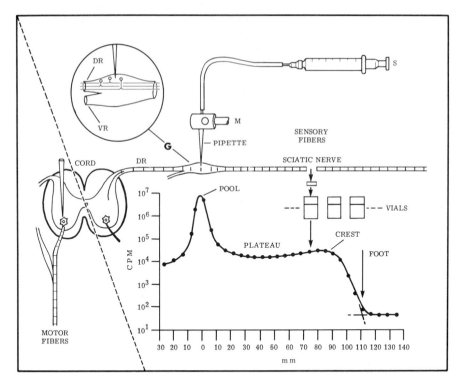

Figure 2-15
Technique of isotope labeling and outflow pattern of fast transport. The L7 dorsal root ganglion (G) is shown with a pipette inserted for injection of ^{3}H-leucine. Insert shows ganglion with T-shaped nerve cells. The dorsal root, ganglion, and sciatic nerve are cut into 5-mm segments. Each segment is placed in a vial and solubilized; scintillation fluid is added and a count is made. The display of activity in counts per minute (CPM) is shown using a logarithmic scale on the ordinate and the distance in millimeters on the abscissa taking zero as the ganglion. A typical pattern for a 6.5-hour downflow is shown. A high level of activity remains in the ganglion and falls distally to a plateau before rising to a crest. A sharp drop is seen at the front of the crest down to the baseline level at its foot. At the left of the dashed line, an injection of the L7 ventral horn is shown for uptake of precursor by the motoneurons. The ventral root and sciatic nerve are cut and measured for outflow as in the case of ganglion injection. (From Ochs, 1965.)

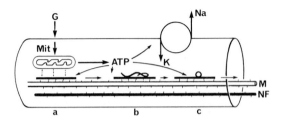

Figure 2-16
Transport filament hypothesis. Glucose (G) enters the fiber, and, after oxidative phosphorylation in the mitochondrion (Mit), ATP is produced. A pool of ATP is shown supplying energy to the sodium pump, which controls the level of Na$^+$ and K$^+$ in the fiber and, as well, movement of the "transport

high levels, and some investigators believe this is their mode of action. A more subtle action on microtubules or the transport mechanism must be considered. The neuropathy resulting from the action of these agents on microtubules in nerve is the chief limitation of the use of vinblastine and other *Vinca* alkaloids in the treatment of leukemia and other neoplasms.

To complete the picture of the movement of materials in nerve fibers: There is, in addition to the anterograde transport described above, a *retrograde transport* of material. An agent which shows retrograde movement is horseradish peroxidase. It is picked up by nerve terminals, carried in the retrograde direction, and accumulated in the cell bodies. In the same way, it appears that viruses and possibly tetanus toxin may pass up the nerve into the nervous system to produce disease.

DEGENERATION AND REGENERATION

When a peripheral nerve is crushed or severed, the part distal to the crush shows the process known as *wallerian degeneration.* After a few days the myelin of the fibers show undulation in its form and a failure to conduct. Over a period of weeks it shows *beading,* then *ovoid formation* with a change to *spheres.* These are later absorbed leaving only Schwann cells and fibrous tissue (Fig. 2-17).

A few days after nerve transection, another process is seen. The central ends of fibers above an interruption sprout buds and a growth of fine fibers, which grow down into the amputated part of the nerve. This process of regeneration of new fibers is similar to the embryonic growth of nerve fibers in that some unknown directive force, a *neurotropic stimulus* (probably chemical in nature), acts on the growing tips of the nerve to direct them to their right "addresses." The Schwann cell material in the degenerated sheaths left in the distal parts of the transected nerve may have such a directive influence on this growth, but little is known on the subject. The surgeon repairing nerve lesions assists regeneration by bringing the cut ends into close apposition and stitching the epineurial sheaths together. Fibrous growth is to be prevented so as not to impede new fiber growth down into the amputated nerve stump. This may cause the outgrowing fibers to turn around, forming a whorl known as a *neuroma,* which may be quite painful (Chap. 3A).

During the several weeks and months during which regeneration of new fibers occurs, the nerve cell body undergoes a series of characteristic changes known as *chromatolysis.* The deep blue staining of the Nissl bodies is decreased or lost and then slowly returns over a period of months. The chromatolytic reaction is due to the dispersion of the Nissl bodies within the soma. Along with this change there is an increased uptake of water and swelling of the cell body with a displacement of the nucleus. The content of ribonucleic acid (RNA) and protein is increased later on in the course of chromatolysis. Various hypotheses have been advanced to account for the phenomenon of chromatolysis. The most likely explanation is that some "signal" substance which normally moves up the fiber by retrograde transport acts to control the level of protein synthesis in the cell. On transection the signal is lost, and the cell undergoes the series of changes characteristic of chromatolysis with an apparent increased protein synthesis.

---

filaments," shown as black bars. The various components bound to it and carried down the fiber include: the mitochondria (a) attaching as indicated by dashed lines to the transport filament and giving rise to a fast forward and retrograde movement (though with a slow net forward movement), soluble protein (b) shown as a folded or globular configuration, as well as other polypeptides and small particulates (c). Thus, a wide range of component types and sizes are carried down the fiber at the same fast rate. Cross-bridges between the transport filament and the microtubules effect the movement when supplied by ATP. (From S. Ochs. *J. Neurobiol.* 2:331–345, 1971.)

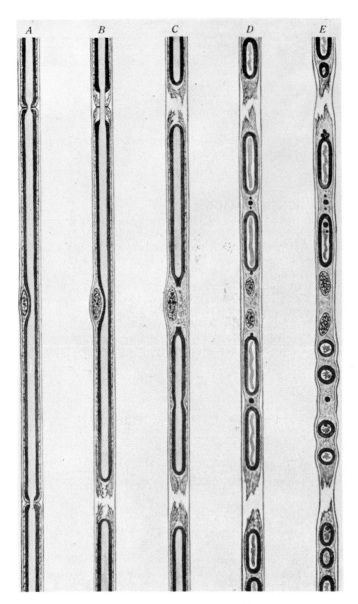

Figure 2-17
Stages of wallerian degeneration of an amputated part of myelinated nerve fiber shown after periods of days and weeks. Normal fiber (A). Early in degeneration (B) there is a retraction of myelin from the nodes. Later (C) long ovoids are formed which still later (D, E) become smaller and round up into vesicular structures. (From Young, 1949.)

REFERENCES

Adelman, W. J., Jr. *Biophysics and Physiology of Excitable Membranes.* New York: Van Nostrand Reinhold, 1971.

Albers, R. W., G. J. Siegel, R. Katzman, and B. W. Agranoff. *Basic Neurochemistry.* Boston: Little, Brown, 1972.

Bodian, D. Introductory survey of neurons. *Cold Spring Harbor Symp. Quant. Biol.* 17:1–13, 1952.

Drachman, D. B. (Ed.) Trophic functions of the neuron. *Ann. N.Y. Acad. Sci.* 228:1–423, 1974.

Erlanger, J., and H. S. Gasser. *Electrical Signs of Nervous Activity.* Philadelphia: University of Pennsylvania Press, 1937 (1968 reprinting).

Gasser, H. S. Pain-producing impulses in peripheral nerves. *Res. Nerv. Ment. Dis. Proc.* 23:44–62, 1943.

Geren, B. B. The formation from the Schwann cell surface of myelin in the peripheral nerves of chick embryos. *Exp. Cell. Res.* 7:558–562, 1954.

Hodgkin, A. L. The Croonian Lecture. Ionic movements and electric activity in giant nerve fibres. *Proc. R. Soc. [Biol.]* 148:1–37, 1958.

Hodgkin, A. L. *Conduction of the Nervous Impulse.* Springfield, Ill.: Thomas, 1964.

Hubbard, J. I. *The Peripheral Nervous System.* New York: Plenum, 1974.

Jacobson, M. *Developmental Neurobiology.* New York: Holt, Rinehart & Winston, 1970.

Katz, B. *Nerve, Muscle and Synapse.* New York: McGraw-Hill, 1966.

Latha, A. (Ed.). *Protein Metabolism of the Nervous System.* New York: Plenum, 1970.

Mitchell, S. W. *Injuries of Nerves and Their Consequences.* New York: Dover, 1965. (Reprinted from Lippincott, 1872 ed.)

Nakamura, Y., S. Nakajima, and H. Grundfest. The action of tetrodotoxin on electrogenic components of squid giant axons. *J. Gen. Physiol.* 48:985–996, 1965.

Narahashi, T. Chemicals as tools in the study of excitable membranes. *Physiol. Rev.* 54:813–889, 1974.

Ochs, S. *Elements of Neurophysiology.* New York: Wiley, 1965.

Ochs, S. Physiology of Nerves. In P. J. Vinken and G. W. Bruyn (Eds.), *Diseases of Nerves (Handbook of Clinical Neurology).* New York: American Elsevier, 1970. Vol. 7, Chap. 3.

Ochs, S. Systems of material transport in nerve fibers (axoplasmic transport) related to nerve function and trophic control. In D. B. Drachman (Ed.), Trophic functions of the neuron. *Ann. N.Y. Acad. Sci.* 228:202–223, 1974.

Ramon y Cajal, S. *Degeneration and Regeneration of the Nervous System* (2 vols., trans. R. M. May, 1928). New York: Hafner, 1968.

Young, J. Z. Factors influencing the regeneration of nerves. *Adv. Surg.* 1:165–220, 1949.

3. Receptors and Effectors
A. Sensation and Neuromuscular Transmission

Sidney Ochs

SENSATION

Sensations are defined as feelings or impressions produced by stimulation of afferent nerves. This definition is extended to include the action of sensory nerves which signal information not consciously registered—for example, the rate of the heartbeat, or impulses from vascular pressure receptors. Sensations may be categorized as *exteroceptive* (those relating to stimuli from the external world), *interoceptive* (those from viscera), or *proprioceptive* (those from somatic structures within the organism).

Sensations are also categorized by the quality of sensation, as *epicritic* and *protopathic*. Epicritic sensations are those readily localized on the surface of the skin. One clinical test for epicritic sensibility is *two-point localization*. The points of a pair of dividers set at various distances are placed simultaneously upon the skin, and the subject is asked whether one or two points are perceived. A close separation can be discriminated over protruding portions of the body (nose, lips, ears, and fingertips). Protopathic sensations are poorly localized. They may be experienced as aches which seem to exist over a wide area or to be poorly localized internally.

According to Müller's *law of specific nerve energies,* when particular nerve fibers are excited in any way, the sensations aroused are those that would be experienced by a normal excitation of the receptors of those fibers. For example, if optic nerve fibers are electrically stimulated, the sensation experienced is one of light. Mechanical stimulation produced by pressure applied to the side of the eyeball also brings about a visual sensation. Similarly, fibers subserving heat perception will, when adequately excited by any means, give rise to the sensation of heat.

Particular sensations can be elicited at discrete points over the skin. The points appear to be associated anatomically with specific receptors for cold, heat, pain, etc. Critical studies of the presumed relation of the various receptor organs identified histologically to the various sensory points within the skin do not show this association in all cases. Over large areas basket-like endings sensitive to touch which surround the hair follicles in hairy portions and free nerve terminals appear to be the main type of receptor organ. Weddell (1961) considers that the pattern of the termination of the free nerve endings can affect the quality of sensation; this depends in part on the overlapping pattern of the nerve fiber terminations in the receptive area. Not only does stimulation of the cornea of the eye elicit pain, but the free nerve endings within the cornea subserve sensations of touch, cold, and heat. The free nerve afferent endings in the skin and other structures are no longer considered to educe only pain. In experiments involving the skin of the cat, cold, heat, and pressure stimulation were effective stimuli in exciting action potential responses from C fibers. The receptors for these stimuli are categorized as noncorpuscular or free nerve endings. The cutaneous *corpuscular* receptors have been categorized as *encapsulated* and *nonencapsulated* receptors. They have classically been associated with specific sensations. Of the corpuscular encapsulated endings we include the *pacinian corpuscle* (touch), *Ruffini ending* (warmth), *Meissner corpuscle* (touch), *Golgi-Mazzoni corpuscle* (pressure), *Krause end-bulb* (cold). Merkel's touch receptor is an example of a nonencapsulated receptor. The significance of these structures in modifying sensory reception will be discussed later.

The Sensory Code

Receptors do not usually respond with a single action potential. When the appropriate sensory terminations are adequately excited, a repetitive discharge is elicited. In general, the rate of discharge is related to the intensity of the stimulation applied.

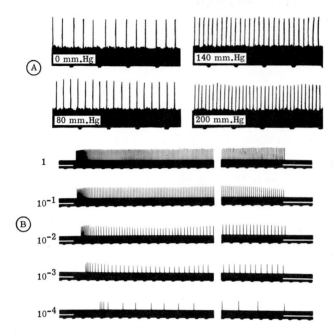

Figure 3A-1
Rate of discharge at different stimulus strengths. In A, a single fiber from a carotid sinus pressure receptor is shown with discharge rates at different levels of intracarotid pressure. As the intrasinus pressure (mm Hg) is increased, the discharge rate increases. In B, the discharge from a single optic receptor of the limulus eye is shown at different intensities of illumination. The filled bar underneath signifies the time of stimulation. As the illumination strength is decreased by factors of 10, the discharge rate is seen to decrease. (A from D. W. Bronk and G. Stella. *Amer. J. Physiol.* 110:711, 1935. B from E. F. MacNichol, Jr. In R. G. Grenell and L. J. Mullins [Eds.]. *Molecular Structure and Functional Activity of Nerve Cells.* Washington, D.C.: American Institute of Biological Sciences, 1956. P. 36.)

Rhythmicity of discharge of sensory nerves upon stimulation was clearly shown in early studies of sensory fibers known to take origin from muscle (Chap. 5). Using small groups of nerves innervating relatively isolated muscles, the experimenters could make recordings from only one or a few afferent nerve fibers. A regular rhythmic discharge was recorded from individual sensory fibers upon stretch of the muscle, the rate depending on the degree of stretch. The same thing is shown for a carotid stretch receptor (Fig. 3A-1A). A similar regular repetitive discharge in afferent fibers upon adequate stimulation was also shown for discharges in limulus optic nerves at different levels of illumination (Fig. 3A-1B).

In both cases the rate of sensory discharge increases approximately with the logarithm of the strength of the stimulus. A similar logarithmic relationship had been found by Fechner and Weber in *psychophysical* studies (see Granit, 1956). When blindfolded subjects were asked to compare weights placed in the palm of the hand, the ability to estimate small increments in weight depended on the weight already present. The relationship of stimulus to sensation is shown by the following expression:

$$\Delta S = \frac{\Delta W}{W}$$

where ΔS = increment of sensation, ΔW = small weight added, and W = weight present. By integrating:

$$S = K_1 \log W + K_2$$

where S refers to the sensation to the total weight, W, and K_1 and K_2 are constants. This logarithmic relationship was generally found to hold for other sensory stimulation—e.g., threshold responses to light or sound of different intensities. The fact that the rate of discharge found in a sensory nerve on stimulation bears a logarithmic relationship to the degree of stimulation suggests that rate is the factor controlling the psychological appreciation of the stimulus. There are, however, deviations from this relation at the extreme ranges. A more general representation of the relationship between stimulation and sensation which holds over a wider range of stimulation is the *power function*, as described by Stevens:

$$\psi = k\phi^n$$

where the sensation (psychological) magnitude ψ is related to the (stimulus) physical magnitude ϕ with a power exponent n which has values from 0.33 to 3.5, depending on the individual sense; k is a constant determined by the choice of units employed.

Adaptation is an important phenomenon relating the nature of the discharge rate seen on continued excitation of sensory inputs. Adaptation is due to processes within the receptors. When action potentials are recorded from the afferent nerves of various receptors during a continued application of a stimulus, the frequency of discharge is seen to decline (Fig. 3A-2). The rate of decline varies according to the type of receptor involved. Adaptation is moderate for stretch receptors in muscle, whereas for touch it is quite rapid. On the other hand, no adaptation is seen in carotid stretch receptors.

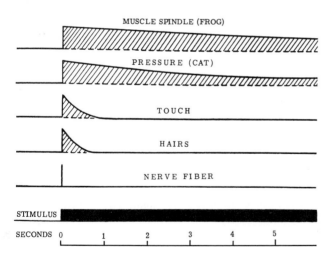

Figure 3A-2
Adaptation in different receptors. Adaptation is shown for various sensory receptors to a constant stimulus represented by a solid black bar in the lower part of the figure. The hatched parts of the various receptors show the rate at which the single units of the receptor are discharging. Adaptation is least for the muscle spindle. For the pressure receptor, adaptation is moderate, and for touch and hair receptors adaptation is rapid. The nerve fiber itself responds with a very brief discharge. (From E. D. Adrian. *The Basis of Sensation.* New York: Norton, 1928.)

Central mechanisms also play a part in the appreciation of sensory input. Stimuli continually present may be ignored. For example, the weight of clothes upon the body is neglected unless attention is drawn to it. Mechanisms which have to do with the attention to stimuli are likewise involved in the adaptive process. Some function of discrimination also exists with regard to the total sensory influx; by this means the field of sensory attention becomes "cleared" and open for new sensory information (see Chap. 9).

Transduction

Microelectrodes have been used to study the mechanism underlying excitation in the single stretch receptor cells of the crustacean (Fig. 3A-3). The afferent nerve of the stretch receptor has its branched dendritic portions wrapped around the central region of the muscular portion of the receptor. When the muscular portion is pulled upon and is elongated, a deformation of the dendritic branches of the sensory stretch receptor occurs. This results in its depolarization and the initiation of repetitive neural discharges. Because the cell body is close to the dendrites, these events can readily be recorded by means of a microelectrode inserted in the cell body. The discharge rate depends on the degree to which the receptor is stretched. The rate of discharge is determined by the depolarization produced by stretching and is known as a *generator potential* (Fig. 3A-4). This depolarization is produced within the deformed dendrites and spreads as a local current (cf. Chap. 2) to the rest of the cell, reaching the initial segment of the axon adjacent above the receptor cell body, the site at which the fiber fires repetitively. The amplitude of the generator potential here determines the rate of discharge.

It has been mentioned that repetitive activity is not usually excited in peripheral axons by a maintained depolarization unless they are subjected to special conditions such as a lower Ca^{++} level (see Chap. 2). The receptor membrane differs from axons in this respect insofar as repetitive activity to a maintained depolarization is its normal mode of behavior. The generator potential is brought about by ionic change subsequent to deformation of the specialized membrane, presumably the mechanical change opening ionic channels so that both Na^+ and K^+ can move through the membrane. This tends to bring the resting membrane potential toward zero (i.e., to depolarize the membrane).

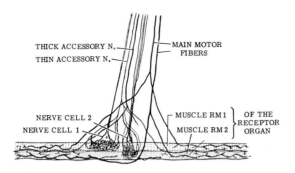

THICK ACCESSORY N.

THIN ACCESSORY N.

MAIN MOTOR FIBERS

NERVE CELL 2

NERVE CELL 1

MUSCLE RM 1

MUSCLE RM 2

OF THE RECEPTOR ORGAN

Figure 3A-3
Crustacean stretch receptors. Two stretch receptors of the crayfish are shown with the dendritic portions of the cells wrapped around the muscular parts of two different receptors. Nerve cell 1 is from the slow-adapting, and nerve cell 2 is from the fast-adapting, receptor. Movements of the muscular portions of the receptor distort the membrane of the dendritic portion of the receptors and cause repetitive discharge in those sensory afferents. Also shown are motor fibers which act in a feedback system to control the level of sensitivity of the stretch receptors. (From J. S. Alexandrowicz. *Q. J. Micr. Sci.* 92:190, 1951.)

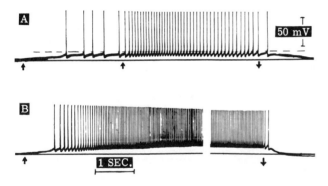

Figure 3A-4
Discharge from a single crustacean stretch receptor fiber. In A is shown the discharge of a single slow-adapting stretch receptor recorded from inside the soma with a microelectrode. A stretch is gradually applied, resulting in depolarization, indicated by the shift in the baseline. The rate of discharge from the receptor increases with the level of depolarization. With a maintained stretch a fairly constant level of discharge is maintained until the stretch is released; then depolarization decreases and the rate of discharge of the receptor falls off. In B, a greater stretch gives rise to a higher level of depolarization and a more rapid rate of discharge. (From C. Eyzaguirre and S. W. Kuffler. *J. Gen. Physiol.* 39:87–119, 1955.)

Two types of stretch receptors are found in the crustacean abdominal segments: the *slow-* and *fast-adapting*. The fast-adapting receptor shows a more rapid adaptation and decrease of rate of discharge with maintained stretch; the slow-adapting cells maintain their firing rates. This difference in behavior is caused in the fast-adapting receptor by the faster fall-off of generator potential. If depolarization is held at a controlled level by a current introduced into the receptor cell body through an electrode, the rate of the repetitive discharge continues undiminished for as long as the depolarization is held constant. A number of studies support the view that the generator potential is the means by which a mechanical change is translated into an electrical correlate; the process is termed *transduction*.

The study of the pacinian corpuscle, the mechanoreceptor subserving touch in the mammal, has complemented and added significantly to our knowledge of receptor function. The onion-like layers around the terminal sensory nerve in the receptor (see Fig. 3A-5A) are not essential to its function. When these tissues are removed by microdissection and the naked central afferent nerve fiber is delicately pressed on by means of a stylus displaced electronically, a rapid repetitive discharge can be excited in the sensory fiber. The action potential discharges are preceded by a generator potential which spreads from the deformed part of the afferent termination to initiate a series of propagated action potentials. These originate either from the nonmyelinated terminals or from the first node of the afferent fiber and are conducted to the central nervous system. The generator potential is brief in duration in this receptor and gives rise to a correspondingly brief repetitive discharge in response to a deformation of the terminal membrane (Fig. 3A-5).

The supporting structures around the receptor terminal serve to modify or limit the range of responses. They may also protect the nerve ending from excessive stimulation but for the most part act as a filter modifying the way in which deformation leads to excitation. This phenomenon is seen in other specialized sensory receptors as well where the secondary structures enable the sensory fibers to respond to selected aspects of the environment (e.g., the auditory hair cells with which the nerve terminal is associated in the ear—Chap. 4B). The static and dynamic hair cell receptors of the semicircular canals, utricle, and saccule are excited by fluid movement or gravity to enable the

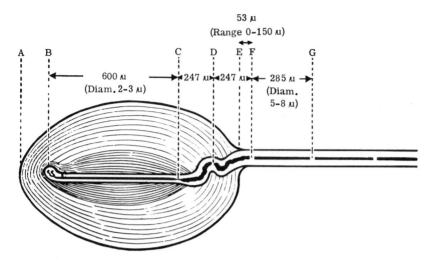

Figure 3A-5
The pacinian corpuscle. The letters C and D indicate the nodes of the sensory fiber within the receptor, F, G nodes outside. The sensory fiber between B and C is surrounded by lamellated structure extended to the corpuscle from A to E. (From T. A. Quilliam and M. Sato. *J. Physiol.* [Lond.] 129:167, 1955.)

sensory organs to signal position and movement of the head in certain directions in space (Chap. 4C). In the retina of the eye, specialized photochemical substances (Chap. 4A) have evolved to enable the optic nerve cells of the retina to respond selectively to light and to colors over a wide range of illumination.

In the specialized sensory organs, secondary cells give rise to electrical potentials in response to physical or chemical stimulation. These *receptor potentials* are similar to the generator potential in that they initiate afferent activity, but they do so in sensory afferent nerve terminals closely apposed to the cells in the receptors. The secondary receptor cells allow a transduction of physical inputs in the ear and eye which approach such high sensitivities that changes in only a few molecules can be perceived.

Unlike the transduction properties of the mechanoreceptors or special sense organs, those of the free nerve terminals of the skin are less amenable to direct study and must be inferred from the discharge of the afferent nerve fibers in response to an appropriate stimulation. By their adaptive properties it is possible to describe the rapidly adapting velocity or acceleration type of receptor compared to the slow-adapting type. The Golgi-Mazzoni corpuscle is, like the pacinian corpuscle, a rapidly adapting type of mechanoreceptor.

Slow-adapting responses are elicited from special domelike pressure receptors seen in certain regions of the hairy skin, those innervated by Merkel cells. These cells continue to discharge on applications of long-maintained pressure, falling off in rate somewhat with time. Another slowly adapting receptor is the Ruffini ending situated intradermally.

Both rapid- and slow-adapting mechanical receptors are found in receptors of the smooth or *glabrous* skin of the palms, soles, and lips.

Pain

Pain has in the past been thought to result from an excessive stimulation of all the various sensory nerves present, but recent experimental evidence shows that there are specific fibers for pain. Pain may range from a mild disagreeableness to an overpowering sensation with all else driven from the mind. Some researchers consider itch and tickle a milder stimulation of the same afferent fibers subserving pain. An argument against this view is that *pruritus* (severe itch) may appear as a separate symptom in certain diseases without being accompanied by the sensation of pain.

Pain is often described by its quality—as dull, burning, sharp, etc.; the terms make a lengthy list. The difference between the epicritic or sharp localizable pain and the duller, less easily localized protopathic pain can be demonstrated by a simple experiment. Touching the skin of the toe with a hot match ember elicits a sharp *first pain* followed after a few moments by a duller *second pain*. This *double-pain* sensation has been interpreted as epicritic pain carried in faster-conducting fibers followed by responses in the slower-conducted protopathic pain fibers. If the skin of the lower leg is similarly excited, the time between the two pain sensations is decreased, and when the knee and then the thigh are stimulated, it is even less. The results are in accord with evidence that slower-conducting C fibers carry the protopathic pain sensations and the delta group of A fibers carry the faster pain sensations.

The technique of studying pain in animals is similar to that for other sensory modalities—i.e., recording from single fibers while applying an adequate stimulation. In the case of pain the adequate stimulus is one that is damaging or *noxious*. By this means it has been appreciated that the pain receptor or *nociceptor* has a relatively high threshold and other characteristics that distinguish it as a specific sensory modality.

At present it is not clear whether the nociceptive fibers are fundamentally *chemo-receptive*—i.e., whether they respond to certain chemical substances released from damaged cells. This was the mechanism originally proposed for histamine or *H substances* or *P factors* and more recently for *bradykinin* and other substances which elicited responses in pain fibers in low concentration. In some experiments the agents are applied to the base of exposed blisters or injected into blisters or into the skin. In similar experiments serotonin, acetylcholine, prostaglandin, etc., have been found effective in eliciting pain and have been considered as possible pain-mediating agents.

Experimental attempts to determine pain thresholds (e.g., by application of graded radiant heat to a blackened spot on the forehead or fingernail of the human subject) have been used in a search for effective analgesic agents. Pain threshold experiments are generally unsuccessful, however, because subjective factors are of overriding importance in human experiments. The central mechanisms responsible for the alleviation of pain by analgesics or for the marked variation with which pain is felt by a given patient are as yet little understood.

An important characteristic of pain relative to central nervous processes is its tendency to *irradiate* and to give rise to *referred pain*. For example, pain due to heart damage may be experienced by the patient as a pain of the left upper arm or a pain passing down the left arm into the hand. Kidney disease may be felt as a back pain and the ureteral dilation and trauma incident to the passing of a kidney stone as a severe pain in the flank. Pain from a stomach ulcer may be referred to the right shoulder. These misdirections of pain sensation appear to be due to the excitation of a common pool of neurons within the spinal cord, brain stem, or cortex acted on by different afferent sources. According to this theory, pain is a functional "spillover" or irradiation of activity in a set of neurons acted on by two different sets of afferent input. Some of these general features led to a *gate* theory of pain in which intermediary cells in the substantia gelatinosa of the cord were considered to receive several incoming sensory afferent inputs, the activity of the larger fiber inputs acting to suppress the pain impulses arising from the smaller pain afferents. Such cells have not, however, been identified, and it is possible that some cells acting in "gating" fashion are present at higher (thalamic or cortical) levels.

Painful states may be long enduring. For example, a painful procedure such as the drilling of teeth may give rise to a feeling of pain long after that stimulus has ceased. The functional change in neurons producing a continued painful state may exist at various levels within the central nervous system.

The neuroma produced at the end of a cut nerve (Chap. 2) is similarly a source of pain. This may also give the sensation that a missing member is still present—a *phantom limb*. In one of the examples of an amputated arm given by Livingstone, the missing arm is experienced as a fist clenched so tightly that it constantly aches. Temporary relief may be experienced by the injection of procaine into the neuroma or its excision. However, procaine is usually not effective for long. Spinal cord section of the afferent tracts may give temporary relief, but another intervention may be required, and in some cases a frontal lobotomy may have to be resorted to for intractable pain when narcotics cannot be used.

The state of the nervous system—in particular the attention and attitude at the time of injury—is most important in determining the degree of pain experienced. It has long been noted in wartime or during a catastrophe that seriously wounded patients may not experience any pain at all at the time of injury, possibly as a result of stress or emotion. Descending influences from the brain acting perhaps on gate cells may determine which types of ascending impulses will reach the brain and consciousness.

NEUROMUSCULAR TRANSMISSION

Transmission between a motor nerve and the muscle fibers it innervates takes place at the *end-plate* (Fig. 3A-6). Transmission at the end-plate is effected by the release of a special chemical agent from the nerve terminal, a *neurotransmitter* which in the vertebrate is known to be *acetylcholine* (ACh). The transmitter released from the nerve endings when the motor nerve is fired moves across the gap between the nerve and muscle membrane in the end-plate to bind to a *receptor protein*. The result is a special potential change in the muscle membrane which then leads to a propagated action potential in the muscle fiber and muscular contraction. The proof that a special

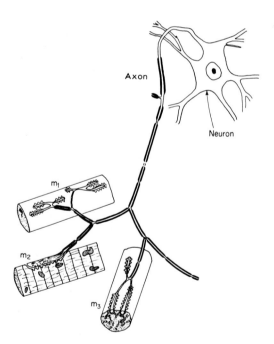

Figure 3A-6
The motor unit. The motoneuron is shown with branches terminating on muscle fibers (m_1, m_2, and m_3), at the end-plates. These are seen as indentations of the surface with folds in the membrane. (Modified from R. Miledi. *Discovery.* Oct. 1962.)

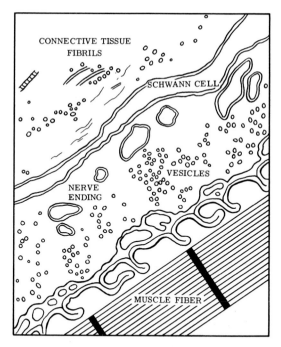

Figure 3A-7
Neuromuscular junction. A longitudinal section of a nerve fiber lying within a channel in the muscle surface is diagrammed from an electron micrograph. The nerve fiber is approximately 1.5 μ in diameter. Within the nerve terminal two types of particulate structure are shown, mitochondria and small vesicles believed to contain the transmitter substance acetylcholine. (From R. Birks, H. E. Huxley, and B. Katz. *J. Physiol.* [Lond.] 150:130, 1960.)

neurotransmitter agent was involved was first given by Loewi. Using an isolated beating frog heart, he found that some substances were released into the fluid medium following a stimulation of the vagal nerve of the heart. This *vagus stuff* caused a slowing in the rate of another similarly isolated frog heart when it was exposed to the fluid containing vagus stuff, much like the response to vagal nerve stimulation itself. The vagus stuff was later identified as ACh. Acetylcholine is labile, and, if blood is present, the enzyme *cholinesterase* in the blood splits the acetylcholine to an inactive form: acetate and choline. If *anticholinesterase agents,* such as *eserine* or *neostigmine,* are present, the enzymatic action of cholinesterase is blocked and ACh can then be detected.

Electron microscopy of the end-plate region shows that there is no merging or continuity of the nerve membrane with the muscle membrane (Fig. 3A-6). The nerve fiber occupies a cleft or channel within the muscle membrane, the space between the two membranes being approximately 300 to 500 Å (Fig. 3A-7). Vesicular bodies approximately 500 Å in diameter are found within the terminal portion of the motor nerve which is believed to contain the transmitter substance ACh. Mitochondria are also seen within the nerve endings. Mitochondria are present wherever metabolic activity occurs and a source of energy is required, usually through the $\sim$P bond of ATP (Chap. 2). Its presence in the terminals is related to the energy needed for the resynthesis of ACh from choline taken up by the terminal. The resynthesized ACh reenters the vesicles of the nerve terminals to again participate in transmission, as will be discussed in the following sections.

The End-Plate Potential

The analysis of neuromuscular transmission was markedly advanced by Claude Bernard's discovery that *curare* blocks at transmission between nerve and muscle. The agent has no effect on excitability when applied directly to the nerve or to the muscle. By *direct* stimulation and recording, propagated action potentials are obtained from either the nerve or muscle membrane, the latter accompanied by normal muscle contractions, with curare present. However, with a blocking amount of curare present, stimulation of the motor nerve to *indirectly excite* the muscle through the nerve does not elicit muscle contractions. The block was inferred to occur at the nerve-muscle junction.

Using the purified form of curare, *d-tubocurarine,* the study of transmission is carried out in vitro with a convenient preparation, the frog sartorius muscle. In a chamber the individual fibers of the isolated muscle can be readily seen, as can the approximate region of the nerve terminals ending on the muscle fibers. The actual end-plates cannot be seen unless a special microscope with Nomarski optics is used. With a microelectrode placed inside a muscle fiber close to an end-plate and a small amount of d-tubocurarine in the bath, sufficient to block indirect excitation of muscle contractions, a special electrical potential is seen remaining (Fig. 3A-8). This depolarization rises about 1 msec after nerve stimulation to reach a peak and then declines over a longer time period of some 15 to 20 msec. The potential is best recorded from the end-plate region and is referred to as the *end-plate potential* (EPP). With increased d-tubocurarine the EPP is diminished and then completely blocked.

The EPP differs from the action potential in several important respects. First, it is not an all-or-none response; it diminishes in amplitude with distance from the region of the end-plate (Fig. 3A-9). Also, unlike the propagated action potential, the EPP can show the property of *summation.* When EPP's are small, their summation may cause the membrane to fire a propagated action potential. While the propagated action potentials are brought about by a brief increased permeability first to Na^+ and then after an interval to K^+, the EPP represents a depolarization brought about by a simultaneous

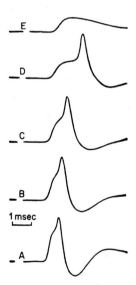

Figure 3A-8
Records taken at the end-plate region. A. Before application of curarine. B, C, and D. During progressive curarization, the diminution of the initial end-plate potential (EPP) and the progressive lengthening of the spike latent period. E. Pure EPP; no spike is set up. (From Kuffler, 1940.)

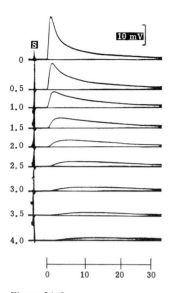

Figure 3A-9
End-plate potential. The end-plate potential (EPP) is recorded with a microelectrode from just under the muscle membrane in the end-plate region. It is a depolarization which occurs after a latency following stimulus at S. The EPP is relatively long lasting and the height decreases with distance, as shown by the numbers indicating in millimeters the distance of the recording micro-electrode from the end-plate. (From P. Fatt and B. Katz. *J. Physiol.* [Lond.] 115:320, 1951.)

increased permeability to both Na^+ and K^+. The extracellular fluid is mainly composed of Na^+ and Cl^-, with small amounts of K^+ and other ions present. When the permeability suddenly increases to Na^+ and K^+ after ACh binds to the receptor, these cations move according to their concentration and electrical gradients. The result is a reduced membrane potential (-12 mv) as it approaches the *equilibrium potential* for the EPP (E_{EPP}). This represents the electrochemical equilibrium for Na^+ and K^+.

The blocking agent d-tubocurarine acts by its binding to the receptor protein; consequently the ACh released from the nerve ending cannot bind to those sites. The resulting block shows competitive inhibition. With increasing curarization the EPP is reduced in size, the extent of reduction depending upon the amount of d-tubocurarine bound to the receptors rending them unavailable to the transmitter ACh. Normally the EPP is relatively large, with an amplitude greater than 50 to 60 mv and able to depolarize the adjoining muscle membrane to excite a propagated action potential in the muscle membrane. When the EPP has been reduced in amplitude below that critical level, it cannot depolarize the muscle membrane to give rise to a propagated action potential. This is the way the block to indirect excitation comes about by d-tubo-curarine.

The permeability increase to Na^+ and K^+ brought about by the action of ACh is relatively brief, of the order of a millisecond or two. The relatively long duration of the EPP is primarily due to the flow of local currents in the adjoining muscle membrane, in the manner previously discussed with respect to local currents and electrotonic potentials in the nerve membrane (Chap. 2). The amplitude of the spread of currents and the EPP decrease with the distance from the end-plate (Fig. 3A-9) are the result of such local current spread. As discussed for local currents, the amplitude of the resulting potential at a distance from the end-plate is determined by the electrical characteristics of the muscle, its internal resistivity and membrane resistance.

The evidence that ACh is in fact the neurotransmitter was substantiated by the study

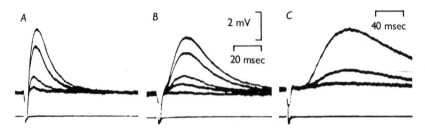

Figure 3A-10
Responses of a junctional sensitive spot to ACh pulses of variable strength delivered at different positions of the pipette. In *A* the pulse duration varied from 1 to 2 msec; in *B*, from 2 to 3 msec; and in *C*, from 3 to 5 msec. The distance from source to receptors was increased from *A* to *C*. In *A* the receptors were located within the source radius; in *B* and *C* the calculated distances were 14 μ and 21 μ respectively. Notice that the time to peak depends on the position of the pipette but not on the pulse strength. Same time calibration for *A* and *B*. (From Feltz and Mallart, 1971.)

of the effect of ACh applied through a micropipette directly to the end-plate region, a technique referred to as *iontophoresis*. The ACh in a microelectrode is ejected from the tip by the passing of an electrical pulse of current. The released ACh causes increased permeability and depolarization in the muscle membrane with an appearance similar to an EPP (Fig. 3A-10). This *ACh potential* can, if it reaches a high enough level, set off a propagated action potential, as does the EPP. When ACh is released from a micro-electrode inserted just under the membrane of the end-plate, it is ineffective. Apparently, only the outer surface of the receptor substance in the membrane of the end-plate is responsive to ACh. The specific cholinesterase acetylcholinesterase (AChE) also present in the muscle membrane at the end-plate acts to terminate the depolarization produced by ACh. It is located adjacent to the receptor protein and seems not to be actually part of the receptor protein itself. When an anticholinesterase agent such as eserine is added to a muscle bath, the EPP is much larger and prolonged in duration.

Choline, succinylcholine, and decamethonium have some similarity in structure to ACh and also act to open the receptor channels. However, they are not acted on by AChE, producing a much longer-lasting depolarization at the end-plate which acts as a block of neurotransmission. The mode of action of these blocking agents therefore differs radically from that of curare, which does not cause a depolarization; hence these compounds are referred to as *depolarizing blocking agents*.

Besides the curariform and the depolarizing groups of neuromuscular blocking agents, other types of neuromuscular block are produced by agents acting on the membrane of the motor endings to interfere with the mechanism of release of the transmitter substance. While the mechanism of release will be discussed in the following section, we may complete this brief survey of blocking agents by noting that *botulinum toxin*, an extremely potent poison, is selectively taken up from the circulation by motor nerve endings, in which it acts to block the release of ACh. *Black widow spider venom* blocks by causing a complete release of all the transmitter with a disappearance of vesicles from the terminal. Death following exposure to any of these agents usually comes about from a paralysis of the respiratory muscles.

MEPP and the Release Mechanism
A continual and irregular discharge of small potentials, several millivolts in amplitude, has been found in the vicinity of the end-plate (Fig. 3A-11). They have a duration and other properties similar to those of the EPP except for their very much smaller size. On this basis they have been termed *miniature end-plate potentials* (MEPP's). The miniature EPP's, like the EPP, are blocked by d-tubocurarine, and their duration is lengthened by the presence of the anti-AChE agents eserine and neostigmine.

The variable latency for each discharge giving rise to an MEPP can be described by a probability curve. Each unit or quantum of the discharge consists of some 10,000 to 20,000 molecules of ACh, the amount apparently contained in one vesicle. This number was determined by using iontophoresis and releasing a known amount of ACh. Comparison of the depolarization produced by the ACh released (Fig. 3A-10) with the depolarization produced by the MEPP made it possible to estimate the number of molecules per quantum in the MEPP.

An attractive theory is that the simultaneous release of a number of quanta from the nerve terminal (each gives rise to an MEPP) brings about an EPP. In support of this concept it has been shown that the rate of discharge of the MEPP is increased greatly by the depolarization of the motor nerve terminals. When the motor fiber is excited, an action potential invades the whole of the nerve terminal in the end-plate region to depolarize it and thus allow the release of a large number of MEPP's. These are sufficiently synchronized to give rise to the EPP.

There is less than a 100 μsec latency for the depolarization when ACh is applied iontophoretically to the end-plate. This is a small fraction of the usual latency, which ranges from approximately 0.5 to 2.5 msec on nerve stimulation. Most of the latter interval is therefore the time required for the release of ACh from the nerve terminal.

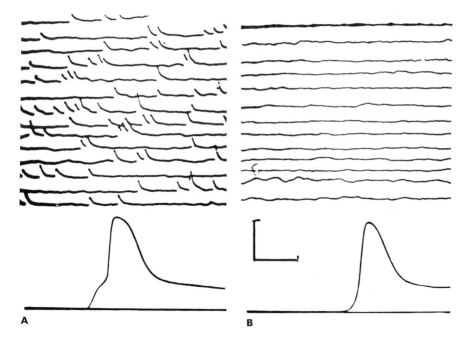

A **B**

Figure 3A-11
Spontaneous miniature end-plate potentials at the myoneural junction. The records were obtained with intracellular recording. *A.* Microelectrode was placed inside a frog muscle fiber at the nerve-muscle junction. *B.* Electrode was placed 2 mm away into the same muscle fiber. The uppermost portions were recorded with slow speed and high amplification (calibrations: 3-6 mv and 47 msec); the lower portions of *A* and *B* show the response to a nerve impulse with fast speed and low-gain recording (calibrations: 50 mv and 2 msec). The stimulus was applied to the nerve at the beginning of the sweep; response *A* (at the end-plate) shows the steplike initial EPP, and response *B* shows the propagated action potential wave, delayed by conduction over a distance of 2 mm. The figure shows that the spontaneous activity is restricted to the myoneural junction. (From del Castillo and Katz, 1956.)

Such release of transmitter also requires that Ca^{++} be present in the medium. The most likely hypothesis is that when the nerve action potential depolarizes the nerve membrane it allows Ca^{++} to enter the terminal and act on the vesicle, which then releases its ACh from the terminal. The need for Ca^{++} to initiate transmitter release was shown by the use of tetrodotoxin (TTX)-treated preparations to block action potentials in the terminal. The latter can then be depolarized. A microelectrode placed on the nerve terminal to depolarize it can initiate a release of transmitter if the medium contains Ca^{++}. This result was shown by timing a pulse of depolarizing current immediately after a pulse to a Ca^{++}-containing micropipette placed on the nerve terminal so that Ca^{++} was present during depolarization (Fig. 3A-12).

The movement of vesicles to the inner wall of the terminal and the discharge of their contents into the synaptic cleft appear to involve fairly complex mechanisms. Some series of internal events are required to account for the temporal changes seen when the nerve is stimulated at high rates. Repetitive stimulation decreases the size of the EPP's. Alternatively, periods of high-frequency stimulation of the motor nerve may be followed by an increase in the size of the EPP for a short time, a *posttetanic potentiation* (PTP). This phenomenon appears to be due to an increased *mobilization* of the transmitter-containing vesicles. Just how the vesicles move to the inner membrane and discharge their quantal contents is a matter of lively debate. There seem to be both

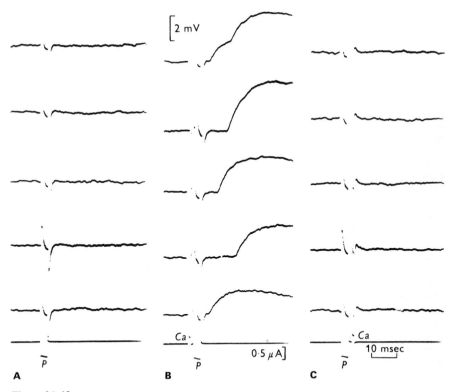

Figure 3A-12
Effect of iontophoretic pulses of calcium (Ca) on end-plate response. Depolarizing pulses (P), and calcium were applied from a twin-barrel micropipette to a small part of the nerve-muscle junction. Intracellular recording from the end-plate region of a muscle fiber. Bottom traces show current pulses through the pipette. A. Depolarizing pulse alone. B. Calcium pulse precedes depolarizing pulse. C. Depolarization precedes calcium pulse. Temperature 4°C. (From Katz and Miledi, 1967.)

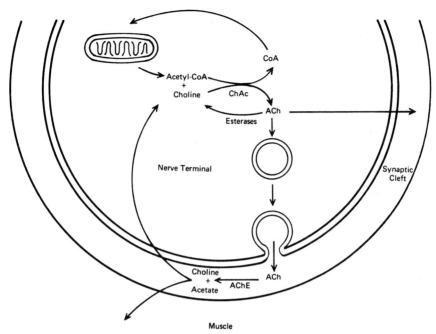

Figure 3A-13
The synthesis, storage, and release of ACh. Arrows indicate various metabolic cycles which interlock in the synthesis of ACh. Release of ACh into the synaptic cleft is indicated in quantal form from a vesicle and in nonquantal form (arrow). Choline uptake from the synaptic cleft by nerve terminals and muscles is indicated by arrows crossing the nerve terminal and muscle boundaries. Choline acetylase (ChAe) is used to resynthesize ACh, which is then packaged into the vesicles shown by arrows, carried to the membrane, and discharged. (From Hubbard, 1974.)

short- and longer-lasting changes related to the mobilization of vesicles. The longer-period changes merge with still longer-lasting processes of transmitter replacement by resynthesis locally in the terminals. Also, over longer time periods a supply of some components is provided by axoplasmic transport (Chap. 2).

Local processes also involve the resynthesis of ACh from choline taken up into nerve terminals by a carrier present in the membrane, with acetate provided as *acetyl-CoA* through metabolism (Fig. 3A-13). By this means the transmitter ACh is resynthesized and it then reenters the vesicles to be used later on in transmission.

Neuromuscular transmission is poor in the muscle disease *myasthenia gravis,* possibly because of a deficiency of transmitter or of components involved in transmitter release or because receptor action is blocked. Recently, a deficiency in the synthesis of receptors was reported present in the myasthenic muscle fibers. Transmission fails in such muscles because the resulting EPP's are too low in amplitude to excite a propagated action potential in the muscle membrane. Anticholinesterase agents such as eserine, neo-stigmine, and similar agents can improve neuromuscular transmission by prolonging the duration of the EPP and increasing its amplitude. It can then attain critical levels of depolarization for muscle action potentials and contraction.

Trophic Influence of Nerve on Muscle
When the motor nerve to a skeletal muscle is transected, the denervated muscle undergoes a series of profound biochemical and functional changes. These take place over a period of some days or weeks. Increased excitability and sensitivity to injected ACh is seen early, the muscle showing *denervation hypersensitivity.* In this state the

muscle shows *fibrillation,* large irregular discharges, and abnormal sensitivity to mechanical stimulation and ACh. Metabolism is abnormal, thus contributing to the slow shrinkage in muscle size over the weeks or months known as *denervation atrophy.*

Recent studies have shown that, whereas normally only the end-plate region is highly sensitive to ACh, the area of sensitivity to ACh spreads slowly outward from the end-plate region of a denervated muscle, so that in a matter of days or weeks, depending on the species, most or all of the membrane becomes sensitive to ACh. This activity was shown by iontophoretic application of ACh at graduated steps all along the length of the fibers. If the muscle becomes reinnervated, the increased sensitivity along the fiber diminishes until only the end-plate as usual shows a high sensitivity to ACh. An interpretation of this phenomenon is that a trophic substance carried down inside the nerve fibers by axoplasmic transport (Chap. 2) passes into the muscle fiber to control the synthesis of the receptor protein within it. In its absence, there is a *derepression* of receptor protein synthesis, and new receptor protein becomes inserted into the muscle membrane outside the end-plate region.

Following nerve interruption or in animals with *botulinum* poisoning where the release of transmitter is blocked as shown by failure of MEPP discharge, the muscle also develops hypersensitization. Thus one possibility is that ACh is the trophic material. However, the failure of ACh applied to denervated muscle to prevent hypersensitization in vitro and other evidence suggests that some as yet unknown trophic substance controls the synthesis of receptor protein in the muscle.

A number of different trophic or control agents appear to be carried from the nerve into the muscle. In the mammal, the white muscles are known to have a faster twitch duration than the red muscles. The slower-contracting red muscles are tonic in function (see Chap. 5), and their longer twitch duration is related to the prolonged contractions required for the maintenance of posture. If the motor nerves to the fast and slow muscles are cut and the nerves united so as to regenerate into the different muscles, the muscles are now innervated by the different nerves. In such *cross-innervations* the fast muscles become slow-contracting and the slow muscles tend to become fast-contracting. This result is explained by the concept that trophic substances carried down within the two types of nerve fiber determine the functional properties of muscles they innervate.

Further study of the trophic control of muscle by nerve is expected to reveal that certain diseases such as muscular dystrophy and other ill-defined muscle diseases may well have their basis in an alteration in the proper supply of trophic substances by their innervating nerves.

REFERENCES

Birks, R., H. E. Huxley, B. Katz.   The fine structure of the neuromuscular junction of the frog. *J. Physiol.* (Lond.) 150:134–144, 1960.

Davis, H.   Some principles of sensory receptor action. *Physiol. Rev.* 41:391–416, 1961.

del Castillo, J., and B. Katz.   Biophysical aspects of neuro-muscular transmission. *Prog. Biophys. Mol. Biol.* 6:121–170, 1956.

Drachman, D. B. (Ed.).   Trophic functions of the neuron. *Ann. N. Y. Acad. Sci.* 228:1–423, 1974.

Eccles, J. C.   *The Physiology of Synapses.* New York: Academic, 1964.

Eccles, J. C.   *The Understanding of the Brain.* New York: McGraw-Hill, 1973.

Feltz, A., and A. Mallart.   An analysis of acetylcholine responses of junctional and extrajunctional receptors of frog muscle fibres. *J. Physiol.* (Lond.) 218:85–100, 1971.

Granit, R.   *Receptors and Sensory Perceptors.* New Haven, Conn.: Yale University Press, 1956.

Granit, R.   *The Basis of Motor Control.* New York: Academic, 1970.

Guth, L.   "Trophic" influence of nerve on muscle. *Physiol. Rev.* 48:645–687, 1969.

Gutmann, E. (Ed.). *The Denervated Muscle.* Prague: Czechoslovakian Academy of Science, 1962.

Gutmann, E., and P. Hnik (Eds.). *The Effect of Use and Disuse on Neuromuscular Functions.* Prague: Czechoslovakian Academy of Science, 1963.

Hubbard, J. I. (Ed.). *The Peripheral Nervous System.* New York: Plenum, 1974.

Hubbard, J. I., R. Llinas, and D. M. J. Quastel. *Electrophysiological Analysis of Synaptic Transmission.* Baltimore: Williams & Wilkins, 1969.

Iggo, A. Cutaneous Receptors. In J. I. Hubbard (Ed.), *The Peripheral Nervous System.* New York: Plenum, 1974.

Katz, B. *Nerve, Muscle, and Synapse.* New York: McGraw-Hill, 1966.

Katz, B., and R. Miledi. Timing of calcium action during neuromuscular transmission. *J. Physiol.* (Lond.) 189:535–544, 1967.

Krnjević, K. Chemical nature of synaptic transmission in vertebrates. *Physiol. Rev.* 54:418–540, 1974.

Kuffler, S. W. Electric potential changes at an isolated nerve-muscle junction. *J. Neurophysiol.* 5:18–26, 1940.

Livingston, W. K. *Pain Mechanisms: A Physiologic Interpretation of Causalgia and Its Related States.* New York: Macmillan, 1947.

Melzack, R., and P. D. Wall. Pain mechanisms: A new theory. *Science* 150:971–979, 1965.

Phillis, J. W. *The Pharmacology of Synapses.* Elmsford, N.Y.: Pergamon, 1970.

Rosenblith, W. A. (Ed.). *Sensory Communication.* New York: Wiley, 1961.

Singer, M., and J. P. Schade. Mechanisms of Neural Regeneration. In M. Singer and J. P. Schade (Eds.), *Progress in Brain Research.* New York: American Elsevier, 1964. Vol. 13.

Sweet, W. H. Pain. In J. Field (Ed.), *Handbook of Physiology.* Washington: American Physiological Society, 1959. Vol. 1, Section 1: Neurophysiology, pp. 459–506.

Thesleff, S. Effects of motor innervation on the chemical sensitivity of the skeletal muscle. *Physiol. Rev.* 40:734–752, 1960.

Weddell, G. Receptors for Somatic Sensation. In M. A. B. Brazier (Ed.), *First Conference on Brain and Behavior.* Washington: American Institute of Biological Sciences, 1961. Pp. 13–48.

3. Receptors and Effectors
B. Effectors: Striated, Smooth, and Cardiac Muscle*

Richard A. Meiss

The task of muscle is to perform some mechanical action in response to a neural or hormonal command. In the more familiar instances, muscle serves as the final link between the voluntary activity of the central nervous system (CNS) and the external environment. Obvious examples include muscles used for locomotion and speech. Equally important, however, are those muscles under hormonal or involuntary nervous control that assist in maintaining a stable internal environment. The heart, the viscera, and the peripheral circulatory system, for example, contain muscle that is not controlled voluntarily.

A consideration of the role of muscle in the total integrated function of the whole body or of a specific organ or system must rest on an understanding of muscle as a contractile system. In this sense, muscle is a type of motor, one that consumes fuel, produces both desired mechanical effects and unwanted thermal and chemical waste products, and is subject to a variety of controls. Muscle may also be considered a form of *transducer* since it changes electrical (nerve) signals into mechanical motion.

In spite of the diversity of muscle types found in the human body, some properties are common to all; the adaptation of a specific muscle to a specific task involves modification and refinement of one or more of these basic properties until it becomes a dominant characteristic. On the basis of their predominant functional and anatomical characteristics, muscles are usually grouped into three major categories. The most familiar group is *skeletal muscle,* so called because it is almost always found attached in some way to skeletal structures. This category is also refered to as *striated* (striped) muscle, a reference to its microscopic structure; it is also called *voluntary* muscle since it is usually (but not always) under the voluntary control of the somatic nervous system. A second category is that of *visceral muscle.* As the name implies, this type is found largely associated with the viscera, or internal organs. It is also categorized structurally as *smooth* muscle because it lacks the microscopic striations of skeletal muscle. Since it is usually (but not always) under involuntary control of the autonomic nervous system (ANS), it is also called *involuntary* muscle. The third category is *heart (cardiac) muscle,* with an obvious anatomical location. While this muscle type has many microscopic similarities to skeletal muscle and some functional similarities to visceral muscle, its mode of control, its unique location, and its vital importance merit putting it into a separate category.

The aim of this chapter is to provide a basic understanding of the shared properties common to muscle in general and then to examine how special modifications of the common properties are used to physiological advantage.

MUSCLE STRUCTURE

Skeletal Muscle

Because they are able only to pull and not to push, all muscles are anatomically located so that some sort of force can be applied to relengthen them after contraction. This universal necessity for providing a lengthening force imposes an important restriction on the possible anatomical arrangements of all muscle types. Hence, skeletal limb muscles are usually arranged around a joint in antagonistic pairs, so that when one muscle shortens, the other is lengthened (Fig. 3B-1). The muscles of the heart, having shortened to pump out blood, are relengthened by the force of new blood entering the

*This chapter is adapted from E. E. Selkurt, *Basic Physiology for the Health Sciences,* Boston: Little, Brown, 1975. Pp. 193–217.

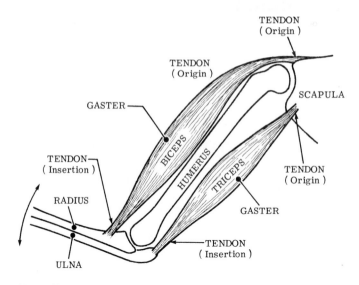

Figure 3B-1
A highly simplified representation of two major muscles of the upper arm. They are arranged antagonistically so that contraction of the biceps will stretch out the triceps, and contraction of the triceps will relengthen the biceps. (In actuality, the biceps has a double origin, as its name implies, and the triceps has a triple origin and a more complex insertion than is shown here.) (From R. A. Meiss. In E. E. Selkurt [Ed.], *Basic Physiology for the Health Sciences.* Boston: Little, Brown, 1975.)

pumping chambers of the heart. Similarly, shortened muscle in the walls of the intestine is relengthened when that segment of gut is actively filled with material moved from another region. In some cases, muscles are arranged to work against elastic tissue structures that cause the relaxed muscle to return to its resting position; the muscles responsible for facial expression are an example.

Skeletal muscles are usually distinct and relatively large structures, and their gross anatomy is straightforward. A whole muscle and the main structures of the muscle cell are shown in Figure 3B-2. The individual muscle cells (called *fibers*) are from 20 to 90 microns (μ) in diameter and may be many centimeters long, depending on the length of the whole muscle (Fig. 3B-2C). The fibers are multinucleate, with the nuclei located at intervals directly beneath the *sarcolemma,* which is a very thin connective tissue sheath enclosing each individual fiber. The sarcolemma also contains the usual *plasma membrane* as one of its components. The cytoplasmic region is largely filled with the contractile substance, the *myofibrils* (Fig. 3B-2D). These bear the characteristic cross-striated pattern, about which more will be said later. The muscle fibers are grouped into bundles called *fasciculi* (Fig. 3B-2B), arranged in an approximately parallel fashion and fastened together side by side with connective tissue called the *perimysium.* Additionally, each fiber is surrounded by a delicate sheet of connective tissue, the *endo-mysium,* which is separate and distinct from the sarcolemma. The large fiber bundles are further bound together to form a complete muscle with its connective tissue covering, the *epimysium.* At both ends, connective structures, the *tendons,* serve to connect the bundled fibers to the skeleton. The end of the muscle attached to the more stationary skeletal part is its *origin,* the end attached to the part to be moved is the *insertion,* and the thickened middle portion is the *belly* or *gaster.*

Many variations on the basic arrangement are found. The *biceps,* for example, is a relatively long muscle with a double origin and a single insertion. Its fibers are arranged to give it a roughly circular cross section. The *diaphragm,* on the other hand, is a flat sheet of roughly parallel fibers that originate on the skeleton but insert on a central

81

SKELETAL MUSCLE

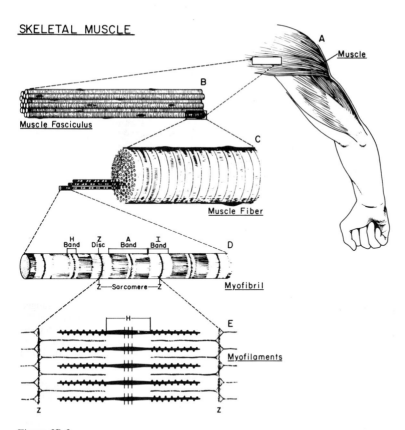

Figure 3B-2
The hierarchy of skeletal muscle organization. Muscle is organized into anatomical units that can be subdivided several times into increasingly smaller functional units. A whole muscle (A) can be subdivided into fiber bundles called fasciculi (B). These can be further broken down into individual muscle fibers (C), which are single muscle cells. The cells contain fibrous structures, the myofibrils (D). Each myofibril is composed of a series arrangement of repeating structures, the sarcomeres. The sarcomere is the smallest level of organization capable of contraction. Every sarcomere is composed of complex protein filaments, the myofilaments (E) (see Fig. 3B-5), which are in turn composed of large molecules of the fibrous protein myosin (thick filaments), or the smaller globulin protein actin (thin filaments), along with other protein components mentioned later in the text. Although muscles throughout the body exhibit a wide range of shapes and sizes, at the ultimate functional level their structures are remarkably similar. (From W. Bloom and D. W. Fawcett. *A Textbook of Histology* [9th ed.]. Philadelphia: Saunders, 1968.)

tendon. Some striated muscles in the tongue and in the upper esophagus neither originate nor insert on the skeleton. The gross structures and connections of other skeletal muscles are similarly tailored to their location and major function.

Smooth Muscle
Smooth muscles existing as separate organs are relatively rare. Smooth muscle usually forms an intimate part of the wall of a hollow tubular or saclike organ; often it forms the largest portion of the wall thickness. The individual smooth muscle cells (Fig. 3B-3) are small, tapered structures 5 to 10 μ in diameter and from 200 to 500 μ long. The nucleus (one per cell) is large and centrally located; the myofibrillar substance shows no periodic cross striations. The cells are grouped by connective tissue into a relatively

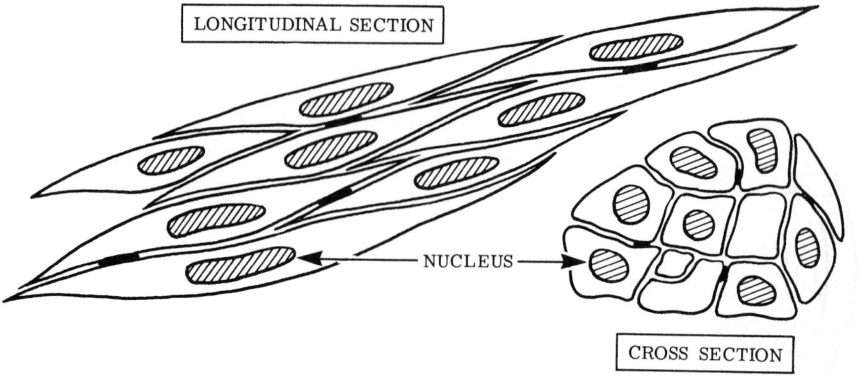

Figure 3B-3
Smooth muscle cells in longitudinal and cross sections. The cells are long and slender, tapering at the ends. In the cross section, some cells are cut so that their nuclei do not appear in the illustration. At many regions throughout the population of cells, adjacent cells are in close contact; this very close proximity aids in cell-to-cell electrical communication. Connective tissue fibers form a network in which the cells are embedded and to which they may also be attached. There are no tendons as such. (From R. A. Meiss. In E. E. Selkurt [Ed.], *Basic Physiology for the Health Sciences.* Boston: Little, Brown, 1975.)

parallel array with a definite directional orientation. In many organs, notably the small intestine (Fig. 3B-4), the smooth muscle is arranged into a circular layer (running around the circumference of the tube) and a longitudinal layer (running along the length of the tube). In other organs, such as the stomach and the uterus, there are three or more layers, with various distinct orientations. Very small blood vessels, on the other hand, may have a single muscle cell completely encircling the structure. A specialized smooth muscle structure, one found at numerous points in the circulatory system and the viscera, is the *sphincter.* This is a region of extra-thick, circularly oriented smooth muscle encircling a tube. Contraction of the sphincter smooth muscle closes off that region of the tube and prevents flow of its contents. Since a sphincter possesses a natural geometrical mechanical advantage, high fluid pressure (such as may be found in the urinary bladder, for example) may be contained with little muscular effort. In the skin, tiny individual smooth muscles are responsible for causing hairs to stand erect and for raising goose pimples. Thus, as was the case for skeletal muscle, the anatomy of smooth muscle has many variations to suit a variety of physiological needs.

Cardiac Muscle

The gross anatomy of cardiac muscle is treated in Chapter 13. It will suffice here to say that the cardiac muscle cells are much shorter (approximately 100μ) and smaller in diameter (9 to 20μ) than skeletal muscle fibers (Fig. 3B-5). The cells are arranged in a branching network with a predominantly longitudinal organization; the outer portions of the cells are covered by a sarcolemma, and the end-to-end connections of the cells are made at the *intercalated disc,* which also functions in the electrical interconnection of the cells. The nuclei are centrally located, and the myofibrillar substance has the cross-striated pattern found in skeletal muscle.

The electron microscope has permitted extensive investigation of the fine structure of muscle in recent years, and a great deal of progress has been made in understanding how the contractile process operates. Here again, studies of skeletal muscle have yielded the clearest picture.

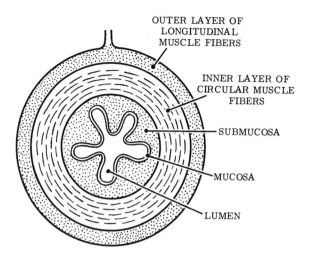

OUTER LAYER OF
LONGITUDINAL
MUSCLE FIBERS

INNER LAYER OF
CIRCULAR MUSCLE
FIBERS

SUBMUCOSA

MUCOSA

LUMEN

Figure 3B-4

A diagrammatic cross-sectional representation of the small intestine. The smooth muscle in this organ is arranged into two layers whose fibers are oriented at right angles to each other. The outer muscle layer is made up of longitudinal fibers; its fibers and fiber bundles run in the direction of the length of the intestine. The fibers of the inner, circular layer run around the intestine. Because the intestine is filled with incompressible fluid, contraction of the longitudinal layer will cause the intestine to become shorter and fatter, while contraction of the circular layer will cause the diameter to decrease and the total length of the intestine to increase, thus stretching the longitudinal muscle fibers. Compare this type of antagonistic relationship with the more direct one shown in Figure 3B-1. (From R. A. Meiss. In E. E. Selkurt [Ed.], *Basic Physiology for the Health Sciences.* Boston: Little, Brown, 1975.)

Fine Muscle Structure

The smallest organized unit of the contractile mechanism of skeletal muscle is the *sarcomere,* as illustrated in Figures 3B-5C, 3B-6, and 3B-8. A single *myofibril* is composed of many sarcomeres arranged both end to end and side by side. The cross striations visible in the light microscope can be associated with the dark and light regions of the sarcomere; the thin dark bands that define the ends of the sarcomere are the *Z lines.* On either side of a given Z line is a light area, the *I band,* and midway between any two Z lines is a dark area, the *A band.* Down the center of the A band is a darker stripe, the *M line,* surrounded on both sides by a lighter *H zone.* Close inspection reveals that the A and I bands are made up of the myofilaments, which are oriented along the longitudinal axis of the muscle—that is, perpendicular to the Z line. The thin filaments of the I band, which attach at one end to the Z line, are seen to penetrate among the thicker filaments of the A band; thus the A band and the I band overlap, and the degree of overlap can vary. The lightness of the H zone is due to the absence of I filaments in this region of the A band; if there is less overlap, the H zone will be wider and the Z lines will be farther apart.

Biochemical and biophysical studies have shown the fibrous protein *myosin* to be the principal constituent of the thick filaments. The individual myosin molecules possess a very long (about 1200 Å) straight tail and a double globular head portion. When packed together to form a thick myofilament, the tails of the myosin molecules lie side by side and form the "backbone" of the myofilament; the globular head portions protrude from the structure at regular intervals to form the cross-bridges seen in Figures 3B-6 and 3B-9. The globular head of the myosin molecules contains the special

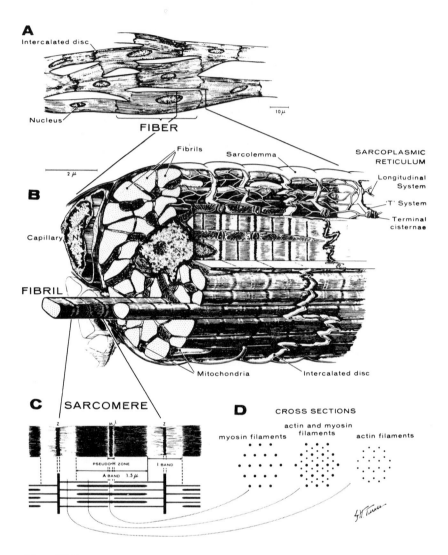

Figure 3B-5
The anatomical organization of heart muscle. As shown in A, the fibers of heart muscle are short and branched and interconnect with one another. Each fiber (cell) has one nucleus. The individual fibers (B) are supplied with many mitochondria and are internally organized into fibrils (or myofibrils). Each fibril in turn is composed of a chain of sarcomeres (C), which are made up of two sets of myofilaments (C, D). Thus, heart muscle shares some of the anatomical features of both skeletal muscle (Figs. 3B-2, 3B-6) and smooth muscle (Fig. 3B-3). (From E. Braunwald, J. Ross, and E. H. Sonnenblick. *Mechanism of Contraction of the Normal and Failing Heart.* Boston: Little, Brown, 1968.)

biochemical and enzymatic properties (to be discussed later) that enable the thick filaments to participate in the contraction process. The myosin tails appear to serve a structural function. Because the myosin molecules pack together in a compact way, the protruding heads extend in all directions from the thick filaments in a spiral fashion that undergoes one complete "turn" every 426 Å; this is the distance between successive

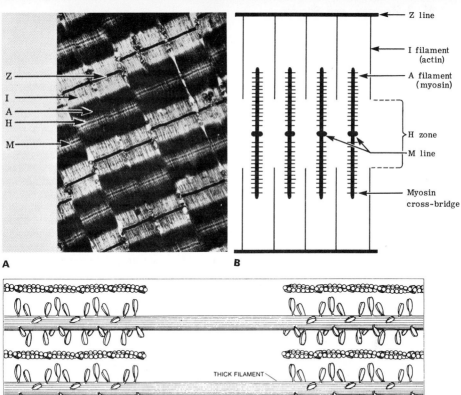

A   B

In figure B, the labels read (top to bottom): Z line, I filament (actin), A filament (myosin), H zone, M line, Myosin cross-bridge. On the left side of A: Z, I, A, H, M.

C

THICK FILAMENT

THIN FILAMENT

BARE ZONE

Figure 3B-6

The ultrastructural organization of skeletal muscle. A. The complex pattern of bands produced by the overlap of the two sets of myofilaments is shown in this electron micrograph. The lightest areas represent the I bands, and the wide dark areas, the A bands. The lighter region at the center of the A band is the H zone; and the center of the H zone is made lighter still by the absence here of *cross-bridges* on the myosin filaments. B. Diagram of a single sarcomere, the fundamental organized unit of the contractile apparatus. The lengths of the filaments are drawn to scale, but the lateral spacings have been exaggerated for clarity. The width of the bridge-free portion of the H zone is constant; the total width of the H zone is measured between the tips of I filaments extending from the opposite Z lines, and this dimension will vary, depending on the degree of myofilament overlap. By convention the sarcomere is considered to run from Z line to Z line. C. The assembly of the various protein components of the myofilaments into a portion of a sarcomere. The helical "bead-chain" structure of the actin polymer is shown in the thin filaments, as is the presence of the molecules of troponin and tropomyosin in the "groove" region of the actin helix (this may be seen more clearly in Fig. 3B-9). The heads of the myosin molecules protrude from the thick filaments, the major portion of which is composed of the myosin tail portions. No heads protrude from the center portion of the thick filaments because this region includes tail portions only; no cross-bridges can be formed in this region of the sarcomere. The M line protein shown in B is not shown here. (A is from H. E. Huxley. Muscle Cells. In T. Brachet and A. E. Mirsky [Eds.], *The Cell* [vol. 4]. New York: Academic, 1960. B is from R. A. Meiss. In E. E. Selkurt [Ed.], *Basic Physiology for the Health Sciences.* Boston: Little, Brown, 1975. C is from J. M. Murray and A. Weber, The cooperative action of muscle proteins. Copyright © 1974 by Scientific American, Inc. All rights reserved.)

cross-bridges as "seen" by any single thin filament. The center region of the thick filament (the pseudo H zone) is devoid of any cross-bridges because this region contains myosin tails only; the myosin molecules are arranged in a "head-out" fashion, starting at the center of the filament. Thus the thick filament is a symmetrical structure, with identical halves extending in either direction from the M line (Fig. 3B-6B). This built-in directionality is important in the contraction process.

The thin filaments are composed of three separate proteins. The largest contributor to the structure is *actin.* This is a globular protein whose relatively small molecules (55 Å) are approximately spherical; they are joined together to form two long bead-chain structures, which in turn wrap about each other to form a helix, which completes a half-turn every seven actin subunits (see Fig. 3B-6C). This type of structure comprises the whole 1-μ length of the thin filament; at one end the thin filament is attached to the Z line protein and then extends through to be a thin filament (1 μ long) of the adjacent sarcomere. Because the two faces of the individual actin molecules are not chemically identical, actin filaments also show a directionality important to the contraction process. That is, the portions of an actin filament on either side of a Z line are oppositely directed along their entire lengths. Stated in another way, the actin filaments inter-digitating with the thick filaments from either end of a single sarcomere have opposite directionalities that are matched to the bidirectional character of the thick filaments.

The two other proteins found on the thin filaments are *troponin* and *tropomyosin.* A pair of the small troponin molecules lie on opposite sides of the actin helix every seven-actin half-turn (Figs. 3B-6C and 3B-9). From each troponin site a long, slender tropomyosin molecule extends for seven actin units along the groove formed by the two strands of the helix. It is the presence of the troponin and the tropomyosin molecules that is responsible for controlling the interaction between the thick and thin filaments during the contraction process.

The precise regularity of the sarcomere in skeletal muscle is also found in cardiac muscle (see Fig. 3B-5C), and the previous description is again applicable. In smooth muscle, however, no such order is apparent. Smooth muscle myofilaments, both thick and thin, are plentiful, but they are not arranged in a precisely interdigitating array. The muscle contains both myosin and actin, which are identified with the thick and thin filaments respectively. No Z lines or sarcomeres as such are present, although scattered plentifully throughout the cytoplasm (myoplasm) are the *dense bodies,* small oblong structures to which myofilaments appear to attach. Thus it appears that, while non-striated (smooth) muscle does not have the precise and esthetically pleasing regularity of skeletal muscle, it nevertheless contains the same essential component parts.

CONTRACTION

The Sliding Filament Theory of Contraction

Skeletal Muscle

The function of muscle is closely related to its observable ultramicroscopic structure. It is not surprising, then, that skeletal muscle, with its beautifully organized ultrastructure, should have provided the major portion of our understanding of the contractile process.

The sarcomere is the fundamental contractile unit of skeletal muscle. A series of interrelated anatomical observations and experimental results have led to the *sliding filament theory* of muscle contraction, based on the properties of the sarcomere. Briefly, the points of evidence are these. When muscles that have been either mechanically stretched out or allowed to contract were fixed and examined under the electron microscope, the distance between Z lines was found to vary directly with the overall muscle length. However, the dimensions of the A bands remained constant, while those of the I bands and the H zone varied. The lengths of both the actin and the myosin

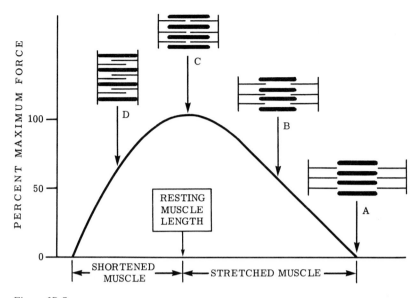

Figure 3B-7
The structural basis for the length-force relationship in skeletal muscle. When the muscle is so highly stretched as to pull the thin filaments completely out from among the thick filaments (A), no force is developed. If partial overlap is allowed (B), less than maximum force is generated. At C there is optimal overlap, and maximum force is developed. In the region between C and A the force is proportional to the degree of thick and thin filament overlap. As the muscle becomes very short (D), thin filaments from opposite sides of the sarcomere interfere with each other, and thick filaments are forced against the Z lines, leading to a rapid fall in force. If the curve shown in this figure were drawn for a single sarcomere, the transitions between the different regions would be abrupt. Because many sarcomeres participate in a contraction and are not all in exact register with each other, the length-tension curve for a whole muscle would be smoothed out as shown. The behavior illustrated is the structural basis for the length-force relationship in skeletal muscle and probably for heart muscle also. It is likely that in the future a similar mechanism will be demonstrated in smooth muscles as well. (Adapted from A. M. Gordon, A. F. Huxley, and F. Julian. *J. Physiol.* [Lond.] 184:170, 1966.)

myofilaments remained constant. Therefore, a change in overall muscle length was directly related to the degree of thick and thin myofilament overlap, with muscle length changes being accompanied by a "sliding" of the array of myofilaments of the I band into or out of the array of A band filaments. Precise studies on single fibers of living muscle showed that the force the muscle could exert was closely related to the degree of myofilament overlap and hence to the number of myosin-filament cross-bridges that could attach to nearby actin filaments (see Fig. 3B-7). When the muscle was stretched so far that there was no myofilament overlap, it could develop no force. On the other hand, when the muscle had contracted to so short a length that the I band filaments coming in from opposite sides of the A band interfered with each other, force was greatly reduced. By these and other means it has been established that the amount of force that skeletal muscle can exert is directly proportional to the degree of thick and thin myofilament overlap, and that the function of the myosin cross-bridges is to slide the thin filaments past the thick filaments in a hand-over-hand sort of way and thereby to produce an overall shortening of the muscle. This in essence is the sliding filament theory; to date, it has survived all experimental tests and appears to approach the truth more closely than do other theories of contraction.

Cardiac Muscle
Because of the structural similarities between cardiac and skeletal muscle, it is reasonable to expect a similarity of contractile mechanisms. Although experimental tests in cardiac muscle have not been so precise, there appears at present to be no reason to expect any fundamentally different contractile mechanism. At present, all evidence indicates that the sliding filament theory is adequate to account for the function of cardiac muscle.

Smooth Muscle
The adequacy of the sliding filament theory is less easy to settle in the case of smooth muscle, because its structure is much less well organized than that of skeletal muscle. However, actin and myosin filaments are present in close proximity to each other, and other biophysical and biochemical criteria are met, so it is entirely possible (and also highly likely) that smooth muscle also contracts by a sliding filament mechanism.

Initiation and Control of Contraction

Control by Muscle Cell Membrane
The greatest degree of control of contractile activity is exerted at the level of the *muscle cell membrane*. It is here that the signal from the central nervous system is translated into a form to which the contractile apparatus can respond. Of the three major types of muscle, skeletal muscle is the one most completely controlled via its cell membrane.

The plasma membrane of a typical skeletal muscle cell (of the *fast*, or *twitch* type) has functional properties similar to those of a nonmyelinated nerve axon (Chap. 2). A *resting potential* of -90 millivolts (mv) (inside negative) is typical, and excitation of the membrane leads to a rapid, overshooting, propagated *action potential* of about 2 msec duration, which is very brief compared with the duration of the contraction it causes. The distribution of Na^+ and K^+ ions across the membrane and the conductance changes that accompany the action potential are similar to those of nerve. A *refractory period* follows the action potential, and the ionic asymmetry of the resting state is maintained by an active sodium pump.

In the case of nerve, the function of membrane activity is to transmit a message down the length of the fiber only; in muscle there are two functions. The first is like that of nerve; the message to contract must be delivered to the whole length of muscle as nearly simultaneously as possible to ensure a coordinated contraction. In addition there must be some communication from the outer membrane of the fiber to its interior in order to activate the contractile apparatus rapidly. Calculations show that a chemical substance released from the periphery of a skeletal muscle fiber cannot diffuse inward nearly quickly enough to account for the rapid onset of contractile activity following an action potential. Muscle must therefore possess a special inwardly conducting system as well as an axially conducting system. The inwardly conducting function has been identified with an anatomical component of the muscle, the *transverse-tubule system* (commonly called the *T tubule system*). This is made up of a network of fine tubes penetrating from the surface membrane deep into the muscle fiber at the level of the Z line of each sarcomere (see Fig. 3B-8). The interior of this network of tubes is continuous with the extracellular space; hence the T tubules may be regarded as an inward extension of the outer cell membrane. As such, they do not disrupt the membrane barrier between the cell interior (cytoplasm) and the extracellular space. It is the function of the T tubule system to conduct the electrical impulse inward. If chemical means are used to destroy the T system while leaving the rest of the muscle intact, the electrical activity of the surface membrane does not penetrate the fiber, and no contraction results in response to an action potential.

As the T tubules penetrate the fiber, they make close contact with yet another membranous component of the muscle interior, *the sarcoplasmic reticulum* (SR). In contrast to the transverse tubules, the SR, a system of *longitudinal tubules* terminating

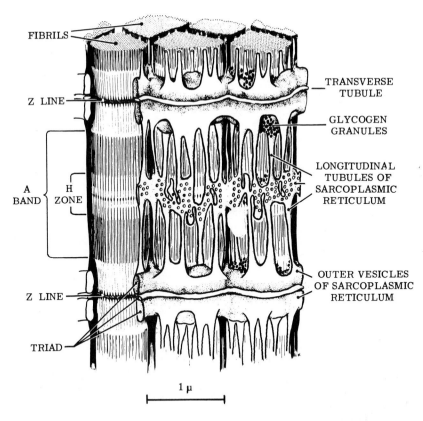

FIBRILS

Z LINE

A BAND

H ZONE

Z LINE

TRIAD

TRANSVERSE TUBULE

GLYCOGEN GRANULES

LONGITUDINAL TUBULES OF SARCOPLASMIC RETICULUM

OUTER VESICLES OF SARCOPLASMIC RETICULUM

1 μ

Figure 3B-8
The internal membrane system of skeletal muscle. Contraction is initiated by a release of stored calcium ions from the longitudinal tubules of the SR. This "trigger" calcium has only a short distance to diffuse in order to reach the myofilaments in the A band, where the ultimate reactions leading to contraction take place. It is by means of this internal membrane system that an action potential on the surface membrane of the fiber rapidly makes its effect felt at the center of the fiber. (After L. D. Peachey. *J. Cell Biol.* 25:209, 1965.)

with *outer vesicles* in each sarcomere, runs parallel to the myofilaments from Z line (where the T tubules are) to Z line. It is at the level of the SR that the final electrically controlled event in the *excitation-contraction coupling* sequence takes place. The SR functions as a storage, release, and uptake site for *calcium ions.* When an impulse from the surface action potential arrives down the T tubules, the SR releases stored calcium ions. These ions diffuse across the short distance from the SR to the cross-bridge region of myofilament overlap, where they provide the signal for the contractile process to begin.

Extensive biochemical studies of the interactions between calcium, adenosine triphosphate (ATP), actin, and myosin have led to a good understanding of the means whereby calcium exercises its control over the contractile process. The key to calcium control has been found to be the presence on the thin filaments of the two additional proteins troponin and tropomyosin.

Figure 3B-9A outlines the basic biochemical steps in the contraction process. The sequence of reactions shown was deduced from studies in which the tails of the myosin

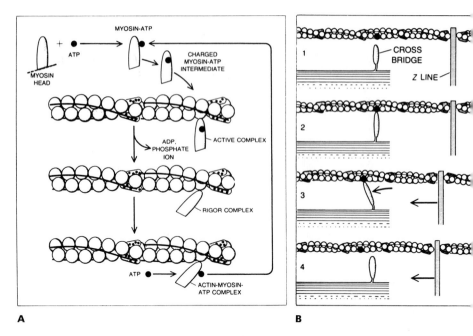

Figure 3B-9
A. The chemical events of the contraction cycle as outlined in the text. In the drawing, the troponin molecule (the oblong structure lying on the actin filament) is fully occupied by calcium ions (small dots), and hence all of its inhibitory activity is suppressed. Thus, as soon as the charged myosin-ATP intermediate is formed, it combines with an actin molecule, and the ATP is split. Sufficient ATP is present to allow the rigor complex to be broken up and the myosin "recycled" through the process. B. The events of the contraction cycle as they occur in intact muscle. Here the myosin head is permanently attached to the thick filament. When ATP is split, the myosin head converts the chemical energy into mechanical energy and undergoes the change in position shown; this action moves the thin filament parallel to the thick filament, and a small shortening of the muscle results. When the myosin head "lets go" of the actin molecule, it returns to its original position and is free to repeat the whole cycle. Many such cycles, when added together, can produce a large movement of the muscle. (From J. M. Murray and A. Weber, The cooperative action of muscle proteins. Copyright © 1974 by Scientific American, Inc. All rights reserved.)

molecules had been chemically removed; only the enzymatically active head portions (which are identified with the cross-bridges of intact muscle) were used. The actin filaments used were intact (but separated from the Z lines) and contained their normal complement of troponin and tropomyosin. An excess amount of calcium, sufficient to activate the process fully, was present. This partly disassembled system was used for experimental convenience and mimics very well the actual process in intact muscle. Figure 3B-9B shows the mechanical counterpart (in intact muscle) of the biochemical reactions.

The reaction sequence begins when a myosin head combines with a molecule of ATP, forming a "charged" myosin-ATP intermediate (step 1). This intermediate species is permitted to interact with an actin molecule (step 2); the actin-myosin combination forms an ATPase system that quickly splits ATP into ADP and an inorganic phosphate ion (step 3). This is the energy-liberating and motion-producing step in the contraction of intact muscle. Following the splitting of the ATP molecule, the myosin head is not free to leave the actin to which it is bound (this association is the *rigor complex*) until it is combined with a new ATP molecule (step 1 again). The myosin head then becomes detached from the actin (step 4) and becomes a new "charged" myosin-ATP intermediate (as in step 1) and may go on to repeat the whole sequence as many times as the calcium concentration (see below) and the ATP supply allow.

Control over this process is exercised by calcium ions; in the absence of calcium the troponin-tropomyosin complex will not allow the myosin-ATP "charged" intermediate to combine with any actin molecule, and the sequence is halted at the end of step 1. The muscle is now ready to contract but is inhibited from further action; this is the normal state of resting muscle. When calcium ions (released from the sarcoplasmic reticulum in intact muscle) combine with troponin, it releases its inhibition of the reaction, and steps 2,3,4, and 1 are free to proceed through as many cycles as possible. The function of the long tropomyosin molecule is to transmit the troponin "inhibition release" message to those actin molecules not directly covered by the troponin molecule itself. Removal of calcium ions from the troponin molecule (as by the action of the sarcoplasmic reticulum in intact muscle) reestablishes the troponin inhibition of the contraction process; no further ATP is split, and the muscle relaxes. Agents such as caffeine cause the sarcoplasmic reticulum to take up calcium ions less readily, and thus the contraction is prolonged. If the ATP supply for the muscle becomes exhausted for some reason, the reaction sequence must stop at the end of step 3. The myosin remains bound to the actin (the rigor complex), and the whole muscle becomes stiff and rigid; this is the cause of the phenomenon of *rigor mortis* that takes place following death, as ATP supply diminishes.

Cardiac muscle also contains a T system and a well-developed SR. In addition, the sarcolemma is able to act as a calcium storage and release site. Because of the smaller size of the cardiac muscle cell, calcium released at the surface membrane can diffuse to a significant proportion of the myofilaments. The essential features of excitation-contraction coupling in cardiac muscle appear not to differ significantly from the scheme that has been worked out for skeletal muscle. This agreement also probably extends to the smooth muscle case as well, although here less is known of the process. In many types of smooth muscle, the SR is almost entirely lacking; the very slow rate of contraction of these muscles suggests that the activator calcium ions must diffuse over rather large distances to reach the bulk of the myofilaments.

Types of Excitation of the Muscle Membrane

The details of the transmission of the nerve signal to the muscle membrane have been given in Chapter 3A. The accounts given there and in this chapter treat principally the *fast* or *twitch* type of muscle fibers, which make up most of the human limb musculature. Another type of specialized skeletal muscle fiber, the *slow* or *tonic* type, is associated with some muscles of the trunk and elsewhere (some of the intercostal muscles, for example). This fiber type, which will not be discussed further, is specialized for slower, more sustained contractions and is characterized by (among other things) a less regularly organized contractile protein arrangement, a greatly reduced internal membrane system (T tubules and SR), and a graded, incremental membrane response to nerve stimulation. Mechanical responses may also be graded by the extent of membrane depolarization. While all skeletal muscles appear to share most of their biochemical and physiological characteristics to a large extent, because of the comparative simplicity, physiological importance, and regularity of structure and behavior of fast skeletal muscle, general discussions (including this one) of skeletal muscle are likely to focus on this type. The reader is referred to the literature (e.g., Close, 1972) for consideration of the other skeletal muscle types.

The basic mechanical response of skeletal muscle, the *twitch*, is a single, brief contraction produced by the arrival of a single nerve action potential (Fig. 3B-10). For a given muscle fiber under constant conditions of temperature, fiber length, and so on, the magnitude of the twitch is a constant, all-or-none property. The duration of a twitch, usually a fraction of a second, is too short to be useful for most tasks. However, if the fiber is restimulated before it has relaxed completely, the second twitch will add its mechanical effect to the first, a process called *mechanical summation*. Repeated stimu-

SKELETAL MUSCLE

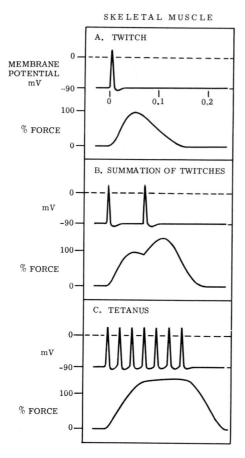

Figure 3B-10
Membrane and contractile events in skeletal muscle. A. The upper trace displays the muscle membrane potential in millivolts. The single action potential shown leads to the single contractile event, a twitch, shown in the lower trace. (The action potential trace time for all the traces is marked off in tenths of a second, and the amplitude of all the contraction tracings is with reference to the peak twitch tension.) It is important to note that the duration of the twitch is much longer than that of the action potential that triggered it. B. Because the refractory period of the muscle cell membrane is over long before the mechanical activity ceases, the muscle may be restimulated while some contractile force is still present. This restimulation leads to mechanical summation of the two twitches, and more force is produced than in a single twitch. C. If action potentials follow each other quickly enough (but not too quickly—why?), no relaxation occurs between successive twitches, and a fused tetanus results. (From R. A. Meiss. In E. E. Selkurt [Ed.], *Basic Physiology for the Health Sciences.* Boston: Little, Brown, 1975.)

lation will produce a state of continued contraction, or *tetanus;* if the interval between stimuli is short enough, the separate twitches will merge into a smooth contraction, a *fused tetanus.* Because of mechanical summation a tetanus produces more force than does a twitch; the usual physiological pattern of activation of a muscle fiber is a tetanus (not necessarily fused) of the appropriate duration.

The motor axons that innervate a muscle undergo varying degrees of branching before they terminate in a number of end-plates. Thus a single axon may have terminals on many separate muscles fibers. A single axon, together with all the muscle fibers it innervates, is a *motor unit.* Since activation of a muscle fiber via the motor end-plate produces an all-or-none contraction (though the maximum force may be increased by tetanic activation), the size of a typical motor unit is related to the fineness of control required of the whole muscle. In muscles in which a high degree of control is exercised over fine movements, the motor axon will control only a few muscle fibers, and hence an action potential from that axon will produce contraction in only a small portion of the whole muscle. Other muscles are specialized for large, rapid movements with little fine control and typically have large motor units, so that a single motor axon will control a relatively large portion of the whole muscle. This variable ratio of muscle fibers to motor axons is the *innervation ratio.*

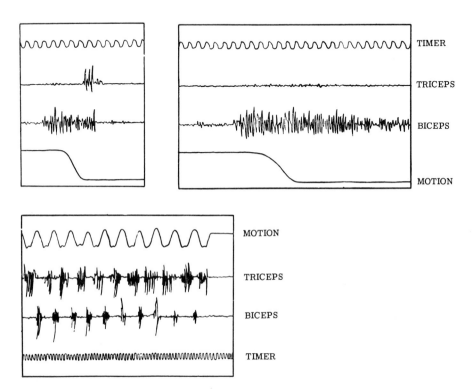

Figure 3B-11
Electromyographs of human muscle. In these actual experimental records the subject had one needle electrode in the biceps (flexor) and one needle electrode in the triceps (extensor) of the upper arm. In the top two panels the four traces (reading downward) represent: time, 0.05 second per cycle; triceps muscle electrical activity; biceps muscle electrical activity; and movement of the forearm. The top left panel is from a rapid, maximal flexion of the arm with no load attached. The biceps is activated prior to, during, and following the rapid movement. At the end of the movement a burst of action potentials shows that the triceps is briefly activated to fulfill its role as antagonist to the biceps and check the movement of the forearm. In the upper right panel, the subject flexed his arm to lift a load that he continued to hold up after the movement stopped. Because the biceps was still activated, action potentials continued; but in this case, the antagonistic action of triceps was not needed, and its electrode showed that it remained inactive. The lower panel shows the antagonistic action more clearly (the positions of the timing and movement traces have been reversed; movement is now at the top). Here, the forearm was moved rapidly back and forth, and the biceps and triceps show electrical activity in an alternate fashion. The anatomical relationships between the antagonistic muscles are presented in Figure 3B-1. (From D. R. Wilkie. *J. Physiol.* (Lond.) 110:249, 1950.)

It is possible for experimental or diagnostic purposes to observe the electrical activity of skeletal muscle in the intact body by applying recording electrodes to the skin above the muscle in question. The pattern produced by the combined action potentials of many motor units is called an *electromyogram* (EMG) (Fig. 3B-11). It may be extended to the observation of a single motor unit by inserting a fine needle electrode, insulated everywhere but at the tip, into the muscle.

Central Nervous System Control

The most highly refined control over muscular movement is exercised by the coordinated activity of the CNS. Sensory receptors located in the muscle and its antagonist and in the joints and tendons, together with visual and aural observation, provide the CNS with a feedback of information concerning the speed, force, position, and so on, of the muscle or limb. This feedback allows the motor signals to the muscle to be modified to produce the desired actions.

The stretch receptors associated with skeletal muscles are of two types; one type, the *tendon receptor,* is found in the tendon region of the muscle (Fig. 3B-12). Much like a cutaneous mechanoreceptor, it has a receptive portion consisting of a profuse network

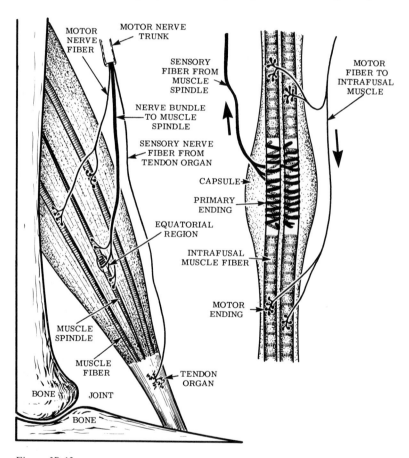

Figure 3B-12
The location and structure of muscle sensory receptors. The inset shows a magnified diagram of the sensory portion of the muscle spindle. (After P. A. Merton, How we control the contraction of our muscles. Copyright © 1972 by Scientific American, Inc. All rights reserved.)

of the fine terminal branches of a myelinated nerve fiber; it is stretched as the muscle exerts force and contracts. Being in the tendon, it is in *series* with the muscle and responds to increasing muscle force with an increase in its firing rate. It is thus suited to monitor the total force exerted by the muscle.

The other muscle receptor, the *muscle spindle,* is more complex in construction and function. It is actually a small group of modified muscle fibers and lies deep within the body of the muscle. Basically (details are simplified here) it is a long, slender structure consisting of a bundle of six or so muscle fibers with a very specialized center (*equatorial*) region (Fig. 3B-12). The equatorial region is devoid of contractile material and is closely wrapped with the terminations of *sensory* nerve fibers. The equatorial region is the actual sensory portion of the spindle. The *intrafusal fibers* (after the Latin *fusus,* meaning "spindle"), muscle fibers connected to the ends of the equatorial region, receive a *motor* innervation. There are two ways in which the equatorial region can be stretched (and thereby stimulated). When the whole resting muscle is stretched, the equatorial regions of all the spindles in the muscle are also stretched, and trains of action potentials (proportional to the stretch magnitude) are set up in the sensory nerve fibers. The equatorial region can also be stretched by a contraction (via their motor axons) of its intrafusal fibers. The sensory result is the same—a train of afferent action potentials—but, surprisingly, the receptor has stimulated itself! Both of these modes of stretch stimulation of the spindle are involved in its normal function.

The normal *extrafusal muscle fibers,* which make up the vast bulk of the muscle, are arranged in parallel with the spindles. Their contraction tends to shorten the spindles and thus reduces the tension to which they can be subjected. Present evidence indicates that, in voluntary muscle contraction, the *intrafusal* fibers are the first ones stimulated by the CNS. Their contraction stretches the equatorial region of the spindle and causes the sensory fibers to send a train of impulses to the central nervous system. In response to this sensory information, reflex connections in the spinal cord cause contraction of the *extrafusal* fibers, and the muscle shortens, tending to collapse the spindle receptors (see Fig. 5-12). Their force signal is thus reduced, and the resultant excitation to the extrafusal fibers is reduced, tending to reduce the total muscle contraction. If motor stimulation to the intrafusal (spindle) fibers is continued, the feedback events just described will continue, with the intrafusal fibers "leading" the sequence and the extrafusal fibers "following," and a sustained contraction will result. This is a biological example of a *servo system,* similar to that found in the power steering system of an automobile.

A simpler manifestation of muscle spindle receptor activity is the familiar knee jerk reflex. Tapping the tendon attached to a resting leg extensor muscle results in a slight stretch to the muscle, which also slightly stretches the spindle receptors. In a simple reflex arc, the extrafusal fibers of the muscle are rapidly stimulated to contract, and the knee jerk results (see Chap. 5 for a detailed account of the reflex processes). Since in this case the intrafusal muscle fibers are not restimulated, the contraction is brief.

Smooth muscle is subject to less precise control by the CNS system, and in this regard it may be divided into two broad categories. *Unitary* smooth muscles include the *iris* (which controls the diameter of the pupil of the eye), the *ciliary muscle* (which controls the focal length of the lens of the eye), the *pilomotor muscles* (which erect the hairs of the skin), and many *blood vessel muscles.* This type of smooth muscle is controlled much as skeletal muscle is, having functional and separate motor units and responding to a single nerve impulse with a single twitch. Fused tetani, associated with a series of action potentials, usually occur. The transfer of the information from nerve to muscle is via a chemical mediator (ACh in some instances), although highly structured motor end-plates are not found.

The visceral smooth muscle, on the other hand, behaves as if all the cells of an extensive region were connected into one large motor unit. In a given region, changes in the membrane potential of any one cell are shared by all the adjacent cells, so the tissue behaves electrically as if all the cell interiors were in direct contact with one another.

Such an arrangement is a *functional syncytium;* this type of muscle tends to have periodic spontaneous variations in membrane potential. When depolarization is sufficient, action potentials and contraction may occur. Myoneural junctions are not present as such, and the transmitter substances (which may have either a stimulating or an inhibiting effect) are released into the extracellular space among the cells. In such muscles the function of nerves is largely to modify and modulate, rather than to initiate, the independent spontaneous and rhythmic control inherent in the muscle cell membranes.

In many smooth muscle tissues, notably the stomach and intestine, periodic fluctuations in the membrane potential ("slow waves") provide the basic timing of the rhythmic contractile activity of the gut musculature. The slow waves do not cause contractions by themselves, but during that portion of the slow wave cycle in which the membrane is most depolarized action potentials ("spikes") can arise and trigger muscle contractions. In this way the periodic muscular contractions are controlled and coordinated by the intrinsic electrical activity of the tissue. Such nerve-independent contractile activity is called *myogenic,* in contrast to the *neurogenic* activity of skeletal muscles. Additionally, there are smooth muscles in the circulatory system and respiratory system whose contraction may be independent of their own cell membrane electrical activity and may instead be controlled by factors such as pH, oxygen tension, and circulating hormones.

While the electrical activity of smooth muscle cells is often complex and varies greatly among the tissue types, some features are held in common with nerve and skeletal and cardiac muscle. The balance between extracellular Na^+ and intracellular K^+ is a primary determinant of the resting potential, which in smooth muscles tends to be smaller in magnitude (less negative; i.e., from -35 to -60 mv) than in other excitable tissues. The resting potential may also be unsteady or undergo rhythmic fluctuations (see above) or show repeated periods of gradual depolarization terminating in action potentials much like the pacemaker activity of some cardiac muscle cells (Chap. 14). Other ions, notably Ca^{++}, may influence the magnitude or stability of the membrane potential. Action potentials in smooth muscles likewise have features in common with those from other excitable tissues. While the spikes are often longer in duration and may not show an overshoot, membrane conductance changes to Na^+ and K^+ (and also Ca^{++}) do appear to be involved. In addition, some of the periodic fluctuations in smooth muscle membrane potentials may be due to the action of spontaneously rhythmic electrogenic pumps (Chap. 1) and not associated with ion conductance changes.

The electrical properties of cardiac muscle lie somewhere between those of striated and smooth muscles. Although the contraction of cardiac muscle fibers is closely controlled by their cell membranes, the electrical activity of the cell membranes is only slightly influenced by the nervous system. Mammalian hearts, under proper conditions, can continue to beat even after they are removed from the body; the beat is myogenic, and its origin lies within the heart itself. It is the special membrane properties of various types of cardiac muscle cells that account for many of the remarkable automatic and rhythmic properties of heart muscle function (see Chap. 14).

THE MECHANICAL FUNCTION OF MUSCLE

Isotonic and Isometric Contraction

One important function of skeletal muscle is to do work in the physical sense—that is, to exert a force that can cause motion of some object (Fig. 3B-13). When the force is constant, as when an item is lifted upward, the contraction is *isotonic* (meaning "same force"). On the other hand, if the object is too heavy to be moved, force will be generated without any change in the muscle length, and the contraction is *isometric* ("same length"). Practically speaking, most contractions are both isotonic and isometric. When we lift a load, at first it does not move (isometric contraction); when sufficient

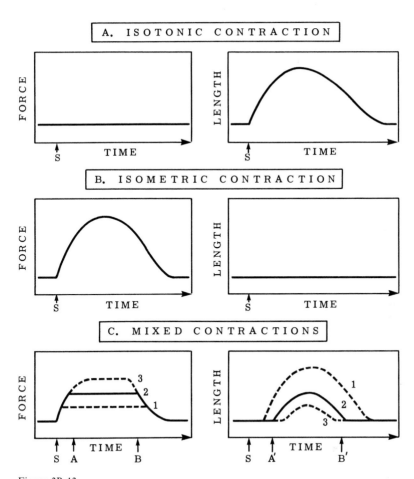

Figure 3B-13

Types of muscular contraction. In all records the stimulus is delivered at the point marked S on the time axis; increasing force and decreasing length (shortening) are plotted as upward movements. A. Isotonic contraction. Here, the stimulated muscle exerts no force; all of its contractile activity is expressed as shortening. B. Isometric contraction. Here, the muscle was prevented from shortening, and contractile activity resulted in the development of force only. C. Mixed contractions. In this case the muscle was presented with a load it could lift. The isometric record is shown at the left. Force increases (following stimulation) until, at point A, the isometric force becomes equal to the load (solid line, trace 2). Between points A and B (while the load is being lifted) the force is constant (isotonic conditions). The length (shortening) record shown at the right shows no movement between S and A′ (isometric conditions), and shortening followed by relaxation between A′ and B′. After B′ the load is no longer being lifted; movement has ceased and relaxation is now isometric. The dotted traces represent different loads on the muscle. If the load is small (trace 1, left), the isometric portion of the contraction is short, because the muscle can soon exert enough force to lift it. Correspondingly, the isotonic portion of the contraction is long (right panel), and the load is lifted quickly and much shortening occurs. If the load is large (trace 3, left), the isometric portion is longer and the isotonic portion is shorter. Less shortening occurs, and the muscle moves at a lower rate of speed. (From R. A. Meiss. In E. E. Selkurt [Ed.], *Basic Physiology for the Health Sciences.* Boston: Little, Brown, 1975.)

force has been developed, the load is lifted. Since it becomes no heavier as it is lifted, force stays constant (isotonic contraction). Sometimes a contraction occurs in which force continually increases and motion continuously occurs; the drawing of a bow and arrow is an example of this *auxotonic* type of contraction. Finally, in many cases, force may be constant with the muscle lengthening, not shortening; lowering an object or descending stairs involves this type of activity.

Length-Force Relationship
In the laboratory, two fundamental mechanical properties of muscle have been revealed by carefully controlling the conditions of contraction. The first of these is the *length-force relationship* (see Fig. 3B-14A, B; see also Fig. 3B-7). If a number of isometric contractions are made with the muscle adjusted to a different length before each contraction, it will be found that there is a length at which the maximum force can be produced. When the muscle is made to be longer or shorter than the optimum length, it will develop less than its maximum force. In muscles with skeletal attachments at both

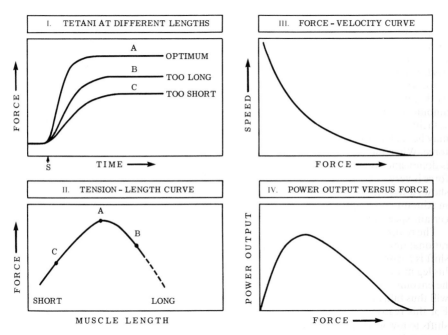

Figure 3B-14
Consequences of changes in experimental conditions. I. Three fused tetani at different resting muscle lengths. At A, the length was optimal for the production of force, while at B and C it was too long or too short respectively. II. The force values from A are plotted against muscle length to form the *tension-* (force)-*length curve*. Compare this with the behavior of muscle shown in Figure 3B-7. The dotted line shows an area of possible damage to the muscle resulting from extreme stretch. III. In Figure 3B-13 it was shown that light loads could be more quickly moved than heavy ones. If the speeds (velocities) of shortening for a large number of different loads (forces) are plotted against the various forces, a characteristic *force-velocity curve* is obtained. This curve is similar in shape for a wide variety of muscles and expresses the general notion that the greater the load, the more slowly it can be lifted. IV. If power (force × velocity) is plotted against the force exerted, the curve shows a distinct optimum force for the production of power. Operation of a muscle at this region of the curve is not an automatic property of the muscle but involves decision-making in the central nervous system. (From R. A. Meiss. In E. E. Selkurt [Ed.], *Basic Physiology for the Health Sciences.* Boston: Little, Brown, 1975.)

ends, this property is largely obscured because the range of length changes the skeleton allows is not very great. Also, the lever action of the skeleton provides a constantly changing mechanical advantage; this can make the force vary with limb position as well as muscle length. For heart muscle, however, the length-force property can be of great importance, since it allows the heart to contract with more force when it has been filled with an increased amount of blood. This property, which is a basis for Starling's law of the heart, will be treated more fully in Chapter 15. As was discussed earlier, the functional basis for the length-force relationship is the change in amount of myofilament overlap at different muscle lengths.

Force-Velocity Relationship
A second fundamental property of muscle is the *force-velocity relationship* (Fig. 3B-14C). It is well known from everyday experience that the heavier a load, the less quickly it can be lifted. Stated another way: the greater the force, the lower the velocity (speed of shortening). When the load (force) is so large that it cannot be lifted, the contraction is isometric (speed = 0). When there is no load at all, the speed of movement is the greatest. Between these two extremes, speed varies with load in a way that is characteristic of the individual muscle; the general form of the curve is similar to that in Figure 3B-14C. The shape of this curve is of some theoretical interest, because it provides information on the way the muscle liberates energy. A further extension of the curve, however, is of practical interest. From elementary physics we know that the term *work* can be defined as a force acting through some distance (work = force × distance). Also, we find that the term *power* is a measure of the rate of doing work (power = work/time). A little more algebraic manipulation leads to the conclusion that the amount of power produced is the product of force and velocity (power = force × velocity). To relate this to the performance of muscle, the force-velocity relationship may be reexpressed as in Figure 3B-14D: force × velocity (= power) is plotted against force. When the force is maximum, power output is zero (since velocity is zero, no work is done), and when the velocity is at its maximum, power output is again very low, since force is very small. Between these two extremes, the curve relating power output to force shows a definite maximum; i.e., there is a combination of force and velocity at which maximum power output occurs. By way of analogy, an automobile engine also has a certain speed of revolution at which its power output is optimum.

The realization that muscles have some of the characteristics of engines can aid in the rational design of efficient muscle-powered machines. A bicycle equipped with a gear shift is a case in point. Let us assume that on flat and level ground the cyclist is moving his leg muscles at such a force and speed that his power output is at its optimum. When he encounters a hill, he is required to exert more force; the speed of his leg movement will thus fall, and he will no longer be working at the optimal point on the power curve. For this reason the bicycle maker has provided a selection of gear ratios. The cyclist shifts to low gear, and the speed of his legs can again increase; the force required of the leg muscles is reduced, and he is again working at the optimal point. (The bicycle is now moving more slowly, however; you can't get something for nothing.) In going down the hill, the situation is reversed; the speed of the leg muscles is now too high, the force too low, and again the optimal conditions are not met. By shifting into high gear he will increase the required force; the leg muscle speed will fall, and conditions can return to optimum. From arguments of this sort, it can be appreciated that there is, for example, an optimal slant to a pedestrian ramp, an optimal weight of boxes for a worker to lift and stack. Successful athletes, whether knowingly or not, employ many such physiologically sound techniques to produce the most from their muscles.

It should be kept in mind, however, that a situation may call for muscular activity to be performed under force-velocity conditions far from the optimum; in such cases the CNS can select the appropriate number of motor units to be activated and determine the duration of activity of each to produce the desired motion. Mechanically optimal performance is not obligatory, but it is useful at times. It should also be noted that under isometric conditions no external work is performed; nevertheless, the contractile apparatus of the muscle is activated and is consuming energy in order to maintain force. In this case, the efficiency of the muscle as a producer of motion is zero. The maximum possible efficiency of the isotonically contracting human muscle is about 25 percent; that is, of the energy consumed by muscle, 25 percent can appear as external work, and 75 percent is wasted as heat. The "waste" heat from inefficient muscular contraction is not always physiologically wasted, however, since muscle plays a large role in the thermal economy of the body (Chap. 29). The efficiency of human muscle is about the same as that of an automobile engine, better than that of a steam engine, and much less than that of an electric motor.

The length-force and force-velocity properties of skeletal muscle are also found in smooth and cardiac muscle. The contraction cycle of cardiac muscle is mechanically quite complex (Chaps. 13, 15), involving a mixture of conditions ranging from nearly isometric to a quasi-isotonic phase during which the afterload is continuously changing. Under controlled conditions, both isolated cardiac and smooth muscle in vitro can be made to perform in much the same way as skeletal muscle, lending support to the notion that the basic contractile mechanisms have much in common. The wide range in the functional capability of the intact heart (Chap. 15) may be more readily appreciated if it is seen largely as the expression of the complex interaction between the force-velocity and length-force properties basic to muscle in general.

Energy Sources

The preceding discussion implies that muscles are motors and that they consume fuel in order to produce work. This is true, but in some respects muscles are more versatile than automobile engines because the body can produce the proper muscle fuel from a variety of foods. The fuel actually utilized in the process of contraction is the ubiquitous biological high-energy compound *adenosine triphosphate* (ATP). During the course of contraction the ATP molecule is degraded to *adenosine diphosphate* (ADP) and *inorganic phosphate* (P_i), and the energy thus liberated is used by the contractile apparatus to produce force and motion. Several biochemical pathways supply the muscle stores of ATP. The most available "ready reserve" source of ATP is the compound *creatine phosphate* (CP). It exists in equilibrium with ATP according to the reaction series:

1. ATP $\rightarrow$ ADP + P_i (liberation of energy for contraction)

2a. ADP + CP $\rightarrow$ ATP + creatine (replenishment of ATP)

2b. creatine + ATP $\rightarrow$ ADP + CP (storage of ATP energy in the CP molecule)

Reaction 1 expresses the liberation of energy for contraction by ATP breakdown. In reaction 2a, the ADP from 1 is rephosphorylated to ATP, using the energy stored in CP. In reaction 2b, ATP produced elsewhere in the muscle is converted to its reserve storage form, CP. The equilibrium constants of reactions 2a and 2b are such that there is approximately 30 times as much CP as ATP. Thus ATP can be quickly supplied to the contractile apparatus for some time even when all non-CP sources are shut off.

The ATP supplied to the CP "reservoir" is produced primarily by two well-known biochemical pathways, one *aerobic* (oxygen-requiring) and one *anaerobic* (oxygen-independent) (Fig. 3B-15). The anaerobic pathway is the sequence of cytoplasmic reactions called *glycolysis*. The aerobic pathway is the mitochondrial reaction sequence variously called the *Krebs cycle, citric acid cycle,* or *tricarboxylic acid* (TCA) *cycle*. The less efficient of these pathways is glycolysis. This series of reactions takes *ADP* and *glucose* (which is stored in muscle largely as its polymer *glycogen*) and from them produces *pyruvic acid* and a small amount of ATP. Anaerobic glycolysis will soon come to a halt if the pyruvic acid is allowed to accumulate. By far the more efficient pathway (in terms of ATP production) is the TCA cycle. In the presence of an adequate supply of oxygen, the TCA cycle reactions utilize the pyruvic acid produced in glycolysis to

MAJOR METABOLIC ENERGY PATHWAYS OF MUSCLES

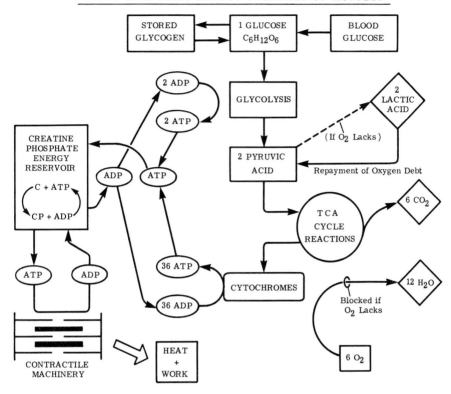

Figure 3B-15
The major metabolic sources of energy for muscular contraction. The number before each of the chemical species represents the net number of molecules of that substance arising from the breakdown of one molecule of glucose. Note that the complete aerobic breakdown of glucose produces 18 times as much ATP as does the glycolytic breakdown alone. The diamond-shaped boxes represent waste products as far as the immediate energy needs of the muscle are concerned. (From R. A. Meiss. In E. E. Selkurt [Ed.], *Basic Physiology for the Health Sciences.* Boston: Little, Brown, 1975.)

rephosphorylate large amounts of ADP to ATP. Water and carbon dioxide are produced as waste products. If, as in the case of heavy exercise, the oxygen supply is inadequate to permit the TCA cycle to operate, the accumulating pyruvic acid may be temporarily disposed of by reducing it to *lactic acid.* At the expense of accumulating lactic acid, then, glycolysis can continue to operate and produce ATP for contraction. The muscle is now said to have acquired an *oxygen debt.* When the oxygen supply is again adequate, some of it must be used to oxidize the accumulated lactic acid back to pyruvic acid and then to CO_2 and H_2O by the TCA cycle. The muscular soreness sometimes experienced after heavy exercise is partially due to the accumulation of lactic acid.

While most skeletal muscles can sustain some degree of oxygen debt, cardiac muscle can sustain very little. The chest pain of *angina pectoris* results when the blood supply of the heart has become insufficient to adjust to the extra oxygen demands of exercise. Any sudden interruption of the normal cardiac blood supply, such as the blockage of a coronary artery by a blood clot (coronary thrombosis), leads rapidly to a depletion of the energy supply of the affected region and may also lead rapidly to death.

FUNCTIONAL ADAPTATIONS OF MUSCLE

The combined mechanical attributes of a muscle at any given time (as described by its length-force characteristic, its force-velocity relationship, and so on) are referred to as its *contractility*. Over a short span of time, the contractility of skeletal muscle is constant except for the property of fatigue, which is a reversible depletion of the metabolic fuel supply brought about by overuse. Certain skeletal muscles, such as those in the arms, become fatigued quickly but are well adapted for performing rapid and forceful movements. Other muscles, notably those of the trunk of the body and the legs, can maintain a force for long periods of time with little fatigue. These *postural* muscles are not so well adapted, however, for rapid movement. Skeletal muscles vary in their ability to sustain an oxygen debt. An adequate supply of oxygen is assured during periods of heavy exercise (see Chap. 29) when the oxygen-carrying blood flow to the muscles is greatly increased through several physiological mechanisms, one of which is the relaxation of the smooth muscle sphincters that constrict the muscle blood vessels.

Certain long-term adaptive changes are found in skeletal muscle. One of these is *hypertrophy*, or an increase in muscle size and force capability. Increased use of a muscle, especially isometric exercise, results in an increase in the cytoplasm and myofilament content, although no new muscle cells are formed. Lack of use of a muscle, such as occurs during long confinement to bed or in a cast, or as a result of nerve paralysis (e.g., poliomyelitis), can result in the opposite process, muscle *atrophy*. Connections with the CNS and some amount of exercise appear necessary to prevent atrophy, which results in an actual loss of contractile protein.

Yet another long-term change is *muscular dystrophy*, a complex of conditions characterized by progressive muscular weakness and a loss of contractile material, with partial replacement by noncontractile tissue. Other muscular disorders include *myotonia*, which is a failure of muscles to relax after a normal contraction because of continued spontaneous muscle membrane electrical activity, and *contracture*, which is a maintained force in the contractile apparatus independent of cell membrane activity. Many disorders of the CNS produce severe muscular symptoms (convulsions, trembling, tetanus) although there is nothing wrong with the muscles themselves.

In contrast to the situation in skeletal muscle, cardiac muscle is subject to many influences that can greatly modify its contractility. This feature is important in enabling the heart to make a wide range of adjustments in its mechanical properties in order to cope with changing physiological demands. Factors that modify the performance of cardiac muscle are called *inotropic agents*. Not all inotropic factors produce a change in contractility, however. For example, the increase in contractile force associated with an increase in resting fiber length (see the preceding section) does not represent a change in contractility, since the basic quantitative properties of the contractile apparatus are little altered. An increase in heart rate, however, increases the force capabilities of the muscle without any change in fiber length. The increased contractility caused by increased heart rate is described by the *force-frequency relationship.* The muscle is said to have been *potentiated* by the rate increase and will now have a force-velocity curve and a force-length property distinctly different from those at the lower rate. A related phenomenon is that of *postextrasystolic potentiation.* If for some reason the heart produces a beat sooner than it should (*extrasystole*), this beat will be followed by a *compensatory pause.* The contraction following the longer-than-normal interval will show increased force; also, if for some reason the heart drops a beat, the next beat will be potentiated. The value of these mechanisms will become apparent in Chapter 15.

Certain drugs and hormones can also modify cardiac contractility. For example, *epinephrine* (Adrenalin) (a hormone released from the adrenal medulla in times of stress) will cause an increase in contractility over and above that due to its rate-increasing effect. For this reason, epinephrine is used as a therapeutic agent. Increasing the extracellular concentration of *calcium ions* has a similar type of effect. The drug *digitalis*, which is administered to failing hearts, produces a strengthened heartbeat without affecting the heart rate.

Finally, as in skeletal muscle, an increased load on the heart muscle for an extended period of time (such as that produced by chronic high blood pressure) can lead to *cardiac hypertrophy* (enlarged heart). Although hypertrophy of the heart represents a useful adaptation to increased stress, the enlarged heart may be subject to disturbances in impulse conduction.

Smooth muscle possesses two mechanical adaptations that suit it well to its physiological roles. First, it is capable of maintaining a force (called *tonus*) for very long periods of time without the consumption of large amounts of energy and consequent fatigue. This property is vital, for example, in keeping sphincters closed over extended periods. The other notable property of smooth muscle is its ability to adjust its resting length over wide ranges without large changes in resting tension. Thus the urinary bladder is able to contain variable volumes of urine without distress, the stomach can expand to accommodate a full meal, and the uterus is able to adjust to the growing fetus.

The hypertrophy that smooth muscle undergoes differs from that of other muscle types in terms of its basic cause. Smooth muscle forms large portions of the target organs of several of the reproductive hormones. At the time of puberty the rising concentration of *estrogens,* for example, induces a marked hypertrophy of the uterine muscle and associated structures. As long as satisfactory hormone levels are maintained, the hypertrophied muscle remains in that state. Withdrawal of hormonal support, as by removal of the ovaries, will result in atrophy of the muscle.

REFERENCES

Carlson, F. D., and D. R. Wilkie. *Muscle Physiology.* Englewood Cliffs, N.J.: Prentice-Hall, 1974.

Close, R. I. Dynamic properties of mammalian skeletal muscles. *Physiol. Rev.* 52:129, 1972.

Hayashi, T. How cells move. *Sci. Am.* March, 1961. P. 184.

Hoyle, G. How is muscle turned on and off? *Sci. Am.* April, 1970. P. 84.

Huxley, H. E. The contraction of muscle. *Sci. Am.* May, 1958. P. 66.

Huxley, H. E. The mechanism of muscular contraction. *Sci. Am.* June, 1965. P. 18.

Katz, B. *Nerve, Muscle and Synapse.* New York: McGraw-Hill, 1966.

Merton, P. A. How we control the contraction of our muscles. *Sci. Am.* May, 1972. P. 30.

Murray, J. M., and A. Weber. The cooperative action of muscle proteins. *Sci. Am.* February, 1974. P. 58.

4. Special Senses
A. The Eye

William McD. Armstrong

The eye is a complex and intricate physiological device. Since it subserves the all-important function of vision, it has been investigated in great detail. Space considerations preclude all but a general survey of the basic principles of the physiology of vision in a textbook of this size. The interested student is referred to the general references listed at the end of the chapter for more detailed accounts of specialized aspects of the subject. In particular, the book by Campbell et al. (1974) gives an up-to-date, readable, and relatively uncomplicated account of physiological optics and its clinical applications. Davson's (1972) monograph is a scholarly summary, within the confines of a single volume, of current knowledge and research in the physiology of vision.

The physiological role of the eye is twofold. First, it is an optical instrument which collects light from objects in the external environment and projects images of them on a light-sensitive organ, the *retina.* Second, it is a sensory receptor which translates these optical images into information that is transmitted to the visual areas of the brain. For convenience, these two functions of the eye will be considered separately. However, it should be remembered that, in the physiological functioning of the eye, they are closely related.

GENERAL STRUCTURE OF THE EYE

Figure 4A-1 represents a horizontal cross section of the right eye. The body, or globe, of the eye has three coats. The outer coat is a layer of connective tissue called the *sclera.* Part of this is seen as the "white" of the eye. The anterior one-sixth of the sclera is a specialized transparent structure, the *cornea,* which forms part of the *dioptric* or light-transmitting apparatus of the eye. The front of the eye, except for the cornea, is covered by a delicate membranous tissue, the *conjunctiva,* which also lines the eyelids.

The middle coat of the eyeball is the *choroid.* This is essentially a layer of highly vascular tissue which is important for the nourishment of the outer layers of the inner coat or *retina.* Anteriorly the middle or vascular coat is continued as the *ciliary body* and *iris* (Fig. 4A-1). The choroid, ciliary body, and iris are known collectively as the *uvea.* The degree of pigmentation of the iris determines the color of the eye.

The major portion of the inner coat of the eye is the *retina.* Anteriorly the retina terminates at the *ora serrata* (Fig. 4A-1). From this the inner coat continues forward as the *ciliary epithelium* which forms part of the ciliary body.

The lens, which together with the cornea forms the dioptric apparatus, is supported by the *ciliary zonule;* this in turn is attached to the *ciliary body.* Between the cornea and the iris lies the *anterior chamber* of the eye. It and the smaller *posterior chamber,* which is bounded anteriorly by the iris and posteriorly by the ciliary zonule, are filled with a transparent aqueous medium, the *aqueous humor.* The main body of the eyeball contains a transparent jelly-like substance, the *vitreous humor.* This essentially aqueous solution has a high viscosity because it contains appreciable quantities of the mucopolysaccharide *hyaluronic acid.* In addition, its structure is maintained by a branching network of fibers, consisting largely of a protein, *vitrosin.*

It is appropriate at this point to consider some of the structural elements in more detail. The cornea is made up of three main regions: an outer region, five or six cells thick, the *epithelium;* a central region, the *stroma,* which accounts for some 90 percent of the total corneal thickness; and finally a single layer of cells, the *endothelium.* Both the

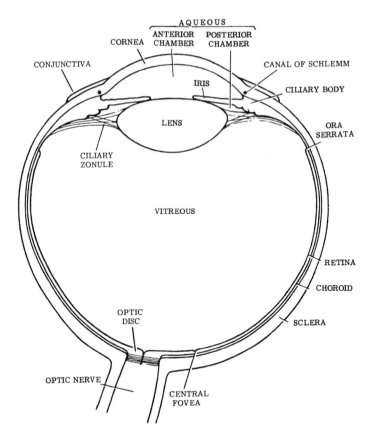

Figure 4A-1
General structure of the eye.

epithelium and the endothelium are typical epithelial sheets (Chap. 1). The stroma is a highly ordered arrangement of parallel collagenous fibers embedded in a relatively structureless ground substance. These fibers tend to have remarkably uniform diameters. In the opaque sclera they are arranged in a more random fashion and have variable diameters. When the cornea swells because of uptake of excess water, it becomes opaque. The opacity seems to be the result of disarrangement of the ordered array of stromal fibers. In the isolated cornea swelling may be induced by cold, metabolic inhibitors, or an increase in *intraocular pressure*. Both the epithelium and the endothelium are permeable to salts, and it appears that, in vivo, K^+, Na^+, and Cl^- ions tend spontaneously to enter the stroma both from the fluid bathing the outer surface (the tears) and from that bathing the inner surface (the aqueous humor). The entry of salt, which is partly, at least, due to Donnan forces (Chap. 1), is accompanied by an osmotic entry of water; if uncompensated, this will lead to corneal swelling and opacity. Thus, the maintenance of corneal transparency requires an outward active transport of ions (with a concomitant osmotic movement of water) across the epithelium or endothelium or both. In the mammalian cornea, the exact nature of the active processes involved is still not completely known (Davson, 1972; Zadunaisky, 1971). In the amphibian cornea an active transport of chloride from stroma to tears is a major factor in the regulation of stromal hydration, and a similar process seems to be a significant component of the mammalian regulatory system (Zadunaisky, 1971).

A peculiarity of the cornea is the fact that it can be readily transplanted from one individual to another of the same species (*homografted*), perhaps because there is a relative lack of antigenic responses in the cornea (Davson, 1972). Whatever the reason, the importance of corneal transplants in the treatment of visual defects due to corneal injury or dysfunction is of course well known.

The cornea and conjunctiva are protected by a fluid film of tears which is secreted continuously by the lacrimal glands, together with mucous and oily secretions from other secretory organs in the conjunctiva and eyelids. Without this protection the cornea would rapidly dehydrate through loss of water by evaporation. There appears to be no secretion of tears during sleep, and secretion decreases markedly with advancing age. Reflex secretion of tears occurs in response to a variety of stimuli—e.g., irritation of the cornea, conjunctiva, or nasal mucosa; thermal stimuli (including "hot" peppery foods) applied to the mouth and tongue; excessively bright lights; vomiting, coughing, and yawning. *Psychic weeping* occurs as a result of emotional upsets. A drainage system consisting of the *canaliculi,* the *lacrimal sac,* and the *nasolacrimal* duct drains off secretions that exceed the loss due to evaporation. The formation of tears is called *lacrimation.* Tears serve to wash away irritating particles and fumes and also contain a bactericidal enzyme, *lysozyme.*

Blinking, a rapid temporary closure of the eyes, occurs reflexly in response to actual or potential mechanical insults. It is effected by contraction of muscles in the upper and lower eyelids. In addition to protecting the eyes from specific hazards, blinking spreads the protective film of lacrimal secretions over the cornea and conjunctiva. During waking hours spontaneous blinking occurs every few seconds. Defective closure of the eyelids or defective lacrimal secretion can give rise to the pathological condition known as *keratitis sicca.* Reflex blinking can be as fast as 0.1 second.

The aqueous humor has an ionic composition basically similar to that of the blood but it has a much lower protein content (about 5 to 15 mg/100 ml in man). Although the details of the process are unknown, it seems well established that aqueous humor is secreted continuously by the cells of the ciliary epithelium into the posterior chamber, whence it flows into the anterior chamber. Because it is produced continuously, the aqueous humor must be drained away continuously if the *intraocular pressure* is to remain normal. The *canal of Schlemm* (Fig. 4A-1) is an important element in the drainage system of the eye. Through this canal fluid drains into veins leaving the eye.

Normal intraocular pressure in the human is about 15 mm Hg (Davson, 1972). There is no significant difference between the sexes nor is there any appreciable correlation between intraocular pressure and age. Chronically elevated intraocular pressure (resulting, for example, from a defect in intraocular drainage) can give rise to *glaucoma.* Glaucoma can cause irreversible damage to certain structures of the eye, including the retina, and is a major cause of blindness. For this reason, the measurement of intraocular pressure is of great importance in ophthalmological practice. It is done with an instrument called a *tonometer.* Clinical tonometers are of two types, *impression tonometers* and *applanation tonometers.* Impression tonometers measure the depth to which a weighted plunger depresses the cornea. Applanation tonometers measure the area of flattening when a metal surface is applied to the cornea with a known force (see Davson, 1972, and Campbell et al., 1974, for further details).

The lens is a transparent biconvex entity consisting of an outer elastic envelope, the *capsule,* which encloses the *lens substance.* The latter is composed of fibers and interstitial cement. In addition, the *epithelium,* a single layer of cells, covers the anterior surface of the lens substance. The tension of the capsule tends to make the lens assume as spherical a conformation as possible. This tendency is counteracted by the pull of the fibers of the ciliary zonules around the lens equator, which help give the lens a more flattened form. With respect to its internal composition the lens behaves as if it were a single cell; i.e., the lens substance has a high K^+ and a low Na^+ and Cl^- content. The maintenance of the internal ionic composition of the lens depends on metabolism; cooling or metabolic inhibitors, for instance, decrease internal K^+ and increase internal

Na^+ and Cl^-. It has been suggested that the lens behaves like other epithelial systems such as the amphibian skin (Chap. 1) in that the active transport of solutes from the interior to the external medium is largely controlled by the epithelium.

The transparency of the lens, like that of the cornea, seems to depend on the maintenance of a very precise arrangement of the fibers within the lens substance, and this in turn probably depends on the maintenance of a normal salt and water content. Lack of transparency in the lens is the condition called *cataract*. It may be localized in spots or may involve the whole of the lens substance. Cataract may result from metabolic or nutritional defects, from trauma (including radiation damage), or simply from age. The last condition (senile cataract) is, unfortunately, rather frequent in occurrence. Abnormalities in calcium metabolism such as may arise from parathyroid-ectomy or rickets can lead to cataract, suggesting a role of calcium in the maintenance of lens transparency.

The structure of the retina is considered below (p. 116).

PHYSICAL PROPERTIES OF LIGHT

The visible light spectrum ranges from 400 to 800 *nanometers* or *millimicra* ($1 \text{ nm} = 1 \text{ m}\mu = 1 \times 10^{-9}$ meter), with blue or violet light having the shortest, and red the longest, wavelength. Immediately adjacent to the visible spectrum are the areas of ultraviolet and infrared radiation. Although not perceived by the eye as such, radiation in these areas is biologically important. For example, ultraviolet radiation is the cause of the tanning and burning effects of the sun on the skin. Infrared radiation has heating effects on the body.

From the time of Newton, scientists have debated whether light is a stream of corpuscles having the properties of matter, or a form of wave reflected or emitted from material objects. Careful experimentation has shown that it actually possesses properties

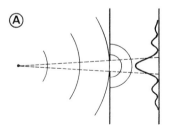

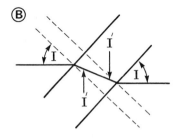

Figure 4A-2
A. Diffraction pattern formed by light rays passing through a narrow slit. Dashed lines indicate the path which light rays would follow if diffraction did not occur. B. Bending of light rays at the points of entrance and emergence from a medium with a higher refractive index. I and I' are angles of incidence and of refraction.

of both matter and waves, and consists of electrical and magnetic energy which travels in packets, or *photons,* at a speed of 186,000 miles per second.

The wavelike properties of light are of particular significance in considering the optical function of the eye. They enable light to be focused, reflected, and refracted like other waves. One important property of wave phenomena is *Huygens' principle,* which states that *any point on a wave front may be regarded as a new source of light.* For example, if light passes through a narrow slit, the rays fan out when they leave the slit as if that slit were a new source of light. If a screen is placed beyond the slit to intercept the rays, they form a central image on the screen with less intense secondary images of the slit on either side of the central image (Fig. 4A-2A). This formation of a *diffraction pattern* is a consequence of the wave nature of light.

Diffraction is of practical importance because it limits the sharpness of the image and consequently the *degree of optical resolution.* For example, if two objects are very close together, their diffraction patterns may overlap and it may be impossible to separate them optically, no matter how much the image is magnified. Ideally, the *limit of resolution* or *resolving power* for any optical instrument which is diffraction limited is about one-half the wavelength of the light it uses.* In practice, resolving power may be considerably less.

REFRACTION AND IMAGE FORMATION

When light rays strike an object, they may be reflected, absorbed, or transmitted through it. Even rays that are transmitted do not escape some alteration. The entering light rays are bent to an extent depending upon the angle at which they strike the surface, and upon the *refractive indices* of the surrounding medium and of the object. The refractive index (n) of any optical medium is formally defined as the *ratio of the velocity of light in a vacuum to its velocity in the given medium.* In practice air rather than a vacuum may be taken for reference purposes since the refractive index of air (1.0003) is close to unity.

As light rays pass from one medium into a second medium which has a higher refractive index (i.e., is more optically dense), the rays are bent toward a perpendicular drawn through the surface between the media. Conversely, as the rays emerge from the more to the less optically dense medium they are bent away from the perpendicular (Fig. 4A-2B). The degree of bending or *refraction* depends upon the refractive indices of the two media and the angle at which the light rays strike the surface between the media. Quantitatively, this is expressed in *Snell's law of refraction,*

$$n \sin I = n' \sin I' \tag{1}$$

where n and n' are the refractive indices of the two media, I is the *angle of incidence,* and I' is the *angle of refraction* (Fig. 4A-2B). One notes from equation 1 that, given constant values of n and n', it is the ratio of the sines of these angles, not the ratio of the angles themselves, that remains constant. Hence, the degree of refraction decreases as I becomes larger. When I is maximal—i.e., 90°—there is no refraction.

The property of refraction is basic to the operation of lenses, as may be seen in Figure 4A-3A. The central rays from the point source of light strike almost perpendicularly to the lens surface and are bent only to a small degree, whereas the peripheral rays strike the surface at an acute angle and bend considerably more. If the surface of the lens is spherical, the transmitted rays intersect at a point on the other side of the lens to form an image of the point source. The lens illustrated is a converging, convex spherical lens.

*This is one reason why the electron microscope is such an important tool in cellular biology. The waves associated with a high-energy electron beam are much shorter than those of visible light. Thus one is permitted to visualize subcellular structures that are far too small to be separated from their surroundings by the ordinary light microscope.

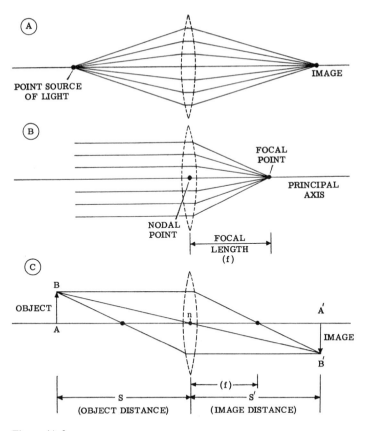

Figure 4A-3
A. Formation of an image by a spherical lens, showing the greater deviation of peripheral rays. B. Illustration of focal length as the distance from the lens at which parallel rays are brought to focus. C. Construction of an image when lens strength, object size, and object distance are known.

The surface of such a lens is a portion of a sphere; i.e., it has the same curvature in all planes. One or both surfaces may be spherical. A lens which causes light to diverge is constructed by making the spherical surfaces concave rather than convex.

Since refraction depends on the angle of incidence, the light rays passing through the peripheral portion of a spherical convex lens do not focus at exactly the same point as those passing through the center. This phenomenon is called *spherical aberration*. An additional limitation is the fact that the refractive index varies with the wavelength of the light used, blue light being bent more than red. Thus, if white light is used, the rays of different wavelength form separate images, and an expanded image is produced. This characteristic is termed *chromatic aberration.** The optical system of the eye possesses both spherical and chromatic aberration to some degree.

GEOMETRICAL OPTICS

Geometrical optics is the study of the properties of lenses and the formation of images. Two important applications of this branch of optics are the determinations of the strength of a lens and of the size and position of the image it forms. When parallel rays

*In high-quality optical instruments, complex combinations of lenses are employed to reduce both spherical and chromatic aberration.

of light fall upon a lens, they come to focus at a point called the *focal point*, as in Figure 4A-3B. The *focal length* is the distance from the focal point to the lens. The focal length is a convenient measure of the strength of a lens. In physiological optics, the lens strength is usually expressed in *diopters*, the reciprocal of the focal length expressed in meters.

Diopters (D) = 1/focal length (meters)

For example, if a lens has a focal length of 0.1 meter, its power is 10 diopters (10 D).

IMAGE FORMATION

The position at which an image is formed by a lens can be determined geometrically by tracing the path of several rays through the lens, as shown in Figure 4A-3C. The determination requires only that object size, object distance, and the focal length of the lens be known. There are three rays whose paths can be readily traced. First, a ray which leaves a point source (the end of the arrow) and passes through the center of the lens is bent as it enters the lens and again as it leaves. Because the two surfaces of the lens are parallel at the center, there is little deviation and the ray passes on, its direction essentially unchanged. Second, a ray which proceeds parallel to the principal axis is bent sufficiently to pass through the focal point on the principal axis on the image side of the lens. As this ray proceeds further, it intersects the ray passing through the center of the lens and an image is formed at the intersection. Third, a ray which passes through the focal point on the object side of the lens is bent just enough to meet the other two rays at their point of intersection.*

If the lens is made stronger or the object is moved farther from the lens, the image is formed closer to the lens. The object distance, image distance, and lens strength are therefore interrelated. This relation is expressed by the *lens formula:*

$$\frac{1}{S} + \frac{1}{S'} = \frac{1}{f} \tag{2}$$

where S is the object distance, S′ is the image distance, and f is the focal length.

Example 1. If an object 20 cm from the lens forms an image 20 mm from the lens, what is the lens strength? First, all values should be expressed in meters. The lens strength can then be obtained in diopters, as follows:

$$\frac{1}{0.2} + \frac{1}{0.02} = \frac{1}{f}$$

$$\frac{1}{f} = 55 \text{ D}$$

Example 2. If an object is placed 40 cm from a lens of 50 D strength, what is the image distance?

$$\frac{1}{0.4} + \frac{1}{S'} = 50$$

Multiplying by 0.4S′:

$$\frac{0.4S'}{0.4} + \frac{0.4S'}{S'} = 50 \times 0.4S'$$

S′ = 0.021 meter

*Note that the image formed by a simple spherical lens is inverted.

If a point source of light is placed at the focal point of a lens, the rays are parallel after passing through the lens and do not come to a focus (or focus only at an infinite distance from the lens). If the source is moved away from the focal point, an image is first formed at a great distance from the lens, gradually moving closer as the source retreats from the lens. When the source is an infinite distance from the lens, the light rays falling on the lens are parallel and the image forms at the focal point. This can be shown by the lens equation. Since S is infinite, the equation becomes

$$\frac{1}{S'} = \frac{1}{f}$$

For any optical system there is a point beyond which the object can be considered for all practical purposes to be at infinity. For the eye this distance is 6 meters (20 feet). Any further movement beyond that point has little effect on the image distance.

Image Size

The size of the image is directly proportional to the size of the object and the ratio of the image and the object distances. For example:

if

AB = object length

and

A'B' = image length

then

$$\frac{A'B'}{AB} = \frac{S'}{S}$$

These relations can be seen from Figure 4A-3C and follow from the fact that ABn and A'B'n are similar triangles.

OPTICAL FUNCTION OF THE EYE

The basic optical principles outlined above can be applied to the function of the eye. The structure of the eye is shown in Figure 4A-1. Light rays striking the eye first enter the *cornea*, which has a high degree of curvature and a refractive index of 1.33. These two factors combine to cause a considerable bending of the light rays as they enter the cornea. The resting eye has a power of 67 D. Most of this optical power (45 D) is in the cornea. When one swims underwater, most of the refractive power of the cornea is lost, since water has a refractive index of 1.33. After transversing the cornea, light passes through the *anterior chamber* (refractive index 1.33) without further alteration but is bent when it enters the *crystalline lens* of the eye. The lens has a graded refractive index which varies from 1.41 at the center to 1.39 at the periphery. This graded index decreases the power of the lens, making it equivalent to one having a uniform index of 1.39. The power of the lens is approximately 20 D at rest but may be increased by as much as 12 D by accommodation.

Accommodation is the process by which the refractive power of the eye is modified for viewing nearby objects. It is produced by a thickening of the center of the lens, mostly at the anterior surface.

As already mentioned, the lens has an elastic capsule which tends to make it assume a spherical shape. However, this tendency is opposed by the tension within the *sclera* or

113

outer coat of the eyeball. The tension is due simply to the intraocular pressure, normally about 15 mm Hg, and is transmitted to the lens by the ring-shaped *ciliary muscle,* which encircles the lens, and by the *suspensory ligaments,* which are attached to the periphery of the lens. The tension applied to the lens by the ligaments flattens it by a stretching process. However, when the ciliary muscle contracts, its diameter decreases, thereby decreasing the tension on the suspensory ligaments and allowing the lens to assume a more spherical shape. Thus accommodation depends upon the elastic properties of the lens capsule. It is illustrated in Figure 4A-4.

Associated with accommodation of the lens is a constriction of a sphincter-like *pupillary muscle* that eliminates rays passing through the peripheral portions of the lens. This action decreases spherical aberration, which might otherwise be a limiting factor in near vision. When one observes a nearby object (accommodation required), the axis of the eye is shifted by the *extrinsic muscles* to train both eyes on the object (see p. 127).

For most purposes the optical properties of the compound lens system of the eye may be considered equivalent to those of a single refractive surface, the *optical center* or *nodal point* of which is situated 5 mm behind the anterior corneal surface and 15 mm in front of the *retina.* This simplified model, called the *reduced eye of Listing,* makes it possible to apply the lens equation to the eye. The normal power of the eye and the power of accommodation may be readily determined. For example, in the normal (*emmetropic*) eye at rest, distant objects are brought to focus on the retina. If the object distance in this case is considered to be infinite, equation 2 becomes

$$\frac{1}{\infty} + \frac{1}{S'} = \frac{1}{f} \quad \text{or} \quad \frac{1}{S'} = \frac{1}{f} \tag{2a}$$

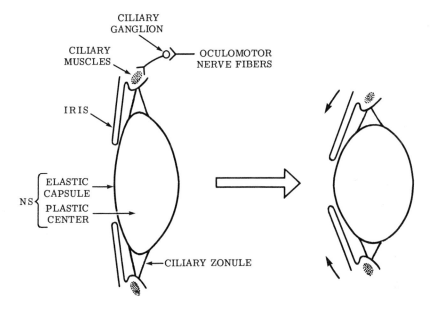

Figure 4A-4
Contraction of ciliary muscles on viewing near object. This causes the lens to become more spherical (accommodation) and the iris to constrict, decreasing the area of its central opening (the pupil).

Table 4A-1. Effect of Age on Amplitude of
Accommodation and Near Point

Age	Amplitude (D)	Near Point (cm)
10	11.3	8.8
20	9.6	10.4
30	7.8	12.8
40	5.4	18.5
50	1.9	52.6
60	1.2	83.3
70	1.0	100.0

Since the image is focused on the retina, $S' = 15$ mm; therefore

$$\frac{1}{.015} = \frac{1}{f}$$

giving 67 D as the optical strength of the resting eye.

The maximal optical strength of the eye can be determined by measuring the shortest distance at which an object may be seen distinctly. This is called the *near point* of vision. In the young adult the distance is about 10 cm. The maximal strength is then

$$\frac{1}{0.10} + \frac{1}{.015} = \frac{1}{f}$$

$$\frac{1}{f} = 77 \text{ D}$$

The difference in strength between the resting and the maximally accommodated eye is the *power of accommodation*. This is about 12 D in children and 10 D in young adults. With increasing age, the elasticity of the lens decreases, thereby reducing the power of accommodation and causing the near point to recede (Table 4A-1). The decrease in accommodation with age is known as *presbyopia*. It is an inevitable result of the aging process. Between ages 40 and 50 the near point normally recedes beyond arm's length, making reading glasses necessary.

OPTICAL ABNORMALITIES

Two of the most common optical defects result from an abnormal size of the eyeball. These conditions are hyperopia, or farsightedness, and myopia, or nearsightedness. *Hyperopia* (or *hypermetropia*) is characterized by an abnormally short eyeball with a decreased distance from lens to retina. Much less frequently it is due to insufficient refractive power of the optical system. This defect causes the image of a distant object to be formed behind the retina in the resting eye (Fig. 4A-5). The hyperopic individual can focus the image on the retina by partially accommodating the lens. This, of course, is not a normal situation and leads to eye fatigue. Because some accommodation is required even for distant viewing, the near point is more distant than normal, and near vision is deficient—hence the descriptive term *farsightedness*. The hyperopic condition can be remedied by placing the appropriate convex spherical lens before the eye (Fig. 4A-5).

Myopia is characterized by an abnormally long eyeball with an increased lens-to-retina distance. Occasionally this condition is produced by an abnormally great curvature of the cornea or lens. The defect causes the image of a distant object to be formed before the retina in the resting eye (Fig. 4A-5). There is no simple physiological way by which the defect can be compensated, since the normal process of accommodation only aggravates the deficiency. Vision of near objects is not impaired, however, and the near point is closer than normal. Myopia can be remedied by placing the appropriate concave spherical lens before the eye (Fig. 4A-5).

Astigmatism is a common optical defect which is most often due to an abnormal curvature of the cornea. Normally, the corneal surface is spherical. In astigmatism the surface is ellipsoid or egg-shaped (Fig. 4A-6), so that the rays traveling in one plane are bent more drastically than those in another. As a result, the rays in one plane may focus on the retina while those in another do not. Astigmatism can be corrected by a *cylindrical* lens. Such a lens may be thought of as a portion of a cylinder which is cut longitudinally. If the lens is viewed from above, the curvature is seen in cross section (Fig. 4A-6). Rays traveling in the horizontal plane also "see" the lens as a curved surface and are bent accordingly. Rays traveling in the vertical plane "see" the lens as a rectangular object and are not bent as they pass through it. By use of a cylindrical lens, the rays not focused on the retina may be brought to focus at that point. Thus it is

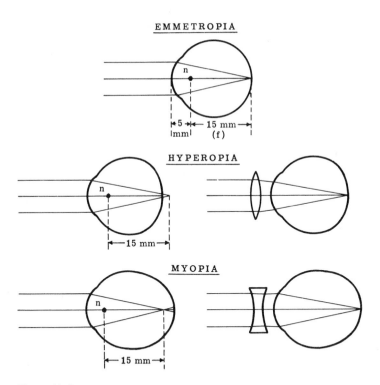

Figure 4A-5
Optical dimensions of the normal and abnormal eye. In the emmetropic eye the nodal point of the lens system is 15 mm before the retina. In the hyperopic eye the nodal point is less than 15 mm before the retina, causing the focal point to fall behind the retina. This defect is corrected by a spherical converging lens placed before the eye. In the myopic eye, rays focus before the retina. This defect is corrected by a spherical diverging lens.

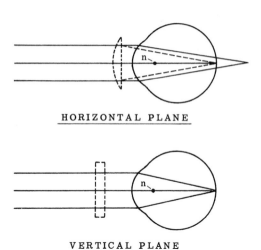

HORIZONTAL PLANE

VERTICAL PLANE

Figure 4A-6
Optic features of astigmatism. In the uncorrected eye, light rays (shown by solid lines) focus behind
the retina in one plane and on the retina in another. The defect is corrected by a cylindrical lens so
placed that the curvature is only in the plane exhibiting the defect.

necessary that the curvature of the correcting lens be placed in the same plane as the
rays having an abnormal focal point. The longitudinal axis of the correcting lens then is
perpendicular to the plane of the astigmatism.

THE OPHTHALMOSCOPE

An ophthalmoscope is a device used for examination of the optical and physical
properties of the eye. It consists essentially of a light source and a mirror or prism which
reflects light into the eye and onto the retina. Some of the light is reflected from the
retina, which becomes, in effect, a new light source. The light passes out of the eye
through the lens and cornea. If the eye is normal and relaxed, the focal point of the lens
and cornea coincides with the position of the retina. Consequently light rays reflected
from the retina are bent just enough to be rendered parallel as they pass through the
lens and leave the eye, whereas if the eye is myopic the rays are convergent.

   The degree of abnormality of the eye may be readily determined by a variety of
means with the ophthalmoscope. For example, if the parallel rays emerging from the
emmetropic eye pass through a +1 diopter lens, an image of the retina is formed 1
meter from the lens. An additional converging lens for the hyperopic eye or diverging
lens for the myopic eye must be used to bring the image to this same point. The strength
of the additional lenses required is a measure of the abnormality that exists. This
method illustrates the principle underlying the use of the ophthalmoscope. At present,
more indirect methods permitting rapid precise measurement of the degree of abnormality are utilized (see Campbell et al., 1974).

THE RETINA

The retina (Fig. 4A-1) is an extremely complex organ consisting largely of nervous
tissue. It is in fact an outgrowth of the central nervous system. Two special areas of its
posterior aspect should be noted: the *optic disc* and the *central fovea* (Fig. 4A-1). The
optic disc is the region where the fibers of the *optic nerve* leave the retina through the
optic foramen (a canal in the bony socket of the eye) and where the *central artery of the*

117

retina (a branch of the ophthalmic artery) and *retinal veins* enter and leave the retinal region. The optic disc contains no visual receptor and is therefore a *physiological blind spot.* The *fovea* is a small depression in the retinal surface, important in relation to visual acuity (p. 119). The fovea and its immediate surroundings form the *yellow spot* or *macula lutea,* so called because of its yellow pigmentation.

The general structure of the nervous elements of the retina is illustrated diagrammatically in Figure 4A-7. Three main layers of cells can be distinguished. These are (going from the inside of the retina toward the outer or vitreous humor side) the *receptor layer,* consisting of *rod cells* (rods) and *cone cells* (cones); a second layer, the *bipolar cells* or *inner nuclear layer;* and a layer of *ganglion cells.* Note that light entering the eyeball must pass through the ganglion layer and bipolar layer before impinging on the photoreceptor layer. In addition to the nerve cells just enumerated, the retina contains numerous *neuroglial* cells which act as supporting and insulating elements.

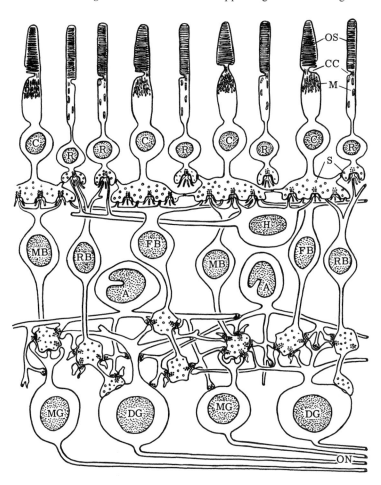

Figure 4A-7
Schematic diagram of the retina showing rod cells (R), cone cells (C) and their specialized structures, the outer segment (OS), connecting cilium (CC), mitochondria (M), and synaptic vesicles (S). Other structures shown are midget (MB), rod (RB), and flat (FB) bipolar cells together with midget (MG) and diffuse (DG) ganglion cells. The axons of the ganglion cells are the fibers of the optic nerve (ON). Horizontal cells (H) and amacrine cells (A) are interneurons.

The rods and cones are so called because of their characteristic shapes. Both have the same general structure, but the rods are usually much thinner than the cones. The light-sensitive pigment is contained in the *outer segment* (OS, Fig. 4A-7). At the other end of these cells are the *synaptic bodies,* which cause the photoreceptor cells to synapse with the bipolar cells. The synaptic regions contain many *synaptic vesicles* (S); these are thought to release chemical transmitters. From the bipolar cells impulses are transmitted to the ganglion cells and hence to the higher centers, the *lateral geniculate body* and the occipital lobe of the cerebral cortex. The axons of the ganglion cells form the fibers of the optic nerve.

The neural pathways in the retina are complex, and their detailed analysis is far from complete. The general pattern may be outlined as follows. The bipolar cells are of two main types: *midget bipolars* (MB, Fig. 4A-7), each of which synapses with one cone; and *diffuse bipolars.* The latter are divided into *rod bipolars* and *flat bipolars* (RB and FB, Fig. 4A-7). Rod bipolars synapse with up to 50 rod receptors. Flat bipolars connect with about 7 cones. Bipolars that synapse with both rods and cones do not seem to occur. Similarly, one finds two types of ganglion cells: *midget ganglion* cells (MG), which synapse with a single midget bipolar, thus providing an exclusive pathway from a single cone to the higher neural centers; and *diffuse ganglion* cells (DG). The latter connect with a number of bipolar cells and may, therefore, respond to stimuli originating from both rods and cones.

Thus the vertical pathways through the retina display both *divergence* (stimulation of several bipolar cells by a single receptor and, in turn, stimulation of several ganglion cells by a single bipolar) and *convergence*—i.e., a single ganglion cell may respond to a number of receptors. The lateral spread of nervous impulses through the body of the retina is thus assured. This spread is greatly assisted by the action of two kinds of interneurons, *horizontal cells* and *amacrine cells.* Overall there would appear to be a great deal of convergence in the human retina since the number of rods (about 125 million) and cones (about 6 million) is much greater than the number of ganglion cells (about 1 million).

SCOTOPIC VISION

The rods contain a reddish purple pigment called *rhodopsin* or *visual purple,* which is bleached upon exposure to light. The bleaching process causes excitation of the receptor

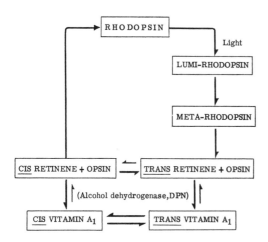

Figure 4A-8
Degradation of rhodopsin when exposed to light. Resynthesis from cis retinene and opsin to rhodopsin occurs in dark.

by means not yet understood. The bleaching and regenerative processes involve changes in the rhodopsin molecule which can be studied in vivo or in vitro. The first stage in the bleaching process is the formation of orange pigments (*lumirhodopsin* followed quickly by *metarhodopsin*), which then split into *retinene* and *opsin* (visual yellow) (Fig. 4A-8). At first, the retinene is in the *trans* isomer form. Some of this retinene is converted to the *cis* form by a photochemical process. Cis retinene in a solution containing opsin is converted in the dark to rhodopsin. Only one of the several cis isomers (11-cis) is the precursor of rhodopsin. The trans retinene which is not isomerized to the cis form is reduced to vitamin A_1. Before the regeneration of rhodopsin can occur, vitamin A_1 must be isomerized from the trans form to the cis form. This action occurs elsewhere in the body by a process not yet understood. Meanwhile, the active form of vitamin A_1 is withdrawn from the blood, to be used in regeneration of visual purple.

Rhodopsin is most readily bleached by light having a wavelength of 500 nm. Its sensitivity falls off on either side of 500 nm; it is about 40 percent as sensitive to light at the blue end of the spectrum. The sensitivity to light in the red region is very low. For this reason, wearing red-tinted glasses allows full adaptation of the rods to night vision even while general illumination is normal. Since some cones are stimulated by red radiation, they do not adapt under these circumstances.

PHOTOPIC VISION

Photopic vision is the term for the visual function performed by the cones. The cones, as a group, are most sensitive to light in the region of 550 nm. The sensitivity of the cones extends throughout the entire visible spectrum but is much less at the two ends of the spectrum than at the center. This is the reason yellow seems much brighter than blue or red. This spectral variation in sensitivity appears to be a summated effect of several different types of cones, each of which responds to a limited portion of the spectrum.

VISUAL ACUITY

The ability of the eye to detect a separation between two adjacent objects depends upon the ability of the retina to perceive a separation between the images that fall upon it. The separation of two adjacent images on the retina depends upon the adequacy of illumination, the fidelity with which the light rays are transmitted by the optical system, and the diffraction pattern of the image on the retina. In addition, the density of the packing of the retinal receptors is a determining factor in the resolution of images. If the separation between images on the retina is to be perceived, a receptor must be present in the intervening space. Since the diffraction pattern tends to diffuse the edges of the image, the separation may not be clear-cut. There may be, in fact, a "gray" zone between two images, which the retina must recognize as a separation.

In viewing nearby objects, the iris of the eye constricts to block peripheral rays and minimize spherical aberration. The visual area of the cortex is apparently able to compensate in some fashion for chromatic aberration. Thus the resolving power of the eye is determined primarily by retinal "grain" or receptor density and the unavoidable diffraction of rays.

The minimum separation between images that can be detected may be conveniently expressed in terms of the visual angle. This is the angle which the separation subtends in the visual field. The minimal visual angle for the normal eye is approximately 1 minute. The images of objects undergoing close scrutiny are projected onto the *macula lutea* or yellow spot, the portion of the retina where the *cones,* the receptors subserving detail vision, are most densely packed. In other portions of the retina the visual angle is greater because the cone population is less dense, and for other reasons which will be discussed shortly.

In clinical studies, visual acuity is determined by use of test charts or letters. The subject is situated 20 feet from the chart so that the emmetropic eye will not be accommodated. The Snellen test chart, which is most often used, is composed of block letters so constructed that the width of the trace and the separation between limbs of the letter are standardized. The capital letter E, for example, is so constructed that the width of the horizontal bars is equal to the separation between them. The chart contains lines of various sizes of type. The test is carried out by having the subject read the smallest type possible from the 20-foot distance, each eye being tested separately. For example, if the subject can read at 20 feet what the normal person can read at that distance, his acuity is expressed as 20/20 and corresponds to a visual angle of 1 minute. If the smallest type the subject can discern at 20 feet corresponds to what the normal individual can read at 40 feet, his acuity is expressed as 20/40 and his visual angle is 2 minutes.

Visual acuity is not uniform over the entire retina. It is greatest in the central region, and particularly in the *fovea,* where the photoreceptors are most closely packed. The central area of the fovea contains only cones, which are the receptors concerned with fine detail and color vision. In addition, the synapses of the cones with the bipolar cells and the connections of the latter with their ganglion cells are of the one-to-one midget type. Other factors contribute to sharp definition of the image in the foveal region: The bipolar and ganglion cell layers are displaced laterally, and the retinal blood vessels largely bypass the area. Thus entering light can impinge more directly on the receptor cells. Furthermore, the yellow pigment of the macula lutea helps to reduce loss of light by reflection. As a result, visual acuity in the central fovea is about twice that just outside the macula lutea and about 40 times that at the retinal border.

As one moves away in a peripheral direction from the fovea, the cones become less abundant and the rods more abundant. Convergence also increases. Vision in the peripheral region of the retina, therefore, is indistinct, but overall light sensitivity is increased since a number of receptors can combine to stimulate a single ganglion cell.

Thus only those images that are projected onto the fovea are perceived clearly and sharply by the brain. However, the peripheral parts of the retina play an important role in *detecting* objects or movement within the visual field. Once an object is detected peripherally, the eyes and/or head are rapidly turned to bring the image onto the fovea and into sharp focus.

Visual acuity and the contrast between an object and its background may be intensified by mechanisms within the retina. When a point source of light is focused on a portion of the retina, some fibers from this area which were previously silent begin to fire. These are called *on* fibers. Other fibers, previously discharging, become silent. These are termed *off* fibers. Still other fibers rapidly adapt to existing light, discharging for a brief period at the beginning and after the end of a light stimulus. These are the *on-off* fibers. Movement of an object across the retinal field leaves a trail of on-off fiber activity in its wake. The fact that the on-off fibers are more numerous than the other two types may account for the great ability of the eye to detect small movements. Normally, the eye does not fix upon an object persistently but scans an area by movements which are rapid and small in amplitude. Apparently this scanning action helps maintain visual acuity, since the image begins to fade if the movements are blocked. The on-off response may be important in this respect, for the scanning tends to create a halo of on-off impulses at the border of the image. There are also lateral neuronal connections in the retina which may be important in visual acuity. When a small area is illuminated, inhibitory impulses sent to surrounding nonilluminated areas tend to intensify the image.

Although visual acuity is lower in the peripheral regions of the retina than in the macula lutea, night vision is better because of the increased population of rods, which are specifically adapted for night vision. The rods are much more sensitive to low levels of illumination than the cones but have a lower level of visual acuity. (Rod vision is about the same throughout the retina and is one-twentieth as acute as the cone vision at the fovea centralis.) The lower visual acuity is due, at least in part, to the convergence of a number of rods upon a single line to the cortex.

ADAPTATION TO ILLUMINATION

If a bright light is shone into the eye, the pupil immediately constricts. This is the *light reflex,* which is initiated at the retina and passes by way of the pretectal region and the ciliary ganglion to cause contraction of the sphincter pupillae muscles. If only one eye is stimulated, the reaction occurs in the pupil illuminated, the *direct light reflex,* and in the opposite pupil was well, the *consensual light reflex.* In tabes dorsalis the light reflex may be lost, while the pupillary reaction during accommodation remains unimpaired. This condition is called the *Argyll Robertson pupil.* The reaction of the pupil to light is usually a temporary phenomenon, the purpose of which is to protect the retina from too intense illumination. As time passes, the retina adapts to the new level of illumination and the pupil returns to its original size. The maximal diameter of the pupil is 8 mm and the minimal is about 1.5 mm. Thus the area, and therefore the amount of light entering, can be changed 30-fold. However, the eye is able to adapt to changes in light intensity of about 10-billion-fold because of changes in the retina.

The reactions of the retina to changes in light intensity may be more readily understood by considering the adaptation to darkness. If an individual enters a dark room, there is a period during which he perceives nothing followed by one in which the objects around him gradually become more discernible. If, during this time, periodic measurements of the light threshold or the just perceptible light stimulus are made, a gradual decrease in this threshold is found, as shown in Figure 4A-9. Both the rods and the cones begin to adapt immediately. The cones initially adapt more rapidly, so that the threshold measured in the early stages of adaptation is essentially that of the cones alone. The cones increase their sensitivity 20-fold to 50-fold in the first five minutes, after which little further adaptation of these receptors occurs. The rods, on the other hand, adapt more slowly but to a much greater degree. After the cones have reached their maximal sensitivity, the threshold of the eye as a whole continues to decrease, because of the adaptation of the rods. Since these adaptations proceed at different rates, the dark adaptation curve for the eye breaks sharply at the point where the rods become more sensitive than the cones. The adaptation of the rods is essentially complete in 25 to 30 minutes, although it may continue slowly for several hours thereafter. In passing

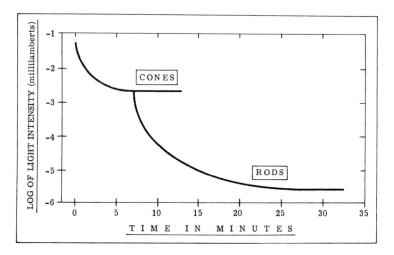

Figure 4A-9
Adaptation of rods and cones to darkness. Change in light threshold is shown on a logarithmic scale.

from a dark area to a light one, the process is reversed. The two-step change in sensitivity is not seen; the photosensitive material in the rods is immediately bleached, and the cones then determine the light threshold.

The process of adaptation is a manifestation of the properties of the photosensitive chemical substance in the light receptor. The nature of this substance has been studied in much more detail in the rods than in the cones.

COLOR VISION

The most widely accepted theory of color vision is the Young-Helmholtz theory. It is based upon the assumption that there are three receptors, red, green, and blue (violet), which subserve color vision. The sensitivity of these receptors, as deduced from indirect evidence, is shown in Figure 4A-10. It can be seen that, although each has its maximum sensitivity in a certain portion of the spectrum, there is considerable overlap between the receptors. The sensation of blue is due to stimulation of the blue receptor alone. The sensation of blue-green is due to stimulation of both the blue and the green receptors. On the other end of the spectrum, the sensation of red is due only to stimulation of the red receptor. However, shifting toward the green portion of the spectrum results in the perception of the sensations of orange and then yellow because of stimulation of both the red and green receptors. Thus the various hues of the spectrum, plus the extra-spectral color purple, may be differentiated on the basis of the three color receptors. If all the receptors are stimulated simultaneously in the correct proportions, the sensation of white is perceived.

For 150 years the Young-Helmholtz theory, first advanced in 1803 by Stephen Young, was without direct substantiation. Indirect evidence from studies of color perception had made it possible to formulate the relationship between the three hypothetical

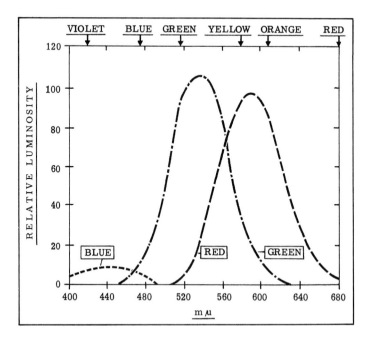

Figure 4A-10
Wavelength sensitivity of the retinal color receptors according to the Young-Helmholtz theory.

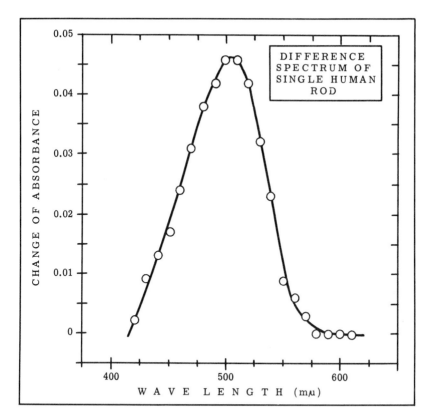

Figure 4A-11
The difference spectrum of a single rod in the human retina (parafoveal region). The spectrum was recorded in a darkened room using very low light intensities to determine the absorption. The experiment was repeated after bleaching with a flash of yellow light. Curve represents difference between the two absorption spectra. (From P. K. Brown and G. Wald. *Science* 144:45–52, Issue 3614, 1964. Copyright 1964 by the American Association for the Advancement of Science.)

receptors shown in Figure 4A-10. However, in 1964 experiments were performed which showed conclusively that such color receptors do exist. Brown and Wald succeeded in measuring the absorption spectrum of single rods and cones in an excised portion of human retina. They found that the rods display an absorption peak at about 505 nm (Fig. 4A-11). They also recorded the absorption of three types of cones: a blue-sensitive cone with an absorption maximum at 450 nm, two green-sensitive cones having absorption maxima at 525 nm, and a red-sensitive cone with its absorption maximum at 555 nm. The results of these remarkable experiments are shown in Figure 4A-12. It should be noted that these absorption spectra agree well with the results obtained in color perception studies.

Equally convincing are the results obtained by MacNichol (1964) on the goldfish retina. Using photometric techniques also, he was able to measure the absorption spectra of more than 100 individual cones and demonstrated that they fell into just three groups, having maxima at 455, 530, and 625 nm. Since this species is known to be color perceptive, the evidence strongly favors the trichromatic theory of color vision. MacNichol's findings also indicate that the mode of excitation of cones strongly

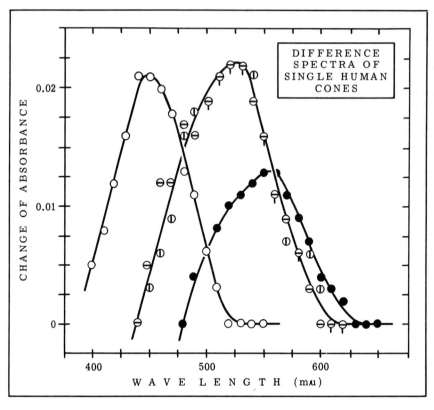

Figure 4A-12
The difference spectra of four cones in the parafoveal region of the retina. Difference spectra represent reduction in light absorption after bleaching with flash of yellow light. The three types of receptors have maximal absorbance at 450 nm (blue sensitive), 525 nm (green sensitive), and 555 nm (red sensitive). (From P. K. Brown and G. Wald. *Science* 144:45–52, Issue 3614, 1964. Copyright 1964 by the American Association for the Advancement of Science.)

resembles that of rods, with a pigment being bleached in the receptor when it is presented with an adequate stimulus.

COLOR BLINDNESS

Color blindness is the inability to perceive a portion of the spectrum or to distinguish between colors recognized by the normal person as being different. The confusion arises between colors that are adjacent in the spectrum. According to the Young-Helmholtz theory, color blindness is due to absence or reduced sensitivity of one or more of the three color receptors. It is present to some degree in 9 percent of males and 2 percent of females.

For example, an individual may be red-blind. This condition is termed *protanopia* and is due to the absence or deficiency of the red receptor. Since the individual has no cones sensitive to red, the red portion of the spectrum is not detected by any color receptor. Apparently the spectral range of the eye is also reduced, since the protanope does not perceive radiation in the red range, particularly beyond 640 nm. The classic example of lack of red perception is the protanope who appeared at a funeral wearing a bright red tie. The protanope has difficulty distinguishing between adjacent colors in the yellow, green, and orange spectral regions. His perception of blue is normal.

Instead of outright lack of a color receptor, some persons may show a color weakness. For example, if this occurs in the red region, the person requires an abnormal amount of red and green to match an intermediate color such as yellow. This condition is known as *protanomaly*. Protanopia and protanomaly are the most common forms of color blindness.

A less common form of color blindness is *deuteranopia*, which is due to lack of the green receptor. The entire spectrum appears to the deuteranope to be composed of yellows and blues. In this instance, the green receptor shifts to the red spectral range and gives the sensation of yellow in the yellow, green, orange, and red ranges. Thus the deuteranope is unable to distinguish between adjacent colors in the region. A weakness of the green receptor is termed *deuteranomaly*. The deuteranope makes many of the same mistakes as the protanope in matching colors. However, the two conditions can be distinguished by the protanope's lack of perception of red.

Tritanopia is a rare form of color blindness which is due to lack of the blue receptor. The spectrum is shortened at the blue end for individuals with this defect.

The types of color blindness just described are due to lack of a single type of receptor. Individuals having them are termed *dichromats*, since their entire spectrum is subjectively composed of two colors or combinations thereof. Normal individuals are *trichromats*. In rare instances, two of the receptors are absent; persons with this defect are *monochromats*. They have no perception of color as such. Apparently they detect only one color and gradations thereof. Total color blindness has also been observed. In this case the cones seemingly do not function at all. The spectral sensitivity curve for persons so afflicted is the same as for persons with scotopic vision (maximum at 500 nm).

VISUAL FIELD

The visual field is the portion of the external environment which is represented on the retina. The extent of this field is limited at most points by anatomical factors such as the supraorbital ridges above and the nose medially. The temporal field is limited only by the orientation of the eye and the sensitivity of the peripheral portions of the retina.

The visual field is also limited to a small extent by the structure of the retina. Since there are no receptors at the exit of the optic nerve from the retina, images falling on this area are not perceived. This is the *blind spot* of the retina. It is situated 10 degrees from the fovea centralis on the nasal side of the retina and encompasses about 3 degrees of visual field.

The extent of the visual field projected onto the retina may be determined by use of the *perimeter*. The perimeter is a metal band shaped as a half-circle, and the subject is situated so that one eye is at its center. If the eye being tested is directed straight ahead at the middle of the band, the two ends of the perimeter are 90 degrees removed from the visual axis. If a small target is moved in from the periphery along the perimeter, the image moves from the peripheral to the central regions of the retina. The subject then indicates when the target is in view. By repeating this test along the horizontal, vertical, and several intervening meridians, one can determine the extent of the visual field. Normally, the visual field for the eye extends 50 degrees upward, 80 degrees downward, 60 degrees nasally, and more than 90 degrees temporally. The field is most extensive for a white target and becomes successively smaller when blue, red, and green targets are used. The extent of the color fields is said to be a relative matter, depending on the brightness of the target.

THE VISUAL PATHWAYS

In man each optic nerve contains about 1 million fibers. In the *optic chiasm* fibers from the nasal region of each retina cross over. Those from the temporal regions do not (Fig.

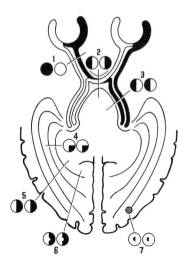

Figure 4A-13
The visual pathways, showing sites of interruption of pathways and the resultant abnormalities in visual fields. (1) Optic nerve—blindness on the side of lesion, with normal contralateral field. (2) Optic chiasm—bitemporal hemianopia. (3) Optic tract—contralateral homonymous hemianopia. (4) Medial fibers of the optic radiation—contralateral inferior homonymous quadrantanopia. (5) Optic radiation in the parietal lobe—contralateral homonymous hemianopia. (6) Optic radiation in the posterior parietal lobe and occipital lobe—contralateral homonymous hemianopia with macular sparing. (7) Tip of the occipital lobe—contralateral homonymous hemianopic scotoma. (From D. O. Harrington. *The Visual Fields* [2nd ed.]. St. Louis: Mosby, 1964.)

4A-13). The nerve trunks that emerge proximally from the optic chiasm are called the *optic tracts*. The visual pathways that carry information from the retina to the cortex are most easily understood by considering the effects on vision of lesions in various parts of them (Fig. 4A-13). First, however, one should note that images formed on the retina are inverted and reversed from left to right (the cortex compensates for this physical inversion). Thus, for example, an object in the lower-left-hand quadrant of the visual field will project an image onto the upper-right-hand quadrant of the retina. Also, images formed on the portions of the visual field of each eye that overlap are projected onto the same area of the cortex.

A lesion of the optic nerve produces blindness in the corresponding eye. A lesion at the optic chiasm involves the fibers from the nasal portion of each retina which cross to join contralateral temporal fibers at this point. The result is *hemianopia,* or blindness in one-half of the visual field of each eye. The blindness is referred to the *visual field* rather than the *retinal field.* Thus, this defect would be termed *bitemporal hemianopia.*

If the lesion occurs in the optic tracts, it interrupts pathways serving the same visual fields in each retina and produces *contralateral homonymous hemianopia.* Since the lesion is on the side opposite the visual field which it affects, it is termed *contralateral.* In the instance shown in Figure 4A-13 the lesion is on the right and the impaired portion of the field is on the left. *Homonymous* refers to the fact that the same side of the visual field is affected in each eye.

The fibers of the optic tract synapse in the *lateral geniculate body* of the thalamus. The latter sends fibers via the *geniculocalcarine tract* to the *visual cortex,* which is located in the occipital lobe of the cerebrum on the same side. A lesion in the geniculocalcarine tract, as in the optic tract, may cause contralateral homonymous hemianopia. If the lesion is less extensive, only one quadrant is obliterated and *contralateral homonymous quadrantanopia* results. If the lesion occurs in the posterior parietal lobe and occipital lobe, a *contralateral homonymous hemianopia with macular sparing* results. The sparing of macular vision occurs because macular fibers terminate before the calcarine cortex. A lesion of the macular field on one side causes *hemianopic scotoma. Scotoma* is a generalized term referring to a blindness or weakness in a portion of the visual field.

BINOCULAR VISION

The extent of the visual field for a single eye has been noted previously. Most of the visual field is the same for both the right and the left eye. If both eyes fix straight ahead upon an object, the visual axes of the eyes are directed toward the object and the centers of the two visual fields coincide. All points within approximately 60 degrees of this center are seen by both eyes.

Each point on the visual cortex within the binocular field receives impulses from a point on each retina. These retinal points are called *corresponding points.* Each gives rise to the same sensation in the cortex. The retina-to-cortex connections are so arranged that when the foveas are trained on the same point, so also are the corresponding points on the two retinas. Thus the same information is sent to the cortex* from all corresponding points on both retinas.

If an object is moved toward an observer, the orientation of the eyes must be shifted so that the foveas remain trained on the object and the corresponding points of the retinas remain matched. This is the act of *convergence,* a reflex rather than a voluntary act. It is accomplished by contraction of an extrinsic striated muscle, the *medial rectus,* which draws the visual axis medially as the object approaches.

Each eye has six extrinsic muscles; they rotate the eyeball on the horizontal, vertical, transverse, and oblique axes. When the eyes are following a moving object, these muscles act in concert to keep the object trained on corresponding points of the retinas. They are capable of moving the eye approximately 50 degrees in any direction from the normal position. If, for some reason, the coordination between these muscles is lost, the images formed in the two eyes no longer fall on corresponding points. A condition known as *diplopia* then exists, in which a double image is projected onto the cortex. If this condition is chronic, one of the images is suppressed and the corresponding eye suffers deterioration of its performance, a condition known as *amblyopia.*

When both eyes fail to focus on the same object, a condition known as *strabismus,* one eye looks directly at the object of interest and the other (the deviating eye) is turned elsewhere. Strabismus can arise from a number of causes: dysfunction of the extrinsic muscles, errors of refraction, and anatomical abnormalities, for example. Depending on the causative agent, strabismus can be corrected by eyeglasses, muscle training exercises, or surgical treatment.

REFERENCES

Brindley, G. S. *Physiology of the Retina and Visual Pathway* (2nd ed.). Baltimore: Williams & Wilkins, 1970.

Brown, P. K., and G. Wald. Visual pigments in single rods and cones of the human retina. *Science* 144:45, 1964.

Campbell, C. J., C. J. Koester, M. C. Rittler, and R. B. Takaberry. *Physiological Optics.* New York: Harper & Row, 1974.

Davson, H. *The Physiology of the Eye* (3rd ed.). New York: Academic, 1972.

MacNichol, E. F., Jr. Retinal mechanisms of color vision. *Vision Res.* 4:119, 1964.

Moses, R. M. *Adler's Physiology of the Eye* (5th ed.). St. Louis: Mosby, 1970.

Zadunaisky, J. A. Electrophysiology and Transparency of the Cornea. In G. Giebisch (Ed.), *Electrophysiology of Epithelial Cells.* New York: Schattauer, 1971. Pp. 225–255.

*Because of the separation of the eyes there are small differences in the shape and pattern of objects in the two visual fields. The closer an object is to the observer, the greater these differences. The observer associates the differences with distance between himself and the object; hence, depth perception depends importantly on these small differences.

4. Special Senses
B. The Ear
Carl F. Rothe

The ear is an organ of exquisite sensitivity and superb design. It is stimulated by energies infinitesimally small, since at threshold it will respond to an energy flux of a millionth of one billionth of a watt per square centimeter. As the function of the ear is described, note the relative ease by which pathological or traumatic changes may occur, and imagine the deleterious effects of calcification of various parts, reduction in compliance, or changes in motion due to inflammation.

Sound is produced and transmitted by the vibratory motion of bodies or air molecules which are displaced from a position of equilibrium in a direction parallel to the direction of propagation, are rapidly restored to this position, overshoot, and again are restored to continue the cycle. Thus sound waves produced by a tuning fork are waves of alternating compression (increased pressure) and expansion (reduced pressure) of air as the tines of the fork swing back and forth. Physiologically, hearing is the subjective interpretation of the sensations produced by vibrations of a frequency and energy adequate to stimulate the auditory apparatus.

Outline of the Hearing Process

Sound enters the external auditory meatus (Fig. 4B-1A) and impinges upon the tympanic membrane or ear drum, putting it into motion. The tympanic membrane in turn is coupled to the auditory ossicles, or bones of the middle ear, which transmit the sound to the inner ear, the cochlea. The inner ear of the living animal is curled in the form of a snail; hence the name. The movement of the auditory ossicles sets into motion the oval window, which separates the aqueous perilymph of the inner ear from the air in the middle ear. The motion of the fluid in the scala vestibuli of the inner ear (the vestibule) causes the basilar membrane to move in a pattern determined by the frequency and intensity of the sound. Movement of fluid in the scala tympani in turn moves the round window. The motion of the basilar membrane stimulates sensory elements in the inner ear (the organ of Corti) so that nerve action potentials are transmitted by the auditory nerve to the auditory cortex. We then perceive these impulses as sound. The vestibular apparatus or labyrinth is part of the inner ear and functions to provide sensory inputs for the maintenance of postural equilibrium (see Chap. 4C).

The *frequency* of sound is given in hertz (Hz) (1 Hz = 1 cycle per second). The psychophysiological appreciation of frequency is pitch. The natural resonant frequency of a system forced to vibrate depends inversely upon the mass in motion (inertia) and directly upon the restoring force (elasticity) acting to bring the body back to equilibrium. The duration of the vibrations depends on the degree of damping (viscous resistance).

The *quality* of a sound refers to the sensation perceived when one hears a mixture of related frequencies. Although musical instruments such as a flute produce a relatively pure sound of only one frequency, most musical sounds are composed of a fundamental frequency plus various amounts of harmonics or overtones which are integral multiples of the fundamental frequency. Nonintegral multiples are also present in some sounds, as from bells. Such mixture of frequencies, regularly repeated, determines the quality of a musical sound and its distinctiveness. *Noise,* on the other hand, is composed of a random mixture of unrelated frequencies. The *duration* of a sound also helps to make it distinctive. Whereas a plucked violin string continues to produce sound for several

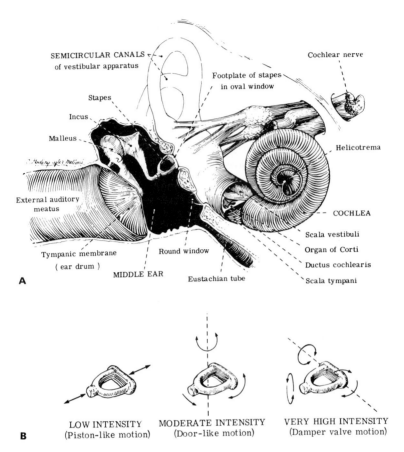

Figure 4B-1
A. Diagram of the human hearing apparatus. B. Modes of vibration of the stapes. At high intensities, near the threshold of feeling, the primary mode of rotation is shifted to a horizontal axis, so that as one edge of the stapes goes in, the other comes out, protecting the inner ear. (A. Redrawn from Melloni in *Dorland's Illustrated Medical Dictionary* [25th ed.]. B. Modified after von Békésy, 1960.)

seconds, the soft tissue of the body effectively dampens the vibrations of such sound-producing organs as the heart.

The *intensity* of a sound is expressed in physical terms as the *amount of energy transmitted per second* through a unit *area* perpendicular to the direction of travel of the wave, i.e., power flux. The usual units are watts per square centimeter. This intensity is dependent on the fluctuation in air pressure. The psychophysiological appreciation of intensity is proportional to the 0.6 power (i.e., approximately the square root) of the sound intensity. It is also related to the frequency. Our ears are most sensitive to sounds with frequencies between 500 and 5000 Hz. If the frequency is increased above or decreased below this level, our ears are less sensitive. This difference in sensitivity is marked, in that it requires about 10,000 times more sound power to hear a 100- or a 15,000-Hz sound than it does to hear a 2000-Hz sound (Fig. 4B-2A). The *pressure* fluctuations must be 100 times as great at 15,000 Hz to be heard as well as a 2000-Hz sound, that is, to sound equally *loud*.

To represent the extreme range of the intensity of sound, the *decibel* (dB) is used. This ratio is somewhat related to perception of intensity. The just noticeable difference of loudness is about 1 dB but ranges between 0.3 and 5.0 dB depending on the frequency and absolute sound level. A decibel is defined as 10 times the logarithm of the ratio of the intensity of the sound in question to the intensity of a reference sound. It is not an absolute unit but always is related to some reference level. In equation form:

$$N_{(dB)} = 10 \log_{10} \frac{I_1}{I_2} = 20 \log_{10} \frac{P_1}{P_2}$$

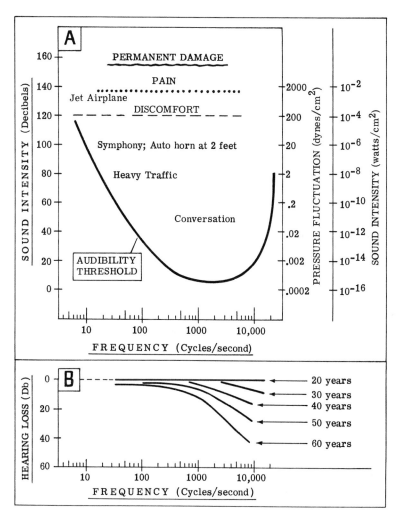

Figure 4B-2
A. Sound intensity as related to audibility at various frequencies, the relative pressure, and energy fluxes. Sound intensity that is uncomfortable or damaging is relatively independent of frequency. B. Presbycusis; progressive loss of hearing ability of men for high frequencies at various ages compared with hearing at 20 years. (Cycles/second = hertz.) (B. Data obtained from C. C. Bunch. *Arch. Otolaryngol.* 9:625, 1929, and from J. C. Webster et al. *J. Acoust. Soc. Am.* 22:473–483, 1950.)

in which I is the intensity in watts per square centimeter and P is pressure fluctuation in dynes per square centimeter. If intensity is in terms of amplitude or pressure change, instead of power flux, the factor is $2 \times 10 = 20$ instead of 10 because for a given system $W = P^2/R$. Here, W is watts of power, P is pressure fluctuation (analogous to voltage in $W = EI = E^2/R$), and R is acoustical resistance. Thus a unit change in power is related to the square (2 times the logarithm) of the potential or pressure fluctuation, assuming a constant resistance. The reference level generally used for intensity is a power flux of 10^{-16} watts/cm^2 and for sound pressure a fluctuation of 0.0002 dynes/cm^2. (Normal atmospheric pressure is about 1 million dynes/cm^2 but is relatively constant, allowing pressure equilibrium throughout the body. However, explosive decompression at high altitude or outer space produces extreme stress on the ear—as well as other structures of the body.) The reference levels, at about 2000 Hz, are about at the threshold of hearing under ideal conditions. Since a dyne of force is roughly equivalent to 1 mg of weight, the usable range of sensitivity to the pressure fluctuations varies between about 0.2 μg/cm^2 at threshold and 2 gm/cm^2 (1.5 mm Hg) as a result, for example, of a sonic boom. This is a ratio of $1:10^7$ or 140 decibels (20×7), since pressure, not power, is the unit considered. The human voice, during normal conversation, develops about 50 microwatts of sound. The sound intensity at the ear of the listener is about 10^{-10} watts/cm^2 (60 dB).

Audiometers are instruments for measuring hearing ability. There are two basic types—the pure-tone audiometer, producing tones of various frequencies and intensity levels, and the speech audiometer, providing for the presentation of speech-testing material by live or recorded voice at different intensities. These instruments are used clinically to measure two basic dimensions of hearing: (1) the threshold of hearing various tones and test words, and (2) how clearly speech can be understood when it is presented at a comfortably loud level. Hearing may be impaired in varying degrees for either one or both of these dimensions.

In the pure-tone audiometer the test frequencies usually are at octave (doubling the frequency) or half-octave intervals, ranging from 125 to 8000 Hz. This frequency range is somewhat less than the capability of the young adult ear, which may perceive frequencies as low as about 10 Hz and as high as about 23,000 Hz without pain. Dogs and some very young children may hear frequencies as high as 40,000 Hz. The thresholds of hearing at the extremes of the frequency range require a high energy flux (Fig. 4B-2A). The intensities on an audiometer are scaled in decibels of hearing threshold level (decibels of *hearing loss* on older audiometers). The zero reference is related to the performance of normal ears rather than to the physicist's standard zero sound pressure level of 0.0002 dyne/cm^2. This threshold reference is more convenient since the ear is not equally responsive at all frequencies. Moreover, the typical normal ear even at the most easily heard frequencies is not quite able to hear sounds as faint as the physicist's zero. The audiometric zero is based on extensive hearing surveys of otologically normal ears of well-motivated young people under ideal conditions. A hearing threshold 25 dB above the ISO 1964 reference level in the range of 500 to 2000 Hz is considered to be a slight hearing impairment.

The ability to hear high-frequency sounds suffers with age, as does the ability to focus for close vision. This poor hearing of the elderly, called *presbycusis,* is particularly obvious in men (Fig. 4B-2B). A man 50 years of age normally has a hearing loss of about 25 dB at 4000 Hz compared with a young adult. This is a ratio of $10^{-25/10}$, which is $10^{-2.5}$, or a loss of $1:300$. Presbycusis is complex, including hair cell degeneration, loss of neuron population, and middle ear impairment (Davis and Silverman, 1970).

The External Ear

The external ear consists of the pinna or auricle, the external auditory meatus, and the tympanic membrane. The *meatus* provides a passage for sound to enter the middle and inner ear. These delicate structures are surrounded and protected by the bone of the skull. The meatus prevents the entrance of large insects and objects which might damage the paper-thin (0.1 mm) ear drum. Furthermore, it acts to keep the air moist and near body temperature ($\pm 0.2\,°C$), essential conditions if the ear drum is to function adequately. The *tympanic membrane* or ear drum is roughly conical in shape, so that a degree of rigidity is provided for coupling to the auditory bones.

The Middle Ear

The middle ear is air-filled and contains the *auditory ossicles.* These bones (Fig. 4B-1A) weigh, in all, about 55 mg. The *malleus,* or hammer, is fastened to the tympanic membrane; the *incus,* or anvil, is firmly attached to the malleus and then acts on the *stapes,* or stirrup, which is attached to the oval window of the inner ear.

The primary function of the middle ear is to efficiently couple the movements of the low-density air to the high-density aqueous medium of the inner ear. If the energy flux is to be transferred efficiently, there must be impedance matching. That is, the relatively high amplitude but low force of movement of the air must be efficiently coupled to the high resistance to movement (inertia) of the fluid of the inner ear, so that the maximum amount of power (force times velocity) is transmitted. The tympanic membrane has an area of about $0.7\ cm^2$ in man. It is coupled through the bones to a much smaller ($0.03\ cm^2$) oval window. The coupling acts to convert the movements of easily compressible air to the higher forces necessary to put into motion the aqueous perilymph of the inner ear. This arrangement is a pressure transformer. It is analogous to holding an automobile door open with one finger while the car is moving, for although there may be a relatively small wind force on each unit area of the door, when the forces are transferred to the one small fingertip a high force is experienced. In addition, there is a small lever action of the ossicular chain of bones (1.3 to 1.0). By these mechanisms the force per unit area is increased about 15 times while the amplitude of vibration is little changed. The minute amounts of energy available are thus efficiently transferred to the inner ear with relatively small loss.

A second important function of the middle ear is to protect the exquisite structure of the inner ear from excessive movements. With low-intensity sound, the motion of the stapes is probably piston-like (Fig. 4B-1B). With a sound of moderate intensity, the axis of rotation of the malleus and incus is such that the stapes rocks about a vertical axis at one edge of the oval window. This motion, though minute in magnitude, is analogous to that of a door opening and closing. Loud, low-pitched sounds cause the axis of rotation to shift so that the major axis of rotation is horizontal, across the oval window. Under these conditions one edge goes in and the other comes out, as diagrammed in Figure 4B-1B. Excessive movements of the fluid within the cochlea are thus prevented by this short-circuiting procedure—a crucial protective mechanism in that the mode of motion can change instantly. It provides a means of protecting the inner ear from transient sounds, such as explosions, which occur much too rapidly for any reflex mechanism.

Another protective device of the middle ear is the reflex action of the muscles, which functions in a manner similar to, and fully as fast as, a blink of the eye. The *tensor tympani* acts to pull the malleus and tympanic membrane into the middle ear, as the name implies. The *stapedius* tends to pull the stapes out of the oval window. The two muscles, acting together, snub low-frequency vibrations and so help protect the inner ear. However, above about 2000 Hz, this mechanism provides scant protection. Furthermore, the response time of this reflex is at least 50 msec, and about one-sixth of a second of a loud, reflex-stimulating tone is required for maximal protection. It is therefore of little value in shielding the inner ear from explosions or blows on the ear. Indeed, protection is not fully adequate, for continued exposure to loud noise is a well-documented cause of hearing loss.

These two protective mechanisms, in conjunction with the characteristics of the inner ear, provide the tremendous dynamic range of the ear, so that we can perceive, without damage, pressures 10 million times the threshold level.

The *eustachian tubes* act to equalize pressures. The body fluids absorb gases, so that the total gas pressure in tissue is about 60 mm Hg below atmospheric pressure. Consequently, air must periodically enter the middle ear if a partial vacuum and impairment of hearing are not to develop. Air enters or leaves by way of the pharyngeal slits of the eustachian tubes when an individual yawns or swallows.

Chronic *otitis media* from infection of the middle ear may so damage the ossicles and their supporting structures that surgical replacement of the ossicles by a single columella between the tympanic membrane and the oval window is required. Since air conduction is impaired, this disorder is an example of *conduction deafness*. A common cause of hearing disability is *otosclerosis*. This disease causes a significant hearing loss in about 1 percent of whites (twice as frequent in females as in males) but is rare among blacks. There is destruction and regrowth of bone about the otic capsule. In most instances otosclerosis produces a primary middle ear lesion. The otosclerotic foci invade the stapes footplate, gradually reducing its mobility. In the early stages of the disease only the low frequencies are affected. However, as the invasion of the footplate continues, the high frequencies are also involved. With complete stapes ankylosis all frequencies are about equally affected, and a hearing loss of about 60 to 65 dB will be present—the maximum loss which can be imposed by a conductive lesion. That part of the impairment from reduced mobility of the stapes may be relieved by surgery, but surgery is not effective at all in reducing impairment attributable to sensory-neural involvement. The surgical procedure of choice used to be fenestration, in which a new oval window was created in the horizontal semicircular canal. Without an impedance-matching transformer system even the best surgical results left a mild hearing deficit of about 25 dB. Later techniques were developed which retained the transformer action of the ossicular chain. At the present time the procedure of choice is stapedectomy with the substitution of a prosthesis.

The Inner Ear

Movement of the stapes causes movement of the perilymph fluid, basilar membrane, and organ of Corti, which in turn triggers neural impulses. Many theories have been proposed to explain frequency analysis by the ear and the triggering of the auditory nerves. The theories presented here are generally accepted and are in accord with most of the data now available.

Place and Volley Theories of Frequency Discrimination

Very-low-frequency pressure waves cause the perilymph to move back and forth through the helicotrema, a minute opening connecting the scala vestibuli to the scala tympani (Fig. 4B-1A). They have little effect on the basilar membrane. At somewhat higher frequencies, for example 30 Hz, the pressure waves tend to short-circuit through the basilar membrane because of the inertia of the fluids; thus they cause it to move back and forth. The movement of the basilar membrane causes distortion of the hair cells, which in turn initiates volleys of neural impulses. Under these conditions frequency discrimination is performed by the cerebral cortex. At these low frequencies the volleys of neural impulses correspond to the fluctuations in pressure of the incoming sound.

At high frequencies perception of sound frequency is based upon the *place* where the maximal movement of the basilar membrane occurs. Although a pattern of the sound pressure changes may still be seen in the action potential pattern to about 3000 Hz, above about 120 Hz the place discrimination process becomes important. The basilar membrane is relatively massive at the distal or apical end (about 0.5 mm wide); here it has a relatively low stiffness, and because of its distance from the stapes a relatively large mass of perilymph is involved in moving the distal end. Thus low frequencies tend to act here. On the other hand, just inside the oval window and stapes the supporting structure is lighter, narrower (0.04 mm), and more rigid. Most important, there is less fluid to move. High frequencies tend to produce the maximal effect here; that is, the amplitude of the oscillation is greater at this place than farther on in the cochlea (Fig. 4B-3).

135

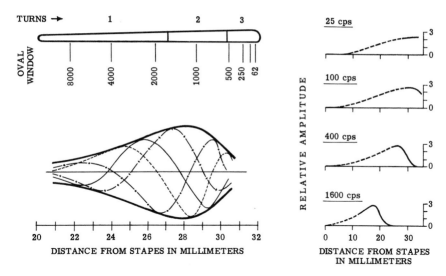

TURNS →

OVAL WINDOW

8000 4000 2000 1000 500 250 62

DISTANCE FROM STAPES IN MILLIMETERS

20   22   24   26   28   30   32

RELATIVE AMPLITUDE

25 cps

100 cps

400 cps

1600 cps

DISTANCE FROM STAPES IN MILLIMETERS

0   10   20   30

Figure 4B-3
(Upper left) Localization of pitch discrimination along the basilar membrane of man is deduced from experiments with guinea pigs. Lesions were produced in the basilar membrane, and electrical audiograms were then made. In man there are three turns in the cochlea, with high frequencies acting near the stapes and oval window. (Lower left) Pattern of motion of a traveling wave with maximal amplitude of the envelope of motion at about 28 mm from the stapes. (Right) Relative amplitude of motion of cochlear partition of a cadaver specimen. (cps = hertz.) (Upper left: After S. S. Stevens et al. *J. Gen. Psychol.* 13:297, 1935. Lower left, right: After G. von Békésy. *Experiments in Hearing.* Copyright 1960, McGraw-Hill Book Company. Used with permission of McGraw-Hill Book Company.)

Von Békésy (1960) has provided much information concerning the dynamic activity of the inner ear. Using a microscope and microtechniques with extreme care, he bored holes in the walls of the cochlear channels, applied minute specks of silver, and with a stroboscope actually measured the displacement at various places along the membrane when sounds of various frequencies were applied to the ears of cadavers (Fig. 4B-3, right). The basilar membrane undulates maximally at a certain place along the membrane when stimulated by a specific frequency. The nature of the undulation is shown in Figure 4B-3. This kind of wave form is called a *traveling wave.* Because of the relatively large mass of fluid and low tension on the basilar membrane, the resonance theory of Helmholtz, which pictured the inner ear as being tuned like a harp, is not adequate.

Studies of inner ears damaged by high-intensity sounds of a specific frequency support the theory that sound frequencies act maximally on specific sites along the basilar membrane. Frequencies above about 10,000 Hz will be discriminated rather poorly because of the closeness of the frequency scale on the basilar membrane (Fig. 4B-3, top left). Thus there is mechanical analysis of frequency by localization of the place of maximal displacement of the basilar membrane. Low frequencies are localized at the far end, and high frequencies act next to the inner ear windows.

The *structure* of the inner ear is complex. A cross-sectional drawing of the cochlea is presented in Figure 4B-4. The total diameter is about 3 mm, and its volume is about 100 μl (the volume of 2 drops of water!). Pressure waves come in along the top, the scala vestibuli, and cause the basilar membrane to move up and down. The lower drawing is still more highly magnified and shows the hair cells and auditory nerves. The *tectorial membrane* is a rather rigid and massive structure which is in contact with the hair cells.

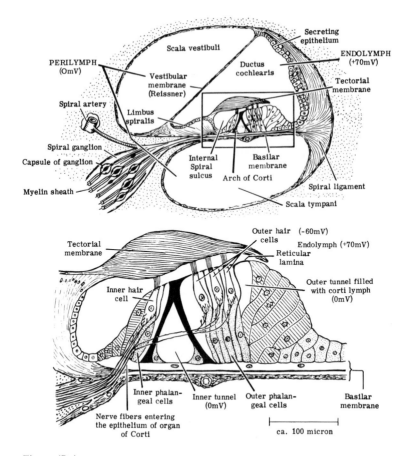

Figure 4B-4
(Upper) Cross section of the human cochlea showing the spiral organ of Corti and the three ducts of the inner ear. (Lower) Basilar membrane with organ of Corti at higher magnification. The periotic spaces are filled with perilymph, and the endotic space is filled with endolymph. (Modified from A. T. Rasmussen. *Outlines of Neuro-anatomy* [3rd ed.]. Dubuque, Iowa: Brown, 1943. Pp. 45, 47.)

Because of the geometry of the attachments of the tectorial membrane and the organ of Corti, sound-induced vibrations produce a shearing action that causes the hairs to distort the cuticular plates of the hair cells. The movement of these hairs somehow excites the auditory nerves. The amount of movement is exceedingly small. A 3000-Hz sound of just threshold intensity moves the tympanic membrane back and forth about 10^{-10} cm. The amplitude of movement is much less than the diameter of a hydrogen molecule (2×10^{-8} cm). The basilar membrane moves only a small fraction as much as this. The auditory apparatus is indeed extremely sensitive. The reason we do not hear blood flowing through the vessels of the tympanic membrane is probably that the otherwise audible higher frequencies are damped so that the flow is steady. It is of significance that there are no blood vessels in the organ of Corti. Its nutrient supply is dependent upon the secretory epithelium of the ductus cochlearis.

Initiation of Neural Impulse
The most recent evidence indicates that neural impulses are triggered in some manner by the development of *receptor potentials* resulting from relative movements of parts of

the organ of Corti. These potentials are detected outside the ear as *cochlear microphonics.*

A unique feature of the inner ear is the steady potential difference across the hair cell membrane available for the development of the receptor potential. The *perilymph,* an ultrafiltrate of plasma, fills the scala vestibuli and the scala tympani and surrounds the otic labyrinth; it is in direct communication with the cerebrospinal fluid and is at the same potential as the rest of the body (Figs. 4B-1, 4). The fluid filling the organ of Corti (the cortilymph) is similar in composition and potential to the perilymph. The *endolymph* fills the interior of the ductus cochlearis (the scala media) located between the basilar membrane and the vestibular membrane. It is probably produced by the secreting epithelium (the stria vascularis) of the ductus cochlearis (Fig. 4B-4). This fluid has, in the scala media, an electrical potential of about 70 mv positive (not negative) with respect to the rest of the body. The mechanism maintaining this potential is not clearly understood. It is highly dependent upon oxidative metabolism and is not particularly sensitive to changes in sodium or other ionic concentrations within the duct. With a few minutes of severe hypoxia, the potential drops to near zero, and if the hypoxia is prolonged, it attains a large negative value. On return of blood flow carrying adequate amounts of oxygen it returns to normal in a matter of seconds. Destruction of the secreting epithelium abolishes the potential. Thus the positive endolymph potential is apparently a secretory potential rather than a diffusion potential. With this positive 50 to 80 mv outside the hair cells, and the usual negative 20 to 80 mv inside, there is a uniquely high (up to 160 mv) potential across the membranes of the hair cells.

Slight movement of the hairs by the relative motion of the tectorial membrane and the basilar membrane probably distorts the hair cell membrane and opens pores to allow a flow of ions to initiate the partial depolarization of the hair cell membrane—the receptor, cochlear, or trigger potential. Present evidence suggests that this potential "acts directly upon the unmyelinated dendrites of the afferent neurons found at the sides and bases of the hair cells" to excite the cochlear nerves (Gulick, 1971).

The receptor potential and cochlear microphonics are not nerve action potentials, for there is no distinct threshold, no refractory period, no all-or-none response, but a potential which is proportional to the displacement of the basilar membrane at moderate sound pressure levels. As a pressure front impinges upon the tympanic membrane, there is a delay of 0.1 msec or less before the initiation of a cochlear microphonic, then a delay of about 0.7 msec before the nerve spikes are seen. The hair cells next to the stapes respond to all frequencies, but in the third turn of the cochlea (Fig. 4B-3), receptor potentials are produced by relatively low frequencies only, supporting the place theory of frequency analysis.

Although sound of a particular frequency sets into motion a relatively large part of the basilar membrane, some mechanism, probably mediated through the auditory cortex, sharpens the sensation so that frequencies differing by 2 to 3 Hz can be distinguished in the range of 60 to 1000 Hz when presented separately. Above 1000 Hz, the ability to discriminate between different frequencies is about 2 per 1000. Thus 8000- and 8016-Hz tones, if heard alternately, will be perceived as just different in frequency. If two tones of similar loudness and nearly the same frequency are heard simultaneously, a phenomenon called *beats* will be heard. There will be a variation in loudness which occurs at a frequency equal to the *difference* of the frequencies of the two tones. Beats between two tones can be detected up to a beat frequency of about six per second. Listening for these beats is the technique used by musicians to tune their instruments. If the beat frequency is greater than about six per second, a sensation of dissonance occurs.

Loudness discrimination is possible since sounds of higher intensity cause a greater movement over a wider area of the basilar membrane than do those of low intensity. By moving more hairs to a greater degree, more hair cells are stimulated to excite more auditory nerves and also to increase the nerve impulse frequency to give the sensation of greater loudness. In addition, the inner hair cells (Fig. 4B-4) have a higher threshold and therefore, when stimulated, may add to the sensation of loudness.

Sensory-neural deafness, involving damage to the inner ear or the auditory nerve, may occur at any age as a result of infections or trauma. Some antibiotics of the streptomycin group are ototoxic and thus cause degeneration of the organ of Corti.

Prolonged exposure over a period of years to high-intensity noise can result in *noise-induced hearing impairment.* Thus hearing conservation measures are recommended if the noise level exceeds 85 dB over the range of 300 to 2400 Hz. At this level of noise, conversation is not possible unless the voice is raised and the people are within a few inches of each other. The availability of high-power audio amplifiers for popular music has made permanent hearing impairment by such "noise pollution" a significant problem for many young people. *Acoustic trauma* to the organ of Corti may be caused by a single high-intensity noise of greater than about 140 dB. When the sound is a pure tone, histologically detectable damage to a localized spot along the basilar membrane occurs. The maximum loss in hearing ability occurs at a frequency about a half-octave above the frequency of the damaging tone, a site nearer the oval window. Hearing impairment by noise is dependent upon not only the intensity of the noise but also its frequency and duration.

Sound intensity of 160 dB, a level attained, for example, close to a jet engine, has a power flux of 1 watt/cm^2. Since the soft tissues of the body absorb sound, particularly that of high frequency, and convert it to heat, there is a significant and dangerous increase in body temperature under these conditions. Furthermore, the pressure fluctuations are of the order of 20 gm/cm^2 (about 15 mm Hg), and delicate tissues such as those of the ear and brain can be literally torn apart by the vibrations. Whereas ultrasound intensities of over 1 watt/cm^2 may cause serious biological damage from excessive temperature increases, pressure variations, or cavitation, levels of 100 mw/cm^2 or less are currently considered safe for diagnostic studies (Ulrich, 1974). However, continuing research may reveal serious dangers from even this low level in some tissues of certain individuals.

High-intensity sound can be used to destroy pathological structures in the brain. Ultrasonic frequencies of about 1 million per second, well beyond the audible range, may be focused to impinge on a small area within the body. When such beams of high-intensity sound are focused on a desired spot in the brain, tissue can be selectively destroyed by the heat generated without serious damage to intervening tissue. The echoes from pulses of ultrasound of *low* intensity are being used to visualize internal structures of the body, such as heart valves, the ventricles of the brain, a fetus.

Olivocochlear Efferent Nerves
In addition to the afferent cochlear nerves, there are efferent nerves—the olivocochlear bundles—going to the hair cells. Their function is not clear. They have vesiculated, densely granulated endings. They appear to be inhibitory, raising the threshold of the auditory nerve fibers, especially to masking noise.

Localization of Source of Sound
The process by which we localize the source of a sound is complex in that several mechanisms are involved. The *time of arrival* of the pressure front (e.g., from a click) provides one mechanism. If the source is to the right of the individual, the sound first stimulates the right ear and then the left ear. By orienting sound sources and having a blindfolded subject indicate the direction of the sound source, or by using two earphones and an electronically produced time-delay for one of the ears, it has been found that the normal individual can distinguish differences in the apparent location of the source when the arrival times of two clicks are spaced as closely as about 50 μsec. A separation of about 1 msec gives a sense of localization to the side of the earliest sounds. If the clicks are more than about 10 msec apart, the sensation is not that of localization but of hearing two separate sounds. In the interpretation of heart sounds, for example, if the aortic and pulmonary valves close within an interval less than 20 msec, the sound will be perceived as a single sound and not as a split sound (Chap. 13).

With low frequencies, there is localization based on the *phase* relationship. This is closely similar to the time-of-arrival mechanism but involves a smoothly changing type of sound (sine wave) rather than sharp waves such as are associated with clicks.

Intensity of sound at each ear provides another important mechanism, particularly at high frequencies. The head tends to shadow the sound, and since we can distinguish a difference in intensity of about 1 dB, this ability furnishes a mechanism for localization of sound. At high frequencies the difference in intensity, because of the shadowing effect, is as high as 30 dB; thus, rather fine localization ($\pm 10°$) is obtainable. This difference in intensity of the high frequencies is of prime importance in the appreciation of stereophonic sound. Interestingly enough, at about 3000 Hz, the frequency of maximal sensitivity of the human ear, both the time of arrival and the intensity mechanism are less than optimal; hence these frequencies are difficult to localize. The human voice, having transient, clicklike sounds and lower frequencies, is easily localized.

REFERENCES

Beranek, L. L. Noise. *Sci. Am.* December, 1966. Pp. 66–76.

Burns, W. *Noise and Man.* Philadelphia: Lippincott, 1969.

Davis, H. Excitation of Auditory Receptors. In J. Field (Ed.), *Handbook of Physiology.* Washington: American Physiological Society, 1959. Section 1: Neurophysiology, vol. 1, pp. 565–584.

Davis, H., and S. R. Silverman (Eds.). *Hearing and Deafness* (3rd ed.). New York: Holt, Rinehart & Winston, 1970.

Eldredge, D. H., and J. D. Miller. Physiology of hearing. *Annu. Rev. Physiol.* 33:281–310, 1971.

Goldstein, M. H., Jr. The Auditory Periphery. In V. B. Mountcastle (Ed.), *Medical Physiology* (13th ed.). St. Louis: Mosby, 1974. Chap. 12.

Grinnell, A. D. Comparative physiology of hearing. *Annu. Rev. Physiol.* 31:545–580, 1969.

Gulick, W. L. *Hearing: Physiology and Psychophysics.* New York: Oxford University Press, 1971.

Jerger, J. (Ed.). *Modern Developments in Audiology* (2nd ed.). New York: Academic, 1973.

Mountcastle, V. B. Central Neural Mechanisms in Hearing. In V. B. Mountcastle (Ed.), *Medical Physiology* (13th ed.). St. Louis: Mosby, 1974. Chap. 13.

Siebert, W. M. Hearing and the Ear. In J. H. U. Brown and D. S. Gann (Eds.), *Engineering Principles in Physiology.* New York: Academic, 1973. Vol. 1, chap. 7.

Ulrich, W. D. Ultrasound dosage for nontherapeutic use on human beings—Extrapolations from a literature survey. *I.E.E.E. Trans. Biomed. Eng.* 21:48–51, 1974.

von Békésy, G. *Experiments in Hearing.* New York: McGraw-Hill, 1960.

4. Special Senses
C. The Ear: Vestibular Apparatus

Julius J. Friedman

Orientation of the body during movement is maintained in part by reflex activity (acceleration reflexes) which originates from the *labyrinths* (vestibular apparatus), the nonacoustical part of the inner ear. The reflexes adjust the position of the head with respect to the horizontal plane and gravity and reorient the body to effect a normal postural alignment. In addition, man uses other sensory information to maintain orientation with respect to the environment. This information includes visual perception, proprioceptive impulses from the body musculature, and external sensations of relative motion and position. The vestibular apparatus consists in part of the *semicircular canals,* which are concerned with *angular* movement of the head. The horizontal (lateral) are the *nongravity* canals, concerned with turning movements around a vertical axis, as in waltzing. The vertical are the *gravity* canals, concerned with movements around a horizontal axis, as in leaping and landing and in protection against falling. The other parts of the vestibular apparatus are the otolithic organs, the *utricle* and *saccule,* which are stimulated by slow tilting (gravity), centrifugal force, and linear acceleration.

STRUCTURE OF THE LABYRINTH

The horizontal (lateral), anterior (superior), and posterior (inferior) membranous semicircular canals are enclosed in the *bony labyrinth* (Fig. 4C-1). These canals are continuous with the utricle, and both are filled with a fluid, *endolymph.* Near the point of junction with the utricle, each canal is equipped with an enlarged region called the *ampulla.* The ampulla contains the receptor organ, *crista ampullaris* (Fig. 4C-2), a ridge of columnar epithelial hair cells possessing hair filaments which are embedded in a gelatinous mass called the *cupula.* The cupula projects from the crista to the opposite wall of the ampulla and represents a movable partition that can be distorted by movements of endolymph within the canals. The hair filaments protruding from each sensory hair cell are composed of 40 to 70 *stereocilia* and one *kinocilium.* The kinocilia of the cells of the horizontal crista are always located at the peripheral margin of the cells on the side toward the utricle, whereas the kinocilia of the cells of the posterior and anterior cristae are on the canal side (Fig. 4C-2). These receptors respond bidirectionally. Displacement of the stereocilia toward the kinocilium depolarizes the cell, increasing its discharge frequency; displacement of the stereocilia away from the kinocilium hyperpolarizes the hair cell, decreasing its discharge rate.

The utricle and saccule are situated anterior to the semicircular canals and communicate with each other through the independent utricular and saccular, and the common endolymphatic, ducts. The receptor organs of the utricle and saccule are called *maculae* (Fig. 4C-2) and are structurally similar to the cristae. However, the stereocilia and the kinocilia of the utricular maculae are oriented in a more complex pattern. Kinocilia of these sensory organs are pointed in both directions, providing both hyperpolarization and depolarization with a displacement in either direction. The depolarization effect must exceed that of hyperpolarization since displacement increases the discharge frequency. Embedded in the hair cell filament membrane of the maculae are numerous small particles of calcium carbonate called *otoliths.* The macula of the utricle is oriented horizontally with the head in the erect position, whereas that of the saccule is oriented vertically.

The nerve endings in the cristae and maculae communicate with the central nervous system via the vestibular ganglion and vestibular branch of the eighth nerve.

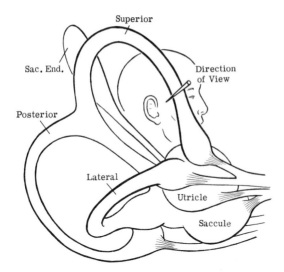

Figure 4C-1
Structural relations of the human vestibular apparatus. (From M. Hardy. *Anat. Rec.* 59:403, 1934.)

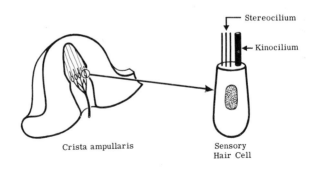

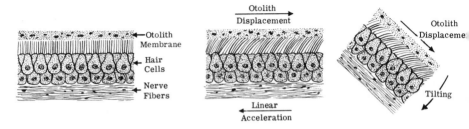

Figure 4C-2
Diagrams of the vestibular sense organs. (Above) Crista ampularis. (From A. Flock and J. Wersall. *J. Cell Biol.* 15:20, 1962.) (Below) Macula. (From J. Fischer. *The Labyrinth.* New York: Grune & Stratton, 1956. P. 17. By permission of the publisher.)

FUNCTION OF THE SEMICIRCULAR CANALS

Mode of Stimulation of the Semicircular Canals

Under resting conditions there is a balanced basal discharge of impulses from the ampullae on both sides of the head. Deviation of either cupula from the resting position will create an imbalance and produce the sensation of rotation. For example, an animal lacking one vestibular apparatus may experience the sensation of rotation whether or not the ampulla is compressed or decompressed, because of the gradient established by a deviation from resting balanced condition. With two vestibular systems present, the intensity of sensation is increased, because of the greater differential between the ampullae on either side.

The crista of a particular canal is stimulated when the head is rotated in the plane of the canal. Because of the inertia of the endolymph, movement of this fluid lags behind movement of the canal throughout the period of rotational acceleration. Since this situation is equivalent to a reverse flow of endolymph, the endolymph is displaced during acceleration in a direction opposite to rotation. Thus, with horizontal rotation to the right, endolymph is displaced to the left, thereby compressing the right and decompressing the left ampullae (Fig. 4C-3). This movement approximates the stereo-

ROTATION	ENDOLYMPH DISPLACEMENT (Post rotary)	VERTIGO	NYSTAGMUS	FALLING
 RIGHT HORIZONTAL	Posterior 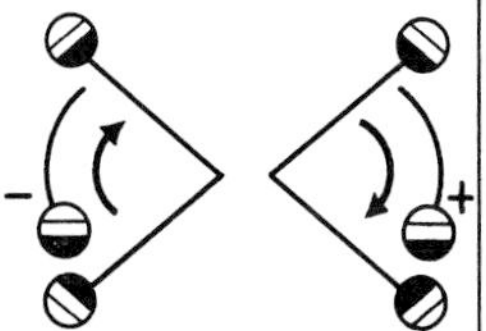 / Anterior	SPINNING LEFT	HORIZONTAL LEFT	TURNING RIGHT
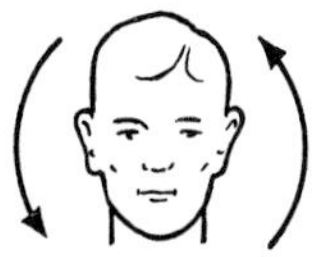 RIGHT TRANSVERSE	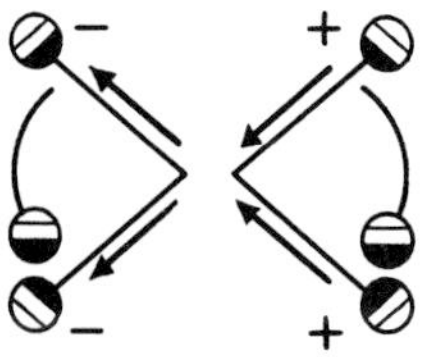	FALLING LEFT	ROTARY LEFT	FALLING RIGHT
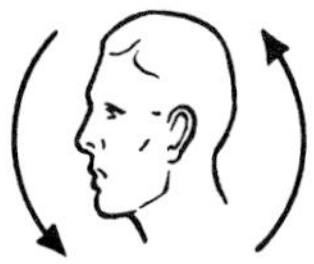 FORWARD MEDIAL	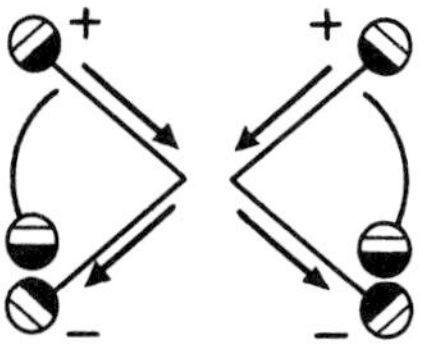	FALLING BACKWARD	VERTICAL UPWARD	FORWARD

* Rotations in the opposite directions will produce reverse endolymph displacement and signs

Figure 4C-3

Effects of rotation in the three main axes on the postrotatory endolymph displacement. The dark segment and thin line of each crista respectively denote the relative position of the kinocilia and stereocilia. Rotations in the opposite directions will produce reverse endolymph displacement and signs.

cilia and kinocilia of the right and separates those of the left cristae. Consequently, the frequency of afferent impulses from the right ampulla increases, whereas that from the left ampulla decreases, so that the sensation of rotation to the right is perceived. It should be noted that the *sensation of rotation is in the direction opposite to the direction of the endolymph displacement.* This is the case with rotation in all planes. As the speed of rotation is held constant and as acceleration becomes zero, the endolymph lag disappears. Because of its elasticity, the cupula returns to the resting position and the discharge of impulses from the ampullae is restored to the basal frequency. In this manner, the sensation of rotation declines as angular acceleration approaches zero. With deceleration, the inertia of the endolymph causes it to be displaced in the direction of rotation, thereby compressing the left and decompressing the right ampullae and producing a postrotatory sensation of rotation in the opposite direction (Fig. 4C-3).

In an ordinary head movement more than one canal is stimulated on each side; the three pairs of canals will always give rise to a combination of endolymphatic streams, the resultant of which will flow in the "plane of rotation." With transverse rotation down and to the right, endolymph is displaced upward and to the left, thereby compressing the left anterior and posterior ampullae and decompressing the right anterior and posterior ampullae. Thus the stereocilia and kinocilia of the right ampullae are brought closer together, and the discharge frequency increases while the opposite events take place in the left cristae. The sensation of rotation down and to the right is thus elicited. With deceleration, the endolymph is displaced in the direction of rotation, compressing the right and decompressing the left anterior and posterior ampullae and causing the postrotatory sensation of rotation up and to the left. Medial rotation forward and down produces endolymph displacement upward and backward, compression of the posterior and decompression of the anterior ampullae, and a sensation of rotation forward and downward. With cessation of rotation, endolymph is displaced in the direction of rotation (forward and down), thereby compressing the anterior and decompressing the posterior ampullae.

Effects of Labyrinth Stimulation

Stimulation of the semicircular canals produces the subjective manifestations of vertigo, nausea, and other autonomic nervous responses, in addition to the objective manifestations associated with changes in tonus of eye muscles (nystagmus) and of the antigravity muscles (falling reaction).

Vertigo

Vertigo is the sensation of spinning with respect to the environment and may cause a person to lose balance and fall. The direction of the sensation of spinning depends on which semicircular canal is stimulated. *In every case, the vertigo is oriented in the direction opposite to the endolymph displacement.* Therefore, the vertigo is in the direction of rotation during the period of rotation, whereas that experienced during the postrotatory state is in the direction opposite to the original rotation.

Autonomic Responses

Autonomic responses resulting from the vertigo include nausea and vomiting, pallor, perspiration, and—with intense stimulation—cardiac, vasomotor, and respiratory influences which can lead to hypotension and hyperpnea.

Alterations in Muscle Tone

Stimulation of the semicircular canals produces an increased muscle tone on the ipsilateral side and a decreased tone on the contralateral side. As a consequence of these changes in muscle tone, one experiences nystagmus, past pointing, and falling reactions. *Nystagmus* is an oscillatory eye movement consisting of a slow and a fast component. The slow component is always in the direction of the flow of the endolymph displacement and is a result of a reflex from the cristae via the vestibular nuclei in the brain stem to the eye muscles. The fast component, the recovery phase of eye rotation, which

145

unfortunately has been designated to define the direction of a nystagmus, is due to an efferent central nervous component of the vestibular reflex from the reticular formation in the brain stem, not from the vestibular nuclei. Rotation in the horizontal plane produces a horizontal nystagmus, medial rotation produces a vertical nystagmus, and transverse rotation produces a rotatory nystagmus. *Past pointing* and *falling reactions* are due to a change in tonus of the antigravity muscles wherein the tone of the muscles increases in the side toward which the endolymph is displaced and decreases on the opposite side. Thus, if the ground gives way under a person's right foot, he and his head lurch to the right, shifting his endolymph to the left. This action reflexly causes an immediate extension of his right arm and leg and flexion of his left arm and leg, accompanied by deviation of his eyes to the left. These are protective righting reflexes.

Caloric Stimulation
Douching the ear with warm or cool water produces convection currents in the endolymph which cause crista stimulation. This approach has an advantage over stimulation by rotation in that the vestibular system of each ear can be tested separately. The technique is used widely in diagnosis of vestibular disorders. Douching each individual ear at suitable intervals with water at 30°C and then at 44°C is a method used to ascertain canal paralysis, paresis, and a phenomenon called *directional preponderance.* The latter means there is a tendency for the nystagmus elicited by both ears to beat more (longer) in one direction (right or left) than in another. This is sometimes indicative of central nervous system disease.

When the head is held backward at 60 degrees, the horizontal canals are brought into vertical position. The douche causes a greater change in the temperature of the endolymph in the part of the canal lying near the external auditory meatus than in the more deeply situated parts. With a cold douche, the currents created move away from the ampulla; with a warm douche, they move toward it. The convection currents thus set up stimulate the crista, and *horizontal nystagmus* and vertigo result. With the head upright, both sets of vertical canals are stimulated, and *rotatory nystagmus* results, away from the cold-syringed ear and toward the warm-syringed ear.

FUNCTIONS OF THE UTRICLE AND SACCULE
The utricle is the most important gravity organ in the labyrinth. Its maculae respond to slow tilting, linear acceleration, and centrifugal force. With the head in the erect position, the utricular maculae are in the horizontal plane, and the otoliths exert uniform pressure down upon the hair cells. In this case there is no distortion of the hair filaments, and the discharges from both right and left utricles are in balance, the hair cells firing at the rate of 10 to 20 per second. Linear acceleration or tilting the head causes the otoliths to be displaced because of inertia and the force of gravity. This displacement produces distortion of the hair filaments and elicits a change in discharge frequency of the hair cells. More specifically, if the left utricle is tilted to the left, the discharge rate will increase, reaching a maximum in the left side down position, whereas a tilt to the right decreases the discharge of the left utricle to a minimum in the left side up position. Complementary changes occur on the opposite side; these amplify the imbalance between the right and the left utricular discharge. The function of the saccule is obscure and still relatively unknown.

REFERENCES
Fischer, J. *The Labyrinth: Physiology and Functional Tests.* New York: Grune & Stratton, 1956.
Gernandt, B. E. Vestibular Mechanism. In J. Field (Ed.), *Handbook of Physiology.* Washington: American Physiological Society, 1959. Section 1: Neurophysiology, vol. 1, pp. 549–564.

5. Reflexes and Reflex Mechanisms

Sidney Ochs

Descartes first clearly defined the basic behavior pattern of the reflex. He gave an example of a foot placed near a fire, which when painfully stimulated is quickly withdrawn. (In modern terms this is called a *flexor withdrawal reflex.*) Involved in reflex activity are: excitation of sensory receptors, conduction over afferent nerve fibers, and finally, after specific central nervous system (CNS) activity, excitation of the motor nerves innervating the musculature to give an appropriate contraction.

Reflex responses can be described as *machine-like* in character, which implies a reproducibility of response. They have also been called *purposeful,* since reflexes are generally of use to the organism. For example, when a beam of light is shone into the eye, a constriction of the pupil occurs, the *pupillary reflex.* This reflex response helps prevent the retina from being subjected to intense illumination. Similarly, reflex control of the muscles of the ear drum acts to prevent excessive sounds from being transmitted to the inner ear. Another example is the *pinna reflex,* found in the cat and the dog. When a foreign object enters the outer ear, a series of reflex ear twitches act to dislodge it. The purposive nature of reflexes is further demonstrated in the spinal frog. Pithing the brain of this animal results in a short period of *spinal shock* (see following section). After recovery, when an irritant acid solution is applied to one flank, the leg on that side is brought up and wiping movements are performed. If the leg is held, the other leg performs the wiping movements. The direction of the limb toward the spot of excitation on the flank is an example of *local sign.* This implies that stimuli which enter the nervous system have sufficient localizing information so that the resulting reflex actions are directed to the appropriate site.

In the last century, the purposiveness of reflexes was taken to indicate that there was some kind of primitive consciousness in the spinal cord that directs reflex activity. Sherrington (1906), however, pointed out that the apparently purposive nature of reflex responses represents a selection, during evolution, of responses which have survival value. Certain reflexes damaging to the organism, rather than protective, may be elicited, indicating the lack of conscious direction.

Some actions are partially voluntary and partially reflex. For example, a mixture of volition and reflex activity is involved in the act of swallowing. Food in the mouth may be voluntarily rejected, but once it has passed beyond the fauces, reflexes coordinated by a swallowing center in the medulla are set in motion.

A practical use made of reflexes is illustrated in anesthesiology. Touching the eyelids causes a reflex closure, the *eyelid reflex,* which is lost with moderately deep anesthesia. Touching the cornea of the eye causes a *corneal reflex;* the lid blinks to cover and protect the cornea. This reflex is diminished or lost when the brain stem has been depressed to a perilous degree. The *pupillary light reflex,* constriction of the pupils to light shone in the eye, is one of the last reflexes to disappear in deep anesthesia, along with respiratory and cardiovascular reflex control mechanisms which have their centers in the medulla. Another reflex of value in judging whether the depth of anesthesia is satisfactory for operative manipulation is the response produced by pinching the skin (one example of *nociceptive,* i.e., injury-provoking, stimulation). This evokes a *flexor withdrawal reflex* in which the limb flexes away from the site where the noxious stimulus is applied. When this reflex is absent, anesthesia is generally considered sufficiently deep to permit operative procedures.

As will be discussed in more detail in a later section, the *stretch reflexes* of skeletal muscle are of great importance for normal posture and locomotion and form our basic concepts of the reflex mechanism. A stretch reflex occurs as a result of a brief extension of the muscle. This is elicited by tapping upon a tendon so that a quick reflex contraction of the muscle, a *phasic* or tendon reflex, occurs. A maintained stretch produces a maintained tonic reflex tension in the muscle. Muscles are termed *flexors* or *extensors* depending on whether they flex or extend a limb at a joint. Again, as will be discussed more fully later on, there is a special relationship of flexor and extensor muscles acting around a joint in a stretch reflex response in that, when one muscle is excited reflexly, the other is inhibited, and vice versa.

The various reflexes are so interrelated in intact animals that the result is one continuous smooth, well-directed behavior pattern, each reflex succeeding and merging with the next in rapid sequence. The interrelation of one reflex with another is demonstrated in locomotion. An animal in which the spinal cord has been cut some weeks previously to allow it to recover its reflex excitability (see below) is suspended in a harness. When the foot is gently but quickly pressed upward, the slight spreading of the toe pads suddenly results in a powerful downward thrust of the leg, the *extensor-thrust* reflex. During locomotion the limb is flexed and brought up from the ground. After the body of the animal has carried it forward, the limb is extended and comes into contact with the surface. With that contact, the extensor thrust is excited and the limb is converted into a rigid column to give a pole-vaulting effect to the body, carrying it forward over the extended limb. The reflex is then quickly inhibited, permitting the leg to flex, and the cycle of flexion and extension is repeated.

The reflex relationship of the two hind limbs is shown in the *crossed reflexes.* When the spinal animal is suspended in a body harness with its legs pendant and a noxious stimulation is applied to one limb, a flexor withdrawal is produced. The contralateral hind limb then may show a *crossed extension.* Forceful flexion of one leg may also cause a reflex extension of the contralateral limb. Some animals do not walk or run but hop or gallop with both hind legs flexing and extending together. Rabbits are hopping animals, and in this species reflex flexing of one hind limb usually causes a *crossed flexion* of the contralateral hind limb rather than the crossed extension seen in the cat and the dog. Another type of response associated with locomotion, the *walking reflex,* may be observed when reflex excitability is high. Noxious stimulation of one leg of a spinal dog to cause flexion may not only produce a crossed extension of the opposite limb but be followed by an extension of the originally reflexly flexed leg. This action proceeds in alternating fashion with a fairly constant rhythm, so that there is a cyclic pattern of flexion and extension of the two hind limbs which may persist for several minutes.

Walking and running are activities of the quadruped requiring all four legs. When one leg is stimulated, not only does the leg of the contralateral side show crossed extension, but the foreleg may also extend on the side on which the hind limb is flexed. The opposite foreleg may become extended. Such patterns are referred to as *reflex figures.* The reflexes elicited by these stimulations are part of patterns within the spinal cord controlling normal locomotory behavior of the four-legged animal.

LOWER AND UPPER MOTOR CONTROL

The basic machinery for the most fundamental types of reflex mechanisms is found within the spinal cord and the brain stem. In *local reflexes,* some of which are described in the preceding section, a complex interplay of control is present. Sherrington used such terms as *prepotency* and *dominance* to describe the interaction and ordering of reflexes. Noxious stimulation is usually prepotent, as demonstrated in competition with the *scratch reflex.* To excite the latter, a stimulus is used to imitate an insect crawling on the skin. This gives rise to a rhythmic scratching movement of the limb, directed toward the stimulus. If the scratch reflex is induced and then followed by a nociceptive stimulation to the skin, the scratch reflex ceases. The more prepotent flexor withdrawal response to the noxious stimulation takes command of the neural channels. Only a relatively few lower motoneuron cells are available to respond to reflex action, and some mechanisms as yet unknown act to switch command from one reflex to the other. Sherrington applied the term *final common path* to the motoneuron cells innervating the muscles. All the varieties of reflex activities and the resulting complex behavior displayed by animals are eventually funneled through the relatively few motoneurons of the final common path.

If, as a result of compression, asphyxiation, or neurotropic diseases such as polio-myelitis, the final common path motoneurons are destroyed, there is a total loss of reflex excitability. It is followed later by an atrophic change of the muscles innervated by these neurons (Chap. 3A). This is part of the condition termed a *lower motor lesion.* If the spinal cord is transected above the level of the motoneurons so that the descending connections from higher motor centers are interrupted, a *higher motor lesion* develops. Immediately upon or after the making the section, a diminution or elimination of motor reflexes occurs with a lessened muscle tone, the state of *spinal shock.* A gradual recovery of tone and reflexes then can take place over a period of minutes, months, or longer depending on the species. Spinal shock is not due to the immediate effects of the lesion,—i.e., a possible stimulation produced in cutting across nerve fibers; if, after recovery from spinal shock, a second cut is made below the level of the first, no further onset of shock ensues. The degree of spinal shock is more profound and the recovery time more prolonged, the higher the animal is in the phylogenetic series. In the frog, much of the reflex function returns within a few minutes. In the dog and cat, reflexes begin to return within several hours, full recovery taking several weeks. In the monkey, months are required for recovery, and in man it may take many months, with only a limited recovery of reflexes. These differences in the time course of recovery from spinal shock in the different species are due to *encephalization,* the greater dependence of the lower motor centers upon higher motor control mechanisms which have developed in the brain of the higher species.

Not only are the somatic reflexes depressed during spinal shock, but one finds a similar depression of the *visceral reflexes.* Some of the higher visceral control mechanisms will be outlined later in the chapter, but because it is an important aspect of reflex loss, mention is made here of the effect of spinal shock on urinary bladder function. A reflex emptying of the bladder normally occurs when it is filled with urine to a certain level and stretch receptors in the musculature of the wall are activated. The reflex mechanisms are localized within the lower spinal cord but can also be controlled in part by higher centers. Voluntary inhibition of the reflex is thus possible. Upon spinal cord transection, the upper influences are interrupted. Also, the threshold for the local spinal reflex becomes greatly elevated. In such an event, the bladder may attain a considerable size before the threshold is reached for reflex bladder emptying. In the clinical management of spinal cases, the bladder must be drained by a catheter or the urine manually expressed at intervals. The procedure is continued for some days or weeks until the threshold falls and reflex emptying occurs. This is the *automatic bladder* condition. Other visceral reflex mechanisms, those of vasomotion and sweating, are also depressed and gradually recover after spinal cord section.

Spinal shock appears to be due to the removal of upper influences which act on lower motor mechanisms in a mainly excitatory fashion. Removal of the excitation results, therefore, in a reduced level of a local excitatory state in the lower motor centers and a depression of spinal reflexes and tone. Possibly spinal shock is also due to the removal of a local inhibition of reflex control.

Later in the course of recovery from spinal shock, somatic and visceral reflexes may increase in their excitability far beyond normal levels, so that a mild stimulation can excite a very great response, a mass discharge. Profuse sweating, flushing, urination, defecation, and flexor or extensor motor spasms of the limbs are seen during a mass discharge in spinal man. It is possible that sprouting of fibers from the entering afferent fibers and interneurons synapsing on motoneurons is responsible for these later manifestations of exaggerated neural activities. Or the membrane properties of the cells themselves may have changed so that their reflex excitation thresholds are abnormally low.

THE FLEXOR REFLEX AND SOME GENERAL PROPERTIES OF REFLEX POOLS

The flexor reflex acts to protect against a potential trauma caused by a painful stimulation, i.e., a nociceptive stimulation. Stimulation of the skin of a limb causes a contraction of the limb muscles with a movement of the limb—or even, if strong enough, the entire body—away from the source of stimulation. Flexor reflexes are widespread, engaging many muscles of a limb and, if strong, may spread to other limbs. For example, a moderate nociceptive stimulation of the lower foot may cause only the foot to be withdrawn. If a stronger stimulation is used, the lower leg is also flexed. With still stronger stimulation, the upper leg is flexed as well. Furthermore, other limbs may enter into the reflex when very strong excitation is used. The *Babinski reflex,* routinely looked for during clinical examination, is in adults an abnormally augmented flexor withdrawal to nociceptive stimulation. The bottom of the foot is stroked along the outer edge. Normally the response is a downward movement of the big toe (plantar flexion). In the Babinski reflex, an upward movement of the big toe takes place, often accompanied by a fanning of the other toes. If reflex excitability is high, the foot may be everted from the site of stimulation and the whole limb withdrawn. The presence of the Babinski reflex has been associated with an upper motor lesion.

A brief electrical stimulus applied to the central end of a cut flexor muscle nerve causes a brief reflex contraction in other flexor muscles of that limb. The extensive distribution of response of flexor muscles in response to stimulation of the nerve of one muscle is brought about by the widespread synaptic connections within the spinal cord. Two classes of afferent fiber synaptic termination and distribution are present in the cord. In one type, afferent fibers synapse directly on the motoneurons in *circumscribed* fashion (Fig. 5-1A). In the other, afferent fibers synapse on interneurons, which in turn have widespread or diffuse spread of synaptic contact with other interneurons that

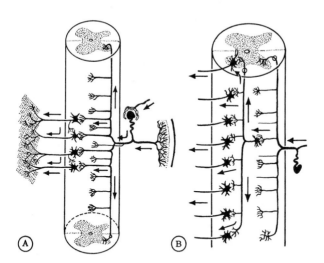

Figure 5-1
Diffuse and circumscribed distribution of afferent fibers terminating in the spinal cord. In A, a circumscribed monosynaptic type of termination is shown. The fibers synapse directly upon the motor horn cells, usually within the spinal cord segment of entry of those fibers. In B, a diffuse transmission within the central nervous system is shown. Sensory fibers terminate upon interneurons within the spinal cord which, by multiple branches, engage a large number of motoneurons that in turn are distributed to many muscles of the peripheral musculature. (From S. Ramón y Cajal. *Histologie du système nerveux.* Madrid: Consejo Superior de Investigaciones Cientificas, 1952. Vol. 1, pp. 531, 532.)

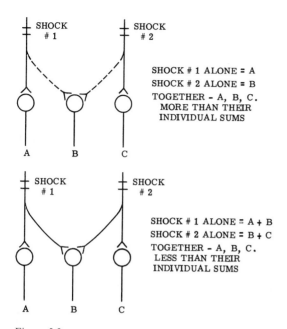

SHOCK # 1 ALONE = A
SHOCK # 2 ALONE = B
TOGETHER - A, B, C.
   MORE THAN THEIR
   INDIVIDUAL SUMS

SHOCK # 1 ALONE = A + B
SHOCK # 2 ALONE = B + C
TOGETHER - A, B, C.
   LESS THAN THEIR
   INDIVIDUAL SUMS

Figure 5-2
Facilitation and occlusion. Excitation of one sensory nerve, # 1, excites a group of neurons, A; and a collateral branch, shown by the dashed line, subliminally excites a group represented by B. Similarly, excitation of sensory nerve # 2 excites neuron group C and subliminally group B. The sum of the two shocks presented separately is A + C. If # 1 and # 2 are excited together, the two subliminal excitations of group B summate, and the resulting discharge of A, B, and C together is more than the sum of the individual shocks. In the lower half of the illustration, occlusion is shown. A shock to # 1 excites neuron groups A and B. A shock to # 2 excites groups B and C. A shock to # 1 and # 2 together, because of their common termination on B, is less than the sum of their individual responses—occlusion.

eventually synapse with motoneurons (Fig. 5-1B). The stretch reflexes are subserved by the circumscribed type of connections, and the flexor withdrawal reflex is controlled by the diffuse type of connectivity.

The entering collateral fibers make synaptic contact either on interneurons or on motoneurons by enlarged endings, the *synaptic boutons*. These are present on nearly all neurons, being particularly dense over the surface of the motoneurons. One class of boutons terminating on the motoneurons acts to excite them, while another set inhibits the discharge of the motoneuron. This result can be inferred from interaction studies made by stimulating two afferent nerves selected from the various nerves innervating the limb muscles. Each afferent nerve eventually activates a number of motoneurons in the pool innervating a part of the peripheral limb musculature to give rise to a flexor reflex. The type of effect observed depends on whether a weak or strong stimulus is used to excite the afferent nerves. With weak stimuli, only a few cells of the motoneuron pool are excited. The cells that do not fire may be *subliminally* excited. When two afferent nerves are stimulated simultaneously, each producing subliminal excitation in a group of neurons common to each, a summation of subliminal excitability occurs and a large response is obtained (Fig. 5-2A). This is called *facilitation*. If the shocks to the afferent nerve inputs are made so strong that a group of cells are excited by each input, the response to stimulation of the two afferent inputs together will be less than the sum of each of the responses elicited individually (Fig. 5-2). This effect demonstrates the phenomenon of *occlusion*.

The presence of multiple synapses over the motoneuron accounts for the variety of complex interactions seen when inputs from different sources are arranged to occur at different times. The spatiotemporal interactions observed with stimuli presented to the same afferent nerve at different times are those of *temporal interaction.* If the strength of the first shock is low in magnitude so that only relatively few cells are fired, with other cells subliminally excited, the later test shocks may show an increased excitatory effect in the motoneuron membrane, *temporal facilitation.* If stronger conditioning shocks are used, the later stimulus shows an inhibition brought about by inhibiting neuron influence, as we will see, which may persist for up to several hundred milliseconds. The long-lasting inhibitions are brought about in part by an effect on the postsynaptic neuronal membrane and in part by interactions through interneurons acting on the motoneurons. The complexities of interaction with this type of reflex activation are simplified by the use of the monosynaptic reflex system described in the following section.

Stretch Myotatic Reflexes—The Monosynaptic Reflex System

A sudden stretch of a limb muscle gives rise to a reflex contraction of that same muscle. The classic example of a *stretch,* or *myotatic, reflex* is the *patellar reflex,* or *knee jerk,* produced when the patella over the knee is given a light tap. The quadriceps muscles are abruptly stretched, thereby exciting stretch *spindle receptors* in the muscle. A volley passes via its afferents to the cord. These impulses excite the motoneurons, producing extension of the lower leg. The reflex nature of the response is shown by cutting the lower lumbar spinal cord dorsal roots through which the afferent fibers from that muscle enter the spinal cord. In this case stretch of the muscle no longer elicits a reflex contraction. The reflex nature of the response to a maintained stretch on the muscle is shown by the increase of tension in the stretched muscle and the drop of tension in it when the dorsal roots are cut. Section of the dorsal root breaks the reflex arc.

Muscles usually operate in pairs for the movement of a limb, so that when one set of muscles, the *agonistic,* is contracting, the opposing *antagonistic* muscles are inhibited and relaxed. This coordinated opposed action of agonist and antagonist muscles is called *reciprocal innervation.* In order to demonstrate the reflex properties of a reciprocally innervated muscle group (e.g., of the anterior tibialis and gastrocnemius muscles), cords are tied to the cut ends of the tendons to stretch them and to record the tension reflexly produced. A stretch of the muscle elicits a reflex contraction in the same muscle and an inhibition of the antagonistic muscle. Reflex inhibition is mediated by inhibitory discharges which act on the motoneurons innervating the antagonistic muscle. This response will be described in the following section.

Both the stretch reflexes excited by a quick stretch of the muscle and the slow myotatic reflexes produced by a maintained stretch are excited by the *spindle receptors* present among the skeletal muscle fibers. A more detailed description of these muscle receptors will be given in a later section of this chapter. The afferent fibers of the spindle receptors enter the cord to synapse directly on the motoneurons with the circumscribed type of connectivity shown in Figure 5-1A. Thus only one synapse between the afferents and motoneurons is involved.

The dorsal root fibers can be arranged so that they may be stimulated with a brief impulse. A reflex discharge is recorded in the ventral roots (Fig. 5-3). An artifact signifies the time of stimulation, and a latency of approximately 1.5 msec occurs before the *monosynaptic response* is recorded in the ventral root (Fig. 5-3). This response is then usually followed by an irregular discharge lasting approximately 15 msec, the *multisynaptic* or *polysynaptic response.* As indicated in Figure 5-3, the monosynaptic response is so termed because it is excited by fibers synapsing directly on the motoneurons. The multisynaptic response occurs after a number of intervening neurons are activated which then terminate on motoneurons.

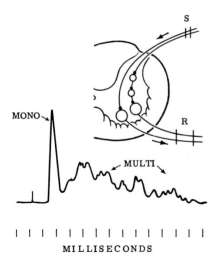

Figure 5-3
Electrical reflexes of the spinal cord. The diagram in the upper portion of this figure shows a stimulus (S) of afferent fibers in the dorsal root which terminate directly on the motoneurons to give the monosynaptic (mono) responses recorded in the ventral root; terminations on interneurons give rise to multisynaptic (multi) discharges. Electrical reflex response patterns are shown in the lower part of the figure.

Evidence that the monosynaptic response involves only a single synapse has been obtained from a knowledge of the conduction time of impulses in afferent and efferent fibers and the time involved in the synaptic transmission at the motoneurons. The delay is similar to the synaptic delay seen at the neuromuscular junction (Chap. 3A). To show this, a brief electrical stimulus is delivered through the tips of the needle electrodes inserted into the ventral horn so as to excite a population of motoneurons (Fig. 5-3). A double response is recorded in the ventral root (Fig. 5-4). The first spike (m) is accounted for by the direct excitation of the motoneurons or of their axons. Little latency is seen other than the time of conduction from the directly stimulated motoneurons to the recording electrodes on the ventral root. The m response is followed after an additional latency of 0.5 to 1.0 msec by an s response, which is due to excitation of afferent fibers synapsing on the motoneurons, the *reflexomotor collaterals of Cajal.* The time between the m and the s responses cannot be reduced below 0.5 msec no matter how strong the stimulation. This represents the synaptic delay time, which ranges from 0.5 to 1.0 msec.

As the strength of a dorsal root stimulus is gradually increased, more dorsal root fibers and more synaptic endings on the motoneuron cells are excited. The size of the resulting monosynaptic response in the ventral root increases in sigmoidal fashion with increasing stimulus strength. When the stimulus is weak, there is a subliminal activation of a large number of fringe cells. When the stimulus strength is increased, a large number of neurons are fired and the fringe group becomes smaller. However, even with a maximal response a significant fringe group exists, as is shown by the phenomenon of *post-tetanic potentiation* (PTP). A PTP is demonstrated by a tetanic stimulation of the dorsal root for several minutes followed by augmented monosynaptic responses. This reaction persists for some minutes before there is a gradual return of responses to control levels. The mechanism responsible for PTP is explained by a change of synaptic transmission, discussed later in this chapter. To anticipate that discussion: The chemical mediator which is considered to underlie synaptic excitation appears to be mobi-

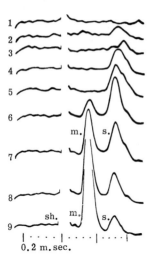

0.2 m. sec.

Figure 5-4

Focal stimulation within the central nervous system. With a stimulating needle electrode in the oculomotor nucleus the motoneurons are excited, as are afferent branches terminating on another group of motoneurons. As the strength of the shock is successively increased (shown in the traces from above downward), a small longer latency response is seen, corresponding to a monosynaptic response with synaptic (s) delay. As the strength is further increased, an earlier m response is seen which is due to a direct excitation of the motoneurons. Notice that as the size of the direct response increases, it occludes and reduces the synaptic response. Even at the strongest shock levels synaptic delay is no shorter than 0.5 msec. Sh is the shock artifact. (After R. Lorente de Nó. *J. Neurophysiol.* 2:409, 1939.)

lized by the tetanic activation with a temporarily increased probability of release of transmitter from the boutons.

The more prolonged electrical discharge appearing after the monosynaptic response, the *multisynaptic response,* is believed to be due to a repetitive activity in chains of interneurons terminating on motoneurons (Fig. 5-3). The motoneurons that give rise to multisynaptic discharges are distributed mainly to flexor muscles, while the monosynaptic response is restricted to the same muscle from which the spindle afferents originate; these muscles are both extensor and flexor. The multisynaptic response is present not only in the ventral root of the spinal cord segment of the stimulated afferent root but also in the ventral roots of a number of segments of the spinal cord, as would be expected from the diffuse type of connectivity (Fig. 5-1B).

A variable factor determining the proportion of monosynaptic to multisynaptic responses is the strength of stimulation. The most excitable afferent fibers, the largest-diameter subgroup of A fibers (IA), take origin from the spindle stretch receptors within the muscles. On activation, these fibers give rise to the monosynaptic response. Fibers of smaller diameter give rise to multisynaptic discharges. By depressing the activity of the spinal cord, either by the use of barbiturate anesthesia or by means of weak shocks to the afferent nerve fibers, one can obtain monosynaptic responses in relative isolation. If the nerve selected for afferent stimulation is purely a skin sensory nerve (e.g., sural nerve), the reflex response in the ventral root is typically only a multisynaptic discharge with little evidence of a monosynaptic discharge.

Facilitation and Inhibition of Monosynaptic Responses

Muscles acting in the same manner at a joint—for example, the lateral and medial gastrocnemius muscles—are *synergists.* Facilitation may be clearly demonstrated in the motoneurons of such muscles by using monosynaptic responses obtained by stimulating the synergistic nerves. For example, assume that the medial gastrocnemius nerve is excited by a *conditioning stimulus* and concurrently or at various intervals thereafter by a *test stimulus* delivered to the central end of the lateral gastrocnemius nerve. The result is that the test response is augmented (Fig. 5-5A). The conditioning stimulus is usually adjusted so that only a few motoneurons fire, giving rise to a just liminal excitation. The test stimulus given concurrently is much augmented. As the time between the two volleys is increased over a period of approximately 15 msec, facilitation gradually falls to zero (Fig. 5-5A).

Because facilitation is seen when the two nerves are stimulated concurrently with no time for an intervening synapse, this is termed a *direct facilitation.* It is brought about by afferent fibers from the one set of stretch receptors in the two muscles synapsing directly

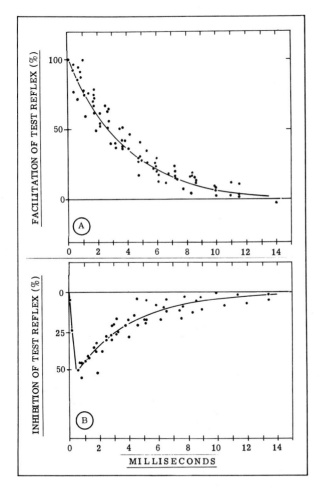

Figure 5-5
Direct facilitation and "direct" inhibition. The monosynaptic response is used as a test response and is preceded at the time intervals shown on the abscissa by a conditioning shock. In A the conditioning shock is a volley in a synergistic nerve. In B the conditioning volley is to an antagonistic nerve. Both facilitatory and inhibitory effects decline gradually to control levels in approximately 15 msec. (From D. P. C. Lloyd. *J. Neurophysiol.* 9:421–438, 1946.)

on synergistic motoneurons. The concurrent excitation in the synergistic nerves causes additional subliminal excitatory activity in the membranes of the fringe group moto-neurons acted on in common by the afferents of the synergists (cf. Fig. 5-2A). Reversing the order of test and conditioning stimulations presented to the synergistic pair of nerves gives rise to the same pattern of facilitation; they show a reciprocal facilitation.

Because of the simplicity of the monosynaptic response, consisting of one presynaptic fiber ending on one postsynaptic neuron, a similar direct system for testing *inhibition* from reciprocally paired inputs of antagonistic muscles might be expected. In this case, however, because a special inhibitory interneuron is interposed (to be more fully discussed in the next section), there is a difference. The inhibition is shown by use of a reciprocal pair of muscle nerves (e.g., anterior tibial and gastrocnemius muscles). A test stimulus delivered to one afferent input is preceded by a conditioning shock to the nerve of the antagonistic muscle, and the test response shows a diminution: i.e., it is inhibited (Fig. 5-5B). When the conditioning and test stimuli are delivered concurrently, very little inhibitory effect may be seen, but with a small increase in the conditioning-test interval the inhibition increases to a peak at an interval of 0.5 msec between the volleys (Fig. 5-5B). As the time between the two stimuli is increased further, the test response gradually regains its normal size over a period of approximately 15 msec. An explana-tion of this difference between facilitation and inhibition requires an understanding of their synaptic mechanisms, described in the next section.

MECHANISMS UNDERLYING SYNAPTIC TRANSMISSION

The preceding discussion dealt with populations of motoneurons involved in the monosynaptic responses. A deeper analysis requires a knowledge of the events occurring in individual motoneurons. Microelectrodes were first successfully used for this purpose by Eccles and his colleagues. These were inserted into the substance of the spinal cord until the entry of the tip of the electrode into the soma of the motoneuron was indicated by the sudden appearance of a negative potential of approximately 70 mv, the resting membrane potential. With microelectrode inside the motoneuron cell body, the stim-ulation of its axon in the ventral root gave rise to a large action potential with overshoot (Fig. 5-6A). The latency of the antidromic response was very short, as expected from the short conduction distance between the ventral root and the cell. When the dorsal root was stimulated to orthodromic activation, a depolarizing potential appeared with a latency of 0.3 to 0.5 msec (Fig. 5-6B). This response is recorded in the cell after it is excited synaptically and thus is referred to as an *excitatory postsynaptic potential* (EPSP). The EPSP lasts approximately 15 msec and has a form similar to the end-plate potential (EPP) recorded in the muscle (Chap. 3A). It increases in size as the volley to the afferent fibers is increased in strength and as more afferent fibers add their synaptic excitation. When the EPSP becomes large enough, a critical level of depolarization is reached and an action potential is excited in the motoneuron (Fig. 5-6C). The action potential propagates down the axon and excites the muscles innervated by that motoneuron.

The EPSP size is related to the number of presynaptic fibers activating bouton endings distributed over the membrane of the motoneuron. Each synapse gives rise to a small current adding to the EPSP. Their addition accounts for the facilitation found in the interaction studies discussed in the preceding section in which the EPSP produced by a conditioning volley summates with a second EPSP produced by a volley in a synergistic nerve. The temporal facilitation is brought about by the summation of EPSP's if these occur in a period of approximately 15 msec, the time course of the EPSP (Fig. 5-7).

When an antagonistic nerve to the motoneuron is stimulated, a smaller hyper-polarizing type of response, an *inhibitory postsynaptic potential* (IPSP), is seen in the motoneuron (Fig. 5-8). The effect of the inhibitory excitation is to reduce membrane potential below the critical level for firing and so to block excitation. As a result of the

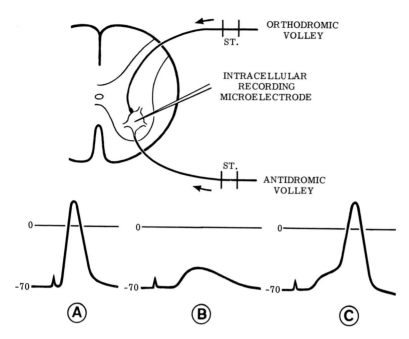

Figure 5-6

Microelectrode recording of responses from motoneurons. A recording microelectrode is shown inside a motoneuron in the top diagram. Stimulation of the cell's fiber in the ventral root gives rise to an antidromic response (A). Stimulation orthodromically with a weak shock gives rise, after a synaptic delay, to an EPSP (B). Stimulation orthodromically with a stronger shock gives rise to EPSP and an action potential (C). The response recorded in the soma upon excitation of the ventral root results from conduction in the reverse direction and it is therefore called an *antidromic response* (Gr. *dromos*, "running"). This is distinguished from propagation in the usual direction, i.e., *orthodromic* excitation via afferent nerve excitation.

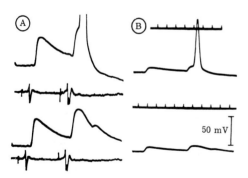

Figure 5-7

Synaptic potentials of motoneurons. Two excitatory postsynaptic potentials (EPSP) recorded from inside a motoneuron in response to two successive weak shocks are shown in A at high and in B at lower amplification. The EPSP's are smaller and longer-lasting depolarizing potentials. The first EPSP of the upper traces is ineffective, but the second, occurring a short time afterward, excites an action potential. In the lower trace, EPSP's only are seen. A small shift in excitability caused the second EPSP to be ineffective. (From Eccles, 1953.)

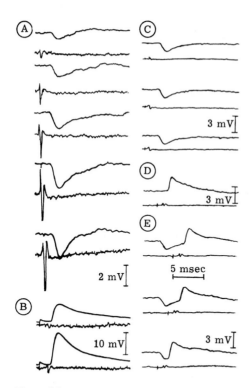

Figure 5-8
Inhibitory postsynaptic potentials. The hyperpolarizing inhibitory postsynaptic potentials (IPSP) recorded with a microelectrode from within motoneurons are shown at increasing strengths of stimulation in A. The action potential of the afferents in the dorsal root volley is shown on the lower traces. As the strength is increased, the size of the IPSP increases. The IPSP's are to be compared with EPSP's excited in these same motoneurons (B). In C, inhibitory volleys are set up, and then (D) an excitatory volley; finally, in E, an excitatory volley follows soon after an inhibitory volley. The IPSP decreases the resultant depolarization brought about by the EPSP to below critical level for firing. (From Eccles, 1953.)

inhibitory action, an EPSP cannot depolarize the cell to the critical level (Fig. 5-8). The IPSP also increases the ionic conductance of the membrane, which further acts to reduce the effectiveness of an EPSP to excite an action potential.

The delay found for the IPSP is greater than that for an EPSP and is due to an additional interneuron on the *inhibitory line* (Fig. 5-9). The afferent fibers have collaterals which synapse onto the inhibitory cells. They have short axons which synapse onto the antagonist motoneurons to give rise in them to an IPSP. The synaptic delay caused by transmission through these intervening cells accounts for the extra time needed for the inhibition to reach a maximum in the interaction studies previously mentioned (Fig. 5-5B).

The action potential of the motoneuron does not normally arise from the cell body membrane. Evidence from a variety of cells indicates that the propagated action potential actually originates from the axon just distal to the cell body, in the non-myelinated *initial segment* (IS). The soma membrane is in fact less excitable than the initial segment, as shown by the occasional failure of an antidromic invasion of the soma, leaving the potential of the initial segment remaining. Other evidence was

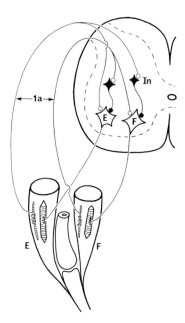

Figure 5-9
Afferent fibers from spindle stretch receptors of extensor muscle (E) are shown terminating on motoneurons, their efferent axons innervating that same muscle. Collateral branches from the afferent fibers branch to terminate on cells in an intermediate group of cord neurons. These are inhibitory cells terminating on flexor (F) motoneurons to inhibit their discharge. Conversely, excitation fibers from flexor muscles terminate onto flexor motoneurons and extensor motoneurons. (Modified from Eccles, 1974.)

obtained from the directly visualized stretch receptors of crustaceans (Chap. 3A). With recording electrodes placed at the soma and at the initial segment, responses were seen to be initiated first at the initial segment. The sequence of events during normal orthodromic excitation is therefore as follows: (1) An EPSP is excited in the soma by synaptic activity, resulting in a flow of current outward through the initial segment of the cell; (2) excitation of an action potential follows at the IS with propagation down the axon to the muscle; (3) at the same time, an action potential is propagated back into the cell body to give rise to a *soma-dendrite action potential.*

The initial segment is the common site on the cell where the depolarizations of the EPSP are tallied up. If a large number of excitatory synapses are fired, a larger flow of outward current passes through the membrane of the initial segment. If at the same time, or just previously, an inhibitory input is excited and IPSP's are elicited, the resulting hyperpolarization of the initial segment also adds up at the initial segment, reducing the level of depolarization below the critical level required for the firing of propagated action potentials.

Synaptic Transmitters
The irreducible latency between the activation of the afferent nerve terminals and the beginning of an EPSP is taken as evidence that a transmitter substance is released from the terminals to cause an excitation of the postsynaptic membrane. However, the transmitter substances giving rise to the EPSP and IPSP are as yet unknown. The transmitters for EPSP and IPSP are known to differ because of the opposite electrical

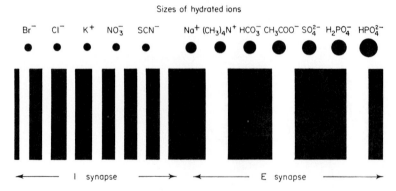

Figure 5-10
The results with ion substitutions using multiple barrel microelectrodes indicate that the smaller-sized ions Br⁻ to SCN⁻ can pass through the inhibitory (I) activated synapses, while after activation by the excitatory (E) transmitter substance, larger ions Na⁺ to HPO⁻ can penetrate. The difference in the resulting equilibrium potentials will determine whether a depolarization or a hyper-polarization PSP will occur. (From Eccles, 1957.)

actions they produce on the postsynaptic membrane of the motoneurons, depolarization and hyperpolarization respectively.

The ionic permeability increases mainly to Na^+ and K^+ in response to excitation, causing the membrane to shift toward a zero equilibrium potential. The inhibitory transmitter substance, on the other hand, produces a more selective permeability increase to K^+ and to Cl^-, with a resultant shift to an equilibrium potential at a higher membrane potential (i.e., from -70 to -80 mv). These different ionic permeability changes during EPSP and IPSP activation were shown by the use of double-barreled microelectrodes inserted into motoneurons. One barrel recorded the EPSP or IPSP; the other was used to pass current and different ions into the cell. By this means it was shown that the inhibitory transmitter permits only ions of smaller overall size to pass through the membrane, apparently because a smaller channel opening is produced in the membrane as compared to the wider channels open by the excitatory transmitter (Fig. 5-10).

Another interesting difference between the action of the two transmitters is that the convulsant agent strychnine blocks the effect of the inhibitory transmitter without affecting excitatory activation. Some recent candidates for transmitter substances are believed to be amino acids. Glutamic acid has been considered as possibly the excita-tory transmitter, and gamma-aminobutyric acid (GABA) and glycine the inhibitory transmitter, with most evidence seeming to favor glycine as the transmitter in the cord.

Renshaw Cell Inhibition

A special type of inhibitory interneuron has been found related to the motoneurons within the ventral horn portion of the spinal cord, the *Renshaw cell* (Fig. 5-11). Collateral fibers from the motoneurons synapse on the Renshaw cells; their axons in turn synapse onto a number of motoneurons in their immediate vicinity to give rise in them to a long-lasting inhibition. The repetitive discharge of the Renshaw cell is responsible for the long-lasting inhibitory effect.

The functional significance of the inhibition brought about by Renshaw cell activity is as yet not completely determined. It may act to sharpen the effect of those motoneurons which have fired by inhibiting other motoneurons in the pool which have not fired. Apparently Renshaw effects are greater on the *tonically* active motoneurons that supply muscles important in maintaining posture for long periods of time than on responses

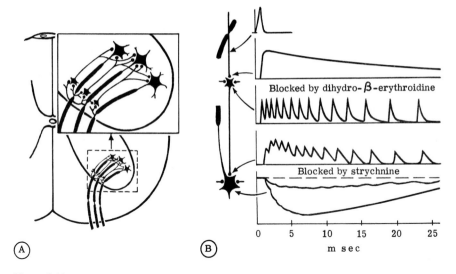

Figure 5-11
Renshaw cell activity of the spinal cord. In A, the ventral horn of the spinal cord is shown, and in the inset an enlarged view. From the axons of three motoneurons, collaterals can be seen taking origin from the first node. These synapse on smaller cells (Renshaw cells), which in turn send axons to the surrounding motoneurons and branches back to the same motoneuron. In B the collateral from the motor horn cell axon is shown at the top synapsing on a Renshaw cell. The Renshaw cell, when it discharges, does so at a very rapid rate and gives rise to a series of discharges which can be blocked by dihydro-β-erythroidine. The Renshaw cell in turn synapses on motoneurons, and at this point inhibitory mediator substance is released in successive summated IPSP's. This IPSP can be blocked by strychnine. (From J. C. Eccles et al. *J. Physiol.* (Lond.) 126:542 and 557, 1954.)

that are rapid (*phasic* responses), in which movement occurs. Phasic changes take place when posture is interrupted, to be followed by another tonic position or a series of movements.

The Renshaw cells have a further significance. The motoneuron collateral fiber ending on the Renshaw cell releases ACh as its transmitter. This action was inferred from *Dale's principle,* which holds that the same transmitter is released at all branches of a neuron. Since the motoneuron releases ACh from its terminals at the neuromuscular junction, on this principle it should also release ACh at the motoneuron collaterals ending in the cord on the Renshaw cells. Injection of substances into the bloodstream that are known to excite or block ACh transmission at the neuromuscular junction should likewise excite or block synaptic activation of the Renshaw cells. The substance dihydro-β-erythroidine, which unlike d-tubocurarine can pass the blood-brain barrier, was found to block Renshaw cell activity (Fig. 5-11). Conversely, ACh and cholinomimetic substances were able to prolong the discharge of the Renshaw cell, as expected from the analogous actions postulated.

The Gamma Loop System

As has already been noted in Chapter 3A, the muscle spindle receptor is excited by stretch to give rise to volleys passing into the spinal cord via the dorsal roots. The large-diameter afferent fibers arise from the central portion of the spindle receptor as annulospiral endings. Elongation of the spindle deforms the sensory terminations of the fibers, thereby exciting discharges with a rate proportional to the degree of stretch (Chap. 3A). Because the spindles are anatomically in parallel with the contracting muscles, as shown in Figure 5-12A, a twitch that contracts the *extrafusal muscle* (muscle

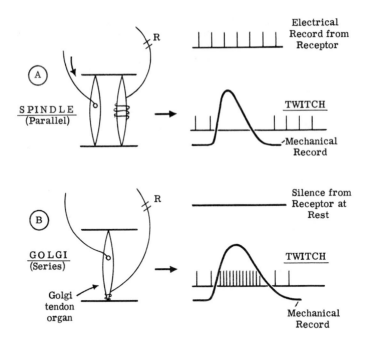

Figure 5-12
Muscle receptor organs. Two main types of muscle receptor organs are pictured, the spindle stretch receptor (A) and the Golgi tendon organ (B). The spindle is placed in parallel with the muscle fibers. When the muscle is caused to contract, the electrical discharge recorded from a spindle receptor shows a decrease, the "pause." The Golgi tendon organ is in series with skeletal muscles, and contraction of the muscle during a twitch causes a speeding up of the discharge rate during the time of the muscle contraction.

fiber other than the spindle) will cause a relaxation of the stretch receptor, *unloading* the tension on it. The result is a diminution or cessation of the discharge in the spindle afferent fibers known as the *pause*. This is seen by recording from single fibers isolated from the dorsal roots. A given fiber is identified as coming from a spindle stretch receptor by the pause (Fig. 5-12A).

The spindle receptor has a complex structure (Fig. 5-13). Small motor fibers synapse on the muscular parts of the spindle receptor, which are present on either side of the central receptor portion. The small motor fibers fall into the gamma range of fiber sizes and are referred to as *gamma motor* or *fusimotor fibers*. When the gamma motor fibers are fired, the muscular portions of the spindle receptor contract and stretch out the central portion of the receptor. This response has the same effect as a pull on the muscle, elongating the spindle receptor, and results in a discharge of the receptor. With an increased excitation of gamma discharge, a smaller stretch of the spindle is required to elicit an afferent discharge. When the gamma system is less active, or if it is stopped altogether, a greater stretch must be placed on the spindle to cause the same afferent discharge. If great enough, gamma activity can overcome the pause, and the stretch receptor can discharge at a fairly high rate even during a twitch response.

Activity in the CNS by exciting or inhibiting the gamma motor fiber discharge can in turn alter the sensitivity of the spindle receptor. An excessive discharge of the gamma neurons appears to underlie the pathological state of *spasticity* through an action of higher brain centers controlling gamma motoneuron activity. This is indicated by the condition of exaggerated hypertonus and increased reflex excitability known as *de-*

cerebrate rigidity. It is produced by transecting the brain stem of a dog or cat at the level of the midbrain between the colliculi. The position of the animal in decerebrate rigidity is one of extreme hypertension of the limbs. The head is arched back and the tail elevated, in a spasm known as *opisthotonos.* The whole posture in this state has been called a caricature of standing. If, in such a decerebrate preparation, the dorsal roots are cut, the limb immediately becomes flaccid, proving that the decerebrate state is produced by reflex activity of stretch reflexes through an augmented gamma activity. The same thing was shown by the early appearance of increased discharges in recordings taken from thin slips from the ventral roots and identified as gamma motor fiber discharge. The discharge increases during the onset of decerebrate rigidity.

If the leg of an animal showing a decerebrate rigidity is forcibly flexed, it resists the applied flexion until at a certain point the resistance suddenly melts away. This phenomenon is known as the *knife-clasp reflex* or the *lengthening reaction.* At some high level of stretch, sensory afferents from the tendon regions of muscles, the *Golgi tendon organs,* reach threshold and they have an inhibitory effect on the motoneurons. This may be a protective reflex. The Golgi tendon organs, unlike the spindles, are placed in a series with respect to the skeletal muscle (Fig. 5-12B). Such afferents may be identified in the following way. A single sensory nerve fiber isolated in the dorsal root is identified

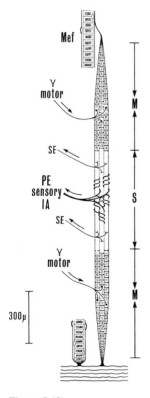

Figure 5-13
The primary sensory (PE) nerve endings of the mammalian spindle terminate in an annulospiral configuration around the center of the receptor. These are type 1A sensory fibers. Secondary receptor (SE) fibers terminate at the sides in the receptor (S) area. In addition, a muscular (M) portion of the receptor is shown containing striped fibers. These are innervated by gamma (γ) motor fibers. Portions of extrafusal muscle fibers (Mef) are shown at the side. (Modified from Barker, 1948.)

as coming from a Golgi tendon organ if it shows an increased repetitive activity during a twitch of the muscle rather than a pause, as in the case of a spindle afferent (Fig. 5-12B). Another identifying characteristic is the lack of effect of increased gamma motor activity on the discharge rate of the Golgi receptor.

Synaptic boutons are found terminating on afferent terminals within the spinal cord. These *presynaptic endings* can inhibit the discharge of the afferent terminal and the cell on which they synapse. The precise relation of presynaptic inhibition to the usual postsynaptic inhibition is a matter of present investigation but seems to act as another means of selective control of afferent input to the nervous system.

Complex interneuronal actions occur in the spinal cord and find expression in the excitation of the motoneurons activating the muscles, the final common path of outflow. In addition to the local influences acting on these systems, described here, upper motor influences descend to and act to control their output. These are discussed in Chapter 8.

REFERENCES

Barker, D.   The innervation of the muscle spindles. *Q. J. Micros. Sci.* 89:143–186, 1948.
Barker, D. (Ed.).   *Symposium on Muscle Receptors.* Hong Kong: Hong Kong University Press, 1962.
Boyd, J. A., C. Eyzaguirre, P. B. C. Mathews, and G. Rushworth.   *The Role of the Gamma System in Movement and Posture.* New York: Association for the Aid of Crippled Children, 1968.
Creed, R. S., D. Denny-Brown, J. C. Eccles, E. G. T. Liddell, and C. S. Sherrington. *Reflex Activity of the Spinal Cord.* London: Oxford University Press, 1932. (Reprinted, 1972.)
DeMyer, W. D.   *Technique of the Neurologic Examination* (2nd ed.). New York: McGraw-Hill, 1974.
Desmedt, J. E. (Ed.).   Human Reflexes. Pathophysiology of Motor Systems. Methodology of Human Reflexes. In *New Developments in Electromyography and Clinical Neurophysiology.* New York: Karger, 1973. Vol. 3.
Eccles, J. C.   *Neurophysiological Basis of Mind.* London: Oxford University Press, 1953.
Eccles, J. C.   *Physiology of Nerve Cells.* Baltimore: Johns Hopkins Press, 1957.
Eccles, J. C.   *The Physiology of Synapses.* New York: Academic, 1964.
Eccles, J. C.   The Dynamic Loop Hypothesis of Movement Control. In K. N. Leibovic (Ed.), *Information Processing in the Nervous System.* New York: Springer-Verlag, 1969.
Eccles, J. C.   *The Understanding of the Brain.* New York: McGraw-Hill, 1974.
Eccles, J. C., R. M. Eccles, and A. Lundberg.   Synaptic actions on motoneurons caused by impulses in Golgi tendon organ afferents. *J. Physiol.* (Lond.) 138:227–252, 1957.
Hunt, C. C., and E. R. Perl.   Spinal reflex mechanisms concerned with skeletal muscle. *Physiol. Rev.* 40:538–579, 1960.
Liddell, E. G. T.   *The Discovery of Reflexes.* New York: Oxford University Press, 1960.
Mathews, P. B. C.   *Mammalian Muscle Receptors and Their Central Actions.* Baltimore: Williams & Wilkins, 1972.
Ochs, S.   *Elements of Neurophysiology.* New York: Wiley, 1965.
Phillis, J. W.   *The Pharmacology of Synapses.* Elmsford, N.Y.: Pergamon, 1970.
Sherrington, C.   *The Integrative Action of the Nervous System.* Cambridge, England: Cambridge University Press, 1906. (Rev. reprint, New Haven, Conn.: Yale University Press, 1947.)
Wiersma, C. A. G.   Neural transmission in invertebrates. *Physiol. Rev.* 33:326–355, 1953.

6. Properties of the Cerebrum and Higher Sensory Function

Sidney Ochs

The term *encephalization* refers to the increased power of command which the more headward part of the central nervous system exerts on the lower reflex mechanisms. An increase in the mass and complexity of brain structure is found in the higher phyla, along with an increased range of function. The cerebral cortex made its appearance late in phylogenetic development and has thus been taken to be the seat of intelligence. But, as will be seen, subcortical centers are also important for the exercise of higher functions.

Clinical observations indicating that certain functions could be localized within the cerebral cortex were confirmed when Fritsch and Hitzig in 1870 showed that electrical stimulation of certain regions of the cortex gave rise to *motor responses* (Chap. 8). Later, various other areas were found to receive *sensory inputs.* Still other areas in the cortex, designated as *associational,* were relegated to the control of higher functions (Chap. 9). However, the concept of strictly separable motor and sensory centers in the cortex has been modified with the realization of the complex interrelations between them for which the term *sensorimotor* is often used. We shall, for didactic reasons, treat them separately. In this chapter the electrical activity of the cerebrum with relation to its function, the changes seen during sleep and wakefulness, sensory reception in the cortex, and general changes including synaptic transmission and its modifications will be discussed. A discussion of motor control from the cerebrum will be reserved for Chapter 8.

THE ELECTROENCEPHALOGRAM

In 1875 Caton first discovered the continuous electrical activity present in the exposed cerebrum of animals. In 1929 Berger, using the string galvanometer, showed that similar electrical activity could be recorded through the intact human skull. Of great importance was his finding that this electrical activity changed with sleep or altered attention. The availability of improved electronic instrumentation since then has allowed the *electroencephalogram* (EEG) to become a valuable clinical tool in the study of epilepsy, brain tumor localization, and the diagnosis of other neurological conditions. It is also of value in relating brain changes to higher function (see Chap. 9).

To record the EEG, electrodes are placed over the surface of the skull in a regular pattern using *monopolar* or *bipolar* recording. Monopolar recording is accomplished with an *active electrode* placed on the scalp and an *indifferent electrode* placed on inactive tissue at a distance. The ear lobes are usually used as the site for the indifferent electrode, because electrical activity from the heart would interfere if it were placed on the body. Bipolar recording is accomplished with two electrodes placed over active sites on the scalp. The potential recorded between them at any instant is the algebraic resultant of voltage appearing under each electrode. Figure 6-1 shows an EEG taken from a human subject. A fairly regular 8 to 12 per second series of waves may be readily distinguished in some portions of the records. These are known as the *alpha waves,* although the term *Berger rhythm* is sometimes used in honor of their discoverer. The alpha rhythm is usually present in an individual who is relaxed, with eyes closed, in a quiet room. If the subject is asked to open his eyes, the alpha rhythm is abruptly blocked and replaced with the smaller-amplitude and faster-frequency *beta* waves. The same blocking of the alpha rhythm can occur if the subject is asked to visualize a scene with

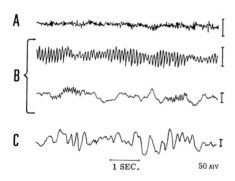

Figure 6-1
EEG during different states of sleep and wakefulness. The EEG of the alert state (A) shows a
low-amplitude, fast *beta* wave type of activity. In the drowsy or light sleep state (B), *alpha* activity is
seen. There may be periods in the drowsy state when a mixture of alpha and slower wave activity is
seen. In deep sleep (C), large-amplitude, slow *delta* waves are predominant. (From W. Penfield and
T. C. Erickson. *Epilepsy and Cerebral Localization.* Springfield, Ill.: Thomas, 1941. P. 401.)

his eyes closed, if a sudden sound is made, or if the subject is otherwise alerted to some
unexpected sensory stimulus.

If the blood supply to the brain is suddenly interrupted, the EEG activity disappears
within 15 to 20 seconds, with loss of consciousness. If the blood supply to the brain is
returned within approximately 5 minutes, recovery is generally complete. (See Metabolism, below.) The correlation of EEG activity to consciousness is clear.

The various rhythms in the EEG may be grouped according to the dominant
frequencies found. The alpha waves range from 8 to 12 cycles per second (cps), the beta
waves from 18 to 30 cps, and the *theta* waves from 4 to 7 cps. Those of still slower
frequencies, below 4 cps, are referred to as *delta* waves. They are usually larger in
amplitude than the alpha waves and are seen often during sleep (see below). The wave
form of the EEG may be difficult to analyze into dominant frequencies by simple
inspection. Automatic frequency analyzers have been used to measure the EEG
spectrum, usually based on Fourier analysis. Frequency analyzers assess the relative
amount of activity in selected divisions of the frequency spectrum. The relative amount
of activity present in the various frequency bands within a short sampling time is then
displayed so that the proportion of alpha, beta, etc., activity can be determined
numerically.

The EEG waves are believed to reflect activity within the dendrites of pyramidal cell
neurons of the cerebral cortex (Fig. 6-2). A large number of the apical dendrites of the
pyramidal cells are found stacked vertically in the cortex, and their finer branches are
numerous in the upper (molecular) layer of the cortex. Many such neuronal elements
must be synchronously activated to produce the EEG voltages recorded. The voltages at
the surface of the cerebral cortex result from the current flow between the active region
of the apical dendrites and the rest of the neuron and are explainable on the basis of
general electrical principles. Polarization differences occurring between the upper and
lower shafts of the apical dendrites or the cell bodies result in the flow of current in the
extraneuronal spaces from polarized to depolarized parts of the cell. Current flow in the
brain occurs in a three-dimensional conducting volume, giving rise to complex electrical
potentials. These can be recorded from the surface of the brain. For purposes of this
discussion it will suffice to consider only the simplest case. Areas of depolarization on
the neuron membrane—for example, the dendrites—are called *sinks.* Current flows into
them from polarized parts of the membrane, referred to as *sources.* An external
electrode records the voltage drop in the electrolyte fluid outside the cells. The size of
the potential is a function of the number of neurons synchronously discharging. The

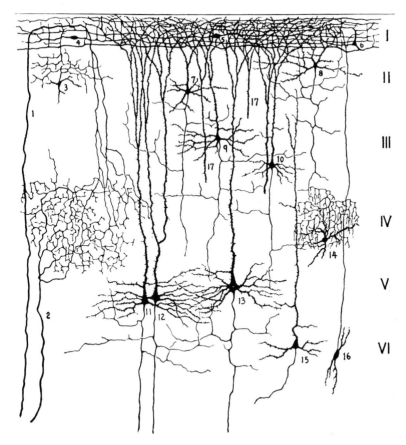

Figure 6-2
Section through the cerebral cortex with characteristic cells through all the layers I–VI. Pyramidal cells 7, 10, 11, 12, 13 have apical dendrites extending up to the molecular (1st) layer, where they arborize profusely. Other cells include short axon cell 3, and Golgi type II cell 14. (From H.-T. Chang. *J. Neurophysiol.* 14:1, 1951.)

depolarized regions must also have similar spatial orientations so that their currents can add together.

A propensity for rhythmic response appears to be a fundamental property of cells in the cerebral cortex. In order to study the origin of rhythmicity, isolated regions or *islands* of cerebral cortex have been prepared. These are made by means of cortical cuts produced some days or weeks previously, leaving the blood supply to the island intact but with all neuronal connections severed. When an island is electrically stimulated, rhythmic responses are excited. Without such stimulation the islands remain electrically silent. An external source of activation, therefore, is considered to *trigger* rhythmic behavior from the cortex. That a trigger or *pacemaker* for the EEG waves is located in the thalamus was shown when electrodes insulated except at their tips were inserted into various nonspecific and interlaminar regions of the thalamus. Stimuli delivered at a rate close to that of the naturally occurring EEG *spindle waves* evoked similar rhythmic waves in the cortex (Fig. 6-3A). Spindle waves normally show *recruitment,* in that the successive waves in the brief train increase in amplitude as more cells add their discharge to the group. Then the cells fall out of the responding group, thus accounting

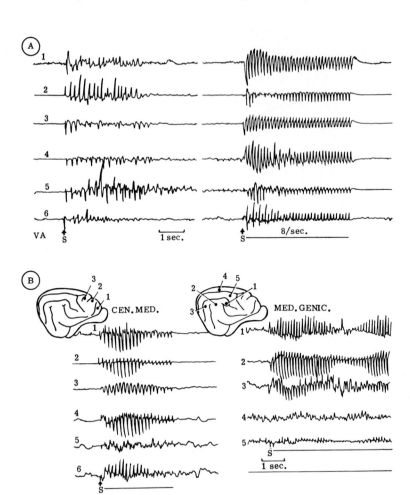

Figure 6-3
Recruiting waves excited by stimulation of various nonspecific thalamic nuclei. In A, to the right, a train of 8-per-second stimuli was initiated. The evoked responses are large and follow the rate of stimulation. To the left, a single stimulus was delivered and a train of repetitive responses was excited. In B, nuclei of the nonspecific group are excited with single shocks, as in A, with trains of stimulation. Recruiting waves are apparent in the various regions of the cat's cortex. (From J. Hanberry and H. H. Jasper. *J. Neurophysiol.* 16:256, 1953.)

for the tapering of the amplitude of the waves in the spindle. Spindles are seen in sleep and in barbiturate anesthesia. A single shock delivered to the thalamus may excite a long train of spindle waves in the cerebral cortex (Fig. 6-3B).

Single-cell studies of thalamic nuclei have shown that a long-lasting hyperpolarization of approximately 100 msec occurs in the neuron following the action potential. This long period of positivity is close to that of waves seen in the EEG and is considered to be the cellular basis for the pacemaker activity of the triggering mechanism in the thalamus. The long-lasting inhibition appears to be similar to that found in the cord when Renshaw cells are activated (Chap. 5).

169

ALERTING AND THE SLEEP-WAKING CYCLE

During sleep a shift to lower-frequency waves is seen in the EEG. In general, the EEG from alert individuals is largely composed of *fast wave (FW) activity* whereas sleeping individuals show a large proportion of delta waves, or *slow wave (SW) activity.*

By transecting the brain stem of cats between the colliculi of the midbrain, a *cerveau isolé* preparation is produced (cut B in Fig. 6-4). The EEG taken from the brain of this preparation shows long runs of slow wave activity, as is typical of a sleeping animal. This cut interrupts the upward influences coming both from the specific sensory afferents and from the *reticular formation.* It was pointed out by comparative anatomists many years ago that the reticular formation, composed of cells with interspersed fibers, might well have an integrative function. Collaterals branching off from the various sensory tracts passing up to the cerebral cortex relay impulses into the reticular formation on their passage through the brain stem. These collateral inputs were demonstrated by recording *evoked responses,* i.e., a grouped discharge of neurons in the reticular formation resulting from various kinds of peripheral sensory stimulation. Evoked responses were elicited in response to sound, skin touch, or a light flash to the eye. The evoked responses recorded in the reticular formation to these different sensory sources showed *occlusion;* that is, a second evoked response was reduced in amplitude if the second input followed too soon after a first such response. Occlusion was also shown with other sensory modalities. Occlusion indicates that the different excitatory inputs on entering the reticular formation synapse onto a common group of cells. In this way the reticular formation reflects the general level of sensory excitation, and its fibers leading to the cerebral cortex have a relation to alert and sleep states, as will be described later.

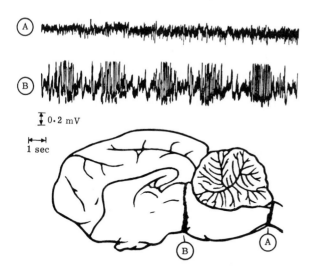

Figure 6-4
EEG patterns after brain stem transections. The medial surface of the brain is represented in the diagram. Cut A produces an *encéphale isolé* preparation. The EEG record taken from this preparation and shown in line A has the low amplitude and fast wave activity characteristic of the alert animal. A transection made through the midcollicular region of the brain stem, cut B, produces a *cerveau isolé* preparation. The EEG pattern (line B) in this case is of the sleep type, with larger-amplitude slow waves. (From F. Bremer. *Bull. Acad. R. Med. Belg.* 2:68, 1937. Modified from M. A. B. Brazier. *The Electrical Activity of the Nervous System.* New York: Macmillan, 1960. P. 212.)

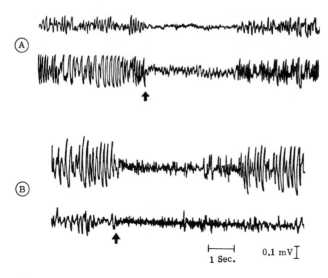

Figure 6-5
Alerting responses in the brain. In A the EEG obtained from the motor cortex and visual cortex is shown. After a sudden sensory stimulation (arrow), which may be a noise, a touch, or a flash of light, the EEG pattern changes to the alert type. Alerting or desynchronization of EEG lasts for a brief time before return of the regular alpha or spindle type of record characteristic of the drowsy animal. In B, a brief electrical stimulation is delivered to the reticular formation, and it has a similar alerting effect on the EEG of the cortex. (From F. Bremer. In G. E. W. Wolstenholme and M. O'Connor [Eds.], *Ciba Foundation Symposium on the Nature of Sleep.* Boston: Little, Brown, 1960. Pp. 32, 38.)

The reticular formation can be selectively destroyed insofar as it is found mainly in the medial portion of the medulla, while the lateral part of the brain stem contains the specific afferent pathways passing upward to the thalamus. Lesions made in the middle part of the medulla, containing the reticular formation, gave rise to a *comatose* state. The EEG recorded from the animals showed the slow, large-amplitude EEG pattern characteristic of deep sleep. When lesions were placed laterally in the brain stem, in the tract containing the ascending specific sensory pathways, the animals remained behaviorally in a waking state and had an EEG pattern typical of an alert animal.

Alerting or *arousal* refers to the change seen in the EEG of a drowsing animal when it is suddenly presented with a brief sensory stimulation—a click of sound, a pinch of the skin, etc. In this event the slow wave activity is abruptly replaced by a low-amplitude pattern of fast EEG activity which may persist for some time after the stimulus (Fig. 6-5A). The pattern of arousal or *desynchronization* of the EEG is found to be widespread in the cortex. A similar period of alerting of the EEG could be obtained following electrical stimulation of the reticular formation through electrodes inserted into its substance (Fig. 6-5B).

When a complete transection is made below the brain stem, an *encéphale isolé* preparation is produced (cut A in Fig. 6-4). The animals may be studied for several hours thereafter. The EEG pattern (EEG record A in Fig. 6-4) is fast; i.e., it is one of wakefulness. The difference between this section and the cerveau isolé is that sensory inputs coming into the brain stem from the trigeminal nerve innervating the head are retained in the encéphale isolé. These inputs into the reticular formation activate it and produce the wakeful EEG state in the cortex. Destruction of the trigeminal nerve afferents in an encéphale isolé preparation promptly brings about the slow wave EEG pattern typical of sleep.

The state of sleep, while generally thought to be due to a diminished corticopetal sensory influence on the basis of such experiments, may also be actively brought about by electrical stimulation of thalamic nuclei of the cat through electrodes chronically implanted in this region. By this means behavioral changes in the animal may be observed. On stimulation of thalamic regions at low frequencies, the animal shows diminished motor activity, seeks out a resting place, curls up, and appears to fall into a natural state of sleep. It may be wakened by brain stem stimulation at another site or when a faster rate of stimulation is used. The systems involved in sleep-waking states will be discussed in the following section.

THE MONOAMINE THEORY OF SLEEP-WAKING

A group of monoamine-containing neurons have recently been shown in the midbrain and diencephalon. These fiber systems are present in the cord, brain stem, and cortex. In the brain stem they have a close relation to the control centers of visceral functions (Chap. 8), and they appear to be involved in the sleep-waking control of the cortex. The cells and their fibers are visualized by freeze-drying a section of the brain and treating it with formaldehyde. A characteristic fluorescence of fibers may then be seen in the fluorescent microscope.

Figure 6-6 shows the cell groups of origin and the terminations of the three main groups of monaminergic fibers in the brain stem. The *noradrenaline* (NA) group of fibers terminates in the cortex and seems to be involved in the sleep-waking cycle as an activator (see below). The *dopaminergic* (DA) system (Fig. 6-7) of dopamine-containing fibers has another pattern of distribution with its fibers terminating in the striatum. This region is involved in motor control (Chap. 8). Some of the other connections of the dopaminergic system are to the cerebellum and to the hypothalamus.

In the monoamine fiber systems, as in the peripheral autonomic nervous system and the chromaffin tissue of the adrenal medulla, a set of related enzymes is found which leads from the amino acid tyrosine to the transmitters dopamine and noradrenaline. Phenylalanine becomes, through related biochemical reactions, yet another transmitter, *5-hydroxytryptamine* (5-HT) or *serotonin*. The distribution of the serotonergic fibers is shown in Figure 6-8, and this system is apparently involved in the sleep-waking cycle.

An uptake system is present in the terminals for these transmitters, a mechanism which may act to terminate their action. A local synthesis of the transmitters also occurs

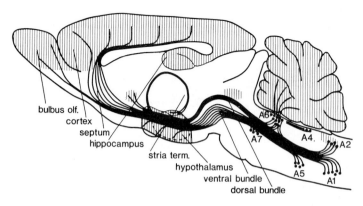

Figure 6-6
Sagittal projection of ascending NA pathways. The striped regions indicate fiber termination areas. Descending NA pathways not shown. (From U. Ungerstedt. *Acta Physiol. Scand.* 367:1–48, 1971.)

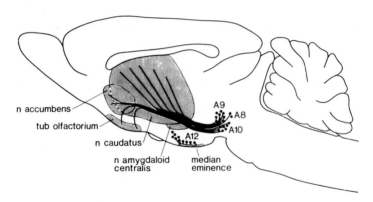

Figure 6-7
Dopamine (DA) pathways shown in sagittal projection. The striped regions show fiber termination areas. (From U. Ungerstedt. *Acta Physiol. Scand.* 367:1–48, 1971.)

in the terminals, and it is of interest that the enzyme required for such synthesis, dopamine-β-hydroxylase, is transported down their fibers to the terminals by axoplasmic transport (Chap. 2). Catabolic enzymes are also involved in the regulation of the level of the monoamine transmitters within the terminals through the enzymes *monoamine oxidase* (MAO) and *catechol-o-methyltransferase* (COMT). It is possible to build up higher levels of monoamines in these fibers by treating animals with antimonoamine oxidase agents, such as isoniazid. These usually give rise to signs of increased behavioral excitability. Conversely, the monoamines may be depleted by the use of

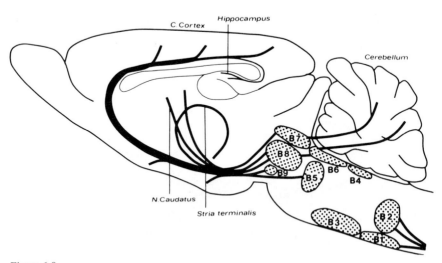

Figure 6-8
Schematic diagram of the central 5-HT cell groups in the lower brain stem. The ascending, descending, and cerebellar 5-HT projections are shown in sagittal section. (From Fuxe and Jonsson, 1974.)

reserpine or other agents which can interfere with the uptake of noradrenaline or 5-HT. This in turn can cause behavioral changes which are mainly in the direction of sedation.

PARADOXICAL (REM) SLEEP

The study of sleep was revolutionized when it was recognized that *rapid eyeball movements* (REM) occur periodically during sleep. REM sleep is associated with an EEG pattern which paradoxically consists of fast low-amplitude waves looking very much like an awake or an alerted EEG. This is termed *paradoxical sleep* (PS) or *fast wave sleep* (FWS). If the subject is awakened during a REM period, he reports a dream. REM periods were found to occur five or six times during a seven-hour sleep session and to occupy approximately 20 percent of the total sleep period in the adult.

REM periods with the same alert-like phases of fast wave EEG have been found during sleep in many species. In the cat they are associated with a marked decrease in the tonus of the neck muscles recorded electromyographically. Using the decreased muscle tone as an index of the paradoxical phase or REM period of sleep, it was shown that the decorticate cat still had such episodes, indicating a lower brain stem origin for the PS periodicity. When lesions were made in the upper pons, these periodic changes were blocked. The rhombencephalon appears, therefore, to be the site of origin of these cycles of the alert-like EEG activity during sleep; hence FWS is at times referred to as *rhombencephalic sleep*. Such active periods are initiated in the pons and cause alerting of the cortex. Judging from loss of neck muscle tonus, a downward discharge of the inhibitory region of the medullary reticular formation to the lower motor center (Chap. 8) also takes place.

The periodic changes occurring during sleep are of great significance. As already mentioned, dreaming is associated with the FW periods, and individuals who are repeatedly wakened during this phase of sleep, and prevented from fulfilling the usual time spent in dreaming, may suffer hallucinations and other behavioral changes. In animals too such deprivation leads to apparently abnormal behavior. The implications for psychiatry are apparent. On the other hand, there are well-authenticated cases of people who require only a few hours or less of sleep and appear to be normal.

In a recent theory of sleep-waking states, Jouvet has proposed that the serotonin system of neurons is involved in slow wave sleep. Depletion of serotonin pharmacologically, by injection of *p*-chlorophenylalanine or by the destruction of the cell bodies of the 5-HT fibers in the raphe nucleus, leads to a constantly wakeful state, an "insomniac" animal. This finding suggests that the system initiates or brings about sleep. Activation of the catecholamine systems is probably related to fast-wave sleep, an activation of the cortex.

Destruction of the dorsal noradrenaline bundle or the groups originating from the A8 cell body of the noradrenaline fibers (Fig. 6-7) leads to a decrease of low-voltage fast activity and a behavior which is sleeplike; i.e., *hypersomnia* results.

The ontogeny of sleep patterns throws some light on the relation of 5-HT and NA to sleep and wakefulness. Newborn animals show mainly REM sleep while the slow wave pattern develops later. The dual system of control of the sleep-waking cycle is shown by the relation of 5-HT and NA levels in neonates and in animals subjected to a variety of brain stem lesions. Slow wave activity is associated with increased 5-HT levels and fast wave activity with relatively more NA.

We have two fundamental problems to contend with in understanding the cyclic shift from sleep to wakefulness and back again. One pertains to the neuronal mechanisms that regulate cyclic changes. No doubt these involve interactions between the monoaminergic neuronal systems described. Another system has also been implicated—the *acetylcholine* neurons which are present within the brain stem and cortex. These may have to do with the initiation of sleep, but it is uncertain how they and the ascending fiber systems in the cerebral cortex act to regulate consciousness. For an understanding of those mechanisms, a knowledge of the action of synaptic networks in the cortex is required. This is discussed in the next section.

SYNAPTIC CONNECTION AND TRANSMITTER CONTENT

Recent new techniques have shown the full extent of the dendritic arborization of pyramidal and stellate cells in the cortex, only partially seen in Golgi preparations. Injection of the cell with Procion yellow or with ^{3}H-glycine, which after incorporation into protein is spread by axoplasmic transport, reveals the great extent of the very fine branches in radioautographs. Synapses on the dendrites have features resembling the presynaptic terminals at the neuromuscular junction (Chap. 3A). The terminal bouton endings typically contain both vesicular structures (approximately 500 Å in diameter) and mitochondria. In the cerebral cortex, these presynaptic terminals may end on special spinelike projections arising from the shafts of the apical dendrites of pyramidal cells. These are the *axoedendritic junctions.* Additionally, typical synaptic junctions are also found on the cell bodies similar to those on the somas of the spinal cord motoneurons (Chap. 5) at the *axosomatic junctions.*

A further advance in study of the synaptic terminals was made possible through the development of the technique of *subcellular fractionation.* The brain is homogenized in an isotonic sucrose solution and subjected to centrifugation at different speeds. One fraction can be selected which contains structures like those of the presynaptic terminals. Vesicles, mitochondria, and, in some cases, part of the postsynaptic membranes torn off in the course of homogenization can be identified. The stems of the fiber apparently are pinched off to form the *synaptosomes,* containing their characteristic components. Upon hyposmotic treatment of the synaptosome, the membranes are ruptured and their vesicles released. Pharmacological and biological studies of the vesicles have shown that ACh, serotonin (5-HT), and NA are present in these synaptic terminals.

A very high concentration of *glutamic acid* and *gamma-aminobutyric acid* (GABA) is characteristic of CNS tissue. By means of microiontophoresis very small amounts of glutamic acid or GABA can be released in the immediate vicinity of neurons in the cortex. As a result, their discharge rate is greatly changed. Glutamic acid excites a rapid increase in the rate of discharge of most neurons, whereas GABA will markedly inhibit the discharge of neurons. The fact that these effects are found with extremely low concentrations of these substances suggests that they may be excitatory and inhibitory transmitters, respectively. Or they may act as *modulators*—i.e., change the excitatory state of cortical neurons in a more general fashion and over a longer period of time than does a neurotransmitter. A series of other possible or putative neurotransmitters is under active investigation.

DIRECTLY EVOKED RESPONSES

The EEG represents the outcome of the complex temporal and spatial interplay of neuronal activity in the cortex. To further the analysis of the electrical responses of the cortex, it is possible to synchronize the discharge of a relatively large population of cortical neurons by directly stimulating the exposed surface of the cortex. When the cortical surface is stimulated by means of a single brief pulse of current, a characteristic surface-negative response is recorded a few millimeters away. A *superficial response* or *direct cortical response* (DCR) is seen as a surface negative wave which increases in amplitude as the strength of stimulation is successively increased (Fig. 6-9). The response is due to an excitation of axons in the molecular (outermost) layer extending tangentially in the cortex and synapsing on the apical dendrites of pyramidal cells, where the response is generated (Fig. 6-10).

The nature of the DCR response is not as yet completely known. It appears from

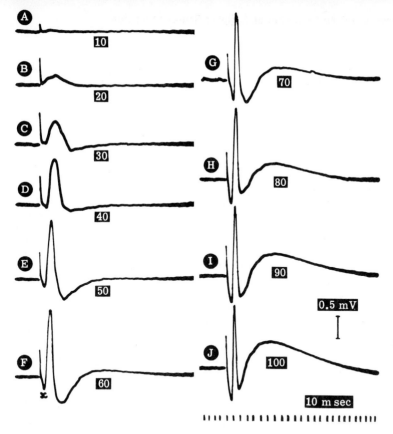

Figure 6-9
Direct cortical responses (dendritic potentials). The responses elicited by direct stimulation of the surface in recording a few millimeters away are shown at successively increasing strengths of stimulation. The response is a slow wave response, broader at the weaker stimulus strength, and at the higher strengths of stimulation is followed by a longer-lasting second negative wave. (From H.-T. Chang. *J. Neurophysiol.* 14:1, 1951.)

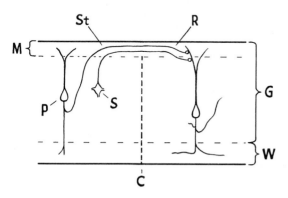

Figure 6-10
Neuronal connections underlying the negative wave of the DCR. Stimulation (St) of axons in the molecular layer (M) inferred by passage over a cut (C) of all other layers. They may arise from pyramidal cell (P) collaterals or stellate cells (S). In the recording site (R) synapse of these axons on apical dendrites of pyramidal cells is shown, the latter generating the response. (From S. Ochs and H. Suzuki. *Electroencephalogr. Clin. Neurophysiol.* 19:230, 1965.)

interaction studies that it represents a graded response in the apical dendrites but does not have the properties expected of an EPSP (cf. Chap. 5). It does underlie in part other evoked responses of the cortex, as is described in the following section.

SENSORY AREAS AND EVOKED RESPONSES

Sensory inputs of various modalities are localized to specified regions in the cerebral cortex. For example, visual input is localized to a part of the occipital cortex, auditory response to part of the temporal lobe, and sensations (touch, pressure, etc.) from the skin to the postcentral cortex of the parietal lobe of the brain.

Nerve fiber pathways from peripheral sensory organs of the eye, ear, and skin have a relay on cells within the various specific thalamic nuclei, and the fibers of these nuclei in turn terminate within the separate sensory regions of the cerebral cortex. Histological studies of the cortex show that these *primary sensory areas* have distinctive cellular features. The visual area of the cortex has a striate appearance which is due to the position of the large number of terminating afferent sensory fibers and small *granular cells* in the fourth layer. These small cells predominate in sensory regions, and sensory cortical regions are therefore often called granular cortex. Because of the variation of fibers and the different cell types found in the different layers of the different cortical areas, these regions have been characterized and given separate letter or number designations. Such *cytoarchitectonic* maps, however, have in the past been carried to the point where too fine an areal division of the cortex was made. Differences in cell shapes are due in part to mechanical distortions produced by the folds of the surface. For these reasons only the distinctly different cytoarchitectural regions are currently considered as having a possible relation to function.

Electrical studies have confirmed the general features of localization of the visual, auditory, and somesthetic receptor areas in the cortex. If the receptor organs are briefly stimulated by a light flash to the eye, a click to the ear, or a touch to the skin, a characteristic electrical response can be detected in the primary sensory area of the cortex to which the fibers of each sense modality project. A more consistent response is seen when a brief single shock is delivered to a sensory projection tract leading to the primary receptor area. This *primary cortical response* consists of a series of small fast waves which continue on to a slower 10- to 20-msec positive wave, followed in turn by a slow negative wave lasting 10 to 20 msec (Fig. 6-11).

The same complex response is seen in each of the corresponding primary sensory areas when its specific sensory relay nuclear group (somesthetic, visual, or auditory) is stimulated. The first of the brief spikelike waves is the sign of activity in the entering specific afferent fibers.

The specific afferent fibers synapse on granular cells in the fourth layer of the cortex. In turn, the granular cells synapse on other neurons and eventually on pyramidal cells to give rise to the slower wave components of the response. The positive portion of the response is due to excitation of elements activated in deeper layers of the cortex, probably on the soma and lower dendrites of pyramidal cells. The negative portion of the sensorily evoked response is due to a later activation and depolarization of the apical dendrites near the surface. It is likely that the same electrogenesis is involved as that described for the DCR; the negative phase of the sensorily evoked response shows occlusive interaction with the DCR, and, as previously mentioned, occlusive interaction indicates that a common element (in this case the apical dendrites) underlies both responses.

Electrical stimulation of the primary sensory cortex of man under local anesthesia gives rise to sensations experienced as either visual, auditory, or somesthetic, depending on the sensory area stimulated: for the visual area, optical; for the auditory area, sound; and for the somesthetic cortex, sensations referred to the skin of the body. Sensory areas relating to olfactory, gustatory, and alimentary sensations are also known. Tumors of the uncus, for example, may be associated with disturbing sensations of smell.

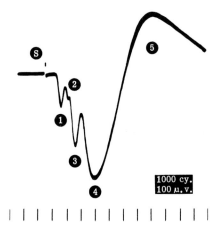

Figure 6-11
Primary cortical response from the visual cortex. A brief electrical stimulation of the afferent pathway leading to a primary sensory area of the cerebral cortex (visual cortex in this case) gives rise to a series of fast spikelike waves (1, 2, 3) inscribed on a slow positive wave (4) which is succeeded by a slow negative wave (5). Spike wave 1—and possibly 2—is due to the activity of the specific afferent endings within the visual cortex. Waves from 3 on are due to intracortical activity. (From L. I. Malis and L. Kruger. *J. Neurophysiol.* 19:175, 1956.)

Within each of the specific sensory regions of the cortex there is an internal spatial organization—a topographical localization of parts of the receptor field to parts of the sensory area, as shown for several species in Figure 6-12. The correspondence of a point within the sensory field to a point of the cortex is referred to as *point-to-point representation.* This correspondence was shown experimentally by the use of a point source of light flashed at a given locus in the visual field of an animal while the cortex was explored with a monopolar electrode. A point on the visual cortex was found where the evoked response had the greatest response amplitude. Reducing the strength of the light and using a small point source allowed the area of the cortex representing the point stimulated in the visual field to be localized to within less than a millimeter.

The topographical relationship within a sensory area may be shown by a figurine, as in Figure 6-13. Such representation more readily shows the extent of the area given over in the cortex to a sensory modality which has a relation to a species' behavior. For example, in the primate, the thumb area in the somesthetic region is large, as is the face region, especially the mouth and tongue. This is also the case in man, and the area of these regions conforms with their importance. Another example is the large extent of the cortical somesthetic representation for the snout in the pig.

Sensory afferent projections to other regions than the primary sensory areas have been found in the cortex. These *secondary sensory areas* (Fig. 6-13) also have a topographical organization with point-to-point representation of the sensory field. The functional relation of the secondary receptor areas to the first sensory areas is at present unknown.

Point-to-point localization does not mean that fibers are activated in a direct line leading from a sensory point within the receptor field to a point in the sensory region of the cortex. Studies of the retina have shown that when a pinpoint source of light falls on the retina, a relatively large number of cells may be discharged. Also, a number of cells in the surrounding area are inhibited. Retinal cells, therefore, show complex and

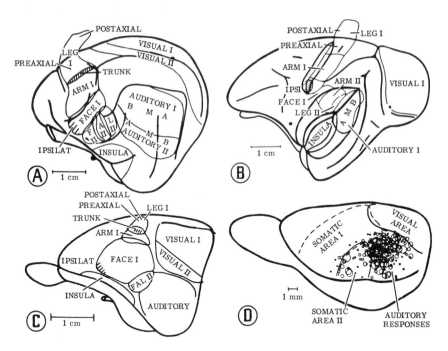

Figure 6-12
Sensory maps of several animal species. A is a sensory map for the cat. Two visual areas are shown, and the somesthetic area I is represented by leg, arm, and face areas. A second somesthetic area II is also shown, and two auditory areas as well are found. B is a corresponding map for the monkey, C for the rabbit, and D for the rat. Notice that in general the relative positions of the main sensory areas have a similar pattern over the surface of the brain. (From C. N. Woolsey and D. H. Le Messurier. *Fed. Proc.* 7:137, 1948.)

Figure 6-13
Figurine representation of the somesthetic area of the dog. The parts of the schematic dog lie over their corresponding tactile areas as mapped by evoked potential techniques. A second somesthetic area is also found which is a smaller duplication of the first, somesthetic area II. Note the greater disproportionate size of the snout region in this species. (From C. N. Woolsey and D. H. Le Messurier. *Fed. Proc.* 7:137, 1948.)

widespread excitatory and inhibitory interactions. The same sort of surrounding *inhibition* (and excitation) occurs at relay stations as the impulses ascend to the cortex. Within the cortex the specific afferent fibers branch profusely and have multiple synaptic terminations on a large number of neurons within the sensory cortex. These afferents thus engage a relatively large volume of cortex, and the "grain" of the cortical region excited by a single afferent fiber would be much too coarse to provide an appreciation of fine detail in the sensory field if no other mechanism of discrimination between points in the visual field were operating. Some additional process must come into play to help "sharpen" the representation of points in the visual field. The sharpening could come about by a surrounding inhibitory effect, such as that described for the Renshaw system in the spinal cord (Chap. 5). Sharpening could also occur by the overlapping of afferent terminations ending with a cortical region so as to bring about the excitation of a much smaller number of cortical cells through the process of occlusion.

Single Cell Sensory Responses—Columnar Arrangement

New principles of organization of neurons in primary sensory regions have been found by recording the activity of single cells within the sensory regions of the cortex. With the tip of a microelectrode close to an active cell, there is a rapid decrement of potential with distance from the discharging surface. Its activity can therefore be recorded in isolation, the more distant cells contributing too small a voltage to be recorded. Unit cell studies have the disadvantage that a statistical survey must be made to determine what a given population of neurons is doing in any small area. For example, during an evoked response to a light flash, some neuronal units increase their rate of firing, others show a decrease, and yet others are not affected at all. These varied types of unit response indicate that much of the function of the primary receptor areas underlying vision is complex. Recording from the cortex reveals that some cortical cells in the visual cortex give rise to a discharge at the onset of light stimulation, an *on response,* and others at the termination of a light stimulation, an *off response.* Still other cells appear to be a combination, the *on-off* cells. Excitation (on response) and inhibition (off response) were discovered to be functionally related, so that in a region of the cortex surrounding

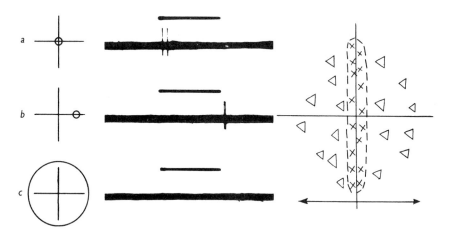

Figure 6-14
Responses of a unit to a stimulation with circular spots of light in the visual field. The receptive area is activated by: a, 1° spot in the center; b, same spot displaced 3° to the right; c, 8° spot covering a large part of the receptive field. (From Hubel and Wiesel, 1959.)

excitatory on units, an inhibitory discharge of off units may be found, and vice versa (Fig. 6-14). In Figure 6-14 action potentials occur at the start of the light, and these are excitatory (X); the response after its termination (Δ) is inhibitory.

Unit studies of neurons in the sensory cortex have revealed an extraordinary kind of *columnar* organization not previously recognized. It was first discovered in the somesthetic sensory area, where individual neurons responded either to movement of hairs of the body surface or to rotation of limb and digit joints. On recording in the cortex downward from the surface, a series of neurons was found to respond successively in the same way to a given stimulus type. The neurons responding to a given type of sensory input were arranged vertically in columns, and the column pattern was seen only when the electrode was passed perpendicularly down from the surface, the tip encountering cell after cell giving rise to the same response—as, for example, to a limb rotation or a touch to a particular place on the skin surface. The regions of the skin surface from which the columns of responding cells can be elicited are generally discrete and located on the body contralateral to the somesthetic cortex. In the secondary sensory areas of the cortex, the peripheral body surface area from which an excitation could elicit responses was more widely spread, and, in addition, the same neurons could be excited from bilateral areas of the body surface.

A similar system of neurons having a columnar organization was found in the visual cortex. Units in the visual cortex responded to slits of light which were oriented at a certain specific angle in the visual field (Fig. 6-15). For those neurons, if the orientation of the slit was varied, the light stimulus brought no response. Recording from cells at successive levels below the surface of the cortex showed that the cells responded only to the slit at a given orientation. Vertical columns of cells nearby responded to slits of other orientations or to other types of visual stimuli—for example, a movement of a slit of light in one certain direction, or different-shaped contours of light stimulation; in the latter case the response is referred to as a *complex type* of cell response.

Cells in the primary visual areas are found connected for binocular vision. A neuron

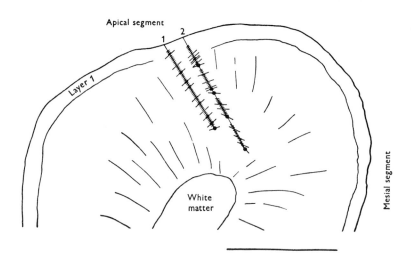

Figure 6-15
Two recording penetrations of microelectrodes (1 and 2) into the visual cortex are shown along with responses from neurons recorded at different distances from the surface. The lines intersecting the tracks represent the orientation of the light slits used for stimulation of the visual field which gave maximal neural responses. Notice the similarity of orientations for the cells along track 1 which is nearly perpendicular to the surface, compared to the changes in orientation of responsivity of the cells along track 2. Horizontal line, 1 mm. (From Hubel and Wiesel, 1963.)

in one visual cortex responding to a given type of slit orientation will often respond to the same stimulus presented in the other visual field.

Use of the simple sensory systems of lower organisms has shown some basic mechanisms involved in vision which may help reveal some principles for study of the complex system met with in higher forms. The *ommatidium* of the arthropod *Limulus* consists of a small cluster of light receptor cells. A number of these form the compound eye, with an *eccentric cell* found in close apposition to them. Discharge in the axon of the eccentric cell has been recorded, the rate of its discharge being determined by the degree of light shone into the ommatidium. There are collateral interactions between the axons of the eccentric cells of the eye, such that discharges in one will inhibit the response in those around it. This is termed *lateral inhibition,* a phenomenon having some relation to the effect of Renshaw cell activity (Chap. 5) in the cord, though without an intervening inhibitory cell.

STEADY POTENTIAL CHANGES AND SPREADING DEPRESSION

The relation of the apical dendrites of pyramidal cells to the production of EEG waves has been discussed in a preceding section. Another cortical activity likely to be related to the dendrites is the *steady potential* (SP). Steady potentials are recorded from the surface of the cerebrum by means of a DC coupled amplifier and for this reason are also called *DC potentials.* Changes in the steady potential have been related to differences in the polarization of the upper and lower portions of the pyramidal cells. The relation of the steady potential to activity was shown by the negative shift in the steady potential when the nonspecific thalamic nuclei were stimulated. The change appears to be due to the termination of axons from the thalamus onto cortical dendrites, which act to depolarize them.

The steady potential is also related to metabolism. If the blood supply to the brain is clamped off or N_2 breathed to initiate anoxia, little change is seen during a latent period of approximately two to five minutes; then there is a rapid negative shift of the steady potential (Fig. 6-16). At the time of the negative shift, referred to as the *asphyxial potential,* the electrical resistance (impedance) of the cortex increases markedly. Both

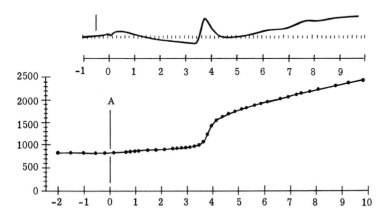

Figure 6-16

Asphyxial changes in the cerebral cortex. In the upper curve, the steady potential is recorded; in the lower curve, the electrical resistivity of the cortex. At A, the brain is asphyxiated by sudden deprivation of its blood supply. A latent period of several minutes ensues before a rapid increase in resistivity is seen. At the same time a negative depolarization of the brain takes place. (From A. van Harreveld and S. Ochs. *Am. J. Physiol.* 187:184, 1956.)

phenomena, the negative change in steady potential and the increased electrical impedance of the cortex, are related to the loss of normal ion permeability of the dendrites, with an increased entry of Na^+, Cl^-, and water from the extracellular space into the dendrites. Evidence for this event was a measured swelling of the apical dendrites when freeze-substitution was used to prepare the tissue for electron micros- copy. The asphyxial potential and ionic changes herald an irreversible damage of cortical neurons. Consequently, it is imperative that oxygenated blood be supplied to the brain before this change takes place. In humans, a permanent loss of consciousness can result if anoxia lasts more than about five minutes at 38°C. The time can be considerably lengthened if the brain temperature is lowered. Advantage is taken of this fact in surgery in which brain circulation must be stopped for longer periods of time, and techniques of lowering body temperature artificially are employed.

Some aspects of the changes seen after asphyxiation are also found to occur in the phenomenon known as *spreading depression* (SD). Spreading depression is elicited by mechanical, chemical, or electrical stimulation of the surface of the brain. It is char- acterized by a depression of the EEG and of evoked responses which spreads through the cortex without regard to the functional area involved—sensory, motor, or associa- tional—at the rate of 2 to 5 mm/per minute. In an area occupied by SD, an increased permeability and ionic changes occur in the neurons, mainly in the apical dendrites of the pyramidal neurons, which cause a negative shift of the steady potential and an increased electrical resistivity (Fig. 6-17). Apparently both K^+ and glutamate are released from the involved cells to act on nearby cells to excite SD in them. The process

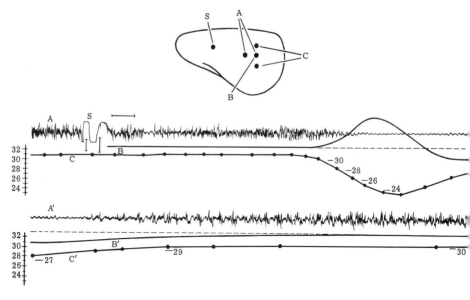

Figure 6-17
Spreading depression of the cerebral cortex. The EEG is shown in line A, the steady potential of the surface of the brain in B, and on line C the electrical conductivity (inverse of resistivity) of the cerebral cortex. These records are continued below as A', B', and C'. The electrode positions on the surface of the brain are shown in the diagram. A brief intense stimulation of the brain is employed at S to start spreading depression. As depression spreads out and occupies the region of the recording electrodes, the EEG shows a diminution, and at the same time a negative variation is seen in the steady potential level of the brain. A decrease in conductivity also occurs at this time. Recovery occurs some minutes later. (From A. van Harreveld and S. Ochs. *Am. J. Physiol.* 189: 159, 1957.)

is repeated, to account for the spread. The apical dendrites of the involved neurons show evidence of swelling, with an entry of Na^+, Cl^-, and water into them, similar to the events occurring during the asphyxial change. However, unlike the asphyxial change, a recovery takes place within a few minutes, the Na^+ pump in the dendrites and cells acting to restore normal ionic conditions. Spreading depression occurs readily in the rat, rabbit, and guinea pig. In higher species it is more readily seen after cortical exposure or an application of Ringer's solution with an augmented amount of K^+ present. It may also occur pathologically in man following a concussion or in some convulsive states.

The term *spreading depression* may be somewhat misleading because a small amount of convulsive spike activity is commonly seen under certain conditions in which a slow spread of a convulsive discharge occurs rather than depression; this is referred to as a *spreading convulsion*. It is due to a "release" of activity of cells in lower cortical layers not invaded by SD and occurring for the same period of time.

METABOLISM

A close dependence of brain function on oxidative metabolism is shown by the rapid alteration of normal function with loss of blood supply. As already noted, consciousness is lost in approximately 10 to 15 seconds. Concomitantly, the EEG changes to an isoelectric pattern; i.e., a loss of rhythmicity is seen. If blood circulation is returned within approximately five minutes, consciousness returns with a return of EEG activity. Similar loss of cortical function due to metabolic interference can come about through a hypoglycemia such as that produced by an overdose of insulin. If prolonged, an irreversible neuronal damage may occur with loss of higher function, as in the case of anoxia.

The maintenance of adequate circulation to the brain is accomplished by a number of important reflex mechanisms outside the brain (Chap. 17). The fine control of circulation within the brain apparently depends not on local neural control but on metabolic mechanisms, primarily by variation in Pco_2 level, which in turn is determined by local metabolism. Increase in Pco_2 increases cerebral blood flow, and vice versa (Fig. 6-18). The use of a Fick technique (Chap. 11) for estimation of cerebral blood flow (CBF) gives values of approximately 60 ml per 100 gm per minute, and this flow does not show much variation with changing conditions such as sleep or waking. Adequate blood flow is maintained in fairly constant fashion as systemic blood pressure varies from approximately 70 to 150 mm Hg and more; i.e., it shows autoregulation. Low blood pressures are dangerous, however, for with decreases to or below a level of approximately 40 to 50 mm Hg, as may occur in shock (Chap. 18C), cerebral blood flow becomes inadequate for proper oxygen uptake and metabolism. Coma and permanent loss of cerebral function may occur if this condition is not soon reversed.

A more subtle alteration of metabolism is possible. As noted earlier, neurons have a high rate of protein synthesis which supplies materials carried outward in the fibers by transport mechanisms necessary to maintain normal nerve function and synaptic transmission. At present not enough is known about the effects of various diseases on this transport function. There is, however, a growing awareness of its role. For example, brain damage can occur as a result of protein deficiencies in the diets of infants, with a resultant subnormal intelligence later in life.

A large portion of the cells in the CNS and in the cortex are neuroglial (glial) cells. The exact function of the glial cells is not understood. According to one theory, the glial cells control the transport of materials from the blood capillaries to the neurons. This role has classically been that of the *blood-brain barrier,* usually ascribed to capillaries, which controls entry of substances to the extracellular space and thence to cells. Part of the evidence for such an intermediating role for glia comes from electron microscopic studies which show that there is very little extracellular space in central nervous tissue, possibly too little for the diffusion of the materials from the blood capillaries to the

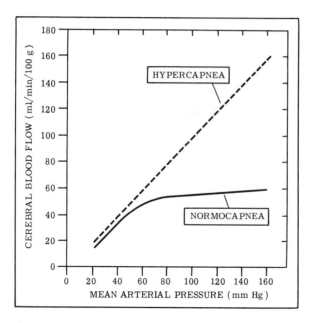

Figure 6-18
Pressure-flow relationship in the cerebral circulation. Autoregulation of blood flow is present at normal arterial blood levels of CO_2 (normocapnea) but is abolished by high arterial blood levels of CO_2 (hypercapnea). Then CBF varies as a linear function of mean arterial blood pressure. (From R. M. Berne and M. N. Levy. *Cardiovascular Physiology.* St. Louis: Mosby, 1967. P. 221.)

neurons. However, the size of the extracellular space is likely to be similar to that found in other tissues—about 20 percent. This has been determined from electrical impedance studies and the increase in impedance upon asphyxiation when Na^+, Cl^-, and water enter the cells with a corresponding reduction of the extracellular space. The seeming lack of an extracellular space noted in the usual electron micrographs (and in turn an anatomical evaluation of the size of the extracellular compartment) is not a settled question. The apparent lack of extracellular space is due to the previously discussed translocation of Na^+, Cl^-, and water into cells during the time the tissue is removed and prepared for electron microscopic examination. An extracellular space with a value closer to that obtained from measurement of cortical impedance has been shown in electron micrographs using freeze-substitution.

REFERENCES

Anderson, P., and S. S. Andersson. *Physiological Basis of the Alpha Rhythm.* New York: Appleton-Century-Crofts, 1968.

Berne, R. M., and M. N. Levy. *Cardiovascular Physiology.* St. Louis: Mosby, 1967. P. 221.

Bureš, J., O. Burešová, and J. Krivanek. *The Mechanism and Applications of Leão's Spreading Depression of Electroencephalographic Activity.* New York: Academic, 1974.

Chang, H.-T. Dendritic potential of cortical neurons produced by direct electrical stimulation of the cerebral cortex. *J. Neurophysiol.* 14:1–21, 1951.

Davson, H. The Blood-Brain Barrier. In G. H. Bourne (Ed.), *The Structure and Function of Nervous Tissue.* New York: Academic, 1972. Vol. 4.

Fuxe, K. and G. Jonsson. Further mapping of central 5-hydroxytryptamine neurons: Studies with the neurotoxic dihydroxytryptamines. *Adv. Biochem. Psychopharmacol.* 10:1–12, 1974.

Hubel, D. H., and T. N. Wiesel. Receptive fields of single neurons in the cat's striate cortex. *J. Physiol.* (Loud.) 148:574–591, 1959.

Hubel, D. H., and T. N. Wiesel. Receptive fields of single neurons in the cat's striate cortex. *J. Physiol.* (Loud.) 160:106–154, 1962.

Hubel, D. H., and T. N. Wiesel. Shape and arrangement of columns in cat's striate cortex. *J. Physiol.* (Loud.) 165:559–568, 1963.

Jasper, H. H., A. A. Ward, Jr., and A. Pope (Eds.). *Basic Mechanisms of the Epilepsies.* Boston: Little, Brown, 1969.

Jouvet, M. The role of monoamines and acetylcholine containing neurons in the regulation of the sleep-waking cycle. *Ergeb. Physiol.* 64:166–307, 1972.

Kety, S. The Cerebral Circulation. In J. Field, H. W. Magoun, and V. E. Hall (Eds.), *Handbook of Physiology.* Washington: American Physiological Society, 1960. Section 1: Neurophysiology, vol. 3, chap. 61.

Krnjević, K. Iontophoretic studies on cortical neurons. *Int. Rev. Neurobiol.* 7:41–98, 1964.

Magoun, H. W. *The Waking Brain* (2nd ed.). Springfield, Ill.: Thomas, 1963.

McIlwain, H. and H. S. Bachelard. *Biochemistry and the Nervous System* (4th ed.). London: Churchill Livingstone, 1971.

Moruzzi, G. The sleep-waking cycle. *Ergeb. Physiol.* 64:1–165, 1972.

Penfield, W. G., and H. H. Jasper. *Epilepsy and the Functional Anatomy of the Human Brain.* Boston: Little, Brown, 1954.

Phillis, J. *The Pharmacology of Synapses.* Elmsford, N.Y.: Pergamon, 1970.

Ross Conference. *Brain Damage in the Fetus and Newborn from Hypoxia or Asphyxia.* Columbus, Ohio: Ross Laboratories, 1967.

Rossi, G. F., and A. Zanchetti. The brain stem reticular formation. *Arch. Ital. Biol.* 95:195–435, 1957.

Resnick, O. The Role of Biogenic Amines in Sleep. In C. D. Clemente, D. P. Purpura, and F. E. Mayer (Eds.), *Sleep and the Maturing Nervous System.* New York and London: Academic, 1972. Chap. 6.

Sholl, D. A. *The Organization of the Cerebral Cortex.* London: Methuen, 1956.

van Harreveld, A. *Brain Tissue Electrolytes.* Washington: Butterworth, 1966.

van Harreveld, A. The Extracellular Space in the Vertebrate Central Nervous System. In G. H. Bourne (Ed.), *The Structure and Function of Nervous Tissue.* New York: Academic, 1972. Vol. 4.

von Bonin, G. *Some Papers on the Cerebral Cortex.* Springfield, Ill.: Thomas, 1960.

7. Regulation of Visceral Function
A. The Autonomic Nervous System

Carl F. Rothe

Homeostasis, the constant and optimal internal environment of the body, is maintained in large part by the actions of the autonomic nervous system, for it is this part of the nervous system that provides a fine control of the visceral or internal functions of the body. It is generally involuntary and acts on the internal effectors such as (1) nonstriated (smooth) muscle, as in the intestine; (2) cardiac muscle; (3) exocrine glands, e.g., the sweat and salivary glands; and (4) some endocrine glands. The endocrine system also participates in the control of the visceral function of the body, but it is generally slower and acts through the release of internal secretions (hormones) which are transported by the cardiovascular system. The autonomic nervous system, along with the endocrine system, thus aids in keeping the internal environment of the body at such a composition and temperature that cellular life may proceed optimally.

CHARACTERISTICS OF THE AUTONOMIC NERVOUS SYSTEM

Anatomically, the autonomic nervous system is the efferent pathway linking the control centers in the brain and the effector organs such as smooth muscle and secretory cells. However, physiologically, control of visceral function must include sensors, afferent pathways, and central control centers as well. The sensory afferents from the viscera in the vagus and splanchnic nerves, for instance, serve both the somatic and the autonomic nervous system. Other sensors, such as those of blood plasma osmolarity and carbon dioxide partial pressure, are located within the cells of the central nervous system itself. The autonomic nervous system differs from the somatic in that the motor neurons which come into immediate functional relationship with the effector cells lie completely outside the central nervous system. These motor neurons are called postganglionic, since they synapse with the preganglionic fibers coming from the spinal cord in ganglia (clumps of neuron cell bodies) located in chains beside the spinal cord or in the organ innervated. The adrenal medullae are an exception, for postganglionic neurons are not present. The adrenal medullary tissue is of the same embryological origin as postganglionic tissue, however, and the gland functions in a manner analogous to that of postganglionic fibers by releasing catecholamines (see below).

The autonomic nervous system may be divided, both functionally and structurally, into the *sympathetic division,* with neurons leaving the spinal cord from the thoracic and lumbar segments (Fig. 7A-1), and the *parasympathetic division,* with neurons leaving from the cranial and sacral segments (Fig. 7A-2).

The Sympathetic Division

The sympathetic division tends to act in a widespread manner to prepare the body for emergencies and vigorous muscular activity. Life is generally possible without the sympathetic division; but the animal must be sheltered, for the sympathectomized animal is much less resistant to extremes of environmental temperature, hypoxia, and other forms of stress. The cardiovascular system is no longer finely adjusted to provide for prolonged or severe exertion. The animal tends to be weak and apathetic.

The synapses are characteristically located either in the sympathetic paravertebral chain on each side of the spinal cord throughout its length or in special ganglia such as the celiac ganglion (*solar plexus*) (Fig. 7A-1). The interconnections between the paravertebral ganglia facilitate the diffuse action of the sympathetic division. In addition, the

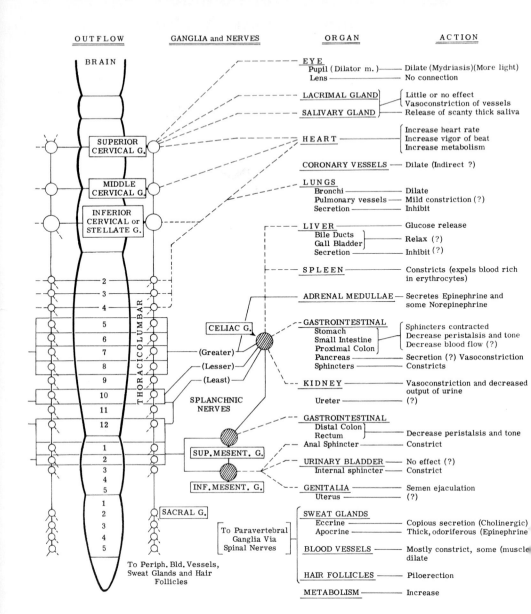

Figure 7A-1
Sympathetic division of the autonomic nervous system. Cholinergic fibers, ——; adrenergic (postganglionic) fibers, ----. The celiac, superior, and inferior mesenteric ganglia are called prevertebral (or paravertebral) ganglia, since the synapses (ganglia) are not in the sympathetic trunk along the spinal column.

189

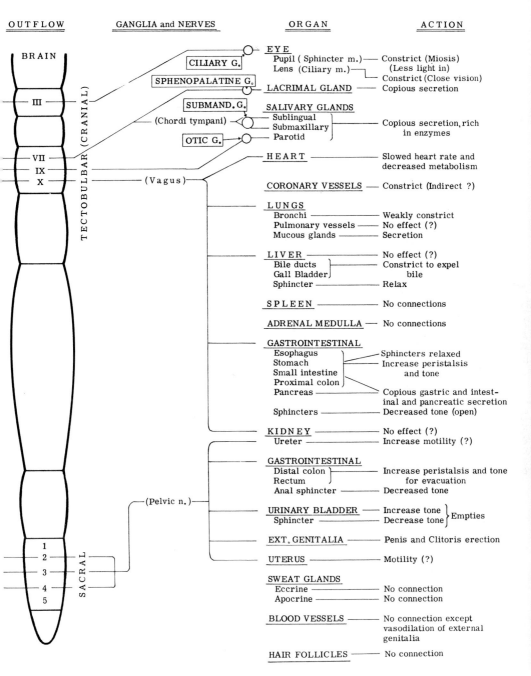

Figure 7A-2
Parasympathetic division of the autonomic nervous system (all nerve fibers are cholinergic).

adrenal medullae release the sympathetic mediators, epinephrine and norepinephrine, into the bloodstream for even more complete and diffuse distribution.

Homeostatic needs are anticipated because of interconnections with the cortex of the brain. For instance, the mere thought of a fight activates the sympathetic division, which in turn increases the activity of the cardiovascular and respiratory systems, curtails gastrointestinal activity, causes the release of glucose from glycogen stores, channels more blood to muscles, and starts sweating for removal of excess heat. The pupils of the eyes dilate, the eyes tend to protrude, and the eyelids widen. In many animals the hairs of the back and tail bristle. All these reactions are in anticipation of vigorous muscular activity. The sympathetic division thus acts in emergencies to adjust the organism to an adverse environment or a rapid change in internal requirements.

Besides participating in massive discharge for emergency situations, the various parts of the sympathetic division function in everyday life in independent and discrete ways. The connections of the sympathetic division with various organs and its actions on them are summarized in Figure 7A-1 and bear careful study.

The Parasympathetic Division
The parasympathetic division acts more discretely on individual organs or regions. The abdominal viscera are innervated by preganglionic neurons which leave the cranial part of the cord and travel via the vagus nerves. The synapses are generally located *within* the innervated organ. The vagi contain most of the parasympathetic efferent fibers, although they consist mainly of sensory afferents which participate in visceral reflexes, such as those from the abdomen and the pressure and chemical receptors located in the aortic arch and the heart. There are also a few somatic efferent fibers. Dogs, in addition, have in the cervical vagus trunks some cervical postganglionic sympathetic nerve fibers to the head. The sacral part of the parasympathetic division is primarily concerned with the emptying of the pelvic organs and with reproductive functions.

Whereas the sympathetic division is mainly involved with homeostasis during voluntary muscular activity, the parasympathetic division is involved with restorative vegetative functions such as digestion and rest. Imagine an old man sleeping after eating. His heart rate is slow, his breathing is noisy because of bronchial constriction, and his pupils are small. Saliva may run from the corners of his mouth, and rumbles from his abdomen reveal much intestinal activity.

The mediator that is released at the parasympathetic nerve endings is the same as in the somatic nervous system and ganglia—acetylcholine. The actions on various organs are presented in Figure 7A-2 and should be studied.

The Hypothalamus
The hypothalamus (Chap. 8) is the center for interrelating the visceral and somatic functions of the body. Not only are motor actions accompanied by widespread and complex visceral responses, but visceral activity modifies somatic reactions. For instance, during digestion the increased blood flow through the intestine tends to reduce the muscular capacity for work. Such interdependence of visceral and somatic functions is implied in the concept that central control of both functions is exerted through neural mechanisms that are located at common levels in the spinal cord, brain stem, diencephalon, and cerebral cortex; and that both have a common sensory inflow. Current research supports the concept of widespread interaction of many levels of the central nervous system in the control of the circulatory system (Smith, 1974).

If the spinal cord is sectioned at about the first thoracic vertebra, there is an immediate depression not only of somatic but also of autonomic reflexes. Blood pressure drops as a result of loss of peripheral vasoconstrictor tone, sweating is absent, and the body temperature is poorly regulated and tends to approach the environmental temperature. The evacuation reflexes for the bladder and bowel are greatly impaired, as are the sexual reflexes. With the passage of time, many of these reflexes return, in part, with the further development of the spinal segmental reflexes; but regulation of blood pressure and of body temperature is severely limited, and micturition, defecation, and sexual reflexes are incomplete.

The decerebrate preparation responds in a similar fashion except that many cardio-vascular centers located in the medulla are intact, and cardiovascular regulation is therefore more nearly normal.

If the hypothalamus is left intact by sectioning the brain above the diencephalon, removing only the cerebral cortex, the animal has normal vegetative regulation of homeostasis, but the adjustments appropriate for the anticipation of muscular activity are missing.

As discussed in Chapters 8 and 9 and other appropriate chapters, the hypothalamus is concerned with the regulation of (1) the cardiovascular system, (2) respiration, (3) body temperature, (4) body water and electrolyte concentrations, (5) food intake and gastric and pancreatic secretion, (6) hypophysial secretion and sexual and maternal behavior, and (7) emotional states.

Reciprocal Action

Most of the visceral organs have a dual, antagonistic innervation, in that the actions of the two divisions are diametrically opposed. For instance, the sympathetic division increases cardiac activity, and the parasympathetic division decreases it. The para-sympathetic division acts to enhance and accelerate visceral functions such as gastro-intestinal motility, while the sympathetic division generally acts to constrict visceral blood vessels and to inhibit visceral functions such as digestion and elimination.

An exception to the generalization of reciprocal action is the control of secretory function of the salivary glands and pancreas; both divisions seem to stimulate secretion. In addition, not all organs have dual innervation. The sympathetic division has exclu-sive innervation of the adrenal medullae, spleen, pilomotor muscles, sweat glands (although the postganglionic fibers are cholinergic), and probably the vasomotor muscles of the blood vessels of the viscera, skin, and skeletal muscle, if not all blood vessels. The parasympathetic division innervates the ciliary and the sphincter muscles of the eye. Functionally, the pupil of the eye has an antagonistic dual innervation; structurally, however, the dilator muscles have sympathetic innervation whereas the constrictor muscles have parasympathetic innervation.

Tone

The autonomic system generally maintains a "tone," a basal level of activity, which then may be either increased or decreased by central control. For example, the flow of blood through an innervated muscle is about one-third to one-half that observed following interruption of sympathetic nerve supply to the muscle, indicating that a sympathetic tone is acting under normal conditions to constrict the vessels. Because of this vaso-motor tone, the blood vessels can be either dilated by a decrease in sympathetic division activity or further constricted by an increase in activity (see Chap. 17).

The parasympathetic division, by way of the vagus, provides a "vagal tone" to the heart. If the vagi of an animal under basal conditions are sectioned, the heart rate increases as a result of removal of this inhibiting effect. Parasympathetic tone tends to increase the intrinsic motility of the intestinal tract. If the lower vagi are sectioned, there is serious and prolonged, but not permanent, gastric and intestinal dysfunction. Generalized sympathetic division activity tends to inhibit intestinal motility.

The presence of dual innervation and the possibility of either increasing or decreasing the tone permit a wide range of control.

MEDIATORS

The chemical mediators released at the nerve endings of the autonomic nerve fibers act on *receptors* to produce, in turn, an effector response. A receptor may be conceived as

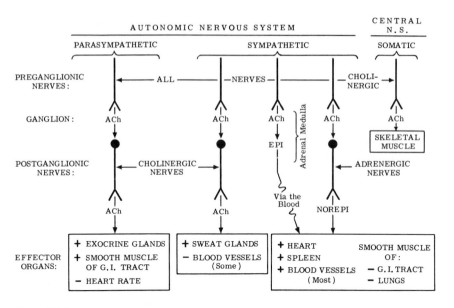

Figure 7A-3
Classification of nerves in terms of transmitter or mediator released and location of cell body and synapse; + means stimulatory, − means inhibitory.

that molecular structure with which a single molecule of mediator or drug interacts. The site may be either an enzyme or the structural configuration of the cell membrane.

The mediator released by the preganglionic fibers of both divisions is acetylcholine. On the other hand, the postganglionic mediator is not the same for all fibers (Fig. 7A-3). The parasympathetic postganglionic fibers are all cholinergic; that is, they release acetylcholine as the transmitter agent. The mediator substance released at most of the postganglionic sympathetic nerve endings is *norepinephrine.* Such fibers are *adrenergic. Epinephrine* and some norepinephrine are released into the blood by the chromaffin cells of the adrenal medullae. Norepinephrine is also called levarterenol, arterenol, or noradrenalin. Epinephrine is norepinephrine plus an N methyl group, and is also called methylarterenol or adrenalin. *Isoproterenol,* a synthetic catecholamine, is norepinephrine with an N isopropyl group. These compounds as a class are called catecholamines. In addition, some postganglionic sympathetic fibers, such as those going to the sweat glands, are cholinergic, releasing acetylcholine. The best evidence for this conclusion is that their activity is blocked by atropine, a drug which blocks the action of acetylcholine on these cells.

The *mode of action* of acetylcholine and the catecholamines on excitable membranes, such as those of the intestinal smooth muscle and the heart, is apparently that of modifying the rate of polarization and depolarization, thus changing the frequency of spontaneous spike discharge which is the basis of the autorhythmicity characteristic of these tissues. Acetylcholine increases and epinephrine decreases the frequency of spontaneous spike discharge of intestinal smooth muscle. The fact that each spike is followed by a slight contraction explains the effect of these mediators on this tissue. On the other hand, acetylcholine slows or stops the heart beat by hyperpolarizing the pacemaker cell membrane and slowing depolarization, both actions being probably due to an increased permeability of the resting membrane to potassium. The net effect is opposite to that of acetylcholine on intestinal smooth muscle or on the neuromuscular junction, for here depolarization is enhanced. Epinephrine accelerates the heart by increasing the rate of depolarization of the pacemaker during diastole. Thus the response of a particular end-organ to autonomic nervous system activity depends upon the individual makeup of that organ.

Epinephrine and norepinephrine generally act in the same manner, in that both tend to constrict the blood vessels, stimulate the heart, and inhibit the viscera. However, the blood vessels of skeletal muscle are constricted by norepinephrine and, under some conditions, dilated by epinephrine. Because epinephrine and norepinephrine show different relative degrees of influence between tissues, two adrenergic receptors are hypothesized: *alpha* and *beta.* Blocking agents such as phentolamine seem to block only the alpha receptors, while drugs such as propranolol block only the beta receptors. The biochemical characteristics of the receptors are still unknown. The *alpha receptor,* when stimulated, generally excites smooth muscle contraction (except for the intestine) and may act via changes in permeability of the cell membrane to change its excitability. The catecholamines may act directly at the actomyosin cross-bridges, for example heart muscle, to increase work output independently of the myosin ATPase, calcium, and sarcomere length (Honig and Takauji, 1974). The *beta receptor,* when stimulated, tends to be associated with increased metabolism, glycogenolysis and lipolysis. (The vigor of cardiac muscle activity is also increased.) Vascular or intestinal smooth muscle relaxation following beta receptor stimulation may be associated with an increased sodium pump activity which stabilizes the membrane by increasing its polarization. The beta receptor seems to be related to cyclic AMP and adenyl cyclase. The complex effect of epinephrine on skeletal muscle blood flow (Fig. 7A-4) seems to be the net result of both alpha and beta receptor stimulation. In the intestine, the alpha receptor response is weak and so relaxation is usually seen. Norepinephrine causes much less beta response than does epinephrine. On the other hand, isoproterenol induces a beta response (increased metabolism, vasodilation, and increased vigor of cardiac contraction).

The mediators must be removed from the site of action if they are to be effective under dynamic conditions. *Acetylcholinesterases* destroy acetylcholine by hydrolysis —normally within less than a minute if acetylcholine is introduced into the bloodstream. The enzymes are also found in high concentration around nerve endings. The catecholamines are relatively stable in blood but are rapidly inactivated in body tissues. The major mechanism for catecholamine inactivation is the active uptake of the catecholamine by the postganglionic sympathetic nerve endings; other mechanisms involve degradation (see, for example, Goodman, 1975).

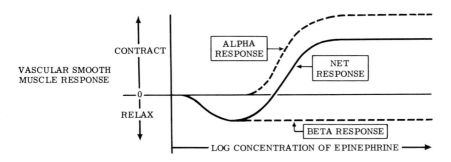

Figure 7A-4
Net effect of various concentrations of epinephrine on the vasculature. (After suggestion of D. O. Allen.)

THE ADRENAL MEDULLAE

The release of epinephrine into the blood provides a mechanism for spreading the effects of generalized sympathetic division activity to all cells of the body. These effects are primarily on metabolic processes. Adrenal medullary secretions are important factors in the homeostatic control of blood glucose concentration and release of free fatty acids from adipose tissue during conditions of stress such as violent muscular activity, hypotension, asphyxia, and hypoxia. Epinephrine secreted by the adrenals facilitates synaptic transmission, increases the heart rate and strength of contraction, and increases metabolism. Although the secretions stimulate the cardiovascular system, the direct adrenergic innervation can produce still higher maximal levels of activity.

The adrenal medullae, like the rest of the sympathetic division, are not essential for life if emergency demands are minimized. In fact, it is difficult to show deficiences under nonstressful conditions following removal of these organs. However, a reduction in glycogenolysis and a reduced stress tolerance are seen. Functionally, the adrenal medullary cells are equivalent to sympathetic postganglionic nerve cells. Their cholinergic preganglionic innervation and their embryological origins are similar. The cells are stimulated to secrete under the same conditions of stress as those affecting the adrenergic postganglionic neurons. In both, the catecholamines are held in minute granules and so are protected from enzymatic degradation. The neuroadrenomedullary and ganglionic junctions respond similarly to ganglionic blocking or facilitatory agents. The one clear difference is the secretion of epinephrine by the medullae and, on the other hand, norepinephrine by the adrenergic fibers. The endocrine secretion of the adrenal medullae supplements the neural action of the sympathetic division activity. In addition to the adrenal medullae, similar chromaffin cells secreting small amounts of epinephrine are located adjacent to the paravertebral chain of sympathetic ganglia, near the carotid sinus, heart, and liver.

Adrenal medullary secretions are controlled by centers located in the posterior hypothalamus. Under normal resting conditions the catecholamine concentration is 0.1 to 1.0 μg per liter of blood. The secretion is about 75 percent epinephrine, although there are marked species differences. At birth, most of the secretion is norepinephrine. With pain, excitement, anxiety, hypoglycemia from insulin injection, cold, hemorrhage, anoxia, or similar stresses, the systemic concentrations increase to as much as 100 μg per liter. The maximal rate of release is about 2 μg per minute per kilogram body weight. There is no clear evidence of separate control of norepinephrine and epinephrine secretion.

Because of the metabolic stimulation and glycogenolytic actions, epinephrine from the adrenal medullae has a particular value in homeostasis in response to decreases in body temperature. Cardiac stimulation and facilitation of synaptic transmission by epinephrine are remarkable during hypothermia.

EFFECTORS—SMOOTH MUSCLE

Smooth muscle is the effector organ of much of the autonomic nervous system. *Multiunit* smooth muscle has innervation similar to that of skeletal muscle. Little or no spontaneous or rhythmic activity is present. The nerve supply is excitatory. Examples include the muscle of the iris, the piloerector and nictitating membrane muscles of animals such as the cat, and probably vascular smooth muscle. *Visceral* smooth muscle, on the other hand, contracts spontaneously and rhythmically in response to the local chemical environment or to stretch. This activity is modified by the innervation from the autonomic nervous system, which either inhibits or excites further activity.

In most, if not all, types of smooth muscle under normal conditions, electrical activity of the muscle membrane precedes contraction (Chap. 3B). A prolonged contraction comes about as a result of repetitive action potentials. The evidence is not certain for all organs because of the minute size of the smooth muscle cells, most of which have diameters of the order of 1 μ or less. In the case of visceral smooth muscle and the heart, the spontaneous contraction results from an unstable resting membrane potential of the pacemaker cell, which upon reaching threshold initiates an all-or-none action potential that is transmitted from cell to cell. Stretching visceral smooth muscle in many cases depolarizes the membrane to initiate an action potential. Sodium and potassium are important in determining the resting membrane potential, while sodium and calcium are involved in the action potential (Chap. 3B). The smooth muscle cell membrane is apparently much more permeable to sodium at rest than is that of neurons or skeletal muscle, because cellular metabolism and the sodium pump activity are important factors in determining visceral smooth muscle tension.

EFFECTS OF THE AUTONOMIC NERVOUS SYSTEM ON SPECIFIC ORGANS

The Eyes

The adjustment of the eyes to varying light conditions and to focusing at various distances is controlled by the autonomic nervous system. Sympathetic activity to the radially oriented muscle of the iris causes it to dilate and to allow more light to enter the eye. Neural impulses of the parasympathetic division make the sphincter muscles of the iris contract, and thereby cause constriction of the pupil. During excitement the pupils of the eye are characteristically dilated (*mydriasis*), whereas poisoning with anti-cholinesterase compounds (see below) causes a characteristic pinpoint pupil (*miosis*).

Focusing of the lens of the eye is controlled almost exclusively by the parasympathetic division. The lens at rest is focused for distant vision and is relatively flat, but excitation by the parasympathetic division causes the ciliary muscles to contract and make the lenses more spherical for accommodation for near vision.

The Heart

As discussed in the chapters on circulation, vigorous muscular activity is accompanied by an increased sympathetic division activity, causing the heart to beat faster and to contract more vigorously so that more blood is pumped. At some later time the parasympathetic division, by slowing the heart rate, serves a restorative function.

Smooth Muscle of the Systemic Blood Vessels

The blood vessels of the abdominal viscera and skin are generally constricted by generalized sympathetic activity. On the other hand, the vessels of the heart and active muscles are dilated either directly or indirectly. Blood vessels other than those to the external genitalia have no parasympathetic innervation. Sympathetic cholinergic nerves for active vasodilation of skeletal muscle act to increase the flow of blood even before contraction starts. The cutaneous vasodilation enhancing heat loss probably results from a decrease in the sympathetic vasoconstrictor tone. Increased blood flow of salivary glands following stimulation via the parasympathetic division is due to the increase in metabolism of the organs and the release of *bradykinin*, a polypeptide which is a potent vasodilator on the one hand and an intestinal and uterine smooth muscle stimulant on the other. The evidence for a *direct* vasodilating action by the parasympathetic or sympathetic division has been uncertain. Much, but not all, of the data can be explained either by a reduction in sympathetic vasoconstrictor tone or by the release of vasodilator materials resulting from the enhanced metabolism brought about by increased activity of the tissues, e.g., heart, muscle, or glands (see Chap. 17).

If the flow of blood to the skin, intestinal tract, and kidneys is reduced (for example, during exercise and hemorrhage), blood flow is available for diversion to active muscle, the heart, and the brain. On the other hand, muscle blood flow is decreased during asphyxia.

Spleen

The contraction of the spleen and subsequent discharge of blood rich in erythrocytes is another effective mechanism for maintenance of homeostasis during physiological stress, such as exercise, hemorrhage, and anoxia. The erythrocyte storage function of the spleen is of minor importance in man.

Respiratory Tract

The autonomic system apparently has a minor effect on the functioning of the lungs. Consistent with the concept of the importance of the sympathetic division for voluntary muscular activity, sympathetic division stimulation causes the bronchi to dilate, thus allowing air to enter more easily. The parasympathetic division acts in an opposite manner, and also stimulates the secretion of mucus.

Exocrine Function

Most of the exocrine glands of the body are stimulated to secrete as a result of parasympathetic division activity. An exception is the *mammary gland.* Prolactin is the primary hormone controlling milk production and its secretion into the alveoli of the breasts. Stimulation of the myoepithelial cells surrounding the alveoli for milk ejection or "let-down" (the exocrine function) is by oxytocin released from the posterior pituitary (the neurohypophysis). Disturbance and other psychogenic factors leading to generalized sympathetic stimulation inhibit the release of oxytocin, depress milk ejection, and so lead to inadequate nursing.

The *nasal* and *lacrimal glands* secrete mucus and tears, thereby providing a protective function. The *salivary glands,* when stimulated by the parasympathetic division, supply a copious secretion for mastication and digestion. The *glands of the stomach* and *pancreas* are stimulated by parasympathetic division impulses and provide an increased secretion of pancreatic and digestive juices for digestion. Not all control of these secretions is neural, however.

The sympathetic division has little or no effect upon exocrine secretion except that its activity generally causes vasoconstriction, which tends to reduce secretion. Exceptions are the effects on the salivary glands and pancreas. Sympathetic stimulation causes a small but definite increase in flow of saliva, which is thicker and more viscous than that following parasympathetic division activity. Splanchnic nerve stimulation (sympathetic) causes some pancreatic secretion and changes in the zymogen granules of the acinar cells of the pancreas.

The *eccrine* type of *sweat gland,* producing copious watery sweat (*hidrosis*), is innervated by sympathetic nerves with no connections from the parasympathetic division. Innervation is by cholinergic fibers, an exception to the general rule that sympathetic postganglionic fibers are adrenergic.

The *apocrine glands,* located primarily in the axillary and pubic regions, secrete a thick material which is decomposed by skin flora to produce characteristic body odors. Although closely related to the eccrine glands, they respond to adrenergic stimulation. In fact, there is evidence that they are not innervated, but respond to circulatory epinephrine only. Whereas the eccrine glands are associated with heat loss, the apocrine glands are associated with fear, anxiety, and vigorous muscular or sexual activity.

Sexual Function

The role of the autonomic nervous system in normal sexual activity is complex. Engorgement of *erectile tissue* of both sexes (penis and clitoris) is by parasympathetic dilation of the arterioles of the tissue. With restricted venous outflow, the increased blood flow engorges the venous sinusoids. Without sexual excitement a sympathetic tone appears to keep blood flow low and the tissue flaccid. In addition, there is parasympathetic stimulation of exocrine glands of the *vagina* which supplies mucus for lubrication and protection. Whereas the autonomic nervous system is clearly involved in psychogenic erection, engorgement of the erectile tissue can also be elicited by physical stimuli involving a simple spinal reflex. In the male, the contraction of the vas deferens and seminal vesicles and *ejaculation* of semen are innervated by the sympathetic division. However, spinal cord transection above the third lumbar segment does not abolish ejaculation. It is not clear whether muscular contractions of the female *orgasm* are mediated by the sympathetics, also. The role of the autonomic nervous system in *parturition* is minimal, although the release of oxytocin from the neurohypophysis is crucial. The sympathetic division is apparently not essential for gestation.

Smooth Muscle of the Gastrointestinal Tract

Although the gastrointestinal system has its own intrinsic neural control, both divisions of the autonomic nervous system affect gastrointestinal activity. Parasympathetic division activity generally increases gastrointestinal motility. The role of the sympathetic division is primarily an inhibiting one seen only in some diseases or during stress.

The esophagus is supplied by parasympathetic nerves. The involuntary parts of the process of swallowing are mediated by the parasympathetic division, which initiates a wave of constriction down the musculature of the esophagus.

Smooth Muscle of the Pelvic Viscera

The parasympathetic system is involved in contraction of the bladder and lower colon for emptying of these organs. A spinal reflex arc, including parasympathetic fibers, activates micturition and defecation. Excitement and generalized sympathetic activity will tend to inhibit those functions. Furthermore, since the external sphincters are under somatic control, elimination can be stopped voluntarily.

RESPONSE TO DENERVATION

Immediately following sectioning of sympathetic or parasympathetic nerves, the denervated organ loses most of its tone and activity, but not permanently. Compensation develops which is primarily an increased sensitivity to circulating chemical agents. Denervation supersensitivity is most pronounced if the postganglionic nerves are severed. This was first studied in connection with *Horner's syndrome,* which results from interruption of the sympathetic division supply to the face. As a result of lack of sympathetic tone the pupils are constricted (miosis), the eyelids droop (*ptosis*), and the face is flushed. With the passage of time, however, these symptoms tend to disappear. If one side of the face has the preganglionic sympathetic supply sectioned and the other side the postganglionic fibers, at first the effects are symmetrical, but with time the side with the postganglionic innervation cut shows a larger pupil (return of sympathetic tone) than does the other side. The discrepancy is intensified when the animal is frightened. These phenomena are explained as follows: With destruction of the postganglionic fibers, the major inactivation mechanism for catecholamine (i.e., uptake by nerve endings) is removed, so the receptors are exposed to stimulating concentrations of blood-borne catecholamines. Following preganglionic section, the neural input is blocked and circulating catecholamines in the vicinity of the receptors are readily inactivated by the postganglionic endings nearby. Some supersensitivity develops but of lesser magnitude. Denervation supersensitivity also follows parasympathetic nerve sectioning. Even skeletal muscle shows increased sensitivity to the mediator after denervation, for sectioning of the motor nerve causes the motor end-plates to become hypersensitive to acetylcholine.

Since preganglionic denervation causes less sensitization than does postganglionic denervation, it provides a surgical tool. For example, *Raynaud's disease* involves painful paroxysmal cutaneous vasospasms (usually in the fingers and toes) following exposure to cold or emotional stress. The vasospasm is so intense that gangrene sometimes results from the lack of circulation. If the postganglionic fibers are sectioned, it is only a matter of time before the blood vessels respond excessively to circulating epinephrine. Preganglionic denervation, however, eliminates the neural constrictor influence without the development of the marked hypersensitivity to the circulating epinephrine.

TRAINING THE AUTONOMIC NERVOUS SYSTEM

With operant conditioning, it is clear that autonomic nervous system responses to various stresses and stimuli can be modified (Smith, 1974). People who are in the early stages of hypertensive disease can perhaps be trained to reduce their blood pressure. Such visceral learning would be of great therapeutic significance and is the subject of continuing research.

DRUGS ENHANCING OR DEPRESSING AUTONOMIC FUNCTIONS

Drugs that mimic or block sympathetic or parasympathetic division activity provide a tool for a better understanding of normal physiological processes and may be used to reverse or counteract some diseases. Pharmacology texts should be consulted for details.

Preganglionic Simulating Drugs

Preganglionic simulating agents (e.g., acetylcholine in very high concentrations and nicotine) mimic the activity of the preganglionic neurons and thereby stimulate postganglionic neurons of both divisions. Since nicotine and "nicotinic" drugs such as carbachol stimulate not only sympathetic and parasympathetic postganglionic fibers but also skeletal muscle, it is thought that the receptor system at the postganglionic neurons is similar to the system at the neuromuscular junction.

Parasympathomimetic or Cholinergic Drugs

Because acetylcholine is so rapidly destroyed in blood after injection, it does not have the same effects throughout the body as parasympathetic division activity does. However, drugs such as methacholine, muscarine, and pilocarpine are less rapidly inactivated and have effects in the body similar to those of the acetylcholine which is released from parasympathetic endings. These drugs also simulate the activity of sympathetic *cholinergic* fibers; e.g., profuse sweating is seen following administration.

The parasympathetic effects may be potentiated if acetylcholine destruction is inhibited. An agent such as eserine or prostigmine inhibits acetylcholine hydrolysis and thus potentiates parasympathetic activity. Because of this anticholinesterase activity, transmission at sympathetic ganglia and neuromuscular junctions is also enhanced. The poisoning action of most organic phosphorus pesticides is by way of anticholinesterase activity.

The action of the alkaloid muscarine on viscera such as cardiac muscle, exocrine glands, and smooth muscle is similar to that of acetylcholine. The effects are termed the *muscarinic actions* of acetylcholine. Muscarine has little effect on skeletal muscle or ganglionic transmission. *Nicotine,* on the other hand, affects autonomic ganglia—both parasympathetic and sympathetic—and skeletal muscle much as acetylcholine does. Thus, the nicotinic actions of acetylcholine may be differentiated from the muscarinic actions.

Sympathomimetic or Adrenergic Drugs

Sympathomimetic or adrenergic drugs mimic the effects of sympathetic division discharge. Drugs such as ephedrine, amphetamine, and isoproterenol have effects similar to the action of epinephrine and norepinephrine, which are natural hormones. They differ, however, in potency at various effector sites, in mode of action, and in duration of their activity.

Blocking Agents

There are two general types of blocking agent used for research and therapy. *Depolarizing agents,* such as nicotine in high concentration, block by maintaining depolarization of the excitable membrane. At first the effector organ is stimulated, but then blockage of transmission occurs. *Antidepolarizers* generally block by competing with acetylcholine for the receptor sites on the postganglionic or postjunction membrane. Hexamethonium and tetraethylammonium act in this manner at the ganglia, d-tubocurarine has a similar blocking action at the skeletal muscle neuromuscular junction, and atropine blocks at the visceral sites.

At the ganglia the nicotinic effects of acetylcholine are blocked by agents such as tetraethylammonium and trimethaphan camphorsulfonate, but atropine or curare is relatively ineffective. Ganglionic blocking agents are used primarily to block sympathetic activity, as in reducing hypertension, but also act on the parasympathetic division ganglia.

At the neuromuscular junction of skeletal muscle the receptor agent for acetylcholine is different from that in smooth muscle, for here atropine is not effective, while d-tubocurarine is. Succinylcholine is an example of a depolarizing blocking agent, whereas curare reversibly binds the receptor.

At the visceral effector organs, the muscarinic effects of acetylcholine are blocked by atropine and scopolamine. Atropine blocks not only the excitatory effects of acetylcholine, such as those on the intestine or exocrine glands, but also the inhibiting effects such as in the heart.

The direct excitatory (alpha) responses of sympathetic effector cells to agents that cause vasoconstriction are blocked by certain drugs—phentolamine or phenoxybenzamine, for example. After administration of these drugs, epinephrine dilates certain blood vessels—the constricting effects are blocked. The drugs do not block the beta receptors, for catecholamines continue to stimulate metabolism and inhibit the gastrointestinal tract.

The beta response to epinephrine is blocked by propranolol. Use of such agents aids in studying the action of catecholamines.

Reserpine appears to block the sympathetic system, but in fact it acts to reduce the amount of catecholamines in the brain and adrenal medullae available for subsequent release after neural stimulation. Indeed, the catecholamines in the heart and blood vessels approach zero in animals treated with high doses of reserpine.

REFERENCES

Ahlquist, R. P. Effects of the Autonomic Drugs on the Circulatory System. In W. Hamilton (Ed.), *Handbook of Physiology.* Washington: American Physiological Society, 1965. Section 2: Circulation, vol. 3, pp. 2457–2476.

Altman, P. L., and D. S. Dittmer (Eds.). *Biology Data Book* (2nd ed.). Bethesda, Md.: Federation of American Societies for Experimental Biology, 1973. Vol. 2, pp. 1150–1180.

Appenzeller, O. *The Autonomic Nervous System.* Amsterdam: North-Holland, 1970.

Burn, J. H. *Autonomic Nervous System* (4th ed.). Oxford, England: Blackwell, 1971.

Goodman, L. S., and A. Gilman (Eds.). *The Pharmacological Basis of Therapeutics* (5th ed.). New York: Macmillan, 1975.

Honig, C. R., and M. Takauji. Mechanochemical coupling at the actomyosin cross-bridges and adrenergic inotropic receptor. *Recent Adv. Cardiac Struct. Metab.* 4:297–303, 1974.

Malmejac, J. Activity of adrenal medulla and its regulation. *Physiol. Rev.* 44:186–218, 1964.

Pick, J. *The Autonomic Nervous System.* Philadelphia: Lippincott, 1970.

Root, W. S., and F. G. Hofmann (Eds.). *Physiological Pharmacology.* New York: Academic, 1967. Vols. 3, 4.

Smith, O. A. Reflex and central mechanisms involved in the control of the heart and circulation. *Annu. Rev. Physiol.* 36:93–123, 1974.

Von Euler, U. S. Autonomic Neuroeffector Transmission. In J. Field (Ed.), *Handbook of Physiology.* Washington: American Physiological Society, 1959. Section 1: Neurophysiology, vol. 1, pp. 215–237.

7. Regulation of Visceral Function
B. Homeostasis and Negative Feedback Control

Carl F. Rothe

Each cell of the body requires an environment that supplies nutrients and removes metabolic wastes. The concept of *a constant and optimal internal environment* as a necessity for normal function was first formulated by Claude Bernard a century ago. In 1929 Cannon further developed the concept of this condition, which he called *homeostasis,* and emphasized the role of the autonomic nervous system. One of the cardinal principles of physiology is that homeostatic mechanisms operate to counteract changes in the internal environment which are induced either by changes in the external environment or by activity of the individual. Thus, disturbances of the internal environment as the result of exercise, nutritional imbalance, trauma, and disease are minimized.

An example of homeostasis is the control of body temperature. If the internal temperature drops, homeostatic mechanisms act to reduce heat loss from the body and to increase heat production. Consequently, these mechanisms limit a decrease in body temperature and so maintain this variable relatively constant. On the other hand, a cold-blooded animal does not possess homeostatic temperature control systems; its body temperature thus tends to be similar to that of its environment.

Homeostatic mechanisms act to minimize the difference between the actual and optimal responses of a system and are therefore biological examples of *negative feedback control.* The level of the controlled variable is sensed, and action is taken which opposes any change from the desired level. If the response increases, a signal is fed back to an effector mechanism in a negative or inhibitory manner so that the subsequent response is reduced. Conversely, a decrease in response elicits a subsequent increase. A familiar example of a negative feedback control system is the thermostatic control of room temperature. This is illustrated in Figure 7B-1A. Room temperature is measured by a temperature-sensing element in a thermostat and compared with a reference (desired or optimal temperature) in such a manner that an *error signal* develops when discrepancies exist. If the room is too hot, the furnace is turned off by the action of the thermostat; if it is too cold, the furnace is turned on. The error signal from the thermostat is used to adjust the system automatically so as to minimize any deviation between the measured temperature and the desired temperature. If the system is properly designed, room temperature can be held nearly constant despite wide fluctuations in outside temperature.

Homeostatic feedback mechanisms of mammals are exceedingly complex and interrelated but are generally amenable to analysis using the approach of engineers. These workers, utilizing the principle of negative feedback control, have made great advances in the design of control systems used in the operation of devices such as automatic airplane pilots, missile guidance systems, computers, and servomechanisms for industrial automation. The primary purpose of using negative feedback in these mechanisms is to provide high accuracy and stability of function in spite of changes in the external environment or changes in the systems themselves. The same general principle acts in mammals to keep constant and optimal such diverse variables as body temperature, blood pressure, blood sugar concentration, electrolyte concentrations, muscle tone, and blood carbon dioxide levels, to mention only a few. The human body has a large number of negative feedback systems, all perfected through evolutionary development. The autonomic nervous system is an important part of most homeostatic mechanisms. To attain a better understanding of the physiology of the normal human, it is necessary to understand the fundamental characteristics of negative feedback control systems.

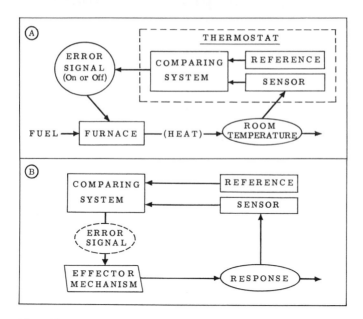

Figure 7B-1
Negative feedback control system. A. Schematic diagram of components of a thermostatically controlled furnace. B. Components of a negative feedback control system.

SYSTEMS

The concept of a system—a set of components which act together, and which can be treated as a whole—has been found to be highly useful for the study of dynamic, complex biomedical situations. With a system, *inputs* are those stimuli, forcings, or disturbances that act on it, while the *output* or response is the consequence of such inputs. Furthermore, the relationship between output and input is the "law" describing the system. It is usually written as a mathematical equation. In some cases, such as that describing the relationship between the input forcing of a transducer and the resulting pen deflection of a recorder, the equation is simple; in others, such as those predicting changes in cardiac function in response to changes in pressure at the pressoreceptors, the relationship is so involved that complex differential equations are required to provide even an approximation. If the purpose of the analysis is to understand and predict the behavior of complex systems, the input-to-output relationship written in mathematical form with carefully defined terms is an effective form for the theory describing the system.

SYSTEMS ANALYSIS, SIMULATION, AND MODELS

As our understanding of biological systems becomes more extensive, it becomes more difficult to organize the information, to check it for consistency, and to predict the consequences of various stimuli or changes of parameters on a system. *Mathematical biology, systems analysis,* and *simulation* are different words for the general approach of (1) describing our understanding through models written in the form of mathematical equations and (2) studying the models with computers to check for internal consistency and for agreement with available experimental data. As an example, both the plasma osmolality and the transmural pressure of the left atrium are important variables in the control of the secretion of antidiuretic hormone, which in turn acts on the kidney in the control of blood volume. What is the relative importance of each, under various conditions? How fast does the system respond? Is there a synergistic response? A threshold? Saturation of response? A mathematical model may provide a concise and accurate summary of thousands of experiments and may permit accurate predictions under a wide variety of conditions.

A *model* is a representation of reality, be it a stereotype, archetype, caricature, syndrome, diagram, prejudice, equation, or verbal description. In comparison to verbal models, mathematical models are more precise, are less ambiguous, and more readily permit quantitative verification. The use of computers facilitates the study of multifactor interaction and dynamic (i.e., transient or moment-by-moment) responses. In the process of developing and testing the model—called *simulation*—new experiments are suggested, for gaps in knowledge often become obvious. Not only do models provide a concise summary of pertinent observations, but for many students they provide a more satisfying description of the system than catalogues of cases, tables of data, or graphs.

CHARACTERISTICS OF HOMEOSTATIC MECHANISMS

The operation of a negative feedback control system may be broken into several separate functions (Fig. 7B-1B). First, the magnitude of the controlled variable must be *sensed* by some means (the sensor). This magnitude must then be *compared* with a reference value which is considered optimal and is often a part of the genetic makeup of the biological organism. The difference between the reference value and the actual value is the *error signal,* which is *fed back* to the *effector mechanism* in such a manner (negative) as to *minimize the error.*

Regulation of the amount of food eaten is an example of a complex biological control system. Hunger is the error signal. The level of intake needed to satisfy the hunger is widely variable, depending in part on the physical activity, psychological status, and general health of the individual. If the food intake is inadequate, the hunger increases and the individual eats more and thereby reduces the hunger; this is a negative feedback system. (Because the reference for hunger is vague and the internal comparator is so easily modified by psychological factors, the control of food intake is imprecise. If food is readily available, there is an all-too-common tendency toward obesity.) Negative feedback control systems also operate at the molecular level, controlling enzymatic reactions. Here all facets may be combined into a system in which an increase in the concentration of the product of a chemical reaction inhibits the reaction itself so that the product concentration is held relatively constant.

The response of a system may be controlled by two general approaches: (1) *Open loop,* or control based on *compensation* for the effect of each source of disturbance. The quantitative effects of all the important disturbances on the system must be determined. The magnitudes of the disturbances are sensed or measured, and the system is then adjusted accordingly. Aiming a gun, after taking into account wind and distance, is an example of this approach. Its effectiveness depends upon precise knowledge of all disturbing variables, both as to their influence and as to their magnitude. In addition, the mechanism itself must be precise and stable, features not characteristic of biological systems. (2) *Closed loop.* In this approach, control is attained by measuring or sensing the response and *feeding back* the deviation from optimal to the effector mechanism. Knowledge of the source and magnitude of the disturbance is immaterial—providing they are not beyond the range of control. The effector mechanism may be unstable and imprecise in constructional details—a hallmark of living organisms. With a closed loop, close and reliable control is attainable. A control system in which the output or response is held constant is often called a *regulator,* while one in which the response is made to follow the pattern of a varying input signal is called a *servomechanism.* Control (in the sense of negative feedback control) implies that the response of the total system will be held near certain desired values even though disturbing inputs to the system fluctuate, or there are changes in the effector system itself.

The Reference Level

The reference level (or set point) is the desired condition. In the control of room temperature, the reference is the temperature setting of the thermostat. In the healthy individual, it is usually the optimal level of the variable and aids in defining what is meant by normal. With a control system, the regulated variable may be changed by changing the reference level. Many pathological conditions can be traced to such a change, although the details of how and why are far from being solved. For example, fever is primarily a change in the reference level for the control of body temperature. Hypertension is possibly a change in the reference level for the control of blood pressure. When the reference is changed, the controlled function fluctuates about the new level.

Comparing the response of the system to the input or reference by subtraction is one way to obtain the error signal. A different approach is to divide the reference signal by a signal representing the response. As applied to an amplifier, this type of control provides an automatic gain control. Although potentially of wide importance in studying biological systems, especially those involving enzymatic actions, this form of description has not been widely exploited as yet.

Since in most biological systems there is no clearly defined arithmetic comparator for the generation of an error signal by subtracting a value representing the response from a reference value, most biological systems function around an *operating point* determined by the characteristics of the chemical and physical structures of the system.

Feedback Factor

The feedback factor has been variously defined, but it generally means the ratio of the response to a disturbing influence of an uncontrolled system to that of the controlled system.

$$\text{Feedback factor} = \frac{\text{Uncontrolled response}}{\text{Controlled response}}$$

It is a measure of the effectiveness of the control system. As an example, the control of body temperature in mammals is excellent; indeed, such animals are called homeotherms. If the body temperature of a group of subjects in a room with the temperature at 40°C was 37.8°C, while that of the same persons under similar conditions but at 2°C was 37.2°C, the feedback factor for these people was

$$\frac{40°C - 2°C \text{ (uncontrolled response to environment)}}{37.8°C - 37.2°C \text{ (controlled response)}} = \frac{38}{0.6} = 63$$

The blood pressure is said to be regulated with a feedback factor of 3. This means that an influence which would cause only a 20 mm Hg change in the normal animal would cause a 60 mm Hg change in the animal without control. Present research suggests that it may be as high as 10. Hypertension might be explained by some factor causing an increase in blood pressure in an animal which has a feedback factor too low to provide adequate compensation, or which has a control system that unfortunately adapts over a long period of time to the pathological pressures.

It is generally quite difficult to obtain a completely uncontrolled response of a biological system since it is hard, and in most cases probably impossible, to remove all the regulatory features without damaging the normal, basic biological mechanisms. The concept of feedback factors is of value, however, in assessing the importance of parts of a biological control system or in testing the adequacy of a homeostatic mechanism. For example, if before anesthetization the change in blood pressure of a dog in response to a temporary loss of blood volume is -2 mm Hg and after anesthetization the response to a similar loss is -20 mm Hg, it can be said that the efficiency of the control of blood pressure is reduced by a factor of 20/2 or 10 by the anesthetic agent. For use as a diagnostic approach to disease, standardized stresses or test situations must be developed.

Modes of Control

Several modes of control are found in biological systems: (1) *Proportional.* In this form the signal (E) driving the effector system is directly proportional to the error (ε) signal ($E = k_1\varepsilon$; where k_1 is a gain or amplification factor). With simple proportional control, the error is relatively large if the disturbance is large. (2) *Integral.* To reduce the steady-state error, the error is integrated (added up as time goes by) to develop an additional signal giving proportional-plus-integral control ($E = k_1\varepsilon + k_2 \int \varepsilon \, dt$). (3) *Derivative.* To anticipate, in effect, the degree of corrective action needed and so to reduce "hunting" (overshooting and undershooting around the desired value), the rate of change of error can also be used with proportional control ($E = k_1\varepsilon + k_3 \, d\varepsilon/dt$). For example, control of blood pressure and body temperature is, in part, rate dependent. Because all physical systems have energy-storing features (e.g., inertia and compliance), the response never exactly follows the stimulus, and so perfect control during transient changes, even with these more complex modes, is impossible. (4) *On-off.* This is the familiar type of control used with automatic appliances such as home furnaces, refrigerators, and pumps. When the controlled variable exceeds set limits, the unit is turned on or off. Though it is discontinuous, this type of control can be highly effective, especially if used as a supplement to open-loop control.

Limits of Control

A control system will perform accurately only within certain limits of change in components or energy supply. If the energy supply is markedly reduced or some component is seriously damaged, the effector mechanism may operate at maximal, yet inadequate, levels. With a serious continuing loss of blood, the blood pressure may be maintained for a while by compensatory mechanisms, but a point will be reached at which the limits of compensation have been exceeded. Then the pressure will fall, often in a fulminating manner. Another example of an overdriven control system is seen when insufficient iodine in the diet limits the production of thyroid hormone. A feedback mechanism involving the pituitary gland causes hypertrophy of the thyroid (simple goiter), which may lead to irreversible damage of this gland if the iodine deficiency is not relieved.

When a biological feedback system is driven to its limits because of malfunction or external stress, the ideal form of treatment is to find and correct the cause of the change rather than attempting to augment compensatory changes which might lead to overloading of the system and possible fulminating failure. Defective function of any link in the homeostatic system is ordinarily reflected in some degree of hyperfunction or hypofunction of other elements in that system. This is the reason symptoms are commonly clustered into characteristic *syndromes.* Because the various parts of a homeostatic system are connected functionally into a closed loop, the chain of cause and effect among symptoms is often difficult to discover unless studied in the light of overall system operation.

Positive Feedback, Response Time, and Stability

If the effect of the feedback is to increase, rather than decrease, the deviation from the desired level, the feedback is said to be positive. Positive feedback may be defined as a situation in which an *increase* in response, when fed back to the control mechanism, causes a still *further increase* in response. Positive feedback occurs if the feedback factor is less than 1—i.e., if the "controlled" response to a disturbing influence is greater than the "uncontrolled" response with no feedback. Positive feedback, if not limited, causes a progressive change in response so that eventually the system operates at a maximum or minimum level—a vicious cycle.

If the thermostat for the furnace were connected in reverse, so that an increase in temperature turned the furnace *on,* the result would be an example of unlimited positive feedback. Biological examples of unlimited positive feedback are relatively rare, since the condition results in extremes of response. The response of a nerve to an adequate stimulus is a good example, however. If the stimulus to nerve action is inadequate, there is no response; but if the stimulus is adequate, once the response of the nerve starts, the process reinforces itself (positive feedback) so that the nerve fiber responds maximally. Only later is the process reversed.

Positive feedback can lead to death. For example, if the heart starts to fail significantly, less blood is pumped, the blood pressure falls, and there is consequently a lesser supply of blood to the heart itself. This weakens it *further,* leading to a further decrease in blood pumped and so, eventually, to death.

The *response time* of a system is a measure of the time required for it to respond to a change in conditions and is defined, for the simplest system, as 63.2 percent $(1 - 1/e)$ of the time taken to achieve the final total change in response. Following a sudden change in conditions, the error signal even in an efficient control system is relatively large at first, but then it declines toward zero as the desired level is again approached.

Because of the interaction between energy-storing devices, oscillations may occur in response to a sudden change in input or to disturbances. The transport time of materials through the vascular system, and the diffusion of products from sites of synthesis are conducive to oscillations. In most control systems, oscillations are a sign of malfunctioning or poor control. The blood pressure sometimes oscillates in an animal with a failing cardiovascular system. Predicting the conditions for system stability (i.e., lack of sustained oscillations) is a major part of the rather complex science of control system design. In biological control systems, some of the components are not even known, so no complete description of the characteristics of the systems can be given. This, then, is an area for further biophysical research.

For accurate control, a high feedback factor is essential so that the effects of changes in the metabolic energy supply, parts of the system, or environment will be reduced toward zero. Such systems, however, tend to be unstable and to oscillate, especially if the feedback system has a relatively long response time (i.e., is sluggish). The response of the somatic nervous system is speedy; therefore, accurate control of posture by rapidly responding muscles is possible. In some disorders, however, the feedback system is sluggish or the feedback factor (gain) is too high, so there is tremor on attempts to make fine movements.

Biological Control Systems

Many biological variables are controlled directly. For example, the amount of carbon dioxide in the arterial blood is controlled by homeostatic mechanisms which regulate the degree of lung ventilation so that the proper amount of carbon dioxide is exhaled. Other variables, such as the amount of oxygen in the arterial blood, tend to be regulated in an indirect manner, for the arterial oxygen tension is dependent largely on the amount of oxygen in the air and the degree of lung ventilation; this in turn is controlled directly by the amount of carbon dioxide in the blood. Finding the primary variables and the interrelationships of these homeostatic mechanisms is one of the challenges of modern physiology.

It must be emphasized that a negative feedback control system is not highly dependent upon the accuracy and stability of its components. An individual's heart may be severely damaged by disease, yet the blood pressure may be normal if the person does not strenuously exert himself. There are marked differences in the dimensions of the organs of people and in their functional reserves between birth and old age, yet these differences do not preclude normal function.

Finally, biological homeostatic mechanisms often possess several feedback systems which act on the same variable, providing constancy and reliability to the organism (though complexity to the physiologist and student). As an example, there are several types of sensors and effector mechanisms for the control of blood pressure.

The autonomic nervous system is an essential part of most mammalian control systems. As better understanding of the homeostatic mechanisms and sites of failure is obtained, more effective therapy for various pathological conditions may be devised.

In conclusion: "The essence of physiology is regulation. It is this concern with 'purposeful' system responses which distinguishes physiology from biophysics and biochemistry. Thus, physiologists study the regulation of breathing, of cardiac output, of blood pressure, of water balance, of body temperature and of a host of other biological phenomena" (Grodins et al., 1954).

REFERENCES

Adolph, E. F. Early concepts of physiological regulations. *Physiol. Rev.* 41:737–770, 1961.

Blesser, W. B. *A Systems Approach to Biomedicine.* New York: McGraw-Hill, 1969.

Brown, J. H. U., and D. S. Gann. *Engineering Principles in Physiology.* New York: Academic, 1973. Vols. 1, 2.

Cannon, W. B. Organization for physiological homeostasis. *Physiol. Rev.* 9:399–431, 1929.

Grodins, F. S., J. S. Gray, K. R. Schroeder, A. L. Norins, and R. W. Jones. Respiratory responses to CO_2 inhalation. A theoretical study of a nonlinear biological regulator. *J. Appl. Physiol.* 7:283–308, 1954.

Jones, R. W. *Principles of Biological Regulation: An Introduction to Feedback Systems.* New York: Academic, 1973.

Milsum, J. H. *Biological Control Systems Analysis.* New York: McGraw-Hill, 1966.

Riggs, D. S. *Control Theory and Physiological Feedback Mechanisms.* Baltimore: Williams & Wilkins, 1970.

Warner, H. R. Simulation as a tool for biological research. *Simulation* 3:57–63, 1964.

Yamamoto, W. S., and J. R. Brobeck. *Physiological Controls and Regulations.* Philadelphia: Saunders, 1965.

Yamamoto, W. S., and E. S. Walton. On the evolution of the physiological model. *Annu. Rev. Biophys. Bioeng.* 4:81–102, 1975.

8. Higher Somatic and Visceral Control

Sidney Ochs

MOTOR AREAS OF THE CEREBRAL CORTEX

In 1870 Fritsch and Hitzig first showed that electrical stimulation of certain parts of the cortex produced movements of specific portions of the peripheral musculature. Since then others have shown, in an increasingly refined way, the correspondence of points within such *motor areas* to the peripheral musculature. In Figure 8-1, the motor area of the human brain is labeled with the body musculature excited from the various cortical sites. A general similarity or homology of motor areas and their muscle control patterns within the cerebral cortex has been found for the different species. A relationship of motor function to the underlying cellular structure was indicated for the area of the brain numbered 4 by Brodmann (Fig. 8-1), which contains large *pyramidal tract* (PT) *cells* also known as *Betz cells.* The large PT cells are stimulated by applied electrical currents to give responses in the different limb muscles. However, there is not an exact correspondence of motor control of a given musculature to a specific PT cell population; the motor area is known to extend beyond area 4. Also, as shown anatomically, axons from a given part of the motor cortex are widely distributed to many motor nuclei of the lower motor centers—although with a predominant convergence onto one group of motoneurons innervating a particular muscle.

With the use of a constant level of anesthesia, and other factors to control the stability of the experimental preparation, a fairly consistent and detailed representation of the peripheral musculature in the motor region may be found. The motor area determined in this way is called the *precentral motor area.* Figure 8-2 shows the result obtained from a *Macaca mulatta* brain, in which a map was constructed from those points of the cortex having the lowest threshold for responses in the various muscles represented. As previously indicated, the correspondence of motor areas with the various cytoarchitectonic regions of the cortex is only approximate, with portions of the motor cortex falling outside area 4. In area 6 of Brodmann, larger muscle groups of the back and the upper limb muscles are represented. From area 4, a control of finer muscles is found, those of the lower arms, the digits, tongue, and face. These areas of finer muscle control are disproportionately large with respect to the control areas of other parts of the musculature. Therefore, lesions of the motor cortex are more destructive to the finer and complex manipulative movements of the digits, while the grosser movements of the axial musculature or of muscles of the upper limbs remain relatively less affected. Recent evidence has shown that, in man and primates in general, pyramidal tract fibers controlling fine movements end directly on spinal motoneurons. However, a large share of the control is apparently also subserved in tracts other than the pyramidal, in the *extrapyramidal tracts.*

A greater susceptibility to disruption of fine control is characteristically seen in patients who have had brain strokes with lesions in the internal capsule interrupting a large part of the fiber downflow from the motor cortex. A patient so affected has a disturbance of posture and movement. The position of the limbs is said to show a reversion to primitive patterns of motor control—flexion of the upper arms and hyperextension of the lower legs with increased reflexes, i.e., *spasticity.* The lower motor centers and brain stem have been *released* by the capsular lesion from upper cortical control. A positive Babinski sign is present with upper motor neuron lesions. But it is in connection with fine complex movements that the disturbance is most apparent. The attempt to move the fingers—opening and closing them or approximating the fingers—is difficult or not possible. The upper limb, however, may be moved. In the lower limb this

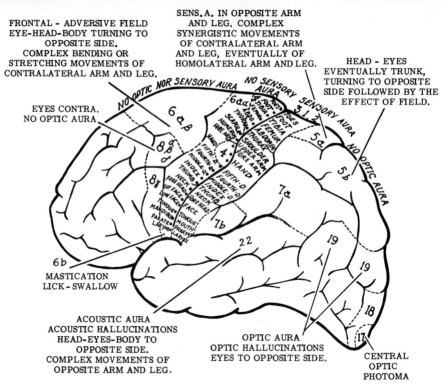

Figure 8-1

Motor areas of the human brain. The motor areas of the brain are in front of the central sulcus. The various parts of the peripheral musculature excited by stimulation in that region are labeled. Various other regions in different areas of the brain also can give motor responses as indicated. Notice the correspondence of the motor area to the sensory regions posterior to the sulcus. (From W. Penfield and T. C. Erickson. *Epilepsy and Cerebral Localization*. Springfield, Ill.: Thomas, 1941. P. 401.)

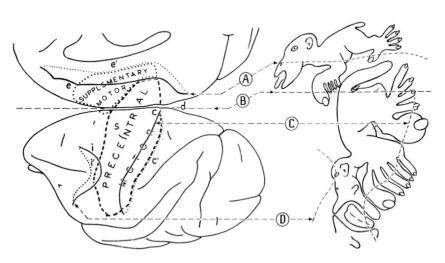

Figure 8-2

Motor areas of the monkey brain. To the left, the precentral motor area is shown with dashed lines, and to the right a map of the body musculature represented within the area. In addition to this primary motor area, a smaller supplementary motor area can be found. Much of it is hidden from pattern of defect gives rise to the spastic gait typical of a stroke patient. The leg is held stiffly extended, and while walking it is swung outward and forward in sticklike fashion. The rhythmic flexion and extension of the limbs at the hip, knee, and ankle joints which are part of the normal walking pattern are lost. The earlier assignation of a coarse control function to area 6 was due in part to an unrecognized involvement with a distinct second motor area, *the supplementary motor area,* which can also be represented by a map. Part of this area is turned around and hidden from view on the medial surface of the hemisphere (Fig. 8-2). The responses to stimulation of the supplementary motor area give rise to the more generalized movements affecting large muscular groups. The efferent pathway from the supplementary area projects separately into the pyramidal tract and does not relay through the precentral motor area of the cortex.

The region from which motor responses are obtained represents a complex interplay of neurons having facilitatory and excitatory effects on the eventual motor outflow in the pyramidal tract. Anesthesia, which might be expected to have an effect on the excitation, causes a decrease in the size of the area from which a motor response of a given muscle is obtained by monopolar anodal excitation. Studies indicate that the organization of the motor cortex consists in widely overlapping areas of muscle "representations" instead of pointlike areas or discrete regions controlling particular muscle groups. The shrinkage of areas of excitability of a given muscle group with increased anesthesia is probably due to the loss of facilitation from surrounding neurons. When local anesthesia is used instead of general anesthesia, or chronically implanted electrodes are employed for stimulation without any anesthesia, motor responses can be elicited over large areas of the cortex.

A technique used to restrict excitation to a small group of pyramidal cells is the insertion of semi-microelectrodes into the cortex to effect a stimulation at various depths in the cortex. The effective current is a cathodal pulse which probably depolarizes cell bodies, or rather the initial segments of the axons of the pyramidal cells. With this technique, excitation was seen at much lower thresholds than with surface monopolar anodal shocks, and small muscle fields were excited from groups of neurons aligned in columnar fashion (Fig. 8-3). In other words, the microelectrode tip, as it was moved down through the depth of the cortex, excited cells having the same motor control. The threshold was lowest at the fifth layer, where the larger pyramidal cells contributing to the corticospinal outflow are found.

Fairly large groups of cells are involved by stimulation of even the small region around the electrode tip. This allows a coordination of excitation to occur via the interconnections between the cells and to spread to a great many neurons.

The various inputs acting on single pyramidal tract (PT) cells have been shown by recording intracellularly from neurons identified as PT cells by antidromic excitation. In this technique the axons in the pyramidal tract are stimulated while recording is taking place from the cell in the cortex. The method is similar to that used to identify the motoneurons of the cord by an antidromic volley (Chap. 5). While recording is being done from an identified PT cell, a stimulation of a primary sensory input to the cortex gives rise to either a discharge or a change in the ongoing discharge rate of that PT cell.

Study of the properties of PT cells by means of microelectrode recording has revealed that the cortex has an inhibitory system similar to that of the Renshaw cell system in the spinal cord. The PT neuron has collaterals synapsing onto cells, giving rise to long-lasting hyperpolarizations with decreased excitability when activated.

The motor cortex is the site of voluntary control of motor behavior. The mechanism of such control is at present unknown, but recent microelectrode studies have shown some of the early events occurring during a voluntary movement. Microelectrodes were chronically implanted into the motor cortex of monkeys that were trained to make a

---

view on the medial aspect of the brain. For both the primary and secondary motor areas the thumb area is relatively very large, as are the areas for the digits of hand and foot. The face is quite large, with a large tongue area. (From Woolsey et al., 1952.)

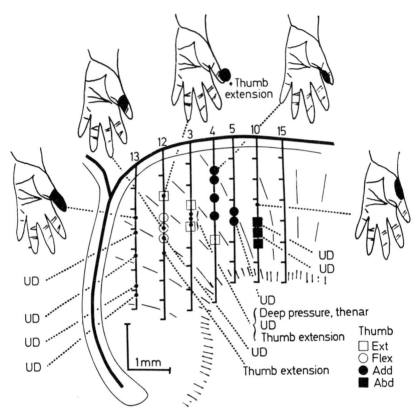

Figure 8-3
Reconstruction of electrode tracks and cell locations. Several electrode penetrations (solid lines, identified by numbers) passed through efferent zones projecting to various thumb muscles. UD represents cells undriven by peripheral stimulation. The peripheral motor effects are indicated by symbols explained in the figure. Cortical spots stimulated without evoking motor effects are shown by small solid lines perpendicular to the lines indicating tracks. Positions of cells encountered are indicated by dots and connected with dotted lines to descriptions of receptive fields and adequate stimuli. (From Rosén and Asanuma, 1972.)

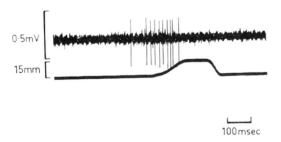

Figure 8-4
Discharge of a precentral neuron in association with one phase of a repetitive movement task. A monkey grasped a horizontal lever and pulled it forward through 15 mm with his right hand. The movement of the lever is indicated in the second trace. The top trace shows the burst of nerve

movement. A discharge of neurons was found to precede the motor response (Fig. 8-4). This presumably represents some part of the integration of command which results in the discharge of the PT cells controlling the peripheral musculature.

Interconnections between associational areas and motor areas permit other parts of the brain to affect motor control. The sensory receptor areas within the cortex have close connections to the motor areas, and electrical stimulation of parts of the cortex outside the "true" motor area may be compounded of sensory and associational cortex excitation which then acts on the motor neurons. A correspondence of the peripheral representation present in the somesthetic sensory cortical areas 1, 2, and 3 with the motor control regions of area 4 can be seen in Figure 8-1. In some species there is more overlap of sensory and musculature representations, and one considers this a *sensorimotor cortex* rather than separate motor and sensory regions. The term *motorsensory* region has also been used to refer to this concept of integration.

RETICULAR FORMATION

The upward spread of sensory activation exerted by the reticular formation on the cerebral cortex was referred to in Chapter 6. Historically, the motor control from the reticular formation acting on lower motor centers was discovered first. Stimulation within certain areas of the reticular formation results in a *facilitatory* effect on lower motor centers, whereas stimulation in other reticular formation areas gives rise to *inhibitory* effects, as is shown in Figure 8-5. While a reflex, the knee jerk, was being elicited continuously at regular intervals, stimulation through electrodes stereotaxically placed in the reticular formation of the lower medial part of the medulla gave rise to a rapid inhibition. On cessation of stimulation, the reflexes recovered within a few seconds. A similar effect of reticular formation stimulation on cortically evoked limb movements was also found from this region. If stimulation was administered to the upper lateral portion of the reticular formation, a facilitation of knee jerk reflexes was found (Fig. 8-6). A similar effect was seen with cortically evoked limb movements. By this means, the presence of *facilitatory and inhibitory regions* within the reticular formation could be plotted (Figs. 8-5 and 8-6). Loss of the normal inhibitory control descending from the cortex and other subcortical centers on the facilitatory regions of the reticular formation centers could result in a *release* of excitatory influences descending to and acting on the centers in the brain stem and spinal cord.

A loss of discharge from the inhibitory reticular formation could also help produce the exaggerated reflex activity associated with spasticity. An important control of the spinal cord gamma motoneurons has been located in the reticular formation, and an exaggerated gamma discharge has been associated with the spastic-like condition of decerebrate rigidity (Chap. 5). As was noted, making a cut between the colliculi of the brain stem produces a decerebrate rigidity preparation. The legs are stiffly extended and the head arched back in a caricature of standing. An explanation of this phenomenon is that the reticular formation inhibition is reduced and an increased excitation of vestibulospinal downflow is involved along with other excitatory downflow from the reticular formation. Not all of this is due to an augmentation of gamma motor activity. Stimulation of the reticular formation can still cause a facilitation or an inhibition of the monosynaptic response. When a monosynaptic reflex is elicited by stimulation of the central end of cut dorsal roots, the gamma motoneurons are excluded from the response. Thus, there is some control over the motoneurons which does not primarily

---

impulses produced by a single neuron in the left precentral gyrus discharging in relation to one performance of this movement task. The cell in the motor cortex began firing about 100 msec before the beginning of the movement. The bursts always began before the movement and always preceded the beginning of electromyographic activity in flexor muscles. Brackets indicate 0.5 mv for cell spikes and 15 mm for movement trace. (From Porter, 1973.)

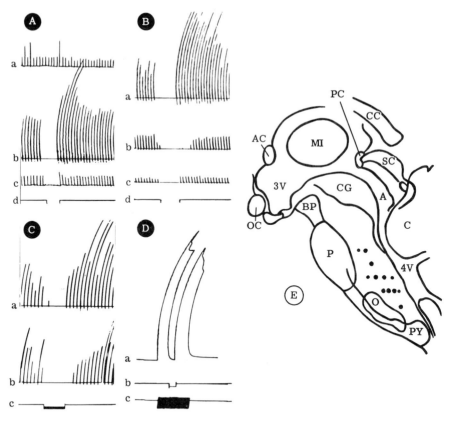

Figure 8-5

Inhibitory regions in the reticular formation. To the right, in E, is a sagittal view of the brain stem, with dots in the lower portion indicating regions effective for causing inhibitory effects. These effects are shown in A and B as an inhibition of the knee jerks at the time of reticular formation stimulation (see lower segment lines). In C and D, stimulation to the inhibitory regions is seen to be effective for inhibiting cortically evoked movements. (From H. W. Magoun and R. Rhines. *J. Neurophysiol.* 9:166, 169, 1946.)

involve the gamma system. The proportion of such motoneuron control varies in different species. In the cat and dog most of the reticulospinal influence on the cord is through interneurons, which eventually act on the motoneurons. In the primates and in man there is a greater synaptic influence from the brain directly on the motoneurons. The synaptic terminations are found on the soma and on the dendrites of the moto-neurons.

The earlier view that the functional control exerted from the reticular formation is "global" in its effect on lower motor centers had to be changed with further study. Mixed facilitatory and inhibitory effects have been found. Under certain experimental conditions—for example by the use of anesthesia—a simplification of the complex control of reflex responses from the reticular formation takes place. This was shown by stimulation of electrodes chronically implanted in the brain stem reticular formation. Without anesthesia a coordinated movement of the limbs occurs rather than generalized facilitation from the "facilitatory," or inhibition from the "inhibitory," areas. Part of the descending motor control exerted by the reticular formation region appears to derive from the *monoamine neurons* discussed in Chapter 6.

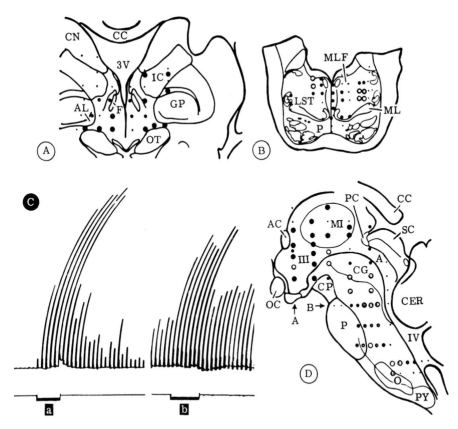

Figure 8-6
Facilitatory regions of the reticular formation. D is a sagittal view of the brain stem. A and B are cross sections of the reticular formation, with dots representing the areas from which facilitatory effects are obtained. In general, these facilitatory regions are higher in the brain stem and more lateral than is the inhibitory region. In C, cortically evoked responses are facilitated during the period of stimulation of the facilitatory region of the reticular formation (signal a on lower line). At the right, a knee jerk reflex is similarly facilitated during the period of stimulation of the reticular formation (signal b on lower line). (From R. Rhines and H. W. Magoun. *J. Neurophysiol.* 9:222, 220, 221, 1946.)

CEREBELLUM

The cerebellum, situated behind the cerebrum, is involved in those mechanisms relating to motor coordination and body equilibrium. An animal with its cerebellum ablated shows an inability to adjust the position of its limbs with respect to a goal. It overshoots or undershoots its goal when, for example, reaching for food: a condition called *dysmetria*. Movements, instead of occurring smoothly, are disjointed or jerky, showing *asynergia*. A cerebellar tremor is also seen during a voluntary movement. The tremor is coarse and becomes greater as the goal is being approached: an *intention* tremor. Decerebellated animals appear weaker (*asthenic*); the weakness of the muscle is described as a diminution of tone, or *hypotonia*.

Removal of different parts of the cerebellum has different effects upon motor behavior. The newer posterior and lateral portions (the *neocerebellum*) are joined with the cortex by extensive afferent and efferent fiber connections. This part of the cere-

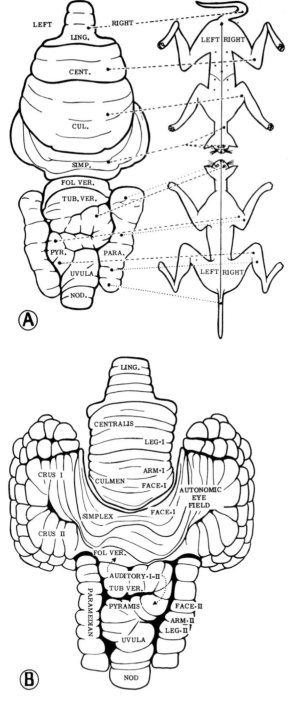

Figure 8-7
Somatotopic localization in the cerebellum. In A the localizations of motor functions found by stimulating areas over the cerebellar surface are shown. In B, the localization within the cerebellum of various sensory inputs using evoked response techniques is shown. (From J. L. Hampson et al. *Association for Research in Nervous and Mental Disease, Proceedings* 30:304, 1952, and from J. L. Hampson, *J. Neurophysiol.* 12:47, 1949.)

bellum is associated with cortically controlled movements. The phylogenetically older portion of the cerebellum (the *paleocerebellum*) is located anteriorly and medially in the cerebellum and is concerned with tonic postural mechanisms as well as with locomotion, which is more automatic in nature. The flocculonodular portions of the cerebellum (part of the *archicerebellum*) are found laterally and are concerned with body equilibrium. Nerve impulses from the vestibular sensory receptor organs (semicircular canals) give information as to the position of the head with respect to spatial orientation (Chap. 4). The afferent fibers of these receptors terminate in the flocculonodular portion of the cerebellum. Ablation of either the vestibular organs or the flocculonodular lobes causes impairment of orientation and loss of equilibrium. A human so damaged falls if asked to close his eyes, because he thereby loses the remaining visual cues for equilibrium. Sense modalities such as vision and sound are represented in topographical fashion, as are sensory inputs from the skin and from muscle receptors (Fig. 8-7).

Analysis of the function of the cerebellum shows that it acts as a feedback control mechanism which smoothes and integrates motor activity initiated in the cerebrum. How the integration of various sensory inputs within the cerebellum comes about is suggested by the recent studies of Eccles, Ito, Llinas, and their colleagues.

The cells of the cerebellar cortex are arranged in three layers. In the outermost or molecular layer are the large cell bodies of the *Purkinje cells*. These have large ramified dendrites which spread fanwise at right angles to the folia of the cerebellum within the molecular layer. Between the dendritic branches are the parallel fibers, which course through the dendrites in the direction of the folia and make synaptic contact with them (Fig. 8-8). The parallel fibers originate from granular cells found in the middle layer of

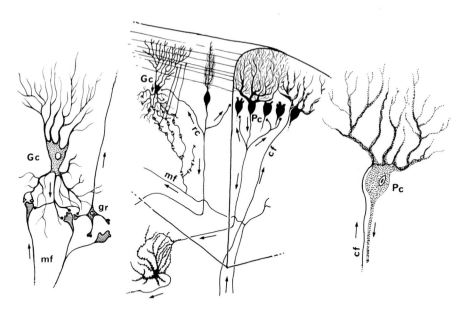

Figure 8-8
Details of cerebellar neurons. Left—afferent mossy fibers (mf) terminate on granule cells (gr) of the granular layer and on Golgi cells (Gc). Right—afferent climbing fibers (cf) terminate on dendrites of Purkinje cells (Pc). Middle—Purkinje cells have an extensive dendritic arborization across the folium in the molecular layer, where parallel fibers thread through and synapse on them. Efferent outflow is from the Purkinje axon. (After C. A. Fox. In E. C. Crosby, T. H. Humphrey, and E. W. Lauer [Eds.], *Correlative Anatomy of the Nervous System*. New York: Macmillan, 1962; and from J. C. Eccles, *Perspect. Biol. Med.* 8:289, 1965. Copyright © 1965, The University of Chicago Press.)

the cerebellum. The deepest layer is the white matter, representing fibers entering and leaving the cerebellar cortex (Fig. 8-9).

There are two sensory inputs to the cerebellar cortex. One, originating from the inferior olivary nucleus of the brain stem, ascends into the molecular layer to intertwine on and synapse on the dendrites of the Purkinje cells as *climbing fiber* afferents. The other afferent input is made up of the *mossy fibers,* representing sensory inputs from the spinal cord, brain stem, and upper motor areas of the brain. These end in bulbous *glomeruli*—synapses at which granule cells are excited at the same time an inhibitory control is exerted by the Golgi cells (Fig. 8-9). The axons of the granule cells ascend to the molecular layer where their fibers divide as the parallel fibers, synapsing on the apical dendrites of the Purkinje cells through which they pass. In addition to these excitatory inputs to the Purkinje cells, there are inhibitory interneurons, the *stellate* and *basket* cells, synapsing on the dendrites and cell body respectively of the Purkinje cell to contribute an inhibitory influence on its outflow (Fig. 8-9). The Purkinje cell

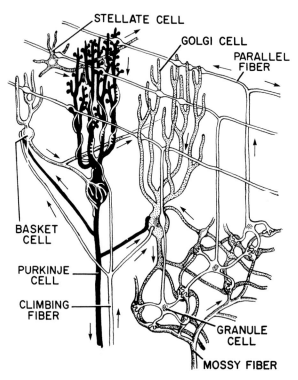

Figure 8-9
Interconnection of neurons in the cortex follows an elaborate but stereotyped pattern. Each Purkinje cell is associated with a single climbing fiber and forms many synaptic junctions with it. The climbing fiber also branches to the basket cells and Golgi cells. Mossy fibers come in contact with the terminal "claws" of granule cell dendrites in a structure called a cerebellar glomerulus. The axons of the granule cells ascend to the molecular layer, where they bifurcate to form parallel fibers. Each parallel fiber comes in contact with many Purkinje cells, but usually it forms only one synapse with each cell. The stellate cells connect the parallel fibers with the dendrites of the Purkinje cell, the basket cells mainly with the Purkinje cell soma. Most Golgi cell dendrites form junctions with the parallel fibers but some join the mossy fibers; Golgi cell axons terminate at the cerebellar glomeruli. Arrows indicate direction of nerve conduction. (From Llinas, 1975.)

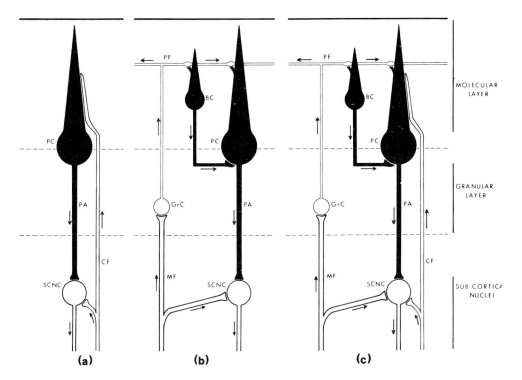

Figure 8-10
Diagram of neuronal pathways in the cerebellum. PC, Purkinje cell; BC, basket cell; PA, Purkinje cell axon; GrC, granule cell; MF, mossy fiber; CF, climbing fiber; SCNC, subcortical nuclear cell. The directions of impulse flow are shown by arrows. (From Eccles, 1969.)

represents the "final common path" from the cerebellum, and its axons are inhibitory to the deeper relay nuclei of the cerebellum (fastigial, interpositus, and dentate). The neurons of these nuclei in turn synapse on other brain stem groups (vestibular, reticular formation, and the red nucleus), which form tracts descending to the lower motor spinal reflex centers (Fig. 8-10). Another major outflow from the dentate nucleus of the cerebellum goes to the ventrolateral nucleus of the thalamus and, after a relay there, to the sensorimotor cortex of the cerebrum to regulate its motor control functions and the pattern of outflow of PT cells.

The interconnections of the neurons of the cerebellum with their fiber outputs passing to the cortex and spinal cord through the deep cerebellar nuclei are part of an inter-connected loop of neurons (Fig. 8-11). Exactly how this network works as a whole to smooth ongoing movement is not yet understood, but the function of such loop connectivities is likely to yield to a systems approach.

BASAL GANGLIA

An important but little understood group of nuclei of the brain stem are the older motor centers, the *basal ganglia*. These include the caudate nucleus, putamen, globus pallidus (the striate body), subthalamic nucleus of Luys, substantia nigra, and red nucleus. Their stimulation might produce a series of complex stereotyped movements—for example, a turning of the body or a sudden arrest of ongoing movement. Inhibitory effects from the

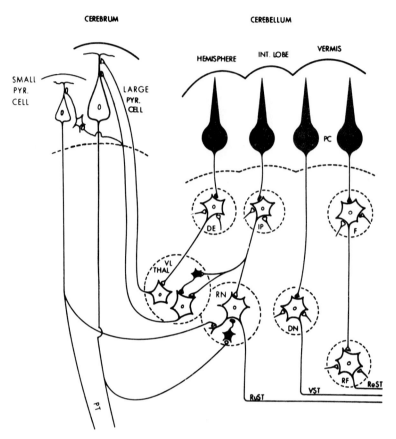

Figure 8-11
Diagram showing the various efferent pathways of Purkinje cells (PC) in the vermis, intermediate lobe, and hemisphere of the cerebellum. All inhibitory cells are shown in black. F, fastigial nucleus; IP, interpositus nucleus; DE, dentate nucleus; RF, reticular formation; DN, Deiters nucleus; RN, red nucleus; VL THAL, ventrolateral nucleus of thalamus; PYR, pyramidal cell; ReST, reticulo-spinal tract; VST, vestibulospinal tract; RuST, rubrospinal tract. The pathways are shown from small and large pyramidal cells to the VL thalamus and red nucleus. (From Eccles, 1969.)

basal ganglia on lower reflexes have been obtained resembling the effect seen on stimulating the inhibitory regions of the reticular formation. *Parkinson's disease* appears to be a disorder of basal ganglia function. It is characterized by an alternating tremor of extensor and flexor muscle groups. Destruction of the efferent outflow of the globus pallidus by electrocoagulation or by focused high-frequency sound waves has in some cases resulted in dramatic relief. Dopaminergic fibers from the globus pallidus termi-nating in the striatum are also important. A deficiency of dopamine was found in the striatum of patients with parkinsonism. L-Dopa is synthesized into the transmitter dopamine in the dopamine neurons (Chap. 6), and on that basis L-dopa was given in large amounts to such patients and benefited a substantial number of them. We can expect that further study will lead to a better understanding of the mechanisms involved in this disease and the means to control the pathological alteration.

HIGHER CENTERS OF VISCERAL CONTROL

In later chapters the physiology of various visceral systems—cardiovascular, respiratory, genitourinary, digestive, and others—is thoroughly dealt with. It will be of value, however, in this section to note briefly the mechanisms of upper visceral control located in the central nervous system and their relation to visceral systems and to the endocrine system of general homeostatic regulation.

The peripheral part of the autonomic nervous system, with its cell bodies located in the intermediary portions of the gray matter of the spinal cord, is discussed in Chapter 7A. This system, like the somatic nervous system, has afferent fibers entering the spinal cord which synapse eventually upon cell bodies within the intermediary parts of the gray matter of the spinal cord, which in turn give rise to efferent fibers emerging from the ventral roots. The autonomic motor nerves are distributed in a more or less diffuse fashion to the heart, various smooth muscles, and glandular structures throughout the body. Like the somatic nervous system, the local cord reflex centers controlling the autonomic visceral functions are in turn controlled by higher brain centers. In spinal shock, transection of the spinal cord gives rise not only to those profound somatic changes characteristic of this state but also to vascular, thermoregulatory, urinary bladder, and other urogenital changes that parallel those described for the somatic nervous system.

Another relationship of higher central nervous system structure to visceral functions is found in the brain regions connected to and regulating the pituitary gland (hypophysis). The relation of the hypothalamus to the hypophysis was suggested by Frölich's study of a syndrome in adolescent boys in which obesity, dwarfing, and sexual infantilism are met (Chap. 30). The syndrome was finally traced to certain portions of the hypothalamus. It is of interest here to describe the control this area of the brain has upon visceral functions—those connected with food and water intake, body fluid and ion regulation, cardiovascular and respiratory control, gastrointestinal regulation, genitourinary function, and sexual and maternal behavior. These visceral functions all relate to the internal economy of the organism or to broader events in the organism's life, such as sex and the struggle for food, and they all have an attending emotional involvement. Emotion is a predominant factor in human behavior, and by inference in the behavior of animals as well. Although closely connected with visceral control, emotional reactions related to the hypothalamus will be discussed separately (Chap. 9).

The hypothalamus may, for present purposes, be considered a controlling station for most of the visceral functions that have been mentioned, serving as a "high level" reflex control system. Other regions of the brain (limbic system) have, as will be noted in the next chapter, important interconnections with the hypothalamus and with the cortex. The limbic system and the neocortex can also therefore exert, at a still higher level, a regulation of visceral functions.

ROLE OF THE HYPOTHALAMUS IN VISCERAL CONTROL

In the preceding chapter, attention was directed to the importance of the hypothalamus as a higher regulating center for the autonomic nervous system. The purpose in this section is to develop this aspect of neural control. The areas to be discussed are (1) cardiovascular control, (2) respiration, (3) temperature regulation, (4) water and electrolyte balance, (5) control of food intake, and (6) sexual and maternal function.

Cardiovascular Control

An animal in a state of spinal shock shows various degrees of loss of vasomotor activity in those parts of the nervous system below the transected level. The peripheral vessels are dilated, blood pressure is reduced, and response to reflex vasomotor activity is lessened. Vasodilation follows from the loss of normal reflex vasomotor tone acting upon the vessels. If the transection of the brain is made forward of the hypothalamus, control of the nervous system is relatively normal insofar as vasomotor activity is concerned, and blood pressure is not affected.

Anesthesia, if very deep, causes a similar state because of its central action with a loss of reflex adjustments. The anesthetic agent chloroform sensitizes the heart to adrenergic action, leading to irregularities of rhythm, extrasystoles, and fibrillation. Cardiovascular control originates from the hypothalamus with a pathway traced from the hypothalamus and descending in the brain stem via sympathetic nerves which exit from the spinal cord at the thoracic levels to innervate the heart. The release of adrenergic transmitter leads to fibrillation if agents such as chloroform are present.

Parts of the brain higher than the hypothalamus also have been shown to have vascular effects. Stimulation of the cingulate cortex and neocortex can give rise to localized vasodilations and constrictions of skin, muscle, and body organs.

Respiration

As indicated in Chapter 21, there is much evidence that a fundamental pacemaker activity underlies the rhythm of respiration. This rhythm is also controlled by higher nervous centers, so that voluntary control is possible. Higher centers in the brain, afferent discharges from the lungs and body, and also the oxygen, carbon dioxide content, and pH of the blood all affect the rhythm of respiration. But the nature of the rhythmic changes in cells leading to periodic discharge is at present little understood. One would like to know what series of molecular events causes the rhythmic changes in those cells. A similar problem relates to the pacemaker cells of the heart (Chap. 14) and EEG activity. The generator potential of sensory nerve endings related to rhythmic discharge (Chap. 3A) is an analogous system; study of it may eventually lead to an understanding of pacemaker activity.

Evidence exists (Chap. 21) that the hypothalamus has an excitatory effect on the inspiratory neurons in the medulla. Higher control by the cerebral cortex is indicated by the modifications of respiration that occur with speech, chewing, swallowing, and smelling. Stimulation of the cingulate cortex and other parts of the limbic system has also been found to affect respiration.

Temperature Regulation

The relation of the hypothalamus to temperature regulation is shown when, upon local heating of small regions within the hypothalamus, peripheral changes are found in the body acting in the direction of removal or loss of body heat from the animal (Chap. 28). Conversely, cooling of the blood passing to the hypothalamus will cause changes in the direction of conserving body heat or increasing heat production by shivering. Shivering results in an increase of body warmth, and electrical stimulation of the posterior hypothalamus is effective in producing shivering. Electrodes placed in the hypothalamus show potential changes indicative of neural activity during heating or cooling of the blood or of the cells in this region. If the outflow from the hypothalamus is cut or the hypothalamus destroyed, heat-regulatory mechanisms are interfered with. In such cases an animal may be changed from a *homeothermic* to a *poikilothermic animal*. In a cold environment the animal will show a reduction in body temperature, with a fall toward that of the external temperature. Conversely, if the animal is placed in a warmer than usual environment, body temperature will rise. In other words, the homeostatic reflex adjustments mediated by the hypothalamus in the homeotherm are interfered with or lost.

Sweating is part of the hypothalamic regulation of body temperature in some species, including man, and operates in conjunction with the aforementioned control of the peripheral vascular bed, which in respect to temperature control provides appropriate vasoconstriction to aid the body's heat retention or vasodilation to assist its heat loss.

Water and Electrolyte Balance

Receptors sensitive to changes in osmotic pressure of the blood, *osmoreceptors,* have been localized within the anterobasal part of the hypothalamus. Injection of hypertonic fluid into the arterial supply of the hypothalamus causes the release of antidiuretic hormone (ADH) from the posterior pituitary (neurohypophysis) after reflex excitation of the osmoreceptors of the hypothalamus (Chap. 30). An increased discharge of neural units within the supraoptic nuclei of the hypothalamus is also found following injection of small amounts of hypertonic fluids into the carotid arteries. Electrical stimulation within the hypothalamus of goats has been shown to lead to excessive thirst and excessive intake of water to the point where death can ensue. From cells in the anterobasal region of the hypothalamus (supraoptic and paraventricular nuclei), tracts pass down into the neurohypophysis. It is believed that ADH is formed in the cells of these hypothalamic nuclei. The ADH moves down inside the fibers of the tracts by axoplasmic transport (Chap. 2), to be stored and later released as required from the pituicytes of the neurohypophysis. This hormone is required for the normal return of water from the kidney tubules to the blood (Chap. 22). In the absence of ADH, polyuria results.

Control of Food Intake

Within the hypothalamus there are areas which on stimulation cause an animal to seek food, and other regions which stop the satiated animal from continuing to eat. Lesions made in the medial part of the hypothalamus will make an animal eat excessively; i.e., it shows *hyperphagia.* Such an animal may more than double its normal weight. On the other hand, lesions in the lateral part of the hypothalamus will cause a reduction of food intake, *hypophagia.* If the lesions are complete, *aphagia* may result. Electrical stimulation of the medial nuclei makes an animal eat less, and stimulation of the lateral hypothalamic nuclei causes increased or continued eating. The medial nuclei are therefore regarded as a *satiety center,* the lateral nuclei as a *hunger center.* These two centers within the hypothalamus receive appropriate sensory inputs when hunger is present or after sufficient food is taken, and together they regulate the intake of food with relation to body needs. Receptors responding to glucose levels seem to be one important factor in the stopping of eating by exciting the satiety center. An increase in the level of food intake over a long enough time causes *obesity.* The deposition of fat in excess amounts in obesity is related to later degenerative changes in the cardiovascular and other systems and is associated with a shorter life span. Obesity may also be brought about through influences from the higher centers (limbic, brain, and cortex) acting upon the food intake mechanisms of the hypothalamus. Some pharmacological agents, such as Dexedrine, appear to act on the hypothalamus to decrease the sensitivity of the food intake center. This adrenergic drug apparently depresses the sensitivity of the food intake center and decreases the desire to eat. Higher centers may also be involved. In the cephalic phase of digestion (Chap. 26), the sight and smell of food stimulate the intake of food, and conditioned reflexes associated with food and eating are of great importance with regard to the level of food intake or the desire to eat.

Regulation of Sexual and Maternal Behavior

The implantation of minute amounts of estrogen within the hypothalamus has shown that this region controls behavior patterns concerned with mating. Other limbic areas controlling overt behavior of this function will be discussed in Chapter 9.

The secretion of gonadotropic hormones by the adenohypophysis in both the male and the female is under the control of the hypothalamus. Lesions placed in the anterior hypothalamus prevent the release of luteinizing hormone, and a condition of constant estrus results. Lesions placed in the posterior tuberal area cause gonadal atrophy. When the adenohypophysis is removed from its normal site in the sella turcica and transplanted to the kidney capsule or anterior chamber of the eye, it is unable to secrete gonadotropins in amounts adequate to maintain gonadal activity. The control of hormonal release which the hypothalamus exerts over the adenohypophysis occurs through passage of hormones (*releasing factors*) in the rich vasculature found around the adenohypophysis, the pituitary portal circulation (Chap. 30).

REFERENCES

Anand, B. K.   Nervous regulation of food intake. *Physiol. Rev.* 41:677–708, 1961.

Asanuma, H., and I. Rosén.   Topographical organization of cortical efferent zones projecting to distal forelimb muscles in the monkey. *Exp. Brain Res.* 14:243–256, 1972.

Brooks, C. McC., J. L. Gilbert, H. A. Levey, and D. R. Curtis.   Humors, Hormones, and Neurosecretion. New York: State University of New York, 1962.

Brooks, V. B.   Information Processing in the Motorsensory Cortex. In K. N. Leibovic (Ed.), *Information Processing in the Central Nervous System.* New York: Springer-Verlag, 1969.

Denny-Brown, D.   *The Cerebral Control of Movement.* Springfield, Ill: Thomas, 1966.

Eccles, J. C.   The Dynamic Loop Hypothesis of Movement Control. In K. N. Leibovic (Ed.), *Information Processing in the Nervous System.* New York: Springer-Verlag, 1969.

Eccles, J. C.   The cerebellum as a computer: Patterns in space and time. *J. Physiol.* (Lond.) 228:1–32, 1973.

Eccles, J. C.   *The Understanding of the Brain.* New York: McGraw-Hill, 1974.

Eccles, J. C., M. Ito, and J. Zentagothai.   *The Cerebellum as a Neuronal Machine.* New York: Springer, 1967.

Evarts, E. V., and Thach, W. T.   Motor mechanisms of the CNS, cerebro-cerebellar inter-relations. *Annu. Rev. Physiol.* 31:451–498, 1969.

Fulton, J. F.   *Physiology of the Nervous System* (3rd ed.). New York: Oxford University Press, 1949.

Granit, R.   *The Basis of Motor Control.* New York: Academic, 1970.

Hardy, J. D.   Physiology of temperature regulation. *Physiol. Rev.* 41:521–606, 1961.

Harris, G. W.   *Neural Control of the Pituitary Gland.* London: Arnold, 1955.

Hoff, E. C., J. F. Kell, Jr., and M. N. Carroll, Jr.   Effects of cortical stimulation and lesions on cardiovascular function. *Physiol. Rev.* 43:68–114, 1963.

Jung, R., and R. Hassler.   The Extrapyramidal Motor System. In J. Field (Ed.), *Handbook of Physiology.* Section 1: Neurophysiology, Washington: American Physiological Society, 1960. Vol. 2, pp. 863–927.

Lassek, A. M.   *The Pyramidal Tract.* Springfield, Ill.: Thomas, 1954.

Liddell, E. G. T., and E. G. Phillips.   Overlapping areas in the motor cortex of the baboon. *J. Physiol.* (Lond.) 122:392–399, 1951.

Llinas, R. R. (Ed.).   *Neurobiology of Cerebellar Evolution and Development.* Chicago: American Medical Association, 1969.

Llinas, R. R.   The cortex of the cerebellum. *Sci. Am.* January, 1975. P. 58.

Magoun, H. W., and R. Rhines.   *Spasticity: The Stretch-Reflex and Extrapyramidal Systems.* Springfield, Ill.: Thomas, 1947.

Marks, J.   *The Treatment of Parkinsonism with L-DOPA.* New York: American Elsevier, 1974.

Penfield, W.   *The Excitable Cortex in Conscious Man.* Springfield, Ill.: Thomas, 1958.

Porter, R.   Functions of the mammalian cerebral cortex in movement. *Prog. Neurobiol.* (Pt. 1) 1:1–51, 1973.

Preston, J. B., and D. G. Whitlock.   Intracellular potentials recorded from motoneurons following precentral gyrus stimulation in primate. *J. Neurophysiol.* 24:91–100, 1961.

Rosén, J., and Asanuma, H.   Peripheral afferent inputs to the forelimb area of the monkey motor cortex: Input-output relations. *Exp. Brain Res.* 14:257–273, 1972.

Snider, R. S., and A. Stowell.   Receiving areas of the tactile, auditory and visual systems in the cerebellum. *J. Neurophysiol.* 7:331–357, 1944.

Spiegel, E. A., and H. T. Wycis. *Stereoencephalotomy.* Vol. II, *Clinical and Physiological Applications.* New York: Grune & Stratton, 1962.

Wiesendanger, M. The pyramidal tract. *Ergeb. Physiol.* 61:72–136, 1969.

Woolsey, C. N., P. G. Settlage, D. R. Meyer, W. Sencer, T. P. Hamuy, and A. M. Travis. Patterns of localization in precentral and "supplementary" motor areas and their relation to the concept of a premotor area. *Association for Research in Nervous and Mental Disease, Proceedings* (1950) 30:238–264, 1952.

9. Higher Nervous Functions

Sidney Ochs

Higher nervous functions involve those complex behaviors resulting from interactions with the environment whereby an animal learns to cope with the many kinds of stimuli or signals that determine success or failure in attaining goals. The field of study of the higher nervous functions is a difficult one. Often, inferences from subjective experience must be used in an attempt to augment limited objective observations made in the laboratory. The subject will be divided into *emotional* and *intellectual* behavior, with full recognition of the difficulty of such a separation.

EMOTION AND THE LIMBIC SYSTEM

Emotion has two aspects: internal awareness and external display. The release of an emotional display appears at times to be almost reflex-like. For example, we may have a feeling of sorrow or depression, but a stronger sorrow may lead to uncontrollable weeping and sobbing. We may be amused or give way to sudden laughter. We may feel pleasure or enter a state of ecstasy. Displeasure and anger in the extreme can turn to rage. These human experiences have some parallels in experimental investigation of the emotions in animals. We ascribe emotional states to those displays in the animal corresponding to the behavior seen in man, usually the sudden reflex-like changes.

The similarity of brain structures in the human and the animal suggests generalizations for the localization of higher nervous function in brain regions with similar form. For the most part the hypothalamus and the *limbic system* are involved. The limbic system is a group of nuclear regions with interconnections found around the brain stem on the medial and ventral aspect of the brain stem (see below). Although the hypothalamus is sometimes included in the limbic system because of important connections made by the fornix carrying impulses from the limbic structures to the hypothalamus, it will be treated separately.

An extreme state of angry behavior known as *sham rage* can be produced in animals by amputation of the brain in front of the hypothalamus. Such an animal shows periodic displays of rage, with claws extended, jaws open, and muscles tense. Upon provocation, or even gentle restraint, the animal growls, thrashes about, lunges, and snaps. The rage display is not directed specifically to the source of provocation because the higher centers necessary to direct an attack are lost. Sympathetic discharge from the hypothalamus is shown by the increased heart rate and blood pressure, pupillary dilation, hair bristling, and salivation (Chap. 7A). An augmented sympathetic discharge is not seen if the hypothalamus has been destroyed. If, in the monkey, a cut is made in the base of the brain just anterior to the hypothalamus, a rage state is produced, but in this case the animal can focus its attacks. The extreme ferocity of this preparation has been appreciated. Some components of a rage reaction can be elicited from yet lower brain stem structures. A chronic decerebrate preparation shows some aspects of ragelike behavior referred to as *pseudoaffective*. However, while some of the components of rage display are evoked from lower brain structures, the more organized response requires the hypothalamus.

Localized electrical stimulation of the hypothalamus by means of chronically implanted electrodes can clearly bring about a display of ragelike behavior, with signs of sympathetic discharge also present. Some experiments show that sham rage elicited by such electrical stimulation of the hypothalamus does not have a subjective component; an animal may purr and even respond to petting between displays and during the stimulation. Stimulation directly in the hypothalamus and conditioning (see following section) would suggest a subjective aspect to the electrically elicited rage. While all of the display may not be conditioned, animals can be conditioned to avoid hypothalamus stimulation by techniques to be described in a later section. The hypothalamus appears to act as a control center for the outward expression of rage responses without necessarily being the place where emotions as such are experienced.

Papez related limbic brain structures (Fig. 9-1) to emotions by theorizing that afferent activity passes from the thalamus to the hypothalamus, then via mamillary bodies to the

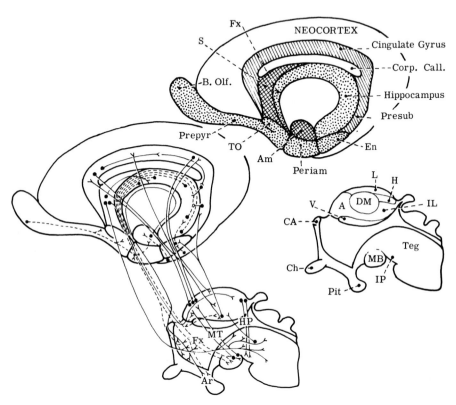

Figure 9-1
Limbic system shown at the right, with upper diagram representing the medial aspect of the brain, lower diagram the medial aspect of the brain stem. The corresponding pair of diagrams at the left show their regional interconnections. Three types of region are distinguished: (1) the *paleocortex*, represented by stippling, including as its chief structure the hippocampus and the olfactory structures, olfactory bulb (B. Olf.), and prepyriformis (prepyr); (2) the *juxtallocortex*, shown by slanted lines, which includes the cingulate cortex and presubiculum (presub); and (3) the *brain stem* structures, cross-hatched in the upper diagram—the septum (S), and amygdaloid complex (Am). These regions are related to one another and to brain stem regions as shown at the left. Important efferent connections are made from the hippocampus to the mamillary bodies in the hypothalamus via the fornix (Fx) and from the mamillary bodies to the anterior thalamic nucleus via the mamillothalamic tract (MT). Connections from the septum to lower brain stem centers are carried by the medial forebrain bundle, which is not labeled. (From J. V. Brady, "The Paleocortex and Behavioral Motivation," in H. F. Harlow and C. N. Woolsey, editors, *Biological and Biochemical Bases of Behavior* [Madison: The University of Wisconsin Press; © 1958 by the Regents of the University of Wisconsin], plate 23.)

anterior thalamic nucleus, and from there to the cingulate cortex. The cingulate cortex was supposed to be the site of the subjective appreciation of the emotional states not only of rage but of sexual and feeding behaviors. This theory has been significant in providing impetus to investigations of the limbic brain structures with regard to their suggested connections with emotions.

The ablation studies of temporal lobes of monkey brain by Klüver and Bucy further directed attention to the relation of limbic brain structures to behavior. In their work, a number of limbic structures were removed, including the amygdaloid nuclei. Monkeys with both temporal lobes ablated showed bizarre behavior patterns. They lost their normal fear of objects such as snakes and other similar genetically determined "fear" objects. They seemed to "explore" their environment orally, putting edible and inedible objects alike into their mouths, no matter how abnormal or repulsive these objects might be with respect to the usual behavior for the animal. They would not necessarily swallow them, spitting out inedible objects and eating food. They appeared to show lack of recent memory by returning soon after to an object previously spit out and again putting it into their mouths. The picking up of any and all objects for oral exploration might suggest a lack of visual discrimination. However, temporal-lobe-ablated animals can learn to discriminate visually between symbols, e.g., a square versus a circle, when these are presented simultaneously in a standard visual discrimination experiment. Apparently the character of the objects is not recognized, or they cannot inhibit oral exploration of them. Animals with lesions in the amygdaloid nuclei of the temporal lobes also showed a steady gain in weight. Their level of food intake seemed to have been raised by the lesion, or they became less active and therefore accumulated fat.

Temporal lobe lesions give rise to peculiar sexual behavior. The animals take other species as sex partners or even inanimate objects and in general show an increase in sexual activity. Other limbic system structures are related to sexual activity. The retrosplenial cortex in the rat is homologous to the cingulate cortex, a limbic brain structure. If it is destroyed, mother rats do not retrieve their pups from danger. Female rats so damaged show an abnormally increased responsiveness to the male. In the male rat, electrical stimulation of cingulate cortical regions can give rise to erection.

The two constellations of behavior (feeding and sex) have been referred to as "preservation of the self and preservation of the species." Obviously, these two biological goals are required for carrying on animal life. Aggressive or placid emotive states connected with hunger and satiation are of biological significance with regard to the organism's drive toward or inattention to those goals.

It has been discovered that there are regions in the brain which when electrically stimulated may produce a "pleasurable" experience in the animal or, more objectively, can be *positively rewarding* for the animal. Electrodes were chronically implanted in either of several subcortical structures of the rat: the hypothalamus and the septum, hippocampus, and other limbic brain regions. Connections from the electrodes were made to an electrical stimulating apparatus which could be controlled by a switch placed inside the animal's box. The animal, when pressing the switch, causes a brief electrical stimulation to be delivered to its own brain structures. Once the animal learns this, it will return to the switch and administer *self-stimulation* by repeatedly pressing the switch. Apparently the self-administered stimulation is so *positively reinforcing* (pleasurable) that the animal may incessantly press the switch for hours on end without stopping. Having learned to self-stimulate, it will endure painful electrical shocks applied through the floor grid in order to get to the lever and administer self-stimulation. It will prefer self-stimulation to other biological goals such as food or mating. Such self-stimulation has been demonstrated in a wide variety of animal species.

There is evidence of some special distribution within the limbic structures of positively rewarding sites. However, other regions, close to these same structures, presumably give rise to painful or unpleasant sensations when electrically excited, as shown by the *aversive* responses of the animal. When the electrodes are inserted into such *negatively reinforcing* sites, an animal will stimulate itself once and not press the switch again. Furthermore, it will exert considerable effort to prevent excitation of negatively rewarding regions.

The positively rewarding sites are related to the medial forebrain bundle found in the lateral hypothalamus. These fibers contain catecholamines and it is likely that the effect of self-stimulation is on these fibers to release transmitters from their terminals. Catecholamines have been implicated in emotional excitatory states, possibly producing pleasurable experiences for the animal.

LOCALIZATION OF LEARNING IN THE BRAIN

In the previous section the relation of parts of the brain to fundamental biological goals which are emotionally charged was emphasized. The alteration of innate behavior with respect to experiences in the environment (i.e., learning) will now be discussed. In their well-known studies, Pavlov and his colleagues investigated what is now known as *classical conditioning.* An animal is conditioned according to the Pavlovian technique by first receiving a sensory stimulus to which it will respond in an unlearned manner—the *unconditioned stimulus.* Food juices placed in the mouth of a hungry dog by means of an implanted cannula represent such a stimulus. This unconditioned food stimulus gives rise to an *unconditioned response,* in this case salivation. The measure of the response is the amount of saliva produced. If, at the time that the unconditioned food stimulus is being presented (or a short time beforehand) a bell is rung, the animal after a number of such pairings will salivate in response to only the ringing of the bell.

The sound of the bell is termed the *conditioned stimulus,* and salivation in response to it is a *conditioned response.* According to Pavlov, some new neural connections are made in the cerebral cortex after conditioning has occurred, so that the conditioned stimulus can take the place of the unconditioned stimulus with which it has been paired. The basis of Pavlov's belief was a report of an inability to condition decorticate dogs. Subsequent investigations have shown that simple types of classical conditioning can occur in the decorticate animal, and this finding directed attention to subcortical regions which may be involved in learning. Stimulation of the centrencephalic system localized in the nonspecific thalamic nuclei causes a petit mal type of widespread convulsive discharge synchronized in both hemispheres with a loss of consciousness (see Chap. 6). On this basis the centrencephalic system was proposed as the "highest" level in Jackson's sense—namely, where more complex functions, in this case higher functions, are controlled or organized.

That some "higher" level of learning is accomplished in subcortical sites is suggested by the use of direct stimulation of subcortical structures as an unconditioned stimulus, and peripheral stimulation or a sensory stimulus, such as a sound, as the conditioned stimulus to cause conditioning. A *deviational response* (a turning of the head) could be obtained from stimulation of limbic structures and was conditioned by sound. From the septum, sexual responses, as the unconditioned responses, could be conditioned to a tone. A more stereotyped behavior such as the *display of rage* obtained from stimulating some subcortical structures could not be conditioned.

In contrast to the simple type of classical conditioning, a greater involvement of the cortex in learning is shown by use of *instrumental* or *operant conditioning.* In instrumental conditioning, an animal makes some response which at first may be accidental or part of its normal repertoire of activity. The experimenter arranges conditions so that when the desired response is made, the animal is immediately rewarded with food or some other reward. It soon learns to discriminate between rewarded responses and other kinds of behavior and will, if sufficiently motivated, consistently perform those responses which have been so *reinforced.* In Skinner's type of experimental arrangement, an animal is placed in a small box with a small bar projecting from one wall. In moving about inside the box, the animal will occasionally strike the bar. If, when a hungry rat strikes the bar, a food pellet is delivered into a food hopper within the box, the animal learns to repetitively strike the bar for food. If reinforcement is removed, the animal gradually decreases its rate of responding until no more responses are made, a phe-nomenon known as *extinction*.

Various schedules of reinforcement have been described. An animal may be required to press the bar a fixed number of times to obtain reinforcement. Or, it may learn that reinforcement will take place only when the bar is pressed after a fixed interval of time. These and other schedules are of interest insofar as they have been shown to be differentially affected by drugs. Cortical ablation or spreading depression present in both cortices will prevent the learning or execution of an instrumental response.

While recent studies have shown the possibility of subcortical conditioning, one should not lose sight of the importance of the cortex for higher functions. It is possible to stimulate the cortex directly for use as a conditioned stimulus in order to elicit a conditioned response. The cortex appears to be critically involved in more complex types of learning tasks. Attention has long been directed to the frontal lobes, an associational region thought to be connected with learning. Animals with frontal lobes ablated show a defect of *delayed response* performance. In a delayed response, an animal is allowed to see a food bait placed under one of two specially marked covers. Then, after a delay, the animal is permitted to choose one or the other of the covers to obtain the bait. Successful performance is measured by the proper choice of the cover over the food bait. Delayed response behavior was greatly interfered with by frontal lobe lesions. Intense electrical stimulation or convulsive activity induced in the frontal cortex was also effective in interrupting delayed responding. Similar disruptive activity induced in other cortical areas was ineffective. Animals with their frontal lobes ablated do not show a loss of ability to discriminate one object from another if both objects are presented simultaneously. An interpretation of this result is that the frontal cortex is related to recent memory—or more likely that frontal-ablated animals are more distractible while they must wait before choosing. They may not have a sufficiently prolonged interest in the test object during the time it is out of sight. Once learning has been achieved, disruption of the activity of the frontal lobes is ineffective in interrupting delayed responding. This finding demonstrates a fundamental difference in the mech-anisms responsible for the *acquisition* of learned behaviors and for their *retention*. Retention (i.e., memory mechanisms) will be discussed in a later section.

EEG CORRELATES OF CONDITIONING

Changes in the electroencephalogram (EEG) have been correlated with a classical conditioning procedure using the alerting reaction (Chap. 6). A light shone in the eye gives rise to the usual period of low-amplitude fast waves (Fig. 9-2B). Also required to show EEG conditioning is a tone signal which does not by itself produce alerting. This result is arranged for by *habituation* of the alerting response to the sound signal. The first presentation of a sound stimulus produces alerting; with repeated presentations there is a diminished alerting response in the EEG. Eventually the stimulus no longer produces alerting, and the animal or human subject has become habituated to that specific stimulus. If the sound is altered to another tone close in frequency to the first, it will produce alerting. In the example of Figure 9-2A, the response of the subject to a sound stimulus had been habituated and it no longer caused alerting. Alerting was caused in Figure 9-2B by presentation of the light signal, and also in Figure 9-2C when, early in the course of conditioning, the light signal was preceded by a tone as a conditioning stimulus. After a number of such pairings, conditioning occurred to the sound stimulus, as shown in Figure 9-2D. Notice that alerting took place in response to the conditioning signal which preceded the onset of the light stimulus.

An interesting type of EEG conditioning is shown by the use of a light flashed at the rate of 7.5 times per second as the unconditioned stimulus, the flashing light exciting a series of evoked waves in the cortex. A tone signal to which the animal was habituated was selected as the conditioned stimulus and was presented a short time before the light stimulus. After a number of such pairings the tone signal elicited conditioned responses

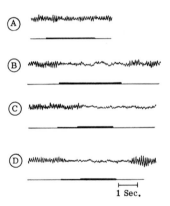

Figure 9-2
Conditioning of the EEG. This EEG of a human shows in A a record of a resting EEG with no alerting response to a tone signal (time of presentation of tone shown by thickening of the line underneath). In B, alerting to a light stimulus is seen (light on at time shown by heavier line underneath). In C, the tone signal precedes the light and no conditioning has yet occurred. After a number of such pairings, conditioning is seen in D when alerting occurs at the onset of the *tone* signal. (From F. Morrell. *Neurology* 6:329, 1956.)

at the same frequency as that of the light flashes used as the unconditioned stimulus. This conditioned response was seen to be widespread in a number of cortical regions. *Discrimination* was shown when, in a later stage of conditioning, the conditioned response became localized to the visual cortex (the unconditioned primary areas).

The limbic cortex, and in particular the hippocampus, has a connection with the learning processes. When the cortex shows the faster, smaller-amplitude waves of alerting, the hippocampus gives rise to large waves at the rate of 5 to 7 per second. During conditioning, EEG changes were registered from the hippocampus via chronically implanted electrodes, and a characteristic shift to a lower frequency was found to take place during learning. The electrical correlate of learning suggests that the hippocampus is involved in an early stage of learning.

MEMORY MECHANISMS

Intellectual ability is based on memories, to which the individual is continually adding. It must be admitted that little is known of the neuronal basis underlying memory, although some general statements may be made. Memories can exist for years, perhaps a lifetime, and thus would appear to depend on some permanent cellular changes. Memories persist after normal cerebral activity has been temporarily stopped, either by a degree of anesthetization sufficient to decrease all EEG activity of the cerebrum or by cooling to a low enough body temperature so that EEG activity is absent. All this supports the conclusion that memory stores are due to some permanent structural change of the neurons rather than being the result of a continuing neuronal activity.

Another general observation concerning memory is that recent memories are less stable than older ones. Clinical experience indicates that a blow upon the head may produce a *retrograde amnesia,* i.e., a loss of memory extending back to include a period of several hours or more before the blow. The susceptibility of recent memories to interruption of neural activity was experimentally shown by application of electrical shock to the heads of rats at different intervals after a conditioning stimulus. Animals electroshocked at too short a time after a learning session do not learn conditioned responses even when numerous conditioning and shock sessions are given. If the interval between the conditioned experience and the administration of the electroshock is lengthened to an hour or more, electroshocked animals are conditioned as well as unshocked animals. The destructive influence of electroshocks on recent memory shows a curve of decreasing effect: It is greatest when shocks are given just after the training session, and the effect diminishes in regular fashion as more time between training sessions and shock is allowed. A similar destructive effect on conditioned learning was found using a quickly induced short period of anesthesia or a rapid cooling of the brain after conditioning sessions. Therefore, two types of memory are distinguished: that level of neuronal activity responsible for recent memories, and molecular changes in neurons responsible for long-term memories.

In an attempt to find the place in the brain where long-term memories are stored, the effects of various cortical lesions on the retention of a learned response were studied. From his work on the rat Lashley concluded that memories or *engrams* could not be specifically located and that the amount of cortex damaged was related to the degree of memory loss. On the other hand, the importance of the temporal lobe with respect to storage and retrieval of memories in man has been noted. Electrical stimulation of the temporal cortex of conscious patients under local anesthesia gave rise to vivid memories. Stimulation of cortical areas outside the temporal lobe did not elicit memories. Penfield considered such responses from the temporal lobe to be support for its *interpretative function.*

The memory responses obtained by electrical stimulation of the temporal region might be a misinterpretation or altered interpretation of present experience with respect to past experiences. Thus, a mother under the influence of electrical stimulation in this region appeared to be seeing her child present in the operating room with all the adjoining sensations and sounds of an actual experience which had occurred previously. It should be noted that such hallucinatory effects may also have taken place in patients as a result of the excitation produced by a tumor in the temporal region. These experiences are related to *auras*—the sensory alterations or hallucinations preceding a generalized epileptic attack.

The fact that electrical stimulation of the temporal cortex can excite a past memory suggests that neuronal activity in the normal temporal lobe cortex is somehow connected with a memory. The memory does not exist in this part of the cortex per se, but it can trigger other regions—e.g., the centrencephalic region in the thalamus. A subcortical focus was indicated by the fact that, after an ablation of the temporal cortex, stimulation of the underlying fibers was also effective in eliciting memories.

The study of learning was further advanced by use of the *split-brain* preparation (Fig. 9-3). To produce a split-brain cat or monkey, the optic chiasm and the corpus callosum tract connecting the two hemispheres as well as the crossing fibers of the anterior commissure are cut in the anteroposterior plane. The visual information input from each eye can therefore pass only to the hemisphere of the corresponding side. With one eye covered and the other eye open, the animal is taught a visual discrimination in the hemisphere on the side on which the eye is open. After training is completed on that side, the eye of the opposite side is uncovered and the eye on the trained side covered. The animal is then unable to perform the visual discrimination to which it had previously been trained. The habit remains localized to the brain on the side on which the eye was open during training. Use of this technique makes it possible for two habits to be laid down, one on each side. Even opposing habits may be trained into the two sides. On one side the animal may learn to respond to a square and not to a circle; on the other side, to a circle and not to a square. With both eyes open the animal responds positively to one or the other symbol. At times one habit dominates, at times the other; the animal does not respond partially to the habits laid down on each side.

If the optic chiasm is cut in the midline as usual but the corpus callosum remains uncut, sensory information is channeled to the other side. When, after an animal was trained with one eye open, the other covered, and then later the previously open eye was closed and the other opened, it readily recognized the visual stimulus and responded to it.

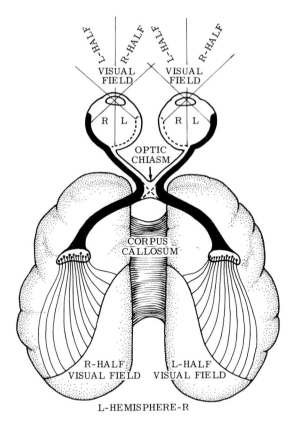

Figure 9-3
The visual input from one eye is restricted to one hemisphere by cutting the optic chiasm in the midline. The corpus callosum is also sectioned to produce a split-brain animal. (From R. W. Sperry. *Science* 133:1749–1757, [June] 1961. Copyright 1961 by the American Association for the Advancement of Science.)

By means of the phenomenon of spreading depression (Chap. 6), one or both cerebral cortices may be functionally incapacitated for up to several hours. During such *temporary decortication* of both hemispheres animals may learn only a simple type of classical conditioned response while the acquisition of a complex response is prevented. With spreading depression present in one hemisphere, an animal will learn to respond with the cortex not occupied by spreading depression. The localization of the learned response to the cortex which was not depressed during the training sessions was shown when later the trained cortex was depressed and the animal then behaved as an untrained or naive animal. Such *lateralization* of the learned response to one hemisphere may remain for days, weeks, or months even though the interhemispheric connections in the corpus callosum might be expected to transfer the habit from the trained side to the other side.

Just such transfer can occur if the animal receives *reinforcement* without depression. The transfer of lateralized conditioned responses was used to great effectiveness by Bureš. Spreading depression was used to lateralize a conditioned response in one hemisphere of rats. Another conditioned response related to the first response was lateralized in the other cortex. After conditioning, the animals could bring these two conditioned responses together to make a chained conditioned response.

235

Attempts have been made to show the sites where memories are stored by means of conditioned responses and placing lesions in various limbic and other subcortical structures. Lesions in the septum and in the pathways connecting the hypothalamus and hippocampus to other regions will produce defects of learning and loss of conditioned responses. A number of subcortical regions have been implicated, but further analysis is required to determine whether memory or motivation is involved in such lesions.

Much is still to be learned about memory. If all memories are present subcortically, does the cortex play a role in retrieving them? Is part of the memory laid down in one site and part in another, and if so how are they gathered together to give rise to a whole memory sequence? Finally, much evidence suggests a molecular basis for permanent memories. Changes in the ribonucleic acid (RNA) composition in neurons, and in turn the synthesis of specific proteins, may be responsible for long-term memories. But how is neuronal activity transformed into specific RNA changes required to synthesize specific proteins? How do such specific proteins later effect a neural discharge of a given pattern? Too little is known as yet to give a satisfactory answer to these questions.

HIGHER FUNCTIONS IN THE HUMAN CEREBRUM
A further advance in the study of brain function was made possible through the split-brain studies carried out by Sperry and his associates on human patients who required surgery for various reasons including severe epilepsy. With the corpus callosum cut and techniques of sensory stimulation arranged so that visual and tactile stimuli could be presented to one eye or hand, a clear difference in function of the two hemispheres was shown. With a word flashed briefly (less than one-tenth second) in the

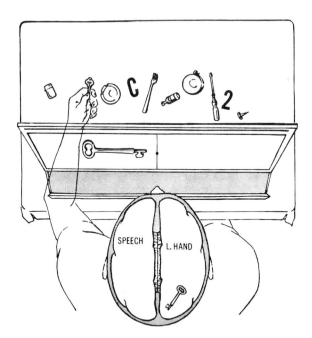

Figure 9-4
Visuotactile associations are made only if visual and tactual stimuli both project to same hemisphere. Subject here denies seeing anything but a flash of light on left. However, if left hand is allowed to "make a guess," correct object is selected or even an object "used with" the projected stimulus if subject is so instructed. (From Sperry, Gazzaniga, and Bogen, 1969.)

left visual field, the normal crossed connections of the optic nerve give rise to an activation of the right visual cortex. In this cortex (right) the split-brain human cannot verbalize what he has seen, but through the control exerted by the right hemisphere over his left hand he can pick out from the group of objects hidden from his sight the object named by the word (Fig. 9-4). In spite of his correct manual selection, the subject cannot

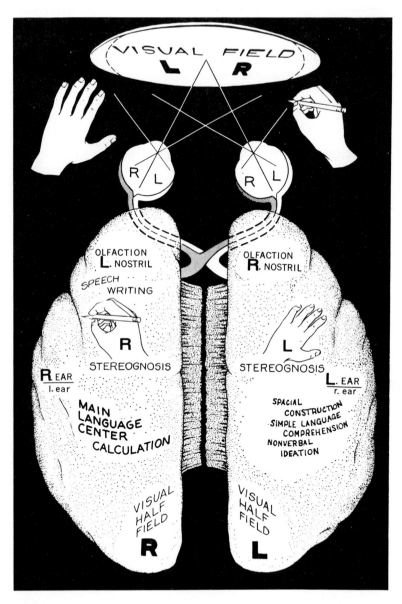

Figure 9-5
Functional lateralization in left and right hemispheres shown often cutting corpus callosum. Language is located in the left and spatial construction and nonverbal ideation mainly in the right hemisphere. (From Sperry, 1970.)

describe what he is doing with his left hand, which is under the control of the right or *minor* hemisphere. The subject may even be able to write out the correct word with his left hand without being subjectively aware of the term. If, similarly, a word is flashed in his right visual field, the object can be named. This is the *major* hemisphere or the hemisphere by which verbal behavior is controlled (Fig. 9-5). The most interesting result found in split-brain patients is that the one hemisphere does not know what is going on in the other hemisphere. There is some degree of subconscious awareness. Certain words or objects presented to the right hemisphere which have embarrassing or emotional value may give rise to behavioral responses, indicating some emotional reaction, but without the subject able to consciously identify its source. Consciousness in the right hemisphere falls far short of that in the dominant (left) hemisphere. Studies of brain lesions have earlier shown that the awareness of speech is perceived in Wernicke's area in the left hemisphere. This is required for conscious awareness of speech. Lesions in the right or minor hemisphere cause defects related to a disturbance of body-sense awareness and spatial relationships in general (Fig. 9-5). Split-brain studies are of necessity limited because the specific conditions which require callosal section in the human are relatively rare, but they are all the more important for a proper under-standing of the functions of the brain and the alterations expected in pathological states.

The studies of callosal section can be related to observations of patients who have lesions in portions of the brain which cause them to have *agnosia* (failure to interpret or recognize objects or speech or parts of their own body) or *apraxia* (failure to execute learned movements).

An interpretation of the effects shows a *disconnection syndrome*. As noted above, Wernicke's area in the left hemisphere (Fig. 9-6) appears to be needed for the compre-hension of spoken language; destruction of this area leads to language agnosia. Upon recognition of a spoken command neural pathways are excited which lead to the left

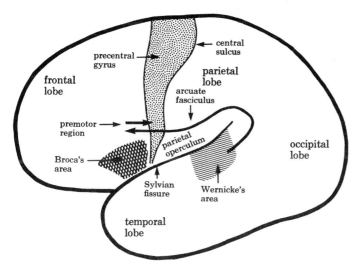

Figure 9-6
Left hemisphere of human showing the pathway used in the performance of motor acts by the right limbs and the cranial muscles in response to a verbal command. The precentral gyrus (stippled) contains the classic motor cortex. At the lower end, near the Sylvian fissure, is the face motor area—the region controlling movements of the muscles of the face, jaw, tongue, palate, and vocal cords. The path from Wernicke's area, where an auditory command is received and interpreted, leads to the premotor region as shown by the curved line. A shorter line gives the path from the precentral gyrus to the pyramidal motor system descending to the brain stem and spinal cord for control of movement on the opposite (right) side. (From Geschwind, 1975.)

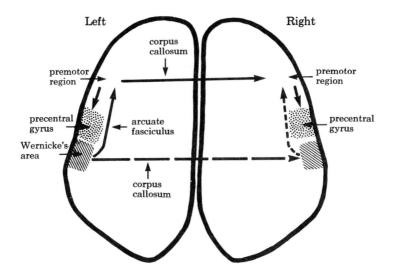

Figure 9-7
Human brain viewed from above showing the intrahemispheric callosal pathways used in carrying out movements with the right or left limbs in response to a verbal command. Solid lines show path from Wernicke's area on the left to the premotor region and from premotor region to precentral gyrus for motor outflow controlling the right side. Another solid line from the premotor region on the left to the premotor area is shown on the right side for control of the right precentral gyrus and movement on the left side. Destruction of the anterior part of the corpus callosum results in the inability to respond with the left arm to a spoken command. Dashed lines show possible alternative paths. (From Geschwind, 1975.)

premotor area. From this region control over the left motor area is possible and through it movement of the *right* hand, normally controlled by the left cortex. To move the *left* hand, excitation of nerve fibers passing via the corpus callosum to the right precentral motor cortex allows for a control of movement through this region (Fig. 9-7). If the corpus callosum is cut through, a spoken command to move the left hand cannot be executed and an apraxia is thereby revealed.

A number of voluntary control movements are bilaterally represented. These include use of the muscles of lips, mouth, larynx, and the axial musculature. For example, a patient with callosal section can move his fascial muscles or walk properly. The difference between unilateral and bilateral control lies in the fact that fine motor control over the fingers in man and primates is a province of the pyramidal tract, which is unilaterally represented in the motor cortex for the most part. The extrapyramidal control in the precentral motor areas subserves the large muscle groups, primarily axial muscles, and these are bilaterally represented in the cortex (cf. Chap. 8 and Fig. 8-2).

REFERENCES

Bureš, J., O. Burešová, and J. Krivanek. *The Mechanism and Applications of Leao's Spreading Depression of Electroencephalographic Activity.* New York: Academic, 1974.

Fulton, J. F. *Frontal Lobotomy and Affective Behavior: A Neurophysiological Analysis.* New York: Norton, 1951.

Gaito, J. (Ed.). *Macromolecules and Behavior* (2nd ed.). New York: Appleton-Century-Crofts, 1971.

Gazzaniga, M. S. *The Bisected Brain.* New York: Appleton-Century-Crofts, 1970.

Geschwind, N. The apraxias: Neural mechanisms of disorders of learned movement. *Am. Sci.* 63:188–195, 1975.

Herrick, C. J. (Memorial Symposium): Neurophysiology of learning and behavior. *Fed. Proc.* 20:601–631, 1961.

Horn, G., and R. A. Hinde. *Short-Term Changes in Neural Activity and Behavior.* Cambridge, England: Cambridge University Press, 1970.

Klüver, H., and P. C. Bucy. Preliminary analysis of functions of the temporal lobes in monkeys. *Arch. Neurol. Psychiatry* 42:979–1000, 1939.

Landauer, T. K. (Ed.). *Readings in Physiological Psychology.* New York: McGraw-Hill, 1967.

Lashley, K. S. In search of the engram. *Symp. Soc. Exp. Biol.* 4:454–482, 1950.

MacLean, P. D. Psychosomatic disease and visceral brain. *Psychosom. Med.* 17:355–366, 1955.

Miller, N. E. Learning of visceral and glandular responses. *Science* 163:434–445, 1969.

Morrell, J. Electrophysiological contributions to the neural basis of learning. *Physiol. Rev.* 41:443–494, 1961.

Mountcastle, V. B. (Ed.). *Interhemispheric Relations and Cerebral Dominance.* Baltimore: Johns Hopkins Press, 1962.

Olds, J. Self-stimulation of the brain. *Science* 127:315–324, 1958.

Papez, J. W. A proposed mechanism of emotion. *Arch. Neurol. Psychiatry* 38:725–745, 1937.

Pavlov, I. P. *Conditioned Reflexes.* New York: Dover, 1960.

Penfield, W. Functional localization in temporal and deep sylvian areas. *Res. Publ. Assoc. Res. Nerv. Ment. Dis.* 36:210–226, 1958.

Sperry, R. W. Perception in the Absence of the Neocortical Commissures. *Res. Publ. Assoc. Res. Nerv. Ment. Dis.* 48:123–138, 1970.

Sperry, R. W., M. S. Gazzaniga, and J. E. Bogen. Interhemispheric Relationships: The Neocortical Commissures; Syndromes of Hemisphere Disconnection. In P. J. Vivken and G. W. Bruyn (Eds.), *Handbook of Clinical Neurology.* Amsterdam: North-Holland, 1969. Vol. 4, chap. 14.

10. Functional Properties of Blood

Julius J. Friedman

Unicellular organisms are in immediate contact with their external environment. Nutrition is derived directly from and excretion occurs directly into the external medium. The organism is able to survive under these conditions because the volume of the extracellular environment greatly exceeds that of the cell. Consequently, acquisitions from and contributions to the surrounding medium produce no significant changes in concentrations of the various constituents of the medium, and the cell can be considered as being exposed to a constant environment.

In complex multicellular organisms, cells neither gain nutrition directly from the external environment nor excrete waste materials directly into it. Instead, the organisms are equipped with specialized tissues which are concerned with the processes of nutrition and excretion. To provide these facilities to the remainder of the cells throughout the organism, a vehicle of communication between them is necessary.

It is the primary function of blood to provide a link between the various organs and cells of the body. Blood maintains a constant cellular environment by circulating through every tissue and continuously delivering nutrients to the tissues and removing waste products and various tissue secretions from them. Blood is a tissue which consists of a fluid plasma in which are suspended a number of formed elements (erythrocytes, leukocytes, and thrombocytes).

PLASMA

Plasma is a pale amber fluid which contains numerous dissolved materials: proteins, carbohydrates, lipids, electrolytes, pigments, hormones, and others. The plasma proteins include *albumin, globulins,* and *fibrinogen.* The total plasma protein concentration is 6 to 8 gm per 100 ml of plasma, with albumin (MW = 69,000) at a concentration of 4 to 6 gm per 100 ml, globulins (α_1, α_2, β_1, β_2, and γ) (MW = 80,000 to 200,000) at 1.5 gm per 100 ml, and fibrinogen (MW = 350,000 to 400,000) at a concentration of 0.2 to 0.4 gm per 100 ml.

The plasma proteins are routinely separated and quantified by means of paper electrophoresis. A normal electrophoretic pattern (Fig. 10-1A) illustrates the differences in mobility of the proteins in an electric field, a characteristic related to molecular size, configuration, and charge. They are formed, in part, in the liver (albumin, fibrinogen, α_1, and α_2 globulins) and in the reticuloendothelial and lymphatic systems (β_1, β_2, and γ globulins). Thus, marked changes in the electrophoretic pattern occur in disorders such as infectious diseases, liver dysfunction, and multiple myeloma (Fig. 10-1C, D). In some renal diseases large quantities of albumin are lost to the urine (Fig. 10-1B).

The plasma proteins contribute significantly to the physical and physiological characteristics of plasma. They determine the specific gravity of plasma, which is 1.028, as well as plasma viscosity. Albumin and the globulins substantially determine the colloidal osmotic pressure of plasma. All the proteins participate in the buffering capacity of blood. Fibrinogen is essential for coagulation. The globulins are important in antibody formation and as hormones and enzymes. They also act as substrate for the release of peptides such as angiotensin and bradykinin.

FORMED ELEMENTS

The formed elements of the blood include erythrocytes, leukocytes, and thrombocytes (platelets).

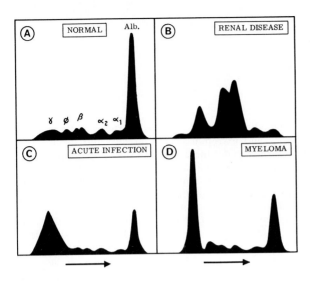

Figure 10-1
Densitometric patterns of paper electrophoresis of human plasma.

Erythrocytes

Erythrocytes, or red blood cells, are anucleated biconcave discs which are formed in bone marrow by means of *erythropoiesis*. They are about 2 μm thick and 8 μm in diameter. The volume of a red cell is about 85 μm^3. By virtue of its biconcave shape, the red cell can undergo marked changes in shape without changes in volume, as occurs with the distortions that take place during passage through the capillaries. It can also endure increases in volume without having marked changes in membrane tension such as occur during osmotic stress. The major function of the red cell is to transport oxygen and carbon dioxide. The red cell must be equipped to release its oxygen rapidly to the tissues and take up carbon dioxide. It contributes significantly, moreover, to the buffering activity of blood. The biconcave shape of the cell is ideally suited to this function, providing the maximum surface area for diffusion for the volume of the cell.

The number of red blood cells in the circulatory system can be estimated by counting those in a small representative sample of blood with the aid of the microscope and the graduated slide called a *hemacytometer*. There are 5.5 to 6.0 million red blood cells per cubic millimeter of blood in the normal male adult and 4.5 to 5.0 million per cubic millimeter of blood in the normal female adult. The number varies with age, activity, climate, and altitude, as well as disease.

The relative red blood cell content of blood is described by the *hematocrit ratio*. This is derived by first rendering a blood sample incoagulable and then centrifuging it in a Wintrobe tube at 1500 g (3000 rpm with an arm 15 cm long) for 30 minutes. A Wintrobe tube is a heavy-walled, small-volume, uniform-diameter tube graduated from 0 to 10 (Fig. 10-2). Since the specific gravity of red cells is about 1.088, compared to 1.028 for plasma, they become packed at the bottom of the tube, and the less dense plasma is situated above them. The white blood cells and platelets possess a specific gravity intermediate between that of plasma and red cells and form the buff-colored layer immediately above the red cell column. The hematocrit ratio is determined simply by measuring the height of the red cell column and dividing this by the height of the total blood column (Fig. 10-2). The white blood cells are not included in the hematocrit ratio. This ratio is usually multiplied by 100 to provide a hematocrit reading in percent. The venous hematocrit of a normal adult determined in this manner is about 45 percent.

However, because the cells are disc-shaped, the small space between adjacent cells is occupied by plasma, which must be subtracted from the initial measurement. It has been estimated, by using radioactive plasma, that approximately 4 to 5 percent of the volume of packed cells in the Wintrobe tube is represented by plasma. Thus, the corrected venous hematocrit reading is approximately 43 percent. Care must be exercised in evaluating changes, since the hematocrit varies with changes in sampling site as well as in many clinical conditions (Fig. 10-2).

The hematocrit is not uniform throughout the body. That of venous blood is slightly greater than that of arterial blood, because of fluid and electrolyte shifts. In addition, the hematocrit varies from tissue to tissue. The splenic hematocrit is approximately 70 percent, whereas that of the kidney is only about 20 percent. The values for other tissues are distributed between these two extremes. Variation in tissue hematocrit has been attributed in part to a phenomenon called *axial streaming*. As blood flows through the peripheral circulation, mostly plasma is "skimmed" off into the capillary-venular circulation from the periphery of the flow stream. Since the capillaries and venules contribute significantly to tissue blood volume, the low hematocrit of this segment of the tissue vasculature imparts a low reading to the remainder of the tissue circulation, and tissue hematocrit is therefore lower than either arterial or venous hematocrit. The most notable exception, the spleen, by virtue of its sinusoidal vasculature, possesses the ability to "filter" red cells from the circulation; thus splenic blood has a high hematocrit.

The most reliable estimate of body hematocrit is derived from independent total

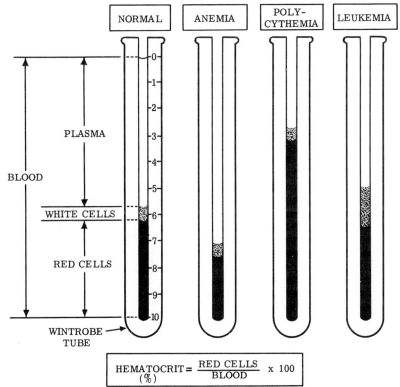

Figure 10-2
Determination of hematocrit ratio.

plasma and total red cell volume determinations. Body hematocrit calculated as the ratio of total red cell volume to total blood volume is about 35 percent. The ratio of the total body hematocrit to the venous hematocrit, which has been referred to as the *F cells ratio,* is regularly less than unity. The average F cells ratio seems to remain fairly constant at 0.91 under a variety of conditions of cardiovascular stress. It does increase during pregnancy and is susceptible to change in severe hemorrhage and transfusion.

Red Blood Cell Production

Red cells are produced in the bone marrow at a normal rate of approximately 400 to 500 ml of packed red cells per month. The rate of red cell production is probably determined by the oxygen tension of the blood, although the precise mechanism is not clear. Any condition that results in a reduction of oxygen supply to the tissues appears to stimulate red blood cell production. Local tissue anoxia apparently leads to the formation of a specific humoral principle called the *erythropoietic stimulating factor.* This factor may then stimulate the bone marrow to increase the production and maturation of red cells. *Erythropoietin,* a glycoprotein (MW = 30,000), is apparently produced in the kidney, and in some obstructive renal disorders its level increases markedly.

Tissue anoxia can develop in a variety of ways. Low tension of atmospheric oxygen, impaired gaseous diffusion between the alveoli and the blood, and altered oxygen uptake by the tissues can lead to local anoxia (Chap. 20). Therefore, erythropoiesis may be stimulated under any of these circumstances. Overstimulation of red cell production can cause the red blood cell population to increase excessively and results in a condition called *polycythemia.* Although the oxygen-carrying capacity of blood is enhanced in polycythemia, the viscosity is also increased, and the heart must perform additional work in order to pump the more viscous blood through the circulatory system. To accommodate the added work load, the heart hypertrophies. Polycythemia and cardiac hypertrophy are most prevalent among people who reside at higher elevations and who are thus chronically exposed to low oxygen tensions.

Red Cell Destruction

The red cell travels about 700 miles during its lifetime and is subjected to considerable mechanical and chemical stress, which eventually leads to its destruction. The life-span of the red blood cell, as determined by radioactive methods, is about 120 days. Approximately 1 percent of the red cells are replaced daily, therefore, and the total red cell volume is replaced every four months. Red cells are primarily destroyed in the spleen. Continuous friction between red cells and the vessel walls and the action of chemical substances in circulation apparently alter the integrity of the red cell membrane, thereby rendering it more susceptible to osmotic stresses. When red blood cells are subjected to progressively increasing osmotic stress by a progressive reduction of the extracellular sodium chloride concentration from the isotonic level, they begin to hemolyze—few at first, then substantially more, and finally the few remaining. The sigmoid character of the degree of hemolysis is thought to represent the hemolytic behavior of cells of different ages. Thus, older cells hemolyze with lower osmotic stress whereas the younger cells are most resistant to osmotic influence. Fragility tests are frequently used to distinguish between hereditary and acquired red blood cell disorders.

The number of red cells in the circulation at any one time is determined by the balance between production and destruction. When production exceeds destruction, the red cell volume and hematocrit rise, and polycythemia results. Conversely, when production is less than destruction, red cell volume and hematocrit fall, and *anemia* results.

Anemia

Anemia is a condition in which either the number of red blood cells per cubic millimeter or the amount of hemoglobin per 100 ml of blood or the red cell fraction of blood (hematocrit) is below normal level. It may result from a number of causes, including hemorrhage (acute and chronic), maturation deficiency, aplastic bone marrow, and enhanced fragility. Each situation influences both size and hemoglobin content of the red cell in a characteristic manner; these two parameters are used to classify anemia. The cell of normal size is *normocytic*. A cell larger than normal is *macrocytic,* while one smaller than normal is *microcytic*. The extent of microcytosis or macrocytosis is reflected in the mean corpuscular volume (MCV)

$$MCV = \frac{\text{Hematocrit ratio} \times 1000}{\text{Red cell count in millions/mm}^3}$$

The normal range of red cell volume is 75 to 95 μm^3. A volume below 75 μm^3 would be microcytic whereas one above 95 μm^3 would be macrocytic.

A cell possessing a normal concentration of hemoglobin is said to be *normochromic*. Red cells with subnormal hemoglobin concentration are *hypochromic,* and those with an above normal concentration are *hyperchromic*.

The degree of hemoglobin alteration is reflected in the mean corpuscular hemoglobin (MCH) determination.

$$MCH = \frac{\text{Hemoglobin (gm/1000 ml)}}{\text{Red cell count in millions/mm}^3}$$

The normal range of hemoglobin per red blood cell is 24 to 34 picograms (pg) with an average hemoglobin content of 30 pg. This, together with an average cell volume of 85 μm^3, provides an average red cell hemoglobin concentration of 35 gm per 100 ml of red cells. With an average hematocrit of 0.45, the blood hemoglobin concentration is about 16 gm/100 ml.

Normocytic, Normochromic Anemia. Although red blood cells can be of normal size and contain the normal concentration of hemoglobin, anemia may exist if the number of red cells in the circulation is below normal. This condition arises with acute hemorrhage. As the result of hemorrhage, both plasma and red cells are lost from the circulation and the red cell volume is reduced. The plasma volume is restored in a short time by compensatory responses (Chap. 12). However, the process of red blood cell production is slow, and the circulation remains deficient in red cells for some time. Since the bulk of the cells present in the circulation during this period are the same ones that were present prior to hemorrhage, the cells are essentially normal. Consequently, the anemia resulting from acute hemorrhage is normocytic and normochromic.

Microcytic, Hypochromic Anemia (Iron Deficiency Anemia). Iron deficiency anemia often occurs in women of childbearing age, during chronic hemorrhage, and in infants.

During chronic hemorrhage, red cell loss from the circulation continues over a period of time, and the replacement of these cells must be maintained throughout the period. The persistent and extended stimulus to erythropoiesis imposes a continuous demand on the iron stores of the body and leads to a condition of iron deficiency. Consequently, the hemoglobin content of newly formed red cells is lower than normal (hypochromic) and the cells are smaller than normal (microcytic).

The iron requirement of infants is so great that nutritional iron deficiency often develops, producing a microcytic, hypochromic anemia.

Macrocytic, Hyperchromic Anemia. In 1929 Castle discovered that a particular anemia could be alleviated by administration of the product formed by incubating meat with gastric juice. This principle was called the *antianemic* or *hematinic factor*. The meat component was labeled the *extrinsic factor* and was later identified as vitamin B_{12}. The gastric juice component was called the *intrinsic factor*. Vitamin B_{12} contained in ingested meat combines with the intrinsic factor to form the hematinic factor, which is absorbed through the intestinal wall and stored in the liver. In the absence of either of these factors, the maturation of red cells is altered and excessively large red cells (macrocytes) are formed in reduced number. They contain a greater than normal amount of hemoglobin, because of their increased size, and therefore are hyperchromic. The clinical condition which characterizes these changes is called *pernicious* or *addisonian anemia.*

Pernicious anemia may occur under any circumstances that interfere with any phase of the incorporation or utilization of the antianemic factor. Consequently, a reduction in extrinsic factor due to dietary deficiency, reduced gastric juice secretion, reduced intestinal absorption, or decreased liver function may lead to pernicious anemia. Because the liver is the storage site of the hematinic factor, either the addition of liver to the diet or administration of liver extract is generally sufficient to alleviate the condition.

Other megaloblastic anemias, which resemble pernicious anemia but have an obscure pathogenesis, occur in tropical sprue and pregnancy, as well as in conditions of folic acid and vitamin C deficiencies, and are referred to as *nutritional anemias.*

Aplastic Anemia. Occasionally a subject with malfunctioning bone marrow is encountered; his capacity to produce normal cells is absent. In acute conditions, such as those due to excessive exposure to radiation, the cells in circulation are essentially those that were present prior to exposure and are therefore normocytic and normochromic. However, since destruction proceeds independently of production, the red cell count progressively falls. Aplastic bone marrow can also occur spontaneously.

Hemolytic Anemia. Anemia may result when enhanced red cell hemolysis takes place because of chemical agents, hereditary defects of cell formation, enhanced reticuloendothelial destruction, and hypersplenism. Venoms, bacterial toxins, and some drugs produce hemolysis by means of lysis of the cell membrane. *Familial hemolytic anemia* is a hereditary condition characterized by the production of small spherical red cells called microcytes or spherocytes. These cells are normochromic but hemolyze readily because their shape imparts great tension to the membrane during the deformations encountered by the cell as it passes through minute vessels. *Sickle cell anemia,* particularly prevalent among Negroes, represents another hereditary abnormality in which the red cells shaped like sickles or crescents are formed during the sickling crisis that occurs as a result of abnormal hemoglobin and therefore are highly fragile. In addition, changes in the nature of the hemoglobin in these cells modify their binding of oxygen and impair gaseous exchange.

Leukocytes

The major function of leukocytes, or white blood cells, is to protect against the invasion of bacteria and infection. These cells are capable of passing through the vascular endothelium and entering the tissue spaces (diapedesis) in accordance with the local needs. They are apparently attracted by chemical substances (chemotaxis) released by the bacteria and can engulf and digest the foreign substances by means of ameboid movement (phagocytosis). Normally there are approximately 5000 to 7000 white blood cells per cubic millimeter of blood. This total is comprised primarily of polymorphonuclear neutrophils (60 percent) and lymphocytes (35 percent), the remaining 5 percent consisting of polymorphonuclear eosinophils, basophils, and monocytes. The leukocytes form part of a broader system which provides protection against infection. This is the *reticuloendothelial system;* it includes all phagocytic cells, fixed and mobile, which are primarily situated in the liver, spleen, lymph nodes, lung, and gastrointestinal tract.

Thrombocytes

Thrombocytes, or platelets, are small, colorless bodies ranging in size from 2 to 4 μm in diameter with a life-span of 7 to 10 days. There are 150,000 to 300,000 platelets per cubic millimeter of blood. Thus, approximately 40,000 platelets per cubic millimeter per day are turned over in the spleen and throughout the vascular system. The number of platelets remains fairly constant for extended periods, suggesting the existence of some regulatory process. Platelets possess dense granules that are believed to be the location of coagulation factors. These cells contain, or have absorbed on them, serotonin, histamine, catecholamines, adenosine, and ribonucleic materials. They also contain a retractile protein, *thrombosthenin*, which is responsible for clot retraction.

Platelets are of great importance in hemostasis. They contribute serotonin and catecholamines for vasoconstriction, undergo transformation to form a platelet plug and establish a matrix for more effective coagulation, and retract to draw the wound margins closer together.

BLOOD VOLUME

The total blood volume in a subject is best determined by estimating both the plasma and the red cell volumes independently and adding them together. These volumes are estimated by means of a dilution principle wherein a known quantity of indicator (Q) is introduced into an unknown volume. Sufficient time is allowed for complete mixing, and the concentration of indicator (C) in a representative sample is determined. The volume of dilution (V) is then calculated as

$$V = \frac{Q}{C}$$

Plasma Volume

Plasma volume (PV) is estimated by injecting into the circulatory system about 5 microcuries of radioactively iodinated (^{125}I) serum albumin (RISA). If albumin distribution were restricted to the vascular system, a sample after 5 to 10 minutes of vascular mixing would be adequate for the determination of indicator concentration. Unfortunately, albumin leaves the circulation slowly (about 10 percent per hour) so that it is necessary to obtain at least two samples—10 and 20 minutes after injection—and extrapolate from these two points to zero time. This procedure provides an estimate of the mixed indicator concentration in the plasma before any extravascular loss occurred.

An average plasma concentration of indicator in a 70-kg male would be about 0.0015 microcurie per milliliter of plasma. Dividing this concentration into the dose administered provides an estimate of about 3200 ml of plasma.

Red Blood Cell Volume

Red blood cell volume (RCV) is generally estimated with red cells labeled with radioactive chromium (^{51}Cr). The procedure employed is as follows: Approximately 20 to 30 ml of blood is withdrawn from the subject and mixed with 5 ml of an anticoagulant acid citrate dextrose (ACD) solution. A dose of 15 to 20 microcuries of sodium chromate (^{51}Cr) is added to the blood, which is gently agitated at room temperature for 30 to 45 minutes. This provides 90 percent binding of ^{51}Cr. Tagging is terminated by adding 50 to 100 mg of ascorbic acid, which reduces the chromate.

The labeled blood is injected intravenously and allowed to mix in the circulation for about 10 minutes. A sample is obtained and corrected for unlabeled ^{51}Cr, and the concentration is determined. Average concentrations are about 0.011 microcurie per milliliter of red cells. Dividing the corrected concentration into the corrected dose provides an estimate of 1800 ml of red cells for a 70-kg adult male.

The times allowed for vascular mixing apply to a subject under reasonably normal cardiovascular conditions. In cases of cardiovascular stress in which tissue blood flow is severely altered, longer mixing times will be necessary.

Total Blood Volume

Total blood volume (BV) is calculated by the expression

$$BV = PV + RCV$$

Thus the total blood volume of a normal 70-kg adult male is about 5000 ml. The blood volume of a normal adult female is about 4500 ml.

It is possible to obtain an estimate of total blood volume from a single determination of plasma or of red cell volume and corrected venous hematocrit according to the expressions

$$BV = \frac{PV}{(100 - \text{hematocrit}) \times F \text{ cells}}$$

$$BV = \frac{RCV}{\text{Hematocrit} \times F \text{ cells}}$$

Total blood volume varies under circumstances involving the integrity of the vascular system, the state of water balance of the individual, and the state of physiological activity. These will be discussed in detail in Chapters 18 and 23.

Changing posture from a lying to a standing position will result in a fall in blood volume due to a reduction in plasma volume. This reflects the change in transcapillary dynamics upon standing, which produces enhanced filtration (see Chap. 12). Exercise produces similar variations in blood volume. Adaptation to low tissue oxygen tensions such as occur at altitude and with physical training increases red cell and blood volume. In pregnancy both plasma and red cell volumes increase.

The blood volume is not uniformly distributed throughout the body. Approximately 80 percent of the total blood volume is contained within the low pressure system (veins, right heart, and lungs). Because of the high distensibility of the pulmonary vasculature, and the large mass of muscle, these two tissues are prominent blood reservoirs which participate in large volume shifts during conditions of circulatory stress. About 50 percent of a 400- to 1000-ml hemorrhage is provided by the intrathoracic compartment. During orthostasis, approximately 700 ml is translocated from the thorax to the legs.

Changes in blood volume distribution are manifested by either passive or active adjustment of the capacitance vasculature. Passive behavior depends on the elastic recoil of the vessel system, whereas active behavior is primarily related to venomotor tone.

BLOOD GROUPS AND TRANSFUSION

In conditions of anemia or hemorrhage, it is often desirable to transfuse additional volumes of blood into the circulation in order to improve the hemodynamics of the system. When this course of treatment is indicated, care must be exercised in selecting the blood to be transfused, since not all blood is compatible. Numerous reactions may arise from incompatibilities in blood. These incompatibilities reflect variations in principles present on the cells and in the plasma of different individuals.

Blood has been classified, according to its antigenic activity, as types O, A, B, and AB. The specific representation in the blood of an individual is determined genetically, so there must be a double representation, one derived from each parent. Consequently, the genotypes of the blood groups are OO, AA, AO, BB, BO, and AB. Present on red cells are specific chemical principles called *agglutinogens,* which determine the antigenic behavior of the cells. When serum possesses a suitable antibody, *agglutinin,* agglutination of the red cells takes place.

Fortunately, when an agglutinogen is present in the blood, the antagonistic agglutinin is absent. Conversely, where the agglutinin is present, the agglutinogen is absent. Type A blood contains agglutinogen A and anti-B or beta agglutinin. Similarly, type B blood contains agglutinogen B and anti-A or alpha agglutinin. Type O blood does not usually contain any agglutinogen, and both alpha and beta agglutinins are present, whereas type AB blood possesses both agglutinogen A and agglutinogen B; consequently no agglutinins are present. The blood groups, their constituents, and their percentage distribution are listed in Table 10-1.

When type A blood is transfused into a type A recipient, no blood reaction occurs because no antagonistic blood principles are present. However, if type A blood is

Table 10-1. Blood Group Constituents

Group	Agglutinogen	Agglutinin	Percent of Population
O	—	α, β	46
A	A	β	42
B	B	α	9
AB	A + B	—	3

transfused into a type B or a type O recipient, two antigen-antibody reactions are possible. The agglutinins of the donor may react with the agglutinogen of the recipient to produce agglutination of recipient cells, or agglutinogens of the donor may react with agglutinins of the recipient to produce agglutination of donor cells. The former reaction is unlikely, because the addition of small quantities of donor agglutinins to a relatively large volume of the recipient's blood provides sufficient dilution to render the donor agglutinins ineffective. Transfusion reaction therefore involves the agglutination of donor cells by recipient agglutinins.

Since type O blood possesses no agglutinogens, type O cells cannot be agglutinated by any of the agglutinins. Consequently type O blood can be transfused into any other type without reaction, and type O persons are referred to as "universal donors." Conversely, type AB blood is without agglutinins and will not agglutinate any type of donor blood. Persons with type AB blood are referred to as "universal recipients."

There are a number of other factors in the blood which can lead to low-level incompatibilities or sensitization. The most prominent of these, the *Rh factor*, is present in approximately 85 percent of the population. When it is present, the blood is Rh-positive; when it is absent, Rh-negative. Antibodies to the Rh factor do not normally exist in the blood but arise following the exposure of individuals with Rh-negative blood to the Rh antigen. An Rh-negative mother bearing an Rh-positive child may become sensitized and develop Rh antibodies. These antibodies can then diffuse into the fetal circulation, to produce an Rh reaction and death (erythroblastosis fetalis). Usually a number of exposures are required to establish an antibody titer sufficiently high to produce fetal death.

Other factors, such as M and N, are occasionally considered in blood typing, particularly in circumstances of disputed parenthood.

HEMOSTASIS

The rupturing of a blood vessel results in blood loss which, if unchecked, would eventually lead to hemorrhagic shock and death. During normal activity, minor accidents frequently occur in which minute blood vessels are ruptured, yet blood loss is minimal. When a blood vessel is ruptured, hemorrhage is limited by a reduction of blood flow to the site of injury by means of local vasoconstriction, and by platelet aggregation, by coagulation, and by clot retraction.

Vasoconstriction

Numerous influences can produce vasoconstriction (Chap. 17). With vascular injury, platelets release serotonin, which may provide the stimulus for constriction of small arteries and arterioles. By reducing blood flow from the severed vessel, which allows the local concentration of the platelet and coagulation factors to increase, vasoconstriction enables platelet aggregation and coagulation to proceed with effectiveness.

Platelet Aggregation

The role of platelets in hemostasis is a very important one. In addition to releasing serotonin, which causes vasoconstriction, the platelets, under the influence of adenosine diphosphate (ADP) released from damaged cells, and from the platelets themselves, undergo transformation from a disc to a sphere and develop increased adhesiveness. This leads to aggregation and formation of the platelet plug. The sphere-shaped platelets grow pseudopodia, which attach to collagen at the site of injury and thereby establish a framework upon which effective coagulation can proceed. It is thought that the prostaglandin PGE_2 may be involved in the platelet transformation, since aspirin, which blocks PGE_2, also blocks the transformation.

Coagulation

Coagulation represents the complex interaction of platelet, plasma, and tissue factors which culminates in the formation of a fibrin meshwork at the site of vessel rupture. While numerous theories have been introduced to account for this phenomenon, there is still considerable uncertainty about it because of the limited availability of purified factors. There are numerous synonyms for the coagulation factors which will be cited, most of them included in Table 10-2.

Coagulation may be considered to proceed in three stages: (I) formation of pro-thrombin-converting enzyme; (II) conversion of prothrombin to thrombin by this agent; and (III) conversion of fibrinogen to the fibrin clot through the action of thrombin. A scheme of coagulation is presented in Figure 10-3.

There is general agreement regarding the reactions of stages II and III. The uncertainty about the coagulation process concerns the reactions of stage I. Therefore, the coagulation sequence will be described in reverse order.

Stage III. Conversion of Fibrinogen to Fibrin

Fibrinogen is converted to fibrin by thrombin. Thrombin partially proteolyzes fibrinogen by splitting off two peptides. The fibrin monomer thus formed is unstable and dissolves easily in urea. However, in the presence of calcium ions and fibrin-stabilizing factor (factor XIII) derived from plasma transglutaminase, fibrin becomes cross-linked to form an insoluble clot.

Stage II. Conversion of Prothrombin to Thrombin

Since blood clots rapidly in the presence of thrombin, significant amounts cannot be present, as such, in the circulation. Instead, it is represented by a precursor, pro-

Table 10-2. Common Synonyms of Blood-Clotting Factors

Factor	Synonym
I	Fibrinogen
II	Prothrombin
III	Thromboplastin (tissue)
IV	Calcium ion
V	Proaccelerin, labile factor, plasma accelerator globulin (AcG)
VI	Discontinued
VII	Proconvertin, stable factor, precursor of serum prothrombin conversion accelerator (pro-SPCA)
VIII	Antihemophilic factor (AHF), antihemophilic globulin (AHG), thromboplastinogen, plasma thromboplastic factor (PTF), platelet cofactor I
IX	Plasma thromboplastic component (PTC), Christmas factor, platelet cofactor II
X	Stuart-Prower factor
XI	Plasma thromboplastin antecedent
XII	Hageman factor
XIII	Fibrin-stabilizing factor

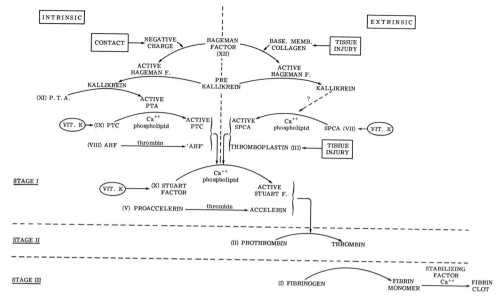

Figure 10-3
Schematic representation of the mechanism of blood coagulation, viewed as a three-stage process.

thrombin. *Prothrombin* is a plasma globulin fraction, with a molecular weight of about 63,000, which is formed in the liver. In the presence of the *converting enzyme, accelerin,* and *calcium ion,* prothrombin is converted to thrombin.

Stage I. Formation of Prothrombin-Converting Enzyme
Coagulation may occur in response to tissue injury or take place without apparent tissue damage, such as intravascularly or in a test tube.

The sequence of events following tissue damage is referred to by some as *extrinsic* coagulation, whereas coagulation without the participation of tissue factors is regarded as *intrinsic,* although at this time the distinction may not be so clear-cut.

Extrinsic Coagulation
The active principle in tissue is *thromboplastin* (III), which is continuously available and released or activated upon damage of tissue. Tissue thromboplastin in the presence of calcium ions and phospholipid apparently acts on or with the precursor substance of serum prothrombin conversion accelerator (pro-SPCA, VII) to activate the Stuart factor (X).

The activated Stuart factor acts with activated proaccelerin (V) to form what may be regarded as the prothrombin-converting enzyme.

Intrinsic Coagulation
Blood obtained free of tissue factors can clot readily, indicating that plasma itself possesses the factors necessary to form a thromboplastic-like material. Plasma thromboplastic activity is initiated by the contact of blood with negatively charged surfaces. This activates the Hageman factor (XII), which converts prekallikrein to kallikrein. Kallikrein then appears to activate plasma thromboplastic antecedent (PTA, XI), which, in the presence of Ca^{++} and phospholipid, activates the plasma thromboplastic component (PTC, IX). PTC, in conjunction with the antihemophilic factor (AHF, VIII) and in the presence of Ca^{++} and phospholipid, activates the Stuart factor.

Clot Retraction

After the clot forms, it retracts and extrudes fluid identical with plasma except for the absence of fibrinogen and factors VII and IX. This fluid is *serum.* Platelets are necessary for clot retraction since the process does not occur in platelet deficiency. The platelets apparently contain a contractile protein, thrombosthenin, which contracts when acted upon with adenosine triphosphate (ATP). The function of clot retraction has been considered to be that of drawing the wound surfaces together, recanalizing thrombosed vessels, or promoting more effective thrombosis.

Fibrinolysis

Blood clots are not permanent. After some time the clot is dissolved or lysed. The process of *lysis* involves the conversion of an inactive precursor, *profibrinolysin* (plasminogen), to the active form *fibrinolysin* (plasmin). It appears that proteolytic fragments from activated Hageman factor convert an activator from its inactive form, and this activator then converts profibrinolysin to fibrinolysin. Once formed, the fibrinolysin feeds back on the Hageman factor to increase the formation of the proteolytic fragments. Fibrinolysin degrades the fibrin meshwork, disrupting the clot.

There is evidence suggesting that a balance may exist between fibrin clot formation and lysis. Interruption of coagulation by heparin administration results in an elevation of profibrinolysin concentration. This mechanism has also been thought responsible for the maintenance of normal capillary flow by preventing microthrombus formation.

Natural Inhibitors of Coagulation

Since tissue breakdown and platelet destruction probably occur under normal circumstances, and since some platelets are no doubt damaged and thromboplastin probably enters the circulation to form thrombin, it would seem possible that coagulation in vivo might occur. Fortunately, inherent inhibitors *antithromboplastin, heparin,* and *antithrombin,* control clotting at every stage. Consequently, when small quantities of thromboplastin are released from tissue or formed in blood, inactivation by antithromboplastin occurs. However, even in situations of low-level release, the concentration of thromboplastin in the plasma may become sufficiently great to activate Stuart factor. Once again coagulation is prevented, this time by virtue of the presence of heparin.

Heparin is a polysaccharide produced primarily in the basophilic mast cells distributed throughout the pericapillary tissue. It is available commercially and serves as one of the major clinical anticoagulants. Heparin prevents coagulation in a variety of ways. Apparently it can inhibit prothrombin conversion, and possibly thrombin. Coagulation is also inhibited by the presence of antithrombin, which prevents the conversion of fibrinogen to fibrin. Antithrombin appears to be a complex of substances which are potentiated by heparin. In addition to this function, antithrombin is believed to clear thrombin from the blood once clotting is completed. Otherwise, it is conceivable that thrombin would progressively accumulate and produce widespread coagulation. With the presence of these anticoagulant factors, obviously coagulation does not normally occur spontaneously but requires adequate tissue damage to take place.

Coagulation Deficiencies

A deficiency of coagulation and a tendency toward bleeding may result from a deficiency of any of the principles cited in Figure 10-3 except the Hageman factor. Deficiencies arise in several ways. A dietary or absorption deficiency of vitamin K will result in a low blood prothrombin level (prothrombinemia). A reduced platelet count (thrombocytopenia) will lead to deficiency of cofactors. Since fibrinogen, prothrombin, and probably other factors in coagulation are formed in the liver, liver disease that alters prothrombin or fibrinogen synthesis will result in coagulation deficiencies. The classical hereditary condition of *hemophilia* results from the absence of antihemophilic factor (AHF), which leads to coagulation deficiency and a bleeding tendency. This effect is an inherited, sex-linked abnormality of males. In each case the replacement of the missing factor is usually sufficient to restore coagulation. Coagulation may also be impeded by excessive concentrations of fibrinolysin such as are seen in fibrinolytic purpura.

Coagulation Excesses

When the concentration of any of the inherent anticoagulants is low, when endothelial damage is great, or when blood is stagnant, coagulation in vivo may occur. A clot which remains fixed in a vessel is called a *thrombus,* and the condition, *thrombosis.* As blood flows by a thrombus, fragments (*emboli*) may break away and circulate throughout the vascular system. An embolus continues to circulate until it reaches a vessel with a cross-sectional area smaller than its own. It then becomes lodged in the vessel, effectively occluding the vessel and depriving the tissue supplied by the vessel of adequate nutrition. When the area supplied is extensive or is situated in a vital organ, death can result. Usually the occluded region, if restricted in size, becomes invaded by a collateral circulation, and this may restore an adequate blood supply. When collateral circulation is inadequate, the occluded area degenerates to form an *infarct.*

Anticoagulant Agents

The most prominent anticoagulant agents administered clinically in circumstances of excessive coagulation in vivo are heparin and *dicumarol,* a drug that acts in the liver to prevent prothrombin formation by antagonizing the action of vitamin K. Because calcium ion is intimately associated with both the formation of the converting enzyme and the conversion of prothrombin, effective removal of calcium ion from the plasma prevents coagulation. However, the use of calcium precipitants or complexing agents is restricted to in vitro anticoagulation because a decreasing of the calcium ion concentration leads to cardiac arrest. *Sodium citrate* is a compound that removes calcium ion from solution by forming a poorly dissociating calcium salt. This substance is used routinely in blood banks as an anticoagulant. *Oxalates* are also occasionally used as in vitro anticoagulants.

Coagulation Tests

Numerous tests are in use to assess the effectiveness of the coagulation system: clotting time, prothrombin consumption time, thromboplastin generation, thrombin generation time, and others. The clotting time is the most comprehensive assay, since it is sensitive to deficiencies of most of the coagulating principles. The other tests are used to identify the coagulation deficiency more specifically. Details of the tests are to be found in any standard hematology text.

REFERENCES

Biggs, R., and R. G. MacFarlane. *Human Blood Coagulation and Its Disorders* (3rd ed.). Philadelphia: Davis, 1962.

MacFarlane, R. G. (Ed.). *The Functions of Blood.* New York: Academic, 1960.

Putnam, F. W. (Ed.). *The Plasma Proteins.* New York: Academic, 1960.

Ratnoff, O. D. *Bleeding Syndromes.* Springfield, Ill.: Thomas, 1960.

Ratnoff, O. D. Some recent advances in the study of hemostasis. *Circ. Res.* 35:1–14, 1974.

Wintrobe, M. M. *Clinical Hematology.* Philadelphia: Lea & Febiger, 1961.

11. Fluid Dynamics

Carl F. Rothe

The rate of blood flow through tissues is a vital determinant of the adequacy of cellular nutrition and waste removal. The physical factors governing this flow are the same as those modifying fluid flow through any living or nonliving system. A thorough under-standing of cardiovascular function therefore requires the application of these principles of fluid dynamics. Once the concepts are understood, the effects of disease on the function of the cardiovascular system can be appreciated more fully, for one may reasonably predict the consequence of various changes. Most of the principles are summarized by the equations given in this chapter. The same equations, with differ-ences in parameters or added terms, are applicable to flow in any system (e.g., air flow in the lungs).

FACTORS INFLUENCING FLOW

The rate of flow of a fluid is proportional to the driving pressure gradient. Expressed mathematically, this is

$$\dot{Q} = G\,\Delta P = \frac{1}{R}\,\Delta P \tag{1}$$

The flow, $\dot{Q}$, is expressed in units of volume per unit of time (e.g., milliliters per minute). The *perfusion pressure,* or *effective pressure gradient* along the vessel in question, $P_1 - P_2$ or ΔP (Fig. 11-1A), is expressed in millimeters of mercury (mm Hg or torr), or newtons per square meter. (The unit of pressure in the System International, the pascal or N/m^2, is rather small, since 1 mm Hg $= 133.3\ N/m^2$.) The *conductance,* G, is the proportionality constant of equation 1 and is expressed in flow units divided by pressure units (e.g., milliliters/minute per millimeter of mercury). The reciprocal of conductance is *resistance,* R. The resistance opposing the flow of fluid cannot be measured directly but is calculated from simultaneous measurements of the flow and pressure gradient using equation 1. It is expressed as the units of pressure divided by the units of flow. The equation $\dot{Q} = P/R$ is fundamentally the same as Ohm's law of electrical circuits: $I = E/R$. There is no commonly accepted resistance unit analogous to the ohm. Note that if the pressure gradient increases, the flow increases, whereas if resistance increases, the flow decreases.

Flow is proportional to the pressure gradient, which in the case of a tissue vascular bed is the arterial minus the venous pressure. Because the venous pressure is much less than the arterial pressure in the systemic circuit, the venous pressure can often be neglected in calculating overall tissue vascular resistance. However, measurement of the resistance to flow through a low-pressure circuit, such as the pulmonary circulation, requires a knowledge of the pressure in the pulmonary veins as well as in the pulmonary artery.

Poiseuille's Equation

Poiseuille's equation describes in more detail the factors determining the resistance to flow (Fig. 11-1A). Poiseuille's equation is

$$\dot{Q} = \Delta P \times \frac{\pi r^4}{8L} \times \frac{1}{\eta} \tag{2}$$

(A) POISEUILLE'S LAW

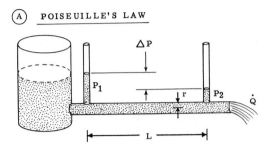

(B) VELOCITY PROFILE OF A FLOWING STREAM

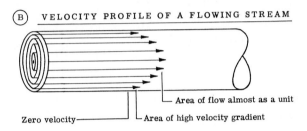

Area of flow almost as a unit

Zero velocity—┘ └Area of high velocity gradient

Figure 11-1
A. Factors in Poiseuille's equation. B. The velocity profile of the flowing stream is parabolic, with the highest velocity in the center, and no flow at the wall.

Here r is the radius of the vessel, L is its length, and η is the coefficient of viscosity. Note that flow will increase if the pressure gradient or radius is increased or if the length or viscosity is decreased.

There are several important consequences from the mathematical derivation of Poiseuille's equation. The primary assumption, attributed to Sir Isaac Newton, is that fluid flows as infinitely thin layers or laminae sliding past each other, giving *laminar* flow. The force per unit area (*shear stress*) tending to impede this sliding is directly proportional to the relative velocity between two adjacent layers (*shear rate*). The shear rate is a velocity gradient and is the longitudinal displacement per second divided by the distance between layers. Hence, the dimensions are: cm/sec per cm or $\sec^{-1}$ (the units of length cancel). The ratio of shear stress to shear rate is called the viscosity of the fluid. Like friction, there is dissipation of energy to heat because of the relative motion of the molecules past each other. If the viscosity is independent of shear rate, the fluid is said to be *newtonian*. Blood and many colloidal suspensions may not be newtonian, for at low shear rates (less than $100 \sec^{-1}$, i.e., 10 cm/sec per millimeter of thickness) blood viscosity becomes sensitive to shear rate, as will be described below.

From the theoretical derivation and from actual measurement, the fluid at the wall of a blood vessel is stationary; the velocity is zero (Fig. 11-1B). However, the velocity gradient at this point is maximum. At the center of the tube the velocity is maximum whereas the velocity gradient is zero.

There are two types of use for an equation such as this: (1) The equation *describes* the characteristic of a physical *system* (Chap. 7B). It is the law relating input to output of the system; i.e., it describes how the variable Q is dependent on the various independent variables ΔP, r, η, and L. The equation is considered valid and the assumptions used in arriving at it are not disproved if, after each of the independent variables and the flow have been measured, the predicted flow equals the measured flow. Statistical analyses are used to estimate whether the discrepancies between measurement and prediction are likely to be real or a result of random variation. (2) The equation may also be used to *estimate* one of the independent variables. Assuming that the equation is valid under the conditions used, if flow and all but one other variable are measured, the value of this other variable may be calculated. One way to determine viscosity is to measure the rate of flow of the fluid through a tube of known dimensions using a known pressure gradient.

Note that the flow is inversely proportional to the viscosity (η) of the fluid. Syrup, for example, is a highly viscous fluid whereas water has a relatively low viscosity. The viscosity of air is much less than that of water but is not negligible; it is about one-thirtieth of that of water at body temperature.

Since flow through a resistance vessel is directly proportional to the fourth power of the radius of that vessel, the resistance is markedly influenced by the radius of minute blood vessels such as the arterioles. For example, at a constant pressure gradient, a 19 percent increase in arteriolar radius permits a 100 percent increase in flow. Physiologically, resistance to the flow of blood is determined primarily by the caliber of the small arteries, arterioles, and precapillary sphincters, which in turn is dependent on changes in the vasomotor tone. This vasomotor tone is the state of tonic contraction of the smooth muscles of these resistance vessels (see Fig. 11-3 and Chaps. 12 and 17).

The *total peripheral resistance* (TPR) is the resistance to flow of blood of the systemic vasculature as a whole. It is the aortic pressure minus the central venous pressure divided by the total flow through the system (the cardiac output). It decreases with exercise and is increased in essential hypertension. Since the circulation consists of many parallel branches, the calculated value of the total peripheral resistance is less than that of any resistance through a particular organ or region. When using total peripheral resistance as a diagnostic tool, one should remember that pathological conditions causing an increase in resistance to flow in one organ may be compensated by opposite changes in other organs, so the total peripheral resistance may remain constant.

Viscosity of Blood

Plasma is about 1.8 times as viscous as water. Whole blood has a variable viscosity of between 2 and 15 times that of water. The causes of this range of blood viscosity are physiologically important.

The *hematocrit ratio* has a marked effect upon the viscosity and flow of whole blood, as may be seen in Figure 11-2A and B. If the hematocrit were zero, the viscosity would be that of plasma, but as the proportion of blood which is cells increases, the viscosity of the blood increases. Beyond the normal hematocrit ratio of about 0.45, the blood becomes much more viscous. Thus, if the concentration of erythrocytes is abnormally large, as in polycythemia, the flow tends to be diminished because of the increased viscosity of the blood. In contrast, in anemia the proportion of cells is so reduced that the flow tends to increase because of the decrease in viscosity.

Temperature has an inverse effect upon viscosity. At 0°C, both blood and water are about 2.5 times more viscous than at 40°C. At body temperature, there is a 2 percent increase in viscosity per 1°C decrease in temperature. This rise in viscosity of cold blood is an important factor in determining blood flow when extremities are cooled by immersion in cold water or when hypothermia is used in surgery.

Fluid velocity may influence blood viscosity, for although blood behaves as a newtonian fluid under normal physiological conditions in all the vasculature, under some pathological conditions of nearly stagnant flow the viscosity of blood increases. Since flow is then not linearly proportional to the driving pressure—that is, the velocity of each layer is not proportional to the force acting on it—the blood is *non-newtonian*. Because the calculated viscosity of blood is different depending upon the shear rate or the diameter of the flow channel (Fig. 11-2A and below), the ratios of shear stress to shear rate are called the *apparent viscosities* under the various conditions. Because the geometry of the minute blood vessel network is so complex that calculation of viscosity from pressure-flow data and the Poiseuille equation is impossible, the viscosity of blood

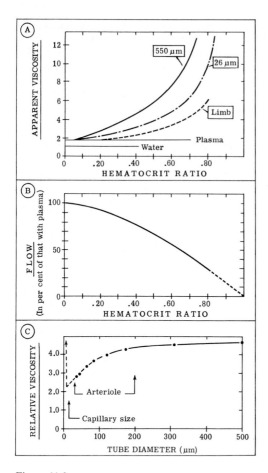

Figure 11-2
A. Effect of the concentration of erythrocytes (hematocrit ratio) on the apparent viscosity of blood perfused through tubing of various diameters and an isolated dog hind limb. There is less effect of hematocrit ratio on viscosity if the blood flows through small, 26 μm diameter tubes than if the flow is through larger, 550 μm glass tubes. (Data from Bayliss in Frey-Wyssling [Ed.], *Deformation and Flow in Biological Systems.* New York: Interscience, 1952.) There is even less effect when blood flows through a hind limb. (Data from Whittaker and Winton. *J. Physiol.* 78:357, 1933.) B. As the hematocrit ratio of blood is increased, the flow of blood through isolated tissue decreases, other factors being constant. (Derived from data of Whittaker and Winton. *J. Physiol.* 78:357, 1933.) C. The effect of tube diameter on the relative viscosity of blood. (Data from Fåhraeus and Lindqvist. *Amer. J. Physiol.* 96:565, 1931.)

is compared to that of saline solution—a fluid of known characteristics. The *relative viscosity* of a fluid is thus the ratio of the flow of (saline solution) water to that of the fluid under the same conditions of tubing size and geometry, temperature and pressure gradient. Such a comparison also provides an estimate of the apparent viscosity.

A major cause of the increased viscosity of blood at abnormally low shear rates is the tendency for erythrocytes to aggregate into rouleaux (i.e., they resemble a stack of coins). Indeed, blood has a very small but finite yield stress. At stresses of less than about 0.1 dyne per square centimeter blood does not flow. Since force is required to break up the rouleaux so that flow can continue through the capillaries, less driving force is available to overcome the viscous friction. As the shear rate is increased, the apparent viscosity decreases toward an asymptote. Fibrinogen is necessary for this anomalous rheological behavior of whole blood. Although normal blood does not undergo rapid aggregation, the tendency toward cell clumping may be important under conditions such as traumatic or burn shock, when the blood tends to *sludge*.

Flow in Minute Vessels

If blood is pumped through very small tubes such as arterioles, the apparent viscosity is decreased as shown in Figure 11-2A and C. Blood thus behaves more like water when flowing through small tubes than when pumped through larger tubes. (With either fluid, however, the flow in small tubes is small!) This phenomenon is called the *Fåhraeus-Lindqvist* effect. It is caused by the erythrocytes flowing through vessels only a few times larger in diameter than the diameter of these particles. In large tubes (>0.5 mm) the viscosity of blood is not dependent upon the size of the tube, but as the tubing diameter is decreased (as in small arteries and arterioles) the apparent viscosity decreases. Finally, however, in the 4 μm diameter capillaries the apparent viscosity is increased, for even though the 8 μm diameter erythrocytes are readily distorted so that the cells can be easily perfused through the tissue, some force is required to distort the cells. If the diameter is yet smaller, as with precapillary sphincter constriction, the erythrocytes occlude the resistance vessel, the apparent viscosity becomes infinite, and the flow ceases (Fig. 11-2C). This phenomenon is superimposed on the general effects of vasoconstriction (Fig. 11-3).

Another aspect of the anomalous viscosity of blood and a partial explanation of the Fåhraeus-Lindqvist effect is the *axial accumulation* and *streaming* of the red blood cells, which leaves a cell-free marginal zone a few micrometers thick along the wall of the

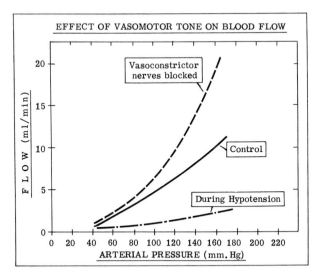

Figure 11-3

At a given pressure, the flow of blood through skeletal muscle is influenced by the vasomotor tone. If impulse conduction along vasoconstrictor nerves is blocked, the flow is increased. The arteriolar diameters become larger—dilate. Conversely, during severe hypotension as a result of hemorrhage, there is an intense, neurally induced vasoconstriction of the blood vessels in vascularly isolated muscles of dogs, so that the flow is reduced even though the perfusion pressure may be normal.

vessel and an increasing concentration of cells toward the center of the flow stream. Erythrocytes tend to migrate toward the middle of the flowing stream in small tubes at moderate velocities. The less viscous plasma is left at the wall, where the greatest velocity gradient exists. Consequently, such blood approaches the viscosity of plasma alone. The forces causing axial accumulation are not well known. Factors of importance include flow rate, cell concentration, and cell shape, size, and deformability (Charm and Kurland, 1974). The reduction in apparent viscosity, in small-diameter tubes in com- parison to large ones, is particularly noticeable with abnormally high hematocrit ratios (Fig. 11-2A). Axial accumulation, by permitting plasma skimming into side branches, is a factor in determining the distribution of erythrocytes and plasma in the capillary bed (Chap. 12).

Continuity Equation

The continuity equation states that the velocity of flow times the cross-sectional area of a vessel is at all points constant:

$$v_1 \times A_1 = v_2 \times A_2 \tag{3}$$

Here v is the velocity and A is the cross-sectional area. As the cross-sectional area of the vascular system increases, the velocity must decrease since the mass of blood moving in the body is constant (see Fig. 16-2). Thus, in a human aorta of about 2 cm^2 in area, the mean velocity is about 50 cm per second, but in the 4 μm diameter capillaries it is only about 0.1 cm per second because the *total* cross-sectional area in man is over 1000 cm^2 (1 square foot).

Equation of Motion

When a pressure or force is applied to a system having mass, it does not start moving instantly. Inertia must be considered. Newton's laws of motion may be applied to a flowing stream to obtain an estimate of the distribution of the force, since the algebraic sum of the forces is zero (d'Alembert's principle). The basic equation applied to a mechanical system, such as a weight suspended on a spring and moving in a viscous liquid, is

$$M\,a + R\,v + S\,x = F \tag{4a}$$

Here F is the applied force; x is the displacement of the spring from the neutral or zero force position; S is the parameter defining the stiffness of the spring (force/unit displacement); v is the velocity or rate of change of position (dx/dt); R is the resistance or coefficient of friction related to the viscosity of the fluid and the geometry of the moving object; a is the acceleration or rate of change of velocity; and M is the mass. The first term (M a) represents the inertial part of the applied force; the second, the frictional or viscous part; and the third, the elastic part.

When applied to a flowing fluid,

$$M\,\ddot{Q} + R\,\dot{Q} + S\,Q = P \tag{4b}$$

where $\ddot{Q}$ is the rate of change of flow or acceleration of the bolus of fluid, $\dot{Q}$ is the flow, Q is the distending volume, and P is the applied pressure. Thus, the pressure developed in the ventricle during systole is distributed into three parts acting to: accelerate the bolus of blood, M $\ddot{Q}$; overcome the resistance to flow through the aortic valve and aorta, R $\dot{Q}$; and distend the aorta, S Q. At the peak of ventricular pressure, the inertial and viscous terms normally represent only about 5 percent each of the total force. Most of the developed pressure is acting to distend the aorta and move the blood through the periphery. The above analysis is only an approximation, for it considers the parameters of mass, viscosity, and stiffness as lumped at discrete points rather than as continuously distributed throughout the system. Nonetheless, consideration of inertia, for example, is crucial in understanding the dynamics of ventricular ejection, closing of the valves, and the resonant behavior of fluid systems. A further consequence of the physics of the system is that energy is dissipated (i.e., lost to the environment) *only* by the frictional term. The energy stored as momentum on acceleration and that used to distend the aorta are available later to drive the blood through the periphery.

Bernoulli's Equation

Rather than considering a balance of forces, as in equations 4a and 4b, one may obtain another useful description of a system by considering a balance of energy. Bernoulli's equation is based upon the principle of conservation of energy and states that the total energy, E, of a laminar (streamline) flow stream *without resistance* is constant and equals the sum of the potential energy of pressure, P V (here, V is volume of fluid); the potential energy due to gravity above a reference plane, M g h, (here, M is mass; g, the acceleration by gravity; and h, the height); and the kinetic energy due to the velocity, v, of the mass which is flowing, $\frac{1}{2}$ M v^2. Thus

$$E = P V + M g h + \tfrac{1}{2} M v^2 \qquad (5)$$

The driving force or *effective pressure gradient* for the propulsion of blood is the total energy gradient—i.e., the difference between the total energy, as expressed in equation 5, at the upstream point of measurement and that at the downstream end. Thus, the effective pressure gradient includes not only the usual *pressure* type of potential energy but also hydrostatic or *gravitational* potential energy and the *inertial* or kinetic energy of the moving mass. The hydrostatic factor is important in understanding the factors determining the flow to and from the legs (Chap. 16). The inertial factor accounts for over half of the driving force in the pulmonary artery or great veins during vigorous exercise. During one part of the ejection phase of the cardiac cycle, the inertial term is great enough so that flow continues even though the pressure gradient is reversed (Chap. 13). When the rate of flow changes, as in the arteries, the resistance to flow, more correctly termed *impedance,* includes this inertial term. With inertance or compliance in the system, the flow pattern does not coincide with that of the pressure but lags behind the pressure if the inertial component predominates, or leads if the arterial compliance (see below) is dominant. The mathematics describing the relationship between the driving arterial pulse and the resulting flow becomes complex (see Attinger, 1968; Noordergraaf, 1969; Bergel et al., 1972; McDonald, 1974).

A consequence of the Bernoulli principle is that, if the velocity of flow increases because of a constriction in a tube or because blood is being rapidly propelled from the heart, more of the total energy is in the form of kinetic energy (velocity) and less is in the form of potential energy (pressure). Under most physiological conditions this factor is minor. However, it does help explain the closure of the valves of the heart, for blood rapidly flowing past partially closed valve leaflets tends to suck them together.

The orientation of a catheter to measure the lateral pressure of a flowing stream is important, for if it is directed upstream against the direction of flow it will give a somewhat higher value than if it is directed at right angles to the flow stream. If the opening is oriented toward the flow stream, as the flow impinges on the cannula some kinetic energy due to the inertia of the blood is converted to potential energy and is added to the lateral pressure. In the larger systemic arteries this "end-pressure" or kinetic factor is minor, but in the pulmonary vessels with high flow velocities and low lateral pressures it becomes significant, for an open-ended catheter pointing into a stream flowing at 100 cm per second will have a pressure 4 mm Hg greater than that measured with a catheter with openings in the side. At 200 cm per second (a value not uncommon in the aorta or pulmonary artery, at the peak of outflow) the difference is 15.9 mm Hg.*

Transmural Pressure

The *transmural pressure*, P_T, is the difference between the pressure inside, P_i, a vessel and the pressure outside, P_o, the vessel:

$$P_T = P_i - P_o \tag{6}$$

Often the outside pressure is atmospheric and so may be considered zero.

The pressure gradient *along* a resistance vessel is a major determinant of flow (equation 1). The transmural pressure *across* a resistance vessel is also a factor determining the resistance to flow, for as the pressure increases, the vessels are passively distended and allow an even higher flow than would be found with constant vessel size (Figs. 11-3 and 11-4B). Contraction of vascular smooth muscle in conjunction with the elastic elements counteracts the distending force and so regulates the flow of blood to the tissue. This phenomenon explains in part the nonlinear pressure-flow characteristics of blood through a living tissue (Fig. 11-3).

Compliance

Hollow organs and tubes show an increase in volume with an increase in transmural pressure. This property is called *distensibility*. Compliance (C) is a general term describing the change in dimension (strain) resulting from a change in transmural pressure (stress). It is usually defined as

$$C = \frac{\Delta V}{\Delta P} \tag{7a}$$

where ΔV is the change in volume in response to a change in pressure, ΔP. The coefficient of stiffness, S, of the third term in equation 4b represents the reciprocal of compliance. The histological structure of most biological blood vessels—and indeed most tissues—is such that with distention, the compliance decreases; i.e., they become stiffer. Furthermore, blood vessels contain a volume at zero transmural pressure. Therefore, it is important to specify the volume and transmural pressure at which a

---

*In working with equations such as these, recall that: Force = mass times acceleration, and weight = mass times the gravitational constant. Thus the M g h in the second term of equation 5 is a force (in newtons) due to gravity (g = 9.806 m/sec²) acting on the mass (in kg). Multiplying by height (in m) gives work in newton-meters [kg · (9.81 m/sec²) · m = N · m], the potential energy available due to height. In the third term ($\frac{1}{2}$ M v²), since a newton is kg · m/sec², one obtains directly the appropriate units

$$\frac{1}{2} \text{ kg} \cdot (\text{m/sec})^2 = \frac{1}{2} (\text{kg} \cdot \text{m/sec}^2) \text{ m} = \frac{1}{2} \text{ N} \cdot \text{m}$$

Note: A newton-meter is a joule of work, and a newton-meter per second is a watt: the rate of doing work or power.

In solving the catheter orientation problem, total work and the height are assumed to be constant, the kinetic energy and potential energy of pressure are set equal to each other (P V = $\frac{1}{2}$ M v²), and the resulting equation is solved for P.

The mass (M) moving is density times volume (ρV). By assuming that this volume is equal to the pressure volume, one obtains

$$P = \frac{\frac{1}{2} M}{V v^2} = \frac{\frac{1}{2} \rho V}{V v^2} = \frac{1}{2} \rho v^2$$

Here, P is the pressure (dyne/cm²) from the conversion of kinetic to potential energy, ρ is the density (gm/cm³), v is velocity (cm/sec), and 1 dyne/cm² = 1 gm · cm/sec² per cm². Note that 1 mm Hg = 1333 dynes/cm².

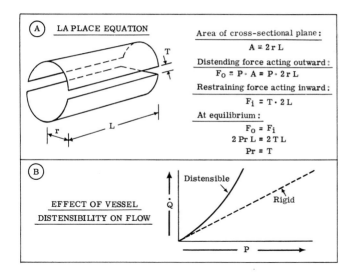

Figure 11-4
A. Derivation of the Laplace equation. B. As the driving pressure in a distensible vessel is increased, the flow increases more rapidly than that in a rigid vessel of the same resting diameter, because both the pressure gradient and the radius are increased by increases in the driving pressure.

compliance measurement is made. *Capacitance,* a term borrowed from electronics, implies a storage function, suggesting the ratio of the total volume contained to the transmural pressure. Some workers equate it with compliance. Because a long vessel would have a larger change in volume for a given pressure increase than a shorter vessel, a *coefficient of volume distensibility* (κ) is defined as the fractional change in volume per unit change in pressure:

$$\kappa = \frac{\Delta V}{V \cdot \Delta P} \tag{7b}$$

The combination of resistance, compliance, and inertance of the blood in the arterial system determines the arterial input impedance, the ratio of pressure gradient to flow at a specified frequency. Because of its compliance, the aortic input impedance for pulsatile frequencies between 1 and 15 Hz is only about 10 percent of that for the steady flows and so acts to reduce the load on the heart, as well as to aid in continuing flow to the peripheral tissues during cardiac diastole.

Laplace Equation

Because blood flow is proportional to the fourth power of the radius of the arterioles and other resistance vessels, small changes in vessel dimensions may have profound effects on blood flow. Therefore, the factors influencing vascular dimensions, such as transmural pressure and wall tension, are important contributors to the state of vascular resistance. The Laplace equation states that the tension (T) in the wall of a vessel required to maintain a given radius is proportional to the product of the *transmural pressure* (P) and the *radius* (r).

$$T = Pr \tag{8}$$

A derivation of the Laplace equation provides a valuable insight into simple mathematical reasoning. If the wall of the vessel is to be stationary, the *distending force* pushing the vessel wall out must be balanced by a *restraining force* produced by the tensions of the fibers in the wall. The distending force is equal to the area of a longitudinal sectional plane times the transmural pressure. In the case of a cylinder, the distending force equals P × 2rL, where P is the pressure in grams per square centimeter, r is the radius, and L is the length of the cylindrical vessel (Fig. 11-4A). The restraining force is equal to T × 2L, where T is the tension per unit length and L is the length. The factor 2 is used because both edges of the vessel are acting to hold it together. (In this simple derivation the walls of the vessel are assumed to be infinitely thin, an assumption which is not valid for calculating wall tensions of small arteries, for example.) Equating these two expressions gives the relationship T = Pr. (For a sphere, the equation becomes T = P r/2.) In a larger vessel the forces in the wall required to keep it from distending at a given pressure must be higher than in a smaller vessel. For the heart to eject blood at a given pressure, the myocardial tension must be greater in a large heart than in a small heart (Chap. 15). In aortic aneurysm the walls of the aorta are damaged, so the diameter becomes much larger than normal. This development is exceedingly serious; the enlarged vessel is much more likely to rupture, because of the high tensions in its walls, even at normal arterial pressures. The capillaries, because of their minute size, have wall tensions of less than 10 mg per centimeter of vessel length and so remain intact in spite of wall thicknesses of about 1 μm.

Turbulence

Blood flow through vascular channels is normally streamline in nature. Under special conditions turbulent fluid flow, with swirls or eddies, may develop. In this event, flow is approximately proportional to the square root of the pressure drop, so a given increase in pressure results in less than a proportionate increase in flow:

$$\dot{Q} \simeq k \sqrt{\Delta P} \tag{9}$$

The reason is that the potential (driving) energy is dissipated in the swirls and eddies as heat. If turbulence is of sufficient magnitude, audible sound is produced like the hiss of air from a compressed-air outlet.

If the ratio of inertial to viscous forces of a flowing stream exceeds a given value, a slight disturbance in a streamline, laminar flow pattern will lead to the development of a randomly oriented turbulent pattern. Turbulence is likely when the ratio, called Reynolds' number (Re), exceeds a value of about 1000 for blood and 2000 for fluids (liquid or gas) without suspended particulates (e.g., erythrocytes). Reynolds' number may be computed for a cylindrical conduit with the following equation:

$$Re = \frac{\rho \, v \, D}{\eta} \tag{10a}$$

where ρ is the density of the fluid in gm/cm³ · g (here g is the gravitational constant, 980 cm/sec²), v is the average velocity in cm/sec, D is the diameter of the vessel in cm, and η is the viscosity in gm · sec/cm². Other factors being constant, an increase in velocity is likely to result in turbulence. Writing the equation in terms of flow, $\dot{Q}$, rather than velocity gives

$$Re = \frac{4 \rho \, \dot{Q}}{\pi \, D \, \eta} \tag{10b}$$

Since blood density is reasonably constant, the factors tending to cause turbulence are an increase in the rate of blood flow, such as occurs in exercise; a decrease in blood viscosity, as seen in anemia; and a decrease in the radius of the blood vessel, as occurs with stenotic valves or coarctation of a blood vessel such as the aorta.

The streamline flow of fluid may also break into organized swirls (called *vortex shedding*) and produce sounds of a rather definite pitch at flow velocities much *lower* than those causing fully developed random turbulence. As a fluid flows past an obstacle in the flow stream, a slight disturbance causes the flow wake to fluctuate at first toward one side. Inertial forces continue the stream into a higher-velocity area, which then forces it back in the opposite direction, where it overshoots and then is forced to swing back to repeat the cycle. The fluttering of a flag in a stiff breeze is produced by the same mechanism. Described another way, the fluid tends to follow the curvature of the obstacle, and so curves in a swirl around it to form a vortex. This is an area of low pressure; hence, the stream tends to flow into that area from the opposite direction. The cycle is repeated indefinitely with swirls of opposite direction being set up, vortices shed and carried downstream. Each time the wake of the flow stream swings back and forth, it changes the lateral pressure on the wall of the vessels.

The studies of Bruns (1959) have shown that the frequency and location of cardiovascular sounds under physiological conditions are explainable by this process. Note that the fluctuating wake has a maximal component of fluctuations at right angles to the flow stream. If lateral fluctuations of pressure from the oscillating flow stream are of sufficient magnitude and frequency, audible sound is produced. In this kind of turbulence, relatively more sound is produced with a given driving pressure than in the randomly oriented flow pattern of classic Reynolds' turbulence. It is probably the source of most of the murmurs heard in the cardiovascular system and can occur under conditions in which Reynolds' number is much less than 2000.

Summary

1. Factors regulating flow are the pressure gradient along the vessel, the viscosity of the fluid, and the physical dimensions of the resistance vessel.
2. An increase in transmural pressure, by distending the resistance blood vessels, tends to increase flow irrespective of its concomitant effect on the driving pressure.
3. With constant pressure gradient, the smooth muscle tension (vasomotor tone) acts to reduce the caliber of the resistance vessels and so reduces flow.
4. The apparent viscosity of blood is increased (hence the flow decreases if other factors remain constant) as a result of increased hematocrit ratio, decreased temperature, turbulence, an increase in minute vessel size up to about 0.5 mm, or erythrocyte aggregation if flow rate is reduced toward zero.

MEASUREMENT OF BLOOD FLOW

Understanding of the performance of the circulatory systems requires a measurement of the flow of blood. In measuring flow, two sources of error must always be considered: errors in the measurement itself, and deleterious effects of the measurement process upon the functioning of the organism. In cardiovascular research, if highly accurate results are to be obtained, measurements must be made without disrupting the organism by concomitant surgery, anesthesia, pain, impeding the normal blood flow, inducing blood clotting, or causing uncompensated loss of blood. Much has been learned about the cardiovascular system, however, without the necessity of placing such stringent requirements on the measuring process.

The simplest and fundamentally the most accurate way to measure blood flow is to collect a known volume of vascular outflow over a measured time interval by means of a graduate and stopwatch. This method, although simple, is obviously of no value for measuring cardiac output because the removal of the blood would seriously affect the animal. However, it is of value in measuring the flow through small amounts of tissue.

The flow of blood from small segments of tissue may be measured by cannulating a vein and counting the number of drops of blood per minute leaving the cannula. If the volume of the drops is known, the flow may be determined. The maximum flow measurable is limited to about 10 ml per minute. Furthermore, the size of the drops is dependent on a multitude of variables; thus, reliable measurement of complex fluids such as blood is difficult.

To measure the pattern of pulsatile blood flow, electromagnetic and ultrasonic flowmeters are available (Geddes and Baker, 1975; Cobbold, 1974). The probes of these instruments can, if desired, be implanted in an experimental animal so that painless measurements can be made at a later date without anesthetics. The artery does not have to be cannulated since the devices detect the flow through the walls of intact blood vessels. Within a few years ultrasonic flowmeters will be commercially available to measure blood flow without penetrating the skin.

The *electromagnetic flowmeter* is based upon the principle of an electric generator. If a conductor is moved through a magnetic field, an electrical potential is developed in the conductor which is proportional to the length and velocity of the conductor. Blood going through a magnetic field forms a single conductor; consequently, electrodes can be placed on both sides of the blood vessel to pick up the potential developed.

The *ultrasonic flowmeter* is based upon the principle that the velocity of sound along a flowing stream in relation to a stationary point is higher going downstream than going upstream. One-half the difference between the two velocities is the stream velocity. By the use of techniques similar to those used for radar and sonar, a pulse of sound is produced and the time required for it to travel downstream is compared with the time for a similar pulse of sound to be transmitted upstream. The Doppler principle of change in frequency following reflection from moving particles (red blood cells) is currently the principle most widely used.

Indicator dilution provides a way to measure blood flow. An indicator is injected into the bloodstream, samples are taken after mixing, and the flow is computed with respect to the degree of dilution of the indicator. The method is widely used for determining cardiac output and is fully as accurate as more direct methods if proper conditions are met.

The *material conservation principle* is used to account for the indicator injected. If a known amount of a substance is injected into a stream flowing into a particular region, it must either (1) leave the region, (2) accumulate in the region, (3) be converted to some other material, or (4) be excreted by some route other than the flow stream.

The *indicator dilution principle* is based on the fact that an amount of a substance (A) mixed uniformly in a volume (V) has a concentration (C) such that by definition of concentration

$$C = \frac{A}{V} \tag{11}$$

This *dilution method* (see also Chaps. 10 and 23) is used to estimate the volumes of various fluid compartments in the body by rearranging the equation to V = A/C. A material is used which equilibrates within the entire fluid compartment in question. Any losses of indicator from the fluid compartment must be accounted for by the material conservation equation.

Indicator dilution can also be used to measure flow. Mean flow ($\overline{Q}$) is volume (V) moved per unit time (t):

$$\overline{Q} = \frac{V}{t} \tag{12}$$

This volume of fluid passing a given point in a unit of time may be estimated by the dilution principle, by substituting V = A/C into equation 12. If the amount of indicator added or removed by an organ is known, and is divided by the change in concentration of a substance as it passes through the organ, the flow can be calculated, as indicated below. The material conservation principle is used to account for the indicator used.

Fick Method (Constant Infusion or Removal)
The blood flow through an organ can be determined by measuring the amount of a substance removed by the organ per minute and dividing by the change in concentration of the substance in the blood as it goes through the organ. For *cardiac output,* for example, measurement of the flow of blood through the heart and thus through the lungs is desired. The transport of oxygen by the lungs provides a convenient indicator agent. The volume of oxygen consumed per minute, A/t, may be measured by means of a spirometer. Oxygen consumption is measured over an interval of 5 to 10 minutes. The concentration of oxygen in the venous blood coming to the right heart and lungs (C_V) and the concentration of oxygen leaving the lungs and left heart in the arterial blood (C_A) are measured to give the change in concentration (ΔC). This change in concentration is the dilution. The *Fick equation* is derived as follows:

$$\dot{A}_a = \dot{A}_v + \dot{A}_r$$

where $\dot{A}_a$ is the amount of indicator (oxygen) leaving via the arteries per unit time, $\dot{A}_v$ is the amount entering via the veins per unit time, and $\dot{A}_r$ is the amount entering the bloodstream via the respiratory system. (It is the rate of oxygen consumption [ml/min] in this case.) Since flow ($\dot{Q}$) times concentration (C) equals amount transported per unit time ($\dot{A}$):

$$\dot{Q}_a\,C_a = \dot{Q}_v\,C_v + \dot{A}_r$$

If we can assume that the venous return ($\dot{Q}_v$) equals the cardiac output ($\dot{Q}_a$) and that the arterial and venous indicator (oxygen) concentrations are constant though different, then

$$\dot{Q}_a\,(C_a - C_v) = \dot{A}_r$$

$$\dot{Q} = \frac{\dot{A}_r}{C_a - C_v} = \frac{A/t}{\Delta C} \tag{13}$$

The technique and calculations are shown schematically in Figure 11-5. The problem to be solved is: If each 100 ml of blood going through the lungs takes up 5 ml oxygen, how much blood must flow per minute to take up 200 ml of oxygen?

Arterial blood samples are obtained by arterial puncture. The major problem with this method is obtaining a valid sample of mixed venous blood. This blood must be representative of all the venous blood. Since the kidneys usually extract only about 2 to 3 volumes of oxygen per hundred volumes of blood, the brain about 6, and the heart and active muscle 10 to 15, the sample must be obtained after mixing. Mixing is adequate only at sites beyond the right ventricle. Cardiac catherization is thus routinely used to obtain these samples. The catheter is inserted into an arm vein and advanced through the veins into the right heart and on through the pulmonary valve into the pulmonary artery. The procedure is safe, but to the patient it is a rather heroic one, so the apprehension tends to give abnormally high values.

The direct Fick is the standard method of determining the cardiac output in man but is subject to errors, particularly during severe exertion. It is a *determination* of mean flow, since the oxygen consumption is averaged over a 5- to 10-minute period. The venous sample should be taken continuously during this period if accurate results are desired. It is not adequate for measuring rapid changes in cardiac output, since it is impossible to measure the true oxygen consumption over short intervals of time because of the variable amount of air held in the lungs from breath to breath during such changes. Repeated determinations using the Fick and dye dilution methods (discussed below) agree within a range of $\pm$ 15 percent. Much of this variation is probably due to true minute-by-minute variation in the cardiac output.

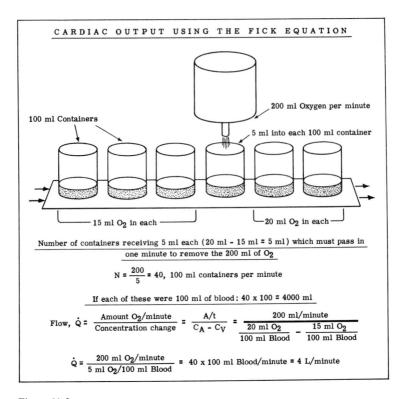

Figure 11-5
Derivation of the Fick equation for the measurement of cardiac output or blood flow.

The indicator dilution method, using the Fick equation, provides a convenient tool for the measurement of mean *organ blood flow* if the indicators used are removed from the blood by these organs. The liver removes bromsulphalein (BSP), and the kidneys remove *p*-aminohippurate (PAH). The brain and heart absorb nitrous oxide to provide arteriovenous (A-V) differences that enable flow to be measured, assuming that the amount of material accumulated is related to the solubility of the gas in the tissue. The pattern of "washout" of radioactive xenon or krypton may also be used to calculate flow.

Stewart-Hamilton Method (Single Injection)
The Stewart-Hamilton method is based upon the rapid injection of a definite, known amount of indicator. If the amount injected and the mean concentration over a given time (the minute-concentration) are measured, the flow can be calculated. It is in many ways more convenient than, and is fully as accurate as, the direct Fick method. In practice, for cardiac outputs, a known amount of an indicator such as indocyanine green or albumin labeled with ^{131}I is injected into a vein. It is diluted and mixed with the blood flowing through the heart and lungs. The resulting concentration of dye in the arterial blood is then measured and the cardiac output calculated.

To understand this procedure, first imagine a flowing stream (Fig. 11-6A). If the total flow containing an added amount of dye (A) is collected over a definite period of time—for example, one minute—and the dye is then mixed throughout the total collected volume to give a concentration (C), the volume can be calculated from the basic dilution equation $V = A/C$. This is the volume per minute—i.e., the rate of flow.

Next, imagine that the dye is uniformly mixed across the flow stream and that a definite proportion (assume one-tenth) of the flow is collected as in Figure 11-6B. Starting before the dye appears and ending after the dye has cleared the sampling site, a volume containing dye is collected for one minute and is thoroughly mixed so that the mean concentration may again be determined. If, in the sample, the proportion of the total amount of dye is exactly equal to the proportion of the total flow collected over one minute (i.e., one-tenth of each), the total flow can again be calculated. Since the

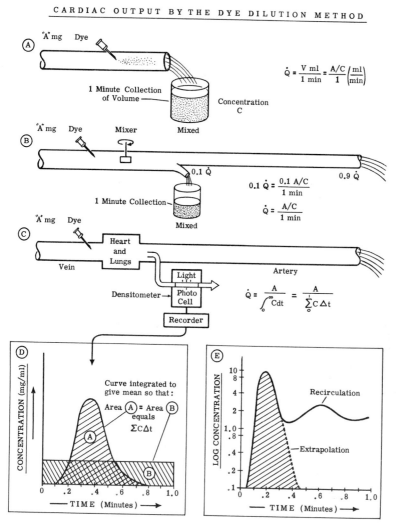

CARDIAC OUTPUT BY THE DYE DILUTION METHOD

$$\dot{Q} = \frac{V\ ml}{1\ min} = \frac{A/C}{1}\left(\frac{ml}{min}\right)$$

$$0.1\ \dot{Q} = \frac{0.1\ A/C}{1\ min}$$

$$\dot{Q} = \frac{A/C}{1\ min}$$

$$\dot{Q} = \frac{A}{\int_0^\infty C\,dt} = \frac{A}{\sum_0^1 C\,\Delta t}$$

Figure 11-6

Derivation of the Stewart-Hamilton indicator dilution equation for the measurement of cardiac output by the dye dilution method. A. Total flow stream collection. B. Collection of a fraction of the flow stream with no recirculation. C. Measuring dynamic concentration changes with a densitometer. D. Record of dye concentration and the derivation of the concentration-time factor, which is usually the mean concentration during 1 minute, i.e., the minute-concentration. E. Effect of recirculation and the method of extrapolation using a semilogarithmic plot of dye concentration.

proportionality constants for the dye and the volume can be canceled (Fig. 11-6B), it is not even necessary to know what proportion of the flow is collected. This proportion must, however, be constant throughout the collection, and the dye must be uniformly mixed across the flow stream.

Finally, instead of collecting a sample over a given time interval and mixing, one may analyze individual samples or make a continuous measure of the indicator concentration with a densitometer (Fig. 11-6C). The mean concentration is mathematically calculated (Fig. 11-6D) over the time interval, usually one minute. The *Stewart-Hamilton indicator dilution equation* is

$$\dot{Q} = \frac{A}{\displaystyle\int_0^\infty Cdt} = \frac{A}{\Sigma C \Delta t} \tag{14}$$

where Δt is the fraction of the unit time interval.

The concentrations at various fractions of a minute are added together to give $\Sigma C \Delta t$, the *minute concentration,* a term somewhat analogous to the *respiratory minute volume.* Usually, after less than a minute the concentration has declined to so near zero that in actuality no further additions are made to the sum, called the integral. Because fractional units of time are used, this is a mean concentration averaged over one unit of time—ordinarily a minute.

In practice, a known amount of dye is injected into the bloodstream. It must mix across the total flow stream and, for cardiac output determinations, does so in the right or the left ventricle. A representative fraction of the cardiac output containing changing concentrations of dye is obtained by taking a continuous sample of arterial blood, as shown schematically in Figure 11-6C. The concentration of dye is analyzed and plotted (solid curve, Fig. 11-6E) over an interval long enough for a complete passage of dye through the circulatory system. In calculating the mean concentration of dye over a certain period (the minute concentration), one must calculate the area under the curve either by measuring with an instrument called a planimeter or by summation of instantaneous values. The latter consists in finding the concentration during the first second and adding it to that at the second second, adding to the third, and so summating throughout the whole minute. Dividing this sum by the number of samples measured during the minute, 60, gives the minute concentration. During the first few samples after the indicator injection, the concentration will be zero. If there were no recirculation of dye, there would be no dye in the last samples and so no additions to be made to the sum. This situation is analogous to the calculation of the daily rainfall over one year. Some days are without rain, and for these zero amount of rain is added in the calculation.

Recirculation is a potentially serious problem associated with this method. Dye in the arterial side rapidly returns through the venous system to recirculate and be diluted a second time. Recirculation must be eliminated in the calculation of the mean concentration. Since the circulation time through the coronary vessels of the heart is about 15 seconds, recirculation usually distorts the curve. The problem is solved by making the assumption that the tail of the curve follows an exponential decay; that is, dye is washed out at a rate proportional to the amount that is present. This assumption is usually valid, so that plotting a logarithm of the concentration will give a straight downslope if plotted against time. Thus the downslope of the curve, as shown in Figure 11-6E, may be extrapolated to some value which can be considered to be of no significance and thus called zero in the summing process. (A logarithmic curve approaches but does not reach zero; hence, an arbitrary baseline of about 1 percent of the peak concentration must be established.) Finally, note that an increase in cardiac output, by diluting the dye to a greater extent, results in a decrease in the concentration-time factor. Special-purpose computers are now available to speed the computation.

In the determination of flow by this technique, several major assumptions are made. Conditions must be such that the assumptions are valid, or appropriate corrections are made. (1) The indicator must be adequately *mixed* within the system. The flow at the point of mixing is the flow that is calculated. This condition usually presents no problem in measuring cardiac output, since the ventricles mix the blood vigorously. It does preclude, however, the possibility of injecting dye in the arch of the aorta and sampling at the femoral artery to obtain accurate information about aortic blood flow. (2) There must be no *loss* (or gain) of indicator. The dye must not be lost in tubing, absorbed on artery walls, or lost by passing through the capillary walls. If indicator is lost, there will be an apparently higher diluting volume, and thus higher indicated flow. (3) In flow determinations, corrections must be made for *recirculation*. There must be enough data on the downward side of the curve to obtain an accurate estimate of the downslope for the extrapolation. Injection of dye near the heart usually solves the problem, but in some clinical conditions, such as heart failure, it may be serious. (4) A *representative* sample must be obtained. That is, the sampling rate should be proportional to the flow at all moments during the curve for the determination. Since the dilution is made over a period of 10 to 20 heart beats, this source of error is not particularly serious. Fluctuations in cardiac output from respiratory efforts do cause variations, however. The same problem of proportionate sampling exists in obtaining a mixed venous sample for the direct Fick method. (5) An *accurate determination* of concentration must be made. If a single calibration factor is used, the calibration curve must be linear and pass through zero. The detector must respond to the indicator in question and not to extraneous indicator—a difficult feat with external counting when radioisotopes are used. Good practice dictates that duplicate or triplicate determinations be made.

Both the dye dilution and Fick methods measure the mean flow over a period of many seconds to several minutes. There are, therefore, variations from one determination to the next because of variations in cardiac output from moment to moment. The respiratory cycle is a particularly important source of this type of error since the cardiac output often fluctuates in response to changes in intrathoracic pressure and respiratory center activity.

A discussion of the factors influencing cardiac output in man is presented in Chapter 15.

REFERENCES

Attinger, E. O. Analysis of pulsatile blood flow. *Adv. Biomed. Eng. Med. Phys.* 1:1–59, 1968.

Bergel, D. H. (Ed.). *Cardiovascular Fluid Dynamics.* New York: Academic, 1972. Vols. 1, 2.

Bloomfield, D. A. (Ed.). *Dye Curves: The Theory and Practice of Indicator Dilution.* Baltimore: University Park Press, 1974.

Brown, J. H. U., and D. S. Gann. *Engineering Principles in Physiology.* New York: Academic, 1973. Vols. 1, 2.

Bruns, D. L. A general theory of the causes of murmurs in the cardiovascular system. *Am. J. Med.* 27:360–374, 1959.

Charm, S. E., and G. S. Kurland. *Blood Flow and Microcirculation.* New York: Wiley, 1974.

Cobbold, R. S. C. *Transducers for Biomedical Measurements: Principles and Applications.* New York: Wiley, 1974.

Eskinazi, S. *Principles of Fluid Mechanics.* Boston: Allyn & Bacon, 1962.

Geddes, L. A., and L. E. Baker. *Principles of Applied Biomedical Instrumentation.* 2d Ed. New York: Wiley, 1975.

Glasser, O. (Ed.). Circulatory System. In *Medical Physics.* Chicago: Year Book, 1960. Vol. 3, pp. 119–193.

Hamilton, W. F. (Ed.). *Handbook of Physiology*. Washington: American Physiological
    Society, 1962, 1963. Section 2: Circulation, vols. 1, 2. See: A. C. Burton, Physical
    Principles of Circulatory Phenomena: The Physical Equilibria of the Heart and
    Blood Vessels, chap. 6, pp. 85–106; K. L. Zierler, Circulation Times and the Theory
    of Indicator-Dilution Methods for Determining Blood Flow and Volume, chap. 18,
    pp. 585–615; and H. D. Green, et al., Resistance (Conductance) and Capacitance
    Phenomena in Terminal Vascular Beds, chap. 28, pp. 935–960.
McDonald, D. A. *Blood Flow in Arteries* (2nd ed.). Baltimore: Williams & Wilkins,
    1974.
Merrill, E. W. Rheology of blood. *Physiol. Rev.* 49:863–888, 1969.
Middleman, S. *Transport Phenomena in the Cardiovascular System.* New York:
    Wiley-Interscience, 1972.
Noordergraaf, A. Hemodynamics. In H. P. Schwann (Ed.), *Biological Engineering.*
    New York: McGraw-Hill, 1969. Pp. 391–545.
Replogle, R. L., H. J. Meiselman, and E. W. Merrill. Clinical implications of blood
    rheology studies. *Circulation* 36:148–160, 1967.
Whitmore, R. L. *Rheology of the Circulation.* Long Island City, N.Y.: Pergamon, 1968.

12. Microcirculation

Julius J. Friedman

The cardiovascular system is organized toward the maintenance of a homeostatic environment for the tissues of an organism. Cardiac and peripherovascular functions are coordinated to transport blood to and from the capillary-venular networks where exchange of nutrients and cell products between blood and tissue fluids takes place. Examination of the anatomical and functional organization of the microcirculation unit should provide an insight into the capability of this system to meet the needs of the tissues.

ANATOMICAL ORGANIZATION

A microcirculatory unit consists of serially and parallel arranged blood vessels including *arterioles, metarterioles, precapillary sphincters, capillaries, arteriovenous anastomoses, venules,* and *collecting venules.*

The precise positioning of these microvascular elements differs in different organs. In the mesoappendix of the rat they are arranged as diagrammed in Figure 12-1. Blood flow enters the microcirculatory unit by means of the *arteriole,* which finally terminates as terminal arterioles or *metarterioles.* The metarteriole differs from the arteriole in that the smooth muscle cells in the vessel wall are discontinuous.

Numerous capillaries originate from single metarterioles and terminal arterioles, and they branch and anastomose repeatedly to produce a great increase in surface area and reduction in blood flow velocity. At the entrance of most capillaries there exist segments of vessels well provided with smooth muscle. These are the *precapillary sphincters,* and they regulate the effective surface area of the *capillary-venular* network. Capillaries come together to form *venules,* which merge further to form *collecting venules* to drain the microcirculatory unit. While it is functionally correct to regard the microcirculation as consisting of units, extensive arteriolar and venular anastomoses provide elaborate collateral circulation. In some tissues the arteriolar system is organized as an arcade network which may act to establish an isopressure perfusion source for all the microcirculatory units.

FUNCTIONAL ORGANIZATION

Functionally, the series and parallel coupled components of the microcirculation may be classified as *resistance vessels, exchange vessels, shunt vessels,* and *capacitance vessels.*

Resistance Vessels

Resistance to flow is manifested by all blood vessels, which may essentially be classified as precapillary and as postcapillary resistance elements. The *precapillary resistance elements* include small arteries, arterioles, and precapillary sphincters. The small arteries and arterioles are the major variable resistance elements of the microvascular unit that determine the extent of total tissue blood flow. The precapillary sphincters, on the other hand, by adjusting the number of capillaries open, determine the distribution of capillary blood flow, the extent of capillary flow velocity, capillary surface area, and the mean extravascular diffusion distances. In short, precapillary sphincter activity exerts a great influence on the nature of transcapillary exchange.

Postcapillary resistance vessels include the muscular venules and small veins. Although they manifest only small changes in resistance, their strategic position,

273

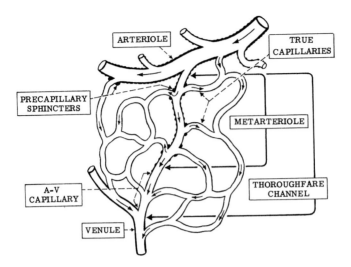

Figure 12-1
Diagram of a microcirculatory unit. (From B. W. Zweifach. In *Factors Regulating Blood Pressure.* New York: Josiah Macy Jr. Foundation, 1949.)

immediately postcapillary, enables them to influence capillary pressure markedly. It is mainly the *ratio* of precapillary to postcapillary resistance that determines capillary hydrostatic pressure.

Exchange Vessels
Exchange between the vascular and extravascular compartments occurs primarily across capillaries and venules. These vessels are uniquely suited for exchange by virtue of their high surface-area-to-volume ratio and their thin wall. Moreover, since exchange vessels are usually within 20 to 50 μm of tissue cells, diffusion distances are minimal. Transvascular exchange is different in different segments of the exchange network in the same tissue as well as in different tissues because permeability of the capillary-venular circuit is not uniform. Within a single tissue the venous end of the exchange vasculature is more permeable to water and to solute than is the arterial end. In different tissues the structure of the capillary wall possesses different characteristics which influence their relative permeabilities.

Shunt Vessels
All elements that bypass the effective exchange circulation of a tissue serve as shunt vessels. They include arteriovenous anastomoses, preferential (thoroughfare) channels, and even capillaries when they receive blood flow greatly in excess of their exchange capacity. Shunts of varying dimensions have been described in most tissues. Except for those in the skin that subserve temperature regulation, their precise functional significance is unclear. When effective shunts exist, some fraction of the total flow to an organ or region will not participate in exchange. With shunting, redistribution of blood flow may occur in the tissue without any apparent change in total tissue blood flow, and variations in total blood flow need not be associated with comparable variations in tissue nutrition.

Capacitance Vessels
Because of their great distensibility, the veins are the major capacitance elements of the vascular system. The distribution of blood volume within the venous circulation is

somewhat uncertain. Current evidence indicates that the venular and small venous division constitutes the major functional capacity of the vascular system. The total system is estimated to contain about 70 percent of the total blood volume.

FUNCTIONAL ACTIVITY OF THE MICROCIRCULATION

Normal Blood Flow
In contrast to the laminar flow seen in small arterial and venous vessels, laminar flow characteristics in the microcirculation are less evident. Red cells flow through capillaries in single file at varying rates and are usually distorted because of a restricted vessel lumen. Arteriolar flow velocity has been estimated at 4.6 mm per second, and venular flow velocity at 2.6 mm per second.

Red blood cell velocity in the capillary is estimated to be about 1 to 2 mm per second. Velocity of flow through a segment of capillary does not necessarily indicate the transit time of red blood cells through the capillary circuit. Capillary flow is intermittent and often reverses direction, so that the effective red cell transit time is probably in excess of the estimated one second.

Intermittency of Blood Flow
Direct examination of the microcirculation reveals that flow is very intermittent. Capillary flow may vary in rate and direction from one moment to the next. The intermittency probably reflects changes in the pressure gradient through a particular vessel due to sphincter action or to transient plugging upstream or down. It does not represent active capillary activity, since capillaries do not contract.

Flow in any single capillary depends primarily upon the state of the precapillary sphincters and metarterioles. These elements exhibit alternate phases of contraction and relaxation with a period of about 30 seconds to several minutes. Such phasic contractile activity has been termed *vasomotion.* Vasomotion determines the distribution of blood flow between various regions of the capillary bed. Metarteriolar vasomotion is primarily concerned with main channel flow. The contraction phase of metarteriolar vasomotion is never intense enough to arrest main channel flow. However, the contraction phase of precapillary sphincter vasomotion is complete, and capillary flow is arrested. With augmented vasomotion, the frequency of contractions and relaxations increases, the contraction phases becoming more prominent. Under these circumstances, capillary blood flow decreases. On the other hand, a reduction in vasomotion involves a reduction in the frequency of the cycle with the dilator phase predominating, and capillary blood flow increases.

Vasomotion is the response of vascular smooth muscle to local chemical, myogenic, and neurogenic influences. However, since vasomotion is seen in denervated regions, the primary influence is probably exerted by chemical and physical influences acting on the vascular musculature. Neurogenic influences may serve to modify the reaction of vascular smooth muscle to the other agents. Consequently, vasomotion is dependent to a large extent upon tissue activity. There is a normal level of vasomotor tone that partially restricts blood flow into the capillary network. As tissue metabolism increases or as a degree of anaerobic metabolism ensues, the enhanced accumulation of metabolic products depresses the peripheral vasculature, thereby reducing vasomotion, and capillary blood flow increases. Conversely, minimal metabolic activity, such as exists in the basal state, provides a low concentration of depressed metabolites, and vasomotion is maintained at a relatively high frequency.

Phasic contractions similar to vasomotion are also seen in arterioles. However, arteriolar vasomotion is probably influenced more by central nervous impulses traveling through the sympathetic outflow to the arterioles than by local humoral influences.

TRANSCAPILLARY MOLECULAR AND FLUID EXCHANGE

Normal tissue function requires that water and soluble material pass between the blood and tissue fluids. This transfer of material across the capillary-venular wall occurs by means of filtration, diffusion, and cytopempsis. The rate and extent of transcapillary movement depend on the nature of the molecule and of the capillary structure.

Capillary Structure

Electron microscopic examination of capillaries shows them to consist of a single layer of *endothelial* cells. At the outer surface of the endothelial cells is an amorphous mucopolysaccharide matrix called the *basement membrane*. It forms a barrier about 500 Å thick which contributes to the permeability characteristics of the capillary. Capillaries have been classified on the basis of the nature of *endothelial perforations* (fenestrations) and the extent of *basement membrane* and *perivascular investment*. Thus, capillaries of muscle, skin, heart, and lung (Fig. 12-2A) have an unperforated endothelium, a prominent continuous basement membrane, and a prominent pericapillary investment. Liver capillaries (Fig. 12-2B) possess large endothelial perforations, a faint discontinuous basement membrane, and no significant pericapillary investment. Intestinal capillaries (Fig. 12-2C) possess thin, fenestrated endothelial walls with a continuous but indistinct basement membrane and only a slight perivascular investment.

Capillary Pores

Transcapillary exchange of lipid-insoluble molecules, such as electrolytes, glucose, and amino acids, is believed to take place through aqueous-filled "pores" which are associated with the intercellular spaces. These spaces (Fig. 12-3) are, on the average, about 200 Å wide with one or two foci of narrowing of the intercellular space where the adjacent endothelial cell membranes become closely approximated and appear to fuse. However, fusion is incomplete, and openings of about 40 Å in radius are present. These small openings are so tiny and dense that they constitute less than 0.1 percent of the capillary surface. Despite the limited area of the openings, they represent sufficient area to provide for very rapid exchange of small molecules (MW < 10,000), primarily by diffusion. In addition to these small pores, there are large pores (radius = 250 Å) which are primarily associated with the venous end of the exchange system. Their contribution to normal transcapillary exchange of small molecules is, however, not very significant on account of their very low density. For each large pore there are approxi-

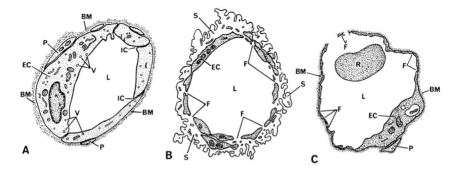

Figure 12-2
Diagrammatic representation of the electron microscopic characteristics of capillaries from different tissues. A. Muscle. B. Liver. C. Intestine. EC = endothelial cell; IC = intercellular gap; L = vessel lumen; V = vesicle; F = fenestration; BM = basement membrane; P = pericyte; R = red cell; S = extravascular space. (After Bennett, Luft, and Hampton, 1959.)

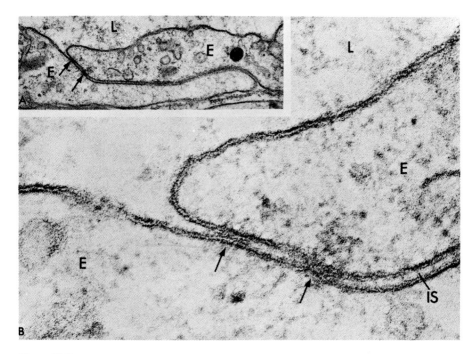

Figure 12-3
The intercellular junction between two endothelial cells. The arrows identify two foci of narrowing. (From R. S. Cotran. The Fine Structure of the Microvasculature in Relation to Normal and Altered Permeability. In E. B. Reeve and A. C. Guyton [Eds.], *Physical Bases of Circulatory Transport: Regulation and Exchange.* Philadelphia: Saunders, 1967.)

mately 30,000 small pores. They may represent one of the routes whereby large molecules, such as plasma albumin, leave the vascular compartment and enter the interstitial space.

The fenestrations, which may be considered as giant "pores," must certainly be prominent elements in capillary permeability. Liver capillaries, with numerous large fenestrations, possess a relatively high permeability. Intestine and kidney capillaries, which have intermediate-size fenestrations, also have intermediate permeabilities. Muscle, skin, heart, and lung capillaries, which have no fenestrations, have low permeabilities.

Cytopempsis (Vesicular Transport)

Electron microscopy reveals the presence of abundant vesicles 500 to 800 Å in diameter in the endothelial membrane. There are also numerous invaginations at each surface of the endothelial membrane. Many vesicles are attached to the surface by necks which can be as large as 400 Å in diameter. These vesicles have been proposed as comprising the "large pore" system cited above. They presumably participate in transcapillary transport by engulfing plasma at the vascular border of the membrane, transporting the molecules contained in the vesicles through the cell cytoplasm by means of brownian movement, and discharging their contents at the interstitial border of the endothelial cell. Vesicles are not very efficient at transporting small molecules, since these are rapidly exchanged across the capillary wall by diffusion. They may be important in the transcapillary movement of large molecules, such as globulins and fibrinogen (molecules too large to transverse the small or large pores), and contribute to some extent to albumin leakage from the vascular system. The size of the vesicles along with their number and rate of movement across the cell can provide a maximum transport capability of about 1.5×10^{-4} ml per second per 50 gm of muscle. Vesicular fluid is probably very close to plasma, so that vesicles are capable of transporting as much as 1 mg of plasma protein per minute per 100 gm of muscle.

Transcapillary Molecular Exchange
(Capillary Permeability)

Transcapillary exchange is a passive process which depends on the diffusion of molecules and on the degree of membrane restriction to particle penetration. The free diffusion of molecules in solution is described by the Fick relationship (Chap. 1). According to this formulation, the rate of diffusion of a quantity of materials (Q) per unit time (t) with a diffusion coefficient (D_s) through a cross-sectional area (A_s) over distance (x) is directly proportional to the concentration gradient (dc/dx).

$$\frac{dQ}{dt} = DA \frac{dc}{dx}$$

Capillary permeability (P_s) may be defined as the degree to which the capillary permits the passage of molecules. The capillary wall is a passive barrier, and the major determinants of molecular penetration are molecular lipid solubility and molecular size or shape and charge. Lipid-soluble molecules utilize the entire endothelial surface for transport in accordance with their oil:water partition coefficients. However, most ionic and molecular constituents of plasma are not lipid soluble, and presumably they must pass through aqueous channels in or between cells in order to enter the extravascular space.

The extent of the gaps and fenestrations and of electrostatic interaction is reflected in the apparent area available for diffusion (A_s) compared to the total area of the membrane (A_m). This may be formulated as follows:

$$P_s = \frac{D_s}{\Delta_x} \times \frac{A_s}{A_m}$$

Capillary permeability to a particular solute differs from the free diffusion of that solute to the degree that the effective or apparent area for diffusion differs from the total area of the membrane modified by any electrostatic interaction that may occur between the charged particles and fixed charges in the capillary membrane. The apparent area for diffusion for all lipid-insoluble molecules is considerably less than the total area of the membrane. Consequently, the rate of capillary penetration is substantially less than that of free diffusion. Since there are relatively few large pores, the area of diffusion for large lipid-insoluble molecules is less than that for smaller molecules, and their capillary permeability is therefore considerably lower than for small molecules. The relationships between molecular size and capillary permeability have been established for muscle and are presented in Table 12-1.

Lipid-soluble molecules such as oxygen and carbon dioxide have an area for diffusion that should be approximately equal to the total endothelial surface. Permeability for these molecules should thus be expected to be approximately equal to the free diffusion coefficient. However, capillary permeability to lipid-soluble molecules is also influenced by the oil:water partition coefficient of each molecule.

Because of variations in the extent of fenestration and the development of the basement membrane of different tissues, it is obvious that capillary permeability is not uniform throughout the body. Capillaries of liver and intestine are highly permeable and restrict protein passage only slightly; those of muscle, skin, heart, and lung, which are poorly permeable, restrict protein passage significantly.

Table 12-1. Permeability of Mammalian Muscle Capillaries to Lipid-Insoluble Molecules

Substance	Molecular Weight (gm)	Diffusion (D) (cm^2/sec)	Molecular Radius (Approx.) (cm)	$A_s{}^*/\Delta x$ (cm)	Permeability (P) cm/sec 100 gm
H_2O	18	3.4×10^{-5}	1.5×10^{-8}	0.55×10^{-5}	28×10^{-5}
NaCl	58	2.0	2.3	0.51	15
Urea	60	1.95	2.6	0.49	14
Glucose	180	0.90	3.7	0.44	6
Sucrose	342	0.70	4.8	0.39	4
Raffinose	504	0.64	5.7	0.34	3
Inulin	5,500	0.24	13.0	0.12	0.3
Myoglobin	17,000	0.17	19.0	0.03	0.1
Serum albumin	67,000	0.085	36.0	0.0007	0.001

*Restricted pore area per unit path length per 100 gm of muscle.
Source: Landis and Pappenheimer (1963).

Further evidence of nonuniform distribution of capillary permeability between tissues has been obtained by examining the escape of different-size molecules of dextran upon their circulation through various tissues. It was found that liver, with a high permeability, passed a greater ratio of large- to small-size molecules than did muscle, with a low permeability. Intestine, with an intermediate permeability, passed some intermediate ratio of small to large dextran molecules.

Blood Flow and Transcapillary Exchange

Effective tissue nutrition depends upon the mutual participation of adequate tissue blood flow, blood flow distribution, and transcapillary exchange. The capillaries are distributed throughout the tissue so that the average diffusion distance from capillary to cell is probably about 10 μm. This short diffusion distance is critical to effective tissue nutrition since the time for transport by diffusion varies inversely with distance. At a distance of 10 μm, approximately 5 seconds are required for oxygen to reach 95 percent of its equilibrium value, whereas at a distance of 100 μm about 10 minutes are necessary.

Figure 12-4 illustrates the influence of tissue blood flow and diffusion distance on the capillary and interstitial Po_2. Blood Po_2 is greatest at the arterial end of the capillary and decreases in an exponential manner as the blood flows through the capillary, reaching its lowest level at the venous end. The Po_2 at any point along the capillary depends upon the relationship between blood flow rate and O_2 extraction by the tissue. O_2 extraction is largely determined by the gradient for Po_2 between the blood and interstitial space and is substantially independent of the time available for exchange since capillary permeability to O_2 is very high. Interstitial Po_2 is determined by the relationship between supply and utilization. Since supply is by way of diffusion and since the effectiveness of diffusive transport diminishes rapidly with distance, the Po_2 at any point in the interstitial space is defined by the level of Po_2 in the adjacent capillary segment and its distance from that capillary.

At normal flow and normal microvascular tone, the Po_2 profile in and around the capillary is as illustrated in Figure 12-4, which reveals that all of the tissue is oxygenated to some extent. As blood flow decreases or as diffusion distance increases because of precapillary vasoconstriction, Po_2 in the tissue can decline to hypoxic and anoxic levels that are incompatible with tissue integrity. In such a case, local factors elicit some degree of precapillary vasodilation which, by increasing the number of open capillaries,

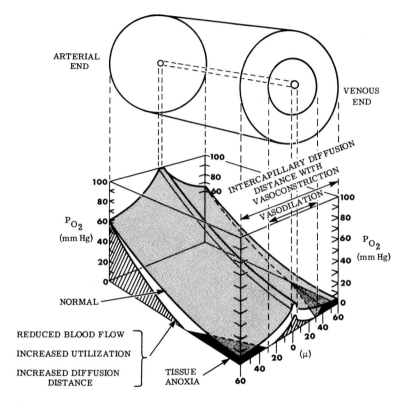

Figure 12-4
The longitudinal and radial concentration gradients for oxygen in and around the capillary during normal and reduced blood flow. (After G. Thews. Gaseous Diffusion in the Lungs and Tissues. In E. B. Reeve and A. C. Guyton [Eds.], *Physical Bases of Circulatory Transport: Regulation and Exchange.* Philadelphia: Saunders, 1967.)

reduces diffusion distance and restores a certain amount of oxygenation to the deprived tissue.

Similar concentration relationships apply to other blood constituents in varying degrees, depending on the capillary permeability to and extravascular distribution of the molecule in question.

Transcapillary Fluid Movement
Net transcapillary fluid movement is determined by the interaction of hydrostatic and osmotic forces acting across the capillary wall, as first described by Starling, and may be formulated as

$$FM = K_f (P_c + \pi_{i.f.} - \pi_p - P_t)$$

FM represents *net fluid movement* and K_f the *capillary filtration coefficient. Capillary hydrostatic* (P_c) and *interstitial fluid colloidal osmotic* ($\pi_{i.f.}$) *pressures* represent the *filtration force* whereas *plasma colloidal osmotic* (π_p) and *tissue hydrostatic* (P_t) *pressures* represent the *absorption force.* FM is positive when the filtration force exceeds the absorption force and fluid is filtered from the circulatory system into the interstitial spaces. Conversely, FM is negative when the filtration force is lower than the absorption force and fluid is absorbed from the interstitial spaces into the circulatory system.

Capillary Filtration Coefficient

The capillary filtration coefficient (K_f) may be defined as the volume of fluid filtered in one minute by 100 gm of tissue for a 1 mm Hg change in capillary pressure. K_f reflects the status of hydraulic conductivity and of surface area of the capillary wall. Capillary membrane hydraulic conductivity is considered to be reasonably constant so that K_f is determined primarily by capillary surface area.

The filtration of water through the capillary wall of the vessels of the human forearm, measured from the slow change in forearm volume which follows an elevation in venous pressure, is approximately 0.0057 ml/min $\times$ 100 gm of tissue per centimeter of water capillary pressure. Fluid filters from kidney, intestine, and liver at rates considerably higher than those of muscle or skin. However, because of the great mass of muscle and skin, the total body filtration rate resulting from elevations of central venous pressure is approximately 0.0061 ml/min $\times$ 100 gm of tissue per centimeter of water venous pressure. In an adult, an increase in central venous pressure of 10 cm HOH will create a loss of 250 ml of fluid from plasma in 10 minutes.

Most of the water is believed to pass across the capillary wall between cell margins and through fenestrations. In muscles it would appear that these comprise approximately 0.1 percent of the total membrane surface area. In kidney, the glomerular filtration rate is considerably greater, so that the effective surface area available for filtration must be greater.

Capillary Hydrostatic Pressure

Under normal circumstances, capillary hydrostatic pressure is probably the major determinant of transcapillary fluid movement, since it may well be the most variable component of the transcapillary forces.

Capillary pressure (Pc) is determined by arterial pressure (Pa), peripheral venous pressure (Pv), precapillary resistance (r_a), and postcapillary resistance (r_v). According to Ohm's law as applied to capillary inflow (Q_i) and outflow (Q_o), when a tissue is in an isovolumetric state (no net filtration or absorption)

$$\frac{Pa - Pc}{r_a} = Q_i = Q_o = \frac{Pc - Pv}{r_v}$$

This expression may be solved for Pc to yield

$$Pc = \frac{r_v}{r_a + r_v} Pa + \frac{r_a}{r_a + r_v} Pv$$

Thus elevations of Pa, Pv, or r_v or reductions of r_a will produce elevations of Pc. Since r_a is usually five to six times greater than r_v, about 85 percent of change in Pv is reflected to the capillaries compared to about 15 percent of change in Pa. This apportionment will be modified by changes in either r_a or r_v.

Pressure in the most distal arterioles of the mesentery is 32 to 40 mm Hg, whereas that in the most proximal venules is 27 to 29 mm Hg (Gore, 1974; Zweifach, 1974). Mean capillary pressure is about 32 mm Hg and appears to be relatively independent of systemic arterial pressure (Zweifach, 1974). The rather flat pressure profile and the relatively constant mean capillary pressure are considered by some to reflect the presence of physiological regulation of pressure. It may be that these pressure characteristics are due to the unique microvascular organization of the mesentery. In muscle, the pressure drops across the capillary are from 34 cm H_2O to 15 cm H_2O, or 19 cm H_2O (Smaje et al., 1970). The greater pressure drop across these capillaries may reflect the smaller diameter and greater length of muscle capillaries.

Table 12-2. Osmotic Pressure of Plasma Proteins

Protein	Molecular Weight	Concentration (gm/100 ml)	Osmotic Pressure (mm Hg)
Albumin	68,000	5.0	20
Globulins	200,000	1.5	5
Fibrinogen	500,000	0.5	1

Plasma Colloidal Osmotic (Oncotic) Pressure

The major contributors to plasma oncotic pressure are the plasma proteins albumin and globulins (Table 12-2). Albumin, the smallest and most concentrated of the plasma proteins, provides the largest contribution. In addition, approximately 30 percent of the total effective osmotic pressure results from the unequal distribution of electrolytes in an ionized colloidal system caused by the Gibbs-Donnan phenomenon. (Chap. 1.)

Tissue Hydrostatic Pressure

Tissue hydrostatic pressure is that which develops in the interstitial compartment outside the capillary wall. It is determined by the volume of fluid in and the distensibility of the interstitial space. Interstitial space distensibility depends on the state of tissue hydration, being only slightly distensible in partially dehydrated tissue and highly distensible in well-hydrated tissue. Attempts to measure tissue pressure have been made with needle micropuncture and by subcutaneous capsule implantation. Micropuncture provides estimates of a positive pressure of 0 to 5 mm Hg, whereas the capsular pressures are found to be about 7 mm Hg negative. It has been suggested that the negative capsular pressure is due to a dynamic osmotic balance that develops across the capsule wall as a result of the formation of a semipermeable membrane about the capsule. Tissue pressure in normally hydrated tissue changes only slightly along the length of the capillary; thus its contribution to normal transcapillary exchange is probably slight. However, in conditions of lymphatic blockage or increased capillary permeability, tissue pressure can achieve high levels and influence transcapillary dynamics significantly.

Interstitial Fluid Colloidal Osmotic (Oncotic) Pressure

Tissue colloidal osmotic pressure is provided by plasma proteins that have passed through the capillary wall into the interstitial fluid. Consequently, this pressure varies in accordance with the permeability of the capillary wall to plasma proteins. Muscle capillaries are poorly permeable to protein, as reflected in the lymph protein concentration of about 1 to 2 gm per 100 ml. In intestine and liver the lymph protein concentration is about 3 to 4 and 5 to 6 gm per 100 ml, respectively.

It is estimated that 1 percent of protein provides about 3 mm Hg osmotic pressure. However, it is difficult to estimate the tissue oncotic pressure from the total protein concentration alone because the interstitial protein is not uniformly distributed. A substantial portion of the interstitial space is occupied by ground substance which consists mainly of hyaluronic acid and chondroitin sulfate. One gram of hyaluronic acid can hold up to 100 ml of water while excluding albumin. Thus, the interstitial protein is distributed in a volume of fluid that is less than the total interstitial fluid volume, and effective interstitial protein concentration and tissue oncotic pressure may be higher than would be anticipated from data of protein content and volume of the interstitial space.

Normal Transcapillary Fluid Movement

Transcapillary exchange of fluid under normal conditions is determined primarily by the relationship of capillary pressure to plasma oncotic pressure. When capillary hydrostatic pressure exceeds plasma oncotic pressure, filtration results. When capillary hydrostatic pressure is less than plasma oncotic pressure, absorption occurs. It was suggested that hydrostatic exceeded oncotic pressure at the arterial end of the capillary, where filtration took place. As blood moved along the exchange vessels, hydrostatic pressure declined progressively until filtration ceased and absorption from the interstitial space to the vascular lumen occurred.

An alternative concept considers the effect of vasomotion on exchange. It was suggested that during the dilator phase of vasomotion, capillary hydrostatic pressure was high and filtration occurred, whereas the constrictor phase of vasomotion produced a reduction in pressure and permitted absorption to take place in the same vessel.

VARIATIONS IN TRANSCAPILLARY MOVEMENT

The net exchange of fluid across the capillary membrane may be influenced by variations in either capillary or tissue hydrostatic pressure, plasma or tissue protein concentration, lymphatic drainage, capillary permeability, or capillary surface area.

Alterations in Capillary Hydrostatic Pressure

Capillary hydrostatic pressure is mainly determined by arterial blood pressure, venous pressure, and precapillary and postcapillary resistance. The effects of changes in these variables on transcapillary fluid exchange are illustrated in Figure 12-5.

The normal balance between filtration and absorption is presented in Figure 12-5A. A reduction in capillary pressure by means of reduced arterial pressure or increased precapillary resistance (Fig. 12-5B) results in reduced filtration and increased absorption. Increasing capillary pressure by increasing arterial pressure or reducing precapillary resistance (Fig. 12-5C) increases filtration and reduces absorption. Marked increases in filtration and reductions in absorption follow elevations of capillary pressure by increasing venous pressure or postcapillary resistance (Fig. 12-5D).

Changes in Plasma Colloidal Osmotic (Oncotic) Pressure

Plasma colloidal osmotic pressure is primarily determined by the plasma albumin concentration. A reduction of plasma albumin concentration, due either to nutritional or to metabolic deficiencies, acts to reduce the colloidal osmotic pressure of plasma. There follows a reduction in the absorption force throughout the length of the capillary, so that filtration increases and absorption decreases (Fig. 12-5F). Under these circumstances there is a net movement of fluid from the circulation into the interstitial space, and circulating plasma volume declines. In conditions of hyperproteinemia or dehydration, the plasma protein concentration is elevated, and consequently plasma oncotic pressure is elevated (Fig. 12-5E). This change reduces the filtration force at the arterial end of the capillary and increases the absorption force at the venous end, so that a net movement of fluid from the interstitial compartment into the circulation takes place and circulating plasma volume tends to rise.

Changes in Tissue Pressure

Either enhanced filtration or reduced lymphatic drainage leads to an increase in interstitial fluid volume and tissue pressure. An elevation in tissue pressure increases the absorption force throughout the capillary. Thus, filtration at the arterial end of the capillary declines while absorption at the venous end increases, and further filtration of fluid from the circulatory system is impeded. For this reason, excessive filtration of fluid from the circulation is somewhat self-limiting. However, before tissue pressure achieves a level of this magnitude, filtration must exceed lymphatic drainage for a sufficient time or by sufficient magnitude. At that point, interstitial fluid accumulation becomes excessive and edema develops.

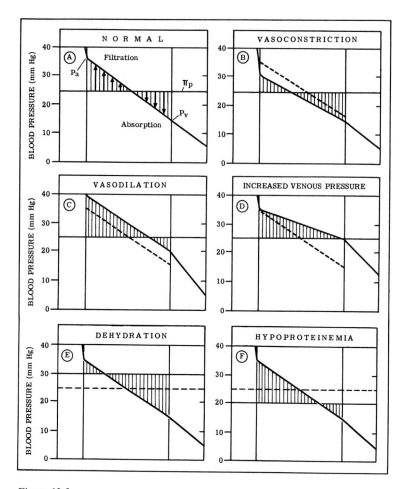

Figure 12-5
Variations in filtration and absorption produced by changes in arterial and venous pressures and resistances and plasma colloidal osmotic pressure. The broken lines represent normal levels.

Changes in Capillary Permeability

The osmotic pressure exerted by the plasma proteins is effective only as long as diffusion of the proteins through the capillary wall is adequately restricted. Under the circumstances of increased capillary permeability, albumin molecules are able to diffuse through the capillary endothelium at a more rapid rate than normal. The result is a reduction of the plasma colloidal osmotic pressure and an increase in tissue colloidal osmotic pressure. Consequently, the absorption force in the circulation is reduced and the relative filtration force is enhanced. In conditions of increased capillary permeability, then, filtration of fluid from the circulation into the extravascular spaces increases markedly and the circulating plasma volume declines.

LYMPH AND LYMPHATICS

Large molecules that reach the interstitial space by filtration, cytopempsis, or cellular metabolism and secretion are not effectively removed by the exchange vessels. If they were to accumulate, they would exert sufficient effective osmotic pressure to upset transcapillary fluid exchange, and excessive fluid would accumulate in the interstitial space. Under normal conditions this event does not occur.

Filtered fluid and other plasma constituents that accumulate in the extravascular spaces are drained and conducted back to the circulatory system by way of the lymphatic network. In addition to this drainage function, the lymphatic system possesses concentrated areas of reticular endothelial cells at various sites which remove bacteria and foreign material from the lymph circulation. This action is one of the major protective mechanisms of the body against the invasion of harmful agents. The lymphatics also serve as a transport system for a number of materials, such as vitamin K and lipids absorbed from the intestine.

The lymphatic network originates in the tissue spaces as very thin, closed endothelial tubes (lymphatic capillaries), which have the same relation to the tissue spaces as have blood capillaries. However, lymphatic capillaries are far more permeable to large particles than are blood capillaries. Although the lymphatic capillaries consist of endothelial cells, their porosity is apparently so great as to provide little or no resistance to the passage of particles into the lymphatic system. The fact that these capillaries are also devoid of any basement membrane may account for their high permeability. The lymphatic capillaries converge on one another to form larger lymphatic vessels which possess valves to ensure a unidirectional flow of the lymph.

Lymph Formation

Lymph is defined as the fluid which returns to the circulation from the tissue space by way of the lymphatics. Since lymph originates from plasma, as the difference between filtration and absorption of fluid across the capillary wall, its composition is very similar to that of plasma, and all influences that modify transcapillary exchange also modify the formation and composition of lymph. Except for variation in protein concentrations and electrolytes associated with the Gibbs-Donnan equilibrium, lymph and plasma are almost identical. The average protein concentration of lymph is approximately 2 percent, compared with 6 percent in plasma. In addition, the concentrations of the different protein fractions in lymph differ from those in plasma. Since albumin is the smallest plasma protein, it permeates the capillary endothelium more easily than does globulin. Consequently slightly more albumin, relative to globulin, accumulates in lymph. Evidence of this differential rate of diffusion is found in the comparison of the ratio of albumin to globulin (A/G) in lymph and in plasma. The plasma A/G ratio is about 1.27, whereas the A/G ratio of lymph is 1.35.

In general, the protein concentration of lymph varies inversely with the rate of formation. With increased filtration from the capillaries, more fluid in relation to protein is cleared from the circulation; and although the quantity of protein filtered increases, the protein concentration in the interstitial compartment is reduced. Since capillary permeability and the rate of filtration vary between different regions of the organism, the protein concentration of lymph from these different regions also varies. The protein concentration of lymph from the liver and intestine may be as high as 5–6 percent, whereas that in lymph from the extremities may be as low as 0.5 percent.

Lymph Flow

The rate of lymph flow in mammals is very low under normal circumstances. Furthermore, it varies between tissues in accordance with local transcapillary dynamics. The flow of lymph in the thoracic duct is approximately 1.38 ml per kilogram per hour. In a 24-hour period this represents a volume of fluid equal to the plasma volume. In addition, 50 to 100 percent of the circulating plasma protein is returned to the circulatory system during the same period by the lymphatics. Therefore, this system is essential in maintaining the circulating blood volume.

Lymph flow is maintained by the influence of active contraction of muscular elements in their walls, and by external forces. Lymph flow is dependent, in part, on the tissue pressure generated by the continuous filtration of fluid from the circulation. Lymph flow is aided by compression of the lymphatic channels as a consequence of contraction of neighboring musculature and is enhanced by the negative intrathoracic pressure. Although increased respiratory or muscular activity enhances lymph flow, the major limitation and therefore the major determinant of the flow is lymph formation. Consequently, any circumstance that increases the rate of filtration of fluid from the capillaries also increases the rate of lymph flow. Raising capillary pressure by either arterial vasodilation or venous constriction enhances the rate of lymph flow.

Intravascular infusions of various solutions enhance the formation of lymph and lymph flow in two ways. First, the solutions increase the volume of fluid in the circulatory system and thereby increase arterial and venous blood pressure; hence capillary pressure is increased and filtration augmented. Moreover, the increased volume of fluid in circulation dilutes the plasma proteins, effectively reducing the plasma colloidal osmotic pressure, thus enhancing filtration and reducing absorption.

Cardiac muscular contraction, the peristaltic action of the intestinal smooth muscle, or voluntary muscular contraction massages the lymphatic channels, and the alternate compression and relaxation exerted by the contracting muscles propel lymph centrally. The valves in the lymphatic channels prevent retrograde flow. Lymph flow is also augmented by increased tissue activity independent of external muscular contraction. The heightened tissue activity results in the accumulation of metabolites, presumably causing the arterioles and precapillary sphincters to dilate and the capillary endothelium to become somewhat more permeable, so that blood flow into the capillary network

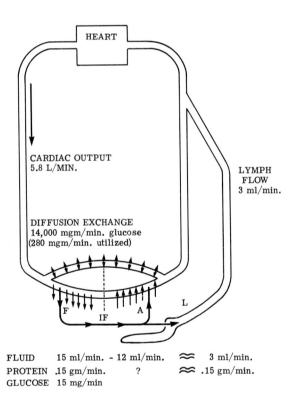

FLUID	15 ml/min. - 12 ml/min.	≈	3 ml/min.
PROTEIN	.15 gm/min.	?	≈ .15 gm/min.
GLUCOSE	15 mg/min		

Figure 12-6
The relationship between transcapillary exchange of fluid and solute and l⸳ Landis and Pappenheimer, 1963.)

increases at the same time that permeability is augmented. Eating results in lymph flows as high as 5.8 ml per kilogram per hour.

The activity of the microcirculation and the relationship between the systemic and lymphatic circulations are illustrated in Figure 12-6. At a cardiac output of 5.8 liters per minute, about 15 ml per minute is filtered across the capillary wall from the circulation and 12 ml per minute reabsorbed. The 3 ml remaining in the interstitial space is returned to the vascular system by way of the lymphatics. This lymphatic contribution does not seem important until one realizes that it amounts to about 4 liters of fluid per day. Clearly, diffusional exchange is the mechanism whereby solute materials such as glucose are supplied to the tissue, for the amount of glucose that could be supplied through filtration is inadequate to satisfy the tissue utilization and is insignificant compared to the diffusional exchange. It can be seen that one of the major functions of the lymphatic system is to return protein to the circulation. The lymphatics handle as much as 200 gm of protein per day, which is almost equal to the total intravascular mass of proteins.

EDEMA

Edema is the condition of excess accumulation of fluids in the tissue spaces. It is of functional significance in that the presence of excess fluid in the interstitial space modifies microvascular flow distribution and retards the exchange of nutrients and metabolites between cells and plasma. Edema results when filtration of fluid from the circulation exceeds the drainage. It may result from any condition that produces sufficiently increased capillary pressure, decreased plasma protein concentration, increased capillary permeability, and reduced lymphatic drainage. When these changes become excessive, fluid accumulates in the extravascular compartments more rapidly than it is removed by lymphatic drainage, and the extravascular compartment swells and becomes congested.

Although elevation in capillary pressure or reduction in plasma oncotic pressure facilitates filtration and may eventually lead to edema, increased capillary permeability and lymphatic blockage cause the most rapid development of severe edema. With increased capillary permeability, plasma proteins permeate the capillary wall more easily and thereby reduce the plasma oncotic pressure, while at the same time they increase the tissue oncotic pressure. Therefore, the accumulation of fluid in the extravascular spaces is doubly augmented.

Capillary permeability increases under a variety of circumstances. Marked increases in capillary volume and pressure created by large intravascular infusions, large doses of vasodepressant drugs, advanced bacterial inflammation, and the production of toxic metabolic products increase the dimensions of the intercellular and intracellular openings. In addition, the toxic metabolites, agents such as histamine, and bacterial toxins appear to act directly upon the capillary wall to increase its permeability. In each of the above conditions lymph flow is greatly enhanced. However, the capacity of lymphatic drainage is limited, and when capillary filtration exceeds lymphatic drainage, edema results.

The most exaggerated form of edema, *elephantiasis* of the limbs and scrotum, is caused by blockage of the lymphatics by the filarial organisms. Under these circumstances, lymphatic drainage from the tissue spaces is retarded or interrupted. Nevertheless, filtration of fluid proceeds, and interstitial fluid accumulation becomes progressive. Ultimately, the accumulation of extravascular fluid becomes so great that tissue pressure rises to a level at which it effectively counteracts the hydrostatic pressure in the capillary. To this extent the condition of edema may be said to be somewhat self-limiting. Unfortunately, because of the great distensibility of the extravascular space and because of the large capacity of body cavities in which the various viscera are contained, this point is not achieved until excessive quantities of fluid have been removed from circulation and circulating blood volume declines significantly. Moreover, on account of the marked diffusion gradient for protein between plasma and interstitial fluid, protein molecules continue to diffuse from the circulation into the interstitial spaces despite a reduction in gross filtration. Tissue oncotic pressure then rises and plasma oncotic pressure is reduced, with the result that edema formation may continue even after hydrostatic pressures have equilibrated. Ultimately, hydrostatic pressure and oncotic pressure equalize across the capillary wall and no further net movement of fluid takes place. The point of equilibrium is seldom, if ever, achieved.

Edema also occurs in renal disease and is an outstanding feature of the nephrotic syndrome. In conditions of nephrosis, excessive protein is lost from the circulation into the urine, thereby reducing the plasma protein concentration. The kidney can also contribute to the formation of edema by means of salt and fluid retention. With this retention, the volume of the circulatory and extravascular systems is increased and capillary pressure is elevated, a condition that facilitates the accumulation of fluid in the extravascular space. A more extensive discussion of fluid balance appears in Chapter 23.

REFERENCES

Bennett, H. S., J. H. Luft, and J. C. Hampton. Morphological characteristics of vertebrate blood capillaries. *Am. J. Physiol.* 196:381–390, 1959.

Gore, R. W. Pressures in cat mesenteric arterioles and capillaries during changes in systemic arterial blood pressure. *Circ. Res.* 34:581–591, 1974.

Intaglietta, M., and B. W. Zweifach. Microcirculatory basis of fluid exchange. *Adv. Biol. Med. Phys.* 15:111–159, 1974.

Johnson, P. C. The Microcirculation and Local and Humoral Control of the Circulation. In A. C. Guyton and C. E. Jones (Eds.), *Physiology, Series One. Cardiovascular Physiology.* London: Butterworth, 1974. Vol. 1.

Karnovsky, M. J. The ultrastructural basis of capillary permeability studied with perodidase as a tracer. *J. Cell Biol.* 35:213–236, 1967.

Landis, E. M., and J. R. Pappenheimer. Exchange of Substances Through the Capillary Walls. In W. F. Hamilton (Ed.), *Handbook of Physiology.* Washington: American Physiological Society, 1963. Section 2: Circulation, vol. 2, pp. 961–1034.

Mayerson, H. S. The Physiologic Importance of Lymph. In W. F. Hamilton (Ed.), *Handbook of Physiology.* Washington: American Physiological Society, 1963. Section 2: Circulation, vol. 2, pp. 1035–1073.

Pappenheimer, J. R. Passage of molecules through capillary walls. *Physiol. Rev.* 33:387–423, 1953.

Renkin, E. M. Transport of large molecules across capillary walls. *Physiologist* 7:13–28, 1964.

Ruszynák, I., M. Foldi, and G. Szabó. *Lymphatics and Lymph Circulation.* Elmsford, N.Y.: Pergamon, 1960.

Smaje, L., B. W. Zweifach, and M. Intaglietta. Micropressures and capillary filtration coefficients in single vessels of the cremaster muscle of the rat. *Microvasc. Res.* 2:96–110, 1970.

Wiederhielm, C. A. Dynamics of transcapillary fluid exchange. *J. Gen. Physiol.* 52:29–63, 1968.

Yoffey, J. M., and F. C. Courtice. *Lymphatics, Lymph and Lymphoid Tissue.* Cambridge, Mass.: Harvard University Press, 1956.

Zweifach, B. W. Quantitative studies of microcirculatory structure and function: I. Analysis of pressure distribution in the terminal vascular bed in cat mesentery. *Circ. Res.* 34:843–857, 1974.

13. Pumping Aspects of Cardiac Activity

Carl F. Rothe
and Ewald E. Selkurt

The function of the circulatory system is to maintain an optimum environment for cellular function. The optimum environment requires control of concentrations of nutritive, hormonal, and waste materials, tensions of respiratory gases, and temperature. Because of continuous cellular activity, an optimum environment may be maintained only by an uninterrupted flow of blood to the tissues to renew nutrients and remove wastes. To maintain this continuous flow, then, is the role of the cardiovascular system, and it is the heart that pumps the blood. Failure of this pump is the prime cause of death among adults. The properties of heart muscle have been discussed in Chapter 3B, and control of cardiac function will be covered in Chapter 15. This chapter deals with the specific anatomical and physiological properties of the heart that fit it for the role of a pump for the circulation of blood.

The arterial blood pressure of a normal person is held relatively constant. The rate of blood flow through individual tissues is controlled, at the tissue, by changes in the caliber of the resistance vessels (the small arteries and arterioles; Chaps. 12 and 17). There is central autonomic nervous system modulation of both cardiac function and the peripheral tissue vasculature (Chaps. 7A and 17). In consideration of normal and pathological function of the cardiovascular system, it is important to note (Chap. 15) that the cardiac output—the flow of blood from the heart—is determined by the input pressure to the heart (the central venous pressure), cardiac contractility, and the arterial pressure load. Furthermore, although the cardiovascular system is a closed circuit, it is made up of distensible vessels which, in conjunction with the blood volume, influence the filling pressure to the heart. Finally, the cardiovascular system is dual, with a systemic (greater) circuit and a pulmonary (lesser) circuit. The flows through these are normally equal, but they have important differences in characteristics, as will be discussed.

There are four general causes of inadequate cardiac function even when the filling pressure is adequate and the outflow load is normal:

1. *Myocardial.* Pumping may be inadequate because the muscle does not contract vigorously enough. A common cause of myocardial failure is myocardial ischemia—an inadequate blood supply to the heart muscle itself (see Chaps. 15 and 18B).
2. *Anatomical.* Abnormalities of structure may be congenital (present at birth) or acquired from disease or accident. Valve damage is a typical example.
3. *Arrhythmia.* Impulse excitation may not be initiated or conducted adequately, or fibrillation occurs in which contractions are chaotic and asynchronized (see Chap. 14).
4. *Pericardial.* Fluids may accumulate between the heart and pericardium (tamponade), or the pericardium may become constricted and so impede adequate filling.

THE FETUS AND NEWBORN

Nutrient and waste exchange for the fetus occurs via the placenta, rather than the lungs and gastrointestinal tract. The "arterialized" umbilical venous blood (Fig. 13-1) goes to the liver, but most passes through the ductus venosus to the vena cava where systemic venous blood is mixed with it. The umbilical flow accounts for about 40 percent of the total cardiac output of the fetus. About half of this rather poorly oxygenated blood

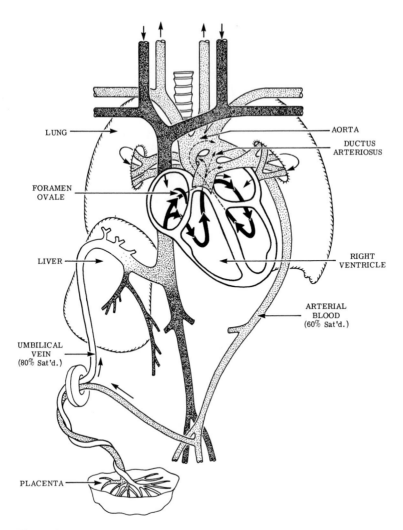

Figure 13-1
Fetal circulation. (Modified from D. Longmore. *The Heart.* New York: McGraw-Hill, 1971.)

crosses from the right to the left atrium through the foramen ovale, bypassing the right ventricle. The pulmonary vascular resistance of the collapsed lung is high, the pulmonary arterial pressure is greater than the aortic pressure, and so most of the right ventricular output passes through the ductus arteriosus into the aorta, bypassing the lungs. As a consequence, both ventricles share the load of the systemic circulation, while only about one-sixth of the total cardiac output goes through the lungs (Fig. 13-2).

Most of the changes from the fetal pattern of blood circulation (Fig. 13-2) to the adult pattern occur at birth when the placenta ceases functioning. The pattern for the neonate (newborn) is similar to that of the adult except for a patent ductus arteriosus. The switchovers at birth are complex:

1. The umbilical blood flow is stopped, and the right atrial pressure decreases.
2. As the lungs inflate, the interstitial tissue, including the pulmonary microvasculature,

is stretched and so the pulmonary vascular resistance is decreased, allowing blood to pass through the lungs more easily. The blood flow through the lungs increases (see Chap. 19). This increased flow into the left atrium causes its pressure to increase.

3. With the reversal of pressure gradient between the right and left atria, the flap valve at the foramen ovale closes, preventing the shunt flow from the right to the left heart.
4. With the decrease in pulmonary arterial pressure, the pressure gradient between the aorta and the pulmonary artery reverses, causing a reversal of the flow through the ductus arteriosus.
5. With respiration and functioning lungs, the oxygen tension of the arterial blood increases. The increased oxygen tension stimulates the smooth muscles of the ductus arteriosus to contract. However, the effect is slow, for flow through the ductus is not restricted until about 10 minutes after birth and requires normally about two days for complete occlusion. The smooth muscle of the ductus arteriosus is almost totally unresponsive to arterial oxygen tension throughout the first two-thirds of gestation, but thereafter it progressively becomes more and more sensitive. The mechanism is unknown, but there seem to be factors in addition to oxygen that elicit ductus arteriosus occlusion in the newly born (Rudolph and Heymann, 1974).

At birth the right ventricle is as large as the left, as might be expected, since the right heart pumps one-half to two-thirds of the total flow at aortic pressures. During the first year the right heart grows somewhat, but the left heart grows rapidly to more than double its weight.

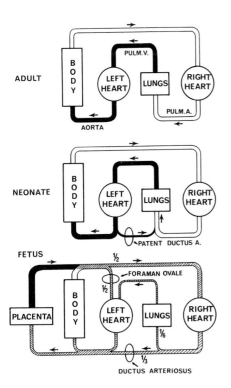

Figure 13-2
Comparison of patterns of fetal, neonatal, and adult circulation.

FUNCTIONAL ANATOMY OF THE HEART

Anatomical drawings, such as those of Netter (1969), should be consulted for details of
the orientation of the great vessels, chambers, and valves of the heart. The dense
connective tissue between the atria above and the ventricles below is the fibrous
skeleton of the heart, and on it are mounted the four heart valves (Fig. 13-3). Through
this skeleton passes the bundle of His, the only structure maintaining the syncytial
pathway between atria and ventricles to provide a conduction path for ventricular
excitation.

Atrial Anatomy

Above the fibrous skeleton sit the cup-shaped right and left atria. The thin-walled atria
provide an easily expandable reservoir for the venous blood returning to the heart while
the ventricles are emptying and the entrance valves are closed.

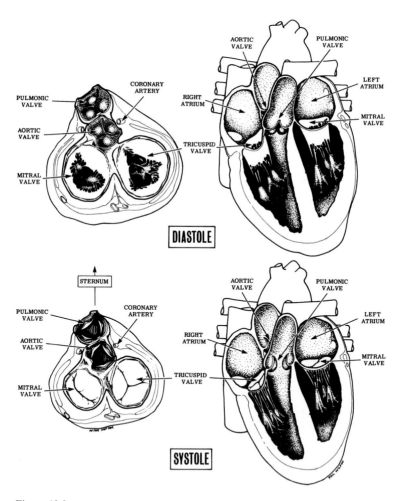

Figure 13-3
Anatomy of the heart and its valves. (Base views redrawn after those of F. H. Netter. *Heart.* CIBA
Collection of Medical Illustrations. Vol. 5. Summit, N.J.: CIBA, 1969.)

Ventricular Anatomy

The heavy-walled ventricles pump blood from the low-pressure venous system into the high-pressure arterial distribution system. The ventricles of the heart are a continuum of interdigitating muscle strands, rather than discrete bands or bundles of muscle. The myocardium contracts in the line of its fibers, which are generally oblique to the base-to-apex axis. The fiber orientation in the free wall tends to be parallel to the base-to-apex axis at both the endocardial (inside) and epicardial (outside) surfaces, but then smoothly but rapidly shifts toward a circumferential orientation in the middle of the wall (see, for example, Brecher and Galletti, 1963; Streeter et al., 1969; and Armour and Randall, 1970, for details and further references). The papillary muscles occupy about 5 to 10 percent of the left ventricle during diastole, but by the end of systole they occupy nearly 30 percent of the remaining chamber volume, tending to obliterate the inflow tract and forming a smooth-walled infra-aortic outflow tract.

The Valves of the Heart

Effective pumping action by the heart demands unidirectional flow from the venous side to the arterial side. This is the function of the four cardiac valves located in the fibrous skeleton of the heart at the entrances and exits of the two ventricular chambers. The *aortic* and *pulmonary* outlet valves, called semilunar valves, located at the exits of the left and right ventricles, are perhaps the simpler pair. They act to prevent the return of blood from the arterial system (Fig. 13-3). Both valves consist of three tough but flexible flaps or half-moon cusps attached symmetrically around the valve rings. A widening or outpouching of the aorta just above the valve ring leads to openings for the two coronary arteries that supply blood to the heart muscle (myocardium). The position of the three aortic valve cusps during their open phase is important, for if the valves were forced back to the aortic wall, blockage of the coronary orifice would occur, causing serious myocardial impairment.

The two atrioventricular (AV) valves (the *tricuspid* on the right side and the *mitral* on the left side) act to prevent the ventricles from ejecting blood back into the atria (Fig. 13-3). Both valves contain two large primary cusps. The tricuspid has an additional small cusp; hence the name. Because the pressure gradient available to fill the heart is low, the large size of the opening is important so as not to impede filling. Furthermore, the total area of the cusps is much greater than that of the orifice they guard. On account of this large size and the thin nature of the AV valves, added support is required during ventricular contraction. Thus, to the lower side of the cusps are attached the chordae tendineae leading to the papillary muscles on the ventricular walls. The papillary muscles function to prevent eversion of the flexible cusps into the atrium. They contract during ventricular ejection to support the valves. If the papillary muscle tension is inadequate during ventricular contraction, they allow bulging into the atria, called eversion, and so an apparent loss of stroke volume.

The valves open passively in diastole whenever the atrial is greater than the ventricular diastolic pressure or, in systole, whenever ventricular pressure exceeds that in the artery. There are two mechanisms involved in valve closure. The most obvious is a reversed pressure gradient causing a backflow of the blood which closes the valve. The other mechanism is based on Bernoulli's principle (Chap. 11). With the jet of high-velocity blood moving through the valve, the lateral pressure is reduced. In addition, eddies of blood swirl in behind (Fig. 13-4A). As a result, the leaflets move toward the center of the flow stream. As filling slows, the momentum of the blood tends to continue the motion to close the valve. The timing of the cardiac cycle is such that normally the valves are about closed when the ventricle contracts. This mechanism markedly reduces the stresses placed upon the extremely thin—but tough—AV valves.

Ventricular Pumping Action

Motion of the heart during contraction is complex (Fig. 13-4B). Not only does the heart become smaller and the apex move upward somewhat, as might be expected to squeeze

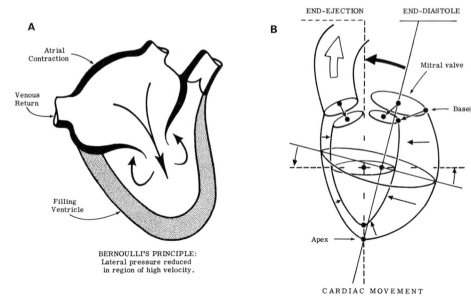

Figure 13-4
A. Effect of Bernoulli's principle and eddies of flowing blood on aid to closure of atrioventricular valves. B. Pattern of movement of the heart during systole. (B modified from J. E. Hinds, et al. *Fed. Proc.* 28:1351, 1969.)

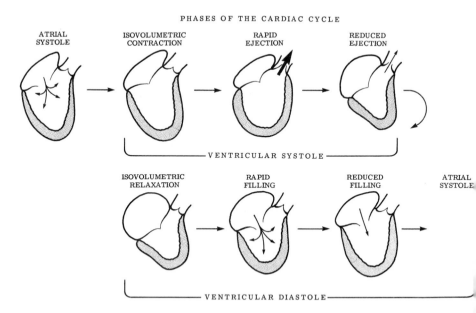

Figure 13-5
Successive phases of the cardiac cycle. (Modified after C. J. Wiggers. *Physiology in Health and Disease.* Philadelphia: Lea & Febiger, 1949. P. 654.)

out blood, but the base moves toward the apex. The reason is that, in accelerating the bolus of blood from the heart into the aorta and pulmonary arteries, there is an opposing force tending to move the heart (and especially the AV valve rings) in the opposite direction, resulting in the relatively stationary apex. As shown in Figure 13-4B, the axis of the heart also shifts. These movements make observation and surgery of the beating heart difficult.

An outline of the sequence of events of the cardiac cycle is shown in Figure 13-5. Note the relation of the phases of systole (contraction) and diastole (filling) to valve action. After atrial contraction, the ventricles contract but the volume is constant (isovolumetric, or isovolumic) because all valves are closed. Ejection follows. During the isovolumetric relaxation phase, the volume contained in the heart does not change, but the pressure rapidly decreases until it is less than that in the atria, allowing the atrioventricular valves to open.

The pattern of left ventricular emptying is shown schematically in Figure 13-4B. For the right heart, emptying is accomplished with a bellows-like action, in addition to simple reduction in diameter and length. The right ventricle has been looked upon as a pocket hanging from the heavier-walled left ventricle (Fig. 13-6). The free wall of the right ventricle is moved toward the interventricular septum by right ventricular muscle contraction, thus simulating a bellows. Another action has been called "left ventricular aid." The curvature of the wall between the two ventricles increases as the left ventricle contracts, pushing the septum into the ventricle and displacing blood from the chamber (Fig. 13-6). The effect of increasing the curvature is analogous to the tendency of snug pants to press tightly when the wearer bends over. The "left ventricular aid" may not be important normally, but in some patients it could mean the difference between life and death, for rather extensive muscular damage can occur in the right ventricular wall without causing death, and in dogs cauterization of the entire right ventricle is not invariably fatal.

When the right and left ventricles are compared as to function, the chief difference is that the left ventricle must pump a volume of blood through the high-resistance systemic (greater) circulation at relatively high pressures, while the right ventricle pumps this same volume through the low-resistance pulmonary (lesser) circulation at pressures that normally are only about one-sixth as high as for the left heart. This difference in function is reflected in the difference in structure, for the right heart has a thinner wall than the left. If the pressure load on the right ventricle is markedly increased over a long period of time, the muscle mass of the right ventricle increases (it hypertrophies) and becomes similar in appearance and function to that of the left heart—a pressure pump.

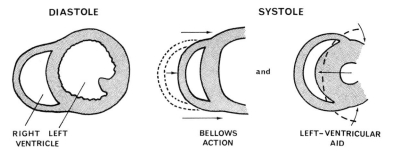

Figure 13-6
Ventricular emptying patterns.

EVENTS OF THE CARDIAC CYCLE

Methodology

The pumping action of the right and left ventricles, examination for possible leaks in the walls between the atria or ventricles, and adequacy of the blood supply to the heart itself (the coronary circulation) are studied by a technique in which a radiopaque (x-ray absorbing) substance is injected into the bloodstream and x-ray motion pictures are taken. This technique is called cinefluorographic angiocardiography—cineangiography, for short (see, for example, Sandler and Alderman, 1974; Verel and Grainger, 1973; Hurst, 1974). For studies of the dynamics of ventricular motion, the pictures are taken from two directions at the rate of at least 60 frames per second. The opaque material in the blood outlines the ventricular chambers and so can be observed in rapid sequence. Other studies have used radiographically opaque metal markers fastened onto the walls of the ventricles of experimental animals so that the motion can be followed. In addition, cardiac function may be studied with echocardiography (see, for example, Fiegenbaum and Chang, 1972). This is a technique based on the transmission of low levels (mW/cm^2) of ultrasound (frequency of 1 to 10 megahertz) into the body and detection of the echos which are reflected from tissue and fluid interfaces. The signal is displayed on an oscilloscope screen. The junction between blood and the wall of the heart, valve leaflets, and the outline of the heart can thus be visualized. Ultrasonic techniques have also been developed to measure the velocity of blood at selected sites (see, for example, Cobbold, 1974).

The complex geometry of the interior of the heart makes computation of stroke volume and residual volume from cineangiograms difficult. The simplest approximation is to consider the left ventricle a cylinder with a cone at the apex end. A better model is to consider the heart an ellipsoid. Computers, used to handle the more complex representations, have improved the accuracy of the determinations. However, the interior of the heart is so complicated and rough because of the papillary muscles that highly accurate estimates of instantaneous volume are not possible.

The events that are to be recorded occupy less than one second in the course of a normal heart beat; significant changes occur in milliseconds. Our current understanding suggests that the reasonable measurement of pressure, flow, and volume requires systems which have a uniform frequency response to at least 25 Hz. Types of transducer-recorder systems needed for these measurements are described by Geddes and Baker (1975), Ray (1974), and Cobbold (1974). Simultaneous recording of many variables gives insight into the chain of events concerned with the opening and closure of the valves and, in turn, filling and emptying of the heart. The classic experiments of Wiggers in open-chest dogs (1952) yielded much valuable information on the events of the cardiac cycle. The modern era in the study of circulatory dynamics in man began about 1941 when Cournand and Ranges demonstrated the practicability and safety of catheterization of the right side of the heart. Their studies were built upon Forssmann's pioneer demonstration in 1929 of the feasibility of catheterizing the human heart; he used himself as the experimental subject. Catheterization of both the right and left sides of the heart are now standard diagnostic procedures. Rushmer (1970) and Hurst (1974) have provided useful presentations of both the normal and pathological patterns.

The Cardiac Cycle

Understanding the normal cause-and-effect relationships between the variables of the heart is the basis for evaluating the significance of pathophysiological situations. To aid in discussing the cycle, it has been divided into the phases indicated in Figures 13-5 and 13-7.

Because a normal heart beats at about 75 beats per minute, a cycle requires about 800 msec and has been arbitrarily divided into sixteen 50-msec intervals. The average values for normal adults at rest are given at the bottom of Figure 13-7 (Table 114 of Altman and Dittmer, 1971). Systole normally requires less than half the cycle time, but

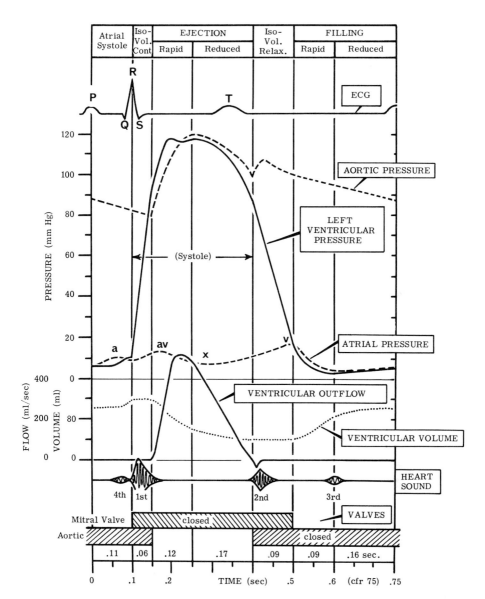

Figure 13-7
Events of the cardiac cycle.

with a tachycardia of 180 beats per minute it occupies more than half the cycle and hence severely limits filling time (see Fig. 15-3). The description that follows is for the left ventricle. Right ventricular events are nearly synchronous and are very similar, except for the level of pressure generated (cf. Fig. 13-9).

1. Atrial Systole
This is the beginning of the sequence of events. A pacemaker cell in the sinoauricular node depolarizes, and there is a spread of excitation across the atria, recognized in the

electrocardiogram as the *P* wave (Fig. 13-7, at the top). Atrial contraction follows, causing a development of pressure in the right atrium. This gives the *a* wave of the atrial pressure and central venous pressure wave form. The increased pressure causes additional ventricular filling. A fourth heart sound is generated, related to the sudden atrial contraction. It can be heard only when the ventricle is distended. Finally, as a result of the filling due to contraction, the ventricular pressure is slightly increased to the value called the left ventricular end-diastolic pressure.

Excitation of the Ventricles. The wave of excitation passes through the base of the heart via the atrioventricular (AV) node, the bundle of His, thence the Purkinje fibers, and finally the myocardium, leading to ventricular depolarization and the QRS complex of the electrocardiogram. As a consequence, ventricular contraction starts within about 10 msec. Then ventricular pressure starts to increase. With increased pressure, there is closure of the atrioventricular valves, and, as the valves are tensed, the first heart sound begins.

2. Isovolumetric Contraction Phase

In response to the contraction of the muscle, the pressure in the ventricle increases very rapidly, but normally there is no change in intraventricular volume because both the valves to the ventricle are closed leak-tight. The rate of change of pressure (dP/dt) of about 1600 mm Hg per second is related to the intrinsic ability of the heart to contract and so provides an estimate of cardiac status compared to normal (Chap. 15). (The value may be remembered by recalling that the pressure increases by about 80 mm Hg—from near zero to near the systolic level—within only one-twentieth of a second.) As the pressure in the ventricle exceeds that of the aorta, the outlet valve opens and flow starts, signaling the end of this phase. Note that the minimum systemic arterial pressure, called the diastolic blood pressure, occurs during early cardiac systole just as the valve opens.

3. Rapid Ejection Phase

The ventricular pressure now approaches its maximum and is greater than the aortic pressure. The magnitude of the difference in pressure is due primarily to the extra force required to accelerate the bolus of blood from the heart. If the outflow valve is stenotic (very small opening), the pressure gradient is increased because of viscous losses (resistance increase). The maximum outflow then follows while the contractile forces of the heart are still high. Blood spurts from the heart with a peak flow of 5 to 10 times the mean flow rate. Because of the high pressure in the ventricle during this phase, unless the papillary muscle tension can perfectly compensate for the pressure there is a tendency for the valves to bulge into the atria early in the ejection phase. The bulge acts as a flow into the atria and thus increases the atrial pressure, giving the *av* wave.* Soon after the increase in atrial pressure comes a decrease in pressure called by some the "x descent." Because of the inertia of the blood acting on the apex to hold it in position, the contracting heart exerts a pull on the valve rings and valves toward the apex. This force tends to move the valves out of the atrium, increase its volume, and so reduce the atrial pressure (Fig. 13-7). The reduced pressure also aids the return of blood to the atrium by providing a larger pressure gradient from the periphery to the heart. The ventricular volume decreases, slowly at first because inertia prevents instantaneous attainment of peak velocities, but then progressively more rapidly until the end of this phase.

Systolic Arterial Pressure. The systolic arterial blood pressure (the maximum value) (Chap. 16) may be used as the arbitrary separation point between the rapid and reduced ejection phases. The peak in arterial pressure follows the peak in flow from the heart

---

*(The term *av*, from atrioventricular valve, is the more appropriate term for this atrial pressure peak. The designation *c* was formerly used because a similar wave occurs in the jugular vein at about the time of the carotid artery systolic pressure peak. The *av* wave occurs much earlier in the heart than the systemic arterial pressure maximum (Fig. 13-7).

because the distensible aorta accumulates some of the outflow, permitting a greater inflow than with a rigid arterial bed, and thereby causes a dampening and lagging of the pressure pattern.

4. Reduced Ejection Phase

During this phase the contractile forces of the heart are decreasing and so is the intraventricular pressure. The aortic pressure is greater by a few millimeters of mercury than the ventricular pressure, but outflow continues because of the momentum of the rapidly moving bolus of blood. Repolarization of the myocardium occurs during the reduced ejection phase, giving the T wave of the ECG. Note that the correlation in time between the electrical and the mechanical events is not exact, and that the membranes of cardiac muscle cells may be electrically repolarized before the mechanical machinery has returned to a relaxed state.

The stroke volume of the average adult is about 80 ml; the end-diastolic volume is about 120 ml. Under normal conditions, a systolic reserve volume of about 40 ml remains in the ventricle at the end of systole. The *ejection fraction* is thus $\frac{2}{3}$, $[80/(40 + 80)]$. The remaining volume provides a reserve which can be utilized if the vigor of contraction is increased or if the aortic pressure is decreased. The ejection fraction increases in exercise and with massive sympathetic activity to about $\frac{3}{4}$ (Chap. 29), and decreases to less than $\frac{1}{2}$ in cardiac failure.

End of Systole. The end of systole occurs at the beginning of the second heart sound (Fig. 13-7). The contractile forces of the muscle are now decreasing very rapidly. There is a reversal of arterial flow toward the valves. This closes the valves and, because ventricular pressure is dropping so fast, the valves are tensed to such an extent that vibrations are set up, giving the second heart sound. Because the valves and aorta are elastic, they will recoil and cause a slight forward flow from the valve area and, as a consequence, a fluctuating pressure in the aorta. The notch in the aortic pressure is called the *incisura,* or *dicrotic notch.* (The dicrotic notch, seen in a peripheral artery pressure wave, is caused by a different mechanism; see Chap. 16.)

5. Isovolumetric Relaxation Phase

Both valves are closed and so, as the ventricular pressure decreases, there is no change in ventricular volume. The pressure decreases almost as rapidly as it had increased during the isovolumetric contraction phase. Because blood has continued to flow into the atria from the venous bed, and the base of the heart is moving back toward its resting position, the atrial pressure increases to a maximum, called the v wave.

6. Rapid Filling Phase

This is a period in which filling of the ventricle occurs quickly. The pressure in the ventricle is now less than in the atria, the AV valves open, inflow begins, and ventricular volume again starts to increase. Filling is rapid because the atrioventricular pressure gradient is relatively high. As the ventricle fills, the ventricular compliance acts to develop more and more back-force to slow the filling, especially at large volumes at which the compliance is reduced (see Fig. 15-5). A third heart sound may sometimes be heard at the end of this phase and is attributable to a rapid decrease in the rate of filling with associated vibrations.

7. Reduced Filling Phase

Demarcation between the rapid and reduced filling phases is arbitrary. Filling slows; indeed, if the diastolic period is abnormally long, a period of quiescence called *diastasis* ensues. During diastole, the aortic outflow to the periphery continues because blood was stored in the elastic aorta, as one might store air in a balloon.

With the basic information available about this marvelously designed system, one should be able to follow the logical sequence of cause and effect, from excitation to contraction, to pressure development, and, hence, to outflow. In studying pathological conditions, it should be possible to reason back to the likely cause of the effect seen, such as a weakness of contraction, a valve problem, or an excitation conduction defect.

Systole/Diastole

The term *diastole* comes from the Greek word meaning "separation" and thus is associated with filling of the ventricles as the walls separate. The end of diastole might be denoted as the time at closure of the AV valve when filling is stopped. The term *systole* comes from the Greek word meaning "a drawing together," or contraction, and thus is associated with the expulsion of blood by shortening of the fibers in the walls of the heart. In practice, however, the beginning of systole is usually said to coincide with the beginning of the first heart sound; the volume of the heart is maximum. (One might quibble by stating that systole starts with an electrical event associated with the Q wave of the electrocardiogram, or the initiation of the muscular contractions, or a development of tension preceding AV valve closure or, even better, the closure of the mitral valve, but all of these are difficult to measure and all occur within about 0.03 second.) The end of systole is even more confusing. By strict denotation, it means the end of contraction. However, the end of contraction, or beginning of relaxation, is highly indeterminate. To be consistent with the definition of diastole—filling—and to leave no undesignated time, one might state that systole ends with the opening of the mitral valve, but clearly the ejection of blood has long since stopped. A functional definition of the end of systole is the beginning of the second heart sound; ventricular volume is at a minimum; outflow has stopped. Thus, ventricular systole is that interval between the beginning of the first and the beginning of the second heart sounds, and diastole is the interval before the occurrence of the next first heart sound.

Heart as an Impulse Force Generator

A muscle twitch tends to be sudden (Chap 3B). The cardiac contraction lasts about one-third of a second, being intermittent between that of skeletal muscle (less than one-tenth of a second) and smooth muscle (more than one second). The cardiovascular system is so designed that normally blood flows relatively smoothly through peripheral capillaries of the body, even though it spurts from the heart. Not only does the suddenness of contraction place high stresses on the valves; it provides the cardiologist with an index of the intrinsic contractile ability of the heart (Chap. 15). It also requires mechanisms for storing the energy available for only a short time so that that energy can then be efficiently used throughout the remainder of the cycle to propel blood to the periphery (Chap. 16).

During the early part of the rapid ejection phase, the intraventricular pressure is appreciably higher than that in the artery, because some of the pressure force is required to accelerate the bolus, in addition to that required to overcome the back-pressure from the aorta and that dissipated in viscous friction (resistance). Under normal conditions, this inertial component may account for only 2 to 10 mm Hg. Under conditions of vigorous contractions, as with heavy exercise or a high level of sympathetic nervous system activity, the acceleration is so great that the pressure difference across the valve due to this factor is of the order of 10 to 20 mm Hg. This phenomenon is a particularly important component of the right ventricular pressure pattern.

With vigorous contraction, the maximum ventricular pressure occurs before the maximum flow, but in some circumstances, especially with a failing heart, maximum ventricular pressure may occur at or even after peak flow. Both the maximum ventricular pressure and the maximum ejection rate occur before the time of maximum arterial blood pressure.

As a consequence of the momentum of the blood leaving the heart, some energy is available to continue flow into the periphery from the aorta during diastole, but most of this kinetic energy of motion is converted to potential energy during the ejection phases by the expansion of the elastic arteries (the Bernoulli principle, Chap. 11). If the aorta is

Table 13-1. Typical Resting Cardiac Variables in the Adult

Variable	Right	Left
Heart rate	70 ± 3	
Atrial pressures (mm Hg)		
Mean	4.5 ± 3	8 ± 3
Ventricular pressures (mm Hg)		
End-diastolic	4.5 ± 4	9.5 ± 3
Peak systolic	26 ± 6	125 ± 15
Maximum rate of change (mm Hg/sec)	250 ± 100	1600 ± 500
Arterial pressures (mm Hg)		
Systolic	25 ± 7	125 ± 15
Mean	14 ± 4	95 ± 10
Diastolic	9 ± 4	80 ± 10
Cardiac output		
Sitting (liters/min)	5.5 ± 1.1	
Supine (liters/min)	6.8 ± 1.5	
Ml/min/kg body weight	90 ± 20	
Cardiac index (liters/min/square meter body surface area)	3.5 ± 0.7	

Mean ± standard deviation.
Primary source: Altman and Dittmer, 1971.

sclerotic (that is, hard and stiff), the pressure must be higher than normal to store a given bolus of blood (Chap. 16). Because the heart ejects less blood if its pressure load is increased (Chap. 15), the heart is less effective as a pump if the aorta is stiffer than normal.

Values of cardiovascular variables in the normal adult are given in Table 13-1.

THE HEART SOUNDS

Auscultation of the Heart

To auscultate means "to listen to," and so the auscultation of the heart means to ascertain its condition by listening to its sounds. The mechanisms of hearing by the human ear (Chap. 4B) and the physics of sound and sound production should be reviewed as background for this section.

Vibrations from the heart are composed of unrelated frequencies (noise), are of brief duration, and have very low frequencies (30 to 250 Hz) and amplitudes so low that the environment must be extremely quiet if one is to make a reasonable examination. Our ears are much less sensitive to sounds of these frequencies than to sounds in the speech range (500 to 2000 Hz; Fig. 13-8). By means of a phonocardiograph, a device consisting of a special microphone, an amplifier, and filters to attenuate extraneous frequencies, sounds of frequencies barely audible to the ear may be converted to a form that can be visualized (Fig. 13-9) (Hurst, 1974; Lewis, 1962; Luisada, 1972). The temporal relationships between the heart sounds and the mechanical events of the cardiac cycle, which are important in interpretation of the significance of sounds and murmurs, may thus be studied (Fig. 13-7).

Origin of Heart Sounds

First Heart Sound

The first heart sound, "lubb," is associated with the following dynamic phenomena: beginning of cardiac muscle contraction, closing of the AV valves, rapid development of

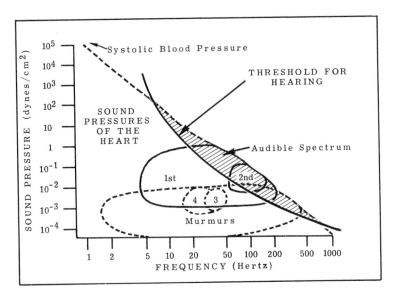

Figure 13-8
Sounds of the heart and threshold for hearing. Note that only a small fraction of the cardiac sounds are audible.

pressure, opening of the semilunar valves, and outflow. Many components of the first heart sound have been proposed (Luisada, 1972). Although some vibrations can be recorded from contracting skeletal and cardiac muscle strips, their intensity is probably not sufficient to contribute to the sound heard on direct auscultation. Because the valve leaflets or cusps have such low mass, closure of valves is also probably silent. However, the sudden acceleration and deceleration of the cardiovascular structures (especially tensing of the valves and myocardium as the result of abrupt pressure change) cause vibrations in the cardiac structures which, when transmitted through the tissues of the body, are picked up from the precordium (chest surface near the heart) by the stethoscope (Luisada et al., 1974). The later part of the first heart sound probably comes from the flow of blood past the valves and into the arteries.

The loudness of the first heart sound depends on (1) the rate of ventricular pressure rise—vigor of ventricular contraction; (2) the stiffness of the ventricle and valve structures; and (3) the positions of the mitral valve leaflets at the beginning of ventricular systole.

With sympathetic nervous system discharge and increased myocardial contractility, the rate of change in ventricular pressure is increased and a louder first heart sound (S1) may be heard. Conversely, in myocardial failure the sounds are reduced. With fibrosis, the AV valves become stiffer, leading to a sound that is louder and somewhat higher in frequency than normal; a "closing snap" may be heard. The sound originating from the right ventricle and tricuspid valve is much less than that from the left ventricle because of the normally lower pressure gradients and rates of change.

If the mitral valve is widely open at the time of ventricular systole, by the time the valve closes, a significant amount of blood will be moving toward the valve, causing some regurgitation and a rapid deceleration on closure. A relatively noisy sound will be produced as the valves are tensed by the then unusually high ventricular pressure. Under normal conditions, the flow of blood past the AV valves tends to cause them to move together in accordance with the Bernoulli principle (Chap. 11 and above), so that they are nearly closed by the time ventricular systole occurs. However, if the PR interval

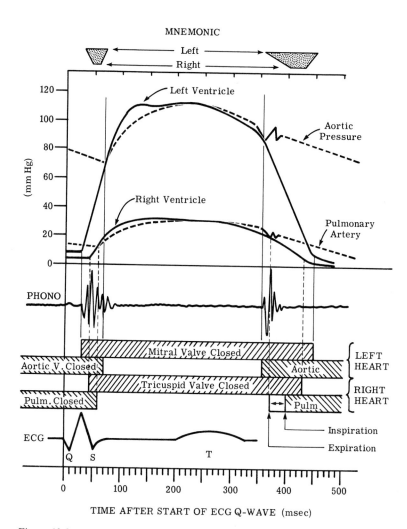

Figure 13-9
Comparison of dynamics of the right and left hearts. (Data in part from Luisada, 1972 and 1974, p. 504, and Hurst, 1974, pp. 184–185.)

(i.e., the time between atrial and ventricular excitation) is shorter than normal (0.16 second), the valves will remain open farther than normal at the onset of ventricular contraction, so that when they finally are closed the sound is louder than normal. Maximum loudness occurs at a P–R interval of about 0.11 second. If the P–R interval is somewhat longer than the normal (e.g., 0.20 second), the first heart sound may be softer than normal because the valves are almost entirely closed when systole starts, since flow from atrial contraction has slowed. However, if the P–R interval is very long (for example, with a first-degree heart block), the first heart sound may again be louder than normal because the valves have time to close and then rebound open.

Second Heart Sound
The second heart sound, "dup," is caused mainly by the tensing of the semilunar valves and the resulting vibrations of the valves, heart, and large arteries. It is normally of

higher frequency and shorter duration than the first heart sound; the following diastolic interval is usually longer than the preceding systolic interval. The sudden, very rapid ventricular relaxation at the end of systole causes a very rapid deceleration of outflow. It appears that the second heart sound starts about 25 msec before the end of outflow and aortic valve closure (Piemme et al., 1966). Blood in the roots of the aorta and pulmonary artery then rushes back toward the ventricular chambers, but this movement is abruptly arrested by closure of the semilunar valves. The momentum of the moving blood stretches the valve cusps, and the recoil causes oscillations to occur in both the arterial and ventricular cavities. Audible splitting of the second heart sound normally occurs during inspiration, as will be explained below.

Third Heart Sound
The third sound, if heard, occurs at the time of transition between rapid filling and reduced filling, when filling slows abruptly, giving a transient soft thud. The mechanism is not clearly understood. The sound is not normally heard in adults but may be heard in many people under age 30; it is of low pitch and intensity. When the third heart sound is of pathological significance, it is associated with a dilated ventricle and high atrial pressure. Filling of the heart until the pericardium is tensed would provide a sudden deceleration that could cause audible vibrations.

Fourth Heart Sound
The fourth sound is almost always inaudible in normal adults and occurs at the time of the peak of atrial contraction. When heard, it is associated with a high atrial pressure, vigorous atrial contraction, and filling of the ventricle. Clinically it occurs when the ventricle is stiff, as in ventricular hypertrophy or ischemia. A fourth heart sound is not heard in mitral or tricuspid stenosis since it originates in the ventricle and is associated with rapid filling.

Synchrony of Contraction of the Right and Left Heart

Although the left ventricle is usually the first to start contracting and the last to start to fill, the differences between the two ventricles are of only a few hundredths of a second. Nonetheless, important clues to disease or malfunction (pathology) are provided by the degree of asynchrony which may be detected when listening to the heart sounds with a stethoscope. The pattern of activity (Fig. 13-9) may be explained by the following sequence: (1) Excitation is initiated in the right atrium and is rapidly conducted to the ventricles. (2) Left ventricular contraction starts first, with closure of the mitral valve occurring within 30 msec of the start of the Q wave of the ECG. The right heart lags, because of the anatomy of the conduction system, by about 15 msec. (3) The pulmonary valve opens (60 msec after the Q wave), and pulmonary flow starts about 10 msec before aortic flow. Because pulmonary arterial diastolic pressure is so much lower than that in the aorta, it takes longer for the left ventricular pressure to develop to the level required to open the aortic valve (Fig. 13-9), even though the rate (dP/dt) is faster in the left than in the right. The isovolumetric period for the right heart is only about 15 msec compared to about 40 msec for the left. (4) Because of the higher pressure in the aorta, left ventricular outflow tends to end first. Thus left ventricular flow *starts last* and *ends first*. Finally, (5) the mitral (left AV) valve opens after the tricuspid (right AV) valve—again, because more time is required for the pressure to drop from aortic pressure to left atrial pressure than for the right ventricular pressure to decrease. Note the mnemonic at the top of Figure 13-9: the isovolumetric phases of the right heart, being shorter, occur *during* the corresponding phases of the left heart.

Split Second Heart Sound

In normal people the second heart sound is audibly split during inspiration because of the difference in time between the aortic and pulmonary valve closings (Fig. 13-10). Our ears are such that sudden sounds (e.g., a click) separated by more than 0.02 second can

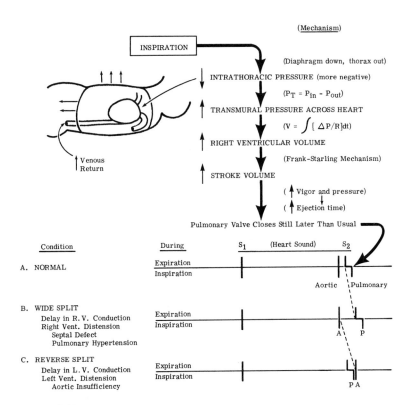

Figure 13-10
Physiology of the split of the second heart sound.

be discerned as separate, or split. Differences in these second heart sounds provide a wealth of information about possible cardiac and lung pathology.

The mechanisms for the delay of the pulmonic valve closure during inspiration are outlined in detail in Figure 13-10. The more negative intrathoracic pressure on inspiration leads to increased right ventricular filling, prolonged ejection and right ventricular systole, delayed valve closure, and hence a delayed P2 (pulmonary second heart sound). With right ventricular conduction defects or right ventricular overload, the split in the second heart sound may be heard even during the expiratory pause—a wide split. On the other hand, left ventricular overload or weakening, or blockage of conduction along the left ventricular branch of the bundle of His may prolong the aortic component so that the split is reversed or paradoxical—that is, is heard during expiration but not during inspiration.

Murmurs

Murmurs are relatively prolonged audible vibrations that originate in either the heart or the great vessels as a result of high-velocity blood movement. Blood flow through vascular channels is normally laminar or streamline. It is silent. Turbulent, noisy flow develops under special circumstances. It may be *random,* with a "whooshing" sound of a wide, uniform-frequency spectrum, or *vortex shedding,* with a predominant frequency or pitch (Bruns, 1959; see also Chap. 11). The factors that determine whether flow is laminar or turbulent are (1) the size and smoothness of the channel, (2) the velocity of flow, (3) the fluid density, and (4) its viscosity. Since blood viscosity and density are reasonably constant in the large vessels, the factors usually of concern are the velocity of flow, the radii of the parts of the cardiovascular system, and the smoothness of the flow channels (see Chap. 11).

It is probable that some turbulence exists normally (e.g., in the aorta or pulmonary artery just beyond the valve), but the noise produced is usually not intense enough to be heard with the stethoscope, or is included with the first heart sound. Such minimal turbulence can be increased and result in an audible benign murmur when the velocity of blood flow is increased, as in exercise, fever, pregnancy, or hyperthyroidism.

When the diameter of a flow channel is reduced by disease or a local obstruction, a murmur may result at normal or even decreased flow rates. Likewise, a murmur is produced by regurgitation of blood through an incompetent valve. The significance of a murmur depends upon (1) the location of its source; (2) when it is heard during the cycle—systolic or diastolic; (3) its quality and pitch; and (4) its intensity.

Systolic Murmurs

Systolic murmurs are those that originate with or after the first heart sound and end with or before the second heart sound.

In *aortic stenosis* (Fig. 13-11), the murmur is proportional to the rate of blood flow and so is loudest in midsystole, giving a diamond shape when visualized with a phonocardiograph. Blood is flowing in the normal direction, but the valve is narrowed to only a small orifice. When stenosis is severe, the ejection of blood is delayed and maximum accentuation occurs later in systole. On catheterization, a large difference in pressure ($\Delta P = 20$ to 100 mm Hg) can be measured across a stenotic aortic valve (Fig. 13-12).

In *mitral insufficiency,* the murmur is associated with regurgitation of blood through an insufficient valve (Fig. 13-11). Whether the valve is insufficient because of disease, dilated annulus, or ruptured chorda tendineae, the sound is similar because the mechanism is similar. When the pressure in the ventricle exceeds that in the corre-

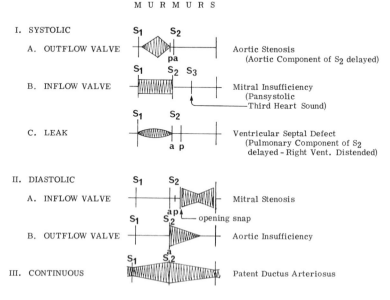

Figure 13-11
Patterns of heart murmurs.

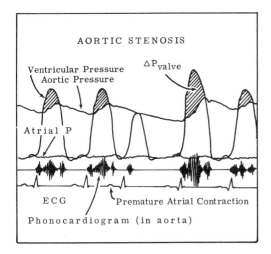

Figure 13-12
Aortic stenosis. (Redrawn from Moscovitz et al., 1963.)

sponding atrium, the mitral and tricuspid valves close, producing the first heart sound, but if closure is not complete, blood is forced back into the atrium under pressure, causing turbulence and a murmur. Usually the murmur begins immediately after the first sound and extends throughout systole with a fairly constant intensity ending with the second sound; it is thus called a *holosystolic* or *pansystolic* murmur because it occurs throughout all of systole.

A *ventricular septal defect* is a hole in the wall between the right and left ventricles. A systolic murmur (Fig. 13-11) will be heard during systole that will be like the murmur in mitral insufficiency, because of the similar mechanism. However, the position of maximum intensity on the precordium will be somewhat different. With the extra filling of the right ventricle—the leak plus the normal venous return—the right ventricular stroke volume will be greater than normal and the pulmonic component of the second heart sound will be delayed, giving a split even during expiration. Valvular regurgitation and septal defects are diagnostically separated by injecting a radiopaque material into the left heart to visualize the leak, called a *shunt*. Measuring the oxygen tension of the blood in the right atrium and right ventricle provides diagnostic information also, since blood leaking from the left side will have a high oxygen tension from passage through the lungs. Therefore, the blood in the right ventricle and pulmonary artery (distal to the shunt) will be better oxygenated than that in the right atrium (proximal to the shunt).

Diastolic Murmurs

In *mitral stenosis* (Fig. 13-11), the murmur occurs during diastole and is proportional to the flow. It thus is loudest early in diastole immediately following the isovolumetric relaxation period when atrial pressure is high, and it also becomes loud during atrial systole ("presystolic"). The murmur is often heralded by a sound, the opening snap of the mitral valve. With mitral stenosis, the pulmonary venous pressure is abnormally high, as is the pulmonary wedge pressure. (This pressure is obtained by floating a small-diameter catheter through the systemic veins, right heart, and pulmonary artery until it is wedged into a small pulmonary artery. Because of the relatively few collateral vessels at this level and low pressure gradient across the pulmonary microvasculature, the pressure beyond where the catheter is wedged in—the *wedge* pressure—has been found to be similar to that of the pulmonary veins and consequently the left atrium on the other side of the pulmonary capillaries. A systemic vein, and thence the pulmonary artery, is much easier and safer to catheterize than the left atrium.)

With *aortic insufficiency* (Fig. 13-11), the murmur is proportional to the flow, starting as soon as the valve partially closes and isovolumetric relaxation starts. It may last throughout diastole but because the arterial blood can leak back into the heart as well as go through the normal pathways, the arterial diastolic pressure is low, and so the murmur usually dies out during diastole. The systolic pressure will be high, from the compensatory mechanisms acting to increase cardiac activity to eject a larger than normal stroke volume.

Continuous Murmurs

Murmurs may continue throughout both systole and diastole—for example, the murmur associated with a patent ductus arteriosus. Since the aortic pressure always exceeds the pulmonic, the flow is continuous, as is the resultant murmur. Furthermore, a valve may be so damaged as to be both stenotic and regurgitant.

Diagnosing possible pathological causes of murmurs requires consideration of numerous possibilities. There are four valves with two possible types of pathology: stenosis or insufficiency, or both. There may be septal defects at either the atrial or ventricular level, or both. There also may be a coarctation (narrowing) of the aorta, or patent ductus arteriosus, or an arteriovenous fistula leading to a murmur. The diagnostician must remember that the heart sounds are associated with sudden changes of motion and subsequent vibrations of elastic structures. Murmurs, on the other hand, are associated with high-velocity squirts of blood.

REFERENCES

Altman, P. L., and D. S. Dittmer. *Biological Handbooks: Respiration and Circulation.* Bethesda, Md.: Federation of American Societies for Experimental Biology, 1971.

Armour, J. A., and W. C. Randall. Structural basis for cardiac function. *Am. J. Physiol.* 218:1517–1523, 1970.

Brecher, G. A., and P. M. Galletti. Functional Anatomy of Cardiac Pumping. In W. F. Hamilton (Ed.), *Handbook of Physiology.* Washington: American Physiological Society, 1963. Section 2: Circulation, vol. 2, pp. 759–798.

Bruns, D. L. A general theory of the causes of murmurs in the cardiovascular system. *Am. J. Med.* 27:360–374, 1959.

Cobbold, R. S. C. *Transducers for Biomedical Measurements.* New York: Wiley, 1974.

Constant, J. *Bedside Cardiology.* Boston: Little, Brown, 1969.

Feigenbaum, H., and S. Chang. *Echocardiography.* Philadelphia: Lea & Febiger, 1972.

Geddes, L. A., and L. E. Baker. *Principles of Applied Biomedical Instrumentation.* 2nd Ed. New York: Wiley, 1975.

Hurst, J. W. (Ed.). *The Heart, Arteries and Veins* (3rd ed.). New York: McGraw-Hill, 1974.

Lewis, D. H. Phonocardiography. In W. F. Hamilton (Ed.), *Handbook of Physiology.* Washington: American Physiological Society, 1962. Section 2: Circulation, vol. 1, pp. 695–734.

Luisada, A. A. *The Sounds of the Normal Heart.* St. Louis: Green, 1972.

Luisada, A. A., D. M. MacCanon, S. Kumar, and L. P. Feigen. Changing views on the mechanism of the first and second heart sounds. *Am. Heart J.* 88:503–514, 1974.

Mirsky, I., D. N. Ghista, and H. Sandler (Eds.). *Cardiac Mechanics.* New York: Wiley, 1974.

Moscovitz, H. L., E. Donoso, I. J. Gelb, and R. J. Wilder. *An Atlas of Hemodynamics of the Cardiovascular System.* New York: Grune & Stratton, 1963.

Netter, F. H. *Heart.* The CIBA Collection of Medical Illustrations. Vol. 5. Summit, N.J.: CIBA, 1969.

Piemme, T. E., G. O. Barnett, and L. Dexter. Relationship of heart sounds to acceleration of blood flow. *Circ. Res.* 18:303–315, 1966.

309

Ray, C. D. (Ed.). *Medical Engineering.* Chicago: Year Book, 1974.

Rudolph, A. M., and H. A. Heymann. Fetal and neonatal circulation and respiration. *Annu. Rev. Physiol.* 36:187–208, 1974.

Rushmer, R. F. *Cardiovascular Dynamics* (3rd ed.). Philadelphia: Saunders, 1970.

Sandler, H., and E. Alderman. Determination of left ventricular size and shape. *Circ. Res.* 34:1–8, 1974.

Streeter, D. D., Jr., H. M. Spotnitz, D. J. Patel, J. Ross, and E. H. Sonnenblick. Fiber orientation in the canine left ventricle during diastole and systole. *Circ. Res.* 24:339–347, 1969.

Verel, D., and R. G. Grainger. *Cardiac Catheterization and Angiocardiography.* 2nd Ed. Edinburgh: Churchill/Livingstone, 1973.

Wiggers, C. J. *Circulatory Dynamics.* New York: Grune & Stratton, 1952.

14. Cardiac Excitation, Conduction, and the Electrocardiogram

Kalman Greenspan

The excitation or stimulus for the heart arises from within the cardiac muscle itself. How and where this excitation occurs and how it is conducted throughout the entire myocardium are discussed in this chapter. The electrical events of the cardiac cycle as manifested in the electrocardiogram and a simplified theory of electrocardiographic interpretation are presented.

RESTING AND ACTION POTENTIAL

The typical transmembrane potential of an inactive myocardial fiber (Fig. 14-1A, B, C) varies between 80 and 100 mv, the interior of the cell being negative with respect to the cell's exterior. When the cell is excited, however, the potential difference is quickly lost and in fact becomes opposite in sign (+20 to +40 mv), indicating that a reversal of polarity has occurred. This sudden initial upstroke (depolarization) is designated as phase 0 and includes the overshoot. Soon after the reversal of polarity the restorative processes start (repolarization), and the potential difference begins to decline toward its resting value. This is characterized by fast repolarization (phase 1), which then slows down and results in a plateau (phase 2). Following the plateau there is once again a fairly rapid wave of repolarization (phase 3), until the resting transmembrane potential or diastolic period is attained (phase 4). The time course of the cardiac action potential differs also according to the fiber type; atrial cells (Fig. 14-1A) show a less prominent plateau. Action potentials obtained from the ventricular myocardium (Fig. 14-1B) and from the Purkinje network (Fig. 14-1C) located in and ramifying throughout these chambers show a more prolonged plateau. However, phase 0 is similar for atrial, ventricular, and Purkinje fibers, being rapid in upstroke.

Quite the contrary is seen in the configuration of the action potentials obtained for the fibers in the sinoatrial (SA) node (Fig. 14-1D) and the atrioventricular (AV) node (Fig. 14-3A). Phase 4 demonstrates a slow continuous diastolic depolarization so that during diastole the SA membrane potential progressively becomes less negative. The magnitude of the transmembrane "resting" potential is about −70 mv. Furthermore, phase 0 is not rapid; the overshoot (phase 1) is much reduced and in many instances is not present. Phases 1 and 2 blend with phase 3, and the entire process of repolarization is slow. Following the repolarization wave there is a progressive decline in the magnitude of the resting potential (an incline in the phase 4 slope) so that on reaching its threshold potential the cell becomes excited in a self-sustaining manner, resulting in full depolarization. All cells potentially capable of developing intrinsic automaticity* exhibit this characteristic sign of slow diastolic depolarization. The time course of these action potentials has great physiological significance, in that cardiac rate and rhythmicity are dependent upon the magnitude of the resting potential, the level of the threshold potential, and the slope or steepness of the diastolic depolarization. In turn, the slow diastolic depolarization is probably responsible for the property of automaticity while the level of membrane potential will determine excitability and conduction. The mechanism responsible for the slow diastolic depolarization has not been clearly defined although there is some evidence of a concomitant progressive increase in membrane

---

*A myogenic property of certain cardiac cells defined as an inherent ability to develop spontaneous depolarization. Such cells are also called pacemaker cells.

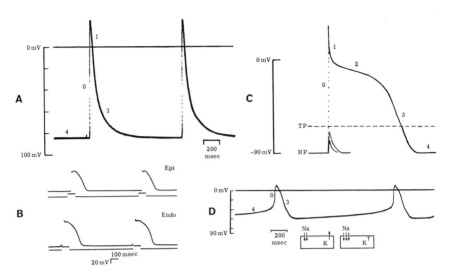

Figure 14-1

Records of membrane potentials obtained from: A. Atrial contractile fiber. B. Ventricular epicardial (Epi) and endocardial (Endo) surfaces. C. Purkinje fiber. D. Sinoatrial pacemaker cell. Note the differences in rapid depolarization (phase 0), the plateau (phase 2), and repolarization (phase 3) between these cardiac cells. C. TP = threshold potential; RP = resting potential. Application of progressively increased cathodal stimuli to cardiac Purkinje fiber. With a threshold stimulus (S_3), the all-or-none action potential is produced. D. In automatic fibers progressive spontaneous self-depolarization occurs during diastole. Excitation occurs when the threshold level of depolarization is reached. Below, in diagrammatic form, are the possible ionic fluxes that could account for the continuous slow diastolic depolarization.

resistance during diastole which may be indicative of a decreased membrane permeability to K^+.

PHYSIOLOGY OF PACEMAKER ACTION

Following depolarization of the primary automatic cell in the sinoatrial node, the current so generated flows longitudinally and lowers the transmembrane potential of adjacent areas. When threshold level is attained, these areas in turn depolarize, and self-sustaining, propagated action potentials spread over the entire myocardium. There is, however, a preferential sequence of activation (Fig. 14-2). Activity is first initiated in the vicinity of the SA node. The velocity of conduction through pacemaker cells is difficult to measure since it increases with distance from the primary pacemaker cell. In the rabbit heart, the earliest activity was detected in the wall of the superior vena cava, a few millimeters away from the crista terminalis, and the activity propagated radially at an extremely low velocity of 0.05 meter per second. As propagation approached the crista terminalis, conduction velocity became increasingly faster. In fact, activity from the crista terminalis spreads into the atrial roof and down to the coronary sinus at velocities ranging from 0.5 to 1.0 meter per second. From the atria, activity spreads through the atrioventricular node, where after some delay the sequence of activation is through the His bundle, bundle branches, peripheral Purkinje fibers, and then ventricular tissue proper.

Of great significance is the delay in the spread of excitation that occurs in the atrioventricular region. Microelectrode studies of rabbit heart have aided immensely in the elucidation of the mechanism of nodal delay. From a functional point of view, the

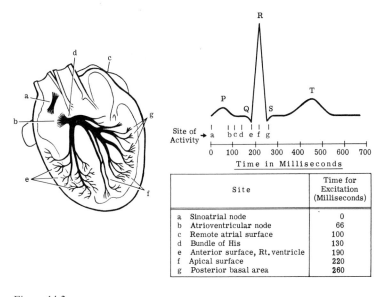

Figure 14-2
Conduction pathways of the heart. The times, in milliseconds, following pacemaker stimulation required for excitation to occur are shown for various regions. An electrocardiogram is shown correlated with the conduction sequence.

AV nodal region can be divided into AN (atrionodal), N (nodal), and NH (nodal-His) zones. Activity approaches the AN boundary region at right angles, and propagation becomes slower. Associated with the reduced conduction velocity is a decrease in the rise time of the action potential when the excitatory wave reaches the narrow, middle N zone; propagation is as slow as that seen in the SA pacemaker (0.05 meter per second). The upstroke of the action potential is slow, and the propagation time consumes about 30 msec of the total AV delay. After the N zone is passed and activity enters the NH zone (Fig. 14-3A), conduction velocity increases (1.0 to 1.5 meters per second) and the action potential starts to assume a configuration almost identical to that seen in the area of the bundle of His (Fig. 14-3B).

It has been postulated that conduction through the AN and NH regions is all-or-none while the mechanism of propagation within the N layer node is one of decremental conduction. Decremental conduction means that, whatever the cause, there is a gradual decrease in the rising velocity and a gradual diminution in the amplitude of the action potential as the activity propagates over a given path. Associated with these phenomena is a decrease in conduction velocity. Accordingly, the N layer is the region where AV block is most likely to occur, since it is in this zone that conduction velocity is at its minimum. If decremental conduction in the N zone is enhanced, as with acetylcholine (ACh) or vagal stimulation, the action potential amplitude attained may not be sufficient to excite the next zone and the propagation wave will not pass to the bundle of His. The result is AV block.

Thus the potential "dam" for excitation lies in the fact that conduction through the N layer is decremental. However, once excitation passes this zone with a high enough amplitude, the wave will reach the bundle of His. As the activity enters the His area and the bundle branches, there is again an increase in conduction velocity (to 3 to 4 meters per second) attributable to the greater diameter of these fibers and their infrequent branching. Action potentials from these cells are characterized by a rapid upstroke, high amplitude, and long duration (Fig. 14-3B). Conduction then slows down (to 1 meter per

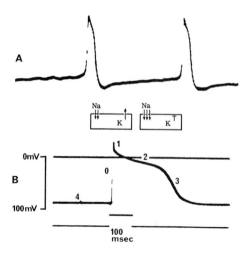

Figure 14-3

A. Record of action potentials obtained from rabbit AV node (NH region). Note the appearance of a spike similar to that in a His-type action potential, but the period of diastole (phase 4) shows an indication of continuous spontaneous depolarization. It may, however, represent a period of hyperpolarization that tends to level off. See text for discussion. Below the potentials are the possible ionic fluxes (decrease in K^+ conductance with an increasing Na^+ influx) that could account for the continuous depolarization. B. Action potential recorded from the bundle of His area. Note that there is no spontaneous self-depolarization; this activity requires a stimulus. The action potential is characterized by a rapid upstroke velocity, a sharp phase 1, and a prolonged plateau (phase 2).

second) as activity enters the ventricular muscle (Fig. 14-1B), presumably because of the smaller fiber diameter and frequent branching.

Generally, impulse transmission in cardiac tissue occurs at different velocities in the various parts of the heart. In any given fiber, however, changes in conduction velocity can be attributed to changes in the resting potential, threshold potential, rate of depolarization, action potential duration, degree of repolarization prior to the arrival of the next impulse, or changes in the core conductor properties of that tissue.

Automaticity in the cardiac cell is demonstrated by a slow continuing depolarization during the diastolic phase (no isoelectric period is observed during phase 4). It can be demonstrated that such automatic cells reside in the SA node (Fig. 14-1D), within the atrium, and in the His-Purkinje system. True atrial and ventricular cells apparently do not meet the electrophysiological and pharmacological criteria for automaticity and hence are not considered as being composed of automatic cells. The cells of the AV node, however, require special consideration. At first glance they seem to show spontaneous depolarization (Fig. 14-3A). But Hoffman and Cranefield (1964) have stated that the AV node does not contain automatic fibers. They base their opinion on the observation that following phase 3 there is a phase of hyperpolarization which rapidly returns to an isoelectric level and remains steady for the duration of phase 4. In addition, a foot appears just prior to the AV nodal action potential upstroke. During rapid heart rates this foot of the nodal action potential merges with the early hyperpolarization, giving the appearance of diastolic depolarization (Fig. 14-3A). Thus, the so-called clinical nodal rhythms probably originate from the transitional fibers located between the NH area and the His area, or from the bundle of His itself. These areas possess automatic cells, and when one of the areas serves as the ventricular pacemaker, electrocardiographic interpretation would be a nodal rhythm.

IONIC BASIS OF ELECTRICAL ACTIVITY

With some reservations, the ionic theory as postulated for the squid axon can be applied to cardiac cells (see Chap. 1). The theory rests essentially upon the concept that any viable membrane possessing the property of excitability can and will alter its permeability to several ionic species. A number of investigators have shown that the intracellular K^+ concentration is 30 to 40 times as much as is present extracellularly. The reverse situation exists for the Na^+ ion. It has also been observed that the membrane at rest is much more permeable to K^+ than to Na^+. Existence of this selective permeability, plus the uneven distribution of these ions across the membrane, allows the cell to act as a battery. The uneven ionic distribution is probably attained through a metabolism-dependent "pump," which actively shifts Na^+ out of the cell interior while transferring the K^+ inside.

Two factors appear to be responsible for, and determine the values of, the cardiac transmembrane resting and action potentials. They are the concentration gradients of K^+ and Na^+, and the relative membrane permeability of these two cations. The equation depicting the contribution of these factors is presented in Chapter 1 (see equation 13). To comprehend the equation fully, one must constantly keep the aspect of permeability in mind. Thus, when the cardiac cell is not excited but at rest, its membrane is relatively more permeable to K^+, and so the transmembrane potential is dominated by the K^+ gradient. With activity, as the heart is depolarized, the permeability to Na^+ increases, permitting some Na^+ to enter the cell and causing further depolarization. This in turn lets the membrane permeability to Na^+ increase further, thus facilitating a greater influx of Na^+. Finally, Na^+ permeability becomes greater than the K^+ permeability, and the membrane potential is hereby determined by the Na^+ gradient. Initially the fast Na^+ channel opens, permitting this cation to enter the cell and causing further depolarization. At some potential level (around -50 mv) a second set of channels opens, permitting the slow inward current of Ca^{++} to occur and complete the depolarization process. At the peak of the action potential, the membrane starts the restorative process, which involves again a reversal of its permeability to Na^+ and K^+. The first step is a decrease in Na^+ permeability which is known as *inactivation* (P_{Na} begins to fall toward its resting value), and soon after P_K begins to rise. Both changes allow the K^+ gradient to dominate, and thus determine the transmembrane potential responsible for repolarization. Membrane conductance during the plateau, or phase 2, of the ventricular and His-Purkinje action potentials remains undefined. It has been suggested, however, that during the plateau P_K and P_{Na} are equal, resulting in a temporary maintenance of the potential at or near the zero level. In addition, there is evidence implicating Ca^{++} as the major cation responsible for the plateau phase. The inward current that flows during phase 2 is in the direction that a Ca^{++} current would take, and the direction is altered by changes in Ca^{++} concentration in the extracellular fluid. Thus, the maintenance of the plateau may be a result of the entrance of Ca^{++} into the cell while cessation of Ca^{++} influx would result in termination of phase 2 (Ca^{++} inactivation).

The ionic theory briefly discussed above postulates that there is an ionic basis for the potential differences existing across the cardiac cell membrane under conditions of rest and activity. In order for a specific ion at any given moment to be responsible for a potential difference, it must build up an electromotive force. This is possible, given two conditions:

1. An uneven distribution of that ion (i.e., causing a gradient to exist across the membrane).
2. Greater permeability for that ion than for any other ion at that given moment.

For the heart, as for other tissues, evidence indicates that during diastole and during repolarization the specific ion responsible for the potential is K^+. At these times the cardiac membrane permeability is much higher for this cation than for any other positively charged ion. Similarly, during the depolarization the Na^+ and Ca^{++} ions determine membrane potential. Hence changes in extracellular ionic concentration of K^+ and Ca^{++} should reflect changes in the amplitude, configuration, and duration of the monophasic action potential. Concomitant changes should occur on the complex of the electrocardiogram. Furthermore, any agent, drug, or hormone which affects membrane permeability will alter its excitability.

EFFECT OF IONS ON THE RESTING POTENTIAL

A prominent effect of increasing the extracellular K^+ concentration is a reduction in the resting potential; the magnitude of the reduction bears an almost linear relation to the amount of K^+ added to the perfusion solution and can be approximately predicted by the Nernst equation (Chap. 1). The depolarization effect of high K^+ concentration has been described for all fiber types of the heart in many species. Structures studied include cat and rabbit atria, and dog SA node, atria, ventricle, and Purkinje fibers, as well as the hearts of cold-blooded animals (frog and turtle). In all these fiber types the initial effect of the reduction of the resting potential by increasing K^+ concentration is an enhancement of excitability. The reason is that the resting potential encroaches upon the threshold potential level so that the stimulus requirement is lower. However, further increase in extracellular K^+ results in a loss of excitability, presumably through inactivation.

An interesting observation was made in relation to the K^+ and Ca^{++} ratio. This ratio appears to be an important component of cardiac function; a similar interaction between K^+ and Ca^{++} exists for the resting potential. Thus, the depolarizing effect of a high concentration of extracellular K^+ can be counteracted by simultaneously raising that of extracellular Ca^{++}, and augmented by lowering the Ca^{++} concentration. With normal K^+ levels the alteration of Ca^{++} to high or low levels does not affect the resting potential. This concept of a K^+ and Ca^{++} antagonism in cardiac tissue is consistent with that observed in other viable cells.

Paradoxically, lowering the extracellular concentration of K^+ results in a resting potential change similar to that of high K^+ concentration. However, this occurs only in mammalian hearts, since in cold-blooded animals the response to severe reduction of extracellular K^+ concentration (0 to 25 percent of normal) is a marked increase in the resting potential, although an increased resting potential in the isolated rabbit heart, perfused with a solution of low K^+ concentration, has been reported. In any event, the depolarization initiated with a low concentration of K^+ in mammalian hearts can be aided by simultaneously raising extracellular Ca^{++} concentration, whereas lowering the concentration of extracellular Ca^{++} in the face of low K^+ concentration will restore the resting potential to its normal value. In fact, it has been observed that ventricular fibrillation induced with a low concentration of K^+ can be terminated if the low-concentrate K^+ perfusate is replaced with a solution low in both K^+ and Ca^{++}.

EFFECTS OF IONS ON THE ACTION POTENTIAL— NONAUTOMATIC CELLS

The change in the resting potential due to an alteration in extracellular K^+ concentration would be expected to affect the contour, amplitude, and duration of the action potential. Thus, following exposure to a high concentration of K^+ there is a reduction in maximal upstroke velocity (Fig. 14-4), amplitude, and action potential duration. These effects have been noted for atria, Purkinje fiber, and ventricular tissue of both mammalian and cold-blooded animals. They may result from the reduced resting potential or perhaps they are due to a direct effect of the added K^+.

The reduced upstroke velocity of phase 0 and the decreased amplitude are apparently

317

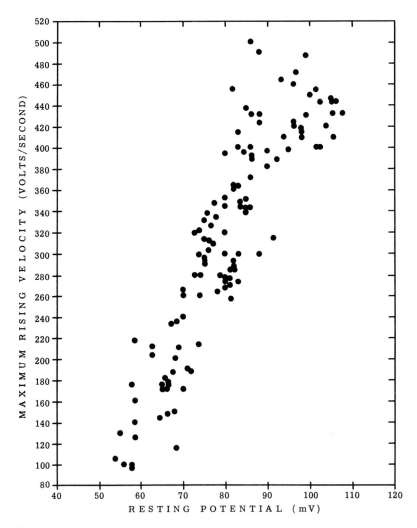

Figure 14-4
Relationship between the transmembrane resting potential and the maximum rising velocity in canine Purkinje cells. As the potential is reduced, the maximal rate of rise decreases. The resting potential in the experiment was reduced by exposing the tissue to a high concentration of K^+.

indirect effects secondary to the reduced resting potential; the observation was made that if the resting potential is restored, as by anodal current, the effect of high-concentrate K^+ on phase 0 and upstroke amplitude is reversed. However, the enhanced repolarization induced by high K^+ concentration is not completely counteracted by anodal current. It appears to be a direct effect of the cation: Infusion of K^+ during the plateau period (phase 2) of the ventricular action potential causes an immediate repolarization. Similar infusion during diastole results only in depolarization without affecting the time course of the action potential. Thus, shortening of the action potential due to high K^+ concentration can occur independently of a reduction in resting potential, and the action of the cation is a direct effect, not a consequence of depolarization. Since during the normal period of repolarization there is an increase in the efflux of K^+, it would seem that the K^+ itself increased K^+ efflux from the cell, presumably by enhancing membrane permeability. Evidence has been advanced that an increase in extracellular K^+ concentration decreases membrane resistance. In fact, this effect is much more pronounced at a membrane potential of -40 mv, the potential level at which the plateau region for sheep Purkinje fiber occurs. Furthermore, both influx and efflux rates of K^+ increase with increasing K^+ concentrations. The above results are highly suggestive that K^+ itself increases membrane conductance or permeability for K^+.

Perfusion of cardiac cells with solutions containing low K^+ concentration affects the action potential configuration in a manner opposite to that seen with high K^+. Thus, with low K^+ concentration the atrial and ventricular action potentials show an increased amplitude and a prolonged duration. Interestingly, continued perfusion with a low-concentrate K^+ solution results in supraventricular and ventricular ectopic beats, ectopic tachycardias, and finally ventricular fibrillation, which are not terminated by replacement in the normal K^+ solution. However, if the low-concentrate K^+ solution is replaced with a control solution prior to the onset of fibrillation, the result is an immediate cardiac standstill lasting 2 to 30 seconds. Subsequently, the heart starts to beat slowly and within two minutes the rate returns to control values. The interesting electrophysiological observation is that, while in the low-concentrate K^+ solution, the ventricular action potential amplitude and duration are increased. Upon the return to the control solution, however, the action potential amplitude is decreased and the duration is shortened, thereby simulating a "high" K^+ effect. In fact, the amplitude of the action potential after the standstill period levels off at a value intermediate between the low-concentrate K^+ solution and the return to the control solution (i.e., smaller than when in the low-concentrate K^+ solution but higher than the original control amplitude).

Since the phase of depolarization is determined by the movement of Na^+, the changes in action potential configuration can be predicted following changes in concentration of this ion in the extracellular components. Thus, a *reduction* in Na^+ concentration results in a decreased upstroke velocity and a decrease in the amplitude of the atrial action potential. Severe reduction of this ion (to 10 percent of normal) leads to a complete loss of excitability. The ventricular muscle behaves somewhat differently in a low-Na^+ medium in that phase 0 is not affected, although the duration is shortened on account of an increase in the steepness of phase 2. *Increasing* the Na^+ concentration has little direct effect on action potential configuration. An indirect effect may be observed as a result of changes in tonicity with the excess Na^+.

Calcium has little effect on the membrane potential and amplitude of the action potential. However, the duration of the action potential is altered by the concentration of Ca^{++} in the medium. In fact, the ventricular action potential following Ca^{++} excess and subsequent intracellular influx becomes markedly prolonged. Inhibition of Ca^{++} influx causes the ventricular cell to resemble an atrial action potential. On the other hand, atrial excitability (unlike ventricular) is markedly decreased by Ca^{++} depletion.

Effect of Temperature on the Action Potential of Nonautomatic Cells

The electrophysiological properties of atrial and ventricular cells are strongly affected by changes in temperature. Temperature affects the action potential phases directly and indirectly via change in heart rate. With hyperthermia, rate is increased and action potential duration decreased. With cooling, depending upon the severity, the heart rate is decreased, the action potential duration markedly prolonged, and the membrane potential reduced; these changes may ultimately lead to a complete loss of excitability. Temperature change can produce the above alterations directly and independently of rate.

FACTORS AFFECTING RATE AND RHYTHM— AUTOMATIC CELLS

The beat of the heart occurs in an extremely regular, continuous, and rhythmic fashion. It is a result of automaticity, a property present in those cells of the heart that exhibit the electrophysiological pattern of slow diastolic depolarization (Fig. 14-1D). This phenomenon confers upon the cells the ability to depolarize spontaneously, independently of the extrinsic and intrinsic nerve supply. The cells of the heart which exhibit automaticity with the most rapid rate of depolarization, and which are responsible for depolarizing the rest of the heart, are located in the sinoatrial node. Although myogenic in origin, the rate and rhythm may change spontaneously, or they may be influenced by temperature, inorganic ions, and neural and humoral activity. Any factor that changes the rate and rhythm of the heart does so by affecting (1) the magnitude of the resting potential, (2) the threshold potential voltage which must be attained, and/or (3) the slope of diastolic depolarization. Study of Figure 14-5 reveals that any alteration either of these potentials or of the slope of depolarization alters the time required for reaching threshold and, hence, the rate of stimulus production.

Effect of ACh and the Vagus

Stimulation of the vagus or administration of ACh reduces the heart rate (bradycardia) and, with sufficient intensity or concentration, may lead to sinus arrest. The effects on the automatic cell of such maneuvers are as follows: (1) Repolarization is enhanced. (2) Hyperpolarization takes place: In the lower diagram of Figure 14-5 the potential goes from *a* to *d*. The magnitude of the hyperpolarization depends upon the initial resting potential value; the smaller the resting potential, the greater the hyperpolarization produced by ACh. (3) There is a shift of the slope of diastolic depolarization: In the

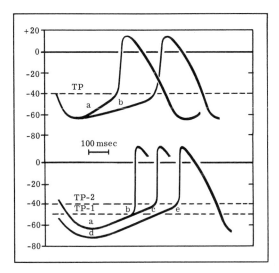

Figure 14-5
Diagram representing the important mechanisms responsible for changes in frequency of automatic fiber. The upper diagram shows a decrease in rate caused by a decrease in the slope of phase 4 from a to b and thus an increase in the time required for the transmembrane potential to reach the threshold potential (TP). The lower diagram shows rate changes associated with a change in the level of the threshold potential from TP-1 to TP-2 and an increase in cycle length from a–b to a–c; also shown is a change in rate due to an increase in resting potential (compare a–c and d–e). (From *Electrophysiology of the Heart,* by B. F. Hoffman and P. Cranefield. Copyright 1960, McGraw-Hill Book Company. Used with permission of McGraw-Hill Book Company.)

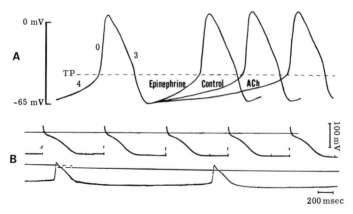

Figure 14-6
A. Epinephrine or sympathetic stimulation increases heart rate by enhancing the slope of diastolic depolarization while parasympathomimetic activity or ACh depresses the steepness of phase 4 depolarization to result in slowing of heart rate. B. Purkinje fiber action potentials (upper strip), which are converted to pacemaker type (lower strip) with influence of epinephrine.

upper diagram of Figure 14-5 the slope goes from a to b. The net effect of cholinergic activity is an increase in the time required for the fiber to reach its threshold potential. The most prominent effect of cholinergic activity is on phase 4 depression; if severe enough, it can result in cardiac standstill. The mode of action of ACh appears to be via an increase in K^+ permeability, which in turn reduces the loss of diastolic membrane potential and thus leads to a reduction in rate (Fig. 14-6A).

Catecholamines and Sympathetic Nerve
The administration of epinephrine or norepinephrine, as well as stimulation of the cardioaccelerator nerve, increases the heart rate (tachycardia). The most striking effect is on the slope of phase 4 (Fig. 14-6; Fig. 14-5, in upper diagram b to a). On latent pacemakers, such as Purkinje fibers, sympathetic activity induces automaticity (Fig. 14-6B) and may lead to multifocal firing and ventricular fibrillation.

Temperature
Cooling of the SA node reduces spontaneous activity (depression of phase 4), and severe cooling may result in complete standstill. Increasing temperatures increase the slope of diastolic depolarization (Fig. 14-6A, similar to that seen with epinephrine). High temperature may also induce Purkinje fiber to become pacemaker cells (as is also seen with catecholamines—Fig. 14-6B).

Effect of Ions on Automatic Cells
Very few studies have been reported on the effect of K^+ upon the electrophysiological properties of cells possessing the property of automaticity. From the available information it appears that a decrease in extracellular K^+ causes an increase in the slope of diastolic depolarization (probably due to a decrease in P_K), a decrease in the maximum diastolic potential, and a reduction in threshold. (See Fig. 14-5, lower diagram; low K^+ results in a shift of d to a; TP-2 to TP-1. In upper diagram slope b shifts to slope a. See also Fig. 14-1D for ionic fluxes which may be exaggerated with a low K^+ environment.) All of these accelerate activity in the normally automatic cells of the heart. SA node tissue perfused with a K^+-free solution develops multifocal activity. In normally quiescent but latent automatic fibers (Purkinje system), low K^+ concentration tends to initiate pacemaker activity and frequently leads to the production of extrasystoles.

Eventually, however, such fibers exposed to an extracellular environment lacking in K^+ will become arrested because of loss of resting potential.

On the other hand, an increase in extracellular K^+ has an opposite effect. The steepness of phase 4 is decreased, probably on account of an increase in membrane permeability to potassium. This would decrease the firing rate of the normal automatic cells. It has been stated, however, that the initial effect of increasing extracellular K^+ is an increased rate of firing from the SA node, which can occur since high K^+ also decreases the magnitude of maximum diastolic potential of the SA nodal cells. The firing rate increases or decreases following high K^+ exposure depending upon whether the cation effect is predominantly on the resting potential or on the diastolic depolarization.

Generally, K^+ affects all the cardiac cells in a similar fashion. However, the sensitivity of the cardiac cells to the cation is markedly different. Within the same heart it appears that the cells of the SA and AV nodes are more resistant to K^+ than are the contractile cells of the myocardium. In turn, the Purkinje fiber network is more sensitive to K^+ than are ventricular cells, while the bundle of His is highly resistant to increases of extracellular K^+. The sensitivity to K^+ also differs within the contractile tissue in that conduction fails earlier in the atrial fibers than in ventricular fibers. Of interest is the high resistance of the SA nodal cells to K^+. Increasing the perfusing solution with K^+ to five times greater than normal results in complete suppression of atrial activity, whereas the SA nodal cells show only a slight decrease in the magnitude of the resting potential and in the slope of diastolic depolarization. Only when the K^+ concentration is markedly increased (to 15 times normal) will the spontaneous activity of the pacemaker become arrested.

Other ions such as Na^+ and Cl^- have little effect on the automatic cells of the SA node unless the ionic changes are severe. Ca^{++}, however, may play an important role in the inherent propensity of automaticity.

THE ELECTROCARDIOGRAM

In any fundamental consideration of the constitution of the electrocardiogram (ECG) due to differences in the basic electrical properties of different cardiac tissue, the registration of the electrocardiographic complex consists of various wave forms. A representative complex is illustrated in Figure 14-7A. The P wave is constituted by the sum of atrial depolarization. The QRS complex reflects ventricular depolarization. The T wave represents ventricular repolarization.

The relative magnitude of these electrocardiographic components reflects the relative mass of tissue represented. For example, the ventricular mass, which is responsible for the relatively large QRS complex, comprises by far the greatest bulk of the heart. The specialized conducting cells and pacemaker cells (SA node, specialized atrial fibers, AV junctional cells, and His-Purkinje cells) are not graphically registered in the conventional ECG because of relatively small total mass and their distance from the conventional recording electrodes. Atrial repolarization is infrequently observed as it usually occurs simultaneously with ventricular depolarization, the wave form of the latter obscuring the atrial T wave.

Normally, the impulse originates in the SA node located in the right atrium at the opening of the great veins, and it traverses through the specialized paths to activate both atria. As a result, we get an inscription on the ECG of the P wave. Hence the P wave represents the electrical record of atrial depolarization. Next, the impulse is delayed for some time in the AV junctional tissue and then is rapidly conducted through the bundle of His, the left and right bundle branches, and the Purkinje system. On the ECG this entire time is recorded merely as a straight line, the so-called isoelectric P–R segment. Finally the myocardium is activated, the impulse going from endocardium to epicardium; on the ECG it is recorded as the QRS complex. The electrical status must then be reestablished, and this process of repolarization gives rise to the T wave on the ECG.

However, the surface ECG does not closely resemble the single fiber action potential, the summated action potentials of which it is comprised. The source of such a marked disparity becomes evident when one considers that the surface ECG represents an algebraic sum of numerous action potentials, oriented in many directions and happening serially as well as simultaneously. If depolarization and repolarization occurred simultaneously in all fibers, the net electrical difference between two recording electrodes would be zero. But it has been amply demonstrated that depolarization of ventricular fibers, initiated via the His-Purkinje conduction system, takes place initially in endocardial fibers and then proceeds radially to the epicardium (Fig. 14-7B). Thus, a QRS dipole vector is established by this temporal sequence of ventricular depolariza-

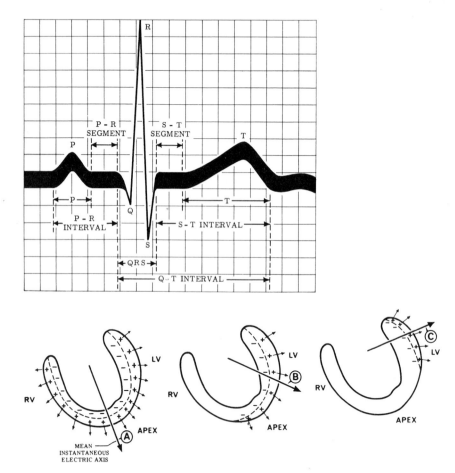

Figure 14-7
(Upper) The normal electrocardiogram, showing the time relationships. The distance between the vertical lines is 40 msec. The distance between the horizontal lines is 0.1 mv. (From G. E. Burch and T. Winsor. *A Primer of Electrocardiography* (6th ed.). Philadelphia: Lea & Febiger, 1972.) (Lower) A. The mean instantaneous electrical axis at the beginning of ventricular depolarization. B. The mean instanteous electrical axis at a later stage of ventricular depolarization. The vector is more to the left and with excitation proceeding toward the epicardium. C. The mean instantaneous axis late in ventricular depolarization. LV = left ventricle; RV = right ventricle. (From Burch and Winsor, 1972.)

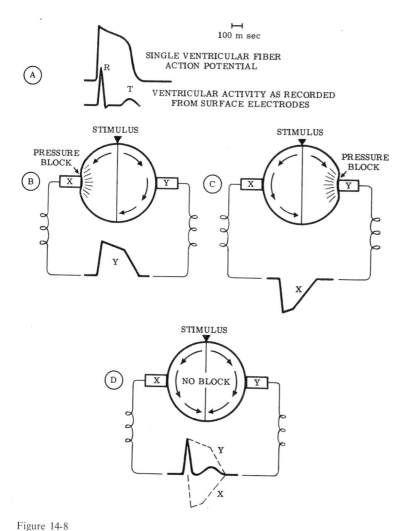

Figure 14-8
A. The action potential of a single ventricular fiber and the electrical activity of the entire ventricle are compared. B. Applying pressure to a hypothetical myocardium with an electrode (X) blocks the conduction of the excitation, and this electrode serves as an indifferent electrode. A monophasic action potential is then recorded under the other electrode (Y). C. The situation is reversed and the direction of the deflection is reversed. If electrodes X and Y were symmetrically influenced by the spread of excitation, no deflection would be recorded, as the upward and downward deflections would cancel each other. D. If some asymmetry existed, the recorded deflection would represent the algebraic sum of the two monophasic potentials. The X electrode is recording from the epicardium and the Y electrode from the endocardium. (From K. Jochim, L. N. Katz, and W. Mayne. *Am. J. Physiol.* 111:177, 1935.)

tion. As schematically illustrated in Figure 14-8, the addition of endocardial and epicardial action potentials yields an isoelectric S–T segment. It would seem reasonable, as well, to expect the T wave to be of equal amplitude and opposite direction to that of the QRS if repolarization occurred in the same sequence as depolarization. As shown in this diagram, however, more rapid repolarization of the epicardium than of the endocardium yields an upright T wave in the presence of an upright QRS complex.

The most conventional method of recording the summation of action potentials involves the use of the three standard leads, as conceived by Einthoven in 1908. Viewing the heart as an electrical dipole situated in the center of an equilateral triangle (see Fig. 14-9), one can perceive the derivation of the three standard leads. Lead I is recorded between the right and left arms, producing an upward deflection if the left arm is positive with respect to the right arm (i.e., when the projection of the dipole vector renders the left arm positive with respect to the right arm). Lead II is recorded between the right arm and left leg, yielding an upward deflection if projection of the cardiac vector upon this lead produces positivity in the left leg. Lead III is recorded between the left arm and left leg, yielding an upward deflection if the left leg is positive with respect to the left arm. As Einthoven's triangle constitutes an electrical network, physical principles require that the sum of all electrical forces equal zero. Thus, Einthoven's law states that lead I voltage plus lead III voltage equals lead II voltage at any instant of the cardiac cycle. (Einthoven's modification of this basic principle involved a reversal of lead II polarity; thus, lead I + lead III = lead II.)

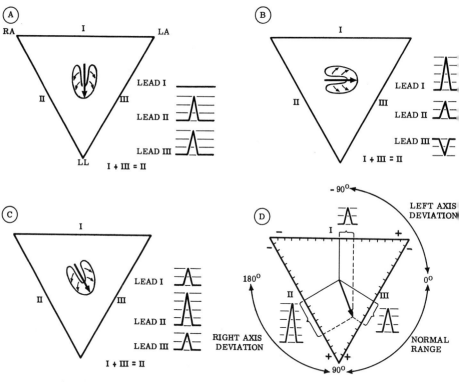

Figure 14-9
Relationship of electrical axis of the heart to the recorded potentials. A. The vector of excitation, or axis, is directly vertical downward. The voltage changes seen at each lead are shown. B. The vector of excitation, or axis, is directly horizontal and to the left arm. C. The vector of excitation, or axis, is downward and toward the left. Lead II, which is most parallel to the axis, has the greatest deflection. D. To determine the mean electrical axis from records obtained for any two leads: Count off from the midpoint of each lead in the appropriate direction the number of millimeters of deflection obtained. Drop perpendiculars from each point. The line drawn from the midpoint of the triangle to the point of intersection of the perpendiculars represents the axis. The arrow head represents its direction, the length of the line represents its magnitude, and the degrees represent its orientation.

So the typical ECG of a cardiac cycle consists of three positive waves, the P, the R, and the T waves. Sometimes one can record a fourth wave called the U wave. The U wave is of unknown origin. It may represent afterpotentials, or it may represent repolarization of the Purkinje network. In any event, besides the positive waves there are two negative waves, the Q wave and the S wave. The Q and S waves with the R wave are collectively referred to as the QRS complex.

It is customary to quantitate these potentials generated from the heart, and we can carry out quantitation from the ECG because of the horizontal and vertical lines that are inscribed on the recording paper. The horizontal and vertical lines are 1 mm apart and represent time and amplitude respectively. Each millimeter of horizontal line is equal to 0.04 second, with every fifth horizontal line being wider and more heavily inscribed so that the distance between two adjacent heavy horizontal lines represents 0.2 second. The vertical lines are also wider after every fifth line, and the machine is so standardized that the addition of 1 mv from the circuit causes the pen to deflect 10 mm or 1 cm. We speak of these waves, intervals, and segments in terms of duration in seconds, and of amplitude in terms of millivolts or millimeters.

To analyze a single cardiac cycle: We first have the P wave, which represents atrial depolarization. The width of the P wave measured from the beginning to the end of the wave should not normally exceed 0.12 second. Its height does not usually exceed 2.5 mm. Note that, although the P wave represents atrial depolarization, we normally do not see an atrial repolarization or auricular T wave (Ta) because it is very small in voltage, and is usually buried in the P-R segment and in the QRS complex. The P-R segment is measured from the end of the P wave to the beginning of the QRS complex, and its duration is about 0.08 second. The measurement of the beginning of the P wave to the beginning of the QRS complex is called the P-R interval or P-Q interval and represents the atrial depolarization time plus conduction through the atrial tissue, the AV node, and the specialized conduction system. Normally this should not exceed 0.22 second.

The P-R interval is followed by an initial downward deflection called the Q wave, then an upward R wave followed by a downward S wave; normally this QRS complex should not exceed 0.1 second. Since the complex represents ventricular depolarization or phase 0, duration of this complex longer than 0.1 second is an indication of intraventricular conduction delay. That portion of the cycle between the end of the QRS and the beginning of the T wave is known as the S-T segment. It represents the depolarized state of the ventricle or phase 2. If one includes the T wave in the measurement, it is called the S-T interval, which represents the time from peak depolarization to complete ventricular repolarization. If one measures from the Q wave to the end of the T wave, it is called the Q-T interval, which normally lasts from 0.36 to an upper limit of 0.4 second, and represents the entire time required for depolarization and repolarization of the ventricular muscle. This time varies with age, sex, and, obviously, heart rate. As a rough index the Q-T interval can be considered normal if its duration is less than half of an R-R interval. Finally, the T wave is equivalent to phase 3 of the action potential and represents ventricular repolarization. Its amplitude and duration depend upon many factors. It is a very labile component and can be altered by emotional factors, smoking, drinking, and position of the heart; the significance of such changes is frequently difficult to evaluate.

We can best account for these phenomena in terms of the migration of a positive and a negative charge, collectively called a dipole. A dipole can be pictured as consisting of two points of opposite electrical charge separated by a very small distance. If the points are connected by a wire, then by convention it is understood that current will flow along the wire from the positive pole (anode) to the negative pole (cathode); actually the electrons flow from the negative to the positive pole. We can then analyze the human ECG in terms of the body surface potentials, the source of which is generated by a series of dipoles emanating from the heart and distributed throughout the body. So we consider the patient as a volume conductor and the electrical impulses originating in the heart as the source of potential difference. Adding and subtracting all of the many dipoles traveling in different directions within the heart at any given moment, we can represent, for practical purposes, the wave of depolarization as a single dipole. This can also be represented as a vector with direction, magnitude, and sense (that is, positivity or negativity). The vector travels in the direction of the depolarization with the arrow indicating the positive field.

The term *cardiac vector* is used to designate all of the electrical events of the heart cycle. It has a known magnitude and direction. The *instantaneous vector* represents the net electrical forces being propagated through the heart at a given instant. A *mean vector* of any given portion of the heart cycle (such as QRS interval) represents the mean direction and magnitude for that period in the heart cycle. The mathematical symbol of a vector is an arrow pointing in the direction of the net potential (that is, polarity), while the length of the arrow indicates the magnitude of the electrical force.

The summation of cardiac action potentials thus yields a net or resultant vector, with force and direction. Since the three standard leads reflect the projection of this cardiac vector upon three separate axes, it is apparent that the cardiac vector can be reconstructed from the standard leads (as shown in Fig. 14-9). The orientation of the QRS axis, in the frontal plane, is usually inferior and somewhat to the left. Change in the mean QRS axis can be produced by mechanical alteration of the cardiac position as well as by a wide variety of cardiac and pulmonary diseases which change the position of the heart. The electrical axis of the P and T waves may be calculated in a similar fashion.

Lead Systems

Actually the ECG which is recorded from the body surface is the result of the summation of all the conducted electrical impulses that are occurring in each cell of the heart. In clinical electrocardiology the electrical potential generated by the heart is recorded with bipolar leads or with unipolar leads.

With bipolar leads, both electrodes are in the electrical field generated by the heart, and the ECG records the difference between two points. The actual potential under either of the electrodes is not given. With unipolar leads, one of the electrodes is so constructed that it is electrically zero. This is called the *indifferent electrode.* The other connection of the unipolar lead sees the potential as absolute and is called the *exploring electrode.*

The important thing to remember in order to understand the clinical ECG is that it is the exploring electrode which determines the polarity of the wave. The ECG is so constructed that when the exploring electrode is in the positive field the ECG records an upright deflection. If the exploring electrode is in the negative field, the ECG registers a downward deflection.

Bipolar Leads

Clinically three bipolar extremity leads are used. We think of the extremities merely as lead wires connected to the body, and the potentials are not altered if the electrodes are moved along the extremities. As stated above, the conventional bipolar limb leads, designated as leads I, II, and III, record differences between right arm, left arm, and left leg. In lead I the connection is between left arm and right arm and the LA lead is the exploring electrode. In lead II the connection is between the right arm and the left leg, and the L electrode is the exploring one. In lead III the connection is between left arm and left leg and again the L is the exploring electrode. The three limb leads are connected in what is known as Einthoven's triangle (Fig. 14-10A). In theory, the heart is located at the center of the triangle, and a line drawn roughly between two electrodes of a lead is known as the *lead axis.* Einthoven deliberately arranged the polarity of lead II so that ventricular activation results in a positive deflection in all three leads of a normal subject. Since the three leads form a closed circuit, the algebraic sum of their potentials at any instant is zero. This has been stated as Einthoven's law, which says that the magnitude of leads I and III is equal to that of lead II. If Einthoven's hypothesis is true,

327

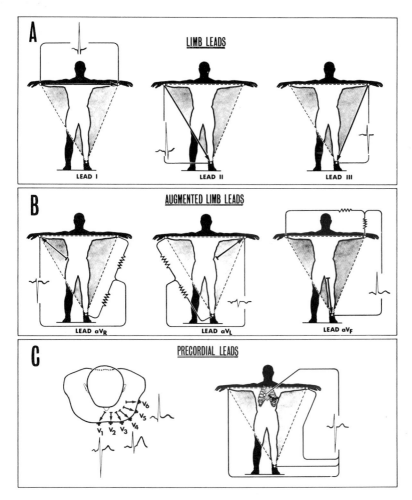

Figure 14-10
Standard limb and augmented unipolar leads. Resistors are used to form an indifferent reference electrode in the unipolar leads. Also shown (C) are the unipolar precordial leads for V_1 to V_6; resistors in the reference leads are omitted. The arrows indicate the vector required for an upward deflection of the recording machine. (From F. L. Abel. In E. E. Selkurt [Ed.], *Basic Physiology for the Health Sciences*. Boston: Little, Brown, 1975. P. 362.)

one should be able to hook up the three extremity leads to an *indifferent electrode,* located so far away from the heart that its potential is considered to be zero. The indifferent electrode constitutes the central terminal of Wilson, while the exploring electrodes are the *unipolar extremity electrodes*—one attached to the right arm, called VR; one to the left arm, VL; and one to the left foot, VF. The V indicates that the Wilson terminal is the reference electrode. In this system the potentials recorded are small, and it was therefore suggested that the potential at an extremity could be obtained simply by disconnecting one extremity electrode from the central terminal and recording the difference in potential between it and the remaining two extremity electrodes. Such a system gives potential amplitudes 1.5 times greater than a V lead amplitude (see below), and hence this system for recording is called the augmented unipolar system with the designations: VR (right arm), VL (left arm), and VF (left foot) (Fig. 14-10B).

The Unipolar Lead System

The unipolar leads include six chest leads called the V leads recorded with the exploring electrode placed at the following positions: V_1 is placed immediately to the right of the sternum at the fourth intercostal space. V_2 is placed left of the sternum also at the fourth intercostal space. V_4 is placed in the fifth intercostal space at the midclavicular line. V_3 is placed between V_2 and V_4. V_5 is placed over the fifth intercostal space in the anterior axillary line, and V_6 too is placed over the fifth intercostal space but at the midaxillary line.

In man, the right ventricle lies anteriorly, the left ventricle posteriorly. The septum is tilted slightly forward apically, and the base-to-apex axis of the ventricle is almost parallel to the diaphragm. The shape of the ventricular complex recorded at the body surface is determined by the pattern of ventricular activation, the particular ECG lead, and the position of the heart within the chest. Again, remember that when depolarization, represented as a dipole or vector, travels *toward* the *exploring* electrode the latter is in the positive electrical field and the ECG registers an upright or positive deflection. When the impulse travels *away* from the exploring electrode, the electrode finds itself in the negative field, and the ECG registers a downward or negative deflection. The initial phase of ventricular activity is usually directed from left to right in the septum, resulting from earlier and/or greater initial left-to-right activity. This activity produces a wave directed to the right, toward the head, and possibly slightly anteriorly. The wave will produce an initial negative deflection in leads I, II, and III which accounts for the Q wave. At the same time, at the precordial leads, the leads on the right side of the chest, V_1 and V_2, face the positive side of the wavefront and so record an upward deflection. The leads on the far left side of the chest, V_5 and V_6, record a negative deflection since they see the negative side of the vector. The leads directly over the heart, V_3 and V_4, usually record biphasic complexes.

Soon after invasion of the septum begins, rapid conduction through the Purkinje system results in an irregular pattern of inside-out spread in the walls. Within the septum, activity is proceeding from left to right, and endocardium to epicardium. In all the limb leads as well as in V_1 and V_6 the potential is near zero. In V_2 and V_5 the deflections are positive because activity proceeds toward the apex and free left wall. At about 40 msec after ventricular activation the overall activity is directed apically, posteriorly, and to the left. At this time all of the standard limb leads record positive deflections, and the leads on the left side of the chest "see" approaching activity while the right side leads "see" negativity (Fig. 14-10C). The overlying lead, V_3, is near zero. Finally, after depolarization of the apical regions is complete, the wave moves forward to the base of the left ventricle, particularly posteriorly, and from apex to base in the septum. This movement causes no potential in lead I and some negativity in II and III; V_R is positive. V_1 and V_2 are returning to zero from negativity, and V_4, V_5, and V_6 to zero from positivity.

To summarize, we can divide ventricular activation into two major phases. The first is initial depolarization of the septum from left to right, down and anteriorly. The second consists in depolarization of the ventricular walls from endocardium to epicardium. The depolarization of the free ventricular wall as registered on an ECG is predominantly that of the left ventricle. Early simultaneous depolarization of the right and left ventricles in opposite directions results in a net "zero" potential. The left ventricle, being about three times as thick as the right ventricle, continues to depolarize after the right ventricle is completely depolarized and generates a strong unopposed vector or electrical field directed to the left (Fig. 14-11).

In the foregoing discussion of the electrocardiogram we have considered all the lead systems. If one emphasizes only the three standard leads, such a limited frame of reference permits an evaluation of the cardiac vector only in relation to the frontal plane (i.e., laterally and superiorly-inferiorly). It should be noted that the mean cardiac vector courses anteriorly, as well as to the left and inferiorly, in the average individual. These anteroposterior forces are projected on the horizontal and sagittal planes but not

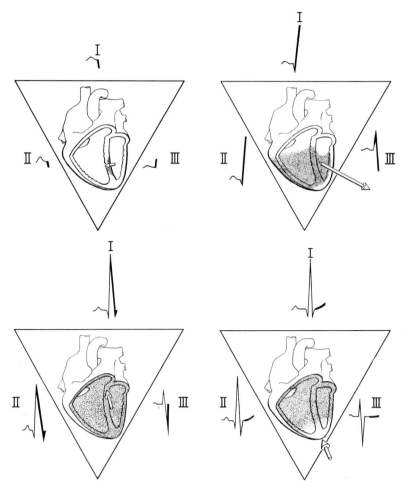

Figure 14-11
Generation of the bipolar ECG signals by the underlying ventricular activity. Progressive depolarization to generate the QRS complex and early repolarization with the beginning of the T wave is
shown (from left to right and from above downward). Stippled areas represent depolarization. The
drawing on the lower right shows the beginning of repolarization of the apex. The resultant
electrical vector (arrow) has components along the Einthoven triangle producing the signals shown
in each limb lead. (From F. L. Abel. In E. E. Selkurt [Ed.], *Basic Physiology for the Health Sciences.*
Boston: Little, Brown, 1975. P. 363.)

on the frontal plane. Conventional electrocardiography provides information regarding
such an anteroposterior component by recording additional leads over the anterior
thorax. Furthermore, besides recording an anteroposterior component, the precordial
leads can disclose the electrical activity of localized areas of the myocardium, to which
they are in closer proximity. This latter capacity is provided by the unipolar character of
the precordial electrodes.

The Normal ECG
As described earlier, the components of the surface electrocardiogram (see Figs. 14-7A
and 14-12) represent circumscribed areas of depolarization and repolarization.
Emanating from the area of the sinoatrial node, or so-called primary pacemaker, the
wave of depolarization first traverses the atria, producing the P wave of the surface
electrocardiogram, of 80 to 100 msec duration. The P wave vector is usually oriented

NORMAL ELECTROGRAM

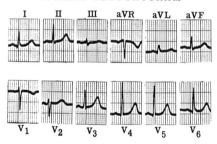

Figure 14-12
A normal human electrocardiogram showing a single complex in each of the 12 leads. Note the R wave amplitude progression in the precordial (V_1-V_6) leads.

toward the left arm and inferiorly, producing an upright deflection in the three standard leads (see Fig. 14-12).

Following the P wave an isoelectric interval occurs, averaging 80 to 100 msec in duration and reflecting the delay in passage through the atrioventricular junctional area. The QRS complex reflects depolarization of the ventricles in a characteristic sequence. The first vector represents depolarization of the interventricular septal fibers, from left to right, often producing a small Q wave (i.e., an initial downward deflection of the QRS complex) in the three standard leads and V_6, and a small R wave in lead V_1. Subsequently, the principal vector becomes manifest, reflecting depolarization of the bulk of the left ventricle, leftward and inferiorly (along the positive axis of the limb leads). This is inscribed as a tall R wave (i.e., an upward deflection of the QRS complex) in leads I, II, III, and the lateral precordial leads V_4, V_5, and V_6. The precordial lead V_3, anatomically positioned to the right of the bulk of ventricular tissue, records a deep S wave as the wave front is moving away from the electrode. (An S wave refers to a downward deflection following an upward deflection of the QRS complex.)

Terminally, the more basal region of the ventricles becomes depolarized, producing a smaller S wave in leads II and III. The entire course of ventricular depolarization, thus completed, normally consumes 80 to 100 msec.

Following ventricular depolarization, the algebraic sum of ventricular repolarization becomes manifest in the electrocardiogram. The S–T segment corresponds approximately to that portion of the single fiber action potential referred to as the plateau or phase 2 (see Fig. 14-8A and compare to Fig. 14-1). The T wave, in contrast, more closely corresponds to the period of rapid repolarization (phase 3) of the single fiber action potential. As illustrated in Figure 14-1B, the epicardial fibers (the last to depolarize) exhibit more rapid repolarization than do the endocardial fibers. This paradoxical sequence results in the inscription of an upright T wave in the presence of a tall R wave, as previously noted.

Determination of Heart Rate

Normally, the paper speed is 25 mm per second. For rapid determination one takes the number of 0.2-second periods between two R waves and divides into 300, because there are 300, 0.2-second intervals in one minute. For more accurate measurements one counts the number of heart cycles (R–R intervals) in a 3-second period and multiplies them by 20 to get the rate per minute.

Lead System and Recording of the ECG

Reading of an ECG. The things to note in attempting to analyze an ECG tracing are the following:

1. First make sure that a 1-mv standardization input has produced a 1-cm deflection.
2. Be sure that each lead recorded has several beats.
3. Determine heart rate. If the heart is not in sinus rhythm, determine the atrial and ventricular rates separately.
4. Note whether the rate is regular or irregular.
5. Measure the P–R interval, Q–T interval, and QRS duration.
6. Note whether the relationships between P, QRS, and T waves are constant or variable.
7. Look at the complexes to see whether their shape and duration are normal.
8. Determine the electrical axis of the heart. This can be computed easily and offers much aid in the diagnosis of ECG abnormalities.

We compute the axis from two limb leads that have been recorded simultaneously. The algebraic sum of the positive and negative peaks in a lead is measured in millimeters and plotted along the proper side of the triangle. A perpendicular is then drawn at the termination of the line. This sequence is repeated for a second lead. The line forming the center of the triangle and the intersection of the two perpendiculars is the mean electrical axis.

In most normal individuals, the axis falls between 0 and +90 degrees. If the axis is greater than +90, there is said to be right axis deviation. If the axis lies in the negative portion of circle, the individual is said to have left axis deviation.

Changes in ECG Configuration

Alterations in repolarization, producing electrocardiographic T wave changes, are the most frequently encountered of ECG abnormalities. While many of these changes are empirically regarded as abnormal, the mechanism by which they occur remains obscure. They are therefore usually referred to as nonspecific T wave changes. Susceptible to a variety of noxious factors, such as ischemia and drugs, or endocrine, metabolic, and electrolyte imbalance, repolarization may be diffusely or locally affected. More rapid repolarization of the endocardium, for example, may result in S–T segment depression and inversion of the T wave (by allowing epicardial repolarization to dominate the summated ECG complex), as can be inferred from Figure 14-8.

Damage to cardiac fibers may impede the course of normal repolarization and produce a so-called injury current, the area involved remaining electrically negative

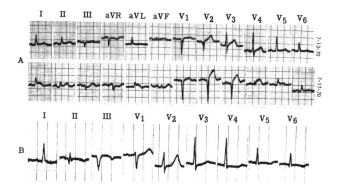

Figure 14-13
A. The ECG recorded on 7-13-70 demonstrates a normal ST–T segment configuration. The ECG recorded on 7-17-70 illustrates the progressive ST–T segment elevation (leads I, aVL, V$_2$, V$_3$, V$_4$, and V$_5$) typical of acute myocardial infarction. B. Healed myocardial infarction is characterized by abnormally wide and deep Q waves (leads II and III).

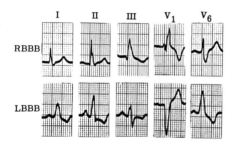

Figure 14-14
Right bundle branch block (RBBB) is characterized by right axis deviation, wide S waves in lead I and V_6, and an RSR' in V_1. Left bundle branch block (LBBB) is typified by broad R waves in lead I and V_6 and absence of the small R wave in V_1.

with respect to the unaffected region. This type of injury results in S–T segment shifts upward in the ECG leads overlying the damaged tissue (Fig. 14-13). As the damaged cells become electrically inactive, the S–T segment subsides. With electrical death of the myocardium, deep wide Q waves develop. Such "electrically silent" areas are manifest by Q waves because the dead myocardium acts as a "window"—the surface electrode records only the residual unopposed forces of depolarization, the way an intracavitary electrode would. Thus, the wave of depolarization travels away from the recording electrode. The changes demonstrated in Figure 14-13A are observed clinically in acute myocardial infarction, where the tissue is ischemic but still viable. The development of Q waves correlates with replacement of the dead tissue with fibrous tissue (Fig. 14-13B).

Conduction block within the Purkinje conducting system of the heart produces characteristic alterations of QRS configuration. Such disturbances in conduction may be due to either "functional" (rate dependent, prolonged refractory period, etc.) or pathological lesions of the Purkinje system. This type of injury or lesion causes conduction of the impulse to be delayed as it courses over longer, more circuitous pathways, and via the more slowly conducting ventricular muscle fibers. Lesions of one or the other of the main bundle branches, for example, yield characteristic electrocardiographic patterns, as illustrated in Figure 14-14. Right bundle-branch block produces alteration of the QRS complex with widening due to late, unopposed depolarization of the right ventricle, represented by a wide terminal S wave in leads I and V_6 and a secondary R wave (R') in V_1. The left bundle-branch block pattern, on the other hand, occurs as the result of a disturbance in the initial course of ventricular depolarization. Normally, septal depolarization takes place from left to right on account of early branching of the left bundle. In the left bundle-branch block, however, the septum is depolarized in the opposite direction with loss of the so-called septal q wave frequently found in leads I, II, and V_6. Late depolarization of the left ventricle then produces a widened, notched R wave in these leads.

Arrhythmias

The concept of arrhythmias implies a disruption, transient or prolonged, in the normal sequence of an electrical activation of the heart. The propensity for creating such a disruption is inherent in cardiac tissue, as is evident from a consideration of the electrophysiological properties of various fiber types. As discussed previously, fibers possessing a capacity for automatic behavior are present in virtually every area of the heart. Specialized fibers are found in the atria; lower AV junctional fibers clearly manifest automaticity; and His-Purkinje fibers pervade the ventricular myocardium.

Normally, the more rapid rate of the sinoatrial pacemaker precludes the expression of activity by these more subordinate pacemakers, which have an inherently slower rate of spontaneous diastolic depolarization. Occasionally, however, a secondary pacemaker

333

(ectopic pacemaker*) depolarizes prematurely and at a time when the surrounding tissue is not refractory, and thus may *initiate* depolarization. An impulse originating from specialized atrial tissue is referred to as an atrial premature excitation (APC). Its ventricular counterpart is known as premature ventricular excitation (VPC).† The atrial ectopic beat (Fig. 14-15E) produces a P wave of atrial depolarization that may differ in amplitude and configuration from the sinus P; the QRS complex approaches the normal configuration (Fig. 14-15A) since the stimulus traverses the AV node and the normal ventricular conduction system. The premature complex, having depolarized the atrium and SA node region, however, is followed by a slight compensatory pause before reestablishment of sinoatrial excitability. Occasional spontaneous activity may originate from the AV junctional–His bundle area as well, producing similar ventricular excita-

*The term *ectopic* refers to an impulse or rhythm initiated by a cardiac pacemaker outside the SA node.
†The terms *APC* and *VPC* have, to date, been used to connote atrial or ventricular premature *contractions.* However, the term *contraction* so used is meaningless when applied to an ECG tracing since the ECG is the recording of the *electrical* events. The latter may or may not lead to the actual mechanical event. Hence *APC* and *VPC* as related to the ECG will here be restricted to defining premature excitations or complexes.

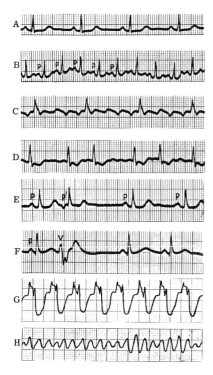

Figure 14-15
A. Normal sinus rhythm. B. Rapid atrial tachycardia. C. Atrial flutter with a 4:1 ventricular response. The baseline demonstrates characteristic sawtooth configuration. D. Atrial fibrillation with an irregular ventricular response. The baseline shows no definitive P waves. E. ECG recording of an atrial premature beat (P′). F. Ventricular premature beat (V), followed by a compensatory pause and two normal sinus beats. G. Ventricular tachycardia characterized by wide aberrant QRS complexes and a rapid rate of 150 per minute. H. Ventricular fibrillation characterized by an undulating baseline without definable QRS complexes.

tion, although atrial depolarization here may be obscured by the QRS. Impulses originating from the above area may activate the atrium in retrograde fashion, resulting in a change of the atrial electrical axis, or they may fail to excite the atrium (retrograde).

The premature ventricular complex, originating from one or the other ventricle (Fig. 14-15F), elicits ventricular depolarization over abnormal pathways, producing a widened QRS complex of unusual configuration. This ectopic complex is not ordinarily associated with atrial depolarization; hence the sinoatrial rhythm is not disturbed. The ventricles are refractory to the next sinoatrial impulse, however, and a so-called *fully compensatory pause* results.

Although such occasional ectopic excitations may be observed in normal individuals, under conditions which enhance the automaticity of subordinate atrial pacemaker fibers ectopic activity may increase in rate sufficiently to usurp the function of the SA node. This circumstance may be evident either as frequent atrial premature impulses interrupting the sinus rhythm or as a persistent atrial rhythm, referred to as *atrial tachycardia*. An arrhythmia of this type may vary in rate from 140 to 240 beats per minute (Fig. 14-15B).

Similarly, latent ventricular pacemaker activity may be enhanced to the point of producing ventricular tachycardia (Fig. 14-15G), by usurping SA dominance (i.e., beating more rapidly than the sinus node, and precluding expression of sinus activity). The rate of such a pacemaker may vary from 100 to 150 beats per minute. This electrocardiographic rhythm is characterized by distorted and prolonged QRS (aberrant) waves, not preceded by atrial deflections. Relatively uncommon, ventricular tachycardia is observed primarily in the presence of severe cardiac disease or digitalis intoxication and is associated with deterioration of cardiac output.

Atrial Flutter and Fibrillation

More rapid ectopic pacemaker activity in the atrium may proceed to the point of atrial flutter (Fig. 14-15C). This is characterized by discrete depolarizations occurring at a rate of 200 to 350 per minute. Slow conduction through the AV node and a longer refractory period of this area do not permit a 1:1 ventricular response. Consequently, AV block of varying degree is produced, the ventricles often responding once to each two or four flutter waves. This type of arrhythmia is most commonly observed in rheumatic and coronary heart disease.

Atrial fibrillation (Fig. 14-15D) has an even more rapid rate (300 to 450 per minute) and, in addition, uncoordinated atrial activation. Atrial contraction is totally ineffective while the ventricular response to such an atrial rhythm is totally irregular (although the ventricular contractions per se are coordinated and effective). The mechanism of arrhythmia is of obscure origin, possibly produced by rapidly firing single or multiple ectopic foci, or by a continuous circus movement* initiated from one focus and perpetuated by a slow conduction velocity (permitting recovery of responsiveness in previously excited areas), a long conduction path, or a short refractory period.

Ventricular Flutter and Fibrillation

Ventricular flutter is an arrhythmia considered by some cardiologists to be a separate entity while others regard it as a phase preceding ventricular fibrillation. Ventricular rhythm during flutter is regular and is characterized by depolarizations that are oscillatory, at a rate of 175 to 250 per minute. The QRS complexes are wide and of large amplitude, and the S–T segment is indistinguishable from the T wave. The ventricular depolarization during flutter is probably from a single ectopic focus. It should be pointed out that ventricular flutter is seldom observed clinically.

Ventricular fibrillation is an arrhythmia (Fig. 14-15H) that is characterized by rapid, uncoordinated ventricular activity and a total irregularity of the QRS complexes. The

---

*Unidirectional transmission of the impulse traversing in a circuitous manner (reentry) resulting in a continuous reexcitation.

condition is incompatible with life, because of the ineffectiveness of ventricular contractions, for it is tantamount to ventricular arrest, with total loss of cardiac output. The mechanism of this type of arrhythmia may be comparable to that underlying atrial fibrillation (although the latter is a *relatively* innocuous rhythm), associated with slowing of conduction velocity and the establishment of a circus movement.

Clinical and experimental experience with exogenous stimulation (i.e., an artificial pacemaker) has provided evidence that ventricular fibrillation may be elicited by a single stimulus at a time when the ventricle is extremely susceptible. This interval is referred to as the vulnerable period and is found toward the end of rapid repolarization (the latter portion of the T wave). These findings have stimulated the development of electrical pacing and defibrillating equipment synchronized to provide an appropriate stimulus well away from the vulnerable period.

Conduction Disturbances

Conduction disturbances may occur in all areas of the heart and may be related to slowing of conduction velocity itself or to an increase in conduction time secondary to a greater tissue mass. The latter results from depolarization traversing a greater mass at the same conduction velocity and may be noted in atrial or ventricular hypertrophy, producing a widening of the P wave or QRS complex respectively.

Conduction disturbances related to a decrease in conduction velocity are perhaps more commonly encountered in clinical practice and most frequently observed in regard to the transmission of depolarizing impulses from atria to ventricles via the AV junctional area. Fibers in the latter region exhibit a resting potential of low magnitude and consequently action potentials of low amplitude. Normally, these low-amplitude action potentials result in slowing of impulse propagation, producing the normal AV conduction delay. This type of conduction, in which action potential amplitude lessens and conduction velocity slows progressively because of successively less adequate action potential generation, is called *decremental conduction*. Thus, when physiological or pathological factors produce a further decrease in action potential amplitude, decremental conduction is enhanced, successive areas of fibers generating lesser and lesser action potentials, to the point of conduction failure. In the latter circumstance the fiber beyond the block remains excitable, but the stimulus is inadequate to effect a response. Illustrations of this phenomenon are found in Figure 14-16, in which AV conduction

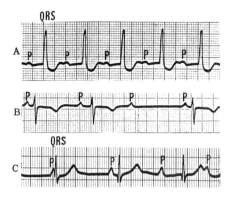

Figure 14-16
A. First-degree block characterized by prolonged P-R intervals. There are no dropped beats. The maximal normal P-R interval is 0.22 second and in this case is 0.31 second. B. Second-degree AV block characterized by a dropped ventricular beat (after the third recorded P). After the dropped QRS complex, P-R is normal. Third-degree heart block or complete AV block. The P waves and QRS complexes are independent of each other. This is characterized by a simultaneous inscription of the P and QRS complexes of the first beat.

delay is seen to produce an increase in AV conduction time without other disturbance (first-degree AV block), intermittent block of atrial impulses (second-degree AV block), and complete cessation of atrioventricular conduction, with the atria and ventricles beating independently of one another (third-degree AV block).

REFERENCES

Brooks, C. McC., B. F. Hoffman, E. E. Suckling, and O. Orias. *Excitability of the Heart.* New York: Grune & Stratton, 1955.

Burch, G. E., and T. Winsor. *A Primer of Electrocardiography* (6th ed.). Philadelphia: Lea & Febiger, 1972.

Fisch, C., K. Greenspan, and S. B. Knoebel. Electrophysiological Basis of Clinical Arrhythmias. In H. I. Russek (Ed.), *Major Advances in Cardiovascular Therapy.* Paul D. White Symposium. Baltimore: Williams & Wilkins, 1973. Pp. 309–319.

Fisch, C., S. B. Knoebel, H. Feigenbaum, and K. Greenspan. Potassium and the monophasic action potential, electrocardiogram, conduction, and arrhythmias. *Prog. Cardiovasc. Dis.* 8:387–418, 1966.

Fishman, A. P. (Ed.). Symposium on the myocardium. Its biochemistry and biophysics. *Circulation* 24 (no. 2, pt. 2): 1961.

Greenspan, K. Mechanisms of Ventricular Fibrillation. In L. S. Dreifus and W. Likoff (Eds.), *Cardiac Arrhythmias: Pathophysiology, Pharmacology, and Treatment.* 25th Hahnemann Symposium. New York: Grune & Stratton, 1973. Pp. 195–202.

Greenspan, K., and E. Steinmetz. Digitalis and Potassium: Pharmacodynamic Relationship. In L. S. Dreifus (Ed.), *Mechanisms and Treatment of Cardiac Arrhythmias.* New York: Grune & Stratton, 1966.

Greenspan, K., E. F. Steinmetz, R. E. Edmands. Clinical and physiological aspects of digitalis and potassium. *Intern. Med. Digest* 1:41, 1966.

Hoffman, B. F. Origin of the Heart Beat. In A. A. Luisada (Ed.), *Cardiology.* New York: McGraw-Hill, 1959. Vol. 1, pt. 2, chap. 4.

Hoffman, B. F., and P. Cranefield. *Electrophysiology of the Heart.* New York: McGraw-Hill, 1960.

Hoffman, B. F., and P. Cranefield. The physiological basis of cardiac arrhythmias. *Am. J. Med.* 37:670–684, 1964.

Katz, L. N., and A. Pick. *Clinical Electrocardiography.* Philadelphia: Lea & Febiger, 1956.

Lipman, B. S., and E. Massie. *Clinical Scalar Electrocardiography* (6th ed.). Chicago: Year Book, 1972.

15. Cardiodynamics

Carl F. Rothe

The function of the heart is to propel enough blood into the aorta to maintain the blood pressure at a level sufficient to assure adequate flow of blood to all tissues. To understand these complex relationships, it is necessary to study the components of the system and the relationships between the following variables (Fig. 15-1): A. The flow (F) of blood through the capillaries supplying the needs of cells depends upon (1) the difference between the systemic arterial blood pressure and venous pressure (ΔP) and (2) the resistance (R) to the flow. This vascular resistance to flow depends on the internal diameter of the arterioles and precapillary sphincters (tissue vasomotor tone), as discussed in Chapters 11, 12, and 17. B. The systemic arterial blood pressure (BP) is the product of (1) cardiac output (CO) and (2) the summed effects of all these peripheral resistance vessels, i.e., the total peripheral resistance (TPR). (These relationships are analogous to Ohm's law for electrical circuits.) C. The cardiac output is the volume of blood pumped by the heart each minute and thus is the product of (1) the volume of each beat (the stroke volume, SV), and (2) the number of heart beats per minute (the heart rate, HR). D. The stroke volume is the difference between (1) the volume of blood in the heart at the beginning of systole (the end-diastolic volume, EDV) and (2) the amount of blood which remains in the ventricles when the valves close at the end of systole (the end-systolic volume, ESV). Four factors, then, are the primary determinants of systemic arterial blood pressure and cardiac output: (1) peripheral resistance to blood flow determined by the tissue vasomotor tone, (2) heart rate, (3) end-diastolic volume, and (4) end-systolic volume (Fig. 15-1).

These factors are closely related. Some are intrinsic, being dependent on the physical characteristics of cardiac muscle and the vasculature, e.g., the Frank-Starling law of the heart (see below). Other interrelationships are influenced by factors extrinsic to a tissue, e.g., neural and hormonal influences such as those from the pressoreceptor reflex acting by way of the autonomic nervous system on all four variables (Chap. 17). In the normal individual, the fine control of cardiovascular function is neural.

The three factors determining cardiac output (heart rate, end-diastolic volume, and end-systolic volume) will now be considered in detail. Factors influencing peripheral resistance are considered in Chapter 17.

HEART RATE

The normal heart rate at rest, in the adult, ranges between 40 and 100 beats per minute and averages about 70. In the newborn, the heart rate is about 135 beats per minute; in the elderly, about 80 beats per minute. The heart of the trained athlete at rest often beats less than 50 times per minute. The heart rate is widely variable, for it is a quickly changeable factor in the maintenance of cardiovascular homeostasis, rather than being controlled.

As shown in Figure 15-2, if the stroke volume is held constant, an increase in heart rate causes a directly proportionate increase in cardiac output (curve 1). However, since an increase in heart rate shortens the filling time between beats, the end-diastolic volume is decreased. (Consider the analogy of a bucket placed under a running faucet. If the time for filling is reduced by half, only half as much water is collected.) Unless the end-systolic volume decreases proportionately to the decrease in end-diastolic volume, there will be a consequent decrease in stroke volume. Indeed, at high heart rates, or in a denervated or isolated heart, the end-systolic volume does not decrease propor-

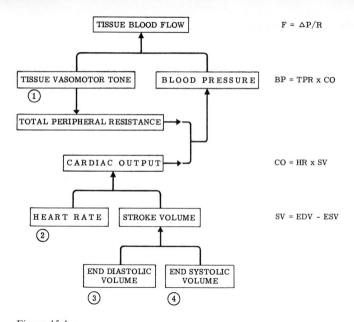

Figure 15-1
Cardiovascular system dynamics. The relationships determining tissue blood flow, blood pressure, cardiac output, and stroke volume. (See page 337 for explanation of the equations.)

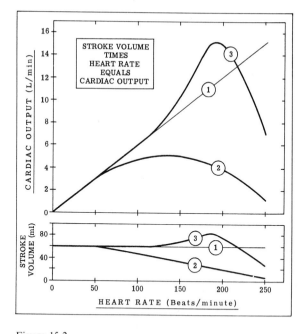

Figure 15-2
The effect of heart rate on cardiac output. Since cardiac output is the product of stroke volume and heart rate, if the stroke remained constant, cardiac output would increase linearly with increases in heart rate (curve 1). If the stroke volume declines because of inadequate filling time or compensatory mechanisms, there is a peak output and then a decrease (curve 2), giving a relatively constant cardiac output as heart rate changes. As in the normal individual, if the stroke volume is held constant or is increased by compensatory mechanisms, such as increased contractility, the output increases to a higher peak before filling time becomes limiting (curve 3).

tionately to the decrease in end-diastolic volume, and the result is a smaller stroke volume (Fig. 15-2, lower curve 2). Without autonomic nervous system control, cardiac performance reaches a peak at moderate heart rates. Any further elevation of heart rate seriously limits filling time and markedly reduces cardiac output (Fig. 15-2, upper curve 2). Because ventricular filling occurs mostly in the first half-second and then tends to plateau after one second, there is little decline in stroke volume until the heart rate exceeds about 60 beats per minute, even in the functionally denervated heart. Over the range of about 60 to 160 beats per minute, cardiac output tends to be independent of heart rate if autonomic nervous system control is blocked, because the stroke volume decreases in proportion to the increase in heart rate above about 60 beats per minute. Furthermore, if the heart is artificially paced over this range and the autonomic reflexes are intact, cardiac output and blood pressure are little changed because of compensatory mechanisms (baroreceptor reflexes, Chap. 17) acting to supplement, if necessary, the reduction in stroke volume as the heart rate increases. Under normal circumstances, such as exercise, mechanisms mediated primarily by the sympathetic division act to maintain the stroke volume at a constant level or even increase it (curve 3) by increasing the vigor of myocardial contraction to decrease the end-systolic volume. Thus, in the normal individual, the output of the heart per minute tends to increase proportionately as the heart rate increases. However, the cardiac output can increase without an increase in heart rate. An increase in heart rate does not *necessarily* mean that the cardiac output is increased, for the stroke volume may be concomitantly reduced.

As heart rate is increased by sympathetic division action, the duration of diastole is decreased proportionally more than that of systole (Fig. 15-3). At rest, the diastolic filling time accounts for about 50 percent of the cycle, but with a doubling of heart rate to 150 beats per minute, the diastolic time accounts for only about 30 percent of the cycle. More important, the actual time available is reduced from about 0.4 second to about 0.1 second, a reduction to 25 percent of the resting value! With even higher heart

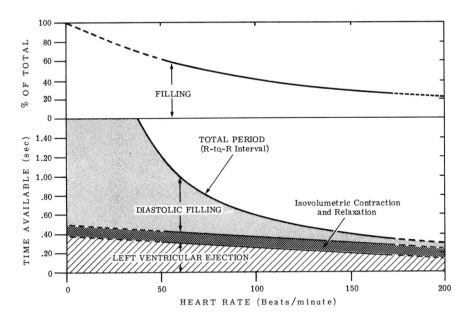

Figure 15-3
Influence of heart rate on time available for left ventricular ejection and filling. Note the marked reduction in filling time at high heart rate. (Values estimated from studies of Weissler et al., *Circulation* 37:152, 1968, and from Jones and Foster, *J. Appl. Physiol.* 19:279, 1964.)

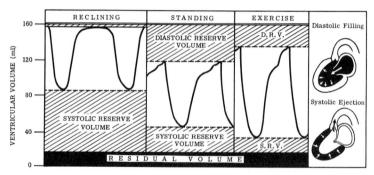

Figure 15-4
Ventricular volume under various conditions. The *diastolic reserve volume* is the difference between the maximum volume and the end-diastolic volume. The *systolic reserve volume* is the difference between the end-systolic volume and the residual volume. The *residual volume* is what would remain in the heart if it ejected against a pressure of zero. (Modified from Rushmer, 1970, p. 81; and Rushmer, *Mod. Concepts Cardiovasc. Dis.* 27:473, 1958.)

rates, the impingement on filling time is even more serious, indicating the need for powerful mechanisms to maintain stroke volume.

The maximum effective heart rate in man is about 180 beats per minute. Severe *tachycardia,* as seen in atrial fibrillation, is also associated with a decrease in cardiac output, because in this case the time for filling is severely limited by the abnormally high heart rate. In *bradycardia,* following complete heart block, the stroke volume may be large, but the cardiac output is severely limited because of the very low heart rate.

Cooling of the pacemaker cells (during surgical hypothermia, for example) causes a profound reduction in heart rate. This might be expected from the marked effect of temperature upon enzymatic reactions. An increase of the body metabolism (during fever, for example) acts to stimulate heart rate. A rise of body temperature of 1°C causes an increase in heart rate of about 12 to 20 beats per minute (7 to 11 beats per minute per 1°F). Generally the heart rate is closely associated with the overall metabolic rate. Any bodily response requiring an increased oxygen supply usually requires an increased cardiac output, which is accomplished as a rule by an increased heart rate. Examples include exercise and digestion. The act of standing upright also elicits an increased heart rate, as part of the homeostatic mechanism maintaining an adequate blood pressure (Chap. 17).

The heart rate is controlled primarily by the autonomic nervous system (Chaps. 7A and 17).

The stroke volume, like the tidal volume in respiration, is not as large as the maximum volume of heart. Consequently, there is usually both a diastolic volume reserve or capacity and a systolic volume reserve (capacity) (Fig. 15-4). Discussion of the various factors influencing these volumes follows.

END-DIASTOLIC VOLUME

If the heart is to be an effective pump, the ventricles must fill before systole. The adequacy of filling is determined by the filling time, the effective filling pressure, the distensibility of the ventricles, and atrial contraction.

The *filling time* is dependent upon the heart rate. As the heart rate increases, the diastolic period is reduced proportionally more than the systolic period, so that available filling time becomes an important factor at high heart rates. Under normal conditions, however, the filling time is adequate.

The *effective filling pressure* is the pressure gradient between the inside of the ventricles and the pressure outside. It is this transmural pressure gradient, across the ventricular walls, that causes the relaxed ventricles to distend during diastole (Fig. 15-5). It is closely dependent upon the rate of venous return as this influences intra-atrial pressure (see below and Chap. 16) and the degree of negative intrathoracic pressure. An increase in the central venous pressure or a more negative thoracic pressure acts to cause an increase in the transmural pressure and thus facilitates filling. Under normal conditions the filling pressure of the heart is adequate and so does not become a limiting factor. However, if the blood volume is reduced—for example, as a result of hemorrhage—or if there is loss of venomotor tone and hence dilation of the capacity vessels (veins), the filling pressure drops and the cardiac output is seriously reduced. Atrial contraction (see below) also contributes to ventricular filling.

Distensibility is the property of a hollow organ that permits it to be distended or expanded (Chap. 11). In Figure 15-5 the lower line is a typical distensibility, or pressure-volume, curve of a relaxed ventricle. A very small pressure is required to cause a marked increase in volume at low volumes. As the ventricle fills, its elastic fibers are stretched, and, because of the nonlinear stress-strain characteristics of these fibers, the ventricle becomes less distensible. Then a unit increase in pressure causes only a small change in volume. Myocardial distensibility does not appear to be significantly modified by sympathetic and parasympathetic division activity. However, it does seem to be decreased in hypertrophied or ischemic hearts—they are stiffer.

The mechanisms of distention of the ventricle during filling are important. During the *rapid-filling phase,* the compression of the elastic cardiac tissue by the previous systole tends to cause a recoil of the tissue. This speeds the filling, just as squeezing a rubber bulb and then releasing it causes the bulb to fill rapidly. The recoil phenomenon is especially important with relatively small end-systolic volumes. Under these conditions

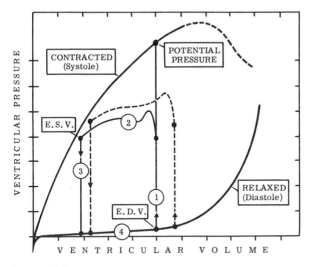

Figure 15-5
Volume-pressure relationships of the heart. 1, isovolumetric contraction phase; 2, ejection; 3, isovolumetric relaxation; 4, filling. A normal cycle is indicated by a solid line. Systole starts at the end-diastolic volume (EDV) and first produces the isovolumetric change in ventricular pressure. It ends at the end-systolic volume (ESV), after which time isovolumetric relaxation occurs. As predicted by Starling's law or heterometric autoregulation (see below), an increase in end-diastolic volume (dotted-line loop) tends to result in an increase in stroke work, since either stroke volume or pressure is increased as a result of the increase in vigor of contraction.

actual ventricular suction (a negative transmural pressure) has been found. The rate of release of the actin-to-myosin bonds (Chap. 3B) and myocardial viscosity (see below) are major factors limiting the rate of rapid relaxation. If the rapid filling is reduced because of low filling pressure or weakened atrial contraction, tachycardia will greatly reduce stroke volume.

Atrial contraction becomes especially important at high heart rates, when the diastolic time is limited. When the atria contract, blood is propelled in both directions: toward the veins and into the ventricles. However, the inertia of the blood tends to act as a valve so that a significant amount of blood is propelled into the ventricles to aid in filling. The cross-sectional area of the AV valve and ventricle is larger and the distance shorter than is the case with the veins leading to the atria; thus the opposing accelerative force is less in this direction than in the retrograde direction. Furthermore, atrial contraction leads to reduced regurgitation by assisting atrioventricular valve closure (Chaps. 11, 13). During exercise, the amount of blood pumped into the ventricles by the atria can account for as much as 40 percent of the stroke volume. But normally, atrial contraction is a minor factor in the filling of the ventricles. Lack of effective atrial contraction, such as occurs with atrial fibrillation, does not jeopardize life but it does limit exertion.

Normally, homeostatic mechanisms compensate for moderate losses of blood volume which tend to decrease the end-diastolic volume. These mechanisms include an increased cardiac contractility and increased venomotor tone (Chap. 17). A human can lose about 1 percent of his body weight of blood without any significant drop in blood pressure. Without the compensatory mechanisms, however, the effects of a decreased filling pressure become highly important. Indeed, a decrease in effective filling pressure (central venous pressure) of only 1 cm of water (about 1 mm Hg) can result in a reduction in cardiac output of nearly 50 percent (Guyton et al., 1973)!

Pericardium

The pericardium provides a limit to the magnitude of the end-diastolic volume. Under normal conditions it is not essential in man. The removal of an adherent, dense, scar-filled pericardium following pericarditis significantly improves the clinical condition of many patients recovering from this disease. The loss of the pericardium does not seriously limit their activity. The function of the pericardium is apparently to check acute overdistention during diastole and to afford a lubricated surface between the beating heart and the lungs. By limiting right ventricular filling, it protects against pulmonary congestion and left ventricular overload. This is important during various cardiac arrhythmias in which prolonged diastolic periods occur. It also aids in keeping the heart and great vessels aligned. Under chronic conditions of myocardial weakness or chronic overloading of the heart, the pericardium slowly distends and cardiac hypertrophy ensues.

If blood or fluids accumulate between the heart and pericardium, filling is impeded. This acute compression of the heart is called *cardiac tamponade*. It is a serious consequence of penetrating wounds of the heart, or of pericarditis, in which there is effusion of fluids from the inflamed membranes.

Cardiac hypertrophy follows a chronically increased load on the heart such as occurs after aortic valvular lesion, mitral insufficiency, or hypertension. The increase in size of the individual muscle fibers (hypertrophy) would be expected to increase the ability of the heart to do the extra work. However, the vascularity of the myocardium usually does not increase proportionately; thus, the 1:1 ratio between the number of capillaries and the number of muscle fibers may decrease as the muscle fibers enlarge and more are formed. In addition, the mean distance which oxygen has to diffuse from the capillaries to the muscle fibers is increased. This development limits the effective oxygen supply, and if the hypertrophy is excessive, it is usually followed by cardiac failure.

The cause of hypertrophy is not clearly known, but it often follows a chronic increase in size of the ventricles (dilation). The dilation would be expected to follow increases in filling pressure as a compensatory mechanism for increased pressure load (e.g., aortic stenosis), regurgitation, or myocardial weakness. Left ventricular heart failure leads to right ventricular hypertrophy because the pulmonary arterial pressure rises as a result of pulmonary congestion and so continuously overloads the right heart (Chap. 18B).

Heterometric Autoregulation—
The Frank-Starling Relationship

If cardiac muscle is stretched, it develops greater contractile tension upon excitation. As Starling stated in 1914, "The law of the heart is therefore the same as that of skeletal muscle, namely, that the mechanical energy set free on passage from the resting to the contracted state depends . . . on the length of the muscle fibers." The vigor of contraction is a function of muscle fiber length. Stated another way, stroke volume tends to be directly proportional to diastolic filling; that is, the ventricle tends to eject whatever volume is put into it. Although many investigators, such as Frank in the 1880s, contributed to the study of this mechanism, Starling's formulation was such that it has become known as *Starling's law of the heart.* Sarnoff (1962) has coined the more descriptive term *heterometric autoregulation.* This implies that the performance of the heart is so regulated as to be an effective mechanism for the homeostatic control of circulatory function, by intrinsic (auto) mechanisms which are not neural or hormonal in origin and which are determined by changes (hetero) in myocardial fiber length (metric).

There is no question concerning the validity of this relationship. However, it is but one factor among many controlling cardiac output. The relationship is described graphically in Figure 15-5. This diagram is similar to diagrams showing the tension developed by contracting skeletal muscle at various initial lengths. At a higher end-diastolic volume (dotted-line loop, Fig. 15-5), the peak isovolumetric pressure (potential pressure with no ejection of blood) which can be developed by the ventricle increases. The increase in end-*diastolic* volume, by causing the myocardial fibers to contract with greater vigor, acts to maintain about the same end-*systolic* volume. Thus the increase in end-diastolic volume is greater than the subsequent increase in end-systolic volume, and there is a net increase in stroke volume following increased cardiac filling at a constant pressure load. The increased vigor of contraction causes more external work to be done by the heart (see also Figs. 15-6B and 15-8).

As another example, if the total peripheral resistance increases so that the arterial pressure is increased (let us assume from 100 to 150 mm Hg), the systolic ejection is limited by the increased pressure head and the end-systolic volume is increased (say from 40 to 50 ml) because the usual amount of blood (80 ml per beat) was not ejected (now only 70 ml, a 10-ml decrease). The stroke volume and cardiac output would then be less, damming up the blood behind the ventricle. With a continuing venous return, the filling pressure therefore increases over a period of several beats, causing the end-diastolic volume to increase by about 10 ml, from 120 to 130 ml. The larger heart then contracts with more vigor and ejects the same volume per beat ($130 - 50 = 80$ ml) as before ($120 - 40 = 80$ ml). Note that the heart is beating with larger end-systolic and end-diastolic volumes. Actually, the stroke volume is not perfectly maintained, as in our example, because the venous inflow is limited to a greater degree at the larger size, on account of the decreased distensibility of the heart at the larger volume. Thus the end-diastolic volume increases only partially from the 120-ml control end-diastolic volume to the 130-ml volume, and so the expected 10-ml increase in end-diastolic volume is not fully achieved.

If the heart is overdistended, the peak pressure tends to decrease (Fig. 15-5), probably because the sarcomeres are extended so much that the actin and myosin myofilaments do not overlap as effectively (see Braunwald et al., 1976). Then the increase in end-diastolic volume is followed by an increased end-systolic volume, and a smaller stroke volume is ejected. This sequence may happen in acute, uncompensated congestive heart failure. However, such a decrease in potential pressure and reduced stroke volume at very high end-diastolic volumes is seldom seen in the normal heart, especially if the pericardium is intact.

From studies with isolated animal hearts, it had long been thought that cardiac output was controlled primarily by changes in stroke volume as determined by the end-diastolic volume and the Frank-Starling relationship following increases in venous return. However, the importance of this phenomenon in the intact unanesthetized man has been seriously questioned, since changes in right atrial pressures have been noted without concomitant changes in cardiac output. Recent evidence from human volunteers shows that if the sympathetic division is blocked by drugs, an increase in blood volume causing an increase in right atrial pressure is, indeed, followed by an increase in cardiac output. By using injections of radiopaque dyes in humans and animals, and special transducers to measure the dimensions of the ventricles in dogs, Rushmer and co-workers (e.g., 1970) have found that the end-diastolic volume is normally near maximum at rest in the reclining person or dog. The normal heart is large in the resting, reclining individual; it apparently almost fills the pericardium (Fig. 15-4). Under these conditions a marked increase in heart size in response to increased loads is not possible; hence, most of an increase in cardiac output must take place by means other than increased end-diastolic volume and heterometric autoregulation. For example, this can be brought about by increased heart rate and decreased end-systolic volume resulting from increased vigor of myocardial contraction (via sympathetic activity).

Standing, excitement, and exertion are often followed by decreases in the diastolic size of the heart as a result of increases in sympathetic division activity (Fig. 15-4). Although the stroke volume increases during maximal exertion, Rushmer (1970) concludes that the stroke volume remains "remarkably" constant over a wide range of voluntary exertion. Normally, the stroke volume increases only if the total oxygen consumption exceeds about eight times the resting rate. This neural control, which minimizes the importance of heterometric autoregulation (Starling's law), had not been seen to such a great degree in earlier animal experiments for the reason that most anesthetics depress the cardiovascular reflexes acting on the heart and cardiovascular system (Chap. 17). Furthermore, the procedure of thoracotomy, almost always used in these studies, removes the negative intrathoracic pressure, with the result that the heart has a lower effective filling pressure. Thus the heart is not adequately filled during diastole. The end-systolic volume of the heart is almost zero under these conditions, and the heart rate is high and relatively constant. It is therefore not surprising that increases in filling pressures distend the heart, giving marked increases in cardiac function explained by Starling's law.

Heterometric autoregulation is without question an important mechanism, in the normal individual, for balancing the outputs of the right and left ventricles. For example, if the right ventricle pumps more blood than the left ventricle, the difference accumulates in the lungs, acts to increase pulmonary venous and left ventricular diastolic pressures, and so increases left ventricular filling. In accordance with Starling's law, the left ventricular output increases and so restores the balance. Heterometric autoregulation is deleterious during changes in body posture in man, for unless other mechanisms act to maintain homeostasis, as the blood pools in the lower extremities there is a profound drop in cardiac output and hence in blood pressure (orthostatic hypotension). Likewise, during thoracic surgery, when the effective filling pressure to the heart is reduced, other mechanisms must provide compensation. The Frank-Starling relationship is of value as a defense against heart failure, for the increase in filling pressure (congestion) within certain limits acts to maintain the cardiac output.

END-SYSTOLIC VOLUME

The end-systolic volume is that volume of blood remaining in the heart at the end of systole at the time when the semilunar valves start to close. At this moment the aortic and ventricular pressure *distending* force is balanced by the *contracting* force produced by the myocardial fibers. The primary factors determining the magnitude of the end-systolic volume are the pressure against which the ventricle is pumping and the strength of the myocardial fiber contraction.

Pressure Load

If the heart is to pump against a higher than previous blood pressure, the myocardial fiber tensions must be greater to expel the same amount of blood—a result of the Laplace relationship. Thus systole will terminate at some higher volume unless the contractility of the myocardium is increased (see below). Conversely, a reduction in systemic blood pressure results in a decrease in end-systolic volume.

There is a complex interaction between filling pressure, aortic (outflow) pressure, and cardiac performance (Fig. 15-6). Although the data shown are from dog experiments, the pattern is applicable to man. Consider point N as the normal (Fig. 15-6A). Increasing ventricular fiber length to point A increases left ventricular (cardiac) output, as predicted by the Frank-Starling relationship. (Ventricular fiber length, ventricular volume, end-diastolic pressure, and mean left atrial pressure are generally closely related.) At a constant filling pressure, an increase in mean arterial pressure (pressure load) results in a decrease in cardiac output, point C. Note that at high arterial pressures (200 mm Hg) increasing the filling pressure has relatively little influence on cardiac output, in sharp contrast to the marked influence of filling pressure on cardiac output at low (50 mm Hg) pressure loads. Note, too, that with abnormally low transmural filling pressures (e.g., 4 mm Hg) the cardiac output is low and relatively independent of pressure load.

The rate of doing work is a better index of cardiac function and is more closely related to cardiac reserves in disease states than is mere flow output. Figure 15-6B is a three-dimensional presentation of left ventricular power output (flow times pressure) as influenced by filling and output pressure—heart rate, and metabolic, neural, and hormonal influences constant. At a given arterial pressure, an increase in filling pressure leads to increased power output (the curve B-N-A), as well as flow output. If the work per beat is used rather than work per minute (power), the relationship is called a *ventricular function curve* (see below).

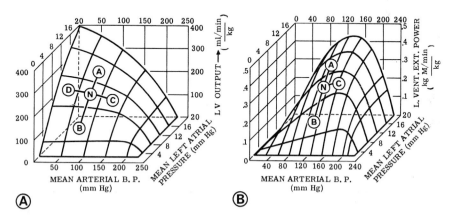

Figure 15-6
Effect of filling pressure (mean left atrial pressure) and output pressure (mean arterial pressure) on: A. cardiac output (left ventricular output), and B. heart work per minute (left ventricular external work). (Modified from K. Sagawa. In E. B. Reeve and A. C. Guyton [Eds.]. *Physical Bases of Circulatory Transport.* Philadelphia: Saunders, 1967.)

Homeometric Autoregulation

In addition to the increased power output of the heart as a result of increased filling —heterometric autoregulation—the physiology of the heart is such that nearly the same stroke volume is ejected even if there is an increased aortic pressure; the work output is increased. This response requires a few beats to develop, but it is not dependent on changes in fiber length (i.e., homeometric) or on neural or hormonal factors (i.e., autoregulation) (Sarnoff and Mitchell, 1962). The relationship is shown in Figure 15-6B, whereby the arterial pressure is increased from 100 mm Hg (point N) to 150 mm Hg (point C). At high mean arterial pressures (greater than 160 mm Hg) a reduction in external power output is seen with additional increases in pressure load. The increased output is not free, for when more work is done by increasing the arterial pressure load, the metabolic energy and hence oxygen requirements are increased. The systolic tension may be the principal determinant of not only homeometric autoregulation but also oxygen consumption (see below).

Finally, since the heart tends to eject the amount of blood returned, it is useful to consider the response of the *entire* cardiopulmonary bed to changes in filling pressure and systemic arterial pressure (Fig. 15-7). In contrast to the responses of the left ventricle alone (Fig. 15-6A), the entire system shows much less dependence upon the systemic arterial pressure because the right heart tends to maintain its flow output as the pulmonary vascular pressures increase somewhat following increased systemic arterial pressure. (Following an increase in systemic arterial pressure there is a consequent increase in both left ventricular end-systolic and then end-diastolic volumes. The increased left ventricular filling pressure is reflected back through the pulmonary bed.)

Myocardial Viscosity
Figure 15-5, uppermost curve, shows the pressure-volume relationship under iso-volumetric, nonshortening conditions. (The curve represents the maximal pressures at various volumes of *isometric* contractions, and each point on it is related to the maximum isometric force [P_0] of a force-velocity curve [Chap. 3B].) The tension developed when strips of myocardium are allowed to shorten is much less than the

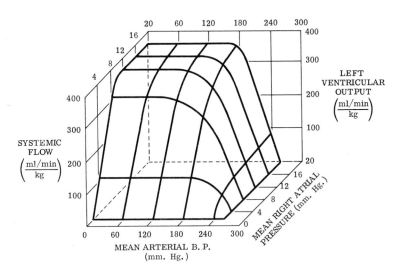

Figure 15-7
Effect of mean *right* ventricular filling pressure and mean systemic arterial pressure on left ventricular pumping capacity with autonomic control blocked. (Modified from C. W. Herndon and K. Sagawa. *Am. J. Physiol.* 217:65–72, 1969.)

tension produced when no shortening is allowed—the force-velocity relationship. This loss of force or tension is similar to that which would be seen if the contractile machinery had to drag through a "viscous medium." The mechanism for the viscous effect is primarily explained by the limited *rate* of formation of the actomyosin cross-bridges, which are assumed to be the cause of the developed tension. The myocardial tension rises rapidly during the isovolumetric contraction phase, but as soon as ventricular ejection starts, the myocardial fibers shorten and the effects of the apparent internal viscosity of the myocardium are seen. As a result, the effective intraventricular pressure developed is less than it would be under isovolumetric conditions. Under normal conditions, myocardial contraction is neither isometric nor isotonic, but *auxotonic;* that is, the contraction is against an increasing load, since the aortic pressure increases during the rapid ejection phase.

Less shortening is required by the large heart to eject a certain stroke volume than by the smaller heart. A change of volume (the amount ejected) of a sphere is related to its diameter cubed, and that of a cylinder or cone to the diameter squared, but the circumference (degree of fiber shortening) is directly proportional to the diameter. This relationship holds true under most conditions, especially for the left ventricle, although the geometric relationships of the two ventricles make a quantitative analysis complex. In summary, with a large heart the velocity of contraction is less, so the tension developed is more; that is, less muscular tension is lost in overcoming viscous forces, and more is available for pumping blood, than when the heart is functioning at a small size.

Elastic Forces

Another factor which must be considered is the development of elastic forces between the fibers and layers of the heart muscles as shortening occurs. Energy is stored during systole because of tissue compression and so less is available for ejecting blood. During diastole there is a facilitation of filling due to elastic recoil from the compression. At normally large ventricular volumes, this factor is minor; consequently, there is more energy available per unit volume for blood ejection.

Three factors, then, permit more effective ventricular function at large diastolic volumes: (1) the relationship described by Starling's law of the heart, (2) minimal myocardial viscosity because less shortening is required, and (3) minimal development of elastic forces in connective tissue. However, these factors are opposed by a fourth, which tends to limit severely the work done by the heart at very large volumes.

Laplace's Equation

According to the Laplace equation (Chap. 11), the tension of the fibers of a sphere or cylinder is directly proportional to the distending pressure times the radius. Thus, the myocardial tension of the contracting ventricles *required* to expel a unit volume of blood against a given pressure must be greater for a larger heart than for a smaller one. This factor tends to compensate, in the small heart, for the loss of effective tension because of the myocardial viscosity, development of elastic tissue compression forces, and reduced vigor resulting from shorter initial lengths of the myocardial fibers.

MYOCARDIAL CONTRACTILITY

A description of cardiac performance is essential for diagnostic evaluation. If performance is inadequate, the physician must be able to distinguish between factors such as inadequate filling or valve malfunction and the intrinsic ability of the heart to contract. *Myocardial contractility* and *vigor of contraction* are vague terms, connoting the tendency to contract (fiber shortening) as well as the force developed by the muscle fibers to expel blood. Thus, the term *contractility* serves a useful purpose by expressing a conceptual focal point between the many influences on the heart and its performance. As emphasized by Rushmer, ambiguity of definition tends to "obscure ignorance and

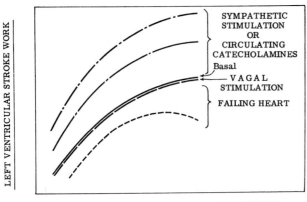

Figure 15-8
Ventricular function curves. Effect of various agents on cardiac function curves relating stroke work (stroke volume times mean systolic blood pressure) and left ventricular end-diastolic pressure. Under normal conditions there is a small sympathetic division activity giving a higher cardiac function curve than the basal level.

impede progress." Much to be preferred is the use of descriptive terms that can be accurately measured in physical terms. Although contractility connotes the ability to contract, the meaning commonly is restricted to the *intrinsic ability of the heart to function other than that induced by changes in presystolic fiber length,* i.e., factors other than those from the Frank-Starling relationship. If the myocardial contractility is increased, either the end-systolic volume will be decreased by the more vigorous contraction (thus increasing the stroke volume) or the same volume of blood will be ejected against a higher pressure by the more forceful beat. In either case, more work will be performed by the heart with each beat.*

One approach to evaluation of myocardial contractility is the determination of a *ventricular function curve* (Figs. 15-6B and 15-8). This is the relationship between ventricular stroke work (stroke volume times the change in pressure) and end-diastolic fiber length or pressure. The latter is often used because it is easier to measure (cardiac catheterization) and because the curvilinear relationship between the ventricular pressure just before systole and the fiber length is remarkably constant for a given heart. Even the mean atrial pressure has been used to derive satisfactory curves. In the usual method of determining the ventricular function curve, transfusion or hemorrhage is used to provide measured changes in the end-diastolic pressure. In addition, cardiac output and heart rate are measured to provide stroke-volume data. The mean aortic pressure and cardiac output furnish sufficient information for evaluating cardiac work, although the integral, during systole, of the instantaneous ventricular pressure times the instantaneous blood outflow provides a more accurate measure. Because such procedures as transfusion and hemorrhage take many seconds, neural and hormonal influences change during the time interval required; in the intact experimental subject, therefore, the curve so produced is not necessarily descriptive of the intrinsic contractility. Although a ventricular function curve provides the current standard of

*Some investigators also exclude from the definition of contractility any changes in power development due to changes in afterload (i.e., arterial blood pressure). The role of heart rate in cardiac power output is not even considered. Not only is it impossible, as yet, to measure contractility satisfactorily; there is no clear and commonly accepted definition as to what criteria the measurement must meet.

reference for myocardial contractility, it is limited, for the work output is a function of the pressure load and consequently the pressure load must be relatively normal throughout the study. Thus the complex surface shown in Figure 15-6B is a more realistic representation of contractility. Furthermore, it should be obvious that the determination of a ventricular function curve is not applicable under clinical conditions. Finally, since the curve is empirical, it cannot be expressed mathematically in a form in which each coefficient of a describing equation has biological meaning.

Four features of ventricular function curves should be considered: (1) Under a given set of conditions, increasing the fiber length leads to greater work output (heterometric autoregulation: the basal curve, Fig. 15-8). (2) If there are extracardiac influences (e.g., sympathetic stimulation), one sees a different curve such that at the same end-diastolic pressure more, or less, work is obtained. (3) Small changes in end-diastolic pressure of the *right* ventricle induce large changes in output of the *left* heart (Fig. 15-7). As Guyton et al. (1973) have emphasized, the right ventricular end-diastolic pressure is dependent upon the factors influencing the flow of blood to the heart (Chap. 16). Because the left ventricle can pump out only what is pumped to it by the right ventricle, and because of the sensitivity of the right ventricle to filling pressure, potent mechanisms must be available to maintain the optimal filling pressure to the heart. (4) It is important to realize that more work than before can usually be obtained from a heart which is weakened (i.e., lower function curve) if the filling pressure is increased at the same time. Thus, a single point on the curve provides only limited information about the condition of the heart. As the filling pressure is increased, a plateau in response is seen. In the failing heart, still further increases in end-diastolic pressure may result in less work than before, even with the same ventricular function curve (Fig. 15-8, bottom curve).

Another approach to evaluating myocardial performance is based on the force-velocity relationship of contracting muscle (Chap. 3B). The computation of the maximum velocity of shortening of the contractile elements (V_{max}) is based on numerous assumptions concerning geometry, uniformity of the ventricles, stiffness of the series elements, and characteristics of parallel elastic elements in conjunction with the peak rate of pressure development in the ventricle. Because some of these assumptions are in doubt, the utility of V_{max} is questionable.

The peak rate of pressure development ([dP/dt]max) is a relatively easily measured and useful index of contractility. The maximum dP/dt usually occurs just as the aortic valve opens (see Fig. 13-7). Measuring rates of pressure rise in the ventricle is easier than developing a ventricular function curve or a force-velocity curve, or even measuring peak rates of flow under various pressure loads. The [dP/dt]max increases in response to positive inotropes and is decreased by negative inotropes. It is markedly depressed in the failing heart. However, the maximum dP/dt changes somewhat with changes in preload (end-diastolic volume) and afterload (arterial pressure). Numerous attempts have been made to compensate for these effects. To date there is no commonly accepted index of myocardial contractility based on the intraventricular pressure pattern.

Yet another approach to estimating changes in contractility in experimental animals is to suture a strain-gauge arch across a slit in the myocardium to obtain an estimate of the tension developed by the myocardium at this site. (Ventricular filling and load must be the same during both the control and experimental periods.)

Factors Influencing Contractility

Myocardial contractility increases markedly in exercise; the heart beats more rapidly and the blood pressure increases. An increase in stroke volume occurs primarily because the end-systolic volume is smaller as a result of the sharply increased myocardial contractility. In the normal individual, autonomic nervous effects obscure Starling's law so that the cardiac output is increased without a definite increase in end-diastolic volume or right atrial pressure (see Fig. 15-9, equilibrium point C, and page 352). The increase in contractility is a result of the *neural influences* on the heart.

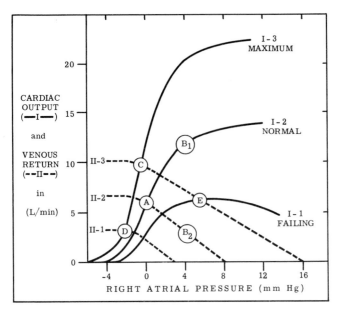

Figure 15-9
Relationship between cardiac function and venous return. *Ventricular function curves,* I, relating atrial pressure to cardiac output: 1, failing heart weakened by negative inotropes or ischemia; 2, normal relationship with moderate sympathetic influences; 3, maximal sympathetic influence. *Venous return curves,* II: 1, after hemorrhage and mean circulatory pressure reduced to 3 mm Hg; 2, normal with mean circulatory pressure of 8 mm Hg; 3, with maximum sympathetic influences or after large transfusion of blood. *Operating points:* A, normal; B_1 and B_2, outflow and inflow after sudden increase in atrial pressure; C, at maximum sympathetic activity; D, following hemorrhage; E, with failing heart showing increased central venous (right atrial) pressure.

Vagal stimulation decreases heart rate and therefore work done per minute. It reduces the strength of atrial contraction and decreases ventricular contractility somewhat (Fig. 15-8). The maximal depressing effect of the parasympathetic division on contractility is very much less than the augmenting effects of the sympathetic division, however.

The health and *metabolic conditions* of the myocardial cells are an obvious factor determining the strength of contraction. Hypoxia, hypercapnia, and acidosis are depressant. Ischemia from obstruction (occlusion) of a coronary artery leads to rapid deterioration of function because of inadequate blood supply.

With a decrease in the time interval between heart beats there is a progressive increase in vigor of contraction (the treppe or staircase phenomenon). Furthermore, although a premature depolarization results in a subsequent weakened contraction, the following beat is then more forceful than normal. This *postextrasystolic* potentiation is independent of ventricular filling. Both phenomena are probably related to the availability and/or transport of calcium between the cell membrane and the contractile machinery (Levy and Martin, 1974). The phenomena are intrinsic to the heart and so are considered by some researchers a form of homeometric autoregulation. Paired electrical stimuli to the weak heart under some conditions may be of significant therapeutic value, but the increased oxygen requirement to do the extra work limits its effectiveness in the most common situation where it might be applied—ischemic heart disease.

Inotropes are agents which cause a change in the contractility of the heart. Negative inotropes act on the heart to weaken the strength of contraction. Prime examples are the bacterial toxins released during many diseases, which reduce the ability of the heart to pump blood. Positive inotropes are agents such as digitalis and *circulating catechol-amines* which act to strengthen the heart beat. The infusion of norepinephrine in order to supplement the endogenously released catecholamines is of real, but limited, clinical value. Excessively large amounts (more than about 1 μg per kilogram of body weight per minute) may produce cardiac or peripheral damage. In last-resort conditions, the infusions of large amounts of norepinephrine in many may only produce more injury.

In summary, there are five classes of influence on the vigor of myocardial contraction:

1. *The metabolic condition* of the cells, which in turn is dependent on an adequate coronary blood flow, oxygen supply, nutrient supply, and lack of toxins.
2. *Heterometric autoregulation,* the Frank-Starling relationship, or the influence of the length of fibers at the beginning of contraction on cardiac performance.
3. *Central nervous system action,* including the effect of circulating catecholamines and the release of the autonomic mediators at the heart.
4. *Homeometric autoregulation,* the added power output of the heart in response to increases in pressure load without changes in end-diastolic fiber length or extra-cardiac influences.
5. *Interval between beats,* the increased vigor of contraction seen with moderate increases in heart rate.

RELATIONSHIP BETWEEN CARDIAC OUTPUT AND VENOUS RETURN

Right atrial pressure is an important variable influencing both cardiac output and venous return. Because the cardiac output must equal the venous return over long periods of time and yet may not be equal on a beat-to-beat basis, a study of the interrelationships helps one's understanding of cardiovascular function (see Guyton, et al., 1973, for details).

The curves labeled I-1 to I-3 of Figure 15-9 indicate the cardiac output (right ventricular outflow) which occurs under various conditions at given right atrial pressures. As filling pressure (right atrial pressure) is increased, cardiac output increases. The relationship is also shown in Figure 15-6A with the added variable of arterial pressure. As the level of sympathetic activity is increased (I-3 representing maximum activity versus I-2, a basal level), the cardiac output at a given atrial pressure (e.g., 2 mm Hg) is increased because contractile ability of the myocardium is increased. (Note: The effective filling pressure is about 5 mm Hg greater than the atrial pressure because of the normal negative 5 mm Hg intrathoracic pressure.)

The curves labeled II-1 to II-3 indicate the venous return (in liters per minute) which occurs at various right atrial pressures (see Chap. 16). Such curves have been obtained experimentally in dogs, although the technique is difficult and rather traumatic. The point at zero venous return defines one end of the curve. It is measured by fibrillating the heart and rapidly pumping blood from the arterial to venous circuits until the pressures are equal. A steady value must be obtained within three to five seconds before significant fluid shifts or autonomic responses have occurred. With the flow being zero, the pressure, called the *mean circulatory pressure,* is the same throughout the entire vasculature. Furthermore, this pressure is the same as that in the peripheral venules or small veins under normal conditions. Here is located the greatest portion of the blood. Setting right atrial pressure at a value less than the peripheral venular pressure, a pressure gradient is obtained, and the resulting flow is measured and plotted. (An external pump is used to maintain the cardiac output at the same rate as the venous return. Reflex compensation is blocked.) At atrial pressures much less than zero, the flow plateaus in value because the vessels entering the chest tend to collapse just outside the chest cavity. Providing a greater negativity does not increase flow, just as sucking harder on a collapsed soda straw does not yield more drink. The slope of the venous return curve represents the resistance and is determined by the geometry of the venous vasculature. The zero-flow pressure intercept is determined by (1) the blood volume and (2) the pressure-volume relationship of the vasculature. Vascular capacity may be reduced to increase the mean circulatory pressure by increasing sympathetic activity which acts on the smooth muscle in the walls of the veins and venules—venomotor tone (Chaps. 16 and 17).

The steady-state cardiac output and venous return must be equal. This condition will occur at the intersection, point A, of the given cardiac output (I-2) and venous return (II-2) curve. If the filling pressure is temporarily higher than the value at the intersection of the curves (e.g., 4 mm Hg), the cardiac output (point B_1) will be greater than the venous return (point B_2). The volume of blood in the atrium will be rapidly decreased by the increased cardiac output and reduced venous return and so will lead to a decrease in atrial pressure until inflow equals outflow. The converse is also true, for if cardiac output is less than venous return, blood accumulates, increasing the atrial pressure and cardiac output and decreasing the rate of venous return. At only one value of atrial pressure—in this case 0—will inflow equal outflow.

Even though cardiac output is greatly affected by a slight change in atrial pressure (e.g., curve I-3 between 0 and 1 mm Hg atrial pressure), a maximal sympathetic effect on both the heart (I-3) and peripheral venous bed (II-3) results in little or no change in atrial pressure, although the operating point is then at a much higher value, point C. Likewise, if only cardiac vigor is increased (curve I-3) without a change in the venous return relationship (curve II-2) there will be a small decrease in atrial pressure and only a slight increase in cardiac output.

Note that massive hemorrhage (curve II-1) greatly reduces the mean circulatory pressure and so venous return. Even with massive sympathetic effects on the heart, the cardiac output is greatly reduced, point D. Because the venous return is reduced, the pressure drop from periphery to heart is also reduced ($\Delta P = R \dot{Q}$). The reduction in right atrial pressure with hemorrhage (from 0 to -2 mm Hg) is less than the reduction in mean circulatory pressure (from 8 to 3 mm Hg).

With cardiac failure (curve I-1), an increased mean circulatory pressure is required to maintain the cardiac output (point E). This increase is obtained by sympathetic discharge or, in the long term, by retention of water, which increases the blood volume.

The factors influencing cardiac output are summarized at the left of Figure 17-1.

Although the stroke volume can be rather accurately predicted from heart rate, arterial pressure load, and filling pressure if the general state of the animal is held constant (Scher et al., 1968), the physiology of the heart under neural control is such that these variables are not likely to be fully independent (orthogonal) as different forms of stress are applied (e.g., exercise versus increased metabolism with fever). Because the sympathetic influences on various parameters of the cardiovascular system are neither proportional to each other nor linear (Chap. 17), a simple model is unlikely.

CARDIAC OUTPUT AND ADAPTATION IN MAN

The cardiac output in normal man is about 5.5 liters per minute. Like many other physical functions of the body, it is associated with metabolism, which in turn is associated with the size of the animal. Therefore a size correction is of value for comparison of various people or animals. Body weight is one factor widely used to determine drug dosage, but the body surface area has been found to be an even better basis for comparison of, for instance, metabolic rate, blood flow, heart rate, size of organs, respiratory rate, and cardiac output. In people of widely different sizes, DuBois and DuBois (in 1916) developed an empirical equation to convert body weight (in kilograms) and height (in centimeters) to body surface area (in square meters):

$$\text{Area} = 71.84 \times \text{weight}^{0.425} \times \text{height}^{0.725}$$

A nomogram is usually used.

In medical terms, an index often means the rate of a physiological function per minute per square meter of body surface area. Thus, the *cardiac index* is defined as the cardiac output per minute per square meter of body surface, and in man it is about 3.5 liters, with values below 2.0 liters or above 5.0 liters definitely abnormal at rest. The determination of the cardiac output in man, as outlined in Chapter 11, provides an important diagnostic tool. If the cardiac index is below about 2 liters, it is a sign of some form of cardiovascular failure. Tachycardia from atrial fibrillation, or bradycardia from atrioventricular block, often results in seriously reduced cardiac output. The cardiac output is reduced in congestive heart failure and does not increase significantly following muscular effort. In patients in mild heart failure, the cardiac output may be within statistically normal limits, but for a given individual it is lower than it would be if the heart of that individual were unimpaired.

Increases in cardiac output demonstrate compensatory mechanisms in the transport of oxygen to the tissues. In anemia, for instance, blood viscosity is decreased and the oxygen-carrying capacity is reduced. Thus a marked increase in cardiac output might be expected because the blood flows more easily; but more importantly, a greater blood flow is induced as a homeostatic adjustment to maintain the same oxygen transport, although the extraction of oxygen may also be increased. In pregnancy, the blood vessels to the placenta and uterus act as an arteriovenous shunt to lower the total peripheral resistance. An adequate blood pressure is maintained by an increased cardiac output (Chap. 17).

Hyperthyroidism (as in thyrotoxicosis), as well as fever or hyperthermia from high environmental temperature, causes an increase in metabolism which requires an increase in oxygen supply if homeostasis is to be maintained. Vessels dilate in response to hypoxia and so permit an increased flow. The cardiac output must increase if the blood pressure is to be maintained. On the other hand, at hypothermic temperatures of 15° to 20°C, the cardiac output is but 5 percent of normal. If the environmental temperature is high, so that the body temperature tends to rise, more blood is shunted to the skin to allow dissipation of more heat. This shunting of blood leads to an increase in the cardiac output.

The most pronounced increase in cardiac output occurs during severe exertion. The cardiac output of a normal individual walking slowly is about 50 percent greater than basal. However, cardiac outputs of about 40 liters per minute have been measured in well-trained athletes during maximal exertion. Since this output is seven times normal, both heart rate and stroke volume were increased.

Hypoxia, emotional excitement, insulin hypoglycemia, cutaneous pain—in fact, almost any mechanism acting to increase sympathetic division activity—increase cardiac output by increasing myocardial contractility and heart rate.

About the only normal cause of a decrease in cardiac output from basal flow is the act of changing from a recumbent to an upright position. On standing, the cardiac output decreases about 25 percent as a result of the redistribution of blood and the tendency toward pooling in the lower extremities. This decrease leads to orthostatic hypotension (low blood pressure on standing erect) unless compensatory mechanisms act to increase heart rate, myocardial contractility, peripheral resistance, and the filling pressure (venomotor tone).

CARDIAC WORK

If blood is to be pumped against a pressure head to the tissues, work must be done by the heart to drive a volume of fluid (V) against a pressure head (P). Thus, force (f in newtons) $\times$ distance (d in meters) = work (W in newton-meters) = pressure (P in newtons per square meter) $\times$ volume (V in meters cubed). Consequently, the effective rate of doing work (power developed) is the integral of instantaneous blood pressure times the instantaneous rate of flow during systole. This may be closely approximated as the product of mean arterial blood pressure times cardiac output.

It is informative to calculate the work done per minute by the left ventricle in the normal man. If he pumps 5.5 liters of blood per minute at a pressure of 95 mm Hg, his left ventricle will do 5.5×10^{-3} cubic meters per minute times 12.66×10^3 newtons per square meter or 69.6 newton-meters or joules of work per minute. (Recall that 1 mm Hg is 133.3 N/m^2.) Since a newton-meter per second is a watt, the useful power output of the left heart is only about 1.2 watts, but nonetheless it is equivalent to lifting three large textbooks (7 kg) 1 meter per minute.

The right heart pumps the same volume as the left heart, but the mean pulmonary arterial pressure is about one-seventh of the mean aortic pressure, and thus the amount of work per minute done by the right heart is one-seventh of that done by the left heart.

In addition to the potential energy, in the form of pressure-volume work, developed by the heart, metabolic energy is also converted to kinetic energy because of the velocity of movement of the blood (see Fig. 11-5). Usually only about 2 percent of the useful work of the heart is in the form of kinetic energy. However, during exercise, when the velocity and rate of flow of blood are higher than normal, the kinetic energy may amount to 25 percent or more of the total useful work. Most of this kinetic energy is reconverted to useful potential energy (pressure times volume) by the distention of the elastic aorta. In effect, the elastic aorta catches the ejected blood and throws it on into the arterial tree.

During exercise, in order to propel more blood to the tissues against the same or higher pressure head, the heart must do more work. The extra work requires an increase in metabolism by the heart. The increase is produced either by a greater utilization of nutrients and oxygen or by an increase in the efficiency of the conversion of these materials to useful work. In man, the biological oxidation of nutrients by 1 ml of oxygen gives about 20.2 joules of energy. Thus the heart would require about 3.5 ml of oxygen per minute to develop the 70 newton-meters per minute of power (1.2 watts) under basal conditions if it were 100 percent efficient in the conversion of oxygen and nutrients to useful work. However, the heart is only about 20 percent efficient. Thus for 70 joules per minute of useful work, about 350 joules per minute of energy must be released. Therefore, the human adult left heart requires about 3.5/0.2 or 18 ml of oxygen per minute at rest and the right about 2.6 ml/per minute. Since our basal requirement is about 250 ml of oxygen every minute, our hearts use about 8 percent of our total oxygen consumption.

CARDIAC EFFICIENCY

Although the oxygen consumption is a good index of the metabolic release of energy under steady-state conditions, the oxygen consumption of the heart is not always closely related to the total useful work done in the intact animal because the efficiency of the heart changes. The oxygen consumption per minute by the heart is roughly proportional to the mean systolic arterial pressure times the duration of systole times the number of beats per minute. Note that the volume of blood moved (i.e., the cardiac output) does not enter this relationship directly. Thus, cardiac oxygen consumption is more closely related to the intraventricular pressure developed than to the external work performed. The relationship may be explained if the following physiological characteristics of muscle are considered: (1) Isometric contraction of muscle requires metabolic energy but results in no useful work since no load is moved. Only tension is developed. The development of a ventricular pressure requires energy even though no blood is ejected and therefore no useful work is done. Useful work is realized only if the heart can eject blood. (2) Heat production by the muscle during contraction is an index of metabolism and is determined primarily by the tension maintained times the duration of maintenance of this tension. However, (3) when a muscle does shorten under a load, some extra energy is required.

An increase in *blood pressure* has little effect on cardiac efficiency. The increased work, when the change is in pressure head or load, with a constant cardiac output and heart rate, requires a *proportionate increase in oxygen,* since pressure is a major factor in the determination of oxygen consumption. This is seen clinically in aortic stenosis. In this condition the resistance to flow of blood out of the heart is greatly increased because of the constricted aortic valve, but the cardiac output is held about normal by various compensatory mechanisms. When the pressure load is high, the oxygen consumption by the heart must be high. Overt myocardial hypoxia and heart failure are seen frequently in aortic stenosis. Hypertension also presents a serious load to the heart and rapidly leads to cardiac failure if the coronary blood flow is restricted by atherosclerosis.

If the *stroke volume* of the heart increases, with blood pressure and heart rate constant, the efficiency of the heart increases to as high as 40 percent, since the greater volume pumped requires but a *small increase in oxygen consumption.* Systole lasts slightly longer, but the effect is relatively small. In experiments with dogs, the work output can be increased about 700 percent by changes in stroke volume with only about 53 percent increase in oxygen utilization. This result is in contrast to increased work by increased pressure, in which case the oxygen consumption parallels the increased work accomplished. Clinically the heart does much work in mitral insufficiency (i.e. blood leaking back to the atrium with each beat) or in intracardiac and extracardiac shunts, which involve a large bypass flow. However, early heart failure is surprisingly rare, considering the enormity of the defect in many cases. Aortic valvular insufficiency is an exception, since there is not only a massive back-and-forth movement of blood through the insufficient valve but also, at the region of the valve, a low blood pressure during most of diastole. Thus the perfusion pressure of the coronary arteries is markedly reduced.

An increase in *heart rate,* blood pressure and cardiac output being constant, acts to decrease the efficiency. An increase in heart rate results in an increase in oxygen consumption without, necessarily, an increase in the amount of work (pressure times volume) done. Since each beat requires a certain amount of energy whether or not useful work is performed, the energy expended increases and the efficiency drops. As a consequence, unless the heart of an athlete who must do much work adapts to provide a larger stroke volume at relatively low heart rates, the athlete will not excel. Likewise the patient at the limit of cardiac reserve is not aided by excitement, which causes an increased heart rate and thereby a decrease in the efficiency. Increases in heart rate can and do increase the cardiac output, but the price is a decrease in efficiency. Since the normal heart has adequate reserves of coronary blood flow, this is not serious.

In the failing heart, for as yet unknown reasons, the efficiency is decreased. In exercise, with a failing heart, the efficiency appears to be decreased even more. Thus adequacy of coronary flow is particularly important under these conditions. Unfortunately, partial or total occlusion of the coronary arteries is commonly the cause of the cardiac failure in the first place.

NUTRITION OF THE HEART

Coronary Blood Flow

The sustained ability of the heart to maintain a high blood flow to the peripheral tissues is limited primarily by the cardiac oxygen supply. There are but three mechanisms by which the supply of oxygen to the myocardial cells may be increased: (1) *Increased oxygen extraction.* This is a minor source of additional oxygen for the myocardium, in contrast to peripheral tissue such as skeletal muscle. In skeletal muscle, the arterial-venous difference in the oxygen content of the blood may be increased markedly by decreasing the oxygen content of the venous blood. The heart normally extracts 70 to 90 percent of the oxygen from the arterial blood, leaving only 2 to 6 volumes per 100 volumes of blood in the coronary sinus venous blood—a low reserve. (2) *Increased myocardial efficiency.* The efficiency of the myocardial utilization of oxygen is determined by the type of work done by the heart and is generally independent of the oxygen needs of this organ. (3) *Increased coronary blood flow.* This is the only remaining possibility for an increased supply of oxygen to the heart. Since the heart can incur only a very limited oxygen debt, an increase in the coronary blood flow by dilation of the coronary arterioles is essential.

Factors Regulating Coronary Blood Flow

Measurement of the coronary circulation is difficult because of the location, short length, and multiplicity of the coronary arteries and veins. Electromagnetic flowmeters are being used to measure pulsatile flow in intact coronary vessels in unanesthetized dogs with the flowmeter chronically implanted, and during cardiac surgery in man. The nitrous oxide method has given reasonably accurate values in unanesthetized humans and dogs. For this method, and the determination of cardiac oxygen consumption, samples of blood are withdrawn from the coronary sinus and aorta by means of indwelling catheters. (See Chap. 11 for a discussion of flow measurements.)

In man, at rest, the coronary blood flow is about 200 ml per minute (about 70 ml per 100 gm of tissue). From data obtained from dogs, the flow rate can increase by at least nine times during severe stress. The myocardium accounts for about 4 percent of the cardiac output and 8 percent of the oxygen consumed by an adult at rest.

Metabolism

Myocardial hypoxia has a pronounced effect on coronary blood flow. Hypoxia can increase the coronary blood flow at least five times. The response to anoxia is rapid, since a five-second occlusion of the coronary artery having little or no effect upon the contractility is followed by a definite increase in flow. This sensitivity and rapidity of response to tissue hypoxia is essential, since the increased oxygen required for increased cardiac work must be supplied promptly by increased coronary flow or there will be incipient cardiac failure. Indeed, the coronary blood flow to the heart is closely and directly correlated with the oxygen consumption by this organ. Although the steps are not completely established, they probably include tissue oxygen tension and adenosine released from the tissue which then acts to dilate the coronary arterioles (see, e.g., Rubio and Berne, 1969; Feigl, 1974). An increase in carbon dioxide tension or a decrease in pH has a similar, but lesser, dilating effect upon the coronary blood vessels. Furthermore, the average coronary blood flow under constant metabolic conditions is relatively constant even if the aortic pressure is varied between 60 and 140 mm Hg, for the coronary bed shows strong autoregulation (see Chap. 17).

Mechanical Regulation of Coronary Flow

Since the coronary arteries are supplied from the aorta and since flow is proportional to the perfusion pressure, the coronary flow is dependent upon the mean systemic arterial blood pressure. The relationship is not simple, for when the heart contracts, the intramural (intramuscular) pressures can and do exceed the arterial pressures and so occlude arteries and stop the blood flow. The normal contracting myocardium tends to impede flow. Although there is usually an inrush of blood very early in systole, during most of systole the high intramural pressure greatly reduces coronary arterial flow so that only 10 to 40 percent of the total flow occurs during this period, even though the aortic pressure is highest then (Fig. 15-10). The venous discharge is hastened during systole because of the squeezing of the veins.

The anatomical arrangement of the arterial tree of the heart is such that if one of the coronary arteries is suddenly occluded, the pressure beyond the occlusion drops to about 30 mm Hg and contractility of the affected myocardium decreases to useless levels. Collateral flow of about 10 percent of normal is present, but unlike skeletal muscle under similar circumstances, the myocardium will require several weeks for the blood flow to increase to adequate amounts. Thus, sudden coronary occlusion often results in a fatal myocardial infarction, whereas gradual occlusion may be followed by the development of an adequate collateral arterial blood supply.

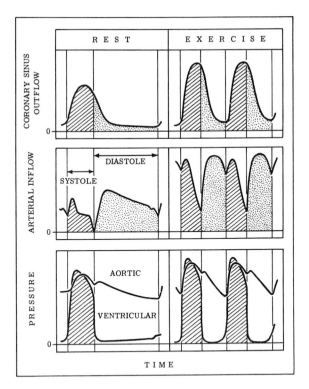

Figure 15-10
Effect of myocardial contraction on coronary venous outflow and arterial inflow at rest and during exercise. Note the hindrance to inflow during systole and the arterial regurgitation or backflow. The sharp rise in inflow during systole may be a rebound filling of the compressed arteries. Pressures given for timing reference. (Modified from D. E. Gregg, E. M. Khouri, and C. R. Rayford. *Circ. Res.* 16:105, 1965; and R. F. Rushmer, *Cardiovascular Dynamics* [2nd ed.]. Philadelphia: Saunders, 1961.)

Neural Control
The innervation of the coronary blood vessels is abundant, but the physiological function of the nerves is not clear, largely because of the difficulty in measuring blood flow under even approximately physiological conditions. Furthermore, it is hard to differentiate the direct neural effect on the resistance vessels and the concomitant effects of changed metabolism as a result of the neural action. Stimulation of the parasympathetic fibers acts to cause a reduction of the coronary blood flow, but in addition there is a decrease in heart rate and thus increased efficiency, decreased metabolism, and so a decreased oxygen consumption. One may therefore expect decreased coronary flow, if the oxygen consumption is the primary regulator of this flow. An increase in the activity of the sympathetic nerves acting on the heart causes an increase in flow. Further, the heart rate increases, the contractility increases, more work is done, and thus the metabolism and oxygen consumption rise. Coronary vasodilation follows. Nonetheless, underlying these metabolic effects is sympathetic coronary alpha-receptor vasoconstriction and parasympathetic vasodilation (Feigl, 1974). The dominant vagal tone in an unanesthetized man acts to keep the heart rate down, the efficiency up, and the coronary blood flow, in consequence, relatively low.

Circulating Epinephrine

Epinephrine causes an increase in coronary blood flow, probably on account of the increased metabolism, as occurs in skeletal muscle (Chap. 17). Since epinephrine increases strength of beat, heart rate, and metabolism, the resulting hypoxia probably elicits the arterial dilation. Angina in susceptible patients, following epinephrine injections or sympathetic nervous system action, has been attributed to the increased metabolism with oxygen demands exceeding the supply. Since the increased blood flow produced by the vasodilating effects of the epinephrine is not adequate to provide for increased oxygen needs, hypoxia and ischemic pain result.

Metabolism of the Heart

A continuous supply of nutrients and oxygen is required by the functioning heart. The primary nutrients during fasting are free fatty acids, but after eating or during exercise the primary nutrients are glucose and lactic acid. In vigorous exercise, the venous oxygen saturation of active muscle is near zero, anaerobic metabolism is taking place, and lactic acid is released. Under these conditions lactate is the primary substrate (see Opie, 1969). It must be remembered, however, that utilization of lactate by the heart requires oxygen. When the heart itself is hypoxic, instead of utilizing lactic acid it produces it as a metabolite and develops a limited oxygen debt. In contrast to skeletal muscle, the mammalian heart can tolerate but little anaerobic metabolism. There is no clinical situation in which the cardiac work capacity is limited by the lack of substrate for energy production.

Electrolytes

In addition to the intermediary metabolism, the heart depends upon an optimal electrolyte environment (Chaps. 3B, 13, and 14). The effect of acids on the heart is complex. In dogs, the lethal pH is about 6.0 and is independent of the anion since both hydrochloric and lactic acids have similar effects. At a critically low pH, there is rather sudden, sharp decrease in contractility of the heart. Cardiac arrest in extreme diastole, similar to that caused by potassium, is seen as a terminal event.

REFERENCES

Berne, R. M., and L. N. Levy. *Cardiovascular Physiology* (2nd ed.). St. Louis: Mosby, 1972.

Braunwald, E., J. Ross, Jr., and E. H. Sonnenblick. *Mechanisms of Contraction of the Normal and Failing Heart* (2nd ed.). Boston: Little, Brown, 1976.

Braunwald, E., E. H. Sonnenblick, P. L. Frommer, and J. Ross, Jr. Paired electrical stimulation of the heart: Physiologic observation and clinical implications. *Adv. Intern. Med.* 13:61–96, 1967.

Feigl, E. O. The Coronary Circulation. In T. C. Ruch et al. (Eds.), *Physiology and Biophysics.* Philadelphia: Saunders, 1974. Vol. 2, chap. 16.

Gregg, D. E., E. M. Khouri, and C. R. Rayford. Systemic and coronary energetics in the resting unanesthetized dog. *Circ. Res.* 16:102–113, 1965.

Guyton, A. C., C. E. Jones, and T. G. Coleman. *Circulatory Physiology: Cardiac Output and Its Regulation* (2nd ed.). Philadelphia: Saunders, 1973.

Levy, M. N., and P. J. Martin. Cardiac Excitation and Contraction. In A. C. Guyton and C. E. Jones (Eds.), *Physiology, Series One. Cardiovascular Physiology.* London: Butterworth, 1974. Vol. 1, chap. 2.

Opie, L. H. Metabolism of the heart in health and disease. Part II. *Am. Heart J.* 77:100–122, 1969.

Pollack, G. H. Maximum velocity as an index of contractility in cardiac muscle. *Circ. Res.* 26:111–127, 1970.

Rubio, R., and R. M. Berne. Release of adenosine by the normal myocardium in dogs and its relationship to the regulation of coronary resistance. *Circ. Res.* 25:407–415, 1969.

Rushmer, R. F. *Cardiovascular Dynamics* (3rd ed.). Philadelphia: Saunders, 1970.

Sarnoff, S. J., and J. H. Mitchell. The Control of the Function of the Heart. In W. F. Hamilton and P. Dow (Eds.), *Handbook of Physiology.* Washington: American Physiological Society, 1962. Section 2: Circulation, vol. 1, chap. 15.

Scher, A. M., A. C. Young, and T. H. Kehl. The regulation of stroke volume in the resting, unanesthetized dog. *Comput. Biomed. Res.* 1:315–336, 1968.

Sonnenblick, E. H., J. Ross, Jr., J. W. Covell, G. A. Kaiser, and E. Braunwald. Velocity of contraction as a determinant of myocardial oxygen consumption. *Am. J. Physiol.* 209:919–927, 1965.

Suga, H., K. Sagawa, and A. A. Shoukas. Load independence of the instantaneous pressure-volume ratio of the canine left ventricle and effects of epinephrine and heart rate on the ratio. *Circ. Res.* 32:314–322, 1973.

16. The Systemic Circulation

Julius J. Friedman

GENERAL CHARACTERISTICS

The systemic circulation includes all the blood vessels that originate from the aorta and terminate at the right atrium. Those lying between the right ventricle and the left atrium constitute the pulmonary circulation (see Chap. 20). The aorta undergoes progressive branching to form, eventually, the capillary-venular network. These vessels then coalesce progressively into the venae cavae. Examination of some of the physical and functional properties of the systemic circulatory system can form a basis for better understanding of the physiological considerations which come later.

Figure 16-1 shows the aorta and large arteries to be heavy-walled vessels containing a relatively large proportion of elastin and collagen in addition to smooth muscle; they are thus equipped to serve as conduits to deliver blood, under high pressure, to the tissues. The terminal arteries and arterioles are especially well endowed with smooth muscle and act as variable resistors to modulate tissue blood flow. The capillaries and venules, with walls essentially one endothelial cell thick, provide a porous barrier between the vascular and extravascular compartments across which exchange of water and solute molecules takes place. The venules, along with small veins, also act as a variable blood reservoir. The large veins and venae cavae are larger in diameter and have thinner walls then their arterial counterparts, which structure is consistent with their function of conducting blood under low pressure from the tissues to the right atrium.

The aorta of an 80-kg man is about 2 cm in diameter and has a cross-sectional area of about 3 cm². As it branches, the cross-sectional area increases slowly but progessively. Figure 16-2 shows that at the level of the arteriole the cross-sectional area increases dramatically to about 800 cm² and at the capillary-venular level reaches about 3500 cm².

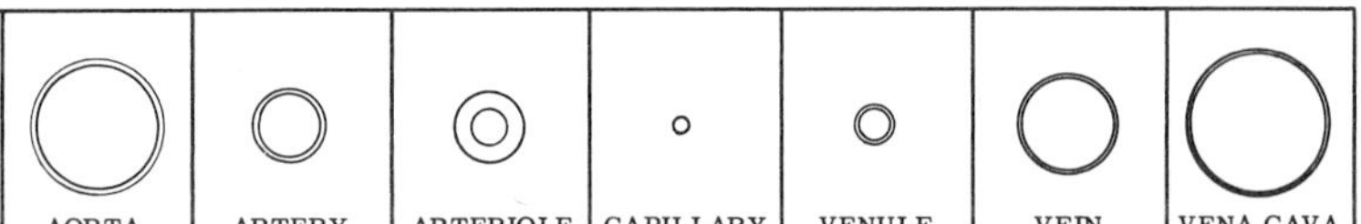

	AORTA	ARTERY	ARTERIOLE	CAPILLARY	VENULE	VEIN	VENA CAVA
DIAMETER	2 cm	4 mm	30 μm	8 μm	40 μm	5 mm	3 cm
WALL THICKNESS	2 mm	1 mm	20 μm	1 μm	2 μm	.5 mm	1.5 mm
WALL THICKNESS / LUMEN RADIUS	$\frac{1}{5}$	$\frac{1}{2}$	>1	$\frac{1}{4}$	$\frac{1}{10}$	$\frac{1}{5}$	$\frac{1}{10}$
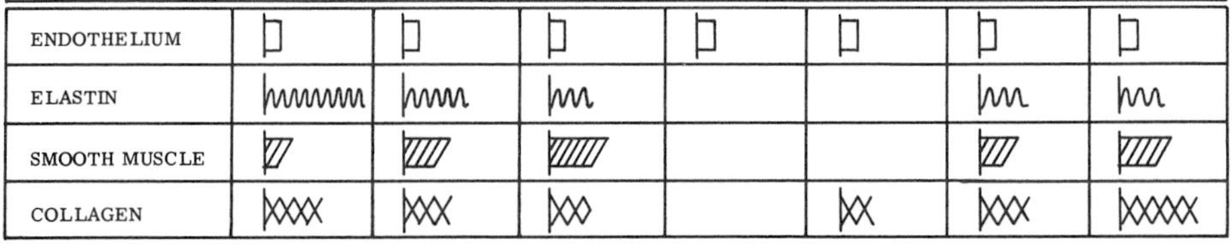 ENDOTHELIUM							
ELASTIN							
SMOOTH MUSCLE							
COLLAGEN							

Figure 16-1
Dimensions and structural constituents of the various vessels of the systemic circulation. (Modified from A. C. Burton. *Physiol. Rev.* 34:619–642, 1954.)

361

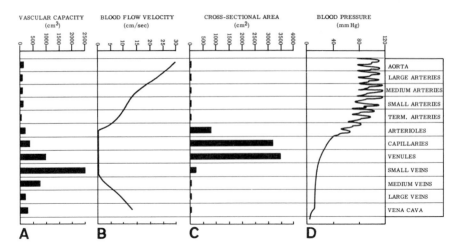

Figure 16-2
Relationship between blood pressure, cross-sectional area, blood flow velocity, and blood volume in the various segments of the systemic vascular system.

This large cross-sectional area then declines abruptly as the venous vessels coalesce to form the 3-cm vena cava with a cross-sectional area of about 7 cm².

The volume of blood flowing through the various vascular segments is the cardiac output; thus, the velocity of blood flow must be inversely related to the cross-sectional area. Figure 16-2 illustrates that the average velocity in the aorta is about 30 cm per second. The velocity falls progressively with branching until it reaches 0.026 cm per second at the capillaries. It then rises slowly and progressively through succeeding segments until it reaches about 14 cm per second in the vena cava. The high cross-sectional area of and the low velocity in the capillaries and venules are particularly well suited to the exchange function of these vessels.

The dimensions of the vascular segments provide them with the vascular capacities shown in Figure 16-2. The arteries contain about 10 percent of the blood volume, the capillaries about 5 percent, the venules and small veins about 54 percent, and the large veins about 21 percent. The heart chambers accommodate the remaining 10 percent. The muscular venules and small veins are, by virtue of their large capacities and smooth muscle, effective reservoirs which can either expand to accommodate larger volumes or contract to mobilize blood for delivery to the heart.

ARTERIAL BLOOD PRESSURES

Clinically, arterial blood pressures are routine measurements that reflect the status of the cardiovascular system. The energy transferred to the arterial system by cardiac ejection generates a pressure pulse (Fig. 16-3). The maximum pressure attained in the arterial system during cardiac ejection is the *systolic pressure* (SP). During the diastolic phase of the cardiac cycle, arterial pressure falls progressively. The minimal pressure achieved during diastole is the *diastolic pressure* (DP). The difference between systolic and diastolic pressures is the *pulse pressure.* The *mean blood pressure* (MBP) is not normally the arithmetic average of systolic and diastolic pressures, because of the unequal distribution of time for systole and diastole. A very crude estimate of mean blood pressure has been made by weighting diastolic pressure by a factor of 2. Thus, MBP = (SP + 2 DP)/3. This formula is based on the consideration that diastole is twice as long as systole. However, as heart rate changes, the duration of diastole changes

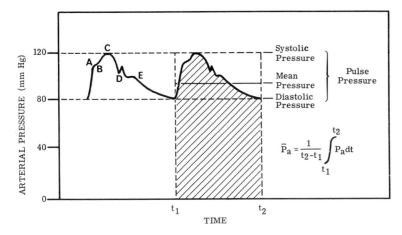

Figure 16-3
Arterial systolic, diastolic, pulse, and mean blood pressures. The mean arterial pressure, $\bar{P}a$, represents the area under the arterial pressure curve (shaded area) divided by the duration of the cardiac cycle ($t_2 - t_1$). The letters A–E refer to clinical designations and are explained in the text.

to a greater extent than does that of systole, and the percentage of the time of the cardiac cycle occupied by systole varies from 30 percent at a heart rate of 40 beats per minute to 70 percent at 180 beats per minute. Therefore, ideally, mean blood pressure should be determined by integrating the pulse contour over the time interval of the cardiac cycle (see Fig. 16-3).

Determinants of Mean and Pulse Pressures

Mean Arterial Pressure
The factors that determine the mean arterial blood pressure can be easily defined by an application of Ohm's law to the cardiovascular system (Chap. 11). This provides the relationship

$$P = Q \times R$$

where P is pressure, Q is blood flow, and R is resistance. In the cardiovascular system, this becomes

$$MBP = CO \times TPR$$

where MBP is mean arterial blood pressure (mm Hg), CO is cardiac output (liters per minute), and TPR is total peripheral resistance (mm Hg per liter per minute). Thus mean blood pressure can be modified by changes in either cardiac output or total peripheral resistance or both.

Pulse Pressure
Direct intra-aortic pressure recorded during the cardiac cycle reveals a number of oscillations, as illustrated in Figure 16-3. The clinician generally refers to these by characteristic labels. With the onset of cardiac ejection, aortic pressure rises very rapidly to an early peak called the percussion wave (A). It dips momentarily to form the anacrotic notch (B) and then rises more slowly forming a second peak, the tidal wave (C). The pressure then falls and the incisura (D) is registered as the aortic valve closes

and oscillates, and a final positive wave, the dicrotic wave (E), is formed as the pressure declines to the level of diastolic pressure. The rising portion of the pulse is called the anacrotic limb and the declining portion the catacrotic limb of the pulse.

Systolic pressure depends on the stroke volume, peak cardiac ejection rate (cardiac contractility), and arterial distensibility. Diastolic pressure depends on the level of systolic pressure, total peripheral resistance, heart rate, and arterial elastic recoil.

Pressure in any distensible vessel is a function of the volume in the vessel relative to its ability to accommodate the volume (i.e., its compliance). The pulse pressure (ΔP) generated during a cardiac cycle is largely determined by the stroke volume (ΔV) and the nature of the arterial compliance (C), and may be expressed as $\Delta P = \Delta V / C$.

Arterial Compliance
Arterial compliance or distensibility is the inverse of elasticity and is determined by the physical properties of the arterial wall. The elastin and smooth muscle in the vessel wall enable it to distend during systole and recoil during diastole. Thus, in accordance with the conservation of energy principle (see Bernoulli equation, Chap. 11), during systole the energy of cardiac ejection is distributed to the wall to produce distention and to the blood to produce pressure and flow. Since only a portion of the energy introduced into the arterial system generates pressure, the level of systolic pressure achieved is reduced. During diastole, as the volume of blood in the arterial system declines and pressure falls, the rate of pressure decline is reduced, and the diastolic pressure is augmented somewhat by the energy of elastic recoil. As the arterial system becomes sclerosed, it becomes stiffer or less compliant. Under this circumstance, less energy is absorbed by the wall during systole, and therefore systolic pressure rises. Moreover, less energy is restored to the blood by elastic recoil during diastole, and diastolic pressure falls so that pulse pressure increases markedly.

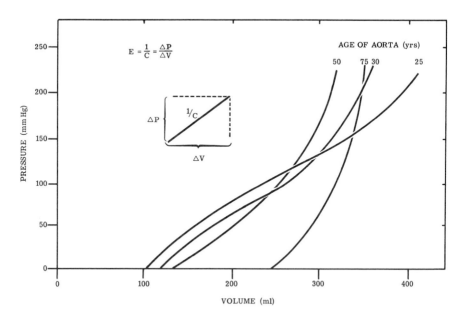

Figure 16-4
The volume-pressure relationship of aortas obtained from patients of different ages. Vessel elasticity (E) is defined as the change in pressure resulting from a change in volume and is the inverse of compliance (C). Thus the slope at any point on these curves represents the inverse of compliance. (After P. Hallock and I. C. Benson. *J. Clin. Invest.* 16:595, 1937.)

The nature of arterial compliance can be better appreciated by examining the volume-pressure relationship (V-P) (Fig. 16-4). The relationship is presented in this manner, rather than the more classic pressure-volume format, since, in the body, cardiac ejection induces a change in volume of the arterial system which gives rise to a change in pressure. The slope of these curves ($\Delta P/\Delta V$) represents the inverse of compliance ($1/C$), which is the elastance (E). Pressure increases with volume, and the extent of increase is directly related to age. This relationship reflects the fact that compliance, and thus distensibility, decreases with increasing age, a consequence of arteriosclerosis. It is also apparent that the curves are not linear, signifying that arterial compliance is not constant.

Residual Arterial Volume

The volume-pressure relationships in Figure 16-4 illustrate the influence of the volume in the arterial system immediately prior to the next cardiac ejection on arterial pressure. This volume, together with arterial compliance, determines the diastolic pressure; it also determines where on the V-P curve, or at what initial compliance, the next pulse will be initiated.

The volume in the arterial system at any moment is determined by the relationship between arterial inflow and outflow. Inflow represents stroke volume; outflow is the blood flow to the tissues. Thus outflow, or arterial runoff, is determined by perfusion pressure and total peripheral resistance. Early in the cardiac cycle, during rapid cardiac ejection, inflow is greater than outflow and arterial volume increases, as does pressure. As systole progresses, the rate of cardiac ejection declines, as do the rates of rise of arterial volume and of pressure. Simultaneously, arterial runoff increases progressively with the increase in arterial pressure. About midway through systole the rate of cardiac ejection and the rate of arterial runoff become equal. At this moment arterial volume is at a maximum, and so is arterial pressure (systolic pressure). From then on, the rate of cardiac ejection is less than the rate of arterial runoff, and arterial volume and arterial pressure decline. When the aortic valve closes, arterial inflow is zero and the arterial blood volume is determined by arterial runoff.

Influence of Stroke Volume

The effect of changes of stroke volume on pulse pressure in a 25-year-old subject is illustrated in Figure 16-5. Normally, the aortic volume at the end of diastole is approximately 180 ml, generating a diastolic pressure of 80 mm Hg. Cardiac ejection of a 70-ml stroke volume causes aortic volume to rise to about 250 ml and generates a systolic pressure of about 120 mm Hg. Increasing stroke volume to 100 ml by means of augmented cardiac contractility would have the effect of increasing aortic volume to about 280 ml and systolic pressure to about 140 mm Hg. Conversely, reducing stroke volume would cause aortic volume and systolic pressure to rise to a lesser extent.

As the individual ages and arterial distensibility declines, less damping of the pulse occurs. The influence of changes in stroke volume in the less compliant 50-year-old aorta is illustrated in Figure 16-6. In this case, if the residual aortic volume were the same as in the 25-year-old aorta, diastolic pressure would be reduced to 50 mm Hg, reflecting the reduction of the contribution to pressure development made by elastic recoil of the arterial wall during diastole. However, at this age residual arterial volume is generally found to be elevated to about 230 ml and diastolic pressure may be elevated to 85 or 90 mm Hg.

A normal stroke volume of 70 ml causes aortic volume to rise to 300 ml, and systolic pressure rises to about 160 mm Hg, or 40 mm Hg beyond that of the more distensible 25-year-old vessel. The augmented systolic pressure reflects the reduction in the damping provided by arterial wall distention. It should be apparent that increases in stroke volume in this subject produce exaggerated increases in systolic pressure, since the aortic volume results in a shift of the volume-pressure relationship to an even less compliant region.

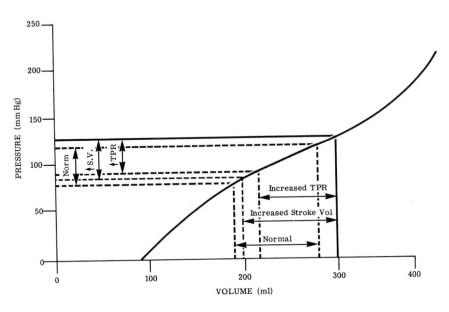

Figure 16-5
The effect of changing stroke volume (SV) and total peripheral resistance (TPR) on the volume-pressure relationship of an aorta from a 25-year-old patient.

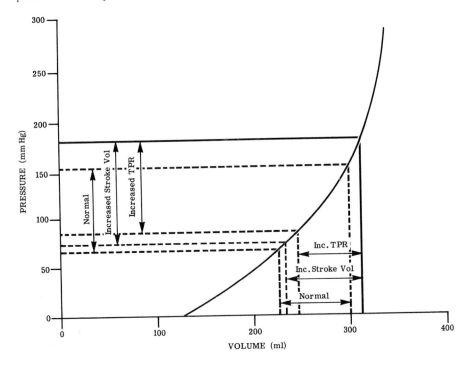

Figure 16-6
The effect of changing stroke volume (SV) and total peripheral resistance (TPR) on the volume-pressure relationship of an aorta from a 50-year-old patient.

Influence of Total Peripheral Resistance

Total peripheral resistance (TPR) is the major determinant of outflow from the arterial system. However, it should be remembered that flow is also proportional to the pressure gradient across the resistance elements. Therefore, systolic pressure, by contributing to the perfusion pressure gradient, also contributes to outflow. Figure 16-5 illustrates the influence of TPR on pulse pressure in the 25-year-old subject.

Increasing TPR, by reducing the rate of runoff, establishes a higher residual arterial volume at the end of diastole and hence a higher diastolic pressure. Now the next cardiac ejection must be delivered into a higher load, and stroke volume will probably decline slightly, as will pulse pressure. Conversely, reducing TPR allows for greater arterial runoff and therefore causes end-diastolic arterial volume and diastolic pressure to decline. In addition, it lessens the load on the heart and permits a greater than normal stroke volume to occur, producing a relative increase in systolic pressure. Thus, reductions in TPR will cause pulse pressure to increase. In the 50-year-old aorta (Fig. 16-6), increasing TPR by increasing end-diastolic arterial volume causes a shift to a steeper, less distensible portion of the V-P curve, and this further reduction in distensibility leads to a further increase in systolic pressure. When distensibility is reduced, then, increasing TPR can increase pulse pressure.

Influence of Heart Rate

Heart rate determines the duration of systole and diastole and thus the time available for cardiac filling and arterial runoff. Increasing heart rate in the 25-year-old person reduces diastole and runoff time, so the residual arterial volume is elevated, as is diastolic pressure. Increasing heart rate also reduces the time for cardiac filling, and stroke volume declines, thereby producing a relatively smaller increase in systolic pressure and pulse pressure decreases. Conversely, reducing heart rate increases the time for arterial runoff and cardiac filling, resulting in a reduction in diastolic pressure and an increase in systolic and thus pulse pressure. In the 50-year-old case, increasing heart rate reduces the time for arterial runoff and thereby increases residual arterial volume. This shifts the V-P relationship to a region of reduced compliance and leads to augmentation of systolic and pulse pressures.

Pathological Pulses

Conditions that alter heart rate, stroke volume, total peripheral resistance, and/or arterial compliance alter systolic and diastolic pressures. Frequently, these changes occur in combination as a result of broad cardiovascular regulatory adjustment to stress. In that event, the relationship of the pulse characteristics to the various determinants is not so clear-cut. In a recent study (Cullen, 1974), in which correlations between systolic and diastolic pressures, heart rate, stroke volume, and total peripheral resistance were examined during hypoxia, anemia, hypercapnia, and anesthesia, positive correlations existed only between the change in systolic pressure and stroke volume, and between the change in systolic pressure and the change in diastolic pressure. No correlation existed between the changes in systolic or diastolic pressure and the changes in heart rate and total peripheral resistance.

A number of clinical disorders generate characteristic pulses. These include aortic insufficiency, arteriosclerosis, aortic stenosis, and hypertension and are illustrated in Figure 16-7.

Aortic Insufficiency

In aortic insufficiency, the aortic valve does not close completely during diastole, and blood regurgitates from the aorta into the left ventricle. Thus, blood leaves the arterial system through two pathways, as peripheral arterial runoff and as backflow into the left ventricle, leading to a reduction in end-diastolic arterial volume and hence diastolic pressure. The regurgitation also leads to an increased left ventricular end-diastolic

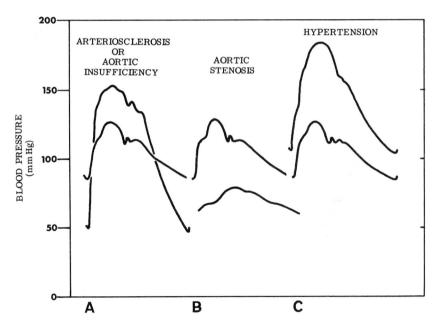

Figure 16-7
Changes in the arterial pulse contour in arteriosclerosis, aortic insufficiency, aortic stenosis, and hypertension.

volume, which produces an increased stroke volume and systolic pressure. In this manner, pulse pressure is markedly increased (Fig. 16-7).

Arteriosclerosis
In arteriosclerosis, sclerotic plaques are deposited on the arterial wall, thereby reducing compliance. As a result, damping is reduced and systolic pressure increases. Elastic recoil is also reduced and diastolic pressure falls. Thus, the pulse resembles that of aortic insufficiency (Fig. 16-7). Differentiation between the two conditions would be made by the diastolic murmur of aortic insufficiency.

Aortic Stenosis
In aortic stenosis, the aortic valve becomes distorted and opens incompletely. The resulting increased resistance to cardiac ejection reduces stroke volume and systolic pressure. The aortic pulse displays a characteristic slow, long anacrotic limb with a prominent anacrotic notch. Diastolic pressure is also reduced, but to a lesser extent than systolic pressure, so that pulse pressure is reduced (Fig. 16-7).

Hypertension
In hypertension, pulse pressure is markedly elevated along with mean blood pressure because systolic pressure rises more than diastolic pressure (Fig. 16-7). The rise in diastolic pressure is associated with an increase in total peripheral resistance, and the greater rise in systolic pressure reflects the superposition of a reduced arterial compliance. Hypertension can occur in a number of conditions the nature of which is discussed in Chapter 18A.

Transformation of Pressure Pulse During Transmission
Following cardiac ejection, the arterial pressure pulse is transmitted throughout the arterial system at a velocity that varies with the stiffness of the arterial wall and the

viscosity of the blood. The stiffer the wall and/or the less viscous the blood, the more rapid the pulse transmission. Normally, the pulse velocity in the aorta is 3 to 5 meters per second, as compared to a mean blood velocity of 0.3 meter per second. The pulse velocity increases to 7 to 9 meters per second in the subclavian and femoral arteries and to 15 to 40 meters per second in the small arteries of the extremities. The increased velocity is due to the lower compliance and tapering of the smaller vessels. In hypertension, or arteriosclerosis, pulse wave velocity may increase to three times normal.

As the arterial pressure pulse travels peripherally, it undergoes marked transformation due to pulse wave reflection and damping. Figure 16-8 shows an anatomical sketch of the complex arterial system which exhibits wave transmission characteristics as though it were a simple pipe with reflecting sites at both ends between which the pulse wave is reflected back and forth. In the normal subject, the wave of pressure emanating from the heart travels out to the periphery of the arterial system to the level of the precapillary resistance section, where it is reflected back. The degree of reflection is directly related to the vascular tone. Thus, during maximal vasoconstriction, reflection is maximal, whereas during maximal vasodilation, reflection may be zero. The reflected wave travels up and down the arterial system until it is damped out. The arterial pressure is augmented at the points of interaction with the reflected waves. Figure 16-8C shows the passage of the impulse generated by ventricular ejection back and forth between the upper and lower reflecting sites at a velocity of 6 meters per second. The amplifications represent the summation of incident and reflected waves. The tidal wave represents reflection from the upper, and the dicrotic wave reflection from the lower, reflecting sites. In addition to reflection, damping of the pressure wave by the arterial wall contributes to pulse modification at the level of the femoral artery, where reflection and damping are prominent. The systolic pressure peaks more sharply and at a higher level than occurs at the base of the aorta. The incisura becomes damped and is replaced by the dicrotic notch, and the catacrotic limb of the pulse is amplified to produce the dicrotic wave.

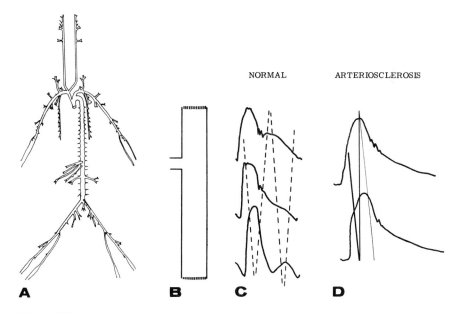

Figure 16-8
The complex arterial tree (A) reflects the pulse wave as though it were a standpipe (B). Transformation of the arterial pressure pulse during peripheral transmission as a result of wave reflection in a normal (C) and a sclerotic (D) arterial system. (After O'Rourke, 1971.)

When the arterial wall is less compliant, as in hypertension or arteriosclerosis, the wave propagation is more rapid (12 to 18 meters per second) and the wave reflection more vigorous to produce an increase in the tidal wave. The increased pulse wave velocity results in increased wave cancellation, which reduces the dicrotic wave. Because of the rapid pulse transmission, the pulses at the base of the aorta and at the femoral artery are similar.

Sphygmomanometry

The use of direct pressure measurements has become somewhat commonplace in the catheterization laboratory. However, routine clinical practice utilizes the simpler indirect method of sphygmomanometry. The basis of this method is illustrated in Figure 16-9. A cuff is placed over the brachial artery and inflated to a level above the expected systolic pressure, thereby occluding the underlying artery and interrupting pulse transmission. The pressure cuff is then slowly deflated at a rate of 2 to 3 mm Hg per beat while one either listens with a stethoscope over the brachial artery for a sound (auscultation) or palpates for a pulse at the radial artery (palpation). As the cuff pressure is lowered, the occluded brachial opens more and more with each pulse to allow blood to spurt through at high velocity. Turbulence is generated and the arterial wall oscillates, producing characteristic sound changes (Korotkoff sounds). The sounds have been divided into several phases, as follows: Phase I, a sharp tapping sound is heard which signals the level of systolic pressure. Phase II, the sound assumes a blowing or swishing quality. Phase III, the sound becomes a soft thud. Phase IV, the sound suddenly becomes softer and develops a muffled quality and then disappears.

Systolic pressure is assigned at the point at which the first sound is heard. There is a great deal of confusion as to when to assign the reading of diastolic pressure. The point of muffling has been recommended by some, whereas others favor the point of silence. Recent studies indicate that the point of muffling provides an overestimate, and the point of silence an underestimate of intravascular diastolic pressure. However, the point of muffling is more consistent. To avoid confusion, diastolic pressure is read at both points. Thus, a normal auscultatory pressure reading would be 120/80/70. The difficulty in performing an accurate indirect pressure estimate concerns the size of the occlusion cuff and the ability to detect the critical sound frequencies generated at the region of occlusion. In 1967 the American Heart Association recommended the use of a cuff 20 percent wider than the limb diameter and containing an inflatable bladder which is placed directly over the artery to be auscultated or palpated and extends halfway around the arm. The recommended bladder appears to be too small and requires excessive inflation to occlude the underlying artery. A bladder that is 30 percent wider and completely circles the limb proves to be more reliable (Steinfeld et al., 1974). When

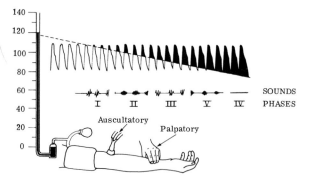

Figure 16-9
Principles of sphygmomanometry illustrating the palpatory and auscultatory techniques for measuring human blood pressure. Included are the Korotkoff sounds and phases.

this is used with an electronic (Doppler) stethoscope, the first sound corresponds to systolic and the silence to diastolic intravascular pressures.

Occasionally, particularly in hypertension, a range of cuff pressure is encountered in which no obvious Korotkoff sound is emitted from the underlying artery. This absence of sound is referred to as the "auscultatory gap." In such cases, it is important either to inflate the cuff to a pressure high enough to exceed the range of the gap or to palpate the radial artery, since the pulse transmission continues within the pressure range of the gap.

Factors That Influence Blood Pressure

Population studies reveal that blood pressure varies with age, sex, weight, race, and socioeconomic status, possibly implicating diet as well as social and economic stresses. In addition, throughout the course of a day, blood pressure varies considerably with physical, mental, and physiological activity, as well as emotional state. In view of this marked variability, diagnoses should not be made on a single blood pressure estimate. Ideally, a series of readings should be made prior to final diagnosis.

Age
Figure 16-10 shows that systolic pressure increases from 110 at age 15 to about 150 to 160 at 65 years of age, or about 1 mm Hg per year. This increased systolic pressure

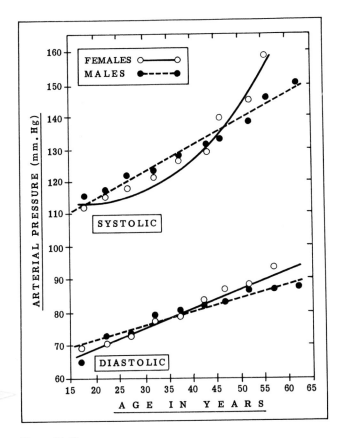

Figure 16-10
Relationship of systolic and diastolic pressures to age and sex. (From J. N. Morris. *Mod. Concepts Cardiovasc. Dis.* 30:635, 1961. By permission of The American Heart Association, Inc.)

probably reflects the progressive reduction in arterial distensibility that accompanies arteriosclerotic vascular changes. Diastolic pressure increases from 70 mm Hg at age 15 to about 90 mm Hg at age 65, or about 0.4 mm Hg per year. The elevation in diastolic pressure probably reflects an increase in total peripheral resistance due possibly either to partial occlusive sclerotic changes in peripheral vessels or to sustained total body autoregulation (Chap. 17) associated with elevated systolic pressure.

Sex
Systolic and diastolic pressures also vary with the sex of the subject. Figure 16-10 shows that they are lower in women under 40 to 50 years of age and higher in women over 50 years than in men. This trend may be due to hormonal changes which take place at menopause.

Weight
Systolic and diastolic pressures are directly related to the weight of the subject. Although pressures in obese people are frequently overestimated because of improper cuff size, this relationship still holds when care is taken to avoid erroneous estimates.

Race and Socioeconomic Status
Systolic and diastolic pressures in blacks are consistently higher than those in whites at all ages and for both sexes. It has been suggested that genetic factors are responsible for this difference. There are, however, data which strongly implicate environmental factors, particularly those associated with the socioeconomic status of the individual. Thus, blacks and whites in rural areas have higher pressures than those in the city. Moreover, blacks and whites in the lower socioeconomic classes have higher pressures than those in higher classes. Furthermore, people living in high-stress areas where social tensions are great routinely have higher pressures. In each case, the blood pressures among the black population are higher than those in the comparable white population. The influence of environment is further exemplified in the high presence of elevated blood pressure in the South, compared to the western mountain states, for whites as well as blacks.

Whether or not these differences are due to genetic, emotional, dietary, or other factors is uncertain. Blacks in higher social and occupational situations have lower pressures than do blacks in lower social and occupational situations, but their pressures are still higher than those of comparable whites. The more affluent blacks may still be subjected to strong emotional stresses.

Posture
When a person stands, gravity acts on venous return to reduce cardiac output and thoracic arterial pressure. Compensatory increases in heart rate and total peripheral resistance cause both systolic and diastolic pressures to rise, but diastolic rises more than does systolic pressure, so that pulse pressure declines.

Exercise
Systolic and diastolic pressures increase with exercise. Systolic pressure increases as a result of increased cardiac contractility. Diastolic pressure may decrease initially because of vasodilation of skeletal muscle vasculature. However, the increased heart rate limits runoff time and causes diastolic pressure to rise.

VENOUS PRESSURE
Pressure in the venous system, as elsewhere in the vascular system, is determined by the volume of blood in the veins relative to their compliance. Venous pressure is low despite a large blood volume because veins are highly compliant. It is this high compliance that accounts for the capacitance function of the venous system.

Venous Pressure Measurement

Venous pressure is quite low compared to arterial pressure and is therefore generally given in centimeters of water, rather than millimeters of mercury. Measurements of low pressures are highly susceptible to the selection of a zero pressure reference point.

In the horizontal position the location of this point is not too critical, since the effect of gravity on the distribution of blood in the cardiovascular system is slight. However, in the standing position gravity acts on the body, with hydrostatic effects on blood pressures.

Guyton et al. (1973) suggest that the point of zero reference, the *physiological reference point,* is located at the base of the tricuspid valve. In a standing position the line passing horizontally through this point is called the *phlebostatic axis* and projects to the surface of the body at the point of intersection of the lateral border of the sternum and the fourth intercostal space. Gauer and Thron (1965), on the other hand, indicate that the location of the reference point—the "hydrostatic indifference point"—depends on the venous distensibility. In the horizontal position it is located 5 cm footward of the diaphragm. In the standing position it moves closer to the lower extremities because of the high venous capacitance in the lower portions of the body. Venous pressure can be measured directly by venopuncture with a needle connected to a suitable manometer. It can also be estimated indirectly by noting the level above the reference point at which engorged veins in a dependent hand collapse as the hand is slowly raised.

The pressure in the thoracic veins poses a special case on account of the negative pressure of the thorax and the cyclical variations associated with respiration. During inspiration, the increasing negative intrathoracic pressure causes the transmural pressure of vessels in the thorax to increase, thereby producing vascular distention which lowers the pressure in the vessel. With passive expiration, intrathoracic pressure increases but still remains negative. The increased pressure is transmitted to the thoracic vessels, and their pressure increases. During forced expiration, intrathoracic pressure can become positive, in which case the transmural pressure of some of the thoracic vessels could become negative, causing collapse.

Effect of Posture

In the recumbent position, gravity acts horizontally on the entire circulatory system so that its influence is minimal. The pressure gradient for venous return is about 15 cm H_2O. The arterial pressure gradient is unaffected (Fig. 16-11). Assumption of the erect position subjects the vascular system to the force of gravity, which acts vertically on the body. According to Pascal's law of hydrostatics, the pressure at the surface of a vertical fluid column is equal to atmospheric pressure. As the point of measurement is moved below the surface, the pressure increases linearly with the height of the fluid column above the point of measurement (Fig. 16-12). If the zero reference pressure, which is situated at the surface, were to be moved down into the fluid column, then all points above it would be at lower pressures and would therefore be negative with respect to the reference point. This is precisely the case in the body. The zero reference point is located either at the base of the tricuspid valve or below the diaphragm. Therefore, the hydrostatic pressures below the reference point are positive, whereas those above the reference point are negative.

The average distance from the level of the right atrium to the foot is approximately 130 cm. Thus, the hydrostatic pressure at the foot is 130 cm H_2O (95 mm Hg). In the reclining position the mean blood pressure in the dorsalis pedis artery is about 105 mm Hg. In the standing position this pressure becomes about 200 mm Hg (105 + 95). The pressure in the corresponding vein would be the sum of the reclining pressure of 20 cm H_2O and the hydrostatic pressure of 100 cm H_2O, or 120 cm H_2O (92 mm Hg). This marked increase in pressure does not affect the perfusion pressure gradient, since arterial and venous pressures are affected equally. Capillary pressure would rise to over 100 mm Hg. However, the capillaries are not endangered by the high pressure because, according to the Laplace relationship, their small radius allows for

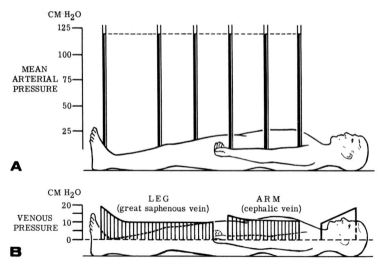

Figure 16-11
Mean arterial and venous pressures in reclining subjects. A. The mean arterial pressure changes little from the aortic arch to the arterial branches (e.g., the radial). B. The peripheral venous pressure diminishes slightly from the periphery toward the heart. (After A. Ochsner, Jr., R. Colp, Jr., and G. E. Burch. *Circulation* 3:674–680, 1951. By permission of The American Heart Association, Inc.)

only a moderate wall tension to develop; moreover, tissue pressure becomes elevated too, although not so rapidly.

The pressure in the vessels above the heart is negative relative to the zero reference point. Thus, the pressure in a cerebral artery situated about 38 cm above the reference point would be about 62 mm Hg (100 − 38) and that in the accompanying cerebral vein would be as low as −28 mm Hg (10 − 38). The negative venous pressure results in a negative transmural pressure, and veins of the head and neck collapse. As blood volume and pressure build up behind the collapsed vessels, they open to allow blood to flow through and then collapse again as pressure falls. Venous flow from the head and neck, then, is intermittent in the erect position. Veins in the *rigid* cranium do not collapse because the hydrostatic influence is equal on all vessels balanced by equal extravascular influence.

Venous Return

Venous return represents the flow of blood from the periphery back to the right atrium. Since the cardiovascular system is a closed tube network, and the system is on the average in a steady state, venous return and cardiac output are equal. In fact, in the short term, venous return substantially determines cardiac output in normal subjects.

Venous return is determined by the pressure gradient for venous flow and the resistance to flow imposed by the large venous system. The pressure gradient is represented by the difference between peripheral venous pressure and right atrial pressure. Peripheral venous pressure is the pressure generated by the flow of blood from the capillaries into the peripheral veins relative to the compliance of the veins. The compliance of the total vascular system, but primarily that of the venules and small veins, relative to the blood volume distending it, is revealed in the *mean systemic pressure* (mean circulatory pressure includes the pulmonary circulation). Mean systemic pressure represents the pressure generated by the total blood volume in the areflexic systemic vascular system when tissue blood flow is zero and the pressure throughout the

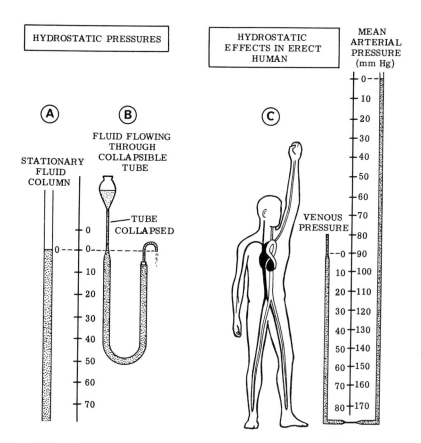

Figure 16-12
The nature and significance of hydrostatic pressures. A. The pressure in a fluid column is dependent upon its specific gravity and the vertical distance from the point of measurement to the meniscus. B. A collapsible tube is distended only so long as the transmural pressure is positive. The transmural pressure is zero in the portion of the tube which is partially collapsed. C. In the erect position, arterial and venous pressures at the ankle are increased by about 85 mm Hg. Above the heart arterial pressure is reduced and effective venous pressure is zero. (After Rushmer, 1970.)

system is equal. Mean systemic pressure varies directly with blood volume and inversely with capacitance of the vascular system. When mean systemic pressure is high, peripheral venous pressure will be high, and thus the pressure head for venous return is determined by mean systemic pressure.

The pressure in the right atrium is normally close to 0 mm Hg. Increasing right atrial pressure above zero, as occurs in conditions of reduced cardiac function, results in reduction of the pressure gradient for venous return. This relationship is illustrated in a venous return curve (Fig. 16-13). As right atrial pressure increases, venous return declines until, at the level where right atrial pressure equals mean systemic pressure, venous return is zero.

Decreasing right atrial pressure below zero would normally be expected to enhance venous return by increasing the pressure gradient. However, venous return does not increase because the negative right atrial pressure is reflected back to the thoracic veins, reducing their transmural pressure; consequently, they collapse. In this situation venous return through the inferior vena cava becomes intermittent.

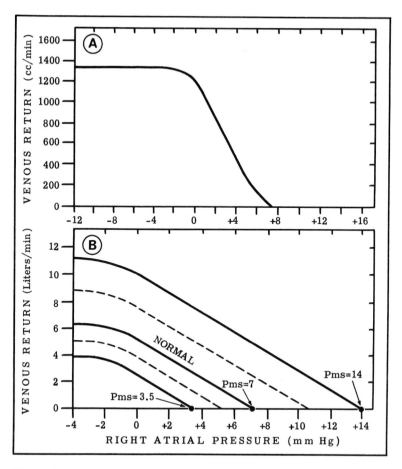

Figure 16-13
A. A normal venous return curve for the dog (from A. C. Guyton et al. *Am. J. Physiol.* 189:609–615, 1957). B. Venous return curve adjusted to values for man, showing the normal curve when mean systemic pressure (Pms) is 7 mm Hg, and showing the effect of altering the mean systemic pressure. (From A. C. Guyton. Circulatory Physiology. In A. C. Guyton (Ed.), *Cardiac Output and Its Regulation.* Philadelphia: Saunders, 1963, P. 201.)

Mechanism for Maintaining Venous Return
Assumption of the erect position increases peripheral venous transmural pressure and thereby results in a shift of blood from the central circulation to the peripheral veins of the lower extremities. This causes venous return to decline, and, in the absence of adequate aids for venous return, syncope would result.

A number of mechanisms exist which act to maintain an adequate flow back to the heart:

Thoracoabdominal Pump. Respiration is an important factor in maintaining venous return. During inspiration, intrathoracic pressure decreases and abdominal pressure increases. The increased negativity in the thorax is transmitted to the right atrium, causing a fall in right atrial pressure. This increases the pressure gradient from the periphery to the right heart, and venous return is increased. The abdominal viscera are part of the peripheral circulation. Increased intra-abdominal pressure during inspiration causes the peripheral venous pressure in this region to increase, further augmenting the pressure gradient for venous return. During expiration the opposite effects occur.

When a person is placed on positive pressure respiration, intrathoracic pressure is elevated to positive values during inspiration and declines to low values during expiration, so that the thoracic pumping action is reversed. Venous return is reduced because in place of the oscillation between high and low negativity in the thorax, which acts to "suck" blood to the heart, there is an oscillation between positivity and low negativity in the thorax that provides a higher average right atrial pressure.

Skeletal Muscle Pump. Another important mechanism for maintaining venous return results from skeletal muscle contraction. The contraction of the skeletal muscles compresses the underlying or adjacent veins, forcing blood from them. The presence of venous valves oriented toward the heart ensures the flow of this displaced blood to the heart, thereby augmenting venous return. The contribution of the muscle pump to the maintenance of venous return and to the maintenance of a reduced peripheral venous pressure is seen during standing. Upon quiet standing, venous pressure in the foot may rise as high as 125 cm H_2O. One single step can translocate enough blood toward the heart to reduce this pressure by 60 to 70 mm Hg. Repeated muscular contraction, such as occurs with walking, can discharge enough blood from the leg veins to maintain this pressure below 40 cm H_2O.

Venomotor Tone. The venous blood reservoir is supplied with smooth muscles that receive sympathetic adrenergic innervation. In conditions in which the cardiac output is reduced, blood pressure will decline, generating a compensatory response through the baroreceptor system (Chap. 17). One component of this compensatory response is an increase in venomotor tone which acts to reduce the capacitance of the venous reservoir, thereby displacing blood toward the right heart.

This mechanism appears to be inoperative during moderate blood volume–vascular capacity discrepancies. A 20 percent reduction in blood volume is required to produce slight venoconstriction in the skin. Hemorrhages greater than 20 percent of the blood volume increase venomotor tone in the skin and muscle circulations.

REFERENCES

Cullen, S. J. Interpretation of blood pressure measurements in anesthesiology. *Anesthesiology* 40:6–12, 1974.

Folkow, B., S. Mellander, and G. Sweden. Veins and venous tone. *Am. Heart J.* 68:397–408, 1964.

Folkow, B., and E. Neil. *Circulation.* New York: Oxford University Press, 1971.

Gauer, O. H., and H. Thron. Postural Changes in the Circulation. In W. F. Hamilton and P. Dow (Eds.), *Handbook of Physiology.* Washington: American Physiological Society, 1965. Section 2: Circulation, vol. 3.

Geddes, L. A. The Direct and Indirect Measurement of Blood Pressure. Chicago: Yearbook Medical Publishers, 1970.

Guyton, A. C. *Textbook of Medical Physiology* (4th ed.). Philadelphia: Saunders, 1971.

Guyton, A. C., C. E. Jones, and T. E. Coleman. *Circulatory Physiology: Cardiac Output and Its Regulation* (2nd ed.). Philadelphia: Saunders, 1973.

Kannel, W. B. Role of blood pressure in cardiovascular morbidity and mortality. *Chest* 65:5–24, 1974.

Kotchen, J. M., T. A. Kotchen, N. C. Schwertman, and L. H. Kuller. Blood pressure distributions of urban adolescents. *Am. J. Epidemiol.* 99:315–324, 1974.

O'Rourke, M. F. The arterial pulse in health and disease. *Am. Heart J.* 82:687–702, 1971.

Rushmer, R. F. *Cardiovascular Dynamics* (3rd ed.). Philadelphia: Saunders, 1970.

Steinfeld, L., H. Alexander, and M. L. Cohen. Updating sphygmomanometry. *Am. J. Cardiol.* 33:107–110, 1974.

17. Control of the Cardiovascular System

Carl F. Rothe and
Julius J. Friedman

If tissue is to receive the oxygen and nutrients required for metabolism and to have waste products removed, the flow of blood through the tissue must be adequate. This flow through each organ is regulated in part by local and in part by central mechanisms which change the vascular resistance to blood flow in accordance with tissue requirements. For the system to be effective, the systemic arterial pressure should be held reasonably constant by homeostatic mechanisms. With a constant blood pressure, the flow of blood through an organ or tissue is inversely related to its vascular resistance (see Chaps. 11 and 12 and Fig. 17-1). To maintain constant systemic blood pressure as the flow through a tissue increases, either the cardiac output must be increased or the flow through some other region of the body must be reduced. Effective treatment of hypertension and many other cardiovascular diseases requires a knowledge of the mechanisms controlling blood pressure.

Figure 17-1 indicates the sequence of cause and effect of the cardiovascular factors determining tissue perfusion. (For simplification, no attempt was made to indicate the many control loops.) Tissue perfusion may be inadequate because of either an inadequate perfusion pressure gradient or excessive resistance due to vasoconstriction (Fig. 17-1). The arterial blood pressure may be inadequate because the cardiac output is low or there is a generalized reduction in total peripheral resistance due to vasodilation. The cardiac output, in turn, is dependent upon stroke volume and heart rate. Stroke volume is dependent upon many factors (Chap. 15), including the effective filling pressure of the right heart—its transmural pressure. Filling pressure is also dependent upon many factors (Chaps. 15 and 16), including the relationship between vascular blood volume and vascular capacitance. Control of blood volume is complex (Chaps. 16, 23, 30). Although the cardiovascular system will function without autonomic system activity, optimal operation, especially during stress, is possible only by the modulation and neurogenic stimulation of the heart and vascular smooth muscle provided by the autonomic nervous system as part of a control system. The effectors for the control of the cardiovascular system are in large measure dependent upon the vigor of contraction of the heart muscle and the level of smooth muscle activity of the arteries and veins.

The organization of our understanding into "simplified" mathematical models has become exceedingly complicated because of the wide variety and closeness of interrelationships of the system (see, for example, Guyton et al., 1972). Such models, however, provide clues to relationships that had not been noted before, suggest experiments to test hypotheses, and are instrumental in helping us understand disease processes. Working with one section at a time in detail or studying the integrated effects of many such blocks together leads to progress.

The homeostatic control of blood pressure involves the integrated activity of the cardiovascular system acting over the short term (minutes), in conjunction with the systems concerned with body fluid and electrolyte balance acting over the long term (days). The nature of body fluid and electrolyte balance and its influence on blood pressure are discussed in Chapters 18A, 22, and 23.

HOMEOSTATIC CONTROL OF BLOOD PRESSURE

The control of cardiovascular activity to maintain constancy of blood pressure is a classic example of a homeostatic, negative feedback system (Chap. 7B). The *effector mechanism* is dual, in that blood pressure is increased by an increase in either cardiac

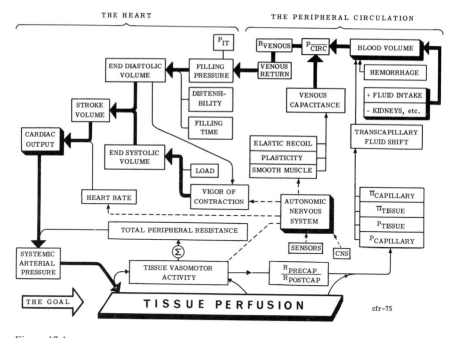

THE HEART   THE PERIPHERAL CIRCULATION

Figure 17-1

Factors influencing cardiovascular function. The sensors influencing the autonomic nervous system include the arterial pressoreceptors, the chemoreceptors, and the atrial pressoreceptors. P_{it} is the intrathoracic pressure. The venous return is in liters per minute. R_{venous} is resistance to venous return. P_{circ} is mean circulatory pressure (the pressure in the small veins). The ratio of vascular resistance (R) between the precapillary (PRECAP) and postcapillary (POSTCAP) segments of the tissue, in conjunction with the rate of blood flow and venous pressure, determines the capillary pressure. The hydrostatic (P) and oncotic (π) pressures influence the rate of transcapillary fluid shifts. The passive elastic recoil of the veins, and so the tendency to redistribute blood from the periphery to the heart, is dependent on the venous transmural pressure, which, in turn, is dependent upon flow and venous resistance.

output or peripheral resistance. The blood pressure is sensed by special organs called *pressoreceptors* (baroreceptors), located primarily in the carotid sinuses and the arch of the aorta. Impulses from these receptors are transmitted via afferent sensory nerves to the *cardiovascular centers,* located in the brain stem. After complex interactions, signals, acting to correct deviations in blood pressure, pass by way of the efferent nerves (the autonomic nervous system) to the peripheral vasculature and to the heart.

The *arterial pressoreceptor reflex* controlling the systemic arterial blood pressure is described schematically in Figure 17-2. An increase in blood pressure stretches the pressoreceptors in the walls of certain arteries, stimulating them to increase the frequency of discharge along afferent nerves. This acts by way of the cardiovascular centers to *inhibit* sympathetic division action on the heart and peripheral vasculature and to *increase* parasympathetic division action on the heart by way of the vagus nerves. Cardiac contractility and heart rate decrease, and the degree of vasoconstrictor tone on both the resistance and capacity vessels is reduced. Thus the tetralogy of reduced cardiac vigor, bradycardia, vasodilation, and venodilation tends to follow an increase in blood pressure. The combination of these responses results in a decrease in blood pressure. This is a negative feedback response, since the reduction in blood pressure is elicited by an increase above normal.

The arterial pressoreceptor reflex is also important in counteracting a decline in blood pressure. When pressure declines, fewer afferent impulses from the pressoreceptors go to the cardiovascular centers. Vagus tone is decreased, and there is less inhibition of the sympathetic division. As a result, cardiac action and vasoconstriction increase. Other names for the arterial pressoreceptor reflex include: carotid sinus reflex, baroreceptor reflex, buffer reflex, and depressor reflex.

The proportion of change of the four effector mechanisms (heart rate, contractility, vasoconstriction, venoconstriction) is not the same under various stresses (see, for example, Korner, 1971 and 1974). With arterial pressure deviations from normal, heart rate and muscle vascular resistance changes predominate. In the conscious dog, carotid sinus nerve stimulation causes bradycardia and vasodilation, especially in the limbs, but little change in kidney or skin vascular resistance or in cardiac output. Because cardiac output is determined by many factors (Fig. 17-1 and Chap. 15), a negligible change in cardiac output does not necessarily mean no change in cardiac vigor of contraction. Indeed, cardiac contractility and venoconstriction (see below) probably also decrease with carotid sinus stimulation (increased arterial blood pressure), but the degree in the normal individual is not clear. Reflex vasoconstriction, of whatever cause, is not uniform throughout the body (Korner, 1974). For example, blood flow through the skin may be reduced with little change in other organs. It is often nonproportional and nonlinear. Under some stressful conditions splanchnic blood flow may be reduced by vaso-constriction with but little change in renal blood flow. With further stress, renal perfusion may be greatly decreased.

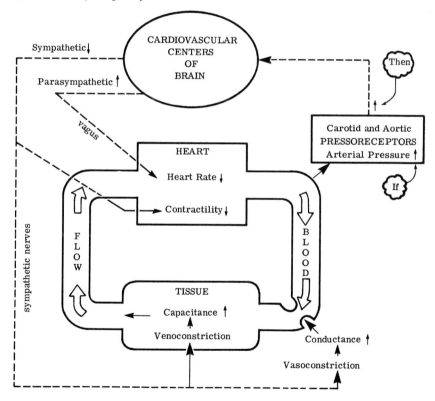

Figure 17-2
Control of systemic arterial blood pressure by the arterial pressoreceptor reflex. Starting the loop at the pressoreceptors, *if* there is an increase in arterial pressure, *then* the firing rate of the presso-receptors increases, with consequences as indicated.

CARDIOVASCULAR CONTROL CENTERS

The neural centers for the control of the cardiovascular system are complex and not completely defined. They are *not* discrete structures but tend to be scattered and intermixed within the neural tissue. The concept of "centers," however, aids in the understanding of the regulatory process. These groups of neurons integrate neural impulses both from peripheral sensors and from the higher brain centers.

Korner (1971) and Sagawa et al. (1974) have provided detailed reviews of the complex field of neural circulatory control. Smith's review (1974) emphasizes the importance of both the central nervous system and spinal centers in the control of cardiovascular function.

Medullary Control

Neurons in the medulla oblongata are responsible for the integration of afferent impulses and the origination of efferent impulses for the homeostatic control of blood pressure. *Vasoconstrictor centers* in the medulla oblongata are responsible for the neurogenic component of basal vasoconstrictor tone. *Cardiostimulator centers* increase cardiac activity. Normally, however, the cardiostimulator centers are relatively quiescent. Electrical stimulation of these *pressor areas* located in the lateral reticular formation causes an increase in blood pressure by increasing vasoconstriction, heart rate, and probably cardiac vigor. The vasoconstrictor neurons can discharge without afferent stimuli, for vasomotor discharge does not cease and the blood pressure may not decrease even if incoming impulses from peripheral pressoreceptors are stopped. There is a continuous "tonic" activity of neuronal discharge *modified* by impulses from the depressor area, chemoreceptors, and higher centers. The medullary respiratory center neurons are intermingled with those of the pressor area. There are not only similarities in cardiovascular and respiratory response to varying conditions but also interactions between the systems both under normal conditions and following various stresses. Arterial pressoreceptor discharge inhibits not only cardiac and vascular activity but also respiration. The respiratory system may influence the cardiovascular system, as seen by fluctuations in arterial blood pressure and a complex pattern of heart rate changes (*sinus arrhythmia*) related to respiratory activity (see below).

Cardioinhibitor and *vasodepressor* areas are located in the caudal and medial reticular formation of the medulla in association with the dorsal nucleus of the vagus. Stimulation of these areas results in a reduction in arterial blood pressure. The response is produced by inhibition of the constrictor tone—not activation of vasodilator fibers—and by vagal slowing of the heart. Like the vasoconstrictor centers, the cardioinhibitor centers are tonically active under basal conditions; cutting the vagus nerves releases the inhibitory tone, and the heart rate increases.

Hypothalamic Control

The hypothalamus, being a generalized center of control of the autonomic nervous system, modifies the activity of the bulbar region. There is probably no tonic activity coming from this region, because section at the level of the pons does not necessarily reduce blood pressure, and the baroreceptor reflex is little influenced by decerebration. However, the suprabulbar centers are of great importance in reflex cardiovascular control (Korner, 1971; Smith, 1974). The cardiovascular adjustments to emotions such as rage are mediated here. Redistribution of blood flow and characteristic patterns of cardiovascular response to exercise are integrated at this level. For example, stimulation of certain highly discrete areas results in a cardiovascular response closely similar to that seen during exercise, including changes in heart rate, blood pressure, cardiac action, and peripheral vascular tone. It has been possible to stimulate sympathetic fibers at this level that influence a particular organ with minimal effects in other organs; the neurogenic outflow, though generally increased, is nonuniformly distributed to all organs, and further, the receptivity of the organs differs.

The hypothalamus is involved in the distribution of blood flow for the control of body temperature. Lesions in this region impair the ability of an animal to protect itself from extreme environmental temperature, because sweating and cutaneous vasodilation or vasoconstriction no longer occur in response to the appropriate conditions. Temperature control is primarily by way of the sympathetic division, since the parasympathetic division has little direct effect except for that on heart rate.

Cerebral Cortical Control
Impulses affecting the cardiovascular system originate in the forebrain (see Smith, 1974). Knowledge of these may have an important bearing on the understanding of psychosomatic medicine. The sympathetic vasodilator outflow to skeletal muscle apparently originates in the cerebral cortex and passes through the hypothalamus and medulla, where the efferent discharge pattern may be modified. Electrical stimulation of the cortex may evoke both autonomic and somatic effects, which are usually anatomically and functionally related.

Spinal Control
Control centers in the spinal cord are apparently of minor importance for the maintenance of an adequate circulation under normal conditions. In cases of circulatory depression the cord is of greater importance. Transection of the thoracic spinal cord causes an immediate and profound fall in blood pressure. However, after a time a certain degree of control of the blood pressure may be regained. Once the blood pressure in a spinal animal has returned to near normal levels, total destruction of the cord results in a permanent reduction in pressure. Thus, there are neurons in the cord responding to pressoreceptor impulses or reduced blood flow and hypoxia which initiate impulses along vasoconstrictor fibers. Stimulation of cutaneous receptors by pain or cold induces segmentally arranged vasoconstriction of the intestinal vessels of spinal animals, a phenomenon indicative of the action of spinal centers on the cardiovascular system.

CARDIOVASCULAR SENSORS AND REFLEX CONTROL OF THE CARDIOVASCULAR SYSTEM
Effective cardiovascular control depends on information provided by sense organs which transmit information to the control centers in the brain.

Stimulation of the various cardiovascular and pulmonary *mechanoreceptors* (stretch, tension, pressure) leads to *reflex inhibition* of the activity of circulatory and respiratory systems. This is a first approximation, but exceptions are few.

Arterial Pressure Receptors
The sensors for the control of the systemic blood pressure by the arterial pressoreceptor reflex are sensitive nerve endings that respond to stretching of the walls of arteries as the transmural pressure is increased. The pressoreceptors are located not only in the carotid sinuses and arch of the aorta but also along the common carotid arteries. The rate of firing of the pressoreceptors at various pressures has been measured (e.g., Fig. 17-3). The carotid artery receptors are effectively quiescent below pressures of about 60 mm Hg, although some may be firing at pressures as low as 30 mm Hg. Furthermore, at threshold there is a minimal firing rate of the order of 20 impulses per second. As the transmural pressure is increased beyond 60 mm Hg, the frequency increases progressively. The *change* in impulse frequency per millimeter mercury pressure change is maximum at about normal blood pressure—the receptors are most sensitive and give the highest gain at this pressure. A plateau is reached at about 160 mm Hg. Additional increases in pressure do not appreciably increase the rate of receptor discharge. Consequently, when the blood pressures decrease below about 60 mm Hg or increase above about 160 mm Hg, little further compensatory response is elicited by this reflex system. The threshold for the aortic pressoreceptors in some species is much higher than that for the carotid receptors, suggesting that their function is primarily to limit

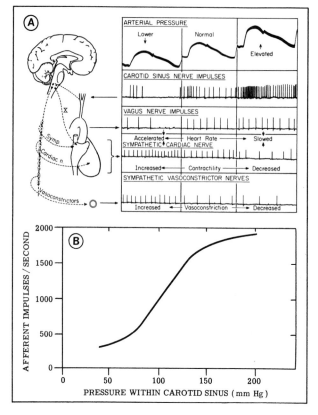

Figure 17-3
Neural relationships of the arterial pressoreceptor reflex. A. Effect of arterial pressure pattern on carotid sinus stretch receptor discharge, and impulse rate of efferent nerves. B. Rate of discharge of afferent nerves at various static sinus pressures. (A from Rushmer, 1970, p. 165. B after W. Kalkoff. *Verh. Dtsch. Ges. Kreislaufforsch.* 23:399, 1957.)

excessively high cardiac and arterial pressures, rather than to elicit reflexes to restore a low blood pressure.

The frequency of discharge of the pressoreceptors is also determined by the *rate* of stretch of the receptors as well as by the average magnitude. Thus, there are bursts of activity with each pulse (Fig. 17-3A). In this manner, the rate of change of pressure (and so the *pulse pressure*) determines, in part, the level of sympathetic division influence on the heart and peripheral vasculature.

The arterial pressoreceptors are not required to maintain a near normal blood pressure and heart rate (Cowley et al., 1973; Sagawa et al., 1974). Although they fire during each heart beat at normal blood pressures and show little adaptation over a period of a few hours, after complete carotid sinus and aortic arch denervation, the blood pressure returns to within about 10 mm Hg of normal in a few days. In contrast to the normal animal, the blood pressure becomes highly variable with activity, but the heart rate is relatively constant.

The role of the pressoreceptors in hypertension is not yet clear. The frequency of impulses is lower than normal in some forms of this disease. However, current evidence suggests that the adaptation of the pressoreceptors over a period of days and weeks is such that the response curve is simply shifted to a higher pressure (to the right, Fig. 17-3B). In most forms of hypertension the pressoreceptors neither attenuate the disease nor cause it (see Chap. 18A).

The surgeon must be cognizant of these receptors, for stimulation of the carotid sinus by probing or pulling can result in profound and sometimes fatal hypotension. Some people have a very sensitive carotid sinus reflex. A tight collar or turning of the neck can stimulate the receptors sufficiently to elicit a response of such magnitude that the blood pressure drops to levels producing cerebral hypoxia and loss of consciousness (syncope).

There are other arterial pressoreceptors located along the thoracic and mesenteric arteries. Their importance in man is not known. Under normal conditions the effects are minimal in the dog. In the cat, however, occlusion of a mesenteric artery results in a marked increase in systemic arterial pressure.

Chemoreceptors

The peripheral chemoreceptors are specialized nerve endings in the carotid and aortic bodies found near the carotid sinus and also in the arch of the aorta. These nerve endings are sensitive to a decrease in oxygen tension, an increase in carbon dioxide tension, or an increase in the hydrogen ion of the blood. Poisons that inhibit oxidative pathways, such as cyanide or fluoride, stimulate the chemoreceptors. Chemoreceptor stimulation causes an increase in pulmonary ventilation (the carotid body is much more important than the aortic body in man) and an increase in blood pressure by peripheral vasoconstriction. Hypoxia generally has little effect on heart rate, in part because peripheral hypoxia causes a slowing of the heart and central (medullary) chemoreceptor activity causes a speeding of the heart.

The carotid body with its glomus cells is one of the most vascular tissues of the body. The tissues weigh only about 2 mg in man, but the blood flow through them is about 20 ml per minute per gram of tissue—4 times that through the thyroid, 30 times that through the brain, and about 200 times that through the body as a whole. In addition, they consume oxygen at the rate of about 90 ml per minute per kilogram of tissue, about three times the rate of the brain. The chemoreceptors respond to decreases in oxygen tension and somewhat to decreases in oxygen content (Paintal, 1973). The high blood flow through this tissue is apparently necessary to supply the metabolic needs of the chemoreceptor cells, for a decrease in blood flow through the carotid and aortic bodies, caused by decreased arterial blood pressure, results in chemoreceptor stimulation. Simultaneous arterial hypoxemia and hypotension are synergistic and thus produce a vigorous response. Although the peripheral arterial chemoreceptors show little activity at normal arterial oxygen and carbon dioxide tensions, they are sensitive to changes from normal. A few minutes after circulatory arrest, however, aortic chemoreceptor activity rapidly declines (Paintal, 1973). If the arterial oxygen availability is very low, their response fails.

The *diving response* is a complex reflex induced by submersion of the nostrils or even by the expectation of diving under water (see, for example, Folkow and Neil, 1971). Extreme bradycardia and peripheral vasoconstriction follow in animals, such as the seal, whale, or duck. The cardiac output is only a small fraction of normal. For skeletal muscle activity oxygen from myoglobin stores and then anaerobic metabolism are utilized. In humans, the response seems to occur in the infant more than the older individual. It is found in expert divers who, on surfacing, experience rapid reversal of the cardiac inhibition and vasoconstriction, leading to a transiently high cardiac output.

Carotid Sinus Reflex

Occlusion of the carotid arteries on the cardiac side of the pressoreceptors and chemoreceptors produces a carotid sinus reflex. The increase in sympathetic division activity follows from both a reduction of stimulation of the arterial pressoreceptors, because the blood pressure within the sinuses is reduced, and a stimulation of chemoreceptors, because the supply of oxygen to these tissues is decreased. These two receptor systems thus help assure an adequate flow of well-oxygenated blood to the central nervous system. Both sets of receptors are important for the homeostatic compensation following hemorrhage. Incidentally, because of cerebral anastomoses and the vertebral arterial supply, the pressure in the sinuses is not reduced to zero following bilateral carotid occlusions.

Cerebral Ischemia

If the flow of blood through the brain is severely restricted, there is a delayed but profound sympathetic division response resulting in intense peripheral vasoconstriction. This ischemic response acts at blood pressures below those which have no *more* effect by way of the arterial pressoreceptor reflex. The brain becomes hypoxic and hypercapnic below blood pressures of about 50 mm Hg. At these perfusion pressures the chemoreceptors along the carotid arteries are of course also stimulated by the stagnant anoxia. In response, there is a marked elevation of systemic blood pressure to as high as 350 mm Hg. The resulting degree of sympathetic vasoconstrictor discharge gives renal vasoconstriction of such a magnitude that urine flow stops. The mechanism of action probably results from an increase in carbon dioxide and a decrease in oxygen tension in the medullary region of the brain. The removal of carbon dioxide is limited by the reduced flow of blood through the brain. Therefore, carbon dioxide accumulates and acts not only on the cardiovascular centers, to cause vasoconstriction, but also on the respiratory centers, to cause a vigorous increase in respiratory activity. (Carbon dioxide apparently produces the effect by changing cerebral intracellular pH.)

The *Cushing reflex* is a corollary of the cerebral ischemic response. If the cerebrospinal fluid pressure is greater than the systemic arterial blood pressure, the flow of blood to the brain is stopped, because the fluid pressure acting in the rigid skull occludes the arteries. Under these conditions the central ischemic response produces an increase in blood pressure by peripheral vasoconstriction. This tends to restore cerebral blood flow. A very high blood pressure following head injury or cerebral vascular accident indicates cerebral hemorrhage and the operation of this mechanism.

With severely depressed cardiovascular function or serious metabolic acidosis, rhythmic variations in arterial blood pressure may be seen. These fluctuations, called *Mayer waves*, have a period of 15 to 60 seconds and are independent of respiration. As the blood pressure declines, central and/or peripheral chemoreceptors excite reflex vasomotor activity, which acts to increase the blood pressure. Chemoreceptor activity then declines, followed by a decrease in blood pressure. The response is oscillatory because of lags in the system. These Mayer waves and Cheyne-Stokes respiration are probably derived from similar mechanisms.

Pulmonary and Cardiac Receptors

Atrial receptors are of two types: Type B receptors respond primarily to atrial volume, and type A receptors respond to contractile tension. The reflex effects of the type A fibers are still confused (Paintal, 1973).

Right atrial and *central vein receptors* respond to distention. If vagal tone is high, giving a relatively slow heart rate, distention of these areas leads to an increased heart rate *(Bainbridge reflex)*—an exception to the generalization that mechanoreceptor stimulation inhibits the cardiovascular system.

Left atrial volume receptors (type B fibers) respond to increased transmural pressure, such as occurs from an increased left atrial volume or a more negative intrathoracic pressure. Impulses transmitted to the brain act via the posterior pituitary to *reduce* antidiuretic hormone (ADH, vasopressin) secretions. An increase in urine flow follows, which reduces blood volume (see Chaps. 23 and 31). Reflex hypotension and bradycardia are also sometimes seen following left atrial distention. In response to hemorrhage and a reduction of left atrial pressure, antidiuretic hormone is released from the hypothalamus to induce water retention. The amounts of vasopressin secreted into the bloodstream are such that there may be a direct pressor effect of importance in the normal control of arterial blood pressure. These receptors may be more important than the arterial pressoreceptors in maintaining cardiovascular homeostasis over the long term following minor hemorrhage (Sagawa et al., 1974).

Ventricular mechanoreceptors respond to ventricular wall tension and distention. There are also receptors (possibly the same nerve endings) that respond to ischemia and drugs such as *Veratrum* alkaloids. Ischemia may sensitize the receptors to distention. The afferent sensory fibers traverse both the vagal and sympathetic tracts. In contrast to the atrial receptors, the ventricular receptors are apparently active only under extreme conditions. The classic response to stimulation is cardiovascular inhibition (bradycardia and hypotension—the Bezold-Jarisch reflex) that tends to protect the ventricle from overload. However, some studies have demonstrated cardiovascular stimulation with cardiac receptor stimulation.

Pulmonary arterial pressoreceptors affect the cardiovascular system in a manner similar to that of the systemic arterial pressoreceptors, but to a much smaller degree; for example, an increased pulmonary arterial pressure induces bradycardia, hypotension, and hypopnea.

Pulmonary congestion, pulmonary edema, pulmonary emboli, and strong irritants stimulate type J (nociceptive) pulmonary alveolar receptors (see Paintal, 1973), giving rise to a sensation of dyspnea, an inhibition of somatic muscle activity, hypotension, and bradycardia. These receptors may be important during maximal exercise or in congestive heart failure.

Sinus arrhythmia is the association of heart rate changes with respiration and is seen in young, normal individuals. With inspiration there is a transient increase in heart rate (tachycardia) followed by bradycardia (slowing). Several mechanisms may be involved: (1) The respiratory center neurons may also act on the cardioexcitatory neurons. (2) The lung stretch receptor afferents not only act to inhibit inspiratory effort but also stimulate the cardioinhibitory (vagal) neurons to slow the heart. (Paintal, 1973, concludes that the pulmonary stretch receptors act to increase the heart rate.) (3) With inspiration and consequent decrease in intrathoracic pressure, the left heart pumps from a lower pressure and the result is some decrease in the systemic arterial pressure. This causes a baroreceptor reflex increase in heart rate. Furthermore, (4) with inspiration there is at first increased filling of the right (Fig. 13-10) and then the left side of the heart, which stimulates the low-pressure receptors. (5) Later, the increased filling results in increased cardiac output (Chap. 15) and so increased arterial blood pressure with consequent stimulation of the arterial pressoreceptors and bradycardia. Because these mechanisms require a relatively fixed time, as the respiration rate increases, the mechanisms overlap. Moreover, expiration tends to elicit a similar heart rate response—transient tachycardia, then bradycardia.

Almost all *sensory nerves* have some connection with the cardiovascular reflex system. Some of the reflexes involve centers in the spinal cord only, since vasoconstrictor and vasodilator reflexes can be elicited in animals with sectioned spinal cords. Generally, a painful stimulus is followed by a rise in blood pressure. The conscious realization of pain, causing anxiety and stimulation of the adrenal medullae, also adds to the response. On the other hand, severe cutaneous pain, painful stimulation of the gastrointestinal or genital tracts, stretching of hollow organs, or the stimulation of deep visceral pain receptors may elicit the opposite response: a fall in blood pressure. However, urinary bladder distention in humans with sectioned spinal cords can cause an increase in systolic arterial pressure to over 300 mm Hg. Emotional stress may be followed by fainting. This is called *vasovagal syncope,* implying a vasodilation of the resistance and possibly the capacitance vessels of the body in addition to a vagal slowing of the heart. Cerebral hypoxia occurs as the blood pressure declines, leading to a loss of consciousness.

Response Time

By recording both receptor activity (e.g., nerve impulse from pressoreceptors) and effector activity (e.g., blood flow through organs such as the kidney or skeletal muscle, heart rate, venoconstriction) we now have a better, but far from complete, understanding of the time course of the various components. The pressoreceptors have a response time of the order of 0.2 second, and within 0.1 second the information is seen in the efferent fibers of both the sympathetic and parasympathetic neurons. Whereas heart rate then changes within less than a second (one heart beat), little change is seen in smooth muscle activity of the resistance vessels for about 3 seconds and in the capacity vessels for about 7 seconds. The complete, steady-state response to a step-change in receptor activity requires at least 20 seconds and often 2 minutes or more. The arterial chemoreceptors also respond rapidly (within less than 1 second) to changes in arterial blood composition, but the final response is delayed depending on the effector system involved. The description of the time course of these reflex responses is complicated by the fact that they are nonlinear (the magnitude of response to an input of a given magnitude is different depending on the initial input level) and they are not symmetrical (the response to an input may vary in time or magnitude depending on whether the receptor input was increased or decreased).

NEURAL CONTROL OF CARDIAC OUTPUT

The primary determinants of cardiac output at a given pressure load are the heart rate, filling pressure, distensibility, and contractility (Chap. 15 and Fig. 17-1). *Neural impulses* from the cardioregulatory centers in the lower brain act to modify these variables.

The *heart rate* is determined primarily by the balance between the inhibitory effects on the pacemaker of acetylcholine released by the vagus nerves of the parasympathetic division, and the excitatory effects of norepinephrine released by the sympathetic nerve endings (Chaps. 7A and 14). The sympathetic and parasympathetic divisions tend to be antagonistic, since an increase in heart rate, caused by an increase in sympathetic division activity, can be reduced to normal by adequate stimulation of the vagus nerves. In fact, massive vagal stimulation stops the heart for many seconds. In the normal resting individual, a continuous vagal tone reduces the heart rate to about 70 beats per minute. The sympathetic outflow is low. The intrinsic heart rate of the adult with all autonomic input blocked is about 100 beats per minute. Blocking of parasympathetic activity with atropine or sectioning of the vagi results in a marked increase in heart rate. During exercise or anesthesia, on the other hand, the sympathetic accelerator fibers exert a stimulating influence on the heart. Sympathectomy by severing the sympathetic outflow in the thorax from T2 to T5 (see Fig. 7A-1) in man is usually followed by a reduced cardiac response to exertion. In dogs with denervated hearts, the cardiac response to moderate exercise is slowed but little reduced—in part because of circulating catecholamines (see below). Maximal exertion is limited, however.

Myocardial contractility is increased by an increase in the rate of discharge of the sympathetic division nerves going to the heart. An increase in myocardial contractility results in an increase in cardiac vigor: The pressure developed by the ventricles increases; the size of the heart, particularly at the end of systole, tends to decrease; and the rates of change of pressure and size increase. The parasympathetic division has little effect upon ventricular contractility, in contrast to its marked effect on heart rate. The atria are richly innervated by the parasympathetic nerve endings, and there appears to be a decrease in atrial contractility as a result of vagal discharge.

Cardiac distensibility is apparently little affected by the autonomic nervous system, although some experiments indicate that sympathetic division activity may increase ventricular distensibility somewhat and so permit easier filling. The rate of relaxation, like the rate of contraction, is increased by positive inotropes.

The *filling pressure* of the right heart is determined primarily by the peripheral factors affecting the return of venous blood and the degree of negative intrathoracic pressure (Chap. 16). The degree of constriction of the venous capacitance vessels is important (see below). The filling of the left ventricle is determined primarily by the action of the right heart, the characteristics of the pulmonary bed, and the left atrium.

Although sympathetic stimulation of the heart appears to be generalized, the discrete regional distribution of sympathetic nerves (Randall et al., 1972) would permit highly localized control—i.e., increased contractility of the left ventricle with little change in the right.

Adrenal Medullary Influence

Catecholamines, released from the adrenal medullae, provide a relatively small, slow-acting, but effective adjunct to the autonomic innervation of the heart. Cardiac output is thereby augmented by means of an increased myocardial vigor and heart rate.

PERIPHERAL CIRCULATION

The peripheral circulation may be considered to be that segment of the circulatory system which is concerned with transport of blood, blood flow distribution, exchange between blood and tissue, and storage of blood.

The various tissue circulations comprising the peripheral circulation are arranged in parallel, except, for example, the portal components of the hepatic circulation. Thus the distribution of the cardiac output to these tissues is primarily determined by the relative resistances of their vascular beds.

The maintenance of normal tissue activity as well as specialized tissue function requires an adequate supply of oxygen and other nutrients. The amount of oxygen available to a tissue is determined by tissue blood flow and the arterial oxygen content. Figure 17-4 illustrates the blood flow and oxygen availability and utilization for the major circulations of the body. The area of the rectangles (blood flow and oxygen content) represent the amount of oxygen made available to the tissues each minute. The crosshatched area represents the quantity of oxygen consumed by the tissue, the number within the crosshatched segment indicating the milliliters of oxygen consumed per minute. Thus, the remaining area of the rectangle represents an oxygen reserve for the tissue.

The level of tissue metabolic activity substantially determines the amount of oxygen consumed. Increased metabolic activity is associated with increased oxygen consumption and is reflected in either a higher blood flow or lower venous oxygen content or both. Except for the heart, brain, and kidney, tissues generally possess a large *effective* reserve of oxygen in the venous blood and can therefore endure modest reductions in blood flow without suffering a severe oxygen deficit. In case of reduced flow more oxygen is extracted from the blood, venous oxygen content decreases, and the arteriovenous oxygen difference increases. The cerebral venous blood normally possesses about 13 ml of oxygen per 100 ml of blood. This does not provide a very adequate reserve relative to the rate of brain metabolism, and even small reductions in blood flow can produce impaired function. Although renal venous oxygen content is high, intrarenal blood flow distribution is such that oxygen extraction does not change appreciably with changing blood flow. Thus, oxygen consumption declines and tissue function suffers with reduced blood flow. Neither blood flow distribution nor oxygen consumption is determined by organ size. The kidneys, which represent 0.5 percent of the body weight, receive 20 percent of the cardiac output, whereas *resting* muscle, which accounts for approximately 40 percent of the body weight, receives only 20 percent of the cardiac output. Tissue blood flow is determined by overall tissue function, which includes normal tissue maintenance as well as specialized tissue function. In this regard, kidney and skin receive blood flow in excess of their oxygen requirement and thus have a low arteriovenous oxygen difference. These flows are related to the respective functions of waste clearance and thermoregulation. The coronary and cerebral circulations, on the other hand, receive relatively high blood flows because of their high metabolic rate and oxygen consumptions. This is reflected in a relatively high arteriovenous oxygen difference.

The parallel architectural arrangement provides a means whereby blood flow to a tissue can vary independently of changes in blood pressure. Figure 17-5 illustrates the

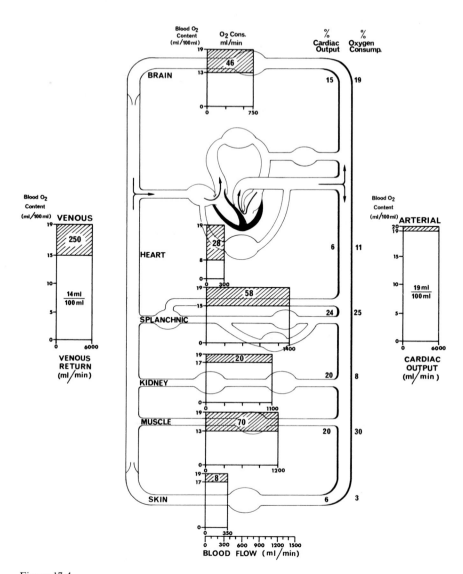

Figure 17-4
Blood flow and oxygen consumption of 70-kg man. The areas of the rectangles represent the amount of oxygen available to the tissues. The crosshatched segments identify the amount utilized by the tissues. The numbers within the crosshatched areas are the milliliters of O_2 consumed per minute. The remaining areas of the rectangles are the O_2 reserves of the tissues. (After Rushmer, 1970.)

nature of the redistribution of the cardiac output to, and the extent of change in oxygen consumption of, skeletal muscle during graded exercise in an average subject and in an athlete. With increasing work of exercise the cardiac output and total body oxygen consumption increase with a progressively greater percentage of the cardiac output distributed to, and a greater percentage of total oxygen consumption accounted for by, the exercising muscle. The figure also illustrates the marked increase in capacity and utilization of oxygen which results from training.

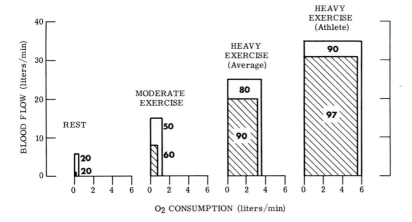

Figure 17-5
The changes in muscle blood flow and O_2 consumption during graded exercise. It appears that the athlete in heavy exercise is a heart-lung-muscle organism. The upper numbers represent percent of cardiac output and the lower numbers the percent of total body oxygen consumption.

The flow of blood through tissues is determined by the arteriovenous pressure gradient, the perfusion pressure, and the resistance to the flow of blood (Chap. 11). Since the pressure gradient, blood viscosity, and length of tissue resistance vessels are relatively constant, the caliber of the resistance vessels is the primary determinant of local blood flow.

The flow of blood through tissues is controlled in part by *central* influences manifested by neural and by humoral mechanisms and in part by *local* mechanisms such as oxygen tension, metabolites, intrinsic reflexes, and autoregulation.

Central and local mechanisms continuously interact to modify the distribution of blood flow. Under some conditions, such as voluntary contraction of muscles, the local and central mechanisms tend to act in the same direction to increase flow. When systemic requirements are great, as occurs during inadequate cardiac function, the central mechanisms will predominate over the local mechanisms for tissues such as skeletal muscle. Conversely, when local factors predominate, the local mechanism may override the central influence.

CENTRAL CONTROL OF THE VASCULATURE
Central control of the peripheral circulation is effected through the activity of the sympathetic division of the autonomic nervous system and through the action of adrenomedullary humoral agents.

Sympathetic Adrenergic Function

Vascular Resistance
The primary mechanism of central control of tissue blood flow is provided by the discharge of *sympathetic vasoconstrictor fibers* causing small artery and arteriolar smooth muscle contraction, thereby decreasing lumen diameter and so increasing resistance to flow. Figure 17-6, curve 1, illustrates the effect of sympathetic adrenergic stimulation on skeletal muscle flow at constant perfusion pressures. These vasoconstrictor fibers apparently provide the *only* neural pathway to the vasculature for the *control* of systemic arterial blood pressure. The resting frequency of nerve fiber discharge to skeletal muscle is about 1 to 3 impulses per second. This provides a basal level of

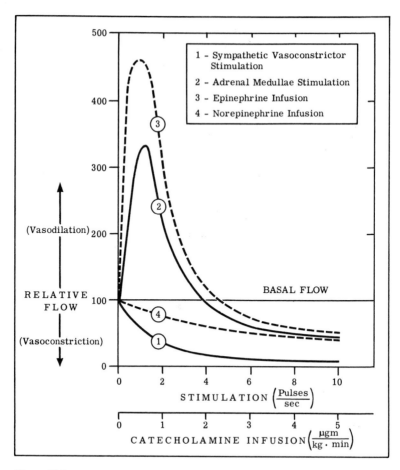

Figure 17-6
The effect of neurogenic and humoral adrenergic influences on skeletal muscle blood flow. (Data from Celander, 1954, p. 48.)

vasoconstrictor tone from which adjustment of vascular resistance can occur for the maintenance of circulatory homeostasis. Maximal resistance responses are developed at a frequency of about 10 impulses per second. Inhibition of this basal vasoconstrictor tone results in vasodilation. Figure 17-7, curve D, shows the relationship between perfusion pressure and flow in resting muscle. At a pressure of 100 mm Hg resting muscle blood flow is about 3 ml per minute per 100 gm, signifying that a high degree of vascular tone is present in this tissue. Maximal sympathetic vasoconstriction (curve E) reduces blood flow at all perfusion pressures. At a pressure of 100 mm Hg the flow is about 0.3 ml per minute per 100 gm tissue. Maximal vasoconstrictor inhibition (curve B) shifts the relationship to the left, and, at the same perfusion pressure of 100 mm Hg, flow is increased about sixfold compared to the normal resting condition. Normally, the tissue vasculature functions well within this range of vascular tone (see Fig. 17-10).

Sympathetic vasoconstriction is not the only determinant of the basal vasculature tone, since an intrinsic level of vascular smooth muscle tone is present after denervation. Blockade of the vascular innervation permits a twofold to fivefold increase in flow

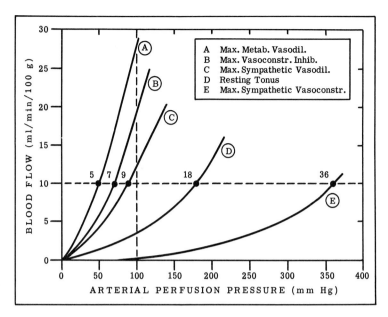

Figure 17-7
The pressure-flow relationship for skeletal muscle. The numbers indicate the peripheral resistance at a flow of 10 ml per minute per 100 gm tissue. (From E. M. Renkin and S. Rosell. *Acta Physiol. Scand.* 54:241, 1962.)

through skeletal muscle (Fig. 17-7, curve B), while skeletal muscular contractions may elicit a further increase to over 20 times the normal flow (Fig. 17-7, curve A).

Figure 17-8 demonstrates that maximal adrenergic stimulation produces constriction and reduced blood flow which is most prominent in skin and kidney, less so in muscle, intestine, and liver, and substantially absent in heart and brain. Control of blood flow in heart and brain is mediated predominantly by local factors (see below).

Figure 17-8 also illustrates the maximal extent of vasodilation which can be achieved in the tissues by means of metabolic or pharmacological procedures. The marked increase in skin blood flow reflects the capability of the arteriovenous anastomosis to distribute blood rapidly to the venous plexus of the skin for the dissipation of body heat during temperature regulation.

Although the sympathetic vasoconstrictor outflow is somewhat diffuse and has a general effect on the overall peripheral resistance, important changes in the distribution of blood flow may be brought about by these fibers. For example, uncomfortable environmental temperatures, hypoxia, digestion, or muscular exertion results in *discrete* cardiovascular adjustments in the distribution of the cardiac output. Sympathetic vasoconstriction during stress, such as severe hemorrhage and hypotension, can markedly reduce the flow of blood through the skin, skeletal muscle, renal, and splanchnic vascular beds. This blood is redistributed to other more immediately vital tissues. The characteristic cold skin, muscular weakness, and anuria so often seen following hemorrhage are due in part to the redistribution. The cutaneous circulation is almost entirely under neural rather than local control.

Vascular Capacitance
Sympathetic adrenergic activity also increases venular and small vein smooth muscle contraction (venomotor tone). Because veins contain blood at zero transmural pressure (the unstressed volume) and because the pressure-volume relationship becomes non-

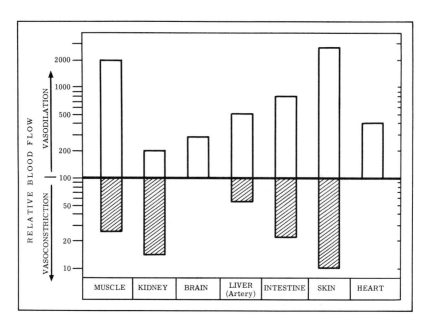

Figure 17-8
The effect of maximal sympathetic adrenergic vasoconstriction and of maximal vasodilation elicited by metabolic or pharmacological procedures on blood flow through the various tissues of the body.

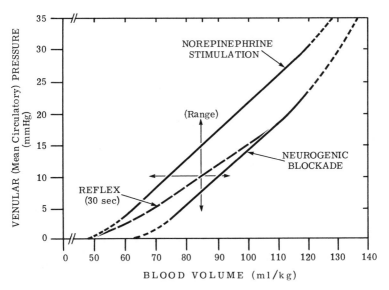

Figure 17-9
Changes in cardiovascular capacitance. The reflex curve is the mean circulatory pressure found 30 seconds after starting a rapid hemorrhage or transfusion to the blood volume indicated. (Modified from Drees and Rothe, 1974.)

linear at high pressure, venous capacitance is difficult to define. *Venous capacitance* is the ability of the veins to store blood and includes the unstressed volume and the compliance ($\Delta V/\Delta P$) over the range of pressures likely to be encountered (Fig. 17-9). The total systemic vascular compliance at normal venular pressures (about 10 mm Hg) is about 3 ml/mm Hg · kg body weight. Within the physiological range it appears to be linear (e.g., Drees and Rothe, 1974), but at high pressures the compliance is much less and so, for example, resists the tendency toward pooling of blood in the lower limbs when standing. Neurogenically induced venoconstriction can act to reduce venous capacitance by either a change in the compliance or a displacement of the entire curve toward a lower volume at the same pressure, or both. As shown in Figure 17-9, small vein pressure (mean circulatory pressure) can be increased to nearly 20 mm Hg by massive sympathetic stimulation at a normal blood volume to increase venous return and therefore cardiac output (Chap. 15). With blockage of venomotor tone, the pressure falls to about 4 mm Hg. When the carotid sinus pressure is decreased about 25 mm Hg, the venoconstrictor response increases mean circulatory pressure by only about 2 mm Hg, but this is enough to account for a 30 percent increase in cardiac output if cardiovascular reflexes are blocked—a highly significant response (Sagawa, 1974). With normal small vein pressure, the equivalent of nearly a liter of blood (10 ml per kilogram) can be compensated for by venoconstriction. With reflexes intact, the apparent compliance measured at 30 seconds after the start of a hemorrhage or transfusion is greater than the maximally constricted or dilated values, for with hemorrhage reflex venoconstriction is induced, while with transfusion the venomotor tone is reduced (Fig. 17-9). However, because of the difficulties in measuring total small vein and venular volume and transmural pressure in relatively intact preparations, our knowledge is uncertain concerning the control of venous capacitance. Gauer et al. (1970) have concluded that active venoconstriction occurs only after a major stress, such as a loss of 15 percent of the blood volume. Other studies (Drees and Rothe, 1974) suggest that changes in venous capacity are linearly related to the degree of hemorrhage.

In addition to the direct neurogenic influence on vascular capacitance, major changes in the blood volume of the venous reservoir occur by means of passive elastic recoil, which follows a reduction in transmural pressure. An increase in precapillary resistance acts to reduce the flow, and thus pressure downstream, and so effectively reduces the transmural pressure.

The magnitude of possible venomotor response is uncertain, but it appears that maximal sympathetic stimulation at a constant blood flow causes less than a 10 percent reduction in contained blood volume of skeletal muscle, about a 20 percent change of skin vascular volume, and up to a 50 percent reduction in the contained volume of the splanchnic bed. A resting venomotor tone similar to the vasoconstrictor tone is normally present. If either is lost, serious hypotension follows, cardiac output being low if the venous tone is abolished and somewhat high if the arteriolar constrictor tone is lost.

Adrenomedullary Influence

Direct neural control of the vascular smooth muscle may be supplemented by circulating catecholamines such as epinephrine and norepinephrine released into the bloodstream through stimulation of the adrenal medulla. The effects on the vasculature are minor in comparison with the direct neural control. Stimulation of the splanchnic innervation to the adrenal medulla over a range of frequencies results in a biphasic change in muscle vascular resistance (Fig. 17-6, curve 2). At frequencies below 3 impulses per second, vasodilation occurs, whereas frequencies from 3 to 10 impulses per second produce progressive vasoconstriction. The basis for this variable effect lies in the fact that the adrenomedullary stimulation liberates primarily epinephrine with some norepinephrine. Norepinephrine produces vasoconstriction by alpha receptor stimulation at all doses (Fig. 17-6, curve 4). Epinephrine in low concentrations (below 2 μg per kilogram-minute) acts on skeletal muscle vessels to cause dilation by stimulation of beta receptors (Fig. 17-6, curve 3, and Fig. 7A-4). At doses greater than 2 μg per kilogram-minute, alpha receptor stimulation predominates, resulting in vasoconstriction.

The precapillary sphincter is considerably more sensitive to circulating catecholamines than are the arterioles, which, in turn, are more sensitive than venules (Korner, 1974).

The dog with a denervated heart is capable of near maximal exercise. However, if circulating catecholamines are also blocked (by propranolol), running performance is drastically reduced. With intact innervation of the heart, the circulating catecholamines add little to exercise performance. Thus, either an intact sympathetic innervation to the heart or an intact sympathoadrenal system permits near maximal exercise performance (Sagawa et al., 1974).

Sympathetic Vasodilator Fibers

Skeletal muscle blood flow increases with the thought of exercise prior to the onset of muscle contraction. This behavior has been attributed to sympathetic vasodilator fibers which originate in the motor cortex and synapse in the hypothalamus and collicular region. Impulses pass through the ventrolateral part of the medulla oblongata to the lateral spinal horns with final distribution from the sympathetic ganglia to the blood vessels. These fibers also innervate the external genitalia. They have no direct effect on the control of blood pressure associated with pressoreceptor activity. The enhanced blood flow to skeletal muscle appears to bypass the capillary exchange network, as evidenced by a reduced oxygen consumption. The functional significance of this system is uncertain although it could establish a blood flow reserve which would be immediately available upon initiation of the metabolic demand of muscle contraction. Figure 17-7, curve C, shows that maximal sympathetic cholinergic vasodilation increases muscle blood flow up to five times the resting level.

Intra-arterial injections of acetylcholine likewise cause vasodilation. Atropine blocks both the vasodilating effect of acetylcholine injected into an artery and the effects of sympathetic vasodilator fiber stimulation. Sectioning of the sympathetic paravertebral ganglia also blocks the response. It is thus probable that acetylcholine is released at postganglionic cholinergic fibers and acts in some manner to cause vascular dilation.

Parasympathetic Vasodilator Action

The erectile tissues of the external genitalia are innervated by cholinergic vasodilator fibers of both the sympathetic and parasympathetic divisions.

Activation of various exocrine glands (e.g., salivary, sweat) by parasympathetic fiber discharge leads to increased blood flow. However, this does not represent direct vascular influence. Instead, increased glandular activity triggers the release of a proteolytic enzyme which acts on a plasma or tissue globulin to form a potent vasodilator, *bradykinin*. This response, following parasympathetic stimulation, may therefore be referred to as secretomotor vasodilation.

There are no known parasympathetic vasoconstrictor fibers that participate in the regulation of blood pressure. Although the vagal slowing of the heart is followed by a vasoconstriction of the coronary vessels, this is probably an indirect consequence of the decreased metabolism.

Dorsal Root Vasodilation

In addition to the autonomic efferent fibers, stimulation of the skin (or the peripheral end of a sectioned dorsal root fiber) often produces a dilation of the adjacent superficial blood vessels. It is probable that this results from impulses arising from receptor sites (pain) and passing up sensory afferent fibers until a branch in the neuron is reached. Impulses then go back over the branch to a vasomotor ending. For this reason they are called *antidromic vasodilator* impulses. The afferent and efferent parts of this *axon reflex* are formed by the branching of a single nerve fiber. Effective stimulation is probably a result of the release of histamine by tissue trauma. Only in areas rich in pain fibers, such as mucous membrane and skin, does vasodilation occur from this mechanism. The axon reflex may be of functional importance in the development of the *flare* following mechanical trauma to the skin and possibly in the development of inflammation about an infected area. The increase in blood flow would presumably aid the healing process.

LOCAL CONTROL OF TISSUE BLOOD FLOW

Passive Response to Changes in Transmural Pressure

In order to appreciate the effectiveness of the local regulation of tissue circulation, one must understand the passive behavior of a vascular bed based on the distensible properties of the walls and the transmural pressure. One can gain some insight into the nature of a vascular bed by examining its *pressure-flow relationship* (Figs. 11-4B, 17-10). For a rigid tube the pressure-flow relationship is essentially linear. That is to say, each elevation in pressure will produce a proportionate elevation in flow. However, blood vessels are distensible tubes, and elevations of pressure in passive vascular beds produce greater than proportionate increases in flow, the extent of which depends on the existing vascular tone (see Fig. 17-7). Ultimately, at high transmural pressures, as the collagen fibers become tensed, the blood vessel will behave like a rigid tube. These passive adjustments in the state of the blood vessels are initiated by changes in transmural pressure. Transmural pressure represents the gradient of pressure across the vessel wall. Consequently, transmural pressure may be *increased* by either an increase in intraluminal pressure or a decrease in extraluminal pressure. Then the vessels will dilate and vascular resistance will decrease. Intraluminal pressure can be increased by greater inflow of blood, upstream vasodilation, or downstream vasoconstriction. In addition, placing the tissue below the heart will increase the intraluminal pressure.

Transmural pressure may be *decreased* by either a decrease in intraluminal pressure or an increase in extraluminal pressure. The radius of the vessel thereby decreases, increasing resistance and consequently reducing blood flow. Intraluminal pressure

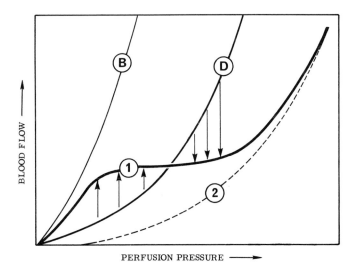

Figure 17-10
Pressure-flow relationships of autoregulating skeletal muscle. B = passive distensible vasculature at very low constant tone (compare with Fig. 17-7, curve B). D = passive distensible vasculature at constant resting tone (compare with Fig. 17-7, curve D). 2 = passive distensible vasculature with increased constant vasculature tone. 1 = Autoregulating vascular bed.

declines as a result of a decrease in blood flow, upstream vasoconstriction, or downstream vasodilation, or of raising the tissue relative to the heart. Extraluminal pressure may increase by means of muscular contraction, abdominal compression, or an increase in tissue pressure resulting from edema. Thus, inflow into skeletal muscle and coronary circulation during contraction is markedly reduced. In the intestine, mechanical compression resulting from peristalsis or distention with luminal contents may also act to retard flow.

In a tissue vascular system, the nonlinear pressure-flow relationship (Figs. 17-7 and 17-10) is also due to recruitment of additional open vascular channels as perfusion pressure is elevated. This passive type of behavior is seen in skin, pulmonary, and hepatoportal systems.

Active Response and Autoregulation of Blood Flow

There are vascular beds, including the cerebral, coronary, muscle, intestine, kidney, and hepatic arterial circulations, in which tissue blood flow is maintained relatively constant over a range of perfusion pressures by means of active local regulation of vascular tone. This phenomenon is called *autoregulation*. It is an intrinsic control mechanism that is independent of neurogenic or systemic humoral factors although it can be modified by them. It represents the local control of blood flow related to the functional needs of the tissue and is seen in Figure 17-10, curve 1, as a deviation from the passive pressure-flow pattern.

Autoregulation is characterized by the phenomenon illustrated in Figure 17-10, curves 1 and D. Elevation of perfusion pressure from the control level results in an initial increase of blood flow followed shortly thereafter by a reduction of flow to about the previous control level. Since flow is thus maintained relatively constant in the presence of an increased perfusion pressure, vascular resistance necessarily increases. Conversely, reduction of perfusion pressure from the control level produces an initial fall in tissue blood flow which is followed within about 30 seconds by an upward adjustment of flow to the control level. In this case a maintained flow during reduced perfusion pressure reflects a reduction in vascular resistance. Thus autoregulation represents the adjustment of vascular resistance to keep blood flow constant over a range of perfusion pressure. The mechanism of autoregulation is complex, based on response of the vascular smooth muscle to changes in transmural pressure, intrinsic metabolic factors, or other local determinants of vascular reactivity. A number of theories have been proposed to account for the phenomenon of autoregulation.

Metabolic Theory

Numerous lines of evidence implicate metabolic factors in the local control of tissue blood flow. Increased metabolic activity is associated with increased tissue blood flow, in what may be considered *function hyperemia*. Enhanced metabolic activity of muscular exercise causes venous blood oxygen saturation to fall and blood flow to increase markedly. Thus, oxygen consumption, the product of blood flow and the arteriovenous oxygen saturation difference, increases dramatically. The hyperemia of muscle during maximal exercise may increase from 1 liter per minute at rest to more than 20 liters per minute (also see Fig. 17-7, curve A).

In metabolic autoregulation, the local concentration of metabolites (of a vasodilator nature) varies inversely with the blood flow through the tissue. For example, the accumulation of vasodilator substances with reduced blood flow tends to dilate the resistance vessels and to restore flow. If flow increases, the metabolites are washed out and flow decreases proportionately.

It is presumed that the hyperemias result from alterations of the concentrations of local "metabolic" factors. The capability of metabolites to produce profound vasodilation is demonstrated by reactive hyperemia. Figure 17-11 shows the level of blood flow that develops when the arterial blood supply to tissues is restored, after having been interrupted for periods of one to four minutes. This manipulation provides an

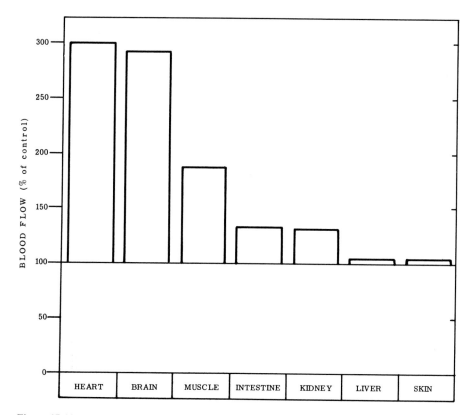

Figure 17-11
The extent of reactive hyperemia following restoration of tissue blood flow after brief arterial occlusion.

indication of the level of metabolic activity and of oxygen reserve in a tissue. Thus, heart and brain, tissues with high metabolic activity and relatively low oxygen reserve, exhibit profound hyperemia, whereas resting muscle, intestine, and kidney display moderate hyperemia reflecting moderate metabolic activity and/or high oxygen reserve. The negligible hyperemia of skin reflects a low metabolic activity with a high oxygen reserve, whereas that of liver represents a low dependency on oxidative metabolism. The precise nature of the metabolic factors responsible for this behavior is uncertain. Many factors have been considered, including oxygen, carbon dioxide, pH, adenosine compounds, histamine, plasma potassium concentration, and plasma osmolality. Moreover, oxygen, potassium, and osmolality have been demonstrated to act in an additive fashion to modify vascular tone (Skinner and Costin, 1971). Figure 17-12 illustrates that reducing oxygen partial pressure in the blood (Po_2) or increasing the arterial carbon dioxide tension (Pco_2) certainly has significant influence on cerebral circulation, possibly through a change in local pH. Adenosine compounds and histamine, although potent vasodilators, achieve inadequate precapillary concentrations to account for the hyperemia. Exercise results in elevated venous plasma potassium concentration, which could produce vasodilation. Moreover, increased plasma potassium together with reduced Po_2 acts synergistically to increase blood flow. Plasma osmolality is also elevated during exercise. It has been proposed that increased plasma osmolality causes

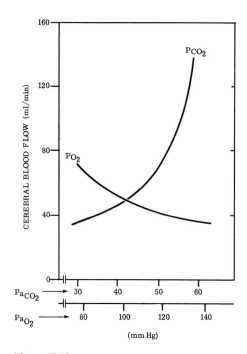

Figure 17-12
The effect of changing P_{O_2} and P_{CO_2} on cerebral blood flow. (After I. M. James et al., *Circ. Res.* 25:77–93, 1969. By permission of The American Heart Association.)

fluid to leave the cells of small arteries and arterioles to reduce their wall thickness and increase their lumen and thus provide for hyperemia. Inasmuch as all these factors are present to some degree during hyperemias, it is possible that they are collectively responsible for the blood flow patterns seen. The relative significance and level of interactions between them probably vary between tissues as well as between species.

Myogenic Theory
Elevating perfusion pressure produces an increase in transmural pressure and therefore wall tension of small arteries and arterioles. Tension on the vascular wall provides the stimulus for vascular smooth muscle contraction, which reduces the diameter, increases the resistance, and impedes flow. The reverse effect occurs when perfusion pressure is reduced. Thus, adjusting vascular resistance in the same direction as the change in perfusion pressure tends to maintain flow at a constant level. Organs in which a myogenic type of autoregulation has been postulated include the kidney, intestine, and hepatic arterial supply.

Other Theories
It has been suggested that, in addition to increasing blood flow, elevating perfusion pressure increases capillary hydrostatic pressure and enhances filtration. The resulting increase in *tissue hydrostatic pressure,* by reducing the transmural pressure, should partially collapse the microcirculation to increase vascular resistance and thereby restore blood flow to control. That this sequence is seen in the encapsulated kidney only at excessive tissue pressures, and in the brain with its rigid cranium only when cerebrospinal pressure exceeds arterial pressure, indicates that it is not a physiological phenomenon.

It has been proposed that autoregulation could represent the participation of *neurogenic reflexes*. However, autoregulation has been demonstrated in isolated denervated preparations under influence of local anesthesia. Consequently, this theory appears to be inappropriate.

Summary of Mechanisms of Autoregulation
A single factor is probably not responsible for the autoregulation observed in various tissues. It is likely that the interaction of multiple factors is responsible for autoregulation in any one tissue, and each factor may have a more prominent effect in one tissue than in another. Thus, the metabolic factor is apparently most notable in highly metabolizing tissues such as exercising skeletal muscle, heart, and brain.

Interaction Between Central and Local Influences

Vascular smooth muscle elements are continuously under central and local influences which may act synergistically or antagonistically to modify vascular tone. The balance between the interactions may be related to the distribution and type of smooth muscle in the various vascular elements. Two types of smooth muscle are present in arterioles. It appears that the smooth muscle at the outer layer of the vessel is of the *multiunit* type and is therefore most responsive to neurogenic influence. The smooth muscle of the inner layer is primarily of the *visceral* type and is most responsive to transmural pressure and local chemical influence. At the level of the arteriole the smooth muscle distribution favors central neurogenic activity, whereas more distally, at the level of the precapillary sphincters, local influences predominate. The smooth muscle of postcapillary resistance vessels seems to be the multiunit type and is therefore primarily under central neurogenic control.

Thus, there exists a framework which can adjust peripherovascular tone to provide for both systemic and local cardiovascular homeostasis. This feature of the control system may be reflected in the effect of sympathetic stimulation on the series resistance vessels in muscle. Sympathetic stimulation produces small artery, arteriolar, and venous vasoconstriction to increase peripheral resistance and venous return and thereby contribute to blood pressure regulation. The reduced tissue blood flow resulting from the vasoconstriction leads to a reduced transmural pressure as well as an accumulation of metabolites at the precapillary sphincter, both of which produce precapillary sphincter dilation. This provides a more uniform distribution of local blood flow and permits a more complete exchange between the vascular and extravascular compartments, despite the reduced total blood flow.

EXTRANEURAL CONTROL OF THE CARDIOVASCULAR SYSTEM

The volume of fluid available to the heart for pumping in relation to the volume of the vascular system provides another factor in the control of cardiovascular function, because an increase in effective blood volume tends to fill the heart more than is normal and so increases cardiac output (see Chap. 15). On the other hand, a decrease in total blood volume tends to decrease cardiac output. However, the passive recoil of the capacitance vessels, associated with precapillary vasoconstriction, provides a powerful extraneural mechanism acting to maintain central venous pressure and cardiac output. The constriction of the resistance vessels results in a decrease in flow, so that the transmural pressure in the capillaries and venules decreases, leading to fluid reabsorption from the extracellular spaces (Chap. 12) and permitting an elastic recoil of the capacitance vessels to redistribute the blood toward the heart. This elastic recoil appears to be a more important mechanism than neurogenic venoconstriction. Also, a decrease in blood volume, by relieving the stretch of left atrial volume, results in fluid retention by renal activity which acts to restore blood volume (Chaps. 23 and 30).

CARDIOVASCULAR SYSTEM RESERVE

The cardiovascular reflexes provide the widely varying flow of blood needed to meet the metabolic demands of the various tissues of the body, but blood flow to the tissue is limited because of the limitations of the cardiovascular system. A person cannot run at maximum speed (high muscle metabolism) after a full meal (splanchnic bed dilation) on a hot day (skin dilation). Heart disease requires the utilization of cardiovascular reserves to provide even the basal blood flow (Chap. 18B).

Control of the cardiovascular system utilizes four primary cardiovascular reserves in the maintenance of circulatory homeostasis: venous oxygen reserve, heart rate reserve, systolic volume reserve, and diastolic volume reserve.

Venous Oxygen Reserve

During normal resting conditions, about 4 ml of oxygen is extracted from the arterial blood per 100 ml of blood flow through the tissue (Fig. 17-4). Thus, a reserve of about 16 volumes of oxygen per 100 volumes of blood (16 volumes percent) is in venous blood. During severe exertion, when the oxygen utilization may be over 10 times the basal level, the extraction of oxygen from the blood is increased to over 12 volumes percent.

A person with heart failure has a decreased cardiac output, but adequate amounts of oxygen can be transported to tissues because most cells can extract a larger than normal fraction of the oxygen from the blood by increasing the oxygen arteriovenous (A-V) difference. Unfortunately, following cardiac failure, the reduced flow to the peripheral tissue is not uniformly distributed. The blood flow through heart muscle and the brain tends to be maintained because there is little neurally induced vasoconstriction in response to the fall in blood pressure following the decrease in cardiac action. Peripheral vasoconstriction causes a disproportionate reduction flow of blood through the kidneys. As renal blood flow is reduced, the oxygen consumption is reduced and function impaired. Thus, although an increase in A-V oxygen difference provides a source of oxygen to tissues, its benefits are limited in the patient with cardiac failure because of the encroachment upon renal function. In the normal individual, however, this reserve provides for a twofold to threefold increase in oxygen utilization by active muscle (Fig. 17-5).

Heart Rate Reserve

During moderate exercise the heart rate is increased by the cardiovascular control system and the stroke volume is held relatively constant. This response provides for an increase in cardiac output. However, the heart rate increase limits the filling time; hence, the venous filling pressure or the effective distensibility of the heart must be increased by homeostatic mechanisms, or the end-systolic volume must be decreased parallel to a decrease in end-diastolic volume, if the stroke volume is to be maintained.

An increase in heart rate does not provide a particularly useful reserve for the cardiac patient, because a rapidly beating heart is less efficient in the use of oxygen than a slowly beating one (Chap. 15). If the heart failure is primarily a result of inadequate coronary blood flow in the first place, this loss of efficiency becomes quite significant. Furthermore, an increase in heart rate tends to limit the coronary blood flow (Chap. 15). Thus, an increase in heart rate is not an efficient means of increasing cardiac output by the failing heart, although it is a highly important mechanism for the normal individual, providing a twofold to threefold increase in cardiac output.

Systolic Volume Reserve

In the normal individual the stroke volume can be increased from about 70 ml to over 100 ml per stroke. Athletes with cardiac outputs of over 30 liters per minute and heart rates of about 200 beats per minute must have stroke volumes of about 150 ml. This increase in stroke volume during *maximal* exertion by well-trained individuals is provided primarily by a decrease in end-systolic volume (see Fig. 15-4). There is a large amount of blood in the heart of a resting individual at the end of systole. It provides a cardiovascular reserve but is available only if the myocardial contractility increases so that the added volume can be ejected. The loss of myocardial contractility in cardiac failure is such that sympathetic activity from the control system is relatively ineffective in utilizing this reserve.

Diastolic Volume Reserve

As the central venous pressure increases in heart failure, the heart becomes distended. The increase in diastolic volume, produced by stretching the myocardial fibers, tends to cause the release of more energy in accordance with Starling's law of the heart, so that more blood is pumped. Furthermore, the heart beats more efficiently because of the decrease in viscous losses and development of internal tensions (Chap. 15). As fluids accumulate, the failing heart becomes larger and more distended. Since even the normal resting reclining individual has a relatively large heart, and therefore a small diastolic volume reserve, there is little reserve for heart failure.

Thus, the person with cardiac failure has severe limits on cardiovascular system reserves. Whereas normally the heart rate is increased and the stroke volume is maintained or increased in response to neural stimuli, the failing heart cannot respond satisfactorily to increased demands, because of myocardial weakness.

REFERENCES

Betz, E.   Cerebral blood flow: Its measurement and regulation. *Physiol. Rev.* 52:595–630, 1972.

Celander, O.   The range of control exercised by the sympathico-adrenal system. *Acta Physiol. Scand.* 32 (Suppl. 116):1–132, 1954.

Chien, S.   Role of the sympathetic nervous system in hemorrhage. *Physiol. Rev.* 47:214–288, 1967.

Cowley, A. W., Jr., J. F. Liard, and A. C. Guyton.   Role of the baroreceptor reflex in daily control of arterial blood pressure and other variables in dogs. *Circ. Res.* 32:564–576, 1973.

Drees, J. A., and C. F. Rothe.   Reflex venoconstriction and capacity vessel pressure-volume relationships in dogs. *Circ. Res.* 34:360–373, 1974.

Fishman, A. P., and D. W. Richards (Eds.).   *Circulation of the Blood: Men and Ideas.* New York: Oxford University Press, 1964. Chap. 7.

Folkow, B., C. Heymans, and E. Neil.   Integrated Aspects of Cardiovascular Regulation. In W. F. Hamilton and P. Dow (Eds.), *Handbook of Physiology.* Washington: American Physiological Society, 1965. Section 2: Circulation, vol. 3, chap. 49, pp. 1787–1824.

Folkow, B., and E. Neil.   *Circulation.* London: Oxford University Press, 1971.

Gauer, O. H., J. P. Henry, and C. Behn.   Regulation of extracellular fluid volume. *Annu. Rev. Physiol.* 32:547–595, 1970.

Green, H. D., C. E. Rapela, and M. C. Conrad.   Resistance (Conductance) and Capacitance Phenomena in Terminal Vascular Beds. In W. F. Hamilton and P. Dow (Eds.), *Handbook of Physiology.* Washington: American Physiological Society, 1963. Section 2: Circulation, vol. 2, chap. 28.

Guyton, A. C., T. G. Coleman, and H. J. Granger.   Circulation: Overall regulation. *Annu. Rev. Physiol.* 34:13–46, 1972.

Johnson, P. C. (Ed.).   *Autoregulation of Blood Flow.* (*Circ. Res.* 15 [Suppl. 1].) New York: American Heart Association, 1964.

Korner, P. I.   Integrative neural cardiovascular control. *Physiol. Rev.* 51:312–367, 1971.

Korner, P. I.   Control of Blood Flow to Special Vascular Areas: Brain, Kidney, Muscle, Skin, Liver and Intestine. In A. C. Guyton and C. E. Jones (Eds.), *Physiology, Series One. Cardiovascular Physiology.* London: Butterworth, 1974. Vol. 1, chap. 4.

Mellander, S.   Comparative studies on the adrenergic neuro-hormonal control of resistance and capacitance blood vessels in the cat. *Acta Physiol. Scand.* 50 (Suppl. 176):1–86, 1960.

Mellander, S., and B. Johanson.   Control of resistance, exchange and capacitance functions in the peripheral circulation. *Pharmacol. Rev.* 20:117–196, 1968.

Paintal, A. S.   Vagal sensory receptors and their reflex effects. *Physiol. Rev.* 53:159–227, 1973.

Pelletier, C. L., and J. T. Shepherd.   Circulatory reflexes from mechanoreceptors in the cardio-aortic area. *Circ. Res.* 33:131–138, 1973.

Randall, W. C., J. A. Armour, W. P. Geis, and D. B. Lippincott.   Regional cardiac distribution of the sympathetic nerves. *Fed. Proc.* 31:1199–1208, 1972.

Rowell, L. B., J.-M. R. Detry, J. R. Blackmon, and C. Wyss.   Importance of the splanchnic vascular bed in human blood pressure regulation. *J. Appl. Physiol.* 32:213–220, 1972.

Rushmer, R. F.   *Cardiovascular Dynamics* (3rd ed.). Philadelphia: Saunders, 1970.

Sagawa, K.   The Circulation and Its Control II: Neural and Humoral Control of the Heart and Vessels. In J. H. U. Brown and D. S. Gann (Eds.), *Engineering Principles in Physiology*. New York: Academic, 1973. Vol. 2, chap. 15.

Sagawa, K., M. Kumada, and L. P. Schramm.   Nervous Control of the Circulation. In A. C. Guyton and C. E. Jones (Eds.), *Physiology, Series One. Cardiovascular Physiology*. London: Butterworth, 1974. Vol. 1, chap. 6.

Scher, A. M.   Control of Arterial Blood Pressure; Control of Cardiac Output. In T. C. Ruch, H. D. Patton, and A. M. Scher (Eds.), *Physiology and Biophysics* (20th ed.). Philadelphia: Saunders, 1974. Chaps. 10, 11.

Skinner, N. S., Jr., and J. C. Costin.   Interactions between oxygen, potassium, and osmolality in regulation of skeletal muscle blood flow. *Circ. Res.* 28 (Suppl. I):I-73 to I-85, 1971.

Smith, O. A.   Reflex and central mechanisms involved in the control of the heart and circulation. *Annu. Rev. Physiol.* 36:93–123, 1974.

18. Pathological Physiology of the Cardiovascular System A. Hypertension

Julius J. Friedman

Arterial hypertension is a condition of sustained elevated systemic arterial blood pressure. There is no clear-cut line of demarcation between normal and hypertensive blood pressures since blood pressure is influenced by many factors. The minimum level of systemic arterial pressure considered to be hypertensive has been arbitrarily set at 140/90 mm Hg. While hypertension usually involves elevations in mean and pulse pressures, the rise in diastolic pressure has been regarded as the critical criterion, suggesting that hypertension involves an increased peripheral resistance. In recent years, greater attention has been directed at the level of systolic pressure because it is thought to be responsible for the development of secondary cardiovascular diseases: arteriosclerosis, congestive heart disease, atherothrombic brain infarction, nephrosclerosis, etc.

Hypertension is a highly prevalent condition. In an extensive study conducted in Framingham, Massachusetts, over an 18-year period involving over 5200 men and women, black and white, between the ages of 30 and 62 years, 18 percent of the men and 16 percent of the women had blood pressures greater than 160/95, and 41 percent of the men and 48 percent of the women had pressures in excess of 140/90. The frequency of hypertension is greater in the older people studied.

Hypertension is a serious cardiovascular disease. It is responsible for approximately 10 to 15 percent of the deaths in people over 50 years of age. Associated with the elevated blood pressure is an increased morbidity as well as mortality. Insurance studies reveal that at age 45, a systolic pressure above 150 mm Hg will reduce the life expectancy of men by 11.5 years and of women by 8.5 years.

The incidence of hypertension is higher in blacks than in whites at every age, and blacks suffer a greater mortality from this disease. Whereas 16 white men in 100,000 die of hypertension annually, 66 black men in 100,000 are so affected, or four times the mortality. The same ratio applies to black women. This higher prevalence of hypertension in blacks has been attributed to genetic factors; however, socioeconomic status must be taken into consideration.

The incidence of cardiovascular disease is greater in hypertensives than in normotensives. Hypertensives suffer from seven times more stroke, four times more congestive heart failure, three times more coronary heart disease, and twice as much occlusive peripherovascular disease.

In approximately 20 percent of the cases of hypertension, the elevated blood pressure is a clinical sign of a specific disease (e.g., renal disease) and not a disease entity in itself. Such conditions of elevated blood pressures are referred to as *secondary hypertension*. In the remaining 80 percent the development of hypertension cannot be attributed to any known origin and probably represents a specific disease state. This form of hypertension is called *primary* or *essential hypertension.*

CLASSIFICATION OF HYPERTENSIVE VASCULAR DISEASE

To gain an insight into the nature of essential hypertension, the various diseases that cause an elevation in blood pressure have been studied both clinically and experimentally. These include cardiovascular, neurogenic, endocrine, and renal disorders.

405

Cardiovascular Hypertension

Systemic arterial blood pressure is determined by cardiac output, total peripheral resistance, and arterial compliance. Consequently a number of cardiovascular changes can result in an elevation of arterial blood pressure. However, elevations of arterial pressure that result from increases in cardiac output alone, or from increases in peripheral resistance not including the renal vasculature, are found to be relatively short term. The elevated pressure increases the urine volume load and produces natriuresis, which in time reduces the extracellular fluid volume and, thus, blood pressure.

Increases in cardiac output can produce sustained elevated arterial pressure by means of autoregulatory increases in peripheral resistance, including the afferent arterioles of the kidney. Guyton (1974) found that total peripheral resistance was elevated within 10 minutes of an increase in cardiac output. In 30 to 60 minutes resistance had increased to three times the increase in cardiac output. Over a period of days and weeks a 1 to 5 percent increase in cardiac output resulted in as much as a 50 percent increase in peripheral resistance.

Neurogenic Hypertension

The centers that regulate and integrate cardiac and vasomotor activity are situated in the brain. Consequently, impairment of cerebral function in such a manner that cardiac and vasomotor activity is enhanced produces an elevation of blood pressure. Cerebral function may be impaired by cerebral ischemia due to either cerebral arterial occlusion or severely increased intracranial pressure, or by tumor development in various regions of the brain. In each case the resulting hypertension may be relieved by appropriate treatment of the organic disturbance.

Another form of neurogenic hypertension may be initiated by carotid sinus baroreceptor denervation. With such denervation the gain of the baroreceptor regulatory loop is reduced, leading to increases in cardiac activity and vasomotor tone. A form of carotid sinus denervation can occur by means of baroreceptor adaptation to a maintained elevation of arterial pressure. In a matter of days following arterial pressure elevation, the baroreceptor discharge frequency approaches the control rate despite the maintained elevation of arterial pressure, and the pressure then becomes regulated at this new level. Restoration of the arterial pressure to the original level will result in a readaptation. The adaptation differs from the "resetting" of the baroreceptors which apparently accompanies the change in physical properties of the baroreceptor vessels following sustained hypertension.

The sympathetic nervous system has been implicated in the development of hypertension. Extensive sympathectomy or ganglionic blocking agents produce a reduction in blood pressure. Whether this sympathetic contribution is a primary one reflecting altered function or a secondary one representing some regulatory activity is not clear.

Endocrine Hypertension

The adrenal medulla contains chromaffin cells which secrete the catecholamines epinephrine and norepinephrine. These cells may develop tumors (pheochromocytomas), which secrete excessive quantities of catecholamines into the circulation, either continuously or periodically; the blood pressure then is elevated by increased cardiac output and peripheral resistance. Surgical removal of the tumorous tissues relieves the condition and usually restores blood pressure to normal levels.

The adrenal cortex has also been implicated; however, the specific role of this tissue and its secretions in the pathogenesis of hypertension remains uncertain. Many cases of hypertension can be relieved by restriction of the sodium content in the diet; a high intake of sodium tends to aggravate the condition. Aldosterone, a substance secreted by the adrenal cortex, stimulates the kidney to retain sodium ion, causing water retention. In conditions of adrenal hyperplasia with primary aldosteronism, hypernatremia with volume expansion develops and the blood pressure becomes elevated. This condition is reversible, and removal of the adrenal gland or administration of spironolactone, an aldosterone antagonist, usually restores the blood pressure and electrolyte pattern to normal levels. This form of hypertension is described as volume-dependent, or low-renin, hypertension and is characterized by hypernatremia and hypokalemia associated with volume expansion, and by minimal sympathetic activity due to reflex inactivation. Volume-dependent hypertension usually responds well to diuretic therapy.

In almost all forms of chronic experimental hypertension, aortic smooth muscle contains an elevated concentration of sodium and potassium ions and an increased volume of water, which may contribute to increased blood pressure in a number of ways. Ion and water accumulation in the walls of resistance vessels could cause swelling of the walls and encroachment upon the lumen, thereby increasing peripheral resistance and blood pressure. The increase in the ratio of vessel wall thickness to lumen diameter results in greater than normal changes in vascular resistance in response to normal degrees of smooth muscle contraction. In addition, alterations of the ionic composition of vascular smooth muscle could, by disturbing the membrane potential, affect the sensitivity of resistance vessels to circulating vasoconstrictor substances and autonomic nervous influences. Also, ionic changes in the muscular elements of the vascular wall may alter the state of actomyosin to produce an increased contractility. Thus, by producing changes in the physical state of resistance vessels and by increasing excitability and contractility of vascular smooth muscle, alterations in electrolyte balance could contribute significantly to the development of hypertension.

Renal Hypertension

Permanent hypertension may be induced by a variety of manipulations that effectively reduce either renal blood flow, functional renal tissue, or tubular sodium delivery. Reduction of renal blood flow by arterial occlusion or by renal compression consistently produces hypertension. The relationship of the kidney to hypertension is independent of nervous activity because complete renal denervation does not prevent the development of hypertension. Therefore, renal hypertension is attributed to a humoral mechanism. As early as 1898 it was recognized that an extract of kidney cortex was capable of producing an increase in blood pressure when injected into a normal subject. The agent responsible for this effect is *renin*.

When renal ischemia exists, the juxtamedullary cells adjacent to the afferent arterioles secrete increased amounts of renin. However, renin itself has no vasoactive properties. It is a proteolytic enzyme which acts upon a plasma protein, alpha-2 globulin, to produce a polypeptide, angiotensin I, which is also vasoinactive. A converting enzyme, present in plasma, then converts angiotensin I to the active form angiotensin II (formerly known as angiotonin or hypertensin). Angiotensin II increases total peripheral resistance through widespread arteriolar vasoconstriction. This agent also produces venoconstriction and reduced vascular capacity, thereby increasing venous return, cardiac output, and blood pressure.

However, in most forms of hypertension, the angiotensin II concentration in plasma is not very great. Unless angiotensin II acts in some indirect manner to increase either the sensitivity or the reactivity of the vascular smooth muscle, or both, its level in the plasma is insufficient to maintain an elevated pressure. Angiotensin II can indirectly influence vascular reactivity as well as the volume of extracellular fluid through its influence on the zona glomerulosa of the adrenal cortex to secrete stimulating aldosterone. Aldosterone acting on the kidney causes sodium retention, which increases extracellular fluid volume. This expanded volume, together with the lessened vascular capacity produced by angiotensin II venoconstriction, results in blood pressure augmentation. In view of the multiple influence of angiotensin II, it is not surprising to find that peptide blocking agents which prevent the conversion of angiotensin I to angiotensin II lower blood pressure in renal hypertension. This form of hypertension is referred to as "high-renin" hypertension and is characterized by reduced extracellular volume due to pressure natriuresis. The condition is characterized by malignant hypertension in which the elevated angiotensin II levels are accompanied by secondary aldosteronism. In high-renin hypertension adrenalectomy, aldosterone antagonists or diuretics are of questionable value.

Self-Perpetuating Nature of Hypertension

Hypertension is characterized by an increase in pulse pressure as well as mean blood pressure. Normally, one would expect an increased peripheral resistance to cause pulse pressure to decline (Chap. 16). The elevated pulse pressure in hypertension is attributed to a reduction of large artery distensibility, which reflects the progressive arteriosclerotic process in hypertension, as well as the nonlinear pressure-volume relationship of the arterial system. The sclerotic process is not restricted to the large arteries. When it involves the small arteries and arterioles, these vessels undergo medial hypertrophy which increases the mass of the vascular smooth muscle causing encroachment on the lumen. Thus, their lumina are permanently reduced and peripheral resistance remains elevated, even after extensive treatment including radical sympathectomy. A similar process is proposed as the basis for the "resetting" of the baroreceptors. In this manner, the condition of hypertension enhances and perpetuates its own development, an example of positive feedback.

ESSENTIAL HYPERTENSION

Essential hypertension refers to the sustained elevation in systemic arterial blood pressure for which there is no discernible origin. The hypertension may be "benign," in that it develops slowly and progressively over many years, or "malignant," in which case it develops rapidly in a brief period of time. Attempts to identify the cause of essential hypertension have been fruitless. After exclusion of hypertension due to renal disease, adrenal dysfunction, or cardiovascular and neurogenic alterations, no consistent explanation is obvious at this time.

Subjects with essential hypertension exhibit an elevated peripheral resistance which is rather uniformly distributed throughout the body and is not a result of changes in central vasomotor activity. Investigators have considered the participation of altered pressure regulation effected by a "resetting" of the baroreceptor mechanism. While the regulatory mechanism of the carotid sinus is active in both normotensive and hypertensive patients, the pressoreceptors of hypertensives fire intermittently, and at lower frequencies, at pressures that elicit continuous, higher-frequency firing in normotensives. Thus, the carotid sinus pressor mechanism may, by becoming adapted (set) to a new, elevated pressure level, perpetuate a higher pressure.

An alternative interpretation of altered pressoreceptor activity concerns the organic alteration of the carotid sinus wall. Sclerosis of the carotid sinus wall reduces its distensibility and limits the deformation of the receptors within the sinus wall. Consequently, larger pressure changes are required to activate the pressoreceptor mechanism to the same extent, and over a wide range the pressure would vary as though the sinus were denervated. Whether this alteration precedes essential hypertension or is caused by hypertension is not clear.

There is evidence from studies of sympathetic blockade that the sympathetic nervous system contributes to the hypertensive state through an increased reactivity of resistance vessels. Patients with essential hypertension exhibit more marked pressor responses to sympathetic influences, both neurogenic and humoral, than do normal subjects. Whether these reactive characteristics are due to altered neurotransmitter function, modification of cyclic nucleotide metabolism, or structural alterations is uncertain. However, the tendency toward hyperactivity is apparently genetically determined, because hypertensive patients almost always possess a family history of the disease.

An increasing body of evidence suggests that the interrelationship between the kidney and the adrenal cortex is significant in severe hypertension. Renal damage releases renin, which results in the formation of angiotensin, and this agent stimulates the secretion of aldosterone by the adrenal glomerulosa zone (Chap. 34). It may be that angiotensin, acting on vessels made hyperreactive by electrolyte alterations, is a significant feature of essential hypertension.

Recently Guyton (1974) has proposed the hypothesis that long-term regulation of arterial pressure is dependent on kidney function, which regulates the extracellular fluid volume and thus pressure by means of the relationship between blood volume and vascular capacity. It is suggested that the baroreceptor system is a rapid-acting short-term regulator of pressure through adjustments of cardiac output and total peripheral resistance.

Hypertension is caused by increased afferent arteriolar resistance which reduces renal blood flow and urine output. The resulting increase in extracellular fluid volume increases blood volume, venous return, cardiac output, and arterial blood pressure. It is stipulated that this feedback loop is designed to restore the urine volume load by increasing blood pressure and renal blood flow; however, when the autoregulation involves the afferent arteriole, a vicious positive feedback cycle is generated and sustained hypertension ensues.

In summary, it is unlikely that any one of the factors cited above is the sole causative agent. In all probability essential hypertension involves the participation of a number of

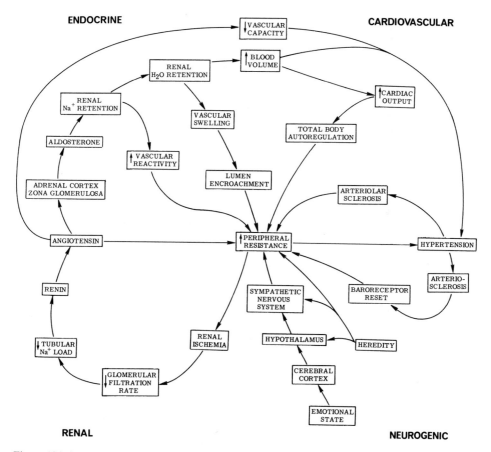

Figure 18A-1
Possible participation of various factors in the development of hypertension.

factors to varying degrees in different cases. This multiple-factor concept is particularly attractive in view of the conflicting experimental evidence and unpredictable pharmacological responses of the disease. Furthermore, there are probably many cases of hypertension classified as "essential" which may be due to other causes. They may represent either diseases not yet identified, an imbalance in the relationship of endogenous vasoactive substances, a modification of the participation of regulatory factors in the maintenance of normal blood pressure, or a vasoactive substance that has not yet been characterized. Figure 18A-1 is a diagram relating the various participating systems to each other.

REFERENCES

Guyton, A. C.   Hypertension: A disease of abnormal control. *Chest* 65:328–338, 1974.

Kannel, W. B.   Role of blood pressure in cardiovascular morbidity and mortality. *Prog. Cardiovasc. Dis.* 17:5–24, 1974.

Laragh, J. H. (Ed.).   Symposium on hypertension: Mechanisms and management. *Am. J. Med.* 55:261–414, 1973.

Page, I. H.   Arterial hypertension in retrospect. *Circ. Res.* 34:133–142, 1974.

Page, I. H., and J. W. McCubbin.   The Physiology of Arterial Hypertension. In W. F. Hamilton and P. Dow (Eds.), *Handbook of Physiology.* Washington: American Physiological Society, 1965. Section 2: Circulation, vol. 3, chap. 61, pp. 2163–2208.

18. Pathological Physiology of the Cardiovascular System
B. Congestive Heart Failure

Ewald E. Selkurt

Heart failure may be simply defined as a state in which the heart fails to maintain an adequate circulation for the needs and demands of the body despite what appears to be satisfactory filling pressure. When this failure is accompanied by an abnormal increase in blood volume and interstitial fluid, the condition is known as *congestive* heart failure. In Chapter 17, four types of cardiovascular reserves were discussed, which should be reviewed in terms of their role in the failing heart. These were (1) the venous blood oxygen reserve, (2) the heart rate reserve, (3) the systolic volume reserve, and (4) the diastolic volume reserve. As was explained, each has limitations as a reserve for the heart in failure. Perhaps the most valuable is the diastolic volume reserve, which acts by way of increased filling pressure. In early stages of heart failure these reserves may be adequate; accordingly, persons at rest or engaged in activity requiring minimal exertion manifest minimal symptomatology. However, increasing venous pressure distends the ventricle beyond the point of critical diastolic stretch, so that ultimately the contractile force declines and the output falls. The diminished cardiac reserve shows up during increased exertion and is evidenced by breathlessness, palpitation (perceptible heart beat), and fatigue with degrees of exercise that were formerly tolerated quite easily. As the condition progresses, these symptoms appear even at rest and are involved in several of the sequelae which result from heart failure and become a part of the overall clinical syndrome.

CAUSES OF HEART FAILURE AND ITS PHYSIOLOGICAL CONSEQUENCES

Circulatory failure occurs for a number of reasons that can be only briefly outlined here. They may be grouped into several categories: First, failure may occur because of *interference with systemic venous return* due to (1) factors usually remote from the heart, as in hemorrhage and shock or the type of peripheral vascular collapse noted in syncope or (2) factors in the heart region which impair venous return, such as pericardial tamponade, constrictive pericarditis, and tricuspid stenosis. Venous congestion is usually a feature of the latter group. Second, failure may result from *interference with the pumping or filling of the ventricle* due to anatomical abnormalities, either inherited or acquired through disease such as rheumatic fever. This type of failure is seen in semilunar valvular stenosis or insufficiency, mitral stenosis or insufficiency, and tricuspid insufficiency, and is also associated with systemic arterial hypertension and pulmonary arterial hypertension. Third, *primary diseases of the myocardium* lead to failure. This happens with myocardial infarcts created by coronary occlusion. Last, *chronic stimulus to high output* of one or both ventricles may ultimately cause failure. This happens with drastic reduction in peripheral vascular resistance, as in beriberi, systemic arteriovenous (A-V) fistula, Paget's disease (in which increased blood flow through the affected bone acts as an A-V fistula), anemia, and some types of congenital cardiovascular malformations—for example, those that have large left-to-right shunts. Despite the high output in the last group of examples, the blood supply to the various tissues and organs is inadequate, particularly in terms of the metabolic needs. On this basis, hyperthyroidism may lead to high-output failure. In all categories, circumstances occur which lead ultimately to failure of cardiac reserves.

Pathological Changes

Sudden occlusion of a branch of coronary artery (done experimentally in dogs) reduces arterial inflow to that region of the heart; after 10 to 15 seconds, the cells become cyanotic, abnormal electrocardiographic changes appear, and the cells cease contracting. Metabolism shifts from aerobic to the anaerobic form.

Characteristic structural changes are seen in severely ischemic myocardial cells after 30 to 40 minutes of coronary occlusion. These changes include virtual depletion of glycogen stores, relaxation of myofibrils, and nuclear and mitochondrial changes (swelling). On reperfusion by release of the occlusion, damaged cells swell enormously as the first phase of development of the irreversible phase of ischemic injury. Thus, the beginning of cell destruction is heralded by the inability of the cell to maintain its normal volume. Impairment of the pump(s) (e.g., sodium) which maintain normal electrolyte balance of the cells is implied.

Ischemic Injury

1. Decrease in arterial flow
2. Hypoxia → anaerobic metabolism
3. Decrease in cellular energy level

   cessation of
   specialized functions

IRREVERSIBILITY
↓
Cell death

The pathological changes characteristic of chronic failure of the heart include hydropic degeneration of the myocardial fibrils followed by necrosis, and fibrosis, particularly when associated with degenerative infections and metabolic heart disease. Ultimately, however, the defect must be sought in the ability of the myocardium to utilize energy and to perform work.

Metabolic Changes

From a biochemical point of view, the heart muscle cell failure could result from abnormalities in *energy production,* in *conservation,* or in *utilization* of the energy. Figure 18B-1 summarizes the steps involved in the transfer of foodstuffs to cardiac work. Substrate metabolism has been discussed in Chapter 15. In so-called low-output failure, resulting from arteriosclerotic hypertensive heart diseases or from valvular defects, some authorities believe there is no defect in energy production or conservation. In a study of congestive heart failure in both man and dog, no significant alteration of oxidative metabolism was found by Olson and Piatnek in 1959. More specifically, the distribution of phosphorus among the high-energy phosphate bonds reveals no significant change from normal. Nor does it occur with thyrotoxicosis (often associated with so-called high-output failure), although in this condition a shift toward utilization of fatty acids from glucose, lactate, and pyruvate has been noted. Such a shift in metabolism could not be a reason for myocardial failure, however.

Schwartz and Lee (1962) have challenged the adequacy of levels of energy-rich phosphate or of oxygen and substrate uptake as indices of the rate of energy liberation or storage. They suggest that normal concentration of high-energy phosphates could exist in the face of reduced rate of energy liberation and/or storage if the rate of energy utilization is also decreased. Such heart muscle preparations, however, may show a definite impairment of mitochondrial oxidative phosphorylation and contractile force. Studying mitochondria from failing guinea pig hearts (produced by aortic constriction), they found them clearly depressed with respect to oxidative phosphorylation. Specifically, phosphorylation associated with oxidation of glutamate, succinate, and

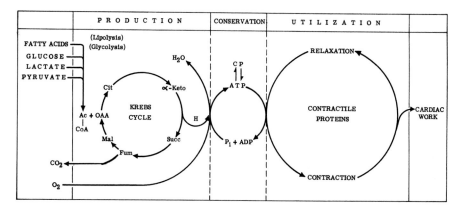

Figure 18B-1

Production. Metabolism of precursors to Ac-CoA (acetyl coenzyme A), which condenses with OAA (oxalacetate) to form citrate. This goes through the steps of the Krebs cycle to yield 8 hydrogen atoms or electrons ($H = H^+ + e$). These represent the energy content of the acetyl, which contributes 4 hydrogen atoms, plus 4 from the water added to the cycle intermediates in effecting the oxidation of the fragment. This step yields "energy-rich" electrons for the transport of oxygen. *Conservation.* This is associated with the electron flow along the hydrogen transport chain, which is depicted as coupling the phosphorylation of adenosine diphosphate (ADP) to the oxidation of the cytochrome enzymes. *Utilization.* The high-energy phosphate bond of adenosine triphosphate (ATP) is channeled into the contractile mechanism and results in the performance of mechanical work. (From R. E. Olson and D. A. Piatnek. *Ann. N.Y. Acad. Sci.* 72:467, 1959.)

α-ketoglutarate was uncoupled. The defect probably resided in the phosphorylation step associated with the electron transfer from cytochrome C to oxygen.

Although several other investigators support the conclusions of Schwartz and Lee, confirmed by the same laboratory (Lindenmayer et al., 1968), that mitochondrial oxidative phosphorylation is impaired in heart failure, this view is not universally accepted. In support of Olson and Piatnek, Stoner et al. (1968) found no impairment in the stressed dog heart (combined tricuspid insufficiency and pulmonary stenosis, pulmonary insufficiency, aortic stenosis, aortic insufficiency, for periods of 332 to 608 days). Chidsey et al. (1966) also found no deficiencies in the energy-transforming capabilities of mitochondria isolated from human myocardial tissues removed at the time of cardiac operations. They concluded that if a biochemical abnormality is responsible for defective ventricular function in the patients, it would appear to involve utilization of energy in the contractile process. Pool et al. (1969), studying right ventricular papillary muscles from cats with heart failure, found the contractile properties severely depressed, to 13 percent of the work of normals, and with this, reduction of use of energy to 7 percent of that used by normal muscle, but with no indication of inefficiency in the direct conversion of chemical energy to mechanical work. Rather, the reduced utilization of ~P was in relation to reduction in contractile element work.

This important controversy will require solution. Although species differences and differences in techniques may be at the basis of the disagreement, very possibly the degree and duration of stress account for the disagreement. Meerson (1962) has defined three significant developmental stages in the failing heart: (1) transient breakdown stage, characterized by contractile insufficiency with deficiency of certain enzymes; (2) protracted stage of relatively stable hyperfunction, characterized by the absence of cardiac insufficiency, by hypertrophy, and by adequate oxidative phosphorylation; and (3) protracted stage of progressing cardiosclerosis and gradual exhaustion of the hypertrophied heart, characterized by disturbances in protein synthesis and sustained hypoxia. It is suggested that stage 3 might be characterized by or caused by an aberration of mitochondrial function(s).

Katz concluded in 1960 that low-output failure does not appear to be related to defects in energy utilization but has suggested rather that muscle tension is not as productive of external work as the normal heart. Olson and Piatnek suggest a defect in the contractile mechanism itself, and this view has been supported by some authorities. They have proposed that an abnormal stable aggregate of myosin is formed in dogs with experimental cardiac failure and that this prevents the formation of actomyosin with normal contractile properties, leading to reduced contractility. But these studies have not been confirmed. Even if changes in the myosin are found, they could be the result, rather than the cause, of myocardial failure. For example, stretching of the sarcomeres as a result of congestive heart failure would decrease ATPase activity which has been found in muscles of failed hearts.

Calcium is important in the contraction and relaxation phases of cardiac action (Chap. 14). Its indispensable role in maintaining the force of cardiac contraction (actomyosin system) is well known. A possible role of calcium in heart failure has been explored. Heart failure can be experimentally produced in laboratory animals by perfusion of the heart with cobalt or nickel ions, which mimics the metabolic and mechanical effects of calcium deficiency, presumably by interfering with the role of Ca^{++} in contraction. Such a failure state can be corrected by addition of calcium to the cardiac perfusion. The rate of uptake of calcium by the sarcoplasmic vesicles of humans with congestive heart failure remains to be determined, however. Studies in this area may become available as heart transplants become more common.

The catecholamine content of the heart is depleted in both experimental heart failure and congestive heart failure in humans. Norepinephrine is synthesized by tyrosine hydroxylase and other enzymes in the heart from tyrosine via dopa and dopamine. Epinephrine cannot be synthesized, however, and that found in the heart is taken up from the circulation. Free catecholamines, derived either from a rapidly turning-over pool of storage particles (by nerve stimulation, or anoxia) or from the circulation, become active in the contraction process at receptor sites on the heart muscle, to play an important role in oxidative metabolism, glycolysis, and contractility of the heart, and to produce the inotropic effect. Nevertheless, the absence of a simple relationship between catecholamine depletion and impaired mechanical function of the heart suggests that depletion is not a basic and primary event in heart failure. This does argue strongly, however, against the use of antiadrenergic drugs in patients with established heart failure.

In certain other pathological conditions the possibility remains that the defect resides in energy production or conservation. For example, in hemorrhagic shock and myocardial infarction, general and local anoxia leads to defective energy production. In anemia, myocardial failure of the high-output type could result because insufficient oxygen is drawn in for substrate metabolism. Likewise, in beriberi heart disease, also associated with high-output failure, a deficiency in thiamine pyrophosphate (cocarboxylase) produces a breakdown of certain decarboxylation reactions that results in interference with normal myocardial energy liberation and conservation.

Hemodynamic Alterations

It has been customary in the past to consider the hemodynamic alterations in terms of two general aspects of heart failure classified as *forward failure* and *backward failure*. Forward failure implies alterations and sequelae which result from reduction in cardiac output and peripheral blood flow. The concept of backward failure is concerned with loss of contractile power and distention of the ventricle with ultimate inability for maintaining cardiac output. Venous pressure rises, and with it serious developments in terms of capillary fluid filtration and the formation of edema fluid (Chap. 12). On a functional basis it is unrealistic to categorize the consequences of heart failure into these two classifications, and the designations should be dropped.

For convenience, our discussion of the hemodynamic changes will begin with a consideration of the consequences of left ventricular failure. Cardiac output is usually less than 4.5 liters per minute. This results in alterations of territorial flow. Renal blood flow in man decreases from a normal of about 1200 ml to as low as 350 ml per minute. The ultimate effects on renal function will be further discussed. Splanchnic blood flow, including blood supply to the liver, is reduced, usually decreasing in proportion to the reduction in cardiac output. Eventually this decline too impairs hepatic function and is also the basis for gastrointestinal manifestations such as anorexia. The flow to the extremities is reduced and may be the basis of fatigue accompanying heart failure. Cerebral blood flow is usually normal, at least in the early phases of heart failure, as is coronary blood flow unless changes in the coronary vasculature have occurred.

When the left ventricle fails to discharge adequately, the output falls and the residual diastolic volume and diastolic pressure rise, as ultimately does the pulmonary venous pressure. If the right ventricle pumps normally, a redistribution of blood can be expected to occur with more in the pulmonary circulation and less in the systemic circulation. The result is pulmonary venous congestion, which leads to a further decrement in left ventricular output. Actually, this is compensated for in several ways. For example, with increased diastolic stretch, the left ventricle pumps somewhat better. Redistribution of the blood volume may in the end lower the output of the right ventricle, which can pump only what it receives.

Pulmonary Congestion

If pulmonary venous pressure rises beyond about 30 mm Hg, edema fluid begins to form and accumulate in the lung tissue and in the alveoli, impairing oxygen exchange. Engorgement of the lungs with blood and the accumulation of edema fluid seriously damage the mechanism of ventilation and increase the work of breathing, leading to undue breathlessness. Fluid accumulation leads to sounds heard by auscultation which are called *rales*. The lung tissue may become tough and lose its resilience because of connective tissue proliferation in the parenchyma. Alveolar membranes become thickened and edematous, further increasing the distance between the alveolar air and blood. *Vital capacity* (Chap. 19) is reduced not only by the extra blood and interstitial fluid in the lungs but also by the usual cardiac hypertrophy which occurs. Finally, there may be accumulation of excessive amounts of mucus and secretions in the small bronchioles and airways, further impeding air exchange. Thus, *dyspnea* is commonly seen. This condition is defined as rapid, shallow breathing of which the patient is conscious, as distinguished from *tachypnea* (rapid breathing) or *hyperpnea* (deep breathing). Dyspnea is not due solely to decreased oxygen and increased carbon dioxide of the blood; it is caused in large part by stimulation of vagus nerve stretch receptors in the lung.

Further manifestations of abnormal respiration are noted, perhaps of more serious consequence. These include *orthopnea, paroxysmal nocturnal dyspnea,* and *Cheyne-Stokes respiration.* Orthopnea is difficulty in breathing in the recumbent position as contrasted with the erect. It probably results from the shifting of blood from the dependent parts of the body to the vessels of the lung, exacerbating the subjective sensations and increasing the respiratory effort, in all likelihood because of enhanced reflex activity from the lung receptors. Paroxysmal nocturnal dyspnea is similar but occurs at night. It is also associated with the redistribution of blood that occurs during recumbency, plus depression of the central nervous system centers during sleep. The subject awakens with a feeling of suffocation, sits up gasping for air, and exhibits excessive pallor of the skin and profuse sweating. The skin may appear cyanotic because of the reduced hemoglobin of the circulation. Coughing and wheezing accompany the attack. The entire episode may last 10 to 20 minutes.

Basically, Cheyne-Stokes breathing presumably results from the depression of the respiratory center which is a consequence of the impaired blood flow following decreased cardiac output. Local vascular disease may contribute to this condition in the elderly. The depressed respiratory center is presumed to be insensitive to normal carbon dioxide tension but may still respond to raised carbon dioxide tension of the blood, and reflexly to anoxia (Chap. 21). With normal Pco_2 and Po_2, breathing stops. During apnea the arterial Po_2 falls and the Pco_2 rises, directly and indirectly stimulating the respiratory center to renew the breathing. However, the hyperpnea leads to a blowoff of carbon dioxide and improvement of the oxygen tension, with the result that apnea occurs again. This sequence is repeated cyclically. A more detailed discussion of periodic breathing is deferred to Chapter 21.

Systemic Congestion

When failure extends to the right side of the heart, other sequelae become manifest. These may be the indirect consequence of left heart failure and the resulting pulmonary congestion and pulmonary artery hypertension, which lead to ultimate impairment of right ventricular function. They may be more directly the consequence of primary lung disease with pulmonary hypertension or pulmonic valvular stenosis. Pulmonary embolism may lead to the symptomatology, as may certain types of atrial septal defects.

The immediate clinical manifestation of this change is elevation of systemic venous pressure (values of 25 to 40 cm H_2O). Venous engorgement is evidenced by the concomitant distention of the jugular veins. These are distended even in the erect position, which is not the case in the normal individual (Chap. 16). The elevation of venous pressure results not only from the inability of the right heart to pump adequately but also, in chronic heart failure, from the increase in blood volume—an increase both of the erythrocytes (hypoxemic stimulation of bone marrow) and of plasma volume (fluid retention). Unquestionable evidence has recently been gained that this is the case by using tagged albumin (^{131}I) and tagged erythrocytes (^{57}Fe and ^{59}Fe) simultaneously. In addition, several investigators believe that there is enhanced venoconstriction, a reflex response to reduced cardiac output. Ganglionic blocking agents (Dibenamine, Arfonad) that decrease sympathetic nerve activity cause significant reductions in venous pressure in the congestive heart failure patient, while only negligible changes in venous pressure occur in the normal subject given the drugs. This finding has been taken as presumptive evidence of increased venomotor activity in the heart patient.

A consequence of the elevated venous pressure is that the liver is engorged and becomes tender and painful in response to pressure. Ultimately, as a combined effect of venous congestion and reduced inflow, this leads to *cirrhosis* (fibrotic degeneration) of the liver, evidenced by increased bilirubin in the blood and jaundice. More specifically, decreased indocyanine green and Bromsulphalein (BSP) clearance result, indicating reduced hepatic blood flow. Splenomegaly is also seen. Peripheral venous congestion (i.e., in the skin) leads to the appearance of cyanosis; this occurs partly as a result of dilation of venules but also because of the probably impaired oxygenation of the arterial blood and greater oxygen extraction by the tissues. Thus, reduced hemoglobin in cutaneous vessels exceeds the 5 gm per 100 ml of blood necessary for the appearance of visible cyanosis.

Mechanism of Peripheral Edema

A number of factors are set into play in congestive heart failure that favor the accumulation of fluid in the interstitial spaces of various parts of the body, leading to edema. This is first manifested by an increase in the subject's weight—a gain of 10 to 20 pounds—before any objective signs are noted. Then the ankles and dependent parts begin to swell, and pitting edema is observed. Pitting edema is detected by a firm finger pressure applied to the edematous region. Fluid is expressed, leaving a depression which slowly refills. In the bedridden patient, sacral edema is prominent. Accumulation of the edema fluid in the lung bed (*hydrothorax*) has been discussed above. Fluid accumulation also occurs in the abdomen. This is called *ascites*. Peripheral edema fluid is low in protein, about 0.5 gm per 100 ml, but ascitic fluid may approach the protein content of plasma—5 to 6 gm per 100 ml.

The production and accumulation of edema fluid in congestive heart failure is a complicated process and is not simply the result of elevation in venous pressure leading to increased capillary filtration pressure. Since intake of fluid is apparently normal, accumulation of fluid must be caused by abnormal retention; hence, the kidney plays a paramount role. An accumulation not only of water but of salt, predominantly sodium chloride, to maintain the isotonicity of the fluid, is involved in this mechanism. Thus, although the first step in the formation of edema fluid is increased capillary filtration, a mechanism for overall accumulation of water and contained salts must be provided.

The mechanism for renal retention of salt and water has not been completely settled. Until the student becomes more familiar with renal physiology (Chap. 22), only brief details can be given. It has been noted that glomerular filtration is usually reduced in congestive heart failure, and some authorities feel that this action may be important for the retention of salt and water. Simply stated, when the load of sodium salts (NaCl and NaHCO$_3$) and water offered to the renal tubular cells is reduced, more complete reabsorption of the smaller load occurs. Although such a mechanism can be experimentally demonstrated, it perhaps oversimplifies the situation in congestive heart failure.

It has been demonstrated that, underlying the overall reduction in renal blood flow in heart failure, redistribution of regional renal blood flow occurs in which outer renal cortical flow is substantially reduced, whereas medullary flow may be little if at all modified. The xenon[133] washout method was employed for this analysis (Kilcoyne et al., 1973). (See Chap. 22 for additional information on regional renal blood flow.) The shift of blood flow from outer cortex to the juxtamedullary nephrons may invoke function of proportionally greater numbers of nephrons having higher filtration-reabsorptive capabilities for sodium than do cortical nephrons, so that, per volume of renal blood flow through this zone, sodium is more avidly retained.

The reduced cortical blood flow due to impaired cardiac output could result in increased renin production by the JGA (juxtaglomerular apparatus) and, in turn, increased angiotensin production. The outer cortex of the kidney has a particularly high renin content and hence might be particularly susceptible to the effects of reduced blood flow. It has been speculated that recirculated angiotensin II, which is a potent renal vasoconstrictor, might promote further reduction in cortical blood flow.

But explanations for sodium retention based purely on renal hemodynamic mechanisms are probably an oversimplification because humoral factors, such as hormones secreted by the adrenal cortex, especially aldosterone, are involved. It is well known that angiotensin stimulates aldosterone release by the adrenal cortical glomerulosa cells (Chap. 34). It has been clearly demonstrated by Leutscher's group (Camargo et al., 1965; Cheville et al., 1966) that there is enhanced production of aldosterone by the adrenal cortex in heart failure patients with edema. In advanced congestive heart failure, aldosterone secretion may be three to four times normal. Plasma aldosterone is elevated also because the liver's capability to metabolize the hormone is impaired by hypoxia, a result of the impaired hepatic blood flow in heart failure (reduced arterial inflow and increased hepatic venous pressure reflecting elevated right atrial pressure). The kidneys too play an important role in the metabolism and excretion of aldosterone and its conjugates, so that when kidney blood flow and filtration rate are reduced, this avenue of metabolism and excretion is damaged and retention of aldosterone is favored. The increased levels of plasma aldosterone promote enhanced reabsorption of sodium by the distal convoluted tubules of the nephron (Chaps. 22 and 34).

However, continued injection of aldosterone in the normal experimental subject, although it may cause transient sodium retention, is followed by later readjustment of renal mechanisms, so that sodium and attendant water is again lost to restore homeostasis. Thus, hyperaldosteronism itself cannot be the sole mechanism involved but is accepted as an important contributing factor. Possible mechanisms for triggering increased release are discussed in Chapter 34. Davis has suggested that an additional factor, or factors, of a humoral or hemodynamic nature, distinct from any identified at present, is involved in congestive heart failure which potentiates the aldosterone influence. The possibility that a sodium-"losing" hormone is involved in the overall homeostasis has been entertained by some investigators.

In conclusion, primary mechanisms involved in congestive heart failure have been presented, and the sequelae of the primary derangements have been discussed. Although the physician is concerned with improving the well-being of his patient by allaying the sequelae—e.g., removing edema fluid by administering diuretics to institute renal loss of salt and water—he continually strives to correct, insofar as possible, the basic derangements. An illustration is his use of drugs such as digitalis to improve the myocardial pumping action.

REFERENCES

Blumgart, H. L. (Ed.). Symposium on congestive heart failure. *Circulation* 21:95–128, 218–255, 424–443, 1960.

Braunwald, E. (Ed.). Symposium on myocardial metabolism. *Circ. Res.* 35 (Suppl. 3): 1–215, 1974.

Burch, G. E., and C. T. Ray. A consideration of the mechanism of congestive heart failure. *Am. Heart J.* 41:918–946, 1951.

Camargo, C. A., A. J. Dowdy, E. W. Hancock, and J. A. Leutscher. Decreased plasma clearance and hepatic extraction of aldosterone in patients with heart failure. *J. Clin. Invest.* 44:356–365, 1965.

Cheville, R. A., J. A. Leutscher, E. W. Hancock, A. J. Dowdy, and G. W. Nokes. Distribution, conjugation and excretion of labeled aldosterone in congestive heart failure and in controls with normal circulation. *J. Clin. Invest.* 45:1302–1316, 1966.

Chidsey, C. A., G. A. Kaiser, E. H. Sonnenblick, J. F. Spann, Jr., and E. Braunwald. Cardiac norepinephrine stores in experimental heart failure in dogs. *J. Clin. Invest.* 43:2386–2393, 1964.

Chidsey, C. A., E. C. Weinbach, P. E. Pool, and A. G. Morrow. Biochemical studies of energy production in the failing human heart. *J. Clin. Invest.* 45:40–50, 1966.

Davis, J. O. The Physiology of Congestive Heart Failure. In W. F. Hamilton and P. Dow (Eds.), *Handbook of Physiology.* Washington: American Physiological Society, 1965. Section 2: Circulation, vol. 3, chap. 59, pp. 2071–2122.

Kilcoyne, M. M., D. H. Schmidt, and P. J. Cannon. Intrarenal blood flow in congestive heart failure. *Circulation* 47:786–797, 1973.

Lindenmayer, G. E., L. A. Sordahl, and A. Schwartz. Reevaluation of oxidative phosphorylation in cardiac mitochondria from normal animals and animals in heart failure. *Circ. Res.* 23:439–450, 1968.

Meerson, F. Z. Compensatory hyperfunction of the heart. *Circ. Res.* 11:250–258, 1962.

Meerson, F. Z. The myocardium in hyperfunction, hypertrophy and heart failure. (Monograph 26, American Heart Association.) *Circ. Res.* 25 (Suppl. 2): 1–163, 1969.

Opie, L. H. Metabolism of the heart in health and disease: Part I. *Am. Heart J.* 76:685–698, 1968; Part II. 77:100–122, 1969; Part III. 77:383–410, 1969.

Pool, P. E., B. M. Chandler, J. F. Spann, Jr., E. H. Sonnenblick, and E. Braunwald. Mechanochemistry of cardiac muscle: IV. Utilization of high-energy phosphates in experimental heart failure in cats. *Circ. Res.* 24:313–320, 1969.

Schwartz, A., and K. S. Lee. Study of heart mitochondria and glycolytic metabolism in experimentally induced cardiac failure. *Circ. Res.* 10:321–332, 1962.

Stoner, C. D., M. M. Ressallat, and H. D. Sirak. Oxidative phosphorylation in mitochondria isolated from chronically stressed dog hearts. *Circ. Res.* 23:87–97, 1968.

Wood, P. H. *Diseases of the Heart and Circulation* (3rd ed.). Philadelphia: Lippincott, 1968.

18. Pathological Physiology of the Cardiovascular System
C. Physiology of Shock

Ewald E. Selkurt

The term *shock* is defined as an abnormal physiological condition resulting from inadequate propulsion of blood into the aorta, and therefore inadequate flow of blood perfusing the capillaries of various tissues and organs. It is typified by deterioration of cellular functions, particularly of vulnerable organs such as the kidney, liver, and heart. It is further characterized by a progressive failure of the circulation eventuating in an *irreversible state.*

CLASSIFICATION OF TYPES OF SHOCK; CAUSES AND MANIFESTATIONS

Circulatory shock describes a syndrome that is characterized by protracted prostration and hypotension, pallor, coldness and moistness of the skin, collapse of superficial veins, reduced sensibilities, and suppression of urine formation. It is important to recognize that all these symptoms may be present without development of the potentially irreversible physiological condition to which the term should be limited. In consequence, some authorities have recommended two subdivisions: primary shock and secondary shock. Primary shock is essentially a bout of acute hypotension. It tends to be transient, and there is usually a natural tendency to recovery. Secondary shock is typified by deterioration of cellular function and progressive failure until an irreversible state is reached, culminating in death. In arriving at a diagnosis, one must therefore relate the existing clinical signs and symptoms to the antecedent events, gauging the probabilities as to whether they indicate development of true shock or are merely a temporary circulatory derangement. Although better clinical criteria could be desired, progressive reduction of arterial pressure and pulse pressure measured at frequent intervals constitutes the best available evidence of the existence of a physiological state of shock. Collapse of the veins is an important sign.

Symptomatology

Clinical symptoms are referable to the skin, the mucous membranes and the countenance, the neuromuscular system, the circulatory and respiratory systems, and the metabolism. The skin is pale and cold because of peripheral vasoconstriction. This is accompanied by sweating because of massive stimulation of the sympathetic division of the autonomic nervous system. The mucous membranes especially tend to be bluish as a result of the enhanced oxygen extraction from the blood (stagnant anoxia). The countenance appears haggard and drawn, in part because of the removal of interstitial fluid from the skin due to hypotension with resultant reduction in capillary hydrostatic pressure which normally opposes the inward-drawing force of the plasma proteins. The reflexes tend to be depressed and responses to noxious stimuli reduced. The blood pressure is decreased, and the pulse is rapid and thready as a result of reduced cardiac output. The respiration is rapid and shallow, on account of both reflex and chemical stimulation. Body temperature may be decreased, partly because of decreased general metabolism resulting from serious reduction in oxygen supply. All these conditions are symptomatic of the loss of circulating blood volume.

Predisposing Causes

In most cases there is a loss of blood or plasma fluid with a consequent prolonged period of hypotension. Some of the more important causes are (1) hemorrhage; (2) extensive edema or serous effusion due to mechanical, chemical, or thermal capillary

419

damage; (3) seepage of wounds or burn surfaces; (4) abstraction of fluids through excessive sweating, vomiting, diarrhea, or drainage of secretions after surgical operations; (5) disturbances of electrolyte and water balance, causing withdrawal of water from the bloodstream; and (6) unusual engorgement of the capillaries with blood, reducing effective blood volume. The last is frequently the causative factor in the shock that occurs during overwhelming infections, such as those from gram-negative bacteria.

STAGES OF SHOCK

Initial or Hypotensive Stage

Primary loss of blood or fluid leads to a disparity between the circulating blood volume and circulatory capacity. This results in reduction of venous return and in an acute decrease in cardiac output, with arterial and pulse pressures and ultimate impairment of circulation and lowered oxygenation of the tissues of the body. Compensatory mechanisms are instituted rapidly, as follows: Hypotension acts via the carotid sinus and aortic arch baroreceptors to cause reflex vasoconstriction and is supplemented by the action of catecholamines released in increased amounts which augment peripheral arteriolar and venous tone. The same reflexes augment the cardiac vigor and rate. The fall in blood pressure also reflexly causes an increase in the depth and rate of respiration, which favors venous return and ultimately cardiac filling—the so-called abdominothoracic pump mechanism discussed in Chapter 16. Increased activity of venopressor mechanisms may contribute to venous return. Further repletion of blood volume results from an increase in plasma volume through the reduction of capillary pressure, which leads to reabsorption of tissue fluid and ultimately plasma replacement. Thus, hemodilution is often noted during this stage.

Increased production and release of aldosterone and antidiuretic hormone (ADH) accompany the bleeding (Fig. 18C-1). These can also be viewed as initial compensatory mechanisms, in the light of favoring salt and water retention by kidney, which is further favored by the reduced glomerular filtration as the result of the hemorrhagic hypotension. The increase in renal sympathetic nerve activity (upper panel, Fig. 18C-1) facilitates afferent arteriolar constriction, reducing glomerular filtration even more (Chap. 22). The ADH increase results from reduced stimulus of the left atrial volume receptors; reduced firing of these receptors (second panel, Fig. 18C-1) releases from inhibition the neurohypophysial secretion of ADH (Chap. 30).

Chien and Usami (1974) have recently confirmed the ADH release in response to hemorrhage and provided convincing evidence that the afferent fibers are in the vagus and carotid sinus nerves. However, extra-vagocarotid factors may play a role in causing ADH release after severe hemorrhagic hypotension.

Progressive or Impending Stage

The arterial pressure, previously stabilized by compensatory mechanisms, may decrease slowly and progressively. There is no further vasoconstriction; in fact, total peripheral resistance (TPR) may begin to lessen. The low arterial pressure together with vasoconstriction exerts additional deleterious effects on the various organs and tissues because of reduction in capillary flow. The generalized tissue hypoxia leads to impairment of enzyme systems and to anaerobic metabolism, with an accumulation of lactates and pyruvates and with failure of resynthesis of organic phosphates. Reduction of alkali reserve occurs, and there is migration of potassium from tissue cells. The vital organs are variously affected. This stage has been called *oligemic shock*.

Organic Alteration in the Progressive Stage

Kidney
Ischemia and the resultant stagnant anoxia result in failure of glomerular filtration, and the reduced blood flow causes eventual tubular damage, evidenced by impaired extraction of para-aminohippuric acid (PAH). These renal alterations lead to anuria and

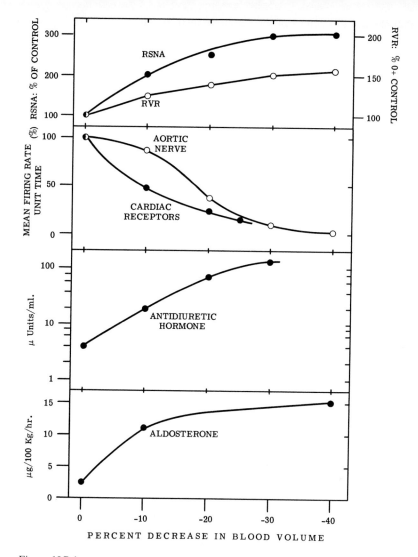

Figure 18C-1

Data showing (from above downward) the measured responses to a graded hemorrhage in terms of the increase of autonomic (sympathetic) drive as the measured change in renal sympathetic nerve activity (RSNA) and in renal vascular resistance (RVR). With loss of the first 10 percent of the blood volume, the sympathetic drive from baroreceptors in the low-pressure system increases sharply as the firing rate of the cardiac (atrial) receptors falls to one-half (second panel). As the loss exceeds 10 percent, there is an increasing influence from the arterial system, as exemplified by the aortic nerve activity. Receptor activity is expressed in percent of resting value: receptor drive is an inverse function of mean firing rate.

Meanwhile, the plasma aldosterone and antidiuretic hormone (ADH) concentrations increase, effecting water and salt retention. Volume-regulating hormones are expressed as micrograms of aldosterone released into the plasma, and microunits per milliliter of ADH. The increase in ADH is presumed to result from cardiac receptors in the left atrium. When stimulated by volume increase, afferent vagal fibers which ascend to the neurohypophysis inhibit ADH release. When atrial volume is decreased, as with hemorrhage, the reverse trend occurs; i.e., ADH output increases. The increase in aldosterone release results, in general terms, from decreased renal blood flow and triggering of the JGA renin-angiotensin mechanism (Chap. 34). (For data sources, see O. H. Gauer, J. P. Henry, and C. Behn. *Annu. Rev. Physiol.* 32:547–598, 1970. Aldosterone was measured in anesthetized dogs, and ADH in unanesthetized dogs—note the logarithmic scale. Atrial [cardiac receptor] firing, RSNA, and RVR activity were measured in anesthetized dogs, aortic nerve activity in the anesthetized cat. RSNA, RVR, and aortic nerve activity data were supplied by W. V. Judy.)

retention of metabolic products (urea, creatinine, uric acid). Furthermore, they contribute to an upset in the acid-base balance in the direction of metabolic acidosis because of failure of the normal mechanisms for hydrogen ion secretion by the kidney, and the accumulation in the blood of phosphates and lactates. When crushing injuries to the limb accompany shock, there is more profound renal injury because the release of myohemoglobin and products of muscle autolysis contribute to renal failure; renal tubules, particularly the distal and collecting, are blocked by debris and hemoglobin deposits. The term *lower nephron nephrosis* has been applied to this situation. A serious consequence of shock is that delayed renal failure may ultimately lead to death in uremia even though the patient may have been treated with transfusion, with adequate recovery of blood pressure. In any event, even after treatment the kidney may for a prolonged period show residual damage characterized by loss of concentrating power, evidenced in excretion of dilute urine of low specific gravity. The loss of concentrating ability is largely due to the loss of the gradient of osmolality in the kidney (Chap. 22), due to "washing out" of the high papillary concentration needed for urinary concentration by the ADH mechanism. The washout is a result of marked reduction or cessation of glomerular filtration, with persistence of blood flow through the medulla (Selkurt, 1969).

The frequent finding of dilute urine of low specific gravity and positive free water clearance ($+CH_2O$, Chap. 22) by the shock kidney invokes the possibility that other factors are involved beyond a simple washout of the gradient. Since a possible deficiency of output of ADH could not be a factor (Selkurt, 1974), an apparent loss of sensitivity of the epithelium of the collecting duct to ADH action is indicated. Strong evidence is accumulating that another category of humoral agents, the prostaglandins (PGE_1 and PGE_2) are involved in the apparent ADH antagonism, as well as in other manifestations of impaired tubular function in shock—e.g., impaired electrolyte reabsorption (Na^+, Ca^{++}, and K^+). It is important to emphasize that these are not direct effects of hypoxia in the tubular epithelium, but rather the consequence of the influence of hypotension and altered renal medullary blood flow on the synthesis of prostaglandins by cells in the medulla which produce them from fatty acid precursors (Chaps. 22 and 30).

The prostaglandins of the renal medulla could play a significant role in altered renal function because of their strategic location. Lipid-containing granules in the cytoplasm of the interstitial cells of the renal medulla and papilla give evidence of the sites of synthesis. These cells lie horizontally in the interstitial space between the vertically oriented loops of Henle and vasa recta (Chap. 22). The granules contain lipids which presumably serve as precursors for synthesis of the prostaglandins. When released, these substances quite possibly diffuse into the neighboring blood vessels to be carried up to the juxtamedullary zone by ascending vasa recta to relax the vascular smooth muscle and sphincters, and thus increase medullary blood flow. This action could favor the washout of the gradient. Ultimately, some prostaglandins leave via the renal vein. Entering the loop of Henle, they could influence tubular cell activity of the ascending limb of Henle and finally the distal convolution and collecting duct, with regard to opposing ADH action.

Interestingly, infusion of PGE directly into the renal artery elicits many of the electrolyte transport and urinary volume changes commonly observed in shock, such as increases in fractional osmolar and electrolyte clearance (Na^+, Ca^{++}, K^+). If the kidney is initially concentrating the urine (Chap. 22), this process may be reversed to production of dilute (hypotonic) urine of relatively greater volume.

The mechanism that triggers synthesis and/or release is still under investigation. One effective means of activating increased appearance in renal vein blood is brief renal ischemia. Renal vasoconstriction by nerve stimulation also enhances release. Strong evidence likewise exists for humoral mediators that promote synthesis and release of prostaglandins. These are catecholamines and angiotensin II. Release of both is stimulated by hemorrhage, as has been discussed earlier (Fig. 18C-1).

Liver

The circulatory anoxia impairs the normal mechanisms of metabolic turnover of such substances as pyruvates, lactates, and amino acids released from traumatized tissue, by disorganizing the hepatic cell enzyme systems. The reason for the accumulation of the amino acids in the blood is impairment of the liver's normal ability to form urea. There may be early glycogenolysis and glycogen depletion and hyperglycemia, but later impairment of gluconeogenesis and ultimate hypoglycemia. In connection with the disorganized enzyme systems, it has been observed that the adenosine triphosphate (ATP) and adenosine diphosphate (ADP) concentrations are reduced in the shock liver.

Cation transport of the liver is impaired in hemorrhagic shock in rats (Sayeed et al., 1974). Thus, sodium-potassium transport is deranged, and sodium accumulation within the liver cells is accompanied by water entry, as cell volume regulatory capability is lost. As a result, the cell swells in an early phase of cell destruction. Failure of ATP and ADP accounts for the membrane electrolyte pump failure, in part. Early in shock (one-half hour of hypotension) the changes could be reversed by reinfusion of shed blood and Ringer's lactate, but not in intermediate (one hour) or late (two hours) shock.

Heart

Eventually the myocardium may become depressed, because of reduced oxygen supply and possibly accumulation of toxic materials. A reduced coronary flow because of hypotension occurs despite compensatory dilation of the vascular supply of the myocardium.

Two types of lesions are characteristic of heart damage in shock (Fig. 18C-2), subendocardial hemorrhage and necrosis, and "zonal lesions" (Hackel et al., 1974). The latter are described as a zone of hypercontraction at the end of the myocyte adjacent to an intercalated disc, accompanied by shortening of the sarcomeres, fragmentation of the Z band, distortion of the myofilaments, and mitochondrial displacement. Such lesions are also seen in certain muscular dystrophies and excessive catecholamine stimulation.

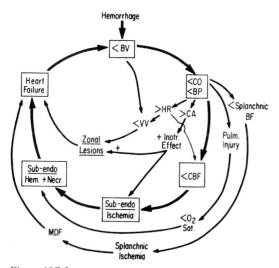

Figure 18C-2

Factors contributing to heart failure in shock. BF = blood flow, CA = catecholamines, Hem. = hemorrhage, Inotr. = inotropic, Necr. = necrosis, Sat. = saturation, Sub-endo = subendocardial, and VV = ventricular volume; BV = blood volume, CO = cardiac output, CBF = coronary blood flow, and MDF = myocardial depressant factor. (From D. B. Hackel, N. B. Ratliff, and E. Mikat. *Circ. Res.* 35:805–811, 1974. By permission of The American Heart Association, Inc.)

Brain

The blood supply to the brain appears to be reasonably well maintained because of compensatory vasodilation, and at the sacrifice of flow to the other organs. For example, the blood flow to the skin, kidneys, and splanchnic bed is diverted by more intense vasoconstriction to the brain, pulmonary, and coronary circulations. Brain samples indicate that the metabolic condition appears to be good in terms of ADP, glycogen, and phosphocreatine contents. However, occasionally death by respiratory failure occurs during this stage, suggesting greater susceptibility of the respiratory centers.

Irreversible Stage

There is further decrease in cardiac output. Blood pressure may be very low, and pulse pressure may be barely detectable. Respiration becomes depressed. Cardiac deceleration may follow, because of default of reflexes or weakening of the myocardium. In some instances the total peripheral resistance begins to decrease, indicating relaxation of compensatory vasoconstriction. This may be the result of failure of the vasomotor centers due to prolonged hypoxia but more probably is caused by the production and release of vasoactive substances in the blood which directly or indirectly depress blood pressure or heart action or both. It is of extreme significance that large infusions of blood and blood substitutes given at this time may prove to be of temporary benefit only. With transfusion, blood pressure and other hemodynamic alterations may be reasonably well restored toward normal for a time, but the restorations are likely not to be maintained. The state is called *normovolemic shock* and may progress into irreversible failure and death.

A major manifestation of the irreversible stage is the observation in experimental animals that the capillaries, particularly in the splanchnic bed, dilate by relaxation of the precapillary sphincters. Thus, there is an eminent possibility that stagnant blood will be impounded in these areas. The capacious pulmonary vascular bed does not appear to be an important site of pooling, however (Abel et al., 1967). Furthermore, release of proteases, histamine, and other substances during the hypotensive phases seems to favor increased capillary permeability, which may allow fluid to escape into the extravascular spaces. There is evidence that fluid moves from the interstitial compartment into cells. This movement may be responsible for the hemoconcentration that is seen in the terminal phases of shock. In the dog especially, the intestinal bed shows marked congestion of the capillaries and venules with extreme extravasation of fluid and even blood into the intestinal lumen; hence, significant amounts of transfused blood and fluid may ultimately be lost into the intestinal lumen and be discharged from the body as bloody diarrhea. However, this mechanism is not important in man and the monkey.

Experimental investigation indicates that myocardial depression contributes to and even accelerates the fatal end: Cardiac output decreases despite terminal elevation of ventricular filling pressure, suggesting weakening of the myocardium; response to equivalent states of diastolic distention is lessened when infusion is given; and the S-T segments of the electrocardiogram are altered.

EXPERIMENTAL PRODUCTION OF SHOCK

The experimental approaches to the study of the basic mechanisms of shock have been many and varied, and space does not permit a detailed description and evaluation of the numerous techniques. Generally, these are designed to simulate the development of shock as it is caused in man. One of the most commonly employed is hemorrhage, illustrated in Figure 18C-3.

The dominant effect of hemorrhage (curve 1) is the reduction in cardiac output that results (curve 3) because of reduced filling pressure (curve 4). Mean blood pressure declines (curve 2), with decrease in pulse pressure, reflecting reduced stroke volume. Reflex stimulation of respiration results (curve 5), supplemented by chemical stimulation from developing acidosis later in the oligemic phase (curve 13). Circulatory

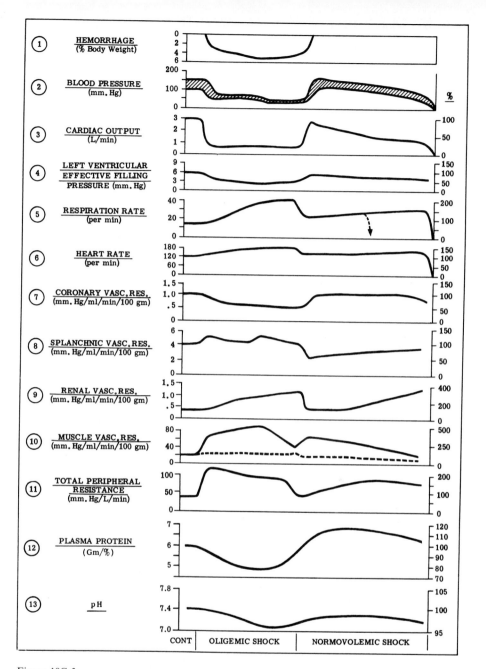

Figure 18C-3

Cardiodynamic trends in experimental hemorrhagic shock in dogs. Values to left are absolute units, to right in percentage change from control (cont) (set at 100 percent).

compensatory mechanisms include increase in heart rate, particularly if the control rate is slow (curve 6), and enhanced vasoconstriction, so that total peripheral resistance increases markedly at first (curve 11). The increase in TPR is both neurogenic and humoral. Regional and organ changes in vascular resistance vary. The resistance actually *decreases* in the coronary circulation (curve 7) and also in the cerebral circulation (not shown); it shows only minor increase in the splanchnic circulation (curve 8) but increases significantly in the kidney (curve 9) and muscles (curve 10; the dashed line shows the relationship with cold block of the nerve). Later, in oligemic shock, TPR declines somewhat. The apparent reduction of plasma protein concentration (curve 12) suggests that a supplementary compensatory mechanism is the influx of interstitial fluid into the capillaries.

Following transfusion, cardiac output and arterial pressure are temporarily restored, then begin a progressive downward trend. Cardiac output declines because of decrease in filling pressure. Although heart rate is restored almost to control, there is a terminal slowing.

Respiratory rate slows somewhat at first, but increases later in normovolemic shock, reflecting in part the underlying disturbance in acid-base balance. Susceptible animals may die abruptly during this phase (see arrow, curve 5) because of respiratory failure. Respiratory changes are, however, variable; a slowing late in normovolemic shock is often observed.

TPR returns toward control following transfusion but then begins a secondary rise. Coronary resistance returns to control but may show a decrease terminally. Renal resistance is temporarily restored toward control level, then progressively increases. Splanchnic resistance shows a phase *below* control which has not been satisfactorily explained, then gradually rises back to the control value. Muscle resistance remains high, although it manifests a downward trend later in normovolemic shock.

Several conclusions can be drawn. Failure of the circulation is not the result of failure of vascular reflex compensatory mechanisms, in view of the well-maintained TPR. Myocardial failure cannot be implicated as an important factor early in normovolemic shock. When extra fluid is given to restore filling pressure, cardiac output is restored. Thus, continued loss of effective circulating blood volume is the reason for the gradual decline in blood pressure. Late in normovolemic shock, despite an attempt to maintain coronary circulation by compensatory dilation during hypotension, flow is impaired because of the reduced cardiac output and blood pressure. Alterations in myocardial metabolism have been demonstrated as a consequence of the impaired flow. Moreover, toxic products of a vasodepressor nature developed in shock probably have an influence on the depression of the myocardium.

Other methods of producing shock include crushing and contusion of muscles, fractures of bones, burns and scalds, intestinal obstruction, and prolonged intestinal exposure and manipulation. Less directly, prolonged ischemia of the limbs, induced by tourniquets and followed by release, or ischemia and release of the intestinal circulation can be used for this purpose. Anoxia of the splanchnic bed, experimentally induced by ischemia, has received considerable attention lately as a method of producing shock, for several reasons. In addition to the products of ischemia (see below), the intestinal bed is the site of toxic amines (e.g., tyramine), which might be released into the general circulation with deleterious effects on the cardiovascular system. Ischemia of the intestine created by clamping of the superior mesenteric artery for an hour or two, then release, leads to irreversible shock with much of the symptomatology characteristic of this condition. The organism becomes particularly susceptible because of concomitant impairment of normal liver function, which therefore cannot detoxify the products that may reach the portal circulation. Experimental evidence exists that vasodepressor substances do pass into the systemic circulation as a consequence of intestinal hypoxia, not only with ischemic shock but during hemorrhagic shock.

Some investigators have emphasized the role of bacterial endotoxins. In their view, bacteria or their products, released from the intestine during shock, are not screened by the impaired liver and reach the systemic circulation to cause cardiovascular collapse. A lipopolysaccharide which is believed to be the endotoxin involved has been isolated from the plasma of the shock animal. Very small amounts of this substance when injected into the experimental animal lead to shock symptoms, particularly if the animal has been weakened by a period of moderate hypotension which by itself is not sufficient to cause shock. During the course of hemorrhagic hypotension, the reticuloendothelial system (RES) becomes depressed, with the result that the antibacterial and antitoxic defense mechanisms of the body are impaired (Blattberg and Levy, 1962). Endotoxins from the normal bacterial flora of the intestine constantly enter the intestinal circulation. Ordinarily, they are inactivated by the RES, principally in the liver. When the RES is severely depressed, the endotoxins invade the general circulation. Endotoxins produce a form of shock resembling in many respects that produced by hemorrhage. Therefore, depression of the RES leads to an intensification of the hemodynamic changes caused by blood loss. Sterilization of the intestine by means of antibiotics significantly reduces the mortality from certain standard shock-provoking procedures, including hemorrhage.

The hemodynamic changes have been studied by intravenous injection of the endotoxins into dogs. Within one minute of injection the endotoxins (derived from *Brucella melitensis* or *Escherichia coli*) cause a sharp fall in arterial blood pressure and an increase in portal venous pressure. The liver and intestines become engorged with blood, suggesting that the fall in arterial pressure and the elevation of portal pressure may result from pooling of blood in the splanchnic venous bed. These changes lead to an impaired venous return to the heart and reduced cardiac output. One consequence of sequestration of blood in the intestine is impaired circulation in the myocardium; thus, myocardial failure may be secondary to the primary event. Additional factors include capillary injury with resultant excessive loss of fluid into interstitial spaces. These changes together are implicated in the progression of shock to irreversibility.

The relationship of endotoxin shock to hemorrhagic shock is not clear at present. Nor is it yet known how the theory of endotoxin shock will fit ultimately into the picture of clinical shock presented by the injured human subject. Unfortunately, the physiology of shock in the experimental animal, such as the dog, differs in significant ways from that in the primate. Much more work needs to be done in the latter species.

THEORIES OF THE MECHANISM OF SHOCK

Several theories of the causes of irreversibility of shock have been based upon findings in experimental animals—dog, rat, and monkey. A key mechanism is fluid loss from the effective circulation. This can follow directly, as by hemorrhage, or indirectly by a loss of fluid resembling plasma, as occurs in burns. There may be local fluid loss at the site of trauma, or it may be caused by prolonged vomiting and diarrhea. The loss of circulating volume triggers a series of mechanisms (described above) which lead to the secondary sequelae of reduced cardiac output, reflex vasoconstriction, and overall impairment of flow in the various tissues and organs, with consequent secondary effects due to deranged metabolism.

Neurogenic Factors

The role of the nervous system has received much attention. The effect of afferent impulses from the sites of trauma, burns, etc., on the vasomotor and respiratory systems has been explored, together with the possibility that failure of the ensuing reflex vasoconstriction accounts for the late irreversibility of shock. Current evidence indicates that total peripheral resistance is well maintained throughout the course of oligemic shock (reduced blood volume) and through most of normovolemic shock (following transfusion). Sympathetic nerve impulse discharge remains high, in the main, and catecholamine output is never decreased (Fig. 18C-4). Thus, progressive fall in blood pressure after transfusion must result in large measure from other factors. Among these, fluid loss has to be given serious consideration.

Changes in Body Fluids

The problem of fluid loss can be considered from the broad point of view of the application of known techniques for the measurement of vascular volume, interstitial volume, and intracellular water during shock, to see whether significant alterations

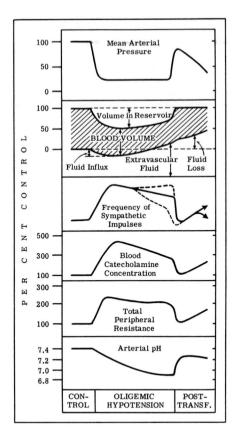

Figure 18C-4

The time course of changes in sympatheticoadrenal activity and circulatory functions in constant-pressure hemorrhagic shock (retransfusion after three to four hours at 40 mm Hg) in the dog. These data were taken from different studies. Because of variations in experimental procedures, the time relationships are not quantitative. Ordinate scales are used only to give an approximate estimation of magnitude of changes and are not exact. In the second panel, the remaining blood volume is indicated by the shaded area. Early in oligemic hypotension a portion of the extravascular fluid enters the circulation and becomes part of the blood volume, whereas the reverse occurs later. The initial, transient increase in sympathetic activity during the first few minutes after hemorrhage is not shown. After prolonged oligemic hypotension and after retransfusion, the changes in frequency of sympathetic impulses may take different time courses. There are no experimental data with simultaneous measurements to show whether or not the different time courses are associated with corresponding variations in other parameters, e.g., total peripheral resistance. Total peripheral resistance generally falls toward the control level after an initial rise. This decrease is parallel to the progressive development of acidosis and gradual decline of sympatheticoadrenal activity. (From S. Chien. *Physiol. Rev.* 47:252, 1967.)

in these volumes indeed occur. Closely related is the consideration of specific organs that might be involved in such disruption by virtue of being particularly susceptible, such as the intestine, or perhaps important because of their mass, such as skeletal muscle.

Many studies have dealt with the changes in plasma volume and red cell mass in the post-transfusion phase of hemorrhagic shock. Most of them have been done on the dog, but more and more data are accumulating for man and other primates. Very typically in the dog, the plasma volume and red cell volume return to normal immediately after transfusion, but both plasma volume and red cell volume later decrease progressively.

During oligemic hypotension, blood is diluted by fluid influx (Fig. 18C-4). This is restored on transfusion, but late in the post-transfusion phase protein concentration and hematocrit value appear to increase, presumably because fluid is lost again. Although plasma loss and hemoconcentration are generally found during the late oligemic period and after retransfusion in the dog, it is important to note that hemoconcentration is not found in man, even after severe hemorrhage. If anything, the findings indicate that in man subjected to hemorrhagic shock there is hemodilution rather than extravasation of plasma fluid. Admittedly, this is often complicated by the therapeutic regimen. In a critical experiment done in another species, the sheep, Gillett and Halmagyi (1966), by the use of ^{51}Cr-labeled cells and ^{131}I-labeled plasma for determination of total blood volume, found it virtually unchanged in irreversible shock. Such findings emphasize that fluid loss alone, when it occurs, need not be the sole factor responsible for circulatory failure in shock.

Other relevant questions pertain to the changes in extracellular and intracellular fluid volumes. Extracellular water (ECW), as measured by distribution of inulin or ^{22}Na or ^{35}S sulfate measurements, interestingly shows a decrease in the late hypotensive and late post-transfusion phases in the splenectomized dogs. The typical finding was a small reduction in ECW in the post-transfusion phase. The magnitude of reduction, regardless of the label used, was quite modest, ranging from about 5 percent to 8 percent. Occasionally, larger changes were observed. For example, in several of the dogs examined by Shizgal et al. (1967), changes were greater than 10 percent and averaged about 20 percent. The question was: What happened to the extracellular fluid water under the circumstances of "irreversible shock"? The possibility that the loss of extracellular fluid water could be accounted for by movement into the cells was suggested by the work of Slonim and Stahl (1968) and Matthews and Douglas (1969). The notion of increased intracellular water volume has led to the speculation that there may have been a failure of the sodium pump as a result of the hypoxia during the oligemic stress, and that now water moved into the cells because of the failure of the pump to move sodium out.

Myocardial Failure

The fact that systemic arterial pressure declines on account of inadequate cardiac output in the phase of deterioration in shock raises the question: Why does the cardiac output decline after the reinfusion of the shed blood? Rothe (1970) has sought an answer by examining ventricular stroke work in dogs during hemorrhage and the post-transfusion phase (Fig. 18C-5). In some animals, massive overtransfusion (twice their original blood volume) was resorted to, supplying an adequate volume and ensuring a good filling pressure. In the animals with more severe stress (hypotension prolonged until a 20 percent to 40 percent uptake of shed blood from the arterial reservoir was indicated), definite evidence of depression of cardiac function was observed in the left ventricle. This persisted even with infusion of 160 ml per kilogram of blood, twice the original bleeding volume.

Humoral Factors

The humoral theories of the mechanism of shock are concerned with the formation of substances influencing the cardiovascular system or release of such substances from the site of injury. Some vasoactive substances may be looked upon as participating in the earlier compensatory mechanisms, for they are vasopressor in action. The increased outpouring of the catecholamines (epinephrine and norepinephrine) has already been cited. Evidence exists that renin or a renin-like substance appears in greater amounts following hemorrhage. Renin release could lead to formation of the pressor substance angiotensin. Increased discharge of vasopressin has been reported. Serotonin (5-hydroxytryptamine), vasoconstrictor in action, may be involved.

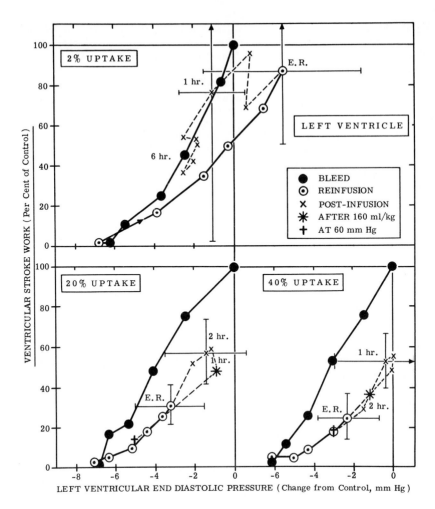

Figure 18C-5
Left ventricular function curves of dogs hemorrhaged to 35 mm Hg until 2, 20, or 40 percent uptake of maximum shed volume of blood. Ninety-five percent confidence intervals given at end of reinfusion (E.R.) and one hour after reinfusion. Animals supported with massive transfusions to maintain central venous pressure within 1 mm Hg of control value. (From C. F. Rothe. *Am. J. Physiol.* 210:1347, 1966.)

Other substances may be toxic to the cardiovascular system and contribute to later circulatory collapse. Some examples are potassium ion, which has a depressing effect on the myocardium, and histamine, which is deleterious to the peripheral circulation. Other products of tissue damage which are vasodepressor in action have been studied. These include the adenosine compounds and vasoactive polypeptides such as bradykinin. Lefer (1970) has proposed a myocardial depressant factor (MDF), probably a polypeptide or glycopeptide of low molecular weight (ca 1000). A possible role is noted in Figure 18C-2.

Cellular aggression (e.g., anoxia) releases cellular proteases (lysosome breakdown) and activator substances which produce in the plasma new active substances: plasma proteases, serotoxin, and plasma kinins. The kinins are formed by the proteolytic action of the proteases on certain plasma protein precursors, splitting off the vasoactive peptides (e.g., bradykinin and MDF). The toxic, shock-producing principle serotoxin is also probably produced by a proteolytic step. It differs from other kinins in that it needs the intermediation of histamine release.

The combined action of proteases and vasodilator kinins may explain the increase in capillary permeability in shock, for the proteases weaken the endothelium by their protein-digesting capabilities, while the kinin dilator action further promotes capillary filtration.

Not only do such products of cellular disorganization, together with substances found in the plasma under their influence, have a local action on the vascular endothelium; they are also transported by the blood to remote structures. There they have similar effects on the vascular endothelium and cause secondary injury to cells. The secondarily injured tissues may be regarded as a new source of pathogenic factors, resulting in a self-perpetuating process, and the final step in the causation of shock. Thus, a positive feedback mechanism is established which may account for irreversible shock. Since the peripheral resistance does not change enough in shock to explain the progressive fall in blood pressure, mechanisms such as the above must be seriously considered. Furthermore, the influences of humoral factors on the capacitance vessels and on the heart remain to be evaluated.

Prostaglandins

Mention has been made of the influences that favor increased synthesis and release of these substances. Among the prominent prostaglandins released by the kidney are the important dilators, PGE_1 and PGE_2. Evidence of increased output of PGE in the hypotensive phases has been adduced (Fig. 18C-6). Arterial concentration rises also, suggesting that the ability of the lung to metabolize prostaglandins is impaired in hemorrhagic hypotension. Markedly reduced pulmonary blood flow would be expected to take out of action the cells that would normally metabolize prostaglandins. Possible shunts open that bypass the pulmonary sites of metabolism. Inasmuch as renal vascular resistance (RVR) progressively increases during hypotension, the obvious conclusion is that the accumulation of pressor substances (cathecholamines, angiotensin, and vasopressin) overrides any vasodepressor action of the prostaglandins during this phase. Upon transfusion, arterial PGE concentration is transiently reduced, but later rises again during normovolemic shock because of increased synthesis and/or reduced metabolism by the lungs. The sharp rise seen late in the post-transfusion phase may be contributory to reduced TPR and the downward trend in mean arterial blood pressure (MABP) noted at this time, heralding the approaching demise of the experimental animal. When the potent prostaglandin synthetase inhibitor indomethacin is administered to the animal, in this phase of shock, concomitant with the reduction in release of PGE, renal blood flow decreases and RVR increases (Selkurt, 1974).

The question may be raised whether increase in circulating PGE is, in fact, detrimental to the animal, by virtue of its potent dilator action. This idea would be in accord with the notion that use of dilator drugs (see below, Treatment of Shock) can be beneficial in promoting better organ and tissue perfusion (if blood volume and cardiac output are adequate). It has recently been demonstrated that PGE_1 improves the coronary flow and myocardial contractility of dogs in hypovolemic shock and hence improves cardiac output (Priano et al., 1974). PGE must be infused into the left ventricle to escape metabolism in the lung, however, for this beneficial effect to be discerned.

TREATMENT OF SHOCK

Comments on the treatment of shock are pertinent here because of the physiological implications. The obvious immediate treatment is transfusion of blood or a suitable substitute, depending on the nature of fluid loss. The value of this procedure depends

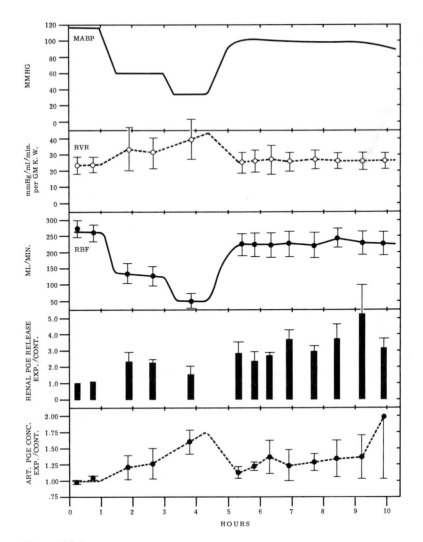

Figure 18C-6
Renal output of total PGE during standardized hemorrhagic shock in the dog (n = 5, mean ± 1 SE). PGE release was calculated as the arteriovenous concentration of PGE × total RBF. RBF was measured by electromagnetic flow probe on the left renal artery. MABP = mean arterial blood pressure; ART = arterial; EXP./CONT. = experimental kidney (hypotension and transfusion) function as compared to starting control values; K. W. = kidney weight. Prostaglandin assay by radioimmune technique. Control PGE values were as follows: arterial PGE concentration, 465 pg/ml (av.); renal release, 8 ng/min/100 gm K. W.

upon the lapse of time since the fluid or blood loss occurred, the quantity transfused, and the character of the transfusate.

The transfusion must be given as soon as possible after the initial blood loss and the development of hypotension, in order to avert irreversible shock. With regard to the quantity infused, the general principle is to give blood until the blood pressure (arterial and venous) and other essential cardiovascular signs (e.g., heart rate) have returned to normal. One of the most reliable of indices that can be rather simply employed is monitoring of the central venous pressure; a catheter in the superior or inferior vena cava is most useful for this purpose.

It was the practice in the recent treatment of battle casualties to transfuse quantities of blood that greatly exceeded the apparent immediate or obvious loss, and results were highly satisfactory. With a weakened myocardium, however, there is some danger of pulmonary congestion if pulmonary venous pressure is raised rapidly by excessive transfusion. It is well to remember that transfusion of 1 liter of blood does not mean that the circulating blood volume is necessarily increased by 1 liter. In the development of irreversible shock, dilated capillary channels might accumulate stagnant blood drawn out of the effective circulation, or rapidly leak out fluid.

The transfusate should conform to a number of requirements. It should be nontoxic and well retained by the vascular system. If a foreign substance, it should not deposit in the tissues and impair their function. It should not hemolyze or agglutinate cells. It should, of course, be isotonic and of proper pH. It should contain colloidal material to provide colloid osmotic pressure similar to that of the plasma protein. Hemorrhage is best treated with transfusion of fresh, properly matched blood. If there has been loss of fluid and electrolytes by vomiting and diarrhea, saline infusion (Ringer's) is the fluid of choice; with burns, plasma, dextran, or PVP (polyvinyl pyrollidone). The latter are synthetic polysaccharide plasma expanders which have proved safer than those formerly used—gelatin, gum acacia, pectin, etc. If anemia supervenes as a sequel to severe burns, blood transfusion is necessary.

The use of vasopressor agents has often been considered in treatment of shock with hypotension. Norepinephrine (levarterenol, Levophed) and metaraminol (Aramine) have been utilized. Use of these drugs in the treatment of hypovolemic (oligemic) shock is of course not justified, in view of the high level of circulating catecholamines. However, they may be helpful allies in the treatment of shock caused by myocardial infarction or bacteremic or hypersensitivity reactions, and they usually have a favorable effect against neurogenic shock.

Under the thesis that tissue and organ blood perfusion is impaired, it appears to be therapeutically advantageous to use vasodilator drugs, coupled with adequate restoration of blood volume, in the treatment of shock. Phenoxybenzamine (Dibenzyline) is the agent for which the largest clinical experience is available. Chlorpromazine, although a somewhat different pharmacological entity, produces equivalent and apparently equally effective peripheral vasodilation.

A limitation on the effectiveness of vasodilator drugs would appear to be the effective circulating blood volume. If the volume is severely depleted, blood perfusion of vital organs (e.g., the brain) might conceivably be further impaired by opening other peripheral vascular beds. Therefore, it cannot be overemphasized that for the most effective use of dilator drugs they must be coupled with adequate blood volume restoration.

The administration of hydrocortisone in pharmacological dosages has been found a useful empirical agent of benefit in shock therapy. Both experimentally and clinically the corticosteroids help fend off irreversibility if given in adequate dosage and as early as possible in the course of the acute disorder. They function by reducing the degree of cellular injury produced by noxious agents (both exogenous and endogenous), by cardiotonic effects, and, interestingly, in this dosage by reducing somewhat the peripheral arterial resistance.

Some investigators believe that constricted venous sphincters (e.g., at the outlet of the liver) are relaxed by corticosteroids, relieving splanchnic pooling and congestion and hence improving circulation. Lefer (1970) takes the view that glucocorticoids in massive doses achieve their beneficial effect by stabilizing the lysosomal enzyme activity and, in turn, minimizing release of proteases which in the final step of plasma interaction lead to production of vasoactive peptides, such as MDF.

Adrenocortical hormones are thought by some workers to provide particular benefit in shock resulting from bacteremia or hypersensitivity reactions when an overwhelming response to inflammation threatens life. They must be used judiciously and with caution. One of them, cortisone, has several undesirable side-effects.

REFERENCES

Abel, F. L., J. A. Waldhausen, W J. Daly, and W. L. Pearce. Pulmonary blood volume in hemorrhagic shock in the dog and primate. *Am. J. Physiol.* 213:1072–1078, 1967.

Blattberg, B., and M. N. Levy. Mechanism of depression of the reticulo-endothelial system in shock. *Am. J. Physiol.* 203:111–113, 1962.

Bock, K. D. (Ed.). *Shock* (Ciba Foundation Symposium). Berlin: Springer, 1962.

Chien, S., and S. Usami. Rate and mechanism of release of antidiuretic hormone after hemorrhage. *Circ. Shock* 1:71–80, 1974.

Gillett, D. J., and D. F. J. Halmagyi. Blood volume in reversible and irreversible posthemorrhagic shock in sheep. *J. Surg. Res.* 6:259–261, 1966.

Green, H. D. (Ed.). *Shock and Circulatory Homeostasis.* Transactions of Five Conferences. New York: Josiah Macy, Jr., Foundation, 1951–1955.

Hackel, D. B., N. B. Ratliff, and E. Mikat. The heart in shock. *Circ. Res.* 35:805–811, 1974.

Hershey, S. G. (Ed.). *Shock.* Boston: Little, Brown, 1964.

Lefer, A. M. Role of myocardial depressant factor in the pathogenesis of hemorrhagic shock. *Fed. Proc.* 29:1836–1847, 1970.

Matthews, R. E., and G. J. Douglas. Sulphur-35 measurement of functional and total extracellular fluid in dogs in hemorrhagic shock. *Surg. Forum* 20:3–5, 1969.

Mills, L. C., and J. H. Moyer (Eds.). *Shock and Hypotension: Pathogenesis and Treatment.* New York: Grune & Stratton, 1965.

Priano, L. L., T. H. Miller, and D. L. Traber. Use of prostaglandin E_1 in treatment of experimental hypovolemic shock. *Circ. Shock* 1:221–230, 1974.

Rothe, C. F. Heart failure and fluid loss in hemorrhagic shock. *Fed. Proc.* 29:1854–1860, 1970.

Sayeed, M. M., M. A. Wurth, I. H. Chaudry, and A. F. Baue. Cation transport in the liver in hemorrhagic shock. *Circ. Shock* 1:195–207, 1974.

Seeley, S. F., and J. R. Weisiger (Eds.). Recent progress and present problems in the field of shock. *Fed. Proc.* 20:Pt. II, 1961.

Selkurt, E. E. Primate kidney function in hemorrhagic shock. *Am. J. Physiol.* 217:955–961, 1969.

Selkurt, E. E. Status of investigative aspects of circulatory shock. *Fed. Proc.* 29:1832–1835, 1970.

Selkurt, E. E. Current status of renal circulation and related nephron function in hemorrhage and experimental shock: I. Vascular mechanisms; II. Neurohumoral and tubular mechanisms. *Circ. Shock* 1:3–15, 89–97, 1974.

Shizgal, H. M., G. A. Lopez, and J. R. Gutelius. Extracellular fluid volume changes following hemorrhagic shock. *Surg. Forum* 18:35–36, 1967.

Slonim, N., and W. M. Stahl, Jr. Sodium and water content of connective versus cellular tissue following hemorrhage. *Surg. Forum* 19:53–54, 1968.

Wiggers, C. J. *Physiology of Shock.* New York: Commonwealth Fund, 1950.

19. The Dynamics of Respiratory Structures

Thomas C. Lloyd, Jr.

Respiration, in the broadest sense, applies both to the processes whereby O_2 and CO_2 are exchanged with the environment and to the utilization of O_2 and production of CO_2 by individual cells. Cellular respiration (i.e., oxidative metabolism) is a subject more appropriately presented in biochemistry texts and will not be considered here. In birds and mammals the exchange of respiratory gases with the environment is a function essentially limited to the lung.

Lung function can be divided into four major divisions for convenience in presentation, but the reader should remember that this separation is arbitrary. Functionally, interactions among the several divisions are profound, and changes in one will result in alteration in some aspect of each of the others. The arbitrary divisions to be used in this text are as follows:

1. Airflow mechanics, which concerns movements of the chest wall, of the lungs, and of air itself.
2. Blood flow mechanics, concerning the way blood is distributed to different parts of the lungs.
3. Diffusion and gas exchange, encompassing transfer of gases through the alveolar membrane, the interacting effects of blood and air flows on the concentrations of O_2, CO_2, and N_2 in blood and alveolar gas, and blood-tissue gas exchanges in other organs.
4. Regulation of respiration, concerned primarily with the control of rate and depth of breathing.

AIRFLOW MECHANICS

The lung, although it contains both muscle and nerve tissue which are in many ways important to the regulation of its function, acts passively as a gas exchanger. As its blood flow is determined by cardiac pumping, so lung airflow is caused by active motion of the chest wall. It is therefore useful to begin with an understanding of the thoracic wall.

Mechanics of the Thorax

The chest wall is a closed container having several notable properties. Of the many thoracic muscles, only four groups (diaphragm, external intercostals, scaleni, and sternocleidomastoids) are important for generating the forces that normally enlarge the volume of the thorax and cause inspiration. Interestingly, none of the thoracic muscles plays a prominent role in expiration, although the internal intercostals may exert a minor expiratory force. In addition to muscles, which supply active forces, the ribs, by acting as rather springy supports, exert passive forces that influence the actions of the respiratory muscles and also help establish the state of inflation of the lungs even in the absence of active muscle tension.

Before proceeding to a more detailed description of the thoracic wall, we should note that a second set of active and passive forces involved with breathing arises from the abdomen. The importance of the abdomen lies in the fact that its upper border, the diaphragm, is the most easily displaced boundary of the thorax. Descent of the diaphragm presses the abdominal viscera against the elastic recoil of the abdominal wall; conversely, contraction of the abdominal muscles displaces viscera cephalad and pushes the diaphragm upward. In fact, contraction of the abdominal wall is the major active expiratory force, though, as will be seen, it is not necessary to use these muscles during quiet breathing. Note also that in a standing person the abdominal viscera can be considered to be suspended from the diaphragm. Their weight causes a downward displacement of the diaphragm and outward movement of the lower abdomen, if the abdominal muscles are relaxed. This gravitational enlargement of the thorax can be overcome passively by lying down, or it can be negated or reversed by the high intra-abdominal pressures attending obesity, fluid accumulation, or pregnancy.

In normal quiet breathing, called *eupnea,* inspiration is caused by contraction of the diaphragm and external intercostal muscles essentially unassisted by other muscles, while expiration occurs passively. At the end of a eupneic expiration the relaxed diaphragm assumes a domelike bulge into the thorax and away from the plane of its attachments to the lower ribs. On inspiration, the dome flattens as the muscles shorten, and the thorax is enlarged by this piston-like descent of its caudal boundary. Because the lung remains in contact with the thoracic wall, it tends to follow the volume expansion of the thorax and in so doing generates a pressure less than atmospheric within the alveolar spaces. The inrush of air through the open airways proceeds because of this pressure gradient. At the end of inspiration, the volume of air that has entered the lungs almost exactly equals the volume change of the thoracic cavity, the small difference being made up by an increase in thoracic blood volume. The lung, however, is an elastic structure which has been stretched by this volume increase. The diaphragm has not only done work moving air but worked against the elastic recoil of the lungs. At the end of inspiration the forces of elastic recoil just equal the muscle force necessary to hold the new volume. If the diaphragm then relaxes, elastic recoil expels air and pulls the diaphragm upward to its initial position. The energy for this passive expiration was stored in the elastic elements of the system during inspiration.

A second effect of diaphragmatic contraction is that it decreases the amount of overlap between lower rib cage and upper abdomen not only by displacing the abdominal viscera anteriorly but also by lifting the rib cage. That is, not only does the center of the dome descend, but its lateral borders ascend and take the rib cage with it. As the ribs move upward, rotations about their spinal attachments cause expansions to occur in both the lateral and anteroposterior directions (see Fig. 19-1).

Greater expiratory flow rates are achieved if expiration is assisted by contraction of the muscles of the abdominal wall. This "active" expiration is used when the demand for breathing exceeds about 40 liters of air per minute, or about 10 times the resting "minute volume." At the higher flow rates inspiration is assisted by contraction of accessory muscles, notably the scaleni and sternocleidomastoids. These muscles, like the external intercostal muscle, bring about inspiration by raising the ribs. During very heavy breathing, muscles of the spine and shoulder girdle also assist in expanding the thorax.

At the end of a quiet expiration all muscle contraction has ceased, and any other forces which remain are balanced so that no motion occurs. The volume of the lungs at this point is named the functional residual capacity (FRC). If, however, the lungs were to be removed from the chest, they would deflate to a much smaller volume by virtue of their elastic recoil property. Furthermore, the isolated rib cage, on account of its own elastic recoil, would spring *outward* from the position of FRC to a position equivalent to about 60 percent of a maximal inspiration. Thus, the relaxation volumes of the rib cage and lungs taken separately differ in opposite directions from the relaxation volume of the combined system. At FRC, the force of inward recoil of the lung when in the chest is exactly offset by the outward recoil force of the rib cage. The action of the diaphragm in quiet breathing is to upset this balance, so that at the end of inspiration the inward-directed force of lung recoil has been increased by the increased stretch, and the outward rib cage recoil force has been decreased by moving toward its relaxation volume. It is this imbalance of forces that causes passive expiration.

With this information it can be understood that eupneic expiration does not simply proceed until the lungs are empty, but proceeds until the declining force of lung recoil is

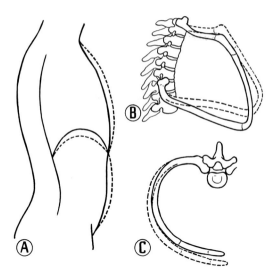

Figure 19-1
A. Changes in diameter of chest and abdomen during breathing. Positions in expiration are shown by solid lines; positions in inspiration, by dashed lines. B. Increase in thoracic diameter due to forward movement of the ribs in inspiration. C. Increase in lateral thoracic diameter in inspiration. (B and C from C. M. Goss [Ed.]. *Gray's Anatomy of the Human Body* [29th ed.]. Philadelphia: Lea & Febiger, 1973. P. 310.)

exactly offset by the increasing opposing force of rib cage recoil. Unlike eupnea, a very large inspiration will cause the thorax to be drawn *above* its resting volume, and, during the first part of expiration from this large inspiration, an inward-directed rib recoil will assist the inward-directed lung recoil until the rib cage relaxation volume is passed. Then further rib movement inward causes a rib cage recoil force which begins to oppose expiration. It should be apparent that anything altering either lung stiffness or chest wall stiffness will thereby alter the FRC, the force available for passive expiration, and the muscular effort required to breathe. (Though it will not be discussed here, a complete analysis of the balance of recoil forces would also require a term for abdominal wall stiffness.)

The Pleural Space
Returning to look at forces present at FRC, and remembering that the lungs and chest wall are really quite separable structures, one might wonder why they do not pull apart from each other, since their respective recoil forces operate to that end. Normally, a thin layer of fluid separates visceral pleura of the lung from parietal pleura of the chest wall, and were one able to insert a measuring device into this fluid-filled space, a pressure about 4 mm Hg below atmospheric pressure would be recorded. This "negative" pressure (according to the standard convention of referring all pressures to local atmospheric pressure) represents the tendency for expansion of the pleural space brought about by the oppositely directed tissue recoil forces. Because of their high intermolecular attractive forces, liquids are essentially unexpandable by the lowering of pressure, so the distance separating the pleural surfaces is determined by the amount of fluid and not by the pleural pressure. However, if a hollow needle is inserted through the chest to connect this narrow space with the atmosphere, air will be drawn inward by the lower pressure and the lung will separate from the chest wall as both lung and wall recoil toward their relaxation volumes. Leaks from the atmosphere into the pleural

space through tears either in the lungs or in the chest wall are not uncommon, and this condition is termed a *pneumothorax*. The presence of substantial volumes of gases or liquids in the pleural space can only result in a diminution of lung volume and is therefore potentially undesirable. Nonetheless, a thin layer of fluid can serve as both a mechanical coupling and a lubricant and thereby play a useful role in the to-and-fro motion of breathing.

The source of pleural fluid is the parietal pleura, which because of its blood supply from the systemic circulation has a capillary pressure high enough to cause a steady loss of fluid by transudation. The visceral pleura, on the other hand, has a larger capillary surface area and is perfused mostly by the lower pressures of the pulmonary circulation, so that fluid absorption is favored. Hence, pleural fluid is a transudate, the volume of which is determined by the balance between formation at the parietal pleura and removal in visceral pleura. The common medical finding of an increased pleural fluid volume (a pleural effusion) can arise from a disturbance in either absorption or transudation. Normally, absorptive forces exceed transudative, and one would expect to find no pleural fluid at all. What in fact happens is that absorption proceeds until the two surfaces contact at numerous points, to leave many intervening microscopic puddles of fluid. Complete absorption is prevented by pleural tissue stiffness, which limits the minimum size of these puddles.

Gases are kept from the pleural space, or are removed if they enter pathologically, by virtue of the fact that, even though the total pressure in the pleural space is slightly subatmospheric, the sum of the partial pressures of gases in pleural capillary blood is always even more subatmospheric. (For further details of gas partial pressures in blood, see Chap. 20.) The effect of this pressure gradient is to transfer gases from the pleural space to blood, preventing the existence of any gas space. Thus, even though lungs and rib cage may tend to separate because of their recoil forces, their surfaces remain near each other because absorption keeps the pleural space free of gases and nearly free of liquid.

Static Mechanical Properties of Lung

It has already been noted that lungs exhibit elastic recoil in much the same way as do stretched springs. For a structure extendable in one dimension, such as a spring, there is a simple relationship between the difference in length (Δl) before and after an additional increment of stretching force (ΔF) is applied: $\Delta l = k \Delta F$. The value of the proportionality constant, k, is determined by elastic properties of the material, the unstretched length of the spring, and its cross-sectional area. A similar expression relating change of volume (ΔV) of a three-dimensional structure to a change in inflating pressure (ΔP) is $\Delta V = C \Delta P$, where C depends upon material properties as well as on initial volume of the structure and on its wall thickness. This relationship between ΔV and ΔP is used by pulmonary physiologists to describe lung distensibility. The constant, C, is called the *compliance*. Normal adult human lungs have a compliance of about 0.2 liter per centimeter of water. Because of the dependence of C on initial volume, a somewhat better measure of lung tissue elasticity can be obtained by measuring not just ΔV but the change of volume with respect to the total volume before ΔV was produced—i.e., $\Delta V/V$, the fractional change in volume. When compliance is calculated on this basis, $C = \Delta V/V \Delta P$, and this is called the *specific compliance*. Specific compliance, although not used as often as compliance, is a better measurement for comparison of distensibilities of lungs of different size, such as those of children and adults.

To determine lung compliance it is necessary to measure the pressure gradient across the lung at times when the only determinant of that pressure is lung recoil—that is, at times of no airflow. In the intact individual that pressure gradient is given by the difference between alveolar space pressure (P_A) and the pleural space pressure (Ppl) when respiration is temporarily arrested. If respiration is stopped with the glottis open, P_A equals the local barometric pressure. At FRC, Ppl is about 4 mm Hg (5 cm H_2O) negative in respect to P_A. Ppl could be measured through a needle inserted into the pleural space through the chest wall, but this procedure is hazardous. The usual procedure is to measure pressure in the esophagus, which is essentially a flaccid tube in the pleural space whose walls are exposed to pleural pressure. The details of the measuring technique are unnecessary here; it is enough to note that when the esophagus is relaxed between peristaltic waves a pressure equivalent to Ppl can be obtained. Volume of air inhaled and exhaled is measured by means of any of several types of devices called spirometers.

If inhalation is done in steps, so that pressure can be determined at each volume plateau, one can obtain a graph of lung volume versus recoil pressure similar to that shown in Figure 19-2. The slope of the graph at any point is the compliance at that lung volume. Note from the figure that a single value for compliance applies to the lung over most states of inflation, but that near maximum and minimum inflation the compliance is less. This pressure-volume curve is characteristic not only of whole lungs but also of any subdivision. Consequently, the variations of compliance with extent of inflation will have important effects on the way an inspired breath is to be distributed throughout the lungs since the degrees of inflation of subsections of lung vary.

In an upright normal human, differences in degrees of inflation of subsections of the lungs always exist because there really is no single value of Ppl for all surfaces of the lung. Just as the pressure is greater at the bottom of a lake than at its top, so the pressure in the lower pleural space is more positive (less negative) than at the top. This is an effect of the weight of the lung, which though it is less dense than water still is able to cause a pleural space pressure difference of about 7.5 cm H_2O between apex and base in an average adult man. Since pressure within alveoli is nominally the same throughout the lungs, the vertical gradient of P_A implies that the elastic recoil pressure (P_A − Ppl) decreases from apex to base. Consequently, the uppermost parts of the lung are more distended than those below, and local compliance varies accordingly. For example, in a normal subject at FRC, the less inflated basal regions have a pressure-volume relationship indicated as region II of Figure 19-2, whereas the more inflated apical regions

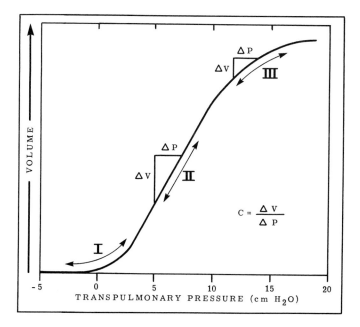

Figure 19-2
Relationship of transpulmonary pressure to lung volume.

behave as indicated by region III. During inspiration of a tidal volume, Ppl changes by an amount which is essentially identical in all regions. This ΔPpl is imposed upon the absolute initial Ppl unique to each region and causes a larger volume of gas to enter the basal regions than those toward the apex. In other words, during normal tidal breathing, the bottom parts of the lungs are less inflated than the tops but receive relatively more of the tidal volume. However, if a large inspiration is made, the relative distributions change. Early in the inspiration of a large breath the largest fraction goes to the basal zones, but these soon become inflated to the point of having their transpulmonary pressures (and compliances) approach those of the apical regions. Subsequent inspired gas is distributed more evenly. An important difference appears if transpulmonary pressures are unusually low at the onset of an inspiration. This can occur in two ways: if a normal lung is made to inspire from an unusually low volume or if the volume at time of inspiration is normal but the lungs are unusually compliant. In either case the same vertical gradient of Ppl continues to be present (because the amount of lung is the same). Therefore, unless there is sufficient recoil to produce pleural pressures at the apex of approximately -7.5 cm H_2O, local Ppl at the base may become positive. For example, if apical Ppl were -5 cm H_2O when P_A was at ambient pressure, basal units would be subjected to a pleural pressure of $+2.5$ cm H_2O acting in the direction to cause collapse. During inspiration under such conditions, no air could enter basal units until regional Ppl had become outward acting. This situation corresponds to region I of Figure 19-2. Hence with either a low initial state of inflation or an increased compliance, basal regions receive less of the inspirate than do those above them, a condition opposite to that of a normal inspiration. Furthermore, basal units may be subjected to collapsing forces during all or part of the respiratory cycle.

Regional tidal volume is not the only determinant of the gas composition change that occurs during breathing. One other factor is the local alveolar volume into which the tidal volume is brought. A given tidal volume causes a greater change in alveolar gas composition if the original volume is small. In other words, regional ventilation depends upon the fractional change of local volume ($\Delta V/V$), and not just the regional tidal volume. Taking this into account, one expects to find the greatest ventilation in regions which are least inflated but which also have the greatest compliance. Illustration is provided in the lower portion of segment II in Figure 19-2, and the implication is that during normal breathing the lower parts of the lung are the best ventilated. The least ventilation would occur in that portion of the lung already most inflated, or that which is nearly collapsed.

Another effect produced by elastic recoil of lung which also influences regional gas composition is that of "interdependence." Imagine a hypothetical sphere of lung tissue embedded in one lobe and receiving air from a single airway, which is partially plugged. Analysis has shown that when the surrounding lung is inflated, it exerts an outward pull on the hypothetical sphere which helps to inflate it. If the sphere itself tended to collapse spontaneously, it would meet a resisting force, depending upon its attachments to surrounding lung. Thus, there is an interdependence between any part of the lung and the lung around it that helps minimize variations in gas exchange caused by local variations in airway resistance or compliance. This interdependence also produces radial traction on larger airways and blood vessels such that their effective extramural pressure may be substantially lower than pleural pressure.

The Role of Alveolar Surface Tension in Pulmonary Mechanics

While compliance is usually determined clinically by measuring V and P over only small volume ranges, the pressure-volume (P-V) relation of excised lungs covering the full volume range from complete lung collapse to full inflation can bring out several important lung characteristics. When lungs are removed from the chest, they collapse to a smaller volume than they could achieve while in situ, but they do not collapse completely. Total spontaneous collapse seems to be prevented because as the lung

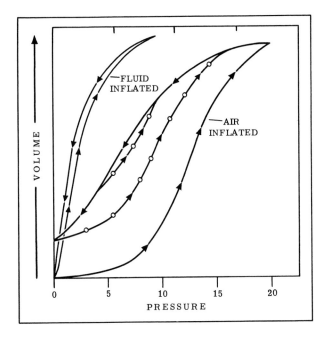

Figure 19-3
Relationship of static recoil pressure to lung volume when excised lungs are inflated from a degassed state with air and with 0.9 percent NaCl solution. Note that when lung is air-inflated, the path of deflation differs from that of inflation, and that deflation to the completely degassed state is not achieved. The curves with open circles are subsequent inflations made in one case from the point of least volume after air inflation. In the other case, the path of a medium-deep inspiration from near mid-capacity is illustrated. All deflations when air-filled take place along the same curve.

collapses the narrowest airways share in the collapse to the point at which their moist walls touch and close off, thereby trapping air in the distal alveoli. Then even suction applied to the trachea will not empty the lung.

However, it is possible to prepare completely degassed lungs by other means. When a degassed excised lung is inflated with air and deflated again by allowing the air to be returned by lung recoil, a P-V curve like that in Figure 19-3 is obtained. Observe that initially there is very little increase in volume until the pressure reaches some critical value, but inflation beyond this is easily achieved. During deflation, pressure again changes more than volume in the first phase, but this relationship gives way to a different slope for most of the expiration curve. The difference between the inflation and deflation curves is prominent, and most of it is in the slopes of the upper and lower quarters of total volume, while slopes of the middle halves are nearly equal. If one remembers that the slope of the curve is equivalent to compliance, it is immediately apparent that both the compliance and the recoil pressure depend on the state of inflation but are even more dependent on the direction of movement. If additional curves are obtained by beginning at lesser states of deflation than the degassed state, it is found that the inflation limbs lie in the area bounded by the loop already seen, whereas all the deflation limbs lie essentially on the same curve. Two inflation-deflation loops from intermediate volumes are included on the figure. Note that the openness of the loop is inversely related to the lung volume at the start of inflation. In striking contrast, if the same lungs are inflated from the degassed state with 0.9 percent NaCl solution instead of air, inflation begins with the first increment of pressure and each volume addition is made with a smaller increment of pressure than was the case with air. Furthermore, the inflation and deflation limbs are virtually identical. If the elastic properties of lung were determined entirely by lung tissue, both liquid and air filling should yield identical results. The significant change brought about by liquid filling, however, was not a change in the tissue but the elimination of the air-fluid interface at the alveolar walls. Thus it is shown that the most important recoil force of lung is not that of tissue but that of surface tension.

Surface tension at an air-liquid interface arises because molecules of the liquid are drawn more into the liquid than into the gas phase. The net result is equivalent to a tension at the surface which tries to decrease surface area. For pure liquids and true solutions the magnitude of this tension is a constant dependent upon the chemical natures of the liquid and gas involved, and the temperature. It can be shown that if the surface is spherical, like a gas bubble in a glass of water, surface tension produces a higher pressure inside the bubble which is related to the bubble size by the expression $\Delta P = 2\gamma/r$. ΔP is the pressure gradient between the inside and outside of the bubble, γ is the surface tension, and r is the bubble radius. It is easy to see (though often hard to believe) that pressures are greater inside small bubbles than large ones. The important corollary is that if one were to attempt to add a small additional volume to a small bubble, a higher pressure would be required to bring about that addition than would be needed to add the same volume to a larger bubble.

This corollary helps explain the inflation limb of the degassed lung P-V curve. Had one looked at the lung surface during that inflation, one would have seen that the larger terminal respiratory units filled with air before the smaller ones, and that none began to fill until some pressure was reached at which the surface tension of the air interfaces at the collapsed airway ends could be overcome. A moment's reflection, however, will reveal that if alveoli really behave like bubbles, the smaller ones should never fill, since to add volume to a larger unit will take less pressure than to add to a smaller, and as the larger units begin to fill they should become even easier to fill! Nevertheless, alveolar expansion does not continue indefinitely, for at some point further expansion begins to stretch relatively uncompliant fibrous tissue elements which heretofore were slack. In this way the most-filled units become constrained and their pressures can be raised without further expansion, allowing smaller units to fill. The progressive opening of smaller and smaller units generates the inflation limb of the P-V curve beyond the sharp upward deflection where the first units open. When all units are air-filled, further inflation only stretches the constraining net of elastin and collagen. Since this net is relatively uncompliant, the pressure-volume curve will have a less steep slope at large volumes, and this change in slope is apparent in both the air- and fluid-filled lungs.

As deflation begins, another property of the alveolar surface becomes apparent, for if ordinary bubbles had been inflated and deflated, the inflation and deflation curves would have coincided. Furthermore, instead of finding that small units deflate first, followed by large ones, as would be expected for soap bubbles, the whole lung has been found to deflate evenly. These and other findings have suggested that alveolar surface tension may be unusual, and many studies of material obtained from lung bear out this suspicion. The alveolar surface tension is not that which would be found if the surface were covered with a pure liquid or a simple solution. Instead, the alveolar liquid layer appears to be covered by an insoluble material, perhaps di-palmitoyl-lecithin, which lowers the surface tension in an unusual way. The source of the material seems to be the alveolar lining cells themselves. This insoluble material, called a *surfactant* (for *sur-face-active* agent) has two important properties. First, its presence lowers alveolar surface tension, with the result that less pressure is required to maintain any given alveolar volume. Second, this material causes the alveolar surface tension to vary as a function of alveolar area, and of the direction of change of area.

Since alveolar surface tension is a force related directly to transpulmonary pressure, and since surface area is a dimension immediately related to alveolar volume, a graph of surface tension of alveolar material versus area of the surface of that material can be used to examine the relation of alveolar surface tension to the state of inflation. Such a

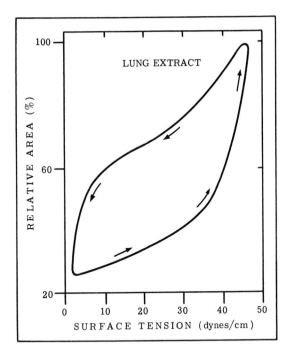

Figure 19-4
Surface tension of an aqueous preparation of lung extract as a function of area of surface. To generate such a curve, a sample of extract is placed on water in a pan which has a movable side wall. The surface tension is continuously measured while the surface area is varied by slowly moving the wall of the pan to and fro.

tension-area curve is shown in Figure 19-4. Note that expansion and contraction of the surface also generate a looped curve, and that the surface tension for any given area is lower during "deflation" than during surface expansion. According to several elegant analyses, it is very likely that during lung deflation, collapse of small units is prevented because, as they begin to collapse, their areas, and hence their surface tensions, fall. Further collapse is minimized, therefore, until the larger units with the still high surface tension have managed to "catch up" in the deflation process. As a separate test of this hypothesis it has been found that if lungs lack surfactant large pressures are needed to fully inflate them, and that during their deflation smaller terminal units collapse completely while large ones remain until last. Thus, the second significant role of surfactant is to allow coexistence of terminal respiratory units of different size, and to ensure that all units can inflate and deflate together.

Airflow Dynamics

The passive pressures described in the preceding section were determined by the recoil properties of the lung, chest, and abdomen and therefore were dependent only on compliances and the lung volume. Pressure gradients due to elastic recoil are indifferent as to whether there is airflow or not. At times of no airflow the transpulmonary pressure (P_L, the pressure gradient between mouth or nose and the pleural space) is determined solely by compliance, but during airflow an additional pressure must be added to achieve flow. That is, at *any* time, $P_L = Pel + Pres$ where Pel is the pressure gradient due to elastic recoil and Pres is the pressure associated with flow resistance. In this section the determinants of Pres will be examined.

Air, like blood, is a fluid, and the concepts of fluid mechanics presented in Chapter 11 will be used again here. As before, resistance (R) can be *defined* as pressure gradient causing flow ÷ flow. Airflow ($\dot{V}$) is the rate of change of lung volume with respect to time and is usually given in liters per second. The units of airflow resistance are centimeters of water per liter per second. Since PL, V, C, and $\dot{V}$ can all be measured or calculated, it is possible to measure PL at several flow rates and lung volumes and calculate the associated Pres from the equation $\Delta Ptp = \Delta V/C + Pres$. It is then found that for airflow rates occurring during rest or mild exertion the relation of Pres to $\dot{V}$ is linear, and the definition of resistance used above is also an adequate mathematical representation of behavior. With higher flow rates the pressure-flow relation deviates from linearity as flows become turbulent, and to describe adequately the Pres-$\dot{V}$ relation it becomes necessary to include another term related to the square of the flow rate, viz: $Pres = k_1\dot{V} + k_2\dot{V}^2$. R has been replaced by k_1 to indicate that k_1 is not entirely determined by the constants and parameters of Poiseuille's equation, which had been the case for lesser flow rates. Furthermore, unlike streamline flow, turbulent flow involves the important factor of fluid density. The constant k_2 contains terms for both density and viscosity, although the latter plays a minor role.

For the remainder of this discussion, flow rates will be considered to be low enough so that turbulent flow is not significant. The reader, however, must remember that at high airflows turbulence will occur and the pressures required may cause an inordinately high energy expenditure for breathing.

Pres is determined by viscous (i.e., frictional) losses from the movements of both air and lung tissue itself. The latter friction arises because lung tissue is deformed by the increase of its contained volume during breathing, much as syrup is deformed when a spilled drop spreads across a table top. This *tissue viscous resistance* amounts to 5 to 10 percent of the total pulmonary resistance. Note carefully that the energy *lost* in viscous deformation of lung is not the same energy accounted for in elastic recoil—the former depends on flow, while the latter depends on position; the former is lost as heat, while the latter is conserved and is available to cause passive expiration.

After tissue viscous resistance has been accounted for, the remainder of pulmonary resistance, related to airway sizes and gas viscosity, can be apportioned among several anatomical divisions. When one is breathing through the nose, airflow resistance of nose and nasopharynx accounts for two-thirds of the total. A switch to mouth breathing causes a substantial fall in total airway resistance. However, with mouth breathing, the resistance of upper airways—those included between the mouth and the intrathoracic portion of the trachea—still accounts for a third of the new total. An average resistance with mouth breathing is 1.5 cm H_2O per liter per second. Surprisingly, the resistance of all airways distal to about the twelfth generation of branching (having diameters of about 1 mm and less) amounts to less than 10 percent of the total. The reason is that even though each individual airway is quite narrow, the large numbers which act in parallel at each order of branching cause the net resistance to be low. Thus the largest fraction of the airway resistance and the greatest pressure gradient occur between trachea and bronchi larger than 2 mm internal diameter. The gas volume contained by the airways in which most of the resistance occurs is less than 3 percent of the total thoracic volume.

It is usually difficult for students to visualize the interrelationships of PL, Pres, Pel, V, and $\dot{V}$ over the course of a breath. The following general statements about those relationships are always true, and it is useful to keep them in mind. (1) Alveolar pressure (PA) is *always* more positive than pleural pressure (Ppl). (2) At any time that $\dot{V} = 0$, PL = Pel. (3) During *expiration* PA is *always* more positive than pressure at the point at which the airways open to the environment (Pao), but during *inspiration* Pao is always more positive than PA. (4) The absolute magnitude of Pel depends on V and is insensitive to $\dot{V}$. (5) The absolute magnitude of Pres depends on $\dot{V}$ and is insensitive to V. (6) Ppl may have *any* value (negative, zero, or positive), depending upon compliance, resistance, muscle strength, flow rate, and flow direction, but conditions 1–5 *always* hold.

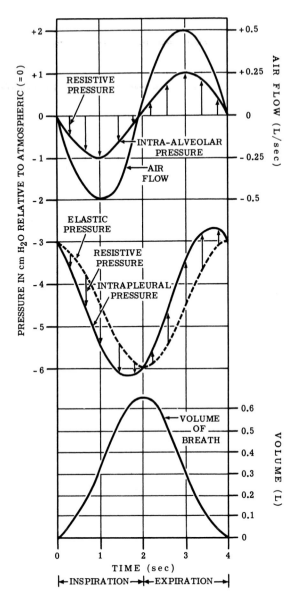

Figure 19-5

Idealized curves of the airflow, tidal volume, and intrapleural pressure which occur during the course of a breath. The two components of the intrapleural pressure, the resistive and elastic pressures, are also shown separately. (After J. F. Perkins. Reproduced from "Respiration" in the *Encyclopaedia Britannica,* 1970.)

Some of the above features can be verified in Figure 19-5. This shows an idealized curve; breathing does not produce exact sine and cosine waves like those shown. However, any actual record can be mathematically handled as the sum of several sine curves, and all conclusions derived from the figure remain valid. Note how the algebraic addition of Pres (shown by the arrows) to Pel yields PL. Because Pel and Pres reach their maximum amplitudes at different times in the cycle, the point of maximum amplitude of PL corresponds in timing with neither of those maxima, nor does it correspond with the greatest displacements of V or V̇. It can be shown that the point of maximum amplitude of PL lies nearer the time of maximum V if Pel is the major determinant of the work of breathing, but maximum PL lies closer to maximum V̇ if Pres is the major load. In eupnea PL and V are nearly "in phase" with each other, indicating that most effort is used to overcome tissue elastic recoil.

Up to now it has been assumed that a single value for airway resistance pertains to all lung volumes and flow rates. This assumption is quite untrue, and variations of resistance are of marked importance, particularly in relation to some lung diseases.

Airway resistance is inversely related to the extent of lung inflation because during expansion the intrapulmonary airways all participate in the volume increase. Though both lengthening and widening occur, changes in radii exert a more profound effect on resistance than do changes in length (see Poiseuille's equation). The graph of airway resistance upon lung volume is approximately hyperbolic, so the greatest changes occur at the smaller lung volumes. An example is shown in Figure 19-6.

Airway resistance can also be increased by contraction of tracheobronchial smooth muscle and by swelling of the mucosal layer. Reflex bronchoconstriction may arise from mechanical or chemical stimulation of a number of receptors within the airways, lung parenchyma, or nasopharynx, besides occurring as part of the reflex response to stimulation of the carotid chemoreceptors or pulmonary vascular stretch receptors. Motor innervation is by the vagus nerve. Though bronchoconstriction from irritant vapors or particulate suspensions such as smoke is usually a reflex response to airway receptor stimulation, a number of chemicals directly alter bronchial muscle tone. Among the more important constrictors are acetylcholine and other drugs known to mimic or enhance the effects of parasympathetic nervous activity, and histamine. Epinephrine and other drugs that mimic the effects of sympathetic nervous activity

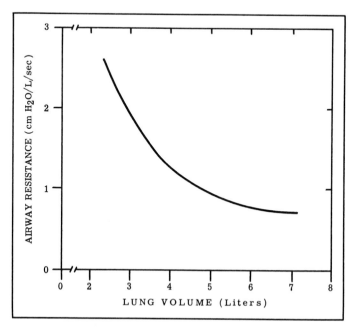

Figure 19-6
The effect of the state of lung inflation upon airway resistance in a normal human subject.

cause relaxation of bronchial muscle. Bronchial muscle tension is also inversely related to the CO_2 concentration of the muscle environment—a characteristic common to most smooth muscle which may be important in homeostatic adjustments of airway resistance (see Chapter 20, Effects of Regional Variation in $\dot{V}A/\dot{Q}$).

Contraction of airway muscle not only narrows the lumen but in cartilaginous airways also stiffens the walls. Bronchial cartilages, when pulled together by the muscles, interlock and overlap much like the plates of an armadillo. This arrangement can help prevent airway collapse caused by negative transmural pressures.

That airways can and do collapse is well known. In fact, airway collapse appears to be the mechanism whereby airflow rates become limited in both normal and diseased lungs. This is a "dynamic" collapse—it occurs during airflow. Because of its importance its mechanism will now be examined in some detail.

Imagine the chest to be only an air-filled cavity connected to the environment by a single airway which is collapsible (Fig. 19-7A). Assume that in order for the cavity to empty at a particular rate, a pressure gradient of 6 units is required for the airway. The energy for this expiratory flow is supplied by a force at the diaphragm equivalent to a pressure in the amount and direction indicated by the arrow. The assumed conditions cause an intraluminal pressure (using the standard reference convention for zero) of $+6$ at the entrance of the airway and 0 at its exit. Since an airway's resistance is distributed

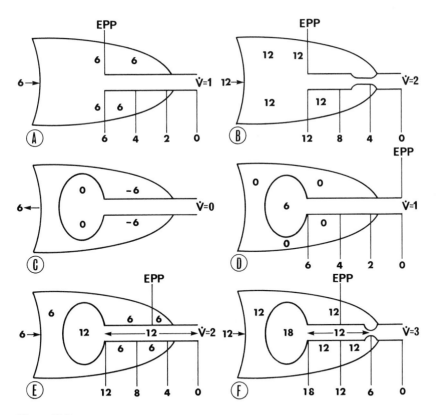

Figure 19-7
Models illustrating forces across airway walls. For details see text. Pressures within spaces are indicated by the numbers therein. Intra-airway pressures are indicated at several points by lines from those points to numbers below. The diaphragmatic pressure equivalent, and its direction, is indicated by arrow.

along its entire length, there is a longitudinal pressure distribution such that intraluminal pressure progressively falls along the airway. At some point the internal pressure and external pressure are equal, and in Figure 19-7A this equal pressure point (EPP) is at the entrance of the airway. However, the pressure on the *outer* wall of the airway is +6 for all parts in the chest and 0 for all parts lying outside. That is, except for the two ends, there is a *positive* (distending) transmural pressure across all the extrathoracic airway and a *negative* (collapsing) transmural pressure gradient across all the intrathoracic airway. Because of the progressive decline of internal pressure down the airway, the greatest negative transmural pressure occurs at the point just before the airway leaves the thorax. If this particular airway wall always collapsed at a transmural pressure of 6 units, no collapse would have occurred under the assumed conditions. If flow is doubled, however, all the pressures will be doubled and the intrathoracic portion will collapse at a point before it leaves the chest (Fig. 19-7B). Further analysis of the dynamics then becomes complicated, but it is sufficient to learn that the collapsed segment acts as a high resistance of variable magnitude which serves to limit the maximum flow. It does so because the harder one tries to force air outward, the greater becomes the resistance of the collapsed segment.

The lung differs importantly from the model described so far in that the lung parenchyma imposes an elastic element between airway and chest cavity. This has effects on airway behavior that can be analyzed by incorporating an elastic balloon to represent the lung in the model of Figure 19-7. Assume that at some time, either with or without airflow, the balloon will have a single volume, and thus a single elastic recoil pressure of 6 units. Pressures in the system under this no-flow condition are given in Figure 19-7C. However, when one assumes air to be flowing out, at the flow rate shown in Figure 19-7A, the pertinent pressures will be those indicated in Figure 19-7D. Under these conditions no external expulsive force is required, and airflow occurs because of elastic recoil. The EPP is at the airway exit, and none of the airway is exposed to a collapsing transmural pressure. If airflow is doubled (Fig. 19-7E), an additional external force must be applied to raise lung pressure sufficiently. But because of the elastic pressure gradient, pressure in the chest cavity is less than it was at the same flow in the absence of the lungs (compare Fig. 19-7B with Fig. 19-7E). The EPP has moved into the chest, but none of the airway collapses because of the requirement for a 6-unit collapsing transmural pressure. Only when flow is increased to three times the initial rate does sufficient collapsing pressure occur on the airway, and with the additional increase in flow, the EPP has descended farther toward the lung (Fig. 19-7F). If still greater expiratory efforts are made, they do not increase the pressure gradient between alveoli and point of collapse but instead move the collapse point closer to the alveoli. This action shortens the segment upstream to the point of collapse and thus lowers its resistance. Because the pressure gradient across that segment is fixed, shortening of the upstream segment may result in increased flow. However, shortening of the upstream segment is accompanied by lengthening of the collapsed segment, which tends to retard increases of flow with added effort. Progressive movement of the collapse point stops when the decreasing resistance of the shortening upstream segment is exactly offset by the increasing resistance of the lengthening collapsed segment. At that degree of effort, maximum expiratory flow is achieved. When the collapse point is fixed, so is the EPP, which is a bit farther upstream to a degree dependent upon collapsibility of the airway.

If the resistance of peripheral airways is increased (equivalent to a narrowing of the tube where it begins in the model lung), pressures in the larger airways will be lower for any given flow than if peripheral airway resistance is low. This effect, like a loss of elastic recoil, will move the EPP toward the alveoli and maximum achievable flow rates will be reduced.

Let us now consider the airway as composed of two segments, an upstream segment from alveoli to EPP and a downstream segment from EPP to airway opening. Flow in each must be equal, since they are in series. At maximum flow the length of each is fixed, since at maximum flow the upstream movement of the EPP has ceased. Note that

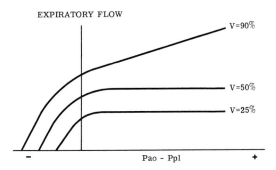

EXPIRATORY FLOW

V=90%

V=50%

V=25%

−　　　　　Pao - Ppl　　　　　+

Figure 19-8
Relationship between expiratory flow velocity and the pressure gradient from pleural space (Ppl) to the end of the airways at the mouth (Pao). The pressure-flow relationships pertaining to three specific lung volumes are shown. Volumes are given as percentage of the vital capacity above residual volume. Static recoil pressure for each volume is indicated by Ppl at the zero-flow point. Pleural pressure rises to zero and becomes positive as expiratory flow velocity is increased. Note that at the two smaller volumes, maximum expiratory flow does not increase beyond a particular point, but that near the maximum inspiratory volume, flow continues to increase with increasing effort.

the pressure gradient for flow in the upstream segment is PA minus Ppl. But this is identically the elastic recoil pressure of the lung at this volume. No greater pressure gradient can be exerted to cause flow along this fixed upstream segment, and since the latter is the sole source of flow into the downstream segment, the elastic recoil is seen to be the principal determinant of maximum expiratory flow. If the lung were inflated to a larger volume, or if it were less compliant, recoil pressure would be greater and thus maximum achievable flow higher. A low recoil pressure, such as would occur if compliance were increased, decreases maximum achievable flow by reducing the driving force. Similarly, an increase in peripheral resistance, because it causes an increased resistance of the upstream segment, will reduce the maximum flow attainable at each lung volume.

Airway collapse does not occur during inspiration. It is a simple task to show that, regardless of airflow, lung volume, or compliance, transmural pressures during inspiration will always be outward directed. Nor does collapse occur during passive expiration.

The above model corresponds with the behavior of lung remarkably well. It is known that pressures around intrapulmonary airways approximate local pleural pressure. Furthermore, studies have shown that even the larger bronchi and trachea collapse at transmural pressures in the range required by theory. The phenomenon of dynamic airway collapse and flow limitation can easily be demonstrated. If one records expiratory flow rate as a function of expiratory effort, one finds that for degrees of inflation less than about 80 percent of total lung capacity the flow at each volume reaches a maximum at a modest effort, and greater effort causes no flow increase. As predicted, maximum achievable flows are greater at large volumes than for smaller volumes. Examples of the pressure-flow relationships at several lung volumes are shown in Figure 19-8.

Clinical Measurement of Lung Mechanics
Measurements of total pulmonary resistance, airway resistance, lung compliance, or flow velocities achievable at graded levels of inflation can be used to assess pulmonary function in patients. While these more elaborate tests may be necessary to estimate pulmonary characteristics independently of some of their interactions, ventilatory

ability is more often estimated from data contained in graphic recordings of single forced expirations. Many different derived values from such records are in use; the choice among them is based on empirical reasons. The ability of the tests to quantify pulmonary function arises from the dependence of airflow on intrinsic pulmonary factors such as resistance, compliance, and lung volume, rather than on muscle strength and subject volition. The interpretation of the tests in terms of specific tissue changes depends on understanding airflow dynamics and the static properties of the system. Some commonly used function tests are the following.

Vital Capacity (VC)
The largest amount of air that can be exhaled in a single breath after a maximum inhalation is known as the vital capacity. The vital capacity is determined by lung size, by subject size and age, and to a large extent by lung and chest wall compliances. It is also influenced by factors that alter thoracic mobility, such as muscle weakness, abdominal fluid, and chest pain. Results are expressed in liters. A graphic display of this and other commonly measured lung volumes is given in Figure 19-9.

Forced Expiratory Vital Capacity (FEV)
When exhalation of a vital-capacity-sized breath is done as rapidly as possible, several useful measures of expiratory flow rates can be obtained from a recording of volume delivered versus time. One of these is the amount exhaled in specific periods. Typically, a person should be able to exhale 70 percent of his VC in the first second (FEV_1) and empty more than 90 percent in three seconds (FEV_3). Diseases that decrease airflow as a result of either increase in resistance or decrease in tissue retractive force will often decrease the rate of delivery of a forced expiratory VC. This is a simple, useful clinical measurement, suitable for following the progress of many cardiopulmonary diseases and for recognizing patients with obstructive airway diseases. Results are expressed either as the percentage of the VC delivered in the chosen time interval or as the absolute volume of gas exhaled in that period. The FEV_3 is now rarely measured, but measurement of the FEV_1 is probably the commonest pulmonary function test.

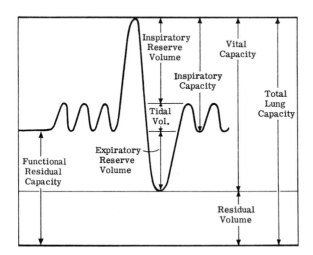

Figure 19-9
Schematic representation of the course of breathing versus time, which includes both tidal breaths and a vital-capacity-sized breath. Note that the residual volume cannot be exhaled. Its size is determined by other methods.

Maximum Expiratory Flow Rate (MEFR)
Some instruments directly measure the greatest instantaneous expiratory flow rate achievable. If this measurement were to be obtained near the onset of a FEV, the result would be highly effort dependent. However, several varieties of instantaneous flow measurement equipment have been developed which can measure the maximum flow rate at any specified state of inflation. Note that these tests measure *instantaneous* flow rates. Results are expressed in liters per second.

Rather than use instantaneous flow rates, or the volume delivered in a specified time interval, some clinical physiologists measure the average rates at which specific segments of the FEV can be delivered. In common use are the mean flow rate at which the middle half of the forced expiratory VC was delivered (called the maximum mid-expiratory flow) and the mean flow rate present between 200 and 1200 ml of exhaled volume. The object of such tests is to measure flow rates over a part of the VC where they are less likely to be effort dependent. In practice, these measurements, like the FEV_1, are obtained from records of single forced VC exhalations. The intercorrelations between all these flow rate measurements are so high that there is probably no reason to measure all in any given subject, or to strongly prefer one measurement to another for clinical use.

Maximum Voluntary Ventilation (MVV)
To measure the greatest minute ventilation of which the subject is capable, he is instructed to breathe as rapidly and deeply as possible, and all his expired air is collected during a short interval, typically 15 seconds. Results are converted to liters per minute. In spite of being influenced by fatigability, coordination, and cooperation, the results correlate extremely well with the flow rate measurements described earlier.

Deposition and Removal of Material from the Airways
About 100 ml of mucus is produced in the tracheobronchial tree of a normal human every 24 hours, and this, along with the particulate and droplet material inhaled, must be continuously removed. The demand for clearance of secretions is greatly increased by hypersecretion with many pulmonary diseases, and by exposure to environments usually laden with airborne suspensions. Cilia in the tracheobronchial tree provide locomotion of the mucus sheet. Ciliary beating is not random but proceeds in an organized wavelike manner. The organization of the ciliary beat can be disturbed by inhaled noxious agents. Individual cilia beat with a quick foreward motion and a slower backward motion, in a fluid sheet that is composed of two layers. The upper mucous layer is stiffer and more viscous than the lower serous layer. During the forward beat, cilia tips approach the upper layer and slide it along, lubricated by the lower layer. During the slower return movement, the rate of movement and curvature of the cilia allow them to return through the lower layer without causing much backward flow. Interference with removal of mucus can, therefore, be caused by alteration or disorganization of ciliary beat, or by changes in the composition and physical properties of the material. Although mucus transport has an important bearing on pathogenesis and management of lung diseases, its physiology has only recently become of general interest.

Aerosol deposition, on the other hand, has been a concern of environmental physiologists for many years. It is known that the level to which a suspended particle will penetrate is determined by its mass and mean radius. Because of the several changes of direction that air takes in descending from mouth or nares to alveoli, large heavy particles tend to impact on the walls rather than make the bends. Smaller and smaller particles descend progressively farther into the lung, until particles of very small size are so light and small that they remain in the gas and never impact on walls but are breathed to and fro, reaching airway walls only by brownian motion. Particles greater than about 10 μm in diameter are removed as air passes through the nose; those between

2 and 10 μm impinge upon the tracheobronchial walls, and particles 0.5 to 2 μm reach the alveolar ducts and alveoli where they may be brought to the walls by gravity or brownian motion. Particles of less than 2 μm are deposited below ciliated epithelium and must be removed by phagocytes, or by the less effective lining fluid transport that exists at those levels. These small particles have the greatest opportunity to induce tissue injury because of their prolonged contact times. There are other important consequences of this deposition pattern. For instance, it is popular to consider an industrial environment hazardous only if the air looks dusty. However, particles that make up visible dusts are large enough to be trapped in upper airways and removed easily, whereas those small enough to enter the periphery of lung (which are usually the most harmful) can be visualized only in unusual circumstances. Another example is the deliberate generation of droplet aerosols for carrying drugs or moisture to peripheral airways; many atomizers and mist generators create particles so large that they barely enter the major bronchi and thus are destined to failure.

Mucus, and the particulate material entrapped in it, can be removed from larger airways by cough. The cough reflex is initiated by receptors in large airways and proceeds in four stages: (1) inspiration, (2) closure of the glottis, (3) compression of the thoracic gas by activation of expiratory muscles against the closed glottis, (4) sudden expulsion of air by opening the glottis. To be expelled, secretions must be moved by cilia into airways in which a large airflow can occur. Cough, therefore, does not clear peripheral airways well. During the explosive decompression, dynamic collapse of the large airways vibrates the walls, and their contact helps to milk secretions upward. Cough is an important way for secretions to pass the larynx, for the vocal cords are not ciliated and secretions accumulate below them until lifted over by coughing. It is not uncommon for segmental bronchi to become completely occluded by secretions. The inspiration preceding cough could not fill lung below those airways were it not for flow into the blocked segment from adjacent segments through collateral ventilatory channels (the pores of Kohn and canals of Lambert). Because the blocked segments can share in the inspired volume, the explosive decompression can unblock their airways. It should also be noted that the collateral channels allow gas to enter lung below blocked airways during normal breathing. Although this gas is essentially inhaled from other alveoli, it does help to maintain some gas exchange in lung below the block and prevents such a region from being a sealed space from which all the gas would be absorbed into capillary blood.

MECHANICS OF THE LUNG CIRCULATION

Bronchial Circulation

The lung, like the liver, has two reasonably separate vascular beds. The smaller of the two is the bronchial vascular system. Since it is a division of the systemic circulation, arterial pressures are four to five times greater than pressures in the pulmonary circulation. The bronchial circulation is the principal blood supply of bronchi and bronchioles, whereas the respiratory bronchioles and more distal lung tissue are nourished by the pulmonary circulation. Bronchial arteries form the vasa vasorum of pulmonary arteries and provide the blood supply of the pulmonary nerves. The bronchial circulation joins the pulmonary circulation in several places, and under some pathological conditions these interconnections become important. The easiest intermingling to confirm experimentally is in the venous bed. Bronchial veins drain from the first one or two divisions of bronchi into the right atrium, but venous drainage from more peripheral airways goes into the left atrium through anastomoses with pulmonary veins. About 1 percent of pulmonary venous flow comes from bronchial veins. There are arterioarterial communications through both capillary and noncapillary networks. The volume of this arterial shunt flow is normally very small, but in congenital absence of part of the pulmonary arterial tree, or following pulmonary arterial occlusions, flow of inadequately oxygenated systemic arterial blood through these anastomoses may become great enough to be of value in gas exchange. Marked enlargement of the bronchial circulation also occurs in bronchi which are chronically infected. Under such conditions bronchial arterial flows as large as 20 percent of the cardiac output have been reported, whereas this circulation normally receives only about 2 percent of the cardiac output.

Pulmonary Circulation

Turning now to the pulmonary circulation we perceive that the most efficient pulmonary gas exchange requires the optimum presentation of CO_2-rich and O_2-poor blood to CO_2-poor and O_2-rich inspired air. An understanding of the way lung blood flow becomes differentially distributed is as important as an understanding of airflow mechanisms, since the two are inextricably linked in gas exchange. Typically the pulmonary circulation has been described as a passive, low-pressure system capable of little regulation. Because of these characteristics pulmonary blood flow distribution can easily be profoundly altered; therefore an understanding of the mechanisms of the alterations is obligatory if one is to comprehend regional variations of gas exchange.

The pulmonary vessels on the arterial side are much more distensible than their systemic counterparts. Their high compliance allows these vessels to have their radii easily changed by variations in transmural forces. In the systemic circulation, except for the heart and other exercising muscles, the intravascular pressure alone is a good approximation of the entire distending force. This is not the case for the lungs, because the low intravascular pressure allows perivascular pressure and the radial pull of extravascular tissue to become significant parts of the total force. Perivascular pressure can be as high as PA or lower than Ppl. Variations of perivascular pressure from main pulmonary arteries to capillaries may be as large as the variation in intravascular pressure, so they play a prominent but often unquantifiable role in establishing the radii of these vessels.

A more readily computed variation in vascular transmural force occurs as a result of the weight of blood. That is, the pressure in a fluid column is higher at the bottom than at the top by an amount $P = \rho gh$, where ρ = blood density, g = gravitational acceleration, and h = column height. If pressure in the main pulmonary artery (with respect to atmospheric pressure) is 20 cm H_2O in an upright individual, the pressure 19 cm further cephalad in an arterial branch of the apical segment will be only 1 cm H_2O. (The densities of blood and water are essentially equal.) Conversely, at any lower level, intravascular pressure is above that at the midlevel. This phenomenon causes a gradation of pulmonary intravascular pressure which tends to distend vessels at the base and collapse vessels at the apex. As would be expected, distribution of blood flow becomes prominently affected by these variations in transmural forces, for they help define the resistance to flow at each horizontal level. Blood flow distributes differently when the person is lying down, because then the gravitational gradient is directed from front to back and acts over a smaller distance. If pulmonary arterial pressure were as high as systemic arterial pressure (about 100 cm H_2O), a difference in height of 10 cm would vary pressures by only 10 percent, whereas with arterial pressures of only 20 cm H_2O, the same difference in height causes a 50 percent change in intravascular pressure. All other things being equal, the proportionally larger stress would produce a proportionally large radial change, so it can be seen that low pulmonary vascular pressures act to augment vascular dimensional instability.

The *perfusion pressure* is the difference in pressure between two longitudinally separated points in a vascular bed. It is usually expressed as the difference between arterial and venous pressures. This gradient is a function of frictional loss between the two measurement sites and also of any differences in kinetic energy between the two sites (see Bernoulli's equation, Chap. 11). The latter effect is ordinarily disregarded, though it is not always correct to disregard it. In vessels having transmural pressures everywhere sufficient to prevent collapse, the downward pull of gravity on arterial blood would be exactly offset by the pull on venous blood. If that were the way the lung vessels behaved, the perfusion gradient for all horizontal sections of the lungs would be given by the difference between pulmonary arterial pressure (Pa) and pulmonary venous pressure (Pv), where both are measured at any single horizontal level. However, in a vascular bed in which collapse can and does occur, any variation of the pressure downstream with respect to the collapse point could not make itself felt upstream beyond the collapse unless the downstream pressure were raised enough to open the collapsed zone. Thus the flow is independent of the downstream pressure as long as the collapsed segment is present.

Such a vascular situation is analogous to the flow over a waterfall. The height of a river below a waterfall has no effect on the flow over the fall, or on the height of the river above the fall, unless the level of the river below the fall becomes as high as the fall itself. The flow of the river is determined by the slope of the river bed above the fall. For this reason the phenomenon of a collapsed vascular segment which determines the blood flow and dissociates the flow from Pv has been called the "vascular waterfall phenomenon." It is familiar to everyone who has attempted to drink through a collapsed straw. In the presence of a collapsed segment the perfusion gradient cannot be Pa minus Pv, but the correct replacement for Pv can easily be chosen after further examination of the collapse mechanism. When collapse occurs, flow stops, and in the absence of flow the pressure everywhere in that vessel upstream of the collapse rises to equal Pa. If Pa is above that necessary to open the collapse, the collapsed zone will open; flow will then resume and the intravascular pressure will fall because of the viscous loss. The collapse is thus reestablished, and the cycle begins again. Actually the "fluttering" of the wall is more apparent than real, and a steady stream seems to be passing through the narrowed portion. In this collapsible system the arterial pressure is working against the pressure that causes collapse, and not against Pv, so the perfusion gradient is the difference between Pa and the pressure necessary to open the vessel. Viscous losses in the segments downstream from the collapse point would not be included in this perfusion gradient.

In lung, capillaries are the vessels one would most expect to collapse because they are least supported by tissue "guy wires." Furthermore, their perivascular pressures, being essentially P_A, are the highest anywhere in the system. It has been found that the pulmonary capillaries collapse when their internal pressures are slightly below P_A. The relevant perfusion gradient for the part of the lung in which collapse occurs is usually taken as Pa minus P_A. It turns out that in a normal standing man there is sometimes an uppermost region of lung where Pa $<$ P_A, and where no perfusion occurs. (This region is known as zone I.) Immediately below zone I there is a region where Pa $>$ P_A but P_A $>$ Pv, so "waterfall" flow occurs and the perfusion gradient is Pa $-$ P_A. This is zone II. Below this (in zone III) Pa $>$ P_A and P_A $<$ Pv, so collapse does not occur and the perfusion gradient is Pa $-$ Pv. Changes in position, P_A, Pv, or Pa will all cause a redistribution of transmural pressures. Since all the above modulating pressures are large in comparison with pressures in the normal pulmonary artery, one would expect them to bring about significant dimensional alterations. A schema of these three zones is shown in Figure 19-10, which also includes a hypothetical plot of flow through each horizontal level as a function of its position.

Increases in Pv cause more of the lung to be in the zone III state and also cause a back pressure which raises Pa and the transmural pressures of all vessels. The increased transmural pressure distends some vessels and opens others which had had enough muscle tone to be closed at lower pressures. Dilation, and "recruitment" of additional parallel paths, lowers the net perfusion gradient, raises Pa somewhat, and increases the number of perfused capillaries.

Increases in P_A cause more of the lung to function in the zone I and zone II conditions. If cardiac output remains constant, the expansion of zone I diverts blood elsewhere, causing concomitant elevation of Pa. This increase in their transmural pressure distends and recruits vessels in zones II and III.

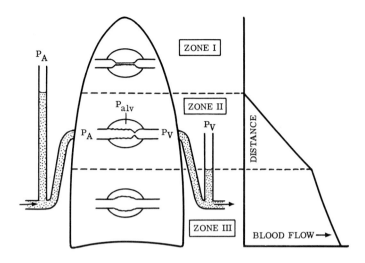

Figure 19-10
Schematic illustration of the three different conditions under which blood vessels exist in the lungs. For details see text. Note that no flow occurs through zone I, and that the way in which flow increases progressively at each lower level in zones II and III is different for each zone. The changes in flow through each zone are caused by gravitationally dependent transmural pressures which alter the length of the collapsed segment in zone II and the radii of distensible vessels in both zones. (From J. B. West, C. T. Dollery, and A. Naimark. *J. Appl. Physiol.* 19:713–724, 1964.)

Like airway caliber, vessel caliber depends upon lung volumes because vessel walls are tethered to the parenchyma. Unlike airways, however, blood vessels do not undergo a progressive decrease in perfusion resistance as progressive lung inflation occurs. Most studies have shown that the relationship of perfusion gradient to volume is U-shaped, and that the point of minimum pressure requirement lies, perhaps fortuitously, at the FRC. The detailed causes of the particular shape for this curve are too complex to be given here but include such things as the changing amount of radial change relative to length change that occurs with each increment of volume; the reapportionment of major sites of pressure drop among arterial, capillary, and venous beds; and the interaction of the (nonlinear) tissue length-tension relation with that characterizing the vessel wall. The effects of volume change are not small; a collapsed lung may require four times the pressure required at FRC to bring about the same blood flow. This factor becomes very important when lungs collapse pathologically and also has a bearing on the way blood is distributed to the fetus, whose lungs are not yet expanded.

Up to this point I have carefully avoided defining pulmonary vascular resistance as pressure gradient ÷ flow ($R = \Delta P / \dot{Q}$). While such a number can tell something about the work of the right ventricle, it is for several reasons a bad way of assessing what the pulmonary vessels are doing. Part of the difficulty lies in defining the pressure gradient. As has been seen, ΔP depends on *either* P_A *or* P_V, and since these are significantly different in intact animals, a correct pressure gradient cannot be selected unless all the perfused lung is in either zone II or zone III, but not some in each. To nearly achieve this end, human subjects are studied in the supine position. In that instance, all of the lung can be considered to be in zone III because left atrial pressure is >0 at mid-chest level.

When the relationship of arterial pressure to flow is obtained for a lung in the zone III state while lung volume and other pressures are held constant, a curvilinear record is

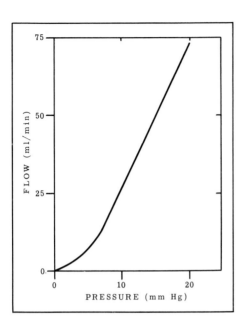

Figure 19-11
Perfusion pressure versus flow obtained with an excised dog lung lobe under zone III conditions using a newtonian fluid as perfusate.

obtained, as shown in Figure 19-11. Since the curve obtained is not a solution set of the equation $P = R\dot{Q}$, a value for pulmonary vascular resistance determined as $P/\dot{Q}$ cannot be used to predict the flow consequent to any pressure gradient except the one at which R was first obtained. The P-$\dot{Q}$ curve deviates from linearity because as arterial pressure is raised it not only causes more flow but also widens vessels and opens channels previously closed. This change in net vascular bed cross-sectional area causes the increase in flow for each increase of pressure to be greater than expected. Widening and recruitment continue only up to a point. Beyond some pressure, all vessels are open, and further dilation is limited by the substantial decrease in wall compliance that occurs when previously slack fibrous tissue begins to be stretched. (In fact, pulmonary vessels are so distensible at low pressures that most vessels which are open at a normal zone III pressure are probably quite close to the greatest diameter that could be achieved even by a several-times increase in transmural pressure.) Once the vessels are fully open and distended, the P-$\dot{Q}$ relation becomes linear. Since vascular dimensions appear to remain constant above that pressure, it would be convenient for any calculated R to be constant also. In addition to having no predictive value, the P/$\dot{Q}$ quotient does not become constant when the P-$\dot{Q}$ curve becomes linear. For this reason some physiologists have defined resistance as the slope of the P-$\dot{Q}$ curve. The choice depends on the use to be made of it. Typically one wants to know whether something done experimentally or therapeutically has altered vessel size. Rather than use either of the above definitions for R, it is better to describe the pulmonary circulation with a P-$\dot{Q}$ curve having at least several points, or by collecting all data when flow is held constant and pressure is the only dependent variable (or vice versa).

Failure to recognize the limitations inherent in assigning a vascular resistance on the basis of the P/$\dot{Q}$ quotient, particularly when blood flow and ventilation both change, has for many years caused profound disagreements about the responses of the pulmonary circulation. Even after recognition of the difficulties it is usually very hard to obtain data unquestionably undisturbed by secondary influences. With this reservation in mind, the presentation of factors altering vascular resistance can continue, but resistance will have to have an intuitive definition.

The part of the pulmonary vascular bed that has the largest share of the resistance is still unresolved. This dilemma arises, to some extent, because the major resistance may vary depending on extravascular conditions. Thus, in zone II the major site is probably the capillaries, but in zone III it is probably the small arteries. The veins are likely always to make up less of the resistance than either of the other beds. A distribution of 45 percent arterial bed, 35 percent capillary bed, and 20 percent venous bed could be considered typical.

The pulmonary circulation differs importantly from the systemic circulation in the amount of vascular smooth muscle present. Small pulmonary arteries, in the range of $100 \ \mu m$ to $1000 \ \mu m$ in diameter, are the vessels most likely to be capable of contracting effectively. The arterioles have too little muscle.

It is known that a number of things can cause vasoconstriction prominent enough to raise pulmonary arterial pressure under conditions in which cardiac output remains constant. Usually the same factors that constrict the systemic circulation will also cause pulmonary vasoconstriction. These vasopressor stimuli include epinephrine and norepinephrine, angiotensin, and increased sympathetic nervous activity. In some pathological conditions several other vasoactive materials may be the causes of pulmonary vasoconstriction. Serotonin, histamine, and various polypeptides may be released either in the general circulation or locally within discrete parts of the pulmonary circulation, and all these bring about vasoconstriction. The absolute changes in pressure caused by maximal pulmonary vasoconstriction under conditions of constant cardiac output are typically in the range of 10 to 30 mm Hg. While these changes may not seem great, the percentage changes in pressure are of about the same magnitude as are caused by intense vasoconstriction in the systemic circulation.

The effects of pulmonary vasoconstriction should be considered from the standpoint of whether vasoconstriction has been generalized or has occurred only in some regions. Generalized vasoconstriction increases right ventricular work load. It also tends to make the distribution of pulmonary blood flow more even. The explanation for the latter is that when the arterial resistance everywhere increases, it becomes the largest component of the total resistance. In that event regional differences among arteriolar, capillary, and venous resistances are of less importance because they are then only small fractions of the total resistance in any region.

Regional vasoconstriction will divert blood flow to other parts of the lungs, and if the region of vasoconstriction is not too large (say, less than half the total circulation), there will be little increase in main pulmonary arterial pressure because the diverted blood opens additional paths. Regional vasoconstriction makes distribution uneven and typically does not result in increased right ventricular work.

One of the best-known stimuli for regional pulmonary vasoconstriction is alveolar hypoxia—that is, a lower than normal O_2 concentration in alveolar gas. In other organs, regional hypoxia causes vasodilation, but in the lungs vasoconstriction occurs. Its exact cause is unknown, but it is not a reflex, nor is it brought about by the usual neuro-mediators. The important factor is the O_2 concentration in the alveolar gas, not the amount of O_2 present in either pulmonary arterial or venous blood. Since a common cause of regional alveolar hypoxia is poor gas exchange resulting from regional impairment of airflow, local vasoconstriction consequent to alveolar hypoxia tends to divert pulmonary blood flow to regions which are best exchanging gas with the outside air. In this way the vascular response to hypoxia tends to enhance the overall efficiency of pulmonary gas exchange. The vasopressor effect of hypoxia is potentiated by acidemia. That potentiation also acts to aid in the diversion of blood to better-ventilated regions, because, as will be discussed in Chapter 20, poor ventilation results not only in low alveolar O_2 concentrations but also in high alveolar CO_2 concentrations. CO_2, acting as an acid anhydride, lowers regional pH, thus in turn not only potentiating hypoxic vasoconstriction but perhaps causing vascular contraction.

If alveolar hypoxia is generalized, or severe enough to cause hypoxemia in the systemic arteries, part of the ensuing pulmonary vasoconstriction will be the result of generalized sympathetic nervous hyperactivity consequent to initiation of chemoreflexes by stimulation of carotid and aortic receptors. Other reflex effects in the pulmonary vessels are much less well established.

The importance of pulmonary vasomotion in adult animals, whether local or general, has not been conclusively decided. Probably the effects of vasomotion, like the passive effects already described, are of little hemodynamic consequence outside the lung. In the lung, however, all these factors act to determine regional gas exchange.

NEONATAL CARDIORESPIRATORY ADAPTATION

The ability of the fetus to convert from placental gas exchange to air breathing is a sine qua non of life. During gestation, respiratory exchange is achieved by circulating fetal systemic blood through the capillary bed immersed in the lakes of maternal blood of the placental sinusoids. The placenta does not provide nearly as complete an exchange or as much functional reserve as will the lungs, and this marginal ability makes any placental maldevelopment or derangement a potentially serious threat to fetal survival. "Fetal distress" from O_2 lack is a common obstetrical complication often requiring premature delivery in order to establish adequate fetal gas exchange.

In the latter part of pregnancy fetal lungs are not collapsed but are somewhat filled with fluid continuously elaborated within the lungs. This is not amniotic fluid, and there seems to be little admixture of amniotic and pulmonary fluids even though fetal respiratory movements sometimes occur. The high fluid viscosity (compared with air) militates against respiratory exchange.

In the last quarter of human gestation, pulmonary surfactant is normally present. This is a critical milestone to have passed, for without surfactant, stable air-filled lungs would be unattainable, and survival of infants born before the appearance of surfactant is unusual.

The conversion to pulmonary gas exchange at birth involves establishment of a gas-filled lung and dissociation of the systemic and pulmonary circulations. Compression of the thorax during delivery expels much of the fluid from conducting airways. The first inspiratory efforts suck the remaining fluid and outside air into alveoli and create the air-liquid interface. The alveolar fluid is apparently absorbed within the next few minutes. Considerable effort is necessary to bring about the first inflation, but once inflation has occurred, the pleural pressure changes required for breathing fall to essentially adult levels if surfactant is present. Most of the lung becomes gas-inflated during the first few inspirations. This enlargement with air does two things: First, it mechanically lowers pulmonary vascular resistance; second, the increased alveolar O_2 removes a potent local stimulus for pulmonary vasoconstriction. The profound fall in pulmonary vascular resistance diverts the right ventricular output into the lungs, accompanied at least temporarily by some blood from the aorta, which now traverses the ductus arteriosus in the opposite direction. The large pulmonary blood flow raises left atrial pressure above that in the right atrium and this closes the flaplike valve at the foramen ovale. With the elimination of right-to-left shunts and appearance of pulmonary O_2 exchange, systemic arterial blood becomes well saturated with O_2. The ductus constricts sufficiently in the presence of the high O_2 content of its blood to prevent flow and complete the separation of the two circulations. If the infant fails to breathe enough to oxygenate alveoli and systemic arterial blood, pulmonary vasoconstriction and ductus arteriosus dilation will occur and may reduce pulmonary perfusion while restoring the fetal right-to-left shunt. Occasionally the ductus fails to close, or atrial septal defects develop such that interatrial shunts remain. These generally divert blood from left to right and so increase pulmonary blood flow. In such instances regression of the pulmonary vascular muscle often does not occur, and persistence of the fetal type of vessels effects the ultimate physiological characteristics of children with congenital heart disease.

REFERENCES

Bouhuys, A. *Breathing—Physiology, Environment and Lung Disease.* New York: Grune & Stratton, 1974.

Campbell, E. J. M. *The Respiratory Muscles and the Mechanics of Breathing.* Chicago: Year Book, 1958.

Dawes, G. S. *Foetal and Neonatal Physiology.* Chicago: Year Book, 1968.

Fenn, W. O., H. Rahn, and P. Dow (Eds.). *Handbook of Physiology.* Washington: American Physiological Society, 1964. Section 3: Respiration, vol. 1.

Macklem, P. T. Airway obstruction and collateral ventilation. *Physiol. Rev.* 51:368–436, 1971.

Mead, J. Mechanical properties of lungs. *Physiol. Rev.* 41:281–330, 1961.

Pattle, R. E. Surface lining of lung alveoli. *Physiol. Rev.* 45:48–79, 1965.

West, J. B. *Ventilation/Blood Flow and Gas Exchange.* Oxford, England: Blackwell, 1965.

20. Respiratory Gas Exchange and Transport

Thomas C. Lloyd, Jr.

Gas exchange in the body occurs by bulk flow of gases, by bulk flow of solutions of gases, and by diffusion of gases through tissues. The preceding chapter discussed some of the determinants of bulk flow but left the story of gas exchange far from complete.

To fully understand gas exchange processes in both lung and systemic tissues it is necessary next to examine some of the physical properties of gases and of solutions of gases in blood. This information can then be used to explore the process of diffusion, by which the internal exchanges occur. With that material in hand, it is possible to amalgamate it with what has been learned about airway and vascular mechanics from Chapter 19 to form a coherent story of pulmonary function. The symbols used to designate the various aspects of pulmonary physiology are listed in Table 20-1.

GASES, AND GASES IN SOLUTION

The composition of a gas mixture can be described by the percentage of each constituent, but composition gradients do not fully determine whether a specified component of a gas mixture will move by diffusion from one place to another. To quantify the tendency for movement, it is necessary to know the total gas pressure, as well as the percentage of the gas represented by any particular molecular type. These two variables are used to calculate an effective pressure for each component of the gas mixture, called the *partial pressure* of that component. The partial pressure (often called the *tension*) of any gas in a gas mixture is the pressure which that gas would exert if *it alone* were present. It is found by multiplying the total gas pressure by the fraction of the composition represented by the gas in question. For example, the partial pressure of O_2 (written Po_2) in air at 1 atmosphere is 0.21×760 mm Hg $= 159.6$ mm Hg. Gas partial pressure is equivalent to voltage in electrical systems. Diffusive gas flow, like flow of electricity, occurs from points of high tension to points of lower tension.

As in meteorology, partial pressures are usually measured in millimeters of mercury. The unit of 1 millimeter of mercury is called a *torr*, in honor of Evangelista Torricelli, the seventeenth-century physicist who invented the barometer. Pressures may also be reported as fractions of a standard atmospheric pressure where 1 atmosphere $=$ 760 torr. To help keep track of the various pressures in the lung, *torr* will be used here for gas partial pressures, but *mm Hg* or *cm H_2O* will be used to quantitate vascular pressures and such pressures as Pel, Pres, or PL.

The concentration of a gas in a mixture may be given as the percentage or as the *mole fraction*, where percentage $= 100 \times$ mole fraction. The sum of the mole fractions of all the constituent gases of a mixture must equal 1, or, alternatively, the sum of all the partial pressures must equal the total gas pressure.

The concept of partial pressure also applies to gases dissolved in liquids, but gas partial pressure is not given simply by the product of hydraulic pressure times the amount of gas dissolved per unit volume. Rather, the partial pressure of a gas in a liquid is equal to that partial pressure in the gas phase *above* the liquid which would be found *under equilibrium conditions*. For instance, in a glass of water in equilibrium with room air at sea level, the Po_2 in both the air and the water is 159.6 torr. Even at the bottom of a lake where the hydraulic pressure might be equivalent to 2 or 3 atmospheres, the Po_2 would be 159.6 torr.

The *amount* of gas dissolved in a liquid at a given temperature is directly related to the gas tension by a coefficient of solubility peculiar to each gas-liquid combination.

461

Table 20-1. Symbols Used in Pulmonary Physiology

Primary Symbols

F	Fractional concentration in gas phase	V	Gas volume
P	Gas pressure in general	$\dot{V}$	Gas volume per unit time
C	Concentration in blood phase	$\dot{Q}$	Blood volume per unit time
S	Saturation	D	Diffusing capacity
R	Respiratory exchange ratio	f	Respiratory frequency

Secondary Symbols

I	Inspired gas	L	Pulmonary
E	Expired gas	b	Blood in general
A	Alveolar gas	a	Arterial
D	Dead space gas	c	Capillary
T	Tidal gas	v	Venous
B	Barometric	$\bar{v}$	Mixed venous blood

STPD	Standard temperature, pressure, dry (0°C, 760 torr)
BTPS	Body temperature, pressure, saturated with water
ATPS	Ambient temperature, pressure, saturated with water

The above symbols are combined to form specific symbols. A dash above a symbol is used to designate a mean value.

Examples

$F_{E_{N_2}}$	Fractional concentration of nitrogen in expired gas
P_B	Barometric pressure
$C_{v_{CO_2}}$	Concentration of carbon dioxide in venous blood
$S_{\bar{v}_{O_2}}$	Saturation of oxygen in mixed venous blood
V_T	Tidal volume
$\dot{V}_{O_2}$	Oxygen uptake
$\dot{V}_A$	Alveolar ventilation
$P_{a_{O_2}}$	Partial pressure of oxygen in arterial blood

(Amount dissolved = constant × partial pressure.) Therefore, knowing the solubility coefficient for the system at hand, one can calculate the partial pressure in a solution by first determining the gas concentration in it. The partial pressure in the solution may be greater or less than the tension of that gas in the gas phase above the solution; this only means that an equilibrium does *not* exist, and more gas must either dissolve or leave solution in accordance with the partial pressure gradient. Temperature affects solubility in such a way that cooling increases the amount dissolvable.

One of the gases present above any solution is the vapor of the solvent itself. The *vapor pressure* of the solvent is determined by its own molecular properties and by temperature, but not by the local barometric pressure. At equilibrium, the partial pressure of solvent in the gas phase above a liquid is equal to the vapor pressure of the solvent. The vapor pressure of water at 37°C is 47 torr. Gas spaces like thoracic airways and alveoli which are fully equilibrated as far as water movement is concerned (i.e., have 100 percent relative humidity) have, therefore, a P_{H_2O} of 47 torr. Since this pressure must appear in the total gas pressure summation, the addition of water vapor to cooler and drier inspired air dilutes the inspired O_2 and N_2 somewhat. Because of the variation of P_{H_2O} with temperature it is usually necessary to correct gas volumes and compositions for this effect. The usual standards are STPD and BTPS. For example, if a gas contains 5 percent of gas x ($Fx = 0.05$) at STPD, its partial pressure will be 0.05 × 760, whereas at BTPS (at sea level) it will be 0.05 × (760 − 47). This correction and the corrections for temperature and pressure on gas volumes ordinarily will not be made in the equations of this text. It must be remembered, however, that in actual practice one must express all measurements under a single condition by use of the gas laws and water vapor corrections. The water vapor pressure in ambient air depends on surface water temperature and also on the relative humidity. The relative humidity indicates the extent to which the liquid and air phases are in equilibrium. At a comfortable 68°F and 50 percent relative humidity, the P_{H_2O} is about 8 torr, but at an unpleasant 90°F and 100 percent humidity, the P_{H_2O} is 36 torr.

The solubility of O_2 in water and in biological fluids other than whole blood is low relative to the amount of O_2 contained in a similar volume of air. For instance, about 0.4 ml O_2 is dissolved in 100 ml of plasma equilibrated with room air, but 100 ml of air contains 50 times that quantity of O_2. Clearly, such liquids would be poor media to transport O_2 to tissues, and all animals more advanced than aquatic forms only a few cells thick have developed compounds which markedly increase the O_2-carrying capacity of their gas transport fluids. It is interesting to note that the type and amount of the specialized material (hemoglobin) in mammals multiplies the O_2-carrying capacity by a factor of about 50, i.e., the content ratio between air and water. The ways in which respiratory gases are carried in solution will next be presented in detail.

Oxygen Transport in Blood

When oxygen diffuses into the plasma from the alveolus, almost all of it finds its way into the red blood cell, where it combines with hemoglobin. Only a small portion of the oxygen remains in the plasma and is carried to the tissues in simple solution. Since the solubility coefficient for O_2 in aqueous solutions such as plasma is 2.44 ml O_2 per 100 ml plasma per atmosphere (760 torr) O_2 pressure, only 0.3 ml of oxygen will be dissolved in 100 ml of plasma with the usual P_{O_2} of 95 torr in arterial blood.

The erythrocyte carries almost all the oxygen (98.5 percent) in the bloodstream by virtue of the capacity of hemoglobin to carry O_2. Hemoglobin is a protein with a molecular weight of 68,000, each molecule having the capacity to combine with four molecules of oxygen. When fully saturated, 1 gm of hemoglobin will hold 1.34 ml of oxygen. Since the normal hemoglobin content of the blood is 15 gm per 100 ml, the oxygen-carrying capacity is approximately 20 ml per 100 ml of blood. This is also expressed as 20 volumes percent.

Unlike the linear relationship between partial pressure and content of gases in true solution, the relationship of P_{O_2} to O_2 content is not a linear one (Fig. 20-1), a fact which has important physiological consequences. Thus there is no single constant usable

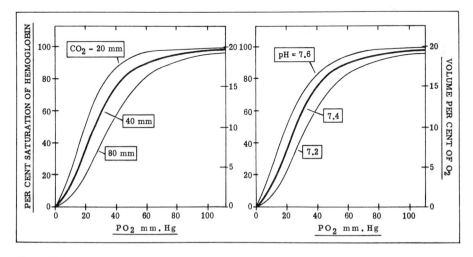

Figure 20-1
The normal oxyhemoglobin dissociation curve. Influences of partial pressure of CO_2 and pH on the curve.

to calculate blood O_2 content from the P_{O_2}. One must always use curves like Figure 20-1 to obtain such information. Close familiarity with this graph is necessary for an understanding of many aspects of gas exchange. The part of the O_2-hemoglobin dissociation curve of greatest importance is the "shoulder" in the P_{O_2} range of 40 to 100 torr. At the top of this shoulder, large changes in P_{O_2} have a small effect on the amount of oxygen carried by the hemoglobin; below that level, small changes in P_{O_2} have a large effect on the oxygen content. It should be noted that the "shoulder" covers the normal range of oxygen tension in the blood under resting conditions. At the upper end of the curve, a decrease in alveolar O_2 tension does not immediately jeopardize the adequacy of the arterial O_2 content. On the other hand, the steep part of the curve enables a lot of O_2 to be delivered from blood to tissues without causing a profound fall in P_{O_2}. This tends to keep the driving force which moves O_2 outward relatively constant along the whole length of a systemic capillary. Conversely, in the pulmonary capillaries a large gradient for O_2 movement into blood is maintained until most of the O_2 has been transferred.

The amount of oxygen combined with hemoglobin also depends upon other factors, such as the P_{CO_2} of the blood (Fig. 20-1). As the carbon dioxide tension increases, the oxygen and the hemoglobin tend to dissociate. This action is an aid to normal function since, in the region of the systemic capillary, the P_{CO_2} is elevated. The elevated carbon dioxide tension helps the blood to unload the oxygen as it passes through the capillary. In the lung, however, the loss of CO_2 by ventilation increases the affinity of hemoglobin for O_2.

The O_2-hemoglobin dissociation curve also depends upon the pH of the blood (Fig. 20-1). An increase in acidity tends to drive oxygen off the hemoglobin molecule. The reason will be considered later, along with the effect of carbon dioxide on the hemoglobin dissociation curve. This effect is also of physiological value since the pH in the systemic capillary region is lower than that of arterial blood, and oxygen is thus released from hemoglobin to diffuse into the tissues.

The temperature of the blood has a significant influence on oxygen carriage, an elevated temperature causing dissociation of oxygen from hemoglobin. This effect may be of value in the muscles, where exercise increases temperature.

Oxygen Extraction in Systemic Capillaries

Under ordinary conditions the arterial blood P_{O_2} is 95 torr and the oxygen content is 20 volumes per 100 ml of blood. When the arterial blood reaches the capillaries, the oxygen in simple solution in the plasma begins to diffuse into the tissue because of the lower P_{O_2} there (< 30 torr). This in turn causes some of the oxygen combined with the hemoglobin to dissociate from it and diffuse first into the plasma and then to the tissues. By the time the blood leaves the capillaries, the P_{O_2} has fallen to 40 torr and the oxygen content to 14 volumes per 100 ml.

The amount of oxygen which leaves the blood depends upon local needs, and conditions in the tissue being perfused. As metabolism increases, the increased O_2 utilization is not matched by an equal increase in capillary blood flow. The amount of oxygen extracted from each milliliter of blood will therefore increase, and it may be expressed in terms of the *oxygen utilization coefficient,* which is the ratio $\dfrac{\text{arteriovenous } O_2 \text{ difference}}{\text{arterial } O_2 \text{ content}}$. Normally this value is $\dfrac{20-14}{20} = 0.3$. In exercise this coefficient may approach 1.0 for the blood flowing through the active muscle and 0.7 to 0.8 for the mixed blood entering the right heart. At the same time, blood flow through the exercising muscle increases greatly, increasing the oxygen available to the tissues in this manner as well.

Carbon Dioxide Transport in Blood

Carbon dioxide is carried by the blood in three forms: (1) as bicarbonate (HCO_3^-), (2) in combination with protein (carbamino), and (3) in simple solution. Among the three forms there is a total of approximately 48 volumes of CO_2 per 100 ml in arterial blood, which is divided as follows: bicarbonate, 43 volumes; carbamino hemoglobin, 2 volumes; carbamino plasma protein, 1 volume; and dissolved carbon dioxide, 2.4 volumes. The CO_2 in normal human arterial blood exerts a partial pressure of 40 torr. The relation between partial pressure and total CO_2 content for arterial blood is shown in Figure 20-2. As P_{CO_2} increases, the total CO_2 content rises in a fashion which is nonlinear. The shape of the curve favors greater CO_2 exchange at lower O_2 tensions.

The total CO_2-blood dissociation curve is made up of several separate curves representing the three forms of CO_2 carriage, as shown in Figure 20-2. The amount of CO_2 in simple solution is a linear function of the CO_2 tension. A small amount of this gas is actually in the form of carbonic acid (H_2CO_3).

The amount carried in the carbamino form (combined with hemoglobin and plasma protein) does not bear a linear relation to CO_2 tension (Fig. 20-2). Also, the O_2 partial

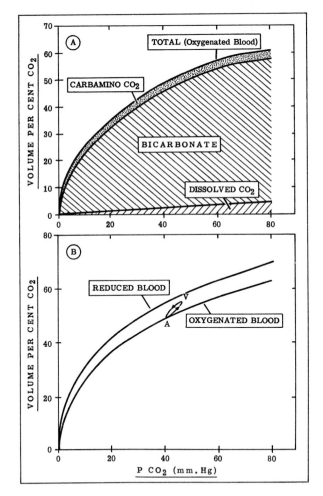

Figure 20-2
A. The CO_2-blood dissociation curve showing the amount of CO_2 present in each of the three forms. B. The dissociation curve of oxygenated and reduced blood. The loop indicates the changes that occur when arterial blood (A) becomes venous blood (V) and when it is again arterialized.

pressure plays an important role here, since saturation of hemoglobin with oxygen drives off carbon dioxide from the molecule. In fact, if it were not for the reduction of the amount of oxyhemoglobin as the blood passes through the capillary, much less CO_2 would be taken up in the carbamino form.

The remainder of the CO_2 is carried in the form of bicarbonate. A solution of bicarbonate will take up appreciable amounts of CO_2 only at CO_2 tensions below 10 torr. However, in the company of other substances, such as hemoglobin, its behavior changes and it readily takes up CO_2 at higher partial pressures. Thus the curve for bicarbonate (Fig. 20-2) is characteristic only for bicarbonate in the blood. The amount of CO_2 which finds its way into the bicarbonate form is dependent upon the O_2 content of the hemoglobin also, as will be seen shortly. The difference in CO_2 dissociation curves of arterial and venous blood reflects the effect of oxygen on bicarbonate and carbamino formation.

The CO_2 tension in the tissues is over 50 torr, while in the arterial blood it is 40 torr, so carbon dioxide will diffuse into the plasma and red cells as the blood passes through the systemic capillaries. As noted earlier, the uptake is aided by the simultaneous outward movement of oxygen from the blood. When the blood leaves the capillary, its CO_2 content has increased from 48 volumes per 100 ml to 54 volumes per 100 ml and the P_{CO_2} from 40 to 46 torr. Since the oxyhemoglobin level has decreased at the same time, the pathway shifts to a new dissociation curve, as shown in Figure 20-2.

The CO_2 entering the blood in the capillary goes into all the three forms, with most (65 percent) becoming bicarbonate, 27 percent forming carbamino compounds, and 8 percent going into solution. The mechanisms by which these reactions occur are outlined in Figure 20-3.

The CO_2 which goes into simple solution in the plasma is slowly converted to carbonic acid, although at equilibrium only a small amount is in acid form. Most of the CO_2 goes into the erythrocyte, where rapid hydration occurs through the catalyzing action of carbonic anhydrase. As the acid is formed, it dissociates to form bicarbonate and hydrogen ions. The bicarbonate concentration rapidly increases in the erythrocyte, and most of it diffuses out into the plasma. To preserve electrical neutrality, chloride

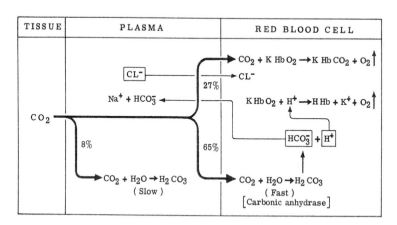

Figure 20-3
Chemical reactions involved in the three principal forms of CO_2 carriage by the blood. A small amount (8 percent) is carried as dissolved CO_2 by the plasma and red blood cell. A larger amount (27 percent) is carried by formation of carbamino compounds (primarily carbamino hemoglobin). The largest fraction (65 percent) is carried by formation of bicarbonate, which occurs within the erythrocyte. The CO_2 entering the cell is hydrolyzed to carbonic acid, which is neutralized by hemoglobin, with the bicarbonate ion diffusing out into the plasma. Simultaneously chloride moves into the cell to preserve electrical neutrality.

ions diffuse into the cell (this is called the *chloride shift*). The hydrogen ions freed by the dissociation of carbonic acid combine with hemoglobin, freeing potassium, which balances the chloride that has diffused into the cell. Another portion of the CO_2 entering the cell combines with the imidazole group of histidine on the globin portion of the hemoglobin molecule. Although this site is different from that of the oxyhemoglobin combination, O_2 and CO_2 do not readily coexist on the hemoglobin molecule. It should be noted that hemoglobin plays an essential role in the bicarbonate and carbamino formation and is therefore indirectly responsible for most of the carriage of CO_2.

In the lung, these processes are reversed; as CO_2 diffuses out of the plasma, bicarbonate is drawn back into the red cell, where it is rapidly broken down to CO_2 through the action of carbonic anhydrase and again released to the plasma to diffuse into the alveoli. At the same time, carbamino CO_2 leaves the hemoglobin molecule, and O_2 joins it.

It might be supposed that movement of CO_2 in and out of the blood would entail considerable alteration in the blood pH. This would be the case were it not for the fact that oxyhemoglobin is more acid than reduced hemoglobin. As the CO_2 enters the blood at the capillary and decreases pH, the hemoglobin at the same time shifts in the alkaline direction with the loss of O_2. The loss of one mole of O_2 causes a loss of hydrogen ion concentration of $0.7M$. Thus, $0.7M$ of CO_2 could be added to the blood for every mole of O_2 lost without change in pH. The change in pH then will depend upon the relative contents of O_2 and CO_2 exchanged at the capillary. Release of acids from the tissues into the bloodstream will also influence blood pH at this point.

The ratio of carbon dioxide produced to oxygen consumed in the tissues under resting, steady-state conditions is termed the *respiratory quotient* (RQ). The ratio varies from 0.7 to 1.0, depending upon diet, and is 0.82 in a fasted human. The change in acidity of the blood as it discharges O_2 and picks up CO_2 in the capillaries will depend upon this ratio: If it is 0.7, there will be no pH change; if 1.0, there will be a considerable change. Under ordinary conditions RQ is 0.82, arterial pH is 7.40, and venous pH is 7.36. It will be shown subsequently (Chap. 24) that these reactions also play an important role in the acid-base balance of the body.

Oxygen and Carbon Dioxide Reciprocity

The reciprocal effects of oxygen and carbon dioxide on gaseous transport seen in Figures 20-1 and 20-2 may be explained on the basis of the chemical reactions described above. For example, increasing the CO_2 level of the blood will increase the hydrogen ion concentration in the red cell and drive O_2 off the hemoglobin molecule. In addition, O_2 and CO_2 compete for hemoglobin directly as oxyhemoglobin and carbamino hemoglobin. The effect of decreased pH on oxygen carriage may also be readily deduced from the chemical reactions which occur in the blood.

Reciprocally, when the P_{O_2} of the blood is increased, CO_2 will be displaced from the hemoglobin and oxyhemoglobin will be formed. The more acidic oxyhemoglobin will provide hydrogen ions to combine with bicarbonate ions, forming carbonic acid, which is then dehydrated to free CO_2. As a result, at any fixed P_{CO_2} level the total amount of CO_2 carried by the blood will decrease as the P_{O_2} is increased.

DIFFUSION

General Concepts

Diffusion is a process by which a difference in concentration (more correctly, a difference in chemical activity) of a substance between two points is evened out by a migration of molecules toward the region of lower activity. Transport by diffusion is an important part of gas exchange. The rate of diffusive movement of gas molecules through a single phase (either liquid or gas) depends on the differences in partial pressure of the gas between any two points and on the distance separating the points.

(For gases, partial pressure gradients are equivalent to activity gradients, whereas concentration gradients are not.) In other words, diffusion depends on the partial pressure gradient per unit length of the diffusion path. Since diffusion is caused by thermal motion of molecules, it will occur at a greater rate at higher temperatures, and at a greater rate for smaller molecules, since thermal agitation depends on molecular weight. If one imagines diffusion to be occurring across a specified area perpendicular to the direction of movement, as in diffusion through a piece of cellophane separating two chambers, the net transport obviously depends on the area across which it can occur. Combining all the above factors,

$$\dot{V} \propto \frac{\Delta PTA}{L\sqrt{M}}$$

where $\dot{V}$ is volume rate of gas flow, ΔP is partial pressure gradient, T is absolute temperature, A is surface area, L is path length, and M is molecular weight.

At the alveolar interface, diffusion occurs from gas to liquid, and the above expression does not fully describe the situation. An additional term must be included for the solubility of the gas in the liquid. As might be expected, diffusivity is directly related to solubility. In the case of gas to liquid diffusion, then:

$$\dot{V} \propto \frac{\Delta PTA\alpha}{L\sqrt{M}}$$

where the new term, α, is solubility. For gases of physiological interest, α lies between 1.0 and 0.01; therefore diffusion rates into liquids are less than diffusion rates in the gas phase alone.

In physiological systems some of the above factors are more important than others. Thus, for the three gases O_2, N_2, and CO_2, differences in molecular weight make very little difference in relative rates of diffusion. Similarly, temperature is essentially constant since even during severe febrile responses the difference in absolute temperature is only about 1 part in 100. On the other hand, solubility and available capillary area play prominent roles in diffusion into blood, making this many times slower than diffusion in the gas phase. In consequence, diffusion from alveolar duct to alveolar interface is an insignificant impediment to gas transport even though the distance of migration may be 10^3 greater than the thickness of the pulmonary membrane. The major impediment is through the tissue phase.

In the lungs one cannot quantitate surface area or path length during life, so a *diffusion coefficient* for lung lacks the meaning of a rigorously derived physical constant. The closest one can approach a diffusion constant for lung is to obtain a *transport factor* or *diffusing capacity* by measuring the rate of uptake or removal of a particular gas and the average partial pressure gradient existing between alveolus and capillary blood which caused that exchange. The diffusing capacity of the lung, D_L, is defined by the relation $D_L = \dot{V}/\Delta P$, and the term D_L contains all the unmeasurable terms and physical constants noted in the proportionality expression given earlier. The dimensions of D_L are milliliters per minute per torr. Note that $1/D_L$ has the dimensions of resistance. After discarding the negligible term for diffusion within the gas phase, the total diffusion barrier to O_2 or CO_2 transfer in the lung is the sum of two major components: (1) diffusion of O_2 through the alveolar and capillary cells, plasma, and red cell, and (2) reaction of the O_2 with hemoglobin. It has been estimated that for O_2 each of these two components of D_L accounts for about half the total, so the rate of uptake of O_2 by blood cannot be considered to be infinitely rapid.

CO_2 is about 20 times as soluble (hence as diffusible) as O_2; therefore it is doubtful that impairment of CO_2 diffusion would ever be critical to survival, since O_2 transfer would be so profoundly altered at that point that life would be unlikely. Actually, even the uptake of O_2 is limited by diffusion only under very unusual circumstances, as will be seen farther on.

Diffusive Transport at the Pulmonary Capillary

The normal adult human lung capillaries contain only about 90 ml of blood, but this is spread across a surface of about 70 square meters. The thickness of the liquid barrier between red cell and alveolar gas is on the order of 1 μm. While this would seem to provide more than enough surface for diffusion, the critical feature is that with a normal resting cardiac output each red cell spends only about 0.75 second in a pulmonary capillary.

Whether an equilibrium is established between gas in the alveolus and the capillary blood depends on the diffusing capacity, but it also depends on the amount of gas which needs to be transferred to reach equilibrium. A gas like N_2, which is less diffusible than O_2, comes into equilibrium very quickly in a unit of blood entering the capillary because it is so (relatively) insoluble that not much must cross the border to establish the same P_{N_2} on both sides. On the other hand, since O_2 and CO_2 have great solubilities in blood, many molecules must be transferred to achieve equality of tensions. It turns out that by the time blood reaches mid-capillary, the P_{O_2} of blood and the P_{O_2} of the alveolus are essentially identical. CO_2, because of its greater diffusivity, reaches equilibrium in a fraction of the time required for O_2, even though about the same number of molecules of CO_2 are transferred. It therefore follows that the blood flow rate could double and O_2 would still reach equilibrium, but at any higher flow with other things unchanged, insufficient time for complete O_2 equilibration may limit oxygenation while not influencing CO_2 removal.

If all pulmonary capillaries were open while the individual was at rest, there could be only a doubling of cardiac output if the condition of reaching equilibrium is to be met. However, the recruitment of additional parallel capillary paths caused by increasing the right ventricular output means that, instead of an increase in capillary flow velocity pari passu with cardiac output, the velocity through any *single* capillary increases only a small amount until all of the available bed is open. Note that the opening of more parallel pathways not only keeps the increase in capillary flow velocity to a minimum but increases the surface area available for diffusion. This property of the pulmonary circulation helps in allowing O_2 uptake to reach about 15 times the resting value without the appearance of a disequilibrium between capillary blood and alveolar gas. Under these conditions, capillary flow velocity will be approximately twice the resting rate, while the diffusing capacity increases about threefold.

Under such high O_2 demands as mentioned above, the blood entering the pulmonary capillary has less O_2 than when at rest, and therefore a lower P_{O_2}. This increase in partial pressure gradient will increase the rate of diffusion in the first part of the capillary. Refer to the hemoglobin dissociation curve (Fig. 20-1) and note that until the "shoulder" around 80 percent saturation is reached, addition of O_2 raises O_2 content more than it raises P_{O_2}. A large partial pressure gradient can be maintained until most of the O_2 has been transferred. The increase in the mean P_{O_2} gradient, as well as the increases in capillary area and flow velocity, is apparently what allows the adequate oxygenation of severe exertion.

O_2 uptake can be limited by diffusion if the capillary bed is reduced by disease or by surgical removal of lung. Diffusion limitations may also follow decreases in the P_{O_2} gradient caused by any mechanism that lowers alveolar P_{O_2}. Less often, diffusivity is compromised by thickening of the alveolar membrane by disease. As one would predict, diffusion impairment influences O_2 uptake during exercise long before interfering with oxygenation during rest.

PULMONARY GAS EXCHANGE

Respiratory Exchange Ratio

In most mammals the alveolar gas composition at 1 atmosphere ambient pressure and a body temperature of 37°C is P_{H_2O} − 47 torr, P_{CO_2} − 40 torr, P_{O_2} − 100 torr, and P_{N_2} − 570 torr. These particular values occur as a consequence of the balance between simultaneous CO_2 delivery and O_2 removal by blood, and CO_2 removal and O_2 delivery by breathing. N_2 is neither used nor produced in the body, so it is a passive filler. The alveolar P_{N_2} can be used as a mirror of the relation between O_2 uptake and CO_2 removal in the following way. If O_2 uptake (in milliliters per minute) is in excess of CO_2 outflow, alveolar N_2 will be more concentrated than is N_2 in air. Conversely, if O_2 uptake is less than the rate of CO_2 removal from pulmonary capillary blood, N_2 will be diluted by the excess of CO_2. Normally the number of moles of O_2 consumed is slightly greater than the amount of CO_2 produced, so the sum of P_{CO_2} and P_{O_2} is slightly less than the P_{O_2} of humidified room air at the same temperature. Since total alveolar pressure equals barometric pressure (the variations consequent to airflow being negligible), the P_{N_2} of average alveolar gas normally exceeds that of the wet warm air by about 10 torr. The relation between CO_2 output and O_2 uptake is expressed by the *respiratory exchange ratio,* abbreviated R. $R = \dot{V}_{CO_2}/\dot{V}_{O_2}$, where $\dot{V}_{CO_2}$ and $\dot{V}_{O_2}$ are the CO_2 lost and the O_2 taken up per unit of time (usually expressed as milliliters per minute STPD). At a steady state, R for the lungs taken as a whole equals the respiratory quotient, the latter being a statement of the metabolic state of the whole animal. For short periods, such as with breath holding or panting, RQ and R may be unequal. R may be appraised either by measurements of the CO_2 and O_2 volumes exchanged or by comparison of alveolar P_{N_2} with P_{N_2} of ambient air. When $R < 1$, alveolar $P_{N_2} >$ ambient P_{N_2}; when $R > 1$, the reverse pertains.

When $R < 1$, exhaled volume is proportionally less than inhaled volume because the volume of O_2 removed from tidal air exceeds the volume of CO_2 added. The converse occurs when $R > 1$. Often it is necessary to know the volume of air inspired per unit of time when the expired volume has been measured. Because the total amount of N_2 remains constant, the fractions of N_2 in inspired and expired air can be used to obtain inspired volume, given expired volume:

$$\dot{V}_I = \dot{V}_E \cdot (F_{E_{N_2}}/F_{I_{N_2}})$$

The $F_{E_{N_2}}/F_{I_{N_2}}$ ratio is also used to "correct" gas concentrations for the effect of differences between O_2 uptake and CO_2 lost. For example, a useful expression for obtaining R (if there is no CO_2 in inspired air) is

$$R = \frac{F_{E_{CO_2}}}{F_{I_{O_2}} \cdot (F_{E_{N_2}}/F_{I_{N_2}}) - F_{E_{O_2}}}$$

In this expression the $F_{E_{N_2}}/F_{I_{N_2}}$ ratio has been used to "correct" $F_{I_{O_2}}$ so that the denominator of the equation is determined only by the amount of O_2 taken up from gas and not by concentration or dilution caused by unequal CO_2 and O_2 fluxes. The usefulness of this equation is that it enables one to calculate R without measurement of gas volumes.

CO_2 Exchange

CO_2 can be removed by breathing only because the act of inspiration continually dilutes alveolar gas, lowers its P_{CO_2}, and thereby establishes a diffusion gradient for movement from the capillary blood. Exhalations can then expel CO_2-laden gas. The balance is such that the alveolar P_{CO_2} at end expiration is normally about 41 torr, and at end inspiration about 38 torr, while the P_{CO_2} of blood entering the capillary is about 46 torr. Under steady-state conditions, O_2 uptake and CO_2 production are constant. CO_2 removal is a function of the rate of removal of alveolar gas (called the alveolar ventilation, $\dot{V}_A$) and the amount of CO_2 in alveolar gas:

$$\dot{V}_{CO_2} = \dot{V}_A \cdot F_{A_{CO_2}}$$

For a given $\dot{V}_{CO_2}$, alveolar ventilation determines $F_{A_{CO_2}}$. If $\dot{V}_A$ is reduced below normal (hypoventilation), $F_{A_{CO_2}}$ rises proportionally. Because alveolar CO_2 is in equilibrium with pulmonary capillary blood, the partial pressure of CO_2 in blood leaving the lung capillary is identical to that in the alveolus, and the P_{CO_2} of systemic arterial blood is essentially the same. Then

$$Pa_{CO_2} = P_B \cdot F_{A_{CO_2}}$$

If, as is usually the case, $\dot{V}_{CO_2}$ has been expressed in milliliters STPD and $\dot{V}_A$ in milliliters BTPS (with the further common assumptions that P_B is 760 torr and the subject's temperature is 37°C), the above equations can be reduced to

$$Pa_{CO_2} = \frac{0.863 \, \dot{V}_{CO_2}}{\dot{V}_A}$$

where the constant, 0.863, includes corrections for pressure, temperature, and water vapor. This often-cited equation shows that Pa_{CO_2} is inversely dependent upon alveolar ventilation when CO_2 production remains constant. Thus, if effective ventilation were halved, a new equilibrium could be achieved by doubling the alveolar P_{CO_2}, since then each half-normal-sized breath would expel the normal number of moles of CO_2. The price paid is an increase in CO_2 stores. This readjustment is often seen in persons with chronic lung disease, but, as will be seen in Chapter 21, to accomplish it means that readjustments in the respiratory control mechanisms must also occur, since these act to hold blood P_{CO_2} quite constant.

If all of each tidal inspiration were equilibrated with capillary blood, the concentration of CO_2 in expired gas would be the same as in alveolar gas. However, part of the inspirate stops within conducting airways far removed from alveoli, and this volume takes no part in gas exchange. The volume of these airways in a normal adult man is between 100 and 200 ml. At end expiration, conducting airways are filled with gas pushed upward from below which has the composition of mixed alveolar gas. On inspiration, this volume is drawn downward, but because it already has alveolar gas composition, it does not alter alveolar gas concentrations. Only the volume inhaled which *exceeds* the volume of the conducting airways becomes a diluent for alveolar gas. The gas in conducting airways is simply shuttled back and forth. The internal volume of these airways is called the *anatomical dead space*. *Dead space ventilation* refers to the movement of gas in and out of some part of the respiratory system in which no actual transmembrane exchange occurs. The presence of a dead space minimizes the effectiveness of the tidal volume.

At the start of expiration, the anatomical dead space contains room air, and this volume is expelled before gas of alveolar composition appears. The composition of expired gas is, therefore, determined by the relative contributions of the alveolar and dead space compartments. Since the number of moles of CO_2 leaving the alveoli must equal the number collected in the expirate, one could calculate the sizes of the dead space and the effective alveolar exchanging volume if one knew the alveolar and mixed-expirate CO_2 concentrations, and the expired volume. A sample calculation brings out these features.

If all the expired air collected from an individual for one minute had a volume of 6.9 liters and a CO_2 content of 4 percent, the subject has expelled 275 ml CO_2 per minute (4% × 6.9 L/min). If his alveolar CO_2 content is determined to be 5.5 percent, the exhaled volume of CO_2 would be contained in 5 liters of alveolar gas (0.275 L/0.055). In this case the subject has had an *alveolar minute ventilation* of 5 liters. The difference between the actual volume exhaled (6.9 liters) and 5 liters expresses dead space ventilation. Since this 1.9 liters was distributed in equal parts to each breath, the dead space volume would be 147 ml if his respiratory rate had been 13 breaths per minute.

In the preceding example it was assumed that all alveoli had a CO_2 content of 5.5 percent. Assume for a moment that only half the alveoli receive any pulmonary blood flow. The remaining half will contain no CO_2 because of the absence of perfusion. If the perfused half has an alveolar CO_2 concentration of 5.5 percent, the average alveolar concentration for the lungs as a whole would be only 2.75 percent and the mixed expirate would be even less than this because of the anatomical dead space gas. In reality there is no way to sample alveolar gas directly. What is done is to use the P_{CO_2} of systemic arterial blood as the average alveolar value. In normal individuals the error is probably less than 2 torr. However, when alveoli receive no perfusion, those alveoli will not be represented in the systemic arterial blood. This means that the alveolar P_{CO_2} assumed by using the P_{CO_2} of arterial blood is weighted in the direction of the alveoli receiving the most blood flow. If alveoli are not perfused, they are physiologically just as dead (from the standpoint of gas exchange) as are conducting airways. When dead space is calculated by assuming the average alveolar P_{CO_2} to be equal to systemic arterial P_{CO_2}, the space so calculated is called the physiological dead space. The above dead space calculation can be expressed in a single equation, which if we include the assumption that Pa_{CO_2} equals PA_{CO_2} is

$$\frac{\dot{V}_D}{\dot{V}_T} = \frac{Pa_{CO_2} - P_{E_{CO_2}}}{Pa_{CO_2}}$$

Therefore, upon measuring arterial P_{CO_2} and the P_{CO_2} of mixed expired air ($P_{E_{CO_2}}$) one can conveniently express the ratio of physiological dead space ($\dot{V}_D$) to total ($\dot{V}_T$) ventilation. This ratio normally is less than 0.3. Carefully note that because of the assumptions for its calculation, the physiological dead space is a *virtual* space and not a real one. Normally, the physiological dead space and anatomical dead space are essentially equal, but in diseases which occlude parts of the lung circulation physiological dead space exceeds anatomical dead space. The air entering and leaving unperfused alveoli represents work done fruitlessly.

So far it has been supposed that there are only two populations of alveoli: those containing CO_2 at a single concentration and those containing no CO_2 at all. But in truth the CO_2 contents of alveoli differ widely in different parts of the lung, because regional ventilation and blood flow vary throughout the lung, as described in Chapter 19. It will now be shown that, rather than just regional ventilation or regional blood flow per se, it is the ratio of ventilation to blood flow ($\dot{V}_A/\dot{Q}$) that determines regional P_{CO_2}. This effect is easiest to see when one looks at extreme values. Assume a mixed systemic venous blood P_{CO_2} of 45 torr, and assume that inspired air has no CO_2. Under normal conditions $\dot{V}_A$ and $\dot{Q}$ are such that PA_{CO_2} is 40 torr. If $\dot{V}_A$ is progressively decreased, PA_{CO_2} approaches 45 torr, since with a $\dot{V}_A$ of zero no CO_2 is lost and PA_{CO_2} equals $P\bar{v}_{CO_2}$. At this extreme the $\dot{V}_A/\dot{Q}$ ratio approaches zero. One could also approach this ratio by increasing $\dot{Q}$ to infinity; once again PA_{CO_2} would approach that of mixed venous blood since there would be too little ventilation to effect significant loss of CO_2 from this infinite volume of blood. At the other extreme, if $\dot{Q}$ approaches zero, the $\dot{V}_A/\dot{Q}$ ratio approaches infinity and PA_{CO_2} approaches that of inspired air, since no CO_2 is being delivered in the absence of blood flow. Similarly, an infinite $\dot{V}_A/\dot{Q}$ ratio can occur if $\dot{V}_A$ approaches infinity, in which event incoming blood is so stripped of its CO_2 by such a high ventilation that its CO_2 tension again would approach zero. There is, therefore, a unique PA_{CO_2} for each $\dot{V}_A/\dot{Q}$ ratio, and the particular PA_{CO_2} lies somewhere between that of inspired gas and that of mixed systemic venous blood.

In Chapter 19 it was mentioned that differences in both ventilation and perfusion occur in the lung as a function of weights of blood and lung tissue. Though in general the bases of the lungs receive both more ventilation and more perfusion than the apices, the relative proportions are such that the $\dot{V}_A/\dot{Q}$ ratio is about 3.5 at the apices and 0.65 at the bases. The equivalent alveolar CO_2 tensions in these regions are 27 and 41 torr, respectively. Intermediate values occur at intermediate locations. The relations of $\dot{V}_A/\dot{Q}$ to alveolar gas tensions and R are shown in Figure 20-4. These are not simple linear relationships because they include combined effects of the hemoglobin dissociation curve, CO_2 blood solubility curve, and (for the gas tensions) influences of R. Note that the range of $\dot{V}_A/\dot{Q}$ in a normal human lung covers the steepest section of the $P_{CO_2} - \dot{V}_A/\dot{Q}$ curve. In other words, the normal range of $\dot{V}_A/\dot{Q}$ is the same range that most influences gas tensions. The overall effective $\dot{V}_A/\dot{Q}$ is about 0.8.

In view of the foregoing, how can it ever be said that there is *an* alveolar P_{CO_2}? This is a difficult problem in pulmonary physiology, for one must usually be content to use an average number and assume that the lungs consist of only one alveolus having a single airway and capillary. The average alveolar P_{CO_2} is most often assumed to be that of mixed pulmonary capillary blood. Systemic arterial blood is considered to have the same P_{CO_2} because there is very little CO_2 added to pulmonary capillary blood (a bit of bronchial and myocardial venous drainage) before it arrives in the systemic arteries. The amount of CO_2 added by the small quantity of bronchial and cardiac venous blood is too small to be detected. Normally, systemic arterial P_{CO_2} is about 40 torr. Note that this is not the average of the extremes of alveolar P_{CO_2} known to be present in a normal lung; the alveoli having the lowest P_{CO_2} contribute the least blood, so their influence is minimized. This effect also makes the average $\dot{V}_A/\dot{Q}$ different from the mean of the extreme values.

Another measure of mean alveolar P_{CO_2} can also be made by assuming it to be equal to P_{CO_2} of the exhaled air which appears after the anatomical dead space has been flushed. This would not yield a value equal to systemic arterial P_{CO_2} if there were a large regional variation in $\dot{V}_A/\dot{Q}$ ratio, but in normal individuals the two are about equal. Expired air measurements have the advantage of not requiring needle puncture of an artery and thus may be more useful in some situations.

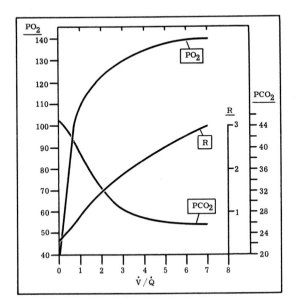

Figure 20-4
The relation of alveolar P_{O_2}, P_{CO_2}, and R to $\dot{V}_A/\dot{Q}$ ratio when inspired air has a normal sea-level composition and mixed systemic venous blood has a P_{O_2} of 40 torr and P_{CO_2} of 45 torr. Partial pressures are in torr. (From data of West, 1971. P. 110.)

O₂ Exchange

Like $P_{A_{CO_2}}$, $P_{A_{O_2}}$ can be expressed as a function of $\dot{V}_A$, but we cannot assume $P_{a_{O_2}} = P_{A_{O_2}}$ as we did with CO_2. Although the exact equation includes a small correction for the effects of R, the following is accurate to within 2 torr during normobaric breathing of ambient air, and for convenience and comparison it is cast in the same units and with the same corrections as the corresponding equation for $P_{A_{CO_2}}$.

$$P_{A_{O_2}} = P_{I_{O_2}} - \frac{0.863 \ \dot{V}_{O_2}}{\dot{V}_A}$$

For reasons soon to be presented, arterial P_{O_2} is a poor measure of mixed alveolar P_{O_2}. Because of this fact, and the questionable applicability of expired air analysis, the average alveolar P_{O_2} is usually calculated from the systemic arterial P_{CO_2} and R. This *alveolar air equation* is easily derived, but the derivation will be omitted here. One can intuitively understand it by remembering that if $R = 1$, the sum of P_{CO_2} and P_{O_2} must equal the P_{O_2} of humidified body-temperature room air. The equation simply does this sum after correcting for the concentration or diluting effects when $R \neq 1$. The calculation of alveolar P_{O_2} is given by

$$P_{A_{O_2}} = P_{I_{O_2}} - P_{A_{CO_2}} \left[F_{I_{O_2}} + \frac{1 - F_{I_{O_2}}}{R} \right]$$

Note from the alveolar air equation that $P_{A_{O_2}}$ is inexorably related to $P_{A_{CO_2}}$, and, since the latter depends upon the $\dot{V}_A/\dot{Q}$ ratio, so must $P_{A_{O_2}}$. As before, $P_{A_{O_2}}$ should vary between P_{O_2} of mixed venous blood and that of room air as $\dot{V}_A/\dot{Q}$ varies from zero to infinity. The variation of $P_{A_{O_2}}$ as a function of $\dot{V}_A/\dot{Q}$ is shown in Figure 20-4. Once again, the steepest part of the curve is that part contained by the range of $\dot{V}_A/\dot{Q}$ ratios in a normal lung. Note that the effect of $\dot{V}_A/\dot{Q}$ on $P_{A_{O_2}}$, in absolute terms, is much greater than the effect on $P_{A_{CO_2}}$. The normal range of $P_{A_{O_2}}$ is from near 132 torr in the apices to about 90 torr at the bases. (See from Figure 20-1 that there will, however, be little difference in O_2 content of blood coming from these two regions.)

In contrast with CO_2, there is usually a difference of about 5 to 10 torr between systemic arterial P_{O_2} and calculated alveolar P_{O_2}. The difference is caused by the addition of small amounts of incompletely oxygenated hemoglobin to pulmonary capillary blood. Because of the shape of the hemoglobin dissociation curve, addition of partially oxygenated blood causes a profound fall in P_{O_2} but only a small change in O_2 content. A sample calculation will bear this out.

Assume a P_{O_2} of 100 torr to exist in all alveoli. Blood passing through those capillaries will emerge with a P_{O_2} of 100 torr, and with its hemoglobin 95 percent saturated (see Figure 20-1). If the total O_2-carrying capacity, when 100 percent saturated, is 20 volumes per 100 ml, blood leaving the hypothetical alveoli will contain 19 volumes per 100 ml. If it is next assumed that, for every 10 volumes of lung capillary blood flow, 1 volume will flow directly from pulmonary artery to pulmonary vein, this "shunt" flow will add less-well-oxygenated blood to that coming from pulmonary capillaries but the composition of the mixture will be predictable. If the shunt blood has a normal systemic venous P_{O_2} of 40 torr, it is 75 percent saturated and therefore contains 15 ml O_2 per 100 ml blood. Since it was initially assumed that the relative flow was 10:1, each liter of capillary blood will be mixed with 100 ml of shunt blood, resulting in a total volume of 1100 ml, which contains 205 ml of O_2 (190 from capillary blood, 15 from shunt flow). This represents 18.6 ml O_2 per 100 ml blood, or 93 percent saturation. Reference again to the dissociation curve shows that 93 percent saturation corresponds to a P_{O_2} of 74 torr. Thus, the addition of the 9 percent shunt of systemic venous blood has reduced the

Po_2 of blood entering the left atrium from 100 to 74 torr, but has changed O_2 content by only 0.4 volume per 100 ml. This effect is particularly apparent if alveolar Po_2 is above the "shoulder" in the dissociation curve at about 50 torr.

It is not necessary to know the cardiac output to calculate the fraction of cardiac output that is shunted around ventilated alveoli. By application of the Fick principle one can derive an equation for the ratio of shunt flow, $\dot{Q}s$, to total flow, $\dot{Q}T$:

$$\frac{\dot{Q}s}{\dot{Q}T} = \frac{C\bar{c}_{O_2} - Ca_{O_2}}{C\bar{c}_{O_2} - C\bar{v}_{O_2}}$$

Here, a, $\bar{v}$, and $\bar{c}$ refer, respectively, to systemic arterial, mixed systemic venous, and mean end-pulmonary capillary O_2 contents per milliliter of blood. Ca_{O_2} and $C\bar{v}_{O_2}$ can be measured directly. $C\bar{c}_{O_2}$ must be calculated from hemoglobin (Hgb) content, and the O_2-Hgb dissociation curve of the subject's blood, by first determining PA_{O_2} with the alveolar air equation and then assuming that this is equal to $P\bar{c}_{O_2}$. If $\dot{Q}s/\dot{Q}T$ is determined while the subject breathes 100% O_2, $PA_{O_2} = PB - PA_{CO_2}$, or $PA_{O_2} \cong (PB - 47) - Pa_{CO_2}$, and it is not necessary to determine R. While breathing O_2, end-capillary blood from all regions of the lung that receive any ventilation at all will be nearly uniform with respect to O_2 content regardless of $\dot{V}A/\dot{Q}$ ratio differences, because PA_{O_2} will be high generally and differ regionally only by the amount of the differences in PA_{CO_2}. Thus, all ventilated regions will have a high, nearly uniform, easily calculable $C\bar{c}_{O_2}$. The shunt fraction determined during O_2 breathing is called the *anatomical shunt* and represents a volume of blood flow passing from right ventricle to left without coming in contact with ventilated spaces. If shunt fraction is calculated when FI_{O_2} is less than 1.0, the effects of regional $\dot{V}A/\dot{Q}$ differences become important and the resultant shunt is called the *physiological shunt*. This will be considered in more detail in the next section.

Effects of Regional Variation in $\dot{V}A/\dot{Q}$

The $\dot{V}A/\dot{Q}$ ratio can be expressed by the following equation:

$$\frac{\dot{V}A}{\dot{Q}} = 0.863 \, R \, \frac{(Ca_{O_2} - C\bar{v}_{O_2})}{PA_{CO_2}}$$

By cross-multiplying, one sees that this is a rearrangement of an equation defining R, with a constant to correct for temperature, pressure, and water vapor. Looking at it in another way, one sees that, for any given capillary blood arteriovenous O_2 content difference and PA_{CO_2}, the $\dot{V}A/\dot{Q}$ ratio is proportional to R. This relationship between R and the $\dot{V}A/\dot{Q}$ ratio is displayed in Figure 20-4. It is important to recall that the relationship shown in the figure pertains to regions as well as to whole lung: Where $\dot{V}A/\dot{Q}$ ratio is high, R is high, and CO_2 loss exceeds O_2 uptake; the PA_{O_2} is high and the PA_{CO_2} is low. We now turn to the problem of determining how much of the lung is functioning at any specific $\dot{V}A/\dot{Q}$ ratio.

In the preceding sections the mechanical causes for $\dot{V}A/\dot{Q}$ ratio variations have been established, and it has been shown that regional gas tensions will vary in a predictable way. The overall $\dot{V}A/\dot{Q}$ ratio (the ratio of alveolar ventilation to pulmonary blood flow) is normally about 0.8, but this tells nothing about the range of regional differences scattered about that mean. Normally, the range of ratio is between the limits of 0.5 and 5.0. The distribution of $\dot{V}A/\dot{Q}$ ratios throughout the lung—that is, the fractions of the total lung volume that receive each specific ratio between highest and lowest—is approximately that of a normal, bell-shaped gaussian curve if one plots the *logarithm* of $\dot{V}A/\dot{Q}$ ratios against the fraction of lung volume (see Fig. 20-5). This is called a *log normal distribution*. Lung diseases may widen this distribution curve without changing the mean; i.e., an unusually large percentage of the lung may have $\dot{V}A/\dot{Q}$ ratios far

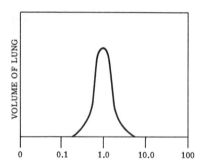

Figure 20-5
Ventilation/perfusion ratio ($\dot{V}_A/\dot{Q}$), plotted on a logarithmic scale, versus relative amount of the lungs which have each $\dot{V}_A/\dot{Q}$ ratio, showing that there is a distribution of ratios around the mean although values near the mean are more common.

above and below the mean ratio. The degree of abnormality could be expressed by the standard deviation of the distribution, a greater standard deviation indicating a wider distribution curve. If the distribution is severely skewed by disease, the change must be expressed in another way. As yet there are no widely applied methods for obtaining distribution curves for individual patients, but suitable techniques are being developed. Characterization of the $\dot{V}_A/\dot{Q}$ ratio distributions will allow much greater insight into the regional mechanical alterations produced by disease, for while one can predict the consequences of regional mechanical alterations, finding them and quantifying the range of variations by measurement of lung mechanics is quite another problem, not yet solved. To satisfy the need to quantify lung abnormality, clinical physiologists resort to simpler models based on the concepts of shunt and dead space to express what is most often an abnormality that is neither an anatomical shunt nor an anatomical dead space, but that can be quantified as if it were. Shunt and dead space will now be reexamined in this regard.

While there is normally about a 2 percent shunt of venous blood into pulmonary veins, it is not necessary that unoxygenated hemoglobin arrive only in this way. Probably, particularly in disease states, blood of less than average alveolar P_{O_2}, yet more than systemic venous P_{O_2}, comes from areas of lung where there is a low $\dot{V}_A/\dot{Q}$ ratio. Note from Figure 20-4 that if $\dot{V}_A/\dot{Q} < 1$, prominent falls in alveolar P_{O_2} will occur. In these cases one can always define a *virtual* lung where all pulmonary end-capillary blood has the average alveolar P_{O_2} calculated by the *alveolar air equation* and where all the difference between that P_{O_2} and the P_{O_2} of systemic arterial blood is ascribed to a *virtual* "shunt" of mixed systemic venous blood. This "physiological shunt" is akin to "physiological dead space," where all the difference between arterial blood and expired gas P_{CO_2} was attributed to the effects of a completely nonexchanging space. It is important to remember that one is calculating a *virtual* shunt or dead space, and, unless real anatomical equivalents are known to exist, it should be presumed that inequalities in $\dot{V}_A/\dot{Q}$ ratio are really determining the partial pressure gradients. That is, decreases in regional perfusion or ventilation are more likely to occur than are complete absences of one or the other, though it is often convenient to express these abnormalities in terms equivalent to complete absences.

A true shunt is essentially a region where $\dot{V}_A/\dot{Q} = 0$: Anatomical dead space is a region where $\dot{V}_A/\dot{Q} =$ infinity. If some part of the lung has a $\dot{V}_A/\dot{Q}$ greater than normal, its effect approaches that of a dead space — it increases the P_{CO_2} difference between arterial blood and expired gas. Similarly, a region of low $\dot{V}_A/\dot{Q}$ approaches in its effect that of an anatomical shunt: It increases P_{O_2} differences between systemic arterial blood and calculated alveolar P_{O_2}. Clinical physiologists commonly calculate physiological dead space and physiological shunt from data obtained while patients breathe room air. This is a simple, but useful, way of expressing the effect of disturbances of the $\dot{V}_A/\dot{Q}$ ratio distribution.

The concept of regional $\dot{V}_A/\dot{Q}$ differences and their effects on systemic arterial blood as being a continuum of effects ranging from the result of a shunt to that of a dead space is a very important one. The normal lung is not a physiologically homogeneous structure. It seems to function like one as far as blood gas contents are concerned because *in the normal ranges of $\dot{V}_A/\dot{Q}$,* blood gas *contents* (not tensions) are rather insensitive to gas tension variations within alveoli. However, pulmonary pathology is especially prone to profoundly distort $\dot{V}_A/\dot{Q}$ ratios in an uneven way throughout the lungs, and the commonest cause of systemic hypoxia of pulmonary origin is alterations in $\dot{V}_A/\dot{Q}$.

It was believed until recently that abnormalities of $\dot{V}_A/\dot{Q}$ ratio distribution, particularly skewness toward lower $\dot{V}_A/\dot{Q}$ ratios, would cause systemic arterial hypoxemia (an increased "physiological shunt") but would not cause systemic arterial hypercapnia. This is not the case. Underventilation of larger parts of the lung results in decreased CO_2 loss from those regions which is not compensated for by the mere displacement of ventilation to other regions. The effect on CO_2 exchange is less apparent than that on O_2 exchange because respiratory control mechanisms do not allow Pa_{CO_2} to rise very much. The impaired CO_2 loss results in CO_2 retention, which greatly stimulates respiration. The increased ventilation to already well-ventilated compartments typically removes this additional CO_2, but at the cost of an overall increase to ventilation and respiratory work. Because little O_2 is added to the capillary blood from regions of a high $\dot{V}_A/\dot{Q}$ ratio as their $P_{A_{O_2}}$ is increased by further increase in ventilation (see Fig. 20-1), the high $P_{A_{O_2}}$ in the overventilated regions does not add enough O_2 to return Pa_{O_2} to normal at the same time that Pa_{CO_2} is returned to normal. If ventilation were not stimulated, Pa_{CO_2} would rise and Pa_{O_2} would fall when regions with abnormally low $\dot{V}_A/\dot{Q}$ ratios develop. Increased Pa_{CO_2} occurs in the presence of $\dot{V}_A/\dot{Q}$ abnormalities if the respiratory control mechanisms are altered, and an increased Pa_{CO_2} is *not* uniquely the consequence of generalized alveolar hypoventilation. Because of its importance it is worth stating again: If overall $\dot{V}_A$ and $\dot{Q}$ *remain normal,* Pa_{CO_2} will *rise* and Pa_{O_2} will *fall* if more of the lung has a lower $\dot{V}_A/\dot{Q}$ ratio than normal.

It might seem that there could be an optimum $\dot{V}_A/\dot{Q}$ ratio for gas transport, and the supposition has been borne out by analysis. Gas exchange, in terms of airflow and blood flow requirements, occurs most efficiently at a $\dot{V}_A/\dot{Q}$ slightly greater than 1.0, which is quite close to the typical ratio of normal lungs taken as a whole. Beyond the limits for $\dot{V}_A/\dot{Q}$ of 0.4 and 4.0, gas exchange becomes much less efficiently accomplished.

As alluded to in Chapter 19, active variations of airway and vascular smooth muscle in response to local P_{O_2} and P_{CO_2} tend to restore toward normal any marked deviations of $\dot{V}_A/\dot{Q}$ away from unity. A fall in alveolar P_{O_2} below about 50 torr, such as would be produced by a decrease in regional ventilation, causes local pulmonary vasoconstriction which directs flow away from that region and so reduces perfusion to bring the $\dot{V}_A/\dot{Q}$ toward normal. In underperfused regions, such as would be caused by vascular changes, the relative hyperventilation leads to a low regional P_{CO_2} which brings about local bronchoconstriction. This increase in local airway resistance (and incidentally a fall in compliance from constriction of alveolar duct muscle) directs airflow elsewhere and returns the $\dot{V}_A/\dot{Q}$ toward normal. Because of the shapes of the O_2 and CO_2 dissociation curves, regulation of $\dot{V}_A/\dot{Q}$ is probably more often important for determining systemic arterial gas *tensions* and the efficiency of exchange than for determining arterial blood gas *contents*. However, when the range of $\dot{V}_A/\dot{Q}$ to be adjusted is expanded to include values found in disease conditions, these regulatory mechanisms significantly decrease systemic hypoxemia and the size of the ventilation wasted in dead space.

EFFECTS OF ABNORMAL CONCENTRATIONS OF RESPIRATORY GASES

Alterations in contents and tensions of the respiratory gases arise in a number of ways. Changes of pulmonary origin, or related to variations in the ambient environment, are always manifested by changes in both peripheral tissues and systemic arterial blood. This is not the case for disturbances such as tissue hypoxia resulting from a fall in peripheral blood flow or from ingestion of agents which interfere with tissue oxidative metabolism, like potassium cyanide or dinitrophenol. In those events, unless respiration also is changed, systemic arterial blood will have normal O_2 and CO_2 contents.

Unlike hypoxia, peripheral hypercapnia is almost always of pulmonary origin, and therefore systemic arterial blood and alveolar P_{CO_2} will be above normal. Hypercapnia may occur secondary to generalized hypoventilation or regional hypoventilation (i.e., broadening or skewness of the $\dot{V}_A/\dot{Q}$ distribution curve to include lower values). Carbon dioxide tensions as high as 90 torr are not infrequently seen in people who continue to survive. The associated P_{O_2} may be in the forties. Were such disturbances to develop very suddenly, collapse or death would usually result, but when they come about gradually, some tolerance and adaptive changes seem to occur.

Hypoxia of systemic arterial blood also can appear from abnormal right-to-left shunts and it is seen at high altitudes in normal subjects. In the case of altitude, the cause is clear. If shunts are the cause, administration of 100 percent O_2 will not cause much increase in arterial blood O_2 content, since all the blood exposed to alveoli is already nearly fully saturated. If subjects who have regional hypoventilation as the cause of hypoxemia breathe 100 percent O_2, the hypoxemia is usually corrected; the poorly ventilated alveoli ventilate enough to wash out their N_2 and replace it with O_2 so that the P_{O_2} becomes much higher than when N_2 was present. Consequently, even under-ventilated units, which still retain their somewhat elevated P_{CO_2}, will now contribute blood with 100 percent O_2 saturation to the systemic circulation.

Rarely, a decreased diffusing capacity causes systemic hypoxemia. This is usually seen only upon exertion, since that puts a greater demand on diffusibility, as noted earlier. In such cases, breathing 100 percent O_2 restores the arterial O_2 content to normal because the gradient for O_2 transport is increased more than tenfold. Unfortunately, most of the conditions that increase the diffusion membrane thickness or decrease the capillary area also alter airways and other blood vessels, so that it is quite difficult to separate the effects of the altered $\dot{V}_A/\dot{Q}$ ratio from the decreased diffusivity. By and large, the $\dot{V}_A/\dot{Q}$ disturbance predominates in these diseases. Arterial hypoxemia may then have two causes. If the $\dot{V}_A/\dot{Q}$ disturbance results in regions of lower than normal $P_{A_{O_2}}$, diffusion impairments may appear which would not have appeared if $P_{A_{O_2}}$ were uniformly high or normal. Systemic hypoxemia is thus the result of the decreased $\dot{V}_A/\dot{Q}$ ratio and also of the failure of capillary blood to come into equilibrium with the low $P_{A_{O_2}}$ of alveoli having the low $\dot{V}_A/\dot{Q}$.

Hypoxia

The effects of hypoxia are insidious. Acute lowering of alveolar P_{O_2} leads progressively to giddiness, lack of judgment, stimulation of respiration and cardiac and vasomotor activity, and finally unconsciousness and death. The first stages are a real threat to aviators and those scuba divers who use rebreathing-type units, for errors in judgment under those conditions can be fatal. Another acute cause of hypoxia is transit to high altitudes on the earth's surface. An acute "altitude sickness" consisting of sleeplessness, nausea, and headache in mild cases, or cerebral and pulmonary edema and retinal hemorrhages in severe cases, is not uncommonly seen when mountaineers quickly ascend to high-altitude base camps. Climbers often progress in stages to allow time for altitude adaptation when making ascents of the highest peaks. Prolonging the ascent time to several days or weeks almost always prevents symptoms, whereas abrupt transport usually causes trouble for new arrivals.

Residents at high altitudes, whether lowlanders who have remained for a time or natives, compensate in part for the decreased O_2 availability by an increase in red cell mass. Natives of Morococha, Peru (a town at an altitude of 15,000 feet), typically have hematocrit readings of 60 percent. Man and most domestic animals seem to develop pulmonary vasoconstriction of varying degree at high altitude. In Morococha Indians, pulmonary arterial pressures were about two to three times higher than were found for the same flows at sea level. The effect is to even out perfusion distribution in a vertical lung, which may increase DL and improve $\dot{V}A/\dot{Q}$ distributions to cause more efficient gas exchange. High-altitude natives and adapted visitors have a decreased ventilatory response to O_2 lack. A high incidence of patent ductus arteriosus has been reported to exist among high-altitude natives, which lends credence to the theory that the ductus normally closes because of the large increase in blood Po_2 that occurs at birth.

Hyperoxia

In recent years it has become clear that one can be exposed to too much O_2. O_2 toxicity is a function of Po_2 and time of exposure. Above 3500 torr, symptoms of central nervous system toxicity develop in man within a few minutes. At these high pressures, muscular twitching progresses rapidly to fully developed convulsions. Even at 1 atmosphere (100 percent O_2 breathing), within 24 hours there are signs of tissue damage in the lung, where exposure to the highest Po_2 is occurring. The changes include edema and hemorrhages of the airway mucosa, and decrease of surfactant, and they have been the cause of complications in patients treated for long periods with 100 percent O_2. It is to be hoped that such treatment is no longer used. Inspired O_2 at 70 percent concentration or less can be tolerated indefinitely, and 70 percent O_2 is adequate to overcome hypoxemia of any conditions that could be significantly aided by 100 percent O_2. The chemical changes of O_2 toxicity are incompletely understood, but the effect seems to be due to the formation of free radical intermediates (hydroxyl and superoxide) in the course of metabolism. In this regard, the toxicity of oxygen resembles a toxic effect of radiation.

A further complication at pressures above 2100 torr is that the high Po_2 enables plasma to carry sufficient O_2 in solution to satisfy resting metabolic needs without recourse to hemoglobin transport. Because hemoglobin does not lose its O_2 in passage through tissue, CO_2 transport is hindered and tissue CO_2 retention occurs. The attendant acidosis may play a role in the toxic changes.

Hypercapnia and Hypocapnia

Increased alveolar (and thus systemic arterial) Pco_2 most commonly occurs as a result of generalized hypoventilation. More insidious exposures are due to defective CO_2 absorbent in closed external breathing systems like scuba units, anesthesia machines, submarines, and space capsules. Acute exposure to very high ambient Pco_2 is a threat to explorers who venture into wet limestone caves—acid groundwater seepage may have liberated CO_2 into a confined space. Industrial exposures also occur among dry-ice workers, miners, and fireman.

The most prominent effect of acute hypercapnia is stimulation of respiration. If hypercapnia results from generalized hypoventilation, this tends to initiate a control mechanism which limits the severity of the underventilation (to be considered further in Chap. 21). In otherwise normal man, respiratory stimulation by CO_2 continues to increase for each increase in Pco_2 as far as it has been tested (about 30 percent CO_2 in inspired gas). Respiratory stimulation is probably caused by not one but several effects, both in the central nervous system and from peripheral chemoreflexes. The predominant mechanism depends on the CO_2 level. Inhalation of CO_2 concentrations above about 15 percent causes generalized convulsions and unconsciousness. Direct effects of CO_2 on vascular smooth muscle tend to produce dilation, but overwhelming effects from central nervous system and reflex stimulation result in hypertension from CO_2 excess, rather than the opposite.

The causes of responses to hypercapnia are not fully understood. Molecular CO_2 itself is apparently responsible for some effects in several ways. In addition, CO_2 hydration leads to an acidosis which elicits some of the response and may become the life-limiting factor.

Hypocapnia occurs as a result of increased ventilation in excess of CO_2 production. This may be the result of simple anxiety or of stimulation by reflexes (including those from O_2 lack). The predominant effects are an increase in blood and tissue alkalinity and a decrease in cerebral blood flow, the latter rising as the typical direct vascular response to P_{CO_2} variation. These changes manifest themselves in dizziness, visual blurring, and neuromuscular hyperexcitability.

Nitrogen Hazards

These are problems peculiar to high ambient pressures, or the abrupt change from high to low ambient pressure. Only two aspects will be described, and those briefly.

First, N_2 is soluble enough so that at high ambient pressure ($>$ 5 atmospheres) it acts as a general anesthetic. The change is not abrupt but progressively follows elevation of pressure. This effect is characteristic of many substances; indeed, it is what allows general inhalation anesthesia at $<$ 1 atmosphere. Narcosis by helium or hydrogen requires greater pressures than is required for N_2, so those gases are used for the O_2 diluent in deep diving, particularly for dives of long duration where there is more time for saturation of body tissues with the inert gas to occur.

The second way in which N_2 becomes important is in decompression. When divers return from depth, or when astronauts ascend to orbit, the N_2 which has been dissolved at the higher partial pressure must be lost. If ascent is done rapidly, the sudden lowering of pressure causes bubble formation in body fluids, which in turn gives rise to several symptoms collectively called "the bends." Small vascular bubbles of N_2 occlude flow to parts of the central nervous system, particularly the spinal cord, and may cause permanent disability unless recompression is quickly accomplished to redissolve them. Bubbles in joint fluid cause local pain, and those in venous blood entering the pulmonary circulation give rise to labored breathing; a condition called "the chokes." These problems are obviated by allowing only gradual return from depth so that N_2 loss can be accomplished without supersaturation and bubble formation. The rate of ascent of astronauts cannot be slowed down economically, so they are depleted of N_2 prior to lift-off by breathing 100 percent O_2 for a short period. This technique could not be used in divers, because O_2 toxicity would supervene before much N_2 was eliminated.

REFERENCES

Fenn, W. O., H. Rahn, and P. Dow (Eds.). *Handbook of Physiology*. Washington: American Physiological Society, 1964. Section 3: Respiration, vol. 1.

Forster, R. E. Exchange of gases between alveolar air and pulmonary capillary blood: Pulmonary diffusing capacity. *Physiol. Rev.* 37:391–452, 1957.

Rahn, H. A concept of mean alveolar air and the ventilation-bloodflow relationship during pulmonary gas exchange. *Am. J. Physiol.* 158:21–30, 1949.

Riley, R. L., and A. Cournand. "Ideal" alveolar air and the analysis of ventilation-perfusion relationships in the lungs. *J. Appl. Physiol.* 1:825–847, 1949.

West, J. B. *Ventilation/Blood Flow and Gas Exchange* (2nd ed.). Oxford, England: Blackwell, 1971.

21. Nervous and Chemical Control of Respiration

Thomas C. Lloyd, Jr.

The lung, like the heart, must undergo cyclic volume changes to perform its function, but it differs from the heart in that its cyclic behavior is not produced by an intrinsic mechanism. Cyclic ventilation requires an interacting control system that includes several organs and several media for information transfer. In addition to genesis of cyclic breathing, ventilation is controlled so as to maintain blood gas composition at or near rather constant levels under a wide range of conditions. While the causes of periodicity will not be ignored, most emphasis in this chapter is placed upon the subsequent control of respiration.

Although the genesis and regulation of breathing have received intense study, they are remarkably poorly understood in their finer details, and they are the subject of considerable controversy. It is the author's intent to minimize controversy by presenting here a "primer" of respiratory control physiology sufficient for use in clinical medicine or for a base from which the advanced student may depart. The neural basis of breathing will be presented first, followed by a discussion of control of breathing by CO_2, O_2, and pH. The chemical effects will be divided further into those occurring in the peripheral chemoreceptors and those occurring by way of central chemosensitive cells.

NEURAL BASIS OF BREATHING

Figure 21-1 portrays a schematic representation of the major neural pathways. The focal point for neural control is the medulla in the region of the obex. Here reside cells that fire during either inspiration or expiration and are labeled singly as inspiratory or expiratory cells of the medullary respiratory center. Although it is convenient to think of these as forming discrete inspiratory and expiratory centers, discrete grouping does not exist. The intermingled inspiratory and expiratory cells communicate with each other and receive inputs from higher regions as well as from peripheral afferent fibers. Given the medullary center and its efferent connections to respiratory muscles, cyclic respiration can occur in the absence of input from both peripheral afferents and higher regions in the central nervous system. Because of the interconnections of inspiratory and expiratory cells, and the properties of these cells, it is possible for them to serve as a self-perpetuating oscillator which sets up a basic respiratory frequency.

The periodic behavior of the medullary center is modified by higher as well as peripheral inputs. Two pontine centers are discernible: the apneustic center, with cells scattered in the caudal two-thirds of the pons, and the pneumotaxic center, whose cells are in the anterior pons. The apneustic center is inspiroexcitatory. Pontine division between the apneustic and pneumotaxic regions causes respiration to be a series of prolonged inspiratory gasps, a pattern called "apneustic breathing." Increased activity of the pneumotaxic center inhibits cells of the apneustic center. The ability of the pneumotaxic center is influenced by activity of the medullary inspiratory cells: Increased inspiratory cell activity increases activity of the pneumotaxic center. This interaction between the medullary and the two pontine centers leads to a more nearly equal inspiratory and expiratory time and a higher respiratory rate than are present if only the apneustic center influences the medullary center. The pneumotaxic center may in part be responsible for alterations of breathing in febrile states.

The medullopontine centers along with peripheral reflexes and control systems mediated by CO_2, O_2, and pH (which act almost entirely at the medullary level) serve to

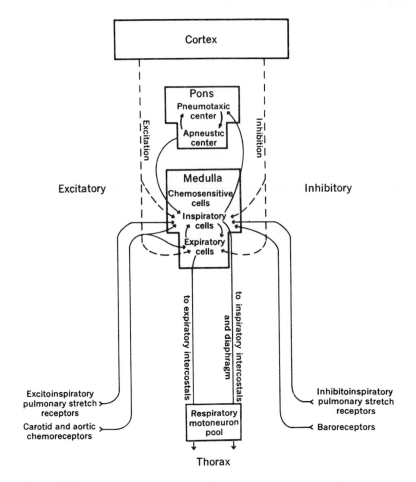

Figure 21-1
Schematic representation of the major central nervous system pathways controlling respiration. (From Christian J. Lambertsen. Neurogenic Factors in Control of Respiration. In Vernon B. Mountcastle [Ed.], *Medical Physiology* [13th ed.]. St. Louis: The C. V. Mosby Co., 1974.)

produce a basic respiratory pattern modified by a number of rather simple feedback loops and forms of information input. These are not adequate, however, for the sophisticated modulation of breathing required for daily living. Cortical and subcortical activity provides both volitional control and less conscious control of respiration during such events as emotional changes, phonation, swallowing, pain, or body movement under a host of conditions. (Consider for a moment why neither hyperventilation nor hypoventilation occurs in the course of a protracted monologue, or why the respiratory rate of a running animal is tightly coupled with the number of steps per minute, and you can begin to appreciate the subtlety and complexity of mammalian respiratory control for which no adequate explanations are forthcoming.)

Important changes in the central control of breathing occur in some diseases. Respiratory arrest following overdoses of hypnotic or sedative drugs, or following medullary ischemia consequent to increased intracranial pressure, is perhaps the most dramatic and life threatening. Medullary and pontine tumors may excite or depress respiration, depending upon their location. Less well understood are several forms of primary hypoventilation, apparently caused by some unknown central change. These manifest as markedly decreased ventilation, especially during sleep, often to such an extent that patients must stay awake and take each breath volitionally. Sleep normally depresses respiration and the respiratory response to chemical stimuli, but in primary hypoventilation this change is far in excess of normal.

Before turning to the chemical control of breathing, which is of great importance in applied physiology, we will consider some ancillary peripheral neural mechanisms, including effects that arise from airway and lung receptors, joint receptors, muscle spindles, and cardiovascular baroreceptors. Not only do these control the pattern of breathing, but several also control the contractile state of airways, larynx, heart, systemic vessels, and skeletal muscle.

Sneezing is a response to activation of nasal receptors whose afferent fibers travel in the trigeminal nerve. Sneezing activates medullary expiratory neurons, but part of the resultant expiratory effects is due to lung reflexes initiated by the antecedent deep inspiration. Reflex bronchoconstriction, laryngeal contraction, and inhibition of skeletal muscle tone are associated with sneezing.

Mechanical irritation of *epipharyngeal receptors* causes bursts of inspiratory activity without facilitation of expiration, leading to a reflex apparently intended to pull obstructing material from the nose into the pharynx where it can be swallowed or coughed up.

Cough can be initiated by stimulation of either laryngeal or tracheal receptors. The cough reflex (see also Chap. 19) includes bronchoconstriction, laryngoconstriction, and variable cardiovascular effects; the first two may enhance expulsion by narrowing and stiffening the airways, thereby increasing expiratory flow velocities and minimizing airway collapse.

Within the epithelium of airways from trachea to respiratory bronchioles are *lung irritant receptors*, whose afferent fibers travel in the vagi. These rapidly adapting receptors respond to chemical irritants, constriction of airway smooth muscle, and mechanical changes of the lung that result in greater pull on, or distortion of, the airway walls. The reflex responses include stimulation of breathing, bronchoconstriction, and laryngeal contraction. These receptors play a major role in an excitoinspiratory response to lung deflation called the Hering-Breuer deflation reflex. At one time this reflex was thought to be a major cause of cyclic respiration, in that deflation induced a subsequent inspiration. It is now known only to be a modifier of the basic cyclic pattern set centrally, but it apparently causes the increased ventilation seen with pneumothorax, pulmonary vascular congestion, or lobar collapse. The important bronchomotor component can be induced not only by chemical and mechanical stimuli but also by allergens, and at least part of allergic bronchospasm arises in this way. Another reflex apparently mediated by lung irritant receptors is "Head's paradoxical reflex," in which a large lung inflation can lead to augmented inspiratory effect. The utility of this response is obscure, but the author has often wondered whether the reflex wasn't the cause of the uncontrollable desire to inhale further near the end of a good yawn, after which effort the event became truly satisfying.

The curious, albeit drowsy, reader may have been provoked at this point to inquire into the meaning and mechanism of yawning. The multiple motor acts are well known, but the causes are many and the mechanisms obscure. Boredom, anxiety, sleepiness, and intracranial disease all cause excessive yawning. Yawning (and its lesser form, sighing) may arise under some conditions from stimulation of inspiration by lung receptors. If nothing else, yawning and sighing perform the important function of intermittently cycling the lung volume over a large range. This has two desirable consequences: First, it increases the concentration of alveolar surfactant in areas where it has decreased consequent to an "aging" of the surface layer that occurs during prolonged duration at a nearly constant lung volume. Second, it pulls open peripheral airways that may have become occluded. Indeed, in prolonged anesthetic sleep, progressive collapse of terminal respiratory units occurs unless patients are forcibly given large tidal volumes every 15 or 20 minutes. In normal sleep, sighing and yawning occur at about this frequency and progressive atelectasis is obviated.

Pulmonary stretch receptors lie among the smooth muscle fibers of the airways. These are low-threshold, slowly adapting receptors which when stimulated by lung inflation inhibit inspiration. They are responsible for the second component of the Hering-Breuer reflex, in which inspiration, particularly if deeper than normal, leads to a prolongation of time before the next breath is taken. The duration of breath holding varies among species. In adult humans it is often difficult to detect, but it is more apparent in the newborn. In rabbits, a forced inspiration may lead to apnea sufficiently prolonged to cause hypoxemia. Stimulation of pulmonary stretch receptors also causes reflex bronchodilation. The responses to stretch receptor stimulation are nearly mirror images of those following stimulation of the irritant receptors.

Receptors within major skeletal joints cause excitation of breathing when stimulated by joint motion. Some investigators consider this reflex a major cause of the hyperpnea of exercise, but not all agree.

The intercostal muscles and diaphragm contain *muscle spindles* that are important in control of breathing. The receptors are attached in series with muscle innervated by gamma efferent fibers. This spindle, in turn, is in parallel with the muscles that are innervated by alpha fibers and that cause most of the movement. If, during muscle stimulation, the alpha-innervated fibers contract and shorten, the stress on the receptor imparted by the contracting gamma-innervated fibers is relieved, and receptor discharge is minimized. If shortening is impaired, as by a high airway resistance, the discharge increases, reflexly facilitating excitation of the alpha efferents. The spindle system serves to maintain tidal volume in the face of increasing mechanical loading because it increases alpha activity until the desired degree of motion is obtained.

Although receptors in the *lung vessels and cardiac chambers* probably influence breathing when stimulated (pulmonary emboli and congestive failure being important examples), the reflex effects are inconsistently reported. The *systemic arterial baroreceptors*, however, are generally found to cause respiratory depression when stimulated by increased pressure. The significance of this in the overall control of respiration is not known.

CHEMICAL CONTROL OF BREATHING

Hypoxemia, hypercapnia, and acidemia stimulate respiration. Of the several stimuli, CO_2 is most important for regulating respiration in normal man. Arterial P_{CO_2} is normally held within 3 torr of the average (40 torr). To demonstrate the control phenomena suitably, each stimulus must be independently varied while respiratory minute volume is measured. Although there is a wide variation among subjects, curves similar to those shown in Figure 21-2 are obtained. Note from the figure that (1) there is little response to hypoxia until Pa_{O_2} falls below 60 torr, (2) there is little response to Pa_{CO_2} at less than 35 torr, (3) the O_2 response depends upon the level of Pa_{CO_2} maintained during the experiment, and (4) the CO_2 response depends upon the level of Pa_{CO_2}. The last two observations are equivalent to saying that the sensitivity to O_2 is set by the Pa_{CO_2}, and vice versa. Hence when both O_2 and CO_2 change, the net result will be more than a simple summation.

If the carotid and aortic chemoreceptors are denervated, the CO_2 response becomes somewhat less and there is no stimulation by hypoxemia. Indeed, severe hypoxia depresses respiration in the absence of the peripheral chemoreceptors. Such experiments show that an important part of the chemical control is mediated elsewhere. A required second site has been found in the medulla just below the floor of the fourth ventricle, but CO_2-sensitive cells are scattered diffusely through the brain stem. We turn now to the roles of peripheral and central chemosensory mechanisms in the control of breathing.

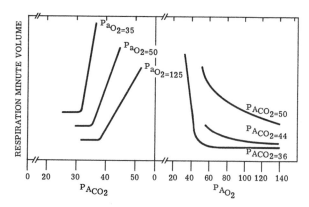

Figure 21-2

Relationships between alveolar gas O_2 and CO_2 tensions and respiratory minute volume. In the left panel, Pa_{CO_2} is varied while Pa_{O_2} is held constant at one of three levels, whereas on the right, Pa_{O_2} is varied while Pa_{CO_2} is held constant at one of three levels. Note that the gas whose tension is held constant not only sets the general level of respiration but also influences the response to changes of composition of the variable gas.

Studies of the peripheral chemoreceptors have shown that changes of Pa_{CO_2} probably induce receptor activation entirely by way of secondary changes in pH. Since molecular CO_2 crosses membranes more rapidly than does H^+, the response to CO_2, although dependent upon the pH change produced, is induced much more rapidly than is the response to changing blood $[H^+]$. The response to hypoxia is more complicated, is less agreed upon, and may include a local neurotransmitter in the chemosensitive tissue. Curiously, the receptor afferent discharge activity does not show the dogleg relationship to hypoxic or hypercapnic stimuli so prominent in the overall response curves of Figure 21-2. Graded responses occur over a much wider band of stimulus values. Why this effect is not reflected in the total response is not understood.

The role of the peripheral chemoreceptors in regulation of breathing remains controversial, particularly with regard to CO_2 and pH effects. With regard to O_2, all would agree that in the adult the chemoreceptors are the sole excitatory input during hypoxemia. Some investigators consider the influence of CO_2, at least in the near normal range, to be only that of setting the sensitivity of the response to hypoxia, and assign the primary site for direct control by CO_2 to the central chemosensitive cells. Others feel that the peripheral chemoreceptors respond importantly to CO_2 changes and that because they are most immediately affected by changes in arterial blood composition they provide the fine control of breathing. An eclectic summary most satisfying to the author is that (1) the peripheral chemoreceptors do provide the most important O_2-related regulatory effect, (2) the sensitivity to hypoxia is significantly determined by arterial blood P_{CO_2}, (3) these receptors may provide for "fine tuning" of the respiratory controller on a moment-to-moment basis in response to small changes in Pa_{CO_2}, and (4) in the overall response to sustained changes in arterial P_{CO_2} or pH the contribution of the peripheral receptors probably accounts for less than 15 percent of the total change in ventilation.

Central chemosensory mechanisms seem to account for most of the responses to changes of blood P_{CO_2} and pH. Studies of the control system have resulted in conflicting proposals. An important discrepancy is among groups who believe that central chemosensitive cells are all protected from an immediate direct influence of blood pH by a blood-brain diffusion barrier and groups who believe there are two central cell populations, one protected by a blood-brain barrier and the other not. Both theories recognize that tissue pH change is the effective stimulus when Pco_2 varies. In the single-center theory, chemosensitive cells are influenced by local pH changes induced by blood Pco_2 changes and by changes of cerebrospinal fluid (CSF) pH. This author, however, subscribes to the less popular dual-center theory because it seems able to incorporate more observations with fewer assumptions and it is easier to understand and use. Both viewpoints are described and contrasted in the suggested readings at the end of the chapter.

The dual-center theory (Fig. 21-3) proposes that two nearly equipotent "centers" exist. Again, "center" has no necessary discrete anatomical meaning and refers only to a population of cells which probably are scattered throughout the brain stem. Diffusion of small charged particles (e.g., H^+) into one center is limited, as is diffusion into most of the brain and CSF, by the blood-brain barrier. This is not an absolute barrier, but a relative one, so that equilibration eventually occurs (with a half-time of 12 to 24 hours). As a result, changes in blood pH affect respiration at the protected center only after a new blood pH has been maintained for several hours. Molecular CO_2, however, crosses the barrier readily and will then influence local pH by its hydration to form carbonic acid. If some of these cells are near the CSF, they can be influenced by CSF pH changes as well, and they would also be affected by brain metabolic acidosis and alkalosis. The second center is not protected from blood pH change by presence of a blood-brain barrier, so that changes in blood pH as well as in Pco_2 will immediately influence respiration. Because a sudden increase in blood pH at a constant Pco_2 is known to cause prompt respiratory stimulation, the dual-center theory can attribute a substantial part of the response to central stimulation, whereas the single-center theory cannot. The dual-center theory also allows CSF pH sometimes to vary as a consequence of respiratory stimulation, rather than always to be the cause of it. The dual-center theory reconciles a number of otherwise conflicting observations, but reconciliation does not, of itself, prove that dual centers exist.

We turn now to some experimental and disease-induced variations of blood gases to see how the chemosensitive control system will act.

Simple hypercapnia at a fixed inspired Po_2 will increase Pa_{CO_2} and lower pH. Thus respiration will be stimulated at three sites: the peripheral chemoreceptors and both

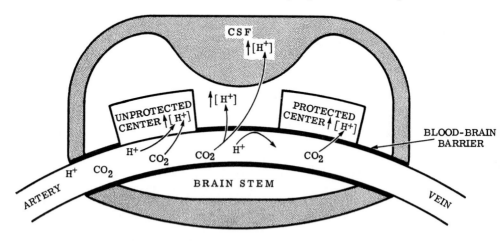

Figure 21-3
Schematic representation of chemosensitive areas in the brain stem. In this dual-center schema, an unprotected center is influenced by both H^+ and CO_2 of blood, whereas other protected chemosensitive cells, along with the remaining brain substance and the CSF, are penetrated immediately by CO_2 but only slowly by H^+ or HCO_3^- because of the blood-brain barrier.

central regions. In time, renal mechanisms will restore pH to normal by increasing plasma and CSF $[HCO_3^-]$. Although this action will reduce ventilation somewhat, it will not fully return ventilation to normal until buffering at the central diffusion-protected center has increased so much that the high Pco_2 no longer causes the tissue pH to be above normal. This buffering amounts to a resetting of the central systems to a new Pa_{CO_2} and ΔCO_2 sensitivity. With the increase in buffering, a given Pa_{CO_2} change will produce a smaller ΔpH at the center; hence less stimulation will occur. Such an adaptation to hypercapnia is seen in patients with chronic obstructive lung disease who retain CO_2 because of limited ventilatory ability.

Addition of fixed acids to blood at first will stimulate respiration centrally at the unprotected chemosensitive cells and, to a lesser degree, at the peripheral chemo-receptors. The result will be a lowering of blood Pco_2 to below normal. Because this lowers stimulation at the protected center, the initial overall response to acidosis is less than if Pa_{CO_2} were held constant. Activity of the protected center returns as CSF pH falls, but this is not an immediate effect because of the barrier. As long as acidosis is present, hyperventilation and hypocapnia will continue. The combination of acidosis, hypocapnia, and hyperventilation is commonly seen in uncontrolled diabetes mellitus or other forms of metabolic acidosis.

In adapting to high altitudes, respiration is stimulated principally by the effect of hypoxia acting at the peripheral chemoreceptors. This hyperventilation results in hypocapnia and alkalosis, which limit the hyperventilation because hypoxic hyper-ventilation withdraws two stimuli (pH, Pco_2) from both the peripheral and the central chemosensitive cells. If Pa_{CO_2} were artificially maintained at the normal level, the hyperventilation of hypoxia would be more intense. Because of the hypocapnia, small changes in Pa_{CO_2} during the first phase of adaptation no longer modulate respiration (see Fig. 21-2). But in the normal adjustment to altitude, respiratory loss of CO_2 and renal compensation restore blood pH to normal within a few days, while reducing blood and tissue buffering abilities. Then, even though the subject remains hypocapnic, he may once again have a normal response to small increases of Pa_{CO_2} because the reduced buffering allows a change in Pco_2 to cause larger changes in tissue pH than were present before compensation. Hereafter CO_2 again becomes important in control of ventilation, but the subject has set his average Pa_{CO_2} at a lower level than it had at sea level. (Compare this process with the adaptation to chronic hypercapnia.)

Patients with chronic obstructive lung disease typically are hypoxic, hypercapnic, and acidotic. The interplay of central factors is considerable, and the attending physician must be aware of them lest his management lead to disaster. Hypoxemia stimulates respiration at the peripheral receptors, but in some patients it may be severe enough to cause central depression, reducing the net effect. The peripheral sensitivity to hypoxia is enhanced by hypercapnia. Acidosis and hypercapnia stimulate both peripherally and centrally, but because of the chronic nature of the illness, increases in buffering capacity mandate a higher Pa_{CO_2} to cause the same effective stimulation. Furthermore, part of the hypercapnia may occur because in the presence of chronic airway obstruction more muscle must be recruited to bring about ventilation, and to bring about a greater effort a greater stimulus is required. In these patients a sudden increase in alveolar Po_2 (such as by giving O_2 to breathe) will withdraw a significant respiratory stimulus and decrease the sensitivity to CO_2, leading to more hypercapnia and acidosis. Similarly, treatment of the acidosis by infusion of alkaline solutions may depress respiration and enhance hypoxemia. Careful use of O_2 and alkalinizing infusions along with mechanically assisted ventilation is often necessary to restore blood gases and buffering capacities to normal.

During muscular exercise ventilation is susceptible to control by all the chemical and neurological mechanisms that have been discussed so far. The physiology of exercise is a field in itself, and an understanding of the adaptations to exercise prepares the student for consideration of adaptive and integrative physiology so much a part of clinical medicine and human engineering. Because of its importance, exercise physiology (including the control of breathing) is given a chapter of its own.

REFERENCES

Mountcastle, V. B. (Ed.).   *Medical Physiology* (13th ed.). St. Louis: Mosby, 1974. Vol. 2.

Fenn, W. O., H. Rahn, and P. Dow (Eds.).   *Handbook of Physiology.* Washington: American Physiological Society, 1964. Section 3: Respiration, vol. 1.

Guyton, A. C., and J. G. Widdicombe (Eds.).   *MTP International Review of Science. Physiology, Series One. Respiratory Physiology.* London: Butterworths, 1974. Vol. 2.

22. Renal Function

Ewald E. Selkurt

The kidney plays a dominant role in maintaining the constancy of the internal environment. It does so in the first instance by the regulation of the water content of the body with the aid of the antidiuretic hormone (ADH) released from the hypothalamo-neurohypophysial system. This regulation is closely integrated with the maintenance of proper salt balance in the body, i.e., the proper proportion of sodium, potassium, calcium, chloride, phosphate, bicarbonate, and sulfate. In the control of electrolyte balance, other endocrine regulations are important (Chap. 34).

A major factor in proper electrolyte balance is the renal adjustment of the acid-base balance. Since the problem is primarily one of ridding the body of the acids created by metabolism, the kidney favors excretion of acid radicals through an ion exchange mechanism whereby the hydrogen ion is secreted in exchange for sodium, making the urine acidic. Second, the kidney forms ammonia, which is secreted by the tubular cells and replaces valuable base. Thus, the renal mechanism for acid-base regulation is essentially a base conservation mechanism. It will be described in greater detail in Chapter 24.

The kidney has the further responsibility of the excretion of waste, such as urea, uric acid, creatinine, and creatine. It must perform this function and others while conserving valuable foodstuffs, such as glucose and amino acids. These must be reabsorbed selectively, while the undesirable waste products are eliminated. Certain additional functions include detoxification, e.g., the conjugation of glycine and benzoic acid to form the more innocuous hippuric acid, which is rapidly secreted by the tubular cells into the urine. Other metabolic functions, such as amino acid oxidation and deamination, also occur in the tubular cells.

In summary, the problem of excretion by the kidney involves filtration at the glomeruli of all but cellular constituents and the plasma proteins; further, it is the function of the tubules to reabsorb selectively the necessary valuable substances and to reject waste products and undesirable excesses of anything taken into the body. In addition, certain substances may be added to the urine by tubular secretion.

FUNCTIONAL ANATOMY OF THE KIDNEY

The functional unit of the kidney is the nephron, and there are approximately $1\frac{1}{4}$ million nephrons per kidney in the human. The nephron is diagrammed in Figure 22-1 with the associated blood supply. It consists of the malpighian corpuscle and the attached tubular system. The malpighian corpuscle includes Bowman's capsule, which consists of an inner *visceral* layer and an outer *parietal* layer. The filtering area of the capsule has been estimated to be about 0.8 square millimeter, which would represent 2 square meters of filtering surface for both kidneys. The visceral layer of Bowman's capsule is in intimate contact with the capillary loops important in filtration; the unit is called the *glomerulus*. Bowman's capsule continues into the *proximal convoluted tubule*. The proximal segment has the widest diameter of any portion of the nephron. It is made of truncated pyramidal cells characterized by a *brush border* on the inner or luminal aspect. At their basal margins they have striations perpendicular to the basement membrane. These striations are related to the distribution of the mitochondria in the cell. The proximal convoluted tubule turns toward the medulla to become the *loop of Henle*. The descending limb is at first of a thickness comparable to that of the proximal convoluted tubule but is then replaced by highly attenuated cells of the thin segment,

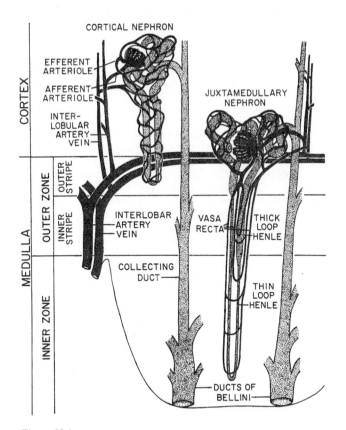

Figure 22-1
Cortical and juxtamedullary nephrons of the human kidney, with representative blood supply. (From *Physiology of the Kidney and Body Fluids*, 3d ed., by R. F. Pitts. P. 8. Copyright © 1974 by Year Book Medical Publishers, Inc., Chicago. Used by permission.)

which are characterized by flattened epithelium. This makes a hairpin turn in the medulla, then turns up toward the cortex into the ascending thick limb of the loop. The thin segment is found only in mammals and in a small percentage of the nephrons of birds. The average length varies considerably in different mammals and, indeed, varies considerably within the human kidney, depending on the location of the nephron (Fig. 22-1).

The ascending limb of the loop of Henle has cells which are at first cuboidal but become more columnar as they approach the cortex. This becomes the *distal convoluted tubule,* the cells of which lack brush borders but do exhibit basal striations. This segment joins the treelike system of *collecting ducts.*

RENAL CIRCULATION

Anatomical Aspects

The renal artery divides into *interlobar arteries,* which subdivide into primary, secondary, and tertiary *arcuate arteries,* from which spring *interlobular arteries.* The *afferent arterioles* arise from these; in the dog, the afferent arterioles usually supply one glomerulus but rarely may branch to supply two to four glomeruli, with a total of 200,000 per kidney. This number compares with estimates ranging from 600,000 to 1,700,000 in each human kidney.

The glomerular capillaries converge in the efferent arterioles, then in a second capillary bed around the convoluted tubules, which drain into the *interlobular veins*. These drain into the *arcuate veins* and *interlobar veins*.

Although Figure 22-1 suggests a one-for-one relationship of the efferent arterioles to the peritubular capillary network, this is an oversimplification. While such relationship may prevail in the superficial cortex, studies by Beeuwkes and Bonventre (1975) in the dog kidney show that it does not apply to mid-cortex and deep cortex. In the bulk of the cortex, proximal tubules were found to be dissociated from the efferent vessels of the same glomerulus. Thus, the blood supply to a given nephron may come from a common capillary bed derived from free anastomoses of the efferent arteriolar branches of other glomeruli. This finding has implications for analysis of single nephron function and its relationship to its peritubular environment.

Arterioarterial and *venovenous* anastomoses have been found in the kidneys of man and dog. Although arteriovenous anastomoses probably exist, their occurrence is infrequent.

Juxtaglomerular Apparatus (JGA)
In the normal human kidney the afferent arteriole, as it approaches the glomerulus, shows a significant increase in the number of cells in the media, thickening the arteriolar wall to form an asymmetrical cap (*polkissen*) on one side of the glomerulus. The polkissen is composed of small, spindle-shaped, afibrillar cells, the juxtaglomerular cells or JG cells. They may have a particular role in regulation of renal blood flow. In rats they show changes in granularity with variations in systemic blood pressure. Some investigators feel that renin, which plays a part in production of angiotensin, the pressor principle in renal hypertension, is formed in these cells.

A portion of the distal convoluted tubule which lies opposite the juxtaglomerular cells shows a distinct modification characterized by condensation of nuclei on one side of the tubule in an epithelial plaque, the *macula densa* and together with the juxtaglomerular cells is called the *juxtaglomerular complex or apparatus*. It is of interest to note that the macula densa is contiguous with both afferent and efferent arterioles.

It has been suggested that the JG cells and macula densa form a regulatory system responding to changes in the composition of the distal tubular fluid, particularly the sodium concentration. This in turn regulates renin production (Nash et al., 1968).

In 1968 Thurau proposed a hypothesis that a local control system regulating the caliber of the afferent arteriole involved angiotensin II production within the JG cells in response to sodium chloride concentration in the distal tubular fluid. The notion has been recently revived (Thurau and Mason, 1974) with presentation of some evidence that renin, converting enzyme, and substrate (alpha-2 globulin) are found in the JGA.

Blood Supply to the Medullary Zone
The glomeruli and associated tubules which lie deep in the cortex adjacent to the medulla (*juxtamedullary glomeruli*) have distinctive modification of their circulation compared with the cortical glomeruli and nephrons. They are typified by long loops of Henle dipping into the medullary zone and papillary portions of the kidney, and have associated modified vascular structures. The efferent arterioles from these glomeruli not only break up into the typical capillary supply to the convoluted portions but also form long hairpin loops of thin-walled blood vessels, the *vasa recta*, which accompany the loops of Henle. Although thin-walled, they have a diameter several times greater than that of the typical peritubular capillary. In man the juxtamedullary glomeruli comprise about 20 percent of the total number of glomeruli in each kidney.

Measurement of Renal Blood Flow

Both direct and indirect methods have been used to measure renal blood flow. Direct methods employ flowmeters of various types, which record either arterial inflow or venous outflow. The indirect methods are an application of the *renal clearance principle*. This is based on the clearance ratio $C = U\dot{V}/P$, in which U is the urinary concentration in milligrams per milliliter, $\dot{V}$ is the minute urine volume, and P is the plasma concentration (usually systemic vein) in milligrams per milliliter. The derivation and meaning of the plasma clearance ratio is explained in greater detail on pages 504–506.

Fick Principle Method

The requisite for the clearance of a substance to measure plasma flow is that it be entirely removed (or nearly so) from the plasma in one transit through the kidney. Clearance is verified by examination of the concentration in arterial plasma (A_c) and in the renal vein (V_c), and by application of the *extraction ratio*

$$E = \frac{A_c - V_c}{A_c}$$

Hence, if V_c equals 0, E will equal 1.0. E is less than unity to the extent that the material is not removed by urinary excretion. It is clear that if the renal clearance $U\dot{V}/P$ is divided by the extraction ratio

$$\frac{U\dot{V}/P}{(A_c - V_c)/A_c} \quad \text{or} \quad \frac{C}{E}$$

the resultant quotient will yield the total plasma flow. The latter is an expression of the Fick principle. For most substances, systemic vein concentration (P) and arterial concentration (A_c) are identical, so that these cancel out to yield the familiar Fick equation:

$$RPF = \frac{U\dot{V}}{A_c - V_c}$$

where RPF is renal plasma flow. In order to obtain total renal blood flow (RBF), the hematocrit measurement of the blood is introduced:

$$RBF = \frac{C}{E} \times \frac{1}{1 - \text{hemat.}}$$

Several substances are so efficiently removed by combined processes of glomerular filtration and active tubular (proximal) transport (secretion) at low plasma concentrations that the renal vein concentration is very low (i.e., extraction nearly complete, and E close to unity). These include Diodrast (D) — E_D is 0.74 — and p-aminohippurate (PAH) — E_{PAH} is 0.80 to 0.85 in the dog and 0.90 in man. Then C_D or C_{PAH} is nearly equivalent to plasma flow. The fact that extraction is not complete is interpreted as indicating that a small fraction of blood does not perfuse excretory tissue: capsule and inert supportive tissue, medullary tissue (loops of Henle, collecting ducts), calycine mucosa, and pelvis. On this basis, Smith has referred to the measured clearance as the "effective" plasma or blood flow.

The Fick principle can be employed with any substance cleared by the kidney which shows a measurable arteriovenous (A-V) difference. Obviously, the smaller the A-V difference, the more prone to error the calculation will be. Thus, phenol red, urea, mannitol, and inulin have been employed, but they have considerably smaller A-V differences than do Diodrast and PAH.

Other Indirect Methods

An application, involving uptake, then washout of radioactive gases krypton (^{85}Kr) and xenon (^{133}Xe) injected into the renal artery, has been used in dog and man. The washout curve can be broken down into different components, signifying differential blood flow rates through different compartments. Following this procedure, computations of cortical blood flow and medullary blood flow have been made (Thorburn et al., 1963; Ladefoged, 1966). Representative values for the human kidney are: cortex, 3.6 to 6.3 ml per minute per gram of cortical tissue; outer medulla, 1.25 ml per minute per gram of tissue.

The distribution of 18 to 36 μm radioactive microspheres (^{85}Sr, ^{145}Ce, ^{169}Y) in the renal substance after intra-arterial injection yields information on regional cortical flow; four zones have been discerned from outer to inner cortex, giving information on juxta-medullary blood flow. The microspheres do not pass the glomeruli and hence are excluded from the medulla (McNay and Abe, 1970). In dogs at normal blood pressure, the zonal flows, outer cortex to inner cortex, averaged 4.18, 6.80, 3.07, and 1.68 ml per gram per minute, respectively. (See below for additional details of regional renal circulation.)

Renal Blood Flow Values

Total renal blood flow averages about 350 ml per minute in the dog (average weight, 15 kg) and about 1200 ml per minute in man, or about one-fifth of the cardiac output. Renal blood flow averages about 3.5 to 4.0 ml per minute per gram of kidney weight.

Oxygen Utilization

The renal venous blood contains considerably more O_2 than does that draining other tissues. The resulting small A-V O_2 difference (1.7 volumes per 100 ml in man and about 3.0 volumes per 100 ml in the dog and cat) remains constant over a wide range of renal blood flows, although it may increase at very low rates of flow. Thus, O_2 consumption (normally 0.06 to 0.10 ml per gram per minute in the dog) is related to flow, so that when flow is reduced, as in shock, the organ ordinarily does not increase its extraction but apparently suffers curtailment of oxidative metabolism.

Warburg kidney tissue slice studies have been done in the guinea pig, dog, and cat. The average Qo_2 values are as follows: in cortex, 21.3 mm^3 per milligram dry tissue per hour; outer medulla, 15.1; inner medulla, 6.2. These findings lead to the conclusion that the structures of the inner medulla (loops of Henle and collecting ducts) do not have important energy-requiring functions, or, alternatively, operate in part anaerobically. This zone is important in the countercurrent urinary concentration process. The blood flow to the medulla has been calculated to be small, perhaps less than 10 percent of the total, 0.7 to 1.0 ml per minute per gram in the outer medulla and only about 0.2 ml per minute per gram in the inner medulla compared with 4.0 in the cortex (Fig. 22-2). The cortex, on the other hand, contains the segments of the nephron involved in active sodium transfer, and it is the sodium-reabsorptive mechanism that largely accounts for the renal O_2 utilization.

The important relationship of renal O_2 utilization of sodium transport is depicted in Figure 22-3. To the left (A), utilization in milliliters of O_2 per minute per 100 gm of kidney weight is related to an experimentally varied range of renal blood flow and related glomerular filtration rate (GFR). Note the inflection and linear increase at about 225 ml per minute per 100 gm of kidney weight, where GFR presumably begins, with attendant filtration and reabsorption of sodium salts by the nephrons. The increase is linear with the progressive increase in RBF and related GFR. To the left of 225 ml per minute, baseline consumption of O_2 by renal tissue metabolism is implied. In B, O_2 utilization is expressed in milliequivalents per minute per 100 gm of kidney weight on the ordinate, as related to absolute sodium reabsorption in milliequivalents per minute per 100 gm of kidney weight.

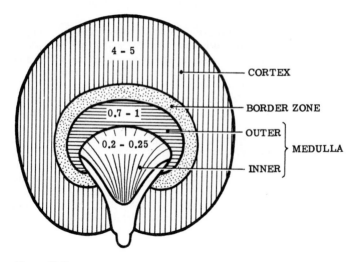

Figure 22-2
Regional blood flow in the kidney. (From K. Thurau. In C. H. Best and N. B. Taylor [Eds.], *The Physiological Basis of Medical Practice* [8th ed.]. Baltimore: Williams & Wilkins, 1966. P. 870. Copyright 1966, The Williams & Wilkins Company, Baltimore, Md. 21202, U.S.A.)

Autonomy of the Renal Circulation

Rein in 1931, in his study of regional blood flow in dogs during changes in systemic arterial pressure, was struck by the relative constancy of the renal blood flow as compared with that in other tissues. This observation has been confirmed many times since. An example of the constancy of renal blood flow with variation in arterial

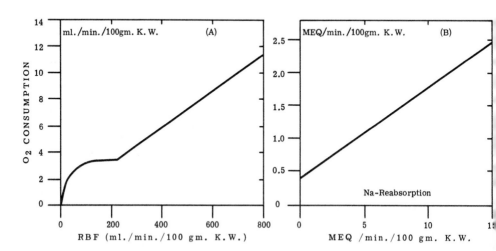

Figure 22-3
A. Relationship of O_2 consumption in milliliters per minute per 100 gm kidney weight (K.W.) to renal blood flow. B. Relationship of O_2 consumption in milliequivalents per minute per 100 gm K.W. to sodium reabsorption. (A from K. Kramer and P. Deetjen. *Pflügers Arch. Ges. Physiol.* 271:787, 1960; B from P. Deetjen and K. Kramer, *Pflügers Arch. Ges. Physiol.* 273:640, 1961.)

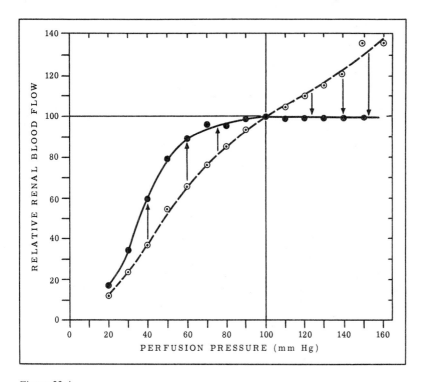

Figure 22-4
Autoregulation of renal blood flow. Transient (open circles) and steady-state (closed circles) effects on blood flow of elevating and reducing perfusion pressure from 100 mm Hg. (From C. F. Rothe, F. D. Nash, and D. E. Thompson. *Am. J. Physiol.* 220:1624, 1971.)

perfusion pressure appears in Figure 22-4. The relative constancy of flow as revealed by this figure is an example of *autoregulation* of the circulation. The mechanism is intrinsic, for it is manifested by the denervated kidney. The constancy of flow is maintained through a range of about 80 to 200 mm Hg of arterial perfusion pressure.

Several hypotheses have been advanced to explain renal autonomy. Of those considered in Chapter 17, the myogenic has been most popular, although the tissue pressure and metabolic theories have received some support. (For a detailed account, see Selkurt, 1963.) Evidence for the myogenic type of response appears to be offered in Figure 22-4, but participation of metabolic factors cannot be ruled out in this type of experiment.

When arterial pressure is suddenly increased, renal blood flow increases, but in 30 to 60 seconds internal adjustments occur, probably in the afferent arterioles, which bring the flow down to the original level despite maintenance of elevated pressure (Fig. 22-4). When pressure is suddenly dropped, an upward adjustment can be observed.

The dynamic reactivity implied in these fairly rapid adjustments corresponds to the type of reactivity anticipated from smooth muscle of the vasculature; hence the term *myogenic.* Isolated arterial segments have been shown to manifest similar responses. Moreover, that a vital phenomenon is involved is supported by the action of a number of agents known to impair smooth muscle activity, which eliminate autoregulation: papaverine, potassium cyanide, procaine, ethyl alcohol, and the anesthetic chloral hydrate. Cooling removes autoregulation, and it is depressed by hemorrhage and anoxia.

The metabolic theory of autoregulation (Chap. 17), as it may apply to the kidney, must take into account two important humoral components: the renin-angiotensin system (Chap. 34) and the prostaglandins. Renin, synthesized in the juxtaglomerular apparatus of the cortical afferent arterioles, converts angiotensinogen of the plasma to angiotensin I; this is converted by plasma-converting enzyme to the potent vasopressor, angiotensin II. Although an intrarenal regulatory role has been proposed for the renin-angiotensin system (Thurau et al., 1968), some difficulties exist in applying this hypothesis to current notions of renal circulatory autoregulation.

The prostaglandins, presumably synthesized in specialized interstitial cells of the renal medulla from fatty acid precursors, include the potent vasodilators PGE and PGA. Reduced renal blood flow, resulting from neurogenic or humoral vasoconstriction, as with hemorrhage (Chap. 18C), or renal ischemia enhances synthesis and release of prostaglandins. The strategic location of the cells of origin to the medullary blood supply (vasa recta) supplies a possible mechanism for intrarenal distribution and circulatory regulation. The importance of these substances for regulating juxtamedullary and medullary blood flow has been recently demonstrated by Kirschenbaum et al. (1974). Presumably, the control points are the arterioles and vasa recta sphincters of the juxtamedullary circulation. In keeping with the autoregulatory role, it is conceivable that initially enhanced RBF, as with sudden increases in arterial perfusion pressure, would "wash out" these dilator substances, eventuating in relative vasoconstriction. It should be added that enhanced production and release of prostaglandins by the kidney does not usually have a systemic influence, since the lungs normally largely metabolize the prostaglandins that arrive in the mixed venous blood (especially PGE and PGF—but PGA, less so).

In summary, it seems likely that any single theory thus far advanced would oversimplify the complex adjustments that the kidney undergoes in its efforts to maintain the constancy of its blood flow. The possibility has been considered that several of the above mechanisms operate together in a composite mechanism.

Factors That Modify the Renal Circulation
Despite the inherent autonomy of its circulation, a number of factors, neurogenic and humoral, are able to decrease or increase kidney blood flow.

Innervation and Neurogenic Regulation
It is generally agreed that the major nerve supply to the kidney has its origin largely from the levels thoracic (T) XII through lumbar (L) II of the sympathetic nervous system in man, and in the dog from T IV to L II, but most abundantly from T X to T XII. In relation to its size, the kidney receives a more profuse and widespread supply than almost any other viscus. The thoracolumbar sympathetic supply to the kidney is a rich source of vasoconstrictor fibers for the kidneys, but presumptive evidence of dilator fibers is at hand (see below). The vagus nerve sends fibers to the kidney, but no evidence exists for vasodilator fibers in this circuit. Hence, the extrinsic vasomotor status of the kidney is maintained largely by variations in vasoconstrictor tone.

Evidence favors the notion that extrinsic regulation is low or absent in the basal state, to be invoked only in emergency states of heightened sympathetic nervous system activity. The principal evidence is that when one kidney is denervated, or denervated and transplanted, function is equal in the experimental and control kidneys in dog and man. This includes concordance of glomerular filtration rate (inulin or creatinine clearance), plasma flow (Diodrast or PAH clearance), diuretic activity, and electrolyte excretion. These findings are in accord with the concept of autonomy of the renal circulation.

Histochemical techniques have recently revealed the presence of both adrenergic and cholinergic fibers in the sympathetic nerve supply to the dog kidney. The cholinergic nonmedullated nerve fibers originate in ganglion cells in the hilus of the kidney and therefore persist after renal denervation (McKenna and Angelakos, 1968). Adrenergic innervation dominates the cortex; some norepinephrine is found in the outer medulla, but little, if any, in the inner medulla. Adrenergic nerve fibers travel along interlobar, arcuate, and interlobular arteries and along afferent arterioles, presumably innervating the smooth muscle cells to provide a vasoconstrictor action. None was seen in association with the glomeruli or efferent arterioles.

The cholinergic fibers, on the other hand, appear to innervate predominantly efferent arterioles, most prominently those of the juxtamedullary nephrons, and the sphincters of the vasa recta (Moffat, 1967). A role in the regulation of the medullary circulation is strongly indicated for these vasodilator fibers. The cholinergic fibers may function to maintain blood flow during reduction of arterial pressure and may participate in alleged renal vasodilator reflexes. In concert with adrenergic constrictor fibers to the afferent arterioles, dilator fibers to efferent arterioles may provide a more versatile and effective control of glomerular filtration pressure, especially in juxtamedullary glomeruli.

Humoral and Pharmacological Factors
The catecholamines L-epinephrine (Arterenol) and Norepinephrine (Levarterenol) are active vasoconstrictors of the renal vasculature. Norepinephrine infused intra-arterially causes rather selective vasoconstriction in the cortex, implying that all the adrenergic terminals are here (Hollenberg et al., 1968). Angiotensin is another important vasoconstrictor. Several well-known and physiologically important humoral agents which cause renal vasodilation are acetylcholine (ACh), bradykinin, serotonin (in low dosage), and prostaglandins (found in high concentration in the renal medulla). When ACh is infused intra-arterially, RBF increases in both cortex and medulla, but mostly in the medulla, implying that the greater proportions of the ACh-sensitive receptors are here.

For the pharmacology of a number of drugs, including anesthetic agents, that exert vasomotor influences on the renal vasculature, see Selkurt (1963).

Response of Renal Blood Flow to Physiological Stress

Exercise
Renal blood flow is decreased in proportion to the severity of exercise. Both the amount of work involved and the duration of the exercise influence the results. As an example, subjects who had run the 440-yard dash at full speed had reductions of 18 to 54 percent below control, which remained for 10 to 40 minutes after exercise. It has been estimated that a saving of 0.5 to 1.0 liter of blood is made available to active tissue by renal vasoconstriction.

In highly conditioned animals (Alaska sled dogs) undergoing heavy exercise, renal blood flow did not decrease (Van Citters and Franklin, 1969). Ultrasonic flowmeters, chronically implanted, were monitored via telemetry. In contrast to exercising humans, the compensatory redistribution of blood flow was not a significant reserve mechanism in these animals during exercise. However, it is well to point out that arterial blood pressure rose during exercise; in the face of constant renal blood flow, increased vascular resistance in this organ is implied. It could be extrinsic or autoregulatory in nature.

Posture and Orthostatic Hypotension
In normal young males, C_{PAH} in the sitting position is 91 percent of that in the supine; in the erect position, it is 85 percent. Motionless standing, or tilting of the subject, leads to progressive venous stagnation, reduced cardiac output, and neurogenic vasoconstriction, until the cerebral circulation becomes inadequate, when syncope occurs. Kidney flow is reduced to about one-half during the compensatory phase.

Renal Hypoxia and Ischemia
Hypoxia due to low O_2 intake or simulated altitude induces at first a slight increase in renal blood flow, but with more severe hypoxia this decreases. Apparently mild hypoxic states, unaccompanied by significant reflex vascular alterations, manifest a slight renal

hyperemia, but more severe hypoxic states or asphyxia triggers vasoconstrictor reflexes in which the kidneys participate.

Acute Renal Ischemia

Very brief periods of ischemia (1 to 5 minutes) result in a flow overshoot (*reactive hyperemia*) after release. Longer periods of ischemia in dogs (20 minutes) produce slight, inconsistent results on renal blood flow. Thirty minutes to two hours of ischemia result in marked, persistent reduction in renal blood flow. Since there is tubular damage, one must employ caution in interpreting clearance methods (e.g., C_{PAH}), although employment of the Fick principle (C_{PAH}/E_{PAH}) serves to correct for effects of tubular damage in the blood flow estimates. Reduction in renal blood flow after two hours of complete renal ischemia has been found to persist for 24 hours to two weeks after release.

Hypercapnia and Acidosis

These conditions cause renal vasoconstriction which is centrally mediated (enhanced activity of the vasoconstrictor centers).

Hemorrhagic Hypotensive Shock

Acute hemorrhage provokes responses in the renal circulation that are typical of general compensatory mechanisms brought into play: reflex vasoconstriction, and shunting of blood to other tissues in order to compensate for blood loss. If blood loss is great enough, there is a shutdown of renal excretory function, which if prolonged might have serious consequences to the organism. Moreover, a prolonged period of anoxic hypotension impairs the function of the tubular epithelium, adding to the problem of shock the probability of renal failure and uremia (see also Chap. 18C).

In *tourniquet* and *traumatic shock,* as well as in *hemorrhagic shock,* blood flow is markedly reduced, as a result of enhanced neurogenic vasoconstrictor activity and of increase in renal vasoconstrictor substances such as epinephrine. Evidently, under these circumstances the renal circulation is subordinate to the welfare of the body as a whole.

Blood Flow in Renal Disease

Relevant information on renal blood flow in acute and chronic nephritis, nephrosis, and hypertensive kidneys has accrued from studies employing C_D and C_{PAH}. Varying degrees of tubular damage, reflected in decrease in *tubular excretory maxima* of Diodrast and PAH (Tm_D and Tm_{PAH}) and extraction (E_D and E_{PAH}) have complicated interpretation. When allowance is made for the influence of impaired tubular function, it is apparent that renal blood flow may actually be elevated in early nephrosis and the acute, inflammatory phase of nephritis. Later, as the disease becomes chronic, blood flow is undeniably reduced to varying degrees and may be very low in chronic nephritis accompanying the resultant disorganization of the kidney vascular pattern. The accompanying renal hypertension seen in nephritis is a manifestation of reduced renal blood flow.

Renal Lymphatic System

The lymphatic circulation is prominent in the cortex, where the lymphatic capillaries begin blindly. Lymphatics within the cortex follow the course of the interlobular vessels and drain in a centripetal fashion toward the arcuate vessels. Some evidence has been presented that lymphatics occur in the medulla of the human kidney, but their presence or absence in other species, including the dog, is controversial. The medullary lymphatics, when they occur, drain the vasa recta system, join the cortical branches at the arcuate level, then pass out with the interlobar vessels toward the renal pelvis. After converging at the hilus of the kidney, the lymphatic vessels course as perivascular channels to the cisterna chyli and the thoracic duct. Lymph flow can be increased by renal venous pressure increment and ureteral blockade.

Slight differences in electrolyte and organic content of renal lymph, as compared to plasma filtrate and thoracic lymph, have suggested a more complex origin than systemic interstitial fluid. Renal lymph involves the tubular reabsorbate and is complicated by the operation of the countercurrent system. The renal lymph protein concentration averages 2.9 gm per 100 ml. Evidently, the medullary renal lymphatics, when present, subscribe an important function for operation of the countercurrent mechanism by draining off excessive protein filtered by the vasa recta, which might otherwise accumulate in the interstitial spaces of the medulla.

THE GLOMERULUS AND ITS FUNCTION

The glomerular capillaries are not simple loops but form a freely branching anastomotic network (Fig. 22-5). More specifically, larger *through channels* exist with an associated capillary network of smaller anastomotic channels. This arrangement may afford a structural basis for the skimming of plasma relatively freed of cells into the network of small capillaries, where the actual filtration proceeds, whereas the greater mass of blood cells directly and rapidly flows through the glomerular lobules to the efferent arterioles as an axial stream. This arrangement facilitates filtration processes by slowing the flow and reducing turbulence, and by possibly exposing more of the plasma which is to be filtered to the effective filtration surface.

The details of the filtering structures have been clarified by the use of the electron microscope (Fig. 22-6). This has revealed that the capillary endothelium, called the *lamina attenuata* or *lamina fenestrae* (0.05 μm thick), has pores 400 to 900 Å in size. The pores are too large to restrain the plasma constituents. Instead, they expose the ultrafiltration membrane, the *glomerular basement membrane* (0.1 μm thick) to the free flow of plasma by removing the endothelial cytoplasmic barrier. Although it exhibits differences in stratification, it is probably a homogeneous layer and appears to be the limiting membrane for restraint of plasma proteins. The visceral layer of Bowman's

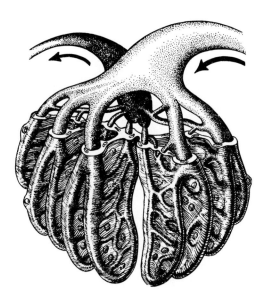

Figure 22-5
Pattern of distribution of *through channels* and *anastomotic channels* in the glomerulus. (From H. Elias et al. *J. Urol.* 83:795, 1960. Copyright © 1960, The Williams & Wilkins Company, Baltimore, Md. 21202, U.S.A.)

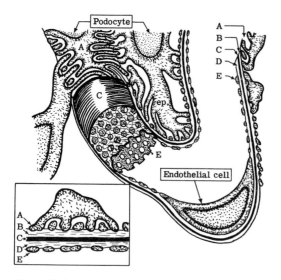

Figure 22-6
Glomerular capillary and associated portions of Bowman's capsule. Epithelial cells (ep.), the *podocytes* of the visceral layer of the capsule, form layer A. Layer C is the lamina densa and with layers B and D (cement layers) forms the basement membrane. Layer E is the capillary endo-thelium (lamina fenestrae). (From D. C. Pease. *J. Histochem. Cytochem.* 3:297, 1955. Copyright © 1955, The Williams & Wilkins Co., Baltimore, Md. 21202, U.S.A.)

capsule is organized into extremely specialized cells, the *podocytes* (foot cells), which pos-sess thousands of foot processes (*pedicels*) resting on the basement membrane. The spacing between the pedicels may be narrow enough (100 Å) to be a limiting dimension in restriction of plasma proteins. It has been designated by some authorities a *slit-pore.*

Dynamics of Glomerular Filtration

The forces involved in glomerular filtration are the same as those that were discussed in Chapter 12 under the consideration of transcapillary molecular exchange. Glomerular filtration rate (GFR) is proportional to what is designated as *net filtration pressure.* This is expressed as:

$$\text{GFR} = K (P'' - P' - \pi'' + \pi')$$
$$60 \quad 10 \quad 30 \quad 0^* \; (\text{mm Hg}) \tag{1}$$

(*assumes no filtration of plasma protein)

K is a proportionality coefficient embodying the volume of fluid filtered per minute per unit area per mm Hg pressure. Net filtration pressure (P_f) equals 20 mm Hg in equation 1. The double primes refer to the blood side of the capillary wall, and the single primes refer to the glomerular capsule fluid side. P denotes hydraulic pressure (P'', glomerular capillary pressure; P', the capsular pressure). π'' denotes colloid osmotic pressure of plasma proteins. The forces shown in mm Hg are values for the human kidney estimated from direct and indirect measurements made in other animals.

The nature of the filtering membranes is important. Chinard (1952), believed that glomerular fluid is formed by diffusion rather than by bulk filtration through pores and felt that most of the capillary surface was involved in the passage of water and most dissolved substances. Another concept, favored by Pappenheimer (1955), and generally

SUBSTANCE	MOL.WT. Grams	DIMENSIONS IN ANGSTROM UNITS		(FILTRATE) (FILTRAND)
		Radius from Diffusion Coefficient	Dimensions from X-Ray Diffraction	
WATER	18	1.0		1.0
UREA	60	1.6		1.0
GLUCOSE	180	3.6		1.0
SUCROSE	342	4.4		1.0
INULIN	5,500	14.8		0.98
MYOGLOBIN	17,000	19.5	←54→ ⊥8⊤	0.75
EGG ALBUMIN	43,500	28.5	←88→ ⊥22⊤	0.22
HEMOGLOBIN	68,000	32.5	←54→ ⊥32⊤	0.03
SERUM ALBUMIN	69,000	35.5	←150→ ⊥36⊤	<0.01

Figure 22-7
Glomerular permeability to molecules of different dimensions. (From *Physiology of the Kidney and Body Fluids*, 3d ed., by R. F. Pitts. P. 52. Copyright © 1974 by Year Book Medical Publishers, Inc., Chicago. Used by permission.)

accepted, assumes that pores exist in the glomerular membranes as straight-bore cylindrical channels about 400 to 600 Å (0.04 to 0.06 μm) in length. He calculated an average pore diameter of 75 to 100 Å, a pore size which would permit only minute quantities of albumin to be filtered. The approximation of the "effective" pore diameter was arrived at by analyzing the glomerular sieving of molecules of various sizes, as illustrated in Figure 22-7. Note that serum albumin is at the critical level of filtration, with an average diameter of 70 to 75 Å based on its diffusion coefficient, and 150 by 36 Å based on x-ray diffraction. The total pore area was calculated by Pappenheimer as about 5 percent of the capillary surface. The membrane resistivity to movement of fluid could thus be described by Poiseuille's equation and characterized as bulk filtration, with fluid and solutes driven by hydrostatic pressure.

Consideration of the filtering membranes suggests that the large pores ("fenestrae") (up to 900 Å in diameter) of the capillary endothelium perform no filtering action for diffusible substances but do in fact freely expose the plasma to the basement membrane, which can be considered as playing a possible role in filtration. The basement membrane has been viewed as a hydrated gel, the supporting structures being protein micelles manifesting a thixotropic behavior, providing a possible sievelike mechanism. Since the membrane is relatively homogeneous and pore-free as viewed by electron microscopy, the concept of bulk filtration through "pores" would of necessity imply a "virtual" pore—i.e., some physical structure operating in a sieving action comparable to a pore.

Another likely site of filtration restraint has recently been emphasized, namely, the slit-pores between the pedicels of the podocytes. Figure 22-8A shows the electron microscope detail of the human kidney, revealing slit-pore membranes between the pedicels or foot processes. The slit-pore has a highly ordered structure, as shown in a tangential section (part B), with an irregular central filament with crossbars to the adjacent foot processes. These irregularly spaced crossbars enclose individual rectangular, electron-lucent spaces with a calculated dimension of 50 × 120 Å. The alternation of crossbars produces a zipper-like appearance for the entire slit. It was calculated that these pores occupied 2.7 percent of the glomerular capillary surface (Schneeberger et al., 1975). Probably the estimate of the slit-pore diaphragm of the human kidney would normally be at the limit of passage of serum albumin, accepting the prolate ellipsoid size of about 150 × 36 Å, which would necessitate an "end-on"

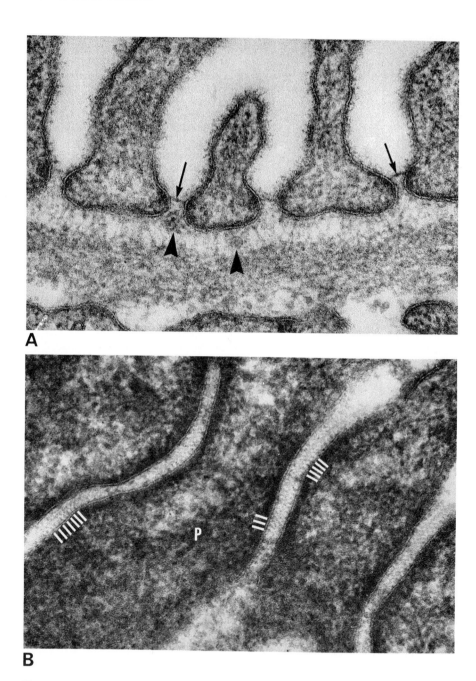

Figure 22-8
A. Human glomerular capillary wall from kidney fixed by perfusion with tannic acid–glutar-aldehyde (×168,000). The slit-pore membrane (arrows) consists of a single electron-dense line in the middle of which is a small black dot measuring 105 Å in diameter. Irregular collections of electron-dense granules present in the lamina rara externa (arrowheads) may represent serum proteins which were incompletely removed by the preliminary washout procedure. The middle layer of the glomerular basement membrane (lamina densa) shows a dense meshwork of irregular

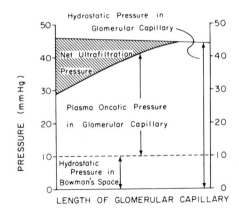

BALANCE OF MEAN VALUES

Hydrostatic pressure in glomerular capillary	45 mmHg
Hydrostatic pressure in Bowman's space	10
Plasma oncotic pressure in glomerular capillary	27
Oncotic pressure of fluid in Bowman's space	0
Net ultrafiltration pressure	8 mmHg

Figure 22-9
Forces involved in glomerular ultrafiltration in rats. As shown, ultrafiltration pressure declines in glomerular capillaries, mainly because plasma oncotic pressure rises. (In contrast, in extrarenal capillaries the decline in ultrafiltration pressure is due mainly to a decrease in intracapillary hydrostatic pressure.) The pattern for the rise in plasma oncotic pressure as a function of capillary length is not known precisely; hence, the mean values for plasma oncotic and net ultrafiltration pressure listed in the table are approximations. It is not yet known at what point in the capillary the sum of the hydrostatic pressure in Bowman's space and of plasma oncotic pressure exactly balances the hydrostatic pressure in the glomerular capillary. If, as shown here, balance occurs before the end of the capillary is reached, ultrafiltration would not take place over the entire length of the glomerular capillary. (From Valtin, 1973. Data from B. M. Brenner, J. L. Troy, and T. M. Daugharty. *J. Clin. Invest.* 50:1776, 1971.)

passage, to leak out, or even the size estimated from physiological experiments indicating a hydrodynamic sphere of 36 Å radius, which would require for passage a hit comparable to a rimless shot in basketball.

The forces involved in glomerular filtration are indicated in Figure 22-9. The figure shows data obtained by direct micropuncture of the glomerular capillaries and pressure measurement with a micropipette transducer in a unique strain of Wistar rats with accessible surface glomeruli. Note the progessive rise of the calculated plasma oncotic pressure along the length of the glomerular capillary, a consequence of ultrafiltration of the plasma, leaving the albumin behind in increasing concentration. Observe that filtration equilibrium is usually attained before the blood leaves the glomerular capillaries (Brenner et al., 1971; Deen et al., 1974). Recently, comparable measurements

---

fibrils ranging from 35 to 110 Å in thickness. The lower layer is the lamina rara interna. B. Tangential view of slit-pore membranes present between epithelial foot processes (P) (× 175,000). Areas in which the crossbars are clearly delineated are indicated by a series of short white lines. In these areas the central filament is also more clearly visible. Photographs supplied through courtesy of Dr. Eveline E. Schneeberger. (From E. E. Schneeberger, R. H. Levey, R. T. McCluskey, and M. J. Karnovsky. *Kidney Int.* 8:48–52, 1975.)

have been made in the anesthetized squirrel monkey (Maddox et al., 1974). The pressure gradient was found to be very similar to that of the rat, when normalized to the same aortic blood pressure. In absolute terms, with a mean aortic pressure of 115 mm Hg, pressure fell across the efferent arterioles to 48.5 mm Hg in the glomerular capillaries, then to about 8 mm Hg in the renal vein across the efferent arterioles. Net filtration pressure (P_f) at the beginning of the glomerular capillaries averaged 12.4 mm Hg $\pm$ 1.9 (SEM).

In summary, it can be stated that the filtration mechanism of the kidney is almost ideal since it provides a very permeable membrane to diffusion and also supplies relatively large forces for ultrafiltration, which are capable of a high degree of regulation by extrinsic and intrinsic (autoregulating) mechanisms.

Micropuncture studies of the capsular space have revealed that the qualifications for filtration are met; although the membranes are largely impermeable to protein, water and solutes pass freely, and the latter are found in approximately equal concentration on both sides of the glomerular membrane, allowing for slight differences in electrolyte concentration due to the Gibbs-Donnan effect (Chap. 1). Thus the classic investigations of Richards (1935) in amphibia and of Walker et al. (1941) in mammals (see Pitts, 1974) have shown that glucose, inorganic phosphate, creatinine, urea, uric acid, sodium and chloride, total osmotic pressure, and pH are approximately the same in the capsular fluid as in the plasma water. When the substance inulin was given, the clearance of which is taken to be a measure of glomerular filtration rate, it was found in equal concentration on both sides of the membrane. Harris et al. (1974) have successfully punctured superficial glomeruli of rats (Munich-Wistar strain) and confirmed that the GF/P (glomerular fluid/plasma) ratio of inulin was 1.00 $\pm$ .01 (SEM).

Regulation of Glomerular Filtration

Glomerular filtration pressure regulation is probably dominated by the influence of adrenergic constrictor activity to the afferent arterioles. This type of control is illustrated for the human kidney in Figure 22-10A, in which line 3 illustrates the influence on GFR of changing vascular resistance in the afferent arterioles alone; this could result from changes in activity of the adrenergic fibers. Since GFR and RBF would change proportionally, the filtration fraction (FF = C_{IN}/RPF) would remain constant (line 3, Fig. 22-10B). Line 1 (A), and curve 1 (B) illustrate the influence of varying efferent arteriolar resistance alone. Here, GFR and FF increase as total RBF is reduced with increased efferent arteriolar resistance, and vice versa. Line 2 in A and curve 2 in B show how proportional changes in both afferent and efferent arteriolar resistance influence GFR and FF: Total GFR remains constant; FF rises with decrease in blood flow and falls with increased RBF.

If cholinergic (dilator) fibers operate simultaneously on the efferent arterioles in concert with constrictor influence on the afferent arterioles, as may be the case in the juxtamedullary glomeruli, a more efficient and versatile mechanism might be provided. Thus, simultaneous afferent arteriolar constriction and efferent dilatation would further lower filtration pressure, and the converse change would result in a higher filtration pressure. A wider range of control might therefore be established.

Figure 22-11 illustrates how various permutations of afferent and efferent arteriolar caliber could affect glomerular capillary pressure.

Glomerular Filtration Rate

Direct methods are not applicable to the measurement of filtration rate in the mammalian kidney, and indirect methods have been devised. These involve an understanding of the *renal plasma clearance concept*. The rate at which a substance (X) is excreted in the urine is the product of its urinary concentration (U_X, mg/ml) and the volume of urine per minute ($\dot{V}$). The rate of excretion ($U_X \cdot \dot{V}$) per minute depends upon the

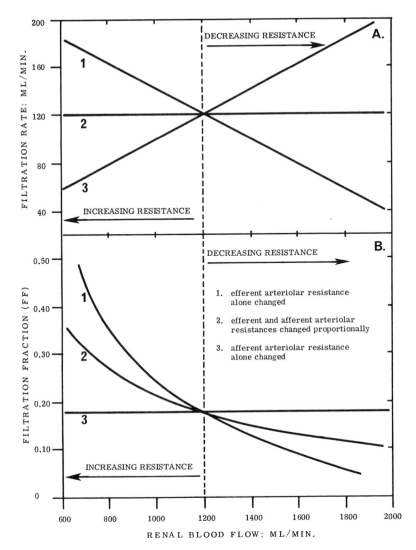

Figure 22-10

A. Relationship of filtration rate to renal blood flow in the human kidney, with various permutations of afferent and efferent arteriolar resistance. B. Relationship of filtration fraction to renal blood flow with corresponding changes in arteriolar resistance.

concentration of X in the plasma (P_X, mg/ml). It is therefore reasonable to relate $U_X \dot{V}$ to P_X, and this relation is called the *clearance ratio:*

$$\frac{U_X \cdot \dot{V}}{P_X} \tag{2}$$

or more generally $U\dot{V}/P$. The calculated value has the dimensions of volume related to time and is in reality the *smallest volume of plasma from which the kidneys can obtain the*

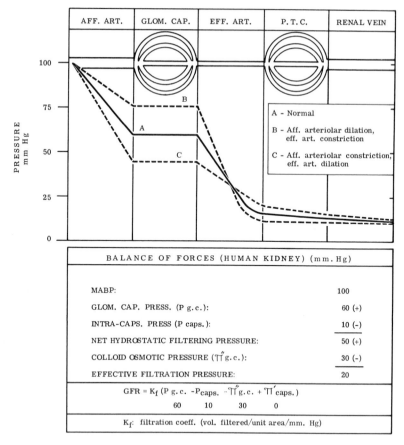

	AFF. ART.	GLOM. CAP.	EFF. ART.	P.T.C.	RENAL VEIN

A - Normal

B - Aff. arteriolar dilation, eff. art. constriction

C - Aff. arteriolar constriction, eff. art. dilation

PRESSURE mm Hg

BALANCE OF FORCES (HUMAN KIDNEY) (mm.Hg)

MABP:	100
GLOM. CAP. PRESS. (P g.c.):	60 (+)
INTRA-CAPS. PRESS (P caps.):	10 (-)
NET HYDROSTATIC FILTERING PRESSURE:	50 (+)
COLLOID OSMOTIC PRESSURE (Π''g.c.):	30 (-)
EFFECTIVE FILTRATION PRESSURE:	20

$$GFR = K_f (P\ g.c.\ -P_{caps.}\ -\Pi''g.c.\ +\Pi'caps.)$$
$$\qquad\qquad 60 \qquad 10 \qquad 30 \qquad 0$$

K_f: filtration coeff. (vol. filtered/unit area/mm. Hg)

Figure 22-11
Balance of forces responsible for ultrafiltration as estimated for the human kidney.

amount of X excreted per minute. It must be understood that the kidneys do not usually clear the plasma completely of a substance in transit through the kidneys, but clear a larger volume incompletely. The clearance therefore is not a real but a *virtual* volume. When substances are being cleared simultaneously, each has its own clearance rate, depending on the amount reabsorbed from the glomerular filtrate or added to it by tubular secretion. The former has the lower clearance, the latter the higher. Those cleared only by glomerular filtration are intermediate, and their clearance, in effect, measures the rate of glomerular filtration in milliliters per minute.

The best-known substance which can be infused into the blood to provide a clearance equal to glomerular filtration rate is *inulin,* a polymer of fructose containing 32 hexose molecules (molecular weight 5200). Strong evidence exists that it is neither reabsorbed nor secreted by the tubular cells (Fig. 22-12), that it is nonmetabolized in the body, that it has no physiological influences, and that it is freely filterable at the glomerular membranes. Several lines of evidence indicate that the clearance of inulin is at the level of *glomerular filtration*—i.e., that it is an index of the total volume of fluid filtered by the glomerular membranes per unit of time. When simultaneous clearances of inulin and other sugars such as glucose, xylose, and fructose are performed, the clearances of the sugars are less than the clearance of inulin. The differences in clearance between these

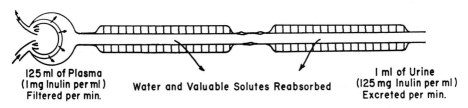

Figure 22-12
Diagram of nephron illustrating requisites of a substance such as inulin for measurement of glomerular filtration rate. (From *Physiology of the Kidney and Body Fluids,* 3d ed., by R. F. Pitts. P. 62. Copyright © 1974 by Year Book Medical Publishers, Inc., Chicago. Used by permission.)

and inulin depend upon their particular reabsorptive mechanisms. However, when the transfer mechanism for sugar is inhibited by the glucoside *phlorizin,* the reabsorption of sugars is blocked and their clearances become identical with that of inulin; i.e., under these circumstances, the sugars are cleared only by the physical process of filtration. Evidence that inulin is not secreted by tubular cells is somewhat more indirect. One factor against this possibility is that inulin is not found in the urine of aglomerular kidneys of certain marine forms even when tremendously high plasma concentrations are attained by intravenous injection of inulin.

The clearance rate of inulin (C_{IN}) in man is 120 to 130 ml per minute. This is taken to be the glomerular filtration rate (also C_F [clearance of filtrate]), or the amount of plasma which is filtered through the glomeruli per minute.* Substances other than inulin have been used to provide an estimate of GFR, the most common of which is creatinine. It has been shown that exogenous (true) creatinine clearance is approximately equal to that of inulin in the dog, rabbit, sheep, seal, frog, and turtle and thus provides a measure of glomerular filtration rate in these animals. However, in addition to being filtered through the glomeruli, creatinine is secreted in part by the tubular cells in man and other primates; hence, exogenous creatinine clearance cannot be used as an estimate of filtration rate.

A creatinine-like chromogen is found in human plasma which gives the same chemical reaction for colorimetric analysis (Jaffe reaction) as does true creatinine, but whose calculated clearance approximates that of inulin. The "noncreatinine chromogen" is apparently not excreted into the urine, so that the plasma level of measured creatinine (true plus false) yields the lower calculated clearance. The so-called endogenous creatinine clearance is widely used in patients as an estimate of GFR. An application appears in Figure 22-13, where the blood urea in developing uremia is related to the endogenous creatinine clearance, a measurement of GFR. Note that GFR needs to be reduced below 60 ml per minute during renal failure before blood urea rises above normal limits (20 to 50 mg per 100 ml) (BUN, about 10 to 25 mg per 100 ml). With a low protein intake, increase occurs only below 30 ml per minute.

---

*Glomerular filtration rate is expressed in terms of milliliters of plasma filtered through glomerular capillaries per minute. This system introduces an ambiguity in terminology, for plasma is 93 to 94 percent water and some 6 to 7 percent protein, and only the aqueous phase is filtered. All of the true crystalloids of plasma are dissolved in the aqueous phase; yet their concentrations are commonly expressed in terms of milligrams per 100 ml (or mEq/L) of plasma, not of plasma water. Thus, all of the crystalloids in 125 ml of plasma are filtered each minute when the rate of glomerular filtration is 125 ml per minute. However, only 0.93×125 ml $= 116$ ml of water is filtered. The student should keep in mind that glomerular filtration rate is defined as *volume of plasma filtered per minute,* not as volume of plasma water, although obviously only the water and its dissolved crystalloids are actually filtered.

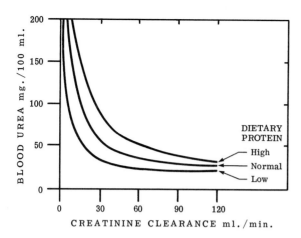

Figure 22-13
The relationship of endogenous creatinine clearance in man to blood urea. (From H. DeWardener. In *The Kidney: An Outline of Normal and Abnormal Structure and Function* [3rd ed.]. Boston: Little, Brown, 1967. P. 28.)

Other substances which have a clearance close to that of inulin in man, dog, and other animals are mannitol and thiosulfate.

A knowledge of the glomerular filtration rate permits quantitation of the amount of any substance freely filtered (C_{IN}, ml/min $\times$ P_X, mg/ml). When dealing with a substance which is reabsorbed, one may subtract one minute's excretion, $U_X \dot{V}$, from the filtered load to find the amount reabsorbed by the tubules in milligrams per minute:

$$T_R = C_{IN} \cdot P_X - U_X \cdot \dot{V} \tag{3}$$

Furthermore, under the same principle, the knowledge of glomerular filtration rate permits the calculation of tubular secretory activity, as follows:

$$T_S = U_X \cdot \dot{V} - C_{IN} \cdot P_X \tag{4}$$

A number of the substances studied for tubular secretion are bound to plasma protein in varying degrees and cannot be freely filtered. The filtered moiety therefore requires correction for the binding. This introduces the factor f, which represents the fraction of the substance not bound by the plasma proteins and therefore free to be filtered. Factor f also depends on the compound (substrate) concentration and plasma protein (albumin) concentration. Factor f may vary with the species (e.g., it is 0.92 for *p*-aminohippurate in the dog and 0.83 in man). Also, it differs for various substances. Thus the f value is 0.40 for phenol red and 0.72 for Diodrast. The estimates are made by dialysis experiments. The completed equation is

$$T_S = U_X \cdot \dot{V} - C_{IN} \cdot P_X \cdot f \tag{5}$$

It should be added that plasma binding does not affect the fraction of clearance that involves tubular secretion. Such mechanisms operate by action of transfer systems which rapidly remove the substances from the plasma water, to establish a gradient whereby the plasma binding is readily broken down, so that the substances continue to be rapidly removed by tubular secretion. However, the diffusion process across the glomerular membranes does not affect the protein binding, and it is the moiety which is free in plasma water that is filtered.

In summary, clearances less than that of inulin represent mechanisms involving filtration and reabsorption; clearances higher than that of inulin represent mechanisms involving glomerular filtration plus some degree of tubular secretion with allowance for possible plasma binding.

TUBULAR REABSORPTION OF ORGANIC SUBSTANCES

Glucose

A classic example of tubular reabsorption is the glucose mechanism. Usually none or only a trace of this hexose appears in the urine. When plasma glucose is elevated in man to about 180 to 200 mg per 100 ml, the so-called *threshold,* glucose appears in the urine. As the concentration is further raised by continued infusion of glucose, the reabsorptive mechanism becomes progressively saturated until the rate of reabsorption becomes constant and maximal. This indication of saturation of the transport system is referred

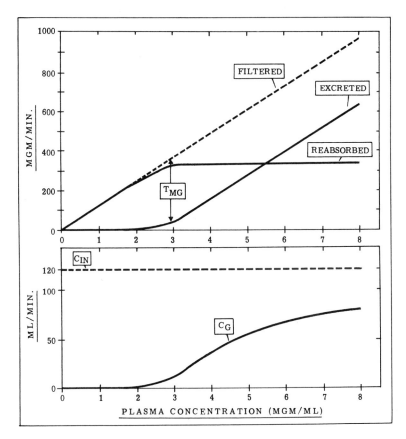

Figure 22-14
Glucose tubular reabsorption and clearance of glucose as related to increase in plasma concentration. C_{IN} = filtration rate; C_G = glucose clearance; T_{M_G} = tubular maximal rate of transport of glucose.

to as the *tubular maximum* or Tm—in this instance Tm_G. In the human, Tm_G has a value of 340 mg per minute; it is about 200 mg per minute in the dog.

Figure 22-14 illustrates the principle of saturation of the reabsorptive system by a progressive elevation of plasma glucose concentration experimentally. It will be noted that $U_G\dot{V}/P$ or C_G increases progressively. This approaches asymptotically the clearance of inulin as the plasma levels increase further beyond the point of saturation of the tubular mechansim. To put it another way, the clearance becomes more and more preponderantly the consequence of glomerular filtration, and the reabsorbed moiety becomes a fractionally smaller part of the total excretory component.

A theory of the kinetics of the glucose transport system is diagrammed in Figure 22-15. A carrier substance, X, is present in the luminal membrane of proximal tubular cells in a fixed and limited amount. The carrier combines reversibly with D-glucose, G, from the tubular fluid, to form the complex $G \cdot X$ within the membrane. The complex moves to the cytoplasmic side of the membrane, where G is split off. From the cytoplasm, G moves out of the basal end of the cell by another step, apparently not rate-limiting, possibly by facilitated diffusion. Carrier X needs to be regenerated via an intermediate step, Y, which handles metabolically abnormal types of sugar (e.g., L-glucose) which are transported to and secreted into the lumen. Then Y is restored to X, involving an enzymatic step and introduction of energy, via adenosine triphosphate (ATP). A sodium dependency, as well as phlorizin sensitivity, has been demonstrated at the carrier site, as in the intestine.

It now seems clear that phlorizin competitively inhibits glucose reabsorption by binding with great affinity at the membrane site which binds or orients glucose (Diedrich, 1963). Phlorizin structurally is glucose with the C_1 hydroxyl replaced with phloretin; hence, competitive binding is not surprising. Other hexoses (galactose, fructose, and, slightly, xylose) compete with glucose for the transport site. The disaccharide sucrose shows very little reabsorption in the in vivo system.

Kleinzeller et al. (1967), using rabbit cortical slices and isolated tubules, have confirmed the notion that the active transport process is localized in the brush border at the luminal face of the cells, analogous to the intestinal glucose transport mechanism. D-Galactose behaved very much the same as D-glucose in their system, and D-fructose and *a*-methyl-D-glucoside were also readily accumulated against a concentration gradient. Considerable specificity was indicated by the fact that other monosaccharides were transported poorly if at all (e.g., D-xylose, D- and L-arabinose, 6-deoxy-D-glucose, and 6-deoxy-D-galactose).

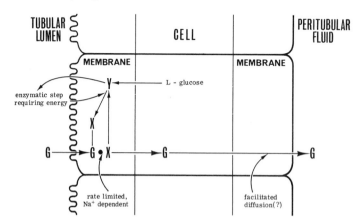

Figure 22-15
Model of glucose transport, illustrating countertransport of L-glucose. (After Huang and Woolsey, 1968.)

Amino Acids

Amino acid clearance is low in man (1 to 8 ml per minute). Much of the earlier characterization of the mechanism has been done in the dog. More recently, rat kidney studies have supplied additional information, utilizing isolated nephrons, kidney slices, and micropuncture techniques. Several amino acids (glycine, arginine, lysine, proline, and hydroxyproline) demonstrate relatively poor reabsorption with small Tm; others, like histidine, methionine, leucine, isoleucine, tryptophan, valine, threonine, and phenylalanine, are so effectively reabsorbed that saturation is not achieved by plasma concentrations which do not cause severe nausea or other physiological disturbances.

There are probably no less than three renal tubular mechanisms for reabsorption of amino acids (Pitts, 1974). One transports lysine, arginine, ornithine, cystine, and possibly histidine; a second handles aspartic and glutamic acid; a third has been shown to involve proline, hydroxyproline, and glycine.

The amino acid transport mechanisms display saturability and substrate specificity, and, in vitro, Michaelis-Menten kinetics. An interesting series of "inborn errors of membrane transport" is manifested by some of the amino acids—cystinuria, hyper-prolinemia, and hydroxyprolinemia, among others. The third is associated with mental deficiency.

Ascorbic Acid

Ascorbic acid has a Tm of 2.0 mg per minute in man (about 1.5 mg per 100 ml of filtrate). In the dog, which is able to synthesize ascorbic acid, reabsorption is 0.5 mg per 100 ml of filtrate. The Tm represents the net activity of a three-component system (filtration, proximal tubular reabsorption, and distal tubular secretion). Distal secretion is promoted by pretreatment with adrenal steroid (deoxycorticosterone acetate), which stimulates distal sodium-potassium exchange. Additional stimulation of distal secretion results from increasing the filtered load of sodium, by alkalinization of the urine with $NaHCO_3$, and administration of acetazolamide. Evidently, a linkage between the ascorbic acid secretory site and the distal sodium-potassium exchange mechanism exists. This may be related to the vitamin's acidic properties.

Urea

Urea, a major product of protein metabolism, is filtered and reabsorbed to varying degrees (40 to 70 percent) throughout the proximal nephron, and again in the medullary collecting tubule and duct (Figs. 22-16 and 22-26). The thick ascending limb, distal convolution, and early collecting tubule are urea impermeable. Urea's reabsorption is inversely related to the rate of urine production because reabsorption is largely a process of back-diffusion in man, dog, rabbit, and chicken. In such a reabsorptive mechanism, the return of urea from the tubular urine, where it is in relatively high concentration on account of water uptake by the nephron, proceeds by the process of diffusion across the tubular epithelium into the peritubular capillaries, where it is in relatively low concentration. With rapid urine flow, the time for diffusion is limited. However, active reabsorptive mechanisms for urea (involving a transfer system and expenditure of energy) operate in the kidneys of elasmobranchii. Protein-depleted rats on a low-protein, high-salt diet show urea/inulin clearance ratios as low as 0.01, with papillary tissue urea concentration about three times that of the final urine, suggesting an adaptive development of an active reabsorption mechanism for urea, necessary to maintain the high papillary solute concentration required for the countercurrent mechanism (Truniger and Schmidt-Nielsen, 1964). On the other hand, in the amphibian (anuran) kidneys the tubules secrete urea by an active process.

Past importance of the urea clearance stems from the fact that it has been used as a clinical indication of glomerular filtration, which it reflects reasonably well at high rates of urine flow, when its reabsorption is minimal. The relationship of the clearance of urea to the rate of urine flow is shown in Figure 22-17. Urea plays an important role in the development and maintenance of the gradient of osmolality, cortex to papilla,

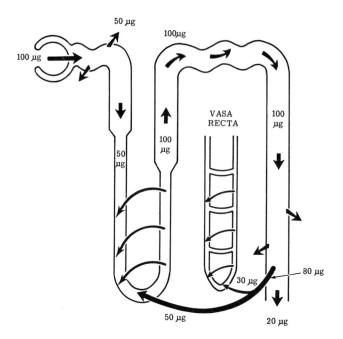

Figure 22-16
The tubular handling of urea. The model shows a representative amount of 100 μg entering the glomerulus. A recent view (see Fig. 22-26) does not accept the recycling of urea in the loop of Henle. (After K. J. Ullrich, K. Kramer, and J. W. Boylan. *Progress in Cardiovascular Disease.* 3:408, 1961. By permission of Grune & Stratton, Inc.)

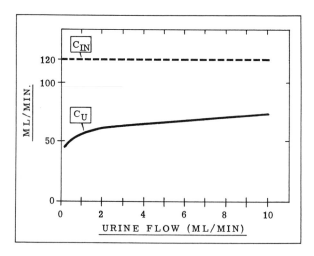

Figure 22-17
Relationship of urea clearance (C_U) to urine flow. C_{IN} = inulin clearance. (From Smith, 1956, P. 79).

needed for urinary concentration. This aspect will be considered in greater detail in a later section.

Uric Acid

In mammals, uric acid appears as a consequence of metabolism of purine bases. In most mammals it is oxidized to allantoic acid, but not in primates or the Dalmatian coach dog. It has been generally assumed that a Tm characterizes its reabsorption in man. But under conditions of injections of a uricosuric drug (sulfinpyrazone) which blocks tubular reabsorption of uric acid, combined with vigorous mannitol diuresis, the ratio of excreted urate over filtered urate exceeds unity and may go up to 1.23 in man, demonstrating tubular secretion. Thus, a three-component system of filtration, reabsorption, and secretion is suggested, so that the amount excreted is normally the *net effect* of these operations. Definite evidence of tubular secretion has also been found in birds and reptiles and in the guinea pig and Dalmatian coach dog. Microperfusion studies in the rat (Lang et al., 1973) have confirmed the three-component nature of the operation and have aided in localization of specific mechanisms. Thus, the clearance is determined by filtration at the glomerulus, proximal tubular secretion, and concomitant reabsorption (bidirectional flux), with secretion predominating early in the proximal segment. Additional reabsorption occurs in the loop of Henle, to account for the normal net reabsorption.

It should be added that other uricosuric agents exist, all of which have been used in the treatment of gout, a disease characterized by deposition of uric acid crystals in the joints. These include cinchophen, salicylate, acetylsalicylic acid, the mercurial Salyrgan, carinamide, and Benemid.

Creatine

Creatine is a product of muscle metabolism which disappears from the urine of humans after adolescence. It is filtered and reabsorbed in concentrations below 0.5 mg per 100 ml. At higher concentrations, reabsorption is incomplete and excretion is enhanced. No Tm has been demonstrated. At higher plasma levels, the creatine/C_{IN} ratio becomes constant at 0.8.

Other Substances

Other substances involved in tubular reabsorption have been demonstrated in dog and man. These include acetoacetic acid, β-hydroxybutyric acid, lactic acid, and guanidoacetic acid. Maximal rates of transport have been demonstrated for β-hydroxybutyric and lactic acids. α-Ketoglutarate, utilized metabolically by the renal tubular cells, is both reabsorbed (Tm limited) and secreted (Balagura and Pitts, 1964). (See also relation to PAH secretory mechanism, pp. 515–516.)

In summary, tubular reabsorption of various substances is limited by a maximal rate, and this varies markedly from one substance to another. In some instances (the sugars and certain amino acids) a transport mechanism may be shared in common and involve the reabsorption of several related substances, but there appear to be many transport systems that operate wholly independently of one another. Besides glucose and the amino acids, these include lactic acid, uric acid, acetoacetic acid, and the vitamin mechanisms.

There is considerable evidence that most of these substances are reabsorbed either in entirety or in part in the proximal convoluted tubule. Information on localization has been obtained by micropuncture of the tubules and by the stop-flow technique. The technique is as follows: The ureter of the experimental animal is clamped during vigorous diuresis, bringing the movement of urine in the nephrons to a halt. Stoppage of the tubular movement of urine for a brief interval (about eight minutes) permits the tubular reabsorptive and secretory processes to proceed to greater completion. Fractional collection of the urine which issues upon release of the ureter permits allocation of observed changes in tubular urine concentration to various segments of the nephron. This is aided by the injection of a marker substance (such as inulin or creatinine) during the period of ureteral blockade to tag the entry of new glomerular fluid or, in effect, to mark the end of the stagnant urine column at the approximate level of the glomerulus. The knowledge of enzymatic and cellular mechanisms involved with these transport systems is rather limited at the present time. Further investigative effort is necessary to clarify the multiplicity of mechanisms involved.

TUBULAR SECRETION
Evidence has accumulated from techniques such as the above that the proximal convoluted tubule is the site of active secretion of some physiologically occurring substances as well as certain foreign substances when injected into the circulation.

Tubular Secretion of Foreign Substances

p-*Aminohippuric Acid (PAH)*
At low concentrations PAH is almost completely cleared from the plasma by a combination of glomerular filtration and efficient tubular secretion. Hence, C_{PAH} measures

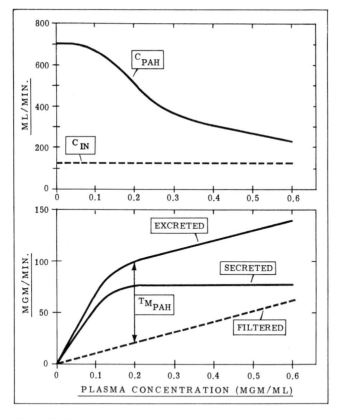

Figure 22-18
The relationship of p-aminohippurate clearance (C_{PAH}) and tubular secretion to increase in plasma concentration. C_{IN} = inulin clearance.

approximately 90 percent of the total plasma flow through the kidneys. In man, the clearance is 600 to 700 ml per minute and corrected for hematocrit gives the *effective renal blood flow.* The proportion of the total plasma passing through the kidneys which is filtered at the glomeruli is called the *filtration fraction.* It can be estimated by the ratio of C_{IN}/C_{PAH}, recognizing that C_{PAH} represents the "effective" plasma flow, not total plasma flow. It has a value of 0.20 in man and 0.32 in the dog.

When plasma levels are elevated to the range of 30 to 50 mg per 100 ml, the secretory mechanism becomes saturated, and the Tm can be discerned. Tm for PAH in man is about 77 mg per minute per 1.73 square meters of surface area (SA). In the dog it is about 33 mg per minute per 1.73 square meters SA. The relationship of the mechanism of urinary excretion as it relates to plasma concentration appears in Figure 22-18. As the tubular secretory mechanism becomes saturated and is exceeded by progressive increments in plasma PAH, C_{PAH} is observed to decrease. The reason is that, with saturation of the tubular transfer system, the total $U\dot{V}$ and the calculated $U\dot{V}/P$ become progressively more a function of glomerular filtration; hence $U\dot{V}/P$ approaches that function asymptotically. The portion of $U\dot{V}$ contributed by tubular secretion becomes a constant but fractionally smaller part of the total $U\dot{V}$.

The mechanism of transfer of PAH and other substances secreted by the tubular cells has been approached by studies involving isolated nephrons and kidney slices. This approach offers the opportunity to study the effect of substances which accelerate or depress accumulation of PAH in the slices as observed in the Warburg apparatus. Taggart (1958) has shown that the accumulation of PAH in tissue slices is dependent on aerobic metabolism by the fact that it is supressed by lack of oxygen, by chilling, or by any metabolic poisons which inhibit respiration. Uptake is also suppressed by 2,4-dinitrophenol and by related compounds, which do not inhibit the uptake of oxygen but uncouple phosphorylation. Accumulation was increased greatly by the addition of acetate to the medium, and to a lesser extent by addition of lactate and pyruvate. Acetate and lactate increase Tm_{PAH} markedly when administered intravenously.

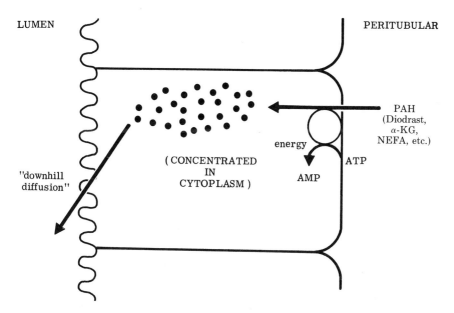

Figure 22-19
Model of PAH transport system in proximal tubular cell. α-KG = α-ketoglutarate; NEFA = nonesterified fatty acids.

The site of secretion of the organic acid PAH is the proximal convolution, and its active transport depends on generation of high-energy phosphate compounds (e.g., ATP) by oxidative metabolism. It is not understood how the energy is coupled to the transport process. Uptake of PAH from the blood side of the tubules probably involves an active pump at the peritubular cell membrane. Foulkes (1963) concluded, from a kinetic analysis in the rabbit kidney, that intracellular PAH is in a free, not bound, form. Transfer from cell to lumen probably involves a facilitated diffusion mechanism, with no energy-requiring step at the luminal membrane. In the mammal, the large volume of glomerular filtrate renders unnecessary an active concentration step at the luminal membrane.

The purpose of the PAH (organic acid) transport system has been speculated upon: (1) It may be to rid the body of foreign organic acids and end products of metabolism, such as glucuronides, sulfate esters, and urates. Interestingly, the liver, ciliary body of the eye, and choroid plexus of the brain actively transport these compounds in an outward direction. (2) This transport system may serve to transport metabolic substrates, such as α-ketoglutarate and nonesterified fatty acids (NEFA) to the site of utilization in the proximal tubular cells (Fig. 22-19).

Other Substances
Additional foreign substances secreted by the tubules are *Diodrast, phenol red,* and *penicillin.* Each has its characteristic Tm; e.g., Diodrast Tm averages 50 mg per minute in man, and Tm for phenol red is 36 mg per minute. The transfer systems for the different substances so far described appear to be mutually interelated, because competitive depression of Tm is noted when they are simultaneously cleared at high plasma levels. Such blocking agents as carinamide and Benemid inhibit the secretion of the above-mentioned substances.

A group of *organic bases* is actively secreted by the proximal tubule: tetraethylammonium (TEA), tetramethylammonium (TMA), tetrabutylammonium (TBA), Darstine (mepiperphenidol), Priscoline (tolazoline), and hexamethonium. The Tm's are small (e.g., Tm_{TEA} is 1.0 to 1.4 mg/min/m² SA in dogs). The mechanism is distinct from that of organic acids such as PAH.

Tubular Secretion of Physiological Substances

Creatinine
Creatinine is derived from creatine in muscles. Exogenous creatinine is cleared by glomerular filtration plus tubular secretion in man and other primates. The Tm is small and about 16 mg per minute. It is also secreted by tubules of certain fish, the alligator, chicken, goat, rat, cat, and guinea pig. Animals from which it is cleared only by glomerular filtration have been cited above.

It has been disclosed that creatinine transport is shared by both the organic acid and organic base transport mechanisms in the monkey and guinea pig (Selkurt et al., 1968; Arendshorst and Selkurt, 1970). This finding may be ascribable to the amphoteric nature of creatinine.

Naturally Occurring Organic Bases
Naturally occurring organic bases include N-methylnicotinamide, guanidine, piperidine, thiamine, histamine, and choline. N-methylnicotinamide, a metabolic derivative of nicotinic acid, has a clearance up to three times GFR, suggesting tubular secretion.

EXCRETION OF ELECTROLYTES
Virtually all segments of the nephron are concerned with the electrolyte economy of the body, to help preserve the proper balance of cations and anions. Reabsorptive mechanisms involve active pumps and energy expenditure.

Electrical Potentials

A knowledge of electrical potentials across the tubular membrane is necessary in order to decide which ions are actively transported. The origin of such potentials was discussed in Chapter 1 and is based upon application of the Nernst equation. The student will find that a brief review of transmembrane potentials is helpful in the understanding of the nature of tubular ion transport. Thus, strong evidence for active transfer of Na^+ is that it moves out of the proximal tubule against a gradient of electrochemical potential.

Cations

Sodium

About 99 percent of filtered Na^+ is reabsorbed by active transport mechanisms (Table 22-1). Estimates vary from 65 to 80 percent reabsorption in the proximal tubules and the remainder in the ascending portion of the loop of Henle, the distal convoluted tubules, and the collecting duct. Reabsorption in the proximal tubule is isosmotic. Reabsorption of NaCl and $NaHCO_3$ here is primary, by an active process; water follows passively and is not influenced by ADH. Another site of NaCl reabsorption occurs in the ascending thick portion of the loop of Henle, significant because this is relatively water-impermeable, and as a consequence the urine becomes hypotonic to plasma at this site.

Evidence of the active nature of sodium reabsorption has been cited in terms of its being pumped against an electrochemical gradient. Other evidence is that the TF/P (tubular fluid/plasma concentration) of Na^+ has been observed to go below 1.0 in the face of brisk osmotic diuresis.

No Tm for sodium has been discerned. In man, about 200 μEq per minute is excreted. Increased loading of the kidney tubules by Na^+ is followed by increased total reabsorption, but with decreased efficiency, since the total load is not quite so effectively reabsorbed. As a result, urinary excretion of Na^+ increases. The hormone aldosterone, secreted by the glomerulosa zone of the adrenal cortex, is necessary for efficient reabsorption. The distal convoluted tubule is the most important site of aldosterone activity, although some evidence suggests a possible proximal tubular action also. (For details of action, see Chap. 34.)

Sodium Transport Mechanisms. Figure 22-20 represents a model of a proximal tubular cell. K^+ content of the cell is high (ca 140 mEq per liter), while Na^+ and Cl^- content is low (10 mEq per liter). The cell is electrically negative with respect to lumen and peritubular fluid. Na^+ can diffuse along an electrochemical gradient from lumen to cell; its concentration within the cell is kept low by a pump which extrudes it into the peritubular fluid. Diffusion of K^+ and Cl^- is the major determinant of the potential across the peritubular membrane. The peritubular membrane pump functions also to return K^+ that has diffused out.

The potential across the luminal cell membrane is produced by a current $\times$ resistance drop as current flows from lumen into cell. The electromotive force (EMF) is created by the diffusion potential of Na^+ and the sodium pump at the peritubular membrane. The circuit is closed by the existence of intercellular channels in parallel with the cellular

Table 22-1. Reabsorption of Sodium and Related Anions

Plasma Conc. (mEq/L)		Donnan Factor	GFR	Filtered (mEq/min)	Excreted (mEq/min)	Reabs. (mEq/min)	Percent Reabs.
Na	140	0.95	125	16.6	0.135	16.465	99.3
Cl	103	1.05	125	13.5	0.135	13.37	99.1
HCO_3	27	1.05	125	3.55	<0.002	3.54	99.9

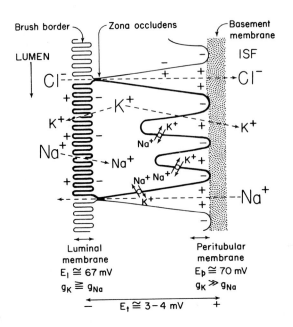

Figure 22-20
Electrolyte transport by the proximal tubular cell. (From L. P. Sullivan. *Physiology of the Kidney.* Philadelphia: Lea & Febiger, 1974. P. 60.)

pathways. The active reabsorption of Na^+ maintains a small electrical potential difference (PD) across the proximal tubular epithelium. The lumen early in the nephron is 3 to 4 mv negative with respect to the surrounding interstitial fluid (Fig. 22-20). This electrical gradient and a small chemical gradient cause passive Cl^- diffusion out across the tubular wall, which has a high conductance to chloride (^{g})(Cl). The diffusion proceeds importantly through the zona occludens. HCO_3^- is removed by a process involving H^+ secretion, which will be discussed more fully in a later section. Present thinking includes the possibility of a passive back-flux of Na^+ into the lumen because of a high conductance of the tubular wall to Na^+ (Fig. 22-20). This may occur through the zona occludens. Nevertheless, the *net* flux of Na^+ out of the tubules is large, as stated above.

Recently, Seely and Chirito (1975) demonstrated that in the rat there is a progressive increase in PD from the initial negative values to positive values (+2 to +4 mv) in later segments of the proximal tubule. This correlates with a rise in tubule fluid-to-plasma [TF/P] chloride ratio to 1.3. The late luminal positivity appears to be a consequence of the development of a chloride diffusion potential. This results from the preferential reabsorption of HCO_3^- with Na^+, along with osmotically obligated water.

At first glance, the late luminal positivity might seem to be inconsistent with a purely passive mechanism for chloride reabsorption in the proximal tubule. Further analysis of the situation (Seely and Chirito, 1975) shows that this is not necessarily the case. The chloride equilibrium potential, E_{Cl} (Chap. 1) between the tubular lumen and the plasma was calculated from the Nernst relationship: $E_{Cl} = 2.303\ RT/F \log P/TF$ (where P/TF is the plasma/lumen chloride concentration ratio). The calculation was done at a number of locations along the tubule, and the results were compared with the measured transtubular potential differences at the same locations. In every instance, the calculated value of E_{Cl} exceeded the corresponding measured potential difference. Thus, there appears to be, at all points along the length of the tubule, a net electrochemical gradient along which chloride ions can diffuse passively in the direction lumen → plasma (cf. Chap. 1). In the opinion of Seely and Chirito, the reason the measured chloride diffusion potential across the tubular wall is significantly less than E_{Cl} is that E_{Cl} is partially "shunted out" by the high passive permeability of the proximal tubule to sodium. (An outwardly directed electrogenic sodium pump in the peritubular membranes of the tubular epithelial cells could also contribute to this effect.) In support of their hypothesis, Seely and Chirito showed that, in the presence of acetazolamide (Diamox), E_{Cl} and the measured transtubular potentials approach the same values at different locations along the proximal tubule.

The potential across the peritubular membrane of the distal convoluted tubule is greater, as is the transtubular potential (-50 mv) (Fig. 22-21). Na^+ still diffuses into the cell from the lumen, but along a less steep gradient, since the luminal fluid content of sodium has been considerably reduced to this point (to about 10 percent of the filtered sodium at the beginning of the distal tubule, and to about 2 percent at the end). Finally, the collecting duct reabsorbs almost all the remaining sodium.

The loop of Henle is also concerned with sodium and chloride transport. Discussion of this will be deferred until a later section, Excretion of Water: Urinary Concentration and Dilution Mechanisms.

Energy Supply of Active Sodium Transport. Extrusion of sodium from the cell against an electrochemical gradient requires expenditure of metabolic energy. The energy for this process is thought to be supplied by the high-energy phosphate bonds in adenosine triphosphate (ATP), which is a product of the cell's metabolism. It will be recalled (Chap. 1) that an enzyme located in the cell membrane can release this energy by catalyzing the splitting of ATP to form ADP and inorganic phosphate.

Potassium

Potassium clearance is usually well below that of inulin, suggesting fairly complete reabsorption. Under certain circumstances, however, entailed with giving large amounts of potassium salts of foreign anions, more potassium is excreted than is filtered. Thus, it

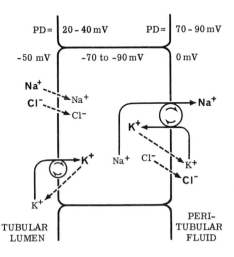

Figure 22-21
Model of distal tubular cell transport mechanisms for Na^+, K^+, and Cl^-. Note that K^+ movement in the luminal membrane has potential of being bidirectional. (From *Physiology of the Kidney and Body Fluids*, 3d ed., by R. F. Pitts. Copyright © 1974 by Year Book Medical Publishers, Inc., Chicago. Used by permission.)

appears that a three-component system operates with proximal tubular reabsorption and distal secretion. This conclusion is supported by micropuncture analysis. K^+ is actively reabsorbed by a pump in the tubular membrane of the proximal tubular cell. Figure 22-21 shows that the theoretical K^+ pump can operate either way in the distal tubule, reabsorbing K^+ under conditions of low intake or secreting K^+ during potassium loading.

The micropuncture studies of Giebisch et al. (1964) show this process more clearly in Figure 22-22. Progressive reabsorption of filtered K^+ is indicated in the proximal tubule. In the distal, the K^+ concentration drops to very low levels with a low-K diet (part A), suggesting continued reabsorption through the collecting duct. In B (normal diet), the rising concentration along the distal segment clearly demonstrates a secretory process.

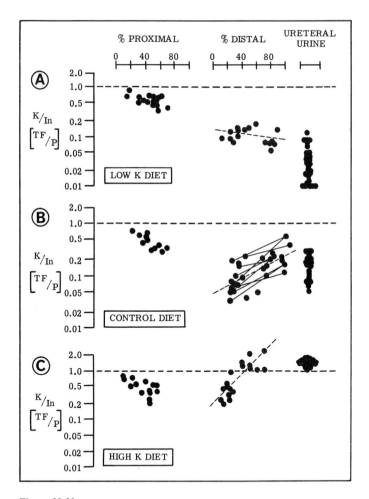

Figure 22-22
Potassium excretion along the nephron of rats on control diet (B), on low-K diet (A), and on high-K, low-Na diet (C). Potassium/inulin [TF/P] concentration ratios given as a function of tubular length. The K/IN [TF/P] ratio corrects for the influence of tubular water reabsorption. (From G. Malnic, R. M. Klose, and G. Giebisch, *Am. J. Physiol.* 206:674, 1964; and from G. Giebisch and E. E. Windhager, *Am. J. Med.* 36:634, 1964.)

This is magnified under the influence of a high-K diet (part C), when the K/IN[TF/P] ratio exceeds unity, indicating net secretion.

The distribution of potassium ions is determined largely by the transtubular potential gradient. In potassium deficiency, the transtubular potential at the secretory site is low, and little potassium is excreted. With a high-K diet, the converse is true.

K^+ secretion is also dependent on the availability of sodium. Although K^+ secretion was originally thought to be effected by a carrier system in exchange for reabsorbed sodium, it has been more recently viewed as an effect of luminal Na^+ concentration on the electrical potential, which in turn influences K^+ secretion. Thus a low sodium concentration is associated with a low potential gradient and a lower rate of K^+ secretion and vice versa (Wright, 1974).

The concomitant depression of reabsorption and increase in excretion of bicarbonate in hyperkalemia (see Chap. 24) results in the delivery of unreabsorbed anions to the secretory site—a condition that, in itself, increases the transtubular potential gradient and potassium secretion.

Calcium

Excretion of calcium is complicated by the fact that a significant part of its plasma content is combined with plasma proteins. However, urinary excretion is nominally low (about 8.5 μEq per minute), suggesting efficient tubular reabsorption of the moiety which is filtered. Parathyroid hormone (PTH) of the parathyroid gland promotes tubular reabsorption of calcium (Chap. 32).

Magnesium

Magnesium not bound by plasma proteins is filtered and reabsorbed. Excretion of about 5 to 6 μEq per minute occurs in man.

Other Electrolytes

In this category can be considered NH_3 and H^+. Ammonium is synthesized in the tubular epithelium from glutamine and other amino acids and is secreted into the tubular urine. It exchanges in this process with Na^+ as a mechanism in acid-base regulation and Na^+ conservation. Finally, the tubule is able to generate and secrete H^+ as a means of acidifying the urine by the aid of the enzyme carbonic anhydrase (further details in Chap. 24).

Anions

Chloride

Because chloride is the chief indifferent anion that accompanies Na^+ through the kidney tubular epithelium, this renal mechanism is conditioned by sodium reabsorption. Thus the Cl^- is pulled by its electrical charge as a result of the active transfer of Na^+. However, there is evidence (Chap. 23) that Cl^- may behave independently of Na^+.

It is now accepted that the thick ascending limb of Henle pumps Cl^- actively (Kokko, 1974; Burg and Stoner, 1974). Recently, Khuri et al. (1974) found intracellular concentrations of Cl^- equal to 42.3 $\pm$ 3.1 mEq per liter in latter distal tubule cells of rat, much higher than a passive equilibrium distribution would predict. Hence, they also postulated an active transport of Cl^- across the luminal cell border, since transport must proceed against both a chemical and an electrical gradient. An electrically neutral NaCl pump involving a double carrier combining with Na^+ and Cl^- simultaneously was considered as a possible alternative to the anion pump.

Bicarbonate

Data on bicarbonate reabsorption are expressed in milliequivalents per 100 ml of glomerular filtrate with an apparent physiological relationship between HCO_3^- reabsorption and filtration rate. As plasma HCO_3^- increases, the quantity filtered increases in direct proportion. Essentially all of the filtered HCO_3^- is reabsorbed until the load exceeds 2.8 mEq per 100 ml of filtrate, when frank excretion of HCO_3^- begins. The quantity reabsorbed remains constant at this value despite further increases in HCO_3^- plasma concentration, the excess being excreted. Thus, there is a grossly maximal reabsorptive capacity which is reached at a filtered load of about 2.8 mEq per 100 ml of glomerular filtrate per minute. The critical value of plasma HCO_3^- required to cause frank excretion is about 27 mEq per liter in man. The role of bicarbonate reabsorption in acid-base regulation will be considered in Chapter 24.

Phosphate
In the plasma, about 80 percent of phosphate occurs as $HPO_4^=$ with about 20 percent as $H_2PO_4^-$. Both are usually combined with Na^+. The ultimate ratio of $H_2PO_4^-/HPO_4^=$ in the urine is determined by the pH. In reabsorption of phosphate there is manifested a Tm which has a magnitude of about 0.13mM per minute. Reabsorption is in the proximal convoluted tubule. Excretion in man is from 7 to 20μM per minute. Excretion is modified by PTH action (Chap. 32), which tends to block reabsorption.

Sulfate
Sulfate is actively reabsorbed in both dog and man and shows a well-defined, although small, Tm. It is 0.05mM per minute in the dog and 0.110mM per minute in man.

EXCRETION OF WATER: URINARY CONCENTRATION AND DILUTION MECHANISMS
The mechanisms that concern tubular reabsorption of water are intimately related to the handling of osmotic constituents, primarily NaCl and urea. Another factor is the action of the ADH which is elaborated by the supraoptic and paraventricular nuclei of the hypothalamus. Finally, the composite mechanisms are integrated in the light of a *countercurrent diffusion multiplier system,* operating particularly in the nephrons which project long loops of Henle and vasa recta into the papillary zones of the renal medulla. This anatomical arangement is especially exemplified by the kidneys of the golden hamster and kangaroo rat but is applicable also to some of the nephrons of the white rat, the dog, and the human kidney.

The basis for this theory is the finding that the osmotic constituents of the kidney are arranged so that they are isosmotic with plasma in the cortex, with the concentration rising three to four times and even more in certain species, at the tip of the papilla. Using a cryoscopic method which depends upon a polarizing microscope to detect tiny crystals of ice formed during the cooling of sections of tissue of 30-μm thickness, Wirz et al. (1951) found that points of equal osmotic pressure form shells concentric to the tip of the papilla and parallel to the interzonal boundary. Uniform distribution of osmotic constituents among the loops of Henle, vasa recta, and collecting tubules has been proved by micropuncture in several animal species. Thus as the osmotic concentration progressively increases from the cortex to the papillary tip, at any given level it is approximately equal in the loop of Henle, the vasa recta, and the collecting tubule.

The arrangement of the loop of Henle system and vasa recta, so that the currents of the fluids in the two limbs of this hairpin flow in opposite directions, has given a foundation for the therory of the *hairpin countercurrent system.* Basically, this operates because of the principle of countercurrent exchange, which has several applications in thermodynamics. The principle of countercurrent exchange has been used by heating engineers in furnace exhaust and air-intake ducts. If the ducts lie side by side, warm air moving from the inside to the outside passes along the duct which brings the cold air from the outside. The heat diffuses across to the incoming cold air, raising its temperature, while the air being exhausted to the outside is cooled. Thus, overall heat conservation results (Fig. 22-23A). The principle is utilized for conservation of heat in the living organism by means of the countercurrent arrangements of the blood vessels in

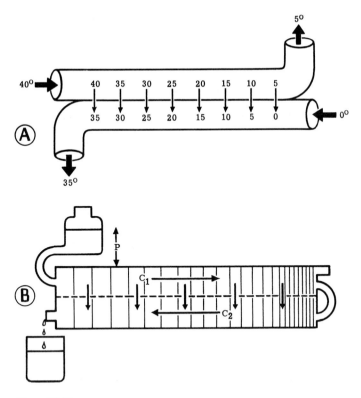

Figure 22-23
A. Diagram of the principle of countercurrent exchange. B. Model illustrating the principle of the countercurrent multiplier. (B from B. Hargitay and W. Kuhn. Z. *Elektrochem.* 55:539, 1951.)

the limbs (see Chap. 28 for further discussion). When the two systems are connected at one end, an opportunity for a *countercurrent multiplier system* is created. This is illustrated by the use of a physical model in Figure 22-23B. Suppose that two parallel circuits are separated by a semipermeable membrane. A solution is driven through the circuit by the hydrostatic pressure (P) of an elevated reservoir. Because the membrane is permeable to water, but not to solute, water is driven across by the hydrostatic force, indicated by the vertical arrows. As fluid leaves the upper tube, the solute (C_1) becomes progressively concentrated toward the end. On making the turn, the highly concentrated solution (C_2) becomes rediluted as it picks up the water flowing through the membrane.

The principle applies to the kidney, but the forces are different, involving changes in osmotic pressure based upon the movement of ions and solutes, rather than hydrostatic pressure forces. The analysis of a model corresponding to the loop of Henle and the collecting duct in Figure 22-24 serves to clarify one possible mechanism. In A, if it is considered that the membrane is unidirectionally permeable (that is, that Na^+ can move from the ascending limb across to the descending limb), the opportunity for countercurrent multiplication exists. In the classic view, the Na^+ from the ascending limb added to the isotonic fluid of the descending limb increases the tonicity of the fluid approaching the hairpin turn. This process continues as the hypertonic fluid rounds the turn. The process is multiplied over and over again, and the Na^+ is, in effect, trapped in relatively high concentration at the tip. Thus, the fluid would be isotonic entering the decending limb, hypertonic at the bend, and hypotonic issuing at the upper end of the ascending limb.

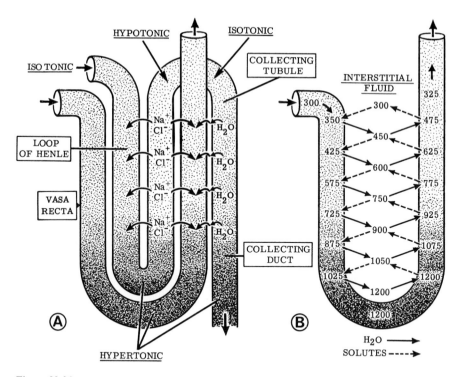

Figure 22-24
A. Relationship of vasa recta to the loop of Henle and the collecting duct, to illustrate role in removal of sodium and water. B. Operation of the countercurrent vascular loops in preserving the osmotic gradient in the renal medulla.

The mechanism for final concentration of the urine becomes more apparent if a tube representing the collecting tubule of the nephron is added to this system. If the upper loop to the right, analogous to the distal convoluted tubule, and the descending tube, analogous to the collecting tubule and duct, are made water-permeable, then urinary concentration becomes possible. It will be seen later that this occurs because of the action of ADH.

Removal of water to the interstitium of the kidney (from the upper right-hand loop analogous to the distal convoluted tubule) establishes isotonicity in this region, but now water is drawn from the collecting tubule and duct to the zone of hyperosmolality established by the countercurrent multiplier system, so that the fluid issuing from the descending right-hand limb, analogous to the collecting duct, becomes hypertonic.

Although this model illustrates the basic principles involved in the countercurrent multiplier as it is utilized in the concentration of the urine, some immediate technical difficulties becomes apparent from the use of the simplified model. One is that the volume of the ascending limb of the loop of Henle and the distal convoluted tubular system would be increased by the return of water from the collecting duct. An extremely important consideration is that interstitial fluid is interposed between the loops of the tubule, and that the vasa recta lie in juxtaposition to the loop of Henle system and the surrounding interstitial fluid. Another important facet is the finding that water leaves the descending limb of the loop of Henle, as indicated by micropuncture studies which show that the inulin U/P ratio increases to the tip of the papilla in the hamster kidney, and probably also in the rat and rabbit, proving that water has left the descending limb of the loop. This would contribute to the operation of the countercurrent multiplier system by producing an increment of osmotic pressure in the descending limb.

Last, consideration should be given to the distribution of osmolality in the vasa recta system. Since their course parallels the loop of Henle system, and they are freely permeable to water and solutes, the vasa recta should manifest the same stratification of osmotic pressure. This then becomes the final important link in the mechanism, for it permits the water taken out of the collecting tubule and duct to be picked up by the plasma moving through the vasa recta system and carried away into the venous channels of the kidney. The force is the colloid osmotic pressure of the plasma protein. The active reabsorption of sodium by the collecting tubular cells may contribute to hyperosmolality of the vasa recta system.

The vasa recta operate also as *countercurrent diffusion exchangers* in terms of preserving the osmotic gradient in the renal medulla (Fig. 22-24B). Blood enters the loop with a concentration of 300 mOsm per liter. As it dips down, water diffuses out, and osmotically active particles (NaCl, urea, ammonia) diffuse in from the medullary interstitium, increasing the concentration in the descending portion of the loop as this process is repeated. Then, as blood traverses the loop and ascends, osmotically active particles diffuse out, and water enters, thus reducing the gradient. The net effect is to trap solutes in the tip of the vascular loop, in equilibrium with the loop of Henle, in a concentration about four times that of the entering blood. Because of the short-circuiting effect of water across the tops of the vascular loops, red cells and plasma proteins are also concentrated, the latter aiding in picking up water reabsorbed from the collecting duct, facilitated by ADH action.

The situation as it exists in the nephron system is illustrated in Figure 22-25, based on tubular puncture techniques. Osmolality values have been arbitrarily assigned as they apply to the human kidney and become the basis for a more extensive recapitulation of the mechanism. The greatest proportion of the filtered sodium and attendant anions is reabsorbed by an active process in the proximal convoluted tubule (A). Water follows passively, not requiring mediation of ADH, so that the proximal tubular fluid remains essentially isosmotic. Micropuncture studies in the hamster and rat have revealed that 67 percent of the filtrate is absorbed in the proximal convolution (Lassiter et al., 1961). The descending limb of the thin segment of the loop of Henle does not transport salt actively. Another site of active NaCl reabsorption is the ascending limb of the loop of Henle, especially the thick portion, which is relatively water-impermeable. According to the newer view, chloride, by an active mechanism, and Na^+ as a result of the electrochemical gradient established, are transported into the interstitium of the outer medulla until a gradient of perhaps 200 mOsm per kilogram of water has been established between the fluid of the the ascending limb and the interstitium. This single effect is multipled as the fluid in the thin descending limb comes into osmotic equilibrium with the interstitial fluid by the diffusion of water out of and possibly the diffusion of some NaCl into the descending loop, thus raising the osmolality of the fluid making the hairpin turn to be presented to the ascending limb. That the descending limb is Na^+-permeable has been questioned recently (Kokko, 1974). Progressive concentration of NaCl and other solutes in the tubular fluid of this segment of the nephron could result largely, if not entirely, from water diffusion into the hypertonic interstitium. In any event, an increasing osmotic gradient is established in the direction of the tip of the papilla (B), and yet at no level is there a large osmotic difference between the luminal and interstitial fluid. An additional 10 to 15 percent of the filtered water is reabsorbed in the pars recta and descending limb of the loop of Henle (passive efflux), and 15 percent in the distal convolution (C), aided by ADH, leaving about 5 percent of the original volume to enter the collecting tubules.

In contrast to the ascending limb, the epithelium of the collecting tubules (D) in the presence of ADH is also believed to be water-permeable, resulting in diffusion of water out of the collecting tubules and ducts into the hyperosmotic medullary interstitium. The volume of fluid remaining in the collecting ducts becomes correspondingly con-

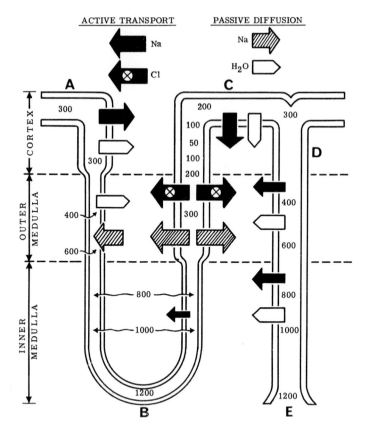

Figure 22-25
Diagram of nephron showing sites of sodium, chloride, and water reabsorption. Figures represent
the distribution of osmolality values. It seems likely that equilibrium (return of osmolality to 300) is
probably not achieved in the distal convoluted tubule of the primate kidney. Although osmolality
values are shown in the simplified diagram to be approximately equal in the thin descending and
thin ascending limbs of corresponding levels of the loop of Henle, the ascending limb values
actually average approximately 100 mOsm lower than the descending, as reported by Jamison et al.,
1967.

centrated (E). The degree to which distal tubular fluid is restored to isotonicity
(300 mOsm per liter) appears to vary with the species. This is achieved in the rat and
hamster kidney before the distal convolution is left, but not in the dog and monkey
kidney, where restoration to isotonicity is delayed until the collecting tubule (D).

For the urine to be significantly concentrated, the flow through the loops of Henle
must exceed considerably the flow through the collecting ducts. This change is accomplished by the action of ADH, which facilitates diffusion of water out of the distal
convolution (C) into the interstitium of the cortex, thereby reducing the volume and
increasing the osmolality to the isosmotic level before the fluid is presented to the
collecting tubules. The rapid flow of blood through the cortex facilitates this removal.

Although the sodium salts are important, urea probably also functions importantly in
this system, from present evidence. It has been proposed that the urea diffuses out of the
medullary collecting tubules and the collecting ducts, to be concentrated in the loop of
Henle by countercurrent exchange, adding to the gradients established in the loop.

Ammonia produced in the proximal tubule also becomes concentrated in the papilla by operation of the medullary countercurrent exchange system.

To work most effectively, the medullary blood flow should be low, and experimental evidence at the present time indicates that it is. Probably the osmotic equilibration of the plasma in the vasa recta with the medullary interstitial fluid is due not only to the diffusion of solute into the descending and out of the ascending limbs but also largely to the diffusion of water in the opposite direction. This short-circuiting across the tops of the vascular loops may be in part the cause of the apparently rich content of cells and plasma protein which has been found in the vasa recta at the tip of the papilla.

The importance of the vasa recta is that the blood entering the medulla picks up not only water that diffuses out from the thin descending limb of the loop of Henle but also that from the collecting ducts. The water movement to the vasa recta system is the result of the favorable gradient established by the electrochemical potential created by the colloid osmotic pressure of the plasma proteins. If indeed, as seems to be the case, the plasma proteins are concentrated in this zone, the greater effectiveness of this mechanism becomes apparent.

The next question to consider is the mechanism in the diluting kidney. This is the kidney that operates in absence of ADH, as occurs following water ingestion. In rats in which the final urine was quite dilute, Wirz (1956) found that the fluid remained dilute throughout the distal convolution and the collecting ducts. Thus an important role of ADH is that it renders the epithelium of the collecting tubules and ducts and perhaps the distal convolution freely permeable to water, so that the osmotic gradients established can operate in the presence of the hormone and are reduced or eliminated in the absence of the hormone. Studies with tritiated water (HTO) show that ADH acts by increasing the permeability of the luminal surface of the cellular membranes of the distal convolution and/or collecting ducts to water diffusion, by enlarging the mean functional diameter of so-called pores within the luminal portion of the membrane of the epithelial cells.

ADH may have an additional role in regulating medullary blood flow. It is well known that ADH is a vasoconstrictor. Reduction in medullary blood flow should favor further increase in osmolality of the papillary tip and thus favor further concentration of the urine. On the contrary, increased blood flow through the medullary circuit would tend to "wash out" the hypertonic zone, thus blunting the capacity for water uptake from the collecting tubule as outlined above.

Passive Equilibration Model of Countercurrent System

The classic description of the countercurrent mechanism has been given above in some detail. The theory was developed largely in the absence of direct measurements of electrolyte and osmotic constituents in the loop of Henle, particularly in the outer medulla, the site of the medullary thick ascending limb (MTAL) of Henle. Recently, a successful technique of perfusing the loop of Henle in vitro (Kokko, 1974; Burg and Stoner, 1974) has permitted measurement of transtubular potentials, net fluid flux, and net movement of solute (e.g., urea, and Na^+ and Cl^-). The investigating groups used the isolated loop of Henle of the New Zealand rabbit. An important unexpected finding was that Cl^- was actively pumped in the MTAL, accompanied passively by Na^+; as expected from this, a positive transtubular potential was found here (+6.7 mv—Kokko, 1974). The correlated demonstration that the thin descending limb (DL) of Henle had a high degree of osmotic water permeability, but a low permeability to NaCl and urea, led to a revision of the classic concept, referred to by Kokko as the "passive equilibration model of the counter-current multiplication system." The model achieves the increasing gradient of osmolality and osmotic equilibration principally by water extracted from the thin DL without requiring net entry of solute (NaCl, urea). The important interstitial osmotic ingredient is urea, diffusing in from the distal collecting tubule and duct (Figs. 22-16 and 22-26). Nevertheless, NaCl attains a high concentration at the bend of the

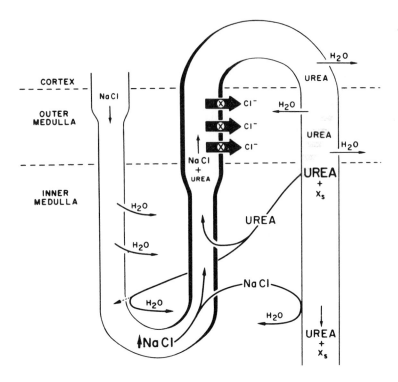

Figure 22-26
Passive equilibration model of the countercurrent system. For explanation, see text. X_S = solutes in excess of urea. (From J. P. Kokko. Reprinted from *Federation Proceedings*, 33:25–30, 1974.)

thin segment because of outward diffusion of water (Fig. 22-26). In the water-impermeable thin ascending limb (AL), NaCl now diffuses outward, down its gradient, to increase the interstitial solute concentration of the inner medulla, along with the urea diffusing into the interstitium from the collecting duct. The active pumping of Cl^-, in the water-impermeable MTAL, accompanied by Na^+, further contributes to the osmotic gradient for H_2O abstraction at this level but leaves behind a hypotonic luminal fluid.

Urea entering the ascending limb of Henle is concentrated in passage through the urea-impermeable distal convoluted tubule and early collecting tubule by water abstraction (favored by ADH action), so that as the tubular fluid enters the inner medulla a more favorable gradient for outward diffusion of urea is created (Fig. 22-26). The final concentration or dilution of the urine depends, as before, on the relative concentration of ADH resulting from the existing state of fluid balance.

In summary, the main features of the "passive equilibration model" which distinguish it from the classic model are as follows: (1) Solute entry (chiefly NaCl) into the thin DL is not required (urea enters slightly at the tip); (2) water abstraction occurs in the relatively solute-impermeable thin DL by water diffusing down its concentration gradient; (3) NaCl diffuses passively out of the thin AL, rather than being actively pumped; (4) the energy-requiring single effect is localized in the MTAL, and is initiated there by the active pumping of Cl^-, with Na^+ following passively; (5) greater emphasis is placed on the urea "recycling" mechanism of the distal nephron and, hence, the important contribution of urea to the mechanism of development of the osmotic gradient.

Critique of the Passive Equilibration Model
Although conceptually satisfying as an energy-conserving biological system, this model presents a difficulty recently pointed out by Marsh and Azen (1975), namely, whether or not the passive mechanism makes a quantitatively significant contribution to the operation of concentrating the urine. Thus, although the mechanism might serve to maintain the gradient of osmolality, it is hard to see how it can develop the gradient fully, or restore it after a washout (e.g., diuresis, hemorrhagic hypotension). These workers reconfirmed in hamsters earlier concepts by providing evidence supporting active transport of NaCl by the thin AL but agreeing that passive transport from the thin AL is not negligible. The concept of passive influx of urea into the thin DL, and probably also entry of NaCl, was supported by their findings.

Pennell et al. (1974) have reexamined the evidence in the rat and likewise concluded that net solute addition contributed substantially to the increase in osmolality of the DL; urea was the principal solute added. Water extraction from the DL also played a significant part (accounting for 60 to 67 percent of the concentration of solutes). Solute entry (urea, electrolytes) accounted for 33 to 40 percent. A preponderant role is played by water abstraction (96 percent) in the Kokko model.

The Concept of Osmolar Clearance and Free Water Clearance

The highest concentration of urine achieved by the human kidney is about 1200 to 1400 mOsm per liter, compared with other mammalian kidneys that can concentrate to higher maxima (e.g., dog, 2500; white rat, 3000; and kangaroo rat, 5000). The lowest concentration observed is about 40 mOsm per liter. In terms of the more commonly employed clinical measurement, the *specific gravity,* this concentration involves a range between about 1.035 and 1.002 for human urine as compared with 1.010 for plasma. Understanding of the mechanism of change in urinary concentration requires the introduction of several concepts which involve the expressions for *free water clearance* and *osmolar clearance.* The volume of water required to contain all the urine solutes in an isosmotic solution with contemporaneous plasma is designated as the osmolar clearance (C_{Osm}). It is given by the equation

$$C_{Osm} = \frac{U_{Osm} \dot{V}}{P_{Osm}} \tag{6}$$

Free water clearance (C_{H_2O}) is the difference between an osmolar clearance and the rate of urine flow:

$$C_{H_2O} = \dot{V} - C_{Osm} \tag{7}$$

When C_{H_2O} is positive, as in water diuresis, the urine is more dilute than plasma; when C_{H_2O} is negative, as in dehydration, the urine is more concentrated than the plasma. Negative C_{H_2O} is more conveniently expressed by the designation Tc_{H_2O}.
To eliminate negative terms, the equation for Tc_{H_2O} can be written

$$Tc_{H_2O} = C_{Osm} - \dot{V} \tag{8}$$

The mechanism of formation of osmotically concentrated urine in hydropenic subjects has been studied by giving additional ADH by Pitressin infusion. The subjects were then given an infusion of hypertonic mannitol solution to produce a progressively increasing osmotic diuresis. Figure 22-27 illustrates the result of plotting of osmolar clearance against urine flow. The dashed line in the figure represents the isosmotic parameter—i.e., the line that would be generated if the osmolar clearance equaled the urine volume at all values, in which case the urine would always be isosmotic with the plasma. The actual line is shifted to the left in the concentrating kidney, indicating that

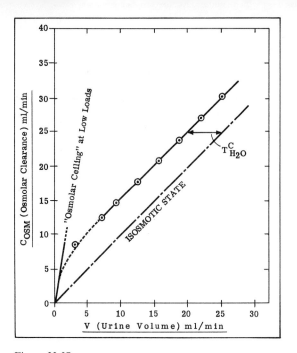

Figure 22-27
Relationship of osmolar clearance (C_{Osm}) to rate of urine flow during osmotic diuresis. (From Smith, 1956. P. 121.)

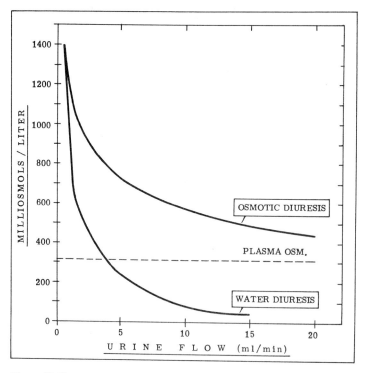

Figure 22-28
Urinary osmolar concentration related to urine flow during osmotic diuresis and water diuresis.

the urine is hypertonic. The horizontal distance between these lines, labeled Tc_{H_2O}, represents the amount of pure water which, if added to the concentrated urine, would restore it to isosmolarity, or conversely, that which must have been removed from the isosmotic tubular fluid to result in the concentrated urine. In this case, the value for Tc_{H_2O} is 5 ml per minute at the higher rates of flow and clearance. The removal of the same volume of pure water from varying volumes of originally isosmotic fluid would lead to a progressive fall in urinary concentration as solute load and flow increase during osmotic diuresis. Thus, removal of 5 ml of pure water from 10 ml of isosmotic tubular fluid would give a final U/P ratio equal to 2, whereas removal of 5 ml from 30 ml of tubular fluid would concentrate the urine to only 1.2, because the amount of solute concentrated in 5 ml of tubular urine would now be dissolved in 25 ml (hence, U is one-fifth of the original concentration instead of twice, and U/P is 1.2 instead of 2.0). Any substance which will institute an osmotic diuresis will produce the same effect. Thus, increasing osmotic diuretic activity would result in increased amounts of isosmotic fluid passing from the proximal system to the loop of Henle and distal system. With removal of a fixed amount of pure water, the U/P ratio would decline (Fig. 22-28).

At very low solute loads and low urine flows, the value for Tc_{H_2O} decreases, the limiting factor being an osmolar ceiling above which the urine cannot be concentrated (at about 1400 mOsm per liter), representing an osmotic U/P ratio of approximately 4.7 (Figure 22-29).

When it is dilute, the urine flow exceeds the osmolar clearance, and the difference becomes the *free water clearance*, C_{H_2O}, which represents the *volume of pure water that originally contained the salt actively reabsorbed in the ascending limb of Henle, the distal convoluted tubule, and the collecting duct.* The capacity of these segments to reabsorb sodium salts, therefore, sets the limit on the amount of water which can be freed and excreted *above that containing the remaining urinary solutes in isotonic solution.*

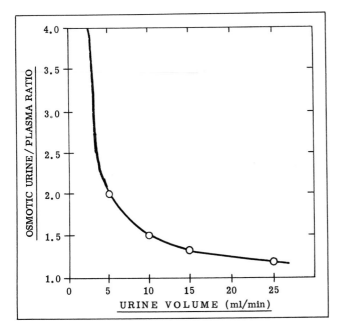

Figure 22-29
Osmolar U/P related to urine volume during osmotic diuresis. (From Smith, 1956. P. 121.)

The role of ADH is primarily to convert the dilute urine of the distal system to isotonicity and then to hypertonicity. The relative quantitative importance of the dilution and concentration processes is illustrated by considering the excretion of 2 liters of isotonic urine in 24 hours. If the subject were to produce maximally diluted urine (absence of ADH), he would excrete 20 liters of urine per 24 hours. Therefore, if ADH were to effect a rise in urinary concentration only to isotonicity, 18 liters of water would be conserved. If the solute were to be excreted in maximally concentrated urine (maximal ADH effect), the urine volume would be reduced to 500 ml, with a further saving of only 1.5 liters. Viewed in this light, it is clear that water conserved by not putting out dilute urine is quantitatively much more important than that conserved in forming concentrated urine.

Discussion of the renal role in body water balance and the control systems related thereto will be deferred until the latter part of Chapter 23, after the facts concerning body water and electrolyte distribution have been considered.

REFERENCES

Arendshorst, W. J., and E. E. Selkurt.   Renal tubular mechanisms for creatinine secretion in the guinea pig. *Am. J. Physiol.* 218:1661–1670, 1970.

Balagura, S., and R. F. Pitts.   Renal handling of α-ketoglutarate by the dog. *Am. J. Physiol.* 207:483–494, 1964.

Beeuwkes, R., and Bonventre, J. V.   Tubular organization and vascular-tubular relations in the dog kidney. *Am. J. Physiol.* 229:695–713, 1975.

Brenner, B. M., J. L. Troy, and T. M. Daugharty.   The dynamics of glomerular ultra-filtration in the rat. *J. Clin. Invest.* 50:1776–1780, 1971.

Burg, M., and L. Stoner.   Sodium transport in the distal nephron. *Fed. Proc.* 33:31–36, 1974.

Chinard, F. P.   Derivation of an expression for the rate of formation of glomerular fluid (GFR). Applicability of certain physical and physico-chemical concepts. *Am. J. Physiol.* 171:578–586, 1952.

Deen, W. M., C. R. Robertson, and B. M. Brenner.   Glomerular ultrafiltration. *Fed. Proc.* 32:14–20, 1974.

Diedrich, D. F.   The comparative effects of some phlorizin analogs on the renal reabsorption of glucose. *Biochim. Biophys. Acta* 71:688–700, 1963.

Foulkes, E. C.   Kinetics of p-aminohippurate secretion in the rabbit. *Am. J. Physiol.* 205:1019–1024, 1963.

Giebisch, G.   Measurements of electrical potential differences on single nephrons of the perfused Necturus kidneys. *J. Gen. Physiol.* 44:659–678, 1961.

Gilmore, P. J.   *Renal Physiology.* Baltimore: Williams & Wilkins, 1972.

Gottschalk, C. W., and M. Mylle.   Micropuncture study of the mammalian urinary concentrating mechanism: Evidence for the countercurrent hypothesis. *Am. J. Physiol.* 196:927–936, 1959.

Harris, C. A., P. G. Baer, E. Chirito, and J. H. Dirks.   Composition of mammalian glomerular filtrate. *Am. J. Physiol.* 227:972–976, 1974.

Hollenberg, N. K., M. Epstein, S. M. Rosen, R. I. Basch, D. E. Oken, and J. P. Merrill.   Acute oliguric renal failure in man: Evidence for preferential renal cortical ischemia. *Medicine* 47:455–474, 1968.

Huang, K. C., and R. L. Woolsey.   Renal tubular secretion of L-glucose. *Am. J. Physiol.* 214:342–347, 1968.

Jamison, R. L., C. M. Bennett, and R. W. Berliner.   Countercurrent multiplication by the thin loops of Henle. *Am. J. Physiol.* 212:357–366, 1967.

Khuri, R. N., S. K. Agulian, and K. Bogharian.   Electrochemical potentials of chloride in distal renal tubule of the rat. *Am. J. Physiol.* 227:1352–1355, 1974.

Kirschenbaum, M. A., N. White, J. H. Stein, and T. F. Ferris.   Redistribution of renal cortical blood flow during inhibition of prostaglandin synthesis. *Am. J. Physiol.* 227:801–805, 1974.

Kleinzeller, A., J. Kolinska, and I. Beneš. Transport of glucose and galactose (and other monosaccharides) in kidney-cortex cells. *Biochem. J.* 104:843–860, 1967.

Kokko, J. P. Membrane characteristics governing salt and water transport in the loop of Henle. *Fed. Proc.* 33:25–30, 1974.

Ladefoged, J. Renal cortical blood flow and split function test in patients with hypertension and renal artery stenosis. *Acta Med. Scand.* 179:641–651, 1966.

Lang, F., R. Greger, and P. Deetjen. Handling of uric acid by the rat kidney. *Pflügers Arch.* 338:295–302, 1973.

Lassiter, W. E., C. W. Gottschalk, and M. Mylle. Micropuncture study of net transtubular movement of water and urea in nondiuretic mammalian kidney. *Am. J. Physiol.* 200:1139–1147, 1961.

Maddox, D. A., W. M. Deen, and B. M. Brenner. Dynamics of glomerular ultrafiltration: VI. Studies in the primate. *Kidney Int.* 5:271–278, 1974.

Marsh, D. J., and S. P. Azen. Mechanism of NaCl reabsorption by hamster thin ascending limb of Henle's loop. *Am. J. Physiol.* 228:71–79, 1975.

McKenna, O., and E. T. Angelakos. Adrenergic innervation of the canine kidney. *Circ. Res.* 22:345–354, 1968; Acetylcholinesterase-containing nerve fibers in the canine kidney. *Circ. Res.* 23:645–651, 1968.

McNay, J. L., and Y. Abe. Pressure-dependent heterogeneity of renal cortical blood flow in dogs. *Circ. Res.* 27:571–587, 1970.

Moffat, D. B. The fine structure of the blood vessels of the renal medulla with particular reference to the control of the medullary circulation. *J. Ultrastruct. Res.* 9:532–545, 1967.

Morel, F., M. Mylle, and C. W. Gottschalk. Tracer microinjection studies of effect of ADH on renal tubular diffusion of water. *Am. J. Physiol.* 209:179–187, 1965.

Nash, F. D., H. H. Rostorfer, M. D. Bailie, R. L. Wathen, and E. G. Schneider. Renin release: Relation to renal sodium load and dissociation from hemodynamic changes. *Circ. Res.* 22:473–487, 1968.

Pappenheimer, J. R. Über die Permeabilität der Glomerulummembrane in der Niere. *Klin. Wochenschr.* 33:362–365, 1955.

Pennell, J. P., F. B. Lacy, and R. L. Jamison. An *in vivo* study of the concentrating process in the descending limb of Henle's loop. *Kidney Int.* 5:337–347, 1974.

Pitts, R. F. *Physiology of the Kidney and Body Fluids* (3rd ed.). Chicago: Year Book, 1974.

Schneeberger, E. E., R. H. Levey, R. T. McCluskey, and M. J. Karnovsky. The isoporous substructure of the human glomerular slit diaphragm. *Kidney Int.* 8:48–52, 1975.

Seely, J. F., and E. Chirito. Studies of the electrical potential difference in rat proximal tubule. *Am. J. Physiol.* 229:72–80, 1975.

Selkurt, E. E. Sodium excretion by the mammalian kidney. *Physiol. Rev.* 34:287–333, 1954.

Selkurt, E. E. Renal Circulation. In W. F. Hamilton and P. E. Dow (Eds.), *Handbook of Physiology.* Washington: American Physiological Society, 1963. Section 2: Circulation, vol. 2, chap. 43, pp. 1485–1489.

Selkurt, E. E., R. L. Wathen, and J. Santos-Martinez. Creatinine excretion in the squirrel monkey. *Am. J. Physiol.* 214:1363–1369, 1968.

Smith, H. W. *The Kidney: Structure and Function in Health and Disease.* New York: Oxford University Press, 1951.

Smith, H. W. *Principles of Renal Physiology.* New York: Oxford University Press, 1956.

Sullivan, P. L. *Physiology of the Kidney.* Philadelphia: Lea & Febiger, 1974.

Taggart, J. V. Mechanisms of renal tubular transport. *Am. J. Med.* 24:774–784, 1958.

Thorburn, G. D., H. H. Kopald, J. A. Herd, M. Hollenberg, C. C. C. O'Morchoe, and A. C. Barger. Intrarenal distribution of nutrient blood flow determined with Krypton[85] in the unanesthetized dog. *Circ. Res.* 13:290–307, 1963.

Thurau, K. and J. Mason.   The Intrarenal Function of the Juxtaglomerular Apparatus. In K. Thurau (Ed.), *Physiology. Series One. Kidney and Urinary Tract Physiology.* Baltimore: University Park Press, 1974. Vol. 6, chap. 11, pp. 357–389.

Thurau, K., H. Valtin, and J. Schnermann.   Kidney. *Ann. Rev. Physiol.* 30:441–524, 1968.

Truniger, B., and B. Schmidt-Nielsen.   Intrarenal distribution of urea and related compounds: Effects of nitrogen intake. *Am. J. Physiol.* 207:971–978, 1964.

Valtin, H.   *Renal Function: Mechanisms Preserving Fluid and Solute Balance in Health.* Boston: Little, Brown, 1973.

Van Citters, R. L., and D. L. Franklin.   Cardiovascular performance of Alaska sled dogs during exercise. *Circ. Res.* 24:33–42, 1969.

Wirz, H.   Der osmotische Druck in den corticalen Tubuli der Rattenniere. *Helv. Physiol. Pharmacol. Acta* 14:353–362, 1956.

Wirz, H., B. Hargitay, and W. Kuhn.   Lokalisation des Konzentrierungsprozesses in der Niere durch direkte Kryoskopie. *Helv. Physiol. Pharmacol. Acta* 9:196–700, 1951.

Wright, F. S.   Potassium Transport by the Renal Tubule. In K. Thurau (Ed.), *Physiology. Series One. Kidney and Urinary Tract Physiology.* Baltimore: University Park Press, 1974. Vol. 6, chap. 3, pp. 79–105.

23. Body Water and Electrolyte Composition and Their Regulation. Micturition

Ewald E. Selkurt

Fluid Volumes

The principle of measurement of fluid volumes was introduced in Chapter 10, where the technique was explained in terms of its application to the determination of plasma volume. The same principle is employed for the measurement of total body fluid and extracellular volumes, based upon the use of a substance which distributes itself throughout the compartment to be measured. The equation is modified from that used for plasma volume, $V = \dfrac{A}{C}$, as follows:

$$V = \frac{A - E}{C}$$

where V is the volume in liters, A is the amount of the substance administered, E is the amount excreted in the urine at the time that C, the concentration per liter, is determined. The introduction of E, a subtraction for the amount excreted in the urine, is necessary because substances commonly used for these calculations are removed from the body by the kidney.

Total Body Water

Total body water (water in the *cellular* plus *intracellular* compartments) is measured as the volume of distribution in the body of an appropriate indicator after a single intravenous injection. The requirement is that the substance must distribute itself uniformly in all the fluid spaces. Apparently, the most reliable and commonly used indicators today are antipyrine and the heavy isotopes, deuterium oxide and tritium oxide. Confirmatory evidence has been supplied by desiccation of human cadaver material to constant weight at a temperature of approximately 105°C. On the basis of determinations with the indicators, total body water ranges from 500 to 600 ml per kilogram of total body weight (50 to 60 percent) for the average adult human subject. Desiccation studies reveal a figure of 640 ml per kilogram. The total body water volume for a 70-kg subject is about 42 liters. Some authorities feel it is more desirable to express the total body water in terms of the lean body mass, since adipose tissue is relatively deficient in water. On this basis, the generally accepted figure is 70 percent of the lean body mass.

The *extracellular* fluid compartment can be directly determined by the use of a substance which distributes itself throughout this particular compartment. Although there is some controversy as to the exact space such a substance measures, at the present time mannitol appears to be the best for measuring this space, and the distribution of sucrose, thiosulfate, and inulin supplies confirmatory evidence. One of the technical problems involved is whether or not the fluid of the connective tissue and bone should be considered part of the extracellular fluid. The tracer substances, mannitol, inulin, sucrose, and others, do not penetrate connective tissue fluid and bone water readily, nor is it likely that they enter the cerebrospinal fluid freely; hence, the space measured may be actually smaller than the true physiological extracellular space. The average figure of 180 ml per kilogram includes the plasma compartment, the interstitial-lymph space, and transcellular fluids. If dense connective tissue and cartilage, and inaccessible bone water, are added, the total becomes 270 ml per kilogram (Table 23-1). *Transcellular* water is that concerned with the transport activity of specialized cells (salivary, pancreas, liver,

Table 23-1. Body Water Distribution in a Healthy Young Adult Male*

Compartment	Percent of Body Weight	Percent of Total Body Water	Liters
Plasma	4.5	7.5	3
Interstitial-lymph	12	20	8.5
Dense connective tissue and cartilage	4.5	7.5	3
Inaccessible bone water	4.5	7.5	3
Transcellular	1.5	2.5	1
Total extracellular	27	45	19
Functional extracellular	21	35	14.5
Total body water	60	100	42
Total intracellular	33	55	23

*All figures rounded to nearest 0.5.
Source: From Edelman and Leibman, 1959.

mucous membrane of respiratory and gastrointestinal tract, cerebrospinal fluid, and ocular fluid). The *intracellular* water can be estimated as the difference between total body water and the total extracellular water. This would yield a volume of 23 liters.

As indicated in Chapter 10, the plasma volume can be measured by the distribution of suitable indicators such as T-1824 or iodinated albumin, giving a plasma volume of about 4.5 percent of body weight, or 3.2 liters for the average-size adult. The breakdown of the various volumes as a percentage of total body water appears in Figure 23-1.

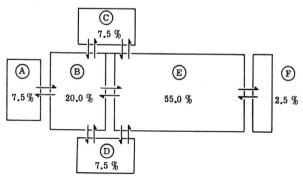

Ⓐ PLASMA WATER – 7.5 % OF BODY WATER

Ⓑ INTERSTITIAL – LYMPH WATER – 20.0 %

Ⓒ DENSE CONNECTIVE TISSUE and CARTILAGE WATER – 7.5 %

Ⓓ BONE WATER – 7.5 %

Ⓔ INTRACELLULAR WATER – 55.0 %

Ⓕ TRANSCELLULAR WATER – 2.5 %

Figure 23-1
Volumes of distribution of water in the body fluid compartments. (From Edelman and Leibman, 1959.)

Table 23-2. Composition of the Body Fluids

Substance	Serum (mEq/L)	Serum Water* (mEq/L)	Interstitial Fluid† (mEq/L)	Intracellular Fluid (mEq/L)
Na^+	138	148	141	10
K^+	4	4.3	4.1	150
Ca^{++}	4	4.3	4.1	—
Mg^{++}	3	3.2	3	40
Cl^-	102	109	115	15
HCO_3^-	26	28	29	10
$PO_4^{\equiv}$	2	2.1	2	100
$SO_4^=$	1	1.1	1.1	20
Organic acids	3	3.2	3.4	—
Protein	15	16	1	60

*Correction to concentration per liter of serum water is based on the value 93 percent.
†Values derived by use of Gibbs-Donnan factor of 0.95 for cations and 1.05 for anions in serum water.

Composition of the Extracellular Fluid

The principal constituents of extracellular fluid are shown in Table 23-2. Note that the concentrations of the interstitial fluid are corrected for the Gibbs-Donnan effect (Chap. 1). Thus, the essential differences between the serum water and interstitial fluid concentrations are the small inequalities resulting from the Gibbs-Donnan equilibrium effect. The exception is the much lower amount of protein which is found in the interstitial fluid. The important cation is Na^+, and the important anions are Cl^- and HCO_3^-. The last column supplies, for comparative purposes, the approximate values found in the intracellular fluid.

Specialized Extracellular Fluids: Aqueous Humor and Cerebrospinal Fluid

The aqueous humor of the eyeball and the cerebrospinal fluid (CSF) are examples of specialized interstitial fluids that cannot be described as simple filtrates from the plasma. Not only is each fluid different in composition from plasma dialysate, but in some respects they differ from each other. In general, CSF has higher Cl^- concentrations than plasma dialysate, while the concentrations of K^+ and Ca^{++} are significantly lower in this fluid. Na^+ is slightly lower in CSF than plasma ($R_{CSF} = 0.98$), but is higher than in the dialysate (0.98 compared to 0.945). The aqueous humor and CSF have similarities (lower urea and glucose concentrations than plasma) but differ from each other in concentrations of Cl^- and K^+. The ionic distribution between plasma and CSF in man and other species is presented in Table 23-3.

Because the solutes are not distributed as one might expect in an ultrafiltrate, the suggestion has been made that these fluids are secretions. Reasons for not viewing the differential concentrations as the result of simple filtration through a highly selective membrane have been advanced (Davson, 1967). Thus, the osmotic pressure required to maintain the difference in concentration of potassium between plasma and CSF can be calculated to be 54 mm Hg. The pressure required to maintain the difference in concentration of protein is 30 mm Hg. Consequently, the total pressure available must be 84 mm Hg, a figure well beyond capillary pressure. In other words, the supply of energy necessary to maintain this difference would be inadequate if the hydrostatic pressure in the capillaries were the sole force involved. Moreover, the concentration of Na^+ and Cl^- is higher in the CSF than in dialysate, as are HCO_3^- and ascorbic acid in

Table 23-3. Distribution of Various Ions Between CSF and Plasma

A. Concentrations of various solutes (mEq/kg H_2O) in plasma and lumbar cerebrospinal fluid of human subjects

Substance	Plasma	CSF	R^*_{CSF}
Na	150.00	147.00	0.98
K	4.63	2.86	0.615
Mg	1.61	2.23	1.39
Ca	4.70	2.28	0.49
Cl	99.00	113.00	1.14
HCO_3	26.80	23.3	0.87
Br	2.45	0.90	0.37
Inorg. P (mg/100 ml)	4.70	3.40	0.725
Osmolality	289.00	289.00	1.00
pH	7.397	7.307	—
P_{CO_2}	41.10	50.50	—

*R represents the ratio: concentration in spinal fluid–water/concentration in plasma–water.
Source: From Davson, 1967, p. 246.

B. Distribution of selected ions between CSF and plasma dialysate (R_{CSF}) in various species

Species	Na Actual	Na Dialysate	K Actual	K Dialysate	Ca Actual	Ca Dialysate	Mg Actual	Mg Dialysate	Cl Actual	Cl Dialysate
Dog	0.98	0.945	0.75	0.96	0.53	0.65	1.33	0.80	1.10	1.04
Goat	0.945		0.69		0.48		0.86		1.15	
Cat	0.97		0.65		0.57		0.99		1.09	
Rabbit	1.00		0.68		0.45		0.92		1.23	

Source: From Pollay, 1974, p. 2064.

the aqueous humor. These facts appear to deny simple filtration. The conclusion that active transport is involved appears warranted.

Formation and Drainage: Aqueous Humor. The ciliary body of the eye is considered the source of the aqueous humor. The highly vascular ciliary processes have been specifically assigned the role of production of fluid. The principal drainage route is the *canal of Schlemm*, which connects via very fine ducts (*collectors*) with the intrascleral venous plexus.

CSF. The CSF is elaborated by the choroid plexuses which project with many folds and villi into the roofs of the third and fourth ventricles, and into the sides of the lateral ventricles. The fluid passes from the lateral ventricles to the third ventricle through the foramen of Monro, then to the fourth ventricle via the aqueduct of Sylvius. It leaves the fourth ventricle through the foramina of Magendie and Luschka, to reach the subarachnoid spaces, here expanding to form the *cisterna magna*. From this region it passes into the dural sinus. Differing from the eye, wherein the canal of Schlemm enters directly via fine ducts into the venous system, the CSF is separated from the blood by the mesothelial lining covering the arachnoid villi.

Transport Systems Operating in Formation of CSF

The major interest in the choroid plexus had centered around the role of this structure in the elaboration of CSF. This crucial function of the plexus clearly has been supported by the demonstrated architecture of its tissue. The most significant finding, however, in the past decade has been the proved ability of plexus tissue to transport bidirectionally a number of biologically important substances. The current knowledge of some of the substances that are transported across the choroidal ependyma is summarized in Figure 23-2.

In the case of the ions, it is believed that actively pumped sodium is the main transported ion responsible for the diffusional transport of water from the blood to CSF across the lamina epithelialis. Magnesium and calcium appear also to be actively transported, as does the potassium ion. The potassium ion transport system is unique in that it saturates at one-half of the blood concentration level. The anion transport systems have been studied in depth and appear to operate in the direction of CSF to blood. This fact explains the efficiency of the cerebrospinal fluid as a "sink" for adjacent brain tissue, since a concentration differential gradient is maintained between these two compartments by the active transport of these anions out of CSF. Although the evidence

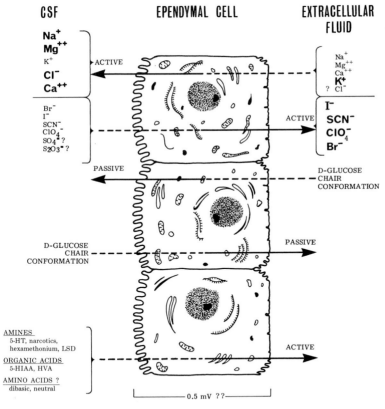

Figure 23-2
Summary of transport systems operating across the choroid plexus. *Active* and *passive* represent active transport and facilitated diffusionary processes, respectively. The size of ion symbols represents usual relative concentration levels in CSF as compared to blood or choroidal extracellular fluid. (From M. Pollay. Reprinted from *Federation Proceedings* 33:2068, 1974.) 5-HT = 5-hydroxytryptamine; HIAA = hydroxyindolacetic acid; HVA = homovanillic acid; LSD = lysergic acid diethylamide.

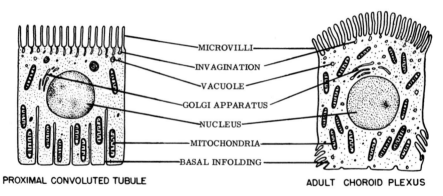

MICROVILLI
INVAGINATION
VACUOLE
GOLGI APPARATUS
NUCLEUS
MITOCHONDRIA
BASAL INFOLDING

PROXIMAL CONVOLUTED TUBULE

ADULT CHOROID PLEXUS

Figure 23-3
Structural similarities of the ependymal epithelial cell of the choroid plexus and a typical proximal convoluted tubular cell.

seems reasonably complete for the bidirectional transport of glucose by facilitated diffusion, the situation for amino acids is unclear at present. The lower part of Figure 23-2 shows a number of other transport systems that are of some pharmacological interest.

The histological similarity of the ependymal cell to the proximal tubular cell of the kidney is worthy of note (Fig. 23-3). Thus, the choroid epithelial cells have a "brush border" and basal infoldings. Cytoplasmic organelles, similar to those in the kidney, may suggest a role in water transport. From a functional aspect, transepithelial potentials are of the same order of magnitude as is seen in the proximal tubular cells.

At this time it can be stated that from both an anatomical and a physiological standpoint the choroid plexus indeed functions as a miniature kidney, and Pollay (1974) has called its secretory product the "neural urine," in the light of the excretory aspects of its function. Clearly, the cerebrospinal fluid is continuous with, and similar in composition to, the interstitial fluid of the brain. Therefore, the blood-CSF barrier (represented by the choroid plexus) and the blood-brain barrier both influence the neural environment.

Composition of Intracellular Fluid

The predominant cations of the intracellular fluid are K^+ and Mg^{++}. The predominant anions are organic $PO_4^{\equiv}$, $SO_4^{=}$, and proteins (see Table 23-2). It should be stated that these are only approximations of the composition of cell fluid and are based on the findings in muscle. Inadequate data are available to form conclusions concerning the complex forms in which the materials exist, the valence of the organic anions, the degree of dissociation of the compounds, and the extent to which the cations are bound, etc. In any event, the osmolar balance between the intracellular and extracellular fluids is operationally equivalent. A bar graph summary of the distribution of cations and anions for the fluids of the body is presented in Figure 23-4.

FLUID EXCHANGE

Exchange Between Extracellular and Intracellular Compartments

Most cell membranes are apparently completely permeable to water, and the total exchange is enormous. The *net exchange* of water is governed by the osmotic pressure changes in the two compartments. When osmotic pressure changes, water moves across

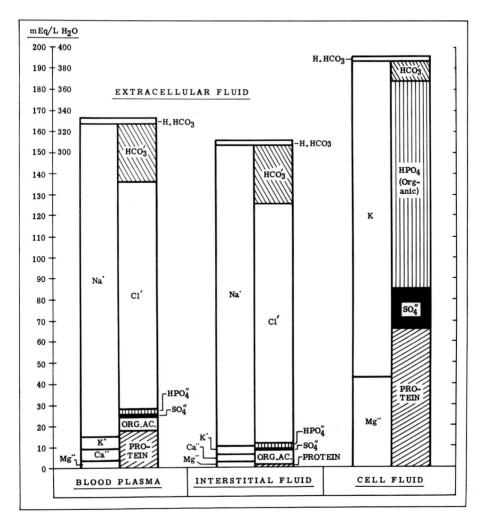

Figure 23-4
Ion distribution in the fluid compartments of the body. (From J. L. Gamble. *Chemical Anatomy, Physiology and Pathology of Extracellular Fluid* [6th ed.]. Cambridge, Mass.: Harvard University Press. Copyright © 1942, 1947, 1954 by the President and Fellows of Harvard College.)

the cellular membrane to maintain an isosmotic state. The situation is diagrammed in Figure 23-5. Part A shows the normal situation—i.e., concentrations of 300 mOsm per liter both within the cell (C) and in the extracellular compartment (E). Part B shows the result of adding 300 mOsm to compartment E. Water moves out of the cell (space C) to dilute the concentration of the materials in E, so the final concentration in each compartment will be somewhat less than 600 mOsm per liter. The net effect will be cellular dehydration. Part C illustrates what will happen if the osmolarity of the extracellular compartment is reduced to 150 mOsm per liter. The cellular fluid is hyperosmolar, and water moves in. Both compartments will become isosmotic, and a value greater than 150 and less than 300 mOsm per liter will be attained, because of the resulting movement of water from compartment E into C. The cell swells.

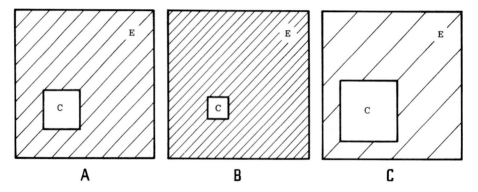

Figure 23-5
Regulation of water exchange between the extracellular and intracellular compartments. The representative volume of extracellular fluid, E, is taken as 1 liter. C = intracellular volume. A. Normal (300 mOsm per liter in both E and C). B. Effect of addition of 300 mOsm to E (initially 300 mOsm per liter). C. The osmolarity of E is reduced to 150 mOsm per liter; C is 300 mOsm per liter. The result is net water gain by the cell as fluid moves into the relatively hyperosmolar zone, C.

Exchange Between Vascular and Interstitial Compartments

When fluid moves between the vascular and interstitial compartments, the total exchange of water and salts in solution is enormous because the capillaries are highly permeable to water and contained solutes, and the area of exchange is great. The total surface area of the capillary bed (muscle) in man has been estimated at 6000 square meters. The limiting factor to diffusion is, in fact, the rate of circulation of the blood through the tissue. However, the net exchange is small because of the normal maintenance of volume of the interstitial and plasma compartments within relatively fixed limits. The protein concentration of the plasma becomes very important in the net exchange of fluid. It will be recalled that the osmotic pressure exerted by the proteins opposes the hydrostatic filtering forces that exist in the capillaries, so that, according to the Starling hypothesis (Chap. 12), the filtration of water is opposed by the retaining action of the plasma protein. Stated simply, the fluid that has been filtered outward at the arteriolar end of the capillaries is returned by the effective gradient of plasma protein at the venular end. The importance of the role of the lymphatics in the overall net water exchange between the capillaries and the interstitial space should be emphasized. The lymphatic system is important in draining off excesses of interstitial fluid.

ELECTROLYTE EXCHANGE

Exchange Between Extracellular and Intracellular Compartments

As was indicated earlier in this book (Chaps. 1, 2, 3, 14, and 22), the exchange of electrolytes between the intracellular and extracellular compartments is a complicated problem. It has been discussed from the standpoint of the function of nerve and muscle cells. The difference in ionic concentration between these two compartments, which reaches tremendous proportions in some instances, must be viewed in terms of mechanisms and operations that require the expenditure of energy. In brief summary, some of the ions under consideration move against the concentration gradient in a transport system that has the following characteristics: (1) a carrier compound which forms a reasonably stable complex with an ion, and (2) a chemical system which can react with this ion carrier compound so as to release the ion from its association with the carrier. Some ions may be held in high concentration in the cell, at least in part, in an intimate and stable relationship with other large complex molecules—for example, the relationship between potassium and myosin in muscle cells.

The thorough treatment of this topic in the above-cited chapters obviates the need for greater development here. The reader is referred to these areas for detailed presentation of the facts.

Exchange Between Vascular and Interstitial Compartments

The movement of electrolytes and other solutes between the vascular compartment and the interstitial compartment is limited only by the movement of the water in which they are in solution. Superimposed upon this is the slight influence exerted by the Gibbs-Donnan effect.

WATER BALANCE

Water balance is achieved between the forces responsible for *intake* of water as manifested by thirst, and the metabolic water and contained water of ingested food-stuffs, and *water loss* through various organs of the body, namely, the skin, the lungs, the gastrointestinal tract, and the kidneys.

Water Intake

Thirst is a subjective sensory impression that results in ingestion of water. Physiological factors that seem to be involved are primarily concerned with increased osmolarity of the extracellular fluid brought about either by salt excess or by water deficit. A decrease in circulating vascular volume may be involved. In addition, influences which are related to habits of salt and water intake have an effect. Basic to stimulation is the sensation of dryness of the mucous membranes of the mouth and pharynx. The central nervous system areas involved are not known precisely, although it is believed that they lie in or near the ventromedial nuclei of the hypothalamus.

Water Loss

Average daily losses are as follows: insensible loss (vaporization through lungs and skin), 800 to 1200 ml; urine, 1500 ml; stool, 100 to 200 ml. In addition, there may be variable loss of water through sensible perspiration (sweat), depending upon environmental conditions and degree of physical activity.

Insensible and Sensible Perspiration

Insensible water loss is modified by the effective osmotic pressure of the body fluids. Thus, there are data suggesting that the rate of insensible loss is inversely related to the concentration of sodium in the extracellular fluids. It follows that the rate of loss of water from the skin and lungs by this mechanism may be related to vapor pressure of the body fluids. Insensible perspiration goes on at a reasonably constant rate when body temperatures remain constant. About half is lost by the skin, half by the lungs.

Sensible perspiration may vary from 0 to 2 liters per hour and more. The output of sensible perspiration is greatly increased in elevated temperatures and with exercise (Chap. 28). Centers regulating sweating are located in the anterolateral portions of the hypothalamus. These respond to elevation in blood temperature or reflex afferent nerve stimulation (e.g., gustatory nerve stimulation). Sensible perspiration is a hypotonic fluid. Average values for the more important solutes appear in Table 23-4.

Gastrointestinal Tract Contribution

Gamble (1958) has estimated that the volume of digestive secretions provided by an average adult in one day is about 8 liters, of which gastric and intestinal mucosal secretions contribute about 5.5 liters. The electrolytic composition of these secretions

Table 23-4. Average Composition of Sweat

Solute	mOsm/liter
Na$^+$	47.9
K$^+$	5.9
Cl$^-$	40.4
NH$_3$	3.5
Urea	8.6

differs considerably in the various regions of the gastrointestinal tract (Chap. 26). However, the average solute concentration in the various secretions is reasonably close to that of the extracellular fluid except for saliva, which is distinctly hypotonic. Although the amount of fluid lost from the gastrointestinal tract is normally small, persistent and excessive vomiting, diarrhea, or drainage from an intestinal fistula very quickly causes serious reduction in volume of the extracellular compartment. In addition, serious disturbance in electrolyte balance is produced. This results from both dehydration and differential electrolyte loss. For example, because potassium is found in these fluids in somewhat higher concentration than in extracellular fluid, potassium deficit would develop. Since disproportionate concentrations of HCO_3^-, Cl^-, and H^+ occur, it is to be expected that disturbances in acid-base balance result from excessive vomiting or loss of intestinal secretions.

Role of the Kidney
The kidney serves as the buffering organ for adjustment of the fluid balance in the sense that it is able to retain water under circumstances of dehydration, or alternatively it can lose large excesses of water when it is necessary to maintain isotonicity and normal volumes in the fluid compartments of the body. Extremes of volumes have been registered between 500 ml for 24 hours to 20 liters or more, the latter in the absence of regulation by antidiuretic hormone (ADH) in diabetes insipidus.

The output of urine is the net result of kidney operation on the original 170 liters of fluid filtered per 24-hour period by the glomeruli. The factors that govern the release of ADH are discussed in greater detail in Chapter 30. In brief, the important controls involve the osmolarity of the extracellular fluid and the volume of the body fluids. Increases in osmolar concentration of the plasma extracellular fluid supplying the zones in the hypothalamus where the osmolar receptors are located cause increased output of ADH and ultimate water conservation by the kidney, according to the findings of Verney (1947). Excessive intake of water causes a reduction in the osmolarity of the extracellular fluid and inhibits the output of ADH by the hypothalamo-neurohypophysial system and leads to water diuresis. Second, changes in volume of the extracellular fluids are perceived by volume receptors and baroreceptors, which cause the requisite correction in total fluid volume, by intermediation of ADH and the kidney.

A related important limiting factor upon the renal regulation of water content of the body is the solute load which must be excreted. The solute load is composed of metabolites, such as urea, creatinine, uric acid, and excess electrolytes. The maximal osmolar concentration of the urine of man, 1400 mOsm per liter, obtains only when the solute load for excretion and obligatory urine volume are both low, the latter being about 0.5 ml per minute. With increasing solute loads, in spite of presumed maximal ADH activity, urine volume increases, and urine osmolarity actually declines progressively. The reason is that the ADH mechanism has attained, in effect, a "ceiling" of operation, so that water is progressively lost with increasing solute loss (see Chap. 22).

SUMMARY OF FACTORS THAT REGULATE URINE VOLUME

Modification of Glomerular Filtration Rate

It is well known that changes in glomerular filtration rate (GFR) may produce parallel changes in urine volume. Changes in GFR can be brought about by changes in systemic arterial blood pressure, which in turn produce corresponding changes in the glomerular filtration pressure. An obvious example is the oliguria or anuria following hemorrhage, under which circumstance arterial pressure and, in turn, glomerular capillary pressure are significantly reduced. Besides the hydrostatic filtering pressure, another important factor is the colloid osmotic pressure which opposes filtration. Hence, if plasma protein concentration is altered, changes in glomerular filtration rate and subsequently changes in urine volume take place. For example, rapid infusion of isotonic salt solution decreases the plasma protein concentration by dilution and increases the effective filtration pressure.

Certain drugs affect glomerular pressure by acting on the afferent or efferent arterioles. As an example, the xanthines (caffeine, theobromine, and theophylline) dilate the afferent arterioles and increase glomerular filtration rate. They also secondarily impair tubular reabsorption of Na^+.

The changes in filtration rate involve changes in solute load to the tubules, which must be considered in explaining the final effect on urine volume. Thus, the extent of solute excretion governs the amount of water that will follow because of osmotic obligation (see below).

Intrarenal Control of Urine Volume by Physical Factors

It will be recalled (Chap. 22) that the bulk reabsorption of water filtered by the glomeruli occurs by passive (osmotic) accompaniment of active sodium reabsorption in the proximal convoluted tubules. A review of the structure and arrangement of the proximal tubular epithelial cells (Fig. 22-20), would be helpful at this time. Attention is particularly directed to the basal infoldings and intercellular channels, and their relationship to the sodium and potassium pumps.

As shown diagrammatically in Figure 23-6, the bulk reabsorptive process begins by the active extrusion of sodium at the basilar end of the cell (process A), causing an increased concentration of Na^+ just outside the cell. Diffusion toward the capillary results, in accordance with the diffusion coefficient, D, along a concentration gradient, ΔC, across the interstitial space, ΔX. Likewise, sodium can diffuse back into the cell ($\Delta X'$), according to its diffusion coefficient in that direction (D'), in part by way of the intercellular space. An expansion of the interstitial space (as from decreased capillary absorption or increased filtration) would increase ΔX, and through operation of Fick's equation would result in a net decrease in the transfer of sodium ($\downarrow dn/dt$). This would cause a further rise of concentration in the interstitium, and finally in the intracellular concentration of sodium, if the transport system operates to maintain a fixed gradient. The increased intracellular concentration of sodium would ultimately result in decreased net passive movement of sodium from the tubular lumen, and overall, decreased net tubular transport of sodium and increased excretion. A possible route for "back-leakage" into the lumen is provided by the intercellular space and the zonula occludens ("tight junction"), which might become leaky with increased volume and pressure in the intercellular space (Diamond, 1974).

Further involvement of physical forces in the movement of salt and water from the tubular epithelium through the interstitium and into the capillaries is illustrated in Figure 23-7. This brings into the picture the Starling forces: the balance of hydrostatic and oncotic forces between the interstitium and the capillary lumen. It is clear from application of facts considered in Chapter 12 that an increase in hydrostatic pressure in

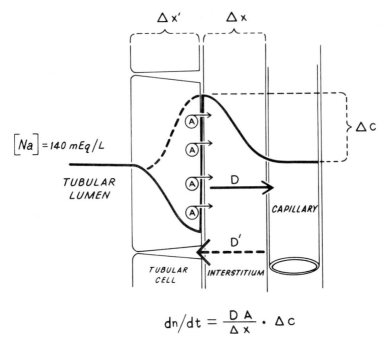

$$dn/dt = \frac{D\,A}{\Delta\,x} \cdot \Delta c$$

Figure 23-6
Possible role of changes in net diffusion of sodium from interstitium to capillary on net transepithelial transport of sodium. The concentration profile for sodium from tubular lumen to
capillary is represented schematically by the heavy line traversing the diagram. (From Earley and
Schrier, 1973.)

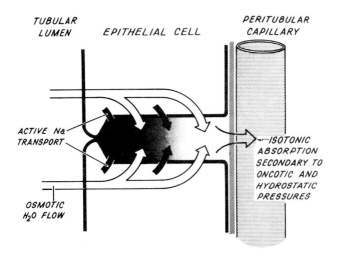

Figure 23-7
Representation of movement of tubular fluid from proximal tubular lumen to peritubular capillary.
This diagram is constructed to conform to the anatomical characteristics of the proximal tubule and
peritubular capillary circulation. Sodium is actively transported into the extracellular spaces

the capillaries would decrease isotonic absorption and tend to impede the osmotic H_2O flow. Ultimately, through mechanisms discussed above in relation to Figure 23-6, uptake of salt and water would be impaired. With reduction of hydrostatic pressure, the reciprocal effect could be anticipated.

Changes in oncotic pressure would similarly influence the mechanism (i.e., dilution of peritubular capillary plasma proteins would tend to impair the osmotic H_2O flow and favor urinary loss); conversely, increase in plasma oncotic pressure would favor uptake and eventuate in decreased urinary excretion of salt and water.

Numerous examples are at hand in which experimental manipulation of the above forces has resulted in changes in electrolyte and water excretion (Earley and Schrier, 1973). Critical experiments would involve alteration of so-called physical factors while GFR and related filtration of water and sodium salts remained constant. These manipulations include changes in renal arterial pressure, assuming that some part of the change in arterial pressure is transmitted to the peritubular capillary bed. It has been demonstrated that increase in renal arterial pressure (RAP) (with constant GFR) increases sodium and water excretion. Agents (e.g., acetylcholine, prostaglandins) which produce renal vasodilation also promote salt and water loss, particularly in conjunction with increased RAP. Decrease in RAP with constant GFR (due to autoregulation) causes reduced urinary output of sodium and water.

Saline infusion would operate in part through this mechanism (note also effect of increase in ΔX, as discussed above) to cause diuresis; dilution of plasma proteins would amplify the effect. Intravenous infusion of hyperoncotic albumin, with increase in peritubular capillary plasma oncotic pressure, causes a net decrease in urine volume and solute excretion, including sodium (Elpers and Selkurt, 1963).

When experimental conditions are set so that GFR remains constant (Fig. 23-8), elevation of renal venous pressure provides an excellent example of operation of physical forces involving isotonic absorption by the peritubular capillaries. Thus, increased venous pressure results in elevated capillary hydrostatic pressure, opposing fluid reabsorption and favoring outward filtration. The mechanisms outlined above operate to decrease net tubular reabsorption of water and sodium, and ultimately result in increased excretion.

Factors Affecting Tubular Reabsorption of Water

The role of ADH in regulation of tubular water handling has been discussed. Control mechanisms are detailed in Chapter 30. Mechanisms that inhibit ADH production and release include reduction of plasma osmotic pressure—for example, by the oral ingestion of large quantities of water or infusion of hypotonic NaCl or isotonic glucose solutions. (When glucose is metabolized, the water of the solution is freed.) Figure 23-9 shows typical water diuresis resulting from ingestion of 1 liter of pure water. Specific gravity of the urine rapidly changes from 1.030 to 1.002 at the peak of diuresis.

Effect of ingestion of 1 liter of isotonic saline is also shown. The diuresis following saline ingestion has been explained by a small increment in GFR. Specific gravity of the urine shows little if any change, for body fluid osmolarity is not altered, and ADH output should not change. In well-hydrated subjects, saline ingestion produces a greater

---

between adjacent epithelial cells, creating a hypertonic pool in this extracellular, intratubular area. Water moves osmotically into the intercellular channel, creating elevated hydrostatic pressure within the intercellular space. The channel is tightly closed at the luminal end of the cell, and the epithelial basement membrane and capillary endothelium have a high hydraulic conductivity. The increased hydrostatic pressure within the channel, together with the oncotic pressure of the peritubular plasma, would drive the column of reabsorbate into the capillary circulation. The continual osmotic flow of water into the channel results in the delivery of virtually isotonic reabsorbate to the capillary circulation. (From Earley and Schrier, 1973.)

diuresis, with decrease in specific gravity of the urine. It is probable that ADH output is decreased in this circumstance by activation of volume receptors located in the left atrium, or by altering of the carotid sinus baroreceptor influence on ADH release (Chap. 30).

Among the factors stimulating ADH secretion and thereby inducing antidiuresis are water deprivation and dehydration, which increase the plasma osmotic pressure, in turn,

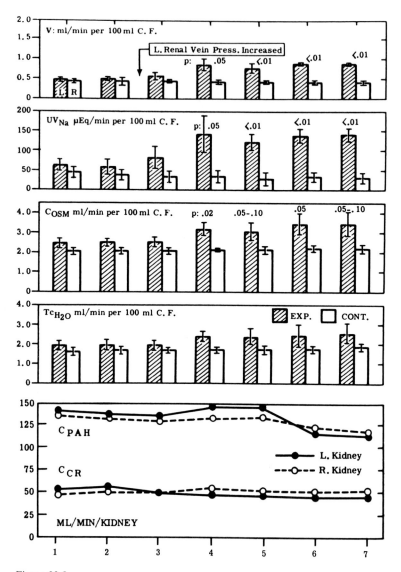

Figure 23-8

After two control periods, renal venous pressure of the left kidney (dog) was increased to 35 mm Hg. Note increase in urine volume, $\dot{V}$; sodium excretion (UV_{Na}); osmolar clearance (C_{Osm}); and negative free water clearance (Tc_{H_2O}). Glomerular filtration rate (C. F., clearance of filtrate), as measured by the clearance of creatinine (C_{CR}), remained constant. (From R. L. Wathen and E. E. Selkurt, *Am. J. Physiol.* 216:1517–1524, 1969.)

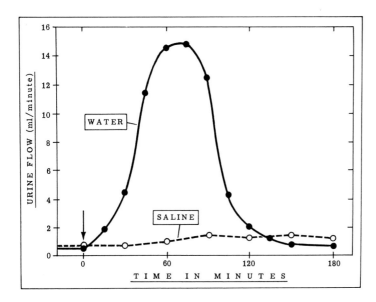

Figure 23-9
Comparison of water diuresis and saline diuresis after ingestion of 1 liter. (From H. W. Smith. *Principles of Renal Physiology.* New York: Oxford University Press, 1956. P. 113.)

acting on the osmolar receptors in the hypothalamus. Also, as the result of reducing plasma volume and, in consequence, stimulation of left atrial volume receptors and carotid sinus baroreceptor activity, ADH output is further enhanced. Pain and emotional stress likewise cause an increase in ADH output. Various drugs such as nicotine, acetylcholine, morphine, and the anesthetic agents barbiturates, ether, and chloroform all act on the hypothalamo-neurohypophysial system to increase output of ADH and lead to antidiuresis.

Another means of tubular regulation of urine volume concerns the role of solutes delivered to the tubular cells, as discussed previously. *Osmotic diuresis* results if the filtered load of a given solute exceeds the reabsorptive capacity for it by the cells at the site of reabsorption in the nephron. The unreabsorbed solutes osmotically retain some of the water filtered at the glomeruli and carry it on through the rest of the system. Common osmotic diuretics include certain organic substances such as mannitol, urea, sucrose, and glucose. Hypertonic NaCl and Na_2SO_4 are others. The acid-forming salts ammonium chloride, ammonium nitrate, and calcium chloride have been used as diuretic agents, alone or in conjunction with others (e.g., mercurials). The ammonium salts act in part by increasing urea formation, which acts as an osmotic diuretic. For example, NH_4Cl is converted to urea and hydrochloric acid and the latter is buffered as follows:

$$HCl + NaHCO_3 \rightarrow NaCl + H_2CO_3$$

H_2CO_3 dissociates to carbon dioxide and water. The resulting acidosis is compensated in part by the release of carbon dioxide by the lungs. There is, however, an accelerated excretion of the residual Cl^- accompanied by Na^+ into the urine, and the loss of NaCl is accompanied by an equivalent loss of water. Oral ingestion of $CaCl_2$ acts by giving up Cl^- to be lost with Na^+; the poorly absorbed Ca^{++} tends to stay in the intestine.

Other Diuretic Agents

Carbonic Anhydrase Inhibitors
Carbonic anhydrase inhibitors are heterocyclic sulfanilamide derivatives of which acetazolamide (Diamox) is the one commonly employed. These compounds are effective inhibitors of carbonic anhydrase in vitro, and the mechanism involved may be related to the mechanism for producing diuresis. It is probable that the drugs reduce the availability of hydrogen ions to both proximal and distal tubular ion exchange mechanisms. The agents promote the renal loss of NaCl and $NaHCO_3$. More indirectly, increased loss of Na^+ favors enhanced K^+ secretion by supplying greater amounts of the ion needed for the exchange mechanism to operate in the distal nephron. Loss of all these ions promotes loss of water by their osmotic diuretic action.

Benzothiadiazine Diuretics
Sulfonamyl benzothiadiazines such as chlorothiazide (Diuril) have a weak carbonic anhydrase–inhibiting action and act chiefly by blocking tubular reabsorption of sodium and chloride ions; potassium loss is also favored. Since they do not interfere with the production of maximally concentrated urine during hydropenia, it has been conjectured that they exert their action predominantly in the distal convoluted tubule.

Aldosterone Antagonists
The antialdosterone substances are spirolactones, the most effective of which is spironolactone (Aldactone). They are structurally similar to aldosterone and act by competitive inhibition, thus blocking aldosterone action and favoring Na^+ loss.

Two Potent Loop of Henle Electrolyte Transport Inhibitors
Ethacrynic acid (Edecrin) is a potent diuretic which acts to inhibit electrolyte transport in the proximal tubule, and throughout the loop of Henle. In hydropenic subjects, the ability to concentrate the urine is lost, and in hydrated patients, the ability to dilute the urine is impaired. Thus, the loop loses its ability to dilute or concentrate, and hence its ability to maintain the gradient of osmolarity in the medulla. Increased Cl^-, K^+, and H^+ excretion also results, because increased amounts of NaCl reach the distal tubule, allowing for greater ion exchange here. Chemically, ethacrynic acid is an unsaturated ketone of aryloxyacetic acid.

Furosemide (Lasix) is a sulfonamide derivative which also acts to block electrolyte transport in the ascending limb of the loop of Henle, as well as proximal sites. If it is confirmed that Cl^- is the ion actively pumped in the ascending limb of the loop, then ethacrynic acid and furosemide presumably act here by blocking Cl^- transport, and sodium is lost with chloride.

Mercurials
Numerous organic mercurial diuretics exist which will not be discussed in individual detail. Diuresis results because of Na^+ and Cl^- loss, being caused by impaired tubular reabsorption. K^+ excretion is variably influenced; if secretion of K^+ is initially low, excretion will be increased, and vice versa.

The discrete mechanism of action of the mercurials has not been settled. Earlier views that the action was due to the affinity of Hg for sulfhydryl groups has been challenged on the basis that there is little good evidence that sulfhydryl-combining enzymes play a significant role in electrolyte transport.

Xanthines
Xanthines also act in part by blocking Na^+ reabsorption.

ELECTROLYTE BALANCE

Sodium Intake

The average adult diet contains 10 to 12 gm of NaCl per day, either in the food (higher in meat diets than in vegetable diets) or added to it. Examples of excessive ingestion of salt or salt craving have been cited in the literature. This is usually associated with deficiency of adrenal cortical secretion, as in Addison's disease. As an extreme example, Strauss (1957) cites the case of a 34-year-old patient with Addison's disease who put an approximately one-eighth-inch layer of salt on his steak, used nearly one-half glass of salt for his tomato juice, put salt on oranges and grapefruit, and even made lemonade with salt. In fact, salt craving is often an early manifestation of Addison's disease that is of considerable diagnostic significance.

However, excesses of sodium (*hypernatremia*) are less common clinically than hyponatremia. Whereas diseases that cause dehydration are usually associated with sodium losses in excess of water losses, excess sodium is almost always associated with an excess accumulation of body water, except in hypernatremia due to cerebral lesions. In cases of intracranial damage, sodium concentration is sacrificed to maintain water volume, and hypernatremia without edema may result.

There will be increased levels of blood sodium in conditions associated with increased adrenocortical activity. High production of aldosterone by the adrenal cortex will favor extraction of excess amounts of sodium from tubular urine by the renal tubular cells (Chap. 22). For the same reason, patients who are on steroid therapy for prolonged periods may have abnormal sodium and water retention (Chap. 34). Although cortisone is a glucocorticoid (not mineralocorticoid), glucocorticoids can cause some sodium retention. This is less a problem with some of the newer steroid preparations than was the case a few years ago, but patients on steroid therapy should nonetheless be observed for signs of sodium and water retention and other untoward reactions.

Aldosterone is normally inactivated by the liver. In some types of liver disease, aldosterone may not be properly degraded, so that there may be abnormal retention of sodium and water (see Chaps. 18B and 34).

After treatment of diabetic coma, when cellular sodium is being returned to the extracellular compartment, there may be a temporary rise in the blood sodium level.

Sodium Loss

Total body sodium is 58 mEq per kilogram of body weight, 25 mEq per kilogram as bone sodium, of which about half is exchangeable (Edelman and Leibman, 1959). The intake of Na^+ is normally quite variable, as is loss by sensible perspiration, so that it falls upon the kidney to regulate the homeostasis of its content in the body. Excesses of Na^+ are excreted by the kidney. Conversely, if the dietary intake is reduced, the urine will become virtually sodium-free in several days in an attempt to maintain sodium balance.

Excessive loss of Na^+ from the extracellular compartment (due to renal reabsorptive impairment in renal disease), gastrointestinal losses (e.g., diarrhea), or excessive sweat loss, leads to low blood Na^+ levels (hyponatremia). The accompanying loss of fluid volume contributes to the observed physiological changes. The symptoms of hyponatremia are weakness, lassitude, apathy, and headache. The decrease in vascular volume leads to faintness, and in extreme cases to orthostatic hypotension, tachycardia, and shock.

Gastrointestinal symptoms associated with sodium loss include anorexia, nausea, and vomiting. Thirst is usually absent. There is a loss of turgor and elasticity of the skin. Normally, when the skin is pinched and picked up, it returns to its previous shape immediately when it is released. When there is a sodium loss and attendant water loss, the fold in the skin may remain for 30 seconds or longer. In severe sodium loss there is confusion, stupor, and eventual coma.

Potassium Intake

The major cation for maintaining proper pH of the intracellular fluid is potassium. All cellular activities involving electrical phenomena, such as skeletal and cardiac muscle contraction and nerve impulse conduction, are dependent on the gradients of K^+ and Na^+ across cell membranes. In considering the shifts of K^+ between the extracellular and intracellular compartments, it is well to keep in mind the quantitative disproportion that exists. Thus a total of 350 mEq in the extracellular fluid compartment contrasts with 3500 mEq in the intracellular fluid (Edelman and Leibman, 1959). Therefore, small losses out of the body cells or small uptakes can cause relatively significant changes in the concentration of potassium in the extracellular fluid. Tissue repair and growth require K^+; conversely, protein breakdown releases K^+ from the cells into the extracellular compartment, from which it is lost in the urine. The transfer of glucose from the extracellular to the intracellular phase requires K^+ by certain cells.

Since there is a narrow range of normal blood levels of K^+ (3.5 to 5.0 mEq per liter), relatively small changes in blood K^+ can cause serious problems. Thus, in *hyperkalemia* (blood K^+ levels of 8 mEq per liter), significant changes occur in the configuration of the electrocardiogram (loss of P wave, changes in the QRS and T waves). Above 11 mEq per liter, ventricular fibrillation is imminent. In addition, neuromuscular symptoms, numbness, tingling, and eventually flaccid muscle paralysis are seen in hyperkalemia.

Increased blood K^+ results from kidney disease, in which excretion of the cation is diminished. In intestinal obstruction, blood K^+ is elevated because 10 percent of the dietary intake of K^+ normally is eliminated in the stool. In Addison's disease, deficiency of aldosterone (Chap. 34) impairs the kidney's ability to excrete K^+, and hyperkalemia is seen; increased loss from tissue cells also occurs.

Potassium Loss

The urine always contains potassium, and if none is administered or ingested in the diet, a deficit of this cation will develop. Ingestion of about 3 gm of KCl a day appears to be adequate under normal circumstances. Normally, about 10 percent of K^+ loss occurs in the stool, the remainder in the kidney. Renal regulation involves the adrenal cortical hormones. Potassium loss may be excessive in certain types of chronic renal diseases associated with polyuria and in the diuretic phase of recovery from acute lower nephron diseases. In a condition called renal tubular acidosis, abnormal amounts of HCO_3^- are excreted in the urine along with accompanying fixed cations, including K^+. In acidosis K^+ is replaced in the cells by Na^+ and H^+. In severe metabolic acidosis, as in diabetic acidosis, K^+ leaves the cells to enter the extracellular fluid compartment and is excreted in the urine. Contributing to cellular loss of potassium is cellular protein catabolism associated with increased gluconeogenesis.

Decreased blood K^+ results from prolonged high gastrointestinal suction. Diuretics may lead to excessive renal excretion of K^+. Increased renal loss also accompanies the excessive administration of bicarbonate. As stated, in alkalosis, K^+ enters the cells, lowering the blood concentration.

Symptoms of hypokalemia tend to be nonspecific. They include malaise, apathy, and intestinal distention leading to paralytic ileus (a result of decreased smooth muscle tone). There is usually postural hypotension, and diastolic blood pressure is low. The terminal event is heart block; the heart stops in *systole*.

Chloride Intake

Chloride intake is governed by the intake of NaCl, as discussed above in relation to sodium.

Although chloride movement across cell membranes is ordinarily viewed as a passive mechanism secondary to that of sodium, some important exceptions occur. The active transport of chloride by the gastric mucosa is an outstanding example. The pumping of Cl^- by the ascending limb of Henle is another important example (Chap. 22). Keynes (1963) has disclosed that the axoplasm of the giant squid axon actively takes up labeled chloride. The uphill inward transport was inhibited by dinitrophenol, and the lack of effect of ouabain indicated that the influx was not linked to transport of cations by the sodium pump. Evidence for an active chloride pump in amphibian cornea was supplied by Zadunaisky (1965), who showed that the short-circuiting current could account for the active transport of Cl^- from the aqueous to the tear side of the cornea. This mechanism appears to have a function in maintaining corneal transparency (by keeping the intracellular ion concentration low and thus preventing swelling).

The most common cause of excess blood chloride (hyperchloremia) is dehydration. Increased amounts of Cl^- are reabsorbed by the kidney with the intensive H_2O reabsorption triggered by dehydration. This favors hyperchloremic acidosis.

Chloride Loss

In general, Cl^- behaves similarly to Na^+ in regard to renal handling and sweat loss. However, there are exceptions. During acid-base alterations, proportions of Cl^- and HCO_3^- associated with Na^+ may be noted in the plasma and in the urine differing from the normal relationship. It has also been observed that mercurial and thiazide diuretics cause greater loss of Cl^- than of Na^+ in the urine.

Abnormal loss of Cl^- because of excessive vomiting or gastric suction may result in alkalosis. When chloride is lost as the hydrochloric acid of the gastric juice, the stomach is obliged to replace the lost HCl; it does so by removing Cl^- from circulating NaCl and combining it with H^+ (see Chap. 24). The remaining Na^+ and OH^- in the bloodstream combine with circulating CO_2 to increase blood levels of sodium bicarbonate ($NaHCO_3$), which causes the alkalosis.

Mercurial and other diuretics (ethacrynic acid, furosemide) interfere with the enzyme systems needed for Cl^- absorption by the kidney. The excreted Cl^- takes Na^+ and H_2O with it; with excessive loss, salt depletion and hypochloremic alkalosis ensue. The alkalosis results because there is more circulating Na^+ than Cl^-, so that with equal loss of these ions into the urine, Na^+ tends to accumulate.

Calcium and Magnesium Balance

Because of their low concentration, the role of calcium and magnesium ions is small in the control of volume and osmolarity of the body fluids, and in acid-base equilibrium. Calcium, a major constituent of bone, is also important because of its influence on cell membrane permeability, neuromuscular excitability, transmission of nerve impulses, blood coagulation, and activation of certain enzyme systems. It is both protein-bound and ionized in the plasma; the extent of ionization is determined by the acid-base equilibrium, being increased in acidosis and decreased in alkalosis. The action of parathyroid hormone in renal regulation of Ca^{++} balance is also related to $PO_4^{\equiv}$ handling. The hormone favors increased tubular Ca^{++} reabsorption, while apparently depressing tubular reabsorption of $PO_4^{\equiv}$, with increased urinary loss. This leads to secondary rise in serum Ca^{++}. Continued mobilization from bone leads to further hypercalcemia and ultimate renal loss by overloading the reabsorptive mechanism (Chap. 32).

Like calcium, magnesium is in either a bound or an ionized phase in the plasma. Its physiological action is noted mostly in connection with the functioning of the neuro-muscular and cardiovascular systems. Thus, magnesium excess results in depression of the central nervous system with loss of the tendon reflexes, drowsiness, and finally coma. An excess may cause bradycardia and depression of conduction through the conducting tissue and myocardium.

DISTURBANCES OF FLUID AND ELECTROLYTE BALANCE

Dehydration

When water loss exceeds intake, the interstitial fluid yields water as long as possible, but eventually desiccation of cells occurs (*dehydration*). Water deficits are incurred in normal individuals under two common conditions: (1) excessive loss of sweat; (2) prolonged deprivation of water.

During preliminary stages of negative water balance, skin and muscle give up fluid first, so that vital organs are protected. Following depletion of interstitial fluid, the plasma water is increasingly removed. The blood becomes more concentrated (*anhydremia*). With prolonged loss or deprivation, or both, intracellular water is also lost.

Characteristic symptoms result from dehydration. Water deficit gives rise to a shrunken appearance of the face and body. The skin loses its elasticity and becomes hard and leathery. There is rapid loss of body weight. When the deficiency reaches such a degree that the water is no longer sufficient for removal of heat of metabolism, high fevers may occur. As the condition worsens, circulatory failure develops. Anuria results, and acid products are retained, leading to acidosis. Cerebral disturbances, excitement, delirium, and coma terminate the episode.

Clinical dehydration may be a consequence of (1) failure of absorption from the alimentary tract (as in pyloric stenosis or high intestinal obstruction); (2) excessive loss from copious sweating, prolonged vomiting, diarrhea, and excessive diuresis; (3) drainage from wounds or burns.

Water loss under these conditions involves also loss of electrolytes, predominantly NaCl. If fluid balance is therefore restored with only pure water, a hypotonic state results in the extracellular compartment. As a consequence, water moves in abnormal amounts into the cells now relatively hyperosmotic, producing symptoms referable to *water intoxication,* among them the well-known heat cramps. Obviously, in treatment, electrolyte (chiefly NaCl) must be given as well as water.

Excess Water Loads

A condition of cellular overhydration may result also from attempts to produce diuresis by forcing hypotonic fluids, particularly in the presence of renal impairment. On the other hand, prolonged and excessive diuresis may result in excessive loss of extracellular solutes, and water will migrate from a zone of relative hypotonicity into the cells. The symptoms of water intoxication which follow are particularly referable to the CNS: salivation, nausea, and vomiting; restlessness, asthenia, muscle tremors, ataxia, and tonic and clonic convulsions that may eventuate in stupor and death.

EDEMA

The fundamental bases for formation of edema fluid have been discussed in Chapter 12 in connection with capillary fluid dynamics. Since edema formation represents an abnormal accumulation of extracellular fluid, particularly that of the interstitial compartment, added facets of the problem of edema are presented here.

Cardiac Edema

This important type of edema was discussed in detail in Chapter 18B.

Cirrhotic Edema

The edema of cirrhosis is localized in the abdominal cavity (ascites). Degeneration of the parenchymal cells of the liver leads to impairment of circulation through the liver, interfering particularly with the low-pressure portal vein system. The resulting portal hypertension favors the exudation and accumulation of fluid in the abdomen, both from the capillaries within the liver and from the capillaries in the intestine. However, portal hypertension is not the only factor in this type of edema. Impaired liver function results in hypoalbuminemia, contributing to the effective outward gradient for filtration of fluid. The increased titer of ADH in the urine of cirrhotics is evidence of greater antidiuretic activity, presumably due to failure of the liver to inactivate this hormone. There is also evidence that aldosterone activity is enhanced under these circumstances, acting to favor salt and water retention. Other circumstances characterized by obstructive conditions in the liver give a similar picture (e.g., sclerosis of the hepatic vessels or compression of the portal vein by tumors, aneurysms, and the acute icteric phase of infectious hepatitis).

Renal Edema

The basic causes of renal edema are (1) decrease in oncotic pressure of the plasma because of loss of albumin via the kidney; (2) coexisting increase in systemic capillary permeability with loss of protein to the interstitial spaces; (3) coexisting congestive heart failure of the hypertensive type when associated with chronic nephritis; and (4) Na^+ retention. In chronic nephrosis, in the nephrotic stage of glomerulonephritis, and in the amyloid kidney, the kidneys excrete 10 to 20 gm of protein a day into the urine because of glomerular damage. Plasma albumin decreases from a normal of about 5 percent to about 2 percent, resulting in a marked reduction of oncotic pressure. The protein content of the edema fluid is low, about 0.1 percent. With acute glomerulonephritis there is widespread capillary injury in all parts of the body; hence, the protein content of the edema fluid may be high (over 1 percent). The edema is a combination of decreased oncotic pressure of the plasma and increased oncotic pressure of the interstitial fluid. Hypertensive heart disease accompanies chronic diffuse glomerulonephritis, pyelonephritis, and polycystic renal disease. With this, the ultimate cardiac decompensation contributes factors which are characteristic of congestive heart failure, including the elevated venous pressure and sodium retention by mechanisms previously described.

Nutritional Edema

Nutritional edema can be caused by faulty metabolism or nutritional deficiency. Although vitamin lack—e.g., lack of vitamin B in beriberi and of vitamin C in scurvy—is contributory, perhaps most important is protein lack in the diet. Inadequate protein intake leads to hypoproteinemia and reduction of the oncotic pressure of the plasma proteins, favoring outward movement of fluid from capillaries into the interstitial spaces. The decrease in tissue pressure which results from the wasting of tissue may favor this outward movement of fluid, and a contributing factor may be the changes in capillary permeability caused by the attendant vitamin deficiency. Some authorities feel that the general reduction in cardiac activity and blood flow impairs kidney function and leads to secondary Na^+ retention as a final contributory factor.

FUNCTION OF URETER AND URINARY BLADDER; MICTURITION

Function of Ureters

Urine collected in the pelvis of the kidney passes through the ureters to the bladder, not only by virtue of forces of gravity in the erect position but also by contraction of muscle layers of the ureter. These muscular contractions are necessary to develop sufficient pressure to overcome the gradually increasing tension in the bladder as urine accumulates. The peristaltic waves observed probably originate in the muscles but are modified by the action of the splanchnic nerves via the renal plexus to the upper portions of the ureters, and via the hypogastric plexus to the lower portions. These are excitatory fibers; inhibitory fibers have been described in the inferior mesenteric plexus. Afferent sensory fibers are present, as evidenced by the excruciating pain felt on passage of calculi through the ureter. The ureters enter the base of the bladder obliquely, thus forming a valvular flap that passively prevents reflux of urine (Waldeyer's sheath).

Bladder Filling and Continence

The bladder consists of a sphere and a cylinder (the neck extending into the urethra) which have special intrinsic properties. The smooth muscle of the bladder can be considered as a sheet of muscle in the form of a reticulum or webwork which extends without interruption down into the urethra as the muscular wall. There is a particularly heavy concentration of elastic connective tissue fibers in the wall of the urethra. The smooth muscle and elastic tissue exert continuous tension in an autonomous manner, with negligible expenditure of energy (Chap. 3B).

When the bladder fundus is being distended with fluid, the smooth muscle fibers of the bladder wall are at first stretched, and caused in turn to contract and increase their tension. Thus, measurement of intravesical (bladder) pressure will demonstrate an initial increase with initial filling of the bladder with fluid; but then, as filling continues, the intravesical pressure will remain approximately constant until bladder capacity is reached (Fig. 23-10). At capacity (250 to 450 ml), the intravesical pressure rises more steeply. The ability of the bladder fundus to maintain a relatively low intravesical pressure is a type of accommodation, which allows easy filling from the ureters.

The urine is kept from leaking out of the bladder, in part, by resistance offered principally by the upper 3 cm of the bladder neck and urethra (Fig. 23-11A). The neck of the bladder can be viewed as an "internal sphincter," although authorities disagree as to whether or not a true sphincter exists here. The urethra aids the sphincteric action by the continuous intrinsic autonomous tension exerted by the smooth muscle and connective tissue in its wall (the "external sphincter"). These tissues keep the urethra compressed, so that its lumen is sufficiently collapsed to prevent urine from flowing out of the bladder fundus under low or moderate pressures. To oppose higher intravesical pressures, the urogenital diaphragm and *levator ani* aid in the act of urinary bladder continence. These muscles increase the efficiency of the urinary sphincter by compressing the urethra circumferentially and elongating it by pulling it cephalad. The striated muscle can act on a voluntary basis, or contraction can take place reflexly as part of postural reflexes. Thus, one can willfully compress the urethra and prevent urinary incontinence when the bladder is full and the patient has an urgent desire to urinate. Or the urinary sphincter can be compressed and elongated reflexly as in assuming the erect posture, coughing, or straining (Fig. 23-11B).

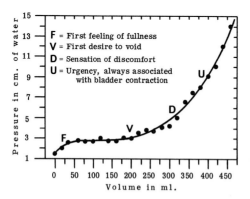

Figure 23-10
The pressure-volume relationship of the normal human bladder during filling. Sharply rising pressures above 400 to 450 ml indicate filling beyond physiological capacity. (Adapted from F. A. Simeone and R. S. Lampson. *Ann. Surg.* 106:413–422, 1937.)

INTRAURETHRAL RESISTANCES

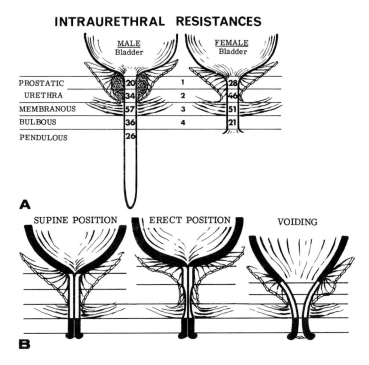

Figure 23-11
A. Comparison of urethra in male and female. The "resistance to pressure" of a given zone is presented in centimeters H_2O. Note that the middle 2 cm of the female urethra corresponds to the distal part of the prostatic urethra and the membranous urethra of the male. This zone is contiguous with the periurethral striated muscle. B. Effect on length of urethra of erect versus supine posture, and effect of voiding when the periurethral striated muscle (shown in section) is relaxed. In addition, the urethrovesical junction is pulled open into a funnel-shaped structure by active contraction of the vesiculourethral smooth muscle sheet. The net result is a functional shortening and widening of the posterior urethra. (Modified from J. Lapides. In M. F. Campbell and J. H. Harrison [Eds.], *Urology*. Philadelphia: Saunders, 1970. Vol. 1, pp. 1348–1349.)

It is important to note that the urinary sphincter is the intact urethra, and that the striated muscles surrounding the urethra are of secondary importance in that they can increase the efficiency of the urinary sphincter but cannot substitute for it (Fig. 23-11B). The striated muscles serve also to interrupt urination rapidly (one to two seconds) when there is an urgent need, by compressing and elongating the urinary sphincter until the vesical smooth muscle stops contracting (10 to 20 seconds).

Micturition
As the volume increases further, the tension starts to rise, in part because tonic contractions of the bladder musculature begin. When bladder volume reaches a capacity of about 150 to 250 ml, stretch and tension (proprioceptive) receptors are stimulated to send out the afferent impulses responsible for the sensation of distention and the desire to urinate. They also cause the act of micturition when spinal reflexes are released from cerebral control. Voluntary control can be exerted until the intravesical pressure increases to about 100 cm H_2O, at which point involuntary micturition begins.

As with the ureter, the extrinsic nerves of the bladder are important in regulating its normal evacuation. They modify changes of tonus, enhance or depress the vigor of rhythmic contractions, and alter the tonus of the external and internal sphincters. This

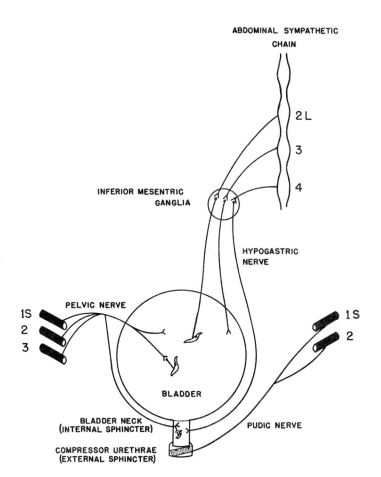

ABDOMINAL SYMPATHETIC
CHAIN

2 L

3

4

INFERIOR MESENTRIC
GANGLIA

HYPOGASTRIC
NERVE

PELVIC NERVE

1S

2

3

1S

2

BLADDER

BLADDER NECK
(INTERNAL SPHINCTER)

PUDIC NERVE

COMPRESSOR URETHRAE
(EXTERNAL SPHINCTER)

Figure 23-12
Innervation of the bladder and urethral sphincters. (After W. S. Root. Physiology of Micturition: A Specific Autonomic Function. In P. Bard [Ed.], *Medical Physiology* [11th ed.]. St. Louis: Mosby, 1961. Modified from J. R. Learmonth. *Brain* 54:147, 1931.)

innervation is by way of the parasympathetic *pelvic* nerves, the sympathetic *hypogastric* nerves, and somatic *pudendal* (pudic) nerves (Fig. 23-12). The pelvic nerves function to maintain tonus of the bladder, and the pudic nerves mediate impulses that contract the striated musculature of the external sphincter. The role of the sympathetic innervation is not so well understood. Afferent fibers run chiefly in the pelvic nerves, although they are also found in the hypogastric and pudendal nerves. They enter the spinal cord at the sacral level, then go on up to the hypothalamus and ultimately to the cortex, where voluntary control resides.

Micturition is normally a voluntary act. Corticospinal impulses sent to the lumbo-sacral region of the cord cause intensive contraction of the bladder, opening of the sphincters, and relaxation of the perineum, and thus lead to voiding (Fig. 23-11B).

The urinary sphincter decreases its resistance during urination by relaxation of the periurethral striated muscle either reflexly or voluntarily. This results in some decrease in length of the urethra and tension of the urethral wall against its lumen. A further decrease in length of the urethra and increase in caliber of the urethral lumen is effected

559

by active contractions of the smooth muscle of the bladder and urethra. When the bladder begins to contract down upon a bolus of urine, the urethrovesical junction and proximal portion of the urinary sphincter are pulled open by active contraction of the muscle sheet which is continuous from bladder into posterior urethra (Fig. 23-11B). The urinary sphincter does not open by passive relaxation but by active contraction of the vesicourethral muscle fibers.

Measurement of intravesical pressure during normal micturition indicates that pressures varying between 25 and 50 cm H_2O are attained at the height of urination. Yet, when the bladder is at rest and storing urine, it may require intravesical pressures of greater than 150 to 250 cm H_2O to overcome the resistance of the posterior urethra or urinary sphincter so that urinary flow occurs. During urination, it is preferable to have low intravesical pressures obtaining, for high pressures, as seen in obstructive uropathy, predispose to dilatation of the urinary tract, infection, and renal deterioration.

In the infant, the normal urinary sphincter can maintain continence between voidings, but it cannot prevent the reflex voiding contractions. Thus, the normal baby will not dribble urine continuously but will wet at intervals during forceful bladder contractions. Voluntary control of urination is gained as the child becomes toilet trained and as the corticoregulatory tract begins to function.

REFERENCES

Burke, S. R. *The Composition and Function of Body Fluids.* St. Louis: Mosby, 1972.

Christensen, H. N. *Body Fluids and Their Neutrality.* New York: Oxford University Press, 1963.

Davis, J. O., C. R. Ayers, and C. C. J. Carpenter. Renal origin of an aldosterone stimulating hormone in dogs with thoracic caval constriction and in sodium depleted dogs. *J. Clin. Invest.* 40:1466–1474, 1961.

Davson, H. *Physiology of the Ocular and Cerebrospinal Fluids.* Boston: Little, Brown, 1956.

Davson, H. *Physiology of the Cerebrospinal Fluid.* London: Churchill, 1967.

Deane, N. *Kidney and Electrolytes: Foundations of Clinical Diagnosis and Physiologic Therapy.* Englewood Cliffs, N.J.: Prentice-Hall, 1966.

Diamond, J. M. Tight and leaky junctions of epithelia: A perspective on kisses in the dark. *Fed. Proc.* 33:2220–2223, 1974.

Dittmer, D. S. (Ed.). *Blood and Other Body Fluids.* Washington: Federation of American Societies for Experimental Biology, 1961.

Earley, L. E., and R. W. Schrier. Intrarenal Control of Sodium Excretion by Hemodynamic and Physical Factors. In J. Orloff and R. W. Berliner (Eds.), *Handbook of Physiology.* Washington: American Physiological Society, 1973. Section 8, Renal Physiology.

Edelman, I. S., and J. Leibman. Anatomy of body water and electrolytes. *Am. J. Med.* 27:256–277, 1959.

Elpers, M. J., and E. E. Selkurt. Effects of albumin infusion on renal function in the dog. *Am. J. Physiol.* 205:153–161, 1963.

Fishman, A. P. (Ed.). *Symposium on Salt and Water Metabolism.* New York: New York Heart Association, 1959.

Gamble, J. L. *Chemical Anatomy, Physiology and Pathology of Extracellular Fluid* (7th ed.). Cambridge, Mass.: Harvard University Press, 1958.

Keynes, R. D. Chloride in the squid giant axon. *J. Physiol.* 69:690–705, 1963.

Maxwell, M. H., and C. R. Kleeman. *Clinical Disorders of Fluid and Electrolyte Metabolism.* New York: McGraw-Hill, 1962.

Meyers, F. H., E. Jawetz, and A. Goldfien. *Review of Medical Pharmacology* (3rd ed.). Los Altos, Calif.: Lange, 1972.

Muntwyler, E. *Water and Electrolyte Metabolism and Acid-Base Balance.* St. Louis: Mosby, 1958.

Pitts, R. F.   *The Physiological Basis of Diuretic Therapy.* Springfield, Ill.: Thomas, 1959.

Pollay, M.   Transport mechanisms in the choroid plexus. *Fed. Proc.* 33:2064–2068, 1974.

Sodeman, W. A.   *Pathologic Physiology.* Philadelphia: Saunders, 1961.

Strauss, M. B.   *Body Water in Man.* Boston: Little, Brown, 1957.

Verney, E. B.   Antidiuretic hormone and the factors which determine its release. *Proc. R. Soc. Lond. [Biol.]* 135:25–106, 1947.

Weldy, N. J.   *Body Fluids and Electrolytes.* St. Louis: Mosby, 1972.

Zadunaisky, J. A.   Chloride active transport in the isolated frog cornea. Abstracts of the 9th Annual Meeting of the Biophysics Society, San Francisco, February, 1965. P. 9.

24. Respiratory and Renal Regulation of Acid-Base Balance

Ewald E. Selkurt

The chemical constituents of the extracellular fluid in which a cell normally lives are regulated within narrow limits to provide an optimal chemical environment for cellular function. One of the most precisely regulated constituents of body fluids is the hydrogen ion (H⁺), and it is the regulation of this chemical species to which the term *acid-base regulation* actually applies.

Deviation from the normal range of hydrogen ion concentration markedly affects many metabolic reactions, accelerating some and depressing others, primarily because of the effect of hydrogen ions on enzyme activity. Therefore, the concentration of hydrogen ions in body fluids must be regulated to an optimal value if homeostasis is to be ensured.

SOURCES OF HYDROGEN IONS

The chief acid product of metabolism is $CO_2(H_2CO_3)$, of which an adult produces about 288 liters per day or 26 Eq, which is equal to 2.6 liters of concentrated HCl. Adults subsisting on a mixed diet produce in addition to CO_2 a substantial quantity of nonvolatile (fixed) acids. For example, 100 gm of protein produces approximately 60 mEq of sulfate by the oxidation of proteins and a quantity of phosphate, requiring 50 mEq of base for its neutralization at pH 7.4. An additional 50 mEq of base is required to neutralize the phosphate from 100 gm of fat (phospholipids). The acids produced are mainly sulfuric and phosphoric but include some hydrochloric, lactic, uric, and β-hydroxybutyric. Approximately one-half of these metabolically produced acids are neutralized by base in the diet, but the remainder must in some fashion be neutralized by the buffer systems of the body.

BUFFER SYSTEMS OF THE BODY

Several buffer systems prevent the organism from being overwhelmed by its own acid products. The one that reacts most readily is the blood buffer system. This depends only upon chemical processes which occur in the blood to attenuate the change in hydrogen ion concentration. Ultimately the buffering capacity of the tissues is brought into play also, since blood pH influences interstitial and intracellular pH as well.

For long-term stability of pH it is necessary to have an effective means of excreting the acids formed in the metabolic processes. This consists in the responses of the respiratory control center and renal tubules to changes in pH and carbon dioxide of the arterial blood. Moreover, these acids must be excreted as acids or as ammonium salts by the kidney since, if they were excreted as neutral salts, the buffers of the body would be rapidly depleted. This possibility becomes increasingly apparent when it is recognized that the total cation available in the blood as buffer salts does not exceed 150 mEq. In the body as a whole, including the protein salts and phosphates of the tissues, the total cation available is 1000 mEq.

The importance of the buffering capacity of the tissues cannot be overemphasized. If acid is produced metabolically, or infused intravenously experimentally, the immediate reaction for neutralization occurs with the blood buffers, and the assumption is made that the acid is largely neutralized by these buffers (in plasma and erythrocytes). However, several groups of investigators have shown that only 15 to 20 percent is neutralized by blood buffers, that a somewhat larger proportion is neutralized by

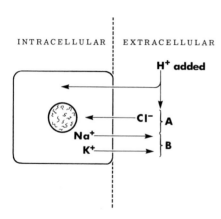

Figure 24-1
Biological (physiological) buffering: diffusion into cells. A. Buffering by diffusion into cells accompanied by ions of opposite charge. B. Buffering of extracellular pH with exchange across cell membrane with ions of like charge.

interstitial fluid, and that a major fraction is neutralized by buffers in tissue cells. Of this, 15 percent of the extracellular hydrogen ions were buffered by exchange with cellular potassium, and 36 percent with cellular sodium (Pitts, 1974). It is probable that some fraction of the sodium ions exchanged for H^+ may actually have come from the apatite crystals of the bone.

This type of buffering, which may be called *biological buffering*, consists in ionic shifts which protect extracellular pH. When acid or base is added to the extracellular space, approximately half of the added ions eventually diffuse into cells where they are probably buffered in the chemical sense. These ions or others that affect acid-base equilibrium (H^+, OH^-, HCO_3^-, etc.) are exchanged across the cell membrane for intracellular ions or are accompanied into cells by ions of opposite charge. For example, if an acid (H^+) is added to the body, some of it will be buffered chemically within the extracellular fluid. Some of the hydrogen ions will also diffuse across cell membranes (Fig. 24-1). Since H^+ is a positively charged ion, this diffusion requires either that a negatively charged ion such as chloride accompany it or that other positively charged ions cross the cell membrane in the opposite direction. Both processes occur, although the movement of cations out of the cell is quantitatively much more important. In any case, extracellular pH is defended by removal of some of the added acid from this compartment. Movements occurring in the opposite direction with alkalosis fall also into this type of buffering.

Blood Buffers

The Brønsted-Lowry concept of acids and bases finds wide acceptance in the consideration of acid-base balance (see Muntwyler, 1968). According to this view, an acid is a proton (H^+) donor, and a base is a proton acceptor. On this basis, the reaction indicating the acidity of an acid, A, takes the form

$$A \rightleftharpoons B + H^+$$

where B is a base since it can accept a proton to form the acid, A, the proton donor. Some examples are given in Table 24-1. Important buffer acids are shown in Table 24-2.

Buffers are substances that tend to stabilize the pH of a solution. A buffer is a mixture of either a weak acid and its conjugate base or a weak base and its conjugate acid. A buffer is effective only when there are appreciable quantities of the alkali salt and either

Table 24-1. Proton Donors and Acceptors

Acid	Proton	Conjugate Base
HCl	$\rightleftharpoons$ H^+	Cl^-
H_2SO_4	$\rightleftharpoons$ H^+	HSO_4^-
NH_4^+	$\rightleftharpoons$ H^+	NH_3

the acid or the base from which it derives. It is most effective when the salt and the acid or the base are present in equal quantities. Buffers are usually thought of in terms of buffer pairs. Acid-base buffer pairs of most importance in the body consist of weak acids associated with a salt of their conjugate base. Examples of common buffer pairs in the blood are $NaHCO_3/H_2CO_3$ and Na_2HPO_4/NaH_2PO_4. The proteins of the blood, hemoglobin, oxyhemoglobin, albumin, and globulin, are also extremely important buffers. Because of the amphoteric nature of proteins, each species represents a buffer pair. The major protein buffer in the blood is hemoglobin because of its high concentration. It is also considerably more important than the bicarbonate and phosphate buffers. The bicarbonate, however, has proved to be an accurate and easily determined index of acid-base balance.

Mechanism of Buffer Action

The mechanism of buffer action may be understood from consideration of the bicarbonate system outlined in equation 1.

$$CO_2 + H_2O \rightleftharpoons H_2CO_3 \rightleftharpoons H^+ + HCO_3^- \tag{1}$$

If hydrogen ions are added to the system, some will combine with bicarbonate ions and drive the reaction to the left. Thus, not all the hydrogen ions added will stay in solution in the ionic form. Conversely, if base is added, some of the H^+ will be removed, shifting the reaction to the right and causing the H_2CO_3 to dissociate into H^+ and HCO_3^-. Thus, there is an inverse relationship between H^+ and HCO_3^- concentration when acid or base is added. This relation may be plotted as shown in Figure 24-2. The pH-bicarbonate diagram used here and in other portions of this chapter was introduced by Davenport (1974). It has proved extremely useful in describing the underlying causes of acid-base disturbances from a limited amount of data. This figure represents the buffer properties of plasma, of which bicarbonate is the principal buffer. The initial point A represents the normal condition of the plasma in arterial blood, namely, a pH of 7.4 and a HCO_3^- concentration of 24mM per liter. When an acid is added to plasma, the pH decreases and, as a result of the reactions described above, the HCO_3^- concentration falls also. When base is added to plasma, both the pH and the HCO_3^- concentration increase. The effects of adding acid and base to blood or to plasma are as shown in

Table 24-2. Most Important Buffer Acids at Physiological pH

Proton Donor		Proton Acceptor		Proton
H_2CO_3	$\rightleftharpoons$	HCO_3^-	+	H^+
$H_2PO_4^-$	$\rightleftharpoons$	$HPO_4^=$	+	H^+
H protein	$\rightleftharpoons$	Proteinate$^-$	+	H^+
$HHbO_2$	$\rightleftharpoons$	HbO_2^-	+	H^+
HHb	$\rightleftharpoons$	Hb^-	+	H^+

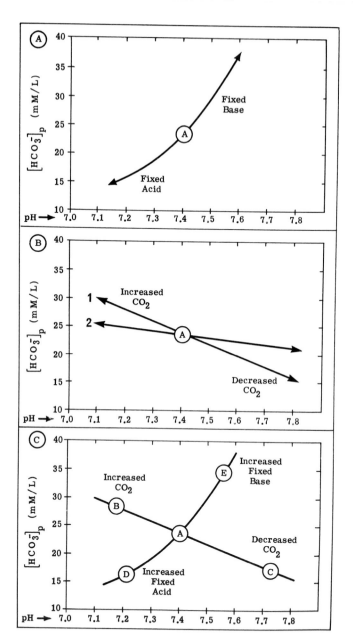

Figure 24-2
A. Effect of fixed acid and fixed base on pH and HCO_3^- content of separated plasma. B. Effect of changes in CO_2 on pH and HCO_3^-. Line 1 is true plasma (plasma equilibrated in presence of red cells and measured anaerobically). Line 2 is separated plasma (removed from red cells). C. Combination of the effects shown in A and B on true plasma. Note increased slope of buffer line with true plasma. (Reprinted from *The ABC of Acid-Base Chemistry* [6th ed.], by H. D. Davenport, by permission of The University of Chicago Press. Copyright 1947, 1949, 1950, 1958, 1969 and 1974, by The University of Chicago.)

Figure 24-2A only when the CO_2 tension is maintained constant. The above effects are seen when nonvolatile acids or bases are added to the blood. These are called *fixed* acids and bases, since they are not removed by respiratory activity. It should be emphasized that the terms *volatile* and *nonvolatile* or *fixed* are useful physiological terms but have no strict meaning in a chemical sense.

The pH of the blood can also be influenced greatly by the CO_2 level. However, the effects of CO_2 on the pH and HCO_3^- concentration are quite different from those of the fixed acids and bases, as may be seen by referring to equation 1. If CO_2 is added or removed, the concentration of H^+ and HCO_3^- changes in the same direction. The behavior of plasma alone in vitro is shown in Figure 24-2B. Thus, the effect of CO_2 can be readily distinguished from that of fixed acid or base simply by noting the shift produced on the graph.

Buffering Effect of Whole Blood

The changes produced in *separated plasma* studied in vitro will differ in some important respects from those seen in plasma which is part of whole blood (*true plasma*) (Fig. 24-2B). When fixed acid or base is added to plasma from whole blood, all the buffers participate in the process of attenuating the pH shift. As a result, the directional changes in the pH-bicarbonate diagram will be the same as in separated plasma, but the magnitude of change in pH will be decreased, per unit of acid or base added. Thus, when a *fixed acid* is added to whole blood, the buffering power of the red blood cell, which is due to the hemoglobin, also tends to attenuate the change in pH. If the amount of acid added were plotted against the pH of the solution, the slope of the buffer line would be steeper for whole blood as compared with separated plasma.

When CO_2 is added to whole blood, there is also a lesser effect on pH than would be the case with separated plasma. In contrast to the effects of fixed acid, the slope of the buffer line (Fig. 24-2C) changes because H^+ and HCO_3^- are formed in equal quantities but some of the H^+ is neutralized by other buffer systems while all the HCO_3^- remains. As a result, in whole blood more bicarbonate ions are formed per unit change in pH than in separated plasma. It should be borne in mind that the concentration of HCO_3^- being considered is that of the plasma portion of the blood only.

The concentration of H_2CO_3 and dissolved CO_2 in the plasma is much smaller than that of $NaHCO_3$. The ratio between these is approximately $1:20$. Ordinarily, this would mean that such a buffer would be rather ineffective, particularly when base is added to the blood. However, since CO_2 is continuously manufactured and to some degree stored in the tissues, an additional supply is readily available. This makes it unnecessary for a large quantity of CO_2 to be present to permit the buffer system to operate effectively.

The curves of Figure 24-2C represent the response of true plasma to fixed acid or base and changes in CO_2 level. The four variations from the normal buffer point represent the four types of acid-base disturbance seen clinically: *respiratory acidosis* (CO_2 excess), *respiratory alkalosis* (CO_2 deficit), *metabolic acidosis* (fixed acid excess), and *metabolic alkalosis* (fixed base excess). By plotting the pH and HCO_3^- levels of a sample of arterial blood on the graph, it is possible to determine whether the subject is normal or in one of these states of acid-base disturbance.

To carry these considerations farther, it is necessary to analyze their quantitative aspects. The pH-bicarbonate diagram is actually a graphic representation of the Henderson-Hasselbalch equation:

$$pH = pK + \log \frac{[HCO_3^-]}{[H_2CO_3]} \tag{2}$$

The pK of the bicarbonate system is known and the pH and HCO_3^- concentration can be measured, but the H_2CO_3 concentration, which is extremely low, cannot be directly measured. Thus the utility of the equation as it now stands is limited. However, reference to equation 1 reveals that H_2CO_3 concentration is directly dependent upon the concentration of dissolved CO_2, which is in turn dependent upon its partial pressure and solubility coefficient of CO_2 in plasma. When these factors are substituted into the equation, the following is obtained:

$$pH = pK + \log \frac{[HCO_3^-]}{a \; P_{CO_2}} \tag{3}$$

where the solubility coefficient, a, = 0.0301 and pK = 6.10.
Substituting, equation 3 becomes:

$$pH = 6.10 + \log \frac{[HCO_3^-]}{0.0301 \; P_{CO_2}} \tag{4}$$

The equation now relates three measurable quantities; if any two are known, the third may be calculated. Nomograms are available to facilitate such calculations.

Abnormal Acid-Base Balance States

Respiratory Acidosis

Since the directional changes on the graph of any form of acid-base disturbance are known, it is possible to determine the type and degree of any abnormality which may exist. Such a representation is shown in Figure 24-3. Point A represents the normal pH and HCO_3^- values. With a disturbance in acid-base balance these values will be shifted according to the type of disturbance. For example, if CO_2 elimination by the lungs is inadequate (e.g., obstructive lung disease, damage to muscles of respiration), the P_{CO_2} of the alveoli and arterial blood will be increased. This situation is called *respiratory acidosis* and is indicated by point B on the diagram. If the deviation from normality is caused only by respiratory acidosis, the point will lie somewhere on the normal buffer line. The change in pH and HCO_3^- produced represents simply the interaction of the increased CO_2 and the blood buffers. As compensatory mechanisms come into play (in this instance the kidney), point B will move away from the normal buffer line in a

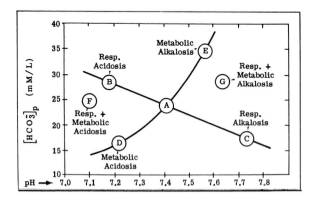

Figure 24-3
Effect of acid-base disturbances on pH and HCO_3^- level of the plasma, plotted on pH-bicarbonate diagram. Note that each type of disturbance occupies a unique place on the diagram. (Reprinted from *The ABC of Acid-Base Chemistry* [6th ed.], by H. D. Davenport, by permission of The University of Chicago Press. Copyright 1947, 1949, 1950, 1958, 1969 and 1974, by The University of Chicago.)

manner which will be described later. The situation illustrated is *uncompensated respiratory acidosis.*

Respiratory Alkalosis

There are also clinical disturbances in which chronic hyperventilation occurs. Examples are hysteria and salicylate poisoning in children. In this instance *respiratory alkalosis* is produced, as represented by point C. Since it lies on the normal buffer line, it is *uncompensated respiratory alkalosis.* Because compensation by the kidney is slow, this uncompensated state will be occasionally observed in the clinic.

Metabolic Acidosis

Metabolic acidosis is produced by conditions such as ketosis induced by diabetes mellitus or excessive loss of alkaline fluids from the lower digestive tract, as occurs in diarrhea. When such a condition occurs, the pH of the blood decreases and the HCO_3^- concentration decreases also (point D). It can therefore be differentiated from respiratory acidosis. If the PCO_2 is not changed as a result of the acidosis, point D will lie on the same PCO_2 isobar as the normal point and the term *uncompensated metabolic acidosis* is used. However, reduction of pH has a stimulatory effect on respiration which will reduce the PCO_2 levels. The effect of this respiratory compensation is to increase pH and reduce HCO_3^- levels, as will be seen later.

Metabolic Alkalosis

Metabolic alkalosis is produced by conditions such as persistent vomiting, which causes loss of a considerable amount of acid. In metabolic alkalosis (point E) the pH and HCO_3^- increase together, enabling it to be distinguished from respiratory alkalosis. If there is no respiratory response to the change in pH, the condition is *uncompensated metabolic alkalosis.*

In some instances there may be a mixture of acid-base abnormalities which can also be detected by use of the pH-bicarbonate diagram. For example, point F represents a combination of metabolic and respiratory acidosis, and point G is metabolic and respiratory alkalosis.

Summary of Changes in Plasma Bicarbonate–Carbonic Acid Buffer System in Disturbances of Neutrality

Figure 24-4 attempts to summarize the primary and secondary changes in the HCO_3^-/H_2CO_3 buffer system with various disturbances in acid-base balance.

In *respiratory acidosis,* the retention of CO_2 and rise in H_2CO_3 are the primary cause. The kidney is the important compensatory organ, secreting H^+ and conserving bicarbonate.

In *respiratory alkalosis* the net loss of carbonic acid by overbreathing is primary, and any tendency to compensate would decrease the bicarbonate ion concentration secondarily. The decrease is achieved by the kidney.

In *metabolic acidosis* the entering hydrogen ion reacts with the bicarbonate ion to produce a primary decrease in its concentration (shown by the heavy arrow). Respiratory activity causes a smaller, secondary decrease in the carbonic acid level (shown by the dotted arrow). Because the numerator is decreased more than the denominator, the pH must be lowered. In this instance the total CO_2 must be distinctly decreased. The total CO_2 content is a commonly employed laboratory determination, in which the serum is acidified and the CO_2 all brought into the gaseous form and measured. The compensatory mechanisms are discussed in greater detail below.

Likewise in *metabolic alkalosis,* the bicarbonate ion is primarily increased, the carbonic acid being secondarily elevated and to a lesser degree. The total effect on the CO_2 content is an elevation. Renal compensation involves loss of HCO_3^-.

Note that the overall direction in the change of the CO_2 content is downward in both metabolic acidosis and respiratory alkalosis and upward in metabolic alkalosis and

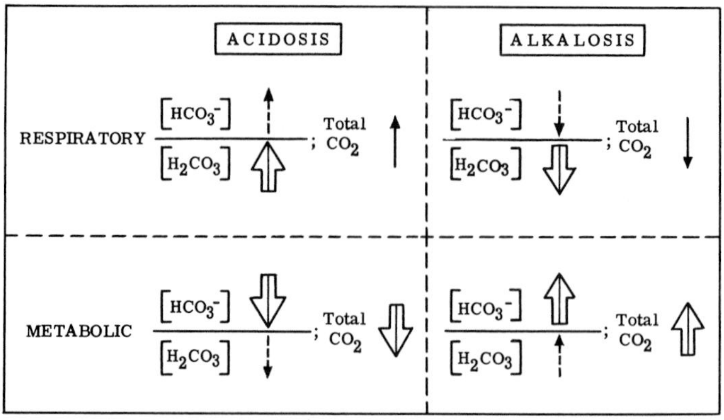

Figure 24-4
Primary and secondary changes in the bicarbonate–carbonic acid buffer system descriptive of the various types of disturbances in acid-base balance. Note that the total CO_2 does not always fall in acidosis or rise in alkalosis. (From H. N. Christensen. *Body Fluids and Their Neutrality.* New York: Oxford University Press, 1963. P. 122.)

respiratory acidosis. The reliability of any association between low CO_2 content and respiratory acidosis is vitiated. Reliance upon the serum CO_2 content (or capacity) alone as an assay of the acid-base balance obviously may lead to mistaken conclusions regarding acid-base balance disturbances of both respiratory and metabolic types.

Respiratory and Renal Compensation in pH-Bicarbonate Diagram

It has already been indicated that the respiratory system acts as a buffer system by responding to changes in blood pH in such a manner as to maintain the pH very nearly constant. This is accomplished only by altering the CO_2 level of the blood. Thus, any change in pH and bicarbonate induced by the respiratory system occurs along a buffer line, whether a respiratory disturbance of acid-base balance or a respiratory compensation for a metabolic disturbance is being considered.

The kidney can influence acid-base balance by conserving hydrogen ions or removing them from the bloodstream. The renal mechanisms that perform this operation have already been considered. The kidney acts to alter the relative amounts of fixed acid or base in the blood. Any changes of this nature in the blood follow the same path as those produced by metabolic acids or bases—that is, up or down a P_{CO_2} isobar.

These compensatory mechanisms may now be considered in terms of their effect on the pH-bicarbonate diagram. The pathways followed by the compensatory mechanisms are shown in Figure 24-5A. When *respiratory acidosis* occurs, for example (point B), renal compensation follows, in the form of an abnormally large excretion of acid by the kidney. With this, HCO_3^- tubular reabsorption is increased. Plasma Cl^- is lost to the extent that HCO_3^- concentration is raised. This acid loss causes movement of the point up the P_{CO_2} isobar toward a normal pH. At point B_1 the pH has been only partially restored to its normal value; at point B_2 it has been fully restored—hence the designations *partially compensated* and *fully compensated respiratory acidosis.* The kidney is able under some circumstances to bring the pH back to 7.4 as indicated by the fully compensated state. This achievement shows a highly effective mechanism of the kidney for hydrogen ion disposal. The response is, however, quite slow. Several days may be required for it to compensate for a chronic condition.

In *respiratory alkalosis* the kidney also acts to compensate for the abnormality, this time by conserving hydrogen ions produced in metabolic processes and by increased excretion of HCO_3^-. With increased HCO_3^- loss, Cl^- is conserved, raising plasma Cl^- concentration to replace lost HCO_3^-. Point C then moves to positions C_1 and C_2 as the degree of compensation increases. Again, it is possible that the compensation may be complete and a normal pH will be restored.

When *metabolic acidosis* occurs (point D), respiratory activity is immediately increased. As a result there is a very rapid partial compensation (point D_1). However, respiratory compensation is never complete enough to bring the pH back to the normal value. Apparently an appreciable pH difference is required to stimulate respiration. It is evident that if pH were returned precisely to its original value the stimulus for increased respiration would cease and the respiratory compensation would disappear. The compensatory patterns of the kidney and the respiratory system therefore differ in the respect that respiratory compensation is never complete. The kidney may later again restore pH to normal.

A similar pattern is seen in *metabolic alkalosis* (points E, E_1). Under the influence of the high pH of the arterial blood, respiratory activity diminishes, allowing CO_2 to accumulate. The compensation therefore occurs along a buffer line which is parallel to but displaced from the normal buffer line. The kidney is again needed for full compensation.

In Figure 24-5B are shown the areas of renal and respiratory compensation. Any set of values of pH and bicarbonate for a given sample of arterial blood must fall within one of these areas or those of combined metabolic and respiratory acidosis or alkalosis.

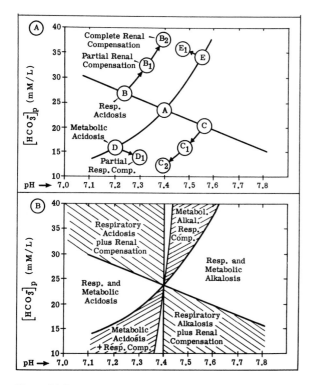

Figure 24-5
A. Effects of compensation on the pH and bicarbonate content. B. Zones of acid-base disturbance and compensation. (Reprinted from *The ABC of Acid-Base Chemistry* [6th ed.], by H. D. Davenport, by permission of The University of Chicago Press. Copyright 1947, 1949, 1950, 1958, 1969 and 1974, by The University of Chicago.)

Thus, the acid-base status of the individual may be determined by plotting the data on the graph and noting the area in which the point falls.

Respiratory Regulation of Body Fluid pH

All of the body's chemical buffers are tied into respiration through the bicarbonate buffer system, as evidenced in the following sequence of reversible reactions:

$$H^+ + HCO_3^- \rightleftarrows H_2CO_3 \rightleftarrows H_2O + CO_2 \tag{5}$$

The conversion of carbonic acid to water and carbon dioxide, and vice versa, is greatly accelerated by the presence of the enzyme *carbonic anhydrase* (CA) within the erythrocyte. An increase in the concentrations of any of the reactants or products drives the reactions in the direction which tends to reduce the increase until equilibrium occurs. For example, an increase in the concentration of hydrogen ions drives the reactions to the right, with the subsequent formation of carbonic acid and a reduction of hydrogen ion concentration toward normal. Conversely, a decrease in the concentration of any of the reactants or products drives the reactions in the direction which tends to increase the concentration toward normal. A decrease in the hydrogen ion concentration, for example, drives the reactions to the left. The subsequent dissociation of carbonic acid elevates hydrogen ion concentration toward normal. Through the elimination of carbon dioxide by the respiratory system, we may effect changes in body fluid pH: an increase in blood P_{CO_2} tending to elevate body fluid hydrogen ion concentration (decrease pH), and a decrease in blood P_{CO_2} tending to lower body fluid hydrogen ion concentration (increase pH).

Alveolar Ventilation and Body Fluid pH

Increases of body fluid hydrogen ion concentration, specifically in the blood and cerebrospinal fluid, result in a reflex acceleration of respiratory rate and an increase in tidal volume (Chap. 21). The net effect is to increase the rate of carbon dioxide elimination from the body, thereby driving the carbonic acid reactions to the right with the subsequent return of hydrogen ion concentration toward normal. Severe decreases in body fluid pH may increase alveolar ventilation to about five times its normal value. Conversely, increases in body fluid pH tend to depress respiratory activity, allowing carbon dioxide and hence hydrogen ions to increase, but the respiratory response is not as marked as it is to a decrease in pH. Severe increases in body fluid pH may decrease alveolar ventilation by 50 percent of normal.

It may be correctly inferred from the foregoing that not only does the pH of body fluids affect respiratory activity but alterations (voluntary or otherwise) in alveolar ventilation have a pronounced effect upon the pH of body fluids. Voluntary apnea (holding one's breath), for example, results in a decrease of pH, while voluntary hyperventilation elevates body fluid pH. Doubling the normal resting rate of alveolar ventilation elevates extracellular pH by about 0.23, whereas reducing the normal resting rate of alveolar ventilation to one-fourth normal reduces extracellular pH by 0.4. As long as the rate at which carbon dioxide is being eliminated from the body keeps up with the rate of carbon dioxide production, there will be no net change in the carbonic acid–bicarbonate ratio of body fluids, and hence under normal circumstances a stable pH is maintained.

Because of the respiratory system's ability to regulate body fluid pH through the elimination of carbon dioxide, it is sometimes referred to as a "physiological buffer system" (Fig. 24-6). As is the case with the body's chemical buffers, the respiratory system cannot eliminate actual hydrogen ions from the body, nor can it eliminate bicarbonate ions. The excretion of carbon dioxide in expired air results in the elimination of potential, not actual, hydrogen ions. The kidneys, however, are able to excrete hydrogen ions as well as effectively regulate the concentration of bicarbonate in body fluids. Although the renal response to changes in body fluid pH is not as quick as the response of either the chemical buffer systems or the respiratory system, the kidneys

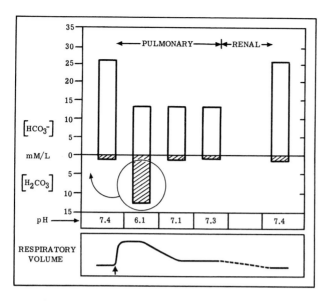

Figure 24-6
Stepwise illustration of the pulmonary response to acid invasion (not carbonic acid). Enough acid is administered at the arrow to cause almost one-half the HCO_3^- to be converted to H_2CO_3. (At the same time other buffer anions also accept H^+.) An explosive increase in respiration results, so that the new-formed H_2CO_3 is swept out as CO_2, and the hypothetical pH of 6.1 (second stage) is never reached. The third stage shows the transient picture when the P_{CO_2} has returned to a normal value. (From H. N. Christensen. *Body Fluids and Their Neutrality.* New York: Oxford University Press, 1963. P. 118.)

play an extremely important role in maintaining acid-base balance because of their ability to selectively conserve or eliminate various acids and bases, thus maintaining effective buffer ratios in body fluids.

Renal Role in Acid-Base Balance
In a quantitative sense, the major role of the kidneys in maintenance of acid-base balance is conservation of circulating stores of bicarbonate. In man, some 5100 mEq of bicarbonate ions are reabsorbed per day from the glomerular filtrate. In exchange, an approximately equivalent number of hydrogen ions are secreted. This is a key mechanism, along with the operation of the lung, in determining the proper ratio of carbonic acid and the bicarbonate ion, and ultimately of all the body buffers. In addition, the kidney generates titratable buffer acid and produces ammonia, these two mechanisms aiding in *base conservation.*

Renal Handling of Bicarbonate
The characteristics of the renal reabsorption and excretion of bicarbonate in man are revealed in Figure 24-7. If the plasma bicarbonate is reduced below about 26 mEq per liter, virtually all of the filtered bicarbonate is reabsorbed and none is excreted; if bicarbonate is infused and plasma concentration is elevated above 28 mEq per liter, a limited but constant amount (2.8 mEq per 100 ml of filtrate) is reabsorbed. All bicarbonate filtered in excess of this is excreted in the urine. Thus, the plasma value of 26 to 28 mEq per liter in man suggests a so-called renal bicarbonate threshold. The term *threshold* is used advisedly, for the statement that bicarbonate is practically absent from the urine at plasma concentrations below 26 mEq per liter is true only so long as the respiratory center is free to regulate pulmonary ventilation in response to the prevailing

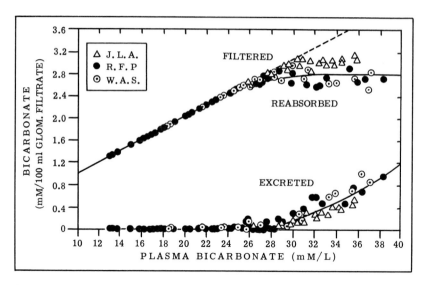

Figure 24-7
Filtration, reabsorption, and excretion of bicarbonate as functions of plasma concentration in normal man. (From R. F. Pitts, J. L. Ayer, and W. A. Schiess. *J. Clin. Invest.* 28:35–49, 1949.)

level of plasma CO_2. During hyperventilation, for instance, the urine may become alkaline and contain considerable amounts of bicarbonate, although the plasma concentration may be well below the so-called normal threshold.

The reason is that, when CO_2 is blown off by the lungs, the cations, such as sodium, are balanced by anionic groups of plasma proteins, which are nonfilterable at the glomerulus. But Na^+ is filtered and picks up HCO_3^- in the tubular urine, and the formed $NaHCO_3$ makes the urine alkaline even though the plasma bicarbonate is relatively low. The bicarbonate is synthesized by the epithelial cells of the tubules, aided by carbonic anhydrase (see below) and is captured by sodium in the fluid passing the cells of origin of HCO_3^- in the tubular lumen.

Proximal Tubular Mechanism

The proximal mechanism is specialized to reabsorb the bulk of the filtered bicarbonate from the tubular urine against a relatively low gradient. The basic elements of the mechanism are illustrated in Figure 24-8. Sodium and bicarbonate ions enter the proximal tubule in the glomerular filtrate. Sodium ions diffuse into the tubular cells down a concentration and electrical gradient and are actively extruded into the peritubular fluid by a pump, which maintains intracellular sodium concentration at a low value. Hydrogen ions move from the interior of the cell to the tubular lumen in exchange for sodium ions against an electrical gradient, necessitating active transport. Whether the passive influx of sodium ions and active efflux of hydrogen ions are carrier linked is not known. However, they are linked operationally. Hydrogen ions associate with bicarbonate ions to form carbonic acid, which decomposes into carbon dioxide and water. The carbon dioxide diffuses into the cell, where it undergoes hydration to form carbonic acid; this reaction is catalyzed by carbonic anhydrase. Subsequent dissociation provides the hydrogen ions, which are exchanged for sodium ions across the luminal membrane, and the bicarbonate ions, which diffuse down a potential gradient into the peritubular fluid. According to this concept, bicarbonate ions are reabsorbed indirectly by conversion within the tubular lumen to carbon dioxide. The key element of the

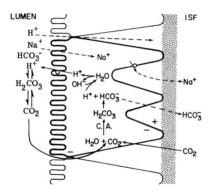

Figure 24-8
Ion exchange mechanism in the proximal tubular cell, emphasizing the role of carbonic anhydrase (CA). ISF = interstitial fluid.

reabsorptive process is the exchange, across the luminal membrane, of intracellular hydrogen ions for sodium ions in the tubular fluid. Some 90 percent of the filtered bicarbonate is reabsorbed in the proximal tubule.

A critical role is played by the enzyme carbonic anhydrase (CA). This is a zinc-containing metalloprotein of molecular weight 30,000. It occurs in high concentration in renal tubules, particularly in the proximal convolutions, where it is found in the brush border as well as in the cytoplasm. The evidence for CA action comes from the fact that certain sulfonamide derivatives which have the ability to inhibit CA block H^+ secretion and produce an excess of sodium in the urine (see Chap. 23). Carbonic anhydrase is not present on the luminal surface of the distal tubules, and none is present in the collecting duct.

The presence of carbonic anhydrase on the brush border of proximal tubules facilitates the bulk reabsorption of bicarbonate, for it reduces by a factor of 10 the gradient against which hydrogen ions must be secreted and correspondingly reduces the energy cost of bicarbonate conservation. It is maintained by some authorities that hydrogen ions are actively secreted by a redox (reduction-oxidation) ion pump. The mechanism is similar to that involved in the secretion of gastric acid.

Distal Tubule and Collecting Duct Mechanisms
The mechanisms for reabsorption of bicarbonate in distal tubules and collecting ducts are qualitatively the same as that in proximal tubules. They differ quantitatively in the following respects: (1) Together they account for only 10 percent of total bicarbonate reabsorption, and (2) they are capable of salvaging essentially all of the bicarbonate remaining in the tubular fluid when plasma concentration is normal or low. To accomplish the latter end, the collecting ducts, specifically, must be able to secrete hydrogen ions against much steeper gradients than either the proximal or the distal tubules. Also, sodium ions may have to be pumped from lumen into cells to achieve their nearly complete removal in states of sodium depletion. Accordingly, a more tightly coupled hydrogen-sodium exchange pump may be present in collecting ducts.

Factors Affecting Bicarbonate Transport
At least four factors influence bicarbonate transport: (1) changes in the carbon dioxide tension of the arterial blood, (2) variations in the body stores of potassium, (3) variations in the plasma level of chloride, and (4) variations in the secretion of adrenocortical hormones.

Effect of Changes in P_{CO_2} *of Arterial Blood.* Figure 24-9 shows the effects of acute variations in arterial P_{CO_2} on bicarbonate reabsorption in the dog. When animals are

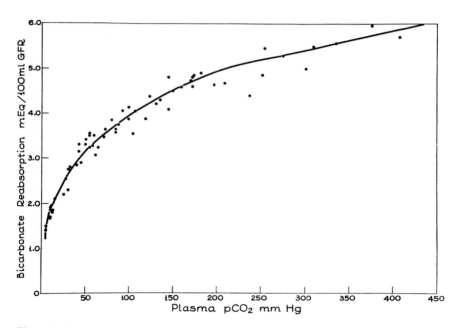

Figure 24-9
The relationship between Pco$_2$ and bicarbonate reabsorption. (From F. C. Rector, D. W. Seldin, A. D. Roberts, and J. S. Smith. *J. Clin. Invest.* 39:1706–1721, 1960.)

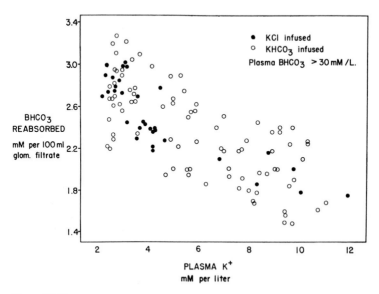

Figure 24-10
The relationship between plasma K$^+$ and bicarbonate reabsorption. (From G. R. Fuller, M. B. MacLeod, and R. F. Pitts. *Am. J. Physiol.* 182:111–118, 1955.)

hyperventilated, so that P_{CO_2} values are reduced to less than the normal 40 mm Hg, the reabsorption of bicarbonate is reduced. When animals breathe carbon dioxide–air mixtures, so that the P_{CO_2} is increased to values above the normal 40 mm Hg, the reabsorption of bicarbonate is increased. At very high levels of P_{CO_2} bicarbonate reabsorption more than doubles. Chronic exposure of animals to high partial pressures of carbon dioxide still further enhances the tubular reabsorption of bicarbonate, the major response occurring within 48 hours.

Variations in Body Stores of Potassium. Figure 24-10 illustrates the effects, in the dog, of the intravenous infusion of potassium salts on the rate of tubular reabsorption of bicarbonate. When the plasma concentration of potassium falls below the normal value of 4 mEq per liter, bicarbonate reabsorption is elevated above the normal value of 2.4 to 2.6 mEq per 100 ml of filtrate. When, in consequence of the infusion of potassium salts, the plasma potassium concentration exceeds 4 mEq per liter, bicarbonate reabsorption is reduced below the normal value. The mechanism of the influence on the bicarbonate reabsorption concerns the intracellular concentration of potassium. When infused, it enters the cells in exchange for H^+, which leaves the cells and is buffered by the plasma HCO_3^-, the concentration of which then falls. Presumably, the renal tubular cells respond in a similar manner, accumulating K^+ from the peritubular fluid and extruding H^+ into it. As a result of reduced intracellular H^+ following potassium infusion, fewer hydrogen ions are secreted into the lumen, and less bicarbonate is reabsorbed. The converse would operate with reduced potassium in the extracellular fluid. It is no longer held that the relationship between hydrogen ion and potassium ion secretion represents competition for a common secretory mechanism.

Hydrogen ion secretion and, thus, bicarbonate reabsorption are increased by hypokalemia and decreased by hyperkalemia along the entire length of the nephron, on account of the changes in cellular hydrogen ion concentration described above. Because filtered potassium is almost entirely reabsorbed in the proximal tubule, no competition between potassium ions and hydrogen ions for secretory transport could occur at this site. The potassium that is excreted is secreted in the distal part of the nephron. The secretory mechanism is passive, and the distribution of potassium ions is determined largely by the transtubular potential gradient (Chap. 22).

Relationship to Plasma Chloride Concentration. When the body is depleted of chloride and hypochloremia develops, plasma bicarbonate increases and bicarbonate reabsorption increases, sustaining a higher plasma concentration of HCO_3^-. Conversely, when body chloride stores are expanded and hyperchloremia develops, plasma bicarbonate decreases. Bicarbonate reabsorption is reduced, and plasma concentration is maintained at a subnormal level. When the chloride deficit is replaced or when the chloride excess is excreted, the plasma level of bicarbonate rapidly returns to normal. The mechanism is not well understood but probably relates to volume of the extracellular compartment. If extracellular volume is expanded by infusion of isotonic solutions, reabsorption of bicarbonate decreases. If extracellular volume is contracted by dialysis, reabsorption of bicarbonate increases.

Influence of Adrenocortical Hormones. In Cushing's syndrome and in hyperaldosteronism the plasma HCO_3^- is moderately elevated. In Addison's disease the plasma concentration of bicarbonate is moderately reduced. The mechanism of action in response to variation in output of the adrenocortical hormones is not known at this time but could be secondary to the hormonal influence on renal potassium handling (Chap. 34).

RENAL MECHANISMS FOR BASE CONSERVATION

Two mechanisms exist for base conservation: (1) *acidification of the urine* and (2) *ammonia synthesis.* In acidification of the urine, the neutral salts of weak buffer acids are converted to acid salts or even free acids. For example, H_3PO_4, NaH_2PO_4, uric acid, etc., can be excreted as titratable acid of the urine, reducing the glomerular filtrate from a pH of 7.4 to urine having a pH as low as 4.5. The general equation for this process is

$$Na_2HPO_4 + HHCO_3 \rightarrow NaH_2PO_4 + NaHCO_3 \tag{6}$$
$$\text{(excr.)}\text{(reabs.)}$$

The other main mechanism is ammonia synthesis. Where there are strong acids (H_2SO_4, HCl), they are finally excreted as ammonium salts.

$$Na_2SO_4 + 2H^+ + 2NH_3 \rightarrow (NH_4)_2SO_4 + 2Na^+ \tag{7}$$
$$\text{(excr.)}\text{(reabs.)}$$

Replacement of sodium by hydrogen ions means that the acid radicals can ultimately be excreted as free acids. There is a limit to the amount of acid that can be excreted in this form, however, because the kidneys cannot excrete urine which is more than about pH 4.5. Excretion of strong acids by this means is accompanied by their full equivalent of base. (Recall that a base is a proton acceptor.) Weaker acids can be excreted in the free form to an extent which is determined by their pK values. Thus on application of the Henderson-Hasselbalch equation:

$$pH = pK + \log \frac{\text{concentration of salt}}{\text{concentration of acid}} \tag{8}$$

Recall also that pK is the pH at which there are equal quantities of acid and the accompanying salt. If pK is the same as the pH of the urine, one-half of the acid will be present in the free form. In the case of a weak acid whose pK is 1 pH unit above the pH of the urine, nine-tenths of the acid will be free and one-tenth combined with base. However, with a strong acid whose pK is 1 pH unit below the pH of the urine, only one-tenth of the total amount of acid can be excreted in the free form, so that nine-tenths of its equivalent of base will have to accompany it. Stronger acids with pK values of more than 1 unit below the minimal urinary pH must always be excreted with more then nine-tenths of the equivalent base.

Those weaker acids which can be excreted to a considerable extent in the free form in the urine within the physiological range of pH are known collectively as the *buffer acids* of the urine. The most important is H_3PO_4, but amino acids and creatinine (pK 4.97) are also in this category. The pK for the dissociation of the second hydrogen ion of phosphoric acid is 6.8; therefore, each equivalent of phosphate in the urine of pH 4.8 requires one equivalent of base to cover it, whereas in the plasma at pH 7.4 it requires nearly two equivalents. Hence, by excreting urine of maximal acidity the kidneys can salvage one equivalent of fixed base for each equivalent of acid phosphate in the urine, as well as additional amounts corresponding to the other buffer acids which are present. The process of acidification of the kidney is reversed when the urine is titrated with alkali back to the pH of the plasma, 7.4, to determine the "titratable acidity." This reflects the amount of base the kidneys have conserved by acidifying the urine. Such base conservation may range from about 30 mEq to as high as 150 mEq per day during diabetic acidosis, when the urine contains large quantities of β-hydroxybutyric acid (pK 4.7), and acetoacetic acid.

Mechanism of Acidification of the Urine

Pitts and Alexander (1945) examined older theories of phosphate and bicarbonate reabsorption with simultaneous measurement of glomerular filtration rate and found that both fell far short of accounting for the possible titratable acidity of the urine, in each case only 20 percent or less, the limitation being the amount of acid in the glomerular filtrate (see Pitts, 1974). To explain the much greater availability of the acid, it was postulated that the tubular cells are able to exchange ions between the tubular urine and the bloodstream. Secretion of H^+ as a basis for an ion exchange mechanism was postulated, and this is now accepted.

Ion Exchange Mechanism

According to the ion exchange mechanism, H^+ is secreted by the tubular cells into the lumen in exchange mainly for Na^+ from the glomerular filtrate. The source of H^+ is the CO_2 of the metabolic activity of the tubular cells, or that brought in by the blood. The mechanism was discussed in relation to the bicarbonate reabsorptive mechanism.

As shown in Figure 24-11, the filtrate is moderately acidified in the kidney of the rat as it flows along the proximal tubule, but the major acidification occurs beyond the distal tubule, i.e., in the collecting duct. This is where the major fraction of the titratable acid represented by buffers of low pK, such as β-hydroxybutyrate (pK_A, 4.7), is formed. In this portion of the nephron a relatively small quantity of H^+ is pumped against a high gradient, whereas in the proximal tubule, a large quantity of H^+ is exchanged against a low gradient. Titratable acid with a high pK, as represented by phosphate, is probably formed chiefly in the proximal tubule, but also in the distal tubule. An example of the ion exchange theory is diagrammed in Figure 24-12. It can be discerned that the $NaHCO_3$ formed in this manner actually represents "new" bicarbonate, as contrasted to "reclaimed HCO_3," as discussed earlier for the bicarbonate reabsorptive mechanism in the proximal tubule.

The overall rate of secretion of H^+ is determined by the following factors: (1) the degree of acidosis, (2) the quantity of buffer acid excreted, and (3) the strength of the buffer—i.e., how strongly it resists giving up Na^+ for H^+ (e.g., phosphate with a pK of 6.8 permits greater secretion of H^+ than creatinine with a pK of 4.97). The significance of the last two factors is illustrated in Figure 24-13.

Ammonia Production

It has been stated that the ability to excrete strong acid in free titratable form is limited because the urine pH cannot go below about 4.5. Another mechanism for conservation

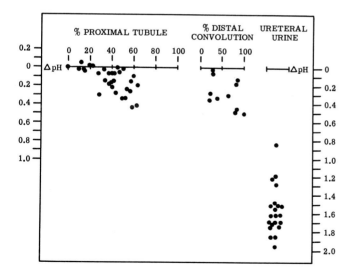

Figure 24-11
Change in pH of tubular fluid along the nephron of the rat. Measurements were made under equilibrium conditions. (From C. W. Gottschalk, W. E. Lassiter, and M. Mylle. *Am. J. Physiol.* 198:581–585, 1960.)

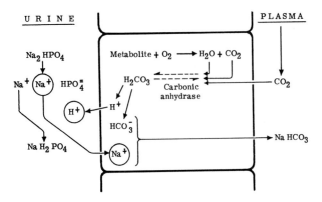

Figure 24-12
The mechanism of formation of titratable acid and conservation of base.

of base is created by secretion of NH_3 into the urine, which binds hydrogen ion and forms the ammonium ion NH_4^+. The ability of the kidney to form ammonia was discovered by Nash and Benedict, who found more NH_3 in the venous blood than in the arterial blood entering the kidney.

According to Van Slyke et al. (1943), the most important mechanism for formation of ammonia is its formation from plasma glutamine with the aid of glutaminase:

$$\text{glutamine} \xrightarrow{\text{(glutaminase)}} \text{glutamic acid} + NH_3$$

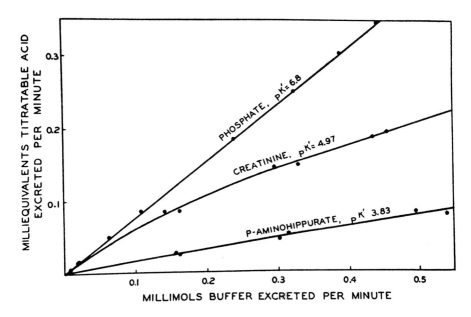

Figure 24-13
The relationship between buffers of different pK′ values and excretion of titratable acid. (From W. A. Schiess, J. L. Ayer, W. D. Lotspeich, and R. F. Pitts. *J. Clin. Invest.* 27:57–64, 1948.)

Thus, glutamine appears to be the immediate precursor of ammonia normally produced by the kidney. Other amino acids (L-asparagine, DL-alanine, L-histidine, L-aspartic acid, glycine, L-leucine, L-methionine, and L-cysteine) when infused into acidotic dogs also increase the secretion of ammonia. These acids are transferred to α-ketoglutarate by transaminases to form glutamate, and this in turn is converted to additional glutamine by the glutamine synthetase system:

α-ketoglutarate $+$ α-amino acid $\rightarrow$ glutamate $+$ α-keto acid
glutamate $+$ ATP $+$ NH_3 $\rightarrow$ glutamine $+$ ADP $+$ PO_4

Additional ammonia can theoretically come from glutamate directly (by undergoing deamination within the tubular cells containing a specific glutamic dehydrogenase), or from the glutamine formed, as above. In summary, glutamine and amino acids contribute amino and amide groups to a nitrogen pool from which ammonia is formed and diffuses into the urine, where it is trapped as ammonium ion (Fig. 24-14). The mechanism of action is as follows: Each molecule of NH_3 (a proton acceptor) secreted binds one hydrogen ion and permits one sodium ion to be reabsorbed. Thus, it is to be expected that the pH of the urine will determine the rate of diffusion of NH_3 into the tubular lumen. The mechanism is understood better by the following series of equations:

$$NH_3 + H^+ \rightleftharpoons NH_4^+ \tag{9}$$

The ionization constant, K_A, is:

$$K_A = \frac{[H^+][NH_3]}{[NH_4^+]} \tag{10}$$

$$pH = pK_A + \log \frac{[NH_3]}{[NH_4^+]} \quad (pK_A = 8.9) \tag{11}$$

It is accepted on good evidence that the cells generally are permeable to the molecular species NH_3 but not the ionic species NH_4^+, perhaps because NH_3 is lipid-soluble, whereas NH_4^+ is water-soluble. Since NH_3 is presumably in the same concentration on both sides of the cell membrane because of its easy diffusibility, the numerator of the last term of the equation can be considered to be a constant; or, to put

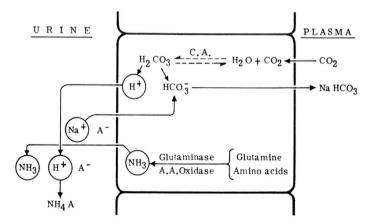

Figure 24-14
The mechanism of production of ammonia by the kidney and its role in base conservation.

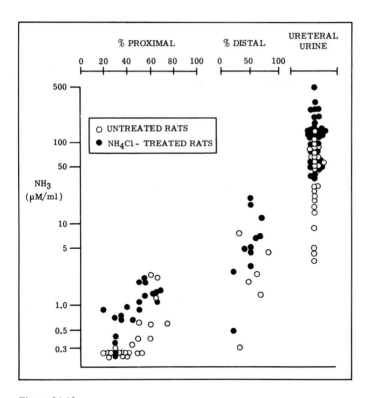

Figure 24-15
Sites of secretion of ammonia in untreated rats as compared to rats made acidotic by NH_4Cl ingestion. (From S. Glabman, R. M. Klose, and G. Giebisch. *Am. J. Physiol.* 205:127–132, 1963.)

it simply, the log of NH_4^+ will vary inversely with the pH. Then, as the urine becomes increasingly acid, the mass law operates to increase the fraction of the total ammonia produced in the distal segment that is captured in the urine as nondiffusible NH_4^+ and excreted as such. The capture of NH_4^+ in the urine serves to neutralize free acid and thus permits a larger quantity of acid radical to be excreted as NH_4^+ salt at a given urine hydrogen ion concentration than would otherwise be possible. Stated another way, the excretion of NH_4^+ salt reduces the quantity of Na^+ which would otherwise accompany the acid radical into the urine. The tubular excretion of NH_4^+ thus serves to conserve an equivalent quantity of Na^+ for the body.

Site of Production of NH_3
Micropuncture data in the rat reveal that the proximal tubule is an important potential source of final urine ammonia, and that increased proximal ammonia addition occurs in response to chronic ammonium chloride acidosis (which leads to metabolic acidosis and increased renal glutaminase activity). Distal tubular ammonia addition is also important in the final ion exchange mechanism (Fig. 24-15). The collecting duct was found to be a significant source of final urine ammonia in the hamster kidney but not in the rat kidney.

Summary of Base Conservation Mechanisms
The micropuncture techniques of Gottschalk et al. (1960) revealed that in the non-diuretic rat the pH of the proximal tubular urine decreased below that of the plasma

(Fig. 24-11). This result is in keeping with the finding, also by micropuncture techniques, that considerable absorption of bicarbonate goes on in the proximal convoluted tubules. The drop of pH in the proximal system is not as great as in the distal nephron (less than one-half pH unit) and is explained in terms of the bulk exchange that proceeds here. In terms of bicarbonate reabsorption in the proximal tubule, the amount of H^+ exchanged here must be considerably greater (3500 mEq per day) than that secreted in the distal convoluted tubule and collecting duct (100 mEq per day) but proceeds on a one-for-one basis ($Na^+ \rightleftharpoons H^+$), so pH does not change significantly in the proximal segment. The reason is that the H^+ which reacts with filtered HCO_3^- is recycled back into the cell and therefore does not ordinarily contribute to the net elimination of acid.

Only that small fraction of the total secreted H^+ which is bound to urinary buffer, either as titratable acid or as NH_4^+, serves to rid the body of metabolically produced acids. The factors influencing the distribution of H^+ between filtered HCO_3^- and non-HCO_3^- buffers determine to what extent the potentially available H^+ will be utilized to conserve filtered HCO_3^- and to what extent to generate new HCO_3^-. If, for example, metabolic acid is administered, $NaHCO_3$ will be decomposed, resulting in a fall in the $[HCO_3^-]$ in glomerular filtrate. Consequently, less H^+ will be utilized in the reabsorption of filtered HCO_3^- and more available for excretion as titratable acid and urinary ammonia. If there is a plentiful supply of filtered buffers (e.g., phosphate, creatinine), titratable acid will be produced. If, on the other hand, there is a paucity of filtered buffers, the removal of H^+ from the renal cell will be restricted and an intracellular acidosis will supervene. The intracellular acidosis will stimulate the formation of new buffer in the form of ammonia, which is then added to the tubular fluid, thus permitting H^+ secretion to return toward its normal rate.

The particular importance of the distal nephron mechanism stems from the fact that the amount of H^+ that can be pumped into the urine is greater than the excretion of base. The base conservation thus facilitated is abetted by capitalizing on the NH_3 production mechanism. However, because ammonia is synthesized in the proximal convolution and pH under certain conditions decreases along this segment, some overlap of function between proximal and distal convolutions can be anticipated.

Summary of the Regulatory Mechanism
Aside from the buffering effects of body cells when hydrogen ion is ingested, there will be buffering by the major plasma buffer system, $H^+ + HCO_3^- \leftrightarrow H_2CO_3 \leftrightarrow H_2O + CO_2$ (Fig. 24-16). If one adds hydrogen ion to this system, the above equation shifts to the right, resulting in a lowering of bicarbonate and increased production of carbon dioxide gas. The CO_2 is then carried to the lungs and expired. This action serves to minimize the change in the ratio of $\dfrac{HCO_3^-}{P_{CO_2}}$, or pH, and is referred to as *compensation*. Despite respiratory compensation, the organism still has excess hydrogen ion; and because the equation has moved to the right, there has been a lowering of serum bicarbonate. The kidney normally, then, has the dual roles of eliminating the excess hydrogen ion and regenerating bicarbonate.

Although the regeneration of bicarbonate is a continuous operation in the nephron, for convenience it can be summarized as a two-step phenomenon.

1. The first step may be regarded as almost complete reabsorption of the filtered bicarbonate taking place in the proximal tubule and *preventing* further reduction in serum bicarbonate. The reabsorption of bicarbonate is not a direct process. The filtered HCO_3^- is converted to H_2O and CO_2 by the H^+ that has been secreted into the lumen. The CO_2 diffuses back into the cell, shifting the reaction toward the formation of H_2CO_3 and HCO_3^-. This intracellularly generated HCO_3^- is delivered to the extracellular fluid with sodium that was reabsorbed in exchange for the secreted hydrogen ion. The result of this phase is the prevention of further lowering of serum bicarbonate and the loss of some hydrogen ion.

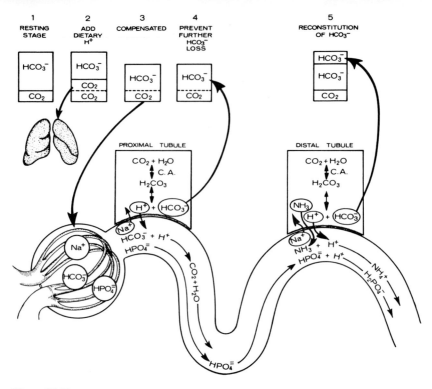

Figure 24-16
Summary of the role of the kidney in regulation of acid-base balance. This is exemplified by the renal handling of the daily dietary acid (hydrogen ion) load of about 50 mEq. (From S. Papper. *Clinical Nephrology*. Boston: Little, Brown, 1971. P. 68.)

2. Although most of the filtered HCO_3^- is converted to H_2O and CO_2, H^+ secretion continues, primarily in the distal nephron. Here the H^+ is buffered by filtered buffers such as $HPO_4^=$ (titratable acid) to form $H_2PO_4^-$, or by ammonia (NH_3) and excreted as ammonium ion (NH_4^+). Ammonia is secreted by the distal tubular cells in response to acidosis. These two essential buffer mechanisms allow for elimination of hydrogen ion without lowering the urinary pH to levels that may damage the tissues. The hydrogen that is secreted is derived from cell H_2CO_3, leaving bicarbonate behind in equimolar amounts. This bicarbonate, along with the sodium reabsorbed in exchange for secreted hydrogen, moves into the extracellular fluid and reconstitutes the serum bicarbonate level. Again, it is as if more bicarbonate is reabsorbed to regenerate the depressed bicarbonate. In renal tubular cells the hydration of CO_2 and consequent dissociation into H^+ and HCO_3^- is catalyzed by carbonic anhydrase.

At the end of this process, the daily dietary acid load is excreted, buffered mostly with ammonia and phosphate, and the serum bicarbonate is reconstituted.

Clinical Abnormalities of Acid-Base Homeostasis

Since the hydrogen ion concentration of the blood ultimately affects the hydrogen ion concentration of all body fluids, and because blood is readily accessible for chemical analysis, disorders of acid-base balance will be described in terms of deviations from the normal pH value of blood.

Acidemia is defined as an acidic condition of the blood signified by a pH value less than 7.35. The physiological processes which are causing acidemia define the term *acidosis* (literally, "a condition of becoming acidic"). *Alkalemia* is defined as an alkaline condition of the blood signified by a pH value greater than 7.45. The physiological processes which are causing the alkalemia define the term *alkalosis* (literally, "a condition of becoming alkalotic"). (See Table 24-3.)

Table 24-3. Examples of Acid-Base Disturbances

Arterial pH	Primary Disturbance	Primary Acid-Base Defect	Example of Blood Acid-Base Values	Compensatory Mechanisms
Normal	None	None	pH 7.35–7.45 P_{CO_2} 35–48 mm Hg $[HCO_3^-]$ 23–28 mEq/L	
Acidemia (low pH)	Respiratory acidosis (chronic obstructive lung disease)	Retention of CO_2	pH 7.30 P_{CO_2} 80 mm Hg $[HCO_3^-]$ 39 mEq/L	Increased respiration; renal retention of bicarbonate, increased excretion of titratable acid and ammonium ions
	Metabolic acidosis (uncontrolled diabetes mellitus)	Excess keto acid production	pH 7.22 P_{CO_2} 30 mm Hg $[HCO_3^-]$ 12 mEq/L	Increased respiration; renal retention of bicarbonate, increased excretion of titratable acid and ammonium ions
Alkalemia (high pH)	Respiratory alkalosis (voluntary hyperventilation)	Excess CO_2 loss	pH 7.62 P_{CO_2} 19 mm Hg $[HCO_3^-]$ 19.5 mEq/L	Decreased respiration; increased renal excretion of bicarbonate
	Metabolic alkalosis (vomiting)	Loss of acid gastric juice	pH 7.57 P_{CO_2} 45 mm Hg $[HCO_3^-]$ 42 mEq/L	Decreased respiration; increased renal excretion of bicarbonate

Source: From R. G. Pflanzer and G. A. Tanner. Acid-Base Regulation. In E. E. Selkurt (Ed.). Basic Physiology for the Health Sciences. Boston: Little, Brown, 1975. P. 535.

Clinical evaluation of the acid-base status of a patient involves the determination of arterial blood pH, P_{CO_2}, and $[HCO_3^-]$. All three variables are mutually related by means of the Henderson-Hasselbalch equation. The application has been discussed earlier.

Clinically, six possible combinations of alkalosis and acidosis may be observed in patients, depending upon etiology and compensatory processes:

1. Respiratory acidosis + metabolic acidosis, arterial pH below 7.4.
2. Respiratory alkalosis + metabolic alkalosis, arterial pH above 7.4.
3. Primary respiratory acidosis + secondary metabolic alkalosis, arterial pH below 7.4.
4. Primary respiratory alkalosis + secondary metabolic acidosis, arterial pH above 7.4.
5. Primary metabolic acidosis + secondary respiratory alkalosis, arterial pH below 7.4.
6. Primary metabolic alkalosis + secondary respiratory acidosis, arterial pH above 7.4.

Physiological compensation for major disturbances of acid-base equilibrium is rarely complete; therefore, the observed values of arterial P_{CO_2} and $[HCO_3^-]$ indicate which component (respiratory or nonrespiratory) is causing acidosis and/or alkalosis, and the value of arterial pH indicates which process is primary and which is secondary or compensatory.

Effects of Acidosis and Alkalosis

When the pH of body fluids falls below normal values, all of the body's systems become affected to varying degrees. Most noticeable is the depressant effect of acidosis upon activity of the central nervous system. When arterial pH falls to near or below 7.0, neuromuscular coordination becomes erratic (e.g., the individual may stagger as though drunk); the condition is compounded by disorientation, and eventually coma ensues followed by death. A classic example is that of the uncontrolled severe sugar diabetic (diabetes mellitus) who, because of excessive metabolism of lipids with resultant elevation of blood keto acids (metabolic acidosis), experiences a fall in arterial pH (diabetic acidemia), loss of neuromuscular coordination and normal orientation (often mistaken for drunkenness, since it is usually combined with excretion of acetone by the lungs), and finally coma.

The principal effects of severe alkalosis are hyperexcitability of both the central nervous system and the peripheral nervous system. An increase in the excitability of peripheral nerves is manifested by tetany of skeletal muscle. Central nervous system involvement is reflected in extreme nervousness, overreaction to normal stimuli, and in some instances (e.g., epilepsy) convulsions.

REFERENCES

Berliner, R. W. Ion Exchange Mechanisms in the Nephron. In A. P. Fishman (Ed.), *Symposium on Salt and Water Metabolism*. New York: New York Heart Association, 1959, P. 892.

Davenport, H. D. *The ABC of Acid-Base Chemistry* (6th ed.). Chicago: University of Chicago Press, 1974.

Dick, D. A. T. *Cell Water*. Washington: Butterworth, 1966.

Gamble, J. L. *Chemical Anatomy, Physiology and Pathology of Extracellular Fluid*. Cambridge, Mass.: Harvard University Press, 1954.

Glabman, S., R. M. Klose, and G. Giebisch. Micropuncture study of the ammonia excretion in the rat. *Am. J. Physiol.* 205:127–132, 1963.

Gottschalk, C. W., W. E. Lassiter, and M. Mylle. Localization of urine acidification in the mammalian kidney. *Am. J. Physiol.* 198:581–585, 1960.

Hayes, C. P., Jr., J. S. Mayson, E. E. Owen, and R. R. Robinson. A micropuncture evaluation of renal ammonia excretion in the rat. *Am. J. Physiol.* 207:77–83, 1964.

Maxwell, M. H., and C. R. Kleeman. *Clinical Disorders of Fluid and Electrolyte Metabolism*. New York: McGraw-Hill, 1962.

Muntwyler, E. *Water and Electrolyte Metabolism and Acid-Base Balance.* St. Louis: Mosby, 1968.

Pitts, R. F. *Physiology of the Kidney and Body Fluids* (3rd ed.). Chicago: Year Book, 1974.

Pitts, R. F., and R. S. Alexander. The nature of the renal tubular mechanism for acidifying the urine. *Am. J. Physiol.* 144:239–254, 1945.

Smith, H. W. *Principles of Renal Physiology.* New York: Oxford University Press, 1956.

Van Slyke, D. D., R. A. Phillips, P. B. Hamilton, R. M. Archibald, P. H. Futcher, and A. Hiller. Glutamine as source material of urinary ammonia. *J. Biol. Chem.* 150:481–482, 1943.

Welt, L. G. *Clinical Disorders of Hydration and Acid-Base Equilibrium* (2nd ed.). Boston: Little, Brown, 1959.

25. Gastrointestinal System
A. Circulation

Ewald E. Selkurt

The act of chewing and swallowing is carried out by the voluntary and involuntary muscles of the upper gastrointestinal (GI) tract. The blood flow through these tissues is closely related to the metabolic requirements of the muscle, as discussed in Chapter 29. When the food reaches the stomach, the digestive process begins; this ultimately involves a number of organs—stomach, small and large intestine, liver, and pancreas—which together comprise what is known as the splanchnic region. While having no digestive function, the spleen is also included in this domain.

General Considerations of the Gastrointestinal Circulation

The above functions require a highly organized vascular bed. After passing through the GI tract circulation the vessels funnel into the liver, which is coupled in series with the GI tract via the portal vein (Fig. 25A-1). The combined vascular beds of the liver, the stomach, and the intestines are called the *splanchnic circulation*. In a 70-kg man, the splanchnic bed totals 3.5 to 4.0 kg. Of this, the liver weighs 1.5 kg and the GI tract almost 2 kg. The spleen weighs 200 gm and the pancreas 60 to 80 gm. The GI tract is about half smooth muscle, half glandular and mucosal tissue.

At rest, the splanchnic vascular bed receives 20 to 25 percent of the cardiac output, 1250 to 1500 ml per minute in a 70-kg man, or 700 ml per minute per square meter surface area.

SPLANCHNIC BLOOD VOLUME

Estimates employing tagged red cells (^{32}P) and labeled albumin (^{131}I) techniques have been variable, but splanchnic blood volume appears to be approximately 25 percent of the total blood volume (Chap. 23). About one-half is in the gastrointestinal tract, and the remainder is equally divided between the liver and spleen.

This relatively high volume, plus rich sympathetic nerve innervation (Chap. 7A), provides an important reservoir depot for "autotransfusion" into the central circulation during hemorrhage. In particular, the spleen serves as a major blood reservoir, perhaps less so in man than other species, such as the dog. The spleen's role will be considered in greater detail in a later section.

HEPATIC CIRCULATION

The liver is the main "chemical factory" of the body, and its O_2 usage is nearly 20 percent of that of the whole body of man at rest. Its O_2 supply, which is largely from the portal vein, is reinforced by the hepatic artery, which provides 30 percent of the blood flow (and 40 to 50 percent of the O_2 used). Portal venous and hepatic arterial blood mix in the hepatic sinusoids, drain into the hepatic veins, then enter the inferior vena cava. The vascular bed of the liver, thus, is coupled both in series (portal vein system) and in parallel (hepatic artery) to the GI tract.

Anatomical Aspects

The anatomical features of the liver should be reviewed before one continues with this section. Certain details of the vasculature are presented here which have a direct bearing on physiological function.

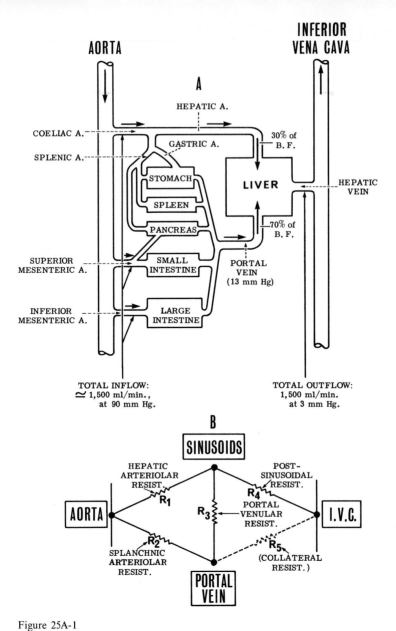

Figure 25A-1

A. The gastrointestinal and hepatic circulation. B. The resistances indicated are the determinants of sinusoidal and portal venous pressures and of flows through the portal vein and hepatic artery. In addition to the hepatic arteriolar resistance (R_1), colic, mesenteric, pancreatic, gastric, and splenic arteriolar (R_2), portal venular (R_3), and postsinusoidal (R_4) resistances; a fifth resistance lying in direct communication between the portal vein and inferior vena cava (the collateral resistance) is shown as a dotted line. It may be seen that the resistance pattern resembles that of the Wheatstone bridge. The total splanchnic resistance (R_T) may be expressed in terms of its constituent resistances as follows (omitting the collateral resistance):

$$R_T = \frac{R_1 R_2 + R_1 R_3}{R_1 + R_2 + R_3} + R_4$$

(After Bradley, 1963.)

588

Hepatic Artery

The role of the hepatic artery is to supply oxygenated blood to the parenchymal cells. This is not an end artery but has many collateral connections. Anastomotic connections with the portal vein occur at several levels, in the interlobular spaces or to terminal branches of the portal vein just before these enter the sinusoids. Terminally, intralobular hepatic arterioles and arterial capillaries empty into the sinusoids lined by the parenchymal cells. These branches supply the interior sinusoids in the central part of the lobule as well as peripheral sinusoids. Other branches supply the capsule, interlobular septa, and tissue of the bile ducts.

Portal Vein

The blood supplied through the portal vein comes from the gastrointestinal tract, pancreas, and spleen. Although functioning primarily to carry digested foodstuffs from the intestine to the liver, the portal vein also supplies oxygen to the liver. When the portal flow is shunted away from the liver to the inferior vena cava (*Eck's fistula*), deranged hepatic function develops because of the impaired circulation. Central portions of the lobule atrophy and become filled with fat.

Sinusoids

Conducting portal vein branches give rise to small distributing veins. These as well as smaller axial portal vein branches give off inlet venules which enter the lobules through holes in the limiting lamina of the hepatic parenchyma. Beyond this, the inlet venules branch into sinusoidal arborizations. The terminal ends of the portal vein branches split directly into sinusoids, the "exchange" vessels of the liver. The sinusoids are slightly ampullate in shape just before they join the draining veins, giving the impression of a sphincter. Also, they appear narrowed at the point of inlet from the venules.

The presumption that sphincters are located at the entry and exit of the sinusoids, permitting regulation of flow and storage of blood, has given rise to the concept of *intermittence of flow*. Such has been observed in the frog and in the rat. Spontaneous local variations in blood flow of the human liver have been observed with a Hensel calorimetry needle. This temperature-recording device, inserted into the liver substance, shows temperature changes that reflect local alterations in blood flow.

Veins

The sinusoids empty into the central vein of the lobule, which connects with the sublobular veins. These in turn converge into the hepatic veins, which discharge into the inferior vena cava. Strong muscular sphincters have been found in the hepatic vein of dogs, but not in the human. They may aid in regulation of hepatic venous outflow.

Methods for Measurement of the Blood Flow of the Liver and Splanchnic Bed

Measurement methods are *direct* (venous outflow, electromagnetic flowmeter, ultrasonic flowmeter, rotameter, bristle flowmeter) or *indirect* utilizing the Fick principle (Chap. 11). An application of the latter method suitable for man utilizes the removal of dyes from the plasma, cleared only by the specific organ being studied—in this case, the liver. Referring to Figure 25A-1, we see that dye extraction by the liver will, in fact, measure the minute blood flow of the entire splanchnic bed. Hence, splanchnic blood flow (SBF) and hepatic blood flow (HBF) are essentially equal.

Dyes used for this purpose are sodium sulfobromophthalein (Bromosulphalein, BSP) and indocyanine green (Cardio-Green). A catheter directed into the hepatic vein is essential for measuring the arteriovenous (A-V) difference. (Systemic venous blood concentration suffices for the arterial concentration of the dye, since these concentrations are essentially in equilibrium.) The rate of dye removal into the bile is determined indirectly (Bradley, 1963) by provisional infusion of dye at varying rates until a constancy of plasma concentration is attained (i.e., dye is being infused at a rate equal to liver removal). (Slight changes in the slope of the dye concentration can be corrected to give the true rate of removal, from a knowledge of the blood volume.) The method assumes removal of the dye only by the liver; extrahepatic removal would lead to obvious overcalculation of splanchnic blood flow. The calculation of blood flow is then made as follows:

$$\text{SBF (or HBF)} = \frac{I}{C_A - C_{HV}} = \frac{R}{C_A - C_{HV}} \times \left(\frac{1}{1 - \text{hemat}}\right)$$

(C_A represents the systemic dye concentration in milligrams per milliliter. C_{HV} represents the hepatic vein concentration. $C_A - C_{HV}$ represents extraction by the liver.) When arterial concentration (C_A) becomes constant, dye infusion rate, I (mg/min), equals removal (R). Since the method analyzes for plasma clearance, total blood flow requires adding the red cell volume. For hematocrit (hemat), see Chapter 10.

In actuality, the method measures "effective" blood flow, in the sense used for the indirect measurement of renal blood flow (Chap. 22), since it measures blood flow to the parenchymal tissue which does the secreting.

This method has been useful in measuring SBF and HBF changes in exercise, hemorrhage, and effects of postural changes and syncope. It has value in estimating hepatic blood flow changes in certain pathophysiological states (e.g., cirrhosis of the liver and congestive heart failure). Generally speaking, HBF is reduced in all of the above circumstances. Obviously, the dye removal method cannot reveal the site or nature of specific hemodynamic changes in individual organs of the combined bed (stomach, small and large intestines, spleen, etc.). Even more specific information is needed for physiological studies of factors influencing flow to various tissues that make up these organs (mucosa, muscle layers). Methods that provide more specific information concerning organ and tissue blood flow will be discussed below in greater detail as individual organs are considered.

Neurogenic Control of Hepatic Circulation
Because of the complexity of the vascular bed of the liver, an analysis of nervous regulation is difficult. By direct observation of liver flow, sympathetic stimulation has been seen to cause constriction of arterioles, sinusoids, and indeed all intrahepatic vessels. It must be kept in mind that some of these changes, such as sinusoidal constriction, may be due to a "passive recoil" effect from arteriolar constriction. Studies using roentgen ray opacity techniques suggest that the constrictor influence upon smaller vessels varies in different regions of the liver. Little is known about the control of the A-V shunts that exist in the liver.

Dilation of the liver vessels apparently cannot be induced by vagal stimulation. It is therefore doubtful that specific vasodilator fibers run to the liver.

It should be remembered that the extrinsic control of the liver circulation adds further complexity, since vasomotor changes in the intestine and spleen modify portal vein inflow, aside from any direct neurogenic or humoral influences on the liver.

Humoral Factors Which Modify Hepatic Blood Flow
Direct injections into the liver circulation or local applications of epinephrine induce a prompt vasoconstrictor response. When low epinephrine concentrations are given intravenously, mostly dilator effects are seen, with evidence of concomitant marked metabolic effects on the liver. It has been speculated that some secondary metabolic by-product may be responsible for the dilating effect. Norepinephrine, whose metabolic influences on the liver are very weak, causes a pure constrictor action. Histamine causes an increase in hepatic blood flow in man, along with a decrease in blood pressure suggesting active dilation of the splanchnic bed. Insulin injection results in a somewhat delayed hepatic hyperemia, probably on a metabolic basis. It suggests an epinephrine response or other compensatory reactions to the insulin-induced hypoglycemia.

Splanchnic Oxygen Utilization

The classic studies of Blalock and Mason in 1936 revealed that hepatic O_2 utilization in unanesthetized dogs averaged 24 ml per minute or 0.045 ml per gram per minute. Average oxygen contents were: hepatic artery, 16.13 volumes per 100 ml; portal vein, 10.51; and hepatic vein, 6.42. Mesenteric utilization (intestine, spleen, pancreas) can be calculated to be 20.4 ml per minute or a total splanchnic utilization of about 44 ml per minute. In man, splanchnic oxygen consumption has been calculated to average about 38 ml per minute per square meter surface area. From these data, O_2 utilization by the liver of man has been computed to be 0.05 ml per gram per minute, a figure close to that of the dog liver.

Following hemorrhage, increased A-V oxygen difference across the splanchnic bed provides a reasonably constant supply of oxygen despite reduction in blood flow. With severe and prolonged hypotension, oxygen utilization of both the liver and intestine is depressed. Under this circumstance, there is a trend for greater contribution of oxygen from the hepatic artery as portal oxygen tension drops to low levels.

STOMACH

Anatomical Aspects

The wall of the stomach is composed of smooth muscle, a submucosal layer containing an intramural nerve cell plexus and glandular structures, and a specialized surface epithelium (mucosa). Branches from the arterial system supplying the stomach penetrate the muscle coat to the submucous layer, where they form a main plexus of arteries, 200 μm in diameter, the loops being connected by cross-anastomosing channels of 150 μm cross section. From this submucosal plexus a rich vascular network supplies the gastric mucosa. Thus, the mucosa literally lies in a vast vascular pool in which there are

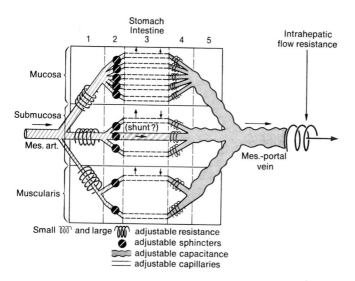

Figure 25A-2

Schematic illustration of the different tissue compartments in the stomach and intestines, with their specialized parallel-coupled vascular circuits and their consecutive sections. (1) Precapillary resistance vessels. (2) "Sphincter" sections. (3) Capillaries. (4) Postcapillary resistance vessels. (5) Capacitance vessels. (From B. Folkow. *Gastroenterology* 52:423–432, 1967. By permission, The Williams & Wilkins Co., Baltimore.)

channels capable of bringing blood to the mucosa or transferring it rapidly to a required point.

In terms of hemodynamic control, the gastric vascular bed should be considered as comprising a set of parallel-coupled circuits, specialized for the requirements of the various tissue compartments. Figure 25A-2 illustrates the numerous control points (*adjustable resistance* and *adjustable sphincters*) regulated by neurogenic and humoral factors, for modifying blood capillary flow and capillary perfusion to the various layers, and illustrates how gastric blood volume can be altered by means of regulation of the adjustable *capacitance vessels*.

Gastric Blood Flow

Delaney and Grim (1965), using the ^{42}K uptake method, determined the distribution of gastric blood flow between the antrum and corpus of the stomach, and to the various tissue layers (mucosa, submucosa, and muscularis) in anesthetized dogs. The ^{42}K uptake method is an application of the Fick principle (Chap. 11), and the organ uptake of the isotope is calculated as a fraction of the cardiac output:

$$\text{Regional flow} = \left(\frac{I_R}{I}\right) \times (\text{cardiac output})$$

For this determination, the animal is killed 30 seconds after the intravenous injection of the ^{42}K, the stomach is removed, and representative tissue samples are digested away in strong acid, then assayed for radioactivity. I_R is that amount recovered in the region, R, as a ratio to the total amount injected, I.

With an average cardiac output of 2.65 liters per minute in the dog, gastric blood flow (GBF) averages 1.85 percent of the cardiac output, or 49 ml per minute. Of this flow, 89 percent goes to the corpus and 11 percent to the antrum. With regard to tissue distribution, the bulk of the flow (72 percent) goes to the mucosa, 13 percent to the submucosa, and 15 percent to the muscularis. Histamine, a potent stimulant of gastric acid secretion, significantly increased GBF, both by increasing cardiac output and by decreasing gastric vascular resistance, along with its excitatory influence on acid secretion. GBF, as percentage of cardiac output (CO), increased from 1.85 to 3.0–4.6 percent. Histamine did not alter the antrum/corpus perfusion ratio. Epinephrine also increased the GBF fraction of CO, with CO also increasing. Hence, epinephrine reduced gastric vascular resistance. On the other hand, norepinephrine increased gastric vascular resistance. It also caused redistribution of blood flow from mucosa to the muscle layers. The other agents did not alter the normal tissue distribution.

Mucosal-Glandular Blood Supply

Another application of the plasma clearance principle has supplied important information on this aspect of the gastric blood supply. Jacobson et al. (1967) have combined the clearance of aminopyrine, which measures mucosal blood flow, with direct measurement of total GBF by electromagnetic flowmeter in unanesthetized dogs to analyze the mechanism of redistribution of blood supply in the stomach (Fig. 25A-3). The clearance (C) of aminopyrine, like renal plasma clearance, is

$$C_{AP} = \frac{G_{AP} \times \dot{V}}{P_{AP}}$$

where G_{AP} is the gastric juice concentration of aminopyrine in milligrams per minute, $\dot{V}$ is the volume of gastric juice secreted per minute, and P_{AP} is the plasma concentration.

The basic assumption is that aminopyrine is cleared solely from the plasma into the stomach. Experimentally, a vagally denervated, fundic (Heidenhain) pouch was used for gastric juice collection. It was proved that 90 percent was recovered in the gastric juice of the pouch.

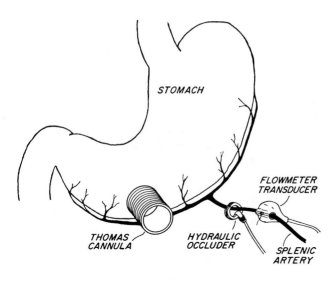

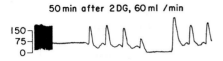

50 min after 2 DG, 60 ml /min

Figure 25A-3
Direct measurement of gastric blood flow (GBF). Upper diagram shows location of electromagnetic blood flow transducer and occlusive device on an artery supplying the stomach. Lower tracing is from a typical recording with this preparation and shows (from left to right) phasic flow at slow paper speed, the integrated flow signal, and then phasic flow at higher paper speed which is reduced to zero flow by applying the occluder. 2DG = 2-deoxy-D-glucose (From Swan and Jacobson, 1967.)

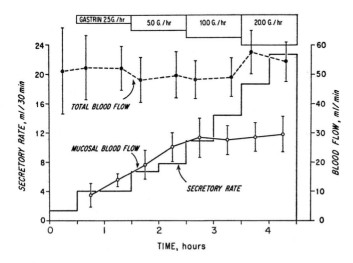

Figure 25A-4
Effects of extracted porcine gastrin on total blood flow, mucosal blood flow (MBF), and secretory rate in the canine Heidenhain pouch (10 exps., 6 dogs). (From Swan and Jacobson, 1967.)

Gastrin and synthetic gastrin-like pentapeptides increased mucosal blood flow (MBF) and secretory rate in a roughly parallel manner (Fig. 25A-4). Importantly, *total* GBF did not change significantly, implying a reduction of blood flow to the smooth muscle and other nonsecretory tissue, as blood flow to the glandular tissue increased. (The mucosal-glandular compartment may have a maximal blood flow of some 300 to 400 ml per minute per 100 gm.)

Another agent that produced similar effects was bethanechol chloride (Urecholine). In stomachs with intact innervation, vagal stimulation also caused an increase in mucosal blood flow, with no change in total GBF. Vagal stimulation was achieved by insulin hypoglycemia; atropine reversed the effect. Histamine increased mucosal blood flow, but also total GBF. Vasopressin and isoproterenol decreased mucosal flow. Epinephrine also decreased MBF, but in view of an overall increase in blood flow (see above), the implication is that blood was shunted from mucosa to the muscle layer.

INTESTINAL TRACT—SMALL INTESTINE

Anatomical Aspects
An outstanding characteristic of the small intestinal vasculature is its *arcade pattern* (cross connections between the peripherally radiating branches). The arcades serve as a blood depot system. The profuse anastomotic connections may have significance in relation to the intestinal movements, as a factor in keeping the blood supply from being seriously interfered with by peristaltic movements. From the arcades arise numerous *vasa recta* which proceed directly to the intestine. Here many anastomoses in the wall permit a separation of 3 to 4 inches from the mesentery, providing protection against distention of the bowel.

Intestinal Blood Flow; Hemodynamic Characteristics
This part of the GI tract is particularly characterized by the large villous surface specialized for absorption and fluid exchange. The total blood flow has been reported in a range of 20 to 80 ml per minute per 100 gm, depending on the species of experimental animal (cat, dog). It can increase to 250 to 300 ml per minute per 100 gm with maximal dilation (Folkow and Neill, 1971).

This vascular bed demonstrates a significant degree of autoregulation (Johnson, 1964). Thus, in a range of arterial pressure from 100 to 35 or 40 mm Hg, the blood flow and capillary pressure change but little.

In meeting the diverse needs of the intestine, a basic constraint is placed upon the system since it is necessary to maintain a reasonably constant capillary pressure. The same is true to some degree of all vascular beds, but the filtration coefficient of intestinal and liver capillaries is quite high; thus small changes in capillary pressure could produce rapid filtration and edema formation. Effective regulation of the peripheral circulation requires mechanisms that provide adequate blood flow to meet the needs while maintaining a nearly constant capillary pressure. Clearly, autoregulation of blood flow in the intestine helps in maintaining the constancy of the capillary pressure.

Local regulation of blood flow (autoregulation) is manifested in the liver in somewhat the same manner as described above for the intestine. However, because of its dual blood supply, the liver possesses the unique feature of reciprocity between portal and hepatic arterial inflow. If portal inflow is reduced, hepatic artery inflow increases, so that a more nearly constant blood supply to the liver tissues is maintained. Apparently, a reduction of portal inflow reduces pressure in the terminal portion of the hepatic arteriolar bed, causing the arterioles to dilate. A myogenic mechanism is probably responsible (Hanson, 1966). While this phenomenon is usually termed reciprocity of flow, it is not truly reciprocal, since a reduction of hepatic arterial inflow does not alter portal flow, which is determined by the resistance vessels of the organs of the digestive tract.

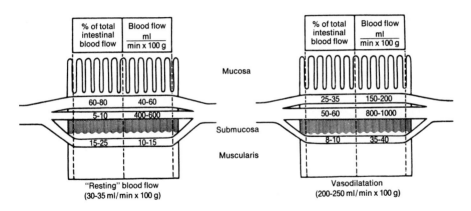

% of total intestinal blood flow	Blood flow $\frac{ml}{min \times 100\,g}$
60-80	40-60
5-10	400-600
15-25	10-15

% of total intestinal blood flow	Blood flow $\frac{ml}{min \times 100\,g}$
25-35	150-200
50-60	800-1000
8-10	35-40

Mucosa

Submucosa

Muscularis

"Resting" blood flow
(30-35 ml/min x 100 g)

Vasodilatation
(200-250 ml/min x 100 g)

Figure 25A-5
Blood flow distribution in the small intestine of the cat at "rest" and intense vasodilatation produced by isopropylnoradrenaline. The three vascular pathways depict in essence the villous mucosal circulation, the submucosal circulation, especially that of the secretory crypts, and the muscularis circulation. Note the huge flows in the submucosa, presumably reflecting the rich vascularization of the secretory crypts. The regional flows were estimated by the wash-out of radioactive krypton (^{85}Kr). (From O. Lundgren. *Acta Physiol. Scand.* [Suppl. 303]: 3–42, 1967.)

Wall Compartment Flow
In the *mucosa* (Fig. 25A-5), there is an especially rich vascularization of the secretory crypts. This zone (secretory crypts) can increase its flow tremendously to values rivaling those in the flow in actively secreting salivary glands (700 ml per minute per 100 gm). In the cat, the absorptive villous portion of the mucosa can attain flows up to 150 to 200 ml per minute per 100 gm (Folkow and Neill, 1971).

Blood Flow During Digestion
In trained, unanesthetized dogs, mesenteric blood flow being measured with chronically installed ultrasonic or electromagnetic flowmeters, presentation and ingestion of food caused an average increase in cardiac output of 62 percent and arterial pressure 33 percent ("anticipation-ingestion phase"). Mesenteric flow began to increase 5 minutes after eating and reached a peak of 132 percent above control in 30 to 90 minutes (average 55) during the "digestion" phase. It gradually returned to control value in 2 to 6 hours postprandially. In the resting dog, a slight compensatory vasoconstriction occurred in the inactive muscle bed (Vatner et al., 1970). Coronary blood flow increased transiently early in the digestion phase (about 5 minutes following eating).

The influence of ingested foodstuffs on intestinal blood flow is probably mediated on a humoral basis, at least in part. Fara et al. (1972) installed intraduodenally in anesthetized cats small volumes of corn oil, L-phenylalanine, or acid (0.127N HCL), and observed selective increase in superior mesenteric blood flow (SMBF) measured with an electromagnetic flow probe, with no change in blood pressure or heart rate.

The changes occurred in association with increases in gallbladder and duodenal motility, and pancreatic enzyme and volume output. It was proposed that the changes were mediated by release of intestinal hormones, such as cholecystokinin (CCK) and secretin (Chap. 26). It was shown that gastrin, a hormone released from the gastric antrum which stimulates secretion of gastric juice, also increased SMBF. Associated was an increase in intestinal oxygen consumption (0.99 ± 0.11 ml per minute per 100 gm, control, to about 1.30) (Fig. 25A-6).

Cross-perfusion experiments appeared to verify the blood-borne nature of the mechanism, for a comparable vasodilation was observed in the recipient cat.

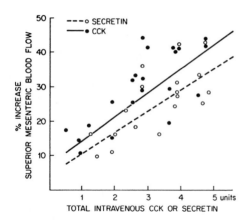

Figure 25A-6
Data points and calculated regression lines for dose-response relationship between amount of intravenous CCK or secretin and percent increase in superior mesenteric blood flow. (From Fara et al., 1972.)

The mesenteric vasodilation could be secondary to the metabolic effect of the intestinal hormones, comparable to that observed in glandular tissue during secretory activity (e.g., salivary glands). This "metabolic" hyperemia may result from activation of local kinins (bradykinin), serotonin, or histamine.

Countercurrent Mechanism of the Villus
The vessels of the villus are arranged in the form of complex hairpin loops (Fig. 25A-7). With this pattern as a base, Lundgren (1967) has presented evidence for a *countercurrent*

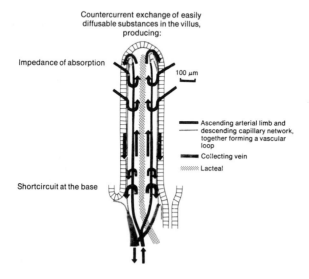

Figure 25A-7
Principles of the countercurrent exchange in the villus hairpin loops of the small intestine. (From B. Folkow. *Gastroenterology*. 52:423–432, 1967. By permission, The Williams & Wilkins Co., Baltimore.)

exchange mechanism in the villi. Sodium, for example, is concentrated at the tip of the villus and falls progressively toward the base of the villus (Haljamäe et al., 1973). Estimated values range from about 700 to 900 mEq per liter at the tip. One view is that actively absorbed sodium is short-circuited from the epithelial capillary network to the central artery, resulting in recirculation and concentration of the sodium toward the villus tip. This process would call for "fenestration" or interendothelial pores of the central arterial vessel. The effect would be multiplied along the villus length, resulting in increasing sodium concentration in the capillaries of the villus and a high concentration at the tip, as in the renal papillae. Alternatively, increase of sodium concentration and osmolarity in the capillaries of the villus could cause transfer of water from the central artery to the capillary network, eventuating finally in a similar sodium multiplication. Both mechanisms may be involved.

Such a "trapping" phenomenon could lead to a "damping" of entrance into the bloodstream (Fig. 25A-7: "impedance of absorption") of other rapidly absorbed materials, particularly small lipid-soluble particles. When such substances are carried with the descending bloodstream of the hairpin loop, they pass very close to the ascending bloodstream and will pass across to it, down their concentration gradient, by cross-diffusion. Consequently, high concentrations will be reached in the upper and outer parts of the villi in the steady-state situation, and they will actively leave the villi by venous drainage only slowly.

Conversely, rapidly diffusing molecules in the arterial bloodstream, such as oxygen, will tend to leave the arterial ascending limbs of the villi and by cross-diffusion become "short-circuited" over to the venous descending limbs (Fig. 25A-7). Thus, the O_2 tension at the tip of the villi will be lower than that at their base. This consequence of the countercurrent system may be one reason for the rapid turnover of epithelial cells at the tip of the villi, where oxygen delivery is probably poor. It is known that there is a steady and considerable production of new epithelial cells from basal parts of the villi, which are gradually displaced toward the tip. There may be other important functional consequences of the hairpin arrangement of the blood flow in the villi that are still unknown. In any case, when the villous bloodstream becomes rapid, as in very intense dilation, less time is available for cross-diffusion, and countercurrent "trapping" may then be less efficient. It is, however, premature to discuss the importance of the villous countercurrent exchange in intestinal absorption in more detail (Folkow and Neill, 1971).

The arrangement of the vascular tree in the intestinal wall also favors the occurrence of "plasma skimming." There is evidence that the blood reaching the villous part of the mucosa has a decidedly lower hematocrit value than does that going to, for example, the muscularis. The functional importance of this arrangement is not known.

INTESTINAL TRACT—LARGE INTESTINE

Essentially the same tissue compartments are present in the large intestine as in the small, but as the colonic mucosa has no villi, its mucosal vessels have no hairpin loops, and its powers of absorption are normally limited to the transference of water and electrolytes. The colonic mucosa produces a mucous secretion.

Flow in Different Wall Compartments

The blood supply of the muscularis shows the same general characteristics as that of the stomach and small intestine. The mucosa is well vascularized, showing a resting blood flow of 30 to 35 ml per minute per 100 gm, with an especially rich supply at the bases of the secretory crypts. This flow is about three times that of the muscularis. Stimulation of the sacral parasympathetic supply causes increased mobility, mucosal secretion, and virtually maximal blood flow of the mucosa (200 to 250 ml per minute per 100 gm).

Table 25A-1. Weight, Share of Cardiac Output, and Blood Flows of the Gastrointestinal Organs in an "Average" 15-kg Dog

Organ	Weight (gm)	Fraction of Cardiac Output (%)	Blood Flow (ml)	Perfusion Rate (ml/min/100 gm)
Stomach	100	1.9	50	50
Intestine	270	6.5	180	70
Colon	50	1.6	40	80
Pancreas	30	0.7	18	60
Gallbladder	2	0.04	1	40

Source: After Delaney and Custer, 1965.

SUMMARY OF BLOOD FLOW OF GASTROINTESTINAL ORGANS

Using the radiorubidium clearance technique, Delaney and Custer (1965) have provided data for distribution of blood to the various parts of the gastrointestinal tract of the dog (Table 25A-1). It is of interest that the duodenum had about the same blood flow as the remainder of the small intestine. Flow to the esophagus was, however, only 20 ml per minute per 100 gm.

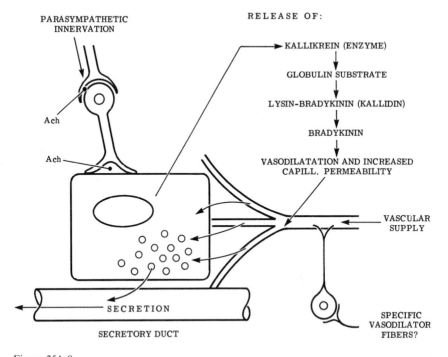

Figure 25A-8
The vasodilatation is due to the presence of kallikrein in the saliva, which acts as an enzyme to split off from a globulin the powerful vasodilator lysin-bradykinin (kallidin). This, in turn, can be rapidly transformed into bradykinin. (From Folkow and Neill, 1971.)

599

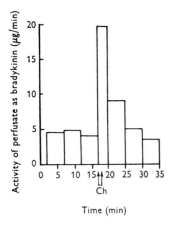

Figure 25A-9
Bradykinin production by the salivary gland during stimulation of the chorda tympani nerve (at Ch). (Modified from G. P. Lewis. *Gastroenterology.* 52:409, 1967. By permission, The Williams & Wilkins Co., Baltimore.)

SALIVARY GLANDS

Salivary glands, when maximally excited, produce nearly their own weight of saliva per minute. Thus, their vascular supply must allow very high rates of blood flow and a large capillary exchange rate. A maximal blood flow of about 700 ml per minute per 100 gm at a pressure head of 100 mm Hg, with a maximum secretion of 70 ml, has been recorded from the cat's salivary gland (cited by Folkow and Neill, 1971).

Parasympathetic nerves (chorda tympani branch of cranial nerve VII) supply and stimulate the salivary glands, and, closely related to this stimulation, hyperemia is induced. This relaxes resistance vessels and precapillary sphincters, and there is an accompanying increase in capillary permeability allowing for huge transfers of fluid and solutes to the secretory gland cell (Fig. 25A-8). Hilton and Lewis in 1956 showed that activation of salivary gland cells (also pancreatic gland cells) released into the tissue spaces a proteolytic enzyme, *kallikrein,* which splits off a vasodilator polypeptide from a plasma and tissue fluid globulin, *kininogen,* which is considered to be a decapeptide and is called *kallidin* (also, lysin-bradykinin).

Plasma and tissue fluid enzymes rapidly transform kallidin into *bradykinin,* a potent vasodilator and an agent for increasing capillary permeability (Fig. 25A-9).

Specific vasodilator fibers may also exist, and both mechanisms may be involved. It may be that the larger resistance vessels are controlled by such fibers and that the kinin mechanism affects mainly the smallest resistance vessels and capillaries.

The sympathetic innervation of the vessels of the salivary glands results in a powerful vasoconstrictor effect but can in some species induce a relatively sparse production of mucus-rich secretion. This type of secretion may also be accompanied by a transient activation of the kinin mechanism.

SPLEEN

Anatomical Aspects

Splenic artery branches pass along the *trabeculae* of the spleen, through the *white pulp,* and break up into arterioles with husklike sheaths (*penicilli*), which constitute the arteriolar stopcocks. These branch into arterial capillaries, which supply the sinusoids (venous sinuses) in the *red pulp* of the spleen. The collecting *venules* converge ultimately into the splenic vein, which drains into the portal vein. Sphincters have been described at the sinusoid outlets, which close during storage phases. The sinusoids are composed of rodlike "barrel stave" endothelial cells, longitudinally arranged but not contiguous, and surrounded by rings or spirals of reticular fibers, reminiscent of barrel hoops, creating a lattice-like structure. A commonly accepted view is that the system is a closed one, but that the latticework is large enough to permit occasional erythrocytes to slip through, as shown by observations on the transilluminated mouse spleen. Thus, the sinusoids represent a mechanism creating a "separatory" circulation, filtering off plasma and storing red cells in high concentration in the dilated sinusoid, and operating in a cycling manner. The separated plasma may return through smaller veins and lymphatics to the general circulation. In a spleen exhibiting a high degree of storage, constant, rapid flow is seen primarily in the capillary shunts directly connecting arterioles with venules. When the splenic capsule contracts under physiological demand, the concentrated cells are discharged directly into the portal venous system. Filtered red cells reenter the venous sinuses by diapedesis, which leads to destruction of fragile erythrocytes. After rupture, the released hemoglobin and stroma are ingested by the reticuloendothelial system of the spleen, or elsewhere in the body.

Other investigators have taken the position that the splenic circulation is an "open" one (whole blood comes into direct contact with splenic cells). Thus, in the mouse, the terminal arterioles are said to end in flared ampullae, the last structures to show signs of endothelial walls. Blood issues forth into the pulp spaces through the ampullae and reenters the vascular system via the open ends of the venous vessels and perforations in the venous walls. However, evidence appears good that the system is closed in man and the rat.

Neurogenic Factors

Nervous regulation is achieved by sympathetic outflow in splanchnic nerve branches accompanying the artery into the spleen, presumably mediated by norepinephrine. Regulation occurs by way of the smooth muscle of the capsule and trabeculae, and also by direct vasomotor action on the arterioles and sphincters. Pressure as high as 100 mm Hg can be developed by splenic contraction (if the vein is occluded). This contraction, combined with constriction of arterioles, forces the blood out. Simultaneous relaxation of the efferent sphincters aids discharge. Discharge is evidently initiated by reflex mechanisms (e.g., via chemoreceptor stimulation during hypoxia or baroreceptor stimulation during hemorrhage).

Since dilator fibers to the spleen have not been demonstrated, filling is taken to be a passive process, achieved by relaxation of arterioles by decrease in splanchnic vasomotor tone and/or by constriction of efferent sphincters and possibly the splenic veins by action of venomotor fibers.

Reservoir Function

In addition to hemorrhage and anoxia, discharge of stored blood is stimulated by exercise, raised external temperature, certain types of anesthesia (e.g., chloroform), certain humoral agents and drugs, and subacutely in estrus, pregnancy, and lactation. Under pentobarbital anesthesia, which apparently promotes greater than normal storage, severe hypoxia caused the spleen of the dog to discharge 16 to 20 percent of the normal circulating blood volume. Because of the high hematocrit, the O_2 capacity of the splenic vein blood can be as much as 80 percent higher than that of arterial blood. During hypoxia, this source may serve as a small emergency store of O_2, for about 30 to 60 ml of O_2 is released with the 100 to 200 ml of blood stored in the canine spleen.

The volume of blood per gram of tissue is high in the spleen compared with other organs: 0.420 ml per gram in dogs as compared with 0.200 in the liver and 0.060 in the bowel.

LYMPHATIC CIRCULATION OF THE GASTROINTESTINAL SYSTEM

Anatomical Aspects

Stomach
The stomach is abundantly supplied with lymphatic vessels. They originate as blind projections among the tubular glands and form a network on the muscularis mucosae. From this plexus many branches pass at right angles through the muscularis mucosae to form a second plexus in the submucosa. From the submucosal collecting branches, large vessels arise and perforate the circular muscle coat to communicate with the inter-muscular plexus, which ultimately drains to lymph nodes on both curvatures of the stomach. These return to the cisterna chyli and thoracic duct.

Small Intestine
The lacteals of the villi represent the blind beginnings of the lymphatic capillary system. The central lacteals at the base of the villi anastomose with the lymphatic capillaries between the glands which form a plexus on the inner surface of the muscularis mucosae. Branches of the plexus, provided with valves, penetrate the muscularis mucosae and form in the submucosa a plexus of larger vessels. This plexus also receives tributaries from the dense network of large, thin-walled lymphatic capillaries surrounding the solitary and aggregated follicles of Peyer's patches.

Large Intestine
The lymphatic vessels of the large intestine and rectum form plexuses in the mucosa, passing eventually through the muscle layer.

Liver
In the liver, the ultimate functional unit, the lobule, is not supplied with lymphatic capillaries. The fluid leaving the liver sinusoids passes through capillary endothelium and, in the lobule, lies between this endothelium and the liver cells. Lymphatic capillaries are found at the periphery of the lobule and carry the highly proteinized liver lymph to collecting trunks which join the thoracic and right lymph ducts.

Spleen
In the spleen, lymphatics are observed only in the capsule and the thickest trabeculae. Fluid which filters through the walls of the capillaries and sinuses must permeate the stroma before reaching the lymphatic vessels.

Functional Aspects: Lymph Flow

Thoracic Duct
In man, lymph has been collected from patients with thoracic duct fistulas. The average flow is in the range 1.0 to 1.6 ml per kilogram per hour (Yoffey and Courtice, 1970). It increases (up to 5.8 ml) with drinking of water and eating.

The relatively large flow of lymph returning to the bloodstream via the thoracic duct is striking and emphasizes the importance of the lymphatic system. In 24 hours it represents a turnover and return by the thoracic duct alone of a volume of fluid roughly equivalent to the plasma volume and from 50 to 100 percent of the total circulating plasma protein. The lymph drainage of the gastrointestinal system contributes notably to the thoracic lymph flow.

Stomach
Absorption from the mucosa is slight. Alcohol, traces of sugar, and possibly a little water are transferred from the gastric lumen, but the lymphatics are normally concerned with

the extravascular circulation of proteins and lymphocytes. Clinically, they are important in the spread of carcinoma of the stomach.

Small Intestine
The flow of lymph from the intestines is variable, being dependent on the activities coincident with digestion and the type of food being absorbed. It is increased during absorption of fluids and foods. The lymph itself is concerned, therefore, with the removal of the materials which have not been taken up by the bloodstream, modified as they may be by the state of digestion at the time. It has a role in transporting vitamin K and long-chain fatty acids.

Under normal circumstances, ingested fat, after passing through the intestinal mucosa, enters the lacteals in the form of chylomicrons which consist mostly of triglycerides and some cholesterol and phospholipid. The lipid, present as lipoproteins, does not alter appreciably during fat absorption. It comes mainly from the capillary filtrate and is small compared with the amount of fat in the form of chylomicrons. Lymph flow is particularly increased during fat absorption.

Large Intestine
The lymphatics normally are concerned with the extravascular circulation of proteins and the migration of lymphocytes. Clinically, their regional drainage is important in the spread of tumors.

Liver
It has been long thought that the liver furnished most of the thoracic duct lymph flow. The accumulated evidence suggests that the contribution is variable, being higher in the cat and dog than in the rat. Collection of lymph from various areas shows that, in the anesthetized dog, permeability of liver blood capillaries to protein is significantly greater than in the intestinal and cervical regions. Furthermore, the liver unquestionably contributes a high proportion of the extravascular protein in thoracic duct lymph (Mayerson, 1962).

REFERENCES

Bradley, S. E.   The Hepatic Circulation.   In W. F. Hamilton and P. E. Dow (Eds.), *Handbook of Physiology*. Washington: American Physiological Society, 1963. Section 2: Circulation, vol. 2, chap. 41, pp. 1387–1438.

Delaney, J. P., and J. Custer.   Gastrointestinal blood flow in the dog. *Circ. Res.* 17:394–402, 1965.

Delaney, J. P., and E. Grim.   Experimentally induced variations in canine gastric blood flow and its distribution. *Am. J. Physiol.* 208:353–358, 1965.

Fara, J. W., E. H. Rubinstein, and R. R. Sonnenschein.   Intestinal hormones in mesenteric vasodilation after intraduodenal agents. *Am. J. Physiol.* 223:1058–1067, 1972.

Folkow, B., and E. Neill.   *Circulation*. New York: Oxford University Press, 1971.

Grim, E.   The Flow of Blood in the Mesenteric Vessels. In W. F. Hamilton and P. E. Dow (Eds.), *Handbook of Physiology*. Washington: American Physiological Society, 1963. Section 2: Circulation, vol. 2, chap. 42, pp. 1439–1456.

Haljamäe, H., M. Jodal, and O. Lundgren.   Countercurrent multiplication of sodium in intestinal villi during absorption of sodium chloride. *Acta Physiol. Scand.* 89:580–593, 1973.

Hanson, K. M., and P. C. Johnson.   Local control of hepatic arterial and portal venous flow in the dog. *Am. J. Physiol.* 211:712–719, 1966.

Hilton, S. M. and G. P. Lewis.   The relationship between glandular activity, bradykinin formation and functional vasodilatation of the sub-mandibular salivary gland. *J. Physiol.* 134:471–483, 1956.

Jacobson, E. D. (Ed.). Symposium on the gastrointestinal circulation. *Gastroenterology* 52:332–471, 1967.

Jacobson, E. D., K. G. Swan, and M. I. Grossman. Blood flow and secretion in the stomach. *Gastroenterology* 52:414–420, 1967.

Johnson, P. C. (Ed.). Autoregulation of blood flow. *Circ. Res.* 15 (Suppl. 1): 1964.

Johnson, P. C., and K. M. Hanson. Capillary filtration in the small intestine of the dog. *Circ. Res.* 19:766–772, 1966.

Lundgren, O. Studies in blood flow distribution and countercurrent exchange in the small intestine. *Acta Physiol. Scand.* (Suppl. 303): 3–42, 1967.

Mayerson, H. S. Lymphatic Vessels and Lymph: Part C. Microscopic Anatomy; Part D. Physiology. In D. I. Abramson (Ed.), *Blood Vessels and Lymphatics.* New York: Academic, 1962. Chap. 23, pp. 708–718.

Selkurt, E. E. Gastrointestinal Circulation: Part D. Physiology. In D. I. Abramson (Ed.), *Blood Vessels and Lymphatics.* New York: Academic, 1962. Chap. 11, pp. 333–340.

Swan, K. G., and E. D. Jacobson. Gastric blood flow and secretion in conscious dogs. *Am. J. Physiol.* 212:891–896, 1967.

Vatner, S. F., D. Franklin, and R. L. Van Citters. Mesenteric vasoactivity associated with eating and digestion in the conscious dog. *Am. J. Physiol.* 219:170–174, 1970.

Yoffey, J. M., and C. F. Courtice. *Lymphatics, Lymph and the Lymphomyeloid Complex.* New York: Academic, 1970.

25. Gastrointestinal System
B. Movements of the Digestive Tract

Leon K. Knoebel

A major component of the gastrointestinal system is the digestive tract, which is essentially a tube about 5 meters in length of variable cross-sectional area running from mouth to anus. The digestive tract includes the mouth, pharynx, esophagus, stomach, and small and large intestines. Other components of the gastrointestinal system, all of which lie outside of the digestive tract, are the salivary glands, pancreas, and biliary system (liver, gallbladder, and bile ducts).

The major function of the gastrointestinal system is to provide nutrients for the cells of the body. These nutrients are derived from the food we eat. Food enters the body from the external environment through the oral cavity and is propelled by muscular contractions through the different regions of the digestive tract. A number of digestive juices are secreted at various points along the route, and the enzymes contained in the secretions catalyze the conversion of complex molecules to simple molecules. These products of digestion and other substances are absorbed into the blood and the lymph, which carry them to the cells of the body. Unabsorbed food residues and a number of waste products are moved to the end of the tract and eliminated from the body.

It is apparent from the above that four major processes occur in the gastrointestinal system: muscular movements, secretion, digestion, and absorption. This chapter discusses the motor functions of the digestive tract, whereas the secretory, digestive, and absorptive functions of the gastrointestinal system are discussed in Chapter 26.

MOTOR FUNCTIONS OF THE DIGESTIVE TRACT

There are basically two types of muscular activity involved in the digestive and absorptive functions of the digestive tract: propulsive movements and mixing movements. In addition, changes occur in the state of the tonus of the musculature of all parts of the digestive tract. In the case of propulsion, the rate of movement of contents through the various regions of the digestive tract depends on the functions served by the different organs of the tract. For example, the passageway from mouth through pharynx and esophagus is simply a conduit for conveying food to the stomach, and transit through these regions is rapid. On the other hand, transit from the stomach and through the small and large intestines is slow, and this is consistent with the time required for the completion of the digestive and absorptive processes that occur in these organs. The mixing movements of the stomach and small intestine promote digestion by mixing the digestive juices with the food that enters from above. In addition, mixing movements facilitate absorption from the small intestine and proximal large intestine by bringing fresh portions of the intestinal contents into contact with the absorbing surfaces.

The muscular movements of the digestive tract are concerned mostly with the activity of the smooth muscle, which extends from the distal esophagus through most of the large intestine. There are three layers of smooth muscle, an outer longitudinal layer running lengthwise along the tract, an inner circular layer, and a small layer, the muscularis mucosa, located between the mucosa and submucosa. It should be noted, however, that skeletal muscle activity is of primary importance at either end of the tract—i.e., the mouth through proximal esophagus at the upper end and the external anal sphincter at the lower end.

CHEWING

When solid food enters the mouth, chewing occurs. This process is important from a number of standpoints. As food is moved about in the mouth, the taste buds are stimulated, and the odors which are released stimulate the olfactory epithelia. These events are significant because much of the satisfaction of eating is derived from these stimuli. Reflex secretion of saliva also occurs during the chewing of food. The food is mixed with saliva, which softens and lubricates the food mass and thereby facilitates swallowing. In addition, chewing reduces the food to a particle size convenient for swallowing.

A crushing force of 100 to 160 pounds can be exerted by the molars and 30 to 80 pounds by the incisors of man. Since a force of 115 to 173 pounds is sufficient to crack hazelnuts, the maximum biting force which can be generated by the muscles of mastication is far greater than that required for the chewing of ordinary food. The force of biting is not the major factor in determining the efficiency of the chewing process. The occlusive contact area between the molars and premolars is much more important in this respect.

Although the act of chewing is under voluntary control, it is also partly reflex in nature. That reflexes can be involved is shown by the fact that an animal decerebrated above the mesencephalon will chew reflexly when food is put in the mouth. The process of mastication is carried out by the combined action of the muscles of the jaws, lips, cheeks, and tongue. The movements of these skeletal muscles are coordinated by impulses over cranial nerves V, VII, IX, X, XI, and XII. Once chewing is accomplished to the satisfaction of an individual, the food mass or bolus is ready for swallowing.

SWALLOWING

Swallowing is the transport of material from the mouth to the stomach. The act of swallowing has been divided into three stages on the basis of the regions through which a bolus passes on its way to the stomach: the mouth, the pharynx, and the esophagus. The forces involved in moving a bolus of food through these parts of the digestive tract are generated by the smoothly coordinated sequence of muscular contractions that occur in response to the stereotyped swallowing reflex.

Swallowing Movements

Many of the events which occur during swallowing can be visualized by means of x-ray motion pictures of a human subject swallowing a radiopaque suspension of barium sulfate. Furthermore, the pressures at various points along the route, both at rest and during a swallow, can be measured by pressure-sensing devices.

Oropharyngeal Stages

The coordinated contractions of a variety of skeletal muscles participate in moving a bolus of food from the mouth through the pharynx into the esophagus (Fig. 25B-1). This series of events occurs in about one second.

During the *first stage* the bolus is passed from the mouth through the isthmus of the fauces into the pharynx. The food mass, either liquid or solid, is rolled toward the back of the tongue, and the front of the tongue is pushed up against the hard palate. Respiration is inhibited briefly. At the same time, the mylohyoid muscles contract rapidly and force the bolus into the pharynx.

In the *second stage,* the bolus is passed through the pharynx into the esophagus, taking about one-fifth of a second. X-ray motion pictures show that a number of events occur simultaneously. The continued contraction of the mylohyoid muscles and the position of the tongue prevent the reentrance of food into the oral cavity. The soft palate is elevated and shuts off the posterior nares. Food is prevented from entering the larynx by the elevation of the larynx and by the approximation of the vocal cords, both of which serve to close the glottis. The epiglottis may or may not be pressed down over the

607

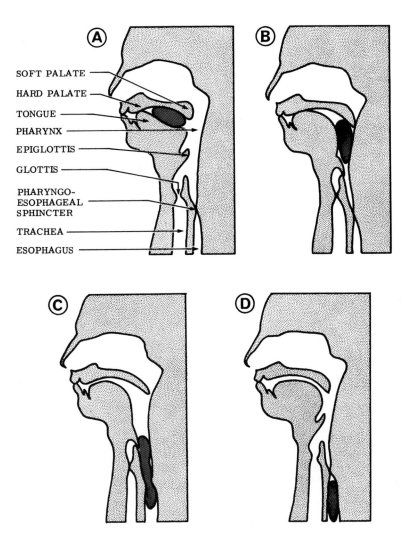

SOFT PALATE
HARD PALATE
TONGUE
PHARYNX
EPIGLOTTIS
GLOTTIS
PHARYNGO-
ESOPHAGEAL
SPHINCTER
TRACHEA
ESOPHAGUS

Figure 25B-1
Passage of a bolus from the mouth through the pharynx and upper esophagus during a swallow.

laryngeal orifice, but even if it is, it probably acts only as an auxiliary mechanism to keep food from entering the respiratory passages, since the epiglottis can be removed without having abnormalities in swallowing result. As these openings close, the pharyngeal constrictors contract and force the bolus into the esophagus. Respiration resumes. As a result of the various muscular activities which occur during this stage, pressure in the pharynx rises from ambient to about 100 cm H_2O (Fig. 25B-2).

Esophageal Stage
The esophagus is a muscular tube which conducts liquids and solids from the pharynx to the stomach during the *third stage* of swallowing. In man, the upper third of the esophagus consists of skeletal muscle, whereas the lower two-thirds is predominately smooth muscle. When the esophagus is in the resting state, the upper part is closed over

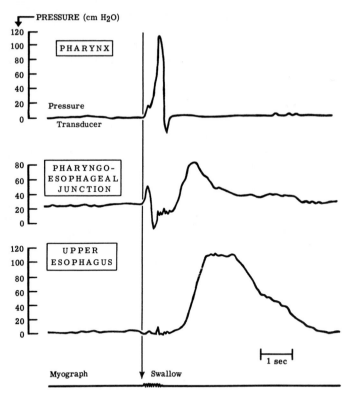

Figure 25B-2
Temporal sequence of pressure changes from rest with swallowing in the pharynx, pharyngoesophageal junction, and upper esophagus. Pressures were measured in the three sites during three different swallows by means of swallowed miniature pressure transducers. The beginning of a swallow is signaled by a myograph, which records action potentials from the jaw muscles. At rest, pressure in the pharyngoesophageal junction is about 40 cm H_2O higher than in the pharynx and upper esophagus. When pressure in the pharynx is high during a swallow, that in the sphincter falls to ambient pressure. Then, when pressure in the pharynx falls to rest value, sphincter pressure rises to about twice its resting level and remains high for about one second before falling to resting pressure once again. Peak pressure in the upper esophagus is attained at the time sphincter pressure is decreasing to resting level. (From E. F. Fyke and C. F. Code. *Gastroenterology* 29:29, 1955. By permission, The Williams & Wilkins Company, Baltimore.)

a distance of 2 to 4 cm by the tonic contraction of a band of skeletal muscle, the cricopharyngeal muscle. This *pharyngoesophageal sphincter* (Fig. 25B-1) exerts a pressure about 20 to 40 cm H_2O above atmospheric pressure, and this zone of high pressure prevents air from entering the esophagus during inspiration. However, almost immediately following the initiation of a swallow, the sphincter relaxes, and pressure in the region drops to atmospheric pressure, allowing a bolus to be forced into the esophagus by the pressure generated in the pharynx. Pressure in the pharyngoesophageal junction then rises to about twice resting level as the result of contraction of the skeletal muscle of this region and prevents reflux of food from esophagus to pharynx. Pressure then gradually subsides to resting level as the muscle relaxes somewhat. The pressures in the pharyngoesophageal junction at rest and during a swallow are shown in Figure 25B-2.

If the bolus is liquid, it is shot through the esophagus by the initial force of swallow-

609

ing and travels by gravity to the stomach in about one second. If semisolid, the bolus is propelled down the esophagus by a type of muscular movement known as *peristalsis.* The main feature of esophageal peristalsis is a contraction (4 to 8 cm in length) of the circular muscle of the esophagus which passes as a wave over the entire esophagus to the stomach at the rate of 2 to 4 cm per second. The wave of contraction, which actually begins in the pharynx, is thought by some researchers to be preceded by a wave of inhibition. However, since the esophagus is normally relaxed at rest, it is difficult to detect any further esophageal relaxation. On the other hand, since the resting tone of the upper sphincter is high, this structure does relax as the wave of inhibition passes over it. The wave of contraction that follows closes the upper sphincter and forces the bolus ahead of it toward the stomach, the transit time generally being four to six seconds. The pressure generated by the contractile component of esophageal peristalsis ranges from 40 to 160 cm H_2O (Fig. 25B-2).

Although there is no well-differentiated muscular structure in the region where the esophagus joins the stomach, a zone of high pressure about 3 to 6 cm in length extends up from the intra-abdominal esophagus through the diaphragmatic hiatus to the intrathoracic esophagus. This is the *gastroesophageal sphincter.* The region is tonically contracted during rest and exerts a pressure of about 10 to 20 cm H_2O. Since pressure in the sphincter is somewhat higher than intragastric pressure, the reflux of gastric contents into the esophagus is prevented. It has been demonstrated recently that one of the properties of the hormone gastrin, which is released from the gastric antral mucosa during the digestion of a meal (Chap. 26), is to increase the tone of the gastroesophageal sphincter. The result presumably is that sphincter pressure increases following the eating of food and thus helps prevent reflux of contents from the full stomach. Figure 25B-3 shows that, almost immediately following the initiation of a swallow, pressure at the gastroesophageal junction drops and remains low during the time a peristaltic wave is

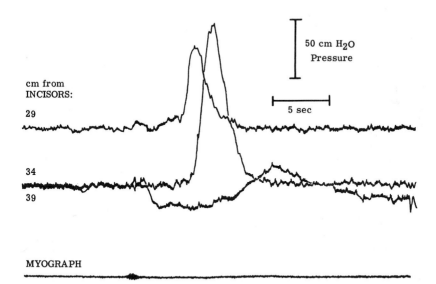

Figure 25B-3
Temporal sequence of pressure changes from rest with swallowing in the lower esophagus and gastroesophageal junction. Pressures were measured simultaneously in the three sites by means of swallowed miniature pressure transducers, the positions of which are given as distances from the incisors. The beginning of a swallow is signaled by a myograph, which records action potentials from the jaw muscles. (From E. F. Fyke, C. F. Code, and J. F. Schlegel. *Gastroenterologia* 86:146, 1956. By permission, S. Karger AG, Basel.)

traversing the lower esophagus. Presumably, the wave of inhibition causes relaxation of the gastroesophageal sphincter so that when a solid bolus is propelled down the esophagus by the wave of contraction, it is easily able to enter the stomach. Occasionally, a liquid bolus can be seen to accumulate momentarily above the sphincter, because a liquid bolus travels so rapidly that it sometimes arrives at the sphincter before relaxation is sufficient to allow entrance of the bolus. Once pressure in the lower esophagus falls to resting level, the pressure in the gastroesophageal junction rises and remains elevated for about 10 seconds before falling to resting level once again.

The Swallowing Reflex

The coordination of swallowing depends on neural mechanisms. The first stage may be initiated voluntarily, but is usually a reflex action. However, there must be some stimulus to the mucous membranes of the mouth to enable one to initiate a swallowing movement. At least a moderate amount of fluid must be present in the mouth, as is shown by the fact that it is virtually impossible to swallow when the mouth is dry.

The remainder of the swallowing response and all other movements of the gastrointestinal tract with the exception of defecation are independent of the will. It is widely held that there is a swallowing center, located in the medulla, which is activated by stimulation of receptors in the mouth and pharynx. This center sends impulses over a number of efferent nerves to the numerous skeletal and smooth muscles involved in the swallowing response, and the complete act of swallowing occurs in the proper sequence. The glossopharyngeal and hypoglossal nerves are mainly concerned with the oropharyngeal stages, whereas the vagus is the important nerve with regard to the esophageal stage.

It can be shown by a simple experiment that the peristaltic wave is not just a conduction of contraction along the muscular wall of the esophagus. If the esophagus is transected, the lower end will still contract at the proper time after a swallow, providing the extrinsic innervation of the esophagus remains intact. Since total esophageal paralysis results following bilateral vagotomy in the neck, the extrinsic nerves which are important in coordinating the orderly progress of the peristaltic wave must be the vagi. A biphasic complex of inhibition and then excitation is probably mediated by the vagi after a swallow. This would result in the passage of a wave of relaxation followed by a wave of contraction over the upper sphincter, esophagus, and gastroesophageal sphincter.

The paralysis resulting from bilateral vagotomy is permanent in the upper esophagus of man, which consists predominately of striated muscle. Weak peristaltic activity does return in the extrinsically denervated distal esophagus, which contains mostly smooth muscle. The myenteric plexus of Auerbach, which contains the parasympathetic postganglionic nerve supply of the esophagus, lies between the longitudinal and circular smooth muscle layers, and it is probable that this cholinergic plexus is largely responsible for the coordination of any peristaltic activity occurring in the distal esophagus both under normal conditions and after vagotomy.

Peristalsis which follows a conscious effort of swallowing is known as *primary peristalsis*. Peristalsis can also be elicited by local stimulation of the esophageal mucosa and without being preceded by a swallowing movement. This is *secondary peristalsis*. The latter type, which like primary peristalsis is reflex and depends on the integrity of the vagi, is important when food remains in the esophagus following the passage of a primary peristaltic wave. Secondary peristalsis facilitates the removal of such residues from the esophagus.

Abnormalities of Swallowing

A condition known as *dysphagia* arises when there are abnormalities of any of the three stages of swallowing. Abnormalities may result from a variety of disorders associated with the nerves and muscles involved in these processes. There are abnormal motility patterns of the esophagus and gastroesophageal junction in *achalasia*. The esophageal

611

musculature may show uncoordinated and spastic contractions, and the gastro-esophageal sphincter often fails to relax following a swallow. Food has difficulty entering the stomach and tends to accumulate above the sphincter, causing dilation of the esophagus. This abnormality is apparently due to a lack of or degeneration of the nerve cells making up Auerbach's plexus.

GASTRIC MOTILITY

Included in the motor functions of the stomach (Fig. 25B-4) are the storage of ingested food for variable periods of time and the discharge of gastric contents into the small intestine at a rate which is optimal for intestinal digestion and absorption. In addition, the mixing movements of this organ aid in the conversion of large, solid food particles to a finely divided, liquid form prior to evacuation.

The Empty Stomach

During periods of fasting, the volume of contents present in the human stomach is only about 50 ml or less. In this circumstance, pressure in the lumen of the resting stomach equals intra-abdominal pressure (5 to 10 cm H_2O), and there is not much tension in the gastric wall. Gastric motor activity is minimal early in fasting, but as a fast is prolonged, gastric contractions become progressively more vigorous. When the stomach is devoid of food, the sensation of hunger is often perceived. As a part of the overall sensation of hunger, some people feel a sense of emptiness accompanied by occasional sharp pangs referred to the abdominal region. Although the matter has been widely debated over the years, the so-called hunger pangs may be directly associated with the more vigorous contractions of the empty stomach. Thus, these gastric contractions may contribute to the sensation of hunger. The control of food intake as related to the sensations of hunger and satiety is discussed in Chapter 8.

Gastric Filling

One function of the stomach is the storage of variable volumes of contents, and it is subserved by the smooth muscle of the fundus and body of the stomach. These regions possess the ability to adapt to the volume of contents they contain in that relatively large volumes can be introduced into them with little increase in intragastric pressure (Fig. 25B-5). The volume adaptation phenomenon has been called *receptive* or *stress relaxation*. A property of smooth muscle which contributes to this volume adaptation is that of plasticity; i.e., smooth muscle fibers, when stretched, either rearrange internally or

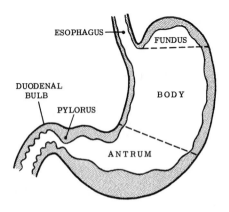

Figure 25B-4
Physiological anatomy of the stomach.

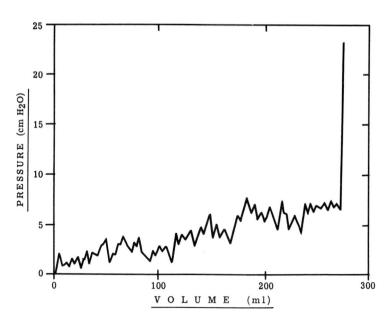

Figure 25B-5
Intragastric pressure during the addition of 2.5 ml of fluid every 2.5 minutes through a pyloric cannula into the cardia-ligated stomach of the rabbit. (From E. G. Grey. *Am. J. Physiol.* 45:276, 1918.)

slide past one another in such a manner that increases in intraluminal pressure are minimized (Chap. 3B). In addition, there is associated with the act of swallowing a reflex inhibition of the tone of the smooth muscle of the body and fundus. That this neural mechanism is mediated via the vagus nerves is shown by the fact that there is relaxation of the vagally innervated body and fundus following a swallow, but not if the vagi to these regions are sectioned.

Movements of the Full Stomach

A certain pressure is maintained on the gastric contents by virtue of a tonic contraction of the musculature of the fundus and body. It might be said that these portions of the stomach have at all times "a grip on their contents." The tonic contraction continually presses on the food mass and aids its delivery to the pyloric antrum.

Repetitive peristaltic contractions of the human stomach occur at the rate of three contractions every minute. Weak peristaltic waves moving at a velocity of 1 to 2 cm per second are first detected in the body of the stomach. However, when they reach the pyloric antrum, the muscle of which is much thicker than that of the upper part of the stomach (Fig. 25B-4), the contractions become much more vigorous and also increase in speed, so that the terminal antrum appears to contract as a unit—the so-called antral systole. Thus, the peristaltic contractions of the pyloric antrum are largely responsible for mixing ingested food with gastric secretions and for generating the forces required for gastric emptying.

Gastric movements have been observed by means of x-ray films, through windows in the abdominal wall of animals, and through fistulas in the stomach of man. The most famous human gastric fistula was that of Alexis St. Martin, a Canadian adventurer, who acquired a gastric fistula in his left side as the result of a gunshot wound. Dr. William Beaumont, a United States Army surgeon, capitalized on this event to keep St. Martin in his employ and for several years made many important observations directly on the human stomach with regard to both motility and secretion. His results, published in 1833, are classic in the field of gastric physiology.

Mechanics of Gastric Emptying

For years it was thought that the state of contraction of the pylorus was the primary factor in the regulation of gastric emptying. However, the currently acepted view is considerably different and minimizes the importance of an anatomical pyloric sphincter. It has been shown that the gastric emptying times of a wide variety of substances are not changed from normal following the surgical removal of the pylorus. Consequently, this structure cannot be of primary importance in the control of gastric evacuation.

Experiments have been performed which have greatly enriched our knowledge of the mechanics of gastric emptying. Gastric motility, activity of the pylorus, motility of the duodenal bulb, and movement of a barium meal through these regions have been followed in the dog by x-ray and fluoroscopic observation. When a peristaltic wave is observed to pass over the pyloric antrum, contents can be seen to move through the relaxed pylorus into the duodenum (Fig. 25B-6). When the contraction wave reaches the pylorus, this structure contracts; then the duodenal bulb contracts. The antrum, pylorus, and bulb all react in the same manner to the passage of a peristaltic wave; the pylorus is no different from the adjacent structures in this respect. Contents continue to escape from the stomach as the gradually contracting pylorus closes, until the increased resistance offered by this musculature prevents further evacuation. Material trapped in the antrum is squirted back forcibly into the body of the stomach as a result of the systolic contraction of the terminal antrum. The pylorus remains contracted somewhat longer than does the duodenal bulb, preventing to some extent regurgitation of duodenal contents into the stomach. The contraction of the duodenal bulb helps to propel contents down the intestine. All these structures finally relax again, and the cycle of emptying is repeated with the coming of the next peristaltic wave.

In addition to these observations, measurements of the pressures developed immediately on each side of the pylorus were made using small pressure transducers placed through antral and duodenal fistulas (Fig. 25B-6). Although absolute pressures may rise 15 to 30 cm H_2O in the antrum and bulb during a cycle of gastric emptying, a pressure gradient of 3 to 4 cm H_2O is maintained between these regions throughout most of the cycle. However, at about the time the pyloric sphincter begins to contract, the antral-duodenal pressure gradient may increase to 20 to 30 cm H_2O. It was concluded that the driving force behind the emptying process is the pressure differential between the antrum and duodenum. If the pressure is higher in the antrum, and high enough to overcome the resistance offered at the pylorus, contents will leave the stomach. Since the muscular activity of the pyloric antrum determines to a large extent the magnitude of

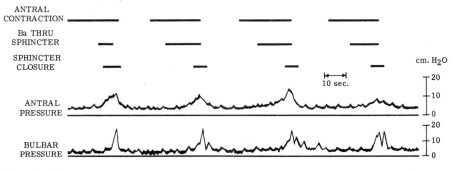

Figure 25B-6
Gastric evacuation-pressure cycle. (From J. M. Werle, D. A. Brody, E. W. Ligon, M. R. Read, and J. P. Quigley. *Am. J. Physiol.* 131:608, 1941.)

this pressure differential, gastric emptying is regulated by mechanisms which control the movements of this region of the stomach.

Control of Gastric Motility

The basic mechanisms controlling the movements of the stomach, small intestine, and large intestine share certain features. These include an autonomous, myogenic control system, the activity of which is modified by impulses over both extrinsic autonomic nerves and neurons located entirely in the wall of the digestive tract as a part of the intrinsic nerve plexuses of this organ system.

Extrinsic Neural Control

The extrinsic innervation of the stomach is by the vagus (parasympathetic) and splanchnic (sympathetic) nerves. These autonomic nerves, in addition to providing afferent fibers from the stomach, serve as the efferent limbs of a variety of reflex arcs which can be initiated from many visceral and somatic receptors throughout the body and can cause either excitation or inhibition of gastric movements. Thus the major function of the extrinsic nerves of the stomach is to correlate activities between the stomach and other parts of the gastrointestinal tract as well as other regions of the body.

The neural control of gastric motility is not completely understood. The results obtained by stimulating the parasympathetic and sympathetic nerve supplies to the stomach have not been uniform. No matter which set of nerves is stimulated, the musculature usually relaxes if it is in a state of contraction, and contracts if the tonus is low. In general, stimulation of the parasympathetic nerves usually produces increased muscular activity, whereas stimulation of sympathetic nerves most often results in inhibition. These generalities can be applied not only to the stomach but also to most of the smooth muscle of the digestive tract.

As will be pointed out below, the vagus nerves play the dominant role in the regulation of gastric motility. Although the sympathetic nerves do participate in the reflex regulation of gastric motor activity, they play only a minor part in the regulation that occurs during the normal course of digestion and absorption.

Intrinsic Neural Control

When the extrinsic nerves of the human stomach are cut, gastric peristalsis is at first very much reduced in strength or even absent, gastric emptying is delayed, and food is retained in the stomach for relatively long periods of time. This is the result of the loss of the excitatory influences that are normally mediated by the vagus nerves. However, after a period of recovery, motility and rate of gastric emptying tend to return to normal. This finding points out a characteristic feature of those parts of the digestive tract having a musculature of the smooth type: They possess a considerable capacity for self-regulation, as demonstrated by the fact that movements are not permanently abolished in the absence of extrinsic innervation. This autonomy probably depends on two factors: (1) the presence of the intrinsic nerve plexuses of Auerbach and Meissner, which are in close association with the smooth muscle of the tract and are primarily cholinergic, and (2) the ability of smooth muscle to contract in the absence of any innervation.

The nerve plexuses contain all components necessary for a local control of gastric movements. A local reflex arc system is subserved by receptors in the gastric wall which are capable of responding to chemical and mechanical intraluminal stimuli, the result being transmission of impulses over afferent fibers through one or more synapses to the effector smooth muscle cells. It might also be visualized that parasympathetic preganglionic fibers synapse on cell bodies somewhere in the nerve net, perhaps on those of the final efferent fibers of the local reflex arc, which then could be considered to represent the parasympathetic postganglionic nerve supply of the muscle fibers. Thus, this would be the point for the vagus nerve to manifest its regulatory influences. Apparently the function of the extrinsic nerves is to modify activity which can be initiated and maintained in the gastric wall itself, and modification can be in the direction of either augmentation or suppression of activity, depending on which extrinsic nerves are involved.

Electrical Activity of Gastric Muscle

Studies of the electrical activity of the stomach have shown that two types of potential changes can be recorded from gastric smooth muscle (Chap. 3B). One wave shape consists of single or multiple short-lived oscillations which represent typical action potentials. The action potential is the electrical event directly associated with contractile activity. Contraction occurs when action potentials are present, but not when they are absent. Action potentials are produced in response to the acetylcholine which is released by parasympathetic postganglionic neurons as a result of their stimulation either by impulses over the vagus nerves or by activation of short reflex arcs through the intrinsic nerve plexuses.

The second potential change is distinct from the action potential in that it is of lesser amplitude and much longer duration. This slow wave represents a relatively small and slow oscillation in "resting" membrane potential. It originates in a pacemaker located in the longitudinal muscle layer somewhere in the upper stomach, spreads to the circular muscle coat, and in man is propagated down to the pylorus at a rate of three waves per minute. Since this potential change shows cyclic activity, it has been called the *basic electrical rhythm* (BER) of the stomach.

The BER is not necessarily associated with contractile activity because, unlike the action potential, it is continually recorded both in the absence and in the presence of contractions. However, there is a relationship between the BER and contraction in that action potentials, when these are recorded, occur at the peak of the slow wave. It would be expected that spiking activity would be most likely at this point because the excitability of the muscle would be highest at this time—i.e., closest to threshold. It is here that any prevailing excitatory influences, such as muscle stretch or release of acetylcholine by nerve endings, would probably cause threshold to be reached. Since the slow wave propagates as a circumferential ring and encompasses a certain population of cells at any one moment, the muscle cells in that ring have an increased probability of spike discharge at about the same time. Thus, a function of the BER is to synchronize spiking activity in a specific region of the stomach, the result being a coordinated and efficient mechanical effort.

Since the BER and gastric peristalsis occur at precisely the same frequency and are propagated at similar velocities, apparently the slow wave is the electrical event that sets the pace for the propagated mechanical event. The frequency of the BER, and thus the rate of peristalsis, is not altered by a variety of excitatory and inhibitory stimuli. On the other hand, these stimuli, by virtue of their ability to enhance or inhibit spiking activity, are of primary importance in determining the strength with which the gastric muscle responds to the exciting signals represented by the BER. For example, if the vagus is stimulated, the vigor of peristaltic contractions is increased, but the rate is unchanged. It might be anticipated that the vagal stimulation induced by feeding increases the strength of gastric contractions, so that these become manifested as typical strong peristaltic contractions.

Regulation of Gastric Emptying

The force with which the stomach contracts to expel its contents into the small intestine is under the combined influence of a variety of excitatory and inhibitory mechanisms. The mechanisms are shown in Figure 25B-7, and this diagram should be referred to during the discussion that follows.

There is a basic pattern of gastric emptying. It has been found that the rate of gastric evacuation of liquid meals decreases with time after ingestion of a meal and that the initial rate of emptying varies directly with the size of the meal; i.e., the greater the volume of material contained in the stomach at any given time, the greater the rate of

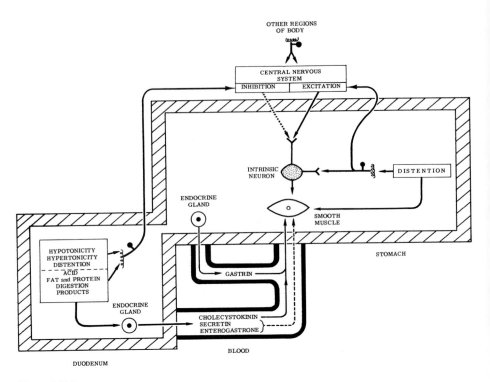

Figure 25B-7
Pathways in the regulation of the gastric evacuation of a meal.
———→ = excitation; – – –› = inhibition.

gastric emptying. A linear relationship is obtained when the square root of the volumes of a meal remaining in the stomach is plotted versus time (Fig. 25B-8). That there may be a physical basis for the square root pattern of gastric emptying is suggested by the fact that the radius of a cylinder varies with the square root of the volume and that circumferential tension is proportional to the radius (law of Laplace).

The relationship between rate of gastric emptying and volume of contents contained in the stomach is related to the degree of distention of the gastric wall. Mechano-receptors in the wall of the stomach are stimulated in proportion to the degree of distention, and impulses are sent over afferent fibers to the brain, which in turn relays impulses back to the stomach by way of the vagi to stimulate contraction. Peristalsis is also augmented by activation of short reflexes mediated through the intrinsic nerve plexuses. These neural influences, along with any contractile activity that is elicited directly in response to stretch of the gastric muscle, are the major mechanisms involved in providing the excitation required to empty the stomach of its contents (Fig. 25B-7). The gastrin which is released from the pyloric antrum during the digestion of a meal (Chap. 26) also augments peristalsis.

Many of the mechanisms that participate in the regulation of gastric emptying are initiated in the duodenum. Gastric contents must come into contact with the duodenal mucosa in order to maintain the normally slow gastric evacuation. If the duodenum is transected close to the pylorus, the stomach empties almost as rapidly as ingested food enters it. It will be obvious from the discussion below that practically any stimulation of the duodenum tends to check gastric emptying. These inhibitory mechanisms are probably utilized to prevent the absorptive powers of the intestinal mucosa from being

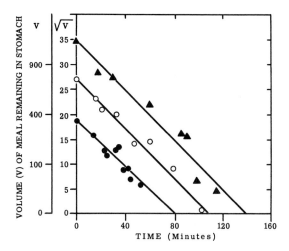

Figure 25B-8
The patterns of gastric emptying of meals of 330, 750, and 1250 ml according to the square root hypothesis. (From A. Hopkins. *J. Physiol.* 182:144–149, 1966.)

overwhelmed by a flood of ingested materials and to prevent undue chemical, mechanical, and osmotic irritation of the duodenum.

Both the chemical and the physical properties of the contents which enter the duodenum have a profound influence on gastric emptying. With regard to the major foods, the presence in the duodenum of the digestion products of carbohydrate, protein, and especially fat impedes gastric emptying. Solutions of pH 3.5 or less greatly retard gastric evacuation. The osmotic pressure of the gastric contents is also an important factor in gastric emptying. Water is evacuated only half as rapidly as equal volumes of isotonic saline. The introduction of hypotonic and hypertonic solutions into the duodenum causes inhibition of gastric motility, suggesting that the osmotic environment most favorable to the duodenum is one close to isotonicity. Carbohydrate and protein probably exert at least part of their inhibitory influences by the formation of osmotically active particles produced by the gastrointestinal digestion of these foods. Distention of the duodenum by increasing intraluminal pressure 10 to 15 mm Hg inhibits gastric emptying. Another factor which has a profound influence on gastric emptying is the consistency of the gastric contents. Dogs fed large chunks of meat take a much longer time to empty the stomach than dogs fed finely ground meat. Liquids are generally evacuated much faster than solids. These observations indicate that the contents of the stomach must be in a finely divided and liquid form prior to evacuation.

Enterogastric Inhibitory Reflex
The gastric inhibitory responses to the duodenal stimuli listed above depend to varying degrees on a neural mechanism called the *enterogastric inhibitory reflex*. The vagus is an important component of this long reflex arc, which is mediated from duodenum to brain to stomach (Fig. 25B-7), since the inhibitory effect is either partially or completely abolished if both vagi are severed. Section of sympathetic nerves has no effect on the reflex. The manner in which vagal inhibition is mediated is unknown, but this may be the result of either inhibition of the vagal nuclei in the medulla by impulses over afferent fibers from the duodenum or the existence of vagal efferent inhibitory fibers. It is also possible that there are a variety of intestinal receptors, such as osmoreceptors, hydrogen ion receptors, chemoreceptors, and mechanoreceptors, that respond to specific intraluminal stimuli to initiate the reflex inhibition of gastric motor activity.

Hormonal Control

The inhibition in response to acid solutions and fat and protein digestion products in the duodenum is not completely removed following vagotomy. However, these stimuli cause two hormones to be released from the duodenal mucosa into the circulation, which carries them to the stomach where they inhibit movements (Fig. 25B-7). One is *secretin,* a well-characterized hormone discussed in detail in Chapter 26, and the other is *enterogastrone,* which has not yet been chemically identified. Another hormone released from the duodenal mucosa in response to these stimuli, and one that stimulates gastric motor activity, is *cholecystokinin,* the biological properties of which will be discussed later in this chapter and in Chapter 26.

Other Gastric Reflexes

Although the regulation of gastric emptying is controlled from the duodenum to a large extent, gastric motility may be stimulated or inhibited reflexly from any sensory region of the body (Fig. 25B-7). Gastric emptying is delayed when the ileum is full (ileogastric reflex) and when the anus is mechanically distended (anogastric reflex). Stimulation of visceral and somatic pain receptors results in inhibition of gastric movements. Various emotional states such as anger, fear, anxiety, and resentment produce changes in gastric motor activity, but the direction of the changes is not always predictable.

Vomiting

The act of vomiting accomplishes the purpose of rapidly emptying the stomach of its contents. Vomiting is generally preceded by profuse secretion of saliva, sweating, rapid heart rate, and a feeling of nausea. A forced inspiration is made, and the glottis and nasal passages are closed by the contraction of the appropriate muscles. The body of the stomach, the gastroesophageal sphincter, and the esophagus relax, and the pyloric antrum contracts. The diaphragm descends on the stomach, and there is a forcible contraction of the abdominal muscles. The pressure generated propels the gastric contents into the esophagus through the pharyngoesophageal sphincter and into the mouth. The major force for vomiting is supplied by the contraction of the skeletal muscle of the diaphragm and abdomen, rather than by contraction of the gastric musculature. If an animal is given curare, an agent that paralyzes striated muscle, vomiting cannot occur.

Vomiting is an extremely complex reflex act and is coordinated by a center located in the medulla. Afferent impulses arrive at this center from many regions of the body. The most potent stimuli arise from the sensory nerve endings of the fauces and the pharynx. Other prominent receptor areas include almost any part of the digestive tract, other abdominal viscera, and the labyrinths during motion sickness. Loss of acid gastric contents by prolonged vomiting can result in profound disturbances in fluid and acid-base balance (Chap. 24).

MOTILITY OF THE SMALL INTESTINE

Ingested food, which is liquefied and partially digested in the stomach, passes into the small intestine, where the major part of digestion and absorption occurs in the duodenum and jejunum. Waste products and food residues are moved into the ileum and then into the colon. Different types of movements which accomplish these functions can be observed and recorded in the small intestine.

Movements of the Small Intestine

The sequence of events in the type of movement most frequently seen in the small intestine and known as *segmenting contractions* is illustrated diagrammatically in Figure 25B-9. Although segmenting contractions probably do help move contents down the

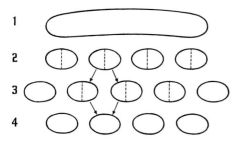

1
2
3
4

Figure 25B-9
Diagram of segmentation movements of the small intestine. An animal is fed a radiopaque meal, and the intestine is observed by means of x-rays. (1) A string of barium-impregnated meal can be seen when the gut is quiet. (2) The circular muscle of the intestine then contracts at a number of places, dividing it into a series of segments. (3) Each segment can be seen to divide, with adjoining halves coming together to form new segments. (4) The original pattern is established once again.

small intestine, this type of activity serves primarily to mix the contents with the digestive juices and to facilitate absorption by bringing fresh portions of contents into contact with the absorbing epithelium. The process can be repeated about 11 times a minute in the upper intestine of man and continues for variable periods of time in a section of intestine.

Another type of movement observed in the small intestine is peristalsis, the function of which is to propel contents along the intestine. Under normal circumstances the wave moves slowly at a rate of 1 to 2 cm per second, and travels only 4 to 5 cm. Thus, this type of motor activity provides the slow transit necessary to permit completion of the absorptive processes. Occasionally peristalsis occurs as a swift movement which sweeps along the entire intestine (peristaltic rush). Such rapidly propagated contractions occur when the excitability of the intestine is high, as, for example, when the intestine is irritated by toxic agents. Diarrhea and malabsorption may result from the rapid transport of contents through the small intestine.

Propulsion of Intestinal Contents
Evidence suggests that there is a one-way transmission mechanism built into the wall of the gut. If a segment of intestine is removed and reversed, the continuity of the gut being maintained, material will accumulate and distend the intestine at this point. The propulsive movements of the reversed segment work in the direction opposite to normal and impede the movement of contents.

One idea that has been advanced to explain this polarity of the intestine is based on the fact that mechanical stimulation of the intestine results in contraction above and relaxation below the point of stimulation. This response, termed the *law of the intestine,* has suggested that peristalsis of the small intestine consists of two components, a wave of inhibition followed by a wave of contraction, the natural consequence of which would be to move contents in a downward direction. However, there is considerable uncertainty regarding the extent to which the law of the intestine contributes to normal intestinal motility.

The *gradient theory* has been advanced in explanation of the polarization of the intestine. This theory is based on studies demonstrating that rhythmic contractions, tonus, excitability, and metabolism of the intestine show the greatest activity in the duodenum but become progressively less active toward the end of the small intestine. The activity gradients presumably account for the normally unidirectional movement of contents through the intestine.

Control of Intestinal Motility

Extrinsic Reflex Control
Stimulation of the vagi generally causes increased intestinal motility, whereas sympathetic (splanchnic) stimulation usually produces inhibition. The importance of the extrinsic nerves in modifying intestinal movements is shown by the fact that intestinal motility can be altered reflexly by stimulation of many sensory areas.

Distention of any part of the intestine inhibits the whole intestine. This effect, known as the *intestinointestinal inhibitory reflex,* does not occur after section of the splanchnic nerves. Intestinal motility is also inhibited when the anorectal region is distended (*anointestinal inhibitory reflex*). Trauma to organs outside the digestive tract, such as irritation of the peritoneum or urinary tract, causes intestinal inhibition. The atonic or flaccid gut, a condition known as *paralytic ileus,* which often follows surgery in these areas is a result of reflex inhibition.

Feeding increases motility of an innervated, isolated loop of intestine, and this effect is abolished by cutting the extrinsic nerves to the intestinal segment. The reflex is probably initiated when food enters either the stomach or the intestine and is termed a *gastrointestinal* or *intestinointestinal excitatory reflex.* It decreases propulsive activity, the reduction in the rate of downward movement of contents presumably being effected by narrowing of the intestinal lumen as a result of increased contraction of the intestinal musculature. This helps allow adequate time for optimal digestion and absorption.

Local Control
Since the small intestine retains normal movements following extrinsic denervation, regulation of intestinal motility must be mainly on a local basis; i.e., the mechanisms which generate and coordinate intestinal movements reside in the wall of the intestine. However, although movements are maintained in the absence of extrinsic innervation, as pointed out above, intestinal motility is influenced by impulses over these nerves.

The segmenting contractions of an extrinsically denervated intestine are not affected by the application of cocaine. Since cocaine paralyzes all nerve structures in the intestinal wall, this type of contraction is initiated in the smooth muscle itself. Peristalsis is eliminated, showing the dependence of this type of muscular activity on the integrity of the intrinsic nerve plexuses of Auerbach and Meissner. It would appear that propagation of contraction depends on a local reflex between the intestinal smooth muscle layers and their intrinsic nerve plexuses.

Local mechanical and chemical stimulation by intestinal contents is probably largely responsible for the initiation and continuance of the movements of the small intestine. Factors that have been shown to increase motility of extrinsically denervated intestine are increased intraluminal pressure; hypotonic, hypertonic, and acid solutions; and products of digestion. These effects are abolished by anesthetization of the mucosa, thus demonstrating the participation of neural elements. The stimuli intensify the slight activity of the resting intestine by way of local reflexes between mucosa and muscle and probably by directly stretching the muscle itself.

The Peristaltic Reflex
Peristalsis first entails contraction of intestinal longitudinal muscle, which shortens the segment of intestine involved, and this is followed shortly by contraction of the circular muscle layer. When hexamethonium, an agent that blocks transmission in nerve ganglia, is applied to the extrinsically denervated gut, longitudinal muscle contraction is not affected, but there is no longer contraction of circular muscle. These results suggest that two neuronal pathways are required for the complete peristaltic reflex. It would appear that one set of neurons without intervening ganglia responds to intraluminal stimuli and initiates the reflex as manifested by contraction of the longitudinal muscle. In addition, another neuronal path with ganglia would subserve the propagated wave over circular muscle.

When the lumen of a segment of intestine exhibiting peristalsis is perfused, the perfusion fluid is found to contain serotonin (5-hydroxytryptamine), and the amount of serotonin released is increased with more vigorous peristalsis. Furthermore, peristaltic activity is increased by adding precursors of serotonin to the perfusion fluid and by the intravenous infusion of serotonin. These observations, plus the fact that the argentaffin cells of the intestinal mucosa contain a relatively high concentration of serotonin, have generated the idea that this substance plays a role in the control of peristalsis. In this regard, it has been shown that it takes less stretch to elicit peristalsis when serotonin is applied to the mucosa, presumably because the threshold of receptors responsive to stretch is reduced. However, the exact role, if any, of serotonin in the peristaltic reflex remains uncertain.

Electrical Activity of Intestinal Muscle
The electrical activity of the small intestine is similar in a number of respects to that of the stomach. The segmenting movements are associated with an intestinal basic electrical rhythm. Slow waves originate in the longitudinal muscle in a pacemaker region close to the entrance of the bile duct, pass to the circular muscle by muscle connections between the two layers, and move down the intestine as a circumferential ring. If a stimulus is present in a certain region of the gut and is sufficient to bring excitability to threshold from the peak of a slow wave as it traverses the region, spikes are fired, and a segmenting contraction results. In man, the duodenal BER is 11 per minute. When on relatively rare occasions every slow wave is accompanied by action potentials, the maximal rate of segmenting contractions is attained and a very rhythmic contraction pattern is observed. There is a gradient from proximal to distal intestine in the frequency of the BER. In the dog, the BER is 17 to 18 per minute in the duodenum, 15 to 16 per minute in the jejunum, and 12 to 14 per minute in the ileum. Thus, it is possible that different regions of the intestine have their own pacemakers.

The frequency of the intestinal BER is not altered by stimulation of extrinsic nerves, fasting, feeding, or a variety of other manipulations. However, the vigor of any contraction which might accompany a slow wave can vary over a considerable range. Vagal stimulation augments segmentation, whereas sympathetic stimulation either depresses or abolishes this movement. These alterations in the strength of the segmenting movements are due to the specific influences of the mediators, acetylcholine and norepinephrine, on spiking activity and once again demonstrate the importance of extrinsic nerves in modifying activity initiated in smooth muscle itself.

Hormonal Control
Evidence is accumulating for the involvement of gastrointestinal hormones in the regulation of intestinal motility. In general, gastrin and cholecystokinin stimulate intestinal movements, whereas secretin is inhibitory.

Ileocecal Sphincter
The terminal ileum receives unabsorbed food residues about three and a half hours after the beginning of gastric evacuation. Contents arrive at a region immediately proximal to the cecum, where the last 2 to 3 cm of the muscular coat is thicker than the rest of the ileum. This is the *ileocecal sphincter,* which is normally closed. A zone of high pressure about 20 cm H_2O above atmospheric pressure and approximately 4 cm in length is present at the ileocecal junction. Distention of the lower ileum produces a drop in pressure in the ileocecal sphincter, whereas distention of the cecum leads to increased pressure in the sphincter. The implication is that, when contents are present in and distend the lower ileum, the sphincter relaxes to allow them to be driven into the colon by the propulsive movements of the distal small intestine, and then contracts, preventing regurgitation into the ileum. The musculature of the ileum is generally not very active, but activity is increased when food enters the stomach (*gastroileal reflex*).

THE GALLBLADDER AND BILE DUCTS

Bile is manufactured and continuously secreted by the liver. This secretion passes from the liver by way of the hepatic duct into the common bile duct, and when the upper intestine is devoid of food, it is diverted by way of the cystic duct into the gallbladder where it is stored for variable periods of time. In addition to acting as a storage organ, the gallbladder provides a safety factor for regulating pressures in the biliary system. However, this organ can be removed without interfering with normal digestion and absorption.

Filling and Evacuation of the Gallbladder

Gallbladder function can be evaluated by watching this organ fill or evacuate its contents. A radiopaque substance, such as tetraiodophenolphthalein, which is secreted by the liver into the bile, is administered. The outline of the gallbladder and bile ducts can then be visualized by means of x-rays. The gallbladder can be seen to fill during periods of fasting. Since relatively little bile enters the duodenum under this circumstance, there must be a mechanism available to prevent the passage of bile into the small intestine and at the same time divert it into the gallbladder. It is widely accepted that there is a sphincter in the region where the common bile duct enters the lumen of the small intestine. If a catheter is inserted into the common bile duct here, the gallbladder will not fill because the sphincter is unable to exert its restraining influence. The resistance offered to the flow of bile by this *sphincter of Oddi* has been investigated by measuring the pressure required to force bile through the region. Such pressure is a measure of the sphincteric resistance, and in fasting, unanesthetized dogs it can be as high as 30 cm H_2O. It is probably this resistance which prevents bile from entering the intestine during fasting.

When food is taken, the gallbladder can be seen to empty over a period of time by means of a series of rather sluggish contractions. The intrabladder pressure rises to about 20 to 30 cm H_2O, and under most circumstances the pressure exerted by the contracting gallbladder is enough to overcome the sphincteric resistance. Further, it is believed that the gallbladder and sphincter act as a functional unit; i.e., when the gallbladder contracts, the sphincter relaxes. The entrance of bile into the gut is also influenced by the state of contraction of the duodenal musculature. During active contraction of the duodenum, the flow of bile is decreased or completely blocked by compression of the duct, but during the relaxation phase, resistance to flow is reduced. Therefore, as the result of duodenal movements, bile may be observed to enter the intestine in squirts.

Regulation of Gallbladder Evacuation

There is a division of opinion regarding the extent of participation of nerves in the control of evacuation of the gallbladder. Many workers feel that contraction of the gallbladder is mediated reflexly, with the efferent limb of the reflex arc being in the vagus. Other investigators believe that evacuation of the gallbladder is only slightly influenced by nervous factors.

That a regulatory mechanism other than a nervous one is involved in contraction of the gallbladder is shown by the fact that normal function is maintained when all extrinsic nerves to this organ are cut. Using dogs in cross-circulation experiments, researchers have shown that the control of gallbladder motility is at least partly hormonal in nature. When acid was introduced into the duodenum of a donor dog, the gallbladder contracted and 6 to 10 minutes later the gallbladder of the recipient dog contracted. Since the only connection between the two dogs was through the circulatory system, a substance must have been transported from the donor through the blood to cause contraction in the recipient. The gallbladder also contracts when extracts of the upper intestinal mucosa are injected intravenously. On the basis of these results it has been postulated that when gastric contents come into contact with the intestinal mucosa, a hormone is released from the mucosa into the blood. This hormone, *cholecystokinin,* travels to the gallbladder and stimulates it to contract. Cholecystokinin, which has been isolated in pure form and has been synthesized, is a 33-amino-acid peptide. Cholecystokinin and pancreozymin (Chap. 26) are the same polypeptide molecule. A variety of substances release cholecystokinin when placed in the duodenum. The most notable are fat and its digestive products. Proteins are effective, but not carbohydrates. Agents that promote emptying of the gallbladder are called *cholecystagogues.*

MOTILITY OF THE COLON; DEFECATION

A firm fecal mass is formed in the colon by the absorption of water from the contents which enter from above. This relatively dehydrated material is stored in the large intestine for variable periods of time and is ultimately evacuated to the outside of the body by defecation.

Colonic Movements

The large intestine of man is inactive for a large proportion of the time. However, when material is present in the proximal colon, segmenting contractions (haustral churnings) occur at a much lower frequency than in the small intestine and produce a limited back-and-forth movement of contents. This is the major type of movement of the large intestine, and by exposing the contents to the mucosa it promotes the absorption of water, thereby facilitating the formation of a firm fecal mass in the proximal colon.

At infrequent intervals of three to four times a day in man, a strong contraction of the proximal colon drives contents into the distal colon, where material accumulates distal to the pelvirectal flexure. These propulsive *mass movements* may represent a powerful segmenting contraction in association with extensive distal relaxation. That they are rather strong contractions is shown by the fact that the pressure in a segment undergoing such a contraction may reach a peak of 100 mm Hg. Peristalsis also aids in moving contents in a downward direction.

It generally takes about 18 hours for contents to reach the distal colon after leaving the small intestine, and they are stored here for varying lengths of time until defecation occurs. This may be 24 hours or longer following the ingestion of food. Although the rectum is normally empty, contents are occasionally shifted into the rectum after one of the mass contractions, and the resultant distention of the region elicits the desire to defecate. The act of defecation is partly voluntary and partly involuntary. The involuntary movements are concerned with smooth muscle; i.e., the distal colon contracts and the internal anal sphincter relaxes. Relaxation of the external anal sphincter, which consists of striated muscle, is voluntary. Other voluntary movements, which can supply one-half of the force involved in evacuation of the rectum, are a contraction of the abdominal muscles and forcible expiration with the glottis closed (straining movements).

Regulation of Colonic Motility

The proximal colon possesses a large degree of autonomy and functions in a relatively normal manner in the absence of its extrinsic motor innervation, which is derived from the vagus nerve. Movements here are probably largely initiated by distention of the colonic walls, but they can also be initiated reflexly, since when food enters the stomach or duodenum a mass contraction often occurs in the proximal colon. These *gastrocolic* and *duodenocolic reflexes* are usually most evident after the first meal of the day and are often followed by the desire to defecate.

The distal colon is somewhat more dependent on its extrinsic nerve supply, and movements in this region, including the act of defecation, disappear after transection of these nerves. However, weak movements do return after a time, and there is a semblance of the act of defecation. Defecation as it normally occurs is under voluntary control. There are subsidiary centers, since a certain region of the medulla can be stimulated to cause defecation. If the spinal cord is transected in the thoracic region, defecation can still occur without voluntary control after spinal shock has passed off. On the other hand, if the sacral cord is destroyed, defecation becomes very imperfect. This portion of the cord presumably serves as a local reflex center for the act of defecation. Distention of the rectum causes impulses to be sent over afferent fibers in the pelvic nerve to the sacral cord, which relays impulses over parasympathetic nerve fibers to the distal colon and anal sphincters. If defecation is inhibited by higher centers, the rectum relaxes, the stimulus of distention disappears, and defecation is postponed. The efferent fibers to the distal colon and internal anal sphincter are in the pelvic nerve, while those to the external anal sphincter are in the pudendal nerve. During periods of nonactivity, the internal and external anal sphincters are maintained tonically contracted by impulses over the lumbar sympathetics and pudendal nerve, respectively.

REFERENCES

Bass, P.   In Vivo Electrical Activity of the Small Bowel. In C. F. Code (Ed.), *Handbook of Physiology*. Washington: American Physiological Society, 1968. Section 6: Alimentary Canal, vol. 4, chap. 100, pp. 2051–2074.

Beaumont, W.   *Experiments and Observations in the Gastric Juice and the Physiology of Digestion*. Plattsburgh, N.Y.: Allen, 1833.

Bortoff, A.   Digestion: Motility. *Annu. Rev. Physiol.* 34:261–290, 1972.

Code, C. F., N. C. Hightower, and C. G. Morlock.   Motility of the alimentary canal in man. *Am. J. Med.* 13:328–351, 1952.

Code, C. F., J. H. Szurszewski, K. A. Kelley, and I. B. Smith.   A Concept of Control of Gastrointestinal Motility. In C. F. Code (Ed.), *Handbook of Physiology*. Washington: American Physiological Society, 1968. Section 6: Alimentary Canal, vol. 5, chap. 139, pp. 2881–2896.

Hightower, N. C., Jr.   Motor Action of the Small Bowel. In C. F. Code (Ed.), *Handbook of Physiology*. Washington: American Physiological Society, 1968. Section 6: Alimentary Canal, vol. 4, chap. 98, pp. 2001–2024.

Hunt, J. N., and M. T. Knox.   Regulation of Gastric Emptying. In C. F. Code (Ed.), *Handbook of Physiology*. Washington: American Physiological Society, 1968. Section 6: Alimentary Canal, vol. 4, chap. 94, pp. 1917–1935.

Ingelfinger, F.   Esophageal motility. *Physiol. Rev.* 38:533–584, 1958.

Kosterlitz, H. W.   Intrinsic and Extrinsic Nervous Control of Motility of the Stomach and Intestine. In C. F. Code (Ed.), *Handbook of Physiology*. Washington: American Physiological Society, 1968. Section 6: Alimentary Canal, vol. 4, chap. 104, pp. 2147–2171.

Magee, D. F.   Physiology of Gallbladder Emptying. In W. Taylor (Ed.), *The Biliary System; A Symposium of the NATO Advanced Study Institute*. Oxford, England: Blackwell, 1965. Chap. 4, pp. 233–247.

Thomas, J. E., and M. J. Baldwin.   Pathways and Mechanisms of Regulation of Gastric Motility. In C. F. Code (Ed.), *Handbook of Physiology*. Washington: American Physiological Society, 1968. Vol. 4, Section 6: Alimentary Canal, chap. 95, pp. 1937–1968.

Truelove, S. C.   Movements of the large intestine. *Physiol. Rev.* 46:457–512, 1966.

Youmans, W. B.   *Nervous and Neurohumoral Regulation of Intestinal Motility*. New York: Interscience, 1949.

26. Secretion and Action of Digestive Juices; Absorption

Leon K. Knoebel

Substances contained in ingested food which are important in the nutrition of the body include carbohydrates, proteins, lipids, vitamins, inorganic salts, and water. Many of the organic constituents of the diet are structurally complex and are not readily absorbed from the digestive tract in their natural states. However, the gastrointestinal system has a variety of exocrine glands which secrete digestive juices into the lumen of the digestive tract, and the enzymes contained in these secretions, along with those in the wall of the small intestine, convert complex organic molecules to smaller molecules. These digestion products are transferred from the intestinal lumen across the wall of the small intestine into the blood and the lymph, which in turn distribute them to the cells of the body. This chapter discusses secretion by the exocrine glands of the gastrointestinal system, describes the actions of the digestive enzymes, and considers the manner in which the end products of digestion and other substances are absorbed into the circulation.

MECHANISM OF SECRETION

The gastrointestinal system contains a variety of secretory cells, and each one produces its own specific type of secretion. Each secretion consists basically of water and inorganic ions, and in addition has one or more organic constituents. The most notable of the organic components are the digestive enzymes of the salivary glands, stomach, and pancreas, the bile salts and bile pigments secreted by the liver, and the mucin which is produced by all regions of the digestive tract.

Plasma is the ultimate source of the digestive secretions in the sense that it supplies the constituents necessary for their elaboration. The function of the secretory cells is to modify plasma in some way to produce a secretion. In the case of the organic constituents, the secretory cells synthesize these from raw materials provided by the plasma, and in some instances the finished products are stored in the cells until an appropriate stimulus causes them to be secreted. Water and some of the inorganic ions of the various secretions are the result of direct movement from plasma through and out of the secretory cells at the time they are stimulated to secrete. On the other hand, certain ions which are present in some secretions are produced by the metabolic activities of the secretory cells, but even in such instances plasma supplies the reactants that participate in the formation of these ions. During passage through the secretory ducts, the composition of the primary cellular secretion may be modified as a result of movement of substances to or from the blood through the duct epithelium.

A number of lines of evidence show that secretion is an active, energy-requiring process and not simply the result of passive movement of water and dissolved materials from blood through glandular epithelium into the lumen of the digestive tract. For example, the classic experiment of Ludwig showed that the secretion of saliva is not simple filtration of fluid from the blood. The duct of the submaxillary gland of the dog was blocked by cannulating the duct and connecting the cannula with a mercury manometer. When the gland was stimulated to secrete, the secretion of saliva continued in spite of the fact that the pressure in the duct rose considerably above arterial pressure. Since filtration is the movement of fluid from a region of high pressure to one of low pressure, the secretion of saliva cannot be produced by this mechanism. The same principle can probably be applied to the secretion of the other digestive juices. Most digestive juices are isosmotic to plasma, but a few are not. For example, parotid saliva is decidedly hypotonic. Furthermore, the concentrations of individual electrolytes in secretions are often quite different from those of plasma. The gastric parietal cell secretes a fluid which is thought to be isosmotic to plasma but which contains hydrogen as its major cation rather than sodium. The acinar cells of the pancreas secrete a juice containing four to five times as much bicarbonate as does plasma. These facts suggest that secretory cells must do considerable work in order to elaborate from plasma a secretion which can have not only a different osmotic activity from that of plasma but a different electrolyte pattern as well. Consequently, it is to be expected that the energy expenditure of an actively secreting gland should be greater than that of a resting gland. That this is the case is indicated by the fact that the oxygen consumption and the utilization of energy-supplying substrates by the digestive glands are increased during secretion.

Although the mechanisms involved in the formation of a secretion are not well understood, it is known that the various types of secretory epithelia are stimulated to secrete by either nervous or humoral mechanisms, and in most instances by both. The exact manner in which nervous and humoral factors act on secretory cells to cause secretion is unknown.

SALIVARY SECRETION

There are three pairs of human salivary glands, the *parotid, sublingual,* and *submaxillary* glands. These glands are composed of secretory cells arranged in acini, the secretions of which empty into a system of small ducts and then into larger excretory ducts that conduct saliva into the oral cavity. There are two types of cells in the acini of the sublingual and submaxillary glands: (1) serous cells, which secrete water, inorganic salts, and the digestive enzyme, *salivary amylase,* and (2) mucous cells, which in addition to water and inorganic salts secrete a number of glycoproteins collectively called *mucin.* Mucin in water produces a solution of high viscosity known as *mucus,* which gives a thick, viscous quality to certain types of saliva. Since the parotid gland contains only serous cells, its secretion is very watery and of relatively low viscosity.

Man secretes between 1.0 and 1.5 liters of saliva every day. The rate of secretion is very low during sleep, increases during the awake state in the absence of apparent stimuli to a level of about 0.5 ml per minute, and under maximal stimulation may attain a level of 7 ml per minute. By virtue of its moistening and lubricating properties, saliva facilitates chewing, swallowing, and speech. Dehydration results in decreased secretion of saliva, and this contributes to the sensation of thirst. Saliva has a cleansing action by diluting noxious substances and flushing them from the mouth. The cleansing action is also important in the prevention of dental caries; in the absence of salivary secretion, the incidence of tooth decay is higher than normal. Salivary amylase, the one digestive enzyme of saliva, initiates the digestion of high-molecular-weight carbohydrates.

Composition and Formation of Saliva

Both the volume and the composition of saliva secreted in response to different stimuli are quite variable and depend on the strength and nature of the stimulus. In general, saliva contains the usual electrolytes of the body fluids, the principal ions being sodium, potassium, chloride, and bicarbonate. At low rates of secretion, the concentrations of all ions except potassium are low (Fig. 26-1), and saliva is decidedly hypotonic. As secretory rate increases, the concentrations of sodium, chloride, and bicarbonate rise, and potassium concentration declines a little. At maximal secretory rates, potassium and bicarbonate concentrations are higher than those of plasma, whereas the concentrations of sodium and chloride are less than the plasma levels of these ions. The overall result of these changes is that, as secretory rate increases, osmolarity increases and at maximal rates of secretion approaches that of plasma. The pH of saliva rises from 6.2 to 7.4 with increasing rates of secretion as the result of the concomitant increase in bicarbonate concentration.

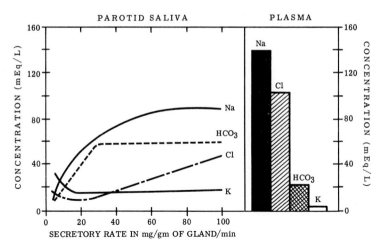

Figure 26-1
Electrolyte composition of human parotid saliva as a function of secretory rate. (From F. Bro-Rasmussen, S. Killmann, and J. H. Thaysen. *Acta Physiol. Scand.* 37:97, 1956.)

A number of theories have been offered in explanation of the changes in osmotic pressure and electrolyte composition of saliva which occur with variations in flow rate. One of them proposes that the primary process in the formation of the secretion of the acinar cells is the active transport of electrolytes, and that this primary secretion is approximately isotonic and has an electrolyte composition similar to that of plasma. It is further proposed that, during passage through the intralobular ducts, the composition of the primary secretion is modified by the transport of electrolytes from duct lumen to blood. The formation of hypotonic saliva would occur at this point, presumably as the result of the absorption of electrolytes in excess of water. This effect would be most marked at low rates of secretion, and both salivary osmolarity and ionic concentrations would be at their lowest values. However, as rate of secretion increases, the duct absorptive mechanism is progressively overwhelmed, so that both osmolarity and the concentration of most ions rise toward plasma levels, reaching maximal values at maximal rates of secretion.

The content of organic matter, the most notable constituents being mucus and amylase, is variable and seems to depend on the type of stimulus applied. For example, dogs with permanent fistulas of the submaxillary gland have been stimulated to secrete reflexly by introducing either meat powder or dilute hydrochloric acid into the mouth. It was found that even though the volumes of secretion, electrolyte concentrations, and blood flows through the gland were the same for both stimuli, the concentration of organic matter in the saliva was considerably greater for meat powder as compared with acid.

It is interesting that a response is often well adapted to the function it must perform. To cite one example, if a dog is given fresh meat, a viscous saliva containing much mucus is secreted which lubricates the bolus, thereby facilitating its passage into the stomach. If the same dog is given dry meat powder, a large volume of a watery secretion is produced and washes the powder out of the mouth.

Reflex Secretion
The secretion of saliva is controlled exclusively by nerve impulses. Saliva is produced in response to impulses acting on salivary centers in the medulla oblongata. These impulses originate mostly from stimulation of sensory nerve endings in the mucous

membranes of the mouth, but they can be initiated from many parts of the body, i.e., from the eyes, the nose, and other regions of the gastrointestinal tract as part of the vomiting reflex. The salivary centers, in turn, send impulses over the parasympathetic and sympathetic nerve supplies of the parotid, submaxillary, and sublingual glands, causing them to secrete their specific juices.

Stimulation of taste buds by introducing substances into the mouth results in the secretion of saliva by decerebrate animals. Higher nerve centers are not required for this secretion of saliva, and such reflexes have been called *unconditioned reflexes*. There are also salivary reflexes which require previous experience and which involve higher nerve centers. These *conditioned reflexes* have been well demonstrated in Pavlov's experiments on salivary secretion in the dog. For example, a bell was rung every time food was brought to a dog, and before long the dog salivated when only the bell was rung. This type of reflex requires that the cerebral cortex be intact and comes into play, for example, when one sees or thinks of appetizing food.

Role of Autonomic Nerves

There has been difference of opinion over the years regarding the precise role of the autonomic nerves in salivary secretion. Contrary to their usual antagonistic actions on effector organs, both parasympathetic stimulation and sympathetic stimulation are excitatory to salivary secretion. When the chorda tympani nerve (parasympathetic) is stimulated, the submaxillary and sublingual glands secrete a large volume of watery saliva which has a low concentration of organic matter. On the other hand, stimulation of the sympathetic nerves to these glands results in the formation of a scanty secretion which is thick and viscous and which has a high concentration of organic matter. The fact that the serous cells in these glands secrete a watery fluid, whereas the mucous cells secrete a highly viscous juice, has suggested to some workers that each type of secretory cell is innervated by only one division of the autonomic nervous system; i.e., the parasympathetic nerves would innervate only the serous cells, and the sympathetic nerves, only the mucous cells.

Other observations are in conflict with this idea. For example, although the parotid gland contains only serous cells, it responds to both parasympathetic and sympathetic stimulation, thereby implying a dual innervation of the cells of this gland. Electrophysiological evidence suggests that the same situation may exist for the cells of the mixed salivary glands. When sympathetic stimulation was superimposed on chorda tympani stimulation, the potential difference measured between these glands and the rest of the body was not changed from that of chorda stimulation alone, as might be anticipated if sympathetic stimulation resulted in the stimulation of an additional cell type. Moreover, when secretory potentials were recorded from the interior of a single cell, potential changes were observed during stimulation of either the chorda tympani or sympathetic nerves.

Although the precise innervation of the salivary secretory cells remains to be defined, the parasympathetic nerves are the important secretory nerves of the salivary glands, because parasympathetic denervation, either by sectioning or by administration of atropine, results in abolition of normal reflex secretion. The role of the sympathetic nerves remains unclear at the present time.

Salivary Digestion

The one digestive enzyme present in saliva is salivary amylase, which in the resting gland is stored in zymogen granules in the serous acinar cells. The action of the enzyme is to degrade complex polysaccharides, such as starch and glycogen, through several intermediate stages (dextrins) to maltose. Small quantities of glucose are also formed. Since salivary amylase does not function in an acid medium such as is present in the stomach, and since food remains in the mouth for such a short time, it might be thought that the action of this enzyme is extremely limited. However, the bolus of food does not disintegrate immediately upon entry into the stomach, and digestion can continue inside the bolus for fairly long periods of time.

GASTRIC SECRETION

Man secretes 2 to 3 liters of gastric juice every day. The gastric mucosa contains a variety of secretory cells, and a number of mechanisms regulate their activities. Gastric juice is most commonly regarded as the juice secreted by the body and the fundus, and at least three types of cell are present in the glands located in these regions of the stomach. The *parietal* or *oxyntic cell* is responsible for the secretion of intrinsic factor and a nearly isosmotic solution of hydrochloric acid, whereas the *chief cell* is the source of pepsin. Cells located in the neck of these glands and in the glands in the cardiac and pyloric regions of the stomach secrete an alkaline fluid which contains a soluble mucus. In addition to the glandular cells, the surface epithelial cells of the stomach secrete an alkaline juice containing an insoluble mucus.

It is impossible to make a single statement concerning the composition of gastric juice, since so many types of secretory epithelia are available for its production. The composition of gastric juice is determined by the relative activities of the different types of secretory cells. These, in turn, depend on the mechanism or combination of mechanisms that stimulate the various cells to secrete their specific juices.

Methods of Study

Surgical procedures have been devised in attempts to obtain pure gastric juice. In 1878 Heidenhain removed a small portion of the greater curvature of the stomach and formed it into a pouch. The secretions of the pouch could be removed through a fistula made by bringing the opening of the pouch through a stab wound in the belly wall. A pouch such as this may be considered to represent a miniature stomach which mirrors the secretory events occurring in the main stomach. Furthermore, the secretions obtained from the pouch are not contaminated with food, saliva, and material regurgitated from the small intestine. The vagal connections to the Heidenhain pouch are completely severed, so results obtained with this preparation may not be representative of secretion as it occurs in the main stomach. However, the pouch is useful when gastric secretion is to be studied in the absence of vagal innervation. In 1902 Pavlov made a pouch to which many vagal connections were maintained intact. Continuity between the pouch and the main stomach was retained by a bridge of tissue through which vagal nerve fibers traveled to the pouch. This pouch secretes reflexly under the proper circumstances. In addition to the pouches of the body and fundus of the stomach just described, both innervated and denervated pouches of the pyloric antrum find wide usage in the study of gastric secretion.

Another preparation is an animal with an esophagostomy. Since ingested material can be drained from the upper esophageal fistula, food can be administered orally without coming into contact with the more distal regions of the digestive tract. On the other hand, food can be introduced through the lower esophageal fistula in order to eliminate stimulation of the oral cavity. A gastric pouch is most often used in conjunction with esophagostomy.

Control of Secretion

Two components of gastric juice that are of considerable importance from a physiological standpoint are hydrochloric acid and pepsin. The rates at which these are secreted generally parallel each other, because secretion of both acid and pepsin is regulated, for the most part, by the same excitatory and inhibitory mechanisms. Both neural and hormonal mechanisms are involved in this regulation.

Excitatory Mechanisms

There are two substances in the body that act directly on the parietal and chief cells to stimulate their secretion. One is acetylcholine, which is released at these cells by the endings of parasympathetic postganglionic fibers present in the intrinsic nerve plexuses of the gastric wall. Since the vagus nerves provide the parasympathetic preganglionic innervation of the stomach, they are the important extrinsic secretory nerves to this organ. The other stimulatory substance is the hormone gastrin, which manifests its excitatory influences following release into the blood from endocrine cells in the mucosa of the pyloric antrum. Acetylcholine and gastrin act synergistically; i.e., the secretory response to both together is greater than to either alone.

Proof of the existence of gastrin has been provided by a variety of experiments. For example, when certain substances are placed in a transplanted denervated pouch of the antrum, there is secretion by a denervated (Heidenhain) pouch of the body of the stomach. Since no nerve connections exist between the two pouches, there must be a humoral link between them. The stimulating agents do not act by being absorbed and transported by the blood, because little secretion occurs when they are injected intravenously. Furthermore, distention of a denervated antral pouch produces a secretory response of a denervated fundic pouch, and no chemical substances are available for absorption under this circumstance. The present concept is that certain stimuli cause the release of gastrin from the mucosa of the pyloric antrum, and this hormone then travels by the blood to excite secretion by the body and fundus of the stomach. Pyloric antral mucosal extracts devoid of histamine (see p. 635) have been prepared and when injected intravenously result in the secretion of gastric juice. More recently, two gastrins have been isolated, their structures determined, and the molecules synthesized. These are heptadecapeptides, the terminal tetrapeptide of which is the active fragment of the total molecule. Even though the potency of the terminal tetrapeptide is only one-fifth that of the whole molecule, it displays all the physiological properties of natural gastrin.

Excitation of gastric secretion by acetylcholine and gastrin has been divided into three phases based on the region in which a stimulus acts to ultimately cause secretion of gastric juice: the *cephalic, gastric,* and *intestinal* phases. Each phase will be discussed in the order in which it occurs following the ingestion of food.

Cephalic Phase. The vagus nerves play the central role in the mediation of the cephalic phase of gastric secretion. Pavlov described the involvement of the vagi in this phase in his experiments using dogs with both an esophageal fistula and a vagally innervated gastric pouch. When these animals are fed, food does not reach the stomach but is returned to the outside through the esophageal fistula (*sham feeding*). As long as the dogs chewed and swallowed food, the Pavlov pouch continued to secrete gastric juice. However, if the vagi were transected just above the stomach, no secretion could be elicited from the pouch under the same experimental conditions—strong evidence that this secretion is produced by reflex action with the vagus nerve supplying the efferent limb of the reflex arcs. The reflexes involved are initiated by excitation of receptors in the head, namely, those associated with the taste, smell, and sight of food. Stimulation of the taste buds involves unconditioned reflexes, but there are also conditioned reflexes. For example, Pavlov's dogs secreted gastric juice when they saw the attendant approach with food or when they heard food being prepared.

Since the mere introduction of solid material into the mouth and the chewing of inert substances do not cause secretion, mechanical stimulation of the oral cavity is not an important factor. Furthermore, the chewing of different foods is not always sufficient to produce a secretion. The secretion produced during the cephalic phase usually fails if the food is not taken with appetite. Consequently, this secretion has been referred to as *appetite juice.* The fact that food must be agreeable to elicit the secretion is supported by the observation that there is little gastric secretion during certain types of emotional upsets, presumably because no food is palatable under these conditions.

The cephalic secretory response is due both to the release of acetylcholine at the site of secretion and to the vagally mediated release of gastrin from the pyloric antrum. The former pathway was demonstrated using dogs equipped with a Pavlov pouch and with the pyloric antrum removed. When the dogs were sham-fed, the vagally innervated pouch secreted acid. Since antrectomy obviated any possible stimulation of gastric secretion by gastrin, this result can be accounted for only on the basis of direct vagal excitation of the secretory cells. On the other hand, the vagal stimulation which results during the cephalic phase does exert part of its effect by liberating gastrin. This response was shown using dogs with an innervated antral pouch and a Heidenhain pouch. When the animals were sham-fed, there was copious secretion of the vagally denervated pouch. Furthermore, denervation of the antral pouch abolished the response.

In addition to the afferent pathways already mentioned, impulses from higher centers in the brain can either excite or inhibit gastric secretion, presumably by influencing the vagal release of acetylcholine and gastrin. The higher centers include the cortex and also the hypothalamus, which among other things is a center for coordinating impulses associated with emotional expression. Various emotional states do produce changes in gastric secretory activity, but the direction of the changes is not always predictable.

The juice secreted during the cephalic phase contains much hydrochloric acid and is rich in pepsin. The same type of juice is obtained when the vagus nerve is stimulated. Since high acidity and high pepsin levels, singly or together, are thought to contribute to ulcer formation, bilateral vagotomy has been used in the treatment of ulcer in an attempt to reduce acid and pepsin secretion in the patients. One way to check whether all vagal fibers to the stomach have been severed is to inject insulin. Insulin reduces the blood glucose level, causing excitation of the vagal nerve centers, which in turn stimulate the stomach to secrete by way of the vagi. If the gastric glands do not secrete following insulin injection, the vagi have been cut.

Gastric Phase. The gastric phase of gastric secretion begins when food enters the stomach and continues for as long as food remains there, which may be three to four hours. In man, a highly acid juice with a high concentration of pepsin is secreted during this phase. As is the case for the cephalic phase, the stimulating actions of acetylcholine and gastrin are responsible for exciting secretion during the gastric phase.

Although gastrin is secreted in response to impulses over the vagus nerve as a part of the cephalic phase, this hormone is also released as the result of local stimulation of the mucosa of the pyloric antrum by substances which either are present in food or result from the digestion of food. The most potent of these substances are ethyl alcohol and the protein digestion products supplied by gastric digestion. Distention of the antrum also results in the release of gastrin. On the other hand, either undigested proteins or fats and carbohydrates in any form are not effective.

Stimulation of gastric secretion by distention of or by the contact of chemical factors with the pyloric antrum persists following extrinsic denervation of the antrum. However, local neural elements are involved in the release of gastrin because, if the antral mucosa of an extrinsically denervated antrum is treated with an anesthetic in order to inactivate all nerves in the antral wall, gastrin is no longer released in response to intraluminal stimuli. Furthermore, that there is at least one synapse in the local neural chain is shown by the fact that the ganglionic blocker, nicotine, abolishes the secretory response. Secretion is also prevented when atropine is used, showing that it is acetylcholine which is released at and stimulates the gastrin-secreting cell. Based on this evidence (Fig. 26-2), the idea has evolved that nerve receptors in the antral mucosa are stimulated by certain chemical agents and distention. Impulses are sent over nerve fibers in the antral wall with at least one intervening synapse, and acetylcholine is ultimately released in the vicinity of the gastrin-producing cells, stimulating the release of the hormone into the blood. It might be further visualized that the vagal preganglionic fibers synapse on the final efferent cholinergic fiber, which could then be considered to represent the parasympathetic postganglionic nerve supply of the gastrin-releasing cells.

Part of the gastric juice secreted during the gastric phase is the result of stimulation of the body and fundus of the stomach. For example, if the body of the stomach is distended in an animal with the antrum removed in order to negate stimulation of secretion by gastrin, secretion of acid and pepsin occurs. This response is very much reduced if the vagi are cut, showing that a long reflex arc over the vagus nerve is primarily responsible for the effect. However, even under the circumstance of vagotomy, a small response remains. Since this is abolished by atropine, there must be, in addition to the vagovagal reflex, a local cholinergic reflex over the nerve nets in the wall of the stomach.

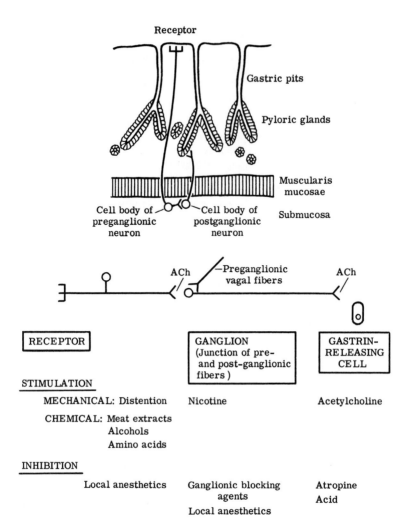

Figure 26-2
Hypothetical mechanism of regulation of release of gastrin and factors that influence it. (From M. I. Grossman. *Handbook of Physiology*. Washington: American Physiological Society, 1967. Section 6: Alimentary Canal, vol. 2, p. 844.)

Figure 26-3 summarizes, among other things, the mechanisms involved in the control of gastric secretion of acid and pepsin as discussed to this point.

Intestinal Phase. The secretion of gastric juice which is initiated during the cephalic phase is maintained during the gastric phase. However, gastric secretion is prolonged by an additional mechanism, the intestinal phase, which accounts for about 10 percent of the acid secreted by the stomach. This phase is demonstrated by the fact that a denervated gastric pouch secretes either when food is introduced directly into the duodenum through a fistula or when the small intestine is distended. It is apparent that stimulation of the intestinal mucosa causes the stomach to secrete in response to humoral stimuli. The mechanism involved is uncertain, but there may be an intestinal gastrin. In addition, cholecystokinin has weak acid-stimulating properties (see below),

633

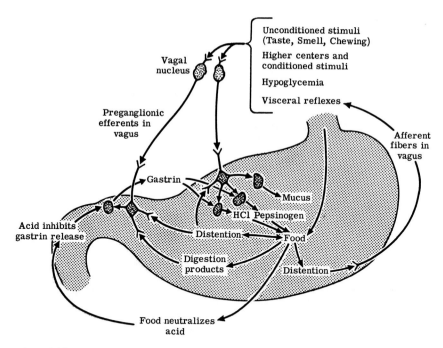

Figure 26-3
The control of gastric secretion. (From *Physiology of the Digestive Tract,* 2nd ed. by Davenport, Horace W. Copyright © 1966, by Year Book Medical Publishers, Inc., Chicago. Used by permission.)

and the release of this hormone from the duodenal mucosa may contribute to the intestinal phase of gastric secretion.

Inhibition of Secretion

When food is ingested, gastric secretion is stimulated by a variety of mechanisms. Nevertheless, there are also inhibitory mechanisms, the function of which is to prevent excessive secretion.

If liver solution at pH 7 is introduced into an antral pouch, a denervated pouch of the body responds to the resultant release of gastrin by secreting acid. On the other hand, if the pH of the liver solution is 2, no response is obtained. It is apparent that acid in contact with the antral mucosa inhibits secretion of acid by the body of the stomach. Furthermore, the extent of this inhibition is proportional to the hydrogen ion concentration of the contents of the antrum. Although it has been suggested that acid inhibition is the result of the release of an inhibitory hormone, the bulk of the evidence is in favor of the idea that acid acts by suppressing the release of gastrin (Fig. 26-3). For example, the gastric secretory responses that are inhibited by bathing the antral mucosa with acid are those known to be elicited specifically by stimuli that release gastrin. On the other hand, there is no inhibition for stimuli that operate by means other than gastrin release. The mechanism by which acid suppresses gastrin release is unknown. However, there is no involvement of nerves in the acid inhibition, because this is not blocked by the application of anesthetics to the antral mucosa. Acid must, therefore, manifest its inhibitory effect distal to the site of acetylcholine release; i.e., the gastrin-releasing cell may be sensitive to hydrogen ions.

An autoregulatory mechanism for the secretion of acid obviously exists, and, since the pH of the contents of the antrum is low at different times during a day, this mechanism must play an important role in regulating gastric acid secretion. For example, during interdigestive periods, the pH of the small volume of contents in the stomach is low, and the release of gastrin is suppressed. When a meal is eaten, the acid which is present is buffered by the constituents of food (Fig. 26-3), antral pH rises, the inhibition is removed, and gastrin is released in response to the usual stimuli. Gastric juice is secreted at a high rate, which continues until the buffering power of whatever food remains in the stomach is exhausted. At this time, pH decreases, and the resultant acidification of the antrum brings the gastric phase to an end.

A variety of substances are known to inhibit gastric secretion when they contact the duodenal mucosa. Examples are fat digestion products, acid, and hypertonic solutions. That a humoral mechanism is involved is demonstrated by the fact that secretion of a denervated pouch in response to gastrin release is inhibited when these substances are introduced into the duodenum. Uncertainty exists regarding the nature of the agent or agents mediating the inhibition. The mechanisms involved are probably the hormonal pathways which are operative under the same circumstances in the inhibition of gastric motility (Fig. 25B-7); i.e., gastric secretion is inhibited by the release from the duodenal mucosa of secretin and enterogastrone.

Although the effect of the intravenous administration of cholecystokinin is to stimulate gastric acid secretion, the actual result of the liberation of this hormone from the intestinal mucosa in response to a meal may be one of inhibition. The possibility is based on the idea that there is competitive inhibition of gastrin by cholecystokinin in the process of gastric secretion of acid. Since both gastrin and cholecystokinin possess the same terminal tetrapeptide, it is reasoned that the two molecules compete for the same acid-stimulating receptor sites on the parietal cell. However, cholecystokinin is a weak stimulant as compared to gastrin, presumably because the remainder of the molecule differs from that of gastrin. When both hormones are released in response to a meal, the overall gastric acid secretory response would be less than that of gastrin alone, because gastrin is unable to manifest its full excitatory effects in the presence of cholecystokinin; i.e., cholecystokinin denies receptor sites to gastrin. The same concept may apply to the roles of these two hormones in the regulation of gastric motility (Chap. 25B).

Secretion of the Parietal Cell
Gastric juice is unique in that it contains a high concentration of a strongly dissociated acid, hydrochloric acid. Although not essential for body function, hydrochloric acid aids in the gastric breakdown of connective tissue and muscle fibers, activates pepsin, provides an optimal low pH for peptic digestion, and kills bacteria that enter the digestive tract with food.

Gastric Acidity
In the study of acid secretion, most work has been done using gastric pouches in attempts to obtain the secretion of the parietal cell in its purest form. It has been concluded that the parietal cell secretes acid at a constant concentration regardless of the rate of secretion. Since parietal cell secretion and plasma are thought to have identical osmotic pressures, the theoretical maximum hydrogen ion concentration obtainable in this secretion would be about 170 mEq per liter. Since the maximal hydrogen ion concentration of parietal cell secretion has been estimated to be about 150 mEq per liter, practically all of the cation of this juice is in the form of hydrogen ion.

The concentration of acid in gastric contents is the result of the interplay of a number of factors. One is the rate of parietal cell secretion, and this depends on the balance between the excitatory and inhibitory mechanisms which influence the parietal cell at any given time. Another factor governing the acidity of gastric contents is the extent to which the acid parietal cell secretion is diluted and neutralized by the secretions of gastric nonparietal cells, saliva, material regurgitated from the duodenum, and food. Of major importance in this regard is the protein of food, which is particularly effective in buffering hydrogen ions secreted by the parietal cell. Consequently, the acidity of gastric contents can vary over a considerable range, the actual acidity depending on the relative amounts of parietal and nonparietal components in the mixture. During periods when there is no food in the stomach, the acidity of the small volume of contents in the stomach is generally high. Following a meal, the rate of parietal cell secretion rises to its highest level. Nevertheless, gastric acidity falls to its lowest value as the result of the predominant effect of dilution and neutralization by food.

Role of Histamine

A powerful stimulus of parietal cell secretion other than acetylcholine and gastrin is histamine, at one time considered the most potent stimulus of acid secretion. It is now recognized that on a molar basis of comparison gastrin is about 500 times more effective than histamine. However, when it is desirable to assess the acid-secreting capacity of the human stomach, frequently histamine or one of its analogues is administered. When histamine is injected subcutaneously or intramuscularly, and if a subject is capable of secreting acid, a juice is obtained which is very high in acid but low in pepsin.

It has been postulated that histamine is the final link in the secretion of hydrochloric acid; i.e., all neural and humoral stimuli may act by causing the release of histamine, and this would be the substance stimulating the parietal cell to secrete. Evidence in favor of this idea includes the following: (1) There are large quantities of histamine in the region of the parietal cells relative to adjacent regions. (2) Vagal stimuli and gastrin reduce the amount of histamine in the gastric mucosa. (3) Histamine appears in gastric juice which is secreted in response to a variety of stimuli. (4) Agents that inhibit the activity of histidine decarboxylase, an enzyme which catalyzes the conversion of histidine to histamine, also inhibit parietal cell secretion in response to diverse stimuli. (5) When diamine oxidase, a histamine-destroying enzyme, is injected into animals which are then vagally stimulated by sham feeding, acid secretion is considerably reduced from control levels. However, since these results are subject to alternative interpretations, not all of them supporting a physiological role of histamine in acid secretion, and since there is a considerable body of evidence in opposition to this idea, the concept that histamine is an obligatory intermediate in the secretion of acid cannot at present be accepted as fact.

Mechanism of Hydrochloric Acid Production

The mechanism of hydrochloric acid production must be extremely efficient because the hydrogen ion concentration of parietal cell secretion is about 2 million times that of plasma. An active transport process such as this requires the expenditure of energy, and it is provided in the form of high-energy phosphate bonds which are derived from oxidative metabolism and aerobic glycolysis of the mucosal cells.

Quantitative data have been obtained in studies in vitro using closed tubes of gastric mucosa of the frog. The tubes were incubated in a saline-bicarbonate solution through which was bubbled a gas mixture consisting of 95 percent oxygen and 5 percent carbon dioxide. When histamine was added to this medium, hydrochloric acid appeared inside the tubes (mucosal surface) and the same number of bicarbonate ions appeared outside the tubes (serosal surface). Furthermore, for every bicarbonate ion and every molecule of acid formed, a molecule of carbon dioxide disappeared from the medium. Although many theories have been proposed in attempts to explain the mechanism of gastric acid secretion, these experiments can be used as the basis for offering a simple working hypothesis in explanation of the formation of hydrochloric acid. The hypothesis is illustrated diagrammatically in Figure 26-4.

Hydrogen ions originate as a result of certain oxidative processes that occur in the parietal cell, and they are actively secreted into the lumen of the stomach. The alkali which is necessarily produced during the secretion of hydrogen ions is represented as hydroxyl ions, and they must be buffered in order to maintain cellular pH constant.

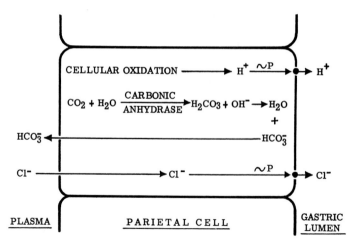

Figure 26-4
A possible mechanism for the secretion of gastric hydrochloric acid.

Carbon dioxide, which is both removed from the plasma and formed by the metabolic processes of the cell, is hydrated to form carbonic acid. The parietal cell is known to contain relatively large amounts of carbonic anhydrase, and this enzyme greatly increases the rate at which the reaction takes place. The hydroxyl ions are buffered in the parietal cell by reacting with carbonic acid with the resultant formation of water and bicarbonate ions. This scheme would explain the production of a bicarbonate ion and the loss of a molecule of carbon dioxide for every hydrogen ion that is produced. In addition, chloride ions are removed from the blood and secreted with hydrogen ions to form hydrochloric acid. Chloride ions do not just move passively along with hydrogen ions, because the surface of the mucosa is negative relative to the serosa. Chloride ions move across the gastric mucosa against a potential difference. In addition, since chloride ions move from a concentration in plasma of about 108mM per liter to a concentration of about 170mM per liter in gastric juice, they are transported against a chemical as well as an electrical gradient. The transport of this ion is therefore an active, energy-requiring process, probably involving a pump distinct from that utilized in the transport of hydrogen ions. In order to maintain electrical neutrality, for all chloride ions which move from plasma into gastric juice, an equivalent number of bicarbonate ions move from the parietal cell into plasma. As a result of the movement of bicarbonate ions, both the blood leaving the stomach and the urine become more alkaline following a meal (*alkaline tide*).

Gastric Digestion

Pepsinogens and Pepsins
There are at least three pepsins, and they are secreted in the form of the inactive precursors, pepsinogens I, II, and III. Pepsinogen I can be recovered from all regions of the stomach and the proximal duodenum, whereas pepsinogens II and III occur only in the oxyntic gland area. Pepsinogen II is synthesized by and stored as granules in the chief cells, but the cellular origin of the other pepsinogens is unknown. An extensively studied pepsinogen is one recovered from hog fundic mucosa; it is a protein having a molecular weight of 42,000. Since the corresponding pepsin has a molecular weight of 34,500, the inactive pepsinogen must be split to form active pepsin. The conversion in vitro of all pepsinogens occurs in the presence of either hydrochloric acid or small

amounts of pepsin. Thus, it seems likely that hydrochloric acid initiates the breakdown of pepsinogen in vivo and that the small amounts of pepsin so liberated then carry the process on autocatalytically. Human pepsins are endopeptidases and split linkages in the interior of both protein and polypeptide molecules, the pH for optimal activity of pepsin I being 3.0 to 3.2 and that of pepsin II being 1.5 to 2.0. As the result of the peptic digestion of protein, some polypeptides along with relatively few amino acids are found in the stomach following ingestion of protein.

Since the most effective stimulus for the secretion of pepsinogen is acetylcholine, the vagal reflexes which occur during the cephalic and gastric phases and result in the release of acetylcholine in the vicinity of the pepsinogen-secreting cells are undoubtedly of considerable importance in determining pepsin levels in gastric contents. Stimulation of pepsinogen secretion by gastrin and histamine is weak compared to that of acetyl-choline. However, there is good correlation between acid and pepsin outputs in man in that most stimuli which increase the secretion of acid also increase secretion of pepsinogen.

Gastric Lipase
A lipase is also present in gastric juice but is of little importance because, in order to act, this enzyme requires a pH close to neutrality. Only small amounts of free fatty acids and partial glycerides are recovered from the contents of the stomach after feeding triglycerides containing long-chain fatty acids.

Secretion of Mucus
All regions of the stomach possess cells that secrete an alkaline fluid containing mucus. Prime stimuli for the secretion of this juice are chemical, mechanical, and thermal irritation of the gastric mucosa. Vagal impulses also stimulate the secretion of mucus.

A layer of mucus, 1.0 to 1.5 mm thick, adheres to the gastric wall and serves as a protective barrier against various forms of irritation. It provides protection against mechanical injury by serving as a lubricant, and against chemical injury by virtue of its neutralizing properties. Mucus holds the alkaline fluid within its gel-like structure, and when acid diffuses into the gel, it can be neutralized before coming into direct contact with the epithelium; 100 ml of mucus neutralizes 40 ml of 0.1 N HCl. When pepsin diffuses into the mucus barrier, this enzyme is inactivated in the medium of high pH, so that the chance of attack on the protein structure of the underlying epithelium is minimized. However, the ultimate barrier is the mucosal cell membrane, the permeability characteristics of which restrict the entrance of even small ions to the interior of the cell. If membrane permeability is increased by toxic agents or mechanical trauma, hydrochloric acid and pepsin enter the cells and cause cellular destruction.

Intrinsic Factor
An important substance which is secreted by the human parietal cell is *intrinsic factor*. This mucoprotein is necessary for the absorption of vitamin B_{12}, which is required for the formation of normal red blood cells. Absorption of vitamin B_{12} by a special transport process in the lower small intestine requires that this vitamin be combined with intrinsic factor. When intrinsic factor is not secreted by the stomach, vitamin B_{12} absorption is defective, and pernicious anemia results.

PANCREATIC SECRETION
In addition to containing islets of alpha and beta endocrine cells which secrete the hormones glucagon and insulin into the circulation (Chap. 31), the pancreas contains exocrine secretory cells arranged in acini that are connected by small intercalated ducts to larger excretory ducts. The excretory ducts converge into one or two main ducts which deliver the exocrine secretion of the pancreas to the duodenum.

The digestion of food, which is initiated in the mouth and continued in the stomach, is carried close to completion in the lumen of the small intestine by the digestive enzymes of the exocrine pancreas. Since the digestive enzymes secreted into the lumen of the small intestine require a pH close to neutrality for optimal activity, the acidity of the contents entering the small intestine from the stomach must be reduced. Reduction may be partly accomplished by the transfer of hydrogen ions across the intestinal membrane, or the exchange of these ions with extraluminal sodium and potassium, or both. However, the alkaline secretion of the exocrine pancreas is of major importance in the neutralization of the acid gastric contents.

Composition of Pancreatic Juice

As is the case for other digestive secretions, the composition of pancreatic juice depends on the type of stimulus applied. One reason is that pancreatic juice consists of two distinct components: (1) an aqueous juice which makes up the largest part of the volume secreted by the pancreas and which contains electrolytes, but very little enzymatic activity, and (2) a juice of very small volume which contains the digestive enzymes of the pancreas. Thus, the total secretion of the pancreas represents a mixture of these two components in varying proportions.

The enzyme juice is elaborated by the acinar cells, which synthesize the digestive enzymes of the pancreas and store them as zymogen granules prior to secretion. The enzyme content of pancreatic juice is variable and depends on the nature of the stimulus. The concentration of proteins in pancreatic juice is an index of enzyme content, which ranges from 0.1 percent to 10 percent in the dog and 0.1 percent to 0.3 percent in man.

It is uncertain which cells are responsible for the secretion of the 1000 to 1500 ml of aqueous juice produced every day. However, the available evidence favors the idea that this water-electrolyte component is secreted by the epithelial cells which line the intercalated ducts. Regardless of the origin of this juice, its composition has been well characterized (Fig. 26-5). The principal cations are sodium and potassium, and these are present in the same concentrations as in plasma at all rates of secretion. The outstanding feature of this juice is that, relative to plasma, it has a high bicarbonate content. The

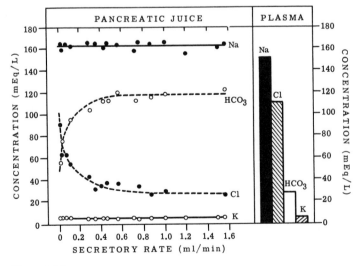

Figure 26-5
Relation between rate of secretion and concentrations of sodium, potassium, chloride, and bicarbonate in the pancreatic juice of the dog (after secretin injection). (From F. Bro-Rasmussen, S. Killmann, and J. H. Thaysen. *Acta Physiol. Scand.* 37:97, 1956.)

bicarbonate concentration increases with increased rates of secretion and ranges from 66 mEq per liter at low to 140 mEq per liter at high secretory rates. The other anion is chloride, the content of which is low relative to that of plasma, but the sum of the chloride and bicarbonate concentrations at any rate of secretion is the same as the sum of these same ions in plasma. This secretion has the same osmotic pressure as that of plasma and, because of its high bicarbonate content, has a pH from 7.6 to 8.2.

Mechanism of Secretion of the Aqueous Component

The bicarbonate of the aqueous juice is derived mostly from plasma, but it is also derived to some extent from carbon dioxide produced as the result of pancreatic cell metabolism. Carbonic anhydrase may play an important role in the formation of that bicarbonate which results from cellular metabolism. For example, the intercalated duct cells contain a high concentration of this enzyme compared to other pancreatic cells. Furthermore, inhibitors of carbonic anhydrase markedly depress pancreatic bicarbonate secretion.

The variations in bicarbonate and chloride concentrations that occur with alterations in rate of secretion are the basis for a number of theories regarding the mechanism of secretion of the aqueous juice. One of them, the *admixture concept,* proposes that pancreatic juice is a mixture of three fluids: an isosmotic chloride solution, an isosmotic bicarbonate solution, and a third fluid containing enzymes, mucus, and chloride. Variations in composition would result from differences in the rates of secretion of these three fluids. The *unicellular theory* suggests that a single cell elaborates the water-electrolyte secretion, and that this cell varies its output of chloride and bicarbonate. And finally, the *exchange theory* proposes that the primary secretion as formed by the intercalated duct cells is an isosmotic solution of bicarbonate. As the bicarbonate solution moves through the duct system, bicarbonate passively exchanges with chloride of the interstitial fluid, the extent to which the exchange occurs being determined by the rate of secretion. The greatest exchange would take place at lower flow rates, because more time would be available for it. In this instance, the concentration of bicarbonate in pancreatic juice as it enters the intestine would be relatively low and that of chloride relatively high.

Control of Secretion

The secretion of pancreatic juice is regulated by both nervous and hormonal mechanisms.

Reflex Regulation

If a dog is sham-fed, the pancreas secretes a small volume of juice which is very rich in enzymes. It has also been claimed that the sight or smell of food results in this same type of secretion. When the vagi are cut or atropine is administered, the juice produced by these stimuli is not obtained. Furthermore, the secretion can be elicited when either the vagi are stimulated or parasympathomimetic drugs, such as acetylcholine, are injected intravenously. These observations are proof for a cephalic phase of pancreatic secretion and show that stimulation of the pancreas occurs reflexly by way of the vagus nerves.

A gastric phase of pancreatic secretion can be demonstrated by distending the body of the stomach. The response of the pancreas is primarily one of increased enzyme output, but there is also a small increase in volume flow. This effect is abolished by vagotomy. The reflex involved is called the *gastropancreatic reflex,* and both afferent and efferent pathways are in the vagus nerve.

Hormonal Regulation

Since there is little impairment of the digestive function of the pancreas when all extrinsic nerves to this organ are cut, other mechanisms must be involved in the regulation of the secretion of pancreatic juice. It is now recognized that hormones play the central role in the control of secretion by the pancreas. In 1901 Bayliss and Starling performed one of the classic experiments in physiology. Their investigations led to the discovery of the first of many hormones now known to be highly important in the regulation of body function. They denervated a segment of jejunum by grossly dissecting the nerves to this section of gut and discovered that a profuse flow of pancreatic juice resulted when dilute hydrochloric acid was introduced into the lumen of the denervated segment. The response obtained was not due to the absorption of hydrochloric acid, since secretion did not occur when this acid was injected intravenously. They concluded that acid in contact with the intestinal mucosa caused the liberation of a substance into the blood and that this substance was transported to the pancreas, stimulating it to secrete. They also found that the same effect could be obtained by injecting intravenously extracts of the intestinal mucosa. The substance which was thought to be present in the mucosal extracts was named *secretin.*

Bayliss and Starling were sharply criticized for suggesting such a drastically different regulatory concept, but time has proved them correct in their conclusions. Unequivocal proof has been supplied in experiments in which the introduction of acid into a denervated jejunal transplant of the dog is followed by profuse secretion of a denervated pancreatic transplant.

Secretin has been isolated in pure form, is a basic linear polypeptide containing 27 amino acids, and has a minimal molecular weight of 3200 to 3500. When highly purified secretin is injected intravenously, a relatively large volume of alkaline, bicarbonate-rich juice with little enzyme activity is secreted; i.e., secretin stimulates secretion of the aqueous component of pancreatic juice. It has been theorized on the basis of this result that pancreatic juice, which is secreted in response to the secretin mechanism, may have its function in the neutralization of acid gastric contents. This argument is supported by the observation that the secretin mechanism responds only to a pH lower than 4.5, and it may be that the mechanism simply guards against high acidities in the gut.

In addition to the neural controls involved in the regulation of the secretion of enzymes and the intestinal hormonal control for the secretion of water and electrolytes, there is an intestinal hormonal mechanism for the secretion of enzymes. Since enzymes are secreted in the absence of pancreatic innervation and since pure secretin preparations produce a juice with low enzymatic activity, another factor must be responsible for stimulating the secretion of enzymes. That an intestinal hormone is involved was suggested by the fact that various foodstuffs elicit a pancreatic secretion of relatively high enzyme content when placed directly into the intestine of dogs with denervated pancreatic transplants. Products of protein digestion, such as proteoses, peptones, and amino acids, are potent in this respect, as are free fatty acids and soaps. Acid is a moderate stimulus, but carbohydrates are not particularly effective. Furthermore, the older and relatively impure secretin preparations stimulate enzyme secretion. The implication is that, in addition to secretin, these impure preparations contain another hormone which is responsible for the secretion of enzymes, and it has been called *pancreozymin.* Cholecystokinin and pancreozymin are the same polypeptide molecule—a chain of 33 amino acids with a terminal tetrapeptide identical to that of gastrin. Since cholecystokinin was discovered first, this name has been given to the polypeptide, but pancreozymin is used to describe its pancreatic-stimulating properties.

The excitatory effects of secretin and pancreozymin are manifested during the intestinal phase of pancreatic secretion. For many years, little emphasis was placed on a gastric phase of pancreatic secretion. However, first the gastropancreatic reflex was demonstrated, and, more recently, evidence has been accumulating for the involvement of gastrin as a stimulator of pancreatic secretion. It is now known that there is a weak stimulation of pancreatic flow and a strong enzyme response under conditions which result in the release of gastrin from the pyloric antrum. A similar result is obtained when pure gastrin is administered intravenously. It is not surprising that this gastric antral hormone stimulates pancreatic enzyme secretion, because it possesses the same terminal tetrapeptide as pancreozymin. Thus, secretion by the pancreas is apparently stimulated by the gastrin which is released as the result of impulses over the vagi during the

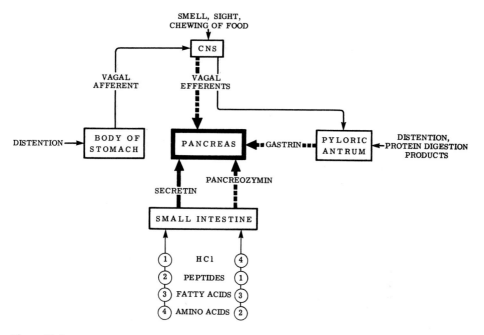

Figure 26-6
Neural and hormonal mechanisms for the secretion of pancreatic juice. The predominant secretory response of the pancreas to neural and hormonal stimuli is indicated by the arrows; dashed arrows, enzyme juice; solid arrow, aqueous juice. The intestinal intraluminal factors listed at the bottom of the diagram are numbered to indicate the relative magnitude of the effect of each of these agents.

cephalic phase and direct stimulation of the antrum during the gastric phase. Since the flow rate that can be attained with gastrin is about one-third that of secretin and the enzyme response is about three-fourths that of pancreozymin, the gastrin mechanism undoubtedly plays a quantitatively important role in the regulation of pancreatic secretion.

Figure 26-6 summarizes the neural and hormonal mechanisms that control secretion by the pancreas.

Pancreatic Digestion
Pancreatic juice is the most versatile and active of the digestive secretions. Its enzymes are capable of almost completing the digestion of all foodstuffs in the absence of the other digestive secretions.

Proteases
Trypsin, a proteolytic enzyme, is secreted as an inactive precursor, *trypsinogen.* Trypsinogen is converted to active trypsin by *enterokinase,* an enzyme which is present in intestinal juice. Small amounts of trypsin are also able to catalyze this conversion.

Trypsin hydrolyzes proteins and the polypeptides which are supplied by peptic digestion of protein. The end products of tryptic digestion are peptides of various sizes and amino acids. The optimal pH for tryptic activity in vitro is 7.8, which is higher than that usually found in the small intestine. Other enzymes secreted into the lumen of the small intestine also show optimal activities at a pH higher than that of intestinal contents. However, this apparently is not a limiting factor, since ingested foods are absorbed efficiently in spite of an environment which is suboptimal for digestion.

There is a trypsin inhibitor present in the pancreas, and it combines with trypsin in the ratio of one molecule of inhibitor to one molecule of trypsin. Since the end product of the combination is enzymatically inactive, the inhibitor prevents the pancreas from digesting itself during those times when small amounts of trypsin are activated in the pancreas. However, if large amounts of trypsin are activated, the inhibitor is overcome and the pancreas may be damaged or destroyed completely.

Another proteolytic enzyme of pancreatic juice is *chymotrypsin.* It is secreted as inactive *chymotrypsinogen* and converted to the active form by trypsin. Except for a few differences, chymotrypsin acts in a manner similar to that of trypsin. Both trypsin and chymotrypsin are endopeptidases and split linkages in the interior of the protein molecule.

The combined activities of pepsin, trypsin, and chymotrypsin result in the degradation of native proteins to polypeptides and amino acids. There are also a number of peptidases present in pancreatic juice. One that has been isolated is *carboxypeptidase.* This enzyme is secreted as *procarboxypeptidase* and is converted to the active form by either trypsin or enterokinase. Carboxypeptidase is an exopeptidase and splits amino acids from peptides with free carboxyl groups, but not from peptides with free amino groups.

Lipase

The pancreas supplies the enzyme *pancreatic lipase,* which is primarily responsible for the gastrointestinal digestion of fats. The action of this enzyme is to split the ester linkages in triglycerides to produce mostly free fatty acids and 2-monoglycerides. The optimum pH for activity of pancreatic lipase depends on the types of fatty acids contained in the triglyceride molecule and ranges from 7 to 9. Although there is an increased loss of carbohydrate and protein in the feces when pancreatic enzyme secretion is deficient, the dominant effect of this condition is *steatorrhea,* the appearance of excessive amounts of undigested fat in the feces; i.e., up to 60 to 70 percent of ingested fat is excreted.

Amylase

The pancreas also secretes an *amylase,* the action of which is similar to that of salivary amylase. Starch and glycogen are degraded to maltose, the optimal activity occurring at pH 7. Some pancreatic amylase escapes into the blood during pancreatic secretion, because stimulation of the pancreas results in a rise in plasma amylase. Plasma amylase levels have been used to evaluate the functional status of the pancreas. For example, trauma to this organ elevates plasma amylase levels, presumably by increasing the escape of pancreatic amylase into the blood. During autolysis of the pancreas by tryptic digestion, both plasma and urinary amylase show marked rises.

Other Pancreatic Enzymes

Additional enzymes secreted by the pancreas include ribonuclease, deoxyribonuclease, elastase, cholesterol esterase, and lecithinase.

BILE SECRETION

Bile is secreted continuously by the parenchymal cells of the liver. It first enters the bile canaliculi and flows from these minute channels through a converging system of small ducts into the hepatic duct and then into the common bile duct, which enters the duodenum at or near the site where the pancreatic duct joins this organ. During periods when there is no food in the upper digestive tract, most of the bile secreted by the liver is diverted into the gallbladder, where it is stored and concentrated by the absorption of fluid. Following a meal, the amount of bile entering the duodenum is increased as the result of contraction of the gallbladder and enhanced secretion by the liver. Bile is the vehicle for the excretion of certain end products of hemoglobin metabolism, the bile pigments, and provides the bile salts, which play an important role in the intestinal absorption of fat.

Although the electrolyte composition of bile is somewhat similar to that of plasma, the process of bile secretion is not simply ultrafiltration from blood, as demonstrated by the fact that secretory pressure can be higher than that in the liver blood sinusoids. In addition, a number of substances, including the bile salts and bile pigments, are transported from blood into bile against large concentration differences. These facts show that secretion of bile by the parenchymal cells of the liver is an active, energy-expending process.

Composition of Bile

The amount of bile secreted by the liver every day in man is 500 to 1000 ml, far greater than the capacity of the gallbladder, which is about 50 ml. However, this inequality of volumes is compensated for by the considerable absorptive capacity of the epithelium of the gallbladder. A comparison of the compositions of bile obtained from the liver and gallbladder as shown in Table 26-1 strongly suggests that water and salts are absorbed by the gallbladder.

The water content of gallbladder bile is much less than that of liver bile, whereas the dissolved materials of gallbladder bile, with the exception of the mineral salts, are much more concentrated than those of the liver. Thus, water and salts are absorbed by the gallbladder, and the other constituents, which are not appreciably absorbed, become concentrated in the process. Normally, the gallbladder concentrates liver bile about tenfold, but if the epithelium is damaged or irritated, the ability to concentrate is decreased or lost completely.

Even though the sum of the concentrations of all of the solutes of bile can be considerably greater than that of plasma, both liver bile and gallbladder bile have the same osmotic pressure as plasma. This apparent discrepancy is reconciled by the fact that certain constituents of bile, namely, the bile salts, bile pigments, lecithin, and cholesterol, form macromolecular aggregates (micelles) that have low osmotic activity. Since the osmotic activity of gallbladder bile is not altered by absorption, there must be an isosmotic absorption of water and electrolytes. The evidence indicates that sodium is actively transported by the gallbladder epithelium and that water follows sodium passively. Sodium transport must be active, because this ion moves against both an electrical and a chemical gradient. That water movement is passive is pointed out by the observations that the flow of water is directly proportional to the transport of sodium and that water does not move in the absence of sodium. As the result of the absorption of electrolytes, the reaction of gallbladder bile (pH 5 to 6) is more acid than that of liver bile (pH 7.4).

Table 26-1. Composition of Human Bile

Constituent	Liver Bile (percent)	Gallbladder Bile (percent)
Water	97.48	83.98
Mucin and pigments	0.53	4.44
Bile salts	0.93	8.70
Fatty acids	0.12	0.85
Cholesterol	0.06	0.87
Lecithin	0.02	0.14
Mineral salts	0.83	1.02

Bile Pigments

The major pigment of bile is *bilirubin,* a water-insoluble end product of the degradation of the porphyrin moiety of hemoglobin by the reticuloendothelial system. The 0.5 to 1.0 gm of bilirubin produced in man every day is extracted by the parenchymal cells of the liver from the blood, where it is maintained in solution by attachment to albumin. This pigment is actively secreted into the bile following its conversion by the liver cells to the water-soluble bilirubin diglucuronide. Intestinal bacteria convert bilirubin into other pigments, and small quantities of these are absorbed from the intestine and excreted by the kidneys. Most of the bile pigments are excreted in the feces, to which they impart a characteristic brown color.

Bile Salts

The parenchymal cells of the liver, in addition to synthesizing the bile salts, also possess an active transport mechanism for their secretion. The transport system enables these cells to secrete bile salts in concentrations many times greater than the concentration of the blood from which they are derived. There is an upper limit to transport capacity at high plasma bile salt concentrations; i.e., the transport system becomes saturated. The

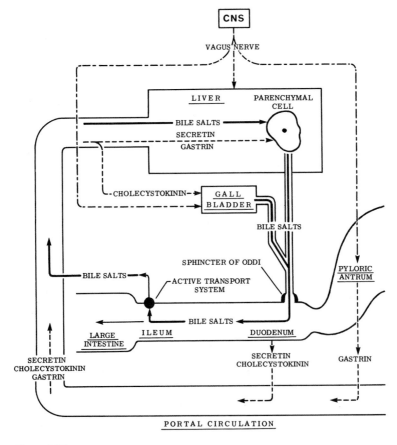

Figure 26-7
Pathways in the regulation of secretion of bile by the liver and contraction of the gallbladder. Solid arrows, enterohepatic circulation of bile salts; dashed arrows, hormonal pathways; dashed and dotted arrows, neural pathways.

bile salts and the bile pigments may be handled by the same transport system, because these molecules compete with one another for secretion.

During interdigestive periods, most of the bile salts of the body pool are stored in the gallbladder. However, when a meal is eaten, contractions of the gallbladder deliver the bile salts into the lumen of the small intestine, where they play an essential role in the absorption of fat. The function of the bile salts in the absorptive process will be discussed later in this chapter. Following their participation in the absorption of fat in the proximal small intestine, the bile salts are absorbed from the terminal ileum by an active process specialized for the transport of these substances. They are then returned by the portal vein to the liver, where they are removed from the blood by the liver cells and resecreted in the bile. The recycling of bile salts between liver and small intestine is called the *enterohepatic circulation of bile salts* (Fig. 26-7).

Since about 4 to 8 gm of bile salts is secreted in response to a meal, and since the total body content of bile salts averages only 3.6 gm, the body pool of bile salts must circulate about once or twice during the digestion of a meal. A total of about 12 to 30 gm of bile salts is secreted every day in response to all food eaten; hence each bile salt molecule circulates from four to nine times, the actual extent of the recirculation depending on the type and amount of food ingested. About 5 percent of the bile salts escapes into the large intestine with each circulation (Fig. 26-7), and the result is a loss of 0.5 to 1.0 gm of bile salts in the feces every day. However, since the liver cells are capable of synthesizing bile salts at rates up to 3 to 5 gm per day, synthesis of new bile salts by the liver easily makes up this loss.

The bile salts form micelles with the lecithin contained in bile. These macromolecular aggregates possess the capacity to dissolve within their micellar structure the highly water-insoluble compound cholesterol, which is also a normal constituent of bile. This is the mechanism by which cholesterol is normally maintained in solution in bile. However, when the ability of the bile salt–lecithin micelles to dissolve cholesterol is exceeded, as, for example, when the liver secretes excessive amounts of cholesterol or when the quantity of bile salts secreted is less than normal, cholesterol precipitates out of solution, small crystals of cholesterol are formed, and these can grow gradually into large gallstones.

Control of Secretion

Any substance that stimulates an increased secretion of bile is a *choleretic,* and the process by which secretion occurs is known as *choleresis.* Chemical, hormonal, and neural factors are operative as choleretics in the regulation of the secretion of bile.

The bile salts, which act directly on the secretory cells of the liver to stimulate secretion of bile in proportion to their concentration in the portal venous blood, are potent chemical choleretics and are the major stimuli for the secretion of bile. Certain studies have quantitated the influence of the bile salts on the rate of flow of water and electrolytes into the bile.

These studies involved the use of trained, unanesthetized dogs with duodenal cannulas installed in such a manner that a catheter could be inserted into the common bile duct for the collection of bile. The bile salt sodium taurocholate was infused at different rates into a vein in order to vary the plasma levels of this salt. As the plasma concentration of taurocholate was increased with higher rates of infusion, the rate of secretion of taurocholate also increased and was about the same as the rate of infusion. Figure 26-8 shows that bile flow increased in direct proportion to the rate of taurocholate secretion. A similar relationship was obtained for the output of chloride and bicarbonate ions. These data suggest that both the flow of bile (water output) and electrolyte output are dependent on the rate at which bile salts are secreted. One idea that attempts to explain these observations is that the active transport of bile salts provides a primary osmotic force for the secretion of bile. It is visualized that, when bile salts are actively transported into the canaliculi, a certain amount of water follows by virtue of osmotic drag. In addition, since there will be a diffusion gradient for inorganic

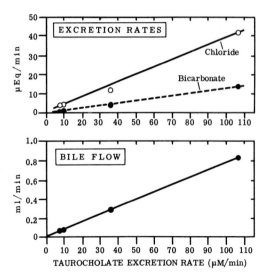

Figure 26-8
Relation between secretion rate of bile salt and output of chloride, bicarbonate, and water (bile flow) in the dog. Sodium taurocholate was infused intravenously at rates of 8, 40, 118, and 8μM per minute (in that order) to obtain the four points. Output of water appeared to be directly proportional to taurocholate secretion rate, as did the output of chloride and bicarbonate ions. (From H. O. Wheeler. *Handbook of Physiology.* Washington: American Physiological Society, 1968. Section 6: Alimentary Canal, vol. 5, p. 2417.)

ions, these move into the canaliculi accompanied by more water, the final result being the formation of a secretion isosmotic to plasma. During periods between meals, the bile salts are sequestered in the gallbladder, the portal venous concentration of these substances is low, and stimulation of bile flow by bile salts is minimal. When food is being digested, the bile salts are circulated between small intestine and liver, the portal venous concentration of bile salts is raised, and the extent to which these substances are actively transported is increased with the result that liver bile flow is stimulated.

Any stimulus that is known to release secretin and cause the pancreas to secrete also stimulates the secretion of bile. Furthermore, intravenous administration of pure secretin increases the flow of bile. However, the role of secretin in the stimulation of secretion by the liver is secondary to that of the bile salts, since secretin produces an increase in bile flow only about one-tenth that of the bile salts. It is of interest that the response of the liver to secretin consists in an increased output of water and electrolytes without any increase in output of bile salts and pigments. This is analogous to secretin stimulation to the pancreas, which has little effect on enzyme output but a decided excitatory influence on the water-electrolyte component of pancreatic secretion. There is also some evidence that, as in the pancreas, the site of action of secretin is the cells lining the small ducts of the biliary system and that, as the secretory rate is increased in response to secretin, bicarbonate concentration rises and chloride concentration falls.

Impulses over the vagi produce an increased liver bile flow. However, it remains to be proved that vagal stimulation has a direct stimulatory effect on the secretory cells of the liver. Since gastrin elicits a small secretory response by the liver, vagal stimulation probably manifests at least part of its excitatory effect through the release of gastrin.

The pathways involved in the excitation of secretion of bile by the liver, as well as those for contraction of the gallbladder, are summarized in Figure 26-7.

SECRETION OF INTESTINAL JUICE
The secretion of intestinal juice is the result of the secretory activities of a variety of cell types. Two types of intestinal glands contribute to this secretion: Brunner's glands, which are present in the duodenum, and the crypts of Lieberkühn, which are scattered through the entire length of the small intestine.

Brunner's Glands
The major contribution to the secretion of the upper duodenum is probably made by the glands of Brunner, which lie in the submucosa and contain only mucous cells. The volume of juice secreted here is low during fasting and increases somewhat following feeding. However, it is known that the upper duodenum possesses a greater resistance to ulceration as compared to the remainder of the small intestine, probably because the mucus secreted by this region is thick and forms a protective coating over the underlying epithelium.

Crypts of Lieberkühn
Four types of cells are present in the tubular glands of the mucosa that constitute the crypts of Lieberkühn. These are (1) undifferentiated cells which migrate up to the villi to become columnar epithelial absorptive cells, (2) mucus-secreting goblet cells, (3) argentaffin cells, which in addition to synthesizing serotonin probably have an endocrine function, and (4) Paneth's cells, the function of which is unknown.

Experiments dealing with the secretion of intestinal juice are complicated by the fact that absorption proceeds at the same time as secretion. Consequently, both the volume and the composition of the juice recovered from the lumen of different regions of the small intestine are the net result of both processes. This fluid is isosmotic to plasma and contains the usual electrolytes of plasma and some organic matter consisting of mucus, enzymes, and cellular debris. The concentrations of sodium, potassium, and calcium are relatively constant throughout the small intestine and similar to those in plasma. Although the total anion concentration is also much the same as in plasma, bicarbonate concentration is lower and chloride concentration is higher in jejunal fluid than in ileal fluid.

The small quantity of enzymes which has been reported to be present in intestinal juice, rather than being a component of an exocrine secretion, is probably derived from epithelial cells that are shed from the intestinal mucosa and disintegrate in the intestinal lumen. This idea is supported by the observation that the amount of enzymes present in juice collected from the intestine is usually related to the content of cellular debris. The process of extrusion of the epithelium is a continuous one, resulting in complete renewal of the intestinal epithelium every one to two days. In any event, only two of the enzymes found in intestinal juice are thought to play a significant role in intraluminal digestive processes. These are enterokinase, which has been mentioned in connection with the activation of trypsin, and intestinal amylase, which is similar in action to salivary and pancreatic amylases.

A proposed function of intestinal secretion is that of providing fluid in large enough quantities to allow efficient intestinal digestion and absorption of food; i.e., the water of intestinal juice serves as a solvent and as a medium of suspension and transport for the solids which are either dissolved or suspended in the contents of the small intestine.

Control of Secretion
The mechanisms involved in the secretion of intestinal juice are not well elucidated. Local mechanical and chemical stimulation of the mucosa of any region of the intestine is an effective stimulus for intestinal secretion. Consequently, it can be concluded that under physiological circumstances local mechanical and chemical stimulation by intestinal contents is important in eliciting the secretion of intestinal juice. Since the same effect can be produced in a segment of gut which does not possess extrinsic

innervation, the response to mechanical stimulation is not necessarily mediated through extrinsic nerves.

Neural Influences

The role of extrinsic nerves in intestinal secretion is uncertain. It has been reported that vagal stimulation increases the rate of secretion of both duodenal juice and that obtained from the rest of the intestine. Extrinsic denervation of an intestinal segment results in a copious secretion, which subsides in a day or two. This *paralytic secretion* of the intestine is also obtained when only the sympathetic nerve supply is cut. This result could be interpreted to mean that the sympathetic nerves send tonic inhibitory impulses to the secretory cells of the intestine. When the sympathetic nerves are severed, the inhibition is removed and the intestinal cells secrete unchecked in response to a cholinergic mechanism until some readjustment of secretory activity occurs. An alternative explanation is that the dilation of blood vessels and resultant increase in blood flow to the intestine which follows section of the sympathetic nerves may augment secretion by providing a greater amount of fluid for secretion.

Humoral Influences

There is evidence that a hormone is involved in regulating the secretion of intestinal juice. When dogs with denervated jejunal transplants are fed, the transplanted segment of intestine responds with an increased secretion. An extract of the mucosa of small intestine has been prepared which produces the same type of secretion when injected intravenously. The substance contained in this extract is called *enterocrinin,* and it is specific for the secretion of intestinal juice; i.e., it does not affect the secretion of pancreatic juice or bile. A hormonal mechanism (duocrinin) has also been postulated for stimulation of secretion of duodenal juice. This idea is based on the fact that a denervated transplant of the Brunner's gland area responds with an increased rate of secretion following feeding. There is also evidence for secretin stimulation of Brunner's glands.

COLONIC SECRETION

The colonic mucosa has numerous tubular glands containing goblet cells which secrete mucus. The fluid secreted by these glands is very viscous and alkaline (pH 8.0 to 8.4). Stimulation of the parasympathetic fibers of the pelvic nerves, chemical irritation, and mechanical irritation increase the volume of colonic secretion. The function of the secretion is to lubricate and facilitate the passage of feces and to protect the mucosa of the colon from mechanical and chemical trauma. The alkaline secretion serves to neutralize irritating acids formed by bacterial action, a process that occurs to a considerable extent in the colon. When there is excessive secretion of mucus by the colon, this substance can constitute the major portion of a bowel movement. This condition is called *mucous colitis* and is thought to be caused by excessive parasympathetic activity, most often attributable to emotional tension.

INTESTINAL ABSORPTION

Absorption has been defined as the passage of substances into the circulation. Absorption from the lumen of the digestive tract consists in the transfer of various digestion products, vitamins, inorganic salts, and water through the barrier imposed by the semipermeable intestinal membrane into the blood and the lymph. One driving force involved in the transport of molecules across this membrane is the difference in concentration of a substance on the two sides of the membrane. The rate of transfer is dependent not only on this diffusion gradient but also on the molecular size and lipid solubility of the substance being transported. Certain chemical reactions which occur in the mucosal cells also supply a driving force for the transport of many substances across the intestinal epithelium. Energy is required for such transport (active), and transfer is impeded when oxidative metabolism of the mucosal cells is inhibited. Chapter 1 supplies a more detailed description of the mechanisms involved in the movement of substances across biological membranes.

Absorption from the mouth, esophagus, and stomach is not important from a nutritional standpoint. The principal site of absorption is the small intestine. Anatomically, the small intestine is well suited to the task, since the surface area available for absorption is greatly increased, not only by the presence of the villi, but also by the microvilli. Practically all substances capable of being absorbed disappear from the lumen of the small intestine by the time mid-jejunum is reached. The ileum does not participate in absorption to a large extent because the more proximal regions of the small intestine are so efficient in this regard that usually only small quantities of absorbable material enter the ileum. However, the distal small intestine has the capacity to absorb, and it does so in situations in which food escapes absorption upstream; i.e., the ileum is a reserve absorptive area. Actually, about 50 percent of the small intestine can be removed without interfering with absorption. It should be recalled, on the other hand, that vitamin B_{12} and the bile salts are absorbed specifically in the terminal ileum, and removal of this region of intestine will result in defective absorption of these substances. The contents recovered from the terminal ileum contain no digestible carbohydrate, very little lipid, and only 15 to 17 percent nitrogen-containing substances. Most of this material can be accounted for as bacteria, desquamated epithelial cells, the remains of various digestive secretions, and undigested and unabsorbed residues of food, such as cellulose and connective tissue. Although the intestinal epithelium does act as a barrier to many materials, it also permits, and in many instances facilitates, the absorption of a large variety of substances. The following discussion will deal with the absorption of carbohydrates, proteins, lipids, water, and electrolytes.

Absorption of Carbohydrates

In terms of calories, carbohydrate makes up approximately half of the American diet. The major digestible carbohydrate of food is the polysaccharide plant starch. This large molecule is composed of straight and branched chains of glucose, and it is reduced by the actions of salivary and pancreatic amylase mostly to the much smaller disaccharide molecule, maltose. Man does not secrete an enzyme capable of digesting cellulose, a plant polysaccharide which also consists of glucose molecules, but joined by linkages different from those of starch. Consequently, this complex carbohydrate is excreted in the feces. Other carbohydrates that are present in the diet in smaller, but significant, quantities are the disaccharides sucrose or table sugar (glucose-fructose) and lactose or milk sugar (glucose-galactose).

It is widely accepted that carbohydrates are absorbed in the form of monosaccharides. This idea is supported by the observation that, following ingestion of a wide variety of carbohydrates, only monosaccharides appear in the portal blood, which is the route of absorption for carbohydrates. The reason polysaccharides and disaccharides are not absorbed as such is that the intestinal epithelium is impermeable to carbohydrates of high molecular weight, and furthermore, the absorptive surface is lacking in special transport systems for these molecules. In any event, the major form in which carbohydrate is presented to the intestinal epithelium for absorption is as disaccharides, which either are present as such in ingested food or result from the amylase digestion of starch. However, the activities of the enzymes that hydrolyze disaccharides are very low in intestinal contents. On the other hand, it has been shown that the brush border of the mucosal cells contains all the disaccharidases that are present in these cells. They include maltase, lactase, and sucrase. Prevailing opinion is that disaccharides, such as maltose, lactose, and sucrose, are hydrolyzed to their constituent monosaccharides on the surface of the brush border of the epithelial cells during the absorptive process.

Evidence accumulated over the years demonstrates that, among the naturally occurring monosaccharides, glucose and galactose are absorbed by an active, energy-

Table 26-2. Absorption of Monosaccharides by Rats*

Sugar	Absorption Rate
Galactose	110
Glucose	100
Fructose	43
Mannose	19
Xylose	15
Arabinose	9

*Values are expressed in relative terms with the absorption of glucose equal to 100.
Source: From C. F. Cori. *J. Biol. Chem.* 66:691, 1925.

requiring process, whereas most of the others are absorbed by passive processes not requiring the expenditure of metabolic energy. In 1925 Cori made observations which showed that the absorption of certain monosaccharides cannot be completely explained on the basis of diffusion. Cori measured the rates of absorption of different mono- saccharides in rats, and the results are summarized in Table 26-2. Rate of diffusion is inversely related to molecular size, and if absorption takes place by diffusion only, the pentoses, xylose and arabinose, should be absorbed more rapidly than the other four monosaccharides, which are hexoses. However, the pentoses are absorbed least rapidly. In addition, even though all four hexoses diffuse at the same rate, there is considerable variation in their rates of absorption. It was concluded from these results that there is a special mechanism in the cells of the intestinal epithelium which increases the rate of absorption of glucose, galactose, and perhaps fructose over the absorption rate of mannose and the pentoses.

More recent work supports and extends the conclusions drawn from the older experiments. Wilson and Wiseman in 1954 used the everted intestinal sac method to study the movement of various sugars across the epithelium of isolated segments of hamster intestine. This method involves the use of small segments of intestine turned inside out, filled with fluid, and tied at both ends. The absorptive mucosal surface is exposed to solutions of known composition. Following a period of incubation, the absorption of specific substances can be measured by determining their concentrations in the mucosal and serosal solutions. Figure 26-9 shows the result of one such experi- ment in which the mucosal and serosal sides of an everted sac of jejunum were simultaneously exposed to glucose solutions of the same concentration. This approach obviates any possibility of a diffusion gradient for glucose. After a period of incubation it was found that glucose moved from the mucosal to the serosal side of the sac against a large concentration difference. The same was true when galactose was used. Since under the conditions of the experiment diffusion cannot be the mechanism involved in the transport of these sugars across the intestinal membrane, some special mechanism must be available. This idea is supported by the fact that the transfer of both glucose and galactose was abolished under anaerobic conditions (Fig. 26-9), thereby demon- strating the importance of oxidative metabolism for the normal functioning of the mechanism. It can be reasoned that certain oxidative metabolic reactions supply energy to facilitate the absorption of glucose and galactose.

There was no net movement of a variety of other monosaccharides in the everted sac studies, showing that no special mechanism is available for transporting these sugars against a concentration difference. The evidence indicates that they are absorbed by passive diffusion. For example, the rates of absorption of these sugars increase in linear relation to intraluminal concentration. This response is typical of a diffusion process; i.e., the greater the concentration difference between lumen and blood, the greater the rate of diffusion, and in direct proportion to the difference. On the other hand, an

651

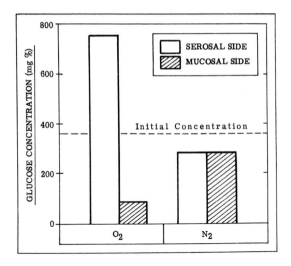

Figure 26-9
The transport of glucose by everted sacs of hamster jejunum under aerobic and anaerobic conditions. (From T. H. Wilson et al. Reprinted from *Federation Proceedings* 19:870–875, 1960.)

intestinal transport maximum can be demonstrated for glucose absorption at high intraluminal glucose concentrations. A transport saturation phenomenon such as this is typical of specialized transport systems.

The relative absorption rate of fructose (Table 26-2) is intermediate between the rates of monosaccharides that are actively absorbed and those of sugars that are absorbed by diffusion. There is an enzyme system in the intestinal mucosa that transforms fructose to glucose. In addition, in certain species the mucosal cells convert fructose to lactate. Since fructose is not transported against a concentration gradient, it is probably absorbed passively. The reason for the rapid rate of absorption of fructose relative to other passively absorbed sugars may be the maintenance between intestinal lumen and mucosal cell of a relatively high diffusion gradient as the result of partial conversion in the cell to glucose and lactic acid.

The process involved in the absorption of glucose and galactose is a clear example of what has been termed an active process; i.e., these sugars move against a concentration difference, the movement depending on certain energy-yielding reactions. The nature of the reactions and the manner in which they participate in the absorption of certain sugars are unknown. It has been suggested that glucose and galactose are phosphorylated in the intestinal epithelial cell and that this is the primary reaction in the active transport of these sugars. However, all attempts to show an alteration in the structure of these sugars during absorption have failed. Present thinking is that a membrane carrier is located on the luminal border of the epithelial cells which transports sugars across the membrane unchanged. The evidence indicates that a common pathway is involved in active sugar transport, because the presence of one actively absorbed sugar inhibits the transport of another actively absorbed sugar; there is competition for the same transport system.

A recent idea is that there is a coupling of the sodium and sugar transport systems of the intestine. It has been shown in studies in vitro that active transport of glucose ceases when the medium is devoid of sodium, and that as the sodium content is increased, absorption of glucose increases. Furthermore, cardiac glycosides which are known to inhibit sodium transport also inhibit glucose absorption. One line of thought is that

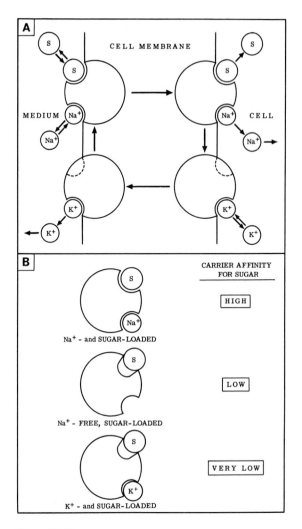

Figure 26-10
A. Model of mobile carrier with two sites, one specific for sugar and one specific for sodium. The influence of intracellular potassium on carrier movement is illustrated. B. Some possible affinities of the carrier for sugar. (From R. K. Crane. *Handbook of Physiology.* Washington: American Physiological Society, 1968. Section 6: Alimentary Canal, vol. 3, pp. 1335–1336.)

sodium increases the affinity of glucose for its carrier and facilitates the subsequent transport of the sugar-carrier complex into the cell (Fig. 26-10).

An assumption is that there is a mobile carrier in the brush border of the epithelial cell and that this carrier has a binding site which has substrate specificity for actively absorbed sugars. It is of interest in this regard that all actively absorbed sugars have certain structural features in common, some of which are lacking in the passively absorbed sugars. These structural characteristics are at least six carbon atoms, a D-pyranose ring structure, and a hydroxyl group at carbon 2. It might be visualized that such a structure is required to "fit" into the binding site. In addition to a binding site for sugar, it is theorized that there is a specific binding site on the same carrier for sodium, and that the carrier affinity for sugar is greatest when sodium is bound to the carrier (Fig. 26-10B). Moreover, when the carrier is loaded with both sugar and sodium, it travels across the membrane, the driving force being the concentration difference which exists for sodium from luminal to interior surface of the brush border membrane (Fig. 26-10A). At this point the carrier arrives in an environment that is favorable for the release of sugar into the cell interior—an environment of high potassium concentration. Potassium is known to compete with sodium for the sodium binding sites, and also, when these binding sites are occupied by potassium, the sugar binding site possesses much less affinity for sugar (Fig. 26-10B). Since potassium concentration is high inside the cell, whereas that of sodium is low, potassium should be able to compete success- fully for the binding sites, thereby displacing sodium and causing sugar to unbind. Once potassium attaches to the sodium binding site, the carrier is driven back to the luminal surface according to the concentration difference for potassium (Fig. 26-10A). Here potassium is replaced by sodium, and more sugar is transported into the cell.

However, in order to ensure continued translocation of the sodium-loaded and sugar-loaded carrier across the brush border membrane and maintain the release of sugar in the cell interior as the result of potassium binding, it is essential that a low intracellular sodium concentration be preserved. This function is fulfilled by an energy-dependent pump which moves sodium from the inside to the exterior of the cell. Since the sodium pump requires energy derived from metabolism for normal function, it would be here that the energy requirement for active sugar transport would be manifested. If sodium pump activity were abolished by anaerobic conditions or metabolic inhibitors, sugar could not accumulate in the cell, because the increased cellular sodium content would not only reduce the sodium gradient across the brush border membrane, thereby decreasing the driving force for movement of the carrier, but also reduce the effectiveness of potassium for competing with sodium for the binding sites.

Absorption of Proteins

Protein which is available for absorption from the small intestine is derived not only from food but also from the desquamated cells and various secretions which enter the lumen of the digestive tract. The protein of endogenous origin comprises a sizable fraction of the total protein presented for digestion and absorption.

The intestine of many newborn mammals can absorb by pinocytosis considerable quantities of intact protein molecules. Although the capacity to absorb proteins is lost shortly after birth, this process is important in a number of species because absorption of antibodies in the colostrum confers passive immunity against infection. From a nutritional standpoint, absorption of native proteins in the adult is insignificant. However, minute amounts of proteins are occasionally absorbed, as shown by the fact that some adults have allergic reactions to various food proteins. When a protein is absorbed unchanged, it sensitizes the individual to future doses of the same protein.

The combined action of the proteolytic enzymes which are secreted into the lumen of the digestive tract is to reduce large protein molecules to small peptides and amino acids. However, proteins are absorbed into the portal blood system mostly as amino acids. In one study when different proteins were fed to dogs, there was a large increase in many free amino acids in the portal vein during absorption. On the other hand, higher products of protein digestion, such as peptides, were not present. It would appear that peptides, even though water-soluble and diffusible, are poorly absorbed. This presumption is substantiated by a study in vitro reporting that when simple dipeptides such as glycylglycine and leucylglycine were put on the mucosal side of rat intestine only small amounts of these dipeptides appeared on the serosal side. Since large amounts of free glycine were present on both sides of the intestine, the inference is that the intestine is far more capable of transporting amino acids than peptides. There are, however, enzymes in the brush borders of the mucosal cells which hydrolyze peptides.

Thus, it is probable that different peptides are hydrolyzed to their constituent amino acids by specific peptidases at the surface of the microvilli, and the amino acids so formed are then transported into the portal blood.

As is the case for certain monosaccharides, many amino acids are subject to selective absorption. The absorption of amino acids has been compared with the absorption of certain compounds, such as polyhydric alcohols and acid amides, which are not normally present in the diet. Although all compounds fed were of approximately the same molecular size, the amino acids were absorbed much more rapidly than were the nonphysiological compounds. The absorption of amino acids cannot be explained solely on the basis of diffusion, since all substances fed had about the same molecular weight and therefore about the same rate of diffusion. These results suggest that the intestinal mucosa is equipped with special mechanisms which facilitate the transport of amino acids across the intestinal membrane. It also appears that the mechanisms involved are selective with respect to the amino acid absorbed. Outright proof of selective transport was provided by experiments in which the absorption of mixtures of the L and D forms of individual amino acids was studied in loops of rat intestine. In all instances the naturally occurring L-amino acid was absorbed much more rapidly than the corresponding D form, although the only difference between the two is their stereoconfiguration. Absolute proof for active transport of certain amino acids has been provided by the observation that they are transported across the intestinal epithelium against a concentration difference, and that this movement is prohibited by anaerobic conditions.

There is probably more than one transport system for amino acids. For example, neutral amino acids compete with one another for absorption, but their absorption is not inhibited by basic amino acids. Furthermore, basic amino acids compete with one another for absorption, and although a few neutral amino acids do inhibit the absorption of some basic amino acids, these results have suggested that there is one transport system for neutral amino acids and another having a primary affinity for basic amino acids. It is probable that other transport systems exist for other groups of amino acids, such as the dicarboxylic amino acids and the imino acids.

Although the nature of the systems which participate in the transport of amino acids is unknown, normal function of the mechanisms involved is dependent on oxidative metabolism because transport is inhibited by anaerobic conditions, cyanide, and dinitrophenol. As is the case for sugars, it has been suggested that membrane carriers are involved in the transfer of amino acids across the intestinal epithelial cell, and there is evidence that the carrier mechanisms are sodium dependent. It is also of interest that glucose and galactose compete with certain amino acids for absorption. The implication is that the carrier mechanisms for sugars and some amino acids are shared. The amino acid molecular configuration which is required for optimal active transport includes the L form of an amino acid, an unsubstituted carboxyl group attached to the alpha-carbon atom, an alpha-amino group, and an alpha-hydrogen atom. Although knowledge of the intestinal transport of amino acids is growing rapidly, the more intimate features of the mechanisms involved remain to be clarified.

Absorption of Lipids

The lipids of the diet are mostly in the form of triglycerides but also include phospholipids, cholesterol, and vitamins A, D, E, and K. Until recently, one difficulty in elucidating the mechanism by which triglycerides are absorbed has been the uncertainty regarding the form in which triglyceride fat enters the epithelial cells from the intestinal lumen. In the case of carbohydrate and protein absorption, only monosaccharides and amino acids can be detected in the portal blood during absorption, and it can be deduced that these are the forms in which carbohydrates and proteins are transported across the intestinal membrane. However, not all products of fat digestion leave the intestinal epithelium unchanged. It is known that considerable quantities of fatty acids with 12 carbon atoms or less are transferred into the portal blood when either fed in the free form or incorporated into triglycerides. On the other hand, the lymph is the major route of absorption for fatty acids possessing 14 or more carbon atoms. Because the triglycerides of the diet are composed chiefly of 16 and 18 carbon atom fatty acids, most of the absorbed fat is transported into the lymph. Since analysis of lymphatic fat of dietary origin shows it to be almost completely in the form of triglycerides, and since pancreatic lipase hydrolyzes ingested triglycerides in the lumen of the intestine, the absorbed digestion products must be resynthesized to triglycerides in the cells of the intestinal epithelium. The resynthesis makes difficult any attempt to determine by analysis of lymphatic fat the form in which triglycerides leave the intestinal lumen to enter the absorptive epithelium.

One problem has been to determine the extent to which triglycerides are hydrolyzed in the intestine prior to absorption. Lipid recovered from intestinal contents of animals fed triglycerides contains diglycerides, monoglycerides, free fatty acids, and un-hydrolyzed triglycerides. The material absorbed could be one of these four lipids, all of them, or some particular combination of the four. Studies in vitro have shown that pancreatic lipase, which acts only at the lipid-water interface, preferentially catalyzes the hydrolysis of the 1- and 3-ester bonds of triglycerides, with the result that there is the consecutive and rapid formation first of 1,2- and 2,3-diglycerides and then of 2-monoglyceride with the liberation of two fatty acids. Since lipase attacks the 2-ester linkage with considerable difficulty, 2-monoglyceride tends to accumulate in the system. If these results can be extended to the intact animal, the inference is that there is relatively little complete hydrolysis of triglycerides to glycerol and fatty acids in the lumen of the intestine. Since hydrolysis to the monoglyceride stage occurs with relative ease, it might be anticipated that a large part of the absorbed fat is in the form of free fatty acids and monoglycerides, and, as will be seen from the following discussion, considerable evidence has accumulated in favor of this idea.

If it is accepted that monoglycerides and free fatty acids are the lipid species that traverse the luminal membrane, the problem remains regarding the mechanism of entrance of these substances into the epithelial cells. Long-chain fatty acids are the most common type of dietary fatty acid, and these, as well as the triglycerides, diglycerides, and monoglycerides which contain them, are water-insoluble. Without the availability of some special mechanism, triglycerides and their digestion products would exist as large oil droplets in the aqueous medium which constitutes intestinal contents, and optimal absorption of fat could not occur. It is known that lipid-soluble substances, by virtue of their solubility in the lipid portion of cell membranes, are able to diffuse through these membranes. However, these substances must be present in some form that assures ready entrance into the cell; i.e., in order to gain access to the surface of a cell, lipids must be in a water-soluble and freely diffusible form. In this regard, studies which have characterized the physical and chemical states of the intraluminal lipids recovered following the feeding of triglycerides have shown that the bile salts play the major role in preparing lipids for absorption by the small intestine (see below).

If intestinal contents are centrifuged at high speed, an oily top phase and a clear bottom phase are obtained. The oil phase is derived from droplets of lipid which are present in intestinal contents, and this phase contains practically all of the intraluminal triglycerides and diglycerides along with some monoglycerides and free fatty acids. Although some of these droplets are quite large, others are part of an extremely stable oil-in-water emulsion consisting of particles having an average diameter of only 5000 Å. This fine emulsion is produced by the combined action of a number of emulsifying agents in the lumen of the small intestine. These include monoglycerides and free fatty acids which result from the lipase hydrolysis of triglycerides, the bile salts, phospholipids and their digestion products, and proteins. The formation of such an emulsion presumably enhances the action of pancreatic lipase by exposing a greater lipid surface area on which this enzyme can act.

The clear phase contains mostly monoglycerides and free fatty acids, and these lipids along with the bile salts are present as water-soluble aggregates of micellar dimensions. At low concentrations, the bile salts form molecular solutions, but as their concentration

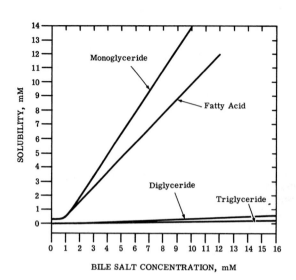

Figure 26-11
Solubility in bile salt solution of the four classes of lipids in the small intestinal lumen during the digestion and absorption of triglycerides. The values were obtained with triolein, diolein, mono-olein, and oleic acid under conditions similar to those present in the jejunal lumen during digestion (pH 6.3; 37°C; Na$^+$, 0.15M). (From A. F. Hofmann. *Handbook of Physiology.* Washington: American Physiological Society, 1968. Section 6: Alimentary Canal, vol. 5, p. 2516.)

is increased, a point, the critical micellar concentration, is reached at which the bile salts aggregate to form micelles. The arrangement of bile salt molecules in a micelle is such that the hydrophilic groups are oriented outward into the water phase and the hydro-phobic groups are oriented inward and grouped together to form a nonpolar core. It is within this nonpolar core that monoglycerides and free fatty acids (and cholesterol and fat-soluble vitamins) are effectively dispersed and solubilized (Fig. 26-11). The product of this solubilization process is a mixed micelle, the average diameter of which is about 40 Å. Since the solubility of triglycerides and diglycerides in bile salt micelles is very low (Fig. 26-11), these lipids are present almost entirely in the oil phase of intestinal contents. Also, since there is a limit to which available bile salt micelles can solubilize monoglycerides and free fatty acids, some of these lipids exist in the oil phase with the triglycerides and diglycerides.

The physical and chemical states of intestinal intraluminal lipids are represented diagrammatically in Figure 26-12. The situation that is visualized is that, when tri-glycerides in the form of large droplets enter the duodenum from the stomach, pancreatic lipolysis is initiated, free fatty acids and monoglycerides are formed, and the presence of these digestion products along with the bile salts and other emulsifying agents results in the formation of a finely dispersed emulsion. However, since free fatty acids, monoglycerides, and bile salts possess the capacity to form mixed micelles, a micellar phase also ensues. Micellar solubilization of monoglycerides and free fatty acids (and cholesterol and fat-soluble vitamins) facilitates entry of these lipids into the absorptive cells by providing an aqueous concentration of intraluminal lipid, thereby creating a gradient for diffusion through the lipid portion of the cell membrane. The monoglyceride and free fatty acid molecules in the micellar aggregates are in physico-chemical equilibrium and rapidly exchange with their corresponding monomers which are dispersed in minute quantities (monoglyceride, 10^{-6}M; free fatty acid, 10^{-5}M) in the surrounding water molecules. Present thinking is that the monomers of monoglyceride and free fatty acid in this molecular solution represent the form in which these lipids penetrate the intestinal brush border. There is an equilibrium between the emulsion and the micellar and monomeric phases in that, as the monomeric and micellar phases are depleted by absorption, they are continually replenished because lipolysis of tri-glycerides and diglycerides of the emulsified fat continues with the production of more monoglycerides and free fatty acids. The process continues until triglyceride digestion is complete. It is of interest that the bile salts, which are required for micelle formation, are not absorbed until they reach the terminal ileum. Thus they apparently perform their fat-dispersing function quite efficiently throughout the entire length of the small intestine in that a minimal amount of bile salts is necessary to prepare a relatively large quantity of fat for absorption.

Two pathways have been defined for the intracellular synthesis of triglycerides from absorbed digestion products. One involves the conversion of fatty acid to triglyceride and the other, monoglyceride to triglyceride—in accord with the concept that tri-glycerides are absorbed as free fatty acids and monoglycerides. It should be noted that the resynthesis of monoglycerides and free fatty acids to triglycerides maintains a concentration gradient for the continued diffusion of these lipids from the lumen of the intestine. The triglycerides delivered into the lymph are aggregated into droplets called *chylomicrons*. Chylomicrons have diameters ranging from 0.05 to 1.0 μm and are stabilized by an enclosure in a layer of phospholipid and protein. The mechanism by which chylomicrons leave the intestinal cell to enter the lacteals is unknown.

Triglycerides containing fatty acids with chain lengths less than 12 carbon atoms are both hydrolyzed and absorbed more rapidly than the longer-chain triglycerides. Thus, there is less interference in the absorption of the shorter-chain triglycerides under

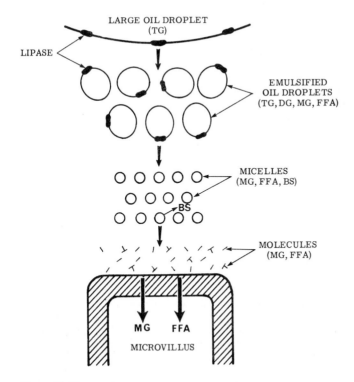

Figure 26-12
Intraluminal and cell penetration phases of triglyceride absorption. TG = triglycerides; DG = diglyerides; MG = monoglycerides; FFA = free fatty acids; BS = bile salts.

conditions of decreased pancreatic lipase and bile salt levels in the intestinal lumen, and it has been considered helpful to feed medium-chain triglycerides (8 to 10 carbon atoms in length) to patients with either pancreatic insufficiency or bile salt deficiency. Fatty acids with less than 12 carbon atoms are absorbed into the portal blood, without being synthesized to triglycerides. This situation may be due to a low activity of the esterifying enzymes of the intestinal mucosa toward these fatty acids, which in turn may be related to the fact that the shorter-chain fatty acids are water-soluble.

Absorption of Water and Electrolytes
Water and electrolytes, which are present in the digestive tract as the result of ingestion and secretion of these substances, are absorbed primarily from the small intestine and to a much lesser extent from the colon. About 1.5 liters of fluid is ingested every day by man and another 5 to 10 liters is secreted by the various regions of the gastrointestinal tract. Obviously any malfunction of the water and electrolyte absorptive mechanisms of the intestine can quickly result in serious depletion of body water and salt.

Small Intestine
Although it has been suggested that water can be actively absorbed from the small intestine, most workers feel that the absorption of water occurs by the simple physical process of osmosis. When solutions of different tonicities are introduced into the lumen of the small intestine, there is a tendency to adjust their osmotic activity to isotonicity in that water is rapidly absorbed from hypotonic solutions, and hypertonic solutions are diluted by the entrance of water into the gut. These results suggest that the movement of water across the intestinal epithelium depends on the solute concentration gradient. Intestinal contents become hypotonic during the absorption of dissolved materials, and water is then driven across the intestinal membrane by the concentration gradient of water which results from the absorption of solute.

The solute which is available in the greatest quantity to account for water absorption is sodium. Studies employing the short-circuit technique (Chap. 1) have shown that the intestinal epithelium possesses a mechanism for the active transport of sodium similar to that which has been postulated to exist in frog skin (Chap. 1). Sodium transport is polarized in that this ion diffuses down an electrochemical gradient from the lumen across the brush border into the intestinal epithelial cell and is then extruded against an electrochemical gradient into the interstitial space by a sodium pump located in the basal margin of the cell. Probably the electrical potential created by the movement of sodium across the intestinal membrane (serosal surface positive to mucosal surface) is responsible for the simultaneous movement of anions.

That the intestine possesses special mechanisms for the transport of certain divalent cations is demonstrated by the observations that both calcium and iron are transferred from mucosa to serosa against a concentration difference, and that such movement is abolished by procedures which interfere with oxidative metabolism. The absorption of calcium requires that this element be in a water-soluble and ionized form in the lumen of the intestine. Carbonate and phosphate ions, especially in alkaline solution, tend to form insoluble salts with calcium and inhibit the absorption of this ion. Insoluble calcium soaps are formed in the presence of large quantities of fatty acids and are excreted in the feces. Both parathyroid hormone and vitamin D are necessary for optimal transport of calcium. A deficiency of dietary vitamin D leads to impaired calcium absorption, and the unabsorbed calcium forms insoluble salts with phosphorus. Since the salts are not readily absorbed, the absorption of phosphorus is also decreased under these conditions. Phosphorus can be rapidly absorbed under normal circumstances, as shown by the fact that radioactive inorganic phosphorus appears in the circulation within five minutes after introduction into the small intestine.

With very few exceptions, the extent to which different substances are absorbed from the digestive tract is not regulated, because few mechanisms are available for this purpose. However, the intestinal absorption of iron represents the absorption of a

659

substance which is regulated in that this cation is absorbed only when the body becomes deficient in this element. Although the average daily intake of iron is about 20 mg, the amount absorbed is only a small fraction (5 percent) of that ingested and is just sufficient to replace the daily loss of iron. In the male, the loss occurs entirely through desquamation of the epithelial cells of the small intestine. The amount of iron absorbed by women is about double that of men because of the additional iron lost during menstrual bleeding. Ferrous iron of the mucosal cells is in equilibrium with circulating iron and with ferritin, a protein-iron complex in the mucosal cells. When the iron of the blood is decreased below normal values, this element is liberated from the mucosal stores of ferritin-iron, and more iron is absorbed from the gut in order to replenish these stores.

Large Intestine
The absorption of ingested carbohydrates, proteins, lipids, and other nutrients is complete by the time material is discharged from the ileum into the proximal colon. This material consists of cellular debris, connective tissue, cellulose, and significant quantities of water and inorganic salts. The large intestine absorbs sodium, chloride, and water from and adds potassium and bicarbonate to its contents. In addition, this organ contains a variety of microorganisms, some of which synthesize certain vitamins. Although these nutrients are absorbed from the large intestine, they provide only a small part of the daily vitamin requirement.

Sodium and water are absorbed from the large intestine by processes similar to those of the small intestine; i.e., sodium is actively transported into the blood, and water follows passively in response to the osmotic gradient created by the removal of sodium from the intraluminal fluid. Water is absorbed to the extent that only about 100 ml of water is lost in the feces every day. Although the colon does absorb water, it is not the water-conserving organ of the digestive tract. About 8000 ml of water is absorbed every day by the small intestine and only about 300 to 400 ml by the colon. It is frequently a misfunction of the small intestine that causes diarrhea. In this instance, large quantities of fluid enter the colon from the small intestine and overwhelm the colonic absorptive mechanism, which is not nearly as efficient as that of the small intestine. The colon has the capacity to absorb 2500 ml of water daily.

REFERENCES
Burgen, A. S. V., and N. G. Emmelin. *Physiology of the Salivary Glands.* Baltimore: Williams & Wilkins, 1961.

Code, C. F. (Ed.). *Handbook of Physiology.* Washington: American Physiological Society, 1967. Section 6: Alimentary Canal, vol. 2, Secretion.

Code, C. F. (Ed.). *Handbook of Physiology.* Washington: American Physiological Society, 1968. Section 6: Alimentary Canal, vol. 3, Intestinal Absorption.

Crane, R. K. Intestinal absorption of sugars. *Physiol. Rev.* 40:789–825, 1960.

Davenport, H. W. *Physiology of the Digestive Tract* (3rd ed.). Chicago: Year Book, 1971.

Florey, H. W., R. D. Wright, and M. A. Jennings. The secretions of the intestine. *Physiol. Rev.* 21:36–69, 1941.

Friedman, M. H. F. (Ed.) *Functions of the Stomach and Intestine.* Baltimore: University Park Press, 1974.

Gregory, R. A. *Secretory Mechanisms of the Gastro-intestinal Tract.* London: Arnold, 1962.

Grossman, M. I. (Ed.). *Gastrin.* Berkeley: University of California Press, 1966.

Hofmann, A. F. Functions of Bile in the Alimentary Canal. In C. F. Code (Ed.), *Handbook of Physiology.* Washington: American Physiological Society, 1968. Section 6: Alimentary Canal, vol. 5, Bile; Digestion; Ruminal Physiology. Pp. 2507–2533.

Hunt, J. N. Gastric emptying and secretion in man. *Physiol. Rev.* 39:491–530, 1959.

James, A. H. *The Physiology of Gastric Digestion.* Monographs of the Physiological Society. London: Arnold, 1957.

Simmonds, W. J. Absorption of Lipids. In E. D. Jacobson and L. L. Shanbour (Eds.), *Physiology. Series One. Gastrointestinal Physiology.* Baltimore: University Park Press, 1974. Vol. 4.

Taylor, W. (Ed.). *The Biliary System; A Symposium of the NATO Advanced Study Institute.* Oxford, England: Blackwell, 1965.

Thomas, J. E. *The External Secretion of the Pancreas.* Springfield, Ill.: Thomas, 1950.

Wilson, T. H. *Intestinal Absorption.* Philadelphia: Saunders, 1962.

27. Energy Metabolism

Leon K. Knoebel

Metabolism is the sum of all transformations of both matter and energy which occur in biological systems. Metabolism endows all cells with the properties of excitability, growth, and reproduction, and for various types of cells makes possible specialized processes such as contraction, conduction, secretion, and absorption. Thus, metabolism is the basis for all physiological phenomena that one can observe or measure. Transformations of matter concern the chemical reactions which occur in the body, and these fall mainly in the province of the biochemist. However, transformations of matter are accompanied by transformations of energy. The following discussion will be confined to this facet of metabolism, energy metabolism.

TRANSFORMATIONS OF ENERGY

A diagrammatic scheme of biological transformations of energy is given in Figure 27-1, and it is suggested that the reader refer to this illustration during the subsequent discussion. Five major forms of energy are encountered in the living organism: chemical, mechanical, osmotic, electrical, and thermal energies. The cells of the body are able to use energy from one source only: the chemical energy which is liberated by chemical reactions. Chemical energy of the body can be transformed into mechanical, osmotic, electrical, and thermal energies, but these transformations are irreversible.

Chemical Energy

The chemical energy which is necessary to sustain the processes of life is provided by the breakdown of carbohydrates, proteins, and lipids contained in the cells of the body. These large organic molecules are oxidized through a series of enzymatically catalyzed reactions to much smaller molecules, principally carbon dioxide and water, and as the reactions proceed, chemical energy is released for the performance of work by the cells. Reactions that involve the degradation of complex molecules to simple molecules with the liberation of chemical energy are designated *catabolic reactions* (catabolism).

The immediate source of chemical energy for the cells is the *metabolic pool*, which can be thought of as existing throughout the extracellular and intracellular fluids of the body. A variety of carbohydrates, proteins, lipids, and the short-chain fragments of catabolism of these molecules are present in the pool, and all can be readily used by cells as a source of energy. Furthermore, there is a free interchange (*dynamic state*) between the cells of the body and the metabolic pool of energy-containing substances in that the chemical energy of certain cells can be mobilized through the metabolic pool and utilized as a source of energy by other cells of the body.

As the chemical energy of the body is constantly being depleted, it must be replenished in order to maintain the energy status of the body at normal levels. The chemical energy of the body is replaced by the chemical energy of ingested food. Since the absorption products of ingested food are identical to some of the substances in the metabolic pool, the body can make free use of these for replenishing cellular energy reserves.

It should be pointed out that the chemical energy released by the catabolism of carbohydrates, proteins, and lipids is not directly used by cells to perform various types of work. Instead, it is first incorporated into high-energy phosphate compounds, the most notable of which is adenosine triphosphate (ATP). It is the hydrolysis of ATP that serves as the immediate source of chemical energy for the performance of work.

661

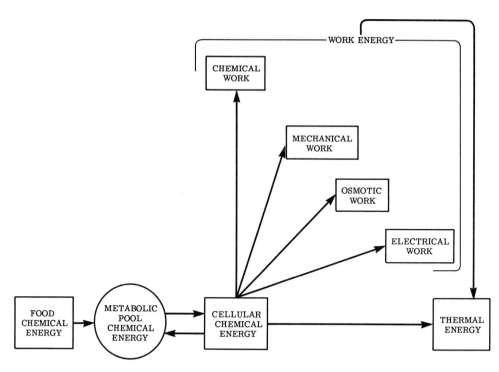

Figure 27-1
Diagram of biological transformations of energy.

Work Energy

Chemical energy is utilized for the purpose of doing work. Energy supplied in the form of work energy enables various types of cells to maintain the processes of life. In the category of work energy are four forms of energy mentioned previously: mechanical, osmotic, electrical, and chemical. The conversion of chemical energy to work in the form of mechanical energy is exemplified by the muscle which shortens and lifts a load. The transformation of chemical energy to work in the form of osmotic energy occurs when a cell maintains a concentration gradient of a substance across its membrane, as, for example, the intestinal epithelial cell which transports glucose from intestinal lumen where the concentration of glucose is low to the cell interior where glucose concentration is relatively high. Both osmotic and electrical work are done at the expense of chemical energy in the maintenance of the unequal distribution of inorganic ions which exists across a nerve or muscle cell membrane. In this instance, not only are individual ionic concentration gradients maintained but, since there is also a separation of charges across the membrane (membrane potential), electrical work is done. Chemical energy can also be used to do chemical work by supplying the energy necessary for synthetic reactions. When complex molecules are synthesized from simple molecules, energy is stored and chemical work is done. *Anabolism* is the term used to designate this type of metabolic activity. Anabolic reactions are always accompanied by catabolic reactions because, in order to synthesize and store energy in a large molecule, chemical energy must be supplied by the breakdown of other molecules; i.e., chemical work is done at the expense of chemical energy.

Thermal Energy

The efficiency of the body in converting chemical energy to work energy is in the neighborhood of only 20 percent. A large part of the chemical energy expended, therefore, must be converted to a form of energy other than work, and it appears as thermal energy or heat. In addition to the thermal energy which is directly liberated by chemical reactions, work energy is also ultimately converted to heat. For example, the heart does mechanical work in pumping blood. However, the work energy of the heart is converted to heat in overcoming friction as blood passes through the circulatory system. The chemical work which is done when a protein is synthesized is lost as heat when that protein is eventually catabolized. Other examples can be cited for other functions. Thus the chemical energy of the body ultimately appears as heat, generated either directly from chemical reactions or indirectly from the dissipation of work energy.

Thermal energy cannot be used for doing work, because cells have no mechanism available for this purpose. Heat derived from metabolic processes may be used to maintain the body temperature at a level which is optimal for the enzymatically regulated reactions occurring in the body. Yet, much of the time more heat is generated than can be used for this purpose, and the body has the problem of getting rid of excess heat (Chap. 28).

ENERGY BALANCE

According to the first law of thermodynamics, energy can be neither created nor destroyed. This law can be applied to living systems in that studies can be made of energy balance as well as of energy transformations; i.e., the total amount of energy taken in by the body must be accounted for by the energy put out by the body. The relationship between the factors involved in the balance between input and output of energy is given in the following equation:

ENERGY INPUT $=$ ENERGY OUTPUT (1)
Chemical Energy of Food $=$ Heat Energy $+$ Work Energy $\pm$ Stored Chemical Energy

All energy of the body is ultimately derived from one source, the chemical energy of food, the intake of which is regulated primarily by hypothalamic centers in the brain (Chap. 8). The major energy output of the body is in the form of heat, the rate of production of which varies from one time to another depending on prevailing circumstances. For example, the heat production of an exercising individual is greater than that of the same person at rest. The eating of food is followed by an increase in the amount of heat produced relative to that which is generated during fasting. Heat production is also influenced by the environmental temperature, circulating levels of a variety of hormones, age, sex, and body size. The role of each of these and additional factors in determining the rate at which heat is produced will be considered in more detail in this and subsequent chapters.

In addition to releasing energy as heat, the body puts energy out in the form of work. When the amount of energy ingested as food is sufficient to balance the amount of energy put out in the form of heat plus work, the chemical energy of the body remains constant. However, this situation is the exception rather than the rule. For example, the chemical energy stores of the body of the growing child increase markedly as body mass is increased over a period of years. The increase in chemical energy stores which is associated with growth is mainly in the form of protein. When an adult becomes obese, the chemical energy stores of the body are increased as the result of an excess deposition of body fat. On the other hand, the chemical energy of the body decreases under conditions of nutritional insufficiency. During early starvation, endogenous chemical energy is lost primarily as the result of the catabolism of fat, and if starvation is extended, protein is also used as an energy source. Even over the course of a day, there are fluctuations in the chemical energy stores of the body of a normal individual, and these are reflected by small changes in body weight.

It is obvious from these considerations that another factor, stored chemical energy, must be included in the energy balance equation in order to provide for changes in the chemical energy of the body. If intake of food energy is greater than the energy put out as heat and work, the body stores of energy increase and storage energy is positive in order to balance the equation. If more energy is released in the form of heat and work than is ingested, the body stores of energy are depleted and storage energy is negative in the equation.

To study the energy balance of a person under the conditions normally encountered in life, three of the variables present in the energy balance equation must be measured so that the fourth can be determined by difference. Although all four variables can be measured, it is not always convenient to measure them. This is especially the case in the clinic, where the facilities are usually inadequate and time precludes such an approach. However, a measurement of energy balance can be simplified by eliminating some of the variables. For example, the energy balance of a subject in the postabsorptive state can be determined, thereby excluding chemical energy derived from food. Moreover, when voluntary movement is restricted, energy in the form of work can be disregarded. With these two variables eliminated, the equation representing energy balance is simplified:

$$- \text{Stored Chemical Energy} = \text{Heat Energy} \qquad (2)$$

This equation simply states that a person in the resting, fasting condition uses a certain amount of stored chemical energy (as shown by the minus sign) with the resultant production of a certain amount of heat. Under these conditions, the chemical energy which is utilized from the metabolic pool is completely converted to heat. Furthermore, since the subject is at rest and postabsorptive (*in the basal state*), the chemical energy is used solely for the purpose of maintaining the vital activities of the body—i.e., for the maintenance of heart action, respiration, etc.

Under the circumstances described above, only two variables are concerned in the energy balance of the body. If one can be determined, the other is known, because they are equal. The rate at which chemical energy is expended by the body is known as *metabolic rate*. Since the rate at which heat is produced is equal to the rate of expenditure of chemical energy, the former is a direct measure of metabolic rate, and metabolic rate is usually expressed in terms of rate of heat production. Metabolic rate can be assessed by determining directly the amount of heat produced per unit time by an individual. However, the more frequently used approach is to calculate rate of heat production from the oxygen consumption of a person—i.e., an indirect determination of heat production. A few of the methods used for these determinations will be considered in the following sections.

ENERGY UNITS

The unit of energy which has been most commonly used in the study of energy metabolism and which will be employed in the following discussion is the kilocalorie (kcal). In terms of heat, 1 kilocalorie is the amount of energy required to raise the temperature of 1 kg of water 1°C. One kilocalorie is equal to 1000 calories (cal). It has recently been suggested that the joule (J) be used as the measure of energy, and for purposes of conversion, 1 kilocalorie is equal to 4187 joules. Rate of energy conversion (power) has been expressed as kilocalories per hour, but it can also be stated in terms of watts (joules per second) by multiplying kilocalories per hour by 1.16.

DIRECT CALORIMETRY

The heat production of man and experimental animals can be determined by *direct calorimetry*. The subject is placed in an insulated chamber (calorimeter) through which cool water is circulated. The rate at which the water flows through the chamber is adjusted so that the temperature of the calorimeter is kept constant. In addition, a meter measures the volume of water which flows through the chamber in a given time, and the temperature of the water entering and leaving the chamber is determined. Thus, a knowledge of the volume of water passing through the chamber and the rise in temperature of the water enables one to calculate in terms of kilocalories the amount of heat transmitted from the subject to the water.

Although this measurement accounts under most circumstances for most of the heat produced by the subject, another avenue of heat loss from the body is represented by the water which leaves the body from the skin and the respiratory membranes in the form of water vapor. For every gram of water vaporized, 0.58 kcal is lost from the body as heat (heat of vaporization). Water vapor produced by the subject is collected in a suitable chemical absorbent and determined by weight. The amount of heat dissipated by evaporation is calculated by multiplying the weight of water vapor by the heat of vaporization. The rate at which heat is lost by evaporation plus the amount of heat absorbed per unit time by the circulating water represents the rate at which the subject produces heat. This method for the assessment of energy exchange is tedious and difficult to perform; although it was used extensively in the past, simpler techniques have now replaced it.

INDIRECT CALORIMETRY

The chemical energy utilized from the body stores is measured by *indirect calorimetry*. By measuring oxygen consumption, carbon dioxide production, and urinary nitrogen excretion, and by knowing certain experimentally predetermined factors, one can determine accurately the type and amount of substances utilized in the metabolic pool. Furthermore, as a measure of metabolic rate, the total amount of heat produced per unit time by the oxidation of these substances can be calculated; i.e., an indirect measurement of heat production is made.

Caloric Value of Foods

Both carbohydrates and fats are completely oxidized to carbon dioxide and water in the body, and the same is true when these substances are combusted in vitro. Since the initial reactants and the final products are the same in both instances, the amounts of energy released as heat in vivo and in vitro are identical. Consequently, it is a relatively simple matter to determine the caloric value of carbohydrates and fats in the body by measuring the heat of combustion of these materials in a bomb calorimeter. A bomb calorimeter is a metal chamber in which is placed a weighed amount of substance. Contact is made between the substance and an iron wire, the chamber is sealed, and oxygen is introduced into the chamber under high pressure. The calorimeter is immersed in a water bath of known volume and temperature, and when an electric current is sent through the wire the material is ignited and complete combustion occurs. Depending on the substance combusted, a certain amount of heat is liberated. The heat of combustion of the substance (kilocalories per gram) can be calculated from the volume of water in the bath and the rise in temperature of the water.

The heat of combustion of various carbohydrates differs. That of glucose is 3.7 kcal per gram, whereas that of starch is 4.2 kcal per gram. However, an average value for carbohydrate is generally taken as 4.1 kcal per gram. The energy content of fat, which is more than double that of carbohydrate, varies, depending on the constituent fatty acids, but an average value of 9.3 kcal per gram is most often used.

Protein differs from carbohydrate and fat in that it is not completely oxidized in the body. Certain end products of protein metabolism appear in the excreta, for the most part in the urine as urea. Urea contains chemical energy, and if it were ignited in a bomb calorimeter, heat would be liberated. Thus the amount of heat which is produced by the oxidation in vivo of 1 gm of protein is less than that which is produced when the same amount of protein is combusted in a bomb calorimeter (5.6 kcal per gram). By taking into account the energy lost in the excreta, it has been calculated that the caloric value of protein in vivo is 4.3 kcal per gram.

Energy Equivalent of Oxygen

In order to calculate heat production from oxygen consumption, it is necessary to relate the number of kilocalories released by the combustion of carbohydrate, fat, and protein to the volumes of oxygen consumed during the oxidation of each of these substances in the body. More specifically, it is necessary to know how many kilocalories of heat are produced when 1 liter of oxygen is used to oxidize each type of substance. This is known as the *energy equivalent of oxygen,* which is expressed as kilocalories per liter of oxygen. Carbohydrate and fat react with definite amounts of oxygen to produce certain quantities of carbon dioxide and water. Using glucose (molecular weight 180) as representative of the oxidation of carbohydrate, one can write the following equation:

$$C_6H_{12}O_6 + 6 O_2 \rightarrow 6 CO_2 + 6 H_2O \tag{3}$$

In this instance, 180 gm of glucose reacts with $6M$ of oxygen or 134.4 liters of oxygen ($6M$ oxygen $\times$ 22.4 liters oxygen/M oxygen). The same volume (134.4 liters) of carbon dioxide is formed. One gram of glucose reacts with 0.75 liter of oxygen (134.4 liters oxygen $\div$ 180 gm glucose) to produce 0.75 liter of carbon dioxide. Since 3.7 kcal is liberated when 1 gm of glucose is oxidized, it follows that the energy equivalent of oxygen is 5 kcal per liter of oxygen (3.7 kcal/gm $\div$ 0.75 liter oxygen/gm).

A similar calculation can be made for a typical fat. The following equation represents the oxidation of a triglyceride (molecular weight 860), the fatty acids of which are oleic, palmitic, and stearic acids:

$$C_{55}H_{104}O_6 + 78 O_2 \rightarrow 55 CO_2 + 52 H_2O \tag{4}$$

In this case, 860 gm of fat combines with 1747 liters of oxygen to form 1232 liters of carbon dioxide. It can be calculated that 1 gm of fat combines with 2.03 liters of oxygen to form 1.43 liters of carbon dioxide. Since 9.3 kcal is liberated by the oxidation of 1 gm of fat, the energy equivalent of oxygen is 4.7 kcal per liter of oxygen (9.3 kcal/gm $\div$ 2.03 liters oxygen/gm).

Determining the energy equivalent of oxygen is more difficult for proteins, since these substances are structurally complex and are incompletely oxidized in the body. However, by using the empirical formula for a typical protein and taking into account the amount of urea that would be formed by the oxidation of this protein, one can calculate the volumes of oxygen consumed and carbon dioxide produced during the oxidation of 1 gm of protein. These values are 0.97 liter of oxygen per gram and 0.78 liter of carbon dioxide per gram. Since 4.3 kcal is released when 1 gm of protein is oxidized, 1 liter of oxygen is equivalent to 4.5 kcal (4.3 kcal/gm $\div$ 0.97 liter oxygen/gm). These data are summarized in Table 27-1.

Respiratory Quotient

It is obvious that there are variations in the volumes of oxygen consumed and carbon dioxide produced by the individual oxidations of 1 gm of carbohydrates, fats, or proteins. Furthermore, the ratio of the volume of carbon dioxide produced to the volume of oxygen consumed during the oxidation of each type of food also varies (Table 27-1). This ratio (vol. CO_2 $\div$ vol. O_2) is called the *respiratory quotient* (RQ). A calculation of the RQ from the rate of oxygen consumption and carbon dioxide

Table 27-1. Metabolic Values for Carbohydrates, Fats, and Proteins

Unit of Measurement	Carbohydrates	Fats	Proteins
Kilocalories per gram	4.1	9.3	4.3
Liters of CO_2 per gram	0.75	1.43	0.78
Liters of O_2 per gram	0.75	2.03	0.97
Respiratory quotient	1.00	0.70	0.80
Kilocalories per liter of O_2	5.0	4.7	4.5

production of a subject furnishes qualitative information regarding the substances utilized in the metabolic pool. If only carbohydrate is oxidized, an RQ of 1.00 is obtained, whereas the specific oxidation of fat results in an RQ of 0.70. Values which fall between these extremes represent the oxidation of various mixtures of carbohydrates, fats, and proteins. This will be discussed in more detail in the following section. The RQ of a subject on an ordinary mixed diet is about 0.85, and that of a fasting subject is about 0.82.

The respiratory quotient is based solely on the rate of exchange of the respiratory gases, since it is calculated from measurements of oxygen consumption and carbon dioxide production. Consequently, this ratio can be affected by factors other than oxidative metabolism. For example, much carbon dioxide is excreted during hyperventilation. This tends to increase the RQ, and values as high as 1.5 to 1.7 are obtained during severe exercise (Chap. 29). On the other hand, carbon dioxide is retained during hypoventilation, and the RQ may fall below 0.70. An RQ obtained under these circumstances simply reflects the state of the respiration and supplies no information about the composition of the metabolic mixture being oxidized. Large amounts of carbon dioxide are also excreted in conditions of metabolic acidosis, and the RQ is high, whereas the opposite effect is obtained during metabolic alkalosis (Chap. 24). In addition, when carbohydrates are transformed to fats in the body, as in the fattening of a farm animal, the RQ can reach 1.4. Carbohydrates are oxygen-rich substances, and fats are relatively deficient in oxygen. Oxygen is liberated from carbohydrates during the conversion to fats, and less oxygen is required from the external environment for oxidative metabolism. The body uses more oxygen than can be accounted for by a measurement of oxygen consumption, and the RQ so obtained is not a true index of oxidative metabolism. Conversely, when fats are converted to carbohydrates, as in hibernating animals, the RQ can fall below 0.70. Enough oxygen must be consumed not only to carry out oxidative processes but also to facilitate the conversion of oxygen-poor fats to oxygen-rich carbohydrates.

It is apparent that an RQ determined under these circumstances is not a reliable measure of oxidative metabolism. A better term to apply to this ratio is *respiratory exchange ratio* (Chap. 20). This has the advantage that the ratio need not always be considered to represent an index of oxidative metabolism; i.e., it is a ratio based simply on the rate of exchange of the respiratory gases. However, if the measurements on which the ratio is based are made under well-controlled conditions, the respiratory exchange ratio can be used to evaluate oxidative processes.

Calculation of Metabolic Rate

It is possible to determine quantitatively the type and amount of substances oxidized in the metabolic pool when oxygen consumption, carbon dioxide production, and urinary nitrogen excretion are measured and when the data given in Table 27-1 are available. Furthermore, the energy liberated as heat by the individual oxidations of carbohydrates, fats, and proteins can be calculated, and from these figures the total amount of heat produced per unit time can be used as a measure of metabolic rate. The principles of indirect calorimetry can be exemplified by considering a resting human subject in the postabsorptive state on whom the following measurements have been made: (1) urinary nitrogen excretion, 0.5 gm per hour; (2) oxygen consumption, 16.0 liters per hour; and (3) carbon dioxide production, 13.5 liters per hour.

Practically all nitrogen which is derived from the oxidation of protein is excreted in the urine. Since the amount of nitrogen contained in a typical protein molecule represents, on the average, 16 percent of the total weight of a protein molecule, the weight of protein oxidized per unit time is determined by multiplying the urinary nitrogen excretion by a factor of 6.25:

$$0.5 \text{ gm N/hour} \times 6.25 \text{ gm protein/gm N} = 3.1 \text{ gm protein/hour} \tag{5}$$

Since the oxidation of 1 gm of protein produced 4.3 kcal,

$$3.1 \text{ gm protein} \times 4.3 \text{ kcal/gm protein} = 13.4 \text{ kcal/hour} \tag{6}$$

The quantity of heat produced by the oxidation of carbohydrate and fat is determined from the oxygen consumption and carbon dioxide production. The total volumes of oxygen consumed and carbon dioxide produced during the oxidation of all three foods are measured. However, it is necessary to determine the amounts of these gases which are involved only in the metabolism of carbohydrates and fats. Thus one must know the volumes of oxygen and carbon dioxide involved in the oxidation of protein, and they are calculated in the following manner:

$$3.1 \text{ gm protein/hour} \times 0.97 \text{ liter } O_2/\text{gm protein} = 3.0 \text{ liter } O_2/\text{hour} \tag{7a}$$
$$3.1 \text{ gm protein/hour} \times 0.78 \text{ liter } CO_2/\text{gm protein} = 2.4 \text{ liter } CO_2/\text{hour} \tag{7b}$$

The total respiratory exchange is then corrected for the amounts of oxygen consumed and carbon dioxide produced by the oxidation of protein:

$$16.0 \text{ liter } O_2/\text{hour} - 3.0 \text{ liter } O_2/\text{hour} = 13.0 \text{ liter } O_2/\text{hour} \tag{8a}$$
$$13.5 \text{ liter } CO_2/\text{hour} - 2.4 \text{ liter } CO_2/\text{hour} = 11.1 \text{ liter } CO_2/\text{hour} \tag{8b}$$

The ratio of the volume of carbon dioxide produced to the volume of oxygen consumed during the oxidation of mixtures of carbohydrates and fats is called the *nonprotein RQ*. This is calculated:

$$11.1 \text{ liter } CO_2/\text{hour} \div 13.0 \text{ liter } O_2/\text{hour} = 0.85 \tag{9}$$

Recall that when pure carbohydrate is oxidized in the body the RQ is 1.00, and when only fat is oxidized the RQ is 0.70. The nonprotein RQ determined in this experiment shows that a mixture of carbohydrate and fat is oxidized. The energy equivalent of oxygen for carbohydrate is 5.0 kcal per liter of oxygen and for fat is 4.7 kcal per liter of oxygen. Every mixture of carbohydrate and fat between the extremes of pure carbohydrate and pure fat has a specific nonprotein RQ and energy equivalent of oxygen. These relationships have been determined and are given in Table 27-2. At the nonprotein RQ (0.85) determined for the subject under consideration, the consumption of 1 liter of oxygen results in the production of 4.862 kcal. Consequently, the heat produced by the oxidation of this particular mixture of carbohydrate and fat is

$$13.0 \text{ liter } O_2/\text{hour} \times 4.862 \text{ kcal/liter } O_2 = 63.2 \text{ kcal/hour} \tag{10}$$

The amount of heat liberated by the oxidation either of carbohydrate or of fat can also be calculated from data supplied in Table 27-2. At a nonprotein RQ of 0.85, carbohydrate supplies 51 percent and fat 49 percent of the heat produced during the oxidation of this mixture of carbohydrate and fat:

$$0.51 \times 63.2 \text{ kcal/hour} = 32.2 \text{ kcal/hour (from carbohydrates)} \quad \text{(11a)}$$
$$0.49 \times 63.2 \text{ kcal/hour} = 31.0 \text{ kcal/hour (from fats)} \quad \text{(11b)}$$

The amounts of carbohydrates and fats that are oxidized can also be calculated from these data:

$$32.2 \text{ kcal/hour} \div 4.1 \text{ kcal/gm carbohydrate} = 7.9 \text{ gm carbohydrate/hour} \quad \text{(12a)}$$
$$31.0 \text{ kcal/hour} \div 9.3 \text{ kcal/gm fat} = 3.3 \text{ gm fat/hour} \quad \text{(12b)}$$

It is apparent that less than half as much fat as carbohydrate need be oxidized to supply approximately the same amount of heat.

Table 27-2. Energy Equivalent of a Liter of Oxygen at Various Nonprotein Respiratory Quotients

Nonprotein Respiratory Quotient	Kilocalories per Liter Oxygen	Kilocalories Derived from	
		Carbohydrate (percent)	Fat (percent)
0.70	4.686	0	100.0
0.71	4.690	1.10	98.9
0.72	4.702	4.76	95.2
0.73	4.714	8.40	91.6
0.74	4.727	12.0	88.0
0.75	4.739	15.6	84.4
0.76	4.751	19.2	80.8
0.77	4.764	22.8	77.2
0.78	4.776	26.3	73.7
0.79	4.788	29.9	70.1
0.80	4.801	33.4	66.6
0.81	4.813	36.9	63.1
0.82	4.825	40.3	59.7
0.83	4.838	43.8	56.2
0.84	4.850	47.2	52.8
0.85	4.862	50.7	49.3
0.86	4.875	54.1	45.9
0.87	4.887	57.5	42.5
0.88	4.899	60.8	39.2
0.89	4.911	64.2	35.8
0.90	4.924	67.5	32.5
0.91	4.936	70.8	29.2
0.92	4.948	74.1	25.9
0.93	4.961	77.4	22.6
0.94	4.973	80.7	19.3
0.95	4.985	84.0	16.0
0.96	4.998	87.2	12.8
0.97	5.010	90.4	9.58
0.98	5.022	93.6	6.37
0.99	5.035	96.8	3.18
1.00	5.047	100.0	0

Source: After G. Lusk. *J. Biol. Chem.* 59:41–42, 1924.

The total heat production is the heat liberated by the oxidation of proteins plus that liberated by the oxidation of carbohydrates and fats:

$$13.4 \, \text{kcal/hour} + 63.2 \, \text{kcal/hour} = 76.6 \, \text{kcal/hour} \tag{13}$$

It can be further calculated that, of the total heat production, 42 percent is derived from carbohydrates, 41 percent from fats, and 17 percent from proteins. Evidently all three major foods are used to supply energy for the maintenance of the vital activities of the body. The total heat production is a measure of metabolic rate; i.e., it is equivalent to the rate at which the chemical energy of the body stores is metabolized.

Determination of Metabolic Rate from Oxygen Consumption

The method of indirect calorimetry described in the preceding sections requires a moderate amount of equipment and considerable time. Since the determination of metabolic rate is a clinical tool of some importance, it has been necessary to devise simpler methods. One means of assessing metabolic rate in the clinic is based on a single measurement, that of oxygen consumption. It was found, using the more complicated techniques of indirect calorimetry, that under standard conditions of resting and fasting the average RQ of a large group of normal subjects was 0.82. On the basis of these determinations, the assumption is now made that the RQ of a subject in the basal state is 0.82. At a nonprotein RQ of 0.82, the energy equivalent of oxygen is 4.825 kcal. Heat production is determined by measuring oxygen consumption over a short interval of time, usually six minutes, converting this value to an hourly basis, and multiplying it by 4.825 kcal per liter of oxygen. This approach considerably simplifies the determination of metabolic rate.

One type of apparatus which is commonly used in the clinic to measure oxygen consumption is shown in Figure 27-2. This apparatus consists of a spirometer bell which is arranged to write on a moving kymograph by means of a pulley system. The bell is filled with oxygen, and during a determination a motor-blower draws oxygen through the inspiratory tube and returns the expired gas mixture through the expiratory tube. The expiratory tube is connected to a canister of soda lime in order to remove all carbon dioxide present in the expired air. As the subject consumes oxygen and as carbon dioxide is removed from the system, individual respiratory excursions and the drop of the bell are recorded. The difference between the height of the bell at the beginning and the end of the experiment is a measure of the volume of oxygen consumed during this time. To determine the distance the bell drops, a line is drawn which best follows either the peaks or the troughs of respiration as recorded, and the vertical distance between the top and the bottom of this sloping line is measured. Since the spirometer is calibrated in terms of volume per unit distance fall of the bell, the total volume of oxygen consumed during the experiment is calculated by multiplying this conversion factor by the distance the bell falls. The oxygen consumption is corrected for water vapor tension, reduced to standard conditions of temperature and pressure, and converted to liters of oxygen consumed per hour. Some subjects find it hard to breathe normally through a mouthpiece and valves, and one serious criticism of this method is that the accuracy of the determination depends on the ability of a subject to breathe regularly. If the rate and/or the amplitude of respiration are irregular, it is difficult to decide precisely where to draw the sloping line.

Obviously, a determination of metabolic rate by this method introduces certain errors into the calculations. Since a nonprotein RQ is used to determine the energy equivalent of oxygen, protein metabolism is neglected. However, the RQ (0.80) and the energy equivalent of oxygen (4.5 kcal per liter of oxygen) for the oxidation of protein are only slightly less than the average values used. Furthermore, protein acounts for a relatively small part of the total energy exchange. Consequently, the error introduced by neglecting protein oxidation is small and has been calculated to be only about 1 percent. In

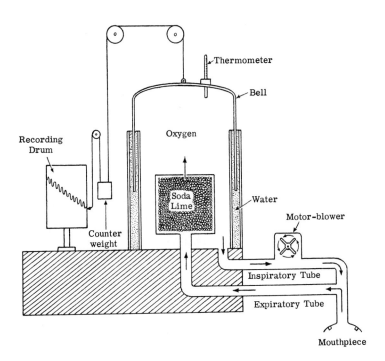

Figure 27-2
Diagram of the Sanborn apparatus for measuring human oxygen consumption.

addition, the assumption is made that the RQ of a subject is normal; but if the RQ is not 0.82, it is not correct to use 4.825 as the energy equivalent of oxygen. Still this use introduces only a small error into the calculations, because the energy equivalent of oxygen varies so slightly between the extreme respiratory quotients of 1.00 and 0.70. By using an RQ of 0.82, which falls midway between the extremes, there can be at most an error of only 3.5 percent.

The errors inherent in the short method are not significant enough to outweigh the advantages gained. Moreover, since a large number of studies have been carried out using this method, good standards are available for comparison. One drawback to the method is that it does not provide information regarding the type and amount of substances oxidized in the body stores. The only information that can be obtained by measuring oxygen consumption is the total heat production of a subject. However, this value can be extremely useful as an aid to diagnosing certain disorders.

BASAL METABOLIC RATE

The methods and calculations of indirect calorimetry described have dealt with the energy exchange of a fasting and resting subject. The term applied to the exchanges of energy that occur under these conditions is *basal metabolism* or *basal metabolic rate* (BMR). Clinically, BMR is determined 12 to 14 hours after the last meal, usually in the morning after at least 8 hours of sleep. There should be no voluntary muscular movement during the test and no muscular exertion within half an hour to an hour prior to the test. In addition, there should be no stress due to extremes of environmental temperature, and the subject should be both mentally and physically at rest. Many factors that produce variable influences on metabolic rate are eliminated by carrying out the determination under these circumstances. The heat production so determined is a measure of the energy exchange required to maintain the vital activities of the body. The basal metabolic state does not represent the minimal functional activity of the body, since energy exchange is about 10 percent lower during sleep. This drop has been attributed to the more complete muscular relaxation achieved during sleep.

Once the basal heat production of a subject has been determined in terms of kilocalories per hour, it is necessary to ascertain whether this value is normal. Consequently, normal values must be available for comparison. If a comparison is to be valid, the heat production of a subject should be compared with the average heat production of a large population of normal subjects having physical and biological characteristics similar to those of the subject. Among the more important factors to consider when making the comparison are the body size, sex, and age of a subject.

Body Size

The total energy exchange of a large animal is greater than that of a small animal. For example, the heat production of an elephant is many times that of a mouse. Likewise, a 21-year-old man weighing 140 kg has a greater total energy exchange than a man of the same age weighing 70 kg. Consequently, when comparisons of metabolic rate are made between individuals of the same or different species, it is essential that there be some basis for comparison which takes body size into consideration. Various criteria of body size that have been considered include body weight, body surface area, and lean body mass.

Metabolic rate is not directly related to body weight, since the caloric output per kilogram of body weight of a small animal is greater than that of a large animal. However, energy exchange is thought to be proportional to certain power functions of body weight ($kcal/kg^{0.67}/hour$ or $kcal/kg^{0.73}/hour$). When heat production is expressed in this manner, large and small animals have approximately the same metabolic rates.

The trend commonly followed in the clinic today is to express metabolic rate in terms of body surface area (kilocalories per square meter per hour). This concept is based on the idea that heat is lost at the surface of the body and that an amount of heat equivalent to what is lost must be produced in order to maintain a constant body temperature; i.e., an animal presumably produces heat in proportion to its surface area. Some relationships between heat production and body surface area are given in Table 27-3. Individuals of the same species (dogs) show considerable variation in body weight, but metabolic rate, when expressed in terms of surface area, is approximately the same for all. A like relationship is evident when different species are compared. Although a hog weighs about 7000 times more than a mouse, the caloric outputs are similar when based on body surface area. The same is true for other species, such as man, dog, and

Table 27-3. Relationship Between Metabolic Rate and Body Surface Area

Dogs		Various Animals		
Body Weight (kg)	$Kcal/m^2/day$	Animal	Body Weight (kg)	$Kcal/m^2/day$
31.20	1036	Hog	128.00	1074
24.00	1112	Man	64.00	1042
19.80	1207	Dog	15.00	1039
18.20	1097	Guinea pig	0.50	1246
9.61	1183	Mouse	0.018	1185
6.50	1153			
3.19	1212			

Source: From M. Rubner. Z. Biol. 19:535, 1883, and 30:73, 1894.

guinea pig. Surface area obviously cannot be measured every time metabolic rate is determined. However, the surface areas of people of many different sizes and shapes have been measured by determining the total area of pieces of paper required to cover each body surface. It has been possible from these measurements to derive an equation that can be used to calculate body surface area from the weight and height of a subject, both of which are easily measured variables. Furthermore, surface area can be conveniently determined from nomograms that have been constructed from this equation.

There has been some opposition to the use of surface area for relating energy exchanges of individuals of different sizes, since heat loss is affected by factors other

Table 27-4. The Mayo Foundation Normal Standards of Basal Metabolic Rate (Kilocalories per Square Meter per Hour)*

Males		Females	
Age	BMR	Age	BMR
6	53.0	6	50.6
7	52.5	$6\frac{1}{2}$	50.2
8	51.8	7	49.1
$8\frac{1}{2}$	51.2	$7\frac{1}{2}$	47.8
9	50.5	8	47.0
$9\frac{1}{2}$	49.4	$8\frac{1}{2}$	46.5
10	48.5	9–10	45.9
$10\frac{1}{2}$	47.7	11	45.3
11	47.2	$11\frac{1}{2}$	44.8
12	46.7	12	44.3
13–15	46.3	$12\frac{1}{2}$	43.6
16	45.7	13	42.9
$16\frac{1}{2}$	45.3	$13\frac{1}{2}$	42.1
17	44.8	14	41.5
$17\frac{1}{2}$	44.0	$14\frac{1}{2}$	40.7
18	43.3	15	40.1
$18\frac{1}{2}$	42.7	$15\frac{1}{2}$	39.4
19	42.3	16	38.9
$19\frac{1}{2}$	42.0	$16\frac{1}{2}$	38.3
20–21	41.4	17	37.8
22–23	40.8	$17\frac{1}{2}$	37.4
24–27	40.2	18–19	36.7
28–29	39.8	20–24	36.2
30–34	39.3	25–44	35.7
35–39	38.7	45–49	34.9
40–44	38.0	50–54	34.0
45–49	37.4	55–59	33.2
50–54	36.7	60–64	32.6
55–59	36.1	65–69	32.3
60–64	35.5		
65–69	34.8		

*The normal limits are usually taken as ± 10 percent, and divergence beyond these limits indicates an abnormal BMR for a subject of this age and sex.
Source: From W. M. Boothby, J. Berkson, and H. L. Dunn. *Am. J. Physiol.* 116:468–484, 1936.

than total surface area. Posture and the insulating effects of hair and clothing influence heat loss to varying degrees by modifying the actual surface exposed to the environment. It has been suggested that a better standard might be the weight of the active tissue of the body, i.e., of those tissues which are actively engaged in metabolic processes and primarily responsible for the production of heat. Since adipose tissue is relatively inactive with respect to energy exchange, lean body mass might serve as a better reference for comparing metabolic rates of individuals of different sizes. Methods for estimating the weight of the body tissues exclusive of adipose tissue have been proposed and are now being evaluated.

Sex
Since the BMR of women is 6 to 10 percent lower than that of men of the same size and age, it is necessary to have standard values for both males and females. The BMR of the female increases markedly during pregnancy as the result of the additional metabolic activity of the fetus.

Age
It is well known that BMR (kilocalories per square meter per hour) is at its peak during the early years and gradually declines throughout the remainder of life (Table 27-4). On the basis of size, the growing animal has a higher metabolic rate than an adult, because the rate of turnover of the body tissues is greater in a growing animal and this is reflected by an increased production of heat. The size of the body increases during growth, and much energy is stored by the synthesis of carbohydrates, fats, and proteins. Since relatively more synthetic activity occurs in a growing animal than in an adult, relatively more energy is liberated as heat by the catabolic reactions which necessarily accompany the synthetic reactions. As a result, standards for energy exchange must be based on age as well as on sex.

Standard Values
Standard normal values based on age, sex, and body surface area have been established for basal metabolic rate. One such set of standard values is given in Table 27-4. When a value in terms of kilocalories per square meter per hour is obtained for a subject, it is then compared with the normal value obtained from individuals of the same age and sex. The usual clinical method of reporting BMR is in terms of the percentage deviation from the standard value. For example, if the BMR of a male subject, age 28, is determined to be 38.3 kcal per square meter per hour and compared with the standard value of 39.8 kcal per square meter per hour:

$$\% \text{ Deviation} = \frac{38.3 - 39.8}{39.8} \times 100 = -3.7\% \tag{14}$$

FACTORS THAT AFFECT METABOLIC RATE

Effect of Food
When a resting subject ingests food, heat production increases over the basal level. The increased production of heat that results from eating is called the *specific dynamic action* (SDA) of foods. Specific dynamic action varies with the type of food ingested. If protein is eaten, the heat production over a period of hours rises above the basal level by 25 to 30 percent of the energy equivalent of the protein ingested. Heat production also increases when either carbohydrates or fats are ingested, but to a lesser extent. For example, suppose that the basal heat production of a subject is 75 kcal per hour. The total heat production of this subject over a period of 4 hours should be 300 kcal. If the subject is fed an amount of protein equivalent to 300 kcal, enough energy should be available in the food to balance that lost as heat over the 4-hour period. However, a

measurement of metabolic rate over the 4 hours following the ingestion of the protein would show that the heat production of the subject is 380 kcal or, on the average, 95 kcal per hour. It can be calculated that 80 more kilocalories were produced during the 4-hour interval than can be accounted for on the basis of the basal heat production. In terms of the energy equivalent of the protein fed (80 kcal ÷ 300 kcal × 100), 27 percent of the energy equivalent of the ingested protein appears as extra heat.

These results show that there cannot be kilocalorie-for-kilocalorie substitution of food energy for energy stored in the body reserves. In the example cited, proteins spare some of the body reserves of energy, but obviously some of the reserve energy is needed to utilize the ingested proteins. Consequently, in order to spare the body stores of energy more completely, the total caloric intake of protein and of other foods should be somewhat greater than the basal energy exchange.

It is not certain why heat production increases following the ingestion of food. The consensus is that the excess heat is the result of intermediary metabolism of absorbed foods. If amino acids are injected intravenously into dogs, heat production increases. However, if the animal is first hepatectomized, heat production increases only slightly. The results suggest that the liver is the major site of the extra heat production, and the importance of the liver with respect to the processes of intermediary metabolism of amino acids and other foods is well established.

Other Factors

A number of factors in addition to those described above influence metabolic rate, and the relationships between some of them and metabolic rate are discussed in detail in other chapters. However, they are briefly summarized here.

Exercise or skeletal muscle activity causes by far the most marked effect on metabolic rate (Chap. 29). Table 27-5 shows the energy expenditure for various activities by a normal 20-year-old male. Metabolic rates up to 350 kcal per square meter per hour can be maintained for long periods of time by a person who is in good physical condition. Higher levels of activity are possible for only relatively short durations.

A variety of hormones influence energy metabolism. Of considerable importance are

Table 27-5. Energy Expenditure for Various Activities

Activity	Kcal/m^2/hr
Rest	
Sleeping	35
Lying awake	40
Sitting upright	50
Light activity	
Writing, clerical work	60
Standing	85
Moderate activity	
Washing, dressing	100
Walking (3 mph)	140
Housework	140
Heavy activity	
Bicycling	250
Swimming	350
Lumbering	350
Skiing	500
Running	600
Shivering	to 250

Source: From A. C. Brown. Energy Metabolism. In T. C. Ruch and H. D. Patton (Eds.), *Physiology and Biophysics* (20th ed.). Philadelphia: Saunders, 1973. Vol. 3, chap. 4.

the catecholamines epinephrine and norepinephrine, which are released by stimulation of the sympathetic nervous system, and thyroxine, the hormone of the thyroid gland. Maximal release of catecholamines can increase metabolic rate 30 to 80 percent by virtue of the stimulatory effects of these substances on the catabolism of carbohydrates and fats. Thyroxine exerts a considerable influence on the rates at which cellular oxidations occur, and excesses or deficits of circulating thyroid hormone modify metabolic rate (Chap. 33). Basal metabolic rate can be 25 to 40 percent lower than normal in hypothyroidism, and 40 to 80 percent higher in hyperthyroidism.

Temperature, both of the environment and of the body, modifies metabolic rate (Chap. 28). One of the responses of the body to a cold environment is shivering, which consists in a series of rapid contractions of skeletal muscle. When the contractions occur at a maximal frequency, a metabolic rate as high as 250 kcal per square meter per hour can be attained (Table 27-5). The heat generated by shivering aids in the maintenance of body temperature at a normal level. Metabolic rate increases with a rise in body temperature and decreases when the temperature of the body falls. For example, the metabolic rate of a person at rest with a 40°C fever would be increased by 33 percent as compared to a normal body temperature of 37°C. The rise is due to a direct effect of temperature on the rates at which the chemical reactions of the body proceed.

The nature of the diet seems to have little influence on metabolic rate, but prolonged malnutrition may cause a decrease of 20 to 30 percent, presumably because of a deficiency of cellular nutrients. Intense mental work does not modify metabolic rate. However, the increased muscular tone and increased sympathetic activity accompanying stressful emotional states can produce increases of 5 to 20 percent.

REFERENCES

Brody, S. *Bioenergetics and Growth.* New York: Reinhold, 1945.

Carlson, L. D., and A. C. L. Hsieh. *Control of Energy Exchange.* New York: Macmillan, 1970.

Clark, W. M. *Topics in Physical Chemistry* (2nd ed.). Baltimore: Williams & Wilkins, 1952.

DuBois, E. F. *Basal Metabolism in Health and Disease* (3rd ed.). Philadelphia: Lea & Febiger, 1936.

DuBois, E. F. Energy metabolism. *Annu. Rev. Physiol.* 16:125–134, 1954.

Garry, R. C. (Ed.). Energy Expenditure in Man (Symposium). *Proc. Nutr. Soc.* 15:72–99, 1956.

Kleiber, M. Body size and metabolic rate. *Physiol. Rev.* 27:511–541, 1947.

Richardson, H. B. The respiratory quotient. *Physiol. Rev.* 9:61–125, 1929.

Swift, R. W., and C. E. French. *Energy Metabolism and Nutrition.* New Brunswick, N. J.: Scarecrow, 1954.

Wilhelmj, C. M. The specific dynamic action of food. *Physiol. Rev.* 15:202–220, 1935.

28. Body Temperature Regulation*

William V. Judy

Man and other *homeothermic* animals are capable of maintaining a near constant internal thermal environment even when exposed to a wide range of climatic conditions. Just as the various physiological systems work together to regulate intracellular and extracellular fluid volume, electrolyte concentration, and pH levels, they also strive to maintain a near constant internal temperature in order to ensure optimal cellular function. Thermal regulation is necessary in warm-blooded animals because cellular biochemical and enzymatic reaction rates are temperature dependent. At the optimal body temperature many cellular enzymatic reaction rates occur between 20 to 50 kcal per mole. If, however, the cellular temperature were to drop by 10°C, these reaction rates would decrease approximately 2.5-fold; that is, metabolic activity in man changes about 25 percent for every centigrade degree change in body temperature. Since cooling body tissues reduces cellular metabolic activity and minimizes tissue hypoxia, induced hypothermia has become an acceptable practice in some types of surgery requiring blood flow occlusion. Body temperature can be lowered several degrees for short intervals (several hours) without serious cellular damage, but it cannot be increased greatly because man regulates his temperature at a point near the maximal tolerable level. Since temperature stimulates biological processes, their influence on cellular activity can lead to deterioration or the burning out of cells. Thus, in order to ensure the optimal cellular thermal environment, there must be some physiological regulatory mechanisms between cellular heat production, heat conductance from the cell, and heat loss or gain by the body surfaces. If heat loss mechanisms become inoperative, body temperature rises, stimulating enzymatic reaction rates and further heat production, and a vicious cycle is created leading quickly to cellular destruction. If man had no means of conserving body heat, his temperature would parallel atmospheric temperature, as occurs in cold-blooded animals (poikilotherms), and when body temperature fell, so would metabolic activity and the functional capabilities.

Homeothermic animals (mammals and birds) are capable of maintaining a near constant internal environment because they have well-developed thermoregulatory systems. Man is a precise temperature regulator compared to hibernating homeotherms because he regulates deep body temperature within a very narrow range at all times. Hibernators show marked seasonal body temperature variations; however, they too regulate their temperature during hibernation to a few degrees above their immediate thermal environments, by as much as 35°C, as a means of slowing down cellular activity to conserve their energy store. Many homeothermic species with different daily and seasonal body temperature fluctuations are found between man and the hibernators, but all have the common capability of regulating their body temperature independent of their external environments, thus establishing the optimal cellular temperature required to meet their metabolic demands.

BODY TEMPERATURE

Oral and Rectal Temperature

The measured and compared indices of body temperature are oral, rectal, and in some cases axillary temperatures. Oral and axillary temperatures are about 0.65°C (1.0°F) lower than rectal temperatures and are subject to even greater variations on account of

*This chapter is enlarged from E. E. Selkurt, *Basic Physiology for the Health Sciences*. Boston: Little, Brown, 1975. Pp. 539–561.

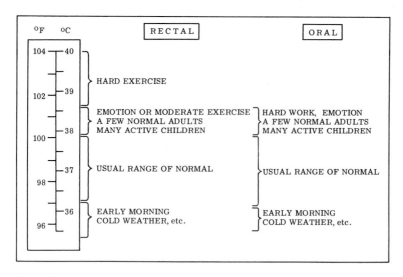

Figure 28-1
An estimate of the ranges in body temperatures found in normal persons. (From Dubois, E. F., *Fever and the Regulation of Body Temperature,* 1948. Courtesy of Charles C Thomas, Publisher, Springfield, Illinois.)

their close proximity to surrounding air (ambient) temperatures. It is difficult to specify one temperature as normal since body temperature varies considerably between individuals (Fig. 28-1). In one study, rectal temperatures in a group of healthy subjects varied from 34.2 to 37.6°C (93.7 to 99.7°F) with a mean of 36.9°C (98.4°F). Rectal temperature, approximately 37°C, refers to deep central body areas such as the brain, heart, lungs, and abdominal organs. This level is considered the normal temperature for most mature individuals, but as shown in Figure 28-1, normal has a wide range. The vital organs produce about 72 percent of total body heat during rest, but only 25 percent during exercise (Fig. 28-2). It is interesting that the vital organs and glands constitute only 8 to 10 percent of total body weight yet produce about three-fourths of the total resting body heat.

Rectal temperature is clinically accepted as being representative of internal temperature; however, variations between 0.1 and 0.9°C may be observed during any measurement, depending on the location of the measuring instrument. In those parts of the rectum adjacent to the veins returning blood from the legs and buttocks, cooler temperatures will be measured because heat from these areas is carried away by the cooler blood. Similarly, in areas away from these great veins the measured temperatures will more realistically reflect the temperature of the viscera. Therefore, body temperature no matter where it is taken (oral, rectal, esophageal, or tympanic membrane) is influenced not only by the metabolic activity of the surrounding tissues but also by the rate at which heat is being conducted away to cooler adjacent tissue and/or fluids. At all times there must be a thermal gradient between deep body or major heat production areas and the skin in order that heat can be easily transported or conducted from internal to external areas.

Skin Temperature
Skin temperatures, unlike deep body temperature, show considerable variations between areas (Table 28-1) and with changes in atmospheric temperatures. Mean skin temperature for the average person at a comfortable room temperature (24 to 25°C) is about 33°C. Surfaces covering the areas of high resting heat production have the highest skin temperatures (34.6°C). Surfaces covering the large muscle masses of the arms and

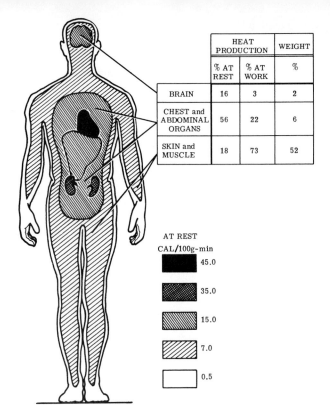

	HEAT PRODUCTION		WEIGHT
	% AT REST	% AT WORK	%
BRAIN	16	3	2
CHEST and ABDOMINAL ORGANS	56	22	6
SKIN and MUSCLE	18	73	52

AT REST
CAL/100g-min

45.0

35.0

15.0

7.0

0.5

Figure 28-2
Heat production at rest and in work of various parts of the body expressed in calories per 100 gm tissue per minute. (Adapted from J. Aschoff and R. Wever. *Naturwissenschaften* 45:477, 1958.)

Table 28-1. Regional Skin Temperatures and Heat Loss for Thermal Comfort at Rest

Region	Area (m²)	Ideal Temp. (°C)	Heat Loss at Ideal Temp. (kcal/m² · hr)
Head	0.20	34.6	20.0
Chest	0.17	34.6	48.3
Abdomen	0.12	34.6	37.5
Back	0.23	34.6	53.9
Buttocks	0.18	34.6	46.2
Thighs	0.33	33.0	36.0
Calves	0.20	30.8	73.0
Feet	0.12	28.6	83.3
Arms	0.10	33.0	84.0
Forearms	0.08	30.8	107.5
Hands	0.07	28.6	228.6
Total body	1.80	33.0 (mean)	59.4

Source: Adapted from D. R. Burton and L. Collier. The development of water conditioned suits. Royal Aircraft Establishment Technical Note. No. Mech. Eng. 400. London: Ministry of Aviation, 1964.

679

legs have a mean temperature of 30.8°C, and areas that cover very little muscle (hands and feet) have the lowest mean resting skin temperature: 28.6°C.

The skin, unlike internal core tissues, can function well at temperature extremes. For example, skin temperature may fluctuate ± 10 to 12°C around its normal mean without damage; however, prolonged skin temperatures as low as 18°C will lead to tissue anoxia, pain, and tissue damage. Sustained high temperatures of 45°C will lead to tissue burns, accumulation of subcutaneous fluid, and pain.

Heat loss from the various body surfaces is also nonuniform; in general, areas with the highest resting temperature have the lowest heat loss per unit surface area. For example, the extremities with their lower skin temperatures have greater heat losses than the warmer trunk and head surfaces. The extremities act as major heat dissipation areas from which heat loss can be increased to maintain thermal comfort by either cutaneous vasodilation or removal of clothing. Heat loss may be decreased from these areas by vasoconstriction and addition of clothing.

Mean Body Temperature

The temperature differences (gradients) between the major internal organ areas, skeletal muscle masses, and the skin have led to the concept that *mean body temperature* ($\bar{T}_b$) is equal to the fractional sums of core temperature, represented by rectal temperature ($\bar{T}_r$), and shell temperature represented by mean skin temperature ($\bar{T}_s$). At a comfortable room temperature (24 to 25°C), two-thirds of the body mass is considered to be at core temperature and one-third at shell temperature (Fig. 28-3); therefore, $\bar{T}_b$ may be expressed by the equation

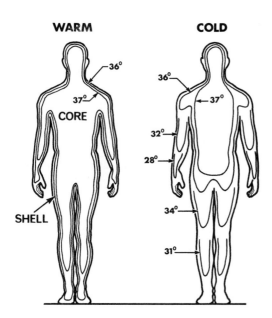

Figure 28-3
Isotherms (surfaces connecting points of equal temperature) in the body. Isotherms in a warm environment (left); in a cold environment (right) in degrees centigrade. The innermost isotherm may be considered the boundary of the body core; the core includes most of the body in hot environments. The outer isotherm is representative of skin or shell temperature. When heat must be conserved, the core contracts to the proportions indicated on the right. In severe cold exposure, the combined effect of vasoconstriction and countercurrent heat exchange results in the pattern of isotherms shown in the limbs. (After J. Aschoff and R. Wever. *Naturwissenschaften* 45:477, 1958.)

$$\overline{T}_b = 0.67\,\overline{T}_r + 0.33\,\overline{T}_s$$

The relative importance of this relationship is that it allows one to estimate body heat lost to or gained from the environment and the rate and amount of heat storage which normally take place in peripheral tissues, or shell. The relationship is, however, subject to error since the proportion of body mass at core and shell temperatures changes as a function of environmental temperatures. At high ambient temperatures the proportion of body mass at core temperature increases, as do mean skin temperature and body heat loss. The result will be a small increase in $\overline{T}_b$ and the peripheral heat stores due to rapid heat transfer from the core to the shell through peripheral vasodilation.

At low ambient temperatures, the proportion of body mass at core temperature is reduced, as are mean skin temperature and heat loss. Mean body temperature will decrease although $\overline{T}_r$ may increase slightly because of peripheral vasoconstriction resulting from direct thermal stimulation or increased sympathetic nerve activity. Thus, the thermal gradients or the isotherms shown in Figure 28-3 are changed by peripheral vasomotor activity, which controls the rate of heat transfer away from vital internal organs during exposure to normal, hypothermic, or hyperthermic conditions.

Factors Influencing Body Temperature
As shown in Figure 28-1 many factors influence body temperature. Most mature individuals show a *diurnal rhythm* (circadian rhythm)—i.e., a 24-hour cycle during which deep body temperature fluctuates ±0.5°C around the person's normal mean temperature (Fig. 28-4). It is usually lowest in the morning during sleep, slightly higher in the early waking hours, and highest in early afternoon or midafternoon during peak activity. The diurnal temperature pattern is probably of a controlled nature, and the depression during sleep appears to be effected through physiological cooling mechanisms and not to be dependent on metabolic depression. These cyclic patterns persist regardless of activity or disease states. They are not well established in small children, and they may be shifted in time or reversed in adults by changing day-night, work-rest

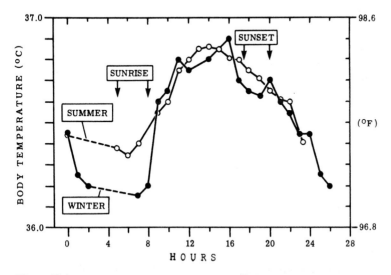

Figure 28-4
Circadian rhythm in body temperature. Body temperature is lowest in early morning, gradually increasing during the day, and reaches a maximum in midafternoon, during peak activity. (From T. Sasaki. *Proc. Soc. Exp. Biol. Med.* 115:1129, 1964.)

schedules or by fast international time zone changes. Therefore, the temperature measured will be influenced by the time of day at which it is taken.

Other factors influence body temperature. Rectal, oral, and skin temperatures all increase during exercise, during prolonged exposure to high ambient temperatures, during emotional stress of pleasure and displeasure, during febrile disease states (fever), and in nonfebrile disease states (hyperthyroidism). Similarly, body and skin temperatures fall during prolonged exposure to severe cold environments, as with frostbite; during prolonged inactivity, as in sleep; and during metabolic disorders such as hypothyroidism (myxedema). Patients with peripheral circulatory obstruction will have marked reduction of muscle and skin temperature in the afflicted area, as well as pain.

Age influences body temperature. Children tend to have higher rectal and oral temperatures (37.5 to 38.0°C) than adults, and their temperatures are more variable since they do not have well-established diurnal patterns. Newborn and premature babies are prisoners of their thermal environments because they do not have well-developed thermoregulatory systems. If they are not protected, their temperature will fluctuate with ambient temperature; therefore, they must be kept in controlled temperature and humidity areas in order to maintain an optimal thermal environment.

Most women show body temperature changes related to the menstrual cycle (Fig. 28-5). A few days before menstruation, core temperature drops 0.6°C, and the new temperature is maintained until just before ovulation, when it drops another 0.2°C. After ovulation, temperature rises to a normal level, where it remains until about the twenty-eighth day, when the onset of menstruation again takes place. If pregnancy occurs, the normal premenstrual temperature drop is inhibited by the continual presence of the hormone progesterone. Temperature also has an influence on conception because high body temperature reduces the viability of sperm; for this reason the male testes, unlike the female ovaries, are located outside the abdominal cavity.

Sex differences have only minor influences on body temperature aside from the apparent menstrual effects. In one study, male and female rectal temperatures were almost identical at an ambient temperature of 22°C. Mean skin temperature was

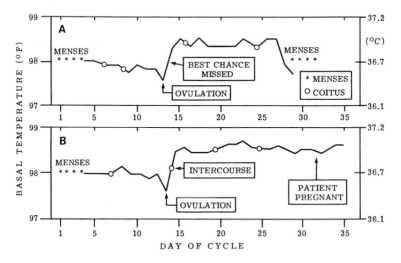

Figure 28-5
Body temperature, recorded daily before arising, of a woman during two menstrual cycles. The second cycle is followed by pregnancy. (From Seller, W. A., *Body Temperature: Its Changes with Environment, Disease and Therapy,* 1952. Courtesy of Charles C Thomas, Publisher, Springfield, Illinois.)

slightly higher for males and heat elimination greater. Rectal and mean skin temperatures increased for both males and females as ambient temperature was raised to 36°C; again rectal temperatures were almost identical, although female mean skin temperature was slightly greater.

PRINCIPLES OF PHYSIOLOGICAL HEAT TRANSFER

Heat is constantly being produced internally, transported to the body surfaces, and exchanged with the environment. Production is through basal metabolic activity, specific dynamic action of food, and muscle contraction (see Chap. 27). Of the total metabolic cost in the basal or active state, 80 percent is heat and only 20 percent is utilized for contractile energy. Basal heat production is about 72 kcal per hour for a 70-kg man and may increase by as much as 20 times during exercise (1440 kcal per hour). Resting man can increase his heat production approximately fivefold through shivering when exposed to cold; however, he cannot willingly decrease or turn off his basal metabolic heat production when exposed to hyperthermic conditions. Thus man has thermoregulatory capabilities which enable him to generate, distribute, and dissipate heat.

To characterize temperature regulation quantitatively, a measure of the change in body heat stores must be utilized. This measurement is the heat capacity or specific heat defined as the ratio of heat supplied (or removed) to the corresponding temperature rise (or decrease) or

$$\text{Specific heat} = \frac{\Delta\text{kcal/kg}}{\Delta T}$$

When 1 kcal is added to 1 kg of H_2O, the temperature is raised 1°C. Thus, the specific heat of water is 1, and the kilocalorie as an energy unit is defined. The average specific heat of the body is about 0.83 because of the high water content with lesser proportions of protein and lipids of lower specific heat. Thus mean body temperature should change 1°C for every 0.83 kcal of heat added or substracted per kilogram body weight from the body heat store. The change in heat content rather than the overall total content is the important factor to the physiologist, and it can be estimated by

$$S \text{ (in kcal/hour)} = \frac{0.83 \, (\overline{T}_{b_1} - \overline{T}_{b_2}) \times \text{body weight}}{\text{time (hours)}}$$

where $\overline{T}_{b_1}$ and $\overline{T}_{b_2}$ are the mean body temperatures at the beginning and the end of the time period.

Body heat stores (body fluids, fat, and muscle) are basically good thermal insulators and poor conductors, therefore providing protective insulation when exposed to cold stress. If, however, man had no effective or rapid means of transferring heat around these insulative stores or to them from internal production sites, they would be detrimental in both normal and hyperthermic environments. To have thermal regulation, there must be a balance between heat production or gain and heat loss. If, for example, all heat loss channels were somehow completely stopped, the basal heat production of a 70-kg man would raise body temperature to the upper (maximal) thermal limit of 40 to 42°C in three to five hours. This increase would require the addition of about 0.83 kcal per kilogram to heat stores each hour. Similarly, a 70-kg man with a metabolic activity of three times basal would have a heat gain of approximately 2.4 kcal per kilogram per hour, which would raise body temperature about 3°C per hour. Considering the stimulating influence of temperature on cellular metabolism, this simple mathematical manipulation becomes even more astounding and shows the importance of internal heat transfer and external heat exchange in the maintenance of thermal equilibrium.

Internal Heat Transfer

Internally produced heat must be readily removed from the cell to prevent excessive heat accumulation. Heat is transferred from the site by *thermal conduction* or *conductance* between adjacent cells and fluids down a thermal gradient, and by *forced convection,* in which heat is conducted across the vessel wall to the vascular fluid and swept away to cooler areas by the circulating blood. Thermal conductance between the core and the shell depends on the steepness of the thermal gradient; however, since the core is a large area of constant temperature, conductance is very slow on the inside and much greater at the core-shell junction, where circulatory adjustments occur readily. If there were no peripheral or skin blood flow, the amount of heat conductance from the internal areas to the periphery would be about 5 to 10 kcal per degree centigrade gradient per hour. Assuming a thermal gradient of 4°C ($\bar{T}_r = 37°C$ and $\bar{T}_s = 33°C$), 20 to 40 kcal per hour would be transported by tissue conductance alone. This value would indicate that anywhere between 38 and 72 kcal per hour would be added to the body heat stores for a 70-kg man in the basal condition (heat production = 72 kcal per hour) and 176 to 196 kcal per hour for a normally active individual. Thus, body tissues are poor thermal conductors and good insulators. For example, the thermal conductance of a 1-cm-thick piece of perfused beefsteak is about equal to that of a 1-cm-thick piece of cork (conductivity = 18 kcal/hour/°C gradient). Internal heat transport by forced convection requires that a thermal gradient exist between circulating blood and the tissues themselves. This is accomplished by having the cooler venous blood return from the periphery or shell area in vessels adjacent to warm arterial blood from the core areas. The circulatory system, then, provides a major heat transport system which maintains a proper cellular thermal as well as fluid, electrolyte, nutrient, and O_2 environment.

External Heat Transfer to the Environment

The major mechanisms of heat exchange between skin and environment are *radiation, conduction, convection,* and *evaporation.* Since radiation, conduction, and convection are technically difficult to separate, they are classified together as *nonevaporative* heat exchange mechanisms. When skin temperatures are greater than the environmental temperature, or that of nearby objects, these pathways are characterized as *heat loss* channels. When skin temperatures are lower than the temperature of the surrounding air or nearby objects, the nonevaporative heat exchange mechanisms become *heat gain* channels through which body temperatures can actually be increased. The relative

Table 28-2. Steady-State Evaporative and Nonevaporative Heat Loss, Rectal and Mean Skin Temperature, and Thermal Conductance at Various Ambient Temperatures

Thermal Regulatory Variables	Calorimeter Temperature (°C)				
	20	25	30	35	40
Evaporative, % of total	17	30	50	93	100
Nonevaporative, % of total	83	70	50	7	0
Rectal temp., °C	36	36.3	36.5	36.8	37
Mean skin temp., °C	27.8	30.0	33.5	35.3	35.7
Thermal conductance, cal/m²/hr/°C	11	16	18	35	39

Source: Adapted from Carlson, Honda, Sasaka, and Judy. *Proc. Soc. Exp. Biol. Med.* 117:327–331, 1964.

685

importance of evaporative and nonevaporative heat exchange mechanisms as means of maintaining deep body temperature varies with the ambient temperature (Table 28-2).

When nude man is exposed to various air temperatures (calorimeter temperature, Table 28-2) between 20 and 40°C, his evaporative heat loss, mean skin temperature, and thermal conductance (heat flow between deep body areas and skin surface) increase greatly, whereas rectal temperature increases only slightly (1°C) and nonevaporative heat loss decreases to zero. These data show that deep body temperature is maintained fairly constant at the various ambient temperatures by changes in the amount of heat transferred from deep body areas to the skin and exchanged with the environment by evaporative and/or nonevaporative mechanisms.

Radiation

Radiation is heat transfer by electromagnetic waves between objects which are not in contact (Fig. 28-6A). All dense objects radiate heat; the greater the temperature difference between two objects, the larger will be the amount of heat radiated from the warmer to the cooler. At low ambient temperatures (20°C, 68°F), radiative heat loss in man may account for 70 percent of total heat production. At high temperatures (35°C,

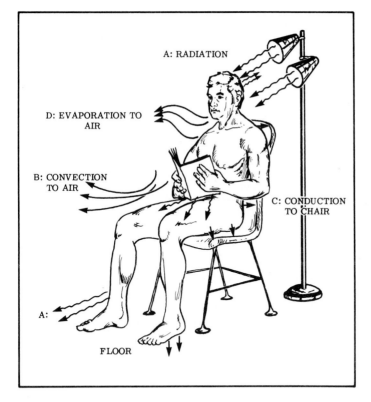

Figure 28-6
Mode of heat transfer with the environment. A. *Radiation*—heat transfer by electromagnetic waves from warm to cooler objects. B. *Convection*—heat transfer by molecules of air or fluid moving between areas of unequal temperature; convective heat exchange is facilitated by air or fluid movements (forced convection). C. *Conduction*—heat exchange between objects in contact; this includes transfer to air. D. *Evaporation*—heat loss by water molecules diffusing from the body surface and by sweating.

95°F), the body may actually gain heat through radiation. The warmth that is felt from a heat lamp or the sun rays is an example of radiant heat. Radiant heat loss is greatly influenced by the total exposed surface area. Normally, not all body surface areas are effective in radiative heat exchange with the environment. Surfaces under the arms, between the fingers, and between the legs radiate heat to the opposing skin surface, and usually there is no net heat loss by the exchange. However, a man who extends his arms and spreads his legs and fingers may increase his effective radiating area from 75 to 85 percent. The spread-eagle stance is commonly observed in many mammals and birds on hot summer days or nights as they increase their effective radiant surface and heat loss. Conversely, men and animals exposed to cooler environments will curl up to reduce radiative heat loss.

Convection
Convection is heat transfer by movement of molecules of a gas or liquid between two locations of different temperatures (Fig. 28-6B). When air temperature is less than that of the skin, heat is transferred from the skin by conduction or radiation to the gas or fluid medium on the skin surface. This medium gains heat, becomes less dense, rises, and is replaced by cooler air in a continuous process. Convective heat loss is aided by air or fluid currents (forced convection). At normal room temperatures and air movements, convective heat loss accounts for only a very small fraction of nonevaporative heat loss; however, forced convection may become quite effective.

Conduction
Conduction is heat exchange in the form of kinetic energy between atoms or molecules of objects in contact (Fig. 28-6C). It includes heat transfer to air or fluid molecules as well as dense objects with which one may be in contact. Actually, little heat is exchanged with the environment by this route unless one is literally in *cold* water. Again, as with convection and radiation, the amount of conductive heat exchange is proportional to the temperature difference between objects.

Evaporation
Evaporation is a constantly operating heat loss mechanism. Small amounts of water are continually being diffused from the skin surface, respiratory passages, and mucous membranes of the mouth. These exchanges are called *insensible* perspiration and account for 20 to 25 percent of basal heat production loss. Profuse sweating becomes the most effective heat loss channel at high ambient temperatures and during heavy work. Under such conditions, the amount of heat lost by evaporation is inversely related to the humidity of the surrounding air, and for sweating to be effective as a cooling mechanism, sweat must be rapidly evaporated.

PARTITIONAL HEAT EXCHANGE
The relationships between body heat production and heat lost to or gained from the environment at any specific environmental temperature are conveniently demonstrated by balance statements. When mean body temperature and the quantity of heat in body heat stores remain constant, man is considered to be at a thermal steady state or equilibrium. Under such conditions, heat sources must equal heat losses:

$$\text{Sources} = \text{losses}$$
$$\text{Metabolism} + \text{heat stores} = \text{evaporative} + \text{nonevaporative losses}$$

If heat sources become greater than heat losses, mean body temperature rises; if losses exceed sources, temperature falls. The relationships between heat sources and losses are different at various ambient temperatures, and the primary mode of thermal regulation changes from one temperature range to another. The modes of temperature regulation are *metabolic, vasomotor,* and *sweating*. Metabolic mechanisms are most effective in cold environments, vasomotor mechanisms in comfortable environments, and sweating in hot environments.

In cool environments, the thermally unprotected man (nude) is subject to body cooling through both evaporative and nonevaporative mechanisms. Evaporative heat loss under these conditions is insensible loss only; it is low, it is fairly constant, and it cannot be greatly decreased to prevent loss of stored heat. Nonevaporative heat loss is the major heat loss channel in a cool environment because mean skin temperature exceeds ambient temperature and the amount of heat loss is proportional to the degree of cold stress (Table 28-2). Thermal steady states may be established by increasing metabolic heat production or decreasing nonevaporative heat loss. Naturally, man protects himself by putting on clothing to reduce nonevaporative heat loss, but the physiological response of the unprotected individual is to increase metabolic heat production and thus replace the heat given up by body heat stores. In addition to basal heat production, cellular metabolic activity may be increased by hormonal (thyroxine, catecholamines), mechanical (muscle activity or shivering), or thermal (direct temperature effects on cell) means.

In warm or hot environments, the primary regulatory mechanism used to maintain thermal equilibrium is evaporative heat loss or *sweating*. When the ambient temperature

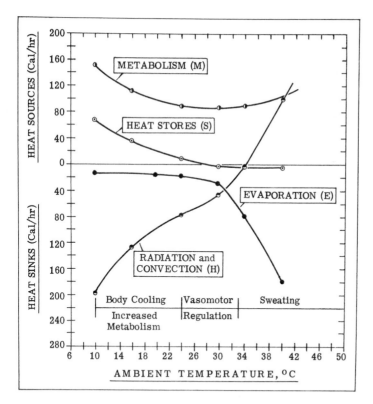

Figure 28-7
Partitional heat exchange of a human subject at different ambient temperatures. All heat sources are above the zero line; all heat sinks are below the line. Note the three modes of thermoregulation (metabolism, vasomotor regulation, and sweating) and their functional temperature ranges. (From R. W. Bullard. Temperature Regulation. In E. E. Selkurt [Ed.], *Physiology* (3rd ed.). Boston: Little, Brown, 1971.)

exceeds mean skin temperature, nonevaporative heat loss mechanisms become heat sources. The problem, therefore, is to prevent body temperature from rising because of heat accumulation. At high temperature, metabolism is increased slightly as a result of direct thermal effects on cellular activity; however, sweating is initiated to maintain thermal equilibrium by effective cooling of heat stores. The sweating rate is directly related to the degree of heat stress, and its effectiveness to cool the body is inversely related to the moisture content of air (humidity).

At comfortable temperatures, thermal equilibrium is achieved primarily through *vasomotor regulation.* Here nonevaporative heat loss is altered by changing the skin to the ambient temperature thermal gradient. This is accomplished by sympathetic nervous system control of skin and extremity blood flow. Thus, thermal equilibrium is maintained without sweating or increasing metabolism.

The partitional heat exchange diagram in Figure 28-7 presents a summary of the avenues of heat loss or gain. The values shown are examples which will change when either exposure time, relative humidity, posture, clothing, or air velocity is varied. The figure indicates the steady-state relationships between heat sources and heat losses (sinks) at various ambient temperatures. It should be noted that balance statements are equal at all points; that is, the algebraic sums of the *sources* above the zero line and *sinks* (losses) below the zero line at any specific temperature are equal.

HEAT TRANSFER WITHIN THE BODY

Internal Heat Distribution
As stated previously, heat is transferred from internal production sites to body surfaces by combined conduction and circulatory-assisted convection (forced convection). Conductive heat transfer between deep body cells is slow and small because the temperature difference (thermal gradient) between adjacent cells is small and body tissues are relatively poor thermal conductors. Circulatory-assisted convection is rapid and is the major means of internal heat transfer. It involves a form of bulk movement of heat by fluids flowing between body areas of unlike temperatures. Heat is conducted to the capillary wall, where convective exchange occurs between the vessel wall and the blood by the action of fluid flow. The circulatory system transports the heat-carrying blood to cooler tissue areas, where the exchange process is reversed and heat is gained by these tissues from the blood.

Effector Systems

Circulation
The circulatory system is essential to body temperature regulation as it is to all body systems. During resting or stress conditions, it distributes heat throughout the body, controls the proportions of body mass at core and shell temperatures, and sets up a countercurrent exchange mechanism in the extremities which reduces the amount of heat available for exchange with the environment. The centrally located vital organ systems (Fig. 28-2) have not only the highest resting metabolic heat production but also the greatest resting blood flow per unit of body weight. This flow acts as an effective cooling mechanism by moving large quantities of centrally produced heat to cooler, less metabolically active peripheral tissues. Thus, an optimal thermal environment is maintained for all cells by keeping most of the body mass at a relatively constant (37°C) temperature.

Effective Body Insulation
The vasomotor responses of systemic arterioles perfusing cutaneous capillaries control the body proportions at core and shell temperatures (Fig. 28-3). The systemic capillaries do not extend into the superficial epidermis; therefore, heat transferred through this

689

outermost layer is by tissue conduction only. At comfortable temperatures, vasomotor activity establishes a thermal steady state without sweating or increasing metabolism. At the upper limits of thermal comfort, cutaneous arterioles dilate, increasing capillary blood flow (100-fold in some areas), and expose large quantities of warm blood to the superficial epidermis. This process reduces the conduction distance between the superficial capillary and the epidermis and increases heat loss by as much as 20-fold. When the cutaneous arterioles are constricted, capillary blood flow and the number of perfused vessels are greatly reduced; conduction distance is increased, and heat loss reduced. By these vasoconstrictor and dilator actions, the circulatory system regulates the amount and rate of heat transferred to the body surface, and the heat loss from the skin, by changing the size of the body shell which separates or insulates deep body core areas from the environment.

Countercurrent Heat Exchange
The route taken by venous blood in the extremities on its return to the heart influences the amount of heat lost from these areas. Venous flow may be shifted from superficial vessels to deep venous plexuses (venae comitantes) that are adjacent to the arterial inflow vessels. The shift sets up a conductive heat exchange between warm arteries and cool veins known as the *countercurrent heat exchange*, which is illustrated for the human arm in Figure 28-8. In essence, cool venous blood is warmed as it returns to the heart, and arterial blood is cooled as it proceeds down the extremities. For example, in

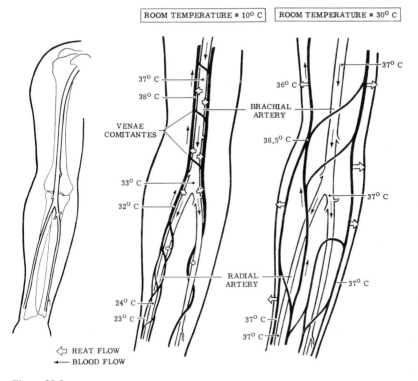

Figure 28-8
The countercurrent heat exchange in the human arm. Venous flow returning to the heart is shifted from superficial vessels at a room temperature of 30°C to deep vessels at 10°C room temperatures. (From R. W. Bullard. Temperature Regulation. In E. E. Selkurt [Ed.], *Physiology* (3rd ed.). Boston: Little, Brown, 1971.)

a cool environment, warm arterial blood (37°C) is pumped down the brachial artery, but before it reaches the radial and ulnar arteries, it is reduced to 33°C; and when it reaches the superficial capillaries, its temperature is 24°C. The temperature decrease is due to conductive heat transfer to adjacent venous vessels as blood returns counter-currently to the heart. Thus, less heat is transported to open cutaneous vessels; the capillary surface thermal gradient is reduced, as is heat exchange with the environment; and the venous blood transports heat back to deep body areas as a means of heat conservation.

At the upper end of vasomotor control, peripheral vasodilation shunts large quantities of blood to the periphery. Cooler venous blood returning to the heart does not gain as much heat because the return flow is directed away from the deep veins. Therefore, vascular control mechanisms initiate blood flow shunting which effectively channels blood through vascular structures in the extremities as required to facilitate heat loss or conservation.

Sweating
When body heat increases because of either muscle activity or decreased nonevapora-tive heat loss, sweating is initiated as a cooling mechanism. Sweat is a true secretion of mostly water and sodium chloride with an osmolar concentration well below that of plasma (Chap. 23). Release of sweat from the subcutaneous glands (eccrine glands) is controlled by the sympathetic postganglionic cholinergic division of the autonomic nervous system (Chap. 7A) and occurs simultaneously in most skin areas. Pumping

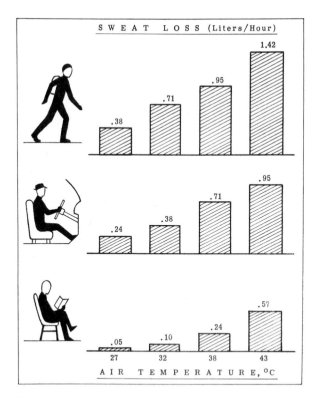

Figure 28-9
Sweating rates for a clothed man in various activities and at various ambient temperatures. (From Adolph et al., 1947.)

action of contractile elements in the gland duct forces sweat to the body surface, and this action increases as both skin and rectal temperature rise.

In the strictest sense, sweating is a type of forced convection, because body heat is conducted to the glandular fluids and transported to the surface by ductile pumping. Pumping rates of 1 to 2 contractions per minute occur at low sweat rates, and 15 to 20 per minute during profuse sweating. This becomes a very effective cooling mechanism when one considers that man has an estimated 2.5 million sweat glands spread under his skin. As a single organ system, they have a combined ductile-surface exposure area of 90 cm^2 (1 square yard), and about 40 ml fluid is required to fill the sweat gland ductile system. Sweating rates for various work levels at different ambient temperatures are shown in Figure 28-9. A man working in severe heat and high humidity can produce 4 liters of sweat per hour for short periods. However, his ability to sweat and maintain high outputs is developed by continuous heat exposure. Individuals who are not accustomed to high heat loads cannot sustain high sweat rates. The sweating mechanism is initiated at lower ambient temperatures in adults (32 to 34°C) than in full-term infants (35 to 37°C), and prematures do not show well-developed sweat patterns but increase their respiratory rates as a means of losing heat.

The effectiveness of sweating as a cooling mechanism is reduced when the air water content (humidity) is high. Sweat must be evaporated easily to be effective in cooling the body surface and enhancing heat loss. Therefore, evaporation is quicker in hot dry climates than in hot humid climates, because the driving force for evaporation is the difference between the vapor tension of water on the skin surface and that in air.

Shivering

Shivering is a compensatory heat production mechanism that is initiated when peripheral vasoconstriction is inadequate to prevent heat loss. When exposed to sustained cold stress, man increases his metabolic heat production by voluntary muscle contraction or involuntary shivering. In voluntary muscle contraction approximately 20 percent of the chemical energy released in the contractile process is converted to work, whereas in shivering almost all of the energy is converted to heat because the muscles do no work. A seminude man will double his heat production during a 60-minute exposure to 5°C (41°F) temperature by shivering; however, the effectiveness of shivering to maintain body heat is very small. At maximum shivering heat production (eightfold), thermal equilibrium in the thermally unprotected man (nude) cannot be attained during prolonged severe cold stress. Newborns and prematures do not demonstrate shivering thermogenesis when exposed to cold, but they become restless and irritable and seem to have an increased sensitivity to cold. They may increase their metabolic heat production (O_2 consumption) by as much as 100 percent over control, which is a much greater value than adults can achieve.

Shivering consists of synchronous contraction and relaxation of small antagonistic muscle groups. The nervous pathways involved are somatic efferent neurons and their proprioceptive and tension feedback system. The amount of heat produced by shivering is influenced by the posterior hypothalamus, and the intensity depends on interactions between cortical, anterior hypothalamic, and cerebellar areas. The rhythmicity is influenced by the cerebellum, and the frequency by interaction in the spinal cord.

THE THERMOREGULATORY SYSTEM

The thermoregulatory system, like other physiological control systems, has at least three major components: sensory receptors, central integrator or controller, and effector organ systems (Fig. 28-10). The sensory thermoreceptors supply skin and deep body temperature information to the central integrator, which compares this information to a standard reference or *set-point* value. On the basis of the difference between the thermoreceptor inputs and the set-point input, the central integrator supplies output information to effector systems controlling heat production or loss, thus regulating body

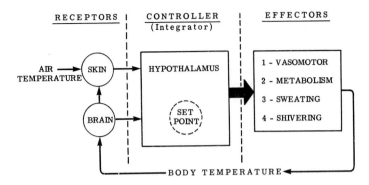

Figure 28-10
Basic components of the thermoregulatory system, showing receptors (skin and central receptors), hypothalamic controller (or integrator) with central set point, and effector systems regulating heat production, conservation, and loss.

temperature around the set-point value. For example, if the peripheral or central thermoreceptor's input information indicates temperature less than that of the set-point, effector heat conservation or production mechanisms are activated. If input information signifies that body temperature is greater than the set-point value, heat loss mechanisms are activated. The information that the receptors transmit to the central nervous system is not temperature per se, but afferent nervous activity (discharge frequency) proportional to the steady state, and to changing temperatures.

Thermoreceptors

Thermoreceptors that monitor skin or shell temperatures are located just beneath the skin. They are classified as cold and warm receptors and are not uniformly distributed over the body. For example, a larger number of cold receptors is found below the skin of the face and hands than below that of the chest and legs. Warm receptors have been histologically defined as the *end-organs of Ruffini,* cold receptors as the *end-bulbs of Krause.* However, current information indicates that the *naked nerve endings* are the major peripheral thermoreceptors. Additional thermal receptors exist in the tongue, in

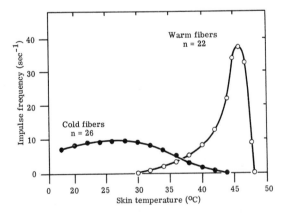

Figure 28-11
Static discharge rate of cutaneous cold and warm fiber populations in the nose of the cat as a function of skin temperature. (From H. Hensel and R. D. Wurster. *Pflügers Arch.* 313:153, 1969.)

693

the respiratory tract, and in deep body areas such as the viscera and spinal cord. Both groups of cutaneous receptors are rapidly adapting, in that their discharge rates are proportional to the rate of temperature change as well as to the steady-state temperature. Figure 28-11 shows the average static discharge rate of cutaneous thermal receptors (warm and cold fiber populations) as a function of skin temperature. These data were taken from the noses of cats at different fixed levels of temperature. Cold fibers are identified by increasing activity as the tissue is cooled, and warm fibers are those whose activity level increases in response to warming. Afferent signals from the cutaneous receptors enter the spinal cord at all levels and ascend to the thalamic portion of the brain through the lateral spinothalamic tract, along with pain fibers.

Central thermoreceptors are found in the hypothalamus, and they are sensitive to local "core" temperature (Fig. 28-12). Cold receptors are located in the preoptic area of the anterior hypothalamus (heat loss center), and heating or stimulation of that area initiates *antirise* responses (vasodilation, sweating, panting, etc.). Warm receptors are found in the posterior hypothalamus (heat production), and local cooling or other stimulation of the area initiates *antidrop* responses (vasoconstriction, epinephrine release, and shivering). It is obvious that the thermoregulatory system utilizes both

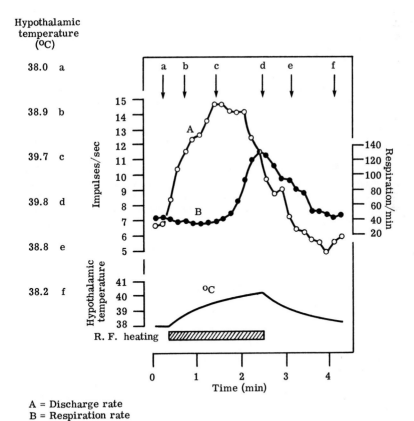

A = Discharge rate
B = Respiration rate

Figure 28-12
Electrical activity of thermal-sensitive units in the preoptic area of the anterior hypothalamus, and respiratory rate in response to local heating and cooling of these central thermal-sensitive areas in anesthetized dogs. (From T. Nakayama et al. *Science* 134:560, 1961.)

central and peripheral temperature information to determine the function of effector systems; however, the relative importance of the receptors in the overall temperature regulation picture is unclear. When the skin is warm, approximately 1°C of local hypothalamic cooling is required to evoke a strong cold-protective response. Since this level of cooling is far greater than that which would occur normally, the peripheral receptors must play a major role in calling forth protective mechanisms. Similarly, local skin temperature changes will greatly alter cutaneous blood flow and sweating rates without measurable change in hypothalamic temperature.

HYPOTHALAMIC INTEGRATION AND TEMPERATURE CONTROL

Experimental evidence points to the *hypothalamus* and its *preoptic areas* as the center of body temperature regulation. These areas, through their various interactions, function as a thermostat operating dually to prevent both excessive body heating and cooling; that is, the mechanisms controlling heat production or conservation and heat loss are regulated by hypothalamic influence over effector systems. The nervous pathways that transmit hypothalamic directions to the various effectors are the somatic and autonomic nervous systems. The sympathetic portion of the autonomic nervous system plays a very large part since it controls vasomotor responses. The somatic nervous system controls voluntary function such as the gross muscle movements required to add or remove clothing and the minute muscle movements associated with muscle tone and shivering.

The overall integrated control of various effector systems is shown in Figure 28-13. This is an expansion of Figure 28-10 showing the peripheral thermoreceptors and their afferent input pathways on the left, the central receptor areas and hypothalamic integrator center in the middle, and the effector systems and their appropriate efferent output pathways on the right. The effector systems are designated according to their *antirise* or *antidrop* effects on body temperature. Antidrop outflow for skin vaso-constriction and piloerection ("gooseflesh") is mediated through sympathetic centers (posterior hypothalamus). The antirise efferent outflow responsible for skin vasodilation acts by an inhibition of sympathetic constrictor activity, possibly through inflow, and is also mediated by the sympathetic system; however, this is a sympathetic cholinergic rather than adrenergic response. Animals (dogs, cats, rabbits, etc.) that do not have sweat glands lose large quantities of heat through the respiratory passages by panting, fluid evaporation from the tongue, and salivation. These mechanisms are para-sympathetic in origin and the only clearly defined parasympathetic contribution to the thermoregulatory process.

The shivering efferent pathways are not completely defined at present. Continuous somatic efferent activity from the central nervous system to the lower motor neurons initiates shivering. Proprioceptive impulses from the muscles that feed back on the spinal cord inhibit the motor input activity, thereby turning shivering off. This mechanism accounts for the shivering rhythm, which may occur at rates of 10 to 20 per second.

The physiological responses to cold and heat also involve the systems that stimulate appetite and thirst, and these drives are initiated through higher center activity. Appetite is related to the antidrop metabolism through the *specific dynamic action* of food. Thirst is related to the antirise sweating and internal heat distribution by bulk fluid flow. Fluid consumption must be stimulated to maintain body fluid levels in the event of high sweat activity. The loss of body fluids (dehydration) causes body temperature to rise, and in severe prolonged dehydration fever is prevalent.

The hypothalamus has direct and indirect influence on the neuroendocrine system. Posterior hypothalamic activity controls the level of autonomic sympathetic activity (cholinergic) to the adrenal medulla and thus the release of epinephrine. Similarly, posterior hypothalamic activity increases sympathetic postganglionic activity and the release of the neurotransmitter norepinephrine. Catecholamines have antidrop vaso-

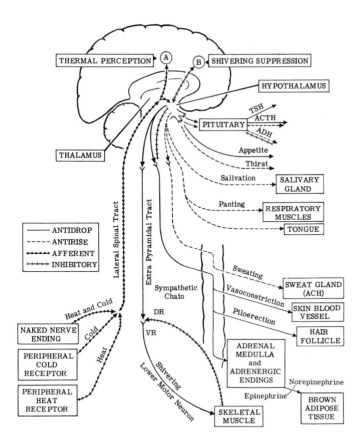

Figure 28-13
Integration of temperature regulation mechanisms. The peripheral thermoreceptors and their afferent pathways are designated on the left; the central thermoreceptors and integrator system in the brain, and the effector systems with the afferent inputs, on the right. DR and VR represent dorsal and ventral root of spinal cord for shivering feedback control. (From A. C. Burton. *J. Appl. Physiol.* 6:65, 1953.)

constrictor and metabolic influences. The calorigenic effects of epinephrine and norepinephrine have been shown to be major stimulants of *nonshivering thermogenesis* in cold-adapted small animals, but the evidence for such action in man is incomplete. Many of the releasing factors that control the synthesis or release of anterior and posterior pituitary hormones are stimulated through hypothalamic activity during various environmental stress conditions. Thyroid-stimulating hormone (TSH) and adrenocorticotropic hormone (ACTH), released from the anterior pituitary, facilitate antidrop metabolism by stimulating thyroxine and glucocorticoid release. Thyroxine's calorigenic activity is synergistic with epinephrine in cold exposure or adaptation. Antidiuretic hormone (ADH) from the posterior pituitary and ACTH from the anterior have antirise influences through their control of fluid and electrolyte reabsorption by the kidney.

In summary, physiological responses to cold or hot stress conditions involve many neuroendocrine hormonal changes. In fact, almost all tissues and organ systems are influenced directly by the environment or indirectly by responses of other systems to environmental changes.

ACCLIMATION TO ENVIRONMENTAL CONDITIONS

Physiological adjustments that take place during continual exposure to hot or cold environments which enhance performance or survival are termed *acclimation.* The two extreme environmental thermal stresses (hot and cold) induce specific physiological and morphological changes in animals; however, the degree of change among animals is not uniform. For example, acclimation to cold is more pronounced in smaller mammals than in larger ones. Man's ability to survive under extreme environmental temperature has been aided by his development of artificial thermoregulatory aids (clothes, shelter, air conditioning, living habits). Nevertheless, considerable physiological changes do occur under some circumstances.

Acclimation to Heat

When man is abruptly exposed to a high heat stress (40°C), his skin temperature, rectal temperature, sweat rate, and heart rate increase. If he tries to work his tolerance and performance are limited. However, after six to nine days of working at high temperatures, he does undergo adaptive changes. Heart rate, rectal temperature, and skin temperature decrease while sweat rate increases (Fig. 28-14). This acclimation is achieved by physiological adjustments in the sweating mechanism. A person who is acclimated to working in hot environments has a lower sweating threshold, sweats at a faster rate with increased evaporation, and has an increased heat conduction from deep body areas to the skin due to lowered skin temperature. As acclimation changes take

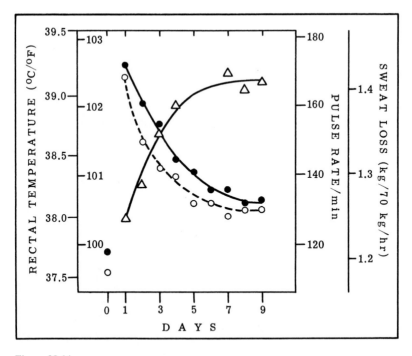

Figure 28-14
Typical average rectal temperatures (●), pulse rates (○), and sweat rates (△) of a group of men during the development of acclimation to heat. On day 0, the men worked for 100 minutes at an energy expenditure of 300 kcal per hour in a cool climate; the exposure was repeated on days 1 to 9, but in a hot climate of 48.9°C. (From Carlson and Hsieh, 1970.)

place, the individual develops a feeling of comfort and well-being instead of heat oppression. The increases in sweating and skin blood flow are the true adaptive responses to heat stress whereas the decreases in heart rate and rectal temperature occur as a result of man's ability to cool his body and to tolerate the work stress. Man cannot, by acclimation, decrease his sweating rate and water consumption.

Cold Acclimation

Acclimation to cold stress by animals is anatomical and physiological. New fur coats are grown, as well as fat layers for insulative protection. Hormonal changes occur that enhance metabolic activity without shivering. This *nonshivering thermogenesis* is moderated through increase in norepinephrine release from sympathetic postganglionic fibers. Thyroxine levels in blood also rise, and thyroxine seems to have a synergistic action with norepinephrine.

Naturally, man does not grow a new fur coat (although his wife may buy one!), and there is evidence that his insulative fat layer decreases; however, he has a higher utilization of thyroxine, and urine levels of norepinephrine increase during prolonged cold exposure, indicating increased production. The aborigine (Australian Bushman) sleeps almost nude, exposing himself to near freezing temperatures, shows no metabolic increase during sleep, but allows his body to cool at night, thus conserving energy. He replenishes his body heat in the early morning by exercising while running naked (streaking) across the Australian bush country. The ama (Japanese and Korean shell divers) show considerable seasonal temperature regulatory variations. They have lower skin blood flow and heat conductance to the skin during winter, and they show what appears to be nonshivering thermogenesis.

FEVER

Fever is an elevation of normal body temperature that is *not* related to work, exposure to hyperthermic conditions, or breakdown of the thermoregulatory system. It involves the mechanisms that establish the central reference or set point and those that regulate temperature around that level.

Common causes of fever are infection, primary neurological disorders, and dehydration. Fever associated with infection may be prolonged or intermittent. The chemical agents that give rise to fever are classified as pyrogens: bacterial pyrogens (endotoxins from gram-negative bacilli) or endogenous pyrogens (leukocyte extracts). The exact mechanism of action of these agents is not known, but indications are that they affect the firing rates of preoptic neurons, which set the temperature reference point at a new high level. The events shown in Figure 28-15 describe the time course of a typical febrile episode. Presence of the febrile agent in the blood shifts the central set point to a higher level. Hypothalamic integrator centers compare peripheral or central thermoreceptor temperature information with the new set-point temperature and register that the body temperature is too low. At the onset of fever, the individual responds physiologically as if he were in the cold. He commonly has a chill and gooseflesh; peripheral vasoconstriction occurs; nonevaporative heat loss is reduced; temperature starts to rise. Feeling cool, the individual may protect himself by turning up the heat, putting on more clothes, etc. Some time later (10 to 60 minutes) shivering may occur with a resultant heat production and temperature rise until it reaches the new set point. Regulation at the new level will persist until the so-called crisis occurs. When the febrile agent is no longer effective or present, the set point is shifted to the lower normal value and the physiological responses are similar to those that take place during heat exposure; i.e., both evaporative and nonevaporative heat loss mechanisms are activated. Peripheral vasodilation and sweating occur, increasing heat loss; body temperature returns to the normal level; the febrile episode is over.

Clinically, the presence of fever is a reliable and common sign of disease, but over prolonged periods it is considered a disadvantage due to adverse effects on the nervous

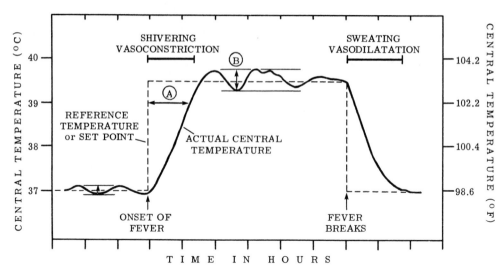

Figure 28-15
Time course of typical febrile episode. The actual body temperature lags behind the rapid shifts in set points (A). Note the regulation is maintained during the fever but is less precise, so that temperature fluctuations are generally greater than normal (B). (From G. Brengelmann and A. C. Brown. Temperature Regulation. In T. C. Ruch and H. D. Patton [Eds.], *Physiology and Biophysics* (19th ed.). Philadelphia: Saunders, 1966.)

and other organ systems. Historically, the crisis was a major recovery sign; today, however, the febrile state is reduced by antipyretic drugs long before the terminal or crisis stage occurs. Before the days of modern drugs, fever therapy was used for treating various diseases. Patients with neurosyphilis were treated by raising body temperature to 40 or 41°C. They were inoculated with tertian malarial parasites or typhoid to raise deep body temperatures. Needless to say, such treatment was not without its own adverse effects, and thanks to the simplest of antipyretic drugs (aspirin) many of the long and sometimes traumatic ordeals experienced by the patient and physician have been alleviated.

DISEASES THAT AFFECT TEMPERATURE REGULATION

Pathophysiological conditions that prevent heat loss or heat production and lead to thermal discomfort influence temperature regulation. Morphological and physiological disorders that prevent the transfer of heat from active cells to the skin, and those that reduce the heat exchange with the environment, reduce man's ability to tolerate heat stress. Similarly, disorders that lower metabolic activity and increase heat exchange with the environment will lower man's tolerance to cold stress. The categories of heat illnesses generally recognized are (1) skin disorders (including sunburn and prickly heat); (2) heat syncope (fainting) resulting from the combined effects of extensive peripheral vasodilation, orthostatic hypotension, and cerebral ischemia; (3) heat stroke, primarily a result of man's inability to sweat, causing brain temperatures to rise to critical levels (41°C) and bringing on unconsciousness, coma, or convulsions—and death if body temperature is not promptly lowered; (4) heat exhaustion and heat cramps, primarily a malfunction related to water and electrolyte metabolism. Sedative and tranquilizing drugs suppress or interfere with temperature regulation in that they cause vasodilation, decreased blood pressure, and lower heart rates. Such effects will predispose an individual to heat syncope and lower his tolerance to cold stress.

Idiopathic malignant hyperthermia during anesthesia, although uncommon, is a striking and highly mortal condition. Patients may react to an anesthetic agent by increasing metabolism, thereby raising body temperature to critical levels. Temperature rises very rapidly (6 to 8°C per hour), and rigorous cooling methods must be used to reduce body heat.

Cold illnesses or disorders influencing man's ability to tolerate cold stress generally involve peripheral circulatory or cardiac defects. Hypersensitivity of the arteries in the fingers to cold (Raynaud's disease) reduces hand and finger blood flow to such critical levels that pain, local asphyxia, and gangrene will occur if the stress is continued. Inflammation of arteries and veins in the extremities (Berger's disease) will result in permanent luminal obstruction, reducing blood flow and heat transfer to and heat loss from the afflicted segment. Cold exposure in some individuals induces *hemagglutination,* i.e., clumping of red blood cells in small superficial vessels which reduces cutaneous blood flow and heat transfer. The incidence of hemagglutination is higher in subjects who have had *frostbite,* and these individuals are hypersensitive to cold. This increased sensitivity to cold is probably due to anatomical rather than physiological changes since naked nerve endings and thermoreceptors are closer to the skin surface.

REFERENCES

Adolph, E. F., et al. *Physiology of Man in the Desert.* New York: Interscience, 1947.

Bullard, R. W. Temperature Regulation. In E. E. Selkurt (Ed.), *Physiology* (3rd ed.). Boston: Little, Brown, 1971.

Burton, A., and O. G. Edholm. *Man in a Cold Environment.* London: Arnold, 1955.

Carlson, L. D., and A. C. L. Hsieh. *Control of Energy Exchange.* London: Macmillan, 1970

Dill, D. B. (Ed.). *Handbook of Physiology.* Washington: American Physiological Society, 1964. Section 4: Adaptation to the Environment.

DuBois, E. F. *Fever and the Regulation of Body Temperature.* Springfield, Ill.: Thomas, 1948.

Greenfield, A. D. M. The Circulation Through the Skin. In W. F. Hamilton and P. E. Dow (Eds.), *Handbook of Physiology.* Washington: American Physiological Society, 1963. Section 2: Circulation, vol. 2, pp. 1325–1353.

Hammel, H. T. Regulation of internal body temperature. *Annu. Rev. Physiol.* 30:641–710, 1968.

Hardy, J. D., A. P. Gagge, and J. A. J. Stolwijk. *Physiological and Behavior Temperature Regulation.* Springfield, Ill.: Thomas, 1970.

Kuno, Y. *Human Perspiration,* Springfield, Ill.: Thomas, 1956.

Selle, W. A. *Body Temperature: Its Changes with Environment, Disease, and Therapy.* Springfield, Ill.: Thomas, 1952.

Strom, G. Central Nervous Regulation of Body Temperature. In J. Field (Ed.), *Handbook of Physiology.* Washington: American Physiological Society, 1960. Section 1: Neurophysiology, vol. 2, pp. 1173–1196.

29. Physiology of Exercise*

William V. Judy

Skeletal muscle contraction is the primary physiological event in exercise, with all other body systems playing a supportive role. The complex events associated with the excitation-contraction coupling mechanism of skeletal muscle were described in Chapter 3, and those concerning the conversion of food substances into chemical energy required by muscle cells to produce mechanical energy and heat were described in Chapter 27. Scores of enzymatic reactions are required in proper sequences before muscle contraction can occur. Activity changes and integration of all body systems occur simultaneously to support active muscle, thereby allowing the continuation of repetitive contraction-relaxation for extended periods. Exercise places demands on all body systems, beginning with conscious or subconscious thought processes, and involves impulses from the central nervous system (CNS) which initiate coordinated muscle, cardiovascular, pulmonary, and other system activity. Greater activity of circulatory and respiratory function is necessary during exercise to provide oxygen and nutrients to active muscle cells and to remove carbon dioxide, other metabolites, and heat. A firm understanding of exercise physiology permits the physician, nurse, and life scientist to evaluate the maximum physical exertion levels of different age groups, and with such knowledge to designate the type and amount of activity patients with cardiac, pulmonary, and other diseases should be allowed during convalescence and rehabilitation.

METABOLIC ASPECTS OF EXERCISE

Energy Source During Exercise

The primary energy source for skeletal muscle contraction is high-energy phosphate units. These are produced from both aerobic and anaerobic metabolism (Fig. 29-1). The aerobic (oxidative) metabolism of glucose or glycogen, lipids, and proteins in the Krebs citric acid cycle produces 38 high-energy units, CO_2, and H_2O, when the electron transport is complete. Anaerobically, these units are produced from glycolysis, which consists in the phosphorylation and breakdown of glycogen to pyruvate or lactate. A net yield of only 2 high-energy units results from a breakdown of each 6-carbon unit compared to the 38 produced aerobically. These units are stored in skeletal muscle as molecules of *adenosine triphosphate* (ATP) and *creatine phosphate* (CP). The actual energy required for the sliding of the thin actin filament over the thick stationary myosin filament is provided by the hydrolysis of ATP. When hydrolysis occurs, mechanical energy of contraction, thermal energy (heat), ADP (adenosine diphosphate), and inorganic phosphate are produced.

Approximately 20 percent of the total energy released is mechanical energy of contraction; the remaining 80 percent is thermal energy. The high thermal energy release during exercise places additional demands on the thermoregulatory system: specifically, the cardiovascular system, which transports heat to the body surface, and the sweat glands, which play a major role in the heat transfer from the body surface to the environment (see Chap. 28).

Continuation of the contractile process requires that the tissue stores of ATP be maintained. These are quickly regenerated from the second high-energy source stored in muscle, CP. This high-energy unit in the presence of ADP yields new ATP units and creatine. As exercise becomes more severe or extended, aerobic metabolism gives way to

---

*This chapter is enlarged from E. E. Selkurt, *Basic Physiology for the Health Sciences.* Boston: Little, Brown, 1975. Pp. 563–587.

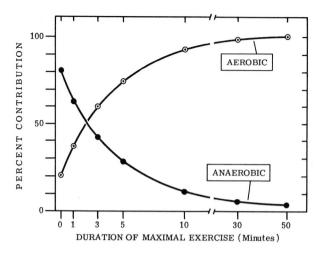

Figure 29-1
The relative contribution of aerobic and anaerobic energy processes during various periods of maximal exercise. Note that the energy for long-term exercise is provided aerobically whereas that required for maximal short-term exercise is provided anaerobically.

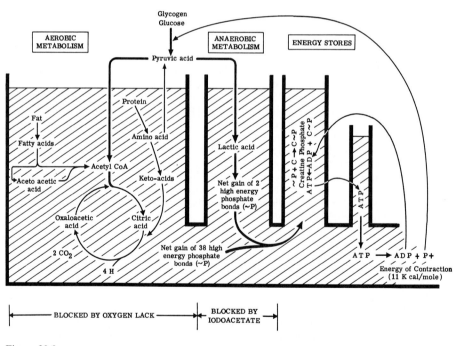

Figure 29-2
Energy sources for muscle contraction. The adenosine triphosphate (ATP) and creatine phosphate (CP) may be considered storage depots for high-energy phosphate bonds. The anaerobic glycolysis and aerobic metabolism function to restore these depots. (From R. W. Bullard. Physiology of Exercise. In E. E. Selkurt [Ed.], *Physiology* (3rd ed.). Boston: Little, Brown, 1971.)

anaerobic conditions because of the lack of available O_2 at the cellular level. In this event, CP is produced anaerobically from glucose and glycogen in addition to pyruvic and lactic acid. Through this pathway tissue ATP stores are replenished anaerobically from the available CP. The CP content of resting human quadriceps muscle is about 8 mg per 100 gram tissue (dry weight), and the ATP content is about 2 mg per 100 gm tissue. During light to moderate exercise the CP content falls to 2 to 4 mg per 100 gm tissue, and during heavy exercise to near zero while the ATP level remains fairly constant. Thus, CP breakdown must quickly serve to replenish tissue ATP stores.

Normally, aerobic ATP production is the major source of mechanical energy during light exercise; however, in short-term maximal exertion or in events in which exhaustion occurs, anaerobic ATP production is the major energy source (Fig. 29-2). In long-term or extended work or exercise such as a cross-country run, aerobic energy production must dominate. In such cases the individual, through training or experience, learns to pace himself so that the cardiopulmonary delivery of O_2 to the active skeletal muscle cells meets the energy demand. A well-conditioned athlete should have considerable metabolic reserve and therefore be able to perform maximally for a short interval through anaerobic energy production. The amount of reserve should determine the duration of the final maximal exertion or "kick" associated with most distance running, provided that sufficient energy (substrate sources) is still available at the cellular level.

Energy Requirements During Exercise

The amount of energy required to do a particular task is proportional to the intensity and duration of the event (Fig. 29-3). Work or exercise classified as *light* to *moderate*,

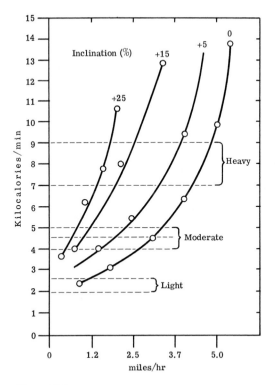

Figure 29-3
Steady-state energy cost for various work loads, and the effect of speed and incline on energy cost of walking on a motor-driven treadmill. (Data from D. B. Dill, *Physiol. Rev.* 16:263, 1936; and from R. Margaria, *Att. des Lancei* 7:299, 1938.)

which includes most everyday occupations except manual labor, requires an energy expenditure of three times the basal (resting) requirements—i.e., about 4 kcal per minute or 1900 kcal in an eight-hour working day. *Hard work,* which includes manual labor (heavy industry, farming, mining, etc.), requires up to 9 kcal per minute or 4300 kcal in an eight-hour workday. Assuming that 500 kcal is expended during sleep and 1400 kcal in the nonworking hours, the total daily energy requirements are about 3800 and 6200 kcal for the moderate and heavy work levels, respectively. Individuals accustomed to hard work can maintain the required energy output for extended periods. Greater activity levels or *maximum* working ability can be sustained for only short intervals before exhaustion occurs. Exhaustion may result from the lack of available energy reserves, insufficient delivery of O_2 to the cellular level, high thermal loads, or accumulation of anaerobic metabolites such as lactic acid in active muscles. Metabolites diffuse from the cellular to extracellular compartments, disturbing the normal homeostatic equilibrium.

The energy (kilocalories) required to do a specific task can be easily equated to the work produced or done in moving a known mass a specific distance. For example, if you are a 50-kg individual who would like to climb a hill 100 meters high to enjoy the view, you would do approximately 5000 kg-meters of work: 5000 kg-meters = 50 kg × 100 meters. Since to do 427 kg-meters of work requires about 1 kcal of energy, you need 11.7 kcal to climb the hill.

$$\frac{5000 \text{ kg-meters}}{427 \text{ kg-meters/kcal}} = 11.7 \text{ kcal}$$

The same task would require about 16.4 kcal for a 70-kg person. Assuming that all this energy was produced aerobically, about 2.5 gm of glucose and 2.3 liters of O_2 would be necessary for the 50-kg person and approximately 3.5 gm glucose and 3.5 liters of O_2 for the 70-kg person (1 gm glucose = 4.7 kcal; 1 liter O_2 = 5.0 kcal). However, since the hydrolysis of ATP to contractile energy is only one-fifth (20 percent) of the total energy production, it will take about five times the original amount of glucose and oxygen to produce the 11.7 and 16.4 kcal of energy, or 58.5 and 72.0 kcal total energy costs, respectively.

The 20 percent mechanical energy produced from the hydrolysis of ATP shows that the *mechanical* or *gross* efficiency (ME) of man is about 20 percent, where efficiency is calculated from the output-to-input energy ratio. Therefore, the percent ME is calculated as

$$\text{Percent ME} = \frac{\text{External work (kcal)}}{\text{Total energy required (kcal)}} \times 100$$

This value is equivalent to that of most man-made combustion engines. Efficiency values up to 30 percent can be obtained in well-conditioned athletes by calculating the *net* and the *absolute* efficiency. The net efficiency takes into consideration the basal metabolic activity

$$\% \text{ E}_{(net)} = \frac{\text{External work (kcal)}}{\text{Total energy (kcal)} - \text{basal energy (kcal)}} \times 100$$

and the absolute efficiency is the energy difference between two measurable work levels

$$\% \text{ E}_{(absol)} = \frac{\text{High work level (kcal)}}{\text{Total kcal (high level)} - \text{total kcal (low level)}} \times 100$$

Exercise or work efficiency varies with the speed of muscle contraction, the load against which the muscle contracts, the fatigue level, training, and the metabolic level being used by the body to perform the work. Calculated efficiency can be improved to a certain extent by training or practice. The trained athlete develops coordinated muscle activity or rhythms that allow him to obtain the maximum functional output from muscle contraction. On the other hand, the untrained individual will waste a lot of energy through unnecessary muscle movements even though he may be in good physical condition. Studies have shown that the trained swimmer may possess an efficiency of at least six times that of an untrained swimmer. Mechanical efficiency is also related to the rate of work or velocity of muscle contraction. Very slow and extremely fast movements are inefficient. For example, climbing a stairs at 50 steps per minute is more efficient for a young person than doing it at a slower or faster rate. The voluntary control of static muscle contraction or "holding back" generates a good deal of thermal energy but not useful work and therefore reduces efficiency. At faster rates of movement, efficiency is reduced because greater forces are required for acceleration and deceleration of the limbs and because the individual is fighting, more and more, the intrinsic or viscous resistance of the contractile and supportive tissue elements. The product of force and velocity is the rate of work or power output; it can be derived from the force-velocity curve for the human forearm as shown in Figure 29-4. The maximum rate of power production that a muscle can sustain is shown with peak power output corresponding to loads of about 30 percent of isometric maximum. Similarly, when man bicycles at a maximum work rate, his leg extensors produce tensions of up to about 30 percent of isometric maximum. The same results have been found for the isolated frog skeletal muscle, and there is a definite inverse relationship between the energy cost of exercise and the mechanical efficiency.

Exertion, as in most sports, requires even greater energy output. Superior athletes can increase their energy output 13 to 14 times their basal metabolic rate. Basal metabolic rate for males is approximately 72 kcal per hour (38 kcal/m^2/hour) and slightly lower for females (Chap. 27). The total energy output for male athletes may exceed 1000 kcal per hour (72 kcal/hour $\times$ 14 = 1008 kcal/hour). Similarly, well-conditioned female athletes (swimmers, for example) can achieve energy outputs of 780 to 870 kcal per hour. Less demanding sports, such as archery, require outputs of only 105 kcal per hour. Obviously, the total energy required to climb the hill previously mentioned (58.5 kcal) now seems minor. However, if you assume that you climbed the 100 meters (328 feet) in eight minutes, it would require about 440 kcal to continue to climb for an hour. These values are based on continuous exercise at fairly constant rates. To be sure, most contact sports do not require sustained higher energy outputs, nor do most track and field events. The actual number of calories used in the performance of a particular event is not very large when compared to a total daily requirement. With the exception of long-distance running (marathon race or cross-country run), the total caloric output for

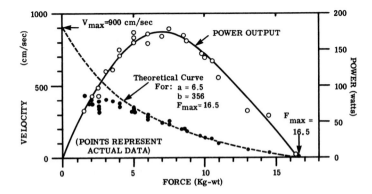

Figure 29-4
Force-velocity and power output relationships of the human forearm muscle. (From R. Meiss, unpublished data.)

most sustained (4 to 10 minutes) athletic events requires only 400 to 800 kcal (200 to 400 kcal/m²). Therefore caloric requirements are not themselves a limiting factor in exercise tolerance. However, a high caloric intake is necessary for training since daily training events may last two to four hours. It takes only about 800 to 1000 kcal for an individual to walk at a rate of 2 miles per hour for six hours (Fig. 29-3; zero incline). Assuming that this is the only caloric utilization in excess of daily requirements, caloric intake should include this amount plus what is needed to repair or return body tissues to some equilibrium condition.

Oxygen Requirements During Exercise

Representative oxygen consumptions are shown in Table 29-1. These are shown for basal conditions, light exercise, and heavy exercise. Note that the basal oxygen consumption (250 ml per minute) is approximately the same for all individuals; for females it is slightly lower. An untrained person can increase oxygen consumption 3-fold (750 ml per minute) during light exercise and 8- to 12-fold (2 to 3 liters per minute) during heavy exercise. Well-trained athletes may increase their oxygen consumption 16 to 20 times (4 to 5 liters per minute), whereas superior individuals or champions may increase theirs 25 times (6.2 liters per minute). Normally, oxygen consumption capacity is expressed in units of body weight (ml O_2/kg/min), and average values of 44 to 51 ml are observed for untrained males and 35 to 43 ml for untrained females. Values between 50 and 60 ml are obtained for average males in good condition. Males normally have aerobic oxygen consumption capacities during exercise about 20 to 25 percent greater than females. The highest measured oxygen consumption on record is 85 ml for a champion cross-country male skier and 65 ml for the champion female skier. Naturally, age is a very important factor in one's exercise tolerance. Oxygen consumption values given thus far have been for 20 to 29-year-old individuals. With age, maximum oxygen consumption decreases for both male and female; for example, the range is 21 to 42 and 25 to 44 ml for 50 to 60-year-olds, females and males respectively.

Similarly, training increases the exercise efficiency and stabilizes the amount of O_2 required to do a specific task. For example, after 2.5 to 3.0 months of training to walk on a treadmill at a rate of 8 miles per hour (zero incline) for 10 to 15 minutes, the mean O_2 consumption for nine men decreased from 47 liters per kilogram per minute to less than 44. The same individuals on an initial exposure to running 7 mph (9 percent grade) had a mean O_2 consumption of about 53 liters per kilogram per minute and they were exhausted after five minutes of effort. However, after two months of training they could run at the same rate and level for 15 minutes without exhaustion because their aerobic metabolic limit was increased to above 56 liters per kilogram per minute. The physiological factors that may limit O_2 consumption in man are (1) the rate of cardiovascular transport to the active tissue, (2) the O_2 utilization by the active cells, and (3) O_2 diffusion capacity in the lungs.

In addition to physical fitness and age, disease and induced inactivity greatly influence man's maximum oxygen consumption capacity. Maximum oxygen consumption is

Table 29-1. Oxygen Consumption (liters/minute)

Individual	Basal	Exercise	
		Light	Heavy
Male athlete	0.25	0.8–1.2	3.5–5.0
Male normal	0.25	0.7–1.0	2.0–3.0
Woman athlete	0.23	0.65–1.0	2.5–3.2
Woman normal	0.23	0.50–.90	1.8–2.2

decreased during long-term inactivity. Young and old patients confined to bed rest or inactivity lose their physical stamina and become easily fatigued with minimal exertion. How quickly they recover depends on the type and severity of their illness, their age, and their willingness to work hard and to endure the extreme exhaustion that accompanies minimal activity after prolonged bed rest. For example, healthy young men confined to bed rest for 20 days showed a 29 percent decrease in maximum oxygen consumption capacity and a 23 percent decrease in the volume of air they could move out of the lungs per minute (maximum expiratory volume). Although these individuals were not ill, they recovered slowly (10 to 20 days), and after 60 days of physical training they were able to increase their oxygen consumption capacities and maximum expiratory volumes 20 and 10 percent respectively above their pre–bed rest control levels. Naturally, patients recovering from long-term illness or even minor surgery would not be able to achieve the same degree of recovery nor would they be able to do so in such a short time. Recovery and the establishment of a high maximum oxygen consumption capacity is a slow process, and the upper limit of oxygen consumption is *one* of the best indices of physical fitness. But having a high maximum oxygen consumption does not necessarily mean that a person will have a high performance or be a superstar.

There is no good correlation between maximum oxygen consumption and performance. Many individuals can, through physical fitness, obtain a high oxygen consumption but have mediocre to poor performances. Other factors, less objectively measured, are also important: motivation, ability, efficiency, and training. During maximum exertion, the main thing, as far as oxygen intake is concerned, is its availability at the cellular level. If the amount of O_2 consumed equals that required for all metabolic activity, then aerobic ATP production is furnishing the necessary energy, and a *pay as you go* condition exists. Such a condition occurs only during light to moderate work. As stated previously, when oxygen delivery to the active cells falls behind the demand, anaerobic ATP production is the major contributor of energy. This can be considered a *buy now, pay later condition,* because the energy stores used must be paid back after the end of exercise. The magnitude of the energy stores used anaerobically that must be paid back is equivalent to the *oxygen debt.*

Oxygen Debt During Exercise
The oxygen debt incurred during moderate to heavy exercise is shown in Figure 29-5. During the initial minutes, oxygen consumption increases abruptly and it plateaus at some steady maximum intake level. After exercise has stopped, consumption declines slowly for several minutes to an hour. The oxygen debt is defined as the oxygen consumed after exercise above the pre-exercise control or basal consumption. In Figure 29-5, area A represents the basal consumption utilizing aerobic energy production; area B represents the incurred debt where the rate of O_2 uptake attains a value appropriate for the level of O_2 expenditure, the so-called steady state. The repaid debt is represented by area D, where the extra oxygen consumed is to repay that incurred in B. Characteristically, the debt is paid rapidly at first and then at a slower rate.

The increased O_2 debt, as measured by the rate of O_2 consumption after stopping an exercise event, probably has three components: (1) a fast component which is paid back in the initial two minutes, probably by means of a small net decrease in O_2 stored in venous blood and muscle hemoglobin (100 ml O_2 in moderate exercise; 250 ml O_2 in exhausting work); (2) the rapid resynthesis of high-energy phosphate bonds of ATP and CP; and (3) the slow removal of lactic acid formed from pyruvate in the aerobic breakdown of glycogen. Although the O_2 debt is generally attributed to the cost of oxidization and reconversion of lactic acid to replenish the tissue stores of high-energy phosphates, there is no good correlation between O_2 debt and blood lactate content until the O_2 consumption reaches 2.5 to 3.0 liters per minute (Fig. 29-6). Note that at low levels of work ($\dot{V}o_2$ less than 2.5 liters per minute) an O_2 debt occurs without a significant increase in blood lactate. This is known as the *alactic O_2 debt,* which is generally attributed to a restoration of the depleted body O_2 stores. With more severe

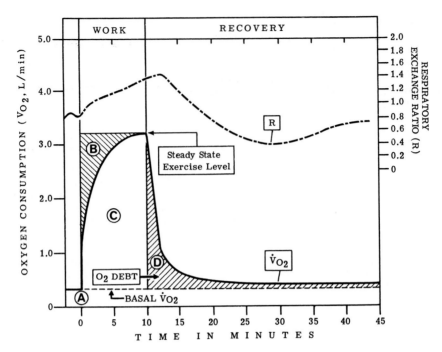

Figure 29-5
Changes in oxygen consumption ($\dot{V}_{O_2}$) and respiratory exchange ratio (R) before, during, and after heavy exercise. Area A represents basal or resting O_2 consumption; area B, the incurred O_2 debt; area C, the period in which the O_2 uptake attains a value appropriate to the O_2 expenditure; and area D, the repaid O_2 debt. (From R. W. Bullard. Physiology of Exercise. In E. E. Selkurt [Ed.], *Physiology* (3rd ed.). Boston: Little, Brown, 1971.)

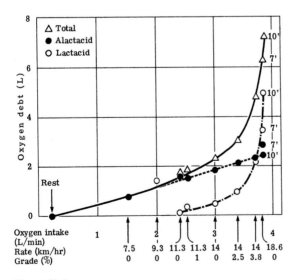

Figure 29-6
Amounts of alactacid and lactacid O_2 debt as a function of O_2 consumption (intake) and work rate on treadmill. (From R. Margaria et al. *Am. J. Physiol.* 106:689, 1933.)

exercise, the O_2 debt and blood lactate levels increase proportionally, and there is a good correlation between alactic O_2 debt and the amount of O_2 required to regenerate the CP store to its original level.

The buildup of an oxygen debt to a critical level appears to be an important but poorly understood factor limiting the duration of heavy exercise and representing one form of exertion. The tolerance of an individual to oxygen lack and his ability to reach a high debt before exercise is stopped is another important criterion of physical fitness or stamina. Untrained individuals can tolerate a maximum of only 10 liters oxygen debt before stopping on account of fatigue, and in some cases nausea and vomiting. The debt ceiling for superior athletes may be as high as 17 to 18 liters. The determinant of the maximum obtainable debt, whether muscle glycogen or creatine phosphate depletion, or other factors, has not been fully evaluated. It is obvious that if the exercise is of long duration (distance running), the individual cannot allow the continuous accumulation of the debt but must work aerobically or at a rate low enough for cardiopulmonary delivery of oxygen to active cells to parallel their oxygen requirement or *cost* (Fig. 29-2). The oxygen cost is that consumed above control or the basal state during exercise plus the oxygen debt. If the total oxygen consumption exceeds that required to perform a particular task in a specific time interval, that task should be easily performed. If the cost is greater than the total consumption, the participant should not be able to finish it within the specific time (running a four-minute mile).

Fuel of Exercise and Respiratory Exchange Ratio

Both carbohydrates and lipids are utilized as fuel during exercise in proportions varying with work intensity, duration, and availability of stored fuels. At low rates of activity, lipid supplies a large portion of the required energy. In heavy exercise (i.e., that consuming above 70 percent of maximum O_2 consumption), the fuel is carbohydrates derived primarily from glycogen stored in skeletal muscle. It appears that the body can substitute carbohydrate for lipid, but lipid cannot be substituted for carbohydrate, particularly at high rates of work. During severe exercise, muscle fatigue is probably due to exhaustion of muscle glycogen stores; however, individuals can continue to exercise at a lower rate by utilizing available lipids. In short-term maximum exertion, such as running the 100-yard dash in less than 10 seconds, the delivery of O_2 by the cardio-pulmonary system to the activity cell is the major limiting factor. In this case a tremendous O_2 debt (5 to 6 liters) is incurred in a short-term effort. The metabolic substrates for aerobic energy production are plentiful. However, their utilization is limited by the available O_2. On the other hand, the availability of tissue oxidative metabolic substrates is probably the major limiting factor in long-term athletic events such as the cross-country and marathon run. In these events athletes through training build up their maximum oxygen consumption capacity and learn to pace themselves so that they primarily work aerobically. Fatigue and exhaustion result from depletion of readily available metabolic substrate stores of carbohydrates or glucose. It is possible to build up or overfill these in the tissue stores by dietary selectivity 10 to 20 hours before the exercise event. For example, many present-day Olympic-caliber distance runners voluntarily increase their carbohydrate consumption by eating spaghetti, crackers, or pizza after their last training event.

Recall that the fuel oxidized during basal steady-state conditions can be determined by the respiratory quotient (volume expired CO_2/O_2 consumed, Chap. 27). During the non-steady-state condition of exercise this ratio is called the *respiratory exchange ratio* (R) and does not indicate the fuels being oxidized. During exercise, R indicates immediate alterations of respiratory gas exchange. Just prior to the onset of exercise some individuals hyperventilate, anticipating the respiratory demands to follow. This action results in a pumping out of CO_2 stores and may temporarily increase R (Fig. 29-5). Further increase in R may be seen during exercise as lactic acid begins to build up in active cells and diffuses into the extracellular spaces. Lactic acid acting as a fixed acid forms carbonic acid, which in turn breaks down into water and carbon dioxide:

Muscle activity

$$\text{Lactic acid} + Na_2CO_3 \rightleftharpoons \text{Na lactates} + H_2CO_3$$
production

$$H_2CO_3 \rightleftharpoons H_2O + (CO_2) \text{ expired}$$

The carbon dioxide is then expired (compensation for metabolic acidosis), thereby raising R. The H^+ content of body fluids increases during exercise to a point where pH may drop 0.6 to 0.7 unit before compensation occurs. Tolerance to low pH seems to be an acquired characteristic developed during training for distance running. Trained individuals can tolerate plasma pH levels of 6.7 to 6.8, normally considered incompatible with cellular function. R may keep increasing for a short time following exertion while O_2 consumption drops rapidly, whereas lactic acid is still diffusing from cells and forming carbon dioxide. As recovery progresses and lactic acid is removed from body fluids, these reactions are reversed, thereby retaining carbon dioxide in bicarbonate formation and replacing of lactate. With retention of carbon dioxide, R values become extremely low and complete recovery is indicated by a return to normal levels.

Body Temperature in Exercise

Body temperature increases during prolonged exercise in response to 10- to 20-fold increase in metabolic activity. Since 80 percent of the total energy expenditure is thermal energy or heat, this energy must be dissipated from the body surface or added to the body heat stores (Chap. 28). Body temperatures up to 39 to 40°C have been observed frequently during long-term exertion, even in cool environments. Deep body temperature (rectal or core temperatures) of champion athletes may reach the fever range (40–41°C or 105–106°F) after prolonged exertion. Temperature increases during exercise are independent of environmental temperature except at extreme ranges; however, they are dependent on the metabolic or work load. Normally, rectal temperature levels off a couple of degrees above normal regardless of the external environment. The temperature rise is not due to a deficiency in the thermoregulatory mechanism except possibly in very hot and humid conditions in which the efficiency of losing heat by sweating is reduced (Chap. 28).

Whether increased body temperature enhances physical performance is not known. Athletes normally warm up before events to increase muscle blood flow, relax the muscles, and prepare them (they hope) for the forthcoming stress by thermal stimulation of cellular metabolism. In addition, heat buildup in active muscle promotes the unloading of oxygen in the tissue by shifting the oxygen-hemoglobin dissociation curve to the right (Chap. 20). Similarly, increased temperature enhances the diffusion rates of respiratory gases, decreases the viscosity of blood, and relaxes vascular smooth muscle; both of the latter will increase muscle blood flow by reducing flow resistance.

An advantage of high body temperature during exercise is that it increases the thermal gradient between the body surface and the environment. This rise reduces the strain on the thermoregulatory system by enhancing convective and radiative heat loss and lowering the heat removal demands placed on the sweating mechanisms. If temperature did not rise during exercise, skin blood flow would have to be greatly increased in order to carry the large amounts of internally produced heat to the body surface. The consequent reduction in muscle blood flow might deprive active cells of required nutrients and oxygen.

Heat is not always beneficial to the athlete; it may be his worst enemy. To compete successfully in hot environments, an athlete must be acclimated to working in heat; therefore, he must have a well-developed sweating mechanism. Recall that the threshold and the rate of sweating are physiological mechanisms that improve with working or training in heat (Chap. 28). Individuals who train or compete in cold climates have less well-developed or responsive sweating mechanisms. Many athletic events have been decided by the differences in heat tolerance of the participants. For example, note the differences between the Minnesota Vikings (trained in the north) and the Miami Dolphins (trained in the south) during a football game in Miami in mid-January with an environmental temperature of 80°F. The northern team, having less conditioned sweat mechanisms, may suffer from overheating since their sweating mechanisms are slow or poorly tuned. On the other hand, the Dolphins, accustomed to working in heat, will have a definite environmental advantage. The combination of high environmental temperature and greatly elevated heat production rapidly brings on the deterioration of performance. Excessive sweating, however, promotes dehydration. This, accompanied by the rapid reduction of intravascular volume of 10 to 15 percent due to increased filtration into the interstitial compartment, could drastically influence tissue perfusion if not corrected by the use of the old-fashioned water bucket or modern Gatorade. Exercise in cold environments (below 20°C) may result in an increase in plasma volume after an extended athletic event. The lack of sweat loss plus metabolic water production during exercise initially produces excessive interstitial fluid, which rapidly shifts to the vascular compartment after exercise.

RESPIRATORY ASPECTS OF EXERCISE

Minute Ventilation and Respiratory Rate

The increased cellular activity during exercise (i.e., increased O_2 consumption and CO_2 production) places extra demands on the respiratory system. Changes occur in the mechanics of breathing, gas exchange, gas transport, and control of respiration. Pulmonary ventilation (minute volume), which is normally about 5 to 6 liters per minute, may exceed 150 liters per minute in severe short-term exercise. The change is attributed to a fourfold increase in respiratory rate (15 to 50 breaths per minute) and a sixfold increase in tidal volume (0.5 to 3.0 liters per breath). Such changes require greater energy output from the respiratory muscles, increasing the work of breathing 100-fold (Fig. 29-7). During exercise, minute volume increases proportionally to the metabolic demand (O_2 consumption and CO_2 production). Figure 29-8 shows the ventilatory response of normal individuals and well-conditioned athletes. Note that athletes have a more efficient ventilation in that they are able to consume more O_2 and remove more CO_2 per unit volume of expired air each minute.

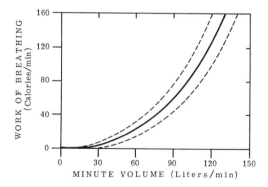

Figure 29-7
The work of breathing as a function of pulmonary ventilation (minute volume). Dashed lines represent estimated range. Under resting conditions with a minute volume of 4 to 5 liters per minute the energy cost for the active portion of breathing is less than 1 percent of resting metabolic activity. During exercise this value increases only to 3 to 5 percent of the maximum aerobic power a person is able to develop. (Redrawn from R. Margaria et al. *J. Appl. Physiol.* 15:354, 1960.)

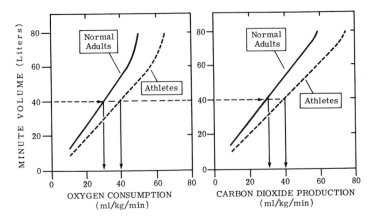

Figure 29-8
Pulmonary ventilation (minute volume) as a function of oxygen consumption and carbon dioxide production during exercise. Note that at a ventilation of 40 liters per minute the athlete consumes approximately 10 ml O_2 per kilogram per minute and produces about 10 ml CO_2 per kilogram per minute (25 percent) more than an untrained adult at the same minute volume. (Redrawn from R. Margaria and P. Cerretelli. In H. B. Falls [Ed.], *Exercise Physiology*. New York: Academic, 1968.)

Similarly, athletes have smaller increases in ventilation rate during exercise than do normal individuals (Fig. 29-9). The rate of breathing increases linearly with pulmonary ventilation; however, athletes have to take fewer breaths to attain the same minute ventilation. It may be concluded that respiratory rate is the major contributor to increased minute volume for the normal individual during exercise whereas tidal volume is the major contributor for the athlete. This difference may be accounted for by two factors: (1) The athlete may have larger upper respiratory airways and therefore lower airflow resistance and greater lung volume change per breath; and (2) he may have greater chest wall and lung compliance and therefore larger lung volume change per unit change in the intrapulmonary pressure (see Chap. 19).

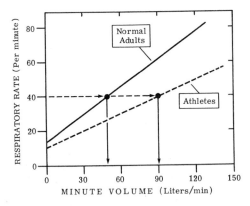

Figure 29-9
Respiratory rate as a function of pulmonary ventilation (minute volume) during exercise. Note that at a rate of 40 breaths per minute athletes have a much greater ventilation (90 to 100 percent) than do untrained adults. (Redrawn from I. Brambilla, P. Cerretelli, and G. Brandi. *Bull. Soc. Ital. Biol. Spec.* 34:1820, 1958.)

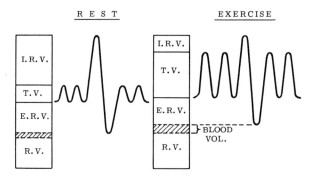

Figure 29-10
Lung volumes at rest and during exercise. IRV = inspiratory reserve volume; TV = tidal volume; ERV = expiratory reserve volume; RV = residual volume; and shaded area = blood volume. Note that the greatest change occurs in TV, which increases at the expense of IRV.

Lung Volume Changes

Lung volume changes during exercise are shown in Figure 29-10; note that lung gas volume decreases slightly because of increased pulmonary blood flow and volume. The major changes occur in tidal volume (TV or V_T) and inspiratory reserve volume (IRV). Tidal volume increases at the expense of IRV, which is correspondingly reduced. The expiratory reserve volume (ERV) and the residual volume (RV) change very little during exercise, provided there is no major change in body position. Changes in body position from standing to supine decrease intrathoracic volume and thereby reduce total lung volume through the influence of reduced gravitational forces on the lung and the abdominal viscera. In prolonged and severe exercise, a slight increase in RV may occur and this may be due in part to more blood in the lung. Pulmonary blood flow increases during exercise, and blood volume distribution changes occur facilitating gas exchange. These aspects will be presented later in the discussion of exercise and ventilation-perfusion.

Forces Involved and the Work of Breathing

In the moving of gas in and out of the lung, work is done to overcome the resistance to air displacement from one position in the lung to another, and to change the position, size, and shape of the lung, rib cage, and abdomen. The driving forces required to do this work are classified as *elastic* (static) and *frictional* (dynamic) (see Chap. 19). The static forces are those required to overcome (1) the elasticity of the chest wall and lung, (2) surface tension of the alveoli, and (3) the gravitational forces on different respiratory structures. The dynamic forces are those required to overcome (1) the viscous resistance of the tissue and (2) the resistance to laminar and turbulent airflow in the pulmonary airways. In healthy individuals, the main opposing forces involved in exercise hyper-ventilation are the dynamic forces required to overcome airflow resistance, and most of the increased work is spent in overcoming this resistance. Recall that the forces required to expand the chest and lung are stored and utilized completely in making passive expiration. During exercise, additional muscle groups are recruited during inspiration and account, in part, for increased work. In normal breathing, the diaphragm, external intercostal muscles, and scaleni muscles (in some individuals) provide the necessary active inspiratory forces, whereas expiration is passive because of the recoiled nature of the lung and thorax. At ventilatory volumes of 50 liters per minute or greater, the sternocleidomastoid, trapezius, and pectoralis muscles of the chest support inspiration; during expiration (at volumes greater than 30 to 40 liters per minute), muscles of the

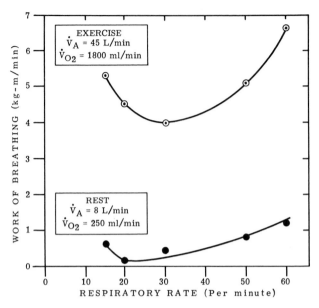

Figure 29-11
Work of breathing as a function of respiratory rate and exercise. $\dot{V}_A$ = alveolar minute ventilation; $\dot{V}_{O_2}$ = O_2 consumption. The minimum point for each condition represents the normal spontaneous rate. Forced hypoventilation or hyperventilation at any exercise level will increase the work of breathing. To convert kilogram-meters per minute to kilocalories (kcal) per minute multiply by 2.34×10^{-3}. (Redrawn from G. Milic-Emili and J. M. Petit. *Arch. Sci. Biol.* (*Bologna*) 43:326, 1959.)

anteroabdominal wall and internal intercostals provide an active expiratory force. Figure 29-11 shows the increased work involved in exercise hyperventilation.

Although there is a very large increase in the work of breathing during exercise, the overall energy required for ventilation is a small fraction of the total energy expenditure. For example, at a ventilation rate of 130 liters per minute, about 0.6 to 0.8 kcal per minute is used for moving air in and out of the lungs. Assuming that this is associated with a maximum aerobic energy production of 13.3 kcal per minute (800 kcal per hour), the work of breathing at this level is only 3 to 5 percent of the total energy expenditure. The percentage can definitely increase in more severe exercise, or where there is some airway obstruction, or where negative pressure breathing occurs. Respiratory rate also affects the work of breathing in that work is elevated at very low and very high rates and attains a minimum value at an intermediate optimal rate. The body tends spontaneously to regulate ventilation at an optimal frequency, which is increased proportionally with ventilation and O_2 consumption (Fig. 29-11).

During exercise, airflow resistance in and out of the lung is reduced by switching from nose to mouth breathing. The size and irregular surfaces of the nasal passages create rather high resistive airflow passages. In addition, as the velocity of airflow in these passages increases, turbulence is induced which also increases airflow resistance. To compensate, an individual who reaches a minute volume of 30 to 40 liters per minute switches from nose to mouth breathing to lower the work of breathing.

Pulmonary Gas Diffusion and Ventilation-Perfusion During Exercise

The diffusion capacity of O_2 and CO_2 across the alveolar-capillary boundaries increases during exercise. Oxygen diffusion capacity rises from an average of about 20 to 25 ml

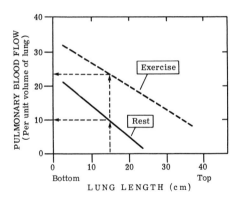

Figure 29-12
Blood flow in different sections of the lung from the apex (top) to the base (bottom), in rest and in exercise. Note that at a 15-cm distance from the bottom, more than twice the resting amount of blood is in the lung during exercise. (Redrawn from J. B. West and C. T. Dollery. *J. Appl. Physiol.* 15:405, 1960.)

per mm Hg pressure gradient at rest to about 80 during exercise. This change is related to the greater number of open capillaries in the lung, increased gas diffusion gradients, and large increases in pulmonary capillary blood volume. Recall that only a fraction of the pulmonary capillaries are open during rest—those in the lower middle or base of the lung (see Chap. 19). Pulmonary blood flow distribution during rest and exercise is shown in Figure 29-12. Note that during exercise the apex of the lung has twice the resting blood flow; therefore, there is a larger lung area for effective gas exchange. During exercise, the ventilation perfusion ratio ($\dot{V}/\dot{Q}$) triples, as does the diffusion capacity, and ventilation and cardiac output increase 15 and 5 times respectively. Although ventilation increases, its distribution in the lung is proportionally equal to that during rest; therefore, with the increased flow and blood volume distribution to the apex a greater gas-blood exchange area occurs, facilitating diffusion (Chap. 20).

Control of Pulmonary Ventilation During Exercise

No single factor has been designated as the primary controlling influence of respiration during exercise. Recall from Chapter 21 that normal respiratory control revolves around the interaction between CNS inspiratory and expiratory centers, pulmonary stretch receptors, and peripheral and central chemoreceptors sensitive to blood gas tension and H^+ concentrations. In exercise, the control mechanisms involved in resting conditions appear to be inadequate. In fact, neither increased arterial CO_2 nor decreased O_2 leads to the high ventilation values observed during exercise. The possibility of other controlling stimuli of a different nature originating from active muscle or from higher CNS areas has been proposed. According to the several theories advanced, the stimuli fall into two main categories, *humoral* stimuli and *neural* stimuli. Figure 29-13 shows the time course of ventilation changes before, during, and after exercise, divided into fast (F, F') and slow (S, S') components. The fast components occur at the onset (F) and at the end of exercise (F'), and the slow components occur after the onset (S) and after stopping (S'). The fast components should realistically be related to nervous stimuli since they occur faster than metabolic changes. The slow components are related to humoral stimuli, which have to be transported to the CNS respiratory centers by the cardiovascular system. The exact nature of both humoral and neural control of ventilation in exercise has not been determined; however, they do not seem to work independently but, occurring together, have a cumulative influence on ventilation.

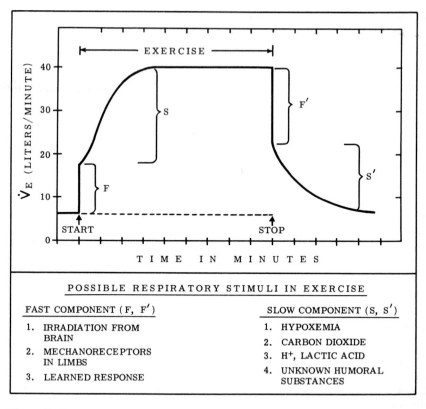

POSSIBLE RESPIRATORY STIMULI IN EXERCISE

FAST COMPONENT (F, F′)	SLOW COMPONENT (S, S′)
1. IRRADIATION FROM BRAIN	1. HYPOXEMIA
2. MECHANORECEPTORS IN LIMBS	2. CARBON DIOXIDE
3. LEARNED RESPONSE	3. H⁺, LACTIC ACID
	4. UNKNOWN HUMORAL SUBSTANCES

Figure 29-13
Ventilatory responses in exercise. (Redrawn from P. Dejours. In J. C. Cunningham and B. B. Lloyd [Eds.], *The Regulation of Human Respiration*. Oxford, England: Blackwell, 1963.)

Humoral Stimuli

Increased Arterial CO_2 Concentration

The CO_2 production is greatly increased in muscle exercise and could lead to an increase in arterial CO_2 concentration of sufficient magnitude to stimulate CNS respiratory centers. Actual measurements of arterial CO_2 concentration indicate that changes in exercise are slight. Small increases often occur during light or moderate exercise, and no change or even a slight decrease may occur during heavy exercise. If CO_2 alone were the primary stimulus, larger increases in blood than those that have been measured would have to take place. For example, inhalation of an 8 percent CO_2, 20 percent O_2, and 72 percent N_2 mixture markedly increases blood CO_2 content and raises ventilation to 100 liters per minute. The striking point is that during exercise trained athletes may attain ventilations of 120 liters per minute without any detectable change in arterial CO_2 content.

Increased H^+ Concentration

The acidification of blood (increased H^+ concentration) due to lactic acid production by active muscle could stimulate respiratory centers and account for exercise hyperventilation. However, only during very strenuous exercise are significant amounts of lactic acid produced, and it has been shown that distance runners (marathon runners) attain only slight increases in blood lactic acid yet have tremendously high ventilation rates. After exercise, lactic acid levels may increase slightly while ventilation rate rapidly falls. It is, therefore, unlikely that blood H^+ concentration alone provides the major stimulus for exercise hyperventilation.

Decreased Arterial O_2 Concentration
Arterial hypoxemia (decreased Po_2) cannot account for more than a few mm Hg changes in arterial oxygen tension (Pa_{O_2}) during exercise since arterial blood O_2 saturation is not appreciably affected in strenuous exercise. Recall that anoxia affects respiration by eliciting chemoreceptor activity from the aortic and carotid bodies (see Chap. 21). The threshold of the receptors is at an arterial blood O_2 tension of approximately 60 mm Hg. Since such Po_2 levels are not reached during strenuous exercise, this mechanism alone cannot provide the necessary stimulation for the hyperventilation.

Circulating Catecholamines
Because the epinephrine and norepinephrine concentrations in blood and urine increase with the intensity of work, and because injected catecholamines produce marked increases in ventilation in resting subjects, these substances may play some role in exercise hyperpnea. The actual mechanism of the response is at present poorly understood but it may involve changes in threshold or sensitivity of chemoreceptors or respiratory centers. It must be concluded that humoral stimuli alone cannot explain the hyperpnea of exercise, but these factors combined with neurogenic stimuli may have a major part in respiratory control during exercise.

Nervous Stimuli

Irradiation of Impulse from Higher CNS
It has been suggested that the sudden increase in ventilation just before and during the onset of exercise is due to impulses arising from the motor cortex. These impulses, which pass through the reticular formation, may irradiate impulses to the respiratory center and evoke a ventilatory increase. There is no doubt that some ventilatory responses are of a volitional nature and learned by experience. They are obvious when one watches a short-distance runner or swimmer hyperventilate just prior to the start of an athletic event. Tests with hypnotized subjects who exercised without full awareness of work intensity have resulted in a lower ventilatory response than would be predicted by metabolic rates.

Peripheral Receptors
Muscle contraction and movement of the extremities may stimulate receptors in joints (proprioceptors) and muscle spindles, which in turn send afferent signals to the respiratory centers. It has been observed that passive limb movement evokes ventilatory response roughly proportional to the number of joints involved in the movement. Such a mechanism may be related to the overall rhythmic pattern or smoothness a superior runner or swimmer adheres to during a race.

Lung Stretch Receptor Activity
Lung stretch receptors when stimulated by filling of the lung send afferent information to the respiratory center, which inhibits inspiration (inhibitoinspiratory reflex) and indirectly activates expiration (see Chap. 21). The excitatoinspiratory reflex may also be activated. The same receptors may play a role in the hyperpnea of exercise by augmenting respiratory rate; however, since voluntary ventilation is not self-perpetuating, this mechanism is not considered a prime stimulus of exercise hyperpnea.

Sensitivity of Respiratory Centers
An increase in respiratory center sensitivity to the amount of CO_2 in the arterial blood could account for exercise hyperpnea, but this would have to be a drastic sensitivity change since arterial CO_2 content in heavy exercise does not seem to decrease much. It

has, however, been observed that if one tries to hold his breath during moderate work, his breaking point occurs at a lower alveolar CO_2 than when at rest.

Body Temperature

Hyperthermia increases ventilation without exercising as a means of losing body heat (see Chap. 28). In exercise, the ventilatory increase is greater than that observed during hyperthermia and occurs before a rectal temperature rises. Since temperature rise increases cellular activity, the hyperthermia of exercise may increase the activity of joint and muscle receptors as well as respiratory center cells.

CARDIOVASCULAR RESPONSE DURING EXERCISE

Microcirculation

Before and during exercise microcirculatory (capillary) changes occur which facilitate the delivery of substances to and from the active skeletal muscle cell. It is estimated that only 20 to 50 percent of the capillaries in resting skeletal muscle are open at any one time. During exercise closed capillaries open as blood flow increases, giving a total estimated capillary surface of 300 to 600 square meters, which considerably shortens the diffusion distance from the capillary to the cell energy unit (mitochondrion). Since the increase in the number of capillary openings and in muscle blood flow is about the same (20 percent of resting value), the velocity of blood flow through the capillaries should not change. Thus transcapillary exchange should be further enhanced by maintenance of the precapillary to postcapillary transit time at approximately one second.

Another factor influencing the delivery of substances to and from the active cell is capillary permeability—i.e., the size and number of membrane pores per unit of capillary surface area. Recall that capillary permeability is not ordinarily affected by the buildup of local vasodilator substances in active muscle (see Chapter 12), but the amount of fluid filtered across the capillary membrane for each mm Hg transcapillary pressure is determined by the total surface area of the open capillaries.

Capillary flow increases during exercise as systolic and mean arterial pressure rises, therefore, increasing blood flow through active muscle. Venous pressure is little affected, and, as a consequence, filtration forces exceed absorption forces (see Chap. 12). This imbalance causes fluid to leave the vascular compartment, upsetting the dynamic equilibrium between outflow and inflow water balance. Mean capillary pressure may rise as much as 10 mm Hg. Assuming a maximum filtration coefficient of 0.04 ml/mm Hg/min/100 gm tissue, circulating blood volume would then be reduced approximately 20 percent (1 liter) in only 10 minutes if the involved muscle mass were 10 kg.

Indeed, intravascular to extravascular fluid shifts occur, as shown by the hemoconcentration (Hct = 56 to 60 percent) and plasma volume reductions of 10 to 15 percent. These observations are more pronounced in long-term exercise events such as distance running and are more obvious in those events performed in cold environments. Naturally, compensatory responses must be activated or circulatory insufficiency would eventually occur.

As fluid enters the interstitial (nonvascular) spaces, the hydrostatic pressure in this compartment rises, and as hemoconcentration occurs in the vascular compartment, plasma colloid osmotic pressure rises. These changes tend to increase reabsorption (Chap. 12) and should therefore strive to maintain effective circulating blood volume; however, complete compensation may not occur because of interstitial protein accumulation and dehydration of 3 to 5 percent of body weight by excessive sweat loss. In addition, fluid return to the systemic circulation is enhanced by the lymphatic circulation, which is aided by the pumping action of contracting and relaxing muscles.

The reduction of circulating blood volume in exercise is mainly compensated for by two mechanisms, one fast and the other slow. The fast mechanism involves the adjustment of the size and volume of the capacitance vessels (venules and veins);

therefore, a reduction in their diameters will greatly increase effective circulating blood volume. These vessels are controlled by changes in vasomotor tone through sympathetic vasoconstrictor activity (see Chap. 7A). Constriction of these vessels during exercise and the force exerted on them by contracting muscle set the volume of the venous compartment and hence venous return and cardiac filling. The slow mechanism (fluid reabsorption) is too slow to aid in early exercise. It obviously has a considerable delay because hemoconcentration of 20 percent may occur early in exercise (10 to 20 minutes).

Cardiac Output

Cardiac output rises during exercise roughly in proportion to the increase in oxygen consumption (Fig. 29-14); however, it does not change to the same extent as pulmonary ventilation and metabolic rate. For example, during moderate to heavy exercise ventilation may change from 8 to 100 liters per minute (12-fold), O_2 consumption from 0.38 to 3.8 liters per minute (10-fold), and cardiac output from 5 to 30 liters per minute (6-fold). The deviation from linearity at the higher O_2 consumption in Figure 29-14 is probably related to thermal regulation at high work levels. As body temperature rises during exercise, skin blood flows must increase so that the high internal heat loads may be transferred to the skin and thus be dissipated (see Chap. 28). Cardiac output during rest and exercise is greatly influenced by body position. Resting output may be 1 to 2 liters per minute greater in the supine position because of the lack of gravitational pooling effects of dependent areas. During heavy exercise, cardiac output changes are less in the supine position but they are nevertheless greater than those for the upright position.

In the transition from rest to exercise cardiac output increases rapidly initially, then plateaus (Fig. 29-15). The rate of initial increase and the peak level are set by the severity of the stress. After cessation of exercise, output decreases gradually toward resting levels, as does oxygen consumption. The recovery similarities for cardiac output and oxygen consumption are undoubtedly related to the incurred O_2 debt. Heart rate and stroke volume influence cardiac output greatly, and the increased output during stress is the product of these two variables.

Stroke Volume
Stroke volume, like cardiac output, depends on cardiac filling; therefore, it is strongly influenced by body position. It is higher in the resting supine position than in standing but changes less during supine exercise. Normal values range from 70 to 150 ml per beat in the resting supine male, depending on body size and whether the individual is an

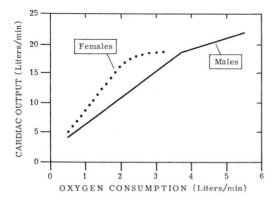

Figure 29-14
Cardiac output at rest (minimal point) and during exercise (increasing O_2 consumption) for males and females. (Redrawn from P. O. Åstrand et al. *J. Appl. Physiol.* 19:268, 1964.)

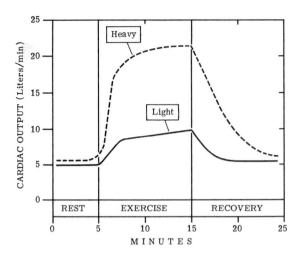

Figure 29-15
Time course of cardiac output changes during and following light and heavy exercise. (From W. V. Judy and F. D. Nash, unpublished data. Measured by impedance cardiograph.)

athlete. Athletes are at the upper range of resting stroke volumes, and females usually have 25 to 30 percent less than males at any position during rest or stress. During exercise in the standing position, stroke volume increases rapidly as venous return is increased by venoconstriction, muscle pumping, and thoracoabdominal pumping mechanisms. It reaches a maximum level (180 to 190 ml per beat) in a short interval (5 to 10 minutes), then plateaus, as does cardiac output. In prolonged exercise events of several hours, stroke volume may decrease 10 to 20 percent below the maximum value; however, cardiac output is maintained fairly constant on account of increases in heart

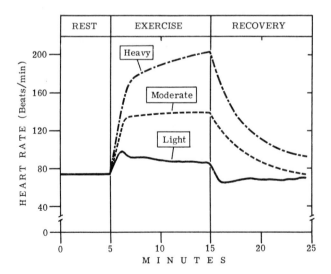

Figure 29-16
Heart rate response during and after light, moderate, and heavy exercise. (From W. V. Judy and F. D. Nash, unpublished data. Measured by impedance cardiograph.)

rate. During supine exercise, stroke volume increases only slightly and is maintained at heart rates of 200 per minute. Evidently the time available for cardiac filling is sufficient at such high rates when the venous return is aided by muscle pumping, thoraco-abdominal pumping, venoconstriction, and reduced gravitational forces on the cardio-vascular system.

Heart Rate

There are little differences between resting supine and standing heart rates or their changes during light or heavy exercise in the respective positions. Athletes have lower resting rates than nonathletes, who do well to increase theirs twofold to threefold before complete exhaustion. Rate may increase in normal individuals from 75 to 100 beats per minute during light exercise, to 130 in moderate exercise, and to near 180 in heavy exercise (Fig. 29-16). Rates of 180 per minute are considered the upper limit for nonathletes, but 200 per minute have often been observed in athletes and children. During light exercise the initial rate increase may be exaggerated, and it is subsequently reduced to a lower steady-state level. In heavy exercise there is a tendency for rate to increase progressively until adequate cardiac output is achieved. After cessation of exercise, rate slowly returns to or below control levels, the rate of return again being proportional to the severity of exercise and the physical condition of the individual.

Blood Pressure in Exercise

Pressures in the systemic arterial system increase during heavy exercise although total peripheral resistance decreases fourfold to fivefold (pressure = cardiac output × total peripheral resistance). Systolic pressure increases more than diastolic and changes at a rate of about 8 mm Hg per 0.5 liter of O_2 uptake (Fig. 29-17). It seldom rises above 180 mm Hg in normotensive individuals but has been observed to rise above 200 mm Hg (50 percent greater than resting) in some athletes. Diastolic pressure does not change much; it may even decrease slightly in light exercise. Since systolic pressure increases more than diastolic, pulse pressure must increase proportionally and mean pressure somewhat less (Fig. 29-17). Venous pressure increases only slightly at the onset but drops to or below resting values during exercise. Therefore, central venous and right

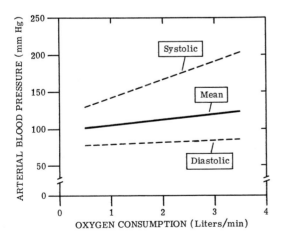

Figure 29-17
Arterial blood pressure during exercise as a function of O_2 consumption. Note the widening of pulse pressure, basically related to a systolic pressure rise and a slight increase in diastolic pressure which reflects the decrease in total peripheral resistance. (Redrawn from S. Bevegaard, A. Holmgren, and B. Johnson. *Acta Physiol. Scand.* 49:279, 1960.)

atrial pressures are not increased, even though venous return and cardiac filling are augmented by the mechanisms previously described. The initial venous pressure rise at the onset of exercise may be due to sympathetic vasoconstrictor activity on the capacitance vessels (veins), causing them to narrow and thus reducing their total volume and increasing blood availability to the heart, or to facilitated venous return due to skeletal muscle and abdominothoracic pumping before the maximum opening of the muscle capillaries.

Pressures in the pulmonary circulation also increase slightly during exercise. Although pulmonary pressures are low compared to those in the systemic arteries (see Chap. 19), small increases may greatly influence pulmonary blood flow distribution by forcing open some of the normally closed capillaries in the middle and upper areas of the lung. Similarly, opening of the capillaries accounts for the marked increase in total lung capillary blood volume during exercise. Volume changes from 60 to 96 ml have been estimated for nonathletes and from 70 to 200 ml for athletes, from rest to exercise respectively. Training or physical conditioning seems to have little influence on various vascular pressures during resting or active conditions.

Regional Blood Flow Distribution

The changes in blood pressure, cardiac output, and cellular oxygen demand during exercise necessitate rather dramatic alterations in blood flow distribution in the body. Although cardiac output increases fivefold to sixfold during maximum exercise, the bulk of this volume flows through active skeletal muscle including the respiratory muscles. Blood flows and the percent of cardiac output for various organs and tissues during rest, light exercise, and heavy exercise are estimated in Table 29-2. Note that brain and kidney flow is maintained during exercise while splanchnic flow is decreased, and skeletal muscle, heart, and skin flows are increased. Obviously, some large vascular beds, as in the viscera, decrease their blood flow so that flow through more vital or active areas may be maintained or increased. Renal flow initially decreases but returns to control during exercise as blood pressure rises, since renal function is damaged by extended periods of low perfusion.

Coronary flow, which is about 250 ml per minute (60 to 70 ml per minute per 100 gm tissue) at rest, may increase four to fivefold or more during exercise. This rise shows the demand of active cardiac cells for oxygen which must be provided to the heart, since myocardial resting extraction of oxygen from capillary blood is 75 to 85 percent, and heart muscle *cannot* utilize the anaerobic glycogenolysis as a major source of energy for the contractile process.

Skin blood flow decreases during the initial moments of exercise, then increases as

Table 29-2. Blood Flow Distribution During Rest and Exercise for a Well-Conditioned Athlete

Area	Rest		Exercise (ml/min, %)			
	ml/min	%	Light		Heavy	
Splanchnic	1,400	24.0	1,100	12.0	300	1.0
Renal	1,100	19.0	1,100	12.0	900	4.0
Brain	750	13.0	750	8.0	750	3.0
Coronary	250	4.0	350	4.0	1,000	4.0
Skeletal muscle	1,200	21.0	4,500	46.0	22,000	85.5
Skin	500	9.0	1,500	14.0	600	2.0
Others	600	10.0	400	4.0	100	0.5
Cardiac output	5,800	100.0	9,700	100.0	25,650	100.0

body temperature rises. In resting adults, total skin flow is about 400 to 500 ml per minute, but when the vessels are fully dilated, as in severe hyperthermia or exercise, it may increase to 3 liters per minute. The vessels of the skin are so constructed and controlled that they play a major role in body temperature regulation (Chap. 28). Blood flow through vascular beds during rest and exercise is influenced by the mechanisms that change the diameters of the resistance vessels (arterioles) and the tightness of the precapillary sphincter. Remember that several mechanisms may be involved: (1) autoregulation, (2) sympathetic tone, (3) circulating hormones, and (4) local metabolites (see Chap. 12). The complex interaction between them makes it difficult to analyze the significance of any one during exercise; however, the dominating factor seems to be sympathetic nervous tone, with some small hormonal and metabolite influence. During rest, there is a high sympathetic tone in muscle arterioles, as signified by the large number of poorly perfused capillaries. Similarly, sympathetic influence on the arterioles of more functional systems such as the heart, kidney, and intestines is small. Therefore, these vessels are open, less resistive, and more easily perfused by the arterial driving pressure.

Circulatory patterns start to change before the onset of exercise in that muscle blood flow increases whereas that to skin, kidney, and intestine decreases. These changes are due to intense sympathetic activity initiated in higher CNS areas which cause constriction of skin, kidney, and intestinal arterioles and dilation of those in muscles (see Chap. 7A). Since the constricted vessels become the most resistive, and the dilated the least resistive, channels to blood flow, pressure created by the pumping action of the heart will force more blood through the less resistive muscle vasculature. This neurogenic mechanism is therefore able to shift or shunt blood flow to more demanding or active areas very quickly. The thought of exercise activates cortical motor areas, which in turn activate hypothalamic and medullary sympathetic constrictor and dilator areas. It has been shown that stimulation of these areas in anesthetized animals increases heart rate and myocardial contractility, constricts resistance vessels in less active organs, dilates vessels in active muscle and heart, and mobilizes extra energy through influences on the liver to release glucose and on metabolism. Inhibition of sympathetic vasoconstrictor activity to resting muscle also enhances blood flow; it is consequently conceivable that sympathetic constrictor activity to muscle during exercise is inhibited, thus allowing flow to increase.

Hormonal control of circulation or blood flow distribution is mainly concerned with epinephrine released from the adrenal gland as a part of the massive sympathetic discharge occurring just before and during exercise. The action of this substance mimics that of sympathetic constrictor influence on less vital systems and dilation of muscle vasculature at low blood levels (see Chap. 7A). The most effective single vasodilator mechanism at the tissue level is that of local metabolites. No one single agent or factor has been confirmed as the primary vasodilator during exercise. Of the many metabolic products, CO_2, lactic acid, histamine, and others have been investigated. Histamine is the most potent vasodilator, but its concentration in muscle decreases during exercise. Several substances may work together to produce local vasodilator control. The local aspects of muscle vasodilation and blood flow distribution during exercise must be emphasized because no evidence exists that metabolites released from contracting muscle circulate to other constricted muscles and organs causing them to vasodilate.

An additional factor influencing muscle blood flow during exercise is the force of muscle contraction against the open blood vessels. Figure 29-18 shows the effects of intermittent contraction on leg blood flow. During rhythmic exercises such as running or walking, rather sharp decreases in flow occur during contraction as the external pressure around the vessels forces blood out of them. During relaxation, equally sharp increases in flow occur since the vascular bed is fully dilated and arterial pressure elevated. Note that the minimal flow during contraction exceeded the controls, and that after exercise ceased, flow was 8-fold to 10-fold greater than control. During recovery, flow gradually decreased over two or more minutes. This sequence shows how dilated the muscle bed

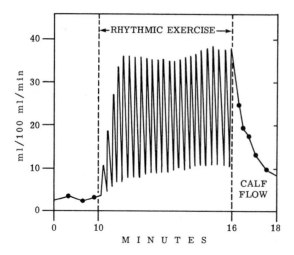

Figure 29-18
Effect of rhythmic muscle contraction on human calf muscle blood flow. (Redrawn from H. Barcroft. *Sympathetic Control of Human Blood Vessels.* London: Arnold, 1953.)

was, the importance of muscle contraction on squeezing blood through the active muscle, and the role of muscle contraction in maintaining venous return.

EXERCISE IN HEALTH AND DISEASE

It is not necessary to have superior muscle strength or physical stamina to be healthy, nor is it true that exercise and physical fitness will protect one from diseases. Most diseases have no human boundaries.

Exercise or physical activity is, however, used in medical science as a diagnostic tool and perhaps more importantly as a therapeutic aid during convalescence and rehabilitation.

Man adapts to his normal daily environment and to the activity therein by building up his muscle strength and exercise stamina according to his daily physical demands. Naturally, then, a wide range of physical fitness can be found among healthy individuals. However, even the most physically fit, such as athletes and astronauts, when placed in a less stressful environment will adapt to the required stress. Bed rest or space flight removes many of the normal daily stresses and leads to physical deconditioning. In addition to the previously stated reduction in maximum oxygen consumption capacity that occurs during bed rest, marked changes occur in skeletal and smooth muscle, bone mineral content, body fluids, and fluid distribution. As little as two days of bed rest or space flight results in an initial diuresis due to vascular fluid displacement to the thoracic cavity; left atrial volume increases, leading to inhibition of antidiuretic hormone release from the posterior pituitary and a resulting reduction in kidney fluid reabsorption. If the condition is prolonged, there is loss of muscle tone and physical stamina, and eventually muscle atrophy occurs. In both bed rest and space flight the lack of gravitational stress on bone induces demineralization, so that calcium is lost; the heart and blood vessels lose their reflex responsiveness, and *cardiovascular deconditioning* is the outcome. Inactivity may lead to constipation, thrombosis due to collapsed venous segments, and eventually psychiatric and neuropsychiatric disorders.

These responses are consistent findings for all age groups. The recovery or readaptation to a more stressful environment may be slow and traumatic. For these reasons, during the last 30 years exercise has become a major therapeutic aid in health and disease. Classically, many physicians advocated complete bed rest for almost all afflictions during convalescence. For countless patients, such as healthy mothers, this was an unnecessary and sometimes harmful kindness. For others, such as those stricken by rheumatic fever, heart attacks, and partial paralysis, it meant a life of inactivity. Today, the physiological and psychological advantages of even the most modest exercise (toe wiggling, coughing) are well documented in the practice of medicine. The medical scientist is aware that the inactivity of bed rest will prolong the recommended therapy and may eventually cause fatal complications. For example, the commonly occurring pulmonary congestion and edema observed in bedridden patients can lead to pneumonia and death. The simple exercises of changing body position and coughing can alleviate this problem. Today, new mothers are up and moving around six to eight hours after child delivery. Patients who have undergone appendectomies and other minor surgery are allowed to sit up as soon as four hours after surgery. In case the patient cannot be gotten up, exercise is still encouraged. Toe wiggling, knee bending, arm stretching, and muscle contraction are encouraged on hourly schedules as preventive as well as therapeutic measures because the sudden exertion of standing or walking after extended inactivity may lead to muscle soreness and fatigue. In many instances this discomfort may reduce the patient's willingness or desire to exercise in order to regain his strength.

Exercise in medicine leading to physical conditioning may take two forms. The first is kinesiological reconditioning—i.e., exercise of a specific muscle group or joint by active or passive movement, massage, or heat in an attempt to restore the function of that unit. The second is physiological reconditioning, in which the entire body is conditioned by graded physical exercise. Here, calisthenics, modified sports, or occupational pursuits are used to restore the physical endurance of the whole body. The growing list of medical specialties, occupational and rehabilitative medicine, physical therapy, occupational therapy, and the appearance of activity rooms and exercise facilities in both old and new medical institutions show the marked awareness of exercise as a major therapeutic tool in health and disease.

RELATION OF METABOLIC, RESPIRATORY, AND CIRCULATORY CHANGES DURING EXERCISE

Exercise, or the ability to perform a specific muscular task, is not simply the end product of the excitation-contraction coupling of the actin and myosin filaments of skeletal muscle. As stated previously, it involves scores of enzymatic reactions occurring in the proper sequence, the central nervous system coordination of all major systems, and local control of microvascular and cellular activity. Therefore, exercise can be described as the sum total of all body systems working together to support active muscle. Psychologically, it starts with the conscious or subconscious thought of having to do a physical task and ends with the final satisfaction or relief that the task is over. The physiological manifestations of exercise discussed in this chapter were primarily concerned with metabolic, respiratory, and circulatory adjustments. Briefly, multiple systems interactions occur; some organs and systems are deactivated or shut down while others are stimulated to increased activity to meet the requirements for continuous muscle contraction. The metabolic increases are supported by similar changes in the systems responsible for oxygen and metabolite transport to and from the active cell. The changes in metabolic activity are closely paralleled by changes in pulmonary ventilation (Fig. 29-8); however, cardiac output does not change in a parallel fashion (Fig. 29-14) but seems to lag behind the metabolic demand. Some investigators feel that the slightly slower response of the cardiovascular system is the weak link in meeting the metabolic demands of active muscle. This is, however, compensated for by several factors which together provide a more efficient oxygen delivery to tissues. The ventilatory increase is supported by an increased oxygen diffusion capacity in the lung, and a similar event occurs at the active cellular level.

In the lung, increased gas and blood volume raises the diffusion capacity by increasing the total capillary-alveolar surface area. Therefore, more red blood cells become oxygen-enriched, and the total oxygen content of blood is greatly increased. At the level of the muscle cells, oxygen delivery is enhanced by the increased metabolism. This lowers the tissue oxygen content, thereby raising the diffusion gradient, which favors quicker and more extensive gas delivery to the cellular mitochondria. In addition, the accumulation of carbon dioxide, lactic acid, heat, and the pH decrease in and around the cell promote the unloading of oxygen by shifting the hemoglobin dissociation curve to the right (see Chap. 20). Another major compensatory factor and perhaps the most important is the pronounced delivery of blood to the active cells by an increase in the number of open capillaries. Arteriolar dilation, along with an increase in total capillary blood volume, notably increases the surface area across which gas can diffuse. At the same time, the distances across which oxygen must travel from the capillary to the cell and carbon dioxide from the cell to the capillary is sharply reduced by the increase in open capillaries. Metabolism, respiration, and circulation tell only part of the story of exercise physiology. It should be obvious that nervous, hormonal, and enzymatic changes are also involved, and that their interaction with metabolism, respiration, and circulation is most important to the total physiological system before, during, and after exercise events. For a more thorough and advanced review of the physiology of exercise, the reader is encouraged to consult the excellent reference list accompanying this chapter.

REFERENCES

Asmussen, E., and M. Nielsen.   Pulmonary ventilation and effects of oxygen breathing in heavy exercise. *Acta Physiol. Scand.* 43:365–378, 1958.

Åstrand, P. O., and K. Rodahl.   *Textbook of Work Physiology.* New York: McGraw-Hill, 1970.

Barcroft, H.   Circulation in Skeletal Muscle. In W. F. Hamilton (Ed.), *Handbook of Physiology.* Washington: American Physiological Society, 1963. Section 2: Circulation, vol. 2, pp. 1353–1368.

Beregard, B. S., and J. T. Shepherd.   Regulation of circulation during exercise in man. *Physiol. Rev.* 47:178–209, 1967.

Bullard, R. W.   Physiology of Exercise. In E. E. Selkurt (Ed.), *Physiology* (3rd ed.). Boston: Little, Brown, 1971.

Chapman, C. B. (Ed.).   Physiology of muscle exercise. *Circ. Res.* 20 (Suppl I): 1967.

Dejours, P.   Control of Respiration in Muscular Exercise. In W. O. Fenn and H. Rann (Eds.), *Handbook of Physiology.* Washington: American Physiological Society, 1964. Section 3: Respiration, vol. 1, pp. 631–648.

Ekelund, L. G.   Exercise. *Annu. Rev. Physiol.* 31:85–116, 1969.

Falls, H. B. (Ed.).   *Exercise Physiology.* New York: Academic, 1968.

Harris, P.   Lactic acid and the phlogiston debt. *Cardiovasc. Res.* 3:381–390, 1969.

Johnson, W. R. (Ed.).   *Science and Medicine of Exercise and Sports.* New York: Harper, 1960.

Kao, F. F.   An Experimental Study of the Pathways Involved in Exercise Hyperpnea Employing Cross Circulation Techniques. In J. C. Cunningham and B. B. Lloyd (Eds.), *The Regulation of Human Respiration.* Oxford, England: Blackwell, 1963.

Wilmore, J. H. (Ed.).   *Exercise and Sports Science Reviews.* New York: Academic, 1973. Vol. 1.

30. The Hypophysis: Neuroendocrine and Endocrine Mechanisms

Ward W. Moore

INTRODUCTION

The degree of complexity of functions observed in man and other vertebrates has been attained in part by the evolution of the two primary integrating systems, the nervous system and the endocrine system. Each participates in the regulation and coordination of the activities of the organism.

The endocrine system is a regulatory system. It functions to maintain the internal environment (the body fluids) at a relatively constant level with respect to volume and concentration, and within the limits of life in the face of changes in the activity of the body. It transmits information by means of chemical messengers, the hormones. These are dispersed without direction throughout the body via the circulatory system, from which they move to act upon genetically conditioned and differentiated cells of the body, the target cells. The effects of the hormones are much more diffuse and relatively slower than are those of the nervous system, which transmits its information rapidly with point-to-point precision via the neurons. The fundamental microscopic anatomy of the glandular elements of the endocrine glands is similar to that of the exocrine glands, with two exceptions. The endocrine gland does not possess a duct system, and each glandular cell of the endocrine system has a surface which abuts a venous sinusoid or a capillary. Thus, the morphology of the system is such that the secretions can be released directly into the circulation.

Emphasis will be placed in this and succeeding chapters on the mechanisms that regulate the rate of secretion of hormones and the role of the hormones in homeostasis. The hormone-secreting elements to be considered are the secretory or glandular cells of the hypothalamus, the hypophysis or pituitary gland, the pancreas, the parathyroid glands and derivatives of the ultimobranchial body, the thyroid gland, the adrenal cortex, the gonads, and the placenta.

Historical Background

Berthold's studies in 1849 were probably the first experimental demonstration of an internal secretion, although they were overlooked as such for nearly half a century. Berthold showed that the grafting of testicular tissue would prevent atrophy of the comb of the capon. Following these studies, nothing approaching the modern concepts of endocrinology was recognized for over 50 years. However, various clinical observations had correlated certain glands with specific disease states and provided the eventual stimulus for the study of the endocrine glands and their functions. In 1902 Bayliss and Starling showed the existence and physiological role of secretin, and the term *hormone* was first used by Starling in 1905. It was Starling who suggested that the constancy of the internal environment, first postulated by Claude Bernard, might be maintained in part through hormonal mechanisms, and he thus linked the function of the endocrine system with Bernard's concept.

The field of endocrinology was advanced on a firm experimental foundation during the first third of this century. During the 1940s endocrinology reached a yet more advanced stage of development. Marked progress in steroid chemistry was stimulated by the clinical studies of Hench and co-workers. New experimental tools and methods such as radioactive isotopes, the phase microscope and electron microscope, tissue culture and perfusion techniques, chromatographic techniques, and electronic recorders were put to use in the study of the hormones, and as a result notable advances in the field have been made during the last two decades. Certain steroids (cortisone and progesterone) were first produced on a mass basis. Experiments utilizing isolated adrenal glands and perfusion of the system with ^{14}C-labeled acetate and cholesterol brought major advances in knowledge of the biosynthesis of the hormonally active steroids. The active substance of the amorphous fraction of the adrenal cortex was finally determined in 1953 by Simpson and Tait to be the steroid aldosterone. Using chromatographic and isotopic techniques, two independent groups isolated a second thyroid hormone. In 1954 Sanger and his associates established the amino acid sequence and full chemical structure of insulin. Du Vigneaud and his associates determined the structure of the active principles of the neurohypophysis, vasopressin and oxytocin, and succeeded in synthesizing the two polypeptides. Many of the larger polypeptide hormones have also been synthesized. The development of techniques utilizing in vitro systems led to improvements concerned with the effects of many hormones on chemical reactions and enzyme systems in many cells. The work pioneered by G. W. Harris laid a solid foundation for our expanding knowledge of the relationships between the central nervous system (CNS) and the anterior pituitary. Recently the radioimmunological techniques introduced by Berson and Yalow for the assay of polypeptide and protein hormones in the plasma, as well as the competitive binding methods for determining plasma steroid concentration, have added much to an understanding of regulation of secretion of the hormones.

Methods of Study
The evidence required to show that an organ functions as an endocrine gland involves the demonstration of specific effects in the absence of the organ, and specific physiological restorative responses following exogenous administration of the hormone by transplantation of the glandular tissue or injection of suitable extracts of the gland. A further step needed to verify the endocrine activity of a gland calls for the administration of extracts of the gland to intact animals with the production of exaggerated effects of the hormone. Other requirements are concerned with evaluation of the chemical and physical characteristics of the hormone, the description of effects of the hormone, and factors which regulate the rate of secretion of the hormone. The endocrinologist also concerns himself with the problem of synthesizing the hormone and determining the mode of action of the hormone. The known hormones of the thyroid, the gonads, the adrenal glands, and the neurohypophysis have been synthesized, and so have many of the larger polypeptide hormones. More and more is being learned about the fundamental action of hormones, including the primary molecular actions.

Functions of Hormones
The effects of hormones fall into three general groups. First, they influence reactions that aid in the maintenance of a constant internal environment. Thus they regulate the rates at which carbohydrates, fats, proteins, electrolytes, and water are deposited in or removed from the tissues of the body. For example, insulin participates in regulating the chemical and/or physical factors which ensure an adequate supply of glucose to most extrahepatic tissues by increasing the permeability of the cell membrane to glucose. Second, the hormones have a morphogenic action; this includes the effects on growth, differentiation, development, maturation, trophic actions, and aging processes. Examples of the morphogenic actions are the effects of the ovarian or testicular hormones (under the influence of the pituitary gland) during the growth and development of the accessory sex organs and secondary sex characteristics at puberty. Finally, the hormones regulate autonomic activity, as well as certain CNS activities and behavioral patterns. For instance, maternal behavioral patterns are linked to the presence of various hormones. Also, the sensitivity of effector cells to epinephrine and norepinephrine is altered by the circulating levels of some hormones.

It must be emphasized that the action of all the hormones is *not* to initiate chemical reactions but to alter the rates of preexisting reactions without contributing significant amounts of either matter or energy to the process. The hormones can function at peak efficiency only when the concentrations of substrate, cofactors, and hydrogen ion, and the temperature are at optimal levels. It follows that, in experimental tests utilizing hormones, the subjects should be in nutritional, temperature, and fluid balance, and as free as possible from psychological stimuli, if representative results are to be expected.

The mechanisms by which hormones exert their control are varied, and recently much information has become available concerning the mechanism of action of hormones. In general, some hormones may be said to act upon genes and alter the rate of protein (enzyme) synthesis, some hormones may act to release small molecules or ions which alter the rate of protein (enzyme) synthesis or activity and/or membrane permeability, and some hormones may act upon membranes and alter their permeability characteristics. In each case the hormones appear to change the activity of rate-limiting steps in a metabolic pathway.

On the basis of experimental studies many hormones have been shown to act as "gene activators" because they alter the rate of deoxyribonucleic acid (DNA)-directed ribonucleic acid (RNA) synthesis and the rate of protein (enzyme) synthesis in target cells. Many of the effects of the steroid hormones and the thyroid hormones on their target cells are probably mediated via this mechanism. A more specific illustration concerns the effect of an adrenal steroid, aldosterone, on sodium reabsorption in the distal renal tubule. It has been proposed that aldosterone becomes concentrated in the nuclei of the cells concerned with sodium reabsorption and promotes messenger RNA (mRNA) synthesis. The latter then serves as a template for the ribosomal synthesis of a specific protein, which increases the permeability of the mucosal surface of the cell to sodium, thus increasing the amount of sodium available for active transport into the extracellular fluid.

The work of Sutherland with 3', 5'-cyclic adenosine monophosphate (cyclic AMP) has added greatly to the understanding of hormone action. Sutherland refers to cyclic AMP as "the 2nd messenger," whereas the hormone is termed the "1st messenger." In this concept many hormones activate an adenyl cyclase at the target cell which produces from adenosine triphosphate (ATP) many "messengers" of cyclic AMP. The cyclic AMP, in turn, may then act to alter the permeability of a membrane, or promote the synthesis or increase the activity of an enzyme, which would alter the rate of physiological response. In many respects cyclic AMP could be considered to amplify the hormonal message. Experimental work has indicated that many hormones, such as vasopressin, glucagon, adrenocorticotropin, luteinizing hormone, thyrotropin, and epinephrine, exert some of their effects by increasing the accumulation of cyclic AMP in their respective target cells.

Recently a family of compounds, the prostaglandins (PG), have been implicated in endocrine functions. They are 20-carbon unsaturated carboxylic acids which are derived from essential fatty acids. The PGs are released from a large variety of tissues in response to many types of stimuli and induce many biological effects. They exert a major action on the adenyl cyclase–cyclic AMP system of numerous cells and may either increase or decrease the activity of this system, depending upon the target cell. By acting thus the PGs may play a role in the transmission of the message of several tropic hormones. It has been proposed that the PGs, in such a situation, might act as the "second messenger" and cyclic AMP would then act as the "third messenger." However, the stimuli which promote the biosynthesis and the release of the PGs remain to be elucidated.

A hormone may also produce an effect by increasing the permeability of a cell to substrate. Insulin has an effect which facilitates the entry of glucose into muscle cells, but the mechanism or mode of action underlying this increase in permeability to glucose is not known. It has been hypothesized that insulin removes a barrier which acts to prevent glucose entry into the cell.

The actions of a hormone upon an end-organ may vary considerably, depending upon the nature of the effector cell. Thus, thyroid-stimulating hormone stimulates the thyroid gland but has little, if any, effect upon the adrenal cortex. Also, the estrogens stimulate growth of the female reproductive tract but have little effect upon skeletal muscle. A hormone through its primary actions may produce a stimulating effect on one type of cell and an inhibitory effect on another type. For example, the hormones of the thyroid gland cause an increase in the oxygen consumption of hepatic tissue but depress the oxygen consumption of the pituitary.

Principles of Hormone Assay

The endocrine glands contain only minute amounts of the hormones they secrete, and the blood contains amounts in even smaller concentration. The physiological activity of the hormones, was in many cases, the only guide to their presence, and these effects formed the basis for bioassay procedures. In such cases, the functional effect of a hormone is used as the basis for the quantitative estimation of the amount of hormone present. The concentration of the steroid and amino acid hormones can be estimated by using chemical techniques, and assay procedures are now available for all of the hormones.

The level of activity of many of the endocrine glands can be estimated by observing the morphology of the glandular cells themselves. Such criteria as size and shape of cells, as well as quantity and characteristics of intracellular granules, supply indices of the level of activity. Similarly, the morphology of cells and functional activity of tissues upon which the hormones act are used to estimate the activity of an endocrine gland.

The hormonal content of each of the endocrine glands can be estimated by means of biological and/or physiochemical techniques. In addition, the level at which a gland is functioning can be assessed by analyzing the body fluids for the secretion of the gland involved. However, only very small amounts of hormone are present in the blood, and small amounts can be recovered in the urine. The extent to which the urinary concentration of hormones and their metabolites reflects the amount in the blood is unknown in many cases. It has become much too convenient to assume that urinary concentration of hormones and their metabolites truly reflects the blood levels. It is the body fluids, exclusive of the urine, from which the hormones must eventually exert their action. Nevertheless, in the absence of adequate techniques for the assay of hormones in the blood, studies on the urinary concentration of hormones have yielded valuable information.

Recently radioimmunological methods have been applied in quantitative assays of protein and polypeptide hormones in the plasma. These methods are sensitive, precise, specific, and accurate. They can be applied to very small quantities of plasma and depend upon highly specific reactions between a protein or polypeptide hormone, the antigen, and its antibody. The assay is based upon competition between unlabeled hormone and radioactively labeled hormone for specific sites on the antibody. Initially radioactively labeled ("hot") hormone is complexed with antibody, and then known amounts of unlabeled ("cold") hormone are added and serve as standards. Some of the "cold" hormone displaces some of the "hot" hormone from the antibody. The free or unbound hormone is then separated from the complexed or antibody-bound hormone, and the amount of radioactivity in both samples is determined. The lower the concentration of "cold" hormone, the more labeled or "hot" hormone will remain bound to the antibody. A standard curve can be prepared by using different concentrations of unlabeled hormone, and unknowns, or plasma samples, can be compared with it. Assays of this type are now available for the determination of plasma levels of insulin, glucagon, parathyroid hormone, calcitonin, growth hormone, adrenocorticotropic hormone, thyroid-stimulating hormone, luteinizing hormone, follicle-stimulating hormone, prolactin, and melanocyte-stimulating hormone, utilizing very small quantities of plasma. Radioimmunoassays based on the competitive binding of proteins with steroids are also available.

Regulation of Endocrine Activity

A variety of stimuli are capable of altering the activity of many of the endocrine glands. Changes in both the external environment and the internal environment precipitate factors that lead to changes in the output of hormones in an attempt to maintain the internal environment within the limits compatible with life. The stimuli that influence the output of a hormone may be transmitted solely via neural channels, solely through humoral influences, or through a combination of both, i.e., neurohumoral channels.

Neural Control of Endocrine Activity

Either directly or indirectly, a normally functioning system of endocrine glands is dependent upon a normal CNS and its many processes. The release of the secretions of two of the endocrine glands, both arising from embryonic neural tissue, is regulated directly by their innervation; i.e., their efferent nerves are secretomotor. These two structures are the adrenal medulla and the neurohypophysis, and both undergo atrophy following denervation (Fig. 30-1).

The anterior pituitary, which has been considered the master gland, is in fact the target of various stimuli, both neural and humoral, which stimulate or inhibit its activity. It now appears certain that the supreme control of the endocrine system resides within the CNS, but this regulation is not expressed via secretomotor fibers to these glands. It is expressed rather through the secretion of mediator substances released by nerves of the hypothalamus which either stimulate or inhibit the release of the respective trophic hormones from the pituitary. The CNS also regulates the function of all the endocrine glands indirectly by exerting its effects via the vasomotor innervation of the various glands. However, it should not be inferred that the rate of hormone secretion by a gland is necessarily related to the rate of blood flow through the gland, for it has been demonstrated that the rate of blood flow through the thyroid gland does not reflect the level of thyroidal activity. It has been shown, nonetheless, that the sympathetic mediators, epinephrine and norepinephrine, increase the responsiveness of the thyroid gland to a constant dose of thyroid-stimulating hormone. Thus, release of the mediators at the vasomotor endings may act at the level of the glandular cells to increase their sensitivity to trophic hormones.

Humoral Control of Endocrine Activity

The endocrine glands should not be considered individually but should be viewed as an integrated system, because a given hormone seldom acts independently and generally

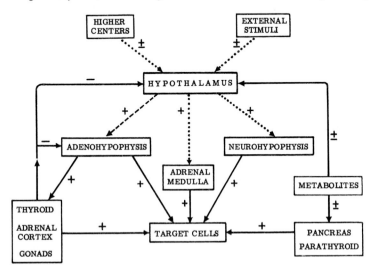

Figure 30-1
Relationships between the central nervous system and the endocrine glands. + = stimulation; − = inhibition; · · · ·> = neural pathway; ---→ = neurohumoral pathway; ——→ = humoral pathway.

affects other endocrine glands and alters their rate of secretion. In this action the hormone may assist or oppose the effect of another hormone. Such interrelationships are illustrated by the following observations. (1) The administration of estrogens inhibits the production of the pituitary gonadotropins, follicle-stimulating and luteinizing hormone. The administration of either of the two gonadotropins increases the weight of the ovaries, but the ovaries do not respond by increasing their output of estrogen. However, the simultaneous administration of the two gonadotropins produces marked ovarian growth and increased estrogen secretion. (2) The two pancreatic hormones, insulin and glucagon, induce opposite effects upon the blood sugar; the former lowers it and the latter elevates it. In addition, the functional state of an endocrine gland may be altered by a hormone secreted by a gland other than the pituitary; e.g., the administration of thyroid hormone may increase the functional activity of the adrenal cortex. Such interrelations between hormones of various glands can account for many of the paradoxical effects observed in experimental and clinical studies.

The pituitary secretes several hormones whose effects may vary considerably. Several endocrine glands are directly affected by the pituitary hormones; these—the gonads, the thyroid, the adrenal cortex—are referred to as the target glands. They can secrete small amounts of their hormones in the absence of the pituitary, but normal trophic hormone secretion by the pituitary is necessary for normal target gland function. An increase in the rate of secretion of a trophic hormone results in an increase in the rate of secretion from the respective target gland. The secretions of the target gland, in turn, tend to produce countereffects which oppose the secretion of its particular trophic hormone by the pituitary and thus inhibit the initial source of stimulus. For example, when there is an increase in the rate of secretion of thyroid-stimulating hormone by the pituitary, the rate of secretion of thyroxine from the target gland, the thyroid, increases. The increasing secretion of thyroxine then inhibits the secretion of thyroid-stimulating hormone. The end result is a balance of the forces involved, and the levels of thyroid function and of heat production by the cells of the body are maintained within relatively narrow limits. This serves to illustrate the concept of the "negative feedback system" of Figure 30-1.

In addition to the trophic hormones, which exert their primary effects upon specific glands, the pituitary gland also secretes a growth hormone which acts through an intermediate on most cells of the body. The hormones of the thyroid gland appear to act upon all cells of the body.

The rate of secretion of some hormones is relatively independent of the pituitary, being regulated by the concentration of nonhormonal metabolites in the blood. An example of this type of regulation concerns secretion of the hormone of the parathyroid gland: A low serum calcium concentration stimulates the secretion of parathyroid hormone, and, conversely, hypercalcemia induces a decrease in the rate of secretion. In each case the calcium acts directly on the gland to alter its secretion rate.

In addition to the effects of the trophic hormones and the reciprocal relations between the target glands and the pituitary, the hormones of the target glands may affect one another. For instance, for normal ovarian function the level of thyroidal activity must be optimal.

THE HYPOPHYSIS (PITUITARY GLAND)

The hypophysis secretes at least nine hormones, each being of polypeptide or protein nature. Some of the pituitary hormones regulate the functional capacities of other endocrine glands, and these are the trophic hormones: adrenocorticotropic hormone (ACTH), thyrotropic hormone (TSH), follicle-stimulating hormone (FSH), luteinizing hormone or interstitial cell–stimulating hormone (LH or ICSH), and prolactin or luteotropic hormone (PRL or LTH). Growth hormone or somatotropic hormone (STH) affects nearly all cell types, whereas the actions of vasopressin (ADH), oxytocin, and melanocyte-stimulating hormone (MSH) are restricted to particular cell types.

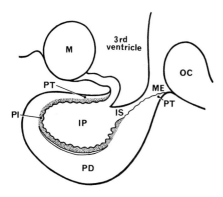

Figure 30-2
Midsagittal section through the hypophysis. The parts of the adenohypophysis or anterior pituitary are the pars distalis (PD), the pars intermedia (PI), and the pars tuberalis (PT). The neuro-hypophysis or posterior pituitary contains the median eminence (ME) of the tuber cinereum, the infundibular stem (IS), and the infundibular process (IP) or neural lobe. The mamillary body (M) and the optic chiasm (OC) are also depicted.

The hypophysis of the adult human weighs about 0.5 gm and is a compound gland of ectodermal origin, arising from two different sources. One part, the *neurohypophysis,* arises from the ventral floor of the diencephalon and remains connected to the hypo-thalamus throughout life by means of its stalk. The glandular portion of the hypophysis, the *adenohypophysis,* stems from oral ectoderm as Rathke's pouch, which is an out-growth from the roof of the mouth. This outgrowth meets the embryonic neural portion of the gland and then loses its connection with the oral epithelium.

The adenohypophysis is described as consisting of three parts: the *pars tuberalis,* the *pars intermedia,* and the *pars distalis* (often referred to as the anterior pituitary or AP). The neurohypophysis is also divided into three parts: the *median eminence of the tuber cinereum,* the *infundibular stem,* and the *infundibular process* (neural or posterior lobe). The median eminence and infundibular stem are collectively referred to as the in-fundibular or *neural stalk,* whereas the *hypophysial stalk* includes the neural stalk plus the sheath portions of the adenohypophysis, the pars tuberalis (Fig. 30-2).

The well-protected hypophysis is one of the most inaccessible organs of the body, being located in the sella turcica, a depression in the sphenoid bone. The gland is encapsulated by the dura mater, but the hypophysial stalk penetrates the dura through the diaphragma sellae. The neurohypophysis is characterized by its rich innervation of hypothalamic origin, whereas the adenohypophysis is characterized by rich vasculari-zation.

The Neurohypophysis
Strictly speaking, the neurohypophysis is not an endocrine organ because it lacks glandular elements and serves merely as a storage place for certain secretions of the hypothalamus. However, it does play an essential role in the release of the stored hormones.

Microscopically, the neurohypophysis is composed of four basis structures: un-myelinated nerve fibers, glial cells, and capillaries surrounded by argyrophilic connec-tive tissue. The nerve fibers have their origin in the cells of the supraoptic and para-ventricular nuclei, and according to an early hypothesis these fibers composed the secretomotor innervation of cells called *pituicytes.* The latter were held to be sites of formation of the posterior lobe hormones, but they are actually glial cells and have no true secretory function. The morphological behavior in vitro of the pituicytes indicates that they are not secretory cells.

It has been observed that the cells of the supraoptic and paraventricular nuclei possess stainable cytoplasmic granules which are regarded as neurosecretory products. These neurosecretory granules have been demonstrated throughout the extent of the supra-opticohypophysial tract, from the nuclear region to where the fibers end within the infundibular process. One can identify, by using a variety of staining and histochemical procedures, the neurosecretory granules along the tracts, and they appear as swellings on the fibers. The neurosecretory granules also appear in the interstitial spaces of the infundibular process. The *Herring bodies* are now considered to represent the swellings or beads on the neurosecretory fibers and therefore the Herring bodies do not lie free in the posterior lobe. The manner in which the neurosecretory material is released into the bloodstream is probably through the process of exocytosis. The chemical composition of the neurosecretory material is protein in nature, and substances that possess all the known biological activities of the posterior lobe hormones have been obtained from extracts of this material.

The above observations, plus the facts that posterior lobe hormones can be extracted from hypothalamic tissues, that the posterior lobe can be depleted of stainable neurosecretory material by procedures which deplete the organ of hormonal activity, and that neurosecretory material piles up at the proximal end of a transected infundibular stem, suggest the following concept: The hormones vasopressin and oxytocin are formed in the cell bodies of the supraoptic and paraventricular nuclei and are transported down the respective nerve tracts to the infundibular process, where they are stored. The terminations of the neurons are regarded as the sites of hormone release. The release of the hormone is apparently triggered by the nerve action potential passing down the hypothalamohypophysial tract. In view of these considerations, it appears proper to refer to this system as the hypothalamo-neurohypophysial system (HNS) and regard it as an organ of internal secretion.

Four different activities have been attributed to the hormones of the HNS. First, in 1895 Oliver and Schafer observed that the intravenous administration of extracts of the whole pituitary resulted in a marked rise in systemic blood pressure (pressor effect). The effect was purely one of vasoconstriction, and Howell showed later that this effect was obtainable if only neural lobe extracts were administered. Second, Sir Henry Dale demonstrated that extracts which caused contraction of the uterus (oxytocic effect) could be prepared from this organ. Third, it was demonstrated that neural lobe extracts were helpful in ameliorating the symptoms of diabetes insipidus (antidiuretic effect). Finally, extracts of the HNS have been shown to induce contractions of the myoepithelial cells of the alveoli of the lactating mammary gland and cause evacuation of milk from the alveoli (milk letdown or draught effect).

The brilliant research of Du Vigneaud and his associates resulted in the isolation of two distinct polypeptides from the neural lobe. Later work led to the synthesis of the two compounds. Each has a molecular weight of about 1000 and contains eight amino acids, six of which are common to both hormones. One compound, vasopressin, possesses a high degree of antidiuretic activity and no oxytocic activity. The other, oxytocin, shows little antidiuretic activity and high oxytocic activity.

It should not be at all surprising, considering the close structural similarity, that the two compounds have overlapping activities. Proteins with molecular weights of 30,000 have been isolated from the posterior pituitary. They are referred to as neurophysins and probably represent storage forms of the hormones. Small amounts of the neurophysins may be released into the plasma.

Vasopressin (ADH)

The most obvious consequence of neurohypophysectomy is the induction of polyuria. This is characterized by the excretion of a high-volume, low-osmolar-concentration urine. As a result of the large volume of water lost, there is an intense thirst and a high water intake (polydipsia). ADH plays an important role in the maintenance of water balance because it regulates the excretion of free water from the kidney. The site of action of ADH on the nephron is discussed in Chapter 22.

Control of ADH Secretion

The rate at which the secretion and release of ADH occur is under nervous control and is subject to changes in the effective osmotic concentration of the extracellular fluid (ECF), the effective volume of the blood, exteroceptive stimuli, and psychic stimuli (Fig. 30-3).

Various drugs such as nicotine, acetylcholine, morphine, barbiturates, ether, and chloroform act on the HNS to increase the output of ADH. On the other hand, alcohol inhibits the secretion of ADH.

Verney has shown that the injection of hypertonic solutions of NaCl, glucose, or sucrose into the carotid artery results in a diminution of water diuresis. Injection of urea (which is ineffective osmotically at the cell) is not an effective inhibitor of water diuresis. Subsequent work has shown that a sustained (about 30 minutes) increase in the effective osmotic pressure of only 2 percent causes an increase in ADH release and a decrease in free water clearance. Similarly, the ingestion of a large volume of water lowers the osmotic concentration, and the secretion of ADH is depressed. It has been suggested that receptors exist which are sensitive to differences in the osmotic concentrations between the ECF and the intracellular fluid (ICF). They are probably located in the anterior hypothalamus near or within the supraoptic nuclei and therefore transmit their stimuli to the neural lobe over this tract. Stimulation of the area around the supraoptic nucleus causes inhibition of water diuresis. Similarly, it has been shown that the injection of hypertonic saline solution increases the electrical activity of single neurons in and adjacent to the supraoptic nuclei.

Superimposed upon the control mechanism whereby changes in the effective osmotic pressure of ECF affect ADH secretion and release by the HNS are alterations induced by nervous reflexes. These alterations may be due to excitation of the HNS by emotional stress or pain.

Studies pioneered by Henry and Gauer and their co-workers have firmly established that "volume" receptors and pressure receptors play a prominent role in the control of

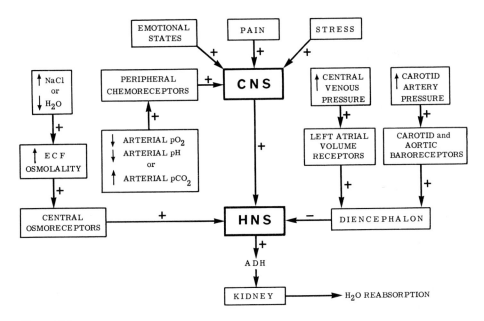

Figure 30-3
Factors that regulate the secretion of ADH by the hypothalamo-neurohypophysial system (HNS).
+ = stimulation; − = inhibition.

vasopressin secretion during isotonic contraction or expansion of the effective blood volume. The best available evidence shows that the monitoring system for such reflexes has an intravascular origin.

It has been shown that distention of the left atrium or the pulmonary vein within the pericardium is invariably followed by a diuresis. The diuresis occurs in spite of the fact that the manipulations markedly diminish the cardiac output. Distention of the pulmonary artery, the right atrium, or the pulmonary vein outside of the pericardium does not lead to either a diuresis or an antidiuresis. Procedures such as negative pressure breathing, the infusion of iso-oncotic or hyperoncotic albumin solutions, isotonic saline infusion, or changing from the sitting to the supine position, each of which results in an increased left atrial volume, lead to a decrease in plasma ADH and an increase in urine volume. These procedures are effective only when the vagus is functional, and associated with each is an increased neuronal activity of afferent fibers running in the vagus trunk. That the volume and not pressure is the stimulus is shown by the fact that bursts of activity in the nerve are minimal during the *a* wave of the atrial pressure cycle, which is a time when the pressure is highest, and the firing rate is greatest during the *v* wave of the atrial pressure cycle, when atrial volume is maximal. Conversely, situations such as venous occlusion of the legs, hemorrhage, orthostasis, positive pressure breathing, and changing from the supine to the sitting position, all of which lead to a decreased left atrial volume, result in increased plasma ADH and reduced urine flow.

The above evidence suggests that activation of the left atrial stretch receptors inhibits the release of ADH from the HNS (Fig. 30-3). Share and his co-workers have shown that the circulating level of ADH increases following isotonic contraction of the blood volume, following bilateral vagotomy, or following carotid occlusion. However, the blood ADH concentration decreases during left atrial distention. They have also shown that carotid sinus baroreceptor activity and carotid body chemoreceptor activity influence ADH secretion. Activity of the baroreceptors appears to be inversely related to ADH secretion, whereas chemoreceptor activity (induced by hypoxia, hypercapnia, or a decrease in pH) appears to be directly related to ADH secretion. The ADH concentration in the blood is elevated in the human in the standing position when compared to that of the individual in the recumbent position. Similarly, the individual acutely exposed to a cold environment shows a decrease in blood ADH, whereas an increase in blood ADH follows acute exposure to a hot environment. Therefore, the rate of receptor discharge from the left atrial receptor is a function of the central blood volume. When the central blood volume is high, ADH secretion is reduced; when low, secretion is increased. Marked increases in plasma ADH may occur in the presence of high plasma osmolar concentration if the mean left atrial pressure and/or volume is elevated. The volume receptor and osmoreceptor regulatory mechanisms provide, in concert, a very precise control over ADH secretion and protect the volume and concentration of the ECF within narrow limits.

Action of ADH

The role of ADH in the regulation of renal tubular water handling and homeostasis of the extracellular fluid has been discussed in Chapter 22. Recent evidence suggests that the action of ADH on the distal renal tubules is mediated through cyclic AMP. However, the nature of the change in permeability to water remains obscure. The distal renal tubule of the human is extremely sensitive to changes in the plasma ADH concentration. The concentrating power of the kidney may be increased from a U_{osm}/P_{osm} ratio of 1 up to a maximum of 4 by increasing the plasma ADH from 1 μU per milliliter to 4 μU per milliliter.

Thus it appears that a type of negative feedback mechanism affords a reciprocity between the kidney, gastrointestinal tract, lungs, skin, and sweat glands and the HNS and ensures a relatively constant extracellular fluid with respect to its effective volume and osmolarity. This system is directly dependent upon its connections with the CNS.

The regulation of the secretion of oxytocin, its role in sperm transport, and its effects upon uterine motility and milk ejection are discussed in Chapter 35.

The Adenohypophysis

The glandular portion of the pituitary is made up of irregular masses and columns of epithelial cells which are supported by a delicate framework of connective tissue. The groups and columns of cells are separated by sinusoids. Two main types of cell are evident. One type does not show any conspicuous stainable cytoplasmic granules, and these cells are referred to as *chromophobes*. The other contains cytoplasmic granules which take up stains readily, and the cells are called *chromophils*. The latter are considered to be secretory in nature. Many cytologists believe them to be daughter cells of the chromophobes, but it is also possible that the chromophobes represent resting cells. The chromophils may be further divided, according to the stainability of their granules, into *alpha cells* (acidophils) and *beta cells* (basophils). The chromophils may also be divided on the basis of their reaction with the periodic acid–Schiff method (PAS stain). This procedure is used in the identification of mucopolysaccharides and mucoprotein, and a positive PAS reaction may be interpreted as the staining of a carbohydrate-containing substance. Additional procedures have shown that there are at least six epithelial cell types in the anterior pituitary.

There is general agreement on the cellular origin of all the anterior pituitary hormones. This agreement has been achieved on the basis of the stainability and cytological appearance of cell types following target gland removal or following excessive (endogenous or exogenous) secretion of target gland hormones, the use of immunological techniques, and functional changes which occur in pituitary tumors.

Hormones of the Adenohypophysis

The anterior pituitary gland could be considered to be six glands in one because its hormones originate from six different types of secretory cells and each type is capable of functioning independently of all the other types. Six hormones have been isolated in a partially purified state from the AP, and all are polypeptide or protein in nature. They are TSH, FSH, LH, PRL, ACTH, and STH. The existence of several other fractions has been postulated, but in all probability the effects of each of the postulated hormones can be accounted for through the effects of the six established hormones.

Partially purified TSH preparations have been obtained, and they have the properties of a relatively low-molecular-weight (10,000) glycoprotein. TSH and other protein hormones are ineffective when administered orally because they are inactivated in the gastrointestinal tract. The cells that secrete TSH are PAS-positive beta cells and are referred to as *thyrotrophs* or *delta cells*. The cells undergo degranulation and become hyperplastic and hypertrophic following thyroidectomy. Following hypophysectomy, the thyroids become atrophic, and all criteria of thyroid function decline rapidly. Any procedure that leads to a marked decrease in the output of thyroid hormone is followed by a marked increase in the secretion and/or release of TSH from the AP and morphological changes in the thyrotrophs. The morphological and functional changes can be prevented by the administration of thyroid hormone. Thus, all available evidence indicates that TSH secretion and/or release is regulated by the level of circulating thyroid hormones. A negative feedback system would then operate between the thyroid and the AP.

Two of the gonadotropic hormones, FSH and LH, are also carbohydrate-containing proteins, and both are secreted by PAS-positive beta cells. The cells that secrete FSH and LH undergo hypertrophy and degranulate following castration. Large cytoplasmic vacuoles also form, and the nucleus is compressed against the cell membrane to form the "signet ring" or castration cell. The cells that secrete FSH are located near the periphery of the AP, whereas those that secrete LH are located centrally. FSH induces ovarian follicular growth and in conjunction with LH leads to estrogen secretion and ovulation in the female. In the male, FSH leads to proliferation of the seminiferous tubules, whereas LH (ICSH) stimulates the Leydig cells of the testes to secrete androgen. A negative feedback regulation also exists between the AP and the gonads.

PRL, another gonadotropin, is a protein hormone with a molecular weight of about 26,000. It is secreted by PAS-negative alpha cells. These cells have a high affinity for azocarmine. The number of carmine-staining cells increases during pregnancy and lactation. PRL plays a fundamental role in the development and secretion of the mammary gland.

The pituitary cells that secrete ACTH are beta cells. This conclusion is based upon marked changes (hyalinization) in the beta cells following spontaneous hyperadrenal states or following the chronic administration of high doses of adrenal steroids. It has also been noted that fluorescent ACTH-induced antibodies combine with ACTH (specific antigen) and localize in a specific type of beta cell. This same cell type also secretes MSH. ACTH is a polypeptide; it contains 39 amino acids, and the full sequence of amino acids is known. This hormone stimulates adrenal cortical growth and hormonogenesis. The relationship between the AP and the adrenal cortex is a reciprocal one in that hormones of the adrenal cortex inhibit the output of ACTH and thus tend to maintain adrenal cortical hormonogenesis at an optimal level.

There is general agreement that the PAS-negative alpha cells secrete STH. Pituitary dwarfism and gigantism and acromegaly correlate well with deficiency and hyperplasia, respectively, of alpha cells. There are practically no alpha cells in the pituitaries of genetic-dwarf mice, and the pituitaries contain only small amounts of STH. The acromegalic has a pituitary alpha cell tumor rich in STH.

The trophic hormones of the AP induce growth and various metabolic changes in the respective target organs, and these changes are not necessarily different from those which take place in tissues in general. Nevertheless, under physiological conditions, the primary effect of the trophic hormones is on the target gland (adrenal cortex, thyroid, gonads) because these tissues possess an unusual sensitivity to the respective trophic hormones. If the trophic hormones are circulating at high levels, such as those after removal of the target gland or after excessive exogenous administration, they may be able to affect other tissues. Thus, the pituitary trophic hormones might be capable of extending their direct effects to tissues other than their specific target organs. For example, ACTH stimulates oxygen consumption and ^{32}P uptake in the adrenal cortex and also increases oxygen consumption in the adrenalectomized animal. The functions and the regulation of the secretion of each of the trophic hormones of adenohypophysial origin will be considered later in the appropriate sections.

The Hypophysial Portal Vessels

For many years it was quite evident that nervous mechanisms could affect the actions of the AP. There was a considerable amount of anatomical evidence against direct neural and vascular connections between the CNS and the AP. Evidence was available which supported the concept of vasomotor control of the AP circulation, and in 1930 Popa and Fielding described what was to be later called the hypophysial portal system. In most species, including man, the hypophysial portal system is arranged as follows: Small branches of the internal carotid and posterior communicating arteries form a rich vascular network within the pars tuberalis and median eminence. A large number of capillary tufts or loops arise from this network and penetrate into the median eminence and form the primary plexus. This plexus comes into intimate contact with neural elements of the hypothalamus. The primary plexus then forms large portal trunks which lie on the anterior or ventral surface of the hypophysial stalk. These vessels break up and distribute the blood into the sinusoids of the AP. Direct observation has revealed that the direction of blood flow in this system is from the median eminence to the AP. The hypophysial portal vessels have a marked capacity to regenerate after transection of the hypophysial stalk, because capillary outgrowths can bridge the gap of a divided stalk within 24 hours. Such regeneration can occur following transplantation of the AP in the subarachnoid space below the median eminence. The functional significance of the hypophysial portal vessels is apparent if one considers that the AP does not possess a direct secretomotor innervation, yet the CNS shows a major regulatory action on the AP. The present concept states that nerves end in and around the primary plexus (within the median eminence) and release specific substances which are transported via the hypophysial portal vessels to the AP. The secretory product of the hypothalamic neuron

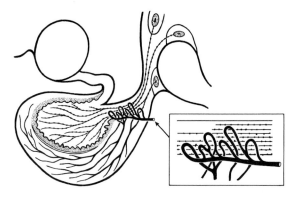

Figure 30-4
Blood supply of the anterior pituitary and innervation of the posterior pituitary. The fibers that innervate the posterior pituitary have their origins in the supraoptic and paraventricular nuclei. A detail of the primary plexus (capillary loops) of the hypophysial portal circulation is shown at the right. Note that the axons of hypothalamic origin terminate in and/or traverse the capillary loops.

then serves as a stimulator or inhibitor of a specific AP activity. Thus the nerve cells of the hypothalamus function as neuronal-hormonal transducers. At the present time seven polypeptide-releasing factors have been isolated from the hypothalamus, the amino acid sequence is known in six of them, and three have been synthesized. Two hypothalamic polypeptides, growth hormone-releasing factor (GRF) and growth hormone-inhibiting factor (GIF) or somatostatin, regulate the secretion of growth hormone. Two, pro-lactin-inhibiting factor (PIF) and prolactin-releasing factor (PRF), regulate prolactin secretion by the AP. Corticotropin-releasing factor (CRF) stimulates ACTH secretion; thyrotropin-releasing factor (TRF) stimulates TSH secretion; and probably a single releasing factor, FSH/LH–RF, stimulates the secretion of FSH and LH in the female. Therefore, the anterior pituitary is as much under hypothalamic control as is the posterior pituitary, but control of the secretions of the AP is not mediated via direct neural connections, as are those of the neural lobe. The control over AP secretions is exerted through the release of specific mediators into the capillary plexus of the portal vessels in the median eminence by hypothalamic neurons (see Fig. 30-4). The regulation of secretion of the trophic hormones of the AP is considered in the appropriate chapters.

Effects of Hypophysectomy
Several notable morphological and functional alterations result from total hypo-physectomy in the young animal: (1) failure of the gonads to mature, with resultant infantile sexual development and sterility; (2) atrophy of the thyroid gland and the characteristics of thyroid insufficiency; (3) atrophy of the adrenal cortex and signs of hypoadrenalism without salt loss; (4) cessation of growth and failure to attain an adult stature; (5) a decided tendency toward hypoglycemia, hypersensitivity to insulin, and a loss of body nitrogen accompanied by diminished fat catabolism. However, if optimal environmental conditions are maintained, removal of the pituitary is not incompatible with life. All parameters of functions of the gonads, thyroid, and adrenal indicate a sharp decline following hypophysectomy in the adult human. Similarly, metabolic derangements coincident with a lack of STH are evident.

Following hypophysectomy or during the development of hypopituitarism the earliest and most frequent deficiencies observed are failure of growth and gonadotropic hormone secretion, and resultant gonadal failure. Evidences of thyroidal insufficiency next appear and are followed by signs of adrenal cortical insufficiency. This sequence of malfunction generally prevails regardless of the species. In several species it has been shown that ablation of increasing amounts of AP tissue results in gonadal, thyroidal, and adrenal cortical failure in that order. The removal of up to 80 percent of AP is compatible with normal gonadal, thyroidal, and adrenal cortical function. Thus, it appears that a large safety factor is present, and the same is true with respect to all the other endocrine glands. Specific deficiencies in the secretion of each of the six hormones of the AP have been described.

The properly treated hypophysectomized human retains a normal appearance, is able to gain weight and attain a positive nitrogen balance, can repair bone, and can be maintained in good health indefinitely. It follows that the administration of extracts of the AP to intact individuals results in marked development of the gonads and accessory sex glands, adrenal and thyroidal hypertrophy with signs of hyperadrenalism and hyperthyroidism, increase in growth rate with the laying down of nitrogen and increase in long bone length, and a tendency toward diabetes with hyperglycemia and glucosuria.

Functional Role of the Adenohypophysis
The adenohypophysis occupies a crucial position in the body economy because it directly influences the output of hormones from the adrenal cortex, thyroid, and gonads through the elaboration of its trophic hormones. It acts directly on all body structures through STH, and it also indirectly affects the output of the hormones from the pancreas and parathyroids as a result of its action, both direct and indirect, on various metabolic pathways.

The effects of an excess or a deficiency of the trophic hormones of the AP are thus mediated through the various target glands. The trophic hormones act only on pre-existing chemical reactions and do not initiate the reactions. However, the trophic hormones are essential for complete morphological and physiological development of their target glands. In general, functional changes in target glands are induced prior to anatomical changes following the administration of hormones.

According to the negative feedback concept of control systems, the trophic hormones are released in amounts varying with the functional state of the individual and not at a constant rate. In all probability, at least two types of negative feedback system operate between the trophic hormone secretion of the AP and secretion of hormones by the target gland. Initially, it was believed that the hormones of the target glands acted directly upon the AP to inhibit the trophic hormone output. In this instance, when target hormone secretion rate was high, trophic hormone secretion would be inhibited, and when target hormone concentration was low, inhibition would be removed and trophic hormone output would be increased, etc. However, it has been shown that various target gland hormones may also act upon specific hypothalamic neurons and indirectly influence the synthesis and secretion of a trophic hormone. Changes in the internal environment and changes in the external environment can affect AP hormone secretion by regulation of the hormone output via the hypothalamus and the hypophysial portal system.

Growth Hormone or Somatotropic Hormone (STH)
As the name implies, growth hormone plays a central role in the growth of the organism. Human growth hormone is a single-chain polypeptide containing 190 amino acid residues. It has a molecular weight of 21,500, and the complete sequence of amino acids has been determined. Growth hormone exerts many biological effects, and it has been only recently that its effects could be studied in man. Early studies concerning the role of growth hormone in human physiology were limited to clinical observations in acromegaly, gigantism, and dwarfism. The hormone affects several parameters of protein, fat, and carbohydrate metabolism, and in contrast to the other adenohypo-physial hormones it does not require a target gland as an intermediary to produce its widespread effects, but acts upon the effector or target cell directly. The plasma level of growth hormone in the adult in a basal state, as determined by radioimmunoassay, is about 2 to 3 ng per milliliter.

Probably the most striking gross effect of STH is that on the skeleton. Hypersecretion of STH or the exogenous administration of STH can lead to gigantism if instituted prior to closure of the epiphysial plates of the long bones. However, the presence of high levels of STH in the adult (after closure of the epiphysial plates) results in acromegaly. The condition is characterized by enlargement of the skeleton, especially the skull, hands and feet, the skin, the subcutaneous tissue, and the viscera; it is generally the result of an acidophilic adenoma. The disease is accompanied by sellar enlargement, gonadal atrophy, changes in the visual fields, and diabetes mellitus.

Growth hormone plays a prominent role in protein metabolism and the regulation of growth. It promotes protein synthesis and nitrogen retention by accelerating the rate of transfer of amino acids from the extracellular to the intracellular compartment and incorporating the transferred amino acids into cell proteins. Some evidence indicates that STH promotes protein synthesis via gene activation because it stimulates the synthesis of messenger ribonucleic acid (RNA), ribosomal RNA, and transfer RNA in liver. This activity is blocked by actinomycin, a substance which forms a complex with deoxyribonucleic acid (DNA), the genetic material which participates in the formation of RNA. The DNA and RNA are required for continued protein synthesis. The actions of growth hormone are mediated by a group of low-molecular-weight polypeptides called *somatomedins*. The somatomedins, presumed to be produced as a result of growth hormone action on the liver, enter the circulation and act on the target cells.

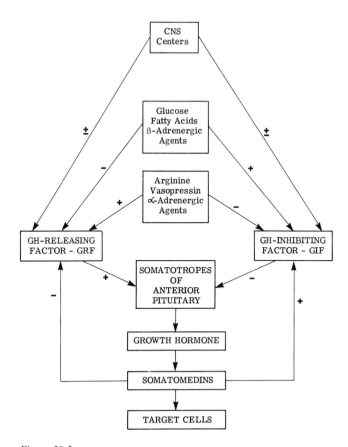

Figure 30-5
Some factors that regulate the secretion of growth hormone by the anterior pituitary. + = stimulation; − = inhibition.

STH also leads to a reduction in the rate of lipid synthesis, promotes the mobilization of fatty acids from adipose tissue, and increases fatty acid oxidation. The first indication that STH played a role in carbohydrate metabolism was the observation that hypophysectomy, when performed on the depancreatized animal, led to the amelioration of the symptoms of diabetes mellitus. In general, growth hormone antagonizes the effects of insulin on both carbohydrate and fat metabolism. Later work showed that the chronic administration of STH could lead to the development of permanent diabetes mellitus.

Recent evidence indicates that the hypothalamus plays an important role in the regulation of growth hormone secretion. Neurons of the ventromedial hypothalamus release two substances into the hypophysial portal circulation that affect growth hormone secretion: growth hormone-releasing factor (GRF), which stimulates growth hormone secretion by the somatotropes; and growth hormone-inhibiting factor (GIF) or somatostatin, which inhibits growth hormone secretion. Somatostatin is a tetradecapeptide and has been synthesized.

Evidence is available indicating that STH may inhibit its own secretion by acting on the hypothalamus and depressing GRF release and presumably also by stimulating somatostatin release. This type of control has been referred to as the "short" feedback mechanism and is probably mediated via one of the somatomedins. Glucose (hyperglycemia), fatty acids, and beta-adrenergic agents each act on the overall system to inhibit growth hormone secretion, while arginine, vasopressin, and alpha-adrenergic agents each stimulate growth hormone secretion. Various areas of the central nervous system may act on the hypothalamic elements to either stimulate or inhibit growth hormone secretion, because several types of stressful stimuli and factors such as exercise, fasting, and sleep stimulate growth hormone secretion. Figure 30-5 illustrates some factors involved in the regulation of the secretion of growth hormone. Secretion of each of the other adenohypophysial hormones is controlled, in part, via the "short" feedback loop as well as via the "long" feedback loop.

Pars Intermedia (Intermediate Lobe)

The pars intermedia, although of the same embryonic origin as the pars distalis, becomes associated anatomically with the neurohypophysis. It secretes the melanocyte-stimulating hormone (MSH), sometimes called intermedin. The hormone β-MSH has a molecular weight of 2177 and is a polypeptide composed of 18 amino acids. The secretion of MSH is regulated by two hypothalamic releasing factors, MSH-releasing factor (MRF) and MSH-inhibiting factor (MIF). This hormone is structurally similar to ACTH in that it shares a common sequence of seven amino acids with ACTH. This similarity accounts for the overlapping actions of the two substances in man. The plasma cortisol concentration plays an important role in MSH release with high levels inhibiting MSH secretion and low levels permitting excessive MSH secretion.

In lower vertebrates, such as the amphibia, MSH causes dispersion of the pigment granules in the chromatophores. The granule dispersion causes the skin to be darkened and thus affords protection to the animal when placed upon a dark background. Conversely, adaptation to a light background is effected by concentration of the pigment granules around the nucleus of the chromatophores following a decrease in MSH secretion. The stimulus for MSH release, the intensity of light as reflected by the environment, is monitored by the retina, and the greater the intensity, the less the MSH released. Thus, impulses transmitted via the optic tract exert an inhibitory effect on MSH release.

In man, MSH stimulates the synthesis of melanin by the melanocytes and thus leads to darkening of the skin. In view of the structural similarity of MSH and ACTH, it is not altogether surprising that ACTH also leads to increasing pigmentation.

Other Hormonal Substances

Several substances have hormonal activity in subprimate species. But since their functions have not been established in man, they will not be discussed in this and following chapters. *Relaxin* is of ovarian and/or uterine origin and relaxes the pubic ligaments in several rodents at the end of pregnancy. *Melatonin* is of pineal origin, and its secretion is controlled by environmental lighting. It affects the state of gonadal maturation in rats and other subprimates. Tumors of the pineal in young boys have been demonstrated to cause precocious puberty. The *prostaglandins* appear to be ubiquitous, and their role in homeostasis remains to be determined.

REFERENCES

Blackard, W. G.   Control of growth hormone secretion in man. *Postgrad. Med. J.* 49:122–126, 1972.

Frohman, L. A., and M. E. Stachura.   Neuropharmacologic control of neuroendocrine function in man. *Metabolism* 24:211–234, 1975.

Gauer, O. H., J. P. Henry, and C. Behn.   The regulation of extracellular fluid volume. *Annu. Rev. Physiol.* 32:547–584, 1970.

Gray, C. H., and A. L. Bacharach.   *Hormones in Blood.* New York: Academic, 1967. Vols. 1, 2.

Guillemin, R., and R. Burgus.   Hormones of the hypothalamus. *Sci. Am.* 227(5):24–33, 1972.

Harris, G. W.   *Neural Control of the Pituitary Gland.* London: Arnold, 1955.

Harris, G. W., and B. T. Donovan.   *The Pituitary Gland.* Berkeley: University of California Press, 1966. Vols. 1, 2, 3.

Hayward, J. N.   Neural control of the posterior pituitary. *Annu. Rev. Physiol.* 37:191–210, 1975.

Jaffee, B. M., and H. R. Behrman.   *Methods of Hormone Radioimmunoassay.* New York: Academic, 1974.

Knobil, E.   The pituitary growth hormone: An adventure in physiology. *Physiologist* 9:25–44, 1966.

McCann, S. M., and J. C. Porter.   Hypothalamic pituitary stimulating and inhibiting hormones. *Physiol. Rev.* 49:240–284, 1969.

Martin, J. B.   Neural regulation of growth hormone secretion. *N. Engl. J. Med.* 288:1384–1393, 1973.

Migley, A. R., G. D. Niswender, and R. W. Rebar.   Principles for the assessment of the reliability of radioimmunoassay methods. *Acta Endocrinol.* (Kobenhavn) 142(Suppl.):163–184, 1969.

Moore, W. W.   Antidiuretic hormone levels in normal subjects. *Fed. Proc.* 30:1387, 1971.

Pastan, I.   Cyclic AMP. *Sci. Am.* 227(2):97–105, 1972.

Ramwell, R. W. (Ed.).   *The Prostaglandins.* New York: Plenum, 1973.

Robinson, A. G.   Isolation, assay, and secretion of individual human neurophysins. *J. Clin. Invest.* 55:360–367, 1975.

Sawin, C. T.   *The Hormones: Endocrine Physiology.* Boston: Little, Brown, 1969.

Segar, W. E., and W. W. Moore.   Regulation of antidiuretic hormone release in man. *J. Clin. Invest.* 47:2143–2151, 1968.

Share, L.   Vasopressin, its bioassay and the physiological control of its release. *Am. J. Med.* 42:701–712, 1967.

Share, L., and C. E. Grosvenor.   The Neurohypophysis. In S. M. McCann (Ed.), *Physiology. Series One.* Baltimore: University Park Press, 1974. Vol. 5, chap. 1, pp. 1–30.

Tepperman, J.   *Metabolic and Endocrine Physiology* (3rd ed.). Chicago: Year Book, 1974.

Turner, C. D., and J. T. Bagnara.   *General Endocrinology* (5th ed.). Philadelphia: Saunders, 1971.

Williams, R. H. (Ed.).   *Textbook of Endocrinology* (5th ed.). Philadelphia: Saunders, 1974.

31. Endocrine Functions of the Pancreas

Ward W. Moore

The relationship between the pancreas and altered carbohydrate metabolism was first shown by the classic experimental work of Von Mehring and Minkowski in 1889. They observed that removal of the pancreas of the dog resulted in the production of all the features of diabetes mellitus. This disease state has been known for centuries, and it is the most frequent, and probably the most studied, metabolic disorder. The endocrine component of the pancreas secretes two hormones, insulin and glucagon, and each functions in the regulation of carbohydrate metabolism. The most prominent pancreatic hormone, *insulin,* was named and its major biological properties were elucidated before Banting and Best obtained the first stable insulin preparation in 1922. The net effects of insulin lead to the storage of potential fuel, because it promotes the synthesis of triglycerides, glycogen, and protein and favors their storage. In 1926 Murlin reported that some of the extracts of pancreatic tissue contained a hyperglycemic principle, in addition to the hypoglycemic material, insulin, and proposed the name *glucagon* for this factor. The net effects of glucagon lead to the mobilization of the primary energy sources, glucose and fatty acids, from their storage sites in liver and adipose tissue.

ANATOMY

The pancreas is derived from gut endoderm and remains connected to the small intestine by means of two ducts (duct of Wirsung and duct of Santorini). Glandular tissue appears during the third month of intrauterine life and develops as side buds from the ducts and duct branches. The acinar tissue remains connected to the duct system and becomes the exocrine portion of the pancreas. Small islets of cells proliferate from the ducts, differentiate, and lose their connection with the duct system and become the endocrine portion of the pancreas. The blood supply of the pancreas is via the left gastric, splenic, and hepatic arteries and the inferior pancreatoduodenal artery. This organ is drained by the splenic and superior mesenteric veins and celiac lymph nodes. It receives a sympathetic innervation via the celiac plexus, and parasympathetic fibers from the vagus. The latter are the only contribution to the endocrine portion of the pancreas.

Langerhans first described small nests or islets of highly vascularized cells which are independent of the duct system of the pancreas. These islets maintain their functional integrity following ligation of the pancreatic ducts, whereas the acinar tissue becomes atrophic. The total volume of the islet tissue makes up about 1 to 3 percent of the entire pancreas. Three cell types have been shown to be present in the islets of Langerhans, and they have been differentiated on the basis of their solubility and affinity for certain stains. The alpha cells contain fine granules which are insoluble in alcohol and which stain bright red with Mallory or Masson staining methods. They are relatively few in number and are the source of glucagon. The beta cells contain small alcohol-soluble granules which stain orange-brown with the above methods. They are the source of insulin and make up about 75 percent of the islet tissue. The delta cells do not contain granules, and their functional significance is unknown, but they may act as mother cells of the alpha and beta cells.

CHEMISTRY OF INSULIN AND GLUCAGON

The fundamental work on the extraction of insulin from pancreatic tissue was done by Banting and Best in 1922. In 1926 Abel and his associates were the first to prepare crystalline insulin, and in 1934 Scott described the crystallization of insulin in the presence of small amounts of metallic ions, such as zinc. Insulin is derived from a single-chain, large-molecular-weight (9000) precursor called proinsulin. Proinsulin is synthesized in the endoplasmic reticulum of beta cells as a single polypeptide chain of 86 residues, and the amino acids are arranged in a folded manner so that two disulfide bridges exist within the molecule. It is transferred to the Golgi complex, packaged in granules, and converted to insulin by proteolytic cleavage. Insulin is complexed with zinc and stored in the granules—beta granules. Secretion of insulin is accomplished by the release of the granules by emiocytosis. Insulin is a simple protein and has a molecular weight of about 6000. In 1954 Sanger and associates established the number and sequence of amino acids composing the two dissimilar polypeptide chains which form the insulin molecule. One chain (chain A) contains 21 amino acids, another (chain B) contains 30 amino acids, and the chains are linked by disulfide (-S-S-) bridges. The biological activity appears to be a function of the entire molecule rather than of specific groupings within the molecule, because potency is lost following slight alterations in the molecular structure. Slight variations in amino acid sequence within the disulfide ring of the A chains do exist between species. Nevertheless, all the insulins cross species barriers with respect to their physiological activities. Insulin is readily inactivated either by acid hydrolysis or by proteolytic enzymes and hence is rendered ineffective in the gastro-intestinal tract and must be administered parenterally.

Insulin is degraded rapidly in the body; less than 0.1 percent may be recovered in the urine following the administration of large doses. The factor responsible for the hepatic inactivation and degradation of insulin is probably a series of proteolytic enzymes which has been termed the insulinase system. The hormone also becomes firmly bound to target tissue such as muscle, adipose tissue, and mammary gland.

The biosynthesis of glucagon, like that of insulin, probably involves a precursor, or pro-glucagon. Pro-glucagon has a molecular weight at least twice that of glucagon, is immunoreactive with glucagon, but lacks the biological activity of glucagon. It is synthesized in endoplasmic reticulum, transported to the Golgi complex, and packaged into alpha cell granules. Secretion is similar to that of insulin. Glucagon is composed of one long chain consisting of 29 amino acid residues and has a molecular weight of 3485. It has been artificially synthesized. Like insulin, it is cleared by proteolytic enzymes, and the integrity of most of the molecule appears to be required for activity, because none of its degradation products possess hyperglycemic activity.

Radioimmunoassays are now available for both insulin and glucagon. With these techniques insulin has been shown to have a half-life of about 30 minutes whereas that of glucagon is about 10 minutes. Normal fasting plasma levels in man have been determined for both insulin and glucagon. The fasting level of insulin is about 8 μU per milliliter with a range of 5 to 20 μU per milliliter, and that of glucagon is approximately 75 pg per milliliter with a range of 35 to 200 pg per milliliter. Small amounts of proinsulin are in the blood, about 0.15 ng per milliliter.

METABOLIC INTERRELATIONSHIPS

The pancreas, via its hormones, plays a prominent role in the regulation of metabolism. A disturbed utilization and regulation of carbohydrate metabolism results from a relative or absolute deficiency in insulin. Similarly, an insulin deficiency causes marked defects in the synthesis, storage, and utilization of fats and proteins. Any essential fault in carbohydrate metabolism necessarily involves the metabolism of protein and fat as well, because the metabolic pathways through which the organism derives energy from food sources are known not to be separate and distinct, but are intermingled. Carbohydrate is the active fuel of the body and is ordinarily the primary source of energy of the cell, but fatty acids can be utilized by several types of cells. Carbohydrate also contributes part of its substance to fatty acid and amino acid formation. Consequently, the proper utilization and formation of protein and fat are dependent upon carbohydrate metabolism. The digestion and absorption of each of these foods has been discussed in Chapter 26.

Carbohydrate Metabolism

Following their absorption from the gastrointestinal tract, the nutrient sugars are transported directly to the liver. This organ converts the sugars to a phosphorylated hexose and acts as a central clearinghouse for their disposition. The hexose phosphate which results from various transformations may be converted to glycogen (glycogenesis) and stored in the liver; it may be broken down (glycolysis) to intermediates, which may be used either as sources of energy or in synthetic processes; or glucose may be released into the general circulation.

Most extrahepatic tissues (brain, muscle, adipose tissue, etc.) utilize glucose for their supply of energy. In order that such tissue may utilize glucose, it must permeate the cell membrane and be trapped by the cell. Immediately after entering a cell, glucose is phosphorylated to glucose-6-phosphate (G-6-P) and thus trapped by the cell, because the reaction is irreversible and the cell membrane is impermeable to phosphate esters. After glucose is trapped in the hepatic cell as G-6-P, it may follow one of four primary avenues (Fig. 31-1). It may be converted under the influence of glycogen synthase I to glycogen and stored. It may be broken down via the hexose monophosphate shunt (HMP). G-6-P may also undergo anaerobic glycolysis to form lactic acid via the Embden-Meyerhof pathway (EMP). Under aerobic conditions, either pyruvate formed via this route undergoes oxidative decarboxylation and irreversibly reacts with co-enzyme A (CoA) to form acetyl CoA, or it may be carboxylated to form malate. Therefore, pyruvate may serve as precursor to both ingredients necessary for the tricarboxylic acid (TCA) or Krebs cycle. The TCA cycle is a major source of CO_2 and contributes to the production of high-energy phosphate bonds via oxidative phosphorylation. G-6-P may also be dephosphorylated by glucose-6-phosphatase (G-6-Pase) and released from the liver cell into the extracellular fluid (ECF) as free glucose.

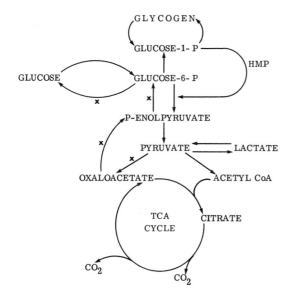

x Gluconeogenesis steps

Figure 31-1
Pathways of carbohydrate metabolism.

The energy for biological work is made available from the degradation and transformation of various substrates. Compounds such as adenosine triphosphate (ATP) which contain high-energy bonds are produced in the process of metabolism of energy-yielding substrates (exergonic processes). The cleavage or hydrolysis of the high-energy bonds is then coupled with energy-requiring processes (endergonic) for biological work—synthetic and secretory processes, muscle contraction, nerve conduction, etc. Compounds such as ATP thus act as the currency which maintains the metabolic economy of the cell. High-energy phosphate bonds may be generated in glycolysis, in the hexose monophosphate shunt, in the formation of acetyl CoA, and in the TCA cycle. Various monographs and biochemistry textbooks contain detailed maps of enzymatic function and biological oxidations.

Acetyl CoA plays an important pivotal role in intermediary metabolism because it is the compound which is common to carbohydrate, protein, and fat metabolism. Furthermore, through this compound the metabolic products of each enter the common pathways that lead to oxidative processes furnishing the organism with energy. When acetyl CoA is oxidized to CO_2 and H_2O, phosphorylation is coupled with oxidation, and 12 moles of ATP are formed per revolution of the TCA cycle (per mole of acetate utilized). Thirty-eight moles of ATP are formed by complete oxidation of 1 mole of glucose by way of the glycolytic and TCA pathways. This amounts to an energy storage of a considerable magnitude, because 1 mole of ATP stores about 11,000 calories of free energy.

Fat Metabolism

A small amount of fat is synthesized in the liver, but most of it is synthesized in adipose tissue. Because 1 gm of fat yields about 9 kcal as compared with 4 kcal per gram of protein or carbohydrate, the fat depots of the body constitute a concentrated store of energy. Formerly, the adipose depots were considered relatively inert, static forms of energy, but newer techniques have shown them to be exceedingly active metabolically. They furnish about 40 percent of the energy content of a normal individual and represent the major source of stored energy.

The fatty acids synthesized in cells derive their carbon from acetyl CoA and malonyl CoA. The synthesis of long-chain fatty acids (lipogenesis) results from carboxylation of acetyl CoA with the formation of malonyl CoA. Repeated condensations of malonyl CoA plus decarboxylation and subsequent reductions in the presence of nicotinamide-adenine dinucleotide phosphate (NADPH) lead to the formation of fatty acids. The free fatty acids then can combine with glycerol and form triglyceride. The degradation of fatty acids (lipolysis) proceeds stepwise by the removal of two carbons at a time and results in the formation of acetyl CoA. Thus the fatty acid is broken down to acetyl CoA units, and these are eventually channeled into the TCA cycle. It is quite evident that any excess of caloric intake, as either carbohydrate or fat, can be channeled via acetyl CoA to result in increased lipogenesis, and that the bulk of storage fat depots must increase if the rate of the lipogenesis exceeds that of lipolysis. Acetyl CoA may also be converted to ketones instead of entering the TCA cycle. The circulating ketone bodies are formed in the liver, and production of them increases as one increases the supply of fatty acids to the liver. Oxidation of the ketone bodies occurs in the extrahepatic tissue, and the circulating level increases when their production exceeds their utilization (Fig. 31-2).

Protein Metabolism

The amino acids absorbed from the gastrointestinal tract either are used by the metabolic mill for the synthesis of protein or contribute their carbon chains for the synthesis of fatty acids, for glucose and glycogen formation (gluconeogenesis), or as an energy source. Eight amino acids must be supplied in the diet of man, because none of these can be synthesized. They are isoleucine, leucine, lysine, methionine, phenylalanine, threonine, tryptophan, and valine. Other amino acids such as alanine, glycine,

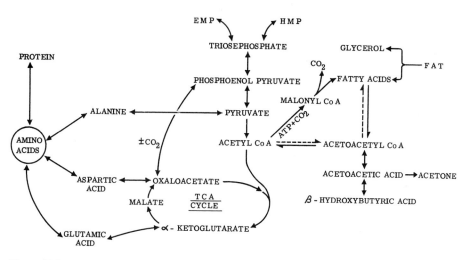

Figure 31-2
Relationship of certain phases of lipid and protein metabolism with carbohydrate metabolism.

glutamic acid, and aspartic acid may be synthesized from specific intermediates of the glycolytic pathway or TCA cycle.

All or a portion of the carbon chain of an amino acid may be converted to glycogen or to ketone bodies. Amino acids that yield pyruvate are channeled into carbohydrate metabolism and classified as glycogenic; those that form acetoacetyl CoA and contribute primarily to reactions of fat metabolism are ketogenic. The great majority of the amino acids are glycogenic.

Man's need for utilizable dietary protein during growth is great, and he has a continuing constant need for protein in order to maintain and repair cells. As indicated in Figures 31-1 and 31-2, liver and muscle glycogen stores may be replaced with glucose derived from protein, and fat may be synthesized and stored in adipose tissue as a result of ingesting protein calories in excess of needs.

ROLE OF INSULIN IN METABOLISM

Numerous effects on the body have been attributed to the administration of insulin. However, many of these alterations in body functions are probably subsidiary to the effect of insulin on glucose transfer from the extracellular fluid into certain cell types. Levine and co-workers (1955) have shown that insulin facilitates the transfer of glucose across the cell membrane. Tissues such as skeletal and cardiac muscle, fibroblasts, mammary gland, the anterior pituitary, and adipose tissue require insulin for the transfer of glucose across their cells membranes, but the permeability of nervous tissue, erythrocytes, intestinal mucosa cells, renal tubules, and hepatic cells to glucose is independent of the action of insulin. Insulin also increases the rate of transfer of amino acids and fatty acids through the cell membranes, particularly in cardiac and skeletal muscle.

Insulin can be considered a potent anabolic agent. It promotes the synthesis of glycogen in both liver and muscle; it promotes the incorporation of amino acids into peptides and proteins in muscle, adipose tissue, and liver; it suppresses lipolysis and promotes triglyceride synthesis in adipose tissue; and it stimulates RNA and DNA synthesis. Thus, overall, insulin promotes the storage of energy and is necessary for normal growth. It is also an effective agent in regulating the blood glucose concentration, being the only hormone to lower the blood glucose; all others tend to elevate it.

Effects of Pancreatectomy

The many and varied effects of insulin on metabolism are best illustrated by the derangements in metabolism which occur in the absence of insulin in diabetes mellitus. Insulin deficiency brings about a decrease in the rate of translocation of glucose from the ECF into the insulin-sensitive cells, a decrease in glucose utilization by cells, and a decrease in glycogen formation. These effects, coupled with increased glycogenolysis and gluconeogenesis, lead to an increased blood glucose level. The oxidation of glucose via the HMP is decreased in relation to the glucose oxidized over the EMP and TCA cycle in liver and adipose tissue. However, to gain entry into the cells where it can be utilized, the concentration of glucose in the ECF must be significantly elevated. When the degree of hyperglycemia reaches a high level, the amount of glucose filtered at the renal glomerulus exceeds the tubular maximum (Tm) for glucose, and glucosuria results. The excessive filtered load of glucose acts as an osmotic diuretic, and the volume of urine excreted increases markedly. Large amounts of Na^+ and Cl^- are also lost as a result of the high rate of urine flow. The ECF volume tends to be depleted, and dehydration, thirst, and polydipsia occur as a consequence of the large amounts of water and salt lost in the urine. The decrease in carbohydrate utilization brings about an essential carbohydrate starvation, and, by an unknown mechanism, centers within the hypothalamus are affected so as to augment the food drive (Chap. 8) and polyphagia results in experimental animals.

The rate of lipogenesis is decreased to about 5 percent of normal. This decline in fatty acid synthesis may be caused by a deficiency in NADPH. However, the decreased lipogenesis may be related to a deficiency in end products of carbohydrate metabolism which are necessary for the formation of fatty acids. The rate of catabolism of fats is increased greatly in insulin deficiency. As the fat stores are mobilized, the fat content of the blood increases. With decreased carbohydrate utilization, most of the energy source of the diabetic must be derived from adipose tissue fat. In this state, lipolysis is increased to such an extent that acetyl CoA is produced in excess of the amount which can be handled by the TCA cycle, and acetoacetyl CoA tends to accumulate in the liver. The excess is channeled at an increasing rate to ketone body (acetoacetic acid, β-hydroxybutyric acid, and acetone) production. When the rate of production of ketone bodies exceeds the capacity of extrahepatic cells to oxidize them, ketonemia and ketonuria result.

The ketone bodies are acids and are buffered by the bicarbonate buffer system, and they markedly decrease the blood pH and therefore lead to an acidosis. Ketones are excreted mainly as sodium and potassium salts, so that the loss of base contributes further to acidosis. Compensatory reactions, such as hyperventilation and the resultant increase in loss of carbon dioxide via the lungs, are triggered by stimulation of the respiratory centers by the lowered pH of the blood. Thus, the HCO_3^-/H_2CO_3 ratio returns toward normal, and the pH is elevated. The respiratory alkalosis may compensate for nearly all the base lost in mild diabetes acidosis, and, along with renal compensations (increase in loss of fixed anions by excreting them with H^+ or NH_4^+), complete compensation may be attained and a normal pH restored. However, in severe insulin insufficiency, uncompensated metabolic acidosis occurs. The presence of an excess of ketone bodies depresses the oxygen consumption of the central nervous system (CNS) and results in depression of CNS activity. If the depression becomes marked, mental disorientation and coma follow.

Protein metabolism is also affected by the lack of insulin. There is an increase in the rate of protein catabolism, and this, coupled with impaired protein synthesis, leads to negative nitrogen balance, impaired growth, delayed healing, and increased susceptibility to infections. The main source of glucose (other than exogenous) excreted in the diabetic is probably protein. In order that protein may be converted to glucose, it must be hydrolyzed to amino acids, which are deaminated and converted to keto acids. The latter may be converted to pyruvate or to ingredients of the TCA cycle. The glycolytic pathway must then be traversed in a reverse direction to form glucose. Because of the increase in protein catabolism and gluconeogenesis, there occur increases in urinary nitrogen and glucose excretion, respectively. Both of the latter contribute further to the osmotic diuresis and the tendency to dehydration.

The effects observed in insulin deficiency appear to result from a diminution in the utilization of glucose for oxidative purposes, lipogenesis, and glycogenesis. The concentration of glucose in the ECF becomes elevated because the rate of transport of the hexose across the cell membrane of most tissues (muscle, adipose tissue) is reduced. A loss of large amounts of available glucose follows because the renal Tm for glucose is exceeded. The relative carbohydrate deficit precipitates lipolysis and proteolysis, which lead to metabolic events whereby the carbon chains of fatty acids and amino acids are utilized as sources of energy. Parts of the substances also may be lost in the urine as ketones, glucose, and nitrogen. The body economy of the diabetic resembles that produced by chronic starvation, and even though the diabetic may avoid acidotic crises, he may succumb eventually to depletion of the body tissues.

Effects of Insulin

Insulin, in contrast to other hormones, does not penetrate the cells it stimulates, and current evidence indicates that it is bound at the cell membrane at specific receptor sites. Insulin receptors are those molecules in the plasma membrane of cells that possess a high degree of affinity and specificity for binding insulin. The manner by which insulin increases the translocation of glucose and certain amino acids from the ECF into sensitive cells does not involve changes in enzymatic activity and remains to be elucidated. Current evidence indicates that insulin alters the activity of membrane-bound enzymes, which release a "second messenger" into the cell interior. The second messenger then carries the insulin signal to influence intracellular enzymatic activity. Two membrane-bound enzyme systems have been implicated in the transmission of the insulin signal: the adenyl cyclase–cyclic AMP system and the Mg^{++}-activated $(Na^+ + K^+)$–ATPase system. Insulin has been shown to suppress adenyl cyclase activity in many cells and promote changes in intracellular enzyme activities which depress lipolytic activity, and to stimulate glycogenic activity. Insulin stimulates membrane ATPase activity and accentuates the accumulation of both K^+ and Mg^{++} in the intracellular compartment of many cell types. It has been proposed that as Mg^{++} accumulates at intracellular loci it acts as a second messenger to influence intracellular enzyme function.

Insulin increases the rate of glucose uptake by muscle cells and adipose tissue and thus induces a hypoglycemia and makes more glucose available for utilization. Insulin decreases G-6-Pase activity and promotes the synthesis of glucokinase in the liver, thereby permitting more G-6-P to enter the glycogenic and glycolytic pathways and allowing less free glucose transfer from the liver to the ECF. The rate of transformation of glycogen synthase from the D to I form, which accelerates the rate of glycogen synthesis, is also stimulated by insulin. In addition to increasing the synthesis of glucokinase, insulin increases the rate of synthesis of phosphofructokinase and pyruvate kinase. It increases HMP activity, and more glucose is oxidized via this pathway. Because of the initial actions, the glucose concentration of the ECF is depressed, the amount of glucose filtered at the renal glomerulus becomes less than the Tm for glucose, glucosuria ceases, and the diuretic effect of glucose is lost. Thus less glucose, water, and salt are lost from the ECF.

The effect of insulin on glucose assimilation in muscle is much more rapid than are its effects on the two hepatic enzymes, glucokinase and G-6-Pase. However, the slower response of these two enzymes to insulin may serve as an important buffer against abrupt changes in blood glucose concentration. Severe hypoglycemia might be expected to be a consequence of an elevated insulin secretion which would follow the ingestion of a high-carbohydrate meal in the normal individual if the activity of the two enzymes paralleled glucose assimilation by extrahepatic tissue.

The administration of insulin to the diabetic returns the rates of lipogenesis and lipolysis to normal levels. The rate of fatty acid synthesis is increased, because the greater assimilation of glucose increases the availability of the intermediates of carbohydrate metabolism which are utilized in fat synthesis. The rate of lipolysis is decreased, thus reducing ketone body production to a level at which extrahepatic tissue oxidation of these substances is equal to or less than their production. Therefore, ketonemia and the resultant ketonuria are inhibited. As the circulating ketone bodies return to normal, the alkali reserve returns to normal, less base is lost via the urine, and electrolyte metabolism returns to normal.

The rates of protein synthesis and gluconeogenesis are returned to normal levels following administration of insulin to the diabetic. The stimulating effect of insulin on protein synthesis is dependent upon adequate glucose metabolism, which supplies the intermediates and cofactors necessary for the process. Insulin also stimulates the transport of amino acids into cells and thus makes them available to enter into synthetic reactions. Decreased gluconeogenesis results in a decrease of urinary nitrogen excretion, and the contribution of protein to the formation of blood glucose declines.

All the metabolic alterations of the diabetic are readily reversed by insulin. These effects of insulin increase the utilization of glucose by peripheral tissue, promote the formation of hepatic and muscle glycogen, stimulate the synthesis of fat and protein, and decrease lipolysis and gluconeogenesis. The administration of excessive amounts of insulin or hypersecretion of insulin causes hypoglycemia. Because the CNS is largely dependent upon blood glucose for its nutrition, this tissue reacts as if it were deprived of oxygen (Chap. 6).

Regulation of Insulin Secretion

The development of an accurate radioimmunoassay for plasma insulin has provided a major advance in the understanding of insulin secretion. Many factors influence the rate of secretion of insulin from the beta cells of the pancreas. One might expect that its secretion would be favored by high ECF concentrations of glucose, amino acids, and fatty acids, metabolic substrates that can be stored by the organisms. Numerous factors, both hormonal and nonhormonal, modulate the secretion of insulin. The transplanted pancreas is capable of maintaining the blood glucose level and insulin level with a high degree of precision. The major physiological stimulus for insulin secretion is glucose. Figure 31-3 depicts the effect of a carbohydrate meal on the plasma insulin concentration in a normal subject. The amount of insulin secreted from an isolated pancreas or isolated pancreatic islet preparations, in vitro, can be increased or decreased by the perfusion of the preparation with blood containing high or low glucose concentrations, respectively. Several other sugars, as well as amino acids and fatty acids, stimulate the secretion of insulin.

Neurogenic mechanisms may also play a role in the regulation of insulin secretion, because stimulation of vagal fibers to the pancreas results in increased insulin secretion. Similarly, acetylcholine and several cholinergic drugs stimulate insulin secretion. The adrenergic agents, norepinephrine and epinephrine, directly inhibit insulin secretion.

Agents such as theophylline and cyclic AMP enhance or promote the release of insulin from the pancreatic beta cells. The compound *alloxan* exerts a selective destruction of the beta cells of the islets; hence its administration leads to diabetes mellitus. The severity of the diabetic state may be of varying degrees. Following pancreatectomy of an alloxanized animal, the insulin requirement is decreased, presumably as a result of the removal of a source of glucagon and pancreatic enzymes which participate in the digestion and absorption of sugars. This method of production of diabetes has advantages over pancreatectomy in that the experimental animal does not lack the exocrine function of the gland. However, alloxan may damage the liver and kidneys if too high a dose is used. The antibiotic streptozotocin also specifically destroys the beta cells and has fewer side-effects than alloxan.

The endocrine portion of the pancreas is independent of direct control by the pituitary gland. However, the endocrine functions of the pancreas can be either directly or indirectly altered by several hormones.

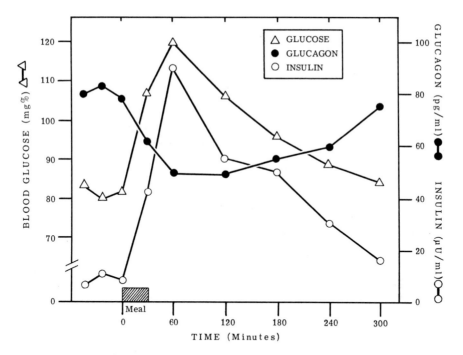

Figure 31-3
Changes in plasma insulin and glucagon concentration in response to a carbohydrate meal.

   The intestinal hormones, secretin and pancreozymin, each stimulate insulin secretion directly. Both pancreatic and intestinal glucagon also stimulate insulin secretion directly. Exogenous somatostatin has a profound inhibitory effect upon insulin secretion. However, it is not known whether or not endogenous somatostatin reaches the beta cells under normal conditions. The prostaglandins also inhibit insulin secretion. Many agents influence the release and/or action of insulin. Some are capable of producing temporary or permanent diabetes by virtue of their ability to antagonize the action of insulin or to destroy the beta cells, the source of insulin.
   The insulin level of the blood may also play a part in the regulation of insulin secretion because, if insulin is perfused prior to and during glucose infusion, degranulation of the beta cells is less than that observed following the infusion of glucose alone, in spite of the presence of marked hyperglycemia in both cases.
   The anterior pituitary gland has indirect influences on insulin secretion. Carbohydrate metabolism is affected by growth hormone, ACTH, and TSH. The effects of the latter two hormones on carbohydrate metabolism are indirect because their influence is mediated via the hormones of their respective target organs.
   Houssay demonstrated that hypophysectomy reduces the effects of pancreatectomy in the dog. Furthermore, the hypophysectomized animal is extremely sensitive to the action of insulin, and growth hormone causes a decided aggravation of diabetes in the hypophysectomized diabetic. In acromegaly and gigantism, there is a strong tendency toward hyperglycemia and the diabetic state. It has been demonstrated that the administration of growth hormone over prolonged periods produces degeneration of the beta cells and results in the production of diabetes. The effect of growth hormone on insulin secretion is indirect and stems from the ability of the hormone to elevate the blood glucose. Growth hormone decreases the peripheral utilization of glucose, increases the release of glucose from the liver, increases the release of a pancreatic hyperglycemic factor, decreases the fixation of insulin to tissues, and thus leads to the diabetic state.

Adrenalectomy ameliorates the symptoms of diabetes whereas the administration of cortisol-like compounds increases the severity of the symptoms. The diabetogenic effect of these hormones results from a marked increase in gluconeogenesis, lipolysis, and ketogenesis and a decrease in the utilization of glucose. Therefore, their administration promotes the secretion of insulin because they produce hyperglycemia.

Diabetes may be produced by the administration of thyroid hormones in the partially depancreatized animal. It results from beta cell degeneration and subsequent decline in insulin secretion. Diabetes increases in severity in hyperthyroidism and decreases in hypothyroidism. The effects of thyroid hormones on carbohydrate metabolism include an accelerative rate of gluconeogenesis and increased glucose oxidation in tissue, both of which contribute to a decrease in the rate of insulin secretion.

Epinephrine and norepinephrine are hyperglycemic agents by virtue of their ability to increase the rate of hepatic and muscle glycogenolysis. The lactic acid produced as a result of muscle glycogenolysis is liberated into the circulation and used in the hepatic synthesis of glucose. Hypoglycemia is a potent stimulus to the sympathetic nervous system with increased catecholamine release. This might afford an important buffer mechanism against acute episodes of hypoglycemia. The catecholamines also increase the release of free fatty acids from adipose tissue. This ability is shared with thyroxine and cortisol, whereas insulin decreases the release of free fatty acids from adipose tissue.

ROLE OF GLUCAGON IN METABOLISM

Glucagon, in contrast to insulin, can be considered a catabolic agent. It acts to decrease energy stores and to increase the supply of glucose and fatty acids for oxidation.

Because glucagon has a rapid hyperglycemic effect, it has been regarded as being primarily involved with the regulation of glucose metabolism. It is a powerful hepatic glycogenolytic agent and increases the rate of gluconeogenesis. These actions are mediated by cyclic AMP. Glucagon also has a lipolytic action in that it stimulates lipase activity in both liver and adipose tissue. Thus, it stimulates free fatty acid release from adipose tissue and increases fatty acid uptake and oxidation in liver and muscle. Being a protein-catabolic agent, glucagon increases the rate of uptake and deamination of amino acids by the liver. However, it depresses the uptake of amino acid by muscle cells. The net outcome of the lipolytic and proteolytic effects of the hormone is to supply more fatty acids and amino acids for the hepatic gluconeogenic pathway.

Glucagon has been shown to have a stimulating effect upon insulin secretion, the adrenal medulla, and cardiac function. It increases the rate of secretion of catecholamines from chromaffin tissue of the adrenal. It also has two important effects upon the heart; they are direct and not mediated through catecholamine release. Glucagon acts directly upon the sinoatrial node and causes an increase in heart rate. In addition to its chronotropic effect, the hormone also exerts a positive inotropic effect upon the myocardium and papillary muscle.

Control of Glucagon Secretion

The plasma glucose concentration appears to be the primary regulator of glucagon secretion. A rise in plasma glucose concentration inhibits the secretion of glucagon, and a fall stimulates secretion. The response of the plasma glucagon concentration to plasma glucose levels in normal subjects is shown in Figure 31-3, which also depicts the relationship between circulating insulin and glucagon levels. A reciprocal relationship exists between the rates of secretion of glucagon and insulin, and both are regulated by the concentration of glucose perfusing the endocrine pancreas. Somatostatin is a potent hypoglycemic agent when administered intravenously. This effect is mediated in part via a marked inhibitory effect on glucagon secretion. However, whether somatostatin plays a significant role in glucagon and insulin secretion is not clear, because it has not been shown that the hypothalamic polypeptide reaches the alpha and beta cells in significant quantities under physiological conditions.

REFERENCES

Bray, G. A., and L. A. Campfield. Metabolic factors in control of energy stores. *Metabolism* 24:99–117, 1975.

Chidedel, E. W., J. Palmer, D. J. Koerker, J. Ensick, M. B. Davidson, and G. J. Goodner. Somatostatin blockade of acute and chronic stimuli of the endocrine pancreas and the consequences of this blockage on glucose homeostasis. *J. Clin. Invest.* 55:754–762, 1975.

Daughaday, W. H., A. C. Herington, and L. S. Phillips. The regulation of growth by endocrines. *Annu. Rev. Physiol.* 37:211–244, 1975.

Krahl, M. E. Endocrine function of the pancreas. *Annu. Rev. Physiol.* 36:331–360, 1974.

Lawrence, A. M. Glucagon. *Annu. Rev. Med.* 20:207–222, 1969.

Lehninger, A. L. *Biochemistry*. New York: Worth, 1975.

Levine, R. Mechanisms of insulin secretion. *N. Engl. J. Med.* 283:522–526, 1970.

Levine, R., and M. S. Goldstein. On the mechanisms of action of insulin. *Recent Prog. Horm. Res.* 11:343–380, 1955.

Mayhew, D. A., P. H. Wright, and J. Ashmore. Regulation of insulin secretion. *Pharmacol. Rev.* 21:183–212, 1969.

Sutherland, E. W., and G. A. Robinson. The role of cyclic AMP in the control of carbohydrate metabolism. *Diabetes* 18:797–819, 1969.

Unger, R. H. Glucagon physiology and pathophysiology. *N. Engl. J. Med.* 285:443–449, 1971.

West, E. S., W R. Todd, H. S. Mason, and J. T. Van Bruggen. *Textbook of Biochemistry* (4th ed.). New York: Macmillan, 1966.

White, A., P. Handler, and E. L. Smith. *Principles of Biochemistry* (rev. ed.). New York: McGraw-Hill, 1973.

Wool, I. G., W. S. Stirewalt, K. Kurihara, R. B. Low, P. Bailey, and D. Oyer. Mode of action of insulin in the regulation of protein synthesis in muscle. *Recent Prog. Horm. Res.* 24:139–213, 1968.

32. Endocrine Control of Calcium Metabolism

Ward W. Moore

Several endocrine glands and their hormones are involved in the regulation of calcium and phosphorus metabolism. Two of them, however, both of branchial origin, secrete hormones that have marked effects on the calcium activity of the extracellular fluid (ECF) and on bone. The parathyroid glands secrete parathyroid hormone in response to a low serum calcium activity or hypocalcemia and induce a hypercalcemic action. The cells, which are derivatives of the ultimobranchial bodies, secrete calcitonin or thyrocalcitonin in response to an elevated serum calcium activity or hypercalcemia. The effect of the actions of this hormone is hypocalcemic in nature. Together, the two hormones exert a precise homeostatic control over the serum calcium.

The functional significance of the parathyroid glands was not realized until about 1909, when MacCullum and Voegtlin demonstrated a lowering of blood calcium in parathyroidectomized animals. Before then, effects of their removal were generally attributed to the removal of thyroidal tissue. The relationship of the parathyroids to calcium and phosphorus metabolism had been realized for several years before their endocrine nature was fully established in 1925 by Collip.

The functional significance of the ultimobranchial glands or their derivative cells has been realized only a short time. Initial observations indicating that thyroparathyroidectomy impaired the control of hypercalcemia were not appreciated. In 1961 Copp and co-workers suggested that the parathyroid glands secreted a rapidly acting hypocalcemic agent and called it calcitonin. However, it was subsequently shown that this agent could be extracted from mammalian thyroid glands but not from parathyroid glands. The name *thyrocalcitonin* was introduced. Recent comparative studies have shown that calcitonin is secreted by the ultimobranchial glands or, in mammals, cells derived from them.

ANATOMY AND BIOCHEMISTRY

The parathyroid glands are small, paired bodies found in the region of the thyroid gland and weigh in total 20 to 40 mg. There are usually four glands, which develop from the third and fourth pairs of branchial pouches. The blood supply to the glands is through the inferior thyroid arteries, although the glands may receive some blood from the superior thyroid arteries. Each gland is surrounded by a connective tissue capsule, and septa divide the glands into lobules. Microscopically, the glands resemble hyperplastic thyroid tissue. Two types of epithelial cells are present: the *chief cells,* the usual source of the hormone of the gland, and the *oxyphil cells,* which have the potential to secrete parathyroid hormone and generally do not appear until after puberty. The functional significance of the oxyphil cells is obscure.

The ultimobranchial glands are derived from the terminal branchial pouch and in nonmammalian vertebrates appear as separate bodies. However, in mammals the cells become embedded in the thyroid. The source of calcitonin is the parafollicular, or "light," or "C" cells. The human parathyroid glands and thymus have been shown to contain biologically active calcitonin. Thus, total thyroidectomy does not remove all the calcitonin-producing cells.

The hormone of the parathyroid glands, *parathyroid hormone* (PTH), is a straight-chain polypeptide containing 84 amino acids and has a molecular weight of about 9600. The biologically active portion of the molecule has a sequence of 34 amino acids, and this portion also has full immunological activity. PTH is derived from a larger protein (molecular weight about 12,000) which is synthesized by the chief cells and termed *pro-parathyroid hormone* (pro-PTH). PTH is formed by cleavage of pro-PTH prior to release into the circulation. Earlier studies of the actions of the hormone were complicated as a result of relatively crude preparations, but recently essentially pure PTH has become available. It is known that the hormone acts at least at three sites, namely, bone, kidney, and gastrointestinal tract, affecting calcium and phosphorus handling by each tissue. *Calcitonin* (CT), contains 32 amino acid residues and has a disulfide bridge between amino acids 1 and 7. Its molecular weight is about 3580, and it has been synthesized. The entire molecule seems to be necessary for complete biological activity. CT has been shown to act on the same tissues that are affected by PTH. Radioimmunoassays are available for both PTH and CT. Normal individuals with serum calcium concentrations between 9 and 10 ng per milliliter have a PTH serum concentration between 200 and 800 pg per milliliter and a CT concentration of 50 to 150 pg per milliliter.

FUNCTIONS OF CALCIUM AND PHOSPHORUS

Calcium is an indispensable mineral constituent of all body tissues and participates in several physiological processes, among which are (1) coagulation of blood, (2) maintenance of cardiac rhythmicity, (3) membrane permeability, (4) neuromuscular excitability, (5) production of milk, and (6) formation of bone and teeth. Primary attention here is focused on this last function, because the other roles of calcium are discussed in preceding chapters.

Calcium is absorbed in the upper intestinal tract, and its absorption is facilitated by the presence of HCl, vitamin D, and protein. Alkali and excessive fat, excessive quantities of phosphate or oxalate, and the presence of chelating agents interfere with the absorption of calcium from the intestinal tract. Most of the calcium which is excreted is lost in the feces, and only a small part is excreted by the kidneys.

Following absorption, the greater part of the calcium is stored in the skeleton, and only a small fraction of the total body calcium is present in other tissue, including the circulating body fluids. The normal serum calcium level is from 9 to 11 mg per 100 ml (5 mEq per liter); about 60 percent of this is ultrafiltrable and diffusible, whereas the remainder is bound to protein. A small fraction of the diffusible calcium is nonionized, but most is ionized.

Phosphorus plays an important role in biological systems because it is involved in the transfer of energy in the intermediary metabolism of foodstuffs; it helps in maintenance of pH of body fluids and is an important constituent of bone. Phosphorus absorption from the gastrointestinal tract is facilitated by acids and an excess of fat and is decreased by high calcium and other cations, and by alkaline salts. About 80 percent of the total body phosphorus is in the skeleton, about 11 percent is in muscle, and the rest is present in the body fluids or distributed in other tissue as organic compounds. The main avenue of phosphate excretion is the kidney, but a small amount appears in the feces. The total phosphorus present in the serum is about 12 mg per 100 ml. This may be divided into three portions: lipid phosphorus (8 mg per 100 ml), ester phosphorus (1 mg per 100 ml), and inorganic phosphorus (3 mg per 100 ml or 2 mEq per liter).

BONE

The manner in which hormones maintain the calcium and phosphorus levels of the internal environment requires a brief discussion of the development and maintenance of bone. Bone has been generally considered an inert tissue. However, studies utilizing isotopes of calcium and phosphorus show that bone is constantly being torn down and remodeled. Bone, in addition to its ground substance, is composed of an organic matrix of connective tissue, collagen, onto which are deposited a complex salt of calcium and phosphate, hydroxyapatite [$3Ca_3(PO_4)_2 \cdot Ca(OH)_2$], and calcium carbonate. More than

98 percent of the calcium of the body and about 80 percent of the body phosphorus are contained in bone.

Three cell types in bone are involved with bone formation or accretion and bone resorption: *osteoblasts, osteoclasts,* and *osteocytes.* The osteoblasts are located on the surface of bone which is forming, and protein (collagen) synthesis is one of the major functions of this cell type. Acid, alkaline, and neutral phosphatases are all associated with the surface of the osteoblasts, and these probably function in the nucleation and calcification of the organic matrix. The role of the osteoblasts appears to be essential in the formation of calcifiable collagen fibers in a suitable ground substance and the alignment of the fibers to form the bone template. The events which initiate calcification of collagen are poorly understood. The mechanisms of nucleation probably involve an enzymatic phosphorylation of collagen. When sufficient concentrations of Ca^{++} and $HPO_4^=$ are present in the ECF, each interacts with specific sites on the collagen fibers and forms the hydroxyapatite crystals. These grow and during the process of growing constitute the exchangeable mineral of bone, and an equilibrium is established between the Ca^{++} and $HPO_4^=$ of the ECF and bone. However, as crystal formation increases, more water is excluded, and because the crystals do not release their ions readily, part of the bone structure becomes a relatively inert, solid, nondiffusible mass. Nearly all (about 99 percent) of the mineral of bone is in this state, the so-called *nonexchangeable bone.* All bone would achieve this state if it were not constantly being remodeled.

The osteoclasts perform a role in processes involved in dissolving bone mineral and the destruction of bone collagen. It is believed that bone resorption results from increased acid phosphatase activity, acid production, and proteolytic activity, which lead to the release of Ca^{++} and $HPO_4^=$ into the ECF and the breakdown of collagen to its constituent amino acids. Hydroxyproline is one of the amino acids unique to collagen, and its rate of excretion in the urine is a reliable index of collagen breakdown. It has been estimated that about 500 mg of calcium from old bone is resorbed under the influence of the osteoclasts each day. Naturally, this old bone must be replaced by an equivalent amount of new bone formation, under osteoblast activity, if equilibrium is to be attained. Conceptually, one may regard the ions of the ECF as being in simple equilibrium with those of exchangeable bone. Thus, the exchangeable bone acts as a buffer in controlling the concentration of Ca^{++} and $HPO_4^=$ in the internal environment.

The osteocyte is required for the homeostatic regulation of bone metabolism. It is capable of promoting both bone formation and bone resorption and is probably very important in the rapid mobilization of calcium induced by PTH.

Vitamin D is essential for normal calcium homeostasis. It is concerned with the formation of bone, and aids in maintaining Ca^{++} levels of the ECF. It reduces the excretion of Ca^{++} in the feces and that of $HPO_4^=$ in the urine. In the absence of vitamin D, the Ca^{++} and $HPO_4^=$ concentrations decline, calcification of collagen ceases, and bone growth stops. The bones readily bend and break and the condition *rickets* develops. Although vitamin D can promote the maximal retention of Ca^{++} and $HPO_4^=$ and maintain the product of the two concentrations above a level at which bone calcification can occur, it cannot do so unless the dietary Ca^{++} is maintained above a critical level.

EFFECTS OF PTH

A considerable amount of controversy has existed concerning the primary effects of PTH. One school contended that the effect of the hormone was primarily on bone; another proposed that the primary site of action was in the kidneys, and that all other effects followed as a consequence of its renal action; another suggested that there are two hormones secreted by the parathyroids, one acting on the kidney, the other on bone. Present data indicate that only one hormone, *parathyroid hormone,* is secreted by the parathyroid gland, and it has three primary sites of action, the aforementioned bone and kidney, and the gastrointestinal tract. The effects of PTH on bone, kidney, and intestine are mediated by stimulation of adenyl cyclase leading to an increase in intracellular cyclic AMP.

The actions of PTH help to maintain the concentration of calcium ion activity in the plasma and protect the organism from hypocalcemia. In addition, PTH permits the extensive remodeling of bone while maintaining a normal level of calcium ion activity. PTH is a potent hypercalcemic agent.

Effect on Bone

Administration of PTH or excessive release of its secretion in parathyroid hyperplasia causes demineralization of bone and a consequent softening of the skeletal system. This demineralization or resorption of bone results from the action of PTH on the osteoclasts and osteocytes. One of the early effects of PTH is a transient hypocalcemia, and this results from an initial uptake of calcium into the skeleton. However, it is followed by the removal of calcium from bone, and the net response to PTH is one of hypercalcemia. Diuresis with increased renal calcium and phosphate loss, the tendency toward increased coagulability of blood, and decreased excitability of nerve result from the elevated Ca^{++} level.

That a primary action of the hormone occurs with bone is shown by the following observations: (1) Transplantation of parathyroid tissue to membranous bone of the skull leads to bone resorption at the site of contact with the transplant, while bone deposition occurs on the opposite surface; (2) the addition of parathyroid tissue or PTH to bone grown in tissue culture leads to increased osteoclast activity and bone resorption; and (3) a prompt fall in the plasma calcium level occurs in the nephrectomized animal following parathyroidectomy. The data indicate that PTH, through its effects on bone cells, increases the rate of bone resorption. PTH induces a variety of changes in the osteoclasts and osteocytes: increases in numbers, enhanced lysosomal activity, enhanced collagenase activity, and enhanced organic acid formation. These activities lead to the destruction of bone matrix. The concentrating hydroxyproline in plasma and urine is a reliable indicator of the rate of destruction of the collagen portion of the matrix. The greater the concentrations, the greater the rate of breakdown of collagen.

Effect on the Kidney

Parathyroid hormone has two renal actions; it induces phosphaturia and decreases calcium excretion. The hormone acts on the renal proximal tubule to reduce the reabsorption of phosphate. This effect of PTH involves a parallel inhibition of sodium and bicarbonate reabsorption in a proximal tubule. The decrease in clearance of calcium induced by PTH probably results from the action of the hormone on the distal nephron. Many factors other than PTH and CT can affect the renal handling of both $HPO_4^=$ and Ca^{++}: dietary intake, plasma concentrations, filtered loads, the renal handling of sodium, adrenal steroids, and growth hormone, etc.

The outstanding metabolic changes that follow parathyroidectomy are *hypocalcemia, hyperphosphatemia,* and *hypocalciuria.* As a result of the fall of Ca^{++} activity in plasma, symptoms attributed to increased neural and muscular excitability are evident. However, the initial change observed following parathyroidectomy is an increase in renal calcium excretion; the subsequent decrease in renal excretion of calcium occurs only after a significant fall in plasma calcium.

Effect on the Gastrointestinal Tract

Experimental studies have indicated that parathyroidectomy results in a 50 percent decrease in the rate of absorption of radiocalcium from the intestine within four hours after the operation. Parathyroid hormone has been shown to stimulate calcium absorption from the intestinal tract in man. This effect of PTH is slow in onset and is dependent upon vitamin D and the dietary intake of both calcium and phosphate. Recently PTH has been shown to increase the uptake of both $HPO_4^=$ and magnesium across the intestinal mucosa. The effects of PTH are diagrammed in Figure 32-1.

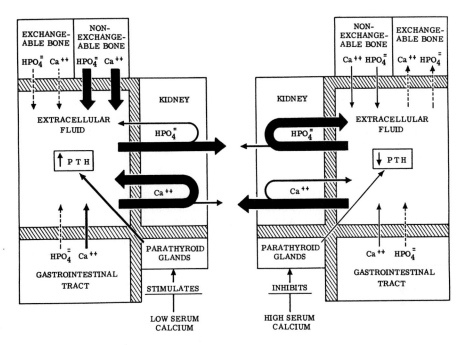

Figure 32-1
Actions and regulation of secretion of PTH. Illustrated are the primary factors regulating PTH secretion and the *early* changes in Ca^{++} and $HPO_4^=$ metabolism over which PTH has a regulatory function. The dashed arrows indicate a lack of PTH effect, and the width of the arrows indicates the magnitude of the response.

EFFECTS OF CALCITONIN

Calcitonin, in contrast to PTH, protects the organism from hypercalcemia, and it reduces the concentration of both calcium and phosphate in the plasma. CT acts on three primary target tissues, bone, kidney, and gut, and induces its effects by increasing the removal of Ca^{++} and $HPO_4^=$ from the ECF and/or by decreasing the rate of entry of the ions into the ECF.

Calcitonin has been shown to inhibit both osteoclastic and osteocytic bone resorption. As a result of its effects on these cells it reduces the net rate of movement of calcium from bone to ECF and induces hypocalcemia. In addition, the rate of collagen destruction is decreased, and the plasma concentration and urinary excretion of hydroxyproline are depressed. In vitro studies have shown that this is a direct effect on bone. As noted above, PTH added to bone in tissue culture leads to a decrease in osteoblast activity and to increased osteocyte and osteoclast activity, hydroxyproline release, and demineralization. However, CT added to PTH-stimulated bone in vitro has no effect on osteoblast activity but reduces osteocyte and osteoclast activity and causes a decrease in the release of calcium. Further, CT inhibits citrate production. In vivo studies on the rate of removal of calcium from PTH-sensitive pools, the urinary excretion of hydroxyproline, and urinary excretion of ^{85}Sr and ^{45}Ca following CT administration indicate that the primary effect of CT is to prevent bone resorption. Some evidence indicates that CT may increase the rate of bone formation or accretion.

The initial effects of CT, in addition to hypocalcemia and hypophosphatemia, are increases in the renal clearances of phosphate and calcium. The increased renal loss of phosphate continues, but the clearance of calcium decreases, and the long-term effects are hyperphosphaturia and hypocalciuria. Micropuncture studies indicate that CT inhibits the reabsorption of phosphate in the proximal tubule and allows increased loss of phosphate in the urine. Using isolated kidney tubules, it has been shown that CT inhibits the extrusion of calcium from the cell. In contrast, PTH increases the influx of calcium into renal cells. Thus the net effect of the two hormones, PTH and CT, on the renal handling of calcium and phosphates is the same. Calcitonin also acts to reduce the reabsorption of sodium, chloride, magnesium, and potassium by cells of the renal proximal tubules. CT administration induces a saline diuresis and a loss of ECF volume and body weight. These losses in sodium and volume result in increases in plasma renin and aldosterone concentration.

Calcitonin depresses the absorption of calcium by the small intestine.

MECHANISM OF ACTION OF PTH AND CT

It has been shown that PTH causes a rapid activation of skeletal and renal adenyl cyclase. In bone the increase in cyclic AMP is assumed to promote the release and synthesis of lysosomal enzymes, which promote osteolysis and calcium mobilization. It has been postulated that the effect of CT is mediated via the same system. The mechanisms of action of PTH and CT on renal tissue are mediated by an increase in intracellular levels of cyclic AMP. The observations that PTH and CT have similar effects on cyclic AMP levels in both bone and renal tissue must indicate that separate receptors exist for the two hormones. The manner in which the effects of PTH and CT are produced is not known.

REGULATION OF SECRETION OF PTH AND CT

The rates of secretion of PTH and CT are independent of the adenohypophysis and the central nervous system. Their secretion rates are controlled by the Ca^{++} concentration in the plasma. Recent studies, involving perfusion of isolated glands, have revealed that the rate of secretion of either hormone can be altered by increasing or decreasing the Ca^{++} concentration of the perfusate. The secretion rates of PTH and CT appear to be reciprocally related.

Studies have shown that there is a simple linear inverse relationship between plasma Ca^{++} and plasma PTH. These studies were conducted over a range of plasma Ca^{++} of 4 to 12 mg per 100 ml (2 to 6 mEq per liter) and they lead to the conclusion that Ca^{++} regulates PTH secretion through a proportional control mechanism. When the concentration of Ca^{++} rises, the glands are inhibited and secrete less hormone, and when the Ca^{++} falls, the glands are stimulated to secrete more hormone. Similarly, high-calcium diets lead to parathyroid hypoplasia, and low-calcium diets result in hypertrophy and hyperplasia of parathyroid glands. Thus, a negative feedback system exists between the parathyroid glands and the circulating Ca^{++} level. Following a depression of Ca^{++} levels, PTH is secreted, and it acts on the intestinal cells to increase calcium absorption; it acts on the kidney tubules to increase reabsorption of Ca^{++} and loss of $HPO_4^=$ from the glomerular filtrate; and it acts upon bone cells to increase osteoclast and osteocyte activity, leading to a release of Ca^{++} from nonexchangeable bone. The hormone also increases the rate of phosphate released by bone, which is offset by increased urinary excretion of phosphate. The final result is an increased plasma Ca^{++} and decreased $HPO_4^=$. When the plasma Ca^{++} concentration reaches a critical level, the parathyroid gland is inhibited and the PTH secretion decreases and the changes are reversed.

PTH is responsible for the hour-to-hour regulation of Ca^{++} concentration of the blood. Studies utilizing isotopic calcium have shown that an amount of calcium equivalent to the total blood calcium is replaced every minute in the young animal. Two mechanisms are responsible for this rapid turnover. The faster mechanism is one of ion exchange between blood and exchangeable bone and is independent of hormonal action. As a result of this exchange, the serum calcium concentration rarely falls below 6 mg per 100 ml, even after parathyroidectomy. The second mechanism is a function of

PTH and involves a calcium-mobilizing effect of the hormone on bone. This effect maintains the serum calcium level at about 10 mg per 100 ml. In addition, an increased rate of bone resorption leads to an increased release of $HPO_4^=$ into serum. If the effect of PTH on bone were the only means of regulating the Ca^{++} level of the serum, the feedback system would lead to wide oscillations in the Ca^{++} level of the serum, for this effect is relatively slow in onset. The effect of PTH on gastrointestinal tract absorption of calcium is more rapid than that on bone and would tend to reduce the oscillations in the serum Ca^{++} level. However, the kidney is capable of responding quickly to PTH and tends to maintain the level of serum Ca^{++} within narrow limits. It also facilitates the excretion of $HPO_4^=$ by inhibiting tubular reabsorption of $HPO_4^=$ and offsets the rise produced by resorption of bone.

A similar relationship exists between plasma Ca^{++} and CT secretion and plasma CT, except the linear relationship is direct. It holds for Ca^{++} levels between 8 and 20 mg per 100 ml (4 to 10 mEq per liter). Thus, an increase in plasma calcium above 8 mg per 100 ml increases CT secretion, and the increase is in direct proportion to the plasma Ca^{++}. A negative feedback system also operates, therefore, between CT secretion and the circulating Ca^{++} level. Following elevation of Ca^{++} levels, CT is secreted and it acts to remove Ca^{++} from the ECF or prevent the addition of Ca^{++} to the ECF via its effects on bone and the kidney. Calcitonin secretion is stimulated following a meal, in effect being probably mediated by the gastrointestinal hormones gastrin, pancreozymin, and enteroglucagon. It has been hypothesized that this effect protects the skeleton from excessive bone resorption during periods of dietary sufficiency.

The two hormones exert a precise homeostatic control over the plasma Ca^{++}. PTH prevents or protects the organism from hypocalcemia, and CT prevents or protects against hypercalcemia.

EFFECTS OF OTHER HORMONES

Several hormones other than PTH and CT influence bone formation and resorption and calcium and phosphorus metabolism. A major effect of growth hormone is to make available an increased amount of calcifiable collagen, and this brings about an increase in the rate of bone formation. It decreases the renal loss of phosphate and increases renal calcium loss. Because of the effects of growth hormone on the plasma Ca^{++} there is a tendency toward hypocalcemia, which leads to more secretion of PTH and thus to a higher rate of bone remodeling.

At puberty the estrogens and androgens promote bone formation and mineralization, and they are required for normal nitrogen, calcium, and phosphorus content in bone of the adult. The thyroid hormones are necessary for normal bone growth and maturation; they increase the rate of turnover of bone and tend to increase the Ca^{++} plasma level. Insulin also appears to function as a growth hormone in bone.

The hormones of the adrenal cortex, particularly cortisol and corticosterone, have a marked effect on bone. They increase the rate of protein breakdown in bone and decrease the rate of collagen synthesis. They thus decrease the total mass of calcified bone. They tend to lower the rate of intestinal absorption of Ca^{++} and increase the loss of renal Ca^{++}. These effects should lead to a rise in PTH secretion.

In summary, negative feedback mechanisms for the regulation of serum calcium and phosphate concentration involve the parathyroids, cells derived from the ultimobranchial body, bone, the gastrointestinal tract, and the kidney. The renal regulator is rapid in response but has limited capacity. The gastrointestinal regulator is slower to respond than is the kidney and also has limited capacity. On the other hand, the bone regulator is slow to respond to PTH but responds rapidly to calcitonin; it is relatively insensitive to PTH but is rather sensitive to the calcitonins; and it has an unlimited capacity for responding to both hormones. Other hormones also play a role in bone and calcium and phosphorus homeostasis.

REFERENCES

Aurbach, G. D., H. T. Keutmann, H. D. Niall, G. W. Tregear, J. L. H. O'Riordan, R. Marcus, S. J. Marx, and J. T. Potts, Jr. Structure, synthesis and mechanism of action of parathyroid hormone. *Recent Prog. Horm. Res.* 28:353–398, 1972.

Bijvoet, O. L. M., J. van der Sluys Veer, H. R. de Vries, and A. T. J. van Koppen. Natriuretic effect of calcitonin in man. *N. Engl. J. Med.* 284:681–688, 1971.

Borle, A. B.   Calcium metabolism at the cellular level. *Fed. Proc.* 32:1944–1950, 1973.

Borle, A. B.   Calcium and phosphate metabolism. *Annu. Rev. Physiol.* 36:361–390, 1974.

Copp, D. H.   Endocrine control of calcium homeostasis. *J. Endocrinol.* 43:137–161, 1969.

Copp, D. H.   Endocrine regulation of calcium metabolism. *Annu. Rev. Physiol.* 32:61–86, 1970.

Gaillard, P.   Parathyroids and Bone in Tissue Culture. In R. O. Greep and R. V. Talmadge (Eds.), *Parathyroids.* Springfield, Ill.: Thomas, 1961. Pp. 20–48.

Heaney, R. P., and T. G. Skillman.   Calcium metabolism in normal human pregnancy. *J. Clin. Endocrinol.* 33:661–670, 1971.

Hirsch, P. F., and P. L. Munson.   Thyrocalcitonin. *Physiol. Rev.* 49:548–622, 1969.

Queener, S. F, and N. H. Bell.   Calcitonin: A general survey. *Metabolism* 24:555–567, 1975.

Swaminathan, R., R. F. L. Bates, S. R. Bloom, P. C. Ganguli, and A. D. Care.   The relationship between food, gastro-intestinal hormones and calcitonin secretion. *J. Endocrinol.* 59:217–230, 1973.

33. Thyroidal Physiology

Ward W. Moore

The thyroid gland plays an important part in a homeostatic control system which aids in maintaining an optimal level of oxidative metabolism and heat production of the organism. Other parts of this regulatory mechanism include the central nervous system (CNS), the adenohypophysis (AP), the general circulation and certain plasma proteins, and the metabolic machinery of all cells of the body. The thyroid gland also has a vital role in the growth, differentiation, and development of the individual. In its role, the thyroid gland excels in three characteristic functions: (1) trapping of iodide, (2) synthesis of organic iodine and the formation of iodotyrosines and iodothyronines, and (3) storage and secretion of hormones which are iodothyronines.

The isolation of the hormone of the thyroid gland, *thyroxine,* by Kendall in 1915, its identification as 3,5,3′,5′-tetraiodothyronine by Harington in 1926, and its synthesis by Harington in 1927 gave great impetus to the study of thyroidal physiology and biochemistry. In 1952, two independent groups, Gross and Pitt-Rivers, and Roche, Michel, and Lissitzky, demonstrated the presence of another iodothyronine in the thyroid gland and blood. This compound has three to four times the biological activity of thyroxine and has been identified as 3,5,3′-triiodothyronine (Fig. 33-1). The extensive use of the radioactive isotopes, electron microscopy, and autoradiography, in conjunction with modern biochemical techniques, has led to important advances in the understanding of thyroid biology.

ANATOMY

Phylogenetically speaking, the thyroid gland is one of the oldest of endocrine glands. The human thyroid first appears in the three-week-old embryo as a ventral diverticulum from the midventral floor of the pharynx between the first and second pharyngeal pouches. By the fifth week, a small hollow sac has developed which remains joined to the pharynx by the thyroglossal duct. The latter atrophies during the sixth week, and with this change the thyroid loses its central cavity and adopts a bilobed form. Discontinuous cavities become manifest within the gland by the eighth week. These represent the beginnings of the follicles of the adult gland; the follicles soon become filled with colloid substance. The human thyroid is made up of two lobes linked together by a thin band of tissue, the isthmus. Normally, the lobes are closely attached to the lateral aspect of the trachea, and the isthmus fits over the anterior surface of the second and third cartilaginous rings. The adult gland weighs about 20 to 30 gm and possesses an enormous potential for growth in goiters weighing up to several hundred grams, which is not rare. The thyroid gland receives its blood supply from paired superior and inferior thyroid arteries arising from the external carotids and subclavians, respectively. The rate of blood flow through the gland is high (5 to 7 ml per gram per minute), more per gram of tissue even than through the kidney. The gland is innervated by laryngeal and pharyngeal branches of the vagus and by cervical sympathetic ganglia. These fibers are probably vasomotor and possibly secretomotor, but the transplanted or denervated gland can function normally. However, various neurohumors can alter the sensitivity of the thyroidal cells to stimuli.

The characteristic and dominant structural feature of the adult thyroid gland is the follicle or acinus, which is lined by epithelium and filled with colloid. The colloid is a viscid homogeneous substance, clear in appearance in the fresh state. The follicle epithelium consists of two cell types: follicular cells and light cells (C cells or para-

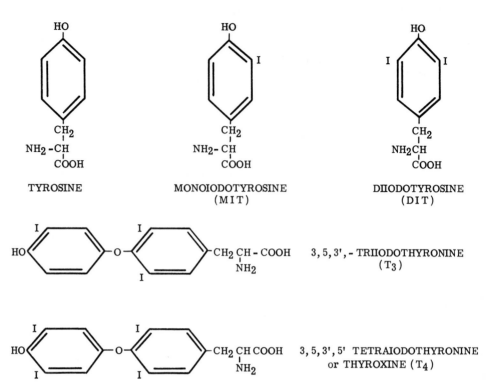

Figure 33-1
Structural formulas of tyrosine, the iodinated tyrosines, and the iodinated thyronines which are present in the thyroglobulin molecule.

follicular cells). The latter type is a derivative of the ultimobranchial bodies, making up only a small portion of the population of epithelial cells, and is a source of calcitonin. The lining epithelium rests on a basement membrane which is surrounded by a rich capillary network. The normal epithelium varies in appearance with the state of activity of the gland; generally speaking, the height of the cells is low when the gland is resting and high when it is active. The ultrastructure of the follicle cells is similar to that observed in other secretory or absorptive cells. In addition, the basal portion of the cells shows an extensive endoplasmic reticulum that contains wide irregular tubules, and numerous microvilli are present on the *apical* portion of the cell and extend into the colloid. The active cell is characterized by an enlarged Golgi apparatus, pseudopodia formation at the apical surface, the appearance of colloid droplets in the apical portion of the cell, and increased numbers of lysosomes. In addition, the active gland typically displays an increase in follicle cell height, a decrease in the amount of colloid, and increased vascularity. The cells lining the follicle have the unique capacity both to release their secretions into the follicular lumen and to permit the passage of the active principles of the cells from colloid to bloodstream.

SYNTHESIS AND STORAGE OF THYROGLOBULIN
It has been shown that the biosynthetic mechanisms involved in the formation of hormones are autonomous. Nevertheless, the rate of formation of the hormones is regulated by extrathyroidal factors. The most important factors are adenohypophysial

thyroid-stimulating hormone (TSH) and the availability of iodide. However, intra-thyroidal mechanisms also regulate hormonogenesis. Several key steps are involved in the synthesis, storage, and secretion of thyroid hormones (see Fig. 33-2).

The thyroid hormones are derived from thyroglobulin (TG), an iodine-containing glycoprotein that has a molecular weight of 670,000. Human thyroglobulin contains about 350 residues of carbohydrate (galactose, mannose, N-acetylglucosamines, sialic acid, and fucose) per mole, and these contribute about 9.7 percent of the total weight of the compound. In addition to containing about 5800 residues of amino acids per mole, thyroglobulin contains four types of iodoamino acids—3-monoiodotyrosine (MIT), 3,5-diiodotyrosine (DIT), 3,5,3'-triiodothyronine (T_3), and 3,5,3',5'-tetraiodothyronine (thyroxine or T_4). The structures of these compounds are shown in Figure 33-1.

The synthesis of thyroglobulin occurs sequentially in three independent stages. The amino acids enter the basal portion of the thyroid cell and are assembled into poly-peptide chains in the rough endoplasmic reticulum. As the protein migrates from the basal to the apical portion of the cell, various carbohydrate moieties are added in the endoplasmic reticulum and Golgi apparatus. Iodination of tyrosyl groups and the formation of the iodothyronines, T_3 and T_4, occur within the matrix of the thyroglobulin molecule near the apical portion of the cell after the incorporation of the amino acids and carbohydrates into the glycoprotein has been completed. The total iodine content of thyroglobulin in general reflects the physiological state of the gland and varies from about 0.2 to 1.1 percent. In goiter the iodine content is generally less than 0.1 percent. Current studies indicate strongly that the protein synthesis necessary for TG synthesis follows the same patterns found in other tissues. Similarly, all the intermediary metabolic processes carried out in the gland are not unique to the gland. The unique-ness of the thyroid gland lies in the fact that in order to synthesize an adequate amount of thyroid hormone it must concentrate iodide (I^-) from the plasma and incorporate it into the existing TG molecule.

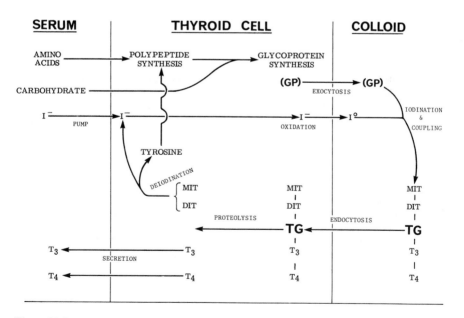

Figure 33-2
Thyroglobulin (TG) synthesis and storage, and secretion of the thyroid hormones triiodothyronine (T_3) and thyroxine (T_4). GP refers to glycoprotein. The iodination of the tyrosyl residues of this molecule probably occurs at the apical surface of the thyroid cell.

Normally, a concentration gradient for iodide (I^-) of about 25 to 1 is maintained between the thyroid and the serum. This is referred to as the thyroid/serum (T/S) ratio. The mechanism by which the thyroid accumulates iodide from the serum (with an iodide concentration of about 0.1 to 0.5 μg per 100 ml) has not been fully elucidated. Iodide is actively transported into the thyroid cell against an electrochemical gradient at the basal membrane, and a potential difference of -40 to -50 mv is maintained across the basal cell membrane. The transport mechanism is a carrier-mediated, energy-requiring process which is related to the function of the Na^+-K^+ dependent adenosine triphosphatase (ATPase) system. The iodide-concentrating mechanism is referred to as the iodide pump. The thyroid gland removes iodide from the blood and clears the plasma of iodide at a rate of about 20 ml per minute.

That the iodide pump is of physiological importance is best illustrated following the administration of thiocyanate, perchlorate, and other similar anions. These anions inhibit the iodide pump, with the result that the thyroid iodide concentration cannot exceed the serum concentration. The mechanism by which the anions interfere with the capacity of the thyroid to concentrate iodide is not known, but it has been postulated that the action of thiocyanate is competitive; i.e., it competes with iodide at the site of the concentrating mechanism. It should be pointed out that other tissues, particularly the salivary glands, gastric mucosa, mammary gland, and placenta, also possess iodide-concentrating mechanisms. Inhibition of the iodide pump in these organs has been observed following treatment with anions of the perchlorate group.

The best-known regulator of the iodide pump is the thyroid-stimulating hormone (TSH). TSH decreases (less negative) the membrane potential of thyroid cells, whereas hypophysectomy increases it. Removal of the adenohypophysis depresses the T/S ratio to 3 to 7, and under certain circumstances the T/S ratio may reach 250 to 300. Thus, the thyroid is able to concentrate iodide in the absence of TSH and therefore retains some degree of autonomous behavior. TSH is the most important factor in regulating the iodide pump, and its action, as well as that of long-acting thyroid stimulator (LATS), is mediated by cyclic adenosine monophosphate (cyclic AMP). Large doses of iodide saturate the pump and reduce the T/S ratio. The iodide-concentrating mechanism of the salivary glands, gastric mucosa, mammary gland, and placenta is not influenced by the removal of the AP or the addition of TSH or dietary iodide.

In addition to the thyroid gland, the kidneys remove iodide from the blood. The thyroid gland of a normal individual accumulates about one-third of a tracer dose of ^{131}I within 24 hours, and most of the remaining iodide is lost in the urine. In the absence of the thyroid, less than 2 percent of a dose of ^{131}I is retained by the body, and nearly all of the remainder is lost via the urine.

Over 90 percent of the thyroidal iodine is organically bound, and therefore it must have appeared as free iodine (I_2), which is a reactive form. Iodide is converted to iodine ($2 I^- \rightarrow I_2$), which can enter into organic combination. Iodine formation from iodide occurs through the action of a peroxidase in the microvilli of the apical portion of the cell. According to some endocrinologists, the presence of a specific oxidase for this reaction is physiologically not necessary.

Agents such as propylthiouracil depress the amount of iodine available for iodination of tyrosyl radicals. Following the use of extremely high doses of propylthiouracil (PTU), only traces of iodotyrosines may still be formed, but significant amounts of iodotyrosines and iodothyronines are still formed after small doses of PTU.

Thyroglobulin acts as a substrate for a series of reactions leading to the formation of the thyroid hormones. The steps involved are the formation of monoiodotyrosine (MIT) and diiodotyrosine (DIT), followed by the oxidative coupling of MIT and DIT with a loss of alanine within the thyroglobulin molecule. Following the formation of elemental iodine the oxidized form is bound to tyrosyl groups which are in peptide linkage within the thyroglobulin molecule. The result is the formation of the iodotyrosines MIT and DIT within the molecule. The formation of T_3 and T_4 is thought to be catalyzed by the activity of a peroxidase or a similar enzyme on the iodotyrosines in peptide chains. It is probable that the phenolic group of DIT moieties of two adjacent thyroglobulin molecules become attached and form a molecule of T_4 on one thyroglobulin molecule and an alanine side chain on the other. T_3 may be formed in a similar manner involving MIT and DIT, or by removal of iodine at the 5' position. In the euthyroid individual each TG molecule contains about 11 molecules of the iodotyrosines, MIT and DIT, and 2 molecules of the iodothyronines, T_3 and T_4. In general, it can be stated that TSH increases the rate of organification of iodine and iodothyronine formation. However, the mechanism or mechanisms by which TSH accelerates organification of iodide are not known. TSH administration decreases the MIT/DIT and T_3/T_4 ratios and increases the absolute amount of T_3 and T_4 in thyroglobulin. It follows that TSH might play a role in stimulating the oxidative coupling of MIT with DIT and DIT with DIT in the thyroglobulin molecule.

It has been indicated that there is an intrathyroidal regulation of the synthesis of iodotyrosines and iodothyronines. This is seen most clearly when an animal is placed upon an iodine-deficient diet. In such a situation, the MIT/DIT and T_3/T_4 ratios within the thyroglobulin molecule are markedly increased, but the concentration of each compound is decreased. If the elevated T_3/T_4 ratio in the gland should be maintained in the serum, an important physiological adjustment in cases of iodine deficiency would be demonstrated. However, it has not been determined whether or not the more potent mixture of T_3 and T_4 is carried over to the serum under these circumstances.

In summary, thyroglobulin (or pre-thyroglobulin) is synthesized in the endoplasmic reticulum and Golgi apparatus. It is transported to the apical cell surface enclosed in vesicles which are formed in the Golgi apparatus, and the vesicles are then moved into the follicle lumen by exocytosis and incorporated into the colloid. The completion of the synthetic process, the formation of the iodoamino acids, occurs in the microvilli at the apical portion of the follicle cell. The hormones are then stored as thyroglobulin in the colloid.

Proteolysis of Thyroglobulins and Secretion of the Hormones

The manner in which the thyroid hormones are stored is unique among the endocrine glands. Thyroxine and triiodothyronine are contained, via peptide linkage, within the large thyroglobulin molecule (molecular weight about 650,000). This is stored in an extracellular site, the follicular colloid, and the rate of its formation and removal is dependent, in part, upon TSH. Recent work, combining electron microscope, biochemical, differential centrifugation, and isotopic techniques, has shed light upon the way T_4 and T_3 are removed from their storage sites and secreted into the plasma. Following TSH administration, droplets or particles of colloid appear in the apex of the follicular cells as a result of an endocytotic process. These vesicles migrate toward the base of the cell and fuse with lysosomes, which have migrated from the base of the cell. The lysosomes contain enzymes that can hydrolyze TG. T_4 and T_3 release occurs by TG lysis within the fused droplet and lysosome or "derived lysosome." MIT and DIT are also released into the thyroid cell by the hydrolysis of TG. The ingestion of colloid by endocytosis occurs at the periphery of the lumen, and the most recently iodinated thyroglobulin is immediately adjacent to the apical cell border. Consequently a higher proportion of recently iodinated thyroglobulin, when compared to older thyroglobulin, should be ingested. Thus, a last-formed, first-secreted or "last come, first served" phenomenon exists.

The iodotyrosines, MIT and DIT, are not normally secreted into the bloodstream because they are deiodinated rapidly by dehalogenases in the thyroid cells. Thus the iodide may be reclaimed immediately by the gland. However, the iodothyronines, T_3 and T_4, are resistant to the dehalogenases and diffuse from the gland as a result of gradients which may be as high as 100 to 1 between the gland and the blood. The diffusion gradient of thyroxine is also aided in consequence of competitive binding favoring certain plasma proteins over the thyroidal proteins. It should be remembered that thyroglobulin does not appear in the blood except under abnormal conditions.

TSH leads to a rapid increase in thyroidal cyclic AMP. The effects of this hormone on thyroid hormone synthesis and secretion are probably mediated via the "2nd messenger." TSH accelerates the rate of proteolysis of thyroglobulin and the rate of release of T_3 and T_4, and following diffusion into the bloodstream each is bound by a specific plasma protein.

Several observations indicate that an intrathyroidal factor participates in the regulation of the release of organic iodine from the gland. When individuals are placed on chronic PTU ingestion, the daily output of hormonal iodine falls in proportion with the store of hormonal iodine in the gland. However, an increase in circulating TSH (which must occur in this instance) should cause an increase in proteolysis of thyroglobulin, with a resultant increase in release of thyroid hormones from the gland. Under this circumstance, one would expect an increase in thyroid hormone output instead of a decline, and it must follow that some mechanism within the thyroid induces a conservation of thyroidal iodine. Apparently the three intrathyroidal factors which regulate thyroid function lead to physiological adjustments which aid in the maintenance of an adequate level of thyroidal iodine stores. At the same time they lead to the secretion of a mixture of T_3 and T_4 of increased potency when challenges that tend to decrease hormonogenesis are put to the thyroid-pituitary-hypothalamus axis.

Transport of Thyroid Hormones in the Circulation

Three plasma proteins readily bind the iodothyronines. A serum globulin, called the thyroxine-binding globulin (TBG), binds approximately 200 μg of thyroxine per liter of serum. The concentration of TBG is elevated by estrogen therapy, during pregnancy, during the neonatal period, and in certain hepatic diseases, but it is depressed by androgen administration. The binding capacity of TBG is increased in hypothyroidism. It appears that TBG accounts for most (about two-thirds) of the transport of thyroxine at physiological concentrations. One of the prealbumins, the thyroxine-binding prealbumin (TBPA), apparently contributes to thyronine binding. Its saturation capacity for thyroxine is greater than that of TBG, but its binding power is less. A serum albumin also reversibly binds thyroxine, but avidity and capacity for thyroxine are much less than those of TBG and TBPA. T_3 is bound by TBG, but not by TBPA or the albumin. It has been shown, by using in vitro systems, that the rate of passage of the iodothyronines from the medium into the tissue cells is inversely related to the concentration of the binding proteins in the medium. This might be an indication that the concentration of TBG and the other binding proteins controls the rate at which the thyroid hormones are taken up by the cells in vivo. An indication of the concentration of the circulating thyroid hormones is the protein-bound iodine (PBI) or the butanol-extractable iodine (BEI) of the serum. These chemical assays entail the determination of the iodine of thyroxine which is combined with TBG. A PBI level of 3.5 to 8.5 μg per 100 ml of serum is said to be compatible with normal thyroidal activity, whereas values below this range are found in hypothyroidism, and those above are associated with hyperthyroid states. However, as noted above, factors other than the thyroid gland can influence the PBI. The normal T_3 concentration in human serum is approximately 200 nanograms (ng) per 100 ml and may reach 800 to 900 ng per 100 ml in untreated hyperthyroidism and fall to 10 to 100 ng per 100 ml in hypothyroid states. Values for T_4 are somewhat higher, with a normal range of about 4 to 10 μg per 100 ml and values of 0.5 to 1.25 in hypothyroid states and 15 to 20 μg per 100 ml in hyperthyroidism. Under ordinary circumstances only small amounts of thyroglobulin (5 to 10 ng per 100 ml) and iodinated tyrosine (0.1 μg per 100 ml) appear in the serum.

It is obvious that there are many points in the synthesis and release processes at which a failure could cause a decrease in thyroid activity, i.e., a deficiency of hormone secreted by the gland and/or delivered to extrathyroidal tissues. Thus, the exogenous supply of iodine may be inadequate, or a deficiency of one of the enzymes needed for trapping, oxidation, coupling, or hydrolysis may result in reduced hormone production. A deficiency of TSH which exerts its effects at several stages in hormonogenesis would cause a deficiency of thyroid hormone. Similarly, an elevated TBG might lead to a decrease in thyroxine uptake by the cells, or an altered extrathyroidal dehalogenase activity might lead to a relative thyroid hormone deficiency.

The primary source of iodine is dietary. However, the iodine of T_3 and T_4 is not completely lost when they are metabolized. In some tissues, notably liver, kidney, heart, brain, and spleen, deiodination of the iodothyronines occurs. The ether linkage within the compounds is broken, with the liberation of MIT and DIT, and further deiodination is accomplished. The iodide released can then be reclaimed by the thyroid. The circulating iodide is about 0.3 μg per 100 ml of serum in the normal individual with a normal iodine intake. The renal tubules reabsorb about 95 percent of the filtered iodide load. The resulting renal plasma clearance of iodide is about 33 ml per minute, and therefore about 150 μg of iodide is lost in the urine per day. Most of the iodine lost from the body fluids is in the form of iodide, but a small part is lost as thyroxine and T_3 in the urine, feces, and sweat. Thyroxine is present in the bile as a glucuronide, and as it passes down the small intestine, part of it is reabsorbed and the remainder appears in the feces. Most of the fecal iodine is in the form of organic iodine.

REGULATION OF THYROID FUNCTION

Many factors influence the rate of thyroid activity, but the common denominator in states of altered thyroid activity is TSH activity. In all instances, the only agent that directly stimulates thyroid growth and thyroid hormonogenesis is TSH, and its activity is mediated by cyclic AMP. Every type of abnormal growth of the thyroid and/or change in hormone output by the gland of a normal individual must be accompanied by a change, either relative or absolute, in TSH output.

The actions of TSH are many. As already stated, it increases the thyroid/serum (T/S) iodide ratio, the rate at which iodothyronines are formed, and the rate at which thyroglobulin is hydrolyzed. In addition, TSH causes an increase in the rate of thyroidal uptake of ^{131}I. It causes an increase in the rate at which organic ^{131}I disappears from the gland, an increase in the rate at which labeled iodothyronines appear in the serum, and an increase in the serum concentration of iodothyronines. It decreases the volume of colloid, causes hypertrophy and hyperplasia of the cells of the thyroidal follicles, and increases the uptake of ^{32}P by the thyroid. Furthermore, TSH induces hypertrophy of retro-orbital tissues, which leads to the condition known as *exophthalmos.* The concentration of TSH in the euthyroid individual varies from 1 to 8 μU per milliliter of plasma with a mean of 3 μU per milliliter, and any maneuver which leads to a decrease in the level of circulating thyroid hormones results in an increased TSH output. In primary hypothyroidism plasma TSH concentrations may reach values exceeding 400 μU per milliliter, and the severity of the defect in T_4 secretion is correlated with the plasma TSH concentration. Conversely, any procedure which elevates the circulating thyroid hormone concentration should decrease TSH secretion. Thus, in the normal individual a balance is established by a negative feedback system that tends to maintain, within certain limits, an optimal level of thyroid activity in the face of altered external or internal environments. Exceptions to these generalizations occur in certain disease states. In Graves' disease (a form of hyperthyroidism) a substance termed long-acting thyroid stimulator (LATS) is found in the serum. LATS is an immunoglobulin, class IgG, which enhances thyroidal iodine uptake and increases thyroid hormone synthesis and secretion. It is distinctly different from TSH and is not dependent upon pituitary function, but, like TSH, its actions are mediated by cyclic AMP.

From the foregoing it follows that thyroidal enlargement (goiter) will result from any condition leading to a marked and sustained imbalance between TSH output and thyroid hormone output, if the imbalance favors TSH output. The reverse is also true, because a chronically low level of thyroidal activity will induce an abnormal growth in the basophils of the adenohypophysis, the so-called *thyrotrophs,* which secrete TSH.

The output of thyroid hormone in "normal" individuals with an iodine-deficient diet is generally normal but is maintained at normal levels only by an increase in TSH output. It is obvious that iodine plays a very important role in the synthesis of thyroxine, since in addition to being part of the thyroxine molecule it decreases the output of thyroid hormone either by inhibiting TSH secretion or by decreasing TSH activity. Therefore, at the outset of iodide deprivation the output of thyroxine is decreased, so a degree of inhibition for the release of TSH is removed. TSH output then rises, increasing the avidity of the thyroid for iodide and making more iodide available for synthesis of thyroid hormone. Thus, a marked sustained iodine deficiency indirectly maintains an increased TSH secretion, which results in hypertrophy and hyperplasia of thyroid cells. The increase in the rate of iodide uptake from the low serum levels can then maintain a level of thyroid hormone production within normal limits, as judged by the observation that all criteria which assess the effects of thyroid hormones appear to be normal; nevertheless, simple goiter is produced. However, if the increase in TSH output cannot restore normal thyroid hormonogenesis over a prolonged period, the gland will eventually become exhausted. On the other hand, if the supply of iodide is increased slightly, the exhaustion is reversible. But if the iodide supply to the gland cannot be increased, the exhaustion is irreversible. It must follow that TSH will lead to thyroid atrophy if the amount of iodide available to the thyroid is inadequate.

The ingestion of large amounts of iodide can also strongly influence thyroidal hormonogenesis. It acts to decrease the T/S iodide ratio through the intrathyroidal mechanism, and by decreasing TSH output or by inhibiting TSH action. It decreases the rate of organification of iodine in the thyroid and may lead to the development of hypothyroidism and goiter. It has no effect upon the rate of release of hormone in the normal individual but decreases the rate of hormone release in the hyperthyroid individual or a normal individual treated with TSH. The manner in which high levels of circulating iodide influence thyroidal physiology is not known.

Effects of Antithyroid Compounds

The antithyroid compounds or goiterogenic drugs are so named because they somehow inhibit the synthesis of thyroid hormones and indirectly induce goiter production by facilitating an increase in TSH output. The first substance observed to interfere with thyroid function was thiocyanate. The administration of thiocyanate and related monovalent anions, such as perchlorate, chlorate, and periodate, inhibits the ability of the thyroid to concentrate iodide. It was shown that the goiterogenic activity of these chemicals could be prevented with simultaneous administration of thyroxine. The anion must therefore act upon some stage of hormonogenesis. It has been demonstrated that the goiter produced as a result of thiocyanate can be reversed by treatment with iodide. The anions inhibit the ability of the thyroid to concentrate iodide. None of the anions completely inhibits movement of iodide into the thyroid, but the iodide which enters by simple diffusion can enter into the synthesis of thyroxine and its intermediates. Therefore (with a maximal thiocyanate effect), one would expect to observe a T/S iodide ratio of 1.0.

Another agent, thiouracil, was shown to produce hyperplastic changes in the thyroid which were accompanied by signs of hypothyroidism. The administration of thyroxine will prevent the goiterogenic effects of thiouracil also, but the administration of iodide will not. Following thiouracil treatment, the T/S ratio increases markedly, and most of the iodine present in the gland is present as iodide. Thiouracil acts on the thyroid by keeping the iodine in the reduced state (I^-), probably by inhibiting peroxidase activity in the thyroid gland. Much work has led to the production of substituted thiouracil and thiocarbamide derivatives with a high degree of therapeutic effect and minimal toxicity for use in hyperthyroid states. The actions of the goiterogenic agents of the thiocyanate and thiouracil type have been used extensively for studying the concentration of iodide by the thyroid and its incorporation into thyroglobulin.

REGULATION OF TSH SECRETION

It has been stated that the thyroid and the adenohypophysis control each other's functions by a feedback mechanism which ensures in the normal individual an adequate, but not excessive, amount of thyroid hormone for the regulation of metabolic processes. Two primary factors participate in the regulation of the rate of TSH secreted by the AP: the hypothalamus, and a direct negative feedback action of the circulating thyroid hormones on the AP.

Several types of experimental procedures have shown that the hypothalamus is intimately concerned with TSH secretion. First, transplantation of the AP to extrasellar sites leads to a reduction in TSH secretion, and thyroid function is decreased, approaching, but not reaching, that observed in the hypophysectomized animal. Second, effective hypophysial stalk section results in a decreased rate of TSH secretion. Third, lesions of the anterior hypothalamus, particularly in the supraoptic region, depress thyroid function—a reflection of decreased TSH secretion. Finally, electrical stimulation of the anterior hypothalamus or rostral portion of the median eminence results in an increase in circulating TSH.

Both hypophysial stalk section and lesions of the anterior hypothalamus lead to a reduction in total TSH content of the AP and a decrease in serum TSH, but the pituitary TSH concentration and the appearance of the thyrotrophs remain normal. Alterations that follow each procedure are: decreases in the T/S iodide ratio, the rate of release of organic ^{131}I, the serum PBI, and thyroid weight; and thyroid function comparable to that observed in the hypophysectomized animal. Another feature observed in these preparations is the failure to respond to stimuli which in the intact animal lead to either an increase or a decrease in TSH secretion. The substance that stimulates TSH synthesis and secretion is thyrotropin-releasing hormone (TRH). TRH is a tripeptide amide which is released by neurons of the anterior hypothalamus into the hypophysial portal vessels, and by activating adenyl cyclase and increasing cyclic AMP in the thyrotrope cell it exerts a tonic stimulatory effect on TSH secretion. The infusion of minute amounts of T_3 or T_4 into the AP results in an inhibition of TSH release by the thyrotrope cell, and because this effect persists after the AP is deprived of its connections to the hypothalamus, it has been deduced that the AP responds directly to an inhibitory influence of the thyroid hormone. Recently it has been shown that this inhibitory action results from blocking the pituitary response to TRH. Various experiments indicate that T_3 and T_4 may inhibit and/or stimulate TRH synthesis and release by acting directly on elements of the CNS, particularly the hypothalamus. Whatever the case, the secretion of TSH is related to the secretion of T_3 and T_4 through a negative feedback system.

The thyrotrope cells of the AP are affected by two chemical stimuli: TRH, which stimulates TSH secretion; and the thyroid hormones, which inhibit it. Present evidence indicates that the principal feedback mechanism of pituitary TSH secretion is a direct one by the thyroid hormones, but for full stimulatory or inhibitory adaptations to external stimuli, the functional integrity of the hypothalamo-hypophysial system must be maintained.

Other neural factors can influence the output of TSH by the AP. The area of the anterior hypothalamus whose neurons release TRH is adjacent to areas containing temperature-sensitive receptors which are involved in integrating mechanisms for the regulation of body temperature (see Chap. 28). Cooling of the temperature-sensitive receptors, in addition to activating heat-production and heat-conserving mechanisms, stimulate TSH secretion and thyroid function. Conversely, fever induced by usual means tends to inhibit TSH secretion and depress thyroid function. Exposure to cold in adults leads to a modest increase in TSH secretion, but in the newborn increases in TSH and thyroid hormone secretion are marked. Thus, peripheral cold receptor activity stimulation can influence TRH and TSH secretion rates. Experimentation has shown that stressful stimuli powerfully inhibit the uptake of ^{131}I and the release of thyroid hormones. The alterations in thyroid function occur prior to the withdrawal of TSH and result from the increased circulating adrenal cortical steroids. These hormones enhance

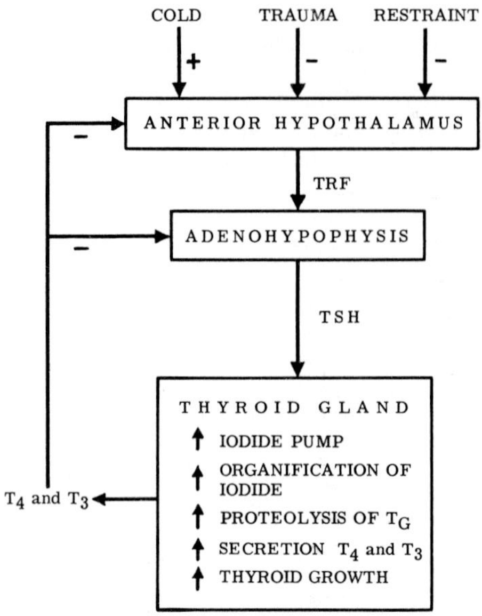

Figure 33-3
Some factors that influence the secretion of TRH, TSH, and the thyroid hormones. + = stimulation; − = inhibition.

urinary iodide loss and slow thyroid circulation, thus contributing to the reduced thyroidal iodide uptake. Chronic stress produces depressed thyroid function, and this is apparently mediated by the action of cortisol to depress the release of TRH. Emotional stress has long been thought to precipitate a marked increase in the rate of TSH release. However, current studies indicate that if emotional stress has any effect upon TSH secretion and thyroid function it is one of inhibition (see Fig. 33-3).

PHYSIOLOGICAL EFFECTS OF THYROID HORMONES

The physiological effects of the thyroid hormones have been extensively studied by observing structural and functional changes which occur in spontaneous and experimental hypothyroidism and hyperthyroidism. The thyroid hormones control many metabolic processes. They influence oxygen consumption and heat production, the metabolism of carbohydrates, fats, proteins, and nucleic acids, and the effect of other hormones. In general, thyroid hormone excess promotes increased rates of turnover of substances because of inefficiency, whereas deficiencies promote reduced turnover rates. Recent evidence suggests that the extrathyroidal conversion of T_4 to T_3, either in the ECF or in effector cells, may be necessary for the metabolic activity of the hormone. This finding would support the concept that T_3 is the primary metabolically active thyroid hormone. The thyroid gland is not indispensable to life, but the presence of adequate amounts of its hormones is necessary for normal heat production, growth, differentiation and development, and well-being of the individual.

Probably the most prominent effect of the thyroid hormones, first noted by Magnus-Levy in 1895, is their effect upon respiratory exchange. His classic discovery that respiratory exchange is decreased in Gull's disease (hypothyroidism) and increased in Graves' disease (hyperthyroidism) has led to the publication of many works concerned with the effects, both pharmacological and physiological, of the thyroid hormones.

The increased oxygen consumption and heat production (calorigenic effect) which follow the administration of the thyroid hormones to animals are marked. The calorigenic effect can also be demonstrated in certain excised tissues obtained from animals with experimentally induced hyperthyroidism. However, when thyroid hormones are added to excised tissues taken from a normal animal, no increase in oxygen consumption results. The respiration of the adenohypophysis, in contrast to other tissues, is not affected by thyroid insufficiency or excess. Similar observations have been made concerning the brain and spleen. The observation that heat production is less in the hypophysectomized individual than in the athyroid individual indicates that the thyroid hormones are not the only hormones involved in the regulation of heat production. Rather it is regulated by a balance of hormonal factors, including growth hormones and the adrenal cortical and medullary hormones, all of which are calorigenic.

It has been estimated that 300 µg of thyroxine daily is required to maintain man in a state of normal metabolic activity under normal circumstances. Disturbances of the heat-regulating mechanism are readily observed in the thyroidectomized individual. The normal person responds to a cool environment by increasing the output of thyroid hormones, stimulating an increase in the basal metabolic rate (BMR), and it has been shown that the thyroxine requirement is inversely related to the environmental temperature. However, the twofold to threefold increase in heat production over the resting level which follows sudden exposure to a cool environmental temperature is not due solely to increased thyroidal activity, for shivering and epinephrine release account for by far the greater part of the increase in heat production. It should be borne in mind that many mechanisms other than hormonal are involved in thermoregulation, and reference should be made to earlier sections (Chap. 6) on energy metabolism and temperature regulation for a discussion of factors which participate in thermoregulation.

The manner in which the thyroid hormones act at the cellular level is not known. Many observations have been made on alterations in the activity and/or concentration of enzyme systems which play a vital role in the reactions concerned with the respiratory activity of cells, and on the release and utilization of energy which might be related to the action of the thyroid hormones. Most of the theories presented for the mode of action of the thyroid hormones have their basis at this level. Since an excess of thyroid hormone uncouples oxidative phosphorylation, cellular respiration and energy are less effectively controlled. These hormones also increase mitochondrial permeability and size. It might be recalled that the mitochondria are regarded as the principal "power plant" of the cell. The thyroid hormones also increase the rate of transfer of transfer-RNA-bound amino acids to microsomal protein, and this property may form the basis on which they function to stimulate metabolic processes. Because an enhanced turnover of nuclear and cytoplasmic RNA precedes the increase in protein anabolic activity and oxygen consumption following the acute administration of thyroxine, it may well act primarily via gene activation.

The effects observed after the administration of thyroid hormones or after thyroidectomy have a considerable latent period. When one considers heat production, for instance, after thyroidectomy, the BMR falls steadily until it reaches a minimum of about −35 to −50 percent in about 40 days. On the other hand, a change in the BMR is not detectable until about 36 hours after a large dose of thyroxine is given to a normal individual, and a maximal effect is not observed until five to seven days later.

Many effects observed in hyperthyroid and hypothyroid states are in all probability results of both direct calorigenic effects and reflex responses to them. Most of these effects change in proportion to the BMR and therefore are assumed to stem from the calorigenic effect of the hormones. Some changes observed in hyperthyroid states are: an increase in fasting nitrogen excretion, increases in the rates at which carbohydrate and fatty acids are absorbed from the gastrointestinal tract, and a diminished serum cholesterol which results from an increased catabolism of the substance. Cardiac output and plasma volume increase, and pulmonary ventilation goes up. Increases in glomerular filtration rate, tubular maxima for para-amino-hyppurate and glucose, and urine volume occur. Many adaptations take place, the increase in thyroid hormone output leading to an increase in oxygen consumption, whereby the organism provides adequate means of satisfying the increased demand for oxygen.

On the other hand, in hypothyroid states varying degrees of anemia and decreased bone marrow activity commonly occur. Obesity is observed, but only if the appetite and caloric intake are not decreased in proportion to the decline in BMR. Sensitivity to external stimuli decreases, and tendon reflex time increases. An accumulation of considerable amounts of water, salts, and nitrogenous material occurs in the extravascular spaces. Certain other changes, opposite to those noted in hyperthyroidism, are also observed (e.g., decreased cardiac output and renal function). In juvenile hypothyroidism sexual development and maturation are retarded and the onset of puberty is delayed. The thyroid hormones are essential for normal reproduction and lactation.

The thyroid hormones also participate in growth, differentiation, and development of cells, organs, and systems, for when the thyroid is absent, or is secreting thyroid hormone (TH) at subnormal rates, several abnormalities in structure and function are noted. The thyroid hormones are probably not required for early fetal growth because the fetus does not synthesize detectable amounts of either TSH or TH until the tenth to eleventh week of gestation and the transplacental transfer of maternal TSH and TH is extremely limited. Thus, it should follow that differentiation and development during this period are independent of the thyroid hormones and are controlled solely by genetic mechanisms. The lack of TH does not interfere with growth prior to birth because birth weights and lengths of athyroid infants at full term are within normal limits. However, the differentiation and maturation of the fetal central nervous system and skeleton do become dependent upon TH prior to birth. Human cretins have delayed appearance of ossification centers, and frequently epiphysial dysgenesis is observed. Similarly, neural development and mental functions are impaired in such individuals. The earlier the onset and the more marked the degree of fetal hypothyroidism, the more marked are the alterations in skeletal and nervous development.

A euthyroid state in postnatal life, in contrast to that in prenatal life, is necessary for normal growth. Postnatal hypothyroidism leads to marked retardation of growth. Skeletal growth is subnormal, and the height and bone age of the cretin lag behind that of the normal. The lagging development of the skeletal system can usually be overcome with replacement therapy. However, if treatment is delayed, a dwarfed stature persists even though accelerated growth rates are attained. An excess of TH in children enhances the rate of linear growth and maturation, and therefore the height and bone age of hyperthyroid children exceed those of normal children of the same age. Large excesses of TH do, however, impair the ultimate linear growth of bone. The mechanism of growth stimulation by TH has not been elucidated. The observation that TH has a mitogenic action suggests a vital role for these hormones in cell division and differentiation. It should be pointed out that the effects of TH on growth are optimal only in the presence of growth hormone.

Probably no tissue suffers more than does nervous tissue from lack of thyroid hormone during fetal and early postnatal life. This deficiency results in a mental deficiency which is usually severe. In adult hypothyroidism the observed mental capacity alteration is manifested as reduced mental alertness. Recent studies concerned with the thyroid hormone and central nervous system structure and function have shown that a deficiency of thyroid hormone results in a delayed appearance and decreased content of myelin in fiber tracts, a decrease in size and number of cortical neurons, and a decreased vascularity. These changes may be reversed if thyroid hormone is given within the third week of postnatal life; if not, irreversible damage results. It has also been observed that brain succinic dehydrogenase decreases in concentration following thyroidectomy of rats at birth. Such activity can be restored to normal with thyroid hormone if treatment is instituted within 10 days. All these observations might provide an explanation for the irreversible retardation of mental development in cretins. On the other hand, the presence of thyroid hormone in excess does not cause growth and development to be accelerated and may even retard development. The mechanism of action of the thyroid hormones in promoting growth, differentiation, and development of the organism is not understood.

REFERENCES

Abuid, J., A. H. Klein, T. P. Foley, Jr., and P. R. Larsen. Total and free triiodo-thyronine and thyroxine in early infancy. *J. Clin. Endocrinol. Metab.* 39:263–268, 1974.

Alexander, W. D., R. M. Harden, and J. Shimmins. Studies of the thyroid iodide "trap" in man. *Recent Prog. Horm. Res.* 25:423–466, 1969.

Björkman, U., R. Ekholm, L.-G. Elmqvist, L. E. Ericson, A. Melander, and S. Smeds. Induced unidirectional transport of protein into the thyroid follicular lumen. *Endocrinology* 95:1506–1517, 1974.

Bray, G. A., and H. S. Jacobs. Thyroid Activity and Other Endocrine Glands. In R. O. Greep and E. B. Astwood (Eds.), *Handbook of Physiology*. Washington: American Physiological Society, 1974. Section 7: Endocrinology, vol. 3, chap. 24, pp. 413–433.

Florsheim, W. R. Control of Thyrotrophin Secretion. In R. O. Greep and E. B. Astwood (Eds.), *Handbook of Physiology*. Washington: American Physiological Society, 1974. Section 7: Endocrinology, vol. 4, part 2, chap. 38, pp. 449–467.

Ghahremani, G. G., P. B. Hoffer, B. E. Oppenheim, and A. Gottschalk. New normal values for thyroid uptake of radioactive iodine, *J.A.M.A.* 217:337–339, 1971.

Hershman, J. M. Clinical application of thyrotropin-releasing hormone, *N. Engl. J. Med.* 290:886–890, 1974.

Hershman, J. M., and J. A. Pittman, Jr. Control of thyrotropin secretion in man. *N. Engl. J. Med.* 285:997–1006, 1971.

Hoch, F. L. Metabolic Effects of Thyroid Hormones. In R. O. Greep and E. B. Astwood (Eds.), *Handbook of Physiology*. Washington: American Physiological Society, 1974. Section 7: Endocrinology, vol. 3, chap. 23, pp. 391–411.

O'Connor, J. F., G. Y. Wu, T. F. Gallagher, and L. Hellman. The 24-hour plasma thyroxin profile in normal man. *J. Clin. Endocrinol. Metab.* 39:765–771, 1974.

Oppenheimer, J. H., D. Koerner, H. L. Schwartz, and M. I. Surks. Specific nuclear triiodothyronine binding sites in rat liver and kidney. *J. Clin. Endocrinol. Metab.* 35:330–333, 1972.

Pittman, C. S., J. B. Chambers, Jr., and V. H. Read. The extrathyroidal conversion rate of thyroxine to triiodothyronine in normal man. *J. Clin. Invest.* 50:1187–1196, 1971.

Snyder, P. J., and R. D. Utiger. Response to thyrotropin releasing hormone (TRH) in normal man. *J. Clin. Endocrinol.* 34:380–385, 1972.

34. The Adrenal Cortex

Ward W. Moore

The adrenal glands consist of two distinct organs, the adrenal cortex and the adrenal medulla. The adrenal cortex is essential to life, and its normal functional integrity is dependent upon the adenohypophysial adrenocorticotropic hormone (ACTH), the hypothalamic corticotropin-releasing factor (CRF), and the renin-angiotensin system. The cells of the adrenal cortex synthesize many substances and may secrete into the circulation several steroidal compounds. Many are metabolically inactive, but some are extremely biologically active and may selectively possess lipolytic, gluconeogenic, protein catabolic, electrolytic, androgenic, estrogenic, and life-sustaining properties. Generally, steroids that have a pronounced effect on intermediary metabolism are referred to as glucocorticoids, while those whose main effect is on salt and water metabolism are called mineralocorticoids. The functional integrity of the adrenal cortex must be normal if the individual is to withstand the stresses of life, such as injury, surgery, and diseases, or any challenge which places an imbalance on homeostatic mechanisms.

ANATOMY

Embryologically, the cortical tissue is first to appear. It becomes evident at about the fifth week of intrauterine life in the human and arises from mesoderm in the urogenital zone. During the seventh week, the mass of presumptive cortical cells is invaded by cells migrating from neural crest material. These cells are surrounded by the mesodermal cells and form the adrenal medulla. The adrenal medulla is regulated directly through its nerve supply, which is part of the thoracolumbar division of the autonomic nervous system. The functional significance of this organ is discussed in Chapter 7. The control of synthesis and secretion of the hormones of the adrenal medulla has recently been reviewed by Fuller (1973).

Normally the paired adrenal glands of the normal adult weigh about 10 gm. Each adrenal receives its blood supply from several small arterial branches which arise from the aorta, the renal arteries, and the phrenic arteries, and has a blood flow of about 5 ml per minute. The venous drainage of each gland is through a single central vein, and the left adrenal vein empties into the renal vein while the right empties directly into the vena cava. The adrenal cortex consists of epithelioid cells arranged in continuous cords or sheets of cells which are separated by capillaries. Structurally, the cortex may be divided into three zones. The *zona glomerulosa,* the outer zone next to the fibrous capsule, has cells arranged in irregular masses. In the middle zone, the *zona fasciculata,* the cords or sheets of cells are straight and radially disposed. In the inner cortical zone, the *zona reticularis,* the cords form an irregular network. The ultrastructure of the human adrenal cortical cells reveals an abundance of mitochondria, and the smooth endoplasmic reticulum is plentiful. These structures are prominent in all steroid-synthesizing cells. All the cortical cells show evidence of accumulation, storage, and secretion of lipid. The chief lipid constituent is cholesterol, which is a precursor of all the adrenal cortical hormones. The fetal adrenal gland may be larger than the normal adult gland, and in relation to fetal size it is immense. The fetal adrenal cortex is composed primarily of large ovoid cells arranged adjacent to the medulla. These cells are referred to as the fetal zone, and they produce a steroidal substance, dehydro-epiandrosterone sulfate, which serves as a precursor for the placental production of estrogens. The importance of this unit, the fetal-placental unit, is discussed in Chapter 35.

779

BIOCHEMISTRY OF ADRENAL STEROIDS

The hormones of the adrenal cortex are closely related structurally to cholesterol. Information regarding the synthesis of the steroids has come from studies using isolated perfused cow and human adrenal glands, the collection of adrenal venous blood from several species, and slices or homogenates of adrenal cortical tissue. Although the results of such experiments (in vitro) do not necessarily reflect synthetic processes (in vivo), these methods afford a very good basis for the understanding of adrenal cortical hormone biosynthesis. Isotopic methods, in particular, have led to an understanding of the intermediates involved in the condensation of acetate molecules to form cholesterol.

Many steroids have been isolated from the adrenal cortex. All are derived from cholesterol, and the major ones of biological importance along with the structural formula of the steroid nucleus, cyclopentanoperhydrophenanthrene, are depicted in Figure 34-1.

The reaction sequence of cortical steroid synthesis (Fig. 34-2) from cholesterol is initiated when cholesterol is cleaved to yield a 21-carbon compound with a ketone group at the 20 position to form Δ^5-pregnenolone. The latter, under the influence of 3β-hydroxydehydrogenase, is converted into progesterone. Progesterone occupies a key position in adrenal cortical hormone biosynthesis, and in the human the greater part of the progesterone is subjected to three successive hydroxylating steps catalyzed by the enzymes 17α-hydroxylase, 21-hydroxylase, and 11β-hydroxylase to form 17α-hydroxyprogesterone, 17α-hydroxydesoxycorticosterone (substance S), and cortisol (hydrocortisone or substance F), respectively. The last, cortisol, is the predominant circulating adrenal steroid in the human. Each of the three cortical zones possesses the 11- and 21-hydroxylases, but the 17α-hydroxylase is present in only the zona fasciculata and the zona reticularis. That amount of progesterone which escapes hydroxylation at the 17α position undergoes hydroxylation at the 21 position to form desoxycorticosterone. This is hydroxylated at the 11 carbon to form corticosterone. The rates of secretion of both cortisol and corticosterone are not constant throughout the day. They are secreted in

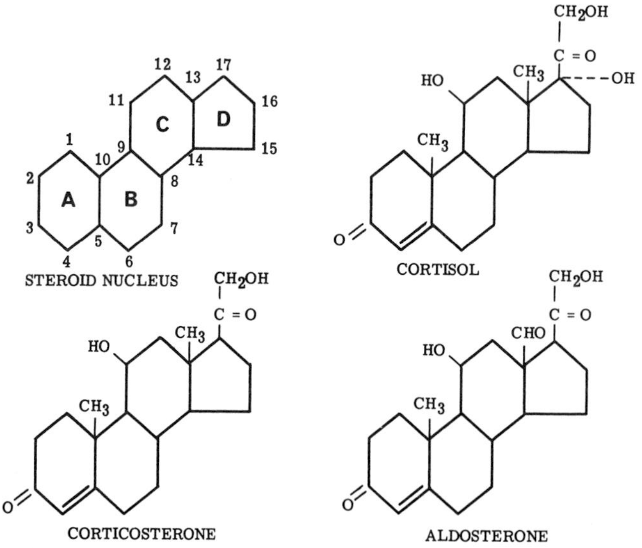

Figure 34-1
Structure of the basic steroid nucleus showing the ring designation and system used for numbering the carbon atoms. Also shown are the three major steroids secreted by the adrenal cortex.

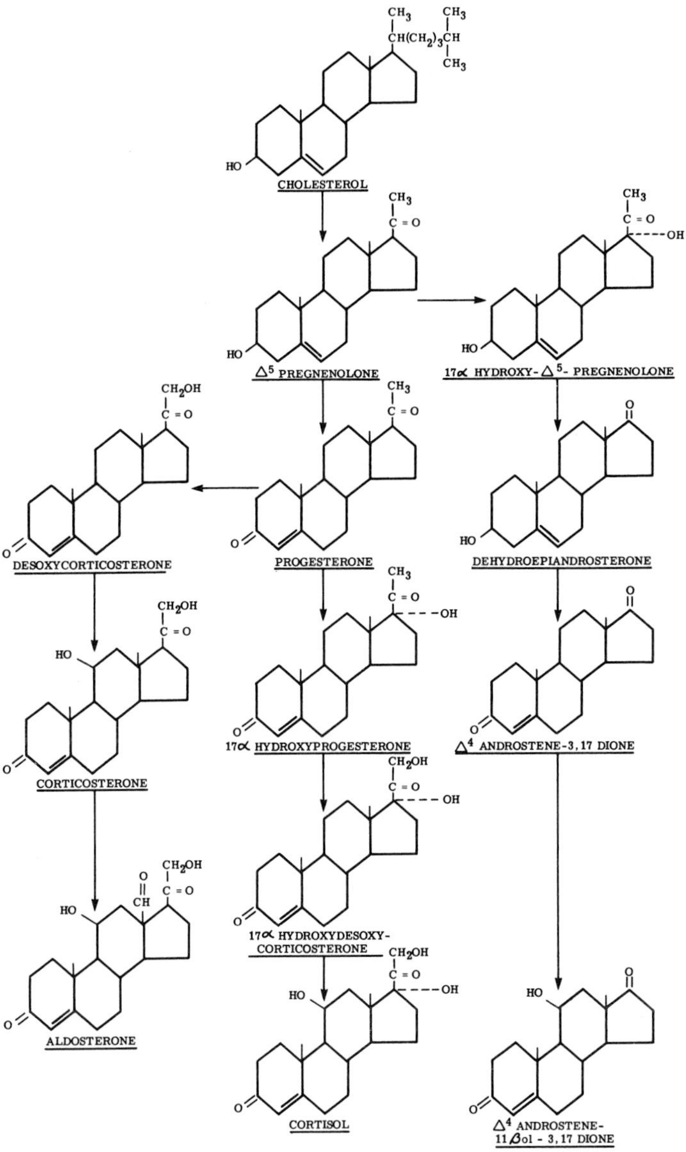

Figure 34-2
Major pathways of hormone biosynthesis in the adrenal cortex of the human.

response to ACTH in a diurnal or circadian manner. The concentration of cortisol in the plasma is much higher in the morning (0600 hours) than in the evening (1800 hours). The diurnal curve of ACTH and cortisol is shown in Figure 34-4. Corticosterone is a direct precursor of aldosterone. It occurs as the result of a substitution of an aldehyde group at the 18 carbon of corticosterone under the influence of "18-aldolase," an enzyme localized only in the zona glomerulosa. The concentration of aldosterone in human peripheral venous plasma is about 0.1 μg per 100 ml.

The secretion of androgenic substances by the adrenal cortex can be of extreme importance in the developing fetus, and one is referred to Chapter 35 for a description of the effects of androgens on differentiation of the fetus.

In the adrenal cortex, Δ^5-pregnenolone plays a pivotal role in the formation of the 19-carbon, androgenic steroids. Δ^5-pregnenolone is hydroxylated at the 17α position to form 17α-hydroxy-Δ^5-pregnenolone, and as a result of the cleavage of the side chain of 17α-hydroxy-Δ^5-pregnenolone, the 19-carbon–17-ketosteroid, dehydroepiandrosterone, is formed. The enzyme, 3β-hydroxydehydrogenase, catalyzes the formation of Δ^4-androstenedione from dehydroepiandrosterone. Δ^4-androstenedione is hydroxylated at the 11β position to form Δ^4-androstene, 11β-ol-3-17-dione. Some Δ^4-androstenedione may be formed in the adrenal cortex directly from 17α-hydroxyprogesterone. The adrenal androgens are biologically very weak but may be converted to the very potent androgenic steroid, testosterone, in many tissues of the body.

The estrogenic steroids secreted by the adrenal cortex generally follow the synthetic pathway which leads from Δ^5-pregnenolone to progesterone, 17α-hydroxyprogesterone, and Δ^4-androstenedione. Δ^4-androstenedione is transformed into estrone following aromatization of ring A and demethylization at carbon 10 leaving an 18-carbon compound. Estradiol can be formed from estrone by transforming the 17-keto group to a 17-hydroxy group. The adrenal cortex normally secretes minimal amounts of estrogenic steroids.

The steroids secreted by the adrenal cortex are excreted mainly as inactive forms, although a small amount of free active steroid may appear in the urine. Cortisol, for instance, is released by the adrenal cortex and is transported in the blood partly bound to an alpha globulin, transcortin, and partly in the free state. It is thought that only the free cortisol is metabolically active. The blood level of cortisol is constantly reduced as a result of degradation to an inactive form in the liver. In this organ and others, the adrenal hormones undergo further changes to form the inactive tetrahydro forms (Fig. 34-3). The active hormones are rendered inactive by saturation of ring A and substitution of hydroxyl groups at carbons 3 and/or 20. The metabolites are rendered water-soluble by conjugation with glucuronic acid in the liver prior to excretion by the kidney. The adrenal cortex of the normal human adult secretes approximately 15 to 30 mg of cortisol, 2 to 5 mg of corticosterone, and 0.05 to 0.20 mg of aldosterone per day.

The major urinary metabolites of the adrenal androgens are dehydroepiandrosterone, the androsterones, and the etiocholanolones, and they are excreted as 17-ketosteroids which have been rendered inactive by saturation of ring A and hydroxyl substitution at carbon 3. These are excreted as sulfates at the rate of about 15 mg per day in the adult male and 10 mg per day in the adult female. The adrenal cortices normally start to contribute to the urinary 17-ketosteroids at an early age in both sexes. The rate of excretion increases with age from about 1 to 2 mg per day at the age of 2 until it reaches the adult level at about 23 to 25 years of age. Part of the adult level is of testicular or ovarian origin, depending upon the sex.

REGULATION OF THE SECRETION OF THE ADRENAL CORTEX

The removal of the anterior pituitary causes notable changes in adrenal cortical function and morphology. Hypophysectomy results in atrophy of the adrenal cortex, a decrease in the level of circulating cortisol and corticosterone to about 10 percent of that seen normally, and a 60 percent decrease in the rate of aldosterone secretion. The level of aldosterone secretion is maintained to such an extent that the animals do not usually die as a result of adrenocortical insufficiency if they are maintained in a very protected environment. Although they show signs of adrenocortical insufficiency, the marked alterations in electrolyte metabolism are absent. Thus, aldosterone secretion rates can be maintained at levels which are compatible with life in the absence of ACTH, but ACTH

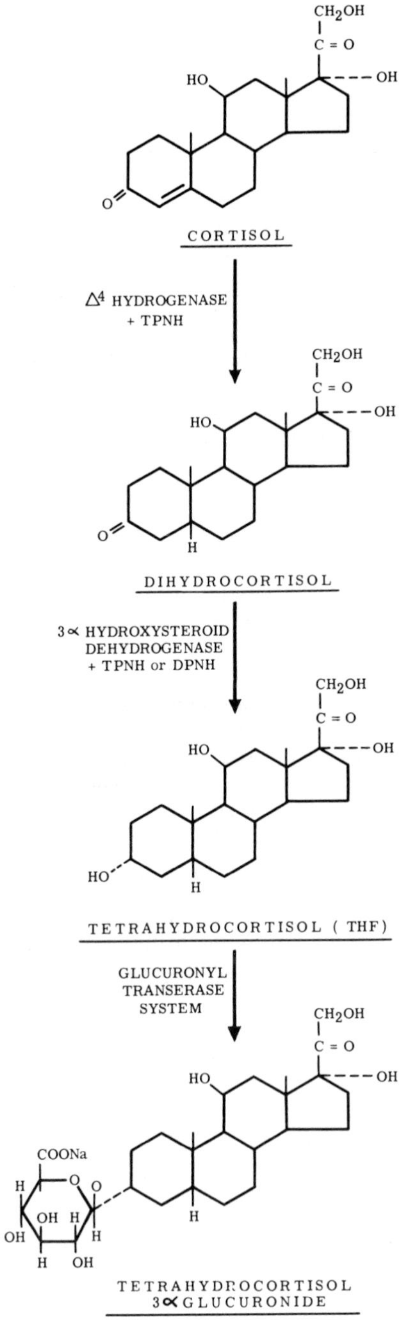

783

Figure 34-3
Major pathway for the inactivation and excretion of cortisol.

is required for maximal aldosterone output. The regulation of aldosterone secretion is discussed later.

ACTH is a polypeptide containing 39 amino acids and has a molecular weight of about 4500. A 24–amino acid fragment of this hormone has the biological activity equal to that of the parent compound, ACTH. ACTH is secreted by a specific basophilic cell (β-R type) of the anterior pituitary (AP). This same cell type secretes beta-melanocyte-stimulating hormone (β-MSH). A radioimmunoassay is available for the determination of the concentration of ACTH in the plasma. There is a sharp diurnal variation in the plasma ACTH. The concentration of ACTH in the plasma during early morning (0600 hours) is about 55 to 65 pg per milliliter in normal undisturbed individuals, whereas similar subjects have a plasma ACTH concentration of about 25 to 35 pg per milliliter at early evening (1800 hours). The plasma cortisol levels are consistent with the plasma ACTH levels, and the concentration of each of these hormones in the plasma is related to the sleep-activity cycle. These changes are shown in Figure 34-4.

Available evidence indicates that ACTH stimulates the secretion of cortisol by affecting enzymatic steps early in the synthetic pathway by accelerating the conversion of cholesterol to Δ^5-pregnenolone, and there appears to be little action of ACTH on the synthesis of steroids on later steps. The early effects of ACTH on adrenal steroid synthesis are probably mediated through cyclic adenosine monophosphate (cyclic AMP). This activates phosphorylase, and glycogen is converted to glucose-1-phosphate and then to glucose-6-phosphate. Increased amounts of generated nicotinamide-adenine dinucleotide phosphate (NADPH) then serve as an energy source for steroid synthesis. The more long-term effects of ACTH on protein synthesis and growth of the gland appear to be dependent upon an increased synthesis of messenger ribonucleic acid (mRNA), increased total RNA content, an increased number of polysomes, and increased mitotic activity.

When ACTH is administered to a normal or hypophysectomized animal, there is a marked increase in the rate of secretion of cortisol, corticosterone, aldosterone, and the adrenal androgens. ACTH also induces an increase in the size of cells in the zona fasciculata and zona reticularis and increases mitotic activity of the cells. The hypertrophic and hyperplastic changes result in an increase in adrenal size. An increase in adrenal blood flow also follows ACTH administration. Further, the hormone acutely depletes the lipid granules of cortical cells and reduces both the cholesterol and ascorbic acid concentrations of the gland.

Changes in the secretion of ACTH by the adenohypophysis follow removal of the adrenal cortex or the administration of large amounts of cortisol. In the former case, an increase in ACTH secretion occurs; in the latter, ACTH secretion may be completely inhibited.

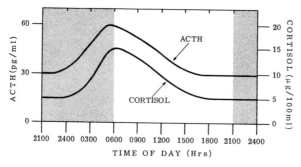

Figure 34-4
Curves showing the diurnal changes in concentrations of ACTH and cortisol in the plasma.

FUNCTIONS OF THE ADRENAL CORTEX

Cortisol, corticosterone, and aldosterone are the physiologically significant steroids secreted by the adrenal cortex of the normal human. Cortisol exerts its effects on the metabolism of protein, carbohydrate, and fat, and on inflammatory and immunological processes, whereas aldosterone is concerned primarily with electrolyte metabolism. The effects of corticosterone are somewhat intermediate between those of cortisol and aldosterone. Steroids that have an androgenic effect alter the metabolism of protein in a manner opposite to that of cortisol.

The effects of the adrenal steroids are best illustrated by the changes that take place in the adrenalectomized animal. Following adrenalectomy, animals invariably die if no supportive means are instituted. The observation that adrenalectomized animals can be maintained indefinitely by a variety of adrenal steroids is taken as proof that the cortical portion of the gland, and not the medullary portion, is essential for life. Symptoms which appear in the adrenalectomized animal include loss of appetite, asthenia, gastrointestinal disturbances, hemoconcentration, reduced blood pressure and body temperature, and renal failure.

Electrolyte Metabolism

A primary abnormality resulting from adrenalectomy is the disturbance in electrolyte metabolism. This alteration depends in part upon the effect of renal handling of sodium, potassium, and water. In the absence of the steroids (aldosterone) a decreased tubular reabsorption of sodium, chloride, and water and an increased reabsorption of potassium occur. Therefore, adrenalectomy results in hyponatremia, hypochloremia, hyperkalemia, and extracellular dehydration. Each abnormality then contributes to hemoconcentration, acidosis, hypotension, decreased glomerular filtration rate, extrarenal uremia, and shock. The administration of aldosterone reverses these changes by virtue of its actions on the distal convoluted tubule to increase the reabsorption of Na^+, along with Cl^- and water, and to decrease the reabsorption of K^+ and increase the urinary loss of H^+.

The effects of aldosterone are mediated through the DNA-directed RNA synthesis of a specific protein which enhances the transcellular movement of sodium from the filtrate into the cells of the distal convoluted tubule. A "sodium pump" then moves the sodium out of the cell on the serosal side into the extracellular fluid (ECF) and conserves sodium. The alterations induce an expansion of the ECF volume, a shift of sodium to the intracellular compartment, and a depletion of potassium in both the extracellular and intracellular compartments. An extracellular metabolic alkalosis which is characterized by increased serum pH and carbon dioxide combining power, mild hypernatremia, and hypochloremia results (Fig. 34-5).

In addition to the renal influences, a deficiency or excess of adrenal steroidal activity exerts a direct effect upon the passage of sodium, potassium, and hydrogen ions across cell membranes. Excessive amounts of aldosterone may lead to an expansion of the ECF volume which is greater than the amount of exogenous water retained. Thus, intracellular fluid (ICF) dehydration must occur. Adrenalectomy or hypoaldosteronism is accompanied by changes which lead to hyponatremia, hypochloremia, hyperkalemia, extracellular dehydration, and circulatory collapse. It should be pointed out that during continued administration of aldosterone to normal subjects an "escape" from the intense sodium-retaining effect is observed within 5 to 10 days even though the steroid administration is continued. Aldosterone also decreases the sodium concentration and increases the potassium concentration in sweat and the secretions of the salivary and intestinal glands. Aldosterone is approximately 25 times more potent than desoxycorticosterone in its sodium-retaining effect, about 800 times more potent than cortisol, and 200 times more potent than corticosterone. The synthetic steroid 9α-fluorocortisol is an extremely potent salt retainer (one-fourth the potency of aldosterone).

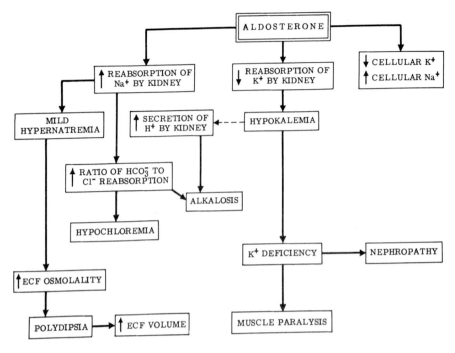

Figure 34-5
Representation of the effects of aldosterone on electrolyte and water balance.

Metabolism of Carbohydrates, Proteins, and Fats

Other defects observed in adrenalectomized or adrenal-insufficient subjects are those concerned with protein, carbohydrate, and fat metabolism. In the adrenalectomized subject, lacking cortisol or corticosterone, there is a tendency for the liver to lose glycogen. The ability of the individual to mobilize the precursors of glycogen from the peripheral tissues is reduced, the rate at which glucose is utilized by the peripheral tissue is elevated, and the deposition of body fat is reduced. Therefore, the adrenalectomized or hypoadrenal individual is prone to develop hypoglycemia and is extremely sensitive to the action of insulin. Such a subject also exhibits an elevated blood nonprotein nitrogen (NPN) which results not from increased protein breakdown but from renal decompensation that develops as a result of circulatory failure.

Cortisol, on the other hand, is diabetogenic. This steroid can reverse all the changes in carbohydrate and fat metabolism observed in adrenal insufficiency. Following the administration of cortisol to a fasting subject, there occurs within hours a marked increase in liver glycogen. This is accompanied by an increase in the blood sugar and a marked increase in the excretion of NPN. The increase in total body carbohydrate in the "steroid diabetic," therefore, must be the result of decreased glucose oxidation and/or an accelerated liver glycogen formation from tissue protein (gluconeogenesis). Various studies have shown that cortisol markedly stimulates protein breakdown in peripheral tissue and the uptake of amino acids by the liver. It has been observed that cortisol has a protein catabolic effect with respect to muscle and lymphatic tissue, and it leads to an overall negative nitrogen balance. However, cortisol does exert a strong protein anabolic effect on hepatic tissue. This differential effect on the liver may explain the diabetogenic effect, when one takes into account the myocytolytic and lymphocytolytic effects induced by cortisol in addition to its inducing an increase in enzyme (protein) synthesis by hepatic cells. One can induce hyperglycemia and glucosuria in intact animals fed a

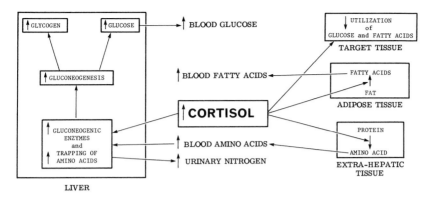

Figure 34-6
Effects of cortisol on carbohydrate, protein, and fat metabolism.

high-carbohydrate diet by the prolonged administration of cortisol-like compounds, and conversely, the metabolic symptoms of diabetes mellitus may be alleviated by adrenalectomy.

Glucocorticoids also mobilize fat from its storage place in adipose tissue and increase the free fatty acid levels in the plasma. Together with greater mobilization of fat, there is a redistribution of fat depots which produces "moon face" (rounding of the cheeks), "buffalo hump" (growth of supraclavicular and upper dorsal fat pads), increased axial fat, and a loss of fat on the extremities. As in the diabetic state, there is a decreased formation of fat from carbohydrate. In general, the glucocorticoids tend to act antagonistically to insulin with respect to its action on carbohydrate and fat metabolism. The androgenic adrenal steroids have a protein anabolic effect, which is opposite to that produced by the glucocorticoids.

Cortisol is the most active naturally occurring steroid in the human with regard to its effects on liver glycogen formation. On a molar basis, cortisone is about 65 percent as effective and corticosterone about 35 percent as effective, whereas aldosterone and desoxycorticosterone show less than 1 percent of the activity of cortisol. The synthetic steroid 9α-fluorocortisol possesses about 10 times the activity of cortisol. The effects of cortisol on protein, carbohydrate, and fat metabolism are illustrated in Figure 34-6. No single mechanism explaining the effects of cortisol on metabolic functions and enzyme systems has been formulated. It is generally accepted that cortisol institutes a "permissive" action on many metabolic functions which permits the effects of other hormones to take place. The mechanisms whereby glucocorticoids exert their effects remain to be elucidated.

Miscellaneous Functions

The glucocorticoids cause a rapid involution of the thymus. They also suppress lymphoid tissue activity and act to induce lymphopenia and eosinopenia. The lymphocytolytic effect of the glucocorticoids interferes with the conversion of lymphocytes into antibody-producing cells and consequently interferes with the immune response. They are thus effective immunosuppressants and are used therapeutically to counteract rejection of organ or tissue transplants in humans. The glucocorticoids profoundly influence the inflammatory response. They decrease vascular permeability, increase capillary resistance, and potentiate the effects of catecholamines. They tend to depress the release of histamine and the migration of inflammatory cells. All of these effects reduce the inflammatory response of many tissues. Local inflammatory reactions to irritating substances are greatly reduced or delayed, hypersensitivity reactions to antigens are suppressed, and healing of wounds is delayed when active steroids are given in excess over a period of hours or days. Therefore, the actions of excess steroids can be deleterious to the combating of infections and the healing of wounds, although alterations of these reactions may be of benefit in certain acute hypersensitivity states.

The adrenal steroids may also have many diverse effects, such as muscular weakness as a result of muscle wasting, increased rate of secretion of gastric HCl and pepsin leading to the development of gastric and duodenal ulcers, hypertension and increased capillary resistance and fragility, increased excitability of brain tissue with euphoria and restlessness, and stimulation of bone marrow. At physiological levels the steroids will prevent the deposition of melanin in the skin of the adrenal-insufficient individual by preventing the release of ACTH, and they are required for normal thyroid, gonadal, and pituitary function.

Role of Adrenal Cortex in Stress

The adrenalectomized subject or adrenal-insufficient individual has little ability to tolerate changes in the internal and external environment, such as cold, heat, infections, trauma, or prolonged exercise. The resistance to environmental change in such subjects is very low, and as a result of the stimulus they will die, whereas the normal individual can tolerate a much greater stimulus. Selye (1956) has shown that when an animal is subjected to a stressor it reacts in a stereotyped manner. A stressor is any stimulus that causes an increase in the secretion rates of corticotropin-releasing factor (CRF), ACTH, and cortisol leading to an increase in the concentration of cortisol in the plasma greater than that normally observed at the same hour of the day in undisturbed normal subjects on a similar activity (sleep-wake) cycle. The entire reaction to the stressor may be termed a stress reaction. If the same stimulus is given to a hypophysectomized or adrenalectomized animal, its ability to survive the injury is minimal. However, survival can be attained if cortisol or ACTH is administered to the hypophysectomized subject. Thus, protection is afforded by glucocorticoids, and activation of the pituitary-adrenal axis must ensue. The adjustments that follow a stimulus must be useful to the subject in the attempt to maintain homeostasis. Many theories have been advanced to suggest how the adrenal steroids act to protect the subject, but none can account for the protection afforded. It should be emphasized that an increase in plasma cortisol concentration is essential for man to withstand severe stress.

REGULATION OF ACTH SECRETION

An intricate homeostatic mechanism regulates the normal secretion of cortisol by the adrenal cortex. The rate of synthesis and secretion of ACTH is regulated directly by a substance, CRF, which is secreted by neurons located in the basal medial hypothalamus. CRF is a polypeptide and is released into the hypophysial portal circulation, through which it gains access to the ACTH-secreting cells of the anterior pituitary, the corticotropes. The regulation of ACTH secretion thus involves factors that regulate the secretion of CRF and effect the response of the corticotropes to CRF. These factors include (1) the neural influences on the CRF-secreting cells, (2) the influence of cortisol on the CRF- and ACTH-secreting elements, and (3) other humoral agents which act upon the hypothalamo-hypophysial unit.

Early studies showed that removal of the anterior pituitary from its vascular connection, the hypophysial portal vessels, to the hypothalamus notably decreased ACTH secretion and abolished the normal stress response. Similarly, destruction of the basal medial hypothalamus abolished all stress responses. Stimulatory neural inputs to the basal medial hypothalamus and the CRF-secreting neurons are several and have been shown to arise in the amygdaloseptal complex and in the medial reticular formation. Lesions of the rostral midbrain and posterior diencephalon depress CRF release and cortisol secretion. The stimuli which activate the CRF-ACTH-cortisol stress response must in part act via the various stimulatory pathways. Inhibitory neural inputs into the CRF-releasing neurons arise in the hippocampus. Electrical stimulation of the posterior hypothalamus decreases CRF and ACTH secretion, and lesions of posterior midbrain increase CRF and ACTH secretion. Stimulation of specific areas of the lateral reticular formation also inhibit CRF and cortisol secretion. Present evidence would indicate that the synapses which stimulate the CRF-secreting neurons are cholinergic in nature, whereas those which inhibit the CRF-secreting neurons are noradrenergic. The circadian rhythm of the CRF-ACTH-cortisol system is probably imposed via neural inputs into the anterior hypothalamus, because lesions in this area abolish the rhythm, but permits the normal stress response of the system to occur. The location of the area that supplies the rhythmic input to the system is not known.

It has been known for many years that a reciprocal relationship exists between the adrenal cortex and the adenohypophysis. Early studies revealed that the systemic administration of cortisol induces a decrease in ACTH secretion and a resultant decrease in cortisol secretion and adrenal atrophy, and that pretreatment with high doses of cortisol prior to exposure to an external or internal stimulus abolishes the normal stress response (e.g., an increase in ACTH secretion). It has also been shown that adrenalectomy or the absence of cortisol results in the secretion of excessive amounts of ACTH. The absence of cortisol likewise leads to excessive β-MSH secretion and high blood levels of β-MSH. Normal blood cortisol levels inhibit β-MSH secretion. More recent studies indicate that the infusion of minute amounts of cortisol solutions or the implantation of cortisol pellets into the anterior pituitary acts to inhibit the release of ACTH under a variety of conditions. By means of in vitro techniques such as the monolayer culture of pituitary cells or isolated incubated glands, it has been demonstrated that the ability of the corticotropes to secrete ACTH in response to CRF is inhibited by adding cortisol to the medium. The steroid also acts on the hypothalamus to decrease CRF secretion, because cortisol administration lowers the content of CRF in the hypothalamus, and the implantation of cortisol in the hypothalamus decreases CRF secretion. Cortisol also decreases CRF secretion by exerting an inhibitory action on the amygdala.

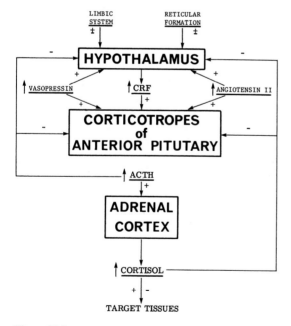

Figure 34-7
Factors influencing the activity of hypothalamo-hypophysial-adrenocortical system. $+$ = stimulation; $-$ = inhibition.

Two other humoral agents, vasopressin and angiotensin II, act on the hypothalamo-hypophysial-adrenal system to increase the activity of the system. Each has been shown not to be the CRF. Each has been shown to stimulate the secretion of CRF directly at the level of the hypothalamus and to stimulate directly the secretion of ACTH at the level of the corticotrope, and vasopressin appears to potentiate the action of CRF on the corticotropes. It should be pointed out that high systemic levels of both vasopressin and angiotensin II are required to induce their CRF- and ACTH-stimulating effects. Figure 34-7 illustrates the factors which regulate the activity of the hypothalamo-hypophysial-adrenocortical system. Psychological factors such as anxiety also affect this system and increase its activity. Similarly, various sedatives, tranquilizers, and anesthetics on initial administration stimulate the stress response but on prolonged administration may suppress it. Each acts on the system at the level of the CNS.

CONTROL OF ALDOSTERONE SECRETION

That the zona glomerulosa does not atrophy following hypophysectomy (short term) was first observed by Houssay. Greep and Deane proposed that this zone secretes an electrolyte-active substance relatively independent of the adenohypophysis. Later, as had been predicted by Greep and Deane on the basis of their morphological work, Ayres demonstrated that aldosterone is preferentially synthesized by the zona glomerulosa.

The rate of secretion of aldosterone is not completely independent of ACTH, because following hypophysectomy it is reduced to one-half the normal rate. Also, varying degrees of atrophy of the zona glomerulosa follow hypophysectomy, and the degree of atrophy is said to be directly related to the time interval between hypophysectomy and autopsy.

The factors that alter aldosterone output are many, but the alterations in sodium balance and in the volume of the body fluids stand out as especially important. Increased output of aldosterone, without a simultaneous rise in cortisol output, has been shown to follow (1) restriction of dietary sodium, (2) high potassium intake, (3) water depletion, (4) thoracic postcaval constriction, and (5) aortic constriction. Hemorrhage, anxiety, surgical stress, and trauma also stimulate aldosterone secretion, but there is a concurrent elevation of cortisol secretion. This would indicate that the hypothalamo-hypophysial activation of ACTH release has occurred.

Davis (1962) first demonstrated that the renin-angiotensin system was an important factor in the regulation of aldosterone secretion by the zona glomerulosa. Hyperaldosteronism resulting from chronic thoracic caval constriction, during chronic sodium depletion, acute blood loss, or acute aortic constriction, is accompanied by a decrease in renal blood flow, an increase in renin content of the kidney, and hypergranulation (with occasional hyperplasia) of the juxtaglomerular cells of the kidney. Nephrectomy in the hypophysectomized animal consistently causes a further marked reduction in the secretion of aldosterone by the zona glomerulosa. Present evidence would indicate that a decrease in renal perfusion pressure or in filtered sodium load acts upon the renal juxtaglomerular complex, and an increase in renin production follows. The renin then acts upon the circulating angiotensinogen to form the decapeptide angiotensin I. The latter, in the presence of chloride and a converting enzyme, is converted to the octapeptide angiotensin II, a potent hypertensive agent. Angiotensin II acts directly on the zona glomerulosa to increase aldosterone secretion. The aldosterone then acts on the nephron and increases sodium reabsorption and, along with it, water reabsorption. The ECF is expanded, renal perfusion pressure is elevated, and the stimulus for renin production is removed. Angiotensin II can stimulate aldosterone secretion in the hypophysectomized, nephrectomized animal without increasing cortisol secretion, and it stimulates the in vitro conversion of cholesterol to aldosterone.

The circulating Na^+ and K^+ exert an effect upon aldosterone secretion. High K^+ and low Na^+ act directly on the zona glomerulosa to increase aldosterone secretion, because,

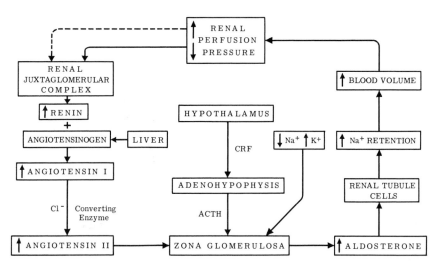

Figure 34-8
Regulation of aldosterone secretion. $\longrightarrow$ = stimulation; $\dashrightarrow$ = inhibition.

in the absence of the adenohypophysis and/or kidneys, aldosterone secretion remains high in either circumstance. Also, using the isolated perfused adrenal gland, lowering the Na^+, or increasing the K^+ concentration of the perfusate provokes a rise in aldosterone secretion.

The factors which regulate aldosterone secretion are shown in Figure 34-8.

ADRENOCORTICAL DISEASES

Observation of the scheme of biosynthetic pathways of adrenocortical hormone leads to the conclusion that many types of functional disorders may be produced as a result either of enzyme deficiency or excess, or of faulty pituitary function. Various forms of congenital adrenal hyperplasia (CAH) (adrenogenital syndrome) are frequently observed in the newborn. CAH occurs as a result of specific enzyme(s) deficiency in biosynthetic pathways of steroids so that there is produced a deficiency of steroids beyond the block and an excessive synthesis and secretion of steroids in front of the block. Most forms of CAH include a deficiency in glucocorticoid secretion and, as a result of the increased rate of ACTH secretion, adrenal stimulations with hypertrophy and, in the most common forms, excessive secretion of adrenal androgens during fetal and postnatal life. In the most marked, but rarest, form cholesterol cannot be converted to Δ^5-pregnenolone, and therefore no adrenal steroidal hormones can be synthesized. In another form the enzyme 3β-hydroxydehydrogenase is lacking, and consequently neither progesterone, 17α-hydroxyprogesterone, nor Δ^4-androstenedione is synthesized. These two forms are not compatible with postnatal life. A condition in which a deficiency in 17α-hydroxylase exists is compatible with life, because the pathways leading to corticosterone and aldosterone are not affected. The two most prevalent forms of CAH involve enzymatic defects in 21-hydroylase and 11-hydroxylase. These blocks inhibit both the glucocorticoid and mineralocorticoid pathways, and adrenal steroidogenesis is shunted to the androgen-producing pathway. Therefore, each of these two conditions is characterized by adrenal insufficiency with excessive androgen secretion and masculinization, whether present in the male or female.

Cushing's syndrome includes those clinical situations that result from the excessive and long-term action of glucocorticoids. Several primary factors may lead to the exposure of individuals to high blood levels of glucocorticoids: (1) excessive ACTH secretion, (2) an adrenal tumor which secretes large amounts of glucocorticoid, (3) excessive ACTH medication, and (4) the administration of excessive amounts of glucocorticoids. Hyperaldosterone states have been described and are associated with hypertension. Primary aldosteronism (Conn's syndrome) results from an adenomatous hyperplasia of the zona glomerulosa, and secondary hyperaldosteronism results from excessive stimulation of the zona glomerulosa by the renin-angiotensin system.

Conditions leading to adrenal insufficiency may arise as a primary dysfunction of the adrenal cortex, as in Addison's disease, in which, in its severest form, one observes changes similar to those seen in adrenalectomy. Adrenal insufficiency may also be the indirect result of a deficiency in ACTH secretion. This is nearly always accompanied by concomitant decreases in the AP hormone secretion.

REFERENCES

Clamon, H. N.   How corticosteroids work. *J. Allergy Clin. Immunol.* 55:145–151, 1975.

Davis, J. O.   The control of aldosterone secretion. *Physiologist* 5:65–86, 1962.

Denton, D. A.   Evolutionary aspects of the emergence of aldosterone secretion and salt appetite. *Physiol. Rev.* 45:245–295, 1965.

Eberlein, W. R., and A. M. Bongiovanni.   Pathophysiology of congenital adrenal hyperplasia. *Metabolism* 9:326–340, 1960.

Edelman, I. S., and G. M. Fimognari.   On the biochemical mechanism of action of aldosterone. *Recent Prog. Horm. Res.* 24:1–44, 1968.

Fuller, R. W.   Control of epinephrine synthesis and secretion. *Fed. Proc.* 32:1772–1781, 1973.

Leung, K., and A. Munck.   Peripheral actions of glucocorticoids. *Annu. Rev. Physiol.* 37:245–272, 1975.

Mulrow, P. J.   The adrenal cortex. *Annu. Rev. Physiol.* 34:409–434, 1972.

Phifer, R. F., D. N. Orth, and S. S. Spicer.   Specific demonstration of the human hypophysial adrenocortico-melanotropic (ACTH/MSH) cell. *J. Clin. Endocrinol. Metab.* 39:684–692, 1974.

Selye, H.   *The Stress of Life.* New York: McGraw-Hill, 1956.

Urquhart, J.   Blood-borne signals: The measuring and modelling of humoral communication and control. *Physiologist* 13:7–41, 1970.

Vander, A. J.   Control of renin release. *Physiol. Rev.* 47:359–382, 1967.

Yates, F. E., and J. W. Moran.   Stimulation and Inhibition of Adrenocorticotropin Release. In R. O. Greep and E. B. Astwood (Eds.), *Handbook of Physiology.* Washington: American Physiological Society, 1974. Section 7: Endocrinology, vol. 4, part 2, chap. 36, pp. 367–404.

35. Physiology of Reproduction

Ward W. Moore

GENETIC BASIS OF SEX

The sex of an individual is the outcome of two distinct processes: sex determination and sex differentiation. Sex determination is regulated by genetic phenomena which are fixed at the time of fertilization and which determine the genotype of an individual. Sex differentiation refers to the course of development of the gonads, genital ducts, external genitalia, and secondary sex characteristics; it results from two developmental processes: the direction of differentiation of the gonads toward ovaries or testes, and the development of the primordial duct systems toward femaleness or maleness. These changes define the somatic sex (phenotypic sex) of the individual. It is generally held that the two processes are predetermined at the time of fertilization by the particular chromosomal pattern of the zygote. Each mature ovum contains 22 chromosomes, termed *autosomes*, plus an X chromosome. Each mature spermatozoon also contains 22 autosomes and has either an X chromosome or a Y chromosome. It is the sex chromosomes, X and Y, that determine the sex of an individual. The male zygote possesses an XY pair of sex chromosomes, while the female zygote possesses the XX pair.

The first specific process of division (meiosis) which leads to the formation of the mature male germ cell involves a division of chromosomes, half of each pair to one daughter cell and half to another. Thus, the two daughter cells bear different sex chromosomes, one an X and the other a Y. In the female, the corresponding division produces two cells, each of which carries an X chromosome. When the ovum of the female and spermatozoon of the male unite, the chances are equal for the formation of an XX zygote (female genotype) or an XY zygote (male genotype). The genotypic or true sex of the zygote is regulated by the components contributed by the male germ cell and is determined at conception. However, the phenotypic or somatic sex may be the outcome of a balance between the influence of the sex hormones and that of an undefined number of genes carried on autosomal chromosomes.

It is now possible to determine the true sex of individuals, varying in age from very young embryos to very old adults, by the use of the so-called *sex chromatin*. This extra mass of chromatin is present in nuclei of most somatic cells of humans and is referred to as the Barr body. The sex chromatin is usually found lying next to the inner surface of the nuclear membrane, is Feulgen-positive, is about 1 μm in diameter, and is found only in genetic females. It is thought to result from the fusing of heteropyknotic portions of the two X chromosomes. The determination, by this technique, of the true sex of an individual has widespread use in clinical medicine as an aid in diagnosis of a variety of conditions.

In a few instances abnormal chromosome combinations occur during the transfer of chromosomes from one cell to the other. In one type the fertilized ovum contains only one X chromosome and no Y chromosome. This will result in an XO genotype, yield a chromatin-negative test, and produce an individual who has either rudimentary gonads or no gonads. The condition is known as Turner's syndrome or ovarian dysgenesis. Another type of abnormal chromosomal complement which occurs not infrequently is the XXY. The individual resulting from this combination has a positive chromatin test, male external genitalia, and abnormal seminiferous tubules with aspermatogenesis. The condition is referred to as Klinefelter's syndrome or seminiferous tubule dysgenesis. Other abnormal XY genotypes include XXX, XYY, and types of mosaicism such as XO/XX and XO/XY.

793

ROLE OF HORMONES IN THE DEVELOPMENT
OF THE GENITAL SYSTEM

The reproductive system passes through a period of early embryonal development during which it is impossible, either grossly or microscopically, to tell the sexes apart. This period is referred to as the indifferent stage. Nalbandov (1964) has stated, "In this stage all *anlagen* for subsequent differentiation into complete male and female systems are present in rudimentary form; the blueprint, as it were, is finished. All the materials for the later elaboration of the fittings and furnishing of structures are present, but no attempt is made to arrange the internal furnishings permanently until final orders are received for the emphasis of either male or female aspects of the different structures."

A bilateral longitudinal ridge, the genital ridge, appears medial to the mesonephros during the fifth week of embryonic life in the human. The surface of this ridge is covered by germinal epithelium and contains the primordial germ cells which have migrated by ameboid movement from yolk-sac endoderm. Specific histological changes occur in this s' ructure (the *indifferent gonad*) during the seventh week, and they indicate that the gonad is committed to either the male or female role. If the gonad is to develop into a testis, the cells of the germinal epithelium become organized within the medulla of the gonad, form the anlagen of the seminiferous tubules, and subsequently join the cords of the mesonephros to form a continuous network of tubules. Mesenchymal cells in the medulla develop into the cells of Leydig, which become apparent at the end of the eighth week. If the gonad is to become an ovary, no changes are apparent until the tenth week. At this time, the cortex of the gonad accumulates nests of cells, and they are differentiated into ovarian follicles, each containing an ovum.

If the ducts are to differentiate in accordance with the sex of the gonads, either the mesonephric system or the müllerian ducts must degenerate. The duct system and the gonads pass through an undifferentiated stage, because the fetus possesses the basic duct systems for both sexes. Only after continued development does one system gain ascendancy and become definite, whereas the other regresses and becomes vestigial. The müllerian ducts begin to degenerate in the male at about the twelfth week, and the mesonephric system develops. In the female the mesonephric system begins to disappear late in the thirteenth week, and the müllerian system becomes dominant. The origins of various parts of the genital system in both sexes are given in Table 35-1.

Removal of the gonads from the early embryo of either sex prior to the development of the genital ducts leads to the development of the female-type internal and external genitalia, regardless of the genetic sex of the fetus. This observation is interpreted as meaning that the testes secrete a substance which leads to the development of the male system, and that in the absence of this substance the genital duct system develops in the direction of the female.

There is little doubt that certain aspects of the development of the reproductive system are effected or facilitated by the sex hormones, the androgens and the estrogens. Both nature and experimentation have shown that while the developing embryo or fetus is in an undifferentiated or differentiating state, the phenotypic sex of the individual may be altered or reversed. The classic papers of Lillie describe a naturally occurring case of sex reversal, the freemartin of cattle. The freemartin is a genetic female and a twin of a genetic male, and the twins possess a common fetal circulation. The genetic female is sterile, has inhibited ovaries and female genital ducts, while the accessory sex ducts and secondary sex characteristics approach those of full maleness. As noted before, the testes differentiate prior to the ovaries and are capable of secreting androgen at an early stage. Thus, androgen acts on the undifferentiated duct system of both twins and causes differentiation of the indifferent duct system in the male direction and regression of the primordial female system.

Varying degrees of masculinization are seen in the newborn human female with congenital adrenal hyperplasia. This is caused by a genetic defect in which simple mendelian recessive genes combine and lead to a deficiency in an enzyme system concerned with adrenal steroid synthesis. The result is excess production of adrenal androgen. The earlier in embryonic life the defect is manifested, the more marked are

Table 35-1. Homologies of the Male and Female Reproductive Systems in the Human

Male	Indifferent Stage	Female
Testis	Indifferent gonad	Ovary
Rete testis		Rete ovarii*
Vas efferens	Mesonephric tubules	Epoophoron*
Paradidymis*		Paroophoron*
Vas aberrans		
Epididymis	Mesonephric duct	Gartner's duct*
Vas deferens		
Ejaculatory duct		
Seminal vesicle		
Prostatic utricle*	Müllerian duct	Hydatid*
		Oviduct
		Uterus
		Vagina
Urethra	Urogenital sinus	Urethra
		Vagina
		Vestibule
		Vestibular glands
Glans penis	Genital tubercle	Glans clitoris
		Corpus clitoris
Raphe penis	Urethral folds	Labia minora
Raphe scrotum		
Scrotum	Labioscrotal swellings	Labia majora

*Rudimentary.

the aberrant changes observed in the newborn female genital system. The newborn male with defects in his adrenal 21- and/or 11-hydroxylases has a high rate of androgen secretion and is precociously and markedly masculinized. The genesis of these disorders is discussed in the preceding chapter.

Under normal circumstances, the effect of the maternal hormones is insufficient to cause profound changes in the fetus. The maternal estrogens and progestin cause cervical enlargement, hypertrophy of the vaginal epithelium, and growth of the mammary gland of the fetus. However, certain synthetic estrogens and progestins, when administered early in pregnancy, may produce marked structural alterations (masculinization) in the female fetus. The placenta is permeable to most steroid hormones, but at birth the infant normally shows little evidence of excessive stimulation by other maternal hormones.

It should be noted that none of the fetal endocrine glands studied has been found to be indispensable for fetal survival; removal of the hypophysis, thyroid, parathyroid, gonads, or adrenals has been accomplished in the fetuses of laboratory animals, and these fetuses survive. Homeostasis of the fetus is generally maintained by way of maternal homeostatic mechanisms, but fetal hormones do play an indispensable role in normal development.

ENDOCRINOLOGY OF THE MALE

The testes of the male have a dual function, one of gamete formation and the other of hormone production. The gametogenesis is dependent to a considerable extent on androgen production by the testes, and both are dependent upon the adenohypophysis (AP). In turn, the AP is dependent upon normal hypothalamic function and intact hypophysial portal vessels. Each is indispensable if the male is to reach full reproductive capacity, but neither is essential for life.

The typical histological picture of the testes is that of seminiferous tubules separated by interstitial connective tissue. This pattern is observed from prenatal stages throughout life with decided variations according to age. At birth, the seminiferous tubules have not yet formed lumina, are small, and contain only spermatogonia. Leydig cells are present at birth, but then regress, and are not again identifiable until the approach of puberty. At this time maturation changes such as thickening of the tunica propria, initiation of spermatogenesis, proliferation of the Leydig cells, and secretion of androgens occur. The changes are dependent on hypophysial gonadotropin production. Thus, from the third month of life, by which time maternal hormone influences have disappeared, until the eighth to the eleventh year, sexual development is largely in abeyance, and individuals of both sexes are in a state of physiological hypogonadism.

During the infantile period, the gonads are potentially able to function in an adult fashion, for they respond to gonadotropins. The secondary sex organs respond either to androgens or to estrogens. Therefore, gross insensitivity of the target organs cannot be responsible for the lack of sexual maturation in childhood. The gonads of immature animals when transplanted to mature animals function as mature gonads; but the gonads of mature animals, when transplanted to immature animals, become quiescent and atrophy.

The newborn adenohypophysis transplanted into the sella turcica of a hypophysectomized mature animal stimulates and repairs gonadal function more rapidly than the pituitary of a mature animal when transplanted into the sella of an infant animal. Thus, the fetal hypophysis contains gonadotropins, and the gland of the prepubertal individual is potentially capable of stimulating the gonads. These observations form the basis of the hypothesis that the secretion of gonadotropins at puberty is dependent upon a neural or neurohumoral phenomenon which is inhibited during childhood. The hypothesis is strengthened by the experimental observations that lesions in the amygdala, stria terminalis, or anterior hypothalamus result in precocious puberty. Similarly, many naturally occurring lesions in the central nervous system (CNS) induce the production of gonadotropins at an early age. These lesions have one characteristic in common, namely, involvement of the hypothalamus. The manner in which the CNS-induced inhibition of gonadotropic secretion is removed at puberty is not known.

Regulation of Testicular Functions

The transition from boyhood to manhood normally begins between the eighth and eleventh years, although it is usually not apparent on gross examination until the tenth to thirteenth year of age. Each stage which occurs with age transition is initiated by the action of gonadotropins, the secretion of which is dependent upon altered CNS activity.

As a result of the increase in follicle-stimulating hormone (FSH) and luteinizing hormone (LH) secretion by the adenohypophysis, the seminiferous tubules are stimulated and the Leydig cells start to proliferate and undergo functional changes. They begin to synthesize androgens or masculinizing steroids, the primary and most potent of which is testosterone.

The testicular androgens are steroidal in nature and are 19-carbon compounds, in contrast to the 21-carbon glucocorticoids and mineralocorticoids and the 18-carbon estrogenic steroids. In the testes, as in the adrenal cortex, the hormones stem from cholesterol, but the primary synthetic pathway in the testes is via Δ^5-pregnenolone, progesterone, 17α-hydroxyprogesterone, and Δ^4-androstenedione. Δ^4-androstenedione is acted on at the 17 position by a reductase to yield testosterone, the main testicular androgen. The secretory rate of testosterone is stimulated by LH. Other androgens which are secreted in small amounts by the Leydig cells are Δ^4-androstenedione, dehydroepiandrosterone, and dihydrotestosterone (DHT). The plasma testosterone is bound to a beta-globulin, with about 2 to 4 percent circulating in the free or unbound state. The testosterone in the serum of the prepubertal boy is generally less than 0.5 ng per milliliter and increases with advancing puberty to an adult level which has a range of 5 to 12 ng per milliliter. Both serum LH and testosterone concentrations increase throughout puberty, the increases in testosterone following those of LH. In several tissues, such as the prostate, testosterone is converted to dihydrotestosterone, and the latter seems to be the effective androgen in this tissue. The major site of inactivation of the testicular androgens is the liver, where they are converted to androsterone, epiandrosterone, and etiocholanolone (Fig. 35-1). These metabolites are conjugated with glucuronide or sulfate, are excreted as water-soluble salts in the urine, and are regarded as 17-ketosteroids (17-KS). The urinary 17-KS excretion is constant and equal in both

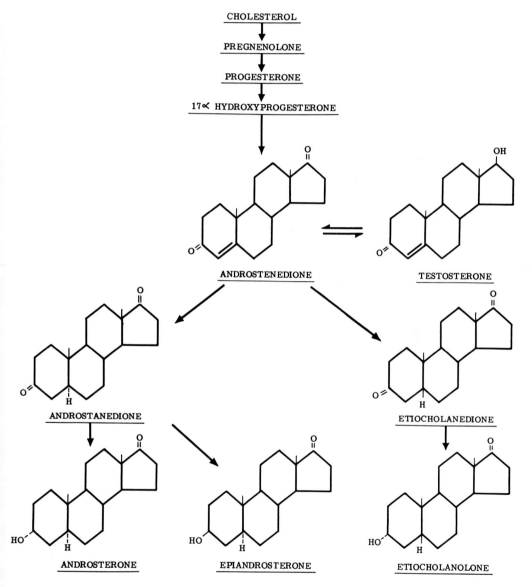

Figure 35-1
Major biosynthetic pathway for the testicular androgen testosterone. The remaining compounds are metabolized and are inactivated by the liver and excreted as 17-ketosteroids in the urine.

boys and girls up to the age of eight or nine years and is of adrenal origin. Of the total of approximately 14 to 16 mg of 17-KS excreted per day by the adult male, about 10 mg is of adrenal origin, because the eunuch excretes only 10 mg per day. Estradiol has been isolated from the human testes. At puberty the testicular secretion of estradiol increases and the urinary excretion of estrogens increases. They both decrease following bilateral orchidectomy and increase following gonadotropin administration.

The mechanism of the initiation and maintenance of gametogenic and secretory activity of the testes is best illustrated in the hypophysectomized male. The seminiferous tubules and Leydig cells are atrophic in such an individual, gametogenesis is halted, and androgen production is minimal.

The administration of LH to the hypophysectomized animal stimulates the Leydig cells to respond by secretion of testosterone. FSH has no demonstrable effect upon the Leydig cells but can cause a significant stimulation of the seminiferous tubules. FSH alone cannot support spermatogenesis and spermiogenesis. In order to maintain complete spermatogenesis, both FSH and LH or testosterone must be administered. It is clear that, in the male, LH is the primary adenohypophysial hormone responsible for stimulation of the seminiferous tubule and Leydig cells. In the latter case, there is direct stimulation, while in the former an indirect action is mediated via the secretion of testosterone by the cells of Leydig. When LH is administered immediately after hypophysectomy, it has the capacity for stimulating the Leydig cells and advancing spermatogenesis and spermiogenesis. That FSH has a physiological role in spermatogenesis is shown when administration is delayed in order to allow posthypophysectomy atrophy of the testes. In this instance, LH cannot act as a complete gonadotropin but

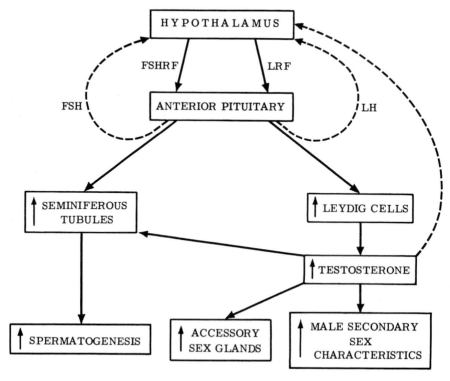

Figure 35-2
Hormonal regulation of reproductive functions in the male. $\longrightarrow$ = stimulation; $-\!-\!-\!\rightarrow$ = inhibition.

functional significance of acid phosphatase in seminal fluid is not known. This enzyme has clinical significance because it enters the bloodstream when malignant growth of the prostate occurs. The secretions of the prostate and seminal vesicles are not necessary for the production of viable sperm because both testicular sperm and sperm taken from the epididymis are viable. However, the viability of such sperm is decreased.

Erection is a vascular phenomenon which is dependent upon the vascular structural pattern of the penis. It is the result of venous constriction and arterial dilation which allows blood to flow under high pressure into the erectile tissue. The erectile tissue is a spongelike system composed of vascular spaces which are relatively collapsed and contain little blood in the flaccid state. However, during erection these spaces become distended with blood, and the pressure approaches that in the carotid artery. Erection may result from either physical or psychic stimuli; the motor pathways are via the sacral component of the craniosacral system, the *nervi erigentes,* and a center for reflex erection exists in the sacral cord. The subsidence of erection can result from stimulation of the sympathetic innervation of the penis.

Ejaculation actually refers to two distinct actions, emission and ejaculation, both of them reflex phenomena. The afferent arc of these reflexes originates in the sense organs of the glans penis, and the impulses are transmitted centrally through the internal pudendal nerves. The first action of ejaculation, which delivers sperm and seminal plasma into the urethra, is the result of contraction of smooth muscles of the genital tract and is called emission. This action includes peristaltic contractions of the testes, epididymis, and vas deferens that cause the expulsion of sperm into the internal urethra. It also includes the contractions of the seminal vesicles, prostate, and bulbourethral glands which result in the movement of their respective secretions into the urethra. Emission is evoked by stimulation of the hypogastric nerves. The second action, ejaculation, causes the expulsion of seminal fluid from the urethra to the exterior as a result of striated muscle contraction (bulbocavernous muscle). This results from increased neural activity transmitted peripherally via the internal pudendal nerves.

The volume of ejaculate is from 2 to 6 ml in man. The ejaculate contains more than 10^7 sperm per milliliter, has a slightly alkaline reaction (pH = 7.1 to 7.5), and has a specific gravity of about 1.028. Coagulation of normal semen occurs promptly after ejaculation. However, it liquefies within several minutes, and the sperm attain full motility.

The life-span of sperm, or the time during which they are able to fertilize ova, is relatively short, about 24 to 36 hours. However, sperm may be frozen to $-169\,°C$, stored, and thawed without impairment of their fertilizing capacity. It has been claimed that sperm concentrations of less than 50×10^6 sperm per milliliter of ejaculate indicate sterility. Out of this number introduced into the female reproductive tract, less than 10^4 sperm reach the oviducts, less than 10^2 reach the vicinity of the ovum, 10^1 may penetrate the zona pellucida of the ovum, but only 1 sperm enters the ovum and accomplishes fertilization. Therefore, the sine qua non for fertility in the male is the number of sperm per ejaculum. Other criteria, such as percentage of abnormal sperm (with respect to morphology) and volume of ejaculate, are also used in assessing semen quality.

ENDOCRINOLOGY OF THE FEMALE

In addition to the production of gametes, the basic function of the ovaries is to secrete hormones which regulate the activity of the female reproductive tract and determine the secondary sex characteristics. The gametogenic and endocrine aspects of ovarian function fluctuate rhythmically during the active reproductive life of the female. The periodic changes in the functional activities of the ovaries are determined by the interrelationships between the activity of the adenohypophysis and the ovaries. The adenohypophysial gonadotropins, FSH and LH, are concerned with ovarian function as well as with testicular function. In addition, a third pituitary hormone, *prolactin* (PRL), is concerned with mammary gland function. Its role in regulating cyclic ovarian activity in the human is not known. The female experiences a shorter reproductive life than does the male. The active reproductive life of the female commences at puberty (10 to 15 years of age) and is maintained for approximately 30 years. Reproductive activity gradually decreases, the ovaries involute, and reproductive cycles cease during the fifth decade of life. This transition is termed the menopause and is followed by hypogonadism for the remainder of the life-span.

Reproductive functions and behavior are markedly influenced by the gonadal hormones. That these are not the only factors involved is emphasized by the adverse effects on the functions of the pituitary-gonadal axis of both female and male of the lack of certain nutritional factors and caloric intake, nongonadal endocrine disorders, changes in the external environment, and CNS and psychological factors.

The reproductive cycle of the female is a complex series of coordinated events involving the CNS, adenohypophysis, ovaries, oviducts, uterus, cervix, and vagina. The most complete studies on the physiology of reproduction have utilized laboratory primates, rodents, and farm animals. Many of those studies have been extended to the human. Although certain quantitative differences exist, the fundamental factors which are known to regulate reproductive phenomena in the rat and bovine probably regulate the same phenomena in the human.

The Ovary

The size and microscopical appearance of the paired ovaries in the adult vary with the period of the reproductive cycle. The most important functional components of the mature ovary (Fig. 35-3) are the *follicles* and *corpora lutea*. Three stages of growth of the follicles can be noted. The immature stage (i.e., the primary follicle) is observed from embryonic to late life of the ovary. This structure makes up the bulk of the population of follicles. In the primary follicle the ovum is surrounded by several layers of granulosa cells and has no vitelline membrane. The secondary follicle is formed as a result of the development of the *zona pellucida,* which is a membrane surrounding the ovum. The

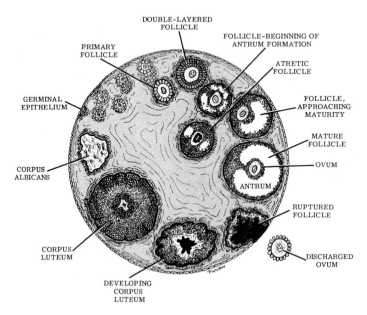

Figure 35-3
Schematic diagram of the human ovary showing various stages of follicular development, ovulation, and formation and regression of the corpus luteum.

tertiary follicle is characterized by the presence of an *antrum,* a fluid-filled space surrounding the ovum. The ovum, enclosed in granulosa cells, is bathed by liquid, the *liquor folliculi.* As the follicle grows, it moves toward the cortex of the ovary, and the antrum enlarges to produce the mature follicle. The mature follicle, called the *graafian follicle,* extends throughout the thickness of the cortex and bulges as a blister on the free surface of the ovary (Fig. 35-3). In addition, the mature ovary contains follicles which exhibit varying degrees of degeneration; these are the *atretic follicles.*

The *corpus luteum* is a temporary endocrine organ that forms following ovulation as a result of luteinization of the *granulosa* and *theca interna* cells. Luteinization involves hypertrophy and hyperplasia of the granulosa and theca interna cells with the development of lipid inclusions within the cells. Immediately after ovulation, the cavity of the follicle becomes filled with blood and lymph. Gradually the space occupied by the fluid is filled by the luteal cells, and a well-developed blood supply is formed within the mass of lutein cells. Seven to eight days after ovulation, regressive changes are seen, such as loss of lipid material and invasion by connective tissue. The span of functional activity of the corpus luteum is approximately 10 to 12 days in the human, after which the gland is nonfunctional and referred to as the *corpus albicans.* During the ensuing weeks, further degenerative changes occur, and the organ is replaced by connective tissue.

Ovarian Hormones

The ovaries secrete three steroid hormones (estrogens, progestins, and androgens). The term *estrogen* refers to any substance which produces vaginal cornification, whereas the term *progestin* refers to any substance which produces secretory changes in the estrogen-primed uterus. These hormones are formed by the follicle and corpus luteum. Cells of the theca interna of the ovarian follicle and thecal lutein cells of the corpus luteum are the source of ovarian estrogens. Progestins are produced by luteinized granulosa cells. The cellular origin of the ovarian androgens is uncertain, although some endocrinologists believe that androgens are produced by cells in the hilus of the ovary. The estrogens and progestins stimulate the growth and development of the female reproductive tract and exert a variety of metabolic effects on other tissues.

The human ovary secretes the estrogen *estradiol* and *estrone;* a third compound with estrogenic activity, *estriol,* is a metabolic end product of estradiol and estrone. It is well established that all the steroid hormones, regardless of their site of origin, are derived from cholesterol. The compound *17α-hydroxyprogesterone* plays a pivotal role in the ovarian synthesis of estradiol and estrone. This becomes apparent if one refers to the pathways of adrenal and testicular steroid biosynthesis. The pathways of ovarian estrogen and progestin biosynthesis are depicted in Figure 35-4. The ovarian estrogens are secreted into the bloodstream, where they combine with plasma protein. However, approximately one-third of the circulating estrogen remains in a free form and is in equilibrium with protein-conjugated estrogens, estroprotein. That the liver inactivates estrogens can be demonstrated by transplanting the ovaries to the spleen, so that the estrogens pass first to the liver. In this instance the estrogens are immediately inactivated and a hypoestrogenism develops. The important role of the liver in estrogen inactivation is further emphasized by the fact that hyperestrogenism accompanies certain types of hepatic disorders. The hyperestrogenism results from failure of hepatic removal of estrogens from the circulation. The estrogens undergo oxidation in the liver to form *estriol,* and the liver conjugates each of the estrogens with sulfate or glucuronide. Large amounts of conjugated estrogens are excreted in the bile, and small amounts appear in the urine. The urinary excretion of estrogens and androgens from birth through life is depicted in Figure 35-5.

Progesterone is secreted into the bloodstream by luteal tissue during its short life-span. The circulating level of progesterone, like that of the estrogens, is very low. It is rapidly metabolized by the liver to *pregnanediol* and *pregnanolone.* Only a small amount (less than 10 percent) of administered progesterone can be recovered in the urine as pregnanediol or pregnanolone. Both appear in the urine as the free compounds, but

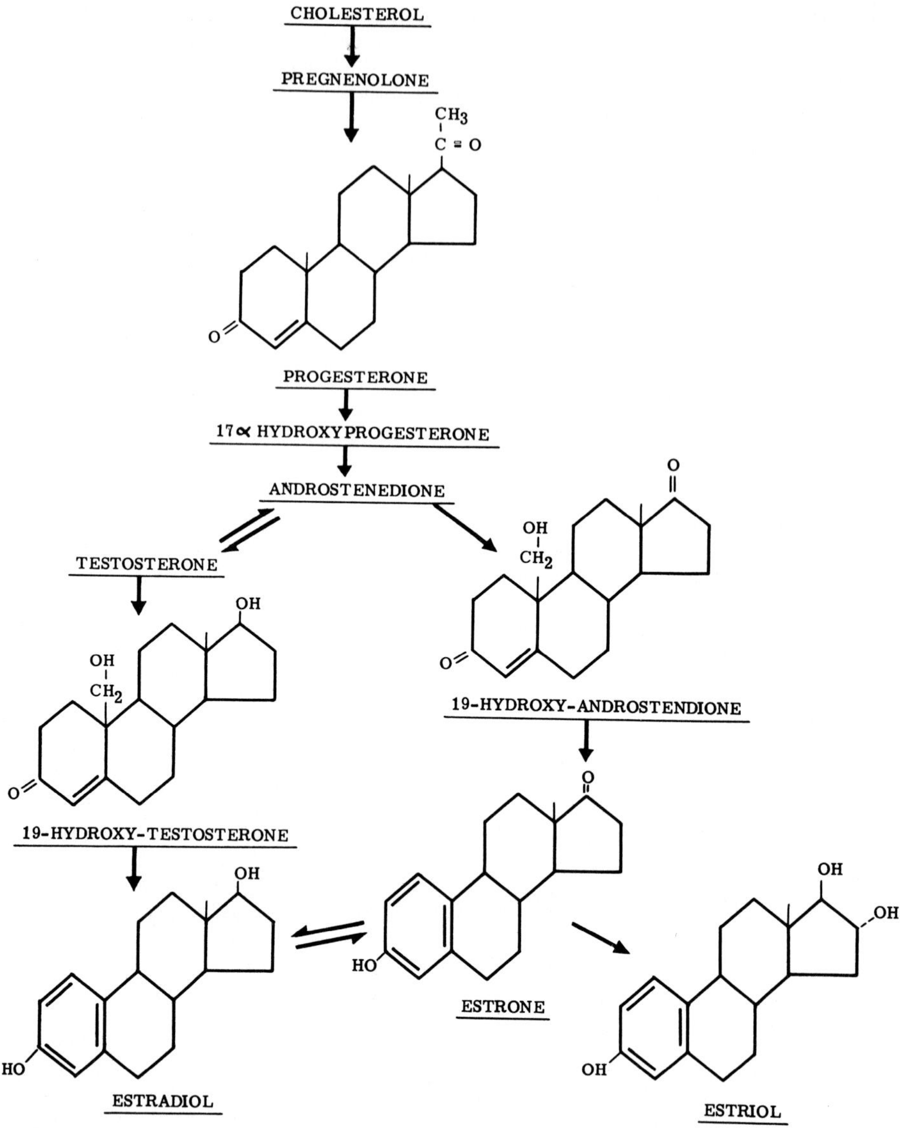

Figure 35-4
Major pathways of the biosynthesis of ovarian progesterone and estradiol.

mostly they are conjugated with glucuronide. The excretion of progesterone metabolites will be considered further when the reproductive cycle, pregnancy, and lactation are discussed.

Effects of Estrogens

The estrogens strikingly influence the cells of the female reproductive tract, and hardly a tissue in the body is not affected by them. Their effects on the genital tract of the female are most dramatically observed when estrogens are administered to the ovariectomized animal.

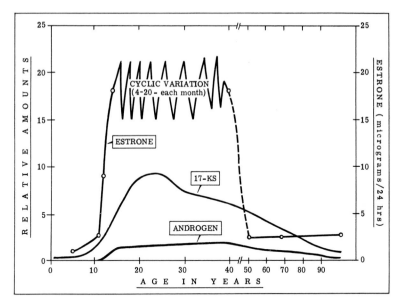

Figure 35-5
Changes in the urinary excretion of estrogens and androgens throughout the life-span of the human female.

Vagina

The vaginal epithelium of either the immature or ovariectomized animal is very thin and shows marked atrophic changes. Estrogen administration brings about a sharp increase in mitotic activity and stratification of the cells of the vaginal epithelium. Cornification of the epithelial cells results from a loss of blood supply to the superficial cells, which occurs because of the rapid growth (increase in thickness) of the vaginal epithelium. Estrogens increase the glycogen and mucopolysaccharide content of the vaginal mucosa, decrease the pH (about 4.5), increase the vascularity of the structure, and produce a slight degree of edema. The appearance of cornified cells in a vaginal smear or lavage is diagnostic of estrogen action.

Uterus

Estrogen administration causes proliferation of the uterine cervical mucosa and the secretion of alkaline mucus from the cervical glands. The estrogens also have a pronounced effect upon the endometrium, inducing endometrial mitoses (hyperplasia), increasing the height of the lining epithelium (hypertrophy), increasing the blood supply and capillary permeability of uterine vessels, and increasing the uterine water and electrolyte content. They also enhance uterine anaerobic and aerobic glycolysis and increase the rate of uterine ribonucleic acid (RNA) and deoxyribonucleic acid (DNA), protein, and glycogen synthesis. Thus, the uterus increases in weight and size on account of marked proliferative changes following estrogen administration. Contrarily, ovariectomy results in definite atrophic changes in all uterine tissue, particularly the endometrium and myometrium.

The smooth muscle is relatively inactive in an ovariectomized animal. Estrogen administration increases myometrial contractility and motility and increases the uterine smooth muscle reaction to oxytocin. These changes result from the ability of estrogens to increase the content of the myometrial contractile substance, actomyosin; thus, a greater isometric tension can be developed. Action potentials are rarely observed in the uterine muscle taken from an ovariectomized animal, but the frequency increases following estrogen administration and brings about greater myometrial activity. It has also been shown that the membrane potential of uterine muscle is elevated by estrogen over that of the castrate. This effect of estrogen is dependent upon the presence of Ca^{++}. Apparently, estrogens set the membrane potential at a critical level so that the myometrial cells show an increased excitability (decreased threshold), increased spontaneous activity, and increased sensitivity (reactivity) to appropriate stimuli.

Estrogen administration also stimulates proliferation of the epithelium and smooth muscle of the uterine tubes or the oviducts; the effects here are similar to those on the body of the uterus. Estrogen administration can lead to failure of implantation of the ovum. This may be due largely to its effects on the musculature of the oviduct. Estrogen stimulates the smooth muscle of the oviduct to increase the peristaltic action in the direction from the uterus to the ovary. This may be of benefit to sperm transport, but if the activity is too great, it prevents the passage of the ovum down the tubes to the uterus.

Metabolism
Estrogen administration results in renal retention of Na^+ and Cl^- and decreased urine volume, with subsequent increases in blood volume, extracellular fluid (ECF) volume, and body weight. The action of estrogen on renal Na^+ handling is independent of changes in renal hemodynamics and the secretions of the adrenal cortex.

The estrogens also promote protein anabolism and affect skeletal growth by increasing osteoblast activity and the rate of deposition of calcium and protein in bone. Bone growth (length) is limited by estrogens because they promote closure of the epiphysial plates, which are the growth centers of bone.

Secondary Sex Characteristics
The secondary sex characteristics of the female are dependent· upon the estrogens. Estrogens act on the mammary gland to induce duct growth and proliferation of stromal tissue. The tendency in the female for adipose tissue to concentrate in the buttocks, hips, thighs, and mammary gland, plus the presence of a uniform layer of subcutaneous fat over the body, leads to a rounded contour with varying convexities and results in a form which is distinctly less angular than that of the male. The female skeleton is smaller than the male, with marked differences in the pelvis, which is adapted in the female to aid in the carriage and delivery of young. The hair of the body is fine, and scalp hair grows faster and is more permanent. Ovariectomy in experimental animals abolishes mating responses and sexual behavior patterns. Libido in the human is relatively unaffected following ovariectomy in the adult but may be decreased if ovariectomy is performed prior to puberty.

Mode of Action
It would be logical to assume that estrogens influence the enzyme systems responsible for protein anabolism, because of their mitogenic and growth-promoting properties. That estrogens act upon energy-supplying systems and provide energy for systems is shown by their stimulating action on oxygen consumption, and oxidation of glucose and pyruvate in endometrial tissue. Studies have shown that the estrogens are involved in stimulating an enzyme which transfers hydrogen ion from reduced nicotinamide-adenine dinucleotide phosphate (NADPH) to nicotinamide-adenine dinucleotide phosphate (NADP). Such a system would accelerate the Krebs cycle and the monophosphate shunt, thereby increasing the amount of energy available for synthetic reactions of the cells. This may be mediated via cyclic adenosine monophosphate (cyclic AMP). Treatment with the estrogens results in activation at the gene level, as evidenced by the observations that they stimulate, in sequence, the rate of synthesis of messenger RNA, transfer RNA, and ribosomal RNA, and this is followed by an increase in the rate of protein synthesis.

Each of the effects of estrogen on the reproductive tract and mammary gland may be considered indicative and diagnostic of puberty, because the estrogen secretion heralds the onset of ovarian endocrine activity. The primary point is that estrogens cause proliferative changes in the female reproductive system. The role of estrogen in cyclic reproductive behavior will be discussed later.

Effects of Progestins

The effects of progesterone, per se, are never seen under physiological conditions and are not normally apparent until after the actions of estrogens have been operative. The effects of estrogens are generalized as being proliferative, whereas progesterone acts upon proliferated tissue to differentiate it into a secretory type of tissue. Progesterone acts on the reproductive tract and mammary gland in such a manner as to prepare the tract for implantation of the fertilized ovum and to maintain gestation and lactation.

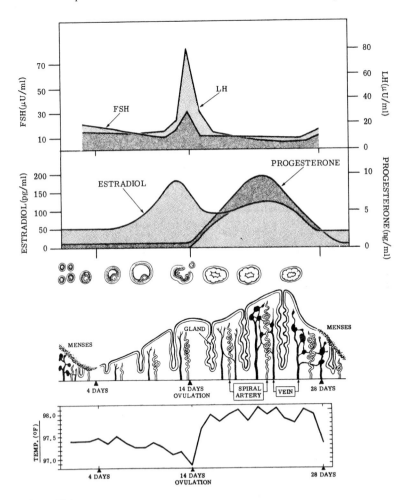

Figure 35-6
Variations in the plasma levels of FSH, LH, estradiol, and progesterone throughout the menstrual cycle in the human. Changes in the ovarian elements, the follicles and corpus luteum, endometrial morphology, and basal body temperature are also shown.

Reproductive Tract
Progesterone transforms the cornified vaginal epithelium to a mucified condition and decreases cervical secretions. The progestational response of the endometrium, which is a result of priming with estrogen followed by progesterone stimulation, is characterized by further endometrial growth, including thickening of the epithelium and accumulation of water. The proliferated endometrium is transformed into the secretory endometrium under the influence of progesterone. These changes include enlargement of endometrial stromal cells and the formation of tortuous glands with marked glycogen deposition. Thus the highly secretory endometrium provides the environment which is necessary for implantation of the blastocyst. That expulsion of the implanted blastocyst does not occur is assured by the effect of progesterone on the myometrium. The membrane potential of uterine smooth muscle is further elevated and results in a decreased excitability of these elements. The hyperpolarized membrane of the progesterone-dominated smooth muscle cell of the uterus shows poor conduction, lowered spontaneous activity, and reduced sensitivity to specific stimuli. It follows that the progesterone-dominated uterus cannot participate in the organized contractions which result in expulsion of the young. Progesterone also results in decreased smooth muscle activity in the oviducts. It acts on the estrogen-primed mammary gland and stimulates lobuloalveolar growth, which is the hallmark of the mature mammary gland.

Body Temperature
Variations in the awakening basal body temperature occur and have been correlated with cyclic ovarian activity. The phase of the cycle during which estrogen is dominant is characterized by a low basal body temperature. At ovulation time, a slight drop in temperature occurs, but when progesterone secretion rises, the body temperature is elevated over that observed during the early stage of the cycle (Fig. 35-6). The basal body temperature remains elevated so long as progesterone secretion is maintained. The mechanism of this action is not understood. It should be obvious, because of the small temperature changes, that even minor infections which elevate the body temperature obscure the cyclic variations in basal body temperature.

Miscellaneous
Progesterone, because of its key position in steroid biosynthetic pathways, might be expected to mimic the effects of a variety of steroids. It possesses adrenal corticoid activity in that it prolongs survival of the adrenalectomized animals and has slight protein catabolic activity. It has androgenic activity and aids in the maintenance of testicular weight in hypophysectomized animals. It also tends to induce a slight hyperventilation, which results in a lowered Pco_2 of arterial blood.

THE REPRODUCTIVE CYCLE

The cyclic variations in reproductive activity occur as the result of neural and hormonal interaction between the central nervous system, the hypothalamus, the anterior pituitary, the ovaries, and the reproductive tract. The normal cycle is characterized by maturation of ovarian follicles, ovulation, and preparation of the reproductive tract for the implantation of a fertilized ovum. If conception and implantation do not occur, these processes are repeated. The hypothalamic substance, gonadotropin-releasing factor (FSH/LH-RF), the pituitary hormones, FSH and LH, and the ovarian hormones, estradiol and progesterone, are all involved in inducing the changes in cyclic activity at all levels of the complex system.

Hormonal Changes and Their Regulation

When one considers the first day of the human menstrual cycle to be coincident with the onset of menstruation, it is clear that the plasma gonadotropic hormone levels during the cycle are characterized by midcycle peaks of both FSH and LH immediately prior to ovulation. The plasma level of LH is relatively constant during the early part of the cycle prior to the peak and slightly higher than that observed during the latter half of the cycle. The plasma LH level tends to decline gradually after ovulation, but as menstruation approaches there is a modest increase in circulating LH. The plasma FSH pattern during the first half of the cycle shows a moderate early elevation followed by a slight decline until the midcycle peak. After the peak the FSH concentration gradually decreases, but it starts to increase a few days prior to menstruation. It has been observed that both FSH and LH plasma levels are not maintained in a steady manner throughout any given day of the cycle; in fact, the plasma concentrations of both are actually maintained by frequent surges of secretion by the pituitary that are much more vigorous during the midcycle peaks. These surges in gonadotropin secretion are in response to the hypothalamic releasing factor, FSH/LH-RF.

The concentration of estradiol in the plasma follows a biphasic pattern with low concentrations during the early and late phases of the cycle. The plasma estradiol level rises rapidly and peaks prior to the surge in FSH and LH secretion. The concentration declines during the time of ovulation and then shows a second gradual rise and fall prior to menstruation. The plasma progesterone concentration, in contrast to the plasma estradiol level, remains very low during the first half of the cycle. It then increases markedly after ovulation and corpus luteum formation, remains high for several days, and decreases to a low level prior to menstruation. The changes in plasma concentration of the four hormones are shown in Figure 35-6. Not included in the figure are the 17α-hydroxyprogesterone levels during the cycle. The plasma concentration of this steroid is very low during the first part of the cycle and rises at midcycle along with FSH and LH. It then follows the same general pattern observed with progesterone. The observation that the ovaries of hypophysectomized animals remain immature, but that gonads of such animals can be restored or maintained with pituitary extracts, particularly the gonadotropins, is taken as proof that the pituitary regulates ovarian function. The pituitary that is removed from its normal site and transplanted to some extrasellar site is incapable of maintaining normal cyclic gonadal function.

FSH, as the name indicates, stimulates growth of the ovarian follicle, which includes hyperplasia of the granulosa cells, enlargement of the antrum, and a marked increase in ovarian weight. These changes occur when FSH is administered to immature hypophysectomized rats. Since changes in the reproductive tract indicative of estrogen secretion do not occur, it is concluded that FSH alone is incapable of stimulating estrogen secretion. If the immature hypophysectomized animal is treated with LH, the ovarian interstitial cells are stimulated, but no estrogen secretion results. However, the combination of FSH and LH notably increases ovarian weight, follicle growth, and estrogen secretion. The rate of follicular growth and estrogen secretion increases under the influence of FSH and LH. The surge of LH release at midcycle induces ovulation and causes luteinization of the granulosa and theca interna cells of the ovulated follicle. LH stimulates the secretion of progesterone and estradiol by the corpus luteum, and present studies indicate that the secretion of small amounts of LH is required for the maintenance of a functional corpus luteum in the human. Once formed, the human corpus luteum appears to have a preset life-span of approximately 12 days.

Various studies indicate that the ovarian steroids, estradiol and progesterone, regulate gonadotropic hormone secretion by the AP by exerting both negative and positive feedback controls over both FSH and LH secretion. The type of feedback control exerted by the steroids appears to be dependent upon the concentrations of the steroids in the plasma. Early investigations have shown that removal of the ovaries results in marked increases in gonadotropic hormone secretion by the AP. Similarly, after the termination of reproductive function at the menopause in the human female, the plasma estradiol concentration is minimal and gonadotropin levels are high. In both examples, the administration of estradiol lowers the plasma FSH and LH concentrations. Thus the negative feedback effect of estradiol can account for the relatively low levels of FSH and LH observed throughout most of the cycle. It has been shown that single or multiple injections of estrogens to anovulatory women can induce the surge of LH secretion and induce ovulation. Progesterone blocks this estrogen-induced surge of LH. On the basis of the evidence, it has been concluded that the rising estradiol levels during the early part of the cycle exert a positive feedback effect and thus trigger massive bursts in LH secretion. It would also appear that as the functional activity of the corpus luteum diminishes, the negative feedback effect of estradiol on FSH and LH is diminished slightly, thereby permitting modest increases in FSH and LH secretion. A new hormonal cycle could then begin. In the event of pregnancy, however, the functional life of the corpus luteum is prolonged, and cyclic activity ceases throughout the duration of the pregnancy.

As a result of observing the effects of emotional states, malnutrition, light, and temperature upon gonadal function, modulation of gonadal activity by the central nervous system has been obvious for many years. It is well known that several species of animal ovulate only after appropriate stimulation, e.g., coitus. These species have been referred to as the induced or reflex ovulators.

There can be little doubt that the functional integrity of the AP-ovary system is dependent upon a control exercised by the nervous system. The nervous control over the AP secretion of FSH and LH is exerted via the secretion of releasing factors by the hypothalamus. It has been suggested that the secretion of FSH and LH is mediated by a single hypothalamic substance: FSH/LH-RF. This is a decapeptide and has been synthesized. The feedback effects of estradiol are directed at the hypothalamus and probably not at the pituitary. Similarly, the two gonadotropins FSH and LH appear to operate through feedback mechanisms at the hypothalamic level to inhibit gonadotropin release.

Available evidence indicates that the hypothalamus acts as a center which integrates brain stem and spinal reflexes into patterns of behavior observed in mating in the female. The activity of the hypothalamus may be modified by influences from the neocortex, rhinencephalon, and brain stem reticular formation.

Cyclic Changes in the Reproductive Tract

The structural and functional changes in the reproductive tract are well known and are repeated without much variation throughout the year. Changes in uterine and vaginal function are directly regulated by the ovarian estrogens and progestins. The changes that occur during the preovulatory stage or early part of the cycle depend upon estrogen secretion, whereas the postovulatory stage is influenced by estrogen and progesterone. Thus, the early part of the cycle is characterized by follicle development and proliferation of the reproductive tract and has been termed the *proliferative* or *follicular* phase of the cycle. During this phase the vaginal epithelium undergoes rapid growth, and the vaginal smear is eventually transformed into one dominated by cornified cells. The uterus begins to accumulate fluid, becomes highly contractile, and undergoes proliferative changes under the influences of estrogen until the time of ovulation. Following ovulation and corpus luteum formation, reproductive tract function is altered by progesterone.

The phase of the cycle which follows ovulation is called the *secretory* or *luteal* phase. During this phase the vagina is infiltrated with leukocytes, and the vaginal smear is characterized by the presence of leukocytes, mucin, and cornified cells. The vaginal epithelium becomes thin, and leukocytes are abundant, so that the vaginal smear contains nucleated epithelial cells and leukocytes. Immediately after ovulation the uterus diminishes in vascularity and contractility. The increasing titer of progesterone transforms the proliferated endometrium into an organ which has highly coiled glands possessing the ability to secrete large amounts of glycogen. Thus, the uterus is prepared for nidation. Following regression of the corpus luteum a new set of follicles is stimulated by the rising levels of FSH, and the cycle is repeated. Cyclic variations in the activity of the fallopian tubes, or oviducts, and in the uterine cervix are also observed, and these changes are dependent upon the circulating estrogens and progesterone.

The reproductive cycle differs among species in several ways. In general, the nonprimate mammalian species exhibit a period of sexual receptivity during a particular phase of the cycle. This period coincides with the time of maximal estrogen secretion prior to ovulation. The period of sexual receptivity, *estrus,* or "heat," is characterized by behavioral changes designed to attract the male at the time when ova are most easily fertilized.

The reproductive cycle in the primates is called the menstrual cycle, and in the normal nonpregnant individual the interval which extends from the onset of a period of uterine bleeding to the onset of the next period of bleeding describes its duration. The mean cycle length is 28 days, but the range is very large, from 20 to 35 days. The menstrual cycle has been divided into three stages, and these divisions are based on the histology of the endometrium.

The first stage, menstruation or *menses,* lasts about 4 to 6 days and is characterized by the occurrence of hemorrhage in the endometrial stroma. The endometrium degenerates and, together with interstitial blood, is sloughed into the uterine lumen and exits through the vagina as menstrual discharge. The second or *proliferative* stage is characterized by repair and proliferation of the endometrium and proceeds under the influence of estrogens during the next 8 to 10 days. The endometrium undergoes marked changes which provide the appropriate environment for the implantation of the fertilized egg during the third or *secretory* stage (about 9 to 10 days' duration). This final stage is dependent upon ovulation and the secretion of progesterone by the corpus luteum. If a fertilized ovum is not available or if implantation does not occur, the corpus luteum and the endometrium degenerate and the cycle is repeated.

Menstruation should not be considered an actively induced process, because it occurs as a result of cessation of stimulation of the endometrium by the ovarian steroid hormones. It is the result of withdrawal of the ovarian steroids, as shown by the following observations. (1) Ovariectomy, if performed during the last two or three weeks of the cycle, precipitates uterine bleeding two to six days postoperatively. (2) The administration of estrogens to ovariectomized individuals results in a proliferative type of endometrium, and the withdrawal of estrogen treatment results in menstruation. (3) The simultaneous administration of estrogen and progesterone to an ovariectomized individual causes the development of a secretory endometrium, and the withdrawal of both steroids results in menstruation within two to three days. (4) The administration of large amounts of estrogen prevents menstruation for long periods of time. Figure 35-6 depicts the changes in the endometrium throughout the menstrual cycle.

Transplantation of the endometrium to the anterior chambers of the eye has enabled investigators to observe directly those factors which are associated with menstruation. The observations have shown that alterations in blood flow through the endometrial vessels are responsible for menstruation. Prior to menstruation (two to six days), the coiled arteries deep in the endometrium offer an increased resistance to blood flow, and because of the inadequate blood flow the endometrium regresses. Immediately prior to menstruation, the coiled arteries constrict further and reduce, to a greater degree, the flow to the endometrium. Then, after varying intervals, the arteries dilate and hemorrhage occurs. The arteries then again constrict, and hemorrhage ceases from the artery involved. It should be remembered that each artery bleeds only once during each cycle; thus each artery dilates, bleeds, then constricts, but not all coiled arteries do so simultaneously, and a small area may be sloughed and repaired before other areas are sloughed. Therefore, as a result of decreasing estrogen and progesterone secretion, the endometrium is deprived of essential materials, and it regresses. The spiral arteries undergo intermittent contraction and cause recurrent ischemia, so that small hemorrhages occur in the tissue; cellular necrosis occurs, and menstruation ensues.

Control of Fertility

Currently a large amount of attention is being given to the study of the control of fertility and family planning. Modern methods of contraception include the use of oral contraceptives ("the pill") and intrauterine devices (IUD). The oral contraceptives in use today consist of synthetic estrogenic and progestational steroids. The combination of synthetic hormones is taken cyclically, one pill daily, for 21 consecutive days of each 28-day menstrual cycle. The steroids act upon the CNS and hypophysis to prevent the midcycle surge of FSH/LH-RF and FSH and LH, respectively. Ovulation and corpus luteum formation are prevented, and ovarian atrophy tends to be produced while one is taking the pill. The endometrium does not undergo atrophy because of the direct stimulating action of the steroids. On withdrawal of the steroid, the endometrium breaks down and menstruation ensues, and a new cycle can be induced. Two general types of methods are used, the *combined* and *sequential.* The combined pill is taken daily and contains both estrogenic and progestational steroids. In the sequential plan, pills containing estrogen only are taken daily for 15 days, followed by 5 pills containing both estrogen and progesterone. These substances act to prevent the surge of LH release and thus ovulation. They also act upon the endometrium and myometrium to create an environment which is hostile to fertilization and implantation. The combined pill has a greater effectiveness in preventing pregnancy than does the sequential pill.

Since primitive times, intrauterine devices have been used in an attempt to control conception. Recently new designs and materials have been employed and are effective. The IUD does not prevent ovulation but probably acts by creating an intrauterine environment that is hostile to fertilization of the ovum and/or implantation of the blastocyst. Well-designed and unbiased studies have shown that both the pill and the IUD are very effective contraceptive agents and that each induces minimal side-effects.

PREGNANCY

Pregnancy may be attained only following a series of events which terminate in the oviducts, where zygote formation takes place as a result of union of an egg and a sperm. The ovum of the human may be viable up to 24 hours after ovulation, but the spermatozoa retain their viability while in the female reproductive tract as long as 36 hours. It has been estimated that in women of proved fertility only one in four conceives when coitus with a fertile male is completed during the midcycle or fertile period of the female. Sperm deposited in the vagina must move or be transported up the reproductive tract to the oviducts. In many species, sperm appear in the oviducts within several minutes after having been deposited in the vagina during midcycle or estrus. The rapid transport of sperm through the uterus and into the oviducts is the result of mechanical propulsion provided by contraction of the uterine musculature. On the basis of work in experimental animals, it has been suggested that the increase in uterine muscle activity is caused by the reflex release of oxytocin from the hypothalamo-neurohypophysial system (HNS).

Fertilization occurs in the ampullary portion of the fallopian tube or oviduct and is generally accomplished within 15 to 20 hours after ovulation. Muscular and ciliary activity of the oviduct propels the zygote toward the uterus, and by the fourth day after ovulation it is in a 12-to 16-cell stage and reaches the uterine cavity. Generally, by the seventh day the cell mass, or blastocyst, makes contact and becomes adhered to the uterus, penetrates the uterine epithelium, and passes through the epithelium. It soon becomes covered by the uterine epithelium.

The mechanism by which the corpus luteum is caused to persist is the result of the action of a hormone secreted by the embryonic trophoblast. This hormone, human chorionic gonadotropin (HCG) is luteotropic and responsible for preventing the demise of the corpus luteum. Because of the persistent luteal synthesis and secretion of estrogen and progesterone, the endometrium fails to undergo its usual menstrual regression, the menstrual period does not occur, and overt cyclic ovarian and uterine activity is absent throughout the pregnancy. Pregnancy can be established and maintained only if adequate amounts of progesterone are secreted prior to and throughout the course of that pregnancy.

The placenta serves as an organ of exchange between the maternal and fetal organisms. It also serves as an endocrine organ, secreting at least four types of hormones: the two gonadotropins, HCG and human placental lactogen (HPL), and estrogen and progesterone. HPL is also called placental somatomammotropin. In addition, other steroidal substances, such as cortisol and aldosterone, are secreted by the placenta. The placenta may also secrete TSH-like and ACTH-like polypeptides. HCG is the first to appear and is generally detected in the plasma about eight days after presumed ovulation. It is a glycoprotein and has the biological activities of LH. HCG maintains the functional activity of the corpus luteum during the early stages of pregnancy and reaches a peak circulating level of approximately 10 to 16 μg per milliliter at about 9 to 11 weeks. The plasma levels of HCG throughout pregnancy are shown in Figure 35-7. The presence of HCG serves as the basis for most of the pregnancy tests. The pregnancy tests based on the presence of HCG are numerous. All are dependent upon the ability of HCG to stimulate gonadal function in a test animal. (1) The Aschheim-Zondek test utilizes the ability of the urinary extract of HCG to induce ovarian hyperemia in the immature mouse ovary. (2) In the Friedman test, urine is injected intravenously in a rabbit and the end point of the assay is the production of ovulation. (3) The Hogben test depends upon the ability of HCG to cause ovulation in the South African clawed toad (*Xenopus laevis*). (4) The Galli Mainini test depends upon the ability of HCG to cause the expulsion of spermatozoa by the male frog.

Also depicted in Figure 35-7 are the plasma levels of HPL. This hormone has been detected in the trophoblast as early as the third week of gestation, can be detected in the plasma by the sixth week, and attains a level of 8 to 12 μg per milliliter prior to parturition. The rising HPL levels parallel the increase in placental weight throughout

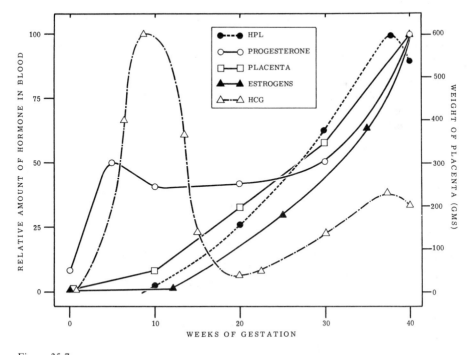

Figure 35-7
Changes in the plasma concentrations of the placental gonadotropins, HCG and HPL, and estrogens and progesterone throughout normal pregnancy in the human. The weight of the placenta is also shown.

the gestation period. HPL is structurally similar to growth hormone and shares most of its biological activities. It also stimulates mammary gland growth and development during pregnancy.

The endocrine function of the ovary can be replaced by the placenta early in the second month of human pregnancy, as shown by the fact that bilateral ovariectomy may be performed during the second month of gestation, yet pregnancy continues to a successful termination at the end of the fortieth week. The levels of urinary estrogens and pregnanediol increase gradually throughout the pregnancy and reach a maximum a few days prior to parturition. The estrogen level increases to about 150 to 175 ng per milliliter of plasma, and the plasma progesterone concentration may increase up to 200 ng per milliliter prior to parturition (Fig. 35-7).

The fetus and the placenta complement each other in the synthesis of certain steroidal metabolites, for neither is capable of completing the entire biosynthetic pathway alone. Many steroidal metabolites produced during pregnancy result from an interplay between the fetus, the placenta, and the mother. Fetal well-being can be assessed by measuring those metabolites which arise primarily from the fetus. Such a metabolite is estriol. The fetal adrenal cortex produces large amounts of the carbon 19 steroids, dehydroepiandrosterone sulfate (DHAS) and 16α-hydroxydehydroepiandrosterone sulfate. The latter is transported to the placenta via the placental arteries and converted to estriol. The mother contributes only small amounts of carbon 18 and carbon 19 to the placental pathway for the synthesis of estriol. This unit is termed the *fetal-placental* unit. Figure 35-8 illustrates the major pathway of estriol synthesis in the fetal-placental unit and the distribution of estriol in the fetus, amniotic fluid, placenta, and mother.

The thyroxine-binding globulin and transcortin levels also increase during pregnancy. The increases in the binding proteins reduces the amount of free circulating T_4 and cortisol, respectively, and indirectly lead to increases in the rates of secretion of TSH and ACTH.

The uterus must serve several related functions during pregnancy. Initially, it must be prepared to receive the fertilized ovum and allow implantation. Once implantation has occurred, it must grow and adapt to the growing products of conception; and it must be prepared to assume delivery at term.

In the nonpregnant state the uterus is thick-walled, weighs about 60 gm, is about 7 cm

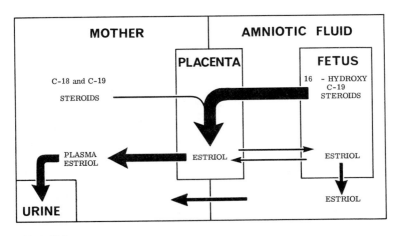

Figure 35-8
Major pathway of estriol synthesis by the fetal-placental unit in the human. The width of the arrow indicates the relative importance of the pathway.

in length, and is contained within the pelvis. At the end of pregnancy it is thin-walled, weighs more than 1 kg, is about 35 cm long, extends up to and elevates the diaphragm, and may contain a total volume up to 5 liters. Thus the pregnant state imposes widespread changes on the maternal organism. Nearly all of them represent adaptive responses which permit the mother to deliver a normal infant without harm to either the mother or child. In order to supply adequate amounts of oxygen and other nutrients to the fetus, the blood flow through the gravid uterus must increase. Studies in the human indicate that uterine blood flow at term in a single pregnancy is about 500 ml per minute, whereas in a twin pregnancy it is over 1000 ml per minute. On the other hand, 24 to 48 hours after delivery the blood flow is decreased to about 205 ml per minute. Vasomotor fibers to the uterus are thoracolumbar in origin; their stimulation results in constriction of the vessels they supply. However, the dominant control over the uterus vessels is exercised by hormones. For instance, the administration of estrogen induces uterine hyperemia prior to any metabolic changes in the uterine tissue. With increased blood flow to the pregnant uterus, various adaptive changes in the cardiovascular system occur, and the uterine blood flow is maintained so that oxygen is supplied in excess of the oxygen requirement of the tissue.

The mother undergoes numerous metabolic changes in response to the demands of the growing fetus and placenta. At the end of an average pregnancy the fetus weighs about 3.5 kg, the placenta weighs 500 to 600 gm, and there is about 1 liter of amniotic fluid. The total blood volume is increased about 1.5 liters in excess of that observed in the nonpregnant state. The increased retention of water, fetal growth, and mechanical factors place marked demands upon the cardiovascular and urinary systems and nutritional status of the mother.

The cardiovascular system adapts by increasing the cardiac output, heart rate, and pulse pressure. The cardiac output, however, is significantly reduced when the subject is in the supine position, because the venous return is reduced as a result of compression of the inferior vena cava and iliac veins by the uterus. An increased femoral venous pressure contributes to the dependent edema frequently observed during the late stages of pregnancy. An increase in cutaneous blood flow is observed, and this contributes to the loss of excess heat generated by the increased metabolic rate.

The glomerular filtration rate (GFR) and renal plasma flow (RPF) increase during pregnancy. Late in pregnancy the GFR, RPF, urine volume, and sodium excretion are markedly affected by posture, with significant decreases in each occurring when a subject in the lateral recumbent position assumes the supine or upright position. The hemodynamic changes are related to the reduced venous return to the heart, while the changes in water and electrolyte excretion are probably related to the decreased GFR and reflex increases in ADH and aldosterone, respectively. As the uterus enlarges during late pregnancy, ureteral compression may occur at the pelvic rim, and the urinary bladder is pushed forward and upward. The latter tends to impair venous and lymphatic drainage.

The metabolic demands placed upon the maternal system to supply the needs for fetal, uterine, placental, and maternal growth are extensive. The metabolic rate and heat production increase about 25 percent as compared to the nonpregnant state. Most of the increase in O_2 consumption is required to meet the demands of the metabolic activity of the products of conception. On the average the total weight gain in pregnancy is in excess of 10 kg. The dietary requirement for protein and total calories for the total pregnancy increases about 10 percent over the requirement in the nonpregnant woman. The requirement for all vitamins increases 10 to 30 percent. Increased amounts of the various minerals—calcium, phosphorus, iodine, iron, and magnesium—for fetal growth and development must be supplied in the mother's diet. The O_2 demand is easily met by the respiratory system. The minute volume increases, and the hyperventilation tends to cause a respiratory alkalosis by lowering the plasma P_{CO_2}. A slight reduction in the plasma bicarbonate compensates for the respiratory alkalosis. The functional residual capacity is reduced as a result of elevation of the diaphragm by the enlarging uterus. All other pulmonary function tests are within normal limits throughout the duration of pregnancy.

Parturition

The average duration of a normal pregnancy in the human is 10 lunar months, 280 days after the onset of the last menstrual flow, or approximately 265 days after ovulation and conception. The termination of a normal pregnancy, parturition or labor, depends upon dilation of the cervix and alterations of the contractile activity of the uterine muscle from a state of quiescence to one of increased activity. Cervical dilatation appears to result from mechanical pressure transmitted by uterine contractions. A variety of factors influence the activity of contractile elements of the myometrial cells, and there is no general agreement as to the relative importance of each in initiating labor. However, once labor begins, whatever the trigger, a series of phenomena carries the delivery process to completion. Some of the factors which influence myometrial activity are progesterone, estrogens, oxytocin, the prostaglandins, the autonomic innervation of the uterus, and mechanical factors. Several experimental observations show that progesterone plays a predominant role in the maintenance of pregnancy, and massive doses of progesterone may maintain pregnancy beyond term. Csapo has advanced the concept that a progesterone block of uterine smooth muscle maintains pregnancy, and that withdrawal of the block is responsible for the onset of parturition. This concept is based on the theory that the placenta gives up the bulk of its progesterone by direct diffusion to the myometrium. Under this circumstance, one would expect that the concentration gradient of progesterone would be greatest at implantation sites rather than at interplacental sites. The latter has been shown to be the case in experimental animals. The concept is further strengthened by observations in women who bear twins in a bicornuate uterus (septum divides uterus into two horns). The twins have separate placentas and can be born several weeks apart. Thus, in the same woman, conditions at the same moment in pregnancy can be appropriate for the maintenance as well as the termination of pregnancy, indicating that local effects of progesterone appear to predominate.

The estrogens increase the synthesis of high-energy phosphate and actomyosin in myometrial cells. They decrease the membrane potential of the contractile cells into the range where the spontaneous discharge of action potentials occurs. The estrogens may also influence the myometrium by making it more sensitive to the actions of oxytocin and the prostaglandins. And they indirectly influence the uterus by increasing the release of oxytocin from the neurohypophysis or the release of prostaglandins from the uterus.

It is well known that the rate of clearance in vivo of oxytocin decreases progressively throughout pregnancy, and that the sensitivity of the myometrium to a constant dose of oxytocin increases up to the thirty-sixth week of pregnancy. There is little doubt that the induction of labor can be accomplished by the administration of oxytocin after progesterone withdrawal and the ascendancy of an estrogen-dominated uterus. During parturition some of the stimuli, in addition to the estrogens, which provoke the release of oxytocin arise primarily from mechanical stimulation of the uterus, cervix, and vagina.

Some prostaglandins are potent stimulators of myometrial activity and have been used clinically in the induction of labor and in therapeutic abortions in women. It has been shown that the uterine decidua is a source of prostaglandins in pregnancy. The concentrations of the prostaglandins in the uterus and in the maternal plasma rise prior to parturition.

Uterine activity in pregnancy can be increased through the sympathetic innervation of the uterus. The action of alpha-adrenergic (excitatory) activity is increased by high estrogen levels and by reduced progesterone levels, and the adrenergic fibers constitute the motor limb of a spinal reflex arc which is stimulated by cervical dilatation.

Alpha-adrenergic activity also increases the sensitivity of the myometrium to oxytocin. The uterine innervation is not essential for normal delivery.

Mechanical stimuli influence the time of parturition. Increasing the volume of the uterus increases myometrial activity. Mechanical stimulation of the cervix and vagina augment uterine contractions, probably by stimulating oxytocin release. Disruption of the relationship between the uterine decidua and the placenta tends to stimulate prostaglandin release.

Labor has been divided into three stages. *Stage one* begins with the first pain and ends when dilatation of the cervix is complete. *Stage two* is the stage of expulsion and ends with the birth of the baby. *Stage three* ends with the delivery of the placenta. The factors that regulate the highly synchronized and integrated sequence of events constituting parturition or labor in the human remain obscure.

LACTATION

Lactation consists of *mammogenesis* (development of the mammary glands), *lactogenesis* (milk secretion), *galactopoiesis* (maintenance of lactation), and *milk ejection.* The most extensive studies have been conducted using dairy animals and laboratory rodents.

Mammary glands are modified sweat glands and are made up of 15 to 25 lobes. The immature gland is comprised of short ducts which radiate from the nipple to each lobe. Each duct has many side branches extending from it to supply the lobules. The estrogens are primarily concerned with growth of the duct system, since, following the administration of estrogen, the duct system becomes extensively arborized. Progesterone acts in conjunction with estrogens to stimulate duct and alveolar growth. Physiological doses of estrogen and progesterone produce full mammary growth in either the castrate male or the female. However, the administration of estrogen plus progesterone to the hypophysectomized animal fails to induce mammary growth. Similarly, the administration of the pituitary hormones, FSH and LH, which stimulate estrogen secretion by the ovary, fails to induce mammary development in the hypophysectomized animal. The full mammary growth observed in late pregnancy is dependent upon several hormones: estrogens, progesterone, prolactin, HPL, growth hormone, and cortisol. Thus, many hormones are required to build up the mammary gland to full functional development. The hormonal environment at the end of pregnancy is such that the mammary gland is morphologically ready to begin lactogenesis.

Some synthesis and storage of milk may begin prior to parturition, but only at or shortly after parturition does a copious secretion and flow of milk occur. The integrity of the hypophysis is necessary for the initiation of milk secretion, because failure of lactation occurs following hypophysectomy, and the administration of extracts of the hypophysis provides a positive lactogenic stimulus. The increased secretion of milk at parturition is probably a result of diminished circulating titers of estrogen and progesterone when the placenta is delivered. The high levels of progesterone during pregnancy tend to inhibit lactogenesis, but the increasing levels of both prolactin and HPL tend to stimulate lactogenesis. Once the placenta is delivered, the check placed on prolactin release and lactogenesis by progesterone is removed and full lactogenesis becomes evident. Many hormones such as growth hormone, T_4, cortisol, estrogen, and insulin are required for milk secretion and its maintenance, but their exact roles in this phenomenon remain obscure. Mechanical factors also play a prominent part in the maintenance of lactation, because as milk production begins, pressure in the glands rises and, if not relieved, reaches a point where milk secretion is retarded. As a result, the mammary glands involute when nursing terminates.

Galactopoiesis requires adequate maternal nutrition and fluid intake. In addition, it requires the permissive effects of numerous hormones. The maintenance of human lactation is influenced by many social and psychological factors. Two neuroendocrine reflexes ensure the secretion of prolactin and oxytocin and aid in the maintenance of lactation.

In the nonpregnant woman the fasting plasma prolactin (PRL) level averages about 10 ng per milliliter and does not vary with the menstrual cycle. Essentially the same concentration is observed in men. During pregnancy the plasma PRL increases, and postpartum levels at the onset of nursing are about three times the mean for nonnursing females. Marked increases in plasma PRL are induced by suckling or manual stimulation of the breast. The response to suckling is prompt (10 minutes), and plasma values as great as 250 to 300 ng per milliliter are attained 30 minutes after the onset of suckling. Spinal cord transection between the last thoracic and first lumbar vertebrae leads to denervation of the caudal nipples of the rat but leaves the more cranial nipples intact. If the cranial nipples are covered so as to force suckling from the denervated nipples, the milk cannot be removed from these glands, and the glands involute. However, if suckling is permitted from two intact nipples, full lactation and milk removal occurs from the denervated nipples. Thus, it appears that suckling induces the release of PRL via a neural pathway.

Suckling also stimulates the release of oxytocin from the neurohypophysis, via a reflex. This reflex, the *milk ejection* reflex, is composed, on the afferent side, of sensory stimuli associated with suckling acting on the receptors in the nipple of the mammary gland. These are conveyed centrally where they facilitate the release of oxytocin from the HNS (Chap. 30). Oxytocin forms the efferent limb of the reflex arc and is carried via the blood to the mammary gland, where it causes contraction of the myoepithelial cells surrounding the alveoli of the mammary gland. The milk is forced out of the alveoli of the mammary gland into the duct system and cisterns, from which it can be removed by suckling. This reflex may be easily conditioned in either experimental animals or the woman. On the other hand, it may be inhibited either by strong emotions or by the administration of epinephrine. As a result of experiments in vitro, the action of epinephrine has been shown to be due to vasoconstriction. The effect of emotional stress on the reflex may act through epinephrine.

In summary, lactation is the result of the action of many hormones and neural elements. The mammary gland develops through the action of estrogen and progesterone, and full secretory function is attained through the action of prolactin, growth hormone, TSH, ACTH, and probably insulin. Milk secretion is maintained only in the animal with intact innervation of the mammary glands and neurohypophysis, and it appears that active stimulation of the mammary gland through suckling participates in the milk letdown reflex by stimulating oxytocin release.

REFERENCES

Assali, N. S. (Ed.). *Biology of Gestation.* New York: Academic, 1968. Vols. 1, 2.

Davidson, J. M., and G. J. Black. Neuroendocrine aspects of male reproduction. *Biol. Reprod.* Suppl. 1: pp. 67–92, 1969.

Diczfalusy, E. Endocrine functions of the human fetoplacental unit. *Fed. Proc.* 23:791–798, 1964.

Fuchs, F., and A. Klopper. *Endocrinology of Pregnancy.* New York: Harper & Row, 1971.

Gier, H. T., and G. B. Marion. Development of mammalian testes and genital ducts. *Biol. Reprod.* Suppl. 1: pp. 1–23, 1969.

Gorski, J., D. Toft, G. Shyamala, D. Smith, and A. Notides. Hormone receptors: Studies on the interaction of estrogens with the uterus. *Recent Prog. Horm. Res.* 24:45–80, 1968.

Lee, P. A., R. B. Jaffe, and A. R. Midgley, Jr. Serum gonadotropin, testosterone and prolactin concentrations throughout puberty in boys: A longitudinal study. *J. Clin. Endocrinol. Metab.* 39:664–672, 1974.

Lloyd, C. W. (Ed.). *Human Reproduction and Sexual Behavior.* Philadelphia: Lea & Febiger, 1964.

McCann, S. M.   Regulation of Secretion of Follicle-Stimulating Hormone and Luteinizing Hormone. In R. O. Greep and E. B. Astwood (Eds.), *Handbook of Physiology*. Washington: American Physiological Society, 1974. Section 7: Endocrinology, vol. 4, part 2, chap. 40, pp. 489–517.

Masters, W. H., and V. E. Johnson. *Human Sexual Response*. Boston: Little, Brown, 1966.

Moore, K. L.   *The Developing Human*. Philadelphia: Saunders, 1973.

Nalbandov, A. V.   *Reproductive Physiology* (2nd ed.). San Francisco: Freeman, 1964.

Neill, J. D.   Prolactin: Its Secretion and Control. In R. O. Greep and E. B. Astwood (Eds.), *Handbook of Physiology*. Washington: American Physiological Society, 1974. Section 7: Endocrinology, vol. 4, part 2, chap. 39, pp. 469–488.

Noel, G. L., H. K. Suh, and A. G. Frantz.   Prolactin release during nursing and breast stimulation in postpartum and nonpostpartum subjects. *J. Clin. Endocrinol. Metab.* 38:413–423, 1974.

Rock, J., C. Garcia, and M. F. Menkin.   A theory of menstruation. *Ann. N.Y. Acad. Sci.* 75:831–839, 1959.

Segal, S. J.   The physiology of human reproduction. *Sci. Am.* 231(3):52–62, 1974.

Steinberger, E.   Hormonal control of mammalian spermatogenesis. *Physiol. Rev.* 51:1–22, 1971.

Stewart-Bentley, M., W. Odell, and R. Horton.   The feedback control of luteinizing hormone in normal adult men. *J. Clin. Endocrinol. Metab.* 38:545–553, 1974.

Vande Wiele, R. L., J. Bogumil, I. Dyrenfurth, M. Ferin, R. Jewelewicz, M. Warren, T. Rizkallah, and G. Mikhail.   Mechanisms regulating the menstrual cycle in women. *Recent Prog. Horm. Res.* 26:63–103, 1970.

Villee, D. B.   Development of endocrine function in the human placenta and fetus. *N. Engl. J. Med.* 281:473–484, 533–541, 1969.

Index